Analysis of Electromagnetic Fields and Waves

Analysis of Electromagnetic Fields and Waves

The Method of Lines

Reinhold Pregla
FernUniversität

with the assistance of
Stefan Helfert

John Wiley & Sons, Ltd

Research Studies Press Limited

Published by John Wiley & Sons Ltd, The Atrium, Southern Gate, Chichester, West Sussex PO19 8SQ, England
Telephone (+44) 1243 779777

Email (for orders and customer service enquiries): cs-books@wiley.co.uk
Visit our Home Page on www.wiley.com

This work is a co-publication between Research Studies Press Limited and John Wiley & Sons, Ltd.

Other Wiley Editorial Offices

John Wiley & Sons Inc., 111 River Street, Hoboken, NJ 07030, USA

Jossey-Bass, 989 Market Street, San Francisco, CA 94103-1741, USA

Wiley-VCH Verlag GmbH, Boschstr. 12, D-69469 Weinheim, Germany

John Wiley & Sons Australia Ltd, 42 McDougall Street, Milton, Queensland 4064, Australia

John Wiley & Sons (Asia) Pte Ltd, 2 Clementi Loop #02-01, Jin Xing Distripark, Singapore 129809

John Wiley & Sons Canada Ltd, 6045 Freemont Blvd, Mississauga, ONT, L5R 4J3

Wiley also publishes its books in a variety of electronic formats. Some content that appears in print may not be available in electronic books.

Library of Congress Cataloging-in-Publication Data

Pregla, Reinhold.
Analysis of electromagnetic fields: the method of lines / Reinhold Pregla; with the assistance of Stefan Helfert.
p. cm.
Includes bibliographical references and index.
ISBN 978-0-470-03360-9 (cloth)
1. Electromagnetic devices–Mathematical models. 2. Electromagnetism–Mathematics.
3. Differential equations, Partial–Numerical solutions. I. Helfert, Stefan. II. Title.
TK7872.M25P72 2008
530.14'1–dc22 2008003751

British Library Cataloguing in Publication Data

A catalogue record for this book is available from the British Library

ISBN 978-0-470-03360-9 (H/B)

Typeset by Sunrise Setting Ltd, Torquay, Devon, UK.
Printed and bound in Great Britain by Antony Rowe Ltd, Chippenham, Wiltshire.

Contents

Preface **xiii**

1 THE METHOD OF LINES **1**
1.1 INTRODUCTION 1
1.2 MOL: FUNDAMENTALS OF DISCRETISATION 5
1.2.1 Qualitative description 5
1.2.2 Quantitative description of the discretisation 7
1.2.3 Numerical example 11

2 BASIC PRINCIPLES OF THE METHOD OF LINES **15**
2.1 INTRODUCTION 15
2.2 BASIC EQUATIONS 16
2.2.1 Anisotropic material parameters 16
2.2.2 Relations between transversal electric and magnetic fields – generalised transmission line (GTL) equations . 19
2.2.3 Relation to the analysis with vector potentials 21
2.2.4 GTL equations for 2D structures 22
2.2.5 Solution of the GTL equations 23
2.2.6 Numerical examples 25
2.3 EIGENMODES IN PLANAR WAVEGUIDE STRUCTURES WITH ANISOTROPIC LAYERS 26
2.3.1 Introduction 26
2.3.2 Analysis equations for eigenmodes in planar structures . 30
2.3.3 Examples of system equations 33
2.3.4 Impedance/admittance transformation in multilayered structures 35
2.3.5 System equation in transformed domain 36
2.3.6 System equation in spatial domain 38
2.3.7 Matrix partition technique: two examples 40
2.3.8 Numerical results 43
2.4 ANALYSIS OF PLANAR CIRCUITS 45
2.4.1 Discretisation of the transmission line equations 45
2.4.2 Determination of the field components 52

2.5 FIELD AND IMPEDANCE/ADMITTANCE TRANSFORMATION . . . 52
2.5.1 Introduction . . . 52
2.5.2 Impedance/admittance transformation in multilayered and multisectioned structures . . . 53
2.5.3 Impedance/admittance transformation with finite differences . . . 61
2.5.4 Stable field transformation through layers and sections . 66

3 ANALYSIS OF RECTANGULAR WAVEGUIDE CIRCUITS 73
3.1 INTRODUCTION . . . 73
3.2 CONCATENATIONS OF WAVEGUIDE SECTIONS . . . 75
3.2.1 LSM and LSE modes in circular waveguide bends . . . 76
3.2.2 LSM and LSE modes in straight waveguides . . . 80
3.2.3 Impedance transformation at waveguide interfaces . . . 82
3.2.4 Numerical results for concatenations . . . 84
3.2.5 Numerical results for waveguide filters . . . 87
3.3 WAVEGUIDE JUNCTIONS . . . 90
3.3.1 E-plane junctions . . . 93
3.3.2 H-plane junctions . . . 96
3.3.3 Algorithm for generalised scattering parameters . . . 98
3.3.4 Special junctions: E-plane 3-port junction . . . 99
3.3.5 Matched E-plane bend . . . 100
3.3.6 Analysis of waveguide bend discontinuities . . . 103
3.3.7 Scattering parameters . . . 110
3.3.8 Numerical results . . . 110
3.4 ANALYSIS OF 3D WAVEGUIDE JUNCTIONS . . . 115
3.4.1 General description . . . 116
3.4.2 Basic equations . . . 117
3.4.3 Discretisation scheme for propagation between A and B 118
3.4.4 Discontinuities . . . 121
3.4.5 Coupling to other ports . . . 122
3.4.6 Impedance/admittance transformation . . . 125
3.4.7 Numerical results . . . 126

4 ANALYSIS OF WAVEGUIDE STRUCTURES IN CYLINDRICAL COORDINATES 131
4.1 INTRODUCTION . . . 131
4.2 GENERALISED TRANSMISSION LINE (GTL) EQUATIONS 132
4.2.1 Material parameters in a cylindrical coordinate system . 132
4.2.2 GTL equations for z-direction . . . 133
4.2.3 GTL equations for ϕ-direction . . . 137

4.2.4 Analysis of circular (coaxial) waveguides with azimuthally-magnetised ferrites and azimuthally-magnetised solid plasma . . . 140
4.2.5 GTL equations for r-direction . . . 144
4.3 DISCRETISATION OF THE FIELDS AND SOLUTIONS . . 150
4.3.1 Equations for propagation in z-direction . . . 150
4.3.2 Equations for propagation in ϕ-direction . . . 153
4.3.3 Solution of the wave equations in z- and ϕ-direction . . 155
4.3.4 Equations for propagation in r-direction . . . 155
4.4 SOLUTION IN RADIAL DIRECTION . . . 155
4.4.1 Discretisation in z-direction – circular dielectric resonators . . . 155
4.4.2 Discretisation in z-direction – propagation in ϕ-direction 162
4.4.3 Discretisation in ϕ-direction – eigenmodes in circular multilayered waveguides . . . 171
4.4.4 Eigenmodes of circular waveguides with magnetised ferrite or plasma – discretisation in r-direction . . . 186
4.4.5 Waveguide bends – discretisation in r-direction . . . 202
4.4.6 Uniaxial anisotropic fibres with circular and noncircular cross-section – discretisation in ϕ-direction . . . 208
4.5 DISCONTINUITIES IN CIRCULAR WAVEGUIDES – ONE-DIMENSIONAL DISCRETISATION IN RADIAL DIRECTION . . . 216
4.5.1 Introduction . . . 216
4.5.2 Basic equations for rotational symmetry . . . 217
4.5.3 Solution of the equations for rotational symmetry . . . 218
4.5.4 Admittance and impedance transformation . . . 219
4.5.5 Open ending circular waveguide . . . 220
4.5.6 Numerical results for discontinuities in circular waveguides . . . 223
4.5.7 Numerical results for coaxial line discontinuities and coaxial filter devices . . . 223
4.5.8 Non-rotational modes in circular waveguides . . . 225
4.5.9 Numerical results and discussion . . . 228
4.6 ANALYSIS OF GENERAL AXIALLY SYMMETRIC ANTENNAS WITH COAXIAL FEED LINES . . . 229
4.6.1 Introduction . . . 229
4.6.2 Theory . . . 230
4.6.3 Regions with crossed lines . . . 239
4.6.4 Two special cases . . . 244
4.6.5 Port relations of section D . . . 247
4.6.6 Numerical results . . . 248
4.6.7 Further structures and remarks . . . 249

4.7 DEVICES IN CYLINDRICAL COORDINATES – TWO-DIMENSIONAL DISCRETISATION . . . 250
4.7.1 Discretisation in r- and ϕ-direction . . . 250
4.7.2 Numerical results . . . 253
4.7.3 Discretisation in r- and z-direction . . . 253
4.7.4 Discretisation in ϕ- and z-direction . . . 254
4.7.5 GTL equations for r-direction . . . 255

5 ANALYSIS OF PERIODIC STRUCTURES 267
5.1 INTRODUCTION . . . 267
5.2 PRINCIPLE BEHAVIOUR OF PERIODIC STRUCTURES . . 269
5.3 GENERAL THEORY OF PERIODIC STRUCTURES . . . 274
5.3.1 Port relations for general two ports . . . 274
5.3.2 Floquet modes for symmetric periods . . . 274
5.3.3 Concatenation of N symmetric periods . . . 280
5.3.4 Floquet modes for unsymmetric periods . . . 281
5.3.5 Some further general relations in periodic structures . . 283
5.4 NUMERICAL RESULTS FOR PERIODIC STRUCTURES IN ONE DIRECTION . . . 286
5.5 ANALYSIS OF PHOTONIC CRYSTALS . . . 291
5.5.1 Determination of band diagrams . . . 291
5.5.2 Waveguide circuits in photonic crystals . . . 297
5.5.3 Numerical results for photonic crystal circuits . . . 299

6 ANALYSIS OF COMPLEX STRUCTURES 311
6.1 LAYERS OF VARIABLE THICKNESS . . . 311
6.1.1 Introduction . . . 311
6.1.2 Matching conditions at curved interfaces . . . 312
6.2 MICROSTRIP SHARP BEND . . . 315
6.3 IMPEDANCE TRANSFORMATION AT DISCONTINUITIES 318
6.3.1 Impedance transformation at concatenated junctions . . 318
6.4 ANALYSIS OF PLANAR WAVEGUIDE JUNCTIONS . . . 320
6.4.1 Main diagonal submatrices . . . 322
6.4.2 Off-diagonal submatrices – coupling to perpendicular ports . . . 323
6.5 NUMERICAL RESULTS . . . 327
6.5.1 Discontinuities in microstrips . . . 328
6.5.2 Waveguide junctions . . . 333

7 PRECISE RESOLUTION WITH AN ENHANCED AND GENERALISED LINE ALGORITHM 345
7.1 INTRODUCTION . . . 345
7.2 CROSSED DISCRETISATION LINES AND CARTESIAN COORDINATES . . . 346
7.2.1 Theoretical background . . . 346

7.2.2 Lines in vertical direction 351
7.2.3 Lines in horizontal direction 357
7.3 SPECIAL STRUCTURES IN CARTESIAN COORDINATES . 361
7.3.1 Groove guide 361
7.3.2 Coplanar waveguide 363
7.4 CROSSED DISCRETISATION LINES AND CYLINDRICAL COORDINATES 366
7.4.1 Principle of analysis 366
7.4.2 General formulas for eigenmode calculation 366
7.4.3 Discretisation lines in radial direction 367
7.4.4 Discretisation lines in azimuthal direction 368
7.4.5 Coupling to neighbouring ports 369
7.4.6 Steps of the analysis procedure 373
7.5 NUMERICAL RESULTS 373

8 WAVEGUIDE STRUCTURES WITH MATERIALS OF GENERAL ANISOTROPY IN ARBITRARY ORTHOGONAL COORDINATE SYSTEMS 377
8.1 GENERALISED TRANSMISSION LINE EQUATIONS 377
8.1.1 Material properties 377
8.1.2 Maxwell's equations in matrix notation 377
8.1.3 Generalised transmission line equations in Cartesian coordinates for general anisotropic material 379
8.1.4 Generalised transmission line equations for general anisotropic material in arbitrary orthogonal coordinates 381
8.1.5 Boundary conditions 383
8.1.6 Interpolation matrices 384
8.2 DISCRETISATION 385
8.2.1 Two-dimensional discretisation 385
8.2.2 One-dimensional discretisation 386
8.3 SOLUTION OF THE DIFFERENTIAL EQUATIONS 388
8.3.1 General solution 388
8.3.2 Field relation between interfaces A and B 389
8.4 ANALYSIS OF WAVEGUIDE JUNCTIONS AND SHARP BENDS WITH GENERAL ANISOTROPIC MATERIAL BY USING ORTHOGONAL PROPAGATING WAVES 389
8.4.1 Introduction 389
8.4.2 Theory 389
8.4.3 Main diagonal submatrices 391
8.4.4 Off-diagonal submatrices – coupling to other ports . . . 393
8.4.5 Steps of the analysis procedure 398
8.5 NUMERICAL RESULTS 398
8.6 ANALYSIS OF WAVEGUIDE STRUCTURES IN SPHERICAL COORDINATES 399

8.6.1 Introduction . . . 399
8.6.2 Generalised transmission line equations in spherical coordinates . . . 400
8.6.3 Analysis of special devices – conformal antennas . . . 408
8.6.4 Analysis of special devices – conical horn antennas . . . 413
8.6.5 Numerical results . . . 419
8.7 ELLIPTICAL COORDINATES . . . 420
8.7.1 GTL equations for z-direction . . . 421
8.7.2 GTL equations for ξ-direction . . . 422
8.7.3 GTL equations for η-direction . . . 423
8.7.4 Hollow waveguides with elliptic cross-section . . . 424

9 SUMMARY AND PROSPECT FOR THE FUTURE 429

A DISCRETISATION SCHEMES AND DIFFERENCE OPERATORS 433
A.1 DETERMINATION OF THE EIGENVALUES AND EIGENVECTORS OF P . . . 433
A.1.1 Calculation of the matrices δ . . . 436
A.1.2 Derivation of the eigenvalues of the Neumann problem from those of the Dirichlet problem . . . 438
A.1.3 The component of ε_r at an abrupt transition . . . 439
A.1.4 Eigenvalues and eigenvectors for periodic boundary conditions . . . 441
A.1.5 Discretisation for non-ideal places of the boundaries . . . 442
A.2 ABSORBING BOUNDARY CONDITIONS (ABCs) . . . 444
A.2.1 Introduction[1] . . . 444
A.2.2 Factorisation of the Helmholtz equation . . . 445
A.2.3 Padé approximation . . . 446
A.2.4 Polynomial approximations . . . 447
A.2.5 Construction of the difference operator for ABCs . . . 449
A.2.6 Special boundary conditions (SBCs) . . . 450
A.2.7 Numerical results . . . 450
A.2.8 ABCs for cylindrical coordinates . . . 453
A.2.9 Periodic boundary conditions . . . 455
A.3 HIGHER-ORDER DIFFERENCE OPERATORS [11] . . . 456
A.3.1 Introduction[3] . . . 456
A.3.2 Theory . . . 457
A.3.3 Numerical results . . . 459
A.4 NON-EQUIDISTANT DISCRETISATION . . . 460
A.4.1 Introduction . . . 460
A.4.2 Theory . . . 460
A.4.3 Interpolation . . . 464
A.4.4 Numerical results . . . 466

A.5 REFLECTIONS IN DISCRETISATION GRIDS 468
A.5.1 Introduction 468
A.5.2 Dispersion relations 468
A.5.3 Reflections at discretisation transitions 471
A.6 FIELD EXTRAPOLATION FOR NEUMANN BOUNDARY CONDITIONS 475
A.7 ABOUT THE NATURE OF THE METHOD OF LINES . . . 476
A.7.1 Introduction 476
A.7.2 Relation between shielded structures and periodic ones . 477
A.7.3 Method of Lines and discrete Fourier transformation . . 478
A.7.4 Discussion 479
A.8 RELATION BETWEEN THE MODE MATCHING METHOD (MMM) AND THE METHOD OF LINES (MoL) FOR INHOMOGENEOUS MEDIA 480
A.9 RECIPROCITY AND ITS CONSEQUENCES 483

B TRANSMISSION LINE EQUATIONS 491
B.1 TRANSMISSION LINE EQUATIONS IN FIELD VECTOR NOTATION 491
B.2 DERIVATION OF THE MULTICONDUCTOR TRANSMISSION LINE EQUATIONS 492

C SCATTERING PARAMETERS 497

D EQUIVALENT CIRCUITS FOR DISCONTINUITIES 499

E APPROXIMATE METALLIC LOSS CALCULATION IN CONFORMAL STRUCTURES 501

Index 503

Preface

This book was written as a contribution to the efficient use of microwave and optical wave technology for a better life and to protect the peace. The author invites all researchers to use the method presented here for these purposes. The author hopes that every user acts in responsibility to God, the creator of our world, and believes that God blesses those activities which do not destroy the Earth and are suitable for helping communication and connecting people.

This book is intended to serve as a reference to practicing engineers, as well as to graduate students in electrical engineering, applied physics and mathematics. The main requirements for the readers of this text are basic knowledge of electromagnetic field theory and wave propagation, of partial differential equations, of linear algebra and of computer languages, like Matlab. All formulas and diagrams were carefully developed. Nevertheless, errors can never be ruled out completely. Therefore, the reader should check the equations before he applies them to his particular problem.

The author would like to thank all his coworkers and PhD students, who cannot all be mentioned because of the long list. They made important contributions to the procedures described in this book, e.g. by verifying the algorithms and applying them to real world problems. Their names are found in the various reference lists. The figures and diagrams in electronic form were made by Georg Schindel. This work is gratefully acknowledged. A special acknowledgement is addressed to Stefan Helfert, for helpful discussions when writing the book and his careful revision of the text.

Finally, I express my deepest gratitude to my wife Renate and our children Anne-Ruth (with fondest memory), Matthias, Kirsten and Corinna, for their patience, sacrifices and prayers.

Acknowledgments
The work on which this book is based has been supported by several grants of the Deutsche Forschungsgemeinschaft and one of the Stiftung Volkswagenwerk.

CHAPTER 1

THE METHOD OF LINES

1.1 INTRODUCTION

The Method of Lines (MoL) is one of the most efficient numerical methods for solving the partial differential equations with which we describe physical phenomena. It has been applied to various problems in theoretical physics [1]. The MoL was developed by mathematicians and the basic principles are described in mathematical books, e.g. in [2,3]. The idea was first applied by the German mathematician Erich Rothe [4] in 1930 to equations of the parabolic type. But it is quite clear that the MoL can be used much more widely. A review from the mathematical point of view is given by O.A. Liskovets [1]. Books which describe the principle of the MoL for use with electromagnetic fields are e.g. [5–8]. The method of lines has certain similarities with the Mode Matching Technique (MMT) and with the Finite Difference Method (FDM). It differs from the latter in the fact that, for a given system of partial differential equations, all but one of the independent variables are discretised to obtain a system of ordinary differential equations. This semi-analytical procedure saves a lot of computing time, because the solution in one coordinate direction is obtained analytically. The method of lines has outstanding properties. It has been shown that:

- The convergence behaviour is monotonic [5]. Hence extrapolation is possible and gives accurate results. This is because the MoL has stationary behaviour.
- The calculation of the field distribution is very accurate because of the relation of the MoL to the Discrete Fourier Transformation (DFT) [5].
- The method of lines does not yield spurious modes, which are known e.g. in the finite element method.
- The relative convergence phenomenon [9] (pp. 603ff), as it sometimes occurs in the Mode Matching Technique as a consequence of the Fourier series truncations, is not observed.

This book describes – based on the author's own and his co-workers' work – how the MoL can be used for the analysis of a wide range of electromagnetic field problems, e.g. of planar and quasi-planar waveguide structures for integrated microwave and optic circuits, rectangular waveguide circuits, circular and conformal antennas, fibre structures and many other problems.

Analysis of Electromagnetic Fields and Waves R. Pregla

This class of waveguide structures can be analysed accurately and nevertheless in an easy way.

However, in spite of the size of this book, there are other problems that can be solved with the MoL which we do not address. We would at least mention some of them here. The modelling of VCSELs (vertical cavity surface emitting lasers) including the electrical and thermal model was described in [10–12]. Particularly, the solution of the heat equation with the method of lines was shown in these papers. The analysis of electrostatic problems was presented in [13]. Non-linear material was taken into account, e.g. in [14]. A BPM (beam propagation method) based on the MoL was given in [7,15]. As will be shown in detail later, we deal in this book with generalised transmission line equations (GTL). These GTL expressions can also be used to develop finite difference BPMs as shown in [16–18].

Now, all these problems deal in the frequency domain (we took the steady state in the electrostatic resp. temperature problems). However, the MoL has also been applied to problems in the time domain [19–23].

Let us begin with the quasi-planar waveguide with constant cross-sections, as shown in Fig. 1.1. It has the following features: The number of substrate layers S_i may be arbitrary, the metallisations M_i can be included in various planes, the boundaries B_i consist of electric or magnetic walls, and the lower and/or upper boundaries B_3 and B_4 respectively may be extended to infinity. The waveguide cross-section in Fig. 1.1 includes common microstrips, microslots and finlines.

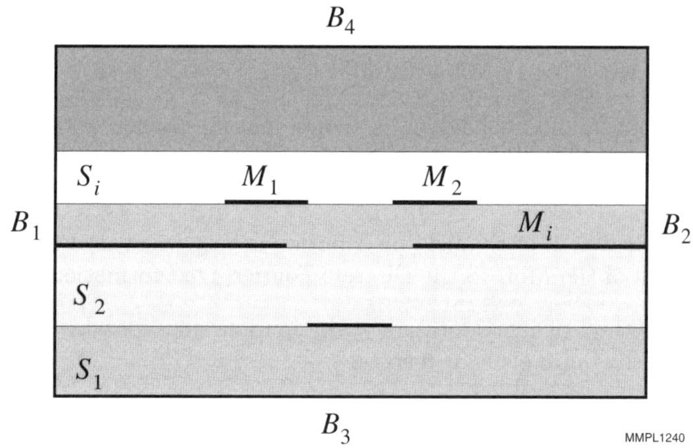

Fig. 1.1 Cross-section of quasi-planar waveguide B_i = boundaries, M_i = metallisations, S_i = substrate layers

Optimum convergence is always assured if the strip edges are located at the right place between the discretisation lines [24]. It should be noted, however, that the convergence of the propagation constant, the characteristic impedance or the resonant frequency does not critically depend on the so-called edge parameters (if the discretisation lines are not positioned on edges). Therefore, the problem of convergence on the whole is not critical. In monolithic integrated circuits the metallisation thickness cannot be neglected compared to the conductor width or slot width. Thus a consideration of the finite metallisation thickness is also presented in this book.

Circuits in integrated optics are basically realised in the same way as in integrated microwave techniques. Instead of the metallisation, dielectric strips with higher dielectric constant are used to bound the light (Fig. 1.2).

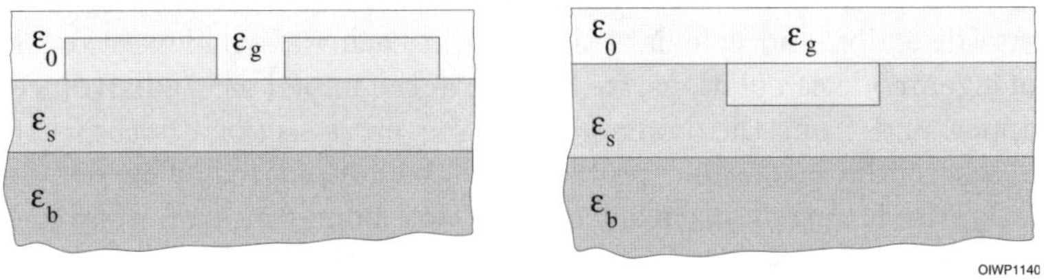

Fig. 1.2 Cross-sections of dielectric waveguides (models for waveguides in the integrated optics)

We will describe how inhomogeneous layers with abrupt transitions in the dielectric constant can be analysed with the MoL. All material parameters can be complex, i.e. losses can be considered. Furthermore, general anisotropic material can also be included.

Complex planar circuits consist of concatenations of various waveguide sections of different cross-sections. Some simple examples are shown in Fig. 1.3.

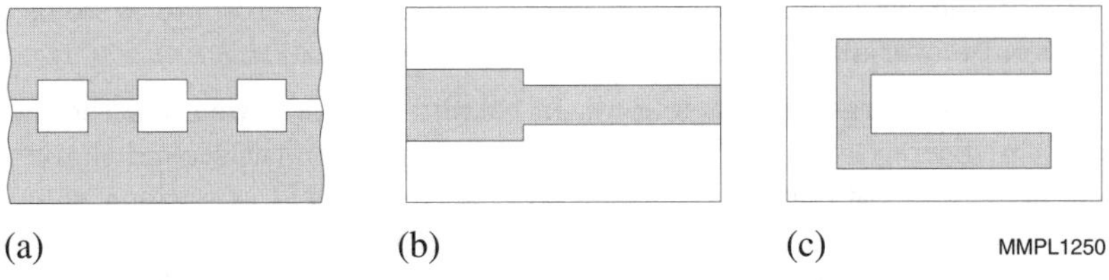

Fig. 1.3 Outlines of longitudinally inhomogeneous waveguides: Examples for (a) periodic structures (b) discontinuities (c) resonators

The MoL, as a special FDM, enables analytic calculation in a specific direction. In this direction, the structures to be analysed can consist of many layers or concatenated sections. To assure numerical stability, we will show that the analysis should be performed by impedance and/or admittance transformation.

The algorithm that we use can be understood as generalised transmission line theory. Therefore, as in transmission line theory we start at the end of the device and transform (analytically) the load impedance/admittance through the sections and between the sections to the input port of the input waveguide. For cases where analytic expressions cannot be given, we have developed algorithms with finite differences. Instead of impedances/admittances we could also use scattering parameters (particularly the reflection coefficient), as shown e.g. in [25]. As is known from transmission line theory (and used in the Smith chart), impedances/admittances and reflection coefficients can be transformed into each other. Therefore, the procedures are equivalent, though different formulas are applied.

If we know the input impedance and the source wave we can go in the opposite direction and calculate the field in each waveguide section. In the case of eigenmode calculations, we transform the impedance/admittance from the upper and from the lower side of the cross-section to the matching plane, where the eigenvalue problem will be formulated. The relations of the discretised tangential fields at the two boundaries of a layer or at the two ports of a section will be written in the form of open circuited or $\boldsymbol{z}$-matrix parameters, or of short circuited or $\boldsymbol{y}$-matrix parameters, as known in scalar form from circuit theory. From these relations, the required impedance/admittance transformation formulas are obtained. The formulas for the impedance/admittance transformation at the (e.g. metallised) interfaces between two layers or concatenation of two different sections have to be developed separately. Here we must take into account the fact that the transverse electric and magnetic field components must be continuous.

The discretisation lines in the MoL can (at least formally) be understood as 'real' lines in multi-conductor transmission line theory. Therefore, the fundamental equations for analysis can be given as generalised transmission line (GTL) equations. These GTL equations are obtained from the Maxwell's equations and can be written in general orthogonal coordinate systems and for general material tensors.

This book is organised as follows. In this first chapter we show the principles of the discretisation and of the analytic solution. In the second chapter we show the analysis of planar structures. We derive the GTL equations, show the solutions and describe particularly the impedance/admittance transformation to obtain stable results. In the following chapters we deal with extensions or special structures. In Chapter 3 we describe rectangular waveguides and how they can be modelled. For an efficient analysis, it is best to use suitable coordinates. Therefore, we introduce cylindrical coordinates in Chapter 4. Of particular interest are periodic structures, which are e.g. used in filters. To study such devices, Floquet's theorem is introduced, as we will show in Chapter 5. Planar structures whose analysis requires an extension of the algorithms presented in the second chapter are presented in Chapter 6. Even more complex devices are given in the seventh chapter, which deals with crossed discretisation lines. Finally, the most general case is described in Chapter 8. Here we introduce arbitrary orthogonal coordinate systems and an arbitrary anisotropy. As special cases, we describe spherical coordinates and elliptical ones in Chapter 8 as well. All chapters give substantial numerical results to show that the algorithms could be used successfully to examine a large variety of devices. The book finishes with several appendices, where some basic algorithms and important properties of the MoL are described.

As mentioned before, in this book we show the MoL as a special transmission line algorithm and want to give a uniform representation. There are of course many papers by other authors that deal with the MoL. Therefore, the reference lists are by no means complete and we want to apologise for being unable to include all of them.

1.2 MOL: FUNDAMENTALS OF DISCRETISATION

1.2.1 Qualitative description

In this chapter we will describe the basic principles of electromagnetic field analysis with the method of lines. We will use a relatively simple waveguide structure to demonstrate how we determine the eigenmodes. For most of the structures that are used in integrated microwave techniques or integrated optics, the electromagnetic fields in waveguide structures can be computed from two independent field components or two potentials. The exception is materials with full anisotropy, for which we need four field components.

Fig. 1.4 shows a simple waveguide structure, with discretisation lines in the z-direction. In most of the cases described in this book, we are going to use the transverse field components (i.e. x and y) for the analysis. In the following chapters we will develop generalised transmission line equations for this purpose. In this section, however, we will remain general and show that the longitudinal components (z-components) could also be used.

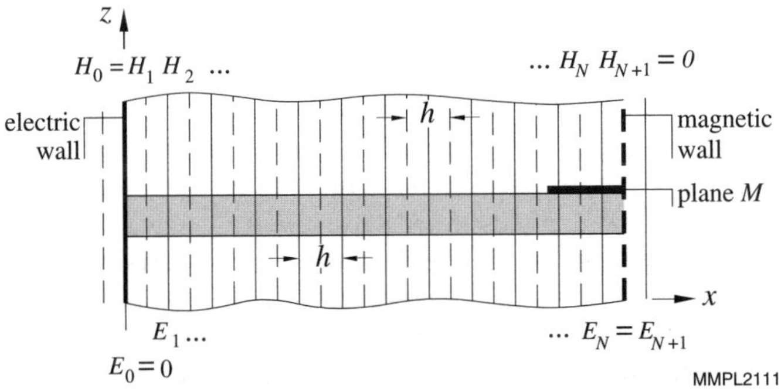

Fig. 1.4 Planar waveguide cross-section (only substrate layer and metal strip) with discretisation lines — lines for E_z, E_y, $\tilde{H}_x$ - - - lines for $\tilde{H}_z$, $\tilde{H}_y$, E_x

We start with suitable components of the electric and/or magnetic fields. Throughout the book we assume a time dependence according to exp (jωt) of the fields. This time dependence will be omitted. From Maxwell's equations we find that in homogeneous sections all components of the electric and magnetic field must fulfil the Helmholtz equation

$$\frac{\partial^2 F}{\partial \overline{x}^2} + \frac{\partial^2 F}{\partial \overline{y}^2} + \frac{\partial^2 F}{\partial \overline{z}^2} + \varepsilon_r F = LF = 0 \qquad F = E_{x,y,z}, H_{x,y,z} \tag{1.1}$$

in each separate homogeneous layer with the individual permittivity ε_r. For F we have to substitute the suitable component. We normalise the coordinates $u = x, y, z$ with the free space wave number $k_0 = \omega\sqrt{\mu_0\varepsilon_0}$ according to $\overline{u} = k_0 u$, and the magnetic field components H_u by the wave impedance $\eta_0 = \sqrt{\mu_0/\varepsilon_0}$ according to $\widetilde{H}_u = \eta_0 H_u$. Moreover, the components tangential to the boundary of $[E]$ (i.e. E_z and E_y) and $[H]$ (i.e. H_z and H_y) must fulfil the following boundary conditions (BCs):

$$\begin{aligned} &\text{electric wall:} \quad E_t = 0 \quad (\text{DC})\,; \quad \frac{\partial \widetilde{H}_t}{\partial n} = 0 \quad (\text{NC}) \\ &\text{magnetic wall:} \quad \widetilde{H}_t = 0 \quad (\text{DC})\,; \quad \frac{\partial E_t}{\partial n} = 0 \quad (\text{NC}) \end{aligned} \tag{1.2}$$

For the components normal to the boundary (here the x-components) we have:

$$\begin{aligned} &\text{electric wall:} \quad \widetilde{H}_n = 0 \quad (\text{DC})\,; \quad \frac{\partial E_n}{\partial n} = 0 \quad (\text{NC}) \\ &\text{magnetic wall:} \quad E_n = 0 \quad (\text{DC})\,; \quad \frac{\partial \widetilde{H}_n}{\partial n} = 0 \quad (\text{NC}) \end{aligned} \tag{1.3}$$

The abbreviations DC and NC stand for Dirichlet and Neumann condition, respectively. The terms on the right $(\partial/\partial n)$ stand for the derivatives in the direction normal to the wall. Here we see the independence of the two components in their dual boundary conditions.

One approach to solving the partial differential eq. (1.1) is to approximate the functions f in a suitable way. This is done in the Mode Matching Technique (MMT) or in the Method of Moments (MoM) in all its variations. Another possibility is to approximate the differential operator L. In the Finite Difference Method the derivatives are substituted by the difference quotient. In the Method of Lines the differential operator is partially substituted by differences, but only as far as is necessary, namely to convert the partial differential equation into a system of ordinary ones. For the determination of the eigenmodes in a waveguide propagating in y-direction with the propagator $\exp(-\mathrm{j}k_y y)$ (Fig. 1.4) we only have to discretise in one direction.

Considering the structure in Fig. 1.4, it becomes clear that the discretisation should be done in the direction parallel to the interfaces of the layers (i.e. in x-direction). The individual layers are homogeneous in y- and, in this chapter, also in x-direction. Generally, the layers may be inhomogeneous in x-direction. This will be described later.

The discretisation of the operator L has the consequence that the fields are considered on straight lines, which are perpendicular to the interfaces of the layers. Due to symmetry, only half of the cross-section has to be considered. In this case a symmetry (i.e. magnetic or electric) wall has to be inserted in the middle of the structure. Fig. 1.4 shows that two separate line systems are used, e.g. for E_z and H_z. There are several reasons for this. First of all, the lateral boundary conditions are immediately fulfilled if the lines are in the right position with respect to the boundaries. In order to fulfil the Dirichlet condition, it is best to put a line on the lateral boundary and to set the corresponding field component to zero. In the subsequent calculation it is not necessary to carry along this component. The Neumann condition is easily satisfied by including the boundary between two lines and enforcing the condition that the components on these two lines are equal to each other.

The shifting of the two line systems has more advantages:

- it allows an optimal edge positioning
- it reduces the discretisation error
- it results in an easy quantitative description.

These advantages will be individually illustrated at convenient points.

1.2.2 Quantitative description of the discretisation

Let the number of E_z and H_z lines in the cross section of Fig. 1.4 be equal to N. The field components E_z and H_z on these lines are combined to column vectors $\mathbf{E}_z$ and $\mathbf{H}_z$ respectively. E_{zi} is the component of the discretised electric field vector on the ith line. Of course, it is a function of y.

In the analysis, we will usually also need other field components. We can e.g. determine E_x and $\widetilde{H}_x$ from the z-components as:

$$\left(\varepsilon_r + \frac{\partial^2}{\partial \overline{z}^2}\right) \begin{bmatrix} E_x \\ \widetilde{H}_x \end{bmatrix} = \begin{bmatrix} \dfrac{\partial^2}{\partial \overline{x} \partial \overline{z}} & -\mathrm{j}\dfrac{\partial}{\partial \overline{y}} \\ \mathrm{j}\varepsilon_r \dfrac{\partial}{\partial \overline{y}} & \dfrac{\partial^2}{\partial \overline{x} \partial \overline{z}} \end{bmatrix} \begin{bmatrix} E_z \\ \widetilde{H}_z \end{bmatrix} \tag{1.4}$$

As previously described, we discretise eqs. (1.1) and (1.4) in x-direction with finite differences. Eq. (1.4) shows that the derivative of E_z with respect to x is needed on an H_z line and that of H_z is needed on an E_z line. This request can be fulfilled by using two shifted discretisation line systems, as shown in Fig. 1.4.

We will show that the approximation has second-order accuracy for this position. Therefore, the derivative obtained by the finite difference is determined exactly in the middle of the lines. For the vector of the discretised field components, we may write in this case:

$$\overline{h}\frac{\partial E_z}{\partial \overline{x}} \longrightarrow \boldsymbol{D}_e\mathbf{E}_z = \boldsymbol{D}_h^{\mathrm{t}}\mathbf{E}_z \qquad \overline{\boldsymbol{D}}_e = \overline{h}^{-1}\boldsymbol{D}_e \tag{1.5}$$

$$\overline{h}\frac{\partial H_z}{\partial \overline{x}} \longrightarrow \boldsymbol{D}_h\mathbf{H}_z = -\boldsymbol{D}_e^{\mathrm{t}}\mathbf{H}_z \qquad \overline{\boldsymbol{D}}_h = \overline{h}^{-1}\boldsymbol{D}_h \tag{1.6}$$

Note: the subscripts e and h correspond to the tangential field components, which are here the z- and y-components (see Fig. 1.4). For other structures, however, differing components might be the tangential ones.

Now, the difference operator for the component E_z with the boundaries in Fig. 1.4 has the following form:

$$\overline{h}\frac{\partial E_z}{\partial \overline{x}} \longrightarrow \underbrace{\begin{array}{c|cccc|c} E_0 & E_1 & \ldots & \ldots & E_{\mathrm{N}} & E_{\mathrm{N}+1} \\ -1 & 1 & & & & \\ & -1 & \ddots & & & \\ & & \ddots & \ddots & & \\ & & & -1 & 1 & \\ \hline & & & & -1 & 1 \end{array}}_{\boldsymbol{D}_e} \tag{1.7}$$

E_0 and $E_{\mathrm{N}+1}$ are positioned outside the computational window. With $E_0 = 0$ at the electric wall on the left side (Dirichlet boundary condition) in Fig. 1.4, the left upper -1 is omitted. Due to the magnetic wall on the right side (Neumann BC), the last row (below the horizontal line) has to be withdrawn. The construction of $\boldsymbol{D}_h$ is analogous, with the conditions exchanged, i.e. Neumann (Dirichlet) BC on the left (right) side. So, we finally obtain:

$$\boldsymbol{D}_e = \begin{bmatrix} 1 & & & \\ -1 & \ddots & & \\ & \ddots & \ddots & \\ & & -1 & 1 \end{bmatrix} \qquad \boldsymbol{D}_h = \begin{bmatrix} -1 & 1 & & \\ & \ddots & \ddots & \\ & & \ddots & 1 \\ & & & -1 \end{bmatrix} \tag{1.8}$$

From these examples it is obvious how difference operators for Dirichlet–Dirichlet or Neumann–Neumann BCs are constructed. The eqs. (1.5) and (1.6) make clear that e.g. the difference operator $\boldsymbol{D}_e$ (that is defined for the representation of E_z in eq. (1.5)) can also be used to approximate the derivative of H_z. This can also be seen in eq. (1.8) and is a further consequence of shifting the line systems. In the difference operator $\boldsymbol{D}$, the lateral boundary conditions are included. Moreover, the second derivatives can also be represented by means of the difference operator $\boldsymbol{D}$. There is:

$$\overline{h}^2 \frac{\partial^2 E_z}{\partial \overline{x}^2} = \overline{h}^2 \frac{\partial}{\partial \overline{x}} \left(\frac{\partial E_z}{\partial \overline{x}} \right) \longrightarrow -\boldsymbol{D}_e^{\mathrm{t}} \boldsymbol{D}_e \mathbf{E}_z = -\boldsymbol{D}_h \boldsymbol{D}_h^{\mathrm{t}} \mathbf{E}_z = -\boldsymbol{P}_e \mathbf{E}_z \tag{1.9}$$

$$\overline{h}^2 \frac{\partial^2 H_z}{\partial \overline{x}^2} = \overline{h}^2 \frac{\partial}{\partial \overline{x}} \left(\frac{\partial H_z}{\partial \overline{x}} \right) \longrightarrow -\boldsymbol{D}_h^{\mathrm{t}} \boldsymbol{D}_h \mathbf{H}_z = -\boldsymbol{D}_e \boldsymbol{D}_e^{\mathrm{t}} \mathbf{H}_z = -\boldsymbol{P}_h \mathbf{H}_z \tag{1.10}$$

Thus the difference operator for the second derivative is obtained as a product of the difference operators for the first derivatives. This can be seen from the chain form of the second derivatives, keeping in mind that the boundary conditions for the outer derivative are dual to those for the inner derivative. The difference operator $\boldsymbol{P}$ can be written in the following way:

$$\boldsymbol{P} = \begin{bmatrix} p_l & -1 & & & \\ -1 & 2 & \ddots & & \\ & \ddots & \ddots & \ddots & \\ & & \ddots & 2 & -1 \\ & & & -1 & p_r \end{bmatrix} \tag{1.11}$$

Therein we have $p_{l,r} = 2$ for the Dirichlet condition on the left resp. right wall and $p_{l,r} = 1$ for the Neumann condition. In our example we chose $\boldsymbol{P}_e = \boldsymbol{P}_{\mathrm{DN}}$ and $\boldsymbol{P}_h = \boldsymbol{P}_{\mathrm{ND}}$.

In the examples so far, only Dirichlet or Neumann boundary conditions have been used. However, with these BCs it is not possible to model radiation. For this we have to introduce absorbing boundary conditions (ABCs), and sometimes periodic boundary conditions (e.g. when examining photonic crystals). All these further BCs will be described in detail in Appendix B.

Now we substitute eq. (1.9) or (1.10) in eq. (1.1). Further, we assume a wave propagating in y-direction according to $\exp(-\mathrm{j}\boldsymbol{k}_y y)$. Then we obtain the following ordinary differential equation system:

$$\frac{\mathrm{d}^2}{\mathrm{d}\overline{y}^2} \mathbf{F}_z + ((\varepsilon_r - \varepsilon_{re})\boldsymbol{I} - \overline{\boldsymbol{P}})\mathbf{F}_z = \mathbf{0} \qquad \overline{\boldsymbol{P}} = \overline{h}^{-2} \boldsymbol{P} = \overline{\boldsymbol{D}}^{\mathrm{t}} \overline{\boldsymbol{D}} \tag{1.12}$$

$\mathbf{F}_z$ is either $\mathbf{E}_z$ or $\mathbf{H}_z$ and $\boldsymbol{P}$ is the corresponding difference matrix. $\boldsymbol{I}$ is the unit matrix in this equation and in the following. We introduced the effective permittivity ε_{re} according to $\varepsilon_{re} = k_y^2/k_0^2$. In eq. (1.12) the fields on three lines are coupled with each other because of the tridiagonal structure of $\boldsymbol{P}$. To decouple these equations, we make a transformation to principal axes. With

$$\mathbf{E}_z = \boldsymbol{T}_e \overline{\mathbf{E}}_z \qquad \widetilde{\mathbf{H}}_z = \boldsymbol{T}_h \overline{\mathbf{H}}_z \tag{1.13}$$

we require that

$$\boldsymbol{T}_e^{\mathrm{t}} \boldsymbol{P}_e \boldsymbol{T}_e = \boldsymbol{\lambda}_e^2 \qquad \boldsymbol{T}_h^{\mathrm{t}} \boldsymbol{P}_h \boldsymbol{T}_h = \boldsymbol{\lambda}_h^2 \tag{1.14}$$

where $\boldsymbol{\lambda}_{e,h}^2$ is a diagonal matrix. $\boldsymbol{\lambda}_{e,h}^2$ is the eigenvalue matrix and $\boldsymbol{T}_{e,h}$ the eigenvector matrix of $\boldsymbol{P}_{e,h}$. In the example above (with the chosen BCs) we have $\boldsymbol{\lambda}_e^2 = \boldsymbol{\lambda}_h^2$. The calculation of these matrices is given in Appendix A for various boundary conditions. As $\boldsymbol{P}_{e,h}$ is a symmetric matrix, $\boldsymbol{T}$ is orthogonal for a suitable normalisation of the eigenvectors:

$$\boldsymbol{T}_{e,h}^{-1} = \boldsymbol{T}_{e,h}^{\mathrm{t}} \tag{1.15}$$

With the abbreviation $\varepsilon_d = \varepsilon_r - \varepsilon_{re}$ we get from eq. (1.12) considering eqs. (1.13) and (1.14):

$$\left(\left(\frac{\mathrm{d}^2}{\mathrm{d}\overline{y}^2} + \varepsilon_d\right)\boldsymbol{I} - \overline{\boldsymbol{\lambda}}^2\right)\overline{\mathbf{F}}_z = \mathbf{0} \qquad \overline{\boldsymbol{\lambda}}^2 = \overline{h}^{-2}\boldsymbol{\lambda}^2 \tag{1.16}$$

By introducing

$$\boldsymbol{\Gamma}_y^2 = \overline{\boldsymbol{\lambda}}^2 - \boldsymbol{\epsilon}_d \qquad \boldsymbol{\epsilon}_d = \varepsilon_d \boldsymbol{I} \tag{1.17}$$

we obtain the general solution for $\overline{\mathbf{F}}_z$

$$\overline{\mathbf{F}}_z = \cosh(\boldsymbol{\Gamma}_y \overline{y})\mathbf{A} + \sinh(\boldsymbol{\Gamma}_y \overline{y})\mathbf{B} \tag{1.18}$$

or in an other form

$$\overline{\mathbf{F}}_z = e^{-\boldsymbol{\Gamma}_y \overline{y}}\overline{\mathbf{F}}_{\mathrm{f}} + e^{\boldsymbol{\Gamma}_y \overline{y}}\overline{\mathbf{F}}_{\mathrm{b}} \tag{1.19}$$

In most cases the components and their derivatives are only needed on the interfaces of the layers. Therefore, we can also give the solution for an arbitrary layer with thickness d (see Fig. 1.5) in the following form:

$$\frac{\mathrm{d}}{\mathrm{d}\overline{y}}\begin{bmatrix}\overline{\mathbf{F}}_{z|y=y_1}\\ \overline{\mathbf{F}}_{z|y=y_2}\end{bmatrix} = \begin{bmatrix}\boldsymbol{\gamma} & \boldsymbol{\alpha}\\ \boldsymbol{\alpha} & \boldsymbol{\gamma}\end{bmatrix}\begin{bmatrix}-\overline{\mathbf{F}}_z(y_1)\\ \overline{\mathbf{F}}_z(y_2)\end{bmatrix} \tag{1.20}$$

where $(\overline{d} = k_0 d)$

$$\boldsymbol{\alpha} = \boldsymbol{\Gamma}_y(\sinh(\boldsymbol{\Gamma}_y \overline{d}))^{-1} \qquad \boldsymbol{\gamma} = \boldsymbol{\Gamma}_y(\tanh(\boldsymbol{\Gamma}_y \overline{d}))^{-1} \tag{1.21}$$

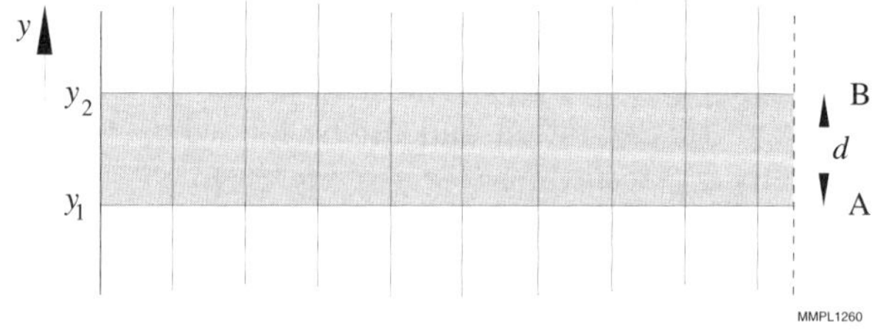

Fig. 1.5 Notation of the interfaces of a layer for the calculation of the field components and their derivatives

1.2.3 Numerical example

As a first simple example, we determine the cut-off wavelength of a rectangular waveguide of width a. In Chapter 3 we will deal with these waveguides in detail. For the H_{n0} (or TE_{n0}) modes, we obtain this wavelength from eq. (1.17) by considering $\boldsymbol{\Gamma}_y^2 = \boldsymbol{0}$ and determining the wavelength for which the condition $\varepsilon_{re} = 0$ holds. From

$$\bar{\lambda}^2 = \varepsilon_r = 1 \quad \bar{h} = \frac{a}{N+1}\frac{2\pi}{\lambda_{cn}}$$

and eqs. (A.4) (A.7) we obtain:

$$\frac{\lambda_{cn}}{a} = \frac{\pi}{(N+1)\sin\dfrac{n\pi}{2(N+1)}}$$

The results for the first two modes ($n = 1$ and $n = 2$) are shown in Fig. 1.6. As is well known, the exact (normalised) values are 2 (H_{10}) and 1 (H_{20}), respectively. As can be seen, both curves approach the analytic value with an increasing number of discretisation lines N for E_z. It should also be stated that the value on the very right could be obtained with only one of these lines. The eigenvectors give the exact values of the eigenmodes in the discretisation points.

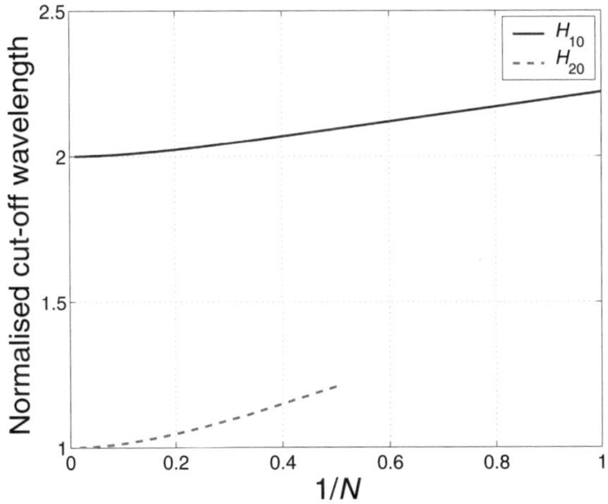

Fig. 1.6 Normalised cut-off wavelength of the H_{10} (or TE_{10}) and H_{20} (or TE_{20}) mode in a rectangular waveguide

References

[1] O. A. Liskovets, 'The Method of Lines, Review', *Differential'nye Uravneniya*, vol. 1, no. 12, pp. 1662–1678, 1965.

[2] W. E. Schiesser, *The Numerical Method of Lines Integration of Partial Differential Equations*, Academic Press, San Diego, USA, 1991.

[3] S. G. Michlin and C. Smolizki, *Approximate Methods for Solution of Differential and Integral Equations*, Teubner, Leipzig, Germany, 1969.

[4] E. Rothe, 'Zweidimensionale parabolische Randwertaufgaben als Grenzfall eindimensionaler Randwertaufgaben', *Math. Anna.*, vol. 102, pp. 650–670, 1930.

[5] R. Pregla and W. Pascher, 'The Method of Lines', in *Numerical Techniques for Microwave and Millimeter Wave Passive Structures*, T. Itoh (Ed.), pp. 381–446. J. Wiley Publ., New York, USA, 1989.

[6] M. N. O. Sadiku, *Numerical Techniques in Electromagnetics*, CRC Press, Boca Raton, London, New York, Washington D.C., 2 edition, 2001.

[7] R. Pregla, 'MoL-BPM Method of Lines Based Beam Propagation Method', in *Methods for Modeling and Simulation of Guided-Wave Optoelectronic Devices (PIER 11)*, W. P. Huang (Ed.), Progress in Electromagnetic Research, pp. 51–102. EMW Publishing, Cambridge, Massachusetts, USA, 1995.

[8] R. Pregla and S. F. Helfert, 'The Method of Lines for the Analysis of Photonic Bandgap Structures', in *Electromagnetic Theory and Applications for Photonic Crystals*, Kiyotoshi Yasumoto (Ed.), pp. 295–350. CRC Press, Boca Raton Fl, London, 2006.

[9] T. Itoh (Ed.), *Numerical Techniques for Microwave and Millimeter Wave Passive Structures*, J. Wiley Publ., New York, USA, 1989.

[10] E. Ahlers, S. Helfert, and R. Pregla, 'Accurate Analysis of Vertical Cavity Surface Emitting Laser Diodes (in German)', in *Deutsche Nationale U.R.S.I Konf.*, Kleinheubach, Oct. 1995, vol. 39, pp. 135–144.

[11] E. Ahlers, S. F. Helfert and R. Pregla, 'Modelling of VCSELs by the Method of Lines', in *OSA Integr. Photo. Resear. Tech. Dig.*, Boston, USA, 1996, vol. 6, pp. 340–343.

[12] O. Conradi, S. Helfert and R. Pregla, 'Comprehensive Modeling of Vertical-Cavity Laser-Diodes by the Method of Lines', *IEEE J. Quantum Electron.*, vol. 37, pp. 928–935, 2001.

[13] R. Pregla, 'The Method of Lines for the Unified Analysis of Microstrip and Dielectric Waveguides', *Electromagnetics*, vol. 15, no. 5, pp. 441–456, 1995.

[14] M. Bertolotti, M. Masciulli and C. Sibilia, 'MoL Numerical Analysis of Nonlinear Planar Waveguide', *J. Lightwave Technol.*, vol. 12, pp. 784–789, 1994.

[15] J. Gerdes and R. Pregla, 'Beam-Propagation Algorithm Based on the Method of Lines', *J. Opt. Soc. Am. B*, vol. 8, no. 2, pp. 389–394, 1991.

[16] R. Pregla, 'Novel FD-BPM for Optical Waveguide Structures with Isotropic or Anisotropic Material', in *European Conference on Integrated Optics and Technical Exhibit*, Torino, Italy, Apr. 1999, pp. 55–58.

[17] S. F. Helfert and R. Pregla, 'A Finite Difference Beam Propagation Algorithm Based on Generalized Transmission Line Equations', *Opt. Quantum Electron.*, vol. 32, pp. 681–690, 2000, special issue on Optical Waveguide Theory and Numerical Modelling.

[18] G. Guekos (Ed.), 'Photonic Devices for Telecommunications', Springer-Verlag, Berlin, Heidelberg, 1999.

[19] S. Nam, S. El–Ghazaly, H. Ling and T. Itoh, 'Time-Domain Method of Lines', *Electron. Lett.*, vol. 24, no. 2, pp. 128–129, 1988.

[20] S. Nam, S. El-Ghazaly, H. Ling and T. Itoh, 'Time-Domain Method of Lines Applied to a Partially Filled Waveguide', in *IEEE MTT-S Int. Symp. Dig.*, 1988, vol. 2, pp. 627–630.

[21] S. Nam, H. Ling and T. Itoh, 'Time-Domain Method of Lines Applied to the Uniform Microstrip Line and its Step Discontinuity', in *IEEE MTT-S Int. Symp. Dig.*, 1989, vol. 3, pp. 997–1000.

[22] S. Nam, H. Ling and T. Itoh, 'Time-Domain Method of Lines Applied to Planar Guided Wave Structures', *IEEE Trans. Microwave Theory Tech.*, vol. 37, no. 5, pp. 897–901, 1989.

[23] S. Nam, H. Ling and T. Itoh, 'Characterization of Uniform Microstrip Line and its Discontinuities Using the Time-Domain Method of Lines', *IEEE Trans. Microwave Theory Tech.*, vol. 37, no. 12, pp. 2051–2057, 1989.

[24] U. Schulz, 'On the Edge Condition with the Method of Lines in Planar Waveguides', *AEÜ*, vol. 34, pp. 176–178, 1980.

[25] S. F. Helfert and R. Pregla, 'The Method of Lines: A Versatile Tool for the Analysis of Waveguide Structures', *Electromagnetics*, vol. 22, pp. 615–637, 2002, Invited paper for the special issue on Optical Wave Propagation in Guiding Structures.

CHAPTER 2

BASIC PRINCIPLES OF THE METHOD OF LINES

2.1 INTRODUCTION

Only in exceptional cases do microwave and millimetre-wave devices have a simple form as integrated circuits. If sections of microstrip lines are used together with coplanar lines, air bridges are necessary. This is also true for microstrip crossings that are very complex.

Various examples of such complex structures are shown in Figs. 2.1 and 2.2. Fig. 2.1 shows various filters. Because of the different propagation paths from input to output, the structure in Fig. 2.1b also allows the formation of phase or group delay characteristics. Fig. 2.2a shows a microstrip crossing. An additional dielectric block is introduced to obtain a mechanically stable circuit. Because of this block, the circuit is no longer planar but quasi-planar. Figs 2.1c and 2.2c show a coplanar line filter and a sketch of a microstrip meander line. Meander lines are particularly applied to group delay equalisation or as delay elements.

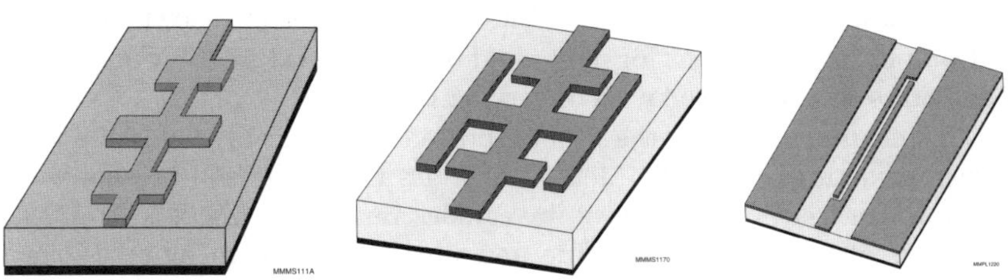

(a) (b) (c)

Fig. 2.1 Various planar microwave filter devices (Reproduced by permission of IGTE (Institut für Grundlagen und Theorie der Elektrotechnik) Graz) (a) microstrip band-stop filter (b) complex microstrip filter (Reproduced by permission of Copernicus Gesellschaft mbH) (c) coplanar line filter (Reproduced by permission of Copernicus Gesellschaft mbH)

In Fig. 2.3 a photo of a microstrip meander line is shown. In this circuit, 114 stubs are coupled with one another. The analysis of each of the circuits must be performed in an adequate way.

In the optics area, we find that planar optical devices are very long. Some examples are shown in Fig. 2.4. The cross-section of integrated circuits for

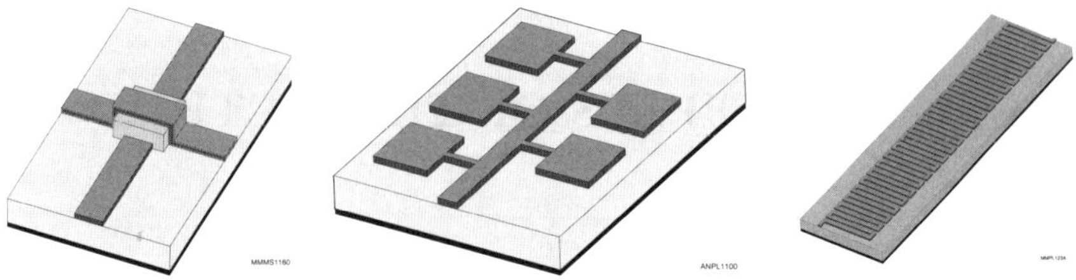

(a) (b) (c)

Fig. 2.2 Complex planar microwave and millimetre-wave devices (a) microstrip crossover (Reproduced by permission of Union Radio-Scientifique Internationale–International Union of Radio Science (URSI)) (b) microstrip patch antenna (Reproduced by permission of IGTE (Institut für Grundlagen und Theorie der Elektrotechnik) Graz) (c) microstrip meander line

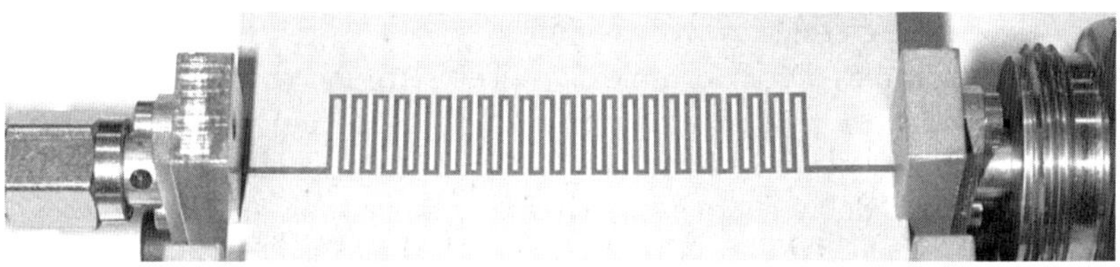

Fig. 2.3 Photo of a microstrip meander line

microwave, millimetre-wave and optical-wave frequencies consist of a number of composite layers. Examples from the microwave area are shown in Fig. 2.5.

Complex cross-sections for integrated optics are sketched in Fig. 2.6. Two of the cross-sections have layers of varying thickness.

This chapter describes the general procedure for a full vectorial analysis of such structures using the Method of Lines (MoL). It will be shown that the MoL can treat these types of waveguides very efficiently. The MoL takes advantage of the layered structure and uses discretisation only as long as necessary. In longitudinal homogeneous waveguides, discretisation is done in one of the transverse coordinate directions, and analytical solution is retained for the other one. The number of layers can be arbitrary. Also, the number of strips or slots may be arbitrary.

2.2 BASIC EQUATIONS

2.2.1 Anisotropic material parameters

In this section we will derive formulas for anisotropic structures. Simple expressions, e.g. for isotropic materials, can be easily deduced from these general ones.

If the main anisotropy axes are not identical with the coordinate axes, the tensor must be modified. We assume that x' and y' are rotated by an angle

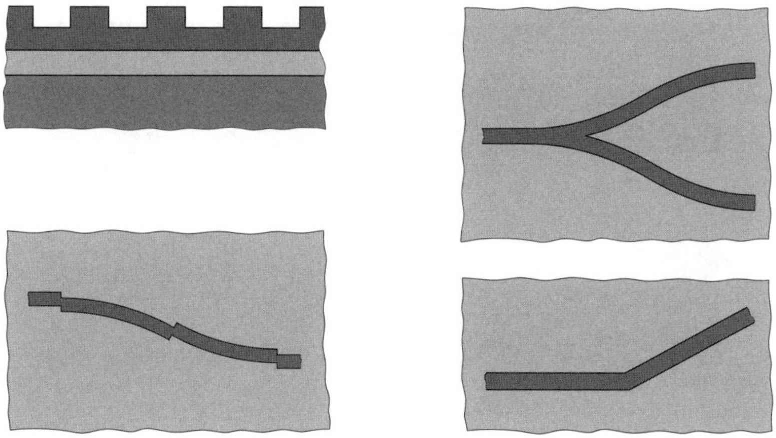

Fig. 2.4 Examples of longitudinally inhomogeneous waveguide structures in integrated optics

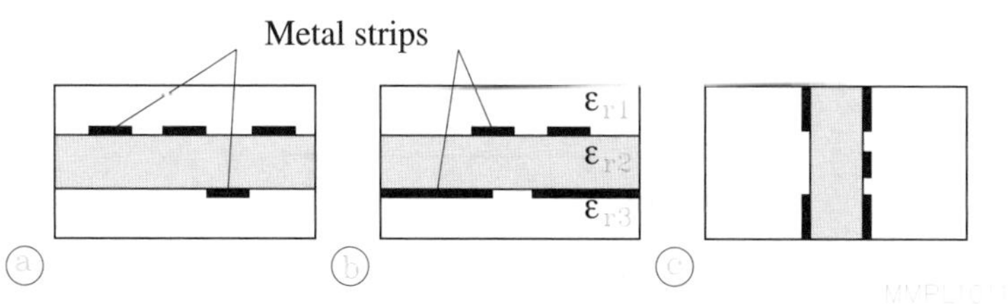

Fig. 2.5 Examples of cross-sections of waveguides for microwave integrated circuits

ϕ with respect to the coordinate axes x and y (see Fig. 2.7) whereas z' and z are identical. The relation of the field components between the two systems is given by:

$$\begin{bmatrix} F_{x'} \\ F_{y'} \end{bmatrix} = \begin{bmatrix} \cos\phi_\circ & \sin\phi_\circ \\ -\sin\phi_\circ & \cos\phi_\circ \end{bmatrix} \begin{bmatrix} F_x \\ F_y \end{bmatrix} \tag{2.1}$$

where F is any of the field components E, D, H or B.

We obtain, e.g. from the dielectric relation:

$$\begin{bmatrix} D_{x'} \\ D_{y'} \end{bmatrix} = \begin{bmatrix} \varepsilon_{11} & \varepsilon_{12} \\ \varepsilon_{21} & \varepsilon_{22} \end{bmatrix} \begin{bmatrix} E_{x'} \\ E_{y'} \end{bmatrix} \tag{2.2}$$

the following expression:

$$\begin{bmatrix} D_x \\ D_y \end{bmatrix} = \begin{bmatrix} \varepsilon_{xx} & \varepsilon_{xy} \\ \varepsilon_{yx} & \varepsilon_{yy} \end{bmatrix} \begin{bmatrix} E_x \\ E_y \end{bmatrix} \tag{2.3}$$

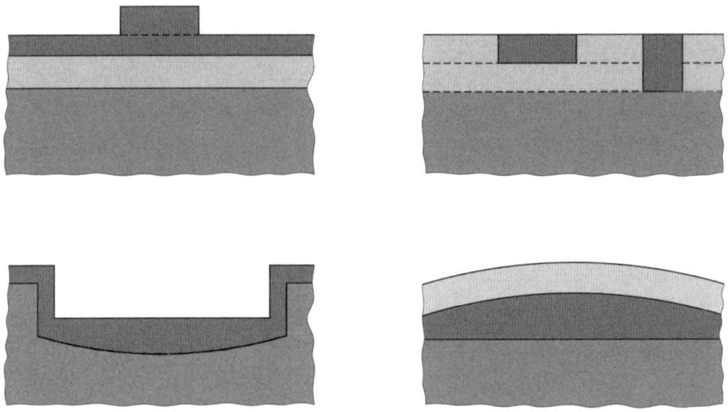

Fig. 2.6 Cross-sections of waveguides for integrated optics

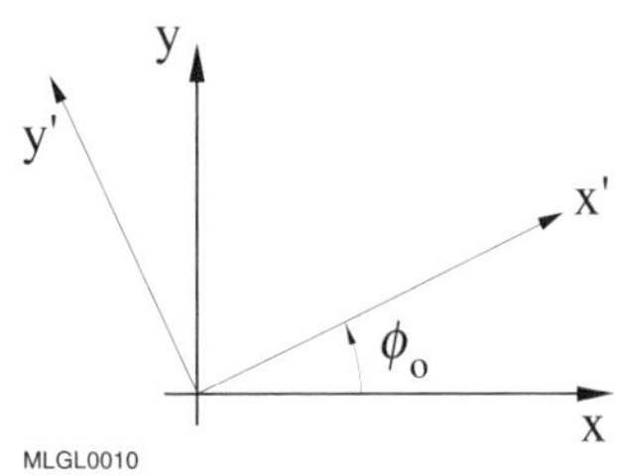

Fig. 2.7 Relation between the anisotropic axes and coordinate system in the cross-section

where we determine the new tensor as:

$$\begin{bmatrix} \varepsilon_{xx} & \varepsilon_{xy} \\ \varepsilon_{yx} & \varepsilon_{yy} \end{bmatrix} = \begin{bmatrix} \cos\phi_\circ & -\sin\phi_\circ \\ \sin\phi_\circ & \cos\phi_\circ \end{bmatrix} \begin{bmatrix} \varepsilon_{11} & \varepsilon_{12} \\ \varepsilon_{21} & \varepsilon_{22} \end{bmatrix} \begin{bmatrix} \cos\phi_\circ & \sin\phi_\circ \\ -\sin\phi_\circ & \cos\phi_\circ \end{bmatrix} \tag{2.4}$$

with:

$$\begin{aligned} \varepsilon_{xx} &= \varepsilon_{11}\cos^2\phi_\circ + \varepsilon_{22}\sin^2\phi_\circ - (\varepsilon_{12}+\varepsilon_{21})\sin\phi_0\cos\phi_\circ \\ \varepsilon_{xy} &= (\varepsilon_{11}-\varepsilon_{22})\sin\phi_\circ\cos\phi_\circ + (\varepsilon_{12}\cos^2\phi_0 - \varepsilon_{21}\sin^2\phi_\circ) \\ \varepsilon_{yx} &= (\varepsilon_{11}-\varepsilon_{22})\sin\phi_\circ\cos\phi_\circ - (\varepsilon_{12}\sin^2\phi_\circ - \varepsilon_{21}\cos^2\phi_\circ) \\ \varepsilon_{yy} &= \varepsilon_{11}\sin^2\phi_\circ + \varepsilon_{22}\cos^2\phi_\circ + (\varepsilon_{12}+\varepsilon_{21})\sin\phi_0\cos\phi_\circ \end{aligned}$$

In case $\varepsilon_{12} = -\varepsilon_{21} = -\mathrm{j}\kappa_\varepsilon$, the last terms in ε_{xx} and ε_{yy} vanish and the terms in brackets of ε_{xy} and ε_{yx} result in $-\mathrm{j}\kappa_\varepsilon$. The calculation of the modified μ-tensor is done analogously. In our further derivations we extend the material parameter matrix to:

$$\overleftrightarrow{\varepsilon}_r = \begin{bmatrix} \varepsilon_{xx} & \varepsilon_{xy} & \\ \varepsilon_{yx} & \varepsilon_{yy} & \\ & & \varepsilon_{zz} \end{bmatrix} \qquad \overleftrightarrow{\mu}_r = \begin{bmatrix} \mu_{xx} & \mu_{xy} & \\ \mu_{yx} & \mu_{yy} & \\ & & \mu_{zz} \end{bmatrix} \tag{2.5}$$

2.2.2 Relations between transversal electric and magnetic fields – generalised transmission line (GTL) equations

In this subsection we will develop the generalised transmission line (GTL) equations that are used for analysis. These GTLs are analogous to the well-known ones for coupled multi-conductor transmission lines (in inhomogeneous media) [1].

$$\frac{\mathrm{d}}{\mathrm{d}z}[U] = -j\omega[L'][I] \qquad \frac{\mathrm{d}}{\mathrm{d}z}[I] = -j\omega[C'][U] \tag{2.6}$$

In [2] these equations were solved by calculating the modal matrices.Here we will proceed in an analogue way.

Eq. (2.6) is a coupled differential equation system for the current and the voltage. We want to derive a (full vectorial) relation between the transverse electric and magnetic field components, which we call generalised transmission line equations.

In this chapter, we restrict the description to permittivity and permeability tensors of the form given in eq. (2.5). The case of arbitrary anisotropic materials (i.e. full tensors) will be described in Chapter 8.

By using the abbreviation $D_{\overline{u}}$, $u = x, y, z$ for $\partial/\partial\overline{u}$, we obtain the following expression from Ampère's law:

$$-\mathrm{j}\begin{bmatrix} \varepsilon_{yy} & \varepsilon_{yx} \\ \varepsilon_{xy} & \varepsilon_{xx} \end{bmatrix}\begin{bmatrix} E_y \\ E_x \end{bmatrix} = \frac{\partial}{\partial\overline{z}}\begin{bmatrix} -\widetilde{H}_x \\ \widetilde{H}_y \end{bmatrix} - \begin{bmatrix} -D_{\overline{x}} \\ D_{\overline{y}} \end{bmatrix}\widetilde{H}_z \tag{2.7}$$

The component $\widetilde{H}_z$ is calculated from the law of induction:

$$\widetilde{H}_z = j\mu_{zz}^{-1}[D_{\overline{x}} - D_{\overline{y}}]\begin{bmatrix} E_y \\ E_x \end{bmatrix} \tag{2.8}$$

By introducing eq. (2.8) into eq. (2.7) and using the following definition for the transversal field vectors:

$$[E_t] = \begin{bmatrix} E_y \\ E_x \end{bmatrix} \qquad [\widetilde{H}_t] = \begin{bmatrix} -\widetilde{H}_x \\ \widetilde{H}_y \end{bmatrix} \tag{2.9}$$

the equation for calculating the transversal electric field components from the transversal magnetic ones takes the form:

$$\frac{\partial}{\partial\overline{z}}[\widetilde{H}_t] = -\mathrm{j}\,[R_{\mathrm{E}}^{z}]\,[E_t] \qquad [R_{\mathrm{E}}^{z}] = \begin{bmatrix} \varepsilon_{yy} + D_{\overline{x}}\mu_{zz}^{-1}D_{\overline{x}} & \varepsilon_{yx} - D_{\overline{x}}\mu_{zz}^{-1}D_{\overline{y}} \\ \varepsilon_{xy} - D_{\overline{y}}\mu_{zz}^{-1}D_{\overline{x}} & \varepsilon_{xx} + D_{\overline{y}}\mu_{zz}^{-1}D_{\overline{y}} \end{bmatrix} \tag{2.10}$$

Note: In the whole book we have to apply operators (like $D_{\overline{y}}$) in a product to all terms on the right of them. Therefore, $D_{\overline{y}}\mu_{zz}^{-1}D_{\overline{x}}$ means the same as $D_{\overline{y}}(\mu_{zz}^{-1}D_{\overline{x}})$. Further, we may not change the order of the terms in such products, i.e. $D_{\overline{x}}\mu_{zz}^{-1}D_{\overline{y}}$ is generally different from $D_{\overline{y}}\mu_{zz}^{-1}D_{\overline{x}}$.

From Faraday's law of induction we obtain:

$$-\mathrm{j}\begin{bmatrix}\mu_{xx} & -\mu_{xy}\\ -\mu_{yx} & \mu_{yy}\end{bmatrix}\begin{bmatrix}-\widetilde{H}_x\\ \widetilde{H}_y\end{bmatrix} = \frac{\partial}{\partial\overline{z}}\begin{bmatrix}E_y\\ E_x\end{bmatrix} - \begin{bmatrix}D_{\overline{y}}\\ D_{\overline{x}}\end{bmatrix}E_z \tag{2.11}$$

We determine the field component E_z from the Ampère's law:

$$E_z = -\mathrm{j}\varepsilon_{zz}^{-1}[D_{\overline{y}} \quad D_{\overline{x}}]\begin{bmatrix}-\widetilde{H}_x\\ \widetilde{H}_y\end{bmatrix} \tag{2.12}$$

By introducing eq. (2.12) into eq. (2.11), the equation for calculating the transverse magnetic field from transverse electric one reads as:

$$\frac{\partial}{\partial\overline{z}}[E_t] = -\mathrm{j}[R_{\mathrm{H}}^z][\widetilde{H}_t] \qquad [R_{\mathrm{H}}^z] = \begin{bmatrix}D_{\overline{y}}\varepsilon_{zz}^{-1}D_{\overline{y}} + \mu_{xx} & D_{\overline{y}}\varepsilon_{zz}^{-1}D_{\overline{x}} - \mu_{xy}\\ D_{\overline{x}}\varepsilon_{zz}^{-1}D_{\overline{y}} - \mu_{yx} & D_{\overline{x}}\varepsilon_{zz}^{-1}D_{\overline{x}} + \mu_{yy}\end{bmatrix} \tag{2.13}$$

Now, we split the matrices $[R_{\mathrm{E,H}}^z]$ into two parts according to:

$$[R_{\mathrm{E}}^z] = [R_{\mathrm{E}}^{z\mathrm{C}}] + [\varepsilon_{rt}^{z\mathrm{C}}] \qquad [R_{\mathrm{H}}^z] = [R_{\mathrm{E}}^{z\mathrm{C}}] + [\mu_{rt}^{z\mathrm{C}}] \tag{2.14}$$

with:

$$[R_{\mathrm{E}}^{z\mathrm{C}}] = \begin{bmatrix}D_{\overline{x}}\mu_{zz}^{-1}D_{\overline{x}} & -D_{\overline{x}}\mu_{zz}^{-1}D_{\overline{y}}\\ -D_{\overline{y}}\mu_{zz}^{-1}D_{\overline{x}} & D_{\overline{y}}\mu_{zz}^{-1}D_{\overline{y}}\end{bmatrix} \qquad [\widehat{\varepsilon}_{rt}^{z\mathrm{C}}] = \begin{bmatrix}\varepsilon_{yy} & \varepsilon_{yx}\\ \varepsilon_{xy} & \varepsilon_{xx}\end{bmatrix} \tag{2.15}$$

$$[R_{\mathrm{H}}^{z\mathrm{C}}] = \begin{bmatrix}D_{\overline{y}}\varepsilon_{zz}^{-1}D_{\overline{y}} & D_{\overline{y}}\varepsilon_{zz}^{-1}D_{\overline{x}}\\ D_{\overline{x}}\varepsilon_{zz}^{-1}D_{\overline{y}} & D_{\overline{x}}\varepsilon_{zz}^{-1}D_{\overline{x}}\end{bmatrix} \qquad [\widehat{\mu}_{rt}^{z\mathrm{C}}] = \begin{bmatrix}\mu_{xx} & -\mu_{xy}\\ -\mu_{yx} & \mu_{yy}\end{bmatrix} \tag{2.16}$$

Combining eqs. (2.10) and (2.13) we obtain:

$$\frac{\partial^2}{\partial\overline{z}^2}[E_t] - [Q_{\mathrm{E}}^z][E_t] = [0] \qquad \frac{\partial^2}{\partial\overline{z}^2}[\widetilde{H}_t] - [Q_{\mathrm{H}}^z][\widetilde{H}_t] = [0] \tag{2.17}$$

Because of $D_{\overline{x}}D_{\overline{y}} = D_{\overline{y}}D_{\overline{x}}$, the products of the matrices $[R_{\mathrm{E}}^{z\mathrm{C}}]$ and $[R_{\mathrm{H}}^{z\mathrm{C}}]$ are always equal to zero.Therefore, we may write for the matrices $[Q_{\mathrm{H}}^z]$ and $[Q_{\mathrm{E}}^z]$:

$$-[Q_{\mathrm{E}}^z] = [R_{\mathrm{H}}^z][R_{\mathrm{E}}^z] = [R_{\mathrm{H}}^{z\mathrm{C}}][\widehat{\varepsilon}_{rt}^{z\mathrm{C}}] + [\mu_{rt}^{z\mathrm{C}}][R_{\mathrm{E}}^{z\mathrm{C}}] + [\mu_{rt}^{z\mathrm{C}}][\widehat{\varepsilon}_{rt}^{z\mathrm{C}}] \tag{2.18}$$

$$-[Q_{\mathrm{H}}^z] = [R_{\mathrm{E}}^z][R_{\mathrm{H}}^z] = [R_{\mathrm{E}}^{z\mathrm{C}}][\mu_{rt}^{z\mathrm{C}}] + [\varepsilon_{rt}^{z\mathrm{C}}][R_{\mathrm{H}}^{\phi\mathrm{C}}] + [\varepsilon_{rt}^{z\mathrm{C}}][\widehat{\mu}_{rt}^{z\mathrm{C}}] \tag{2.19}$$

Now the submatrices of $[Q_{\mathrm{E,H}}^z]$:

$$[Q_{\mathrm{E,H}}^z] = \begin{bmatrix}Q_{\mathrm{E,H11}}^z & Q_{\mathrm{E,H12}}^z\\ Q_{\mathrm{E,H21}}^z & Q_{\mathrm{E,H22}}^z\end{bmatrix} \tag{2.20}$$

can easily be determined. The formulation on the right-hand side of eqs. (2.18) and (2.19) should be used in computer codes to determine these submatrices.

2.2.3 Relation to the analysis with vector potentials

In a previous form the MoL analysis was performed by the use of the special Hertzian vector potential. To obtain this special Hertzian vector potential a second component in the vector potential of Collin [3] was introduced first in [4]. This led to two coupled scalar wave equations (Sturm–Liouville differential equation) which could be solved with the MoL in a straightforward way.

Here we will show that the analysis using vector potentials [5], [6] in the case of isotropic materials can be transformed to the analysis described here. Using the definition:

$$[\phi] = \begin{bmatrix} \phi_h \\ \phi_e \end{bmatrix} \tag{2.21}$$

we obtain from eq. (2.8) in [5] for $\mu_r = 1$:

$$[\widetilde{H}_t] = \mathrm{j}\frac{\partial}{\partial \overline{z}}[\phi] \qquad [E_t] = [R^{\overline{z}}_{\mathrm{H}}][\phi] \qquad \longrightarrow \qquad \frac{\partial}{\partial \overline{z}}[E_t] = -\mathrm{j}[R^{\overline{z}}_{\mathrm{H}}][\widetilde{H}_t] \tag{2.22}$$

We obtained the third equation by differentiating the second equation with respect to $\overline{z}$ and introducing the first equation. The coupled wave eqs. (2.5) and (2.6) in [5] can be written with supervectors according to the definition in (2.21):

$$\frac{\partial^2}{\partial \overline{z}^2}[\phi] - [Q^{\overline{z}}_{\mathrm{H}}][\phi] = [0] \qquad \longrightarrow \qquad \frac{\partial^2}{\partial \overline{z}^2}[\widetilde{H}_t] - [Q^{\overline{z}}_{\mathrm{H}}][\widetilde{H}_t] = [0] \tag{2.23}$$

Differentiating the left equation with respect to $\overline{z}$ and introducing the first equation in (2.22), we obtain the equation on the right side, which is identical to the second equation in (2.17).

Differentiating the first equation in (2.22) with respect to $\overline{z}$ and introducing $\frac{\partial^2}{\partial \overline{z}^2}[\phi]$ from the first equation in (2.23), we obtain the equation on the left hand:

$$\frac{\partial}{\partial \overline{z}}[\widetilde{H}_t] = \mathrm{j}[Q^{\overline{z}}_{\mathrm{H}}][\phi] \qquad \longrightarrow \qquad \frac{\partial}{\partial \overline{z}}[\widetilde{H}_t] = -\mathrm{j}[R^{\overline{z}}_{\mathrm{E}}][E_t] \tag{2.24}$$

Knowing that $[Q^{\overline{z}}_{\mathrm{H}}]$ can be split according to $[Q^{\overline{z}}_{\mathrm{H}}] = -[R^{\overline{z}}_{\mathrm{E}}][R^{\overline{z}}_{\mathrm{H}}]$, we can introduce $[\phi]$ from the second equation in (2.22) and obtain the result in the equation on the right side.

The GTL equations have the advantage of allowing us to immediately obtain the wave equation for the transverse electrical fields. Further, we see that we start with the wave equation for the magnetic or the electric field and obtain the same expressions later on.

By using GTL equations we can very easily take into account anisotropic material as well. Furthermore, the modal matrices obtained from GTL equations lead to simpler analysis equations. Therefore, the description with GTL equations is very flexible.

2.2.4 GTL equations for 2D structures

To show the fundamentals of the Method of Lines, we will begin with the two-dimensional case. In this way all main principles can be explained. When we look at more complex problems, only the numerics that are involved will become more complicated, not the fundamental principles.

To develop equations for 2D structures we must consider materials with diagonal tensors (i.e. $\nu_{xy} = \nu_{yx} = 0$ in eq. (2.5)). We assume that there is no variation in y-direction. The expression $D_y = 0$ is introduced into the operators $[R_{\mathrm{E}}^{z}]$ and $[R_{\mathrm{H}}^{z}]$ in eqs. (2.10), (2.13). We see that the off-diagonal elements are zero. Therefore, we obtain two decoupled polarisations that can be treated independently.

As usual in the MoL, the operators are now discretised with finite differences. The positions of the discretisation lines in a 2D-structure are shown in Fig. 2.8. As mentioned before, the fulfilment of the boundary conditions suggests determining the field components on different positions, which are shifted by half a discretisation distance. This is shown in Fig. 2.27 for a 2D-discretisation. Assuming $\partial/\partial y = 0$, we see that the •- and the □-lines, respectively the ○- and the ◇-lines, become identical. They become the straight (•- and □-lines) and dashed lines (○- and ◇-lines) in Fig. 2.8. We then compute the corresponding field components at the positions as stated below. Due to the discretisation, the operators become operator matrices and we combine the discretised fields in vectors. Then we obtain:

- TE$_z$ modes with the components E_y, H_x, H_z computed on the bullets • in Fig. 2.27(or straight lines in Fig. 2.8). The field vectors for the GTL equations are given by $\mathbf{E} = \mathbf{E}_y$ and $\mathbf{H} = -\mathbf{H}_x$. The third field component is obtained from $\mathbf{H}_z = \mathrm{j}\boldsymbol{\mu}_{zz}^{-1}\overline{\boldsymbol{D}}_{\overline{x}}^{\bullet}\mathbf{E}_y$. The operator matrices for this case are:

$$\boldsymbol{R}_{\mathrm{E}}^{\bullet} = \boldsymbol{\epsilon}_{yy} - \overline{\boldsymbol{D}}_{\overline{x}}^{\bullet\mathrm{t}}\boldsymbol{\mu}_{zz}^{-1}\overline{\boldsymbol{D}}_{\overline{x}}^{\bullet} \qquad \boldsymbol{R}_{\mathrm{H}}^{\bullet} = \boldsymbol{\mu}_{xx} \tag{2.25}$$

  and:

$$\boldsymbol{Q}_{\mathrm{E}}^{\bullet} = \boldsymbol{R}_{\mathrm{H}}^{\bullet}\boldsymbol{R}_{\mathrm{E}}^{\bullet} = \boldsymbol{\mu}_{xx}\boldsymbol{\epsilon}_{yy} - \boldsymbol{\mu}_{xx}\overline{\boldsymbol{D}}_{\overline{x}}^{\bullet\mathrm{t}}\boldsymbol{\mu}_{zz}^{-1}\overline{\boldsymbol{D}}_{\overline{x}}^{\bullet} \tag{2.26}$$

$$\boldsymbol{Q}_{\mathrm{H}}^{\bullet} = \boldsymbol{R}_{\mathrm{E}}^{\bullet}\boldsymbol{R}_{\mathrm{H}}^{\bullet} = \boldsymbol{\mu}_{xx}\boldsymbol{\epsilon}_{yy} - \overline{\boldsymbol{D}}_{\overline{x}}^{\bullet\mathrm{t}}\boldsymbol{\mu}_{zz}^{-1}\overline{\boldsymbol{D}}_{\overline{x}}^{\bullet}\boldsymbol{\mu}_{xx} \tag{2.27}$$

- TM$_z$-modes with the components H_y, E_x, E_z determined on the circles ○ in Fig. 2.27 (or dashed lines in Fig. 2.8). The field vectors for the GTL equations are given by $\mathbf{E} = \mathbf{E}_x$ and $\mathbf{H} = \mathbf{H}_y$. The third field component is obtained from $\mathbf{E}_z = -\mathrm{j}\boldsymbol{\varepsilon}_{zz}^{-1}\overline{\boldsymbol{D}}_{\overline{x}}^{\circ}\mathbf{H}_y$. Here we have the following operator matrices:

$$\boldsymbol{R}_{\mathrm{E}}^{\circ} = \boldsymbol{\epsilon}_{xx} \qquad \boldsymbol{R}_{\mathrm{H}}^{\circ} = \boldsymbol{\mu}_{yy} - \overline{\boldsymbol{D}}_{\overline{x}}^{\circ\mathrm{t}}\boldsymbol{\epsilon}_{zz}^{-1}\overline{\boldsymbol{D}}_{\overline{x}}^{\circ} \tag{2.28}$$

and:

$$\boldsymbol{Q}_{\mathrm{H}}^{\circ} = \boldsymbol{R}_{\mathrm{E}}^{\circ}\boldsymbol{R}_{\mathrm{H}}^{\circ} = \boldsymbol{\epsilon}_{xx}\boldsymbol{\mu}_{yy} - \boldsymbol{\epsilon}_{xx}\overline{\boldsymbol{D}}_{\overline{x}}^{\circ t}\boldsymbol{\epsilon}_{zz}^{-1}\overline{\boldsymbol{D}}_{\overline{x}}^{\circ} \tag{2.29}$$

$$\boldsymbol{Q}_{\mathrm{E}}^{\circ} = \boldsymbol{R}_{\mathrm{H}}^{\circ}\boldsymbol{R}_{\mathrm{E}}^{\circ} = \boldsymbol{\epsilon}_{xx}\boldsymbol{\mu}_{yy} - \overline{\boldsymbol{D}}_{\overline{x}}^{\circ t}\boldsymbol{\epsilon}_{zz}^{-1}\overline{\boldsymbol{D}}_{\overline{x}}^{\circ}\boldsymbol{\epsilon}_{xx} \tag{2.30}$$

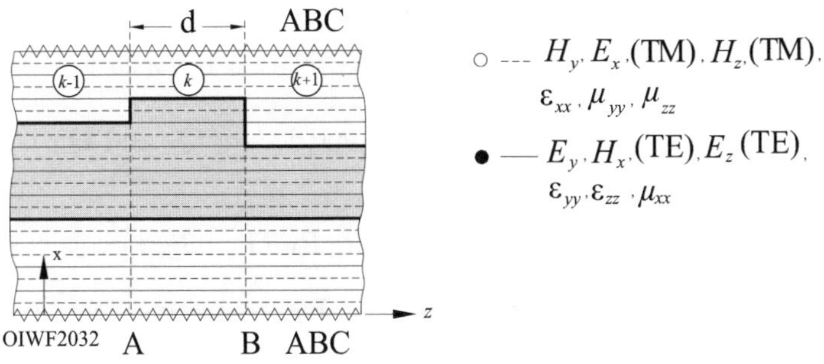

Fig. 2.8 Discretisation of a 2D-waveguide structure

2.2.5 Solution of the GTL equations

Unless otherwise stated, the formulas that we derive here are valid for the TE polarisation as well as for the TM case. Furthermore, when we come to describe the full 3D case we may use the same expressions.

We would like to solve eq. (2.17). After discretisation, we obtain:

$$\frac{\mathrm{d}^2}{\mathrm{d}\overline{z}^2}\mathbf{H} - \boldsymbol{Q}_{\mathrm{H}}\mathbf{H} = \mathbf{0} \qquad \frac{\mathrm{d}^2}{\mathrm{d}\overline{z}^2}\mathbf{E} - \boldsymbol{Q}_{\mathrm{E}}\mathbf{E} = \mathbf{0} \tag{2.31}$$

For this purpose we transform the magnetic and electric field according to:

$$\mathbf{H} = \boldsymbol{T}_{\mathrm{H}}\overline{\mathbf{H}} \qquad \mathbf{E} = \boldsymbol{T}_{\mathrm{E}}\overline{\mathbf{E}} \tag{2.32}$$

Therefore the eqs. (2.17) in discretised form are converted to:

$$\frac{\mathrm{d}^2}{\mathrm{d}\overline{z}^2}\overline{\mathbf{H}} - \boldsymbol{\Gamma}_{\mathrm{H}}^2\overline{\mathbf{H}} = \mathbf{0} \qquad \frac{\mathrm{d}^2}{\mathrm{d}\overline{z}^2}\overline{\mathbf{E}} - \boldsymbol{\Gamma}_{\mathrm{E}}^2\overline{\mathbf{E}} = \mathbf{0} \tag{2.33}$$

with:

$$\boldsymbol{T}_{\mathrm{H}}^{-1}\boldsymbol{Q}_{\mathrm{H}}\boldsymbol{T}_{\mathrm{H}} = \boldsymbol{\Gamma}_{\mathrm{H}}^2 \qquad \boldsymbol{T}_{\mathrm{E}}^{-1}\boldsymbol{Q}_{\mathrm{E}}\boldsymbol{T}_{\mathrm{E}} = \boldsymbol{\Gamma}_{\mathrm{E}}^2 \tag{2.34}$$

The equations (2.26)–(2.30) (in discretised form) or (2.18)–(2.19) show that the matrices $\boldsymbol{Q}_{\mathrm{E}}$ and $\boldsymbol{Q}_{\mathrm{H}}$ are determined as products of $\boldsymbol{R}_{\mathrm{E}}$ and $\boldsymbol{R}_{\mathrm{H}}$. As we know (see e.g. [7]), a product of two square matrices results in identical eigenvalues if the order of the product is reversed. Also, the corresponding eigenvectors are related. We will use this characteristic later. Therefore, both equations (2.34) result in the same propagation constants:

$$\boldsymbol{\Gamma}_{\mathrm{E}}^2 = \boldsymbol{\Gamma}_{\mathrm{H}}^2 = \boldsymbol{\Gamma}^2 \tag{2.35}$$

In what follows, we show the steps of the analysis for the cases where we start with the electric or the magnetic field in parallel. However, when applying the MoL, we will only solve one of the eigenvalue/eigenvector problems in eq. (2.34). The second eigenvector (or transformation) matrix is determined due to the above-mentioned relation between the eigenvector products.

This relation between the eigenvector matrices of $\boldsymbol{Q}_{\mathrm{E}}$ and $\boldsymbol{Q}_{\mathrm{H}}$ is obtained from[1] ($\boldsymbol{\Gamma}^2 = -\boldsymbol{\beta}^2$):

$$\boldsymbol{R}_{\mathrm{E}}\boldsymbol{R}_{\mathrm{H}}\boldsymbol{T}_{\mathrm{H}} = \boldsymbol{T}_{\mathrm{H}}\boldsymbol{\beta}_{\mathrm{H}}^2 \qquad \boldsymbol{R}_{\mathrm{H}}\boldsymbol{R}_{\mathrm{E}}\boldsymbol{T}_{\mathrm{E}} = \boldsymbol{T}_{\mathrm{E}}\boldsymbol{\beta}_{\mathrm{E}}^2 \tag{2.36}$$

Multiplying these equations by $\boldsymbol{R}_{\mathrm{H}}$ and $\boldsymbol{R}_{\mathrm{E}}$, respectively, gives:

$$\boldsymbol{R}_{\mathrm{H}}\boldsymbol{R}_{\mathrm{E}}(\boldsymbol{R}_{\mathrm{H}}\boldsymbol{T}_{\mathrm{H}}) = (\boldsymbol{R}_{\mathrm{H}}\boldsymbol{T}_{\mathrm{H}})\boldsymbol{\beta}_{\mathrm{H}}^2 \qquad \boldsymbol{R}_{\mathrm{E}}\boldsymbol{R}_{\mathrm{H}}(\boldsymbol{R}_{\mathrm{E}}\boldsymbol{T}_{\mathrm{E}}) = (\boldsymbol{R}_{\mathrm{E}}\boldsymbol{T}_{\mathrm{E}})\boldsymbol{\beta}_{\mathrm{E}}^2 \tag{2.37}$$

Since the product $\boldsymbol{R}_{\mathrm{E}}\boldsymbol{R}_{\mathrm{H}}$ ($\boldsymbol{R}_{\mathrm{H}}\boldsymbol{R}_{\mathrm{E}}$) gives the operator matrix $\boldsymbol{Q}_{\mathrm{H}}$ ($\boldsymbol{Q}_{\mathrm{E}}$), we see that we can in principle determine $\boldsymbol{T}_{\mathrm{H}}$ ($\boldsymbol{T}_{\mathrm{E}}$) from the product $\boldsymbol{R}_{\mathrm{E}}\boldsymbol{T}_{\mathrm{E}}$ ($\boldsymbol{R}_{\mathrm{H}}\boldsymbol{T}_{\mathrm{H}}$). The amplitudes of the eigenvector matrices can be chosen arbitrarily. The electric and magnetic field is, however, coupled by the GTL equations. We choose the following normalisation:

$$\boldsymbol{T}_{\mathrm{E}} = \boldsymbol{R}_{\mathrm{H}}\boldsymbol{T}_{\mathrm{H}}\boldsymbol{\beta}^{-1} \qquad \boldsymbol{T}_{\mathrm{H}} = \boldsymbol{R}_{\mathrm{E}}\boldsymbol{T}_{\mathrm{E}}\boldsymbol{\beta}^{-1} \tag{2.38}$$

The last relation can also be written as:

$$\boldsymbol{T}_{\mathrm{E}}^{-1}\boldsymbol{R}_{\mathrm{H}}\boldsymbol{T}_{\mathrm{H}} = \boldsymbol{\beta} \qquad \boldsymbol{T}_{\mathrm{H}}^{-1}\boldsymbol{R}_{\mathrm{E}}\boldsymbol{T}_{\mathrm{E}} = \boldsymbol{\beta} \tag{2.39}$$

These expressions are introduced into the GTL equations, and we obtain:

$$\frac{\mathrm{d}}{\mathrm{d}\overline{z}}\overline{\mathbf{E}} = -\mathrm{j}\boldsymbol{\beta}\overline{\mathbf{H}} \qquad \frac{\mathrm{d}}{\mathrm{d}\overline{z}}\overline{\mathbf{H}} = -\mathrm{j}\boldsymbol{\beta}\overline{\mathbf{E}} \tag{2.40}$$

The combination of these two equations again results in eq. (2.33).

The general solution of eq. (2.33) can be given easily as:

$$\overline{\mathbf{F}} = e^{-\boldsymbol{\Gamma}\overline{z}}\mathbf{A} + e^{-\boldsymbol{\Gamma}\overline{z}}\mathbf{B} \tag{2.41}$$

Using the first part of the general solution in eq. (2.41) (i.e. the forward propagating fields $\overline{\boldsymbol{F}}_{\mathrm{f}}$), we can define wave impedance/admittance matrices. From:

$$\overline{\mathbf{F}}_{\mathrm{f}} = e^{-\mathrm{j}\boldsymbol{\beta}\overline{z}}\mathbf{A} \tag{2.42}$$

[1] When determining β from β^2 (i.e. computing the square root), we must take care to choose the correct sign, because β is the propagation constant of the forward propagating mode.

we obtain:

$$\frac{\mathrm{d}}{\mathrm{d}\bar{z}}\overline{\mathbf{F}}_{\mathrm{f}} = -\mathrm{j}\boldsymbol{\beta}\overline{\mathbf{F}}_{\mathrm{f}} \tag{2.43}$$

Introducing this relation into eq. (2.40) results in:

$$-\mathrm{j}\boldsymbol{\beta}\overline{\mathbf{H}}_{\mathrm{f}} = -\mathrm{j}\boldsymbol{\beta}\overline{\mathbf{E}}_{\mathrm{f}} \qquad -\mathrm{j}\boldsymbol{\beta}\overline{\mathbf{E}}_{\mathrm{f}} = -\mathrm{j}\boldsymbol{\beta}\overline{\mathbf{H}}_{\mathrm{f}} \tag{2.44}$$

So we have:

$$\overline{\mathbf{Z}}_0 = \boldsymbol{I} \qquad \overline{\mathbf{Y}}_0 = \boldsymbol{I} \tag{2.45}$$

Due to the normalisation introduced in eq. (2.38), the characteristic impedance/admittance matrices are equal to unit matrices. However, to be in accordance with the literature, we will keep the wave impedance/admittance in the expressions.[2] Introducing eq. (2.40) into an expression analogous to eq. (1.20) results in:

$$\begin{bmatrix} \overline{\mathbf{H}}_{\mathrm{A}} \\ -\overline{\mathbf{H}}_{\mathrm{B}} \end{bmatrix} = \begin{bmatrix} \overline{\mathbf{y}}_1 & \overline{\mathbf{y}}_2 \\ \overline{\mathbf{y}}_2 & \overline{\mathbf{y}}_1 \end{bmatrix} \begin{bmatrix} \overline{\mathbf{E}}_{\mathrm{A}} \\ \overline{\mathbf{E}}_{\mathrm{B}} \end{bmatrix} \qquad \begin{aligned} \overline{\mathbf{y}}_1 &= \overline{\mathbf{Y}}_0(\tanh(\boldsymbol{\Gamma}\bar{d}))^{-1} \\ \overline{\mathbf{y}}_2 &= -\overline{\mathbf{Y}}_0(\sinh(\boldsymbol{\Gamma}\bar{d}))^{-1} \end{aligned} \tag{2.46}$$

The inversion results in:

$$\begin{bmatrix} \overline{\mathbf{E}}_{\mathrm{A}} \\ \overline{\mathbf{E}}_{\mathrm{B}} \end{bmatrix} = \begin{bmatrix} \overline{\mathbf{z}}_1 & \overline{\mathbf{z}}_2 \\ \overline{\mathbf{z}}_2 & \overline{\mathbf{z}}_1 \end{bmatrix} \begin{bmatrix} \overline{\mathbf{H}}_{\mathrm{A}} \\ -\overline{\mathbf{H}}_{\mathrm{B}} \end{bmatrix} \qquad \begin{aligned} \overline{\mathbf{z}}_1 &= \overline{\mathbf{Z}}_0(\tanh(\boldsymbol{\Gamma}\bar{d}))^{-1} \\ \overline{\mathbf{z}}_2 &= \overline{\mathbf{Z}}_0(\sinh(\boldsymbol{\Gamma}\bar{d}))^{-1} \end{aligned} \tag{2.47}$$

Later on we will also need transfer matrix relations, which are given by:

$$\begin{bmatrix} \overline{\mathbf{E}}_{\mathrm{A}} \\ \overline{\mathbf{H}}_{\mathrm{A}} \end{bmatrix} = \begin{bmatrix} \overline{\mathbf{V}} & \overline{\mathbf{Z}} \\ \overline{\mathbf{Y}} & \overline{\mathbf{V}} \end{bmatrix} \begin{bmatrix} \overline{\mathbf{E}}_{\mathrm{B}} \\ \overline{\mathbf{H}}_{\mathrm{B}} \end{bmatrix} \qquad \begin{bmatrix} \overline{\mathbf{E}}_{\mathrm{B}} \\ \overline{\mathbf{H}}_{\mathrm{B}} \end{bmatrix} = \begin{bmatrix} \overline{\mathbf{V}} & -\overline{\mathbf{Z}} \\ -\overline{\mathbf{Y}} & \overline{\mathbf{V}} \end{bmatrix} \begin{bmatrix} \overline{\mathbf{E}}_{\mathrm{A}} \\ \overline{\mathbf{H}}_{\mathrm{A}} \end{bmatrix} \tag{2.48}$$

with:

$$\overline{\boldsymbol{V}} = \sinh(\boldsymbol{\Gamma}\bar{d}) \qquad \overline{\boldsymbol{Z}} = \overline{\mathbf{Z}}_0\sinh(\boldsymbol{\Gamma}\bar{d}) \qquad \overline{\boldsymbol{Y}} = \overline{\mathbf{Y}}_0\sinh(\boldsymbol{\Gamma}\bar{d}) \tag{2.49}$$

With these relations, the impedance/admittance transformation through sections may be performed according to Chapter 5. The impedance/admittance transformation through section interfaces is also described there.

2.2.6 Numerical examples

Let us now give some examples, using the expressions developed in this section. Fig. 2.10 shows the normalised effective index of a slab waveguide (see Fig. 2.9) for the TE polarisation. As can be seen, by decreasing the discretisation distance (h_x) we approach the analytic value.

As a second example, we examine the propagation of two eigenmodes in a slab waveguide in Fig. 2.11. The interference can be seen clearly. Finally,

[2]Introducing the normalisation in eq. (2.38) is not mandatory. E.g. without the factor $\boldsymbol{\beta}$ we obtained a different expression for the wave admittance/impedance. Now the following formula can be used for this case as well.

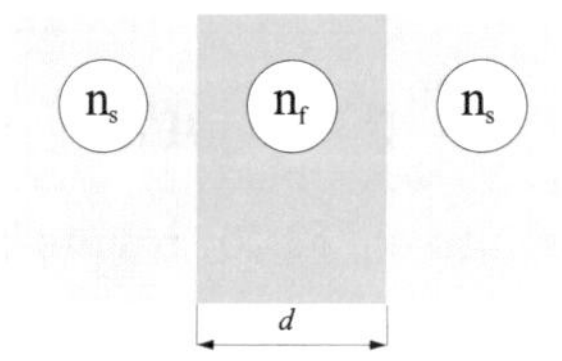

Fig. 2.9 Slab waveguide; the structure is infinite in vertical direction

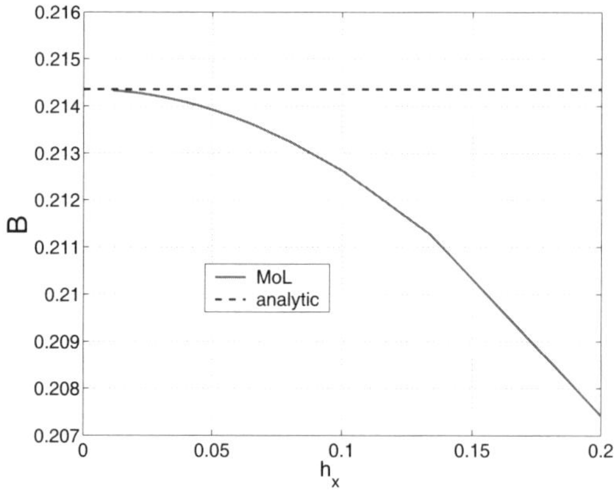

Fig. 2.10 Normalised effective index $B = (n_{\text{eff}}^2 - n_{\text{S}}^2)/(n_{\text{f}}^2 - n_{\text{S}}^2)$ vs. disretisation distance h_x of a slab waveguide (see Fig. 2.9), with $n_f = 3.24$, $n_s = 3.17$, $d = 0.4$ μm, wavelength $\lambda = 1.55$ μm

we examine an abrupt end of a waveguide in air region. Fig. 2.12 shows the magnetic field distribution. In the waveguide section ($z < 0$) we have reflections, as can be seen by the standing wave pattern. In the air region we have a broadening of the beam and radiation. It should also be noted that no reflections at the lateral boundaries are visible, because absorbing boundary conditions (for details see Section A.2 in Appendix A) were introduced.

2.3 EIGENMODES IN PLANAR WAVEGUIDE STRUCTURES WITH ANISOTROPIC LAYERS

2.3.1 Introduction

A second important problem that can be solved with a one-dimensional discretisation is the determination of the mode characteristics in the waveguides, i.e. the determination of eigenmodes. The waveguide structures used in integrated microwave, millimetre-wave and optical circuits consist of multilayered configurations. A general model of a waveguide cross-section is shown in Fig. 2.13. In the $\pm z$ direction, the structure may be infinite or a closed electric or magnetic wall. If the structure is of the microstrip type,

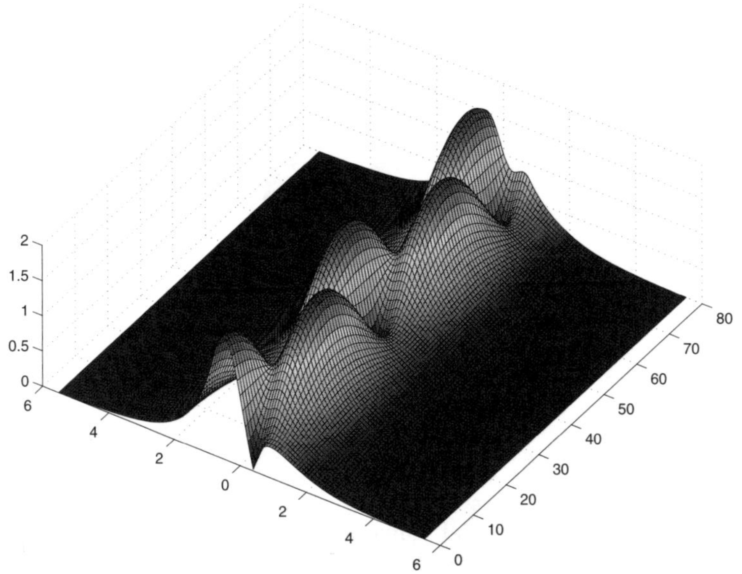

Fig. 2.11 Independent propagation of two guided modes in a slab waveguide (see Fig. 2.9), data: $n_f = 3.24$, $n_s = 3.17$, $d = 1.6\ \mu\text{m}$, wavelength $\lambda = 1.55\ \mu\text{m}$

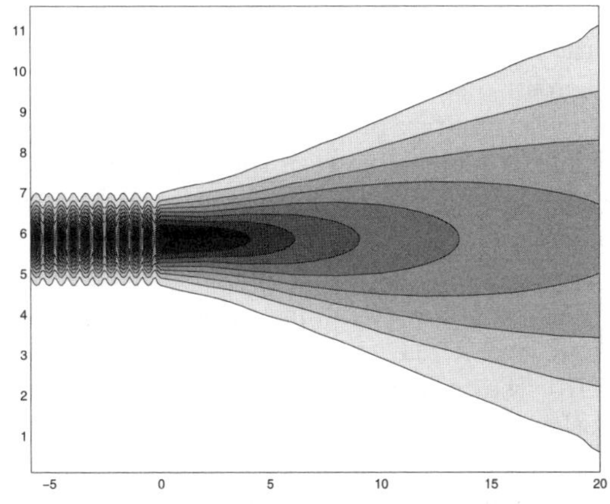

Fig. 2.12 Concatenation of a waveguide and air region, magnetic field distribution, data of the slab waveguide: n_f=1.4, n_s=1; d = 1.6 μm, wavelength λ = 1.55 μm (Reproduced by permission of Elsevier)

the ground metallisation acts as a boundary. In the x-direction, electric or magnetic walls may be introduced as well.

The permittivity of the layers might be either a continuous or step-wise function of the lateral coordinate, and the layers may differ in width.

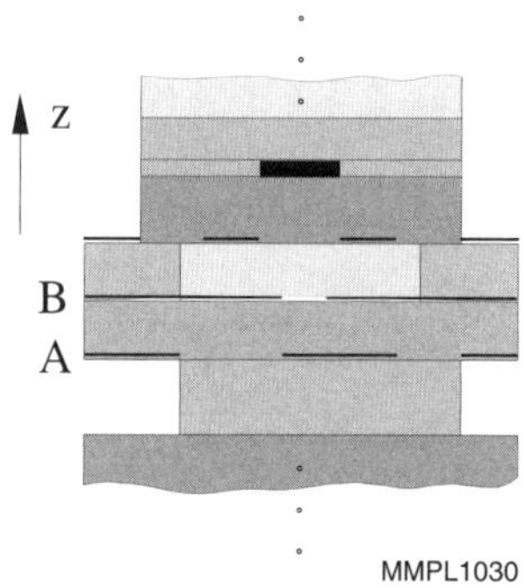

Fig. 2.13 General cross-section of a multilayered waveguide (Reproduced by permission of Elsevier)

Metallic strips are embedded at different interfaces. The metallisations are either infinitely thin or of finite thickness. Lossy materials are modelled by complex permittivities. Therefore, lossy strips are modelled as inhomogeneous layers with the thickness of the metallisation. The permittivity of the metal is given by $\varepsilon_r = \mathrm{j}\kappa_c/(\omega\varepsilon_0)$, where κ_c is the conductivity of the metal [10].

Usually the layers have equal width. However, it is also possible to model structures where the width of the layers is varied. This was first described for the analysis of microstrip structures with finite metallisation thickness [8]–[10]. The transformation matrix partition technique was introduced there to describe the fields in the regions between the strips. In this section, the GTL equations will be adapted to be best suited for eigenmode calculation. For that reason, the transformation matrix partition technique will be generalised so that this technique can also be used for in homogeneous layers. Furthermore, with the help of this partition technique, a general impedance/admittance transformation from layer to layer will be developed. Therefore, in a plane (e.g. A in Fig. 2.13), the impedances/admittances on the upper side of all the layers above the plane and on the lower side of all the layers below the plane can be calculated. With these impedances/admittances, an indirect eigenvalue problem for the field in that plane can be formulated easily. The chosen position of the plane is arbitrary. Therefore, the eigenmode connected with any plane can be calculated correctly. If we formulate the eigenvalue problem for the field in plane B of Fig. 2.13, we can determine the mode for the slot in this plane. It is clear that the strips in the neighbouring interfaces have an influence on this mode. However, this influence must be considered in a proper way. The computation of the field distribution is a difficult numerical task. It will be shown how this problem can be solved in an accurate way. The procedure will be evaluated by numerical results for an inset dielectric guide (IDG) and a groove guide.

The following steps are essential in the determination of eigenmodes with the MoL in case of layered structures:

- Partitioning of the cross-section into suitable layers according to the model in Fig. 2.14.
- Discretisation of the GTL equations in one coordinate direction (x-direction.
- Transformation to obtain decoupled ordinary differential equations for the transverse fields.
- Solution of the equations and determination of the impedance/admittance matrix parameters in each layer.
- Transformation of the impedance/admittance matrices through the multilayered structure, i.e.
    - through the layers
    - through the metallisation between the layers.
- Determination of the propagation constant and field distribution as the solution of an indirect eigenvalue problem in a chosen plane.
- Determination of the fields in the whole structure.

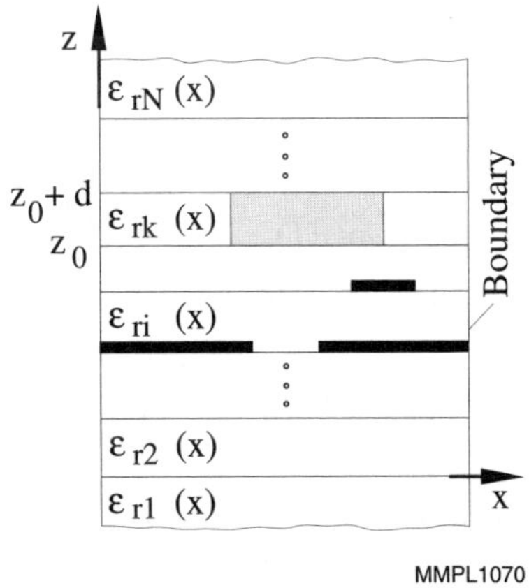

Fig. 2.14 General analysis model for multilayered waveguide structures (Reproduced by permission of Taylor & Francis)

2.3.2 Analysis equations for eigenmodes in planar structures

From the general equations (2.10) and (2.13) we will now derive the analysis equations for eigenmodes in planar structures. The analytical solution will be performed in z-direction, assuming a mode propagation in y-direction (perpendicular to the cross-section, cf. Fig. 2.14), according to $\exp(-\mathrm{j}\sqrt{\varepsilon_{re}}\overline{y})$. Therefore, we replace the operator $D_{\overline{y}}$ by $D_{\overline{y}} = -\mathrm{j}\sqrt{\varepsilon_{re}}$. Alternatively, the analytical solution could be performed in y-direction, assuming a mode propagating in z-direction (cf. Fig. 2.16). In the first case we obtain:

$$[R_{\mathrm{E}}^{z}] = \begin{bmatrix} D_{\overline{x}}\mu_{zz}^{-1}D_{\overline{x}} + \varepsilon_{yy} & \mathrm{j}\sqrt{\varepsilon_{re}}D_{\overline{x}}\mu_{zz}^{-1} + \varepsilon_{yx} \\ \mathrm{j}\sqrt{\epsilon_{re}}\mu_{zz}^{-1}D_{\overline{x}} + \varepsilon_{xy} & -\epsilon_{re}\mu_{zz}^{-1} + \varepsilon_{xx} \end{bmatrix} \tag{2.50}$$

$$[R_{\mathrm{H}}^{z}] = \begin{bmatrix} -\varepsilon_{re}\varepsilon_{zz}^{-1} + \mu_{xx} & -\mathrm{j}\sqrt{\varepsilon_{re}}\varepsilon_{zz}^{-1}D_{\overline{x}} - \mu_{xy} \\ -\mathrm{j}\sqrt{\varepsilon_{re}}D_{\overline{x}}\varepsilon_{zz}^{-1} - \mu_{yx} & D_{\overline{x}}\varepsilon_{zz}^{-1}D_{\overline{x}} + \mu_{yy} \end{bmatrix} \tag{2.51}$$

Because the 'off diagonal' submatrices are complex, we rewrite the eqs. (2.10) and (2.13) by redefining the field vectors according to:

$$\frac{\partial}{\partial \overline{z}}\begin{bmatrix} -\mathrm{j}E_y \\ E_x \end{bmatrix} = -[R_{\mathrm{H}}^{z}]\begin{bmatrix} H_x \\ -\mathrm{j}H_y \end{bmatrix} \qquad \frac{\partial}{\partial \overline{z}}\begin{bmatrix} H_x \\ -\mathrm{j}H_y \end{bmatrix} = -[R_{\mathrm{E}}^{z}]\begin{bmatrix} -\mathrm{j}E_y \\ E_x \end{bmatrix} \tag{2.52}$$

Now the matrices $R_{\mathrm{E,H}}^{z}$ are given by:

$$[R_{\mathrm{E}}^{z}] = \begin{bmatrix} \varepsilon_{yy} + D_{\overline{x}}\mu_{zz}^{-1}D_{\overline{x}} & \sqrt{\varepsilon_{re}}D_{\overline{x}}\mu_{zz}^{-1} - \mathrm{j}\varepsilon_{yx} \\ -\sqrt{\varepsilon_{re}}\mu_{zz}^{-1}D_{\overline{x}} + \mathrm{j}\varepsilon_{xy} & \varepsilon_{xx} - \varepsilon_{re}\mu_{zz}^{-1} \end{bmatrix} \tag{2.53}$$

$$[R_{\mathrm{H}}^{z}] = \begin{bmatrix} \varepsilon_{re}\varepsilon_{zz}^{-1} - \mu_{xx} & \sqrt{\varepsilon_{re}}\varepsilon_{zz}^{-1}D_{\overline{x}} - \mathrm{j}\mu_{xy} \\ -\sqrt{\varepsilon_{re}}D_{\overline{x}}\varepsilon_{zz}^{-1} + \mathrm{j}\mu_{yx} & -D_{\overline{x}}\varepsilon_{zz}^{-1}D_{\overline{x}} - \mu_{yy} \end{bmatrix} \tag{2.54}$$

For diagonal material tensors without losses, the matrices are real. We obtain in discretised form (see Fig. 2.15):

$$\widehat{\boldsymbol{R}}_{\mathrm{E}}^{z} = \begin{bmatrix} \boldsymbol{\epsilon}_{yy} - \overline{\boldsymbol{D}}_{e}^{\mathrm{t}}\boldsymbol{\mu}_{zz}^{-1}\overline{\boldsymbol{D}}_{e} & \sqrt{\varepsilon_{re}}\overline{\boldsymbol{D}}_{h}\boldsymbol{\mu}_{zz}^{-1} - \mathrm{j}\boldsymbol{\epsilon}_{yx} \\ -\sqrt{\varepsilon_{re}}\boldsymbol{\mu}_{zz}^{-1}\overline{\boldsymbol{D}}_{e} + \mathrm{j}\boldsymbol{\epsilon}_{xy} & \boldsymbol{\epsilon}_{xx} - \varepsilon_{re}\boldsymbol{\mu}_{zz}^{-1} \end{bmatrix} \tag{2.55}$$

$$\widehat{\boldsymbol{R}}_{\mathrm{H}}^{z} = \begin{bmatrix} \varepsilon_{re}\boldsymbol{\epsilon}_{zz}^{-1} - \boldsymbol{\mu}_{xx} & \sqrt{\varepsilon_{re}}\boldsymbol{\epsilon}_{zz}^{-1}\overline{\boldsymbol{D}}_{h} - \mathrm{j}\boldsymbol{\mu}_{xy} \\ -\sqrt{\varepsilon_{re}}\overline{\boldsymbol{D}}_{e}\boldsymbol{\epsilon}_{zz}^{-1} + \mathrm{j}\boldsymbol{\mu}_{yx} & \overline{\boldsymbol{D}}_{h}^{\mathrm{t}}\boldsymbol{\epsilon}_{zz}^{-1}\overline{\boldsymbol{D}}_{h} - \boldsymbol{\mu}_{yy} \end{bmatrix} \tag{2.56}$$

The indices e and h at $\overline{\boldsymbol{D}} = \overline{h}^{-1}\boldsymbol{D}$ correspond to the transverse fields, i.e. E_y or E_z and H_y or H_z, respectively. $\overline{h} = k_o h$ is the normalised discretisation distance. The field components E_z and H_z are given by:

$$\mathbf{E}_z = \boldsymbol{\varepsilon}_{zz}^{-1}[\sqrt{\varepsilon_{re}}\boldsymbol{I} \;\; \overline{\boldsymbol{D}}_h]\begin{bmatrix} \widetilde{\mathbf{H}}_x \\ -\mathrm{j}\widetilde{\mathbf{H}}_y \end{bmatrix} \qquad \widetilde{\mathbf{H}}_z = -\boldsymbol{\mu}_{zz}^{-1}[\overline{\boldsymbol{D}}_e \;\; \sqrt{\varepsilon_{re}}\boldsymbol{I}]\begin{bmatrix} -\mathrm{j}\mathbf{E}_y \\ \mathbf{E}_x \end{bmatrix} \tag{2.57}$$

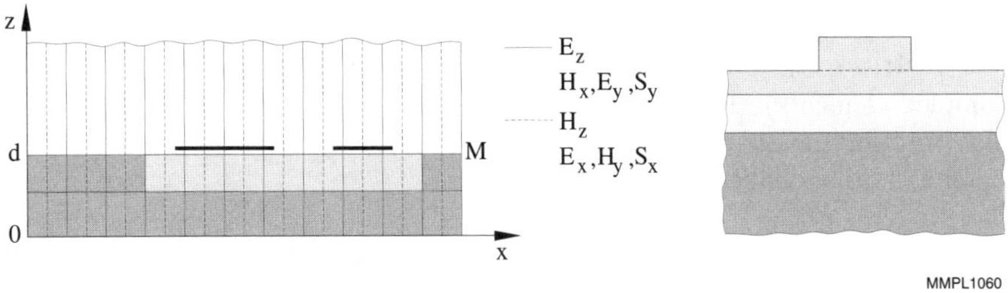

Fig. 2.15 Cross-sections of waveguides for microwave integrated circuits (Reproduced by permission of Taylor & Francis)

Sometimes the coordinate z is parallel to the waveguide axes and the coordinate y is perpendicular to the layers. The equations for this case are derived by coordinate transformation: $x_o \to x$, $y_o \to -z$, $z_o \to y$. The subscript o is used for 'old'. The field components are transformed according to: $F_{x_o} \to F_x$, $F_{y_o} \to -F_z$, $F_{z_o} \to F_y$. Furthermore, we have to change $\sqrt{\varepsilon_{re}} \to -\sqrt{\varepsilon_{re}}$.

Now we assume isotropic dielectric material. R_{E}^{z} and R_{H}^{z} simplify to:

$$\widehat{\boldsymbol{R}}_{\mathrm{E}}^{z} = \begin{bmatrix} \boldsymbol{\epsilon}_e - \overline{\boldsymbol{D}}_e^{\mathrm{t}} \overline{\boldsymbol{D}}_e & \sqrt{\varepsilon_{re}}\,\overline{\boldsymbol{D}}_h \\ -\sqrt{\varepsilon_{re}}\,\overline{\boldsymbol{D}}_e & \boldsymbol{\epsilon}_h - \varepsilon_{re}\boldsymbol{I}_h \end{bmatrix} \tag{2.58}$$

$$\widehat{\boldsymbol{R}}_{\mathrm{H}}^{z} = \begin{bmatrix} \varepsilon_{re}\boldsymbol{\epsilon}_e^{-1} - \boldsymbol{I}_e & \sqrt{\varepsilon_{re}}\,\boldsymbol{\epsilon}_e^{-1}\overline{\boldsymbol{D}}_h \\ -\sqrt{\varepsilon_{re}}\,\overline{\boldsymbol{D}}_e\boldsymbol{\epsilon}_e^{-1} & \overline{\boldsymbol{D}}_h^{\mathrm{t}}\boldsymbol{\epsilon}_e^{-1}\overline{\boldsymbol{D}}_h - \boldsymbol{I}_h \end{bmatrix} \tag{2.59}$$

and we have $\overline{\boldsymbol{D}}_{\mathrm{h}} = -\overline{\boldsymbol{D}}_{\mathrm{e}}^{\mathrm{t}}$.

As in the last section, we show the solution starting with the wave equation for the electric or magnetic field in parallel. However, we should state that only one of these is used in the actual calculations, while the other field is later determined using the GTL equations.

In this case, the discretised wave equations have the form:

$$\frac{\mathrm{d}^2}{\mathrm{d}\overline{z}^2}\widehat{\mathbf{E}} - \widehat{\boldsymbol{Q}}_{\mathrm{E}}^{z}\widehat{\mathbf{E}} = \mathbf{0} \qquad \frac{\mathrm{d}^2}{\mathrm{d}\overline{z}^2}\widehat{\mathbf{H}} - \widehat{\boldsymbol{Q}}_{\mathrm{H}}^{z}\widehat{\mathbf{H}} = \mathbf{0} \qquad \widehat{\boldsymbol{Q}}_{\mathrm{E}}^{z} = \widehat{\boldsymbol{R}}_{\mathrm{H}}^{z}\widehat{\boldsymbol{R}}_{\mathrm{E}}^{z} \qquad \widehat{\boldsymbol{Q}}_{\mathrm{H}}^{z} = \widehat{\boldsymbol{R}}_{\mathrm{E}}^{z}\widehat{\boldsymbol{R}}_{\mathrm{H}}^{z} \tag{2.60}$$

where:

$$\widehat{\boldsymbol{Q}}_{\mathrm{E}}^{z} = \begin{bmatrix} \varepsilon_{re}\boldsymbol{I}_e - \boldsymbol{\epsilon}_e + \overline{\boldsymbol{D}}_e^{\mathrm{t}}\overline{\boldsymbol{D}}_e & \sqrt{\varepsilon_{re}}(\overline{\boldsymbol{D}}_e^{\mathrm{t}} - \boldsymbol{\epsilon}_e^{-1}\overline{\boldsymbol{D}}_e^{\mathrm{t}}\boldsymbol{\epsilon}_h) \\ \boldsymbol{0} & \varepsilon_{re}\boldsymbol{I}_h - \boldsymbol{\epsilon}_h + \overline{\boldsymbol{D}}_h^{\mathrm{t}}\boldsymbol{\epsilon}_e^{-1}\overline{\boldsymbol{D}}_h\boldsymbol{\epsilon}_h \end{bmatrix} = \begin{bmatrix} \boldsymbol{Q}_e & \boldsymbol{Q}_{eh}^{\mathrm{E}} \\ \boldsymbol{0} & \boldsymbol{Q}_h^{\mathrm{E}} \end{bmatrix} \tag{2.61}$$

$$\widehat{\boldsymbol{Q}}_{\mathrm{H}}^{z} = \begin{bmatrix} \varepsilon_{re}\boldsymbol{I}_e - \boldsymbol{\epsilon}_e + \overline{\boldsymbol{D}}_e^{\mathrm{t}}\overline{\boldsymbol{D}}_e & \boldsymbol{0} \\ \sqrt{\varepsilon_{re}}(\overline{\boldsymbol{D}}_e - \boldsymbol{\epsilon}_h\overline{\boldsymbol{D}}_e\boldsymbol{\epsilon}_e^{-1}) & \varepsilon_{re}\boldsymbol{I}_h - \boldsymbol{\epsilon}_h + \boldsymbol{\epsilon}_h\overline{\boldsymbol{D}}_h^{\mathrm{t}}\boldsymbol{\epsilon}_e^{-1}\overline{\boldsymbol{D}}_h \end{bmatrix} = \begin{bmatrix} \boldsymbol{Q}_e & \boldsymbol{0} \\ \boldsymbol{Q}_{he}^{\mathrm{H}} & \boldsymbol{Q}_h^{\mathrm{H}} \end{bmatrix} \tag{2.62}$$

The hat ($\widehat{\ }$) on the quantities indicates supervectors and supermatrices. The matrices $\widehat{\boldsymbol{R}}_{\mathrm{E}}$ and $\widehat{\boldsymbol{R}}_{\mathrm{H}}$ are symmetric (but not Hermitian in the case of complex elements): $\widehat{\boldsymbol{R}}_{\mathrm{E,H}} = \widehat{\boldsymbol{R}}^{\mathrm{t}}_{\mathrm{E,H}}$. Therefore, we have $\widehat{\boldsymbol{Q}}_{\mathrm{H}} = \widehat{\boldsymbol{Q}}^{\mathrm{t}}_{\mathrm{E}}$, which is also true for the submatrices. Eq. (2.60) is solved by transformation to the main axis (diagonalisation):

$$\widehat{\mathbf{E}} = \widehat{\boldsymbol{T}}_{\mathrm{E}}\widehat{\overline{\mathbf{E}}} \qquad \widehat{\mathbf{H}} = \widehat{\boldsymbol{T}}_{\mathrm{H}}\widehat{\overline{\mathbf{H}}} \qquad \widehat{\boldsymbol{T}}_{\mathrm{E}}^{-1}\widehat{\boldsymbol{Q}}_{\mathrm{E}}\widehat{\boldsymbol{T}}_{\mathrm{E}} = \widehat{\boldsymbol{T}}_{\mathrm{H}}^{-1}\widehat{\boldsymbol{Q}}_{\mathrm{H}}\widehat{\boldsymbol{T}}_{\mathrm{H}} = \widehat{\boldsymbol{\Gamma}}^2 \tag{2.63}$$

If matrices $\widehat{\boldsymbol{Q}}_{\mathrm{E,H}}$ have the form given in eqs. (2.61) and (2.62), we can write for the eigenvalues and eigenvectors:

$$\widehat{\boldsymbol{\Gamma}}^2 = \begin{bmatrix} \boldsymbol{\Gamma}_e^2 & \boldsymbol{0} \\ \boldsymbol{0} & \boldsymbol{\Gamma}_h^2 \end{bmatrix} \qquad \widehat{\boldsymbol{T}}_{\mathrm{H}} = \begin{bmatrix} \boldsymbol{T}_e & \boldsymbol{0} \\ \boldsymbol{T}_{he} & \boldsymbol{T}_h^{\mathrm{H}} \end{bmatrix} \qquad \widehat{\boldsymbol{T}}_{\mathrm{E}} = \begin{bmatrix} \boldsymbol{T}_e & \boldsymbol{T}_{eh} \\ \boldsymbol{0} & \boldsymbol{T}_h^{\mathrm{E}} \end{bmatrix} \tag{2.64}$$

From eq. (2.60), considering eq. (2.63), we obtain uncoupled equations:

$$\frac{\mathrm{d}^2}{\mathrm{d}\overline{y}^2}\widehat{\overline{\mathbf{E}}} - \widehat{\boldsymbol{\Gamma}}^2\widehat{\overline{\mathbf{E}}} = \mathbf{0} \qquad \frac{\mathrm{d}}{\mathrm{d}\overline{y}^2}\widehat{\overline{\mathbf{H}}} - \widehat{\boldsymbol{\Gamma}}^2\widehat{\overline{\mathbf{H}}} = \mathbf{0} \tag{2.65}$$

where $\boldsymbol{\Gamma}_h^2$, $\boldsymbol{T}_h^{\mathrm{E,H}}$ and $\boldsymbol{\Gamma}_e^2$, $\boldsymbol{T}_e$ are the eigenvalues/eigenvectors of $\boldsymbol{Q}_h^{\mathrm{E,H}}$ and $\boldsymbol{Q}_e$, respectively. The eigenvector submatrices $\boldsymbol{T}_{he}$ and $\boldsymbol{T}_{eh}$ are obtained from these as the solutions of the following system of equations:

$$\boldsymbol{Q}_h^{\mathrm{H}}\boldsymbol{T}_{he} - \boldsymbol{T}_{he}\boldsymbol{\Gamma}_e^2 = -\boldsymbol{Q}_{he}^{\mathrm{H}}\boldsymbol{T}_e \qquad \boldsymbol{Q}_e\boldsymbol{T}_{eh} - \boldsymbol{T}_{eh}\boldsymbol{\Gamma}_h^2 = -\boldsymbol{Q}_{eh}^{\mathrm{E}}\boldsymbol{T}_h^{\mathrm{E}} \tag{2.66}$$

The kth column vectors of $\boldsymbol{T}_{he}$ and $\boldsymbol{T}_{eh}$, respectively, are given from the equations:

$$(\boldsymbol{Q}_h^{\mathrm{H}} - \boldsymbol{\Gamma}_{ek}^2\boldsymbol{I})\boldsymbol{T}_{he}^{(k)} = -\boldsymbol{Q}_{he}^{\mathrm{H}}\boldsymbol{T}_e^{(k)} \qquad (\boldsymbol{Q}_e - \boldsymbol{\Gamma}_{hk}^2\boldsymbol{I})\boldsymbol{T}_{eh}^{(k)} = -\boldsymbol{Q}_{eh}^{\mathrm{E}}\boldsymbol{T}_h^{(k)} \tag{2.67}$$

The inversion of the matrices $\widehat{\boldsymbol{T}}$ results in:

$$\widehat{\boldsymbol{T}}_{\mathrm{H}}^{-1} = \begin{bmatrix} \boldsymbol{T}_e^{-1} & \boldsymbol{0} \\ -\boldsymbol{T}_h^{\mathrm{H}-1}\boldsymbol{T}_{he}\boldsymbol{T}_e^{-1} & \boldsymbol{T}_h^{\mathrm{H}-1} \end{bmatrix} \qquad \widehat{\boldsymbol{T}}_{\mathrm{E}}^{-1} = \begin{bmatrix} \boldsymbol{T}_e^{-1} & -\boldsymbol{T}_e^{-1}\boldsymbol{T}_{eh}\boldsymbol{T}_h^{\mathrm{E}-1} \\ \boldsymbol{0} & \boldsymbol{T}_h^{\mathrm{E}-1} \end{bmatrix} \tag{2.68}$$

For general matrices $\widehat{\boldsymbol{Q}}_{\mathrm{E,H}}$, the eigenvalues and eigenvectors $\widehat{\boldsymbol{\Gamma}}^2_{\mathrm{E,H}}$ and $\widehat{\boldsymbol{T}}_{\mathrm{E,H}}$ must be determined numerically. The general solution of eq. (2.65) is given by ($\mathbf{F} = \mathbf{E}, \mathbf{H}$):

$$\widehat{\overline{\mathbf{F}}} = \cosh(\widehat{\boldsymbol{\Gamma}}(\overline{z} - \overline{z}_{\mathrm{o}}))\mathbf{A} + \sinh(\widehat{\boldsymbol{\Gamma}}(\overline{z} - \overline{z}_{\mathrm{o}}))\mathbf{B} \tag{2.69}$$

What we would like to obtain from eq. (2.69) is a relation between the transverse electric and magnetic fields at interfaces A and B. In the generalised transmission line equations, the derivative of the transverse E field is proportional to the transverse H field and vice versa. Therefore, we have to introduce the GTL equations into eq. (2.69) as in eq. (1.20). To do this in an adequate way, we determine only one of the transformation matrices $\widehat{\boldsymbol{T}}_{\mathrm{E}}$ or $\widehat{\boldsymbol{T}}_{\mathrm{H}}$ by the procedure described above or by solving one of the eigenvalue problems directly. Because the $\boldsymbol{Q}_{\mathrm{E,H}}$ matrices are products of the $\boldsymbol{R}_{\mathrm{E,H}}$ matrices, the transformation matrices for the electric resp. magnetic field can be determined from one another in the following way:

$$\widehat{\boldsymbol{T}}_{\mathrm{H}} = \boldsymbol{R}_{\mathrm{E}} \widehat{\boldsymbol{T}}_{\mathrm{E}} \widehat{\boldsymbol{\Gamma}}^{-1} \quad \text{or} \quad \widehat{\boldsymbol{T}}_{\mathrm{E}} = \boldsymbol{R}_{\mathrm{H}} \widehat{\boldsymbol{T}}_{\mathrm{H}} \widehat{\boldsymbol{\Gamma}}^{-1} \tag{2.70}$$

The multiplication with $\boldsymbol{\Gamma}^{-1}$ was introduced for normalisation, leading to a unit matrix for the wave impedance/admittance matrix (see also the last section).

With the relations given in eq. (2.70), we obtain for the GTL equations in transformed domain:

$$\frac{\mathrm{d}}{\mathrm{d}\overline{y}}\widehat{\overline{\mathbf{E}}} = -\widehat{\boldsymbol{\Gamma}}\widehat{\overline{\mathbf{H}}} \qquad \frac{\mathrm{d}}{\mathrm{d}\overline{y}}\widehat{\overline{\mathbf{H}}} = -\widehat{\boldsymbol{\Gamma}}\widehat{\overline{\mathbf{E}}} \tag{2.71}$$

Now, by introducing these expressions into an expression analogous to eq. (1.20), we obtain for the kth layer:

$$\begin{bmatrix} \widehat{\overline{\mathbf{H}}}_{\mathrm{A}}^{(k)} \\ -\widehat{\overline{\mathbf{H}}}_{\mathrm{B}}^{(k)} \end{bmatrix} = \begin{bmatrix} \widehat{\overline{\mathbf{y}}}_1 & \widehat{\overline{\mathbf{y}}}_2 \\ \widehat{\overline{\mathbf{y}}}_2 & \widehat{\overline{\mathbf{y}}}_1 \end{bmatrix}_{|k} \begin{bmatrix} \widehat{\overline{\mathbf{E}}}_{\mathrm{A}}^{(k)} \\ \widehat{\overline{\mathbf{E}}}_{\mathrm{B}}^{(k)} \end{bmatrix} \qquad \begin{aligned} \widehat{\overline{\mathbf{y}}}_1 &= \widehat{\overline{\mathbf{Y}}}_0/\tanh(\widehat{\boldsymbol{\Gamma}}\overline{d}) \\ \widehat{\overline{\mathbf{y}}}_2 &= -\widehat{\overline{\mathbf{Y}}}_0/\sinh(\widehat{\boldsymbol{\Gamma}}\overline{d}) \end{aligned} \tag{2.72}$$

$$\begin{bmatrix} \widehat{\overline{\mathbf{E}}}_{\mathrm{A}}^{(k)} \\ \widehat{\overline{\mathbf{E}}}_{\mathrm{B}}^{(k)} \end{bmatrix} = \begin{bmatrix} \widehat{\overline{\mathbf{z}}}_1 & \widehat{\overline{\mathbf{z}}}_2 \\ \widehat{\overline{\mathbf{z}}}_2 & \widehat{\overline{\mathbf{z}}}_1 \end{bmatrix}_{|k} \begin{bmatrix} \widehat{\overline{\mathbf{H}}}_{\mathrm{A}}^{(k)} \\ -\widehat{\overline{\mathbf{H}}}_{\mathrm{B}}^{(k)} \end{bmatrix} \qquad \begin{aligned} \widehat{\overline{\mathbf{z}}}_1 &= \widehat{\overline{\mathbf{Z}}}_0/\tanh(\widehat{\boldsymbol{\Gamma}}\overline{d}) \\ \widehat{\overline{\mathbf{z}}}_2 &= \widehat{\overline{\mathbf{Z}}}_0/\sinh(\widehat{\boldsymbol{\Gamma}}\overline{d}) \end{aligned} \tag{2.73}$$

As mentioned before, $\widehat{\overline{\mathbf{Y}}}_0$ $\widehat{\overline{\mathbf{Z}}}_0$ are unit matrices here. However, the equations (2.72) and (2.73) can also be used in cases of different normalisations leading to other expressions. Therefore, we kept $\widehat{\overline{\mathbf{Y}}}_0$ and $\widehat{\overline{\mathbf{Z}}}_0$ here.

The formulas for the relation of the fields in planes A and B are independent of the coordinate system.

2.3.3 Examples of system equations

Let us show the procedure for the analysis of the structure shown in Fig. 2.16.

We start with a complete metallisation in plane A ($z = 0$), leading to $\widehat{\overline{\mathbf{E}}}_{\mathrm{A}} = \mathbf{0}$. Therefore, we can write in transformed domain:

$$\widehat{\overline{\mathbf{H}}}_{\mathrm{M}}^{d} = -\widehat{\overline{\mathbf{Y}}}_{\mathrm{M}}^{d}\widehat{\overline{\mathbf{E}}}_{\mathrm{M}} = -\widehat{\overline{\mathbf{y}}}_1^{d}\widehat{\overline{\mathbf{E}}}_{\mathrm{M}} \tag{2.74}$$

The subscript M (instead of B) is used for the plane of metallisation (lower side). The superscript d denotes the (inhomogeneous) dielectric layer.

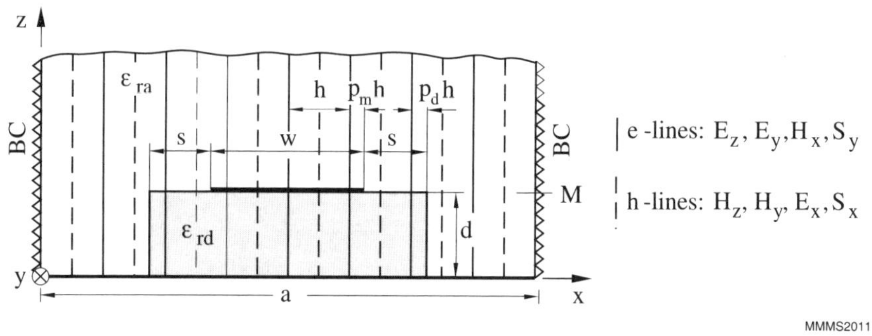

Fig. 2.16 Cross-section of a microstrip with finite substrate width (Reproduced by permission of Taylor & Francis)

The layer above the strip (indicated with the superscript a) may be of finite or infinite thickness. We start in the upper plane of this layer and assume complete metallisation as well, i.e. $\widehat{\overline{\mathbf{E}}}_B = 0$. By using eq. (2.72) we obtain in the plane of the strip (upper side) in transformed domain (note that the transformation matrices for both layers are different!):

$$\widehat{\overline{\mathbf{H}}}^a_{\mathrm{M}} = \widehat{\overline{\mathbf{Y}}}^a_{\mathrm{M}} \widehat{\overline{\mathbf{E}}}_{\mathrm{M}} = \widehat{\overline{\boldsymbol{y}}}^a_1 \widehat{\overline{\mathbf{E}}}_{\mathrm{M}} \tag{2.75}$$

If the upper layer has an infinite height, we obtain (independent of the boundary condition at the top) the wave admittance as the relation between the electric and magnetic field (which is a unit matrix in our case), i.e. $\widehat{\overline{\boldsymbol{y}}}^a_1 = \widehat{\boldsymbol{I}}$, because $\tanh(\widehat{\boldsymbol{\Gamma}}\overline{d})$ approaches a unit matrix. After inverse transformation (to obtain the equations in spatial domain), the difference of the two equations yields:

$$\widehat{\mathbf{H}}^a_{\mathrm{M}} - \widehat{\mathbf{H}}^d_{\mathrm{M}} = \widehat{\mathbf{S}}_{\mathrm{M}} = (\widehat{\mathbf{Y}}^a_{\mathrm{M}} + \widehat{\mathbf{Y}}^d_{\mathrm{M}})\widehat{\mathbf{E}}_{\mathrm{M}} \tag{2.76}$$

We assume that the electric field is equal in the plane M on both sides of the strip. $\widehat{\mathbf{S}}_{\mathrm{M}}$ is a supervector containing the current densities on the strip:

$$\widehat{\mathbf{S}}_{\mathrm{M}} = \eta_0[\mathbf{S}^t_y,\, \mathrm{j}\mathbf{S}^t_x]^t \tag{2.77}$$

From eq. (2.76) we obtain:

$$(\widehat{\mathbf{Y}}^a_{\mathrm{M}} + \widehat{\mathbf{Y}}^d_{\mathrm{M}})^{-1}\widehat{\mathbf{S}}_{\mathrm{M}} = \widehat{\mathbf{E}}_{\mathrm{M}} \tag{2.78}$$

Now a further condition in the plane M must be introduced. The components of $\widehat{\mathbf{E}}_{\mathrm{M}}$ on the strip and the components of $\widehat{\mathbf{S}}_{\mathrm{M}}$ outside the strip must be zero. Therefore eq. (2.78) splits into two equation systems (see Section 2.3.7):

$$(\mathbf{Y}^a_{\mathrm{M}} + \mathbf{Y}^d_{\mathrm{M}})^{-1}_{\mathrm{red}}\widehat{\mathbf{S}}_{\mathrm{MS}} = \mathbf{0} \qquad (\mathbf{Y}^a_{\mathrm{M}} + \mathbf{Y}^d_{\mathrm{M}})^{-1}[\widehat{\mathbf{0}}^{\mathrm{t}}_l, \widehat{\mathbf{S}}^{\mathrm{t}}_{\mathrm{MS}}, \widehat{\mathbf{0}}^{\mathrm{t}}_r]^{\mathrm{t}} = \widehat{\mathbf{E}}_{\mathrm{M}} \tag{2.79}$$

The first equation is an indirect eigenvalue problem, and the second one allows the determination of $\widehat{\mathbf{E}}_{\mathrm{M}}$ from $\widehat{\mathbf{S}}_{\mathrm{MS}}$, the supervectors of the currents on the strip.

$\widehat{\mathbf{0}}_l$ and $\widehat{\mathbf{0}}_r$ are the subvectors indicating the regions without metal, i.e. the part where the current densities are zero. This expression should be understood symbolically, because all the vectors contain two components.

2.3.4 Impedance/admittance transformation in multilayered structures

Now we assume an arbitrary cross-section, as shown in Fig. 2.13, and use a procedure for impedance/admittance transformation through the layers and interfaces. The general case for cross-sections with different width will be described in Section 2.5.2.

If we have a metallic shielding in the interface 0 (the lower plane of the cross-section), we start as in the previous section (see eq. (2.74)). In case of a magnetic shielding, we obtain the analogous result.

$$\begin{aligned} &\text{metallic shielding:} \quad \widehat{\widehat{\mathbf{H}}}_{\mathrm{B1}} = -\widehat{\widehat{\mathbf{Y}}}_{\mathrm{B1}}\widehat{\widehat{\mathbf{E}}}_{\mathrm{B1}} = -\widehat{\widehat{\mathbf{y}}}_1\widehat{\widehat{\mathbf{E}}}_{\mathrm{B1}} \\ &\text{magnetic shielding:} \quad \widehat{\widehat{\mathbf{E}}}_{\mathrm{B1}} = -\widehat{\widehat{\mathbf{Z}}}_{\mathrm{B1}}\widehat{\widehat{\mathbf{H}}}_{\mathrm{B1}} = -\widehat{\widehat{\mathbf{z}}}_1\widehat{\widehat{\mathbf{H}}}_{\mathrm{B1}} \end{aligned} \tag{2.80}$$

These values are valid for the upper side of the lowest layer. The procedure of impedance/admittance transformation through a general layer k is obtained from the eqs. (2.69) and (1.20). In the layer k we use the definition:

$$\overline{\mathbf{H}}^{(k)}_{\mathrm{A,B}} = -\overline{\mathbf{Y}}^{(k)}_{\mathrm{A,B}}\overline{\mathbf{E}}^{(k)}_{\mathrm{A,B}} \qquad \overline{\mathbf{E}}^{(k)}_{\mathrm{A,B}} = -\overline{\mathbf{Z}}^{(k)}_{\mathrm{A,B}}\overline{\mathbf{H}}^{(k)}_{\mathrm{A,B}} \tag{2.81}$$

(We are looking in the $-z$-direction. Therefore, we have introduced the minus sign into the definition.) We obtain for the impedances/admittances $\overline{\mathbf{Z}}^{(k)}_{\mathrm{B}}$, $\overline{\mathbf{Y}}^{(k)}_{\mathrm{B}}$ from s $\overline{\mathbf{Z}}^{(k)}_{\mathrm{A}}$, $\overline{\mathbf{Y}}^{(k)}_{\mathrm{A}}$:

$$\begin{aligned} \overline{\mathbf{Z}}^{(k)}_{\mathrm{B}} &= \overline{\mathbf{z}}^{(k)}_1 - \overline{\mathbf{z}}^{(k)}_2(\overline{\mathbf{z}}^{(k)}_1 + \overline{\mathbf{Z}}^{(k)}_{\mathrm{A}})^{-1}\overline{\mathbf{z}}^{(k)}_2 \\ \overline{\mathbf{Y}}^{(k)}_{\mathrm{B}} &= \overline{\mathbf{y}}^{(k)}_1 - \overline{\mathbf{y}}^{(k)}_2(\overline{\mathbf{y}}^{(k)}_1 + \overline{\mathbf{Y}}^{(k)}_{\mathrm{A}})^{-1}\overline{\mathbf{y}}^{(k)}_2 \end{aligned} \tag{2.82}$$

$\overline{\mathbf{Y}}^{(k)}_{\mathrm{C}}$, C = A,B can be represented in the following way:

$$\overline{\mathbf{Y}}^{(k)}_{\mathrm{C}} = \begin{bmatrix} \overline{y}_{11} & \overline{y}_{12} \\ \overline{y}_{21} & \overline{y}_{22} \end{bmatrix}_{|k} \tag{2.83}$$

At an interface we have the following relation between the fields in transformed domain:

$$\overline{\mathbf{E}}_{\mathrm{k}} = \boldsymbol{T}^{(\mathrm{k})-1}_{\mathrm{E}}\boldsymbol{T}^{(\mathrm{k}+1)}_{\mathrm{E}}\overline{\mathbf{E}}_{\mathrm{k}+1} \qquad \overline{\mathbf{H}}_{\mathrm{k}} = \boldsymbol{T}^{(\mathrm{k})-1}_{\mathrm{H}}\boldsymbol{T}^{(\mathrm{k}+1)}_{\mathrm{H}}\overline{\mathbf{H}}_{\mathrm{k}+1} \tag{2.84}$$

Therefore, we obtain for the transformation of the impedance/admittance at the interface between layer k and layer $k+1$:

$$\overline{\mathbf{Y}}_{\mathrm{A}}^{(k+1)} = \boldsymbol{T}_{\mathrm{H}}^{(k+1)-1}\boldsymbol{T}_{\mathrm{H}}^{(k)}\overline{\mathbf{Y}}_{\mathrm{B}}^{(k)}\boldsymbol{T}_{\mathrm{E}}^{(k)-1}\boldsymbol{T}_{\mathrm{E}}^{(k+1)} \tag{2.85}$$

$$\overline{\mathbf{Z}}_{\mathrm{A}}^{(k+1)} = \boldsymbol{T}_{\mathrm{E}}^{(k+1)-1}\boldsymbol{T}_{\mathrm{E}}^{(k)}\overline{\mathbf{Z}}_{\mathrm{B}}^{(k)}\boldsymbol{T}_{\mathrm{H}}^{(k)-1}\boldsymbol{T}_{\mathrm{H}}^{(k+1)} \tag{2.86}$$

If we have different material parameters in the two layers, the transformation matrices are different, even in case of homogeneous layers. It is assumed that the interface contains no metallisation. If there is a metallisation we have to use the general procedure described in Chapter 5.

In a similar way, we can start from the upper cover and transform downwards. If we start counting the layers at the top with 1, we obtain for the lower side of layer l:

$$\begin{aligned} &\text{metallic shielding:} \quad \widehat{\overline{\mathbf{H}}}_{\mathrm{A1}} = \widehat{\overline{\mathbf{Y}}}_{\mathrm{A1}}\widehat{\overline{\mathbf{E}}}_{\mathrm{A1}} = \widehat{\overline{\boldsymbol{y}}}_{1}\widehat{\overline{\mathbf{E}}}_{\mathrm{A1}} \\ &\text{magnetic shielding:} \quad \widehat{\overline{\mathbf{E}}}_{\mathrm{A1}} = \widehat{\overline{\mathbf{Z}}}_{\mathrm{A1}}\widehat{\overline{\mathbf{H}}}_{\mathrm{A1}} = \widehat{\overline{\boldsymbol{z}}}_{1}\widehat{\overline{\mathbf{H}}}_{\mathrm{A1}} \end{aligned} \tag{2.87}$$

By using the definition:

$$\overline{\mathbf{H}}_{\mathrm{A,B}}^{(l)} = \overline{\mathbf{Y}}_{\mathrm{A,B}}^{(l)}\overline{\mathbf{E}}_{\mathrm{A,B}}^{(l)} \qquad \overline{\mathbf{E}}_{\mathrm{A,B}}^{(l)} = \overline{\mathbf{Z}}_{\mathrm{A,B}}^{(l)}\overline{\mathbf{H}}_{\mathrm{A,B}}^{(l)} \tag{2.88}$$

we obtain in the layer l for the impedances/admittances $\overline{\mathbf{Z}}_{\mathrm{A}}^{(l)}$, $\overline{\mathbf{Y}}_{\mathrm{A}}^{(l)}$ from $\overline{\mathbf{Z}}_{\mathrm{B}}^{(l)}$, $\overline{\mathbf{Y}}_{\mathrm{B}}^{(l)}$:

$$\begin{aligned} \overline{\mathbf{Z}}_{\mathrm{A}}^{(l)} &= \overline{\boldsymbol{z}}_{1}^{(l)} - \overline{\boldsymbol{z}}_{2}^{(l)}(\overline{\boldsymbol{z}}_{1}^{(l)} + \overline{\mathbf{Z}}_{\mathrm{B}}^{(l)})^{-1}\overline{\boldsymbol{z}}_{2}^{(l)} \\ \overline{\mathbf{Y}}_{\mathrm{A}}^{(l)} &= \overline{\boldsymbol{y}}_{1}^{(l)} - \overline{\boldsymbol{y}}_{2}^{(l)}(\overline{\boldsymbol{y}}_{1}^{(k)} + \overline{\mathbf{Y}}_{\mathrm{B}}^{(l)})^{-1}\overline{\boldsymbol{y}}_{2}^{(l)} \end{aligned} \tag{2.89}$$

The transformation between the layers occurs analogously to the previous case.

2.3.5 System equation in transformed domain

In the next step we want to derive an algebraic relation, which connects the tangential electric fields with the currents at the interfaces with metallisation (the subscript M in the following basic equations shall mark this association).

$$\overline{\mathbf{S}}_{\mathrm{M}} = f(\overline{\mathbf{E}}_{\mathrm{M}}) \quad \text{or} \quad \overline{\mathbf{E}}_{\mathrm{M}} = f(\overline{\mathbf{S}}_{\mathrm{M}}) \tag{2.90}$$

Let e.g. the interface on the upper side of layer k (counted from the bottom) be identical to the interface on the lower side of layer l (counted from the top) and let a metallisation only be at this interface (now marked M, see e.g. Fig. 1.4 with $k = 2, l = 1$). At this interface the matching equations hold:

$$\overline{\mathbf{E}}_{k} = \overline{\mathbf{E}}_{l} = \overline{\mathbf{E}}_{\mathrm{M}} \qquad \overline{\mathbf{H}}_{l} - \overline{\mathbf{H}}_{k} = \overline{\mathbf{S}}_{\mathrm{M}} \tag{2.91}$$

With the eqs. (2.81) and (2.88) we get the system equations in the common transformed domain:

$$(\overline{\mathbf{Y}}_{\mathrm{B}}^{(k)} + \overline{\mathbf{Y}}_{\mathrm{A}}^{(l)})\overline{\mathbf{E}}_{\mathrm{M}} = \overline{\mathbf{S}}_{\mathrm{M}} \tag{2.92}$$

or:

$$(\overline{\mathbf{Y}}_{\mathrm{B}}^{(k)} + \overline{\mathbf{Y}}_{\mathrm{A}}^{(l)})^{-1}\overline{\mathbf{S}}_{\mathrm{M}} = \overline{\mathbf{E}}_{\mathrm{M}} \tag{2.93}$$

As another example, we look at the waveguide in Fig. 2.17. The layers have the same width. We assume that all quantities are recalculated into the transformed domain as for the layer II. At the interfaces A and B, relations according to eq. (2.81) and eq. (2.88) hold, namely:

$$\overline{\mathbf{H}}_{\mathrm{B^I}}^{\mathrm{I}} = -\overline{\mathbf{Y}}_{\mathrm{B^I}}^{\mathrm{I}}\overline{\mathbf{E}}_{\mathrm{A}} \qquad \overline{\mathbf{H}}_{\mathrm{A^{III}}}^{\mathrm{III}} = \overline{\mathbf{Y}}_{\mathrm{A^{III}}}^{\mathrm{III}}\overline{\mathbf{E}}_{\mathrm{B}} \tag{2.94}$$

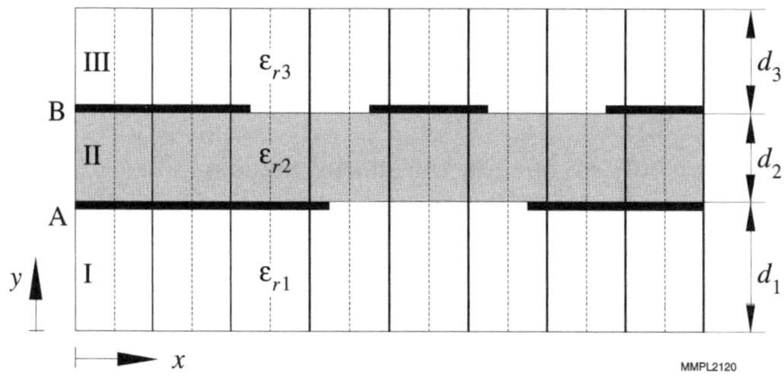

Fig. 2.17 Quasi-planar waveguide with 2 interfaces with metallisation (model for a finline)

(Note that interface A is identical with interface $\mathrm{B^I}$ and interface B is identical with interface $\mathrm{A^{III}}$.) Moreover, an equation according to (2.72) holds for the tangential field components at the interfaces A and B of the layer II:

$$\begin{bmatrix} \overline{\mathbf{H}}_{\mathrm{A}}^{\mathrm{II}} \\ -\overline{\mathbf{H}}_{\mathrm{B}}^{\mathrm{II}} \end{bmatrix} = \begin{bmatrix} \overline{\mathbf{y}}_1^{\mathrm{II}} & \overline{\mathbf{y}}_2^{\mathrm{II}} \\ \overline{\mathbf{y}}_2^{\mathrm{II}} & \overline{\mathbf{y}}_1^{\mathrm{II}} \end{bmatrix} \begin{bmatrix} \overline{\mathbf{E}}_{\mathrm{A}} \\ \overline{\mathbf{E}}_{\mathrm{B}} \end{bmatrix} \tag{2.95}$$

In the formulation of eqs. (2.94) and (2.95), by using the same $\overline{\mathbf{E}}_{\mathrm{A}}$ and $\overline{\mathbf{E}}_{\mathrm{B}}$ respectively in both systems of equations, the continuity condition for the tangential electric field is already fulfilled. For the magnetic field holds:

$$\begin{aligned} \overline{\mathbf{H}}_{\mathrm{A}}^{\mathrm{II}} - \overline{\mathbf{H}}_{\mathrm{B^I}}^{\mathrm{I}} &= \overline{\mathbf{S}}_{\mathrm{A}} \\ \overline{\mathbf{H}}_{\mathrm{A^{III}}}^{\mathrm{III}} - \overline{\mathbf{H}}_{\mathrm{B}}^{\mathrm{II}} &= \overline{\mathbf{S}}_{\mathrm{B}} \end{aligned} \tag{2.96}$$

The combination of the equations (2.94) to (2.96) leads to the following system equation in the transform domain:

$$\begin{bmatrix} \overline{\mathbf{y}}_1^{\mathrm{II}} + \overline{\mathbf{Y}}_{\mathrm{B^I}}^{\mathrm{I}} & \overline{\mathbf{y}}_2^{\mathrm{II}} \\ \overline{\mathbf{y}}_2^{\mathrm{II}} & \overline{\mathbf{y}}_1^{\mathrm{II}} + \overline{\mathbf{Y}}_{\mathrm{A^{III}}}^{\mathrm{III}} \end{bmatrix} \begin{bmatrix} \overline{\mathbf{E}}_{\mathrm{A}} \\ \overline{\mathbf{E}}_{\mathrm{B}} \end{bmatrix} = \begin{bmatrix} \overline{\mathbf{S}}_{\mathrm{A}} \\ \overline{\mathbf{S}}_{\mathrm{B}} \end{bmatrix} \tag{2.97}$$

This equation should be converted depending on the width of the metallisations compared with the width of the slots. If e.g. at interface B only the metal strip in the middle exists, i.e. the width of the two outer metallisations is zero, then a conversion into the form:

$$\begin{bmatrix} \overline{\boldsymbol{Y}}_{\mathrm{A}} & \overline{\boldsymbol{V}}_{\mathrm{AB}} \\ \overline{\boldsymbol{V}}_{\mathrm{BA}} & \overline{\boldsymbol{Z}}_{\mathrm{B}} \end{bmatrix} \begin{bmatrix} \overline{\mathbf{E}}_{\mathrm{A}} \\ \overline{\mathbf{S}}_{\mathrm{B}} \end{bmatrix} = \begin{bmatrix} \overline{\mathbf{S}}_{\mathrm{A}} \\ \overline{\mathbf{E}}_{\mathrm{B}} \end{bmatrix} \tag{2.98}$$

should take place. The reason for this form of representation will be made clear by the ideas put forward in the next section. The equations (2.97) and (2.98) will allow us to formulate eigenmode equations with the electric fields in plane A and B or electric field in plane A and currents in plane B as eigenvectors. In the next chapter we will demonstrate how for the field components in a chosen layer, the eigenmode equations can be formulated.

2.3.6 System equation in spatial domain

The interface conditions in the interfaces with metallisation still have to be fulfilled. This can only be done in the spatial domain. The tangential electric field components on the metallic strips and the electric current densities in the slots must be zero.

$$E_{xi}, E_{yi} = 0 \qquad S_{xk}, S_{yk} = 0 \tag{2.99}$$

We take the index i for the lines on the metallisation and k outside (i.e. the slot). For the interface M in Fig. 1.4, the quantities $\mathbf{E}$ and $\mathbf{S}$ can be calculated in the following way:

$$\begin{aligned} \mathbf{E}_{x\mathrm{M}} &= \begin{bmatrix} \mathbf{E}_{xs} \\ \mathbf{0} \end{bmatrix} \qquad \mathbf{E}_{y\mathrm{M}} = \begin{bmatrix} \mathbf{E}_{ys} \\ \mathbf{0} \end{bmatrix} \\ \mathbf{S}_{x\mathrm{M}} &= \begin{bmatrix} \mathbf{0} \\ \mathbf{S}_{xm} \end{bmatrix} \qquad \mathbf{S}_{y\mathrm{M}} = \begin{bmatrix} \mathbf{0} \\ \mathbf{S}_{ym} \end{bmatrix} \end{aligned} \tag{2.100}$$

The subscript s ($\mathbf{E}_{xs}$ and $\mathbf{E}_{ys}$) stands for 'slot' and the subscript m in $\mathbf{S}_{xm}$ and $\mathbf{S}_{ym}$ for 'metallisation'. These conditions must be substituted in eqs. (2.92) or (2.93), which have to be transformed back to spatial domain. We assume that the width of the metallisation is smaller than the width of the slot, therefore we use eq. (2.93). As the matrices have a structure as shown in eq. (2.83), we write for eq. (2.93):

$$\begin{bmatrix} \overline{\boldsymbol{Z}}_{11} & \overline{\boldsymbol{Z}}_{12} \\ \overline{\boldsymbol{Z}}_{21} & \overline{\boldsymbol{Z}}_{22} \end{bmatrix} \begin{bmatrix} \overline{\mathbf{S}}_{y\mathrm{M}} \\ \mathrm{j}\overline{\mathbf{S}}_{x\mathrm{M}} \end{bmatrix} = \begin{bmatrix} -\mathrm{j}\overline{\mathbf{E}}_{y\mathrm{M}} \\ \overline{\mathbf{E}}_{x\mathrm{M}} \end{bmatrix} \tag{2.101}$$

The quantities $\overline{\mathbf{S}}_{\mathrm{M}}$ must now be transformed back with $\boldsymbol{T}_{\mathrm{H}}$ and the quantities $\overline{\mathbf{E}}_{\mathrm{M}}$ with $\boldsymbol{T}_{\mathrm{E}}$. Thus the inverse transformation of eq. (2.101) in the spatial domain is done as follows:

$$\widehat{\boldsymbol{T}}_{\mathrm{E}} \begin{bmatrix} \overline{\boldsymbol{Z}}_{11} & \overline{\boldsymbol{Z}}_{12} \\ \overline{\boldsymbol{Z}}_{21} & \overline{\boldsymbol{Z}}_{22} \end{bmatrix} \widehat{\boldsymbol{T}}_{\mathrm{H}}^{-1} \begin{bmatrix} \mathbf{S}_{y\mathrm{M}} \\ \mathrm{j}\mathbf{S}_{x\mathrm{M}} \end{bmatrix} = \begin{bmatrix} -\mathrm{j}\mathbf{E}_{y\mathrm{M}} \\ \mathbf{E}_{x\mathrm{M}} \end{bmatrix} \tag{2.102}$$

or:

$$\begin{bmatrix} \boldsymbol{Z}_{11} & \boldsymbol{Z}_{12} \\ \boldsymbol{Z}_{21} & \boldsymbol{Z}_{22} \end{bmatrix} \begin{bmatrix} \mathbf{S}_{y\mathrm{M}} \\ \mathrm{j}\mathbf{S}_{x\mathrm{M}} \end{bmatrix} = \begin{bmatrix} -\mathrm{j}\mathbf{E}_{y\mathrm{M}} \\ \mathbf{E}_{x\mathrm{M}} \end{bmatrix} \tag{2.103}$$

Now we introduce the conditions according to eq. (2.100) into eq. (2.103) and obtain:

$$\begin{bmatrix} \boldsymbol{Z}_{11} & \boldsymbol{Z}_{12} \\ \boldsymbol{Z}_{12}^{\mathrm{t}} & \boldsymbol{Z}_{22} \end{bmatrix} \begin{bmatrix} \mathbf{0} \\ \mathbf{S}_{ym} \\ \mathbf{0} \\ \mathrm{j}\mathbf{S}_{xm} \end{bmatrix} = \begin{bmatrix} -\mathrm{j}\mathbf{E}_{ys} \\ \mathbf{0} \\ \mathbf{E}_{xs} \\ \mathbf{0} \end{bmatrix} \tag{2.104}$$

The columns of the submatrices $\boldsymbol{Z}_{ik}$ that are multiplied by zeros from the current vector have no contribution to the right side. Hence it is not necessary to perform those calculations. For this reason, we rewrite eq. (2.104) in a convenient way. In the first step we omit the columns of $\boldsymbol{Z}_{ik}$ that would be multiplied by zero, yielding the so-called reduced matrix. In the second step we partition each of the rectangular submatrices in an upper (superscript u) and a lower (superscript l) submatrix, according to the partition of the vector on the right-hand side. We obtain from eq. (2.104):

$$\begin{bmatrix} \boldsymbol{Z}_{11}^{ru} & \boldsymbol{Z}_{12}^{ru} \\ \boldsymbol{Z}_{11}^{rl} & \boldsymbol{Z}_{12}^{rl} \\ \boldsymbol{Z}_{21}^{ru} & \boldsymbol{Z}_{22}^{ru} \\ \boldsymbol{Z}_{21}^{rl} & \boldsymbol{Z}_{22}^{rl} \end{bmatrix} \begin{bmatrix} \mathbf{S}_{ym} \\ \mathrm{j}\mathbf{S}_{x\mathrm{m}} \end{bmatrix} = \begin{bmatrix} -\mathrm{j}\mathbf{E}_{ys} \\ \mathbf{0} \\ \mathbf{E}_{xs} \\ \mathbf{0} \end{bmatrix} \tag{2.105}$$

The superscript r means 'reduced' in connection with the first step. Now, the system of equations (2.105) is divided into the two systems:

$$\begin{bmatrix} \boldsymbol{Z}_{11}^{rl} & \boldsymbol{Z}_{12}^{rl} \\ \boldsymbol{Z}_{12}^{rl\mathrm{j}} & \boldsymbol{Z}_{22}^{rl} \end{bmatrix} \begin{bmatrix} \mathbf{S}_{ym} \\ \mathrm{j}\mathbf{S}_{xm} \end{bmatrix} = \begin{bmatrix} \mathbf{0} \\ \mathbf{0} \end{bmatrix} \tag{2.106}$$

and:

$$\begin{bmatrix} \boldsymbol{Z}_{11}^{ru} & \boldsymbol{Z}_{12}^{ru} \\ \boldsymbol{Z}_{21}^{ru} & \boldsymbol{Z}_{22}^{ru} \end{bmatrix} \begin{bmatrix} \mathbf{S}_{ym} \\ \mathrm{j}\mathbf{S}_{xm} \end{bmatrix} = \begin{bmatrix} -\mathrm{j}\mathbf{E}_{ys} \\ \mathbf{E}_{xs} \end{bmatrix} \tag{2.107}$$

System (2.106) is an indirect eigenvalue system. The elements $(\boldsymbol{Z}_{ij}^{rl})_{lk}$ of the system matrix contain the normalised propagation constant ($\varepsilon_{re} = (k_z/k_0)^2$). The eigenvalue $\boldsymbol{\varepsilon}_{re}$ must be varied until the determinant of this system matrix vanishes.

The current vector $[\mathbf{S}_{ym}^{\mathrm{t}}, \mathrm{j}\mathbf{S}_{xm}^{\mathrm{t}}]^{\mathrm{t}}$ is afterwards determined as an eigenvector. If $\boldsymbol{\varepsilon}_{re}$ and the current vector are evaluated, the field vector $[-\mathrm{j}\mathbf{E}_{ys}^{\mathrm{t}}, \mathbf{E}_{xs}^{\mathrm{t}}]^{\mathrm{t}}$ can be calculated with eq. (2.107). Then all other field components in the various interfaces can be determined by means of the equations given in Chapter 3. They must be transformed back to spatial domain first. The field quantities can also be determined inside the layers, as the transformed field quantities vary in y-direction according to the solution given in eq. (2.78).

The system matrix in eq. (2.106) is smaller than the matrix in eq. (2.104); considerably smaller in the case of narrow strip widths compared with the slot widths. Often a few lines on the strip are sufficient for very accurate results. If the width of the strip is greater than the slot width, eq. (2.92) should be transformed back into spatial domain instead of eq. (2.93), and should be reduced by an approach analogous to the above one. Instead of eq. (2.106), we obtain an indirect eigenvalue system for the vector $[-\mathrm{j}\mathbf{E}_{ys}^{\mathrm{t}}, \mathbf{E}_{xs}^{\mathrm{t}}]^{\mathrm{t}}$ and a second system of equations to determine the vector $[\mathbf{S}_{ym}^{\mathrm{t}}, \mathrm{j}\mathbf{S}_{xm}^{\mathrm{t}}]^{\mathrm{t}}$. This corresponds to the fact that the analysis of a time-invariant network can be either done by the node or by the mesh analysis, or by a mixed analysis. A mixed analysis should be used in the case of waveguides as shown in Fig. 2.17. Here, we have a strip in interface B and the slot in interface A under the assumption of the depicted proportions. Whereas in interface A we have less lines in the slot region, the number of lines in interface B is smaller in the region of the strip than in rest of the interface. Thus eq. (2.98) should be transformed back and then reduced corresponding to those components which are different from zero in the subvectors $\mathbf{E}_{\mathrm{A}}$ and $\mathbf{S}_{\mathrm{B}}$. Finally, the system is partitioned into two systems again, according to the parts on the right-hand side, which are either zero or not zero. In Section 2.3.7 we show how the reduction may be done by a partition of the transformation matrices.

2.3.7 Matrix partition technique: two examples

In this subsection we will describe the transformation matrix partition technique (TMPT) and demonstrate its use in an easy formulation of eigenvalue problems. This will be explained by two examples: the inset dielectric guide [11] and the groove guide.

2.3.7.1 The inset dielectric guide

The inset dielectric guide (IDG) proposed by Rozzi [11] (see Fig. 2.18) consists of a groove (region I) in a metallic block which is filled with a dielectric. The region above the groove (region II) is the open space. The field is concentrated mainly in the dielectric. Therefore, we may use special ABCs or even Dirichlet boundary conditions at the sides of region II. However, the boundaries must be positioned far enough from the groove. Since the regions I and II differ in width, the transformation matrices are of different size, too. The fields in plane M in the individual transformed domains are obtained by using eq. (2.72) as:

$$\widehat{\overline{\mathbf{H}}}^{\mathrm{I}} = -\overline{\boldsymbol{y}}_1^{\mathrm{I}}\widehat{\overline{\mathbf{E}}}^{\mathrm{I}} \qquad \widehat{\overline{\mathbf{H}}}^{\mathrm{II}} = \overline{\boldsymbol{y}}_1^{\mathrm{II}}\widehat{\overline{\mathbf{E}}}^{\mathrm{II}} \tag{2.108}$$

In region I we are looking in $-z$-direction and in region II in $+z$-direction. The fields in the spatial domain are obtained by an inverse transformation. For this purpose we partition each of the two submatrices $\boldsymbol{T}_h$ and $\boldsymbol{T}_e$ in $\widehat{\boldsymbol{T}}_{\mathrm{II}}$ into two parts ($\boldsymbol{T}_{he} = 0$ because the region II is homogeneous, see eq. (2.64)). The transformation matrix partition is shown in Fig. 2.19 (the subscripts e

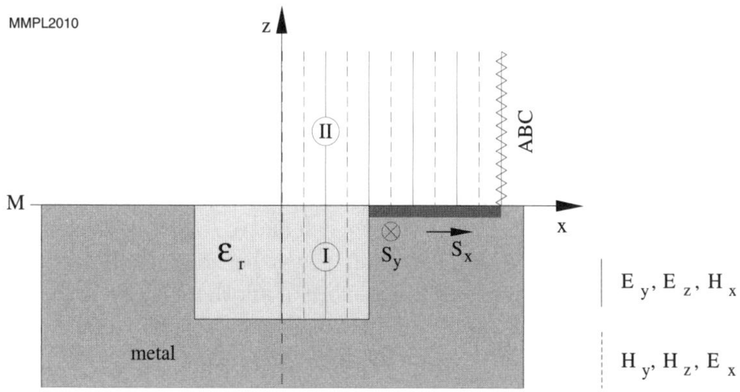

Fig. 2.18 Cross-section of an inset dielectric waveguide (Reproduced by permission of Elsevier)

or h are not written here). The part $\boldsymbol{T}_{\text{IIc}}$ belongs to the part of the region II above the region I. The part $\boldsymbol{T}_{\text{IIm}}$ belongs to the part of region II above metal block. If the inverse $\boldsymbol{T}_{\text{II}}^{-1}$ is given as $\boldsymbol{T}_{\text{II}}^{t}$, the parts of $\boldsymbol{T}_{\text{II}}^{-1}$ can immediately be given by transposing the parts of $\boldsymbol{T}_{\text{II}}$. Otherwise the inversion must be performed first. Now we obtain:

$$\widehat{\mathbf{H}}^{\text{I}} = -\widehat{\boldsymbol{T}}_{\text{IH}}\, \overline{\boldsymbol{y}}_1^{\text{I}}\, \widehat{\boldsymbol{T}}_{\text{IE}}^{-1}\widehat{\mathbf{E}}^{\text{I}} \tag{2.109}$$

$$\widehat{\mathbf{H}}_{\text{c}}^{\text{II}} = \widehat{\boldsymbol{T}}_{\text{IIHc}}\overline{\boldsymbol{y}}_1^{\text{II}}\widehat{\boldsymbol{T}}_{\text{IIEc}}^{-1}\widehat{\mathbf{E}}_{\text{c}}^{\text{II}} \tag{2.110}$$

$$\widehat{\mathbf{S}} = \widehat{\boldsymbol{T}}_{\text{IIHm}}\overline{\boldsymbol{y}}_1^{\text{II}}\widehat{\boldsymbol{T}}_{\text{IIEc}}^{-1}\widehat{\mathbf{E}}_{\text{c}}^{\text{II}} \tag{2.111}$$

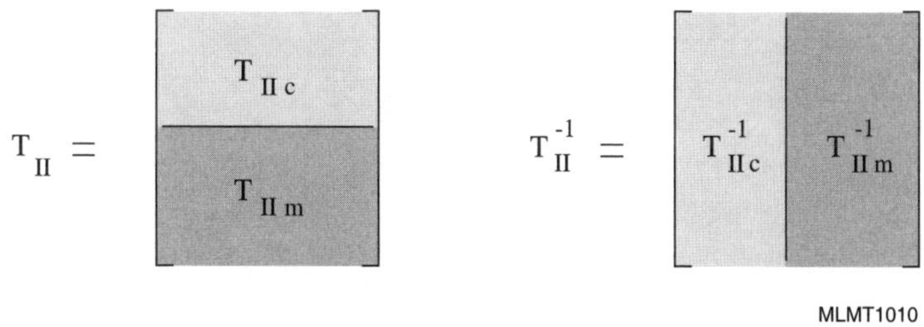

Fig. 2.19 Transformation matrix partition in region II of Figure 2.18 (Reproduced by permission of Elsevier)

A homogeneous dielectric has been assumed in region I. Matching the tangential field components $\mathbf{E}^{\text{I}} = \mathbf{E}_{\text{c}}^{\text{II}}$ and $\mathbf{H}^{\text{I}} = \mathbf{H}_{\text{c}}^{\text{II}}$ yields:

$$(\widehat{\boldsymbol{T}}_{\text{IIHc}}\overline{\boldsymbol{y}}_1^{\text{II}}\widehat{\boldsymbol{T}}_{\text{IIEc}}^{-1} + \widehat{\boldsymbol{T}}_{\text{IH}}\overline{\boldsymbol{y}}_1^{\text{I}}\widehat{\boldsymbol{T}}_{\text{IE}}^{-1})\widehat{\mathbf{E}}^{\text{I}} = \mathbf{0} \tag{2.112}$$

This equation is obtained by subtracting eq. (2.109) from eq. (2.110). After solving this eigenvalue equation for $\widehat{\mathbf{E}}^{\mathrm{I}}$, the current density $\widehat{\mathbf{S}}$ according to eq. (2.111) can be calculated.

2.3.7.2 The groove guide

As a second example, the dispersion diagram and the characteristic impedance of the groove guide (Fig. 2.20) will be calculated. All layers are assumed homogeneous. Region I is smaller than the other regions II and III. The transformation matrix in region I is therefore of another size than the matrices of regions II and III. The system equation in the transformed domain for regions II and III can be written in the following form:

$$\begin{bmatrix} \overline{\boldsymbol{Z}}_{\mathrm{II}} & \overline{\boldsymbol{V}}_{\mathrm{IIa}} \\ \overline{\boldsymbol{V}}_{\mathrm{IIb}} & \overline{\boldsymbol{Y}}_{\mathrm{II}} \end{bmatrix} \begin{bmatrix} \widehat{\overline{\mathbf{S}}}_{\mathrm{B}} \\ \widehat{\overline{\mathbf{E}}}_{\mathrm{A}}^{\mathrm{II}} \end{bmatrix} = \begin{bmatrix} \widehat{\overline{\mathbf{E}}}_{\mathrm{B}} \\ \widehat{\overline{\mathbf{H}}}_{\mathrm{A}}^{\mathrm{II}} \end{bmatrix} \tag{2.113}$$

where $\widehat{\overline{\mathbf{E}}}_{\mathrm{B}} = \overline{\mathbf{E}}_{\mathrm{B}}^{\mathrm{II}} = \overline{\mathbf{E}}_{\mathrm{B}}^{\mathrm{III}}$. $\widehat{\overline{\mathbf{S}}}_{\mathrm{B}}$ is here defined by $\widehat{\overline{\mathbf{S}}}_{\mathrm{B}} = \overline{\mathbf{H}}_{\mathrm{B}}^{\mathrm{II}} - \overline{\mathbf{H}}_{\mathrm{B}}^{\mathrm{III}}$. It is the current density distribution in transformed domain in plane B. The matrices $\overline{\boldsymbol{Z}}_{\mathrm{II}}, \overline{\boldsymbol{Y}}_{\mathrm{II}}, \overline{\boldsymbol{V}}_{\mathrm{II}}$ in eq. (2.113) are determined in the following way:

$$\begin{aligned} \overline{\boldsymbol{Z}}_{\mathrm{II}} &= (\overline{\boldsymbol{y}}_1^{\mathrm{II}} + \overline{\boldsymbol{y}}^{\mathrm{III}})^{-1} & \qquad \overline{\boldsymbol{V}}_{\mathrm{IIa}} &= \overline{\boldsymbol{Z}}_{\mathrm{II}} \overline{\boldsymbol{y}}_2^{\mathrm{II}} \\ \overline{\boldsymbol{Y}}_{\mathrm{II}} &= \overline{\boldsymbol{y}}_1^{\mathrm{II}} + \overline{\boldsymbol{y}}_2^{\mathrm{II}} \overline{\boldsymbol{Z}}_{\mathrm{II}} \overline{\boldsymbol{y}}_2^{\mathrm{II}} & \qquad \overline{\boldsymbol{V}}_{\mathrm{IIb}} &= \overline{\boldsymbol{y}}_2^{\mathrm{II}} \overline{\boldsymbol{Z}}_{\mathrm{II}} = \overline{\boldsymbol{V}}_{\mathrm{IIa}}^{t} \end{aligned} \tag{2.114}$$

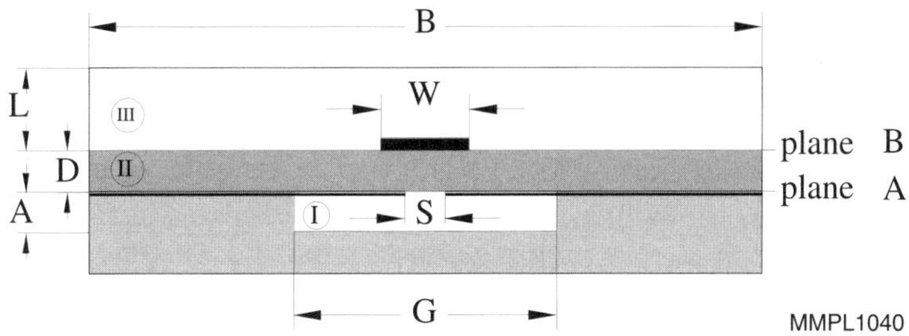

Fig. 2.20 Cross-section of a groove guide (Reproduced by permission of Elsevier)

In region II at interface A, the fields in the spatial domain are given by:

$$\mathbf{H}_{\mathrm{A}}^{\mathrm{II}} = \widehat{\boldsymbol{T}}_{\mathrm{IIH}} \overline{\mathbf{H}}_{\mathrm{A}}^{\mathrm{II}} = \begin{bmatrix} \widehat{\boldsymbol{T}}_{\mathrm{IIHm_{Al}}} \overline{\mathbf{H}}_{\mathrm{A}}^{\mathrm{II}} & = & -\widehat{\mathbf{S}}_{\mathrm{A}l}^{\mathrm{II}} \\ \widehat{\boldsymbol{T}}_{\mathrm{IIHc_A}} \overline{\mathbf{H}}_{\mathrm{A}}^{\mathrm{II}} & = & \widehat{\mathbf{H}}_{sl} \\ \widehat{\boldsymbol{T}}_{\mathrm{IIHm_{Ar}}} \overline{\mathbf{H}}_{\mathrm{A}}^{\mathrm{II}} & = & -\widehat{\mathbf{S}}_{\mathrm{A}r}^{\mathrm{II}} \end{bmatrix} \qquad \widehat{\mathbf{E}}_{\mathrm{A}}^{\mathrm{II}} = \widehat{\boldsymbol{T}}_{\mathrm{IIE}} \widehat{\overline{\mathbf{E}}}_{A} = \begin{bmatrix} \mathbf{0} \\ \widehat{\mathbf{E}}_{sl} \\ \mathbf{0} \end{bmatrix} \tag{2.115}$$

The partition of $\widehat{\boldsymbol{T}}_{\mathrm{IIE}}$ is done analogous to $\widehat{\boldsymbol{T}}_{\mathrm{IIH}}$. $\widehat{\mathbf{S}}_{\mathrm{A}l}^{\mathrm{II}}$ and $\widehat{\mathbf{S}}_{\mathrm{A}r}^{\mathrm{II}}$ are the current densities on the metallisation on the left and right side of the slot.

The fields in the slot $\widehat{\mathbf{H}}_{sl}$ and $\widehat{\mathbf{E}}_{sl}$ are related on the groove side, too.

$$\widehat{\mathbf{H}}_{sl} = -\widehat{\boldsymbol{T}}_{\mathrm{IHc}}\overline{\boldsymbol{y}}_{\mathrm{I}}\widehat{\boldsymbol{T}}_{\mathrm{IEc}}^{-1}\widehat{\mathbf{E}}_{sl} = -\mathbf{Y}_{\mathrm{I}}^{r}\widehat{\mathbf{E}}_{sl} \tag{2.116}$$

$\mathbf{Y}_{\mathrm{I}}^{r}$ is the slot admittance looking into the groove side. Now, the total reduced system takes the form:

$$\begin{bmatrix} \boldsymbol{Z}_{\mathrm{II}}^{r} & \mathbf{V}_{\mathrm{IIa}}^{r} \\ \mathbf{V}_{\mathrm{IIb}}^{r} & \mathbf{Y}_{\mathrm{II}}^{r} + \mathbf{Y}_{\mathrm{I}}^{r} \end{bmatrix} \begin{bmatrix} \widehat{\mathbf{S}}_{\mathrm{B}} \\ \widehat{\mathbf{E}}_{sl} \end{bmatrix} = \begin{bmatrix} \mathbf{0} \\ \mathbf{0} \end{bmatrix} \tag{2.117}$$

with:

$$\begin{aligned} \boldsymbol{Z}_{\mathrm{II}}^{r} &= \widehat{\boldsymbol{T}}_{\mathrm{IIEm_B}}\overline{\boldsymbol{Z}}_{\mathrm{II}}\widehat{\boldsymbol{T}}_{\mathrm{IIHm_B}}^{-1} & \mathbf{V}_{\mathrm{IIa}}^{r} &= \widehat{\boldsymbol{T}}_{\mathrm{IIEm_B}}\overline{\boldsymbol{V}}_{\mathrm{IIa}}\widehat{\boldsymbol{T}}_{\mathrm{IIEc_A}}^{-1} \\ \mathbf{Y}_{\mathrm{II}}^{r} &= \widehat{\boldsymbol{T}}_{\mathrm{IIHc_A}}\overline{\boldsymbol{Y}}_{\mathrm{II}}\widehat{\boldsymbol{T}}_{\mathrm{IIEc_A}}^{-1} & \mathbf{V}_{\mathrm{IIb}}^{r} &= \widehat{\boldsymbol{T}}_{\mathrm{IIHc_A}}\overline{\boldsymbol{V}}_{\mathrm{IIb}}\widehat{\boldsymbol{T}}_{\mathrm{IIHm_B}}^{-1} \end{aligned} \tag{2.118}$$

The transformation matrix part $\boldsymbol{T}_{\mathrm{IIm_B}}$ belongs to the strip in plane B, whereas the transformation matrix parts $\boldsymbol{T}_{\mathrm{Ic}}$ and $\boldsymbol{T}_{\mathrm{IIc_A}}$ belong to the slot in plane A. Eq. (2.117) is an indirect eigenvalue problem for the coupled supervectors of the current densities on the strip in plane B and the tangential electric field in the slot in plane A.

The general procedure is described in Chapter 5.

2.3.8 Numerical results

To verify the algorithm presented in previous subsections, some planar microwave and optical structures were analysed. One important advantage of the MoL is the monotonic convergence behaviour [8]. This is because the MoL has stationary properties [12]. As a first result, in Fig. 2.21 dispersion diagrams for the fundamental and higher order even modes in the inset dielectric guide are given and compared with results from [13]. An excellent agreement is seen. The dispersion diagram for the quasi-TEM mode of the groove guide shown in Fig. 2.20 is plotted in Fig. 2.22a, parameterised with the slot size S. Fig. 2.22b shows the characteristic impedance Z_0 calculated from the transmitted power and the current on the strip vs. frequency. The diagram shows that Z_0 is nearly independent of frequency around 5 GHz. As we have seen, the impedance can be controlled by the slot width S. Furthermore, our investigations have shown that the impedance is relatively insensitive to the groove width G.

The next results are for the special microstrip structure in Fig. 2.16 with finite substrate. The results for the microstrip with $s = 0$ are given in Table I and compared with those of [14], obtained with the spectral domain MoM. The difference is less than 0.2%.

Fig. 2.23 shows the MoL results for ε_{re} as a function of s/d, in comparison with those of [14] and [15]. Results from [14] at $s/d = 0.0$ and $s/d = 0.5$ are very close to those of the MoL. The differences between these and the results of [15] are much higher. The reason for this may be the non-equidistant discretisation in [15], with discretisation distances growing too

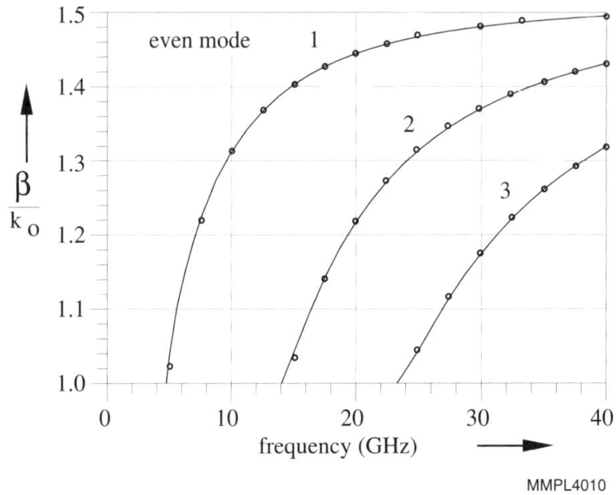

Fig. 2.21 Dispersion diagram for the fundamental and higher order even modes in the inset dielectric guide (• results from [13]) (Reproduced by permission of Elsevier)

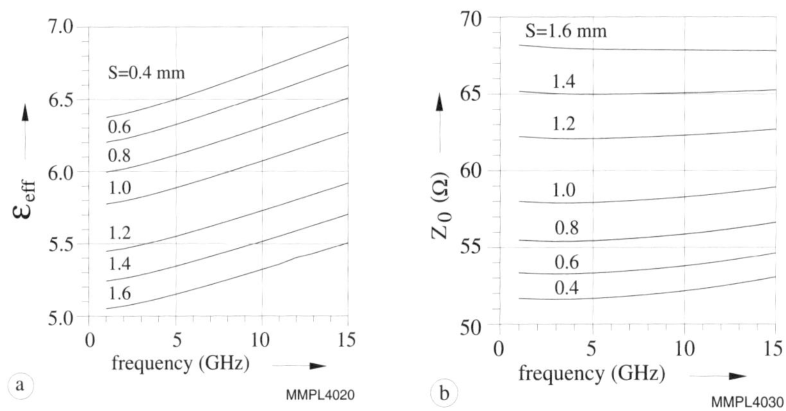

Fig. 2.22 Dispersion diagram (a) and characteristic impedance (Reproduced by permission of Elsevier) (b) for the quasi-TEM mode in the groove guide of Fig. 2.20 (Reproduced by permission of Taylor & Francis)

fast. In Fig. 2.24 the current distribution on the microstrip is sketched and compared with the results of G. Hanson [14]. In principle, the runs are analogous and the values are nearly equal. The S_x component is lower by about two orders of magnitude. The main difference in this component is about 1% of the minimum value of the S_y component. The MoL results are nearly independent of the number of lines on the strip. This is because the MoL is related to the discrete Fourier transformation, which gives exact results at the discretisation points (see appendices). In Fig. 2.25, dispersion diagrams

Table I Results for the microstrip in Fig. 2.3. $\varepsilon_{rd} = 9.7$, $\varepsilon_{ra} = 1.0$, $s = 0$ [MoM]: results from spectral domain MoM (G. Hanson [14])

d/λ_0	0.01	0.05	0.1	0.15
ε_{re} [MoL]	4.0018	4.5942	5.6655	6.6492
ε_{re} [MoM]	4.0009	4.5911	5.6620	6.6484

are given for a rib waveguide and compared with the results given in [16] by a previous MoL algorithm, and those obtained by the mode matching technique [17]. As can be seen, the results obtained by the algorithm described here are practically identical with those from the previous algorithm. The results from the mode matching technique are slightly different.

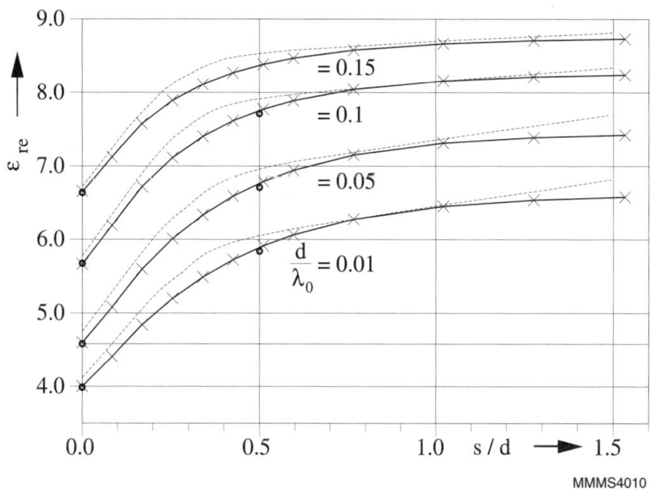

Fig. 2.23 Effective permittivity ε_{re} for the fundamental mode (x) in the waveguide of Fig. 2.16. Results for comparison: • from [14] - - - from [15] (Reproduced by permission of Taylor & Francis)

The corresponding fields for the fundamental mode HE_{00} in the middle of the slab layer (layer below the rib) are presented in Fig. 2.26 ($k_0 d = 6$).

2.4 ANALYSIS OF PLANAR CIRCUITS

2.4.1 Discretisation of the transmission line equations

This section can now be viewed as an extension of Section 2.2.4, where we examined 2D structures with a one-dimensional discretisation. Here, we will show the general 3D case, where we must discretise in two dimensions. As before, the transmission line equations have to be discretised. Fig. 2.27

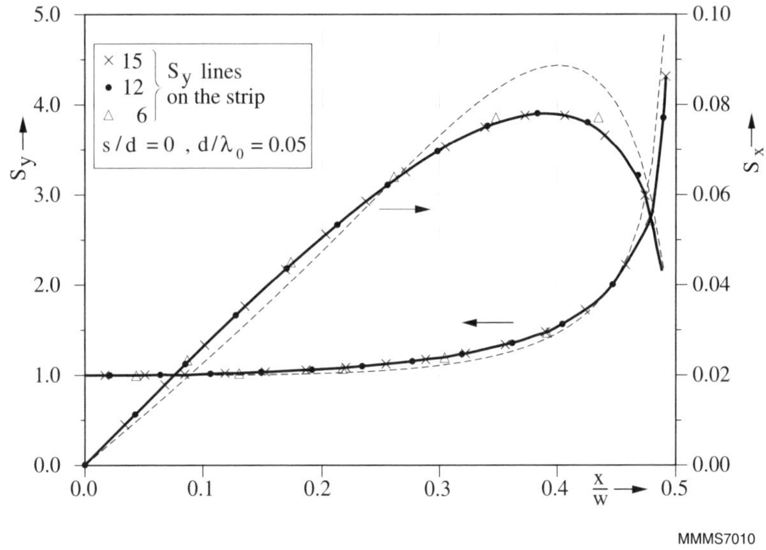

Fig. 2.24 Current density distribution on the right half of the strip - - - [14] (Reproduced by permission of Taylor & Francis)

shows the cross-section of a planar waveguide with the adequate discretisation points.

Usually, the field components are discretised at different positions. However, $\mathbf{E}_x$ is determined at the same place as $\mathbf{H}_y$, and $\mathbf{E}_y$ at the same one as $\mathbf{H}_x$. These are the field components with which the two parts of the Poynting vector in z-direction – the assumed direction of wave propagation – are calculated. Now we collect the discretised field components in column vectors. We order these components starting with the adequate point in the left upper point and going downwards in the first column. Coming down to the last point in the first column, we continue with the highest point in the second column and so on. As we can see, the values in the columns from the left to right are put below each other. Each of the column vectors may be represented by a boldface capital letter of the field component and by a subscript i for the ith column. We mark the collection of all these vectors in the total column vector with a hat ($\hat{\ }$).

If we have N_x° columns of $\circ$ discretisation points and N_y° points in each column, the total number of points is $\mathrm{N}_y^{\circ}\mathrm{N}_x^{\circ}$. In case of ideal metal in the cross-section, the total number of discretisation points is reduced by the number of points in the metal area. In case of lossy metals, we consider these metals as dielectrics with complex permittivity. The numbers of $\bullet$ columns and rows are $\mathrm{N}_x^{\bullet}$ and $\mathrm{N}_y^{\bullet}$, respectively. The values of the permittivities and permeabilities are collected in the same order as the field components, not in column vectors

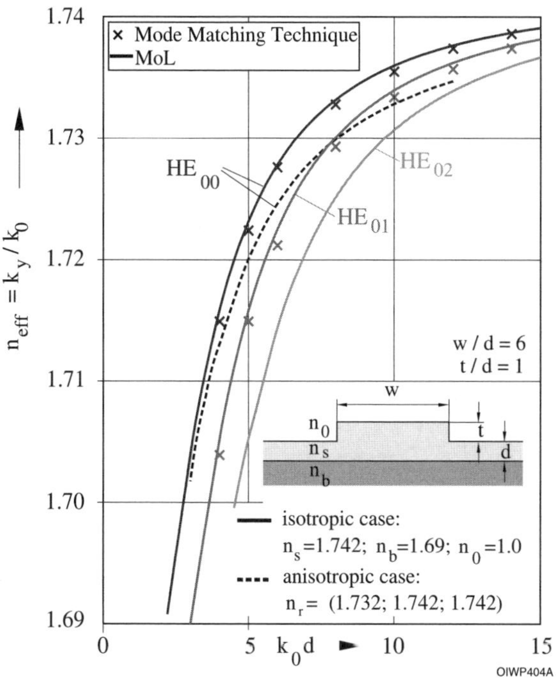

Fig. 2.25 Dispersion diagrams for the fundamental and the two first higher order TE polarised modes (HE_{00} and $\mathrm{HE}_{01}, \mathrm{HE}_{02}$); mode matching technique: [17]

but on the main diagonal of a diagonal matrix. Each of the components of the tensor is discretised on a different permittivity or permeability point. Therefore, we have five different permittivity and permeability matrices each. However, Fig. 2.27 shows that e.g. ε_{rx} is discretised on the same points as $\mathbf{H}_y$. Even in the case of isotropic materials, we need to have three different matrices for each of the parameters. So there is no big difference between the algorithms for isotropic and anisotropic materials; in particular the numerical effort is identical.

Next, the differential operators are replaced by central differences. All the central differences in one row or column are collected to difference operators (matrices) $\boldsymbol{D}_x^{\circ,\bullet}$ and $\boldsymbol{D}_y^{\circ,\bullet}$, respectively. The collection of all rows and columns is marked by a hat ($\widehat{\ }$).

Introducing the supervectors of the discretised field components according to:

$$\widehat{\mathbf{E}} = [\widehat{\mathbf{E}}_y^{\mathrm{t}}, \widehat{\mathbf{E}}_x^{\mathrm{t}}]^{\mathrm{t}} \qquad \widehat{\mathbf{H}} = [-\widehat{\mathbf{H}}_x^{\mathrm{t}}, \widehat{\mathbf{H}}_y^{\mathrm{t}}]^{\mathrm{t}} \tag{2.119}$$

we write for the GTL eqs. (2.10) and (2.13) in discretised form:

$$\frac{\mathrm{d}}{\mathrm{d}\overline{z}}\widehat{\mathbf{H}} = -\mathrm{j}\widehat{\boldsymbol{R}}_{\mathrm{E}}^{z}\widehat{\mathbf{E}} \qquad \frac{\mathrm{d}}{\mathrm{d}\overline{z}}\widehat{\mathbf{E}} = -\mathrm{j}\widehat{\boldsymbol{R}}_{\mathrm{H}}^{z}\widehat{\mathbf{H}} \tag{2.120}$$

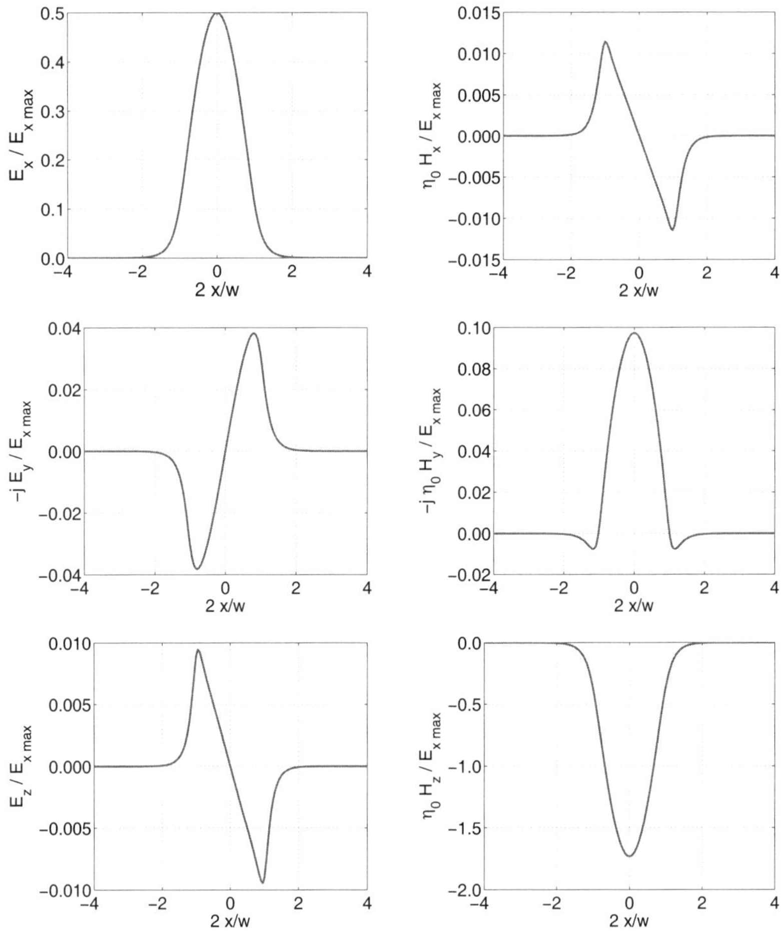

Fig. 2.26 Field distribution of HE_{00} mode at the half height of the slab layer (see Fig. 2.25; $k_0 d = 6$), normalised with the maximum of E_x

Eq. (2.120) show the analogy between the MoL and the transmission line theory: the lines used in the MoL can be represented as transmission lines.

Fig. 2.27 shows e.g. that the permeability $\widehat{\epsilon}_{xy}$ is determined at the same position as E_y. However, not the same this is when we compute the magnetic field component H_y. Looking at eq. (2.10), we see that the operator product $[R_{\mathrm{E}}^{z}][E_t]$ contains the expression $\varepsilon_{xy} E_y$ to calculate H_y. Since the discretisation points are shifted in this case, we must 'move' the results of the product to obtain it at the correct position. For this purpose, we introduce

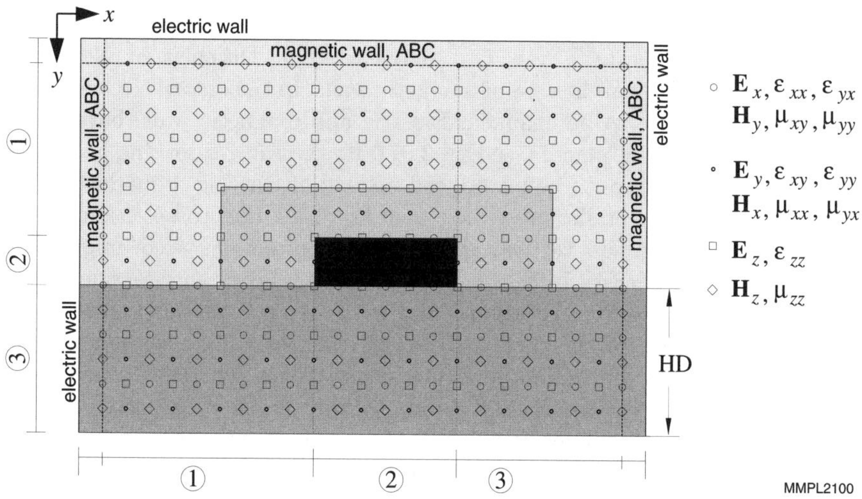

Fig. 2.27 Cross-section of a general planar structure with discretisation points (R. Pregla, 'Efficient and Accurate Modeling of Planar Anisotropic Microwave Structures by the Method of Lines', *IEEE Trans. Microwave Theory Tech.* vol. 50, pp. 1469–1479. © 2002 Institute of Electrical and Electronics Engineers (IEEE))

interpolation matrices, e.g. $\widehat{\boldsymbol{M}}_{xy}^{\diamond\circ} = \widehat{\boldsymbol{M}}_{x}^{\diamond}\widehat{\boldsymbol{M}}_{y}^{\circ}$:

$$\widehat{\boldsymbol{R}}_{\mathrm{E}}^{z} = \begin{bmatrix} \widehat{\boldsymbol{\epsilon}}_{yy} - \widehat{\overline{\boldsymbol{D}}}_{\overline{x}}^{\bullet\mathrm{t}}\,\widehat{\boldsymbol{\mu}}_{zz}^{-1}\widehat{\overline{\boldsymbol{D}}}_{\overline{x}}^{\bullet} & \widehat{\boldsymbol{M}}_{xy}^{\diamond\circ}\widehat{\boldsymbol{\epsilon}}_{yx} + \widehat{\overline{\boldsymbol{D}}}_{\overline{x}}^{\bullet\mathrm{t}}\,\widehat{\boldsymbol{\mu}}_{zz}^{-1}\widehat{\overline{\boldsymbol{D}}}_{\overline{y}}^{\circ} \\ \widehat{\boldsymbol{M}}_{yx}^{\diamond\bullet}\widehat{\boldsymbol{\epsilon}}_{xy} + \widehat{\overline{\boldsymbol{D}}}_{\overline{y}}^{\circ\mathrm{t}}\,\widehat{\boldsymbol{\mu}}_{zz}^{-1}\widehat{\overline{\boldsymbol{D}}}_{\overline{x}}^{\bullet} & \widehat{\boldsymbol{\epsilon}}_{xx} - \widehat{\overline{\boldsymbol{D}}}_{\overline{y}}^{\circ\mathrm{t}}\,\widehat{\boldsymbol{\mu}}_{zz}^{-1}\widehat{\overline{\boldsymbol{D}}}_{\overline{y}}^{\circ} \end{bmatrix} \tag{2.121}$$

$$\widehat{\boldsymbol{R}}_{\mathrm{H}}^{z} = \begin{bmatrix} \widehat{\boldsymbol{\mu}}_{xx} - \widehat{\overline{\boldsymbol{D}}}_{\overline{y}}^{\bullet\mathrm{t}}\,\widehat{\boldsymbol{\epsilon}}_{zz}^{-1}\widehat{\overline{\boldsymbol{D}}}_{\overline{y}}^{\bullet} & \widehat{\boldsymbol{M}}_{yx}^{\square\diamond}\widehat{\boldsymbol{\mu}}_{xy} - \widehat{\overline{\boldsymbol{D}}}_{\overline{y}}^{\bullet\mathrm{t}}\,\widehat{\boldsymbol{\epsilon}}_{zz}^{-1}\widehat{\overline{\boldsymbol{D}}}_{\overline{x}}^{\circ} \\ \widehat{\boldsymbol{M}}_{xy}^{\square}\widehat{\boldsymbol{\mu}}_{yx} - \widehat{\overline{\boldsymbol{D}}}_{\overline{x}}^{\circ\mathrm{t}}\,\widehat{\boldsymbol{\epsilon}}_{zz}^{-1}\widehat{\overline{\boldsymbol{D}}}_{\overline{y}}^{\bullet} & \widehat{\boldsymbol{\mu}}_{yy} - \widehat{\overline{\boldsymbol{D}}}_{\overline{x}}^{\circ\mathrm{t}}\,\widehat{\boldsymbol{\epsilon}}_{zz}^{-1}\widehat{\overline{\boldsymbol{D}}}_{\overline{x}}^{\circ} \end{bmatrix} \tag{2.122}$$

The order of the interpolation may also be changed.

If we have diagonal material tensors, no losses and lossless boundaries (i.e. Neumann or Dirichlet), the matrices $\widehat{\boldsymbol{R}}$ are real and symmetric, i.e. the following relations hold:

$$\widehat{\boldsymbol{R}}_{\mathrm{E,H}} = \widehat{\boldsymbol{R}}_{\mathrm{E,H}}^{\mathrm{t}} \tag{2.123}$$

However, if these matrices become complex due to losses or because of absorbing boundaries, they are not Hermitian.

The difference operators can be constructed in the following way. In case of an inhomogeneous cross-section without metallic subsections, the difference operators can be described by Kronecker products of the difference operators for the rows ($\overline{\boldsymbol{D}}_{x}^{\circ,\bullet}$) or columns ($\overline{\boldsymbol{D}}_{y}^{\circ,\bullet}$) and unit matrices with the size of the number of rows or columns. So we obtain:

$$\widehat{\overline{\boldsymbol{D}}}_{\overline{y}}^{\circ,\bullet} = \boldsymbol{I}_{x}^{\circ,\bullet} \otimes \overline{\boldsymbol{D}}_{\overline{y}}^{\circ,\bullet} \quad \widehat{\overline{\boldsymbol{D}}}_{\overline{x}}^{\circ,\bullet} = \overline{\boldsymbol{D}}_{\overline{x}}^{\circ,\bullet} \otimes \boldsymbol{I}_{y}^{\circ,\bullet} \tag{2.124}$$

$\boldsymbol{I}_x^{\circ,\bullet}$ and $\boldsymbol{I}_y^{\circ,\bullet}$ being unit matrices of size $\mathrm{N}_x^{\circ,\bullet}$ and $\mathrm{N}_y^{\circ,\bullet}$, respectively. The difference operators $\overline{\boldsymbol{D}}_{\overline{x}}^{\circ,\bullet}$ and $\overline{\boldsymbol{D}}_{\overline{y}}^{\circ,\bullet}$ have to fulfil Dirichlet or Neumann boundary conditions in the case of magnetic walls. In the case of electric walls, the conditions have to be changed. Electric and magnetic walls have to be positioned at different places (see Fig. 2.27). If necessary, absorbing boundary conditions (ABC) have to be introduced instead of Dirichlet boundaries.

In the case of a metallic subsection as in Fig. 2.27, the cross-section should be divided with lines in vertical and horizontal directions according to the metallisation boundaries. In Fig. 2.27 we thus obtain in both directions three different subregions. First we assume that we have discretisation points inside and on the surface of the metal. However, the discretised components on these points should be removed in our calculations. Therefore, we reduce the total number of discretisation points to that outside the metal. Let us assume that in our example the number N_x° of $\circ$ columns consists of N_{x1}°, N_{x2}° and N_{x3}° columns in the three subregions $1, 2, 3$ in x-direction, respectively. Analogously, the number N_y° of $\circ$ rows in y-direction is N_{y1}°, N_{y2}° and N_{y3}° in the three subregions in y-direction. Let us define unit matrices $\boldsymbol{I}_{x1}^\circ$, $\boldsymbol{I}_{x2}^\circ$, $\boldsymbol{I}_{x3}^\circ$, $\boldsymbol{I}_{y1}^\circ$, $\boldsymbol{I}_{y2}^\circ$, and $\boldsymbol{I}_{y3}^\circ$ of the order N_{x1}°, N_{x2}° etc. The complete vector $\mathbf{F}_\mathrm{c}^\circ$ of the field quantity $\mathbf{F}$ discretised on $\circ$ points can now be reduced to the vector $\mathbf{F}^\circ$ with components only outside the metal by:

$$\mathbf{F}^\circ = \boldsymbol{J}_\circ^\mathrm{t}\mathbf{F}_\mathrm{c}^\circ \tag{2.125}$$

The matrix $\boldsymbol{J}_\circ^\mathrm{t}$ is obtained from the matrix:

$$\boldsymbol{I}_x^\circ \otimes \boldsymbol{I}_y^\circ = \mathrm{diag}\,(\boldsymbol{I}_{x1}^\circ, \boldsymbol{I}_{x2}^\circ, \boldsymbol{I}_{x3}^\circ) \otimes \mathrm{diag}\,(\boldsymbol{I}_{y1}^\circ, \boldsymbol{I}_{y2}^\circ, \boldsymbol{I}_{y3}^\circ) \tag{2.126}$$

as:

$$\boldsymbol{J}_\circ^\mathrm{t} = \begin{bmatrix} \boldsymbol{I}_{x1}^\circ \otimes \boldsymbol{I}_y^\circ & & \\ & \boldsymbol{I}_{x2}^\circ \otimes \begin{bmatrix} \boldsymbol{I}_{y1}^\circ & \\ & \boldsymbol{I}_{y3}^\circ \end{bmatrix} & \\ & & \boldsymbol{I}_{x3}^\circ \otimes \boldsymbol{I}_y^\circ \end{bmatrix} \tag{2.127}$$

In the middle expression we have discarded those rows of $\boldsymbol{I}_y^\circ$ which are also rows of $\boldsymbol{I}_{y2}^\circ$. The matrices $\boldsymbol{J}_\bullet^\mathrm{t}$, $\boldsymbol{J}_\Box^\mathrm{t}$ and $\boldsymbol{J}_\Diamond^\mathrm{t}$ can be constructed likewise. Now the reduced difference operators can be obtained in the following way:

$$\begin{aligned} \widehat{\boldsymbol{D}}_{\overline{x}r}^\circ &= \boldsymbol{J}_\Box^\mathrm{t}\widehat{\boldsymbol{D}}_{\overline{x}}^\circ\boldsymbol{J}_\circ & \widehat{\boldsymbol{D}}_{\overline{x}r}^\bullet &= \boldsymbol{J}_\Diamond^\mathrm{t}\widehat{\boldsymbol{D}}_{\overline{x}}^\bullet\boldsymbol{J}_\bullet \\ \widehat{\boldsymbol{D}}_{\overline{y}r}^\Diamond &= \boldsymbol{J}_\Diamond^\mathrm{t}\widehat{\boldsymbol{D}}_{\overline{y}}^\circ\boldsymbol{J}_\circ & \widehat{\boldsymbol{D}}_{\overline{y}r}^\bullet &= \boldsymbol{J}_\Box^\mathrm{t}\widehat{\boldsymbol{D}}_{\overline{y}}^\bullet\boldsymbol{J}_\bullet \end{aligned} \tag{2.128}$$

The reduction of the interpolation matrices is done in an analogue way. The diagonal matrices for the material parameters have to be reduced according to:

$$\begin{aligned} \boldsymbol{\epsilon}_{xxr} &= \boldsymbol{J}_\circ^\mathrm{t}\boldsymbol{\epsilon}_{xx}\boldsymbol{J}_\circ & \boldsymbol{\mu}_{xxr} &= \boldsymbol{J}_\bullet^\mathrm{t}\boldsymbol{\mu}_{xx}\boldsymbol{J}_\bullet \\ \boldsymbol{\epsilon}_{yyr} &= \boldsymbol{J}_\bullet^\mathrm{t}\boldsymbol{\epsilon}_{yy}\boldsymbol{J}_\bullet & \boldsymbol{\mu}_{yyr} &= \boldsymbol{J}_\circ^\mathrm{t}\boldsymbol{\mu}_{yy}\boldsymbol{J}_\circ \\ \boldsymbol{\epsilon}_{zzr} &= \boldsymbol{J}_\Box^\mathrm{t}\boldsymbol{\epsilon}_{zz}\boldsymbol{J}_\Box & \boldsymbol{\mu}_{zzr} &= \boldsymbol{J}_\Diamond^\mathrm{t}\boldsymbol{\mu}_{zz}\boldsymbol{J}_\Diamond \end{aligned} \tag{2.129}$$

The reduction of the off-diagonal material parameters is done analogously, where e.g. $\boldsymbol{\epsilon}_{yx}$ ($\boldsymbol{\epsilon}_{xy}$) corresponds to $\boldsymbol{\epsilon}_{xx}$ ($\boldsymbol{\epsilon}_{yy}$).

The absorbing boundary conditions can be realised by replacing the difference matrices in eqs. (2.121) and (2.122) in the following way:

$$\begin{aligned} \widehat{\boldsymbol{D}}_{\overline{x}}^{\circ} &\longrightarrow \widehat{\boldsymbol{D}}_{\overline{x}}^{\diamond a} = \boldsymbol{D}_{\overline{x}}^{a} \otimes \boldsymbol{I}_{y}^{\circ} \\ -\widehat{\boldsymbol{D}}_{\overline{x}}^{\bullet \mathrm{t}} &\longrightarrow \widehat{\boldsymbol{D}}_{\overline{x}}^{\bullet a} = \boldsymbol{D}_{\overline{x}}^{a} \otimes \boldsymbol{I}_{y}^{\bullet} \\ \widehat{\boldsymbol{D}}_{\overline{y}}^{\bullet} &\longrightarrow \widehat{\boldsymbol{D}}_{\overline{y}}^{\bullet a} = \boldsymbol{I}_{x}^{\bullet} \otimes \boldsymbol{D}_{\overline{y}}^{a} \\ -\widehat{\boldsymbol{D}}_{\overline{y}}^{\circ \mathrm{t}} &\longrightarrow \widehat{\boldsymbol{D}}_{\overline{y}}^{\circ a} = \boldsymbol{I}_{x}^{\circ} \otimes \boldsymbol{D}_{\overline{y}}^{a} \end{aligned} \tag{2.130}$$

where:

$$\boldsymbol{D}_{\overline{x}}^{\circ a} = \boldsymbol{D}_{\overline{x}}^{a} \qquad \boldsymbol{D}_{\overline{y}}^{\bullet a} = \boldsymbol{D}_{\overline{y}}^{a} \tag{2.131}$$

It can be seen that the relation $\widehat{\boldsymbol{D}}_{\overline{x}}^{\circ a}\widehat{\boldsymbol{D}}_{\overline{y}}^{\circ a} = \widehat{\boldsymbol{D}}_{\overline{y}}^{\bullet a}\widehat{\boldsymbol{D}}_{\overline{x}}^{\bullet a}$ holds.

As mentioned before, the ABCs are introduced in the matrices with Dirichlet boundary conditions. We could also determine the latter operators $\boldsymbol{D}_{\overline{x}}$ and $\boldsymbol{D}_{\overline{y}}$ from $\boldsymbol{D}_{\overline{x}}^{a}$ and $\boldsymbol{D}_{\overline{y}}^{a}$, respectively. In this case, we have to use for the parameters $a = 1$ and $b = c = 0$ (see Appendix B).

The remaining difference operators in eqs. (2.121) and (2.122) have to be chosen as follows:

$$\begin{array}{lcrcl} \widehat{\boldsymbol{D}}_{\overline{x}}^{\circ \mathrm{t}} & \rightarrow & \widehat{\boldsymbol{D}}_{\overline{x}}^{\circ \mathrm{t}} & = & \boldsymbol{D}_{\overline{x}}^{\mathrm{t}} \otimes \boldsymbol{I}_{y}^{\circ} \\ \widehat{\boldsymbol{D}}_{\overline{y}}^{\circ} & \rightarrow & -\widehat{\boldsymbol{D}}_{\overline{y}}^{\circ \mathrm{t}} & = & -\boldsymbol{I}_{x}^{\circ} \otimes \boldsymbol{D}_{\overline{y}}^{\mathrm{t}} \\ \widehat{\boldsymbol{D}}_{\overline{y}}^{\bullet \mathrm{t}} & \rightarrow & \widehat{\boldsymbol{D}}_{\overline{y}}^{\bullet \mathrm{t}} & = & \boldsymbol{I}_{x}^{\bullet} \otimes \boldsymbol{D}_{\overline{y}}^{\mathrm{t}} \\ \widehat{\boldsymbol{D}}_{\overline{x}}^{\bullet} & \rightarrow & -\widehat{\boldsymbol{D}}_{\overline{x}}^{\bullet \mathrm{t}} & = & -\boldsymbol{D}_{\overline{x}}^{\mathrm{t}} \otimes \boldsymbol{I}_{y}^{\bullet} \end{array} \tag{2.132}$$

Now, the two matrices $\widehat{\boldsymbol{R}}_{\mathrm{E,H}}^{z}$ are given by:

$$\widehat{\boldsymbol{R}}_{\mathrm{E}}^{z} = \begin{bmatrix} \widehat{\boldsymbol{\epsilon}}_{yy} - \widehat{\boldsymbol{D}}_{\overline{x}}^{\bullet a}\widehat{\boldsymbol{\mu}}_{zz}^{-1}\widehat{\boldsymbol{D}}_{\overline{x}}^{\bullet \mathrm{t}} & \widehat{\boldsymbol{M}}_{xy}^{\diamond \circ}\widehat{\boldsymbol{\epsilon}}_{yx} + \widehat{\boldsymbol{D}}_{\overline{x}}^{\bullet a}\widehat{\boldsymbol{\mu}}_{zz}^{-1}\widehat{\boldsymbol{D}}_{\overline{y}}^{\circ \mathrm{t}} \\ \widehat{\boldsymbol{M}}_{yx}^{\diamond \bullet}\widehat{\boldsymbol{\epsilon}}_{xy} + \widehat{\boldsymbol{D}}_{\overline{y}}^{\circ a}\widehat{\boldsymbol{\mu}}_{zz}^{-1}\widehat{\boldsymbol{D}}_{\overline{x}}^{\bullet \mathrm{t}} & \widehat{\boldsymbol{\epsilon}}_{xx} - \widehat{\boldsymbol{D}}_{\overline{y}}^{\circ a}\widehat{\boldsymbol{\mu}}_{zz}^{-1}\widehat{\boldsymbol{D}}_{\overline{y}}^{\circ \mathrm{t}} \end{bmatrix} \tag{2.133}$$

$$\widehat{\boldsymbol{R}}_{\mathrm{H}}^{z} = \begin{bmatrix} \widehat{\boldsymbol{\mu}}_{xx} - \widehat{\boldsymbol{D}}_{\overline{y}}^{\bullet \mathrm{t}}\widehat{\boldsymbol{\epsilon}}_{zz}^{-1}\widehat{\boldsymbol{D}}_{\overline{y}}^{\bullet a} & \widehat{\boldsymbol{M}}_{yx}^{\square \circ}\widehat{\boldsymbol{\mu}}_{xy} - \widehat{\boldsymbol{D}}_{\overline{y}}^{\bullet \mathrm{t}}\widehat{\boldsymbol{\epsilon}}_{zz}^{-1}\widehat{\boldsymbol{D}}_{\overline{x}}^{\circ a} \\ \widehat{\boldsymbol{M}}_{xy}^{\square \bullet}\widehat{\boldsymbol{\mu}}_{yx} - \widehat{\boldsymbol{D}}_{\overline{x}}^{\diamond \mathrm{t}}\widehat{\boldsymbol{\epsilon}}_{zz}^{-1}\widehat{\boldsymbol{D}}_{\overline{y}}^{\bullet a} & \widehat{\boldsymbol{\mu}}_{yy} - \widehat{\boldsymbol{D}}_{\overline{x}}^{\circ \mathrm{t}}\widehat{\boldsymbol{\epsilon}}_{zz}^{-1}\widehat{\boldsymbol{D}}_{\overline{x}}^{\circ a} \end{bmatrix} \tag{2.134}$$

As mentioned before, the matrices are not Hermitian.

Alternatively, the absorbing boundary conditions could also be introduced in $\widehat{\boldsymbol{D}}_{\overline{x}}^{\bullet a}$ and $\widehat{\boldsymbol{D}}_{\overline{y}}^{\circ a}$.

Combining eq. (2.120) we obtain:

$$\frac{\mathrm{d}^2}{\mathrm{d}\overline{z}^2}\widehat{\mathbf{E}} - \widehat{\boldsymbol{Q}}_{\mathrm{E}}^{z}\widehat{\mathbf{E}} = \mathbf{0} \qquad \frac{\mathrm{d}^2}{\mathrm{d}\overline{z}^2}\widehat{\mathbf{H}} - \widehat{\boldsymbol{Q}}_{\mathrm{H}}^{z}\widehat{\mathbf{H}} = \mathbf{0} \tag{2.135}$$

with the $\boldsymbol{Q}$ matrices:

$$\widehat{\boldsymbol{Q}}_{\mathrm{E}}^{z} = -\widehat{\boldsymbol{R}}_{\mathrm{H}}^{z}\widehat{\boldsymbol{R}}_{\mathrm{E}}^{z} \qquad \widehat{\boldsymbol{Q}}_{\mathrm{H}}^{z} = -\widehat{\boldsymbol{R}}_{\mathrm{E}}^{z}\widehat{\boldsymbol{R}}_{\mathrm{H}}^{z} \tag{2.136}$$

If eq. (2.123) is valid, the two $\widehat{\boldsymbol{Q}}$ matrices are transposed of each other:

$$\widehat{\boldsymbol{Q}}_{\mathrm{H}} = \widehat{\boldsymbol{Q}}_{\mathrm{E}}^{\mathrm{t}} \tag{2.137}$$

Now the solution of eq. (2.135) occurs in exactly the same way as described for the 2D case. Therefore, we will not go into details but just add a few comments. Since we are dealing with a two-dimensional discretisation, we must replace all vectors and matrices with supervectors (supermatrices), indicated by a hat ($\widehat{}$). Obviously, the numerical effort for solving in 3D with a two-dimensional discretisation is much higher than before. However, formally we arrive at the same expressions.

2.4.2 Determination of the field components

After determining the field distribution of the eigenmodes (i.e. computing the transformation matrices $\widehat{\boldsymbol{T}}_{\mathrm{E}}$ and $\widehat{\boldsymbol{T}}_{\mathrm{H}}$), they may be partitioned according to:

$$\widehat{\boldsymbol{T}}_{\mathrm{E}} = \begin{bmatrix} \widehat{\boldsymbol{T}}_{\mathrm{E}_y} \\ \widehat{\boldsymbol{T}}_{\mathrm{E}_x} \end{bmatrix} \qquad \widehat{\boldsymbol{T}}_{\mathrm{H}} = \begin{bmatrix} \widehat{\boldsymbol{T}}_{\mathrm{H}_x} \\ \widehat{\boldsymbol{T}}_{\mathrm{H}_y} \end{bmatrix} \tag{2.138}$$

The number of rows in each part is equal to the number of discretisation points for the field components $\widehat{\mathbf{E}}_y$ and $\widehat{\mathbf{E}}_x$ or $\widehat{\mathbf{H}}_x$ and $\widehat{\mathbf{H}}_y$. The fields in the original domain are obtained from the ones in transformed domain by:

$$\widehat{\mathbf{E}}_x = \widehat{\boldsymbol{T}}_{\mathrm{E}_x} \widehat{\overline{\mathbf{E}}} \quad \widehat{\mathbf{E}}_y = \widehat{\boldsymbol{T}}_{\mathrm{E}_y} \widehat{\overline{\mathbf{E}}} \quad \widehat{\mathbf{H}}_x = -\widehat{\boldsymbol{T}}_{\mathrm{H}_x} \widehat{\overline{\mathbf{H}}} \quad \widehat{\mathbf{H}}_y = \widehat{\boldsymbol{T}}_{\mathrm{H}_y} \widehat{\overline{\mathbf{H}}} \tag{2.139}$$

The matrices $\widehat{\boldsymbol{T}}_{E_x}$, $\widehat{\boldsymbol{T}}_{\mathrm{E}_y}$, $\widehat{\boldsymbol{T}}_{\mathrm{H}_x}$, $\widehat{\boldsymbol{T}}_{\mathrm{H}_y}$ may be further partitioned. As can be seen in Fig. 2.27, we have N_x° columns of $\circ$ discretisation points. Therefore, we may partition $\widehat{\boldsymbol{T}}_{\mathrm{E}_x}$ in N_x° submatrices where the number of rows in each of the submatrices is equal to N_y° – the number of discretisation points in each column. If we name the submatrix with the nth column of $\boldsymbol{T}_{\mathrm{E}_x}$ as $\boldsymbol{T}_{\mathrm{E}_{xn}}$, the subvector of the field components $\mathbf{E}_x$ in this nth $\circ$ column is given by:

$$\mathbf{E}_{xn} = \boldsymbol{T}_{\mathrm{E}_{xn}} \widehat{\overline{\mathbf{E}}}, \quad n \in \{1, \mathrm{N}_x^\circ\} \tag{2.140}$$

In an analogous way, the other field components can be given in their columns.

2.5 FIELD AND IMPEDANCE/ADMITTANCE TRANSFORMATION

2.5.1 Introduction

In this section we describe the procedure for stable and accurate numerical field calculations. In transmission line problems we transform the admittances or impedances from the output to the input. With the input impedance/admittance we obtain the current and the voltage at the input.

Then we transform these quantities back to the output. Currents and voltages can also be determined at arbitrary places on the transmission line.

Instead of currents and voltages in our multilayered waveguides (or multisectioned device structures), we use the transverse electric and magnetic field components. Furthermore, we consider more than just one mode. The number of modes that we take into account is equal to the number of lines. For analysis of such structures we use the concept of impedance and admittance matrix transformation. We have to transform these matrices from one side of a layer/section to the other, and from one side of an interface between two layers/sections to the other. We may even have metallisation in the interface. The lower side of each layer (input side of a section) will be denoted by the subscript A and the upper side (output side of a section) by the subscript B (see e.g. A and B in Fig. 2.28).

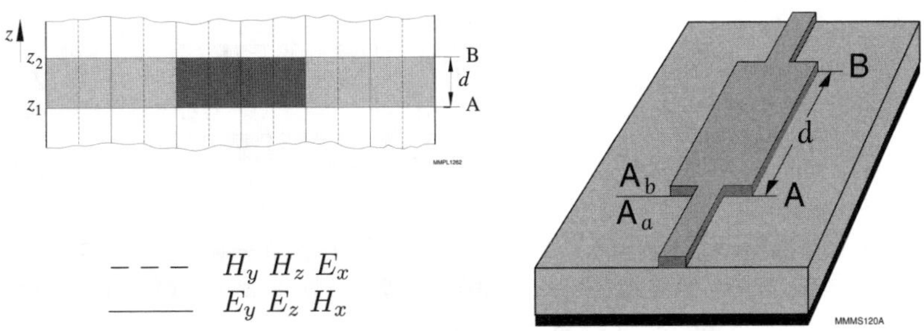

Fig. 2.28 General inhomogeneous anisotropic layer with discretisation lines and waveguide section (Left: R. Pregla, 'Modeling of Optical Waveguide Structures with General Anisotropy in Arbitrary Orthogonal Coordinate Systems', *IEEE J. of Sel. Topics in Quantum Electronics* vol. 8, pp. 1217–1224. © 2002 Institute of Electrical and Electronics Engineers (IEEE))

The principles of the impedance/admittance transformation are the same for all the cases described previously (2D–3D). However, in those cases where we explicitly have supervectors, we will mark them and the corresponding matrices with a hat ($\hat{}$). In all other cases we can have either supervectors or 'normal' ones.

2.5.2 Impedance/admittance transformation in multilayered and multisectioned structures

2.5.2.1 Field relation between interfaces A and B

Consider Fig. 2.29, where we have the cross-section of a waveguide or the concatenation of longitudinal sections in a circuit. Let us examine the layer (or section) k with the planes A and B.

The relation of the transversal fields in these two planes can be written in the following form (for details of special coordinate system and structures see

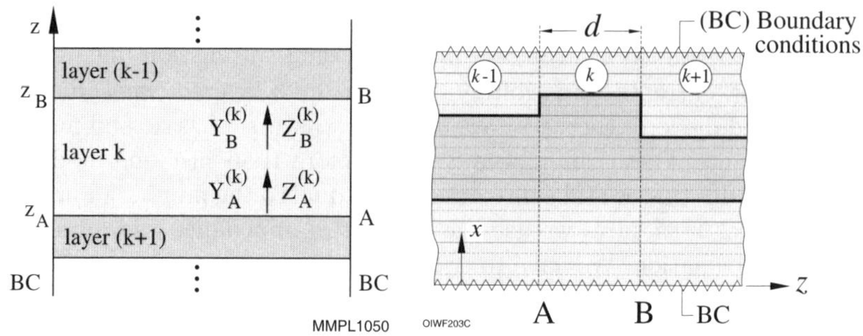

Fig. 2.29 Definitions in a layer and in a waveguide section

the corresponding chapters):

$$\begin{bmatrix} \overline{\mathbf{E}}_{\mathrm{A}} \\ \overline{\mathbf{E}}_{\mathrm{B}} \end{bmatrix} = \begin{bmatrix} \overline{\mathbf{z}}_{11} & \overline{\mathbf{z}}_{12} \\ \overline{\mathbf{z}}_{21} & \overline{\mathbf{z}}_{22} \end{bmatrix} \begin{bmatrix} \overline{\mathbf{H}}_{\mathrm{A}} \\ -\overline{\mathbf{H}}_{\mathrm{B}} \end{bmatrix} = \overline{\mathbf{z}}_{\mathrm{AB}} \begin{bmatrix} \overline{\mathbf{H}}_{\mathrm{A}} \\ -\overline{\mathbf{H}}_{\mathrm{B}} \end{bmatrix} \tag{2.141}$$

$$\begin{bmatrix} \overline{\mathbf{H}}_{\mathrm{A}} \\ -\overline{\mathbf{H}}_{\mathrm{B}} \end{bmatrix} = \begin{bmatrix} \overline{\boldsymbol{y}}_{11} & \overline{\boldsymbol{y}}_{12} \\ \overline{\boldsymbol{y}}_{21} & \overline{\boldsymbol{y}}_{22} \end{bmatrix} \begin{bmatrix} \overline{\mathbf{E}}_{\mathrm{A}} \\ \overline{\mathbf{E}}_{\mathrm{B}} \end{bmatrix} = \overline{\boldsymbol{y}}_{\mathrm{AB}} \begin{bmatrix} \overline{\mathbf{E}}_{\mathrm{A}} \\ \overline{\mathbf{E}}_{\mathrm{B}} \end{bmatrix} \tag{2.142}$$

In many cases we have $\overline{\mathbf{z}}_{11} = \overline{\mathbf{z}}_{22} = \overline{\mathbf{z}}_1$, $\overline{\mathbf{z}}_{12} = \overline{\mathbf{z}}_{21} = \overline{\mathbf{z}}_2$, $\overline{\boldsymbol{y}}_{11} = \overline{\boldsymbol{y}}_{22} = \overline{\boldsymbol{y}}_1$ and $\overline{\boldsymbol{y}}_{12} = \overline{\boldsymbol{y}}_{21} = \overline{\boldsymbol{y}}_2$ [see e.g. eqs. (2.46), (2.47)]. The relation of the tangential fields in the particular planes is given by:

$$\overline{\mathbf{H}}^{(k)}_{\mathrm{A,B}} = \overline{\mathbf{Y}}^{(k)}_{\mathrm{A,B}} \overline{\mathbf{E}}^{(k)}_{\mathrm{A,B}} \qquad \overline{\mathbf{E}}^{(k)}_{\mathrm{A,B}} = \overline{\mathbf{Z}}^{(k)}_{\mathrm{A,B}} \overline{\mathbf{H}}^{(k)}_{\mathrm{A,B}} \tag{2.143}$$

$\overline{\mathbf{Y}}_{\mathrm{A,B}}$ and $\overline{\mathbf{Z}}_{\mathrm{A,B}}$ are the generalised admittances and impedances, respectively. Inserting eq. (2.143) into eqs. (2.141) and (2.142), we obtain the impedance and admittance transformation formulas, respectively.

$$\overline{\mathbf{Z}}^{(k)}_{\mathrm{A}} = \overline{\mathbf{z}}^{(k)}_{11} - \overline{\mathbf{z}}^{(k)}_{12} (\overline{\mathbf{z}}^{(k)}_{22} + \overline{\mathbf{Z}}^{(k)}_{\mathrm{B}})^{-1} \overline{\mathbf{z}}^{(k)}_{21} \tag{2.144}$$

$$\overline{\mathbf{Y}}^{(k)}_{\mathrm{A}} = \overline{\boldsymbol{y}}^{(k)}_{11} - \overline{\boldsymbol{y}}^{(k)}_{12} (\overline{\boldsymbol{y}}^{(k)}_{22} + \overline{\mathbf{Y}}^{(k)}_{\mathrm{B}})^{-1} \overline{\boldsymbol{y}}^{(k)}_{21} \tag{2.145}$$

For the transformation in the opposite direction we obtain:

$$-\overline{\mathbf{Z}}^{(k)}_{\mathrm{B}} = \overline{\mathbf{z}}^{(k)}_{22} - \overline{\mathbf{z}}^{(k)}_{21} (\overline{\mathbf{z}}^{(k)}_{11} + (-\overline{\mathbf{Z}}^{(k)}_{\mathrm{A}}))^{-1} \overline{\mathbf{z}}^{(k)}_{12} \tag{2.146}$$

$$-\overline{\mathbf{Y}}^{(k)}_{\mathrm{B}} = \overline{\boldsymbol{y}}^{(k)}_{22} - \overline{\boldsymbol{y}}^{(k)}_{21} (\overline{\boldsymbol{y}}^{(k)}_{11} + (-\overline{\mathbf{Y}}^{(k)}_{\mathrm{A}}))^{-1} \overline{\boldsymbol{y}}^{(k)}_{12} \tag{2.147}$$

As can be seen, we have analogue formulas for the $\pm z$-direction. To obtain the latter expressions from the former ones we just have to introduce a negative impedance/admittance as a relation between the electric and magnetic fields.

In each plane we have two input impedance/admittance matrices: one when looking in $+z$-direction and the other when looking in $-z$-direction.

The impedance/admittance transformation formulas (2.144)–(2.147) are numerically stable even for very thick layers. This can easily be shown for isotropic structures in Cartesian coordinates. With increasing layer thickness d_k, the parameters $\overline{\mathbf{y}}_1$ and $\overline{\mathbf{z}}_1$ approach a constant value ($\tanh(\boldsymbol{\Gamma}\overline{d}_k) \rightarrow \boldsymbol{I}$), whereas $\overline{\mathbf{y}}_2$ and $\overline{\mathbf{z}}_2$ tend to zero ($1/\sinh(\boldsymbol{\Gamma}\overline{d}_k) \rightarrow \boldsymbol{0}$). Therefore, we obtain from eqs (2.144) and (2.145):

$$\begin{aligned} \lim_{d_k\to\infty} \overline{\mathbf{Z}}_{\mathrm{A}}^{(k)} &= \lim_{d_k\to\infty} (\overline{\mathbf{z}}_1^{(k)}) = \overline{\mathbf{Z}}_o^{(k)} \\ \lim_{d_k\to\infty} \overline{\mathbf{Y}}_{\mathrm{A}}^{(k)} &= \lim_{d_k\to\infty} (\overline{\mathbf{y}}_1^{(k)}) = \overline{\mathbf{Y}}_o^{(k)} \end{aligned} \tag{2.148}$$

The quantities $\overline{\mathbf{Z}}_o$ and $\overline{\mathbf{Y}}_o = \overline{\mathbf{Z}}_o^{-1}$ in eq. (2.148) are the generalised wave impedances and admittances, respectively, which are equal to the unit matrix $\boldsymbol{I}$ with the normalisation introduced in eq. (2.38).

For very thin layers or very short sections, the above formulas (2.141) or (2.142) should be rewritten:

$$\begin{bmatrix} \overline{\mathbf{E}}_{\mathrm{A}}^{(k)} \\ \overline{\mathbf{H}}_{\mathrm{A}}^{(k)} \end{bmatrix} = \begin{bmatrix} \overline{\boldsymbol{V}}_{\mathrm{E}} & \overline{\boldsymbol{Z}} \\ \overline{\boldsymbol{Y}} & \overline{\boldsymbol{V}}_{\mathrm{H}} \end{bmatrix}_{|k} \begin{bmatrix} \overline{\mathbf{E}}_{\mathrm{B}}^{(k)} \\ \overline{\mathbf{H}}_{\mathrm{B}}^{(k)} \end{bmatrix} = \begin{bmatrix} \cosh(\boldsymbol{\Gamma}\overline{d}) & \overline{\mathbf{Z}}_0 \sinh(\boldsymbol{\Gamma}\overline{d}) \\ \overline{\mathbf{Y}}_0 \sinh(\boldsymbol{\Gamma}\overline{d}) & \cosh(\boldsymbol{\Gamma}\overline{d}) \end{bmatrix}_{|k} \begin{bmatrix} \overline{\mathbf{E}}_{\mathrm{B}}^{(k)} \\ \overline{\mathbf{H}}_{\mathrm{B}}^{(k)} \end{bmatrix} \tag{2.149}$$

The second part is valid for Cartesian coordinates with isotropic materials. For $d_k \rightarrow 0$ we obtain: $\overline{\boldsymbol{Z}} \rightarrow \mathbf{0}$, $\overline{\boldsymbol{Y}} \rightarrow \mathbf{0}$, $\overline{\boldsymbol{V}} \rightarrow \boldsymbol{I}$. The impedance and admittance transformation formulas from (2.149) are:

$$\overline{\mathbf{Z}}_{\mathrm{A}}^{(k)} = (\overline{\boldsymbol{V}}_{\mathrm{E}}^{(k)} \overline{\mathbf{Z}}_{\mathrm{B}}^{(k)} + \overline{\boldsymbol{Z}}^{(k)})(\overline{\boldsymbol{V}}_{\mathrm{H}}^{(k)} + \overline{\boldsymbol{Y}}^{(k)} \overline{\mathbf{Z}}_{\mathrm{B}}^{(k)})^{-1} \tag{2.150}$$

$$\overline{\mathbf{Y}}_{\mathrm{A}}^{(k)} = (\overline{\boldsymbol{Y}}^{(k)} + \overline{\boldsymbol{V}}_{\mathrm{H}}^{(k)} \overline{\mathbf{Y}}_{\mathrm{B}}^{(k)})(\overline{\boldsymbol{V}}_{\mathrm{E}}^{(k)} + \overline{\boldsymbol{Z}}^{(k)} \overline{\mathbf{Y}}_{\mathrm{B}}^{(k)})^{-1} \tag{2.151}$$

It can easily be seen that for small lengths ($d_k \rightarrow 0$) the impedance (admittance) $\overline{\mathbf{Z}}_{\mathrm{A}}^{(k)}$ ($\overline{\mathbf{Y}}_{\mathrm{A}}^{(k)}$) in eq. (2.150) and eq. (2.151) approaches $\overline{\mathbf{Z}}_{\mathrm{B}}^{(k)}$ ($\overline{\mathbf{Y}}_{\mathrm{B}}^{(k)}$).

If we have only one mode, these equations reduce to the well-known impedance transformation formulas from transmission line theory. The admittance transformation formula in the form of eq. (2.145) was first derived by Rogge [18] for the isotropic case in Cartesian coordinates. It was applied to the analysis of diffused optical waveguides, where the variation of permittivity in the vertical direction was modelled by up to 80 layers with different permittivity distributions in the lateral direction.

2.5.2.2 *Impedance transformation at interfaces*

(a) *Interfaces without metallisation*

Now we would like to transform the impedances/admittances at interfaces from one layer to the other or from one section to the other. Due to the continuity of the transverse field components, the impedances (or admittances) have to be continuous as well. The fields in the original domain are computed by those in the transformed domain by a multiplication with the transformation matrices; see eq. (2.32). Since we have individual transformation matrices $\boldsymbol{T}_{\mathrm{E,H}}$ in each homogeneous layer (or section), we must transform the admittances/impedances back to the interfaces between the layers. At the interface of the layers (or sections) $k-1$ and k, the corresponding transformation equations are:

$$\overline{\boldsymbol{Z}}_{\mathrm{B}}^{(k)} = \boldsymbol{T}_{\mathrm{E}k}^{-1}\boldsymbol{T}_{\mathrm{E}k-1}\overline{\boldsymbol{Z}}_{\mathrm{A}}^{(k-1)}\boldsymbol{T}_{\mathrm{H}k-1}^{-1}\boldsymbol{T}_{\mathrm{H}k} \tag{2.152}$$

$$\overline{\boldsymbol{Y}}_{\mathrm{B}}^{(k)} = \boldsymbol{T}_{\mathrm{H}k}^{-1}\boldsymbol{T}_{\mathrm{H}k-1}\overline{\boldsymbol{Y}}_{\mathrm{A}}^{(k-1)}\boldsymbol{T}_{\mathrm{E}k-1}^{-1}\boldsymbol{T}_{\mathrm{E}k} \tag{2.153}$$

So far we have assumed layers of equal width and with no metallisation in the interface.

(b) *Interfaces with metallisation*
The question is how the formulas (2.152) and (2.153) must be modified if we have metallisations of ideal conductivity at the interfaces of the layers, or if the layers or the metallisation are of different width. Fig. 2.30 shows such interfaces of planar and circular layers. The concatenation of waveguide sections with different cross-sections is presented in Fig. 2.31. Generally, the field matching at the interfaces (or ports) of the layers (or sections) must be performed in the original (or spatial) domain. For this purpose, we introduce the special matrix $\boldsymbol{J}$. This is a unit matrix which we split in the metal area (see below).

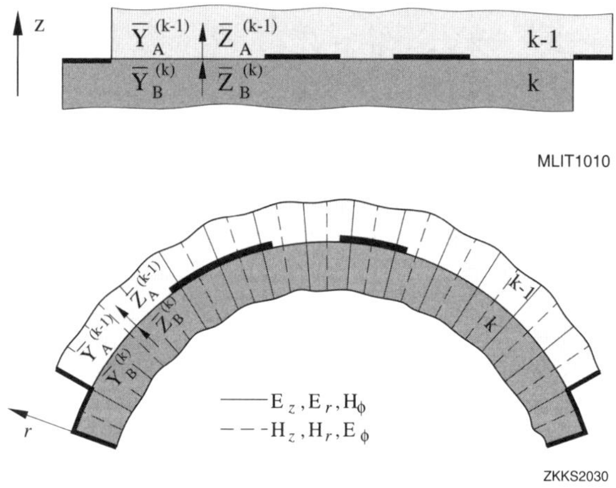

(a)

(b)

Fig. 2.30 General layer interfaces between layers of different widths ((a) Reproduced by permission of Elsevier)

Alternatively, if we use quantities in a transformed domain, the field matching can easily be done by using the so-called transformation matrix

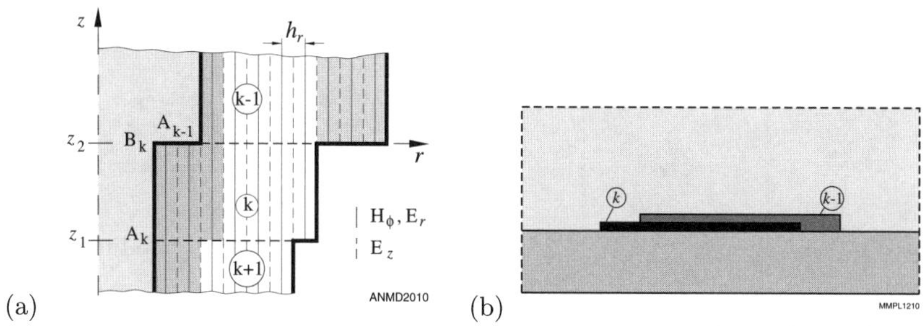

Fig. 2.31 Concatenation of waveguide sections with different cross-sections: (a) coaxial waveguide sections (b) planar waveguide sections

partition technique. Both procedures are in close connection. Here, we perform this transformation in the original or spatial domain, where $\boldsymbol{Z} = \boldsymbol{T}_{\mathrm{E}}\overline{\boldsymbol{Z}}\boldsymbol{T}_{\mathrm{H}}^{-1}$ and $\boldsymbol{Y} = \boldsymbol{T}_{\mathrm{H}}\overline{\boldsymbol{Y}}\boldsymbol{T}_{\mathrm{E}}^{-1}$.

We will explain the procedure with the help of the concatenation of coaxial waveguide sections $k-1$ and k in Fig. 2.31a. Due to the azimuthal symmetry, the vectors $\mathbf{E}$ and $\mathbf{H}$ contain only one component. Some of the discretisation lines in the sections below and above the interface B_k-A_{k-1} end on metal and the other ones (in the concatenated region or in the slots) are common to both sections. We label the first part with m (for 'metal') and the second part with c (for 'common'). To take into account the metallisation, we start with a unit matrix $\boldsymbol{J}$ whose size is equal to the number of discretisation lines in the corresponding section (or layer). Then we split this matrix into two parts, where $\boldsymbol{J}^{\mathrm{c}}$ contains only the rows for the lines crossing the slots outside the metal and $\boldsymbol{J}^{\mathrm{m}}$ contains the remaining rows. Both matrices have rectangular form.

This dividing is equivalent to the partition of the transformation matrices. The partition of the transformation matrix $\boldsymbol{T}$ for the layer k on the upper interface in Fig. 2.31a is sketched in Fig. 2.32. The two parts of the field $\mathbf{F}_k$ with $\mathbf{F}_k = \mathbf{E}_{\mathrm{B_k}}$ or $\mathbf{F}_k = \mathbf{H}_{\mathrm{B_k}}$ (section connected to dielectric area or to metal) are given by:

$$\begin{aligned}
\mathbf{F}_k^{\mathrm{m}} &= \boldsymbol{J}_k^{\mathrm{m}}\mathbf{F}_k = \boldsymbol{J}_k^{\mathrm{m}}\boldsymbol{T}_k\overline{\mathbf{F}}_k = \boldsymbol{T}_k^{\mathrm{m}}\overline{\mathbf{F}}_k \qquad & \boldsymbol{T}_k^{\mathrm{m}} &= \boldsymbol{J}_k^{\mathrm{m}}\boldsymbol{T}_k \qquad & \boldsymbol{T}_k^{-1^{\mathrm{m}}} &= \boldsymbol{T}_k^{-1}\boldsymbol{J}_k^{\mathrm{mt}} \\
\mathbf{F}_k^{\mathrm{c}} &= \boldsymbol{J}_k^{\mathrm{c}}\mathbf{F}_k = \boldsymbol{J}_k^{\mathrm{c}}\boldsymbol{T}_k\overline{\mathbf{F}}_k = \boldsymbol{T}_k^{\mathrm{c}}\overline{\mathbf{F}}_k & \boldsymbol{T}_k^{\mathrm{c}} &= \boldsymbol{J}_k^{\mathrm{c}}\boldsymbol{T}_k & \boldsymbol{T}_k^{-1^{\mathrm{c}}} &= \boldsymbol{T}_k^{-1}\boldsymbol{J}_k^{\mathrm{ct}}
\end{aligned} \tag{2.154}$$

The partitioning should be understood symbolically. If we have various metal regions, we must do this portioning for all of them. Further, if the field contains more than one component and/or in the case of a 2D-discretisation, the metal points are usually not on successive positions. Hence, the portioning has to be done at various positions in the matrix.

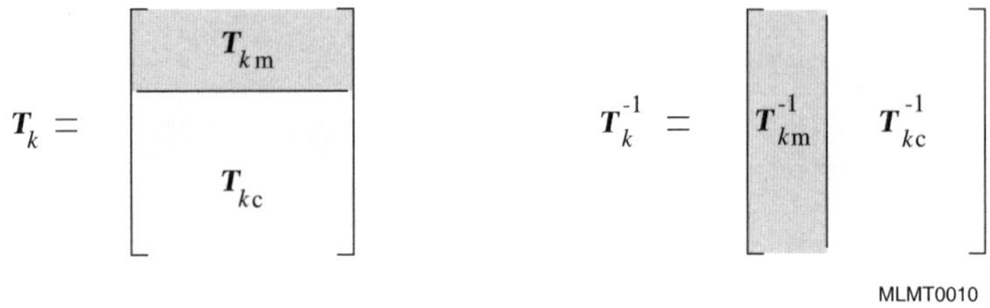

Fig. 2.32 Transformation matrix partition in region k of Figure 2.31a

Let us now derive the formulas for the impedance transformation at interfaces. Matching the tangential electric field results in:

$$\boldsymbol{J}_k^{\mathrm{m}}\mathbf{E}_k = \mathbf{0}_k \tag{2.155}$$

$$\boldsymbol{J}_k^{\mathrm{c}}\mathbf{E}_k = \boldsymbol{J}_{k-1}^{\mathrm{c}}\mathbf{E}_{k-1} \tag{2.156}$$

$$\mathbf{0}_{k-1} = \boldsymbol{J}_{k-1}^{\mathrm{m}}\mathbf{E}_{k-1} \tag{2.157}$$

To explain again: the matrix $\boldsymbol{J}_k^{\mathrm{c}}(\boldsymbol{J}_{k-1}^{\mathrm{c}})$ reduces the field $\boldsymbol{E}_k$ $(\boldsymbol{E}_{k-1})$ in the original domain of layer k $(k-1)$ to that part common to the field of layer/section $k-1$ (k). $\boldsymbol{J}_k^{\mathrm{m}}\boldsymbol{E}_k(\boldsymbol{J}_{k-1}^{\mathrm{m}}\boldsymbol{E}_{k-1})$ is the field part of the metallic front end of layer/section $k-1$ (k). As mentioned before, the dividing of the matrix in Fig. 2.32 should be understood symbolically. Therefore, it is usually more complicated than shown there.

Each part consists of rows which belong to different parts of the interface, e.g. in the case of Fig. 2.30b, the matrix $\boldsymbol{J}_{k-1}^{\mathrm{m}}$ consists of four parts and the matrix $\boldsymbol{J}_{k-1}^{\mathrm{c}}$ consists of three parts, because we have four metallic and three common parts at the interface from layer k to layer $k-1$. Now, we obtain for the tangential electric field transformation between the two sides:

$$\mathbf{E}_k = \begin{bmatrix} \boldsymbol{J}_{k-1}^{\mathrm{c}}\mathbf{E}_{k-1} \\ \boldsymbol{0}_k \end{bmatrix} = \boldsymbol{J}_k^{\mathrm{ct}}\boldsymbol{J}_{k-1}^{\mathrm{c}}\mathbf{E}_{k-1} \qquad \mathbf{E}_{k-1} = \begin{bmatrix} \boldsymbol{J}_k^{\mathrm{c}}\mathbf{E}_k \\ \boldsymbol{0}_{k-1} \end{bmatrix} = \boldsymbol{J}_{k-1}^{\mathrm{ct}}\boldsymbol{J}_k^{\mathrm{c}}\mathbf{E}_k \tag{2.158}$$

Matching the tangential magnetic field results in:

$$\boldsymbol{J}_k^{\mathrm{c}}\mathbf{H}_k = \boldsymbol{J}_{k-1}^{\mathrm{c}}\mathbf{H}_{k-1} = \boldsymbol{J}_{k-1}^{\mathrm{c}}\mathbf{Y}_{k-1}\mathbf{E}_{k-1} = \boldsymbol{J}_{k-1}^{\mathrm{c}}\mathbf{Y}_{k-1}\boldsymbol{J}_{k-1}^{\mathrm{ct}}\boldsymbol{J}_k^{\mathrm{c}}\mathbf{E}_k \tag{2.159}$$

$$\boldsymbol{J}_k^{\mathrm{m}}\mathbf{H}_k = \mathbf{S}_k \qquad \boldsymbol{J}_{k-1}^{\mathrm{m}}\mathbf{H}_{k-1} = \mathbf{S}_{k-1} \tag{2.160}$$

$\mathbf{S}_{k,k-1}$ is the current distribution on the metallisations (we do not need these equations for the matching process and they are only given for completeness). Eq. (2.159) may be rewritten and combined with eq. (2.155) (we have not used this equation in the matching process yet):

$$\begin{bmatrix} \boldsymbol{J}_k^{\mathrm{c}} \\ \boldsymbol{J}_k^{\mathrm{m}} \end{bmatrix} \mathbf{E}_k = \begin{bmatrix} (\boldsymbol{J}_{k-1}^{\mathrm{c}}\mathbf{Y}_{k-1}\boldsymbol{J}_{k-1}^{\mathrm{ct}})^{-1}\boldsymbol{J}_k^{\mathrm{c}}\mathbf{H}_k \\ \mathbf{0}_k \end{bmatrix} \tag{2.161}$$

As our final result we obtain:

$$\mathbf{E}_k = \boldsymbol{J}_k^{\mathrm{ct}}(\boldsymbol{J}_{k-1}^{\mathrm{c}}\boldsymbol{Y}_{k-1}\boldsymbol{J}_{k-1}^{\mathrm{ct}})^{-1}\boldsymbol{J}_k^{\mathrm{c}}\mathbf{H}_k = \boldsymbol{Z}_k\mathbf{H}_k \tag{2.162}$$

or:

$$\boxed{\boldsymbol{Z}_k = \boldsymbol{J}_k^{\mathrm{ct}}(\boldsymbol{J}_{k-1}^{\mathrm{c}}\boldsymbol{Y}_{k-1}\boldsymbol{J}_{k-1}^{\mathrm{ct}})^{-1}\boldsymbol{J}_k^{\mathrm{c}}} \tag{2.163}$$

Alternatively, the result may also be obtained by using partitioned transformation matrices. We write the result as impedance $\overline{\boldsymbol{Z}}_{\mathrm{B}k}$ for the structures shown in Fig. 2.30:

$$\boxed{\overline{\boldsymbol{Z}}_{\mathrm{B}k} = \boldsymbol{T}_{\mathrm{E}k,\mathrm{c}}^{-1}(\boldsymbol{T}_{\mathrm{H}k-1,\mathrm{c}}\overline{\boldsymbol{Y}}_{\mathrm{A},k-1}\boldsymbol{T}_{\mathrm{E}k-1,\mathrm{c}}^{-1})^{-1}\boldsymbol{T}_{\mathrm{H}k,\mathrm{c}}} \tag{2.164}$$

Two typical special cases will be considered briefly.

1. The layer $k-1$ is smaller than the layer k, and the aperture of the layer $k-1$ is enclosed completely within the aperture of the layer k. In this case the matrix $\boldsymbol{T}_{k-1}$ needs not be partitioned. Therefore, we obtain:

$$\overline{\boldsymbol{Z}}_{\mathrm{B}k} = \boldsymbol{T}_{\mathrm{E}k,\mathrm{c}}^{-1}\boldsymbol{T}_{\mathrm{E}k-1}\overline{\boldsymbol{Z}}_{\mathrm{A},k-1}\boldsymbol{T}_{\mathrm{H}k-1}^{-1}\boldsymbol{T}_{\mathrm{H}k,\mathrm{c}} \qquad \boldsymbol{Z}_k = \boldsymbol{J}_k^{\mathrm{ct}}\boldsymbol{Z}_{k-1}\boldsymbol{J}_k^{\mathrm{c}} \tag{2.165}$$

2. In the second case we have the opposite situation. Now the layer k is smaller and its aperture is totally contained in the aperture of the layer $k-1$. In this case the matrix $\boldsymbol{T}_k$ needs not be partitioned. By inverting eq. (2.164) we obtain:

$$\overline{\boldsymbol{Y}}_{\mathrm{B}k} = \boldsymbol{T}_{\mathrm{H}k}^{-1}\boldsymbol{T}_{\mathrm{H}k-1,\mathrm{c}}\overline{\boldsymbol{Y}}_{\mathrm{A},k-1}\boldsymbol{T}_{\mathrm{E}k-1,\mathrm{c}}^{-1}\boldsymbol{T}_{\mathrm{E}k} \qquad \boldsymbol{Y}_k = \boldsymbol{J}_{k-1}^{\mathrm{c}}\boldsymbol{Y}_{k-1}\boldsymbol{J}_{k-1}^{\mathrm{ct}} \tag{2.166}$$

2.5.2.3 Finite metallisation thickness

Next we describe the impedance transformation in the case of finite metallisation thickness, as shown in Fig. 2.33. The layer with the metallisation is labelled l. The transformation will be done in two steps. First we transform the impedance at the interface between layer $k-1$ and l by using:

$$\overline{\boldsymbol{Z}}_{\mathrm{B}l} = \boldsymbol{T}_{\mathrm{El}}^{-1}(\boldsymbol{T}_{\mathrm{Hk-1,c}}\overline{\boldsymbol{Y}}_{\mathrm{A},k-1}\boldsymbol{T}_{\mathrm{Ek-1,c}}^{-1})^{-1}\boldsymbol{T}_{\mathrm{Hl}} \tag{2.167}$$

The matrices $\boldsymbol{T}_{\mathrm{E,Hl}}$ are block diagonal matrices:

$$\boldsymbol{T}_{\mathrm{E,Hl}} = \mathrm{Diag}(\boldsymbol{T}_{\mathrm{E,Hl1}}, \boldsymbol{T}_{\mathrm{E,Hl2}}, \ldots, \boldsymbol{T}_{\mathrm{E,HlL}}) \tag{2.168}$$

where the matrices $\boldsymbol{T}_{\mathrm{E,Hl}i}, i = 1, \ldots, L$ are the transformation matrices for the ith dielectric intermediate region (counting e.g. from left to right) between the metallisations in the layer l. Each of these matrices is quadratic. Therefore, we obtain for the admittance matrix (see eq. (2.166)):

$$\overline{\boldsymbol{Y}}_{\mathrm{B}l} = \boldsymbol{T}_{\mathrm{Hl}}^{-1}\boldsymbol{T}_{\mathrm{Hk-1,c}}\overline{\boldsymbol{Y}}_{\mathrm{A},k-1}\boldsymbol{T}_{\mathrm{Ek-1,c}}^{-1}\boldsymbol{T}_{\mathrm{El}} \qquad \overline{\boldsymbol{Z}}_{\mathrm{B}l} = \overline{\boldsymbol{Y}}_{\mathrm{B}l}^{-1} \tag{2.169}$$

(a)

(b)

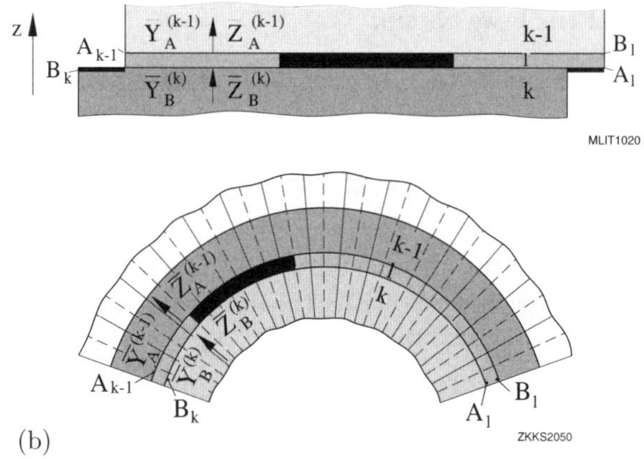

Fig. 2.33 Metallization of finite thickness at the interfaces ((a) Reproduced by permission of Elsevier)

In the next step we transform the impedance $\overline{\boldsymbol{Z}}_{\mathrm{B}l}$ (or admittance $\overline{\boldsymbol{Y}}_{\mathrm{Bl}}$) of the upper side of the layer l to the lower one by:

$$\begin{aligned}\overline{\boldsymbol{Y}}_{\mathrm{A}l} &= \overline{\boldsymbol{y}}_{11}^{l} - \overline{\boldsymbol{y}}_{12}^{l}(\overline{\boldsymbol{y}}_{22}^{l} + \overline{\boldsymbol{Y}}_{\mathrm{B}l})^{-1}\overline{\boldsymbol{y}}_{21}^{l} \\ \overline{\boldsymbol{Z}}_{\mathrm{A}l} &= \overline{\boldsymbol{z}}_{11}^{l} - \overline{\boldsymbol{z}}_{12}^{l}(\overline{\boldsymbol{z}}_{22}^{l} + \overline{\boldsymbol{Z}}_{\mathrm{B}l})^{-1}\overline{\boldsymbol{z}}_{21}^{l}\end{aligned} \tag{2.170}$$

where we combine all parameters $\overline{\boldsymbol{z}}_{jk}^{l}$ and $\overline{\boldsymbol{y}}_{jk}^{l}$ (with $j, k = 1, 2$) analogously to eq. (2.168). We have e.g.:

$$\overline{\boldsymbol{y}}_{11}^{l} = \mathrm{Diag}(\overline{\boldsymbol{y}}_{11i}^{l}) \qquad \overline{\boldsymbol{z}}_{11}^{l} = \mathrm{Diag}(\overline{\boldsymbol{z}}_{11i}^{l}) \tag{2.171}$$

with $i = 1, 2, \ldots, L$.

For small thickness of layer l, we should apply transfer matrix expressions instead of the eqs. (2.170). We obtain in this case:

$$\begin{aligned}\overline{\boldsymbol{Y}}_{\mathrm{A}l} &= (\overline{\boldsymbol{Y}}_{l} + \overline{\boldsymbol{V}}_{\mathrm{H}l}\overline{\boldsymbol{Y}}_{\mathrm{B}l})(\overline{\boldsymbol{V}}_{\mathrm{E}l} + \overline{\boldsymbol{Z}}_{l}\overline{\boldsymbol{Y}}_{\mathrm{B}l})^{-1} \\ \overline{\boldsymbol{Z}}_{\mathrm{A}l} &= (\overline{\boldsymbol{Z}}_{l} + \overline{\boldsymbol{V}}_{\mathrm{E}l}\overline{\boldsymbol{Y}}_{\mathrm{B}l})(\overline{\boldsymbol{V}}_{\mathrm{H}l} + \overline{\boldsymbol{Y}}_{l}\overline{\boldsymbol{Z}}_{\mathrm{B}l})^{-1}\end{aligned} \tag{2.172}$$

where we have, analogous to eq. (2.168):

$$\overline{\boldsymbol{Y}}_{l} = \mathrm{Diag}(\overline{\boldsymbol{Y}}_{li}) \qquad \overline{\boldsymbol{V}}_{\mathrm{E,Hl}} = \mathrm{Diag}(\overline{\boldsymbol{V}}_{\mathrm{E,Hli}}) \qquad \overline{\boldsymbol{Z}}_{l} = \mathrm{Diag}(\overline{\boldsymbol{Z}}_{li})$$

with $i = 1, 2, \ldots, L$. For zero thickness of layer l, we can clearly see that the relation $\overline{\boldsymbol{Y}}_{\mathrm{A}l} = \overline{\boldsymbol{Y}}_{\mathrm{B}l}$ holds. This is true because $\boldsymbol{V}_{\mathrm{E,Hl}} \to \boldsymbol{I}$, i.e. it approaches a unit matrix, whereas the other submatrices become zero-matrices. Now we can transform the admittance/impedance at the interface between the layers l and k and obtain:

$$\overline{\boldsymbol{Z}}_{\mathrm{B}k} = \boldsymbol{T}_{\mathrm{Ek,c}}^{-1}(\boldsymbol{T}_{\mathrm{Hl}}\overline{\boldsymbol{Y}}_{\mathrm{A}l}\boldsymbol{T}_{\mathrm{El}}^{-1})^{-1}\boldsymbol{T}_{\mathrm{Hk,c}} = \boldsymbol{T}_{\mathrm{Ek,c}}^{-1}\boldsymbol{T}_{\mathrm{El}}\overline{\boldsymbol{Z}}_{\mathrm{A}l}\boldsymbol{T}_{\mathrm{Hl}}^{-1}\boldsymbol{T}_{\mathrm{Hk,c}} \tag{2.173}$$

For a layer with zero thickness and $\overline{\boldsymbol{Y}}_{\mathrm{B}l} = \overline{\boldsymbol{Y}}_{\mathrm{A}l}$ we can introduce eq. (2.169) into eq. (2.173) and obtain as a result eq. (2.164), as expected. The described procedure allows the analysis of more complicated structures, like those sketched in Fig. 2.34. This can be done by segmentation of the structure in suitable layers. Arbitrarily-shaped metallic insets can be examined in this way, too.

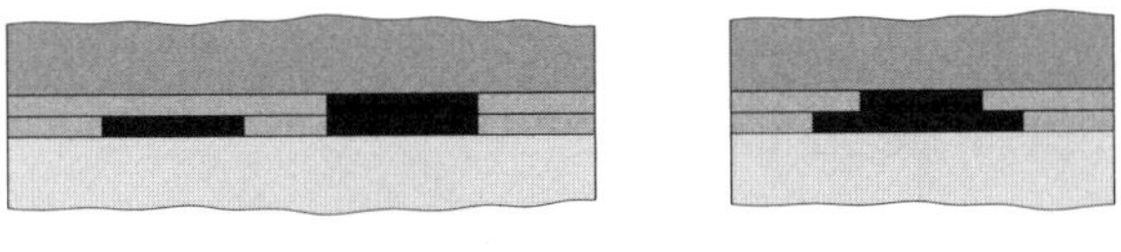

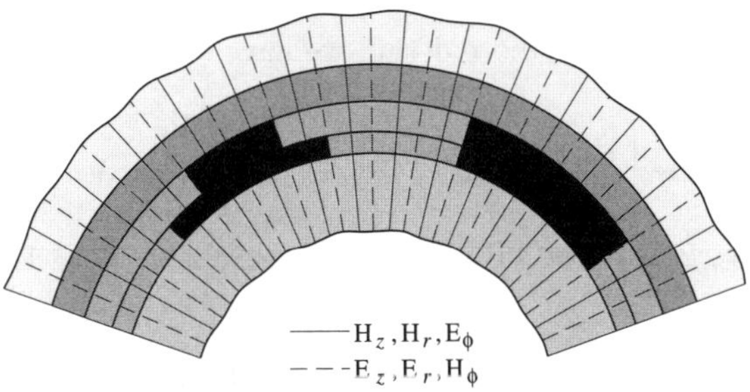

Fig. 2.34 Cross-sections of waveguide structures with different heights of the metallisation (Top: Reproduced by permission of Elsevier)

2.5.3 Impedance/admittance transformation with finite differences

2.5.3.1 Linear field interpolation

For some geometries and coordinate systems, the combined GTL equations cannot be solved analytically. In this case we can use a finite difference solution, as was described in [19].[3] By combining the GTL equations to a first-order differential equation, we obtain:

$$\frac{\mathrm{d}}{\mathrm{d}\overline{u}}\widehat{\mathbf{F}} = \widehat{\boldsymbol{Q}}\widehat{\mathbf{F}} \qquad \widehat{\mathbf{F}} = [\mathbf{E}^{\mathrm{t}}, \mathbf{H}^{\mathrm{t}}]^{\mathrm{t}} \qquad \widehat{\boldsymbol{Q}} = -\begin{bmatrix} \boldsymbol{S}_{\mathrm{E}} & \mathrm{j}\boldsymbol{R}_{\mathrm{H}} \\ \mathrm{j}\boldsymbol{R}_{\mathrm{E}} & \boldsymbol{S}_{\mathrm{H}} \end{bmatrix} \tag{2.174}$$

u is the coordinate for which we would like to solve the equation. The matrix $\widehat{\boldsymbol{Q}}$ is given here in a general form and can depend on u. In most cases, the submatrices $\boldsymbol{S}_{\mathrm{E,H}}$ are equal to zero. If we want to determine eigenmodes we have matrices $\boldsymbol{R}^{e}_{\mathrm{E,H}}$ instead of $\mathrm{j}\boldsymbol{R}_{\mathrm{E,H}}$.

[3]Though we do not show it in this book, we would like to mention that such discrete impedance transformations could also be applied to problems with crossed discretisation lines, as are presented in Chapter 7.

Now we divide a layer (or section) in sublayers (or subsections), with subports A and B. For small distances $\Delta\overline{u}$ between the subports A and B, we can apply a linear field interpolation. By using finite differences, we obtain:

$$\widehat{\mathbf{F}}_{\mathrm{B}} - \widehat{\mathbf{F}}_{\mathrm{A}} = \widehat{\widehat{\boldsymbol{Q}}}(\widehat{\mathbf{F}}_{\mathrm{B}} + \widehat{\mathbf{F}}_{\mathrm{A}}) \tag{2.175}$$

with:

$$\widehat{\widehat{\boldsymbol{Q}}} = 0.5\Delta\overline{u}\widehat{\boldsymbol{Q}}(\overline{u}^{\mathrm{m}}) \qquad \overline{u}^{\mathrm{m}} = 0.5(\overline{u}_{\mathrm{A}} + \overline{u}_{\mathrm{B}}) \tag{2.176}$$

From eq. (2.174) we obtain:

$$\widehat{\mathbf{F}}_{\mathrm{A}} = (\widehat{\boldsymbol{I}} + \widehat{\widehat{\boldsymbol{Q}}})^{-1}(\widehat{\boldsymbol{I}} - \widehat{\widehat{\boldsymbol{Q}}})\widehat{\mathbf{F}}_{\mathrm{B}} = \widehat{\mathbf{V}}_{\mathrm{AB}}\widehat{\mathbf{F}}_{\mathrm{B}} \tag{2.177}$$

If we write the terms in more detail, we have:

$$\begin{bmatrix}\mathbf{E}_{\mathrm{A}}\\ \mathbf{H}_{\mathrm{A}}\end{bmatrix} = \widehat{\mathbf{V}}_{\mathrm{AB}}\begin{bmatrix}\mathbf{E}_{\mathrm{B}}\\ \mathbf{H}_{\mathrm{B}}\end{bmatrix} \qquad \widehat{\mathbf{V}}_{\mathrm{AB}} = \begin{bmatrix}\mathbf{V}_{\mathrm{E}} & \boldsymbol{Z}_{\mathrm{H}}\\ \mathbf{Y}_{\mathrm{E}} & \mathbf{V}_{\mathrm{H}}\end{bmatrix} \tag{2.178}$$

$\mathbf{V}_{\mathrm{E}}$, $\mathbf{V}_{\mathrm{H}}$, $\mathbf{Y}_{\mathrm{E}}$, $\boldsymbol{Z}_{\mathrm{H}}$ are the submatrices of the matrix $\widehat{\mathbf{V}}_{\mathrm{AB}}$. Eq. (2.178) may be rewritten as:

$$\begin{bmatrix}\mathbf{E}_{\mathrm{A}}\\ \mathbf{E}_{\mathrm{B}}\end{bmatrix} = \begin{bmatrix}\boldsymbol{z}_{11} & \boldsymbol{z}_{12}\\ \boldsymbol{z}_{21} & \boldsymbol{z}_{22}\end{bmatrix}\begin{bmatrix}\mathbf{H}_{\mathrm{A}}\\ -\mathbf{H}_{\mathrm{B}}\end{bmatrix} \qquad \begin{bmatrix}\mathbf{H}_{\mathrm{A}}\\ -\mathbf{H}_{\mathrm{B}}\end{bmatrix} = \begin{bmatrix}\boldsymbol{y}_{11} & \boldsymbol{y}_{12}\\ \boldsymbol{y}_{21} & \boldsymbol{y}_{22}\end{bmatrix}\begin{bmatrix}\mathbf{E}_{\mathrm{A}}\\ \mathbf{E}_{\mathrm{B}}\end{bmatrix} \tag{2.179}$$

with:

$$\begin{aligned}
\boldsymbol{z}_{11} &= \mathbf{V}_{\mathrm{E}}\mathbf{Y}_{\mathrm{E}}^{-1} & \boldsymbol{y}_{11} &= \mathbf{V}_{\mathrm{H}}\boldsymbol{Z}_{\mathrm{H}}^{-1}\\
\boldsymbol{z}_{12} &= \mathbf{V}_{\mathrm{E}}\mathbf{Y}_{\mathrm{E}}^{-1}\mathbf{V}_{\mathrm{H}} - \boldsymbol{Z}_{\mathrm{H}} & \boldsymbol{y}_{12} &= \mathbf{Y}_{\mathrm{E}} - \mathbf{V}_{\mathrm{H}}\boldsymbol{Z}_{\mathrm{H}}^{-1}\mathbf{V}_{\mathrm{E}}\\
\boldsymbol{z}_{21} &= \mathbf{Y}_{\mathrm{E}}^{-1} & \boldsymbol{y}_{21} &= -\boldsymbol{Z}_{\mathrm{H}}^{-1}\\
\boldsymbol{z}_{22} &= \mathbf{Y}_{\mathrm{E}}^{-1}\mathbf{V}_{\mathrm{H}} & \boldsymbol{y}_{22} &= \boldsymbol{Z}_{\mathrm{H}}^{-1}\mathbf{V}_{\mathrm{E}}
\end{aligned} \tag{2.180}$$

Now the impedance and admittance transformation formulas between the ports A and B are analogous to the well-known cases. By using the definitions $\mathbf{E}_{\mathrm{A,B}} = \boldsymbol{Z}_{\mathrm{A,B}}\mathbf{H}_{\mathrm{A,B}}$ and $\mathbf{H}_{\mathrm{A,B}} = \mathbf{Y}_{\mathrm{A,B}}\mathbf{E}_{\mathrm{A,B}}$, we have:

$$\mathbf{Y}_{\mathrm{A}} = \boldsymbol{y}_{11} - \boldsymbol{y}_{12}(\boldsymbol{y}_{22} + \mathbf{Y}_{\mathrm{B}})^{-1}\boldsymbol{y}_{21} \tag{2.181}$$

$$\boldsymbol{Z}_{\mathrm{A}} = \boldsymbol{z}_{11} - \boldsymbol{z}_{12}(\boldsymbol{z}_{22} + \boldsymbol{Z}_{\mathrm{B}})^{-1}\boldsymbol{z}_{21} \tag{2.182}$$

We showed in the case of the analytic algorithm that the field and the impedances/admittances can also be transformed in the opposite ('$+u$')-direction. The expressions were similar to those used in the '$-u$'- direction. The same is true for the impedance/admittance transformation with finite differences.

The transfer matrix relation for the $+u$-direction can be given as:

$$\begin{bmatrix}\mathbf{E}_{\mathrm{B}}\\ \mathbf{H}_{\mathrm{B}}\end{bmatrix} = \widehat{\mathbf{V}}_{\mathrm{BA}}\begin{bmatrix}\mathbf{E}_{\mathrm{A}}\\ \mathbf{H}_{\mathrm{A}}\end{bmatrix} \qquad \widehat{\mathbf{V}}_{\mathrm{BA}} = \begin{bmatrix}\mathbf{V}_{\mathrm{E}}' & \boldsymbol{Z}_{\mathrm{H}}'\\ \mathbf{Y}_{\mathrm{E}}' & \mathbf{V}_{\mathrm{H}}'\end{bmatrix} \tag{2.183}$$

With this expression we can determine the submatrices $\mathbf{z}_{ik}$- and $\boldsymbol{y}_{ik}$ in eq. (2.179) as:

$$\begin{aligned}
\mathbf{z}_{11} &= -\mathbf{Y}_{\mathrm{E}}'^{-1}\mathbf{V}_{\mathrm{H}}' & \boldsymbol{y}_{11} &= -\boldsymbol{Z}_{\mathrm{H}}'^{-1}\mathbf{V}_{\mathrm{E}}' \\
\mathbf{z}_{12} &= -\mathbf{Y}_{\mathrm{E}}'^{-1} & \boldsymbol{y}_{12} &= \boldsymbol{Z}_{\mathrm{H}}'^{-1} \\
\mathbf{z}_{21} &= -\mathbf{V}_{\mathrm{E}}'\mathbf{Y}_{\mathrm{E}}'^{-1}\mathbf{V}_{\mathrm{H}}' + \boldsymbol{Z}_{\mathrm{H}}' & \boldsymbol{y}_{21} &= -\mathbf{Y}_{\mathrm{E}}' + \mathbf{V}_{\mathrm{H}}'\boldsymbol{Z}_{\mathrm{H}}'^{-1}\mathbf{V}_{\mathrm{E}}' \\
\mathbf{z}_{22} &= -\mathbf{V}_{\mathrm{E}}'\mathbf{Y}_{\mathrm{E}}'^{-1} & \boldsymbol{y}_{22} &= -\mathbf{V}_{\mathrm{H}}'\boldsymbol{Z}_{\mathrm{H}}'^{-1}
\end{aligned} \tag{2.184}$$

It should be emphasised that the submatrices $\mathbf{V}_{\mathrm{E}}'$, $\mathbf{V}_{\mathrm{H}}'$, $\mathbf{Y}_{\mathrm{E}}'$, $\boldsymbol{Z}_{\mathrm{H}}'$ which we used here are determined from the matrix $\widehat{\mathbf{V}}_{\mathrm{BA}}$ and are different to those in eq. (2.178). For the impedance/admittance transformation from A to B we have:

$$\begin{aligned}
-\mathbf{Y}_{\mathrm{B}} &= \boldsymbol{y}_{22} - \boldsymbol{y}_{21}\left(\boldsymbol{y}_{11} + (-\mathbf{Y}_{\mathrm{A}})\right)^{-1}\boldsymbol{y}_{12} \\
-\boldsymbol{Z}_{\mathrm{B}} &= \mathbf{z}_{22} - \mathbf{z}_{21}\left(\mathbf{z}_{11} + (-\boldsymbol{Z}_{\mathrm{A}})\right)^{-1}\mathbf{z}_{12}
\end{aligned} \tag{2.185}$$

2.5.3.2 Quadratic field interpolation

In eq. (2.175) we used a linear interpolation for the right side of eq. (2.174). Instead, we can also introduce a quadratic interpolation. For $\mathbf{F}$ on the right side of eq. (2.174), we have in this case (see Subsection A.4.3 about interpolation):

$$\mathbf{F}(\overline{u}^{\mathrm{m}}) = -\frac{1}{8}\mathbf{F}_{\mathrm{O}} + \frac{3}{4}\mathbf{F}_{\mathrm{A}} + \frac{3}{8}\mathbf{F}_{\mathrm{B}} \qquad \mathbf{F}(\overline{u}^{\mathrm{m}}) = \frac{3}{8}\mathbf{F}_{\mathrm{A}} + \frac{3}{4}\mathbf{F}_{\mathrm{B}} - \frac{1}{8}\mathbf{F}_{\mathrm{C}} \tag{2.186}$$

The field $\mathbf{F}_{\mathrm{O,C}}$ is determined at the positions $\overline{u}_{\mathrm{O}} = \overline{u}_{\mathrm{A}} - \Delta\overline{u}$ and $\overline{u}_{\mathrm{C}} = \overline{u}_{\mathrm{B}} + \Delta\overline{u}$. The point O is positioned on the left side of A and the point C is on the right side of B. Both formulas, however, are used for the same point, $\overline{u}^{\mathrm{m}}$.

We would like to use these two different formulas for eq. (2.174) at two different points, $\overline{u}_1^{\mathrm{m}} = 0.5(\overline{u}_1 + \overline{u}_2)$ and $\overline{u}_2^{\mathrm{m}} = 0.5(\overline{u}_2 + \overline{u}_3)$. We obtain:

$$\mathbf{F}_2 - \mathbf{F}_1 = \widehat{\widehat{\boldsymbol{Q}}}_1\left(\frac{3}{8}\mathbf{F}_1 + \frac{3}{4}\mathbf{F}_2 - \frac{1}{8}\mathbf{F}_3\right) \qquad \widehat{\widehat{\boldsymbol{Q}}}_1 = \Delta\overline{u}\widehat{\boldsymbol{Q}}(\overline{u}_1^{\mathrm{m}}) \tag{2.187}$$

$$\mathbf{F}_3 - \mathbf{F}_2 = \widehat{\widehat{\boldsymbol{Q}}}_2\left(\frac{3}{4}\mathbf{F}_2 + \frac{3}{8}\mathbf{F}_3 - \frac{1}{8}\mathbf{F}_1\right) \qquad \widehat{\widehat{\boldsymbol{Q}}}_2 = \Delta\overline{u}\widehat{\boldsymbol{Q}}(\overline{u}_2^{\mathrm{m}}) \tag{2.188}$$

If the field is described by a linear function, e.g. $\mathbf{F}_3 = \mathbf{F}_2 + (\mathbf{F}_2 - \mathbf{F}_1)$ in the first equation and $\mathbf{F}_1 = \mathbf{F}_2 - (\mathbf{F}_3 - \mathbf{F}_2)$ in the second one, we obtain again the result given in eq. (2.175). Knowing e.g. the field at plane (or cross-section) 1 (or A), we can now determine the fields at planes (cross-sections) 2 (or B), and 3 (or C) by:

$$\begin{bmatrix} \boldsymbol{I} - \frac{3}{4}\widehat{\widehat{\boldsymbol{Q}}}_1 & \frac{1}{8}\widehat{\widehat{\boldsymbol{Q}}}_1 \\ \boldsymbol{I} + \frac{3}{4}\widehat{\widehat{\boldsymbol{Q}}}_2 & \frac{3}{8}\widehat{\widehat{\boldsymbol{Q}}}_2 - \boldsymbol{I} \end{bmatrix}\begin{bmatrix}\mathbf{F}_2 \\ \mathbf{F}_3\end{bmatrix} = \begin{bmatrix}\boldsymbol{I} + \frac{3}{8}\widehat{\widehat{\boldsymbol{Q}}}_1 \\ \frac{1}{8}\widehat{\widehat{\boldsymbol{Q}}}_2\end{bmatrix}\mathbf{F}_1 \tag{2.189}$$

2.5.3.3 Field and impedance/admittance transformation in propagation direction with quadratic field interpolation

The field in plane (or cross-section) 2 is given from the system of eqs. (2.189) by:

$$\begin{aligned}\mathbf{F}_2 &= \widehat{\mathbf{V}}_1\mathbf{F}_1\\ \widehat{\mathbf{V}}_1 &= \left[\left(\boldsymbol{I}-\frac{3}{4}\widehat{\widehat{\boldsymbol{Q}}}_1\right)+\frac{1}{8}\widehat{\widehat{\boldsymbol{Q}}}_1\left(\boldsymbol{I}-\frac{3}{8}\widehat{\widehat{\boldsymbol{Q}}}_2\right)^{-1}\left(\boldsymbol{I}+\frac{3}{4}\widehat{\widehat{\boldsymbol{Q}}}_2\right)\right]^{-1}\\ &\quad\times\left[\left(\boldsymbol{I}+\frac{3}{8}\widehat{\widehat{\boldsymbol{Q}}}_1\right)+\frac{1}{64}\widehat{\widehat{\boldsymbol{Q}}}_1\left(\boldsymbol{I}-\frac{3}{8}\widehat{\widehat{\boldsymbol{Q}}}_2\right)^{-1}\widehat{\widehat{\boldsymbol{Q}}}_2\right]\end{aligned} \tag{2.190}$$

The fields at plane (or cross-section) 3 at $\overline{u}_3 = \overline{u}_2 + \Delta\overline{u}$ can then be calculated from eq. (2.188) by:

$$\left(\boldsymbol{I}-\frac{3}{8}\widehat{\widehat{\boldsymbol{Q}}}_2\right)\mathbf{F}_3 = \left(\boldsymbol{I}+\frac{3}{4}\widehat{\widehat{\boldsymbol{Q}}}_2\right)\mathbf{F}_2 - \frac{1}{8}\widehat{\widehat{\boldsymbol{Q}}}_2\mathbf{F}_1 \tag{2.191}$$

The fields in plane (or cross-section) 4 ($\overline{u}_4 = \overline{u}_3 + \Delta\overline{u}$) can be obtained by an analogous formula. This procedure can be continued with further steps $\Delta\overline{u}$ until the upper side of the layer or the end of the section is reached. The general formula is:

$$\left(\boldsymbol{I}-\frac{3}{8}\widehat{\widehat{\boldsymbol{Q}}}_k\right)\mathbf{F}_{k+1} = \left(\boldsymbol{I}+\frac{3}{4}\widehat{\widehat{\boldsymbol{Q}}}_k\right)\mathbf{F}_k - \frac{1}{8}\widehat{\widehat{\boldsymbol{Q}}}_k\mathbf{F}_{k-1} \qquad k \geq 2 \tag{2.192}$$

or:

$$\begin{aligned}\mathbf{F}_{k+1} &= \widehat{\mathbf{V}}_k\mathbf{F}_k\\ \widehat{\mathbf{V}}_k &= \left(\boldsymbol{I}-\frac{3}{8}\widehat{\widehat{\boldsymbol{Q}}}_k\right)^{-1}\left(\boldsymbol{I}+\frac{3}{4}\widehat{\widehat{\boldsymbol{Q}}}_k-\frac{1}{8}\widehat{\widehat{\boldsymbol{Q}}}_k\widehat{\mathbf{V}}_{k-1}^{-1}\right) \qquad k \geq 2\end{aligned} \tag{2.193}$$

where:

$$\widehat{\widehat{\boldsymbol{Q}}}_k = \Delta\overline{u}\widehat{\boldsymbol{Q}}(\overline{u}_k^{\mathrm{m}}) \qquad \overline{u}_k^{\mathrm{m}} = 0.5(\overline{u}_k + \overline{u}_{k+1}) \qquad \overline{u}_1 \equiv \overline{u}_{\mathrm{A}} \tag{2.194}$$

For $k = 1$ see eq. (2.190).

The short-circuit admittance and open-circuit impedance matrices in the sublayer k for impedance/admittance transformation are calculated from eqs. (2.183) and (2.184) by using $\widehat{\mathbf{V}}_{\mathrm{BA}} = \widehat{\mathbf{V}}_k$. The impedance or admittance transformation is performed as before, with eq. (2.185), which was also used in the case of linear field interpolation. The field transformation formulas can also be used for beam propagation.

2.5.3.4 Field and impedance/admittance transformation opposite to direction of propagation with quadratic field interpolation

Now we start on the upper side of the layer (or the end of the section). Instead of eqs. (2.187) and (2.188) we have:

$$\mathbf{F}_{\mathrm{N}} - \mathbf{F}_{\mathrm{N-1}} = \widehat{\widehat{\boldsymbol{Q}}}_{\mathrm{N}} \left(\frac{3}{8}\mathbf{F}_{\mathrm{N}} + \frac{3}{4}\mathbf{F}_{\mathrm{N-1}} - \frac{1}{8}\mathbf{F}_{\mathrm{N-2}} \right) \tag{2.195}$$

$$\mathbf{F}_{\mathrm{N-1}} - \mathbf{F}_{\mathrm{N-2}} = \widehat{\widehat{\boldsymbol{Q}}}_{\mathrm{N-1}} \left(\frac{3}{4}\mathbf{F}_{\mathrm{N-1}} + \frac{3}{8}\mathbf{F}_{\mathrm{N-2}} - \frac{1}{8}\mathbf{F}_{\mathrm{N}} \right) \tag{2.196}$$

$$\widehat{\widehat{\boldsymbol{Q}}}_{\mathrm{N}} = \Delta\overline{u}\widehat{\boldsymbol{Q}}(\overline{u}_{\mathrm{N}}^{\mathrm{m}}) \qquad \widehat{\widehat{\boldsymbol{Q}}}_{\mathrm{N-1}} = \Delta\overline{u}\widehat{\boldsymbol{Q}}(\overline{u}_{\mathrm{N-1}}^{\mathrm{m}}) \tag{2.197}$$

From eq. (2.196) we obtain $\mathbf{F}_{\mathrm{N-2}}$ as:

$$\mathbf{F}_{\mathrm{N-2}} = \left(\boldsymbol{I} + \frac{3}{8}\widehat{\widehat{\boldsymbol{Q}}}_{\mathrm{N-1}} \right)^{-1} \left[\left(\boldsymbol{I} - \frac{3}{4}\widehat{\widehat{\boldsymbol{Q}}}_{\mathrm{N-1}} \right) \mathbf{F}_{\mathrm{N-1}} + \frac{1}{8}\widehat{\widehat{\boldsymbol{Q}}}_{\mathrm{N-1}}\mathbf{F}_{\mathrm{N}} \right] \tag{2.198}$$

This expression is introduced into eq. (2.195), resulting in:

$$\mathbf{F}_{\mathrm{N-1}} = \widehat{\mathbf{V}}_{\mathrm{N}}\mathbf{F}_{\mathrm{N}} \tag{2.199}$$

with:

$$\begin{aligned} \widehat{\mathbf{V}}_{\mathrm{N}} = & \left[\boldsymbol{I} + \frac{3}{4}\widehat{\widehat{\boldsymbol{Q}}}_{\mathrm{N}} - \frac{1}{8}\widehat{\widehat{\boldsymbol{Q}}}_{\mathrm{N}} \left(\boldsymbol{I} + \frac{3}{8}\widehat{\widehat{\boldsymbol{Q}}}_{\mathrm{N-1}} \right)^{-1} \left(\boldsymbol{I} - \frac{3}{4}\widehat{\widehat{\boldsymbol{Q}}}_{\mathrm{N-1}} \right) \right]^{-1} \\ & \times \left[\boldsymbol{I} - \frac{3}{8}\widehat{\widehat{\boldsymbol{Q}}}_{\mathrm{N}} + \frac{1}{64}\widehat{\widehat{\boldsymbol{Q}}}_{\mathrm{N}} \left(\boldsymbol{I} + \frac{3}{8}\widehat{\widehat{\boldsymbol{Q}}}_{\mathrm{N-1}} \right)^{-1} \widehat{\widehat{\boldsymbol{Q}}}_{\mathrm{N-1}} \right] \end{aligned} \tag{2.200}$$

Interpreting $\mathbf{F}_{\mathrm{N-1}}$ as $\mathbf{F}_{\mathrm{A}}$ and $\mathbf{F}_{\mathrm{N}}$ as $\mathbf{F}_{\mathrm{B}}$ we can continue with eq. (2.178). By using eqs. (2.182) and (2.185) we obtain the impedance (admittance) $\mathbf{Z}_{\mathrm{N-1}}(\mathbf{Y}_{\mathrm{N-1}})$ from the known ones $\mathbf{Z}_{\mathrm{N}}(\mathbf{Y}_{\mathrm{N}})$.

By introducing $\mathbf{F}_{\mathrm{N}} = \widehat{\mathbf{V}}_{\mathrm{N}}^{-1}\mathbf{F}_{\mathrm{N-1}}$ from eq. (2.199) into eq. (2.198), we obtain:

$$\mathbf{F}_{\mathrm{N-2}} = \widehat{\mathbf{V}}_{\mathrm{N-1}}\mathbf{F}_{\mathrm{N-1}} \tag{2.201}$$

where:

$$\widehat{\mathbf{V}}_{\mathrm{N-1}} = \left(\boldsymbol{I} + \frac{3}{8}\widehat{\widehat{\boldsymbol{Q}}}_{\mathrm{N-1}} \right)^{-1} \left(\boldsymbol{I} - \frac{3}{4}\widehat{\widehat{\boldsymbol{Q}}}_{\mathrm{N-1}} + \frac{1}{8}\widehat{\widehat{\boldsymbol{Q}}}_{\mathrm{N-1}}\widehat{\mathbf{V}}_{\mathrm{N}}^{-1} \right) \tag{2.202}$$

Therefore, the general field transformation occurs according to:

$$\mathbf{F}_{k-1} = \widehat{\mathbf{V}}_k\mathbf{F}_k \tag{2.203}$$

with:

$$\widehat{\mathbf{V}}_k = \left(\boldsymbol{I} + \frac{3}{8}\widehat{\widehat{\boldsymbol{Q}}}_k \right)^{-1} \left(\boldsymbol{I} - \frac{3}{4}\widehat{\widehat{\boldsymbol{Q}}}_k + \frac{1}{8}\widehat{\widehat{\boldsymbol{Q}}}_k\widehat{\mathbf{V}}_{k+1}^{-1} \right) \qquad k \leq \mathrm{N} - 1 \tag{2.204}$$

Again, interpreting $\mathbf{F}_{k-1}$ as $\mathbf{F}_{\mathrm{A}}$ and $\mathbf{F}_k$ as $\mathbf{F}_{\mathrm{B}}$, we can continue with eq. (2.178) by using $\widehat{\boldsymbol{V}}_{\mathrm{AB}} = \widehat{\boldsymbol{V}}_k$. With eqs. (2.181) and (2.182) we obtain the impedance (admittance) $\boldsymbol{Z}_{k-1}(\boldsymbol{Y}_{k-1})$ from the known one $\boldsymbol{Z}_k(\boldsymbol{Y}_k)$.

Eq. (2.174) has some variations. E.g. if the left side is written as $\overline{u}\, \mathrm{d}\widehat{\mathbf{F}}/\mathrm{d}\overline{u}$, the finite difference equivalent would be:

$$\overline{u}\frac{\mathrm{d}}{\mathrm{d}\overline{u}}\widehat{\mathbf{F}} \quad \longrightarrow \quad \overline{u}^{\mathrm{m}}(\widehat{\mathbf{F}}_{\mathrm{B}} - \widehat{\mathbf{F}}_{\mathrm{A}})/\Delta\overline{u} \tag{2.205}$$

with $\overline{u}^{\mathrm{m}}$ being an intermediate point (usually in the middle). In the case of a different factor in front of the derivative (e.g. $\overline{u}^2$), we proceed analogously.

Furthermore, it should be mentioned that in some cases the inversion of the matrix expressions can be simplified. For $\boldsymbol{S}_{\mathrm{E,H}} = \boldsymbol{0}$, the expressions $\boldsymbol{I} + a\widehat{\widehat{\boldsymbol{Q}}}_k$ can be written in a particular form with unit matrices on the main diagonal blocks.

$$\begin{bmatrix} \boldsymbol{I} & \boldsymbol{A} \\ \boldsymbol{B} & \boldsymbol{I} \end{bmatrix} \longrightarrow \begin{bmatrix} \boldsymbol{I} & \boldsymbol{A} \\ \boldsymbol{B} & \boldsymbol{I} \end{bmatrix}^{-1} = \begin{bmatrix} \boldsymbol{I} & -\boldsymbol{A} \\ -\boldsymbol{B} & \boldsymbol{I} \end{bmatrix} \begin{bmatrix} (\boldsymbol{I} - \boldsymbol{A}\boldsymbol{B})^{-1} & \boldsymbol{0} \\ \boldsymbol{0} & (\boldsymbol{I} - \boldsymbol{B}\boldsymbol{A})^{-1} \end{bmatrix} \tag{2.206}$$

Therefore, the inverse can be computed with the formulas shown above. (The order of the matrix product on the right side can also be exchanged.)

To check the impedance/admittance transformation formulas with finite differences, we recalculated the value n_{eff} for the HE_{00} mode of the structure in Fig. 2.35a. The normalised frequency range was $k_0 d = 5 \ldots 6$. We divided the layer with the rib into sublayers as shown in Fig. 2.35b, and used the FD formulas. The results in the frequency range between $k_0 d = 5$ and $k_0 d = 6$ shown in Fig. 2.36.

Fig. 2.37 shows the difference between the results obtained with the FD impedance/admittance transformation and those shown in Fig. 2.35a as a function of the number of sublayers (1/number). As normalised frequency we chose $k_0 d = 6$.

We can see that the value n_{eff} converges to the value obtained with analytical impedance/admittance transformation. The error for the linear field approximation is more than two times larger than the error for quadratic field approximation.

2.5.4 Stable field transformation through layers and sections

In this section we would like to describe the numerical stable field transformation through layers and sections. Again, the derived expressions can be used in the 2D as well as in the 3D case. Therefore, $\mathbf{F}$ can represent fields with either two components (i.e be a supervector) or just one. As before, we will therefore not mark the occurring matrices and vectors with a hat ($\widehat{\ }$). The solutions of the wave equations for Cartesian coordinates can be written as:

$$\overline{\mathbf{F}}^{(k)} = \overline{\mathbf{F}}_f^{(k)} + \overline{\mathbf{F}}_b^{(k)} = e^{-\boldsymbol{\Gamma}\overline{z}}\, \overline{\mathbf{F}}_{f0}^{(k)} + e^{\boldsymbol{\Gamma}\overline{z}}\, \overline{\mathbf{F}}_{b0}^{(k)} \tag{2.207}$$

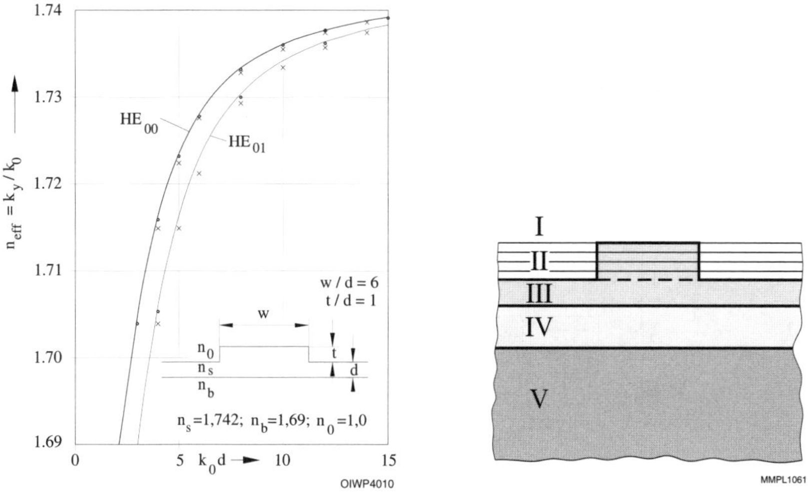

(a) (b)

Fig. 2.35 (a) Dispersion curves for a rib waveguide structure MoL [16] • MoL × mode matching technique [17] (Reproduced by permission of Taylor & Francis); (b) subdivision of the rib layer (Reproduced by permission of Springer Netherlands)

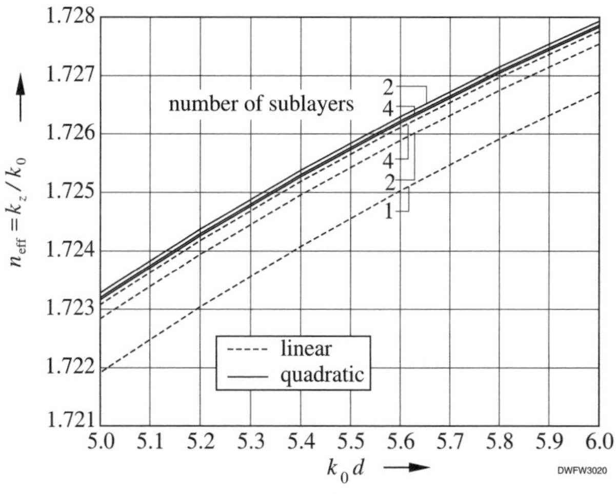

Fig. 2.36 Results for the dispersion of the HE_{00} mode of the structure in Fig. 2.35 (Reproduced by permission of Springer Netherlands)

where the subscripts f and b stand for 'forward' and 'backward' propagation, respectively. Now, we split the fields $\overline{\boldsymbol{H}}^{(k)}$ and $\overline{\boldsymbol{E}}^{(k)}$ into forward and backward

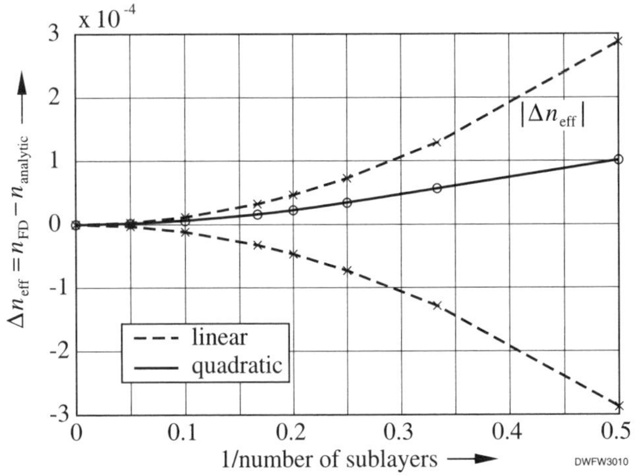

Fig. 2.37 Convergence of the n_{eff} value for the HE_{00} mode at $k_0 d = 6$ using linear and quadratic FD impedance/admittance transformation formulas in the rib layer of the structure in Fig. 2.35 (Reproduced by permission of Springer Netherlands)

propagating parts:

$$\overline{\mathbf{H}}^{(k)} = \overline{\mathbf{H}}_f^{(k)} + \overline{\mathbf{H}}_b^{(k)} \quad \overline{\mathbf{E}}^{(k)} = \overline{\mathbf{E}}_f^{(k)} + \overline{\mathbf{E}}_b^{(k)} \tag{2.208}$$

The forward and backward parts are connected by the characteristic impedance $\overline{\boldsymbol{Z}}_o^{(k)}$:

$$\overline{\mathbf{E}}_f^{(k)} = \overline{\boldsymbol{Z}}_o^{(k)}\overline{\mathbf{H}}_f^{(k)} \quad \overline{\mathbf{E}}_b^{(k)} = -\overline{\boldsymbol{Z}}_o^{(k)}\overline{\mathbf{H}}_b^{(k)} \tag{2.209}$$

Using the normalisation of the modal matrices described previously, we have $\overline{\boldsymbol{Z}}_o = \boldsymbol{I}$. Now we can determine the forward and backward parts from the total fields as:

$$\begin{aligned}
\overline{\mathbf{E}}_f^{(k)} &= \frac{1}{2}(\overline{\mathbf{E}}^{(k)} + \overline{\boldsymbol{Z}}_o^{(k)}\overline{\mathbf{H}}^{(k)}) \qquad \overline{\boldsymbol{E}}_b = \frac{1}{2}(\overline{\mathbf{E}}^{(k)} - \overline{\boldsymbol{Z}}_o^{(k)}\overline{\mathbf{H}}^{(k)}) \\
\overline{\mathbf{H}}_f^{(k)} &= \frac{1}{2}(\overline{\mathbf{H}}^{(k)} + \overline{\mathbf{Y}}_o^{(k)}\overline{\mathbf{E}}^{(k)}) \qquad \overline{\boldsymbol{H}}_b = \frac{1}{2}(\overline{\mathbf{H}}^{(k)} - \overline{\mathbf{Y}}_o^{(k)}\overline{\mathbf{E}}^{(k)})
\end{aligned} \tag{2.210}$$

or by introduction of eq. (2.143) (generalised for an arbitrary plane):

$$\begin{aligned}
\overline{\mathbf{E}}_f^{(k)} &= \frac{1}{2}(\overline{\boldsymbol{I}} + \overline{\boldsymbol{Z}}_o^{(k)}\overline{\mathbf{Z}}^{(k)-1})\overline{\mathbf{E}}^{(k)} \qquad \overline{\mathbf{E}}_b^{(k)} = \frac{1}{2}(\overline{\boldsymbol{I}} - \overline{\boldsymbol{Z}}_o^{(k)}\overline{\mathbf{Z}}^{(k)-1})\overline{\mathbf{E}}^{(k)} \\
\overline{\mathbf{H}}_f^{(k)} &= \frac{1}{2}(\overline{\boldsymbol{I}} + \overline{\mathbf{Y}}_o^{(k)}\overline{\mathbf{Y}}^{(k)-1})\overline{\mathbf{H}}^{(k)} \qquad \overline{\mathbf{H}}_b^{(k)} = \frac{1}{2}(\overline{\boldsymbol{I}} - \overline{\mathbf{Y}}_o^{(k)}\overline{\mathbf{Y}}^{(k)-1})\overline{\mathbf{H}}^{(k)}
\end{aligned} \tag{2.211}$$

With the formulas obtained we have the basic equations for numerical stable field transformation from plane to plane or interface to interface.

Remember that we have obtained accurate impedances and/or admittances in each interface/plane before.

Now we assume that we know the field (e.g. from solution of the eigenvalue problem) for interface A. The problem is to obtain the accurate fields in plane B. To avoid numerical instabilities, we should use the following algorithm:

- With the help of eq. (2.210) or (2.211) we calculate the vectors $\overline{\mathbf{E}}_{fA}^{(k)}$ and $\overline{\mathbf{H}}_{fA}^{(k)}$ from $\overline{\mathbf{E}}_{\mathrm{A}}^{(k)}$ and $\overline{\mathbf{H}}_{\mathrm{A}}^{(k)}$.

- From the vector $\overline{\mathbf{E}}_{fA}^{(k)}$ ($\overline{\mathbf{H}}_{f\mathrm{A}}^{(k)}$) we obtain $\overline{\boldsymbol{E}}_{f\mathrm{B}}^{(k)}$ ($\overline{\mathbf{H}}_{f\mathrm{B}}^{(k)}$) as:

$$\overline{\mathbf{E}}_{f\mathrm{B}}^{(k)} = \mathrm{e}^{-\boldsymbol{\Gamma}\bar{d}_k}\overline{\mathbf{E}}_{f\mathrm{A}}^{(k)} \qquad \overline{\mathbf{H}}_{f\mathrm{B}}^{(k)} = \mathrm{e}^{-\boldsymbol{\Gamma}\bar{d}_k}\overline{\mathbf{H}}_{f\mathrm{A}}^{(k)} \tag{2.212}$$

  This transform is very stable because the exponential function decreases with increasing $\bar{d}_k$ and we avoid the explicit use of the exponential increasing terms that cause the numerical problems.

- At the interface B we use again eq. (2.211) and obtain:

$$\overline{\mathbf{E}}_{\mathrm{B}}^{(k)} = 2(\overline{\boldsymbol{I}} + \overline{\mathbf{Z}}_{o}^{(k)}\overline{\mathbf{Z}}_{\mathrm{B}}^{(k)-1})^{-1}\overline{\mathbf{E}}_{f\mathrm{B}}^{(k)} \tag{2.213}$$

  $\overline{\boldsymbol{H}}_{\mathrm{B}}^{(k)}$ can then be computed with eq. (2.143).

- If necessary, the backward components in layer k are given by:

$$\overline{\mathbf{E}}_{b\mathrm{B}}^{(k)} = \overline{\mathbf{E}}_{\mathrm{B}}^{(k)} - \overline{\mathbf{E}}_{f\mathrm{B}}^{(k)} \qquad \overline{\mathbf{H}}_{b\mathrm{B}}^{(k)} = \overline{\mathbf{H}}_{\mathrm{B}}^{(k)} - \overline{\mathbf{H}}_{f\mathrm{B}}^{(k)} \tag{2.214}$$

  The fields in a plane z between the interfaces $\mathrm{A}(z = z_{\mathrm{A}})$ and $\mathrm{B}(z = z_{\mathrm{B}})$ may now be calculated as:

$$\begin{aligned}\overline{\mathbf{E}}^{(k)}(z) &= \mathrm{e}^{-\boldsymbol{\Gamma}(\bar{z}-\bar{z}_{\mathrm{A}})}\overline{\mathbf{E}}_{fA}^{(k)} + \mathrm{e}^{-\boldsymbol{\Gamma}(\bar{z}_{\mathrm{B}}-\bar{z})}\overline{\mathbf{E}}_{b\mathrm{B}}^{(k)} \\ \overline{\mathbf{H}}^{(k)}(z) &= \mathrm{e}^{-\boldsymbol{\Gamma}(\bar{z}-\bar{z}_{\mathrm{A}})}\overline{\mathbf{H}}_{fA}^{(k)} + \mathrm{e}^{-\boldsymbol{\Gamma}(\bar{z}_{\mathrm{B}}-\bar{z})}\overline{\mathbf{H}}_{b\mathrm{B}}^{(k)}\end{aligned} \tag{2.215}$$

  Also, the terms with the backward propagating parts have a stable form.

- If plane B is an interface between layer (or section) k and the one labelled $k+1$ we can determine the (transformed) fields in this plane. Due to the continuity of the transverse components we can then compute the transformed fields at the input of this section (layer) $k+1$ and following the field everywhere within it. By repeating this procedure (computing the field at the interfaces, determining the fields in the remainder of the section) we are able to calculate the field distribution in the whole structure. As mentioned before, in the described algorithm, we avoid the evaluation of exponential increasing terms. Therefore, numerical problems do not occur.

References

[1] L. B. Felsen and N. Marcuvitz (Eds.), *Radiation and Scattering of Waves*, IEEE Press, New York, USA, 1996.

[2] V. K. Tripathi, 'On the Analysis of Symmetrical Three-Line Microstrip Circuits', *IEEE Trans. Microwave Theory Tech.*, vol. MTT-25, no. 9, pp. 726–729, Sep. 1977.

[3] R. E. Collin, *Field Theory of Guided Waves*, McGraw-Hill, New York, USA, 1960.

[4] R. Pregla, J. Gerdes, E. Ahlers and S. Helfert, 'MoL-BPM Algorithms for Waveguide Bends and Vectorial Fields', in *OSA Integr. Photo. Resear. Tech. Dig.*, New Orleans, USA, 1992, vol. 10, pp. 32–33.

[5] R. Pregla, 'MoL–BPM Method of Lines Based Beam Propagation Method', in *Methods for Modeling and Simulation of Guided-Wave Optoelectronic Devices (PIER 11)*, W. P. Huang (Ed.), Progress in Electromagnetic Research, pp. 51–102. EMW Publishing, Cambridge, Massachusetts, USA, 1995.

[6] R. Pregla, 'The Method of Lines for the Unified Analysis of Microstrip and Dielectric Waveguides', *Electromagnetics*, vol. 15, no. 5, pp. 441–456, 1995.

[7] R. Zurmühl and S. Falk, *Matrizen und ihre Anwendungen, Teil 1*, Springer Verlag, Berlin, Germany, 5th edition, 1984.

[8] R. Pregla and W. Pascher, 'The Method of Lines', in *Numerical Techniques for Microwave and Millimeter Wave Passive Structures*, T. Itoh (Ed.), pp. 381–446. J. Wiley Publ., New York, USA, 1989.

[9] F. J. Schmückle and R. Pregla, 'The Method of Lines for the Analysis of Planar Waveguides with Finite Metallization Thickness', *IEEE Trans. Microwave Theory Tech.*, vol. MTT-39, pp. 101–107, 1991.

[10] F. J. Schmückle and R. Pregla, 'The Method of Lines for the Analysis of Lossy Planar Waveguides', *IEEE Trans. Microwave Theory Tech.*, vol. MMT-38, pp. 1473–1479, 1990.

[11] T. Rozzi and S. J. Hedges, 'Rigorous Analysis and Network Modelling of the Inset Dielectric Guide', *IEEE Trans. Microwave Theory Tech.*, vol. MTT-35, pp. 823–834, Sep. 1987.

[12] W. Hong and W.-X. Zhang, 'On the Equivalence Between the Method of Lines and the Variational Method', *AEÜ*, vol. 45, pp. 198–201, 1991.

[13] Z. Ma, T. Ishikawa and E. Yamashita, 'An Efficient Analysis Approach for Inset Dielectric Guide (IDG) Structures and Its Variations', *IEEE Microwave Guided Wave Lett.*, vol. 5, pp. 117–118, Apr. 1995.

[14] George W. Hanson, 'Propagation Characteristics of Microstrip Transmission Line on an Anisotropic Material Ridge', *IEEE Trans. Microwave Theory Tech.*, vol. 43, no. 11, pp. 2608–2613, 1995.

[15] Xian Hua Yang and Lotfollah Shafai, 'Full Wave Approach for the Analysis of Open Planar Waveguides With Finite Width Dielectric Layers and Ground Planes', *IEEE Trans. Microwave Theory Tech.*, vol. MTT-42, no. 1, pp. 142–149, 1994.

[16] U. Rogge and R. Pregla, 'Method of Lines for the Analysis of Strip-Loaded Optical Waveguides', *J. Opt. Soc. Am. B*, vol. 8, no. 2, pp. 459–463, Feb. 1991.

[17] K. Yasuura and K. Shimohara, 'Numerical Analysis of a Thin-Film Waveguide by Mode-Matching Method', *Opt. Quantum Electron.*, vol. 70, pp. 183–191, 1980.

[18] U. Rogge and R. Pregla, 'Method of Lines for the Analysis of Dielectric Waveguides', *J. Lightwave Technol.*, vol. 11, no. 12, pp. 2015–2020, 1993.

[19] R. Pregla, 'Modeling of Optical Waveguides and Devices by Combination of the Method of Lines and Finite Differences of Second Order Accuracy', *Opt. Quantum Electron.*, vol. 38, no. 1–3, pp. 3–17, 2006, Special Issue on Optical Waveguide Theory and Numerical Modelling.

Further Reading

[20] S. B. Worm, 'Full-Wave Analysis of Discontinuities in Planar Waveguides by the Method of lines Using a Source Approach', *IEEE Trans. Microwave Theory Tech.*, vol. MTT-38, pp. 1510–1514, 1990.

[21] R. Pregla, 'The Analysis of Wave Propagation in General Anisotropic Multilayered Waveguides by the Method of Lines', in *U.R.S.I Intern. Symp. Electromagn. Theo.*, Thessaloniki, Greece, May 1998, pp. 51–53.

[22] R. Pregla, 'A Generalized Algorithm for Analysis of Planar Multilayered Anisotropic Waveguide Structures by the Method of Lines', *AEÜ*, vol. 52, no. 2, pp. 94–98, 1998.

[23] Y. Chen and B. Beker, 'The Method of Lines Analysis of Double-Layered or Suspended Bianisotropic Biaxial Substrates', *IEEE Trans. Microwave Theory Tech.*, vol. MTT-42, no. 5, pp. 917–920, May 1994.

[24] Y. Chen and B. Beker, 'Study of Microstrip Step Discontinuities on Bianisotropic Substrates Using the Method of Lines and Transverse Resonance Technique', *IEEE Trans. Microwave Theory Tech.*, vol. 42, no. 10, pp. 1945–1950, Oct. 1994.

[25] S. B. Worm, *Analysis of Planar Microwave Structures with Arbitrary Contour* (in German), PhD thesis, FernUniversität – Hagen, 1983.

[26] S. B. Worm and R. Pregla, 'Hybrid Mode Analysis of Arbitrarily Shaped Planar Microwave Structures by the Method of Lines', *IEEE Trans. Microwave Theory Tech.*, vol. MTT-32, pp. 191–196, 1984.

[27] J. Machac and W. Menzel, 'On the Design of Waveguide-to-Microstrip and Waveguide-to-Coplanar Line Transitions', in *European Microwave Conference*, Madrid , Spain, 1993, pp. 615–616.

[28] F. Alimenti, P. Mezzanotte, L. Roselli and R. Sorrentino, 'A Revised Formulation of Modal Absorbing and Matched Modal Source Boundary Conditions for the Efficient FDTD Analysis of Waveguide Structures', *IEEE Trans. Microwave Theory Tech.*, vol. MTT-48, no. 1, pp. 50–59, 2000.

[29] I. Wolff, G. Kompa and R. Mehran, 'Calculation Method for Microstrip Discontinuities and T-Junctions', *Electron. Lett.*, vol. 8, pp. 177–179, 1972.

[30] R. Mehran, *Grundelemente des rechnergesttzten Entwurfs von Mikrostreifenleitungs-Schaltungen*, Wolff, Aachen, Germany, 1984.

[31] Z. Ma and E. Yamashita, 'Port Reflection Coefficient Method for Solving Multi-Port Microwave Network Problems', *IEEE Trans. Microwave Theory Tech.*, vol. MTT-43, no. 2, pp. 331–337, Feb. 1995.

[32] E. M. Sich and R. H. Macphie, 'The Conservation of Complex Power Technique and E-Plane Step-Diaphragm Junction Discontinuities', *IEEE Trans. Microwave Theory Tech.*, vol. MTT-30, pp. 198–201, 1982.

[33] M. Koshiba, M. Sato and M. Suzuki-M, 'Application of Finite-Element Method to E-Plane Waveguide Discontinuities', *Trans. IECE Japan*, vol. E66, no. 7, pp. 457–548, 1983.

CHAPTER 3

ANALYSIS OF RECTANGULAR WAVEGUIDE CIRCUITS

3.1 INTRODUCTION

Rectangular waveguide circuits are essential parts of microwave devices. In Section 1.2.3 we showed how the cut-off wavelength of the fundamental (H_{10}) mode and of the H_{20} mode are determined. Here, we show a detailed analysis of these waveguide structures. Generally, they consist of concatenations of different waveguide sections and junctions. In Fig. 3.1 such concatenations of straight and/or curved waveguides are drawn. The impedance transformer or the filter structure shown in Fig. 3.2 can be seen as concatenation of straight waveguide sections and steps. E-plane and H-plane junctions are shown in Figs. 3.3 and 3.4, respectively.

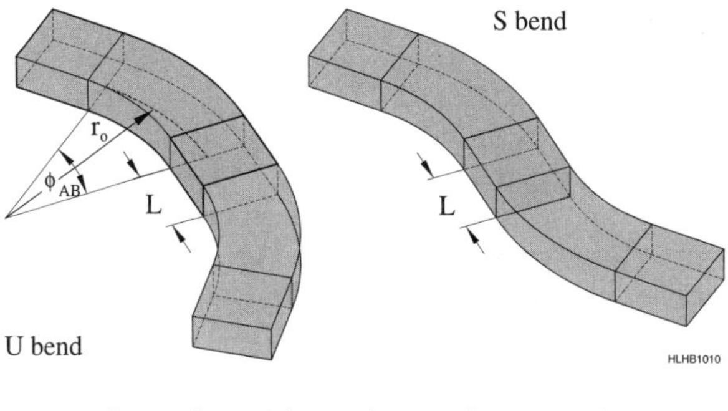

Fig. 3.1 Concatenations of straight and curved waveguide sections (R. Pregla, 'Concatenations of Waveguide Sections', April 1997. The Institution of Engineering and Technology (IET))

This chapter aims at presenting how the overall performance of concatenations of waveguide sections can be analysed. First it will shown how the modes of the curved sections can be determined without using Bessel functions. For this purpose a radial discretisation is used. This can also be applied to the analysis of optical bends (see Section 4.4.5 and [1]). There, the MoL was extended to analysis of structures with inhomogeneous radial layers. Even in the case of homogenous layers the proposed discretisation procedure differs substantially from that given in [2]. To construct the difference operators for the derivatives in the radial direction in the wave equations, we use two different diagonal matrices of the discretised radii. Only in this way can the

Analysis of Electromagnetic Fields and Waves R. Pregla

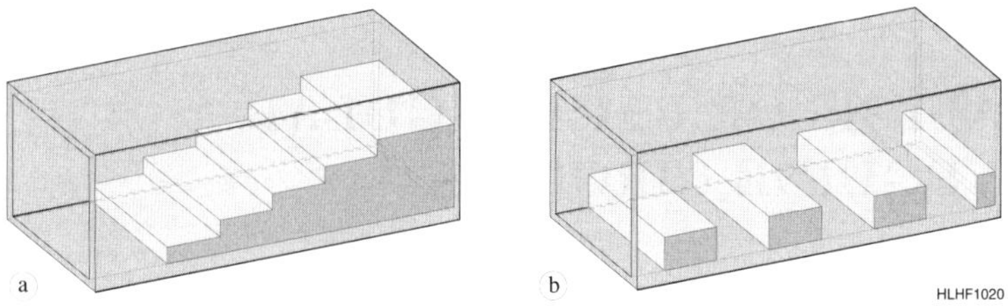

Fig. 3.2 Longitudinal section of an impedance transformer (a) and a filter structure (b) (R. Pregla, 'Concatenations of Waveguide Sections', April 1997. The Institution of Engineering and Technology (IET))

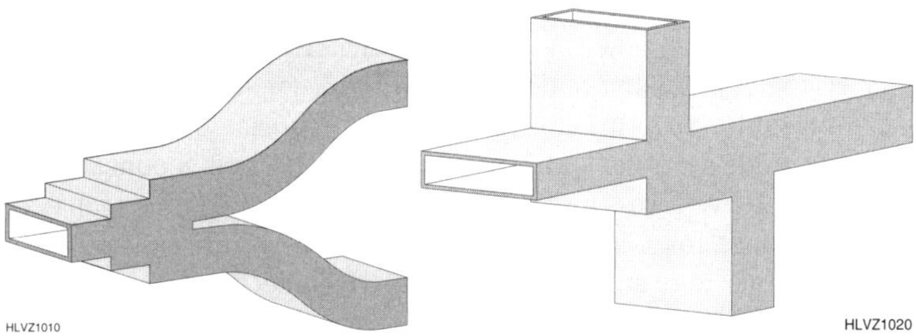

Fig. 3.3 Waveguide E-plane junctions (Reproduced by permission of Elsevier)

field components be given accurately. This is equivalent to modelling dielectric steps in the MoL [4]. After showing the solution of the wave equation in homogeneous sections, formulas for the impedance transformation from the end of a section to its input and vice versa are given. For the interfaces of the different waveguide sections, adequate impedance/admittance transformation formulas are developed. The formulas are general and valid even in the case of steps in the cross-section of concatenation. The very powerful concept of transformation matrix partition technique proposed in [5] is used for this purpose. The formulas for straight sections are obtained from those of curved ones by setting the radius of curvature to infinity. It should be mentioned that concatenations of straight and curved waveguide sections were analysed previously by Tripathi et. al. [6] using the Method of Moments, and by Guglielmi et. al. [7] using a multimode network representation. The algorithm described in this chapter is very compact and allows the analysis of a wide class of structures. The cross-sections of the waveguides can even have different dimensions.

Two different E-plane junctions are drawn in Fig. 3.3. Such structures can be described by rectangular 4-ports concatenated with waveguide section. For the second structure in Fig. 3.3 there are two possibilities for analysis; these are shown in Fig. 3.5. The subdivision in Fig. 3.5a results in one 4-port junction J connected with four waveguides. The subdivision in Fig. 3.5b leads to two 3-port junctions J_1 and J_2 concatenated by a waveguide section C.

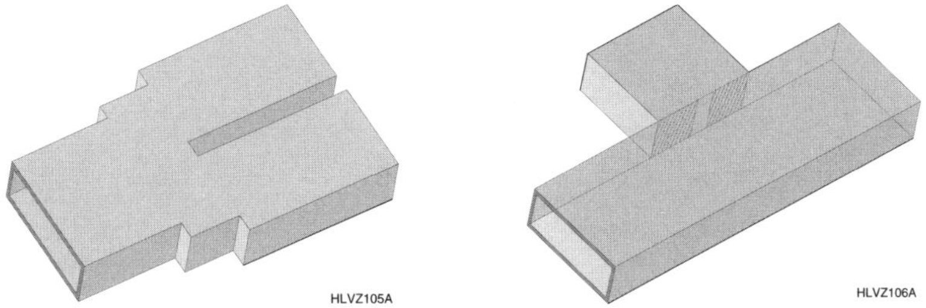

Fig. 3.4 Waveguide H-plane junctions

The waveguide section C may also have a different height (marked by the dark shades). The subdivision in Fig. 3.5b requires a slightly higher programming effort but leads to results with higher accuracy. This subdivision shows that it is very easy to analyse concatenations of junctions with the proposed algorithm. The 3-ports J_1 and J_2 can be replaced by N-ports ($N \geq 4$). It should be mentioned that in one plane there can be more than one port, which is the case in the junction region J in Fig. 3.5c, which is a longitudinal section of the right E-plane junction in Fig. 3.3. The separation of the ports in one plane can easily be performed by the transformation matrix partition. We assume that the permittivity in the junction region is constant. Each inner side of the rectangular junction builds one inner port. On each of the outer sides one or more waveguides can be connected. Therefore, very complex structures can be analysed. To analyse waveguide junction regions J we use the concept of crossed discretisation lines that was introduced in [19]. This concept will be further developed and highly improved by introducing general field relations for the inner side ports of the junctions. To describe the connection of these ports to the connected waveguides, we use the impedance/admittance transfer concept demonstrated in [8], which is further developed here for junctions. The input impedance of the fourth port can be calculated from the known impedances at the three other ports by the 4-port field relations. The numerical results of the algorithm reported here are compared with the results of other papers. The algorithm described in this chapter is very compact and allows the analysis of a wide class of structures. Its implementation in computer codes is very easy.

Now, due to the metal walls, we must introduce Dirichlet conditions for the transverse electric fields (electric walls). By using Neumann walls instead, we are also able to model microstrips (magnetic wall model). The algorithms themselves are very similar. This was e.g. used to examine microstrip bends in [14].

3.2 CONCATENATIONS OF WAVEGUIDE SECTIONS

In this section we will first describe LSE and LSM eigenmodes in circular waveguide bends and straight waveguides. We will then derive impedance/

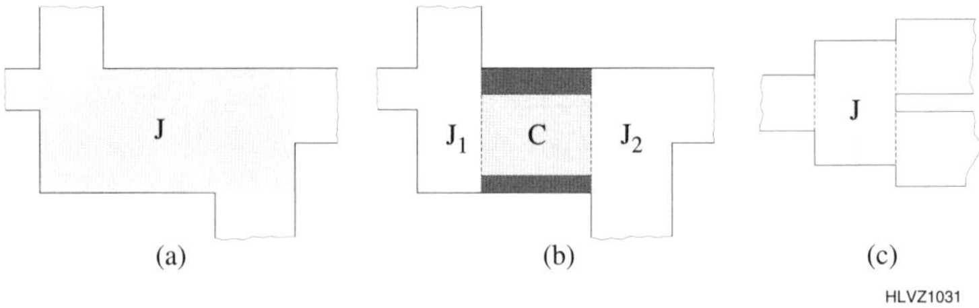

Fig. 3.5 Longitudinal sections of the E-plane junctions in Fig. 3.3, with alternative subdivisions for analysis of the left one (Reproduced by permission of Elsevier)

admittance transformation formulas for homogeneous sections and for interfaces.

3.2.1 LSM and LSE modes in circular waveguide bends

The bends to be analysed are shown in Fig. 3.6. The cylindrical coordinate system that is used is also presented. In the H(E)-plane bend all field components are obtained from $E_z(H_z)$. Therefore, in the H(E)-plane bend only LSM or TM_z (LSE or TE_z) modes (fields) are under consideration. As a consequence, the structures should not change in z-direction. In both cases the fundamental and most interesting TE_{10} mode with respect to the propagation direction in the rectangular waveguide is examined.

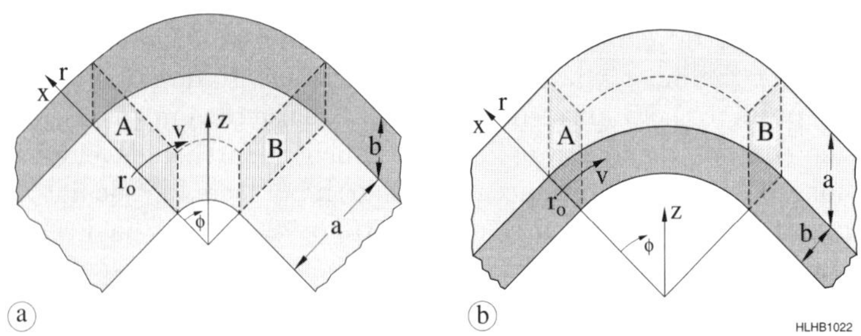

Fig. 3.6 H-plane (a) and E-plane (b) bend of a rectangular waveguide

3.2.1.1 Basic equations

The general equations are derived in Section 4.2.3. We summarise here the special procedure. To avoid the computation of Bessel functions, the discretisation is performed in radial direction (Fig. 3.7). Then, the analytical solutions are obtained on lines in azimuthal direction ϕ. If the permittivity ε_r is constant, we can write for the z dependence of the field components E_z

and H_z:

$$E_z = \cos(k_z z) E_{cz} \quad H_z = \sin(k_z z) H_{sz} \tag{3.1}$$

where the subscripts c and s stand for 'cosine' and 'sine' respectively.

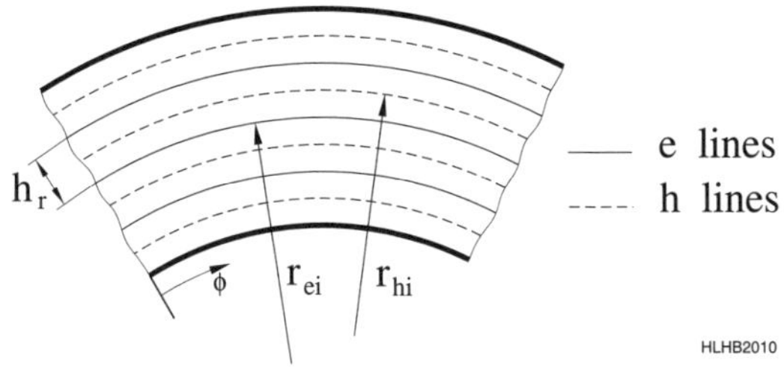

Fig. 3.7 Discretisation lines in azimuthal direction

To obtain symmetrical matrices in the analysis algorithm the following normalisations/abbreviations are introduced:

1. The waveguide dimensions and the coordinates r and z are normalised with the free space wave number k_0 according to: $\overline{r}_0 = k_0 r, \overline{a} = k_0 a$, $\overline{b} = k_0 b$. r_0 is the radius of the curvature in the middle of the waveguide of the bend (see Fig. 3.6).

2. Instead of the azimuthal coordinate ϕ we use the coordinate $v = \phi r_0$, or in normalised form: $\overline{v} = k_0 v$.

3. The field components E_{cz} and H_{sz} are normalised with $\sqrt{r_n}$ ($r_n = \overline{r}/\overline{r}_0 = r/r_0$) according to:

$$E_{zn} = \sqrt{r_n}^{-1} E_{cz} \quad H_{zn} = \sqrt{r_n}^{-1} H_{sz} \tag{3.2}$$

4. The remaining field components $F_\xi = E_\xi(H_\xi)$ ($\xi = r$ or ϕ) are normalised according to:

$$F_{\xi n} = \sqrt{r_n} F_\xi \tag{3.3}$$

5. We introduce the following abbreviations:

$$\partial^2/\partial\overline{z}^2 = -\overline{k}_z^2 \qquad \varepsilon_q = \varepsilon_r - \overline{k}_z^2 \qquad D_{\overline{r}n} = \sqrt{r_n}\frac{\partial}{\partial\overline{r}}\sqrt{r_n} \tag{3.4}$$

Moreover, we have $\overline{k}_z = n\pi/\overline{b}$ with $n = 0$ for the fundamental mode (TE_{10} mode in direction of propagation) in the H-bend and $\overline{k}_z = m\pi/\overline{a}$ with $m = 1$ for the fundamental TE_{10} mode (quasi-TE_{10} mode with respect to the direction of propagation) in case of E-bend. We obtain the following equations for the normalised field components from Section 4.2.3:

H-bend: TM$_z$ modes: $\widetilde{H}_{zn} = 0$ **E-bend: TE$_z$ modes** : $E_{zn} = 0$

$$\begin{aligned}
\frac{\partial E_{zn}}{\partial \overline{v}} &= -\mathrm{j} R^e_{\mathrm{H}} \widetilde{H}_{rn} \quad R^e_{\mathrm{H}} = \varepsilon_r^{-1}\varepsilon_q & \frac{\partial(-\widetilde{H}_{zn})}{\partial \overline{v}} &= -\mathrm{j} R^h_{\mathrm{E}} E_{rn} \quad R^h_{\mathrm{E}} = \varepsilon_q \\
\frac{\partial \widetilde{H}_{rn}}{\partial \overline{v}} &= -\mathrm{j} R^e_{\mathrm{E}} E_{zn} & \frac{\partial E_{rn}}{\partial \overline{v}} &= -\mathrm{j} R^h_{\mathrm{H}} (-\widetilde{H}_{zn}) \\
R^e_{\mathrm{E}} &= \varepsilon_r(\varepsilon_q^{-1} D_{\overline{r}n} D_{\overline{r}n} + \overline{r}_n^2) & R^h_{\mathrm{H}} &= \varepsilon_q^{-1} D_{\overline{r}n} D_{\overline{r}n} + \overline{r}_n^2 \\
E_{\phi n} &= \mathrm{j}\overline{k}_z \varepsilon_r^{-1} \widetilde{H}_{rn} & \widetilde{H}_{\phi n} &= \mathrm{j}\overline{k}_z E_{rn} \\
\varepsilon_q E_{rn} &= -\overline{k}_z D_{\overline{r}n} E_{zn} & \varepsilon_q \widetilde{H}_{rn} &= -\overline{k}_z D_{\overline{r}n}(-\widetilde{H}_{zn}) \\
\varepsilon_q \widetilde{H}_{\phi n} &= -\mathrm{j}\varepsilon_r D_{\overline{r}n} E_{zn} & \varepsilon_q E_{\phi n} &= -\mathrm{j} D_{\overline{r}n}(-\widetilde{H}_{zn})
\end{aligned} \tag{3.5}$$

Please note the symmetry of the equations on the left and right sides. (Instead of ε_r (left side), we have μ_r on the right side, with $\mu_r = 1$.) In the first three rows we see the GTL equations and the structure of the operators. The field components that are used in the GTL equations $E_{gn} = E_{zn}(E_{rn})$ and $H_{gn} = H_{rn}(-H_{zn})$, $g = r, z$ satisfy the wave equations:

$$\frac{\mathrm{d}^2 E_{gn}}{\mathrm{d}\overline{v}^2} - Q^f_{\mathrm{E}} E_{gn} = 0 \qquad \frac{\mathrm{d}^2 H_{gn}}{\mathrm{d}\overline{v}^2} - Q^f_{\mathrm{H}} H_{gn} = 0 \qquad f = e, h \tag{3.6}$$

with:

$$Q^f_{\mathrm{E}} = -R^f_{\mathrm{H}} R^f_{\mathrm{E}} \qquad Q^f_{\mathrm{H}} = -R^f_{\mathrm{E}} R^f_{\mathrm{H}}$$

Note that the superscript $f = e$ (h) is for the components of the TM$_z$ (TE$_z$) modes.

3.2.1.2 Discretisation

We discretise in r-direction. The discretisation of E_{zn} and H_{rn} ($\widetilde{H}_{zn}$ and E_{rn}) is performed on the e-line (h-line) system and indicated by the superscript e (h) (see Fig. 3.7):

$$\begin{aligned}
E_{zn} &\rightarrow \mathbf{E}_{zn} & \widetilde{H}_{zn} &\rightarrow \mathbf{H}_{zn} \\
\widetilde{H}_{rn} &\rightarrow \mathbf{H}_{rn} & E_{rn} &\rightarrow \mathbf{E}_{rn} \\
r_n &\rightarrow \mathbf{r}_e, \mathbf{r}_h & r_n &\rightarrow \mathbf{r}_h, \mathbf{r}_e \\
\overline{h}_r D_{\overline{r}n} &\rightarrow \sqrt{\mathbf{r}_h}\boldsymbol{D}_e\sqrt{\mathbf{r}_e} = \boldsymbol{D}_{en} & \overline{h}_r D_{\overline{r}n} &\rightarrow \sqrt{\mathbf{r}_e}\boldsymbol{D}_h\sqrt{\mathbf{r}_h} = \boldsymbol{D}_{hn}
\end{aligned} \tag{3.7}$$

For $\overline{h}_r^{-1}\boldsymbol{D}_{fn}$ we use the abbreviation $\overline{\boldsymbol{D}}_{fn}$. $\mathbf{E}_{zn}$ and $\mathbf{H}_{rn}$ ($\mathbf{H}_{zn}$ and $\mathbf{E}_{rn}$) are vectors whose components give the fields on the $e(h)$-lines. $\mathbf{r}_e$ and $\mathbf{r}_h$ are diagonal matrices of the discretised radii (see Fig. 3.7). $\boldsymbol{D}_e$ and $\boldsymbol{D}_h$ are difference matrices (see Section 1.2.2) that have to fulfil Dirichlet ($\boldsymbol{D}_e$)

or Neumann ($\boldsymbol{D}_h$) boundary conditions. Furthermore, the following relation holds: $\boldsymbol{D}_{hn} = -\boldsymbol{D}_{en}^{\mathrm{t}}$. The superscript t denotes the transposed matrix. As before, we obtain the approximation of the second derivative by multiplying the discretised first derivative with the (negative and transposed) operator matrix from the left. Since the difference operators are not diagonal we must transform the fields according to:

$$\mathbf{E}_{gn} = \boldsymbol{T}_{\mathrm{E}}^{f}\overline{\mathbf{E}}_{gn} \qquad \mathbf{H}_{gn} = \boldsymbol{T}_{\mathrm{H}}^{f}\overline{\mathbf{H}}_{gn} \tag{3.8}$$

with:

$$\boldsymbol{T}_{\mathrm{E}}^{f^{-1}}\boldsymbol{Q}_{\mathrm{E}}^{f}\boldsymbol{T}_{\mathrm{E}}^{f} = \boldsymbol{\Gamma}_{f}^{2} \qquad \boldsymbol{T}_{\mathrm{H}}^{f^{-1}}\boldsymbol{Q}_{\mathrm{H}}^{f}\boldsymbol{T}_{\mathrm{H}}^{f} = \boldsymbol{\Gamma}_{f}^{2} \tag{3.9}$$

$\boldsymbol{\Gamma}_f^2 = -\boldsymbol{\beta}_f^2$ is a diagonal matrix with the eigenvalues of the above matrix. Further, we have $f = e(h)$ for $\mathbf{E}_{zn}, \mathbf{H}_{rn}(\mathbf{H}_{zn}, \mathbf{E}_{rn})$, respectively.

The following relation between the modal matrices will be used:

$$\boldsymbol{T}_{\mathrm{H}}^{f} = \boldsymbol{R}_{\mathrm{E}}^{f}\boldsymbol{T}_{\mathrm{E}}^{f}\boldsymbol{\beta}_{f}^{-1} \qquad \boldsymbol{T}_{\mathrm{E}}^{f} = \boldsymbol{R}_{\mathrm{H}}^{f}\boldsymbol{T}_{\mathrm{H}}^{f}\boldsymbol{\beta}_{f}^{-1} \tag{3.10}$$

The discretised and transformed wave equations are:

$$\frac{\mathrm{d}^2\overline{\mathbf{E}}_{gn}}{\mathrm{d}\overline{v}^2} - \boldsymbol{\Gamma}_f^2\overline{\mathbf{E}}_{gn} = \mathbf{0} \qquad \frac{\mathrm{d}^2\overline{\mathbf{H}}_{gn}}{\mathrm{d}\overline{v}^2} - \boldsymbol{\Gamma}_f^2\overline{\mathbf{H}}_{gn} = \mathbf{0} \tag{3.11}$$

with the well-known general solutions (see e.g. eq. (2.41)). The remaining field components may be calculated from eq. (3.5). They are obtained as:

H-bend: TM$_z$ modes: $\mathbf{H}_{zn} = \mathbf{0}$

$$\begin{aligned}
\frac{\mathrm{d}}{\mathrm{d}\overline{v}}\overline{\mathbf{E}}_{zn} &= -\mathrm{j}\boldsymbol{\beta}_e\overline{\mathbf{H}}_{rn}\\
\frac{\mathrm{d}}{\mathrm{d}\overline{v}}\overline{\mathbf{H}}_{rn} &= -\mathrm{j}\boldsymbol{\beta}_e\overline{\mathbf{E}}_{zn}\\
\mathbf{E}_{\phi n} &= \mathrm{j}\overline{k}_z\varepsilon_r^{-1}\boldsymbol{T}_{\mathrm{H}}^{e}\overline{\mathbf{H}}_{rn}\\
\mathbf{E}_{rn} &= -\overline{k}_z\varepsilon_q^{-1}\overline{\boldsymbol{D}}_{en}\boldsymbol{T}_{\mathrm{E}}^{e}\overline{\mathbf{E}}_{zn}\\
\mathbf{H}_{\phi n} &= -\mathrm{j}\varepsilon_r\varepsilon_q^{-1}\overline{\boldsymbol{D}}_{en}\boldsymbol{T}_{\mathrm{E}}^{e}\overline{\mathbf{E}}_{zn}
\end{aligned}$$

E-bend: TE$_z$ modes: $\mathbf{E}_{zn} = \mathbf{0}$

$$\begin{aligned}
\frac{\mathrm{d}}{\mathrm{d}\overline{v}}(-\overline{\mathbf{H}}_{zn}) &= -\mathrm{j}\boldsymbol{\beta}_h\overline{\mathbf{E}}_{rn}\\
\frac{\mathrm{d}}{\mathrm{d}\overline{v}}\overline{\mathbf{E}}_{rn} &= -\mathrm{j}\boldsymbol{\beta}_h(-\overline{\mathbf{H}}_{zn})\\
\mathbf{H}_{\phi n} &= \mathrm{j}\overline{k}_z\boldsymbol{T}_{\mathrm{E}}^{h}\overline{\mathbf{E}}_{rn}\\
\mathbf{H}_{rn} &= \overline{k}_z\varepsilon_q^{-1}\overline{\boldsymbol{D}}_{hn}\boldsymbol{T}_{\mathrm{H}}^{h}\overline{\mathbf{H}}_{zn}\\
\mathbf{E}_{\phi n} &= \mathrm{j}\varepsilon_q^{-1}\overline{\boldsymbol{D}}_{hn}\boldsymbol{T}_{\mathrm{H}}^{h}\overline{\mathbf{H}}_{zn}
\end{aligned} \tag{3.12}$$

The overlined $\boldsymbol{D}$ matrices were divided by $\overline{h}_r$. $\mathbf{E}_{rn}$ and $\mathbf{H}_{\phi n}$ ($\mathbf{E}_{\phi n}$ and $\mathbf{H}_{rn}$) in TM$_z$ (TE$_z$) case are discretised on $h(e)$-lines.

3.2.1.3 Field transformation

We can write for the relation between the main tangential field components in the planes (or cross-sections) A and B of a waveguide (see Fig. 3.6):

$$\begin{bmatrix}\overline{\mathbf{E}}_{zn}^{\mathrm{A}}\\ \overline{\mathbf{E}}_{zn}^{\mathrm{B}}\end{bmatrix} = \begin{bmatrix}\overline{\mathbf{z}}_1 & \overline{\mathbf{z}}_2\\ \overline{\mathbf{z}}_2 & \overline{\mathbf{z}}_1\end{bmatrix}_{|e}\begin{bmatrix}\overline{\mathbf{H}}_{rn}^{\mathrm{A}}\\ -\overline{\mathbf{H}}_{rn}^{\mathrm{B}}\end{bmatrix} \qquad \begin{bmatrix}\overline{\mathbf{E}}_{rn}^{\mathrm{A}}\\ \overline{\mathbf{E}}_{rn}^{\mathrm{B}}\end{bmatrix} = \begin{bmatrix}\overline{\mathbf{z}}_1 & \overline{\mathbf{z}}_2\\ \overline{\mathbf{z}}_2 & \overline{\mathbf{z}}_1\end{bmatrix}_{|h}\begin{bmatrix}(-\overline{\mathbf{H}}_{zn}^{\mathrm{A}})\\ -(-\overline{\mathbf{H}}_{zn}^{\mathrm{B}})\end{bmatrix} \tag{3.13}$$

with:

$$\overline{\mathbf{z}}_1 = \overline{\mathbf{Z}}_{0f}(\tanh \boldsymbol{\phi}_f)^{-1} \quad \overline{\mathbf{z}}_2 = \overline{\mathbf{Z}}_{0f}(\sinh \boldsymbol{\phi}_f)^{-1} \quad \boldsymbol{\phi}_f = \boldsymbol{\Gamma}_f \overline{L}_v \tag{3.14}$$

and $\overline{L}_v = \phi_{\mathrm{AB}}\overline{r}_o$ as normalised length of the arc between the planes A and B. Note that the fields in the H(E)-bend are labelled with the subscript $e(h)$. These subscripts are omitted for brevity. The inversion of the above relations yields:

$$\begin{bmatrix} \overline{\mathbf{H}}_{rn}^{\mathrm{A}} \\ -\overline{\mathbf{H}}_{rn}^{\mathrm{B}} \end{bmatrix} = \begin{bmatrix} \overline{\mathbf{y}}_1 & \overline{\mathbf{y}}_2 \\ \overline{\mathbf{y}}_2 & \overline{\mathbf{y}}_1 \end{bmatrix}_{|e} \begin{bmatrix} \overline{\mathbf{E}}_{zn}^{\mathrm{A}} \\ \overline{\mathbf{E}}_{zn}^{\mathrm{B}} \end{bmatrix} \qquad \begin{bmatrix} (-\overline{\mathbf{H}}_{zn}^{\mathrm{A}}) \\ -(-\overline{\mathbf{H}}_{zn}^{\mathrm{B}}) \end{bmatrix} = \begin{bmatrix} \overline{\mathbf{y}}_1 & \overline{\mathbf{y}}_2 \\ \overline{\mathbf{y}}_2 & \overline{\mathbf{y}}_1 \end{bmatrix}_{|h} \begin{bmatrix} \overline{\mathbf{E}}_{rn}^{\mathrm{A}} \\ \overline{\mathbf{E}}_{rn}^{\mathrm{B}} \end{bmatrix} \tag{3.15}$$

where:

$$\overline{\mathbf{y}}_1 = \overline{\mathbf{Y}}_{0f}(\tanh \boldsymbol{\phi}_f)^{-1} \quad \overline{\mathbf{y}}_2 = -\overline{\mathbf{Y}}_{0f}(\sinh \boldsymbol{\phi}_f)^{-1} \tag{3.16}$$

Due to the normalisation in eq. (3.10), the characteristic impedance and admittance are unit matrices: $\overline{\mathbf{Z}}_{0f} = \overline{\mathbf{Y}}_{0f} = \boldsymbol{I}$ $(f = e, h)$. The impedance and admittance transformation through sections is described in Section 2.5.

3.2.2 LSM and LSE modes in straight waveguides

The formulas for straight waveguides are obtained from those for curved waveguides by letting the radii r, r_o approach infinity. Their ratio (the normalized radius r_n) is then equal to 1 and the diagonal matrices $\mathbf{r}_e$ and $\mathbf{r}_h$ are replaced by identity matrices. Therefore, these quantities could be omitted from the formulas. The normalised fields are then equal to the original ones. In what follows, we replace the radial coordinate r with u. Again, v is the direction of propagation. The term $\boldsymbol{\phi}$ in eq. (3.16) is given as $\boldsymbol{\phi} = \mathbf{k}_v \overline{L}$ (see Fig. 3.1).

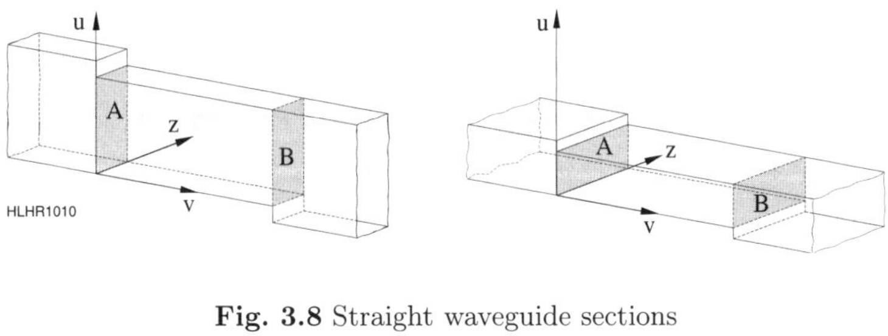

Fig. 3.8 Straight waveguide sections

We use here u, v and z for the Cartesian coordinates instead of x, y and z. In this way, we can choose x or y as the direction of propagation v. This allows a compact description, particularly for junctions.

3.2.2.1 Field components in transformed domain

From eq. (3.5) we obtain immediately the following expressions for the matrices $\boldsymbol{R}^f_{\mathrm{E,H}}$:

$$\boldsymbol{R}^e_{\mathrm{H}} = {\varepsilon_r}^{-1}\varepsilon_q \boldsymbol{I}_u \qquad \boldsymbol{R}^e_{\mathrm{E}} = \varepsilon_r \boldsymbol{I}_u - \varepsilon_r \varepsilon_q^{-1} \boldsymbol{D}^{et}_{\overline{u}} \boldsymbol{D}^e_{\overline{u}} \tag{3.17}$$

$$\boldsymbol{R}^h_{\mathrm{E}} = \varepsilon_q \boldsymbol{I}_u \qquad \boldsymbol{R}^h_{\mathrm{H}} = \boldsymbol{I}_u - \varepsilon_q^{-1} \boldsymbol{D}^{ht}_{\overline{u}} \boldsymbol{D}^h_{\overline{u}} \tag{3.18}$$

With these matrices the eigenvalue problems are again given by the eqs. (3.9) and (3.10). For the field components in transformed domain we obtain:

LSM or TM$_z$ modes **LSE or TE$_z$ modes**

$$\begin{aligned}
&\frac{\mathrm{d}}{\mathrm{d}\overline{v}}\overline{\mathbf{E}}_z = -\mathrm{j}\boldsymbol{\beta}_e \overline{\mathbf{H}}_u && \frac{\mathrm{d}}{\mathrm{d}\overline{v}}(-\overline{\mathbf{H}}_z) = -\mathrm{j}\boldsymbol{\beta}_h \overline{\mathbf{E}}_u \\
&\frac{\mathrm{d}}{\mathrm{d}\overline{v}}\overline{\mathbf{H}}_u = -\mathrm{j}\boldsymbol{\beta}_e \overline{\mathbf{E}}_z && \frac{\mathrm{d}}{\mathrm{d}\overline{v}}\overline{\mathbf{E}}_u = -\mathrm{j}\boldsymbol{\beta}_h (-\overline{\mathbf{H}}_z) \\
&\mathbf{E}_u = -\overline{k}_z \varepsilon_q^{-1} \overline{\boldsymbol{D}}^e_{\overline{u}} \boldsymbol{T}^e_{\mathrm{E}} \overline{\mathbf{E}}_z && \mathbf{H}_u = \overline{k}_z \varepsilon_q^{-1} \overline{\boldsymbol{D}}^h_{\overline{u}} \boldsymbol{T}^h_{\mathrm{H}} \overline{\mathbf{H}}_z \\
&\mathbf{E}_v = \mathrm{j}\overline{k}_z \varepsilon_r^{-1} \boldsymbol{T}^e_{\mathrm{H}} \overline{\mathbf{H}}_u && \mathbf{H}_v = \mathrm{j}\overline{k}_z \boldsymbol{T}^h_{\mathrm{E}} \overline{\mathbf{E}}_u \\
&\mathbf{H}_v = -\mathrm{j}\varepsilon_r \varepsilon_q^{-1} \overline{\boldsymbol{D}}^e_{\overline{u}} \boldsymbol{T}^e_{\mathrm{E}} \overline{\mathbf{E}}_z && \mathbf{E}_v = \mathrm{j}\varepsilon_q^{-1} \overline{\boldsymbol{D}}^h_{\overline{u}} \boldsymbol{T}^h_{\mathrm{H}} \overline{\mathbf{H}}_z
\end{aligned} \tag{3.19}$$

For the TM$_z$-polarisation we discretise $\boldsymbol{E}_u$ and $\mathbf{H}_v$ in u-direction on h-lines. In the TE$_z$ case, $\boldsymbol{E}_v$ and $\mathbf{H}_u$ are discretised in u-direction on e-lines.

With this modification eq. (3.11) also holds for straight waveguides. Therefore, we can write the following expression for the fields between the planes (cross-sections) A and B of a waveguide (see Fig. 3.8):

$$\begin{bmatrix} \overline{\mathbf{E}}_z^{\mathrm{A}} \\ \overline{\mathbf{E}}_z^{\mathrm{B}} \end{bmatrix} = \begin{bmatrix} \overline{\mathbf{z}}_1 & \overline{\mathbf{z}}_2 \\ \overline{\mathbf{z}}_2 & \overline{\mathbf{z}}_1 \end{bmatrix}_{|e} \begin{bmatrix} \overline{\mathbf{H}}_u^{\mathrm{A}} \\ -\overline{\mathbf{H}}_u^{\mathrm{B}} \end{bmatrix} \qquad \begin{bmatrix} \overline{\mathbf{E}}_u^{\mathrm{A}} \\ \overline{\mathbf{E}}_u^{\mathrm{B}} \end{bmatrix} = \begin{bmatrix} \overline{\mathbf{z}}_1 & \overline{\mathbf{z}}_2 \\ \overline{\mathbf{z}}_2 & \overline{\mathbf{z}}_1 \end{bmatrix}_{|h} \begin{bmatrix} (-\overline{\mathbf{H}}_z^{\mathrm{A}}) \\ -(-\overline{\mathbf{H}}_z^{\mathrm{B}}) \end{bmatrix} \tag{3.20}$$

$$\begin{bmatrix} \overline{\mathbf{H}}_u^{\mathrm{A}} \\ -\overline{\mathbf{H}}_u^{\mathrm{B}} \end{bmatrix} = \begin{bmatrix} \overline{\mathbf{y}}_1 & \overline{\mathbf{y}}_2 \\ \overline{\mathbf{y}}_2 & \overline{\mathbf{y}}_1 \end{bmatrix}_{|e} \begin{bmatrix} \overline{\mathbf{E}}_z^{\mathrm{A}} \\ \overline{\mathbf{E}}_z^{\mathrm{B}} \end{bmatrix} \qquad \begin{bmatrix} (-\overline{\mathbf{H}}_z^{\mathrm{A}}) \\ -(-\overline{\mathbf{H}}_z^{\mathrm{B}}) \end{bmatrix} = \begin{bmatrix} \overline{\mathbf{y}}_1 & \overline{\mathbf{y}}_2 \\ \overline{\mathbf{y}}_2 & \overline{\mathbf{y}}_1 \end{bmatrix}_{|h} \begin{bmatrix} \overline{\mathbf{E}}_u^{\mathrm{A}} \\ \overline{\mathbf{E}}_u^{\mathrm{B}} \end{bmatrix} \tag{3.21}$$

where:

$$\overline{\mathbf{z}}_1 = \boldsymbol{I}_f (\tanh(\boldsymbol{\Gamma}_f \overline{L}_v))^{-1} = \overline{\mathbf{y}}_1 \quad \overline{\mathbf{z}}_2 = \boldsymbol{I}_f (\sinh(\boldsymbol{\Gamma}_f \overline{L}_v))^{-1} = -\overline{\mathbf{y}}_2 \tag{3.22}$$

$\overline{L}_v$ is the normalised length of the waveguide between planes A and B. Again, we have unit matrices ($\boldsymbol{I}_f$) for the characteristic impedance/admittance. For each of the cross-sections A, B we may write:

$$\overline{\mathbf{E}}_z^{\mathrm{A,B}} = \overline{\boldsymbol{Z}}_{\mathrm{A,B}} \overline{\mathbf{H}}_u^{\mathrm{A,B}} \qquad \overline{\mathbf{H}}_z^{\mathrm{A,B}} = \overline{\boldsymbol{Y}}_{\mathrm{A,B}} \overline{\mathbf{E}}_u^{\mathrm{A,B}} \tag{3.23}$$

in case of TM$_z$ modes and:

$$\overline{\mathbf{E}}_u^{\mathrm{A,B}} = \overline{\boldsymbol{Z}}_{\mathrm{A,B}} (-\overline{\mathbf{H}}_z^{\mathrm{A,B}}) \qquad -\overline{\mathbf{H}}_z^{\mathrm{A,B}} = \overline{\boldsymbol{Y}}_{\mathrm{A,B}} \overline{\mathbf{E}}_u^{\mathrm{A,B}} \tag{3.24}$$

for the TE_z modes. The impedances and admittances transformation from the cross-section B to the cross-section A is performed as described in Section 2.5 and is therefore not repeated here.

3.2.3 Impedance transformation at waveguide interfaces

At the interfaces between different waveguide sections (see e.g. Fig. 3.9) the tangential fields must be matched. This must be done in the original domain and can easily be performed e.g. with the concept of transformation matrix partition as described in Section 2.5.

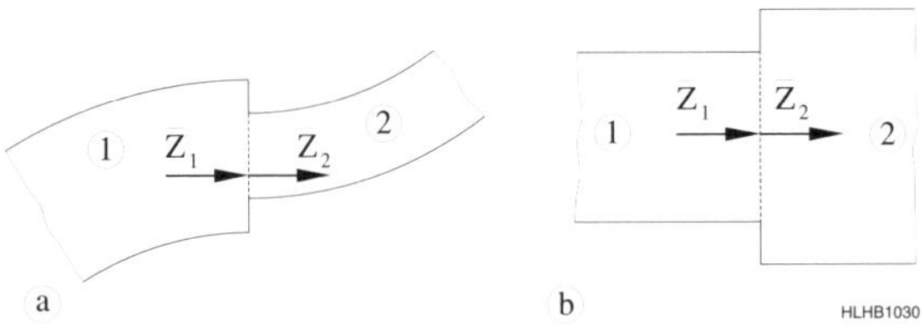

Fig. 3.9 Two typical transitions between waveguide sections

We will give the general formulas for the structure sketched in Fig. 3.10, which shows two waveguide sections with different cross-sections and opposite curvatures. All special cases can easily be obtained from this general one. If one (or both) of the waveguide sections is a straight one, the corresponding $\boldsymbol{r}$ matrices are replaced by unit matrices. During the discretisation process the discretised field components were put into vectors in the order of increasing radii. Now the r coordinate directions of the sections (see Fig. 3.10) are oppositely directed. Therefore, the order of the components is reversed. To get the components at the same position, we introduce the exchange matrix $\boldsymbol{J}_{ex}$, which contains ones on the secondary diagonal and zeros everywhere else. $\boldsymbol{J}_{ex}$ changes the order of the field components in one of the regions. For the matching purpose we divide the transformation matrices into submatrices corresponding to the different parts (common interface with the other waveguide section (subscript c) or metallisation (subscript m)). The partition of the transformation matrix $\boldsymbol{T}_1$ into two parts is shown in Fig. 3.11 (the subscripts E or H and the superscripts e or h were omitted). The submatrix $\boldsymbol{T}_{1\mathrm{c}}$ corresponds to the part of region 1 that is common with region 2. We have $\boldsymbol{T}_{1\mathrm{m}}$ at the part with the metallisation. If $\boldsymbol{T}_1^{\mathrm{t}}$ is the inverse of $\boldsymbol{T}_1$ we can immediately also give the submatrices of $\boldsymbol{T}^{-1}$. Otherwise the inversion must be performed before the partitioning. The dividing of the transformation matrix $\boldsymbol{T}_2$ (section 2) is shown in Fig. 3.12. Here we have three parts. The middle submatrix $\boldsymbol{T}_{2\mathrm{c}}$ of $\boldsymbol{T}_2$ corresponds to the part of region 2 that is common with region 1. The two other parts $\boldsymbol{T}_{2m_1}$ and $\boldsymbol{T}_{2m_2}$ are for the metallisation and can be combined to a single block $\boldsymbol{T}_{2\mathrm{m}}$.

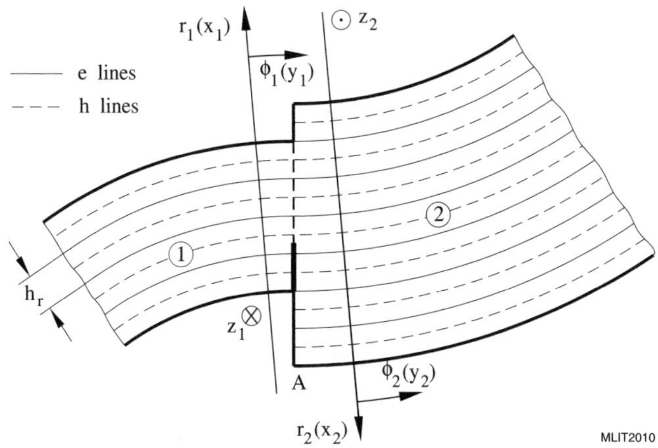

Fig. 3.10 Concatenations of two curved waveguide sections of opposite curvature and different cross-sections

In Figs. 3.11 and 3.12 we see also that the two matrices $\boldsymbol{T}_1$ and $\boldsymbol{T}_2$ have different sizes, whereas the number of rows of the submatrices $\boldsymbol{T}_1^{\mathrm{c}}$ and $\boldsymbol{T}_2^{\mathrm{c}}$ is equal.

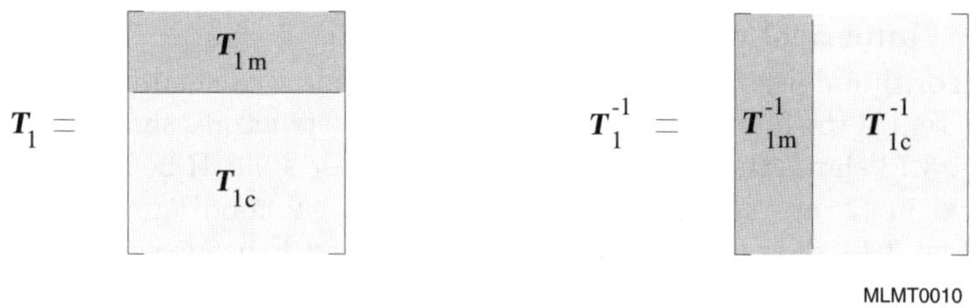

Fig. 3.11 Transformation matrix partition in regions 1 and 2 of Fig. 3.10. The subscripts E or H and the superscripts e or h were omitted

3.2.3.1 H-plane discontinuities

Now, assume we know $\overline{\mathbf{Y}}_2 = \overline{\mathbf{Z}}_2^{-1}$ (see Fig. 3.9). The impedance matrix $\overline{\mathbf{Z}}_1$ (defined as $\overline{\mathbf{E}}_{zn1} = \overline{\mathbf{Z}}_1 \overline{\mathbf{H}}_{rn1}$) is determined as:

$$\overline{\mathbf{Z}}_1 = (\mathbf{w}_1 \boldsymbol{T}_{\mathrm{E}1})_{\mathrm{c}}^{-1}((\boldsymbol{J}_{\mathrm{ex}} \mathbf{w}_2^{-1} \boldsymbol{T}_{\mathrm{H}2})_{\mathrm{c}} \overline{\mathbf{Y}}_2 (\boldsymbol{J}_{\mathrm{ex}} \mathbf{w}_2 \boldsymbol{T}_{\mathrm{E}2})_{\mathrm{c}}^{-1})^{-1} (\mathbf{w}_1^{-1} \boldsymbol{T}_{\mathrm{H}1})_{\mathrm{c}} \quad (3.25)$$

where the abbreviation $\mathbf{w} = \sqrt{\boldsymbol{r}_e}$ was introduced (the subscripts 1 or 2 are omitted for the sake of brevity). We see in eq. (3.25) that the matrix partition described above is actually performed for matrix products of $\boldsymbol{T}$ with $\mathbf{w}$, $\mathbf{w}^{-1}$ and $\boldsymbol{J}_{ex}$.

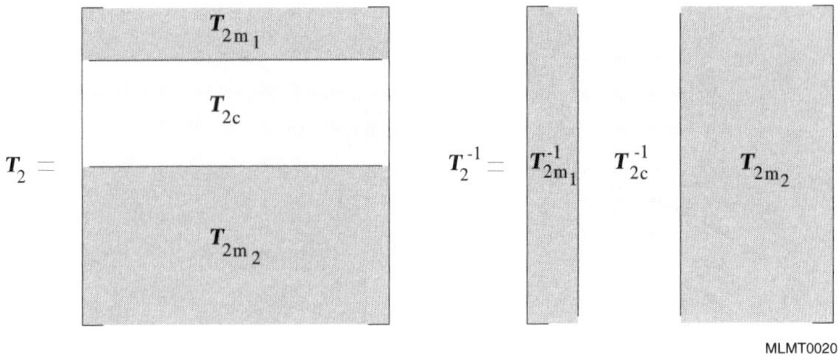

Fig. 3.12 Transformation matrix partition in region 2 of Fig. 3.10. The subscripts E or H and the superscripts e or h were omitted

3.2.3.2 E-plane discontinuities

In the case of an E-plane discontinuity we can determine the impedance $\overline{\boldsymbol{Z}}_1$ (defined as $\overline{\mathbf{E}}_{rn1} = \overline{\boldsymbol{Z}}_1(-\overline{\mathbf{H}}_{zn1})$) from a known one $\overline{\boldsymbol{Z}}_2 = \overline{\mathbf{Y}}_2^{-1}$ as:

$$\overline{\boldsymbol{Z}}_1 = (\mathbf{w}_1^{-1}\boldsymbol{T}_{\mathrm{E1}})_{\mathrm{c}}^{-1}((\boldsymbol{J}_{\mathrm{ex}}\mathbf{w}_2\boldsymbol{T}_{\mathrm{H2}})_{\mathrm{c}}\overline{\mathbf{Y}}_2(\boldsymbol{J}_{\mathrm{ex}}\mathbf{w}_2^{-1}\boldsymbol{T}_{\mathrm{E2}})_{\mathrm{c}}^{-1})^{-1}(\mathbf{w}_1\boldsymbol{T}_{\mathrm{H1}})_{\mathrm{c}} \quad (3.26)$$

where the abbreviation $\mathbf{w} = \sqrt{\boldsymbol{r}_h}$ was introduced. Again, we see that the partition has to be done for the matrix products in eq. (3.26).

3.2.4 Numerical results for concatenations

The algorithm developed in this chapter was used to calculate the scattering parameters of the fundamental mode for different structures shown in Fig. 3.1.

Fig. 3.13 shows the reflection coefficient S_{11} for a 90° H bend. The results reported in [7] are also plotted for comparison. A good agreement can be seen. Fig. 3.14 shows the results for two cascaded E-bend sections sketched in Fig. 3.1. The scattering parameters S_{11} for S and U bends are completely different. The results were compared with those published in [6] and a very good agreement is seen.

Fig. 3.15 shows the convergence behaviour for the U bend at 12 GHz. Because of the monotonic convergence behaviour (the MoL has stationary properties) we may extrapolate the 'exact' value with a quadratic polynomial.

Fig. 3.16 shows results for an S bend with an intermediate straight section. As we know, at discontinuities (here at the interface between the bend and the straight waveguide) higher order modes are excited. If the length of this straight section is large enough, these higher-order modes decrease in the straight section to negligible values decreased to on the output of the first bend. Therefore, the scattering parameters S_{11} for S and U bends are nearly equal in this case. By introducing diaphragms into the bends, the reflections at concatenations with straight waveguides can be reduced. Fig. 3.17 shows the result for a 90° bend with one diaphragm in its middle and Fig. 3.18 for a 180° bend with two diaphragms.

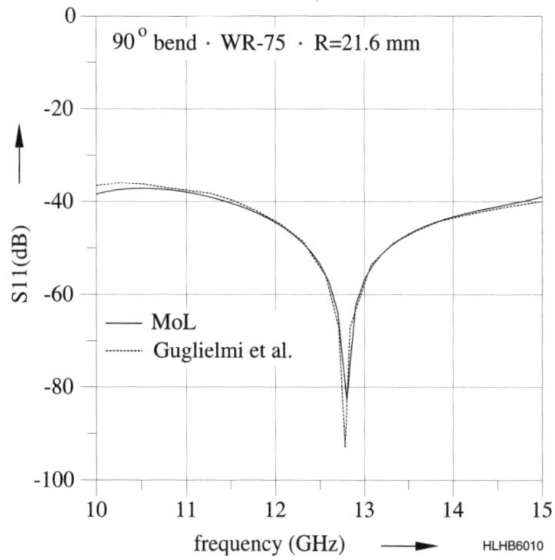

Fig. 3.13 Reflection coefficient of an H-plane bend as a function of frequency (R. Pregla, 'Concatenations of Waveguide Sections', April 1997. The Institution of Engineering and Technology (IET))

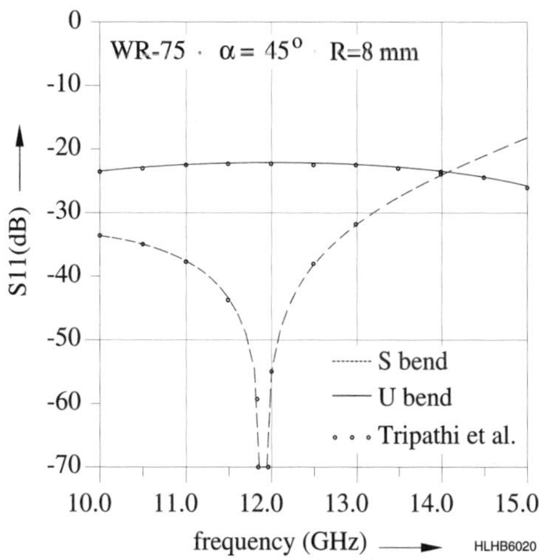

Fig. 3.14 Concatenations of straight and curved waveguide sections (see Fig. 3.1a). Intermediate straight section length $L = 0$ (R. Pregla, 'Concatenations of Waveguide Sections', April 1997. The Institution of Engineering and Technology (IET))

When modelling such structures, the convergence behaviour depends strongly on the position of the discretisation lines relative to the edge. This is shown in Fig. 3.19, where we examined various positions of the discretisation lines (ph is the distance between the edge and an h-line). We clearly see the best behaviour for $p = 0.75$.

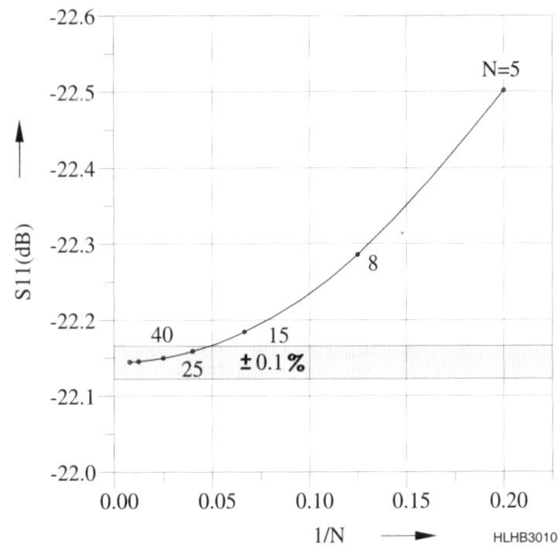

Fig. 3.15 Convergence behaviour of the reflection coefficient as a function of 1/N (N: number of discretisation lines) (R. Pregla, 'Concatenations of Waveguide Sections', April 1997. The Institution of Engineering and Technology (IET))

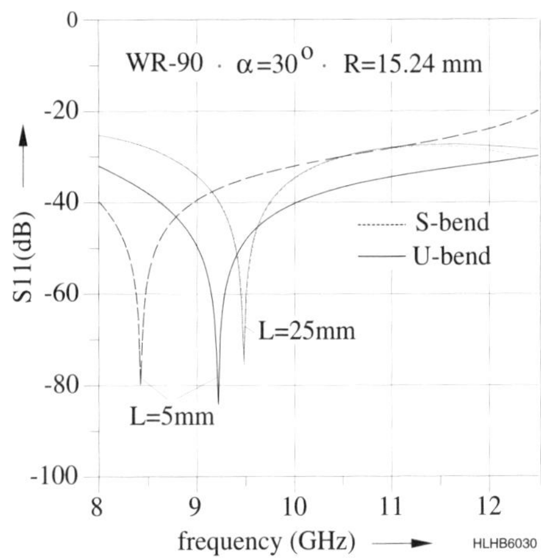

Fig. 3.16 Concatenations of straight and curved waveguide sections (cf. Fig. 3.1b) (R. Pregla, 'Concatenations of Waveguide Sections', April 1997. The Institution of Engineering and Technology (IET))

A waveguide with a step and a diaphragm in the cross-section was examined in Fig. 3.20. The normalised equivalent shunt admittance versus normalised height of the diaphragm is shown. The input impedance for $\overline{\mathbf{Z}}_{o2} = \boldsymbol{I}$ is given by:

$$\overline{\boldsymbol{Z}}_1 = \boldsymbol{T}_{\mathrm{E1c}}^{-1}(\boldsymbol{T}_{\mathrm{H2c}}\boldsymbol{T}_{\mathrm{E2c}}^{-1})^{-1}\boldsymbol{T}_{\mathrm{H1c}} \tag{3.27}$$

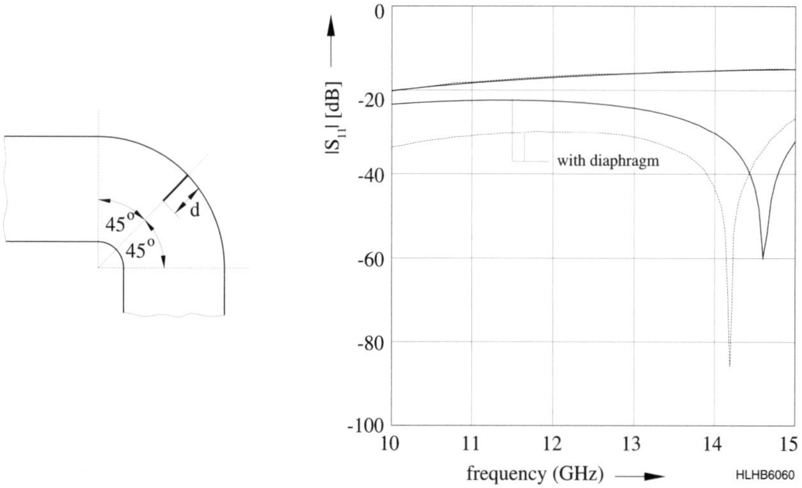

Fig. 3.17 Reflection coefficient as a function of frequency. 90° E bend in WR75 waveguide, $d = 2.43$ mm, - - - [9]

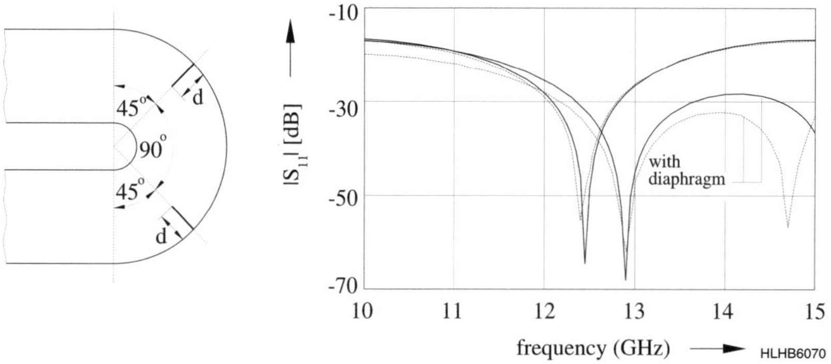

Fig. 3.18 Reflection coefficient as a function of frequency. 180° E bend in WR75 waveguide, $d = 2.43$ mm, - - - [9]

3.2.5 Numerical results for waveguide filters

Waveguide filters can be analysed in the same way as concatenations of waveguides. Fig. 3.22 shows the scattering parameters S_{11}, S_{21} for the waveguide filter [10], [11] sketched in Fig. 3.21. The filter consists of directly coupled resonators. For comparison, the computed and measured results from [10] are shown in the diagram, too.

The computed results in [10] were obtained by using equivalent shunt inductances for the diaphragm. Therefore, the results here show the validity of the approximation used in [10].

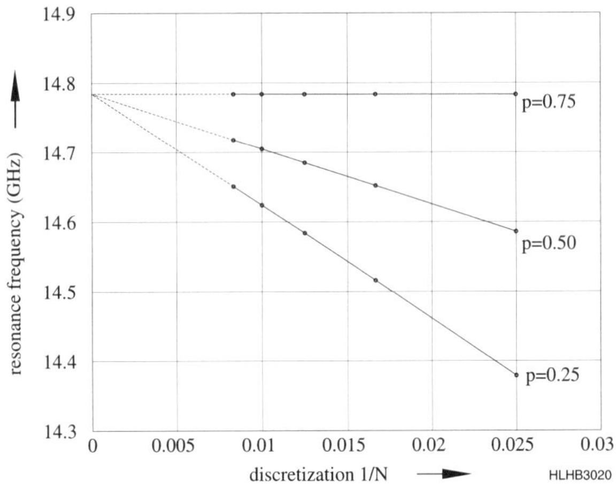

Fig. 3.19 Convergence behaviour in the resonance frequency as a function of the number of discretisation lines N and the position of edges between the discretisation lines: bend in Fig. 3.17 with $d = 0.25\ b$

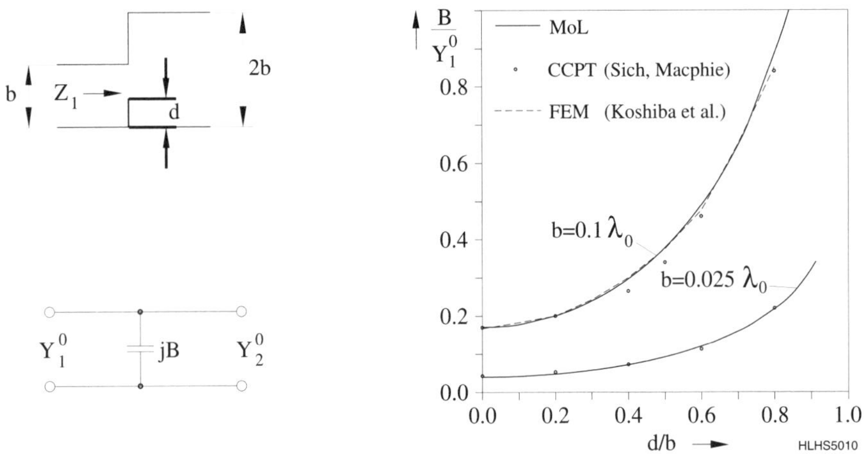

Fig. 3.20 Step in a parallel plate waveguide with diaphragm. Normalised equivalent shunt admittance B/Y_1 as a function of normalised diaphragm height d/b (R. Pregla, 'Concatenations of Waveguide Sections', April 1997. The Institution of Engineering and Technology (IET))

An example of a millimetre wave component is instructive. Fig. 3.23 shows an E-plane filter [11], for which the necessary analysis equations are taken from the equations above. In Fig. 3.24 the scattering parameters are plotted

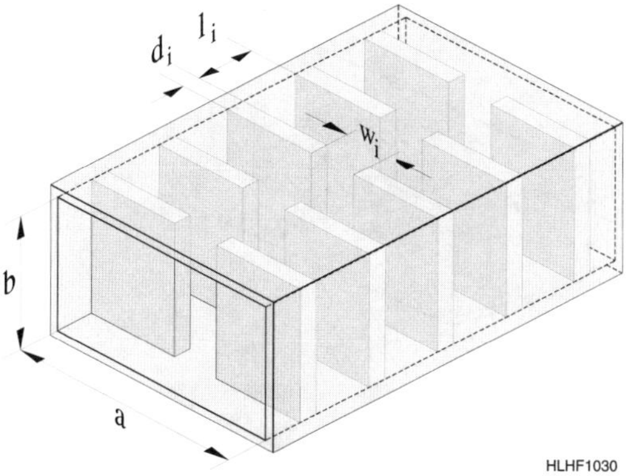

Fig. 3.21 Waveguide filter of direct coupled resonators. $a = 2b = 3.7592$ mm; $l_1 = l_4 = 2.73$ mm; $l_2 = l_3 = 3.02$ mm $d_1 = d_5 = 0.56$ mm; $d_2 = d_4 = 0.52$ mm; $d_3 = 0.63$ mm; $w_1 = w_5 = 1.88$ mm; $w_2 = w_4 = w_3 = 1.18$ mm (R. Pregla, 'Concatenations of Waveguide Sections', April 1997. The Institution of Engineering and Technology (IET))

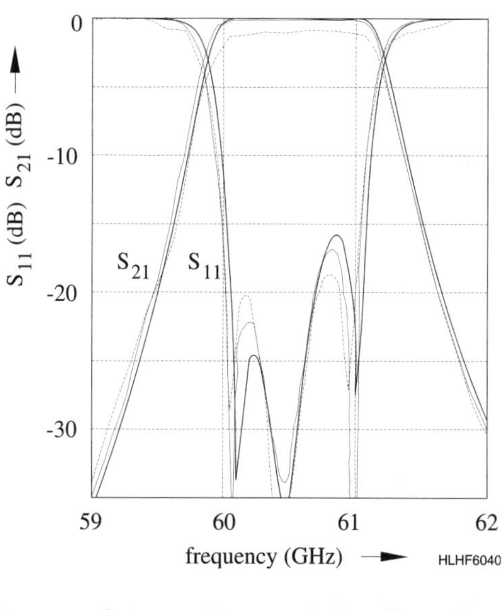

Fig. 3.22 Reflectance and transmittance of the filter shown in Fig. 3.21 — computed results by MoL — computed and - - - measured results in [10] (R. Pregla, 'Concatenations of Waveguide Sections', April 1997. The Institution of Engineering and Technology (IET))

vs. frequency and compared with calculated and measured values reported in [11] and [12].

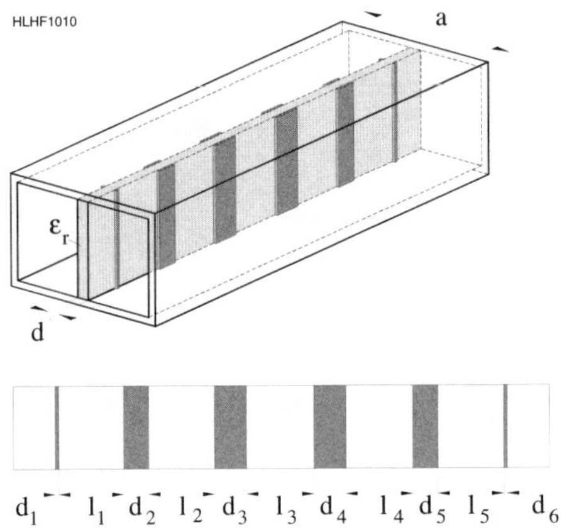

Fig. 3.23 E-plane filter: $a = 1.651$ mm, $d = 0.0762$ mm, $\varepsilon_r = 2.1$ $d_1 = d_6 = 0.03175$ mm, $d_2 = d_5 = 0.3365$ mm, $d_3 = d_4 = 0.887$ mm $l_1 = l_5 = 0.887$ mm, $l_2 = l_4 = 0.8989$ mm, $l_3 = 0.8992$ mm

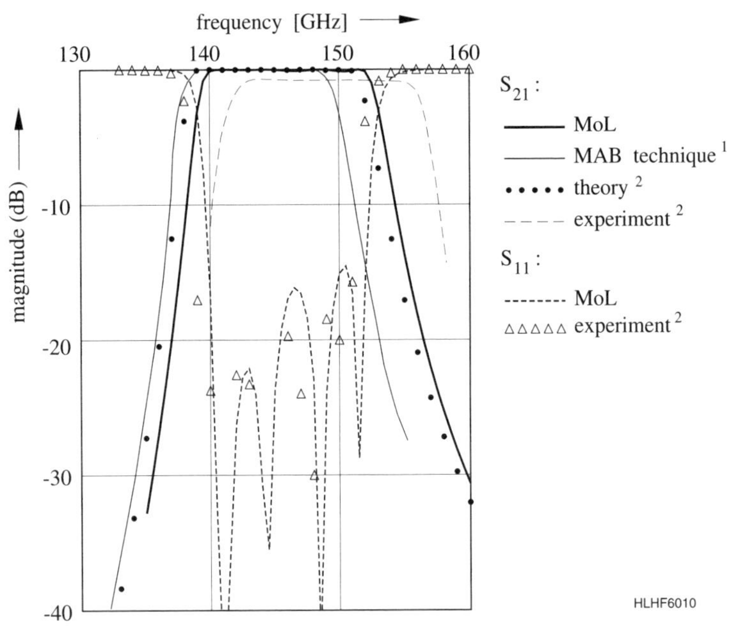

Fig. 3.24 Scattering parameters S_{11} and S_{21} for the filter in Fig. 3.23. Comparison with results from [11][1] and [12][2]

3.3 WAVEGUIDE JUNCTIONS

To analyse waveguide junctions (see the example of an E-plane junction in Fig. 3.3) we use the concept of crossed discretisation lines. General waveguide junctions are drawn in Figs. 3.25 and 3.26.

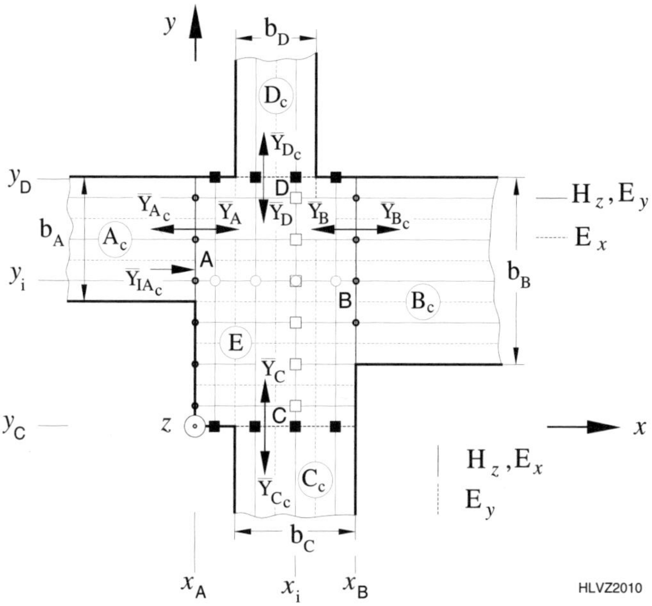

Fig. 3.25 General E-plane junction with crossed discretisation lines (Reproduced by permission of Elsevier)

The four surfaces of region E can be considered as four generalised ports. Our goal is to obtain a general relation between the fields at these four ports A, B, C and D, i.e. in the inner side of region E. Therefore, the metallic parts of the ports are not yet relevant. They must be taken into account when the junction is connected with the waveguides A_c, B_c, C_c and D_c. By extending the two-port description with $\boldsymbol{z}$ and $\boldsymbol{y}$ matrices to a four port relation, which we need for the E-plane and H-plane junctions, the relation of the fields at the four ports can be written as follows:

$$\begin{bmatrix} \widehat{\mathbf{H}}_{\mathrm{AB}} \\ \widehat{\mathbf{H}}_{\mathrm{CD}} \end{bmatrix} = \begin{bmatrix} \widehat{\boldsymbol{y}}_{\mathrm{AB}}^{\mathrm{AB}} & \widehat{\boldsymbol{y}}_{\mathrm{AB}}^{\mathrm{CD}} \\ \widehat{\boldsymbol{y}}_{\mathrm{CD}}^{\mathrm{AB}} & \widehat{\boldsymbol{y}}_{\mathrm{CD}}^{\mathrm{CD}} \end{bmatrix} \begin{bmatrix} \widehat{\mathbf{E}}_{\mathrm{AB}} \\ \widehat{\mathbf{E}}_{\mathrm{CD}} \end{bmatrix} \tag{3.28}$$

The vectors $\widehat{\mathbf{F}}_{\mathrm{UV}}$ (U = A,C and V = B,D) of the discretised fields (where $\mathbf{F}$ is equivalent to $\mathbf{E}$ or $\mathbf{H}$) are supervectors that contain the fields on opposite ports:

$$\widehat{\mathbf{E}}_{\mathrm{UV}} = \left[\mathbf{E}_{\mathrm{U}}^{\mathrm{t}}, \mathbf{E}_{\mathrm{V}}^{\mathrm{t}}\right]^{\mathrm{t}} \quad \widehat{\mathbf{H}}_{\mathrm{UV}} = \left[\mathbf{H}_{\mathrm{U}}^{\mathrm{t}}, -\mathbf{H}_{\mathrm{V}}^{\mathrm{t}}\right]^{\mathrm{t}} \tag{3.29}$$

All the submatrices are obtained by short circuiting the ports alternately. By short circuiting the ports C and D the submatrices:

$$\widehat{\boldsymbol{y}}_{\mathrm{AB}}^{\mathrm{AB}} \qquad \widehat{\boldsymbol{y}}_{\mathrm{CD}}^{\mathrm{AB}}$$

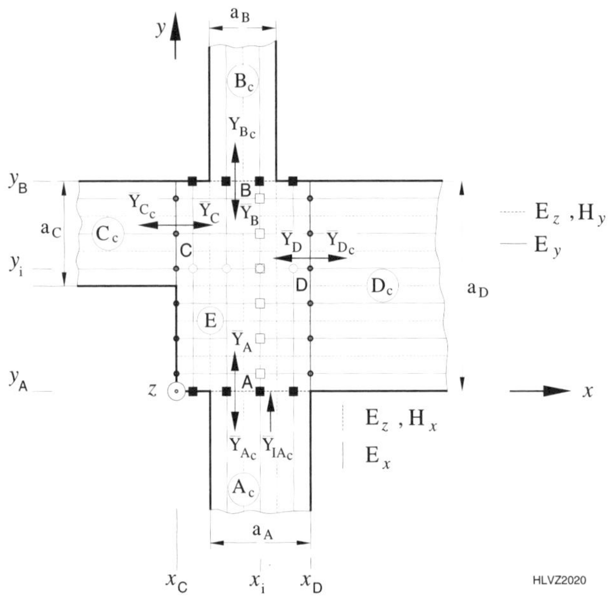

Fig. 3.26 General H-plane junction with crossed discretisation lines (Reproduced by permission of Elsevier)

can be calculated. For the E-plane (H-plane) junction these matrices are related to the discretisation in y- (x)-direction. The fields in planes A and B are described (as usual) by the equations for the lines in x-direction. We also have to determine the field parts at the short-circuited ports C and D or, to be accurate, on the vertical lines in the case of the E-plane junction. The other submatrices

$$\widehat{\boldsymbol{y}}_{\mathrm{CD}}^{\mathrm{CD}} \qquad \widehat{\boldsymbol{y}}_{\mathrm{AB}}^{\mathrm{CD}}$$

can be computed by an analogous procedure by short circuiting ports A and B. For the E-plane (H-plane) junction, these matrices are related to the discretisation in x- (y)-direction.

By inverting eq. (3.28) and taking the fields in transformed domain, we may write:

$$\begin{bmatrix} \overline{\mathbf{E}}_{\mathrm{AB}} \\ \overline{\mathbf{E}}_{\mathrm{CD}} \end{bmatrix} = \begin{bmatrix} \widehat{\overline{\mathbf{z}}}_{\mathrm{AB}}^{\mathrm{AB}} & \widehat{\overline{\mathbf{z}}}_{\mathrm{CD}}^{\mathrm{AB}} \\ \widehat{\overline{\mathbf{z}}}_{\mathrm{AB}}^{\mathrm{CD}} & \widehat{\overline{\mathbf{z}}}_{\mathrm{CD}}^{\mathrm{CD}} \end{bmatrix} \begin{bmatrix} \overline{\mathbf{H}}_{\mathrm{AB}} \\ -\overline{\mathbf{H}}_{\mathrm{CD}} \end{bmatrix} \tag{3.30}$$

We assume constant material parameters in region E. All waveguides should have the same dimensions in z-direction (see Fig. 3.25 for the coordinate system) and no steps in this direction. The more general case will be described in Section 3.4, where we deal with 3D junctions.

3.3.1 E-plane junctions

The fields in E-plane junctions are described by the LSE_z or TE_z modes (see Section 3.2.2). Let us now examine the structure shown in Fig. 3.25, where we have the connection of four waveguides. The waveguide region E is bounded by the planes A and B at x_A and x_B in the horizontal direction and the planes C and D positioned at y_C and y_D in the vertical direction. As we can see, the heights b of the connecting waveguides may be different. The width a, however, must be constant. These four connections may be seen as four generalised ports.

The magnetic field in region E is a superposition of the two parts:

$$H_z^\mathrm{E} = H_{zx}^\mathrm{E} + H_{zy}^\mathrm{E} \tag{3.31}$$

The second subscripts x, y denote the direction of the wave propagation, with respect to the discretisation lines.

Let us now take a look at the magnetic fields in planes A and B. We want to describe them as functions of the electric ones. We have contributions from the electric fields from all four ports (A to D). The dependence of the fields at ports A and B is described by the analytical expressions that we use for 2-ports where the lines are directed in x-direction. Additionally, we must express the parts of the magnetic fields that are caused by the electric ones at ports C and D.

The same holds for planes C and D, i.e. we must determine the magnetic fields from the electric ones at these ports, and additionally, from the electric fields at ports A and B. In the following we omit the superscript E for brevity, because we only derive the relations for this region.

3.3.1.1 Magnetic fields from discretisation in y-direction

In this section we want to determine the magnetic fields at ports A to D, where we consider the discretisation in y-direction and analytic expressions in x-direction.

The relation between the magnetic and electric fields at ports A and B can be given by the expressions determined in Section 3.2.2 for the TE_z modes. We have to replace v with x, u with y and z with $-z$. The field component in z-direction H_z has to be changed to $-H_z$. From eq. (3.21) we therefore obtain:

$$\begin{bmatrix} \overline{\mathbf{H}}_{zyA} \\ -\overline{\mathbf{H}}_{zyB} \end{bmatrix} = \begin{bmatrix} \overline{\mathbf{y}}_1 & \overline{\mathbf{y}}_2 \\ \overline{\mathbf{y}}_2 & \overline{\mathbf{y}}_1 \end{bmatrix}_{|\mathrm{AB}} \begin{bmatrix} \overline{\mathbf{E}}_{yyA} \\ \overline{\mathbf{E}}_{yyB} \end{bmatrix} = \hat{\overline{\mathbf{y}}}_{\mathrm{AB}}^{\mathrm{AB}} \begin{bmatrix} \overline{\mathbf{E}}_{yyA} \\ \overline{\mathbf{E}}_{yyB} \end{bmatrix} \tag{3.32}$$

where:

$$\overline{\mathbf{y}}_1 = \boldsymbol{I}(\tanh(\boldsymbol{\Gamma}_x \overline{d}_\mathrm{AB}))^{-1} \overline{\mathbf{y}}_2 = -\boldsymbol{I}(\sinh(\boldsymbol{\Gamma}_x \overline{d}_\mathrm{AB}))^{-1} \tag{3.33}$$

and $\overline{d}_\mathrm{AB} = k_0(x_\mathrm{B} - x_\mathrm{A})$. $\boldsymbol{\Gamma}_x$ is the diagonal matrix of propagation constants in x-direction. Since the characteristic admittance is a unit matrix (see Section 3.2.2), we write $\boldsymbol{I}$ here. The supermatrix of $\overline{\mathbf{y}}_{1,2}$ parameter matrices is abbreviated by $\widehat{\overline{\mathbf{y}}}^{\mathrm{AB}}_{\mathrm{AB}}$. Because of the analytic expressions, we can easily write the dependence of $\mathbf{H}_{zy}$ on x:

$$\overline{\mathbf{H}}_{zy}(x) = \frac{\sinh(\boldsymbol{\Gamma}_x(\overline{d}_{\mathrm{AB}} - \overline{x}))}{\sinh(\boldsymbol{\Gamma}_x\overline{d}_{\mathrm{AB}})}\overline{\mathbf{H}}_{zy\mathrm{A}} + \frac{\sinh(\boldsymbol{\Gamma}_x\overline{x})}{\sinh(\boldsymbol{\Gamma}_x\overline{d}_{\mathrm{AB}})}\overline{\mathbf{H}}_{zy\mathrm{B}} \tag{3.34}$$

Note: the arguments of the hyperbolic sine function are diagonal matrices. Since such diagonal matrices can be treated like scalars, we have written fractions here. This equation must now be evaluated at the x values of the discretisation lines in y-direction (i.e. the lines perpendicular to the ones we have already examined). At any such position x_i (see Fig. 3.25) we may write:

$$\overline{\mathbf{H}}_{zy}(x_i) = \boldsymbol{\Lambda}^{\mathrm{d}}_{\mathrm{A}i}\overline{\mathbf{H}}_{zy\mathrm{A}} + \boldsymbol{\Lambda}^{\mathrm{d}}_{\mathrm{B}i}\overline{\mathbf{H}}_{zy\mathrm{B}} \tag{3.35}$$

with diagonal matrices $\mathbf{\Lambda}^{\mathrm{d}}_i$ given by the expressions:

$$\boldsymbol{\Lambda}^{\mathrm{d}}_{\mathrm{A}i} = \sinh(\boldsymbol{\Gamma}_x(\overline{d}_{\mathrm{AB}} - \overline{x}_i))(\sinh(\boldsymbol{\Gamma}_x\overline{d}_{\mathrm{AB}}))^{-1} \tag{3.36}$$

$$\boldsymbol{\Lambda}^{\mathrm{d}}_{\mathrm{B}i} = \sinh(\boldsymbol{\Gamma}_x\overline{x}_i)(\sinh(\boldsymbol{\Gamma}_x\overline{d}_{\mathrm{AB}}))^{-1} \tag{3.37}$$

By multiplying with the transformation matrix $\boldsymbol{T}_{\mathrm{Hy}}$ we obtain: $\mathbf{H}_{zy}(x_i) = \boldsymbol{T}_{\mathrm{Hy}}\overline{\mathbf{H}}_{zy}(x_i)$. These are the values H_{zy} in the points marked by squares □ in Fig. 3.25. $\boldsymbol{T}_{\mathrm{Hy}}$ is the transformation matrix for the discretisation in y-direction (propagation in x-direction). Therefore, $\boldsymbol{T}_{\mathrm{Hy}}$ and $\boldsymbol{\Gamma}_x$ were obtained from the corresponding eigenvalue/eigenvector problem.

Next, we want to determine the values of the field at the boundaries C and D with $y = y_{\mathrm{C}}$ and $y = y_{\mathrm{D}}$. Since we have Neumann conditions, the fields have not been computed directly at the boundaries, but must be extrapolated from points next to the boundaries. This can easily be done with the help of the transformation matrix $\boldsymbol{T}_{\mathrm{Hy}}$. For the position x_i on the left (R ≡ C, $y = y_{\mathrm{C}}$) or right (R ≡ D, $y = y_{\mathrm{D}}$) boundary (marked by ■ at the surfaces C and D), we can write:

$$H_{zy\mathrm{R}}(x_i) = \mathbf{T}^{\mathrm{H}}_{y\Delta\mathrm{R}}\overline{\mathbf{H}}_{zy}(x_i) = \mathbf{T}^{\mathrm{H}}_{y\Delta\mathrm{R}}(\boldsymbol{\Lambda}^{\mathrm{d}}_{\mathrm{Ai}}\overline{\mathbf{H}}_{zy\mathrm{A}} + \boldsymbol{\Lambda}^{\mathrm{d}}_{\mathrm{Bi}}\overline{\mathbf{H}}_{zy\mathrm{B}}) \tag{3.38}$$

R ≡ C(D) denotes the lower (upper) surface of region E. $\mathbf{T}^{\mathrm{H}}_{y\Delta\mathrm{R}}$ is a row vector constructed from weighted first (last) rows of the matrix $\boldsymbol{T}_{\mathrm{Hy}}$. It is constructed as:

$$\mathbf{T}^{\mathrm{H}}_{y\Delta\mathrm{C}} = 0.125(15\mathbf{T}_{\mathrm{Hy1}} - 10\mathbf{T}_{\mathrm{Hy2}} + 3\mathbf{T}_{\mathrm{Hy3}}) \tag{3.39}$$

or:

$$\mathbf{T}^{\mathrm{H}}_{y\Delta\mathrm{D}} = 0.125(15\mathbf{T}_{\mathrm{Hy,N}} - 10\mathbf{T}_{\mathrm{Hy,N-1}} + 3\mathbf{T}_{\mathrm{Hy,N-2}}) \tag{3.40}$$

$\mathbf{T}_{\mathrm{Hy1}}, \mathbf{T}_{\mathrm{Hy2}}$ and $\mathbf{T}_{\mathrm{Hy3}}$ are the first three and $\mathbf{T}_{\mathrm{HyN}}, \mathbf{T}_{\mathrm{Hy,N-1}}$ and $\mathbf{T}_{\mathrm{Hy,N-2}}$ the last three row vectors of the transformation matrix $\boldsymbol{T}_{\mathrm{Hy}}$.

Since we have a vector-matrix product in eq. (3.38), we may change the order of the terms, if we also reshape these terms. $\mathbf{T}^{\mathrm{H}}_{y\Delta\mathrm{R}}$ should be changed to a diagonal matrix $\boldsymbol{T}^{\mathrm{Hd}}_{y\Delta\mathrm{R}}$ and $\boldsymbol{\Lambda}^{\mathrm{d}}_i$ to a row vector $\boldsymbol{\Lambda}_i$. By doing this, we can combine eq. (3.38) for all values $x_i(i = 1, 2 \ldots \mathrm{N}^{\mathrm{E}}_x)$ in one single equation:

$$\mathbf{H}_{zy\mathrm{R}} = \boldsymbol{\Lambda}_{\mathrm{A}} \boldsymbol{T}^{\mathrm{Hd}}_{y\Delta\mathrm{R}} \overline{\mathbf{H}}_{zy\mathrm{A}} + \boldsymbol{\Lambda}_{\mathrm{B}} \boldsymbol{T}^{\mathrm{Hd}}_{y\Delta\mathrm{R}} \overline{\mathbf{H}}_{zy\mathrm{B}} \tag{3.41}$$

$\boldsymbol{\Lambda}_{\mathrm{A}}$ and $\boldsymbol{\Lambda}_{\mathrm{B}}$ are full matrices. Their components are determined in the following way:

$$(\boldsymbol{\Lambda}_{\mathrm{A}})_{ik} = \sinh(\boldsymbol{\Gamma}_{xk}(\overline{d}_{\mathrm{AB}} - \overline{x}_i))(\sinh(\boldsymbol{\Gamma}_{xk}\overline{d}_{\mathrm{AB}}))^{-1} \tag{3.42}$$

$$(\boldsymbol{\Lambda}_{\mathrm{B}})_{ik} = \sinh(\boldsymbol{\Gamma}_{xk}\overline{x}_i)(\sinh(\boldsymbol{\Gamma}_{xk}\overline{d}_{\mathrm{AB}}))^{-1} \tag{3.43}$$

$\boldsymbol{\Gamma}_{xk}$ is the kth component of the diagonal propagation matrix $\boldsymbol{\Gamma}_x$ which we had determined after discretisation in y-direction. At ports C and D, we have waveguides (and therefore also discretisation lines) directed in y-direction. Therefore, we would like to transform the expression in eq. (3.41) into the domain of this waveguide. To do this we multiply with the inverse of the transformation matrix $\boldsymbol{T}_{\mathrm{Hx}}$. ($\boldsymbol{T}_{\mathrm{Hx}}$ is obtained after discretisation of region C (D) and determining the eigenvalues/eigenvectors, as before.) Then we obtain:

$$\overline{\mathbf{H}}_{zy\mathrm{R}} = \boldsymbol{T}^{-1}_{\mathrm{Hx}} \boldsymbol{\Lambda}_{\mathrm{A}} \boldsymbol{T}^{\mathrm{Hd}}_{y\Delta\mathrm{R}} \overline{\mathbf{H}}_{zy\mathrm{A}} + \boldsymbol{T}^{-1}_{\mathrm{Hx}} \boldsymbol{\Lambda}_{\mathrm{B}} \boldsymbol{T}^{\mathrm{Hd}}_{y\Delta\mathrm{R}} \overline{\mathbf{H}}_{zy\mathrm{B}} \tag{3.44}$$

Now the relation between the magnetic fields at the four ports A, B, C and D associated with the discretisation lines in x-direction can be expressed in the following matrix form:

$$\begin{bmatrix} (-\overline{\mathbf{H}}_{zy\mathrm{C}}) \\ -(-\overline{\mathbf{H}}_{zy\mathrm{D}}) \end{bmatrix} = \begin{bmatrix} \overline{\boldsymbol{V}}_{\mathrm{CA}} & \overline{\boldsymbol{V}}_{\mathrm{CB}} \\ \overline{\boldsymbol{V}}_{\mathrm{DA}} & \overline{\boldsymbol{V}}_{\mathrm{DB}} \end{bmatrix} \begin{bmatrix} \overline{\mathbf{H}}_{zy\mathrm{A}} \\ -\overline{\mathbf{H}}_{zy\mathrm{B}} \end{bmatrix} = \widehat{\overline{\boldsymbol{V}}}^{\mathrm{AB}}_{\mathrm{CD}} \begin{bmatrix} \overline{\mathbf{H}}_{zy\mathrm{A}} \\ -\overline{\mathbf{H}}_{zy\mathrm{B}} \end{bmatrix} \tag{3.45}$$

with:

$$\begin{aligned} \overline{\boldsymbol{V}}_{\mathrm{CA}} &= -\boldsymbol{T}^{-1}_{\mathrm{Hx}} \boldsymbol{\Lambda}_{\mathrm{A}} \boldsymbol{T}^{\mathrm{Hd}}_{y\Delta\mathrm{C}} & \overline{\boldsymbol{V}}_{\mathrm{CB}} &= \boldsymbol{T}^{-1}_{\mathrm{Hx}} \boldsymbol{\Lambda}_{\mathrm{B}} \boldsymbol{T}^{\mathrm{Hd}}_{y\Delta\mathrm{C}} \\ \overline{\boldsymbol{V}}_{\mathrm{DA}} &= \boldsymbol{T}^{-1}_{\mathrm{Hx}} \boldsymbol{\Lambda}_{\mathrm{A}} \boldsymbol{T}^{\mathrm{Hd}}_{y\Delta\mathrm{D}} & \overline{\boldsymbol{V}}_{\mathrm{DB}} &= -\boldsymbol{T}^{-1}_{\mathrm{Hx}} \boldsymbol{\Lambda}_{\mathrm{B}} \boldsymbol{T}^{\mathrm{Hd}}_{y\Delta\mathrm{D}} \end{aligned} \tag{3.46}$$

The field $\overline{\mathbf{H}}_{zy}$ does not cause an E_x component at ports C and D because we have introduced electric walls (i.e. Dirichlet boundary conditions for E_x) there. Now using eq. (3.32) we can write, instead of eq. (3.45):

$$\begin{bmatrix} (-\overline{\mathbf{H}}_{zy\mathrm{C}}) \\ -(-\overline{\mathbf{H}}_{zy\mathrm{D}}) \end{bmatrix} = \widehat{\overline{\boldsymbol{V}}}^{\mathrm{AB}}_{\mathrm{CD}} \begin{bmatrix} \overline{\mathbf{H}}_{zy\mathrm{A}} \\ -\overline{\mathbf{H}}_{zy\mathrm{B}} \end{bmatrix} = \widehat{\overline{\boldsymbol{V}}}^{\mathrm{AB}}_{\mathrm{CD}} \widehat{\overline{\boldsymbol{y}}}^{\mathrm{AB}}_{\mathrm{AB}} \begin{bmatrix} \overline{\mathbf{E}}_{yy\mathrm{A}} \\ \overline{\mathbf{E}}_{yy\mathrm{B}} \end{bmatrix} \tag{3.47}$$

Therefore, the left off-diagonal matrix $\widehat{\overline{\boldsymbol{y}}}^{\mathrm{AB}}_{\mathrm{CD}}$ in eq. (3.28) is given by:

$$\widehat{\overline{\boldsymbol{y}}}^{\mathrm{AB}}_{\mathrm{CD}} = \widehat{\overline{\boldsymbol{V}}}^{\mathrm{AB}}_{\mathrm{CD}} \widehat{\overline{\boldsymbol{y}}}^{\mathrm{AB}}_{\mathrm{AB}} \tag{3.48}$$

3.3.1.2 Fields from discretisation in x-direction

In an analogous manner we can determine the submatrices of the right column in eq. (3.28). Here, we short circuit ports A and B by using discretisation lines in y-direction, and obtain:

$$\widehat{\overline{\mathbf{y}}}_{\mathrm{CD}}^{\mathrm{CD}} \qquad \widehat{\overline{\mathbf{y}}}_{\mathrm{AB}}^{\mathrm{CD}} = \widehat{\overline{\mathbf{V}}}_{\mathrm{AB}}^{\mathrm{CD}} \widehat{\overline{\mathbf{y}}}_{\mathrm{CD}}^{\mathrm{CD}} \tag{3.49}$$

3.3.2 H-plane junctions

A general waveguide H-plane junction is drawn in Fig. 3.26. All waveguides should have the same height b in z-direction, whereas the widths a may be different. There is no step in z-direction. The fields are described by the TM$_z$ modes of Section 3.2.2. As shown in Fig. 3.26, region E has interfaces with waveguides where the waves propagate in x- and y-propagation direction. The discretisation lines are orientated in the same direction. We assume that the permittivity in region E is constant. To describe the fields in region E we use linear field combinations. Again, the two field parts are discretised on different discretisation line systems perpendicular to each other. The general region E with crossed lines is sketched in Fig. 3.26. In vertical direction, this region is bounded by planes A and B at y_{A} and y_{B}, and in horizontal direction, by planes C and D at x_{C} and x_{D}, respectively.

3.3.2.1 Field part related to discretisation in x-direction

For the field part with discretisation in x-direction and with discretisation lines and propagation in y-direction we can adapt the solution obtained in Section 3.2.2 for the TM$_z$ modes. We have to substitute u with x, v by y, while z remains z. Therefore, we obtain from eq. (3.21):

$$\begin{bmatrix} \overline{\mathbf{H}}_{xx\mathrm{A}} \\ -\overline{\mathbf{H}}_{xx\mathrm{B}} \end{bmatrix} = \begin{bmatrix} \overline{\mathbf{y}}_1 & \overline{\mathbf{y}}_2 \\ \overline{\mathbf{y}}_2 & \overline{\mathbf{y}}_1 \end{bmatrix}_{|\mathrm{AB}} \begin{bmatrix} \mathbf{E}_{zx\mathrm{A}} \\ \mathbf{E}_{zx\mathrm{B}} \end{bmatrix} = \widehat{\overline{\mathbf{y}}}_{\mathrm{AB}}^{\mathrm{AB}} \begin{bmatrix} \mathbf{E}_{zx\mathrm{A}} \\ \mathbf{E}_{zx\mathrm{B}} \end{bmatrix} \tag{3.50}$$

with:

$$\overline{\mathbf{y}}_1 = \boldsymbol{I}(\tanh(\boldsymbol{\Gamma}_y^e \overline{d}_{\mathrm{AB}}))^{-1} \quad \overline{\mathbf{y}}_2 = -\boldsymbol{I}(\sinh(\boldsymbol{\Gamma}_y^e \overline{d}_{\mathrm{AB}}))^{-1} \tag{3.51}$$

and $\overline{d}_{\mathrm{AB}} = k_0(y_{\mathrm{B}} - y_{\mathrm{A}})$. $\boldsymbol{\Gamma}_y^e$ is the diagonal matrix of propagation constants in y: direction for LSM modes and $\boldsymbol{I}$ the unit matrix of the associated characteristic admittances. Again, the main diagonal submatrix is obtained from the previous analysis.

The coupling from ports A and B to ports C and D is given in this case through the magnetic field component $\mathbf{H}_{yx}$. (The second subscript denotes the direction of discretisation.) The relation between $\mathbf{E}_{zx}$ and $\overline{\mathbf{H}}_{yx}$ is given by:

$$\mathbf{H}_{yx} = -j\varepsilon_r \varepsilon_q^{-1} \overline{\boldsymbol{D}}_{\overline{x}}^e \boldsymbol{T}_{\mathrm{Ex}}^e \overline{\mathbf{E}}_{zx} = \mathbf{S}_x \overline{\mathbf{E}}_{zx} \tag{3.52}$$

The non-quadratic matrix $\boldsymbol{S}_x$ can be seen as a transformation matrix. The component $\mathbf{E}_x$ has normal direction at ports C and D, while $\mathbf{E}_y$ is zero at these ports. At ports A and B the latter component is matched if $\mathbf{E}_z$ is matched. For the y dependence of the electric field $\mathbf{E}_{zx}$ we may write:

$$\overline{\mathbf{E}}_{zx}(y) = \frac{\sinh(\boldsymbol{\Gamma}_y^e(\overline{d}_{\mathrm{AB}} - \overline{y}))}{\sinh(\boldsymbol{\Gamma}_y^e\overline{d}_{\mathrm{AB}})}\overline{\mathbf{E}}_{zx\mathrm{A}} + \frac{\sinh(\boldsymbol{\Gamma}_y^e\overline{y})}{\sinh(\boldsymbol{\Gamma}_y^e\overline{d}_{\mathrm{AB}})}\overline{\mathbf{E}}_{zx\mathrm{B}} \tag{3.53}$$

This equation must now be evaluated at particular positions on the discretisation lines in y-direction. At such a position y_i (see Fig. 3.26), we may write:

$$\overline{\mathbf{E}}_{zx}(y_i) = \boldsymbol{\Lambda}_{\mathrm{A}i}^{\mathrm{d}}\overline{\mathbf{E}}_{zx\mathrm{A}} + \boldsymbol{\Lambda}_{\mathrm{B}i}^{\mathrm{d}}\overline{\mathbf{E}}_{zx\mathrm{B}} \tag{3.54}$$

with diagonal matrices $\boldsymbol{\Lambda}_i^{\mathrm{d}}$ given by the expressions:

$$\boldsymbol{\Lambda}_{\mathrm{A}i}^{\mathrm{d}} = \sinh(\boldsymbol{\Gamma}_y^e(\overline{d}_{\mathrm{AB}} - \overline{y}_i))(\sinh(\boldsymbol{\Gamma}_y^e\overline{d}_{\mathrm{AB}}))^{-1} \tag{3.55}$$

$$\boldsymbol{\Lambda}_{\mathrm{B}i}^{\mathrm{d}} = \sinh(\boldsymbol{\Gamma}_y^e\overline{y}_i)(\sinh(\boldsymbol{\Gamma}_y^e\overline{d}_{\mathrm{AB}}))^{-1} \tag{3.56}$$

The vector $\mathbf{H}_{yx}(y_i) = \mathbf{S}_x\overline{\mathbf{E}}_{zx}(y_i)$ gives the values H_{yx} in the points marked by circles ∘ in Fig. 3.26.

For the matching procedure we need the values of the field at the boundaries C and D with $x = x_{\mathrm{C}}$ and $x = x_{\mathrm{D}}$. Because of the Neumann conditions for H_y we must extrapolate. We take points that are close to the boundary to obtain the ones on the boundary. This can easily be done with the help of the 'transformation' matrix $\mathbf{S}_x$. We obtain at position y_i on the left ($\mathrm{R} \equiv \mathrm{C}$, $x = x_{\mathrm{C}}$) or right ($\mathrm{R} \equiv \mathrm{D}$, $x = x_{\mathrm{D}}$) boundary (marked by • at the surfaces C and D):

$$H_{yx\mathrm{R}}(y_i) = \mathbf{S}_{x\Delta\mathrm{R}}\overline{\mathbf{H}}_{yx}(y_i) = \mathbf{S}_{x\Delta\mathrm{R}}(\boldsymbol{\Lambda}_{\mathrm{Ai}}^{\mathrm{d}}\overline{\mathbf{E}}_{zx\mathrm{A}} + \boldsymbol{\Lambda}_{\mathrm{Bi}}^{\mathrm{d}}\overline{\mathbf{E}}_{zx\mathrm{B}}) \tag{3.57}$$

$\mathrm{R} \equiv \mathrm{C(D)}$ denotes the left (right) surface of region E. $\mathbf{S}_{x\Delta\mathrm{R}}$ is a row vector constructed analogously to $\mathbf{T}_{y\Delta\mathrm{R}}^{\mathrm{H}}$ in Section 3.3.1.1. As before, we would like to change the order of $\mathbf{S}_{x\Delta\mathrm{R}}$ and $\boldsymbol{\Lambda}_i^{\mathrm{d}}$ to get compact expressions. For this purpose, we reshape these quantities to a suitable matrix form. $\mathbf{S}_{x\Delta\mathrm{R}}$ is changed to a diagonal matrix $\mathbf{S}_{x\Delta\mathrm{R}}^{\mathrm{d}}$ and $\boldsymbol{\Lambda}_i^{\mathrm{d}}$ becomes a row vector $\boldsymbol{\Lambda}_i$. By doing this, we can write for all values y_i ($i = 1, 2, \ldots, \mathrm{N}_y^{\mathrm{E}}$):

$$\mathbf{H}_{yx\mathrm{R}} = \boldsymbol{\Lambda}_{\mathrm{A}}\mathbf{S}_{x\Delta\mathrm{R}}^{\mathrm{d}}\overline{\mathbf{E}}_{zx\mathrm{A}} + \boldsymbol{\Lambda}_{\mathrm{B}}\mathbf{S}_{x\Delta\mathrm{R}}^{\mathrm{d}}\overline{\mathbf{E}}_{zx\mathrm{B}} \tag{3.58}$$

The full matrices $\boldsymbol{\Lambda}_{\mathrm{A}}$ and $\boldsymbol{\Lambda}_{\mathrm{B}}$ are given by:

$$(\boldsymbol{\Lambda}_{\mathrm{A}})_{ik} = \sinh(\boldsymbol{\Gamma}_{yk}^e(\overline{d}_{\mathrm{AB}} - \overline{y}_i))(\sinh(\boldsymbol{\Gamma}_{yk}^e\overline{d}_{\mathrm{AB}}))^{-1} \tag{3.59}$$

$$(\boldsymbol{\Lambda}_{\mathrm{B}})_{ik} = \sinh(\boldsymbol{\Gamma}_{yk}^e\overline{y}_i)(\sinh(\boldsymbol{\Gamma}_{yk}^e\overline{d}_{\mathrm{AB}}))^{-1} \tag{3.60}$$

In transformed domain we therefore have:

$$\overline{\mathbf{H}}_{yx\mathrm{R}} = \boldsymbol{T}_{\mathrm{Hy}}^{-1}\mathbf{E}_{zx\mathrm{R}} = \boldsymbol{T}_{\mathrm{Hy}}^{-1}\boldsymbol{\Lambda}_{\mathrm{A}}\mathbf{S}_{x\Delta\mathrm{R}}^{\mathrm{d}}\overline{\mathbf{E}}_{zx\mathrm{A}} + \boldsymbol{T}_{\mathrm{Hy}}^{-1}\boldsymbol{\Lambda}_{\mathrm{B}}\mathbf{S}_{x\Delta\mathrm{R}}^{\mathrm{d}}\overline{\mathbf{E}}_{zx\mathrm{B}} \tag{3.61}$$

We are now in a position to give the expressions for the coupling from ports A and B to ports C and D. With discretisation lines orientated in y-direction we may write the following formula with the supermatrix $\overline{\boldsymbol{y}}_{\mathrm{CD}}^{\mathrm{AB}}$:

$$\begin{bmatrix} (-\overline{\mathbf{H}}_{yx\mathrm{C}}) \\ -(-\overline{\mathbf{H}}_{yx\mathrm{D}}) \end{bmatrix} = \begin{bmatrix} \overline{\boldsymbol{y}}_{\mathrm{CA}} & \overline{\boldsymbol{y}}_{\mathrm{CB}} \\ \overline{\boldsymbol{y}}_{\mathrm{DA}} & \overline{\boldsymbol{y}}_{\mathrm{DB}} \end{bmatrix} \begin{bmatrix} \overline{\mathbf{E}}_{zx\mathrm{A}} \\ \overline{\mathbf{E}}_{zx\mathrm{B}} \end{bmatrix} = \overline{\boldsymbol{y}}_{\mathrm{CD}}^{\mathrm{AB}} \begin{bmatrix} \overline{\mathbf{E}}_{zx\mathrm{A}} \\ \overline{\mathbf{E}}_{zx\mathrm{B}} \end{bmatrix} \tag{3.62}$$

The submatrices are determined as:

$$\begin{aligned} \overline{\boldsymbol{y}}_{\mathrm{CA}} &= -\boldsymbol{T}_{\mathrm{Hy}}^{-1}\boldsymbol{\Lambda}_{\mathrm{A}}\boldsymbol{S}_{x\Delta\mathrm{C}}^{\mathrm{d}} & \overline{\boldsymbol{y}}_{\mathrm{CB}} &= -\boldsymbol{T}_{\mathrm{Hy}}^{-1}\boldsymbol{\Lambda}_{\mathrm{B}}\boldsymbol{S}_{x\Delta\mathrm{C}}^{\mathrm{d}} \\ \overline{\boldsymbol{y}}_{\mathrm{DA}} &= \boldsymbol{T}_{\mathrm{Hy}}^{-1}\boldsymbol{\Lambda}_{\mathrm{A}}\boldsymbol{S}_{x\Delta\mathrm{D}}^{\mathrm{d}} & \overline{\boldsymbol{y}}_{\mathrm{DB}} &= \boldsymbol{T}_{\mathrm{Hy}}^{-1}\boldsymbol{\Lambda}_{\mathrm{B}}\boldsymbol{S}_{x\Delta\mathrm{D}}^{\mathrm{d}} \end{aligned} \tag{3.63}$$

3.3.2.2 Fields from discretisation in y-direction

Analogously, we obtain the submatrices of the right column in eq. (3.28):

$$\widehat{\overline{\boldsymbol{y}}}_{\mathrm{CD}}^{\mathrm{CD}} \qquad \widehat{\overline{\boldsymbol{y}}}_{\mathrm{AB}}^{\mathrm{CD}} = \widehat{\overline{\boldsymbol{V}}}_{\mathrm{AB}}^{\mathrm{CD}}\widehat{\overline{\boldsymbol{y}}}_{\mathrm{CD}}^{\mathrm{CD}} \tag{3.64}$$

Here, we short circuit ports A and B and use discretisation lines in x-direction.

3.3.3 Algorithm for generalised scattering parameters

Knowing the sources and the loads at the ports, all the fields at the ports can be computed. Let us define the admittance matrices $\overline{\boldsymbol{Z}}_{\mathrm{A}}$, $\overline{\boldsymbol{Z}}_{\mathrm{B}}$, $\overline{\boldsymbol{Z}}_{\mathrm{C}}$ and $\overline{\boldsymbol{Z}}_{\mathrm{D}}$ at the inner side of the ports (see Figs. 3.25 and 3.26) by:

$$\begin{aligned} \overline{\mathbf{E}}_{\mathrm{A}} &= \overline{\boldsymbol{Z}}_{\mathrm{A}}\overline{\mathbf{H}}_{\mathrm{A}} & \overline{\mathbf{E}}_{\mathrm{B}} &= \overline{\boldsymbol{Z}}_{\mathrm{B}}(-\overline{\mathbf{H}}_{\mathrm{B}}) \\ \overline{\mathbf{E}}_{\mathrm{C}} &= \overline{\boldsymbol{Z}}_{\mathrm{C}}\overline{\mathbf{H}}_{\mathrm{C}} & \overline{\mathbf{E}}_{\mathrm{D}} &= \overline{\boldsymbol{Z}}_{\mathrm{D}}(-\overline{\mathbf{H}}_{\mathrm{D}}) \end{aligned} \tag{3.65}$$

If e.g. the connecting waveguide $\mathrm{B_c}$ is treated as a load, with input impedance $\overline{\boldsymbol{Z}}_{\mathrm{B_c}} = \overline{\boldsymbol{Y}}_{\mathrm{B_c}}^{-1}$, then we obtain with eq. (2.164) in Chapter 2.5:

$$\overline{\boldsymbol{Z}}_{\mathrm{B}} = -(\boldsymbol{T}_{\mathrm{EB}})_{\mathrm{c}}^{-1}((\boldsymbol{T}_{\mathrm{HB_c}})_{\mathrm{c}}\overline{\boldsymbol{Y}}_{\mathrm{B_c}}(\boldsymbol{T}_{\mathrm{HB_c}})_{\mathrm{c}}^{-1})^{-1}(\boldsymbol{T}_{\mathrm{EB}})_{\mathrm{c}} \quad \overline{\boldsymbol{Y}}_{\mathrm{B}} = \overline{\boldsymbol{Z}}_{\mathrm{B}}^{-1} \tag{3.66}$$

The minus sign is required because the impedances are defined for opposite directions. In the following, we assume that the source is connected to the waveguide $\mathrm{A_c}$. The input impedance matrix $\overline{\boldsymbol{Z}}_{\mathrm{A}}$ has to be determined from (3.30)[1] by the algorithm described in the following. We first combine the port impedance matrices according to:

$$\overline{\boldsymbol{Z}}_{\mathrm{AB}} = \mathrm{Diag}(\overline{\boldsymbol{Z}}_{\mathrm{A}}, \overline{\boldsymbol{Z}}_{\mathrm{B}}) \qquad \overline{\boldsymbol{Z}}_{\mathrm{CD}} = \mathrm{Diag}(\overline{\boldsymbol{Z}}_{\mathrm{C}}, \overline{\boldsymbol{Z}}_{\mathrm{D}}) \tag{3.67}$$

Instead of eq. (3.30), we can therefore write:

$$\mathbf{0} = \begin{bmatrix} \overline{\boldsymbol{z}}_{\mathrm{AB}}^{\mathrm{AB}} - \overline{\boldsymbol{Z}}_{\mathrm{AB}} & \overline{\boldsymbol{z}}_{\mathrm{CD}}^{\mathrm{AB}} \\ \overline{\boldsymbol{z}}_{\mathrm{AB}}^{\mathrm{CD}} & \overline{\boldsymbol{z}}_{\mathrm{CD}}^{\mathrm{CD}} + \overline{\boldsymbol{Z}}_{\mathrm{CD}} \end{bmatrix} \begin{bmatrix} \widehat{\overline{\mathbf{H}}}_{\mathrm{AB}} \\ -\widehat{\overline{\mathbf{H}}}_{\mathrm{CD}} \end{bmatrix} \tag{3.68}$$

[1]We may write such an expression also for H bends.

Introducing the second equation of this system into the first results in:

$$\overline{\mathbf{Z}}_{\mathrm{AB}}\overline{\mathbf{H}}_{\mathrm{AB}} = (\overline{\mathbf{z}}_{\mathrm{AB}}^{\mathrm{AB}} - \overline{\mathbf{z}}_{\mathrm{CD}}^{\mathrm{AB}}(\overline{\mathbf{z}}_{\mathrm{CD}}^{\mathrm{CD}} + \overline{\mathbf{Z}}_{\mathrm{CD}})^{-1}\overline{\mathbf{z}}_{\mathrm{AB}}^{\mathrm{CD}})\overline{\mathbf{H}}_{\mathrm{AB}} \tag{3.69}$$

If we replace the term in parentheses a matrix, we can write:

$$\begin{bmatrix} \overline{\mathbf{Z}}_{\mathrm{A}}\overline{\mathbf{H}}_{\mathrm{A}} \\ -\overline{\mathbf{Z}}_{\mathrm{B}}\overline{\mathbf{H}}_{\mathrm{B}} \end{bmatrix} = \begin{bmatrix} \overline{\mathbf{z}}_{\mathrm{AA}}^{\mathrm{n}} & \overline{\mathbf{z}}_{\mathrm{AB}}^{\mathrm{n}} \\ \overline{\mathbf{z}}_{\mathrm{BA}}^{\mathrm{n}} & \overline{\mathbf{z}}_{\mathrm{BB}}^{\mathrm{n}} \end{bmatrix} \begin{bmatrix} \overline{\mathbf{H}}_{\mathrm{A}} \\ -\overline{\mathbf{H}}_{\mathrm{B}} \end{bmatrix} \tag{3.70}$$

Then, we obtain for the impedance $\overline{\mathbf{Z}}_{\mathrm{A}}$:

$$\overline{\mathbf{Z}}_{\mathrm{A}} = \overline{\mathbf{z}}_{\mathrm{AA}}^{\mathrm{n}} - \overline{\mathbf{z}}_{\mathrm{AB}}^{\mathrm{n}}(\overline{\mathbf{z}}_{\mathrm{BB}}^{\mathrm{n}} + (-\overline{\mathbf{Z}}_{\mathrm{B}}))^{-1}\overline{\mathbf{z}}_{\mathrm{BA}}^{\mathrm{n}} \tag{3.71}$$

The last equation again has the known form. From $\overline{\mathbf{Y}}_{\mathrm{A}} = \overline{\mathbf{Z}}_{\mathrm{A}}^{-1}$, the input impedance is calculated using eq. (2.164) in Chapter 2.5 according to:

$$\overline{\mathbf{Z}}_{\mathrm{IA_c}} = (\boldsymbol{T}_{\mathrm{EA_c}})_{\mathrm{c}}^{-1}((\boldsymbol{T}_{\mathrm{HA}})_{\mathrm{c}}\overline{\mathbf{Y}}_{\mathrm{A}}(\boldsymbol{T}_{\mathrm{EA}})_{\mathrm{c}}^{-1})^{-1}(\boldsymbol{T}_{\mathrm{HA_c}})_{\mathrm{c}} \quad \overline{\mathbf{Y}}_{\mathrm{IA_c}} = \overline{\mathbf{Z}}_{\mathrm{IA_c}}^{-1} = -\overline{\mathbf{Y}}_{\mathrm{A_c}} \tag{3.72}$$

With the input impedance matrix $\overline{\mathbf{Y}}_{\mathrm{IA_c}}$ and the source, the fields $\overline{\mathbf{E}}_{\mathrm{A_c}}$ and $\overline{\mathbf{H}}_{\mathrm{A_c}}$ can be calculated as described in Section 2.5.4. From $\overline{\mathbf{E}}_{\mathrm{A_c}}$ and $\overline{\mathbf{H}}_{\mathrm{A_c}}$ we obtain $\overline{\mathbf{E}}_{\mathrm{A}}$ and $\overline{\mathbf{H}}_{\mathrm{A}}$ according to the equations in Section 3.2.3. With $\overline{\mathbf{E}}_{\mathrm{A}}$ and $\overline{\mathbf{H}}_{\mathrm{A}}$ we obtain $\overline{\mathbf{E}}_{\mathrm{B}}$ and $\overline{\mathbf{H}}_{\mathrm{B}}$ with eqs. (3.70) and (3.65). Then $\overline{\mathbf{E}}_{\mathrm{CD}}$ and $\overline{\mathbf{H}}_{\mathrm{CD}}$ may be obtained from eqs. (3.68) and (3.65) and, in the next step, the fields in the connecting guides. By using different source modes and repeating the procedure for the other ports, the resulting generalised scattering matrix can be determined.

3.3.4 Special junctions: E-plane 3-port junction

In this subsection we would like to specify the analysis procedure for the asymmetric junction in Fig. 3.27. Introducing $\mathbf{E}_{\mathrm{C}} = 0$ into eq. (3.28) results in:

$$\begin{bmatrix} \overline{\mathbf{H}}_{\mathrm{A}} \\ -\overline{\mathbf{H}}_{\mathrm{B}} \\ \overline{\mathbf{H}}_{\mathrm{D}} \end{bmatrix} = \begin{bmatrix} \begin{bmatrix} \overline{\boldsymbol{y}}_1 & \overline{\boldsymbol{y}}_2 \\ \overline{\boldsymbol{y}}_2 & \overline{\boldsymbol{y}}_1 \end{bmatrix}_{\mathrm{AB}} & \begin{matrix} \overline{\boldsymbol{y}}_{\mathrm{AD}} \\ \overline{\boldsymbol{y}}_{\mathrm{BD}} \end{matrix} \\ \overline{\boldsymbol{y}}_{\mathrm{DA}} \quad \overline{\boldsymbol{y}}_{\mathrm{DB}} & \overline{\boldsymbol{y}}_{1\mathrm{CD}} \end{bmatrix} \begin{bmatrix} \overline{\mathbf{E}}_{\mathrm{A}} \\ \overline{\mathbf{E}}_{\mathrm{B}} \\ \overline{\mathbf{E}}_{\mathrm{D}} \end{bmatrix} \tag{3.73}$$

By taking into account $\overline{\mathbf{H}}_{\mathrm{D}} = -\overline{\mathbf{Y}}_{\mathrm{D}}\overline{\mathbf{E}}_{\mathrm{D}}$ (for definition, see Fig. 3.25) we obtain from the third equation:

$$\overline{\mathbf{E}}_{\mathrm{D}} = -(\overline{\boldsymbol{y}}_{1\mathrm{CD}} + \overline{\mathbf{Y}}_{\mathrm{D}})^{-1}[\overline{\boldsymbol{y}}_{\mathrm{DA}} \quad \overline{\boldsymbol{y}}_{\mathrm{DB}}] \begin{bmatrix} \overline{\mathbf{E}}_{\mathrm{A}} \\ \overline{\mathbf{E}}_{\mathrm{B}} \end{bmatrix} = \begin{bmatrix} \overline{\mathbf{V}}_{\mathrm{DA}}^{n} & \overline{\mathbf{V}}_{\mathrm{DB}}^{n} \end{bmatrix} \begin{bmatrix} \overline{\mathbf{E}}_{\mathrm{A}} \\ \overline{\mathbf{E}}_{\mathrm{B}} \end{bmatrix} \tag{3.74}$$

For infinite long waveguide at port D, and without step, we have $\overline{\mathbf{Y}}_{\mathrm{D}} = -\overline{\mathbf{Y}}_{0\mathrm{D_c}} = -\boldsymbol{I}$). Now eq. (3.74) has to be introduced into the first two parts of eq. (3.73):

$$\begin{bmatrix} \overline{\mathbf{H}}_{\mathrm{A}} \\ -\overline{\mathbf{H}}_{\mathrm{B}} \end{bmatrix} = \left[\begin{bmatrix} \overline{\boldsymbol{y}}_1 & \overline{\boldsymbol{y}}_2 \\ \overline{\boldsymbol{y}}_2 & \overline{\boldsymbol{y}}_1 \end{bmatrix}_{\mathrm{AB}} + \begin{bmatrix} \overline{\boldsymbol{y}}_{\mathrm{AD}} \\ \overline{\boldsymbol{y}}_{\mathrm{BD}} \end{bmatrix} \begin{bmatrix} \overline{\mathbf{V}}_{\mathrm{DA}}^{n} & \overline{\mathbf{V}}_{\mathrm{DB}}^{n} \end{bmatrix} \right] \begin{bmatrix} \overline{\mathbf{E}}_{\mathrm{A}} \\ \overline{\mathbf{E}}_{\mathrm{B}} \end{bmatrix} \tag{3.75}$$

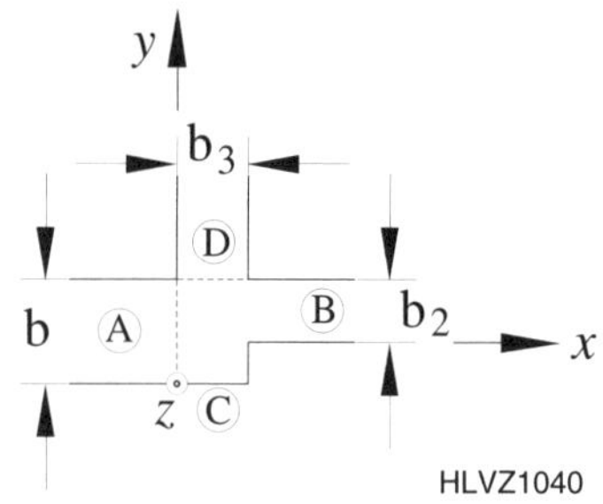

Fig. 3.27 Asymmetric E-plane T-junction

In shorter form, we may write:

$$\begin{bmatrix} \overline{\mathbf{H}}_{\mathrm{A}} \\ -\overline{\mathbf{H}}_{\mathrm{B}} \end{bmatrix} = \begin{bmatrix} \overline{\boldsymbol{y}}_{\mathrm{AA}} & \overline{\boldsymbol{y}}_{\mathrm{AB}} \\ \overline{\boldsymbol{y}}_{\mathrm{BA}} & \overline{\boldsymbol{y}}_{\mathrm{BB}} \end{bmatrix} \begin{bmatrix} \overline{\mathbf{E}}_{\mathrm{A}} \\ \overline{\mathbf{E}}_{\mathrm{B}} \end{bmatrix} \tag{3.76}$$

With the load impedance $\overline{\mathbf{Z}}_{\mathrm{B}}$ according to $\overline{\mathbf{E}}_{\mathrm{B}} = -\overline{\mathbf{Z}}_{\mathrm{B}}\overline{\mathbf{H}}_{\mathrm{B}}$ we obtain:

$$\boxed{\overline{\mathbf{Y}}_{\mathrm{A}} = \overline{\boldsymbol{y}}_{\mathrm{AA}} - \overline{\boldsymbol{y}}_{\mathrm{AB}}(\overline{\boldsymbol{y}}_{\mathrm{BB}} - \overline{\mathbf{Z}}_{\mathrm{B}}^{-1})^{-1}\overline{\boldsymbol{y}}_{\mathrm{BA}}} \tag{3.77}$$

By introducing $\mathbf{E}_{\mathrm{B}} = \mathbf{0}$ into system (3.73) we obtain the equations for the E-plane corner according to:

$$\begin{bmatrix} \overline{\mathbf{H}}_{\mathrm{A}} \\ \overline{\mathbf{H}}_{\mathrm{D}} \end{bmatrix} = \begin{bmatrix} \overline{\boldsymbol{y}}_{1\mathrm{AB}} & \overline{\boldsymbol{y}}_{\mathrm{AD}} \\ \overline{\boldsymbol{y}}_{\mathrm{DA}} & \overline{\boldsymbol{y}}_{1\mathrm{CD}} \end{bmatrix} \begin{bmatrix} \overline{\mathbf{E}}_{\mathrm{A}} \\ \overline{\mathbf{E}}_{\mathrm{D}} \end{bmatrix} \tag{3.78}$$

The input impedance can be calculated analogously to eq. (3.77).

3.3.5 Matched E-plane bend

In this section we would like to show how the concept of crossed lines can be used even in the case of inhomogeneous common regions. As an example, we will analyse the matched E-plane bend shown in Fig. 3.28.

3.3.5.1 Fields on discretisation lines in x-direction

Due to the inhomogeneity of the common regions in x-direction, we must divide the structure into two subregions I and II. These subregions are separated by plane F, as shown in Fig. 3.28a. Owing to the short circuit in plane B, the magnetic field at the horizontal lines in subregion I can be written as follows:

$$\overline{\mathbf{H}}^{\mathrm{I}}_{zy}(x) = \frac{\cosh(\boldsymbol{\Gamma}^{\mathrm{I}}_{x}(\bar{b}_{\mathrm{D}} - \overline{x}))}{\cosh(\boldsymbol{\Gamma}^{\mathrm{I}}_{x}\bar{d}_{x})}\overline{\mathbf{H}}^{\mathrm{I}}_{zy\mathrm{F}} \tag{3.79}$$

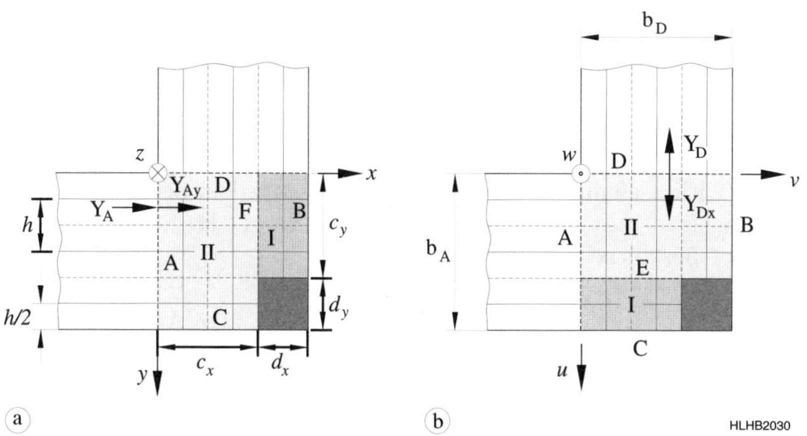

Fig. 3.28 Matched E-plane bend with discretisation lines (Reproduced by permission of Elsevier)

Note: as before, we have diagonal matrices in numerator and denominator, therefore we may treat these matrices as scalars. The impedance in plane F is given as:

$$\overline{\mathbf{Z}}^{\mathrm{I}}_{\mathrm{F}} = \boldsymbol{I}(\tanh(\boldsymbol{\Gamma}^{\mathrm{I}}_{x}\overline{d}_{x}))^{-1} \tag{3.80}$$

In subregion II we adapt eq. (3.32) and obtain:

$$\begin{bmatrix} \overline{\mathbf{E}}^{\mathrm{II}}_{yy\mathrm{A}} \\ \overline{\mathbf{E}}^{\mathrm{II}}_{yy\mathrm{F}} \end{bmatrix} = \begin{bmatrix} \overline{\mathbf{z}}_1 & \overline{\mathbf{z}}_2 \\ \overline{\mathbf{z}}_2 & \overline{\mathbf{z}}_1 \end{bmatrix}_{|\mathrm{II}} \begin{bmatrix} \overline{\mathbf{H}}^{\mathrm{II}}_{zy\mathrm{A}} \\ -\overline{\mathbf{H}}^{\mathrm{II}}_{zy\mathrm{F}} \end{bmatrix} \tag{3.81}$$

With:

$$\overline{\mathbf{E}}^{\mathrm{II}}_{yy\mathrm{F}} = \overline{\mathbf{Z}}^{\mathrm{II}}_{\mathrm{F}}\overline{\mathbf{H}}^{\mathrm{II}}_{zy\mathrm{F}} \tag{3.82}$$

where (see eq. (3.26)):

$$\overline{\mathbf{Z}}^{\mathrm{II}}_{\mathrm{F}} = (\boldsymbol{T}^{\mathrm{II}}_{\mathrm{Ey}})^{-1}\boldsymbol{T}^{\mathrm{I}}_{\mathrm{Ey}}\overline{\mathbf{Z}}^{\mathrm{I}}_{\mathrm{F}}(\boldsymbol{T}^{\mathrm{I}}_{\mathrm{Hy}})^{-1}(\boldsymbol{T}^{\mathrm{II}}_{\mathrm{Hy}})_c \tag{3.83}$$

we obtain from the second expression in eq. (3.81):

$$\overline{\mathbf{H}}^{\mathrm{II}}_{zy\mathrm{F}} = (\overline{\mathbf{z}}^{\mathrm{II}}_1 + \overline{\mathbf{Z}}^{\mathrm{II}}_{\mathrm{F}})^{-1}\overline{\mathbf{z}}^{\mathrm{II}}_2\overline{\mathbf{H}}_{zy\mathrm{A}} = \overline{\mathbf{V}}^{\mathrm{II}}_{\mathrm{FA}}\overline{\mathbf{H}}_{zy\mathrm{A}} \tag{3.84}$$

The input impedance for this field part in plane A is therefore given as:

$$\overline{\mathbf{Z}}_{\mathrm{Ay}} = \overline{\mathbf{z}}^{\mathrm{II}}_1 - \overline{\mathbf{z}}^{\mathrm{II}}_2(\overline{\mathbf{z}}^{\mathrm{II}}_1 + \overline{\mathbf{Z}}^{\mathrm{II}}_{\mathrm{F}})^{-1}\overline{\mathbf{z}}^{\mathrm{II}}_2 = \overline{\mathbf{Y}}^{-1}_{\mathrm{Ay}} \tag{3.85}$$

The magnetic field vector $\overline{\mathbf{H}}^{\mathrm{I}}_{zy\mathrm{F}}$ as a function of $\overline{\mathbf{H}}^{\mathrm{II}}_{zy\mathrm{F}}$ or $\overline{\mathbf{H}}_{zy\mathrm{A}}$ is obtained with the help of eq. (3.84):

$$\overline{\mathbf{H}}^{\mathrm{I}}_{zy\mathrm{F}} = (\boldsymbol{T}^{\mathrm{I}}_{\mathrm{Hy}})^{-1}(\boldsymbol{T}^{\mathrm{II}}_{\mathrm{Hy}})_c\overline{\mathbf{H}}^{\mathrm{II}}_{zy\mathrm{F}} = (\boldsymbol{T}^{\mathrm{I}}_{\mathrm{Hy}})^{-1}(\boldsymbol{T}^{\mathrm{II}}_{\mathrm{Hy}})_c\overline{\boldsymbol{V}}^{\mathrm{II}}_{\mathrm{FA}}\overline{\mathbf{H}}_{zy\mathrm{A}} = \overline{\boldsymbol{V}}^{\mathrm{I}}_{\mathrm{FA}}\overline{\mathbf{H}}_{zy\mathrm{A}} \tag{3.86}$$

Now, with a similar procedure to that described in Section 3.3.1.1, we obtain the field part $\overline{\mathbf{H}}_{zy\mathrm{D}}$ in plane D from $\overline{\mathbf{H}}_{zy\mathrm{A}}$. $\overline{\mathbf{H}}_{zy\mathrm{D}}$ consists of two subvectors due to the two subregions:

$$\overline{\mathbf{H}}_{zy\mathrm{D}} = \overline{\boldsymbol{V}}_{\mathrm{DA}}\overline{\mathbf{H}}_{zy\mathrm{A}} \qquad \overline{\boldsymbol{V}}_{\mathrm{DA}} = (\boldsymbol{T}^{\mathrm{I}}_{\mathrm{Hx}})^{-1}\begin{bmatrix} \boldsymbol{\Lambda}_{\mathrm{A}}\boldsymbol{T}^{\mathrm{HIId}}_{y\Delta\mathrm{D}} + \boldsymbol{\Lambda}^{\mathrm{II}}_{\mathrm{F}}\boldsymbol{T}^{\mathrm{HIId}}_{y\Delta\mathrm{D}}\overline{\boldsymbol{V}}^{\mathrm{II}}_{\mathrm{FA}} \\ \boldsymbol{\Lambda}^{\mathrm{I}}_{\mathrm{F}}\boldsymbol{T}^{\mathrm{HId}}_{y\Delta\mathrm{D}}\overline{\boldsymbol{V}}^{\mathrm{I}}_{\mathrm{FA}} \end{bmatrix} \tag{3.87}$$

where:

$$(\boldsymbol{\Lambda}_{\mathrm{A}})_{ik} = \sinh\big(\boldsymbol{\Gamma}^{\mathrm{II}}_{xk}(\bar{b}_{\mathrm{D}} - \bar{d}_x - \bar{x}_i)\big)\big(\sinh\big(\boldsymbol{\Gamma}^{\mathrm{II}}_{xk}(\bar{b}_{\mathrm{D}} - \bar{d}_x)\big)\big)^{-1} \tag{3.88}$$

$$(\boldsymbol{\Lambda}^{\mathrm{II}}_{\mathrm{F}})_{ik} = \sinh\big(\boldsymbol{\Gamma}^{\mathrm{II}}_{xk}\bar{x}_i\big)\big(\sinh\big(\boldsymbol{\Gamma}^{\mathrm{II}}_{xk}(\bar{b}_{\mathrm{D}} - \bar{d}_x)\big)\big)^{-1} \tag{3.89}$$

$$(\boldsymbol{\Lambda}^{\mathrm{I}}_{\mathrm{F}})_{ik} = \cosh\big(\boldsymbol{\Gamma}^{\mathrm{I}}_{xk}(\bar{b}_{\mathrm{D}} - \bar{x}_i)\big)\big(\cosh\big(\boldsymbol{\Gamma}^{\mathrm{I}}_{xk}\bar{d}_x\big)\big)^{-1} \tag{3.90}$$

3.3.5.2 Fields on discretisation lines in y-direction

Now we divide the common region in y-direction into two subregions I and II. These are separated by plane E, as shown in Fig. 3.28b. Using the coordinates u, v and w instead of x, y and z we can directly adapt all the equations derived in the previous subsection. Then we can go back immediately to the coordinates x and y. Therefore, we obtain:

$$\overline{\mathbf{H}}_{zx\mathrm{A}} = \overline{\boldsymbol{V}}_{\mathrm{AD}}\overline{\mathbf{H}}_{zx\mathrm{D}} \qquad \overline{\boldsymbol{V}}_{\mathrm{AD}} = \boldsymbol{T}^{-1}_{\mathrm{Hy}}\begin{bmatrix} \boldsymbol{\Lambda}_{\mathrm{D}}\boldsymbol{T}^{\mathrm{HIId}}_{x\Delta\mathrm{A}} + \boldsymbol{\Lambda}^{\mathrm{II}}_{\mathrm{E}}\boldsymbol{T}^{\mathrm{HIId}}_{x\Delta\mathrm{A}}\overline{\boldsymbol{V}}^{\mathrm{II}}_{\mathrm{ED}} \\ \boldsymbol{\Lambda}^{\mathrm{I}}_{\mathrm{E}}\boldsymbol{T}^{\mathrm{HId}}_{x\Delta\mathrm{A}}\overline{\boldsymbol{V}}^{\mathrm{I}}_{\mathrm{ED}} \end{bmatrix} \tag{3.91}$$

where:

$$(\boldsymbol{\Lambda}_{\mathrm{D}})_{ik} = \sinh\big(\boldsymbol{\Gamma}^{\mathrm{II}}_{yk}(\bar{b}_{\mathrm{A}} - \bar{d}_y - \bar{y}_i)\big)\big(\sinh\big(\boldsymbol{\Gamma}^{\mathrm{II}}_{yk}(\bar{b}_{\mathrm{A}} - \bar{d}_y)\big)\big)^{-1} \tag{3.92}$$

$$(\boldsymbol{\Lambda}^{\mathrm{II}}_{\mathrm{E}})_{ik} = \sinh\big(\boldsymbol{\Gamma}^{\mathrm{II}}_{yk}\bar{y}_i\big)\big(\sinh\big(\boldsymbol{\Gamma}^{\mathrm{II}}_{yk}(\bar{b}_{\mathrm{A}} - \bar{d}_y)\big)\big)^{-1} \tag{3.93}$$

$$(\boldsymbol{\Lambda}^{\mathrm{I}}_{\mathrm{E}})_{ik} = \cosh\big(\boldsymbol{\Gamma}^{\mathrm{I}}_{yk}(\bar{b}_{\mathrm{A}} - \bar{y}_i)\big)\big(\cosh\big(\boldsymbol{\Gamma}^{\mathrm{I}}_{yk}\bar{d}_y\big)\big)^{-1} \tag{3.94}$$

These equations are written in such a form that the components of $\overline{\mathbf{H}}_{zx\mathrm{A}}$ have the same order as those of $\overline{\mathbf{H}}_{zy\mathrm{A}}$.

The input impedance $\overline{\mathbf{Z}}_{\mathrm{Dx}}(=\overline{\boldsymbol{Y}}^{-1}_{\mathrm{Dx}})$ for this part of the field in plane D can be calculated by a formula analogous to eq. (3.85). The equations for $\overline{\boldsymbol{V}}^{\mathrm{I}}_{\mathrm{ED}}$ and $\overline{\boldsymbol{V}}^{\mathrm{II}}_{\mathrm{ED}}$ are also similar to those for $\overline{\boldsymbol{V}}^{\mathrm{I}}_{\mathrm{FA}}$ and $\overline{\boldsymbol{V}}^{\mathrm{II}}_{\mathrm{FA}}$, respectively. The expressions for $\overline{\boldsymbol{V}}^{\mathrm{I,II}}_{\mathrm{ED}}$ and $\overline{\boldsymbol{Z}}^{\mathrm{I,II}}_{\mathrm{ED}}$ are constructed like eqs. (3.82)–(3.87).

3.3.5.3 System equation

The total magnetic field at ports A and D consists of contributions from both discretisation directions and is obtained by superposition. Therefore, we have:

$$\begin{bmatrix} \overline{\mathbf{H}}_{z\mathrm{A}} \\ \overline{\mathbf{H}}_{z\mathrm{D}} \end{bmatrix} = \begin{bmatrix} \boldsymbol{I}_x & \overline{\boldsymbol{V}}_{\mathrm{AD}} \\ \overline{\boldsymbol{V}}_{\mathrm{DA}} & \boldsymbol{I}_y \end{bmatrix} \begin{bmatrix} \overline{\mathbf{H}}_{zy\mathrm{A}} \\ \overline{\mathbf{H}}_{zx\mathrm{D}} \end{bmatrix} \tag{3.95}$$

To determine the relation of the magnetic fields on the crossed discretisation lines we introduced electric walls at the other port. Therefore, the total electric field only has contributions from one part of the magnetic field. We have:

$$\overline{\mathbf{E}}_{y\mathrm{A}} = \overline{\boldsymbol{Z}}_{\mathrm{A}y}\overline{\mathbf{H}}_{zy\mathrm{A}} \quad \overline{\mathbf{E}}_{x\mathrm{D}} = \overline{\boldsymbol{Z}}_{\mathrm{D}x}(-\overline{\mathbf{H}}_{zx\mathrm{D}}) \tag{3.96}$$

Therefore, we obtain:

$$\begin{bmatrix} \overline{\mathbf{H}}_{z\mathrm{A}} \\ \overline{\mathbf{H}}_{z\mathrm{D}} \end{bmatrix} = \begin{bmatrix} \overline{\boldsymbol{Y}}_{\mathrm{Ay}} & -\overline{\boldsymbol{V}}_{\mathrm{AD}}\overline{\boldsymbol{Y}}_{\mathrm{Dx}} \\ \overline{\boldsymbol{V}}_{\mathrm{DA}}\overline{\boldsymbol{Y}}_{\mathrm{Ay}} & -\overline{\boldsymbol{Y}}_{\mathrm{Dx}} \end{bmatrix} \begin{bmatrix} \overline{\mathbf{E}}_{y\mathrm{A}} \\ \overline{\mathbf{E}}_{x\mathrm{D}} \end{bmatrix} \tag{3.97}$$

We can now determine a relation between input admittance in plane A and load admittance in plane D: $\overline{\mathbf{H}}_{z\mathrm{A}} = \overline{\mathbf{Y}}_{\mathrm{A}}\overline{\mathbf{E}}_{y\mathrm{A}}$, $\overline{\mathbf{H}}_{z\mathrm{D}} = \overline{\mathbf{Y}}_{\mathrm{D}}\overline{\mathbf{E}}_{\mathrm{xD}}$. The latter is the input admittance of the straight waveguide section D, which can be e.g. the characteristic admittance matrix $\overline{\mathbf{Y}}_{0\mathrm{D}}$. We obtain:

$$\overline{\boldsymbol{Y}}_{\mathrm{A}} = \overline{\boldsymbol{Y}}_{\mathrm{A}y} - \overline{\boldsymbol{V}}_{\mathrm{AD}}\overline{\boldsymbol{Y}}_{\mathrm{Dx}}(\overline{\boldsymbol{Y}}_{\mathrm{D}} + \overline{\boldsymbol{Y}}_{\mathrm{Dx}})^{-1}\overline{\boldsymbol{V}}_{\mathrm{DA}}\overline{\boldsymbol{Y}}_{\mathrm{A}y} \tag{3.98}$$

3.3.6 Analysis of waveguide bend discontinuities

In Section 3.3.3 we analysed 90° waveguide bends. Now we would like to examine waveguide bends with a bend angle α less then 90° (see Fig. 3.29). Again we have to distinguish between bends in E-plane and bends in H-plane (see Figs. 3.30 and 3.31).

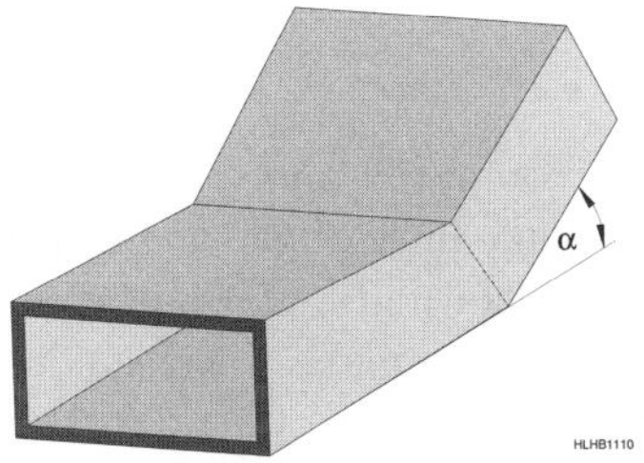

Fig. 3.29 E-plane bend discontinuity (Reproduced by permission of IGTE (Institut für Grundlagen und Theorie der Elektrotechnik) Graz)

For the analysis, the cross-section and the fields are discretised in the plane of the bend. As before, the field is described by trigonometric functions in the perpendicular direction (vertical direction in Fig. 3.29). In the direction of

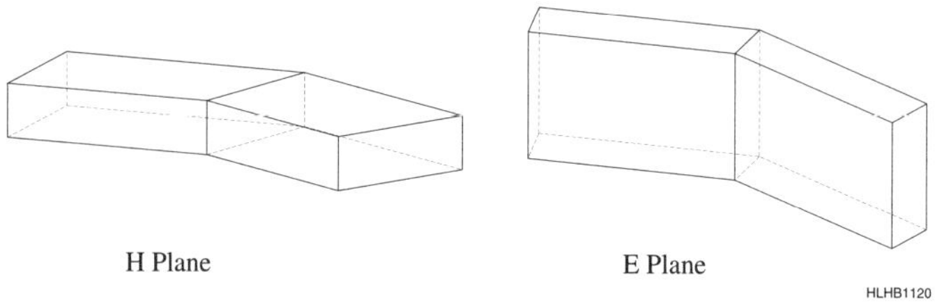

Fig. 3.30 Sharp rectangular waveguide bends of arbitrary angles (Reproduced by permission of IGTE (Institut für Grundlagen und Theorie der Elektrotechnik) Graz)

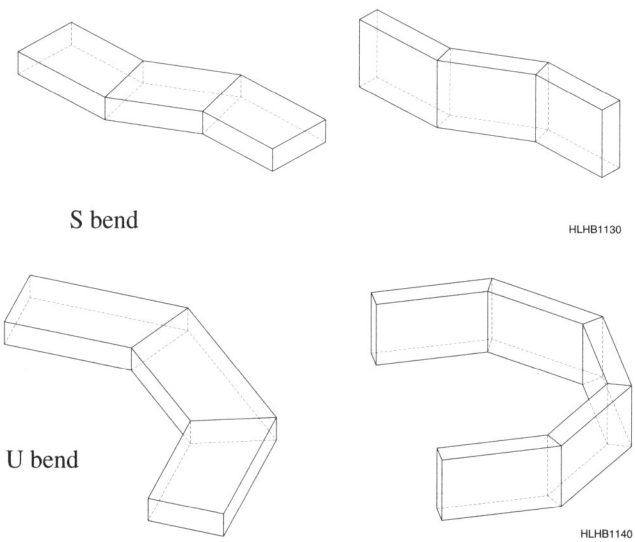

Fig. 3.31 Concatenations of sharp rectangular waveguide bends (Reproduced by permission of IGTE (Institut für Grundlagen und Theorie der Elektrotechnik) Graz)

propagation, the fields are computed analytically on discretisation lines. To model the sharp bends, we take lines of varying length (see Fig. 3.33).

We derive general field transformation formulas and impedance/admittance transformation formulas for rectangular waveguides by using generalised transmission line equations. In the case of bend sections with tilted ports, the impedance/admittance transformation formulas are obtained from transfer matrix expressions. This procedure is similar to that given in [13] for posts in waveguides and for microstrip bends [14]. Very simple formulas for transformation in the bend cross-section are obtained from the matching process of tangential field components. With these impedance transformation formulas, concatenations of sharp bends of arbitrary angles can also be analysed. Numerical results for scattering parameters are presented for a

WR90 waveguide E bend. From these results we derive the elements of an equivalent circuit in π form.

3.3.6.1 Analysis equations

For as long as it is possible, we would like to analyse E and H bends in parallel. The corresponding waveguides, with their coordinate systems, are sketched in Fig. 3.32.

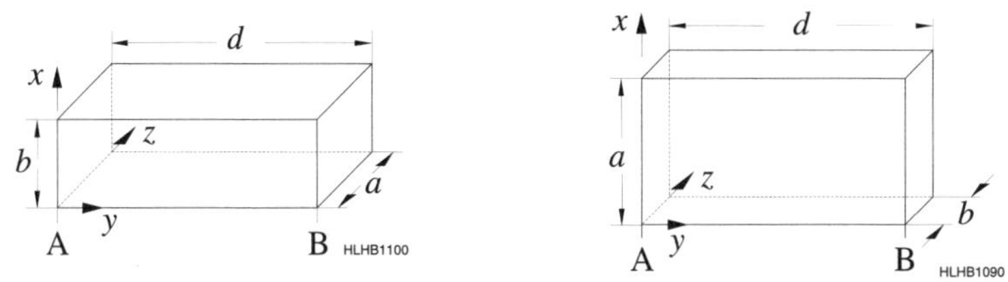

Fig. 3.32 Rectangular waveguides for H bends and E bends with coordinates (Reproduced by permission of IGTE (Institut für Grundlagen und Theorie der Elektrotechnik) Graz)

The fields in discretised form are given by eq. (3.19) (see Section 3.2.2) with $u = x$, $v = y$. For the H_{10} mode we have:

$$\begin{array}{ll} \textbf{H bend: } H_z = 0 & \textbf{E bend: } E_z = 0 \\ \overline{k}_z = 0 \rightarrow \varepsilon_q = \varepsilon_r & \overline{k}_z = \dfrac{\pi}{\overline{a}} \rightarrow \varepsilon_q = \varepsilon_r - \left(\dfrac{\pi}{\overline{a}}\right)^2 \end{array} \tag{3.99}$$

General field transformation equations of a straight section with parallel ports A and B (length d) can be given by transfer matrix relations:

$$\begin{bmatrix} \overline{\mathbf{E}}_{\mathrm{A}} \\ \overline{\mathbf{H}}_{\mathrm{A}} \end{bmatrix} = \begin{bmatrix} \cosh(\boldsymbol{\Gamma}\overline{d}) & \overline{\mathbf{Z}}_0 \sinh(\boldsymbol{\Gamma}\overline{d}) \\ \overline{\mathbf{Y}}_0 \sinh(\boldsymbol{\Gamma}\overline{d}) & \cosh(\boldsymbol{\Gamma}\overline{d}) \end{bmatrix} \begin{bmatrix} \overline{\mathbf{E}}_{\mathrm{B}} \\ \overline{\mathbf{H}}_{\mathrm{B}} \end{bmatrix} \tag{3.100}$$

with $\overline{\mathbf{Z}}_0 = \boldsymbol{I}$, $\overline{\mathbf{Y}}_0 = \boldsymbol{I}$, where $\boldsymbol{I}$ is the unit matrix and where we have:

$$\begin{array}{llll} \text{H bend:} & \overline{\mathbf{E}}_{\mathrm{A,B}} = \overline{\mathbf{E}}_{z\mathrm{A,B}} & \overline{\mathbf{H}}_{\mathrm{A,B}} = \overline{\mathbf{H}}_{x\mathrm{A,B}} \\ \text{E bend:} & \overline{\mathbf{E}}_{\mathrm{A,B}} = \overline{\mathbf{E}}_{x\mathrm{A,B}} & \overline{\mathbf{H}}_{\mathrm{A,B}} = -\overline{\mathbf{H}}_{z\mathrm{A,B}} \end{array} \tag{3.101}$$

Now we assume a straight waveguide section with nonparallel ports (e.g. section II in Fig. 3.33, between ports B and E). We obtain the fields in plane E as function of the fields in plane B:

$$\begin{bmatrix} \overline{\mathbf{E}}^{\mathrm{II}}_{\mathrm{E}} \\ \overline{\mathbf{H}}^{\mathrm{II}}_{\mathrm{E}} \end{bmatrix} = \begin{bmatrix} \boldsymbol{T}_{\mathrm{E}} \bullet \boldsymbol{S}_{\mathrm{C}} & \boldsymbol{T}_{\mathrm{E}} \bullet \boldsymbol{S}_{\mathrm{S}} \\ \boldsymbol{T}_{\mathrm{H}} \bullet \boldsymbol{S}_{\mathrm{S}} & \boldsymbol{T}_{\mathrm{H}} \bullet \boldsymbol{S}_{\mathrm{C}} \end{bmatrix} \begin{bmatrix} \overline{\mathbf{E}}_{\mathrm{B}} \\ \overline{\mathbf{H}}_{\mathrm{B}} \end{bmatrix} = \begin{bmatrix} \overline{\mathbf{V}}^{\mathrm{II}}_{11} & \overline{\mathbf{Z}}^{\mathrm{II}}_{12} \\ \overline{\mathbf{Y}}^{\mathrm{II}}_{21} & \overline{\mathbf{V}}^{\mathrm{II}}_{22} \end{bmatrix} \begin{bmatrix} \overline{\mathbf{E}}_{\mathrm{B}} \\ \overline{\mathbf{H}}_{\mathrm{B}} \end{bmatrix} \tag{3.102}$$

The unit matrices for the wave impedances/admittances are omitted. The multiplication sign $\bullet$ means that the matrices $\boldsymbol{T}_{\mathrm{E,H}}$ and $\boldsymbol{S}$ have to be

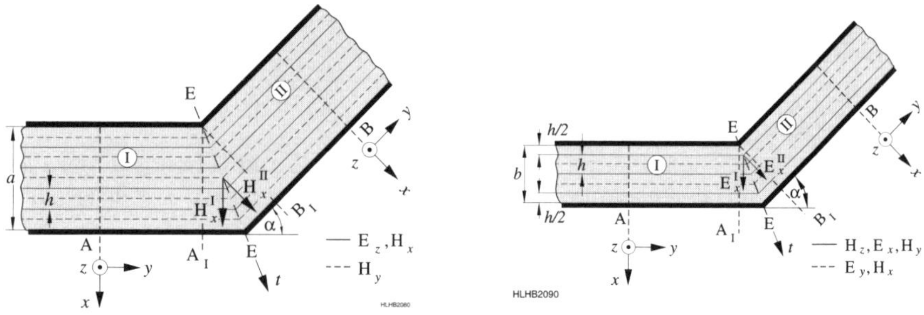

Fig. 3.33 Discretisation of the H bend and E bend (Reproduced by permission of IGTE (Institut für Grundlagen und Theorie der Elektrotechnik) Graz)

multiplied element by element (array multiplication). $\boldsymbol{T}$ is the transformation matrix and the matrices $\boldsymbol{S}_\mathrm{C}$ and $\boldsymbol{S}_\mathrm{S}$ are constructed from $\cosh(\boldsymbol{\Gamma}\overline{d})$ and $\sinh(\boldsymbol{\Gamma}\overline{d})$, respectively. We must take into account that the distance between planes B and E depends on the position of the lines. We label these lengths as $\overline{d}_i$. By collecting all $\overline{d}_i$ in a column vector $\overline{\mathbf{d}}$, we obtain:

$$\boldsymbol{S}_\mathrm{C} = \cosh(\overline{\mathbf{d}}\boldsymbol{\Gamma}^r) \quad \boldsymbol{S}_\mathrm{S} = \sinh(\overline{\mathbf{d}}\boldsymbol{\Gamma}^r) \tag{3.103}$$

$\boldsymbol{\Gamma}^r$ is a row vector that contains the diagonal elements of the (diagonal) matrix $\boldsymbol{\Gamma}$. Therefore, the matrices $\boldsymbol{S}_\mathrm{C}$ and $\boldsymbol{S}_\mathrm{S}$ are full matrices.

To determine the fields in plane E from the fields in plane A (section I), we have to go in negative directions. The expressions are analogous to eq. (3.102). We obtain:

$$\begin{bmatrix} \mathbf{E}_\mathrm{E}^\mathrm{I} \\ \mathbf{H}_\mathrm{E}^\mathrm{I} \end{bmatrix} = \begin{bmatrix} \boldsymbol{T}_\mathrm{E} \bullet \boldsymbol{S}_\mathrm{C} & -\boldsymbol{T}_\mathrm{E} \bullet \boldsymbol{S}_\mathrm{S} \\ -\boldsymbol{T}_\mathrm{H} \bullet \boldsymbol{S}_\mathrm{S} & \boldsymbol{T}_\mathrm{H} \bullet \boldsymbol{S}_\mathrm{C} \end{bmatrix} \begin{bmatrix} \overline{\mathbf{E}}_\mathrm{A} \\ \overline{\mathbf{H}}_\mathrm{A} \end{bmatrix} = \begin{bmatrix} \overline{\mathbf{V}}_{11}^\mathrm{I} & -\overline{\boldsymbol{Z}}_{12}^\mathrm{I} \\ -\overline{\mathbf{Y}}_{21}^\mathrm{I} & \overline{\mathbf{V}}_{22}^\mathrm{I} \end{bmatrix} \begin{bmatrix} \overline{\mathbf{E}}_\mathrm{B} \\ \overline{\mathbf{H}}_\mathrm{B} \end{bmatrix} \tag{3.104}$$

Next, we have to derive equations to match the fields at plane E. Here we must distinguish between the two cases of an E-plane and an H-plane bend.

H bend:

The tangential electric field in plane E is given by $\mathbf{E}_{t\mathrm{E}} = \mathbf{E}_{z\mathrm{E}}$ and must be equal on both sides of that plane. Therefore, matching the tangential electric field results in:

$$\mathbf{E}_{z\mathrm{E}}^\mathrm{I} = \mathbf{E}_{z\mathrm{E}}^\mathrm{II} = \mathbf{E}_\mathrm{E} \tag{3.105}$$

The tangential magnetic field in plane E is given by the equations:

$$\mathbf{H}_{t\mathrm{E}}^\mathrm{I} = \mathbf{H}_{x\mathrm{E}}^\mathrm{I} \cos\frac{\alpha}{2} + \mathbf{H}_{y\mathrm{E}}^\mathrm{I} \sin\frac{\alpha}{2} \tag{3.106}$$

$$\mathbf{H}_{t\mathrm{E}}^\mathrm{II} = \mathbf{H}_{x\mathrm{E}}^\mathrm{II} \cos\frac{\alpha}{2} - \mathbf{H}_{y\mathrm{E}}^\mathrm{II} \sin\frac{\alpha}{2} \tag{3.107}$$

Matching the tangential magnetic field in plane E results in:

$$\mathbf{H}^{\mathrm{I}}_{x\mathrm{E}} = \mathbf{H}^{\mathrm{II}}_{x\mathrm{E}} - (\mathbf{H}^{\mathrm{I}}_{y\mathrm{E}} + \mathbf{H}^{\mathrm{II}}_{y\mathrm{E}}) \tan \frac{\alpha}{2} \tag{3.108}$$

The components $\mathbf{H}_{y\mathrm{E}}$ are obtained with eq. (3.19) according to:

$$\mathbf{H}^{\mathrm{I,II}}_{y\mathrm{E}} = -\mathrm{j}\boldsymbol{M}^{e}_{x}\overline{\boldsymbol{D}}^{e}_{\overline{x}}\mathbf{E}_{\mathrm{E}} \tag{3.109}$$

$\overline{\boldsymbol{D}}^{e}_{x}$ has to fulfil Dirichlet boundary conditions. The interpolation matrix $\boldsymbol{M}_e$ is required to shift the fields to the original discretisation line system (see Appendix B). With the definition:

$$\mathbf{Y}_e = 2\mathrm{j} \tan \frac{\alpha}{2} \boldsymbol{M}^{e}_{x}\overline{\boldsymbol{D}}^{e}_{\overline{x}} \tag{3.110}$$

we obtain from eq. (3.108):

$$\mathbf{H}^{\mathrm{I}}_{x\mathrm{E}} = \mathbf{H}^{\mathrm{II}}_{x\mathrm{E}} + \mathbf{Y}_e\mathbf{E}_{\mathrm{E}} \tag{3.111}$$

E bend:

The tangential magnetic field in plane E contains the component $\mathbf{H}_z$ and an orthogonal component in t-direction (see Fig. 3.33). For the component H_z[2] we can simply write:

$$-\mathbf{H}^{\mathrm{I}}_{z\mathrm{E}} = -\mathbf{H}^{\mathrm{II}}_{z\mathrm{E}} = \mathbf{H}_{\mathrm{E}} \tag{3.112}$$

The t component of the magnetic field is determined as:

$$\mathbf{H}^{\mathrm{I}}_{\mathrm{tE}} = -\mathrm{j}k_z \left(\mathbf{E}^{\mathrm{I}}_{y\mathrm{E}} \cos \frac{\alpha}{2} - \mathbf{E}^{\mathrm{I}}_{x} \sin \frac{\alpha}{2} \right) = -\mathrm{j}k_z \mathbf{E}^{\mathrm{I}}_{\mathrm{nE}} \tag{3.113}$$

$$\mathbf{H}^{\mathrm{II}}_{\mathrm{tE}} = -\mathrm{j}k_z \left(\mathbf{E}^{\mathrm{II}}_{y\mathrm{E}} \cos \frac{\alpha}{2} + \mathbf{E}^{\mathrm{II}}_{x} \sin \frac{\alpha}{2} \right) = -\mathrm{j}k_z \mathbf{E}^{\mathrm{II}}_{\mathrm{nE}} \tag{3.114}$$

The magnetic components in t-direction are proportional to the normal electric field components in plane E. Therefore, if $\mathbf{H}_z$ and the tangential electric components $\mathrm{E}^{\mathrm{I,II}}_{\mathrm{tE}}$ are matched, $\mathrm{E}^{\mathrm{I,II}}_{\mathrm{nE}}$ and $\mathrm{H}^{\mathrm{I,II}}_{\mathrm{tE}}$ are matched at the same time.

The tangential electric field in plane E is given by the equations:

$$\mathbf{E}^{\mathrm{I}}_{\mathrm{tE}} = \mathbf{E}^{\mathrm{I}}_{x\mathrm{E}} \cos \frac{\alpha}{2} + \mathbf{E}^{\mathrm{I}}_{y\mathrm{E}} \sin \frac{\alpha}{2} \tag{3.115}$$

$$\mathbf{E}^{\mathrm{II}}_{\mathrm{tE}} = \mathbf{E}^{\mathrm{II}}_{x\mathrm{E}} \cos \frac{\alpha}{2} - \mathbf{E}^{\mathrm{II}}_{y\mathrm{E}} \sin \frac{\alpha}{2} \tag{3.116}$$

Matching the tangential electric field results in:

$$\mathbf{E}^{\mathrm{I}}_{x\mathrm{E}} = \mathbf{E}^{\mathrm{II}}_{x\mathrm{E}} - (\mathbf{E}^{\mathrm{I}}_{y\mathrm{E}} + \mathbf{E}^{\mathrm{II}}_{y\mathrm{E}}) \tan \frac{\alpha}{2} \tag{3.117}$$

[2]Analogously to eq. (3.101), we introduce $-\mathbf{H}_{z\mathrm{E}} = \mathbf{H}_{\mathrm{E}}$.

The components of $\mathbf{E}_y$ are obtained from eq. (3.19) according to:

$$\mathbf{E}_{y\mathrm{E}}^{\mathrm{I,II}} = \mathrm{j}\varepsilon_{\mathrm{q}}^{-1}\boldsymbol{M}_x^h\overline{\boldsymbol{D}}_x^h\mathbf{H}_{z\mathrm{E}} \tag{3.118}$$

$\overline{\boldsymbol{D}}_x^h$ has to fulfil Neumann boundary conditions. $\boldsymbol{M}_x^h$ is required to shift the result to the original discretisation line system (see Appendix B). With the definition:

$$\boldsymbol{Z}_h = 2\mathrm{j}\varepsilon_{\mathrm{q}}^{-1}\tan\frac{\alpha}{2}\boldsymbol{M}_x^h\overline{\boldsymbol{D}}_x^h \tag{3.119}$$

we obtain from eq. (3.117):

$$\mathbf{E}_{x\mathrm{E}}^{\mathrm{I}} = \mathbf{E}_{x\mathrm{E}}^{\mathrm{II}} + \boldsymbol{Z}_h\mathbf{H}_{\mathrm{E}} \tag{3.120}$$

3.3.6.2 Impedance transformation

The next steps of the analysis are done with the admittance/impedance transformation concept. We have to distinguish between sections I and II and the transition from II to I in plane E.

H bend:

We start with the load admittance in plane B:

$$\overline{\mathbf{H}}_{\mathrm{B}} = \overline{\boldsymbol{Y}}_{\mathrm{B}}\overline{\mathbf{E}}_{\mathrm{B}} \tag{3.121}$$

If the waveguide II is infinitely long then we have $\overline{\boldsymbol{Y}}_{\mathrm{B}} = \overline{\boldsymbol{Y}}_0 = \boldsymbol{I}$. Introducing this relation into eq. (3.102) gives:

$$\mathbf{E}_{\mathrm{E}}^{\mathrm{II}} = (\mathbf{V}_{11}^{\mathrm{II}} + \boldsymbol{Z}_{12}^{\mathrm{II}}\overline{\boldsymbol{Y}}_{\mathrm{B}})\overline{\mathbf{E}}_{\mathrm{B}} \quad \mathbf{H}_{\mathrm{E}}^{\mathrm{II}} = (\mathbf{Y}_{21}^{\mathrm{II}} + \mathbf{V}_{22}^{\mathrm{II}}\overline{\boldsymbol{Y}}_{\mathrm{B}})\overline{\mathbf{E}}_{\mathrm{B}} \tag{3.122}$$

Therefore, the impedance and admittance matrices in plane E are given by:

$$\mathbf{E}_{\mathrm{E}}^{\mathrm{II}} = \mathbf{Z}_{\mathrm{E}}^{\mathrm{II}}\mathbf{H}_{\mathrm{E}}^{\mathrm{II}} \qquad \boldsymbol{Z}_{\mathrm{E}}^{\mathrm{II}} = (\mathbf{V}_{11}^{\mathrm{II}} + \boldsymbol{Z}_{12}^{\mathrm{II}}\overline{\boldsymbol{Y}}_{\mathrm{B}})(\mathbf{Y}_{21}^{\mathrm{II}} + \mathbf{V}_{22}^{\mathrm{II}}\overline{\boldsymbol{Y}}_{\mathrm{B}})^{-1} \tag{3.123}$$

$$\mathbf{H}_{\mathrm{E}}^{\mathrm{II}} = \mathbf{Y}_{\mathrm{E}}^{\mathrm{II}}\mathbf{E}_{\mathrm{E}}^{\mathrm{II}} \qquad \mathbf{Y}_{\mathrm{E}}^{\mathrm{II}} = (\mathbf{Y}_{21}^{\mathrm{II}} + \mathbf{V}_{22}^{\mathrm{II}}\overline{\boldsymbol{Y}}_{\mathrm{B}})(\mathbf{V}_{11}^{\mathrm{II}} + \boldsymbol{Z}_{12}^{\mathrm{II}}\overline{\boldsymbol{Y}}_{\mathrm{B}})^{-1} \tag{3.124}$$

Since the matrices $\boldsymbol{S}_{\mathrm{C}}$ and $\boldsymbol{S}_{\mathrm{S}}$ contain hyperbolic sine and cosine functions, the quantities $\boldsymbol{V}_{11}, \boldsymbol{V}_{22}, \boldsymbol{Z}_{12}, \boldsymbol{Y}_{21}$ increase exponentially. Hence, eqs. (3.123) and (3.124) can be numerically unstable. To obtain a more stable algorithm we use the short-circuit matrix parameters. These $\boldsymbol{y}$ parameters are given by:

$$\begin{aligned} \tilde{\boldsymbol{y}}_{11}^{\mathrm{II}} &= \mathbf{V}_{22}^{\mathrm{II}}\boldsymbol{Z}_{12}^{\mathrm{II}-1} & \qquad \tilde{\boldsymbol{y}}_{12}^{\mathrm{II}} &= \mathbf{Y}_{21}^{\mathrm{II}} - \mathbf{V}_{22}^{\mathrm{II}}\boldsymbol{Z}_{12}^{\mathrm{II}-1}\mathbf{V}_{11}^{\mathrm{II}} \\ \tilde{\boldsymbol{y}}_{22}^{\mathrm{II}} &= \boldsymbol{Z}_{12}^{\mathrm{II}-1}\mathbf{V}_{11}^{\mathrm{II}} & \qquad \tilde{\boldsymbol{y}}_{21}^{\mathrm{II}} &= -\boldsymbol{Z}_{12}^{\mathrm{II}-1} \end{aligned} \tag{3.125}$$

The parameters are marked with a tilde to distinguish them from those which relate to the field quantities in transformed domain. Here, the parameters connect the fields in transformed domain (plane B) with those in original domain (plane E). With these matrices, the input admittance in plane E of section II can also be calculated by:

$$\mathbf{Y}_{\mathrm{E}}^{\mathrm{II}} = \tilde{\boldsymbol{y}}_{11}^{\mathrm{II}} - \tilde{\boldsymbol{y}}_{12}^{\mathrm{II}}(\tilde{\boldsymbol{y}}_{22}^{\mathrm{II}} + \overline{\mathbf{Y}}_{\mathrm{B}})^{-1}\tilde{\boldsymbol{y}}_{21}^{\mathrm{II}} \tag{3.126}$$

Using this input admittance $\mathbf{Y}_{\mathrm{E}}^{\mathrm{II}}$ we can calculate the load admittance of section I by introducing $\mathbf{H}_{x\mathrm{E}}^{\mathrm{II}} = \mathbf{Y}_{\mathrm{E}}^{\mathrm{II}}\mathbf{E}_{\mathrm{E}}$ into eq. (3.111), and obtain:

$$\mathbf{Y}_{\mathrm{E}}^{\mathrm{I}} = \mathbf{Y}_{\mathrm{E}}^{\mathrm{II}} + \mathbf{Y}_{\mathrm{e}} \tag{3.127}$$

The transformation of the admittance from plane E to plane A (see Fig. 3.33) is performed using a formula analogous to eq. (3.126). To reduce numerical problems we move ports A and B towards the discontinuity (positions $\mathrm{A_I}$ and $\mathrm{B_I}$).

E bend:

We start with the load impedance in plane B according to:

$$\overline{\mathbf{E}}_{\mathrm{B}} = \overline{\mathbf{Z}}_{\mathrm{B}}\overline{\mathbf{H}}_{\mathrm{B}} \tag{3.128}$$

In case of an infinitely long waveguide II, we have $\overline{\mathbf{Z}}_{\mathrm{B}} = \overline{\mathbf{Z}}_0 = \boldsymbol{I}$. Introducing this relation into eq. (3.102) results in:

$$\mathbf{E}_{\mathrm{E}}^{\mathrm{II}} = (\mathbf{V}_{11}^{\mathrm{II}}\overline{\mathbf{Z}}_{\mathrm{B}} + \mathbf{Z}_{12}^{\mathrm{II}})\overline{\mathbf{H}}_{\mathrm{B}} \qquad \mathbf{H}_{\mathrm{E}}^{\mathrm{II}} = (\mathbf{Y}_{21}^{\mathrm{II}}\overline{\mathbf{Z}}_{\mathrm{B}} + \mathbf{V}_{22}^{\mathrm{II}})\overline{\mathbf{H}}_{\mathrm{B}} \tag{3.129}$$

The impedance and admittance matrices in plane E are given by:

$$\mathbf{Z}_{\mathrm{E}}^{\mathrm{II}} = (\mathbf{V}_{11}^{\mathrm{II}}\overline{\mathbf{Z}}_{\mathrm{B}} + \mathbf{Z}_{12}^{\mathrm{II}})(\mathbf{Y}_{21}^{\mathrm{II}}\overline{\mathbf{Z}}_{\mathrm{B}} + \mathbf{V}_{22}^{\mathrm{II}})^{-1} \tag{3.130}$$

$$\mathbf{Y}_{\mathrm{E}}^{\mathrm{II}} = (\mathbf{Y}_{21}^{\mathrm{II}}\overline{\mathbf{Z}}_{\mathrm{B}} + \mathbf{V}_{22}^{\mathrm{II}})(\mathbf{V}_{11}^{\mathrm{II}}\overline{\mathbf{Z}}_{\mathrm{B}} + \mathbf{Z}_{12}^{\mathrm{II}})^{-1} \tag{3.131}$$

Like eqs. (3.123) and (3.124), eqs. (3.130) and (3.131) are numerically unstable due to the form of the matrices $\mathbf{S}_{\mathrm{C}}$ and $\mathbf{S}_{\mathrm{S}}$. A more stable algorithm is obtained by using the open-circuit matrix parameters. These z parameters are given by:

$$\begin{aligned} \tilde{\boldsymbol{z}}_{11}^{\mathrm{II}} &= \mathbf{V}_{11}^{\mathrm{II}}\mathbf{Y}_{21}^{\mathrm{II}-1} & \tilde{\boldsymbol{z}}_{12}^{\mathrm{II}} &= \mathbf{V}_{11}^{\mathrm{II}}\mathbf{Y}_{21}^{\mathrm{II}-1}\mathbf{V}_{22}^{\mathrm{II}} - \mathbf{Z}_{12}^{\mathrm{II}} \\ \tilde{\boldsymbol{z}}_{22}^{\mathrm{II}} &= \mathbf{Y}_{21}^{\mathrm{II}-1}\mathbf{V}_{22}^{\mathrm{II}} & \tilde{\boldsymbol{z}}_{21}^{\mathrm{II}} &= \mathbf{Y}_{21}^{\mathrm{II}-1} \end{aligned} \tag{3.132}$$

With these matrices we determine the input impedance in plane E of section II as:

$$\mathbf{Z}_{\mathrm{E}}^{\mathrm{II}} = \tilde{\boldsymbol{z}}_{11}^{\mathrm{II}} - \tilde{\boldsymbol{z}}_{12}^{\mathrm{II}}(\tilde{\boldsymbol{z}}_{22}^{\mathrm{II}} + \overline{\mathbf{Z}}_{\mathrm{B}})^{-1}\tilde{\boldsymbol{z}}_{21}^{\mathrm{II}} \tag{3.133}$$

By introducing $\mathbf{E}_{x\mathrm{E}}^{\mathrm{II}} = \mathbf{Z}_{\mathrm{E}}^{\mathrm{II}}\mathbf{H}_{\mathrm{E}}$ into eq. (3.120) we obtain for the load impedance of section I in plane E:

$$\mathbf{Z}_{\mathrm{E}}^{\mathrm{I}} = \mathbf{Z}_{\mathrm{E}}^{\mathrm{II}} + \boldsymbol{Z}_h \tag{3.134}$$

Similar to the H bend case, we transform the impedance matrix from plane E to plane A, using expressions analogous to eq. (3.133). As before, planes A and B should be moved towards the discontinuity ($\mathrm{A_I}$ and $\mathrm{B_I}$ in Fig. 3.33) to minimise numerical problems.

3.3.7 Scattering parameters

Starting at the output of a concatenated structure, we can calculate the input impedance by using the impedance transformation formulas developed above. Usually, the load impedance is the wave impedance $\overline{\boldsymbol{Z}}_0^{\text{out}} = \overline{\boldsymbol{Z}}_h^{\text{out}}(\overline{\boldsymbol{Z}}_e^{\text{out}}) = \boldsymbol{I}$ of the output waveguide.

At the input of the device we have a feeding waveguide with the wave impedance $\overline{\boldsymbol{Z}}_0^{\text{in}}$ or admittance $\overline{\mathbf{Y}}_0^{\text{in}}$. From the transformation procedure we compute the input admittance $\overline{\mathbf{Y}}_A$. By dividing the electric field into a forward-travelling part and a backward-travelling part we obtain at the input of the device:

$$\overline{\mathbf{H}}_{\text{A}} = \overline{\mathbf{Y}}_{\text{A}}(\overline{\mathbf{E}}_{\text{Af}} + \overline{\mathbf{E}}_{\text{Ab}}) = \overline{\mathbf{Y}}_0^{\text{in}}(\overline{\mathbf{E}}_{\text{Af}} - \overline{\mathbf{E}}_{\text{Ab}}) \tag{3.135}$$

Usually, a forward-propagating field $\overline{\mathbf{E}}_{\text{Af}}$ is injected at the input port. From eq. (3.135) we are able to determine the reflected part. We can therefore write for the generalised scattering parameter $\boldsymbol{S}_{11}$:

$$\boldsymbol{S}_{11} = (\overline{\mathbf{Y}}_0^{\text{in}} + \overline{\mathbf{Y}}_{\text{A}})^{-1}(\overline{\mathbf{Y}}_0^{\text{in}} - \overline{\mathbf{Y}}_{\text{A}}) \tag{3.136}$$

The first element gives the reflection coefficient of the fundamental mode, if the modes have been ordered properly. Knowing the injected and reflected field at the input, we are able to determine the field distribution in the whole structure, and with this the remaining scattering parameters.

For monomode, lossless devices, the absolute value of the transmittance S_{21} can also be calculated from:

$$|S_{11}|^2 + |S_{21}|^2 = 1 \tag{3.137}$$

3.3.8 Numerical results

3.3.8.1 Numerical results for rectangular waveguide bends

Results for an E bend are shown in Fig. 3.34. The scattering parameters are drawn as a function of frequency, with the bend angle α as a parameter. The number of discretisation lines is 20. The results are in good agreement with those presented in [15]. From these parameters, the elements of an equivalent circuit are calculated. These are sketched in Fig. 3.35. Appendix D shows how these elements can be determined. Fig. 3.35 is a plot of the normalized shunt and series admittance parameterised with $2b/\lambda_g$ vs. the bend angle α. λ_g is the guide wavelength, defined as $\lambda_g = 2\pi/\gamma_0 k_0$, where $\gamma_0 = \sqrt{\varepsilon_q}$ is the normalised propagation constant.

3.3.8.2 Numerical results for sharp waveguide bends and junctions

To verify the results when analysing junctions, the scattering parameters and the resulting equivalent circuit parameters for two exemplary structures have been computed and compared with results of the mode matching

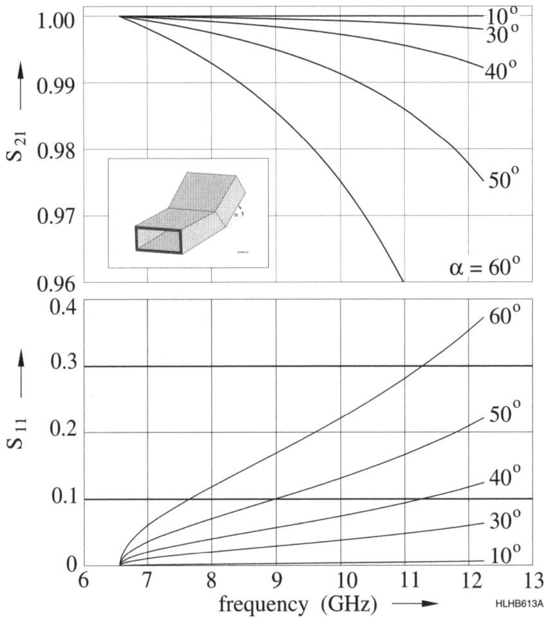

Fig. 3.34 Scattering parameters S_{11} and S_{21} as a function of frequency with bend angle α as parameter (Reproduced by permission of IGTE (Institut für Grundlagen und Theorie der Elektrotechnik) Graz)

technique (MMT). Fig. 3.36 shows the equivalent circuit parameters for a symmetric rectangular E-plane corner [16]. The equivalent circuit is valid for the fundamental mode only. The network parameters are computed from the scattering coefficients S_{11} and S_{21} by the following relations:

$$\mathrm{j}B_1 Z_0 = \frac{2}{1 + S_{11} + S_{21}} - 1 \qquad \mathrm{j}B_2 Z_0 = \frac{2S_{21}}{(1 + S_{11})^2 - S_{21}^2} \tag{3.138}$$

where $B_1 \equiv B_\mathrm{P}$ and $B_2 \equiv -B_\mathrm{S}$ in Fig. 3.35 and:

$$\frac{X}{Z_0} = -\left(2B_1 Z_0 + \frac{(B_1 Z_0)^2 + 1}{B_2 Z_0}\right)^{-1} \quad \frac{d}{b} = 1 + \frac{1}{\gamma_0 \bar{b}} \tan^{-1}(B_1 Z_0 + 2B_2 Z_0)^{-1} \tag{3.139}$$

As before, λ_g is the guide wavelength and Z_0 is the characteristic impedance for H_{10} mode.

In Fig. 3.36 the normalised shunt reactance and its location d/b are presented as a function of normalised frequency. Both curves coincide exactly with the reference values.

As an example for cascaded discontinuities, the scattering coefficients of an asymmetric E-plane T-junction are presented in Fig. 3.37 as a function of frequency. A comparison with MMT results [17] shows a very good agreement.

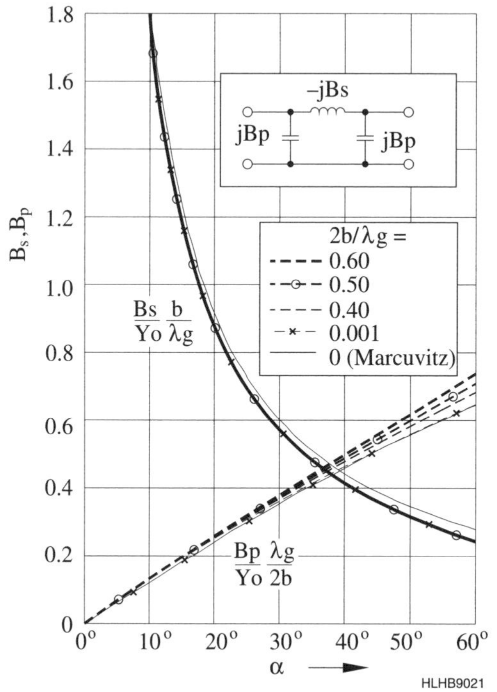

Fig. 3.35 Equivalent circuit parameters for E-plane bends of arbitrary angle (Reproduced by permission of IGTE (Institut für Grundlagen und Theorie der Elektrotechnik) Graz)

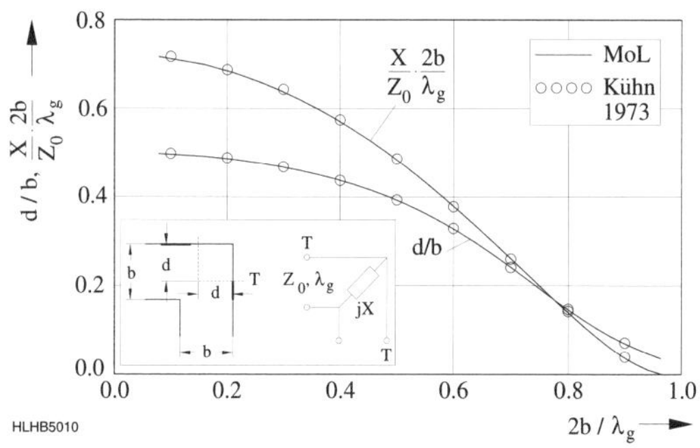

Fig. 3.36 Normalised shunt reactance $X/Z_0 \cdot 2b/\lambda_g$ and its location d/b as a function of normalised frequency. (λ_g guide wavelength, Z_0 characteristic impedance) ∘ ∘ ∘ MMT [16]

Results for a second E-plane T-junction are presented in Fig. 3.38. The scattering parameters are shown as a function of frequency. The results

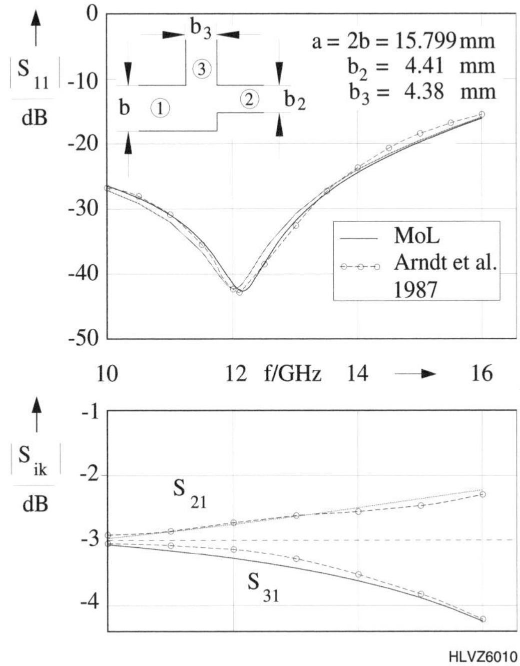

Fig. 3.37 Asymmetric E-plane T-junction. Magnitude of the scattering coefficients as a function of frequency ◦ ◦ ◦ MMT [17] (W. Pascher and R. Pregla, 'Analysis of Rectangular Waveguide Junctions by the Method of Lines', *IEEE Trans. Microwave Theory Tech.*. vol. MTT-43, no. 12, pp. 2649–2653. © 1995 Institute of Electrical and Electronics Engineers (IEEE))

were compared with those obtained by the port reflection coefficient method (PRCM) [18]. Again, a good agreement was found. The junction was analysed, with $N_1 = 120$ and $N_2 = 92$ discretisation lines in waveguide 1 and waveguide 2, respectively. The number of discretisation lines N_3 in waveguide 3 was chosen as $N_3 = N_1 b_3 / b_1 = 63$.

By introducing a metal block of height $t = 0.931$ mm in the junction region, the input port 1 can be matched over a wide frequency band. The power balance (eq. (3.137)) was investigated and the following result was found:

$$||S_{11}|^2 + |S_{21}|^2 + |S_{31}|^2 - 1| < 10^{-6}$$

The next example shows an E-plane corner with matching elements (see Fig. 3.39). Results for an unmatched E-plane corner are presented in [3]. Also, convergence diagrams are given and a comparison is made with the results of [16]. In contrast to Section 3.3 a further matching step is introduced into the junction region. However, no further formulas need be developed for the analysis. The junction region is the same as in Fig. 3.28. The step from

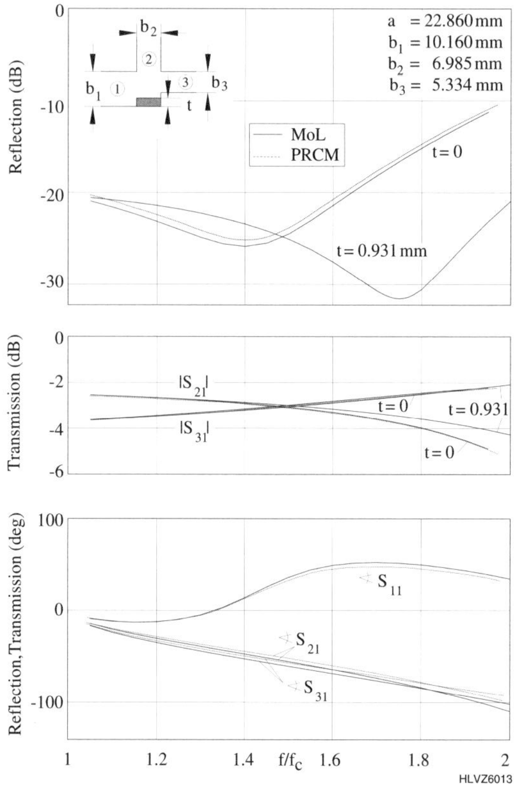

Fig. 3.38 Asymmetric E-plane T-junction. Magnitude and phase of the scattering coefficients as a function of frequency. The results are compared with those reported in [18] (Reproduced by permission of Elsevier)

dimension c to dimension b is modelled with an impedance or admittance transformation (eq. (3.65)). This simple procedure cannot be used in the algorithm described previously [3]. The absolute value of S_{11} (input reflection coefficient) versus frequency is plotted in Fig. 3.41. Curve 1 was obtained for one-step matching ($b = c$).

The optimal solution (i.e. minimum reflection for the frequency range up to 12.4 GHz) was obtained for $d/b = 61/120$. This result is compatible with the well-known recipe that the insert should be chosen as $d \approx b/2$. The absolute value of S_{11} decreases drastically if two matching steps are performed. For the curve labelled 2 in Fig. 3.41, the ratio $c/b = 109/120$ was chosen, with d being the same as before. Curves 3 and 4 were obtained for U bends and S bends, respectively (see Fig. 3.40), where we chose the same dimensions as for curve 2. The separation distance was $e = 15$ mm. These concatenations were analysed in the same way as the concatenation of circular

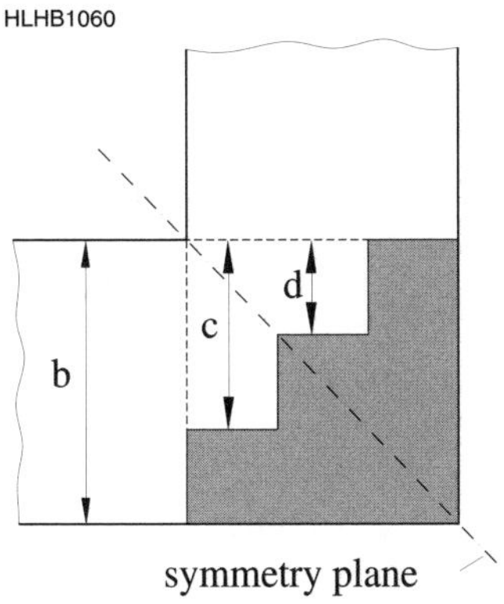

Fig. 3.39 Analysed E-plane corner in WR90 waveguide (Reproduced by permission of Elsevier)

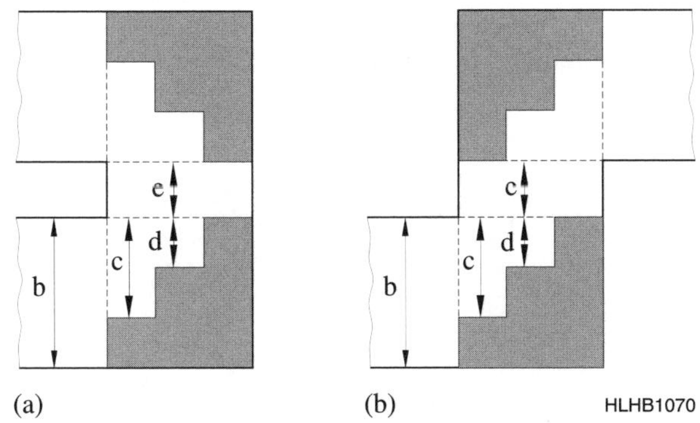

Fig. 3.40 Concatenations of two E-plane corners: (a) E-plane U corner (b) E-plane S corner (Reproduced by permission of Elsevier)

bends described in Section 3.2. No additional formulas need be developed. It should be mentioned that the numerical effort for analysing unsymmetric structures is only slightly higher than that for symmetrical ones.

3.4 ANALYSIS OF 3D WAVEGUIDE JUNCTIONS

The structures described in this section can be understood as a generalisation of the waveguide junctions that were presented in Section 3.3. An example is shown in Fig. 3.42. As we can see, we have E-plane and H-plane bends at the same time. Again, we will use crossed discretisation lines, but this time in three directions. Further, we have to perform a 2D discretisation here.

As before, this will be done by general short- or open-circuit matrices, analogous to the short- or open-circuit parameters in circuit theory. Here, we will use a six-port description.

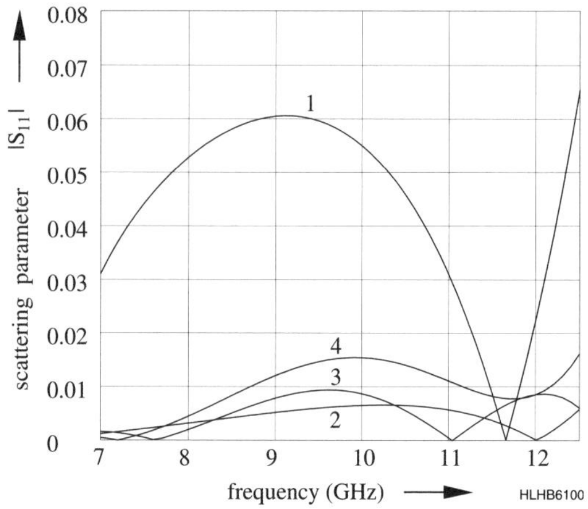

Fig. 3.41 Matched E-plane corner. Magnitude of the scattering coefficients S_{11} as a function of frequency. 1: One-step matching, 2: two-step matching 3: S-bend corner, 4: U-bend corner (Reproduced by permission of Elsevier)

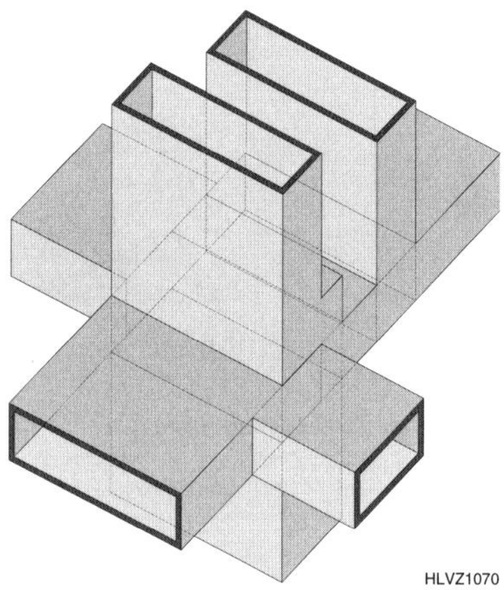

Fig. 3.42 Example of a 3D waveguide junction (Reproduced by permission of Copernicus Gesellschaft mbH)

3.4.1 General description

For the analysis, we extract the connecting waveguides from the structures and obtain a reduced junction. For the structure shown in Fig. 3.42 this reduced junction is sketched in Fig. 3.43a. We now define an inner junction that is identical to the reduced junction but remove all the planar metallisations (see Fig. 3.43b). Only the edges are left metallised, like thin wires. The six surrounding planes are the six generalised ports of the junction. The term 'generalised' means that there exist various modes on each of the ports. Most of them are non-propagating (evanescent). Therefore we have various subports a at each port. The number of these subports corresponds to the number of modes. By extending the description to the E-plane and H-plane junctions, the relation of the fields at the generalised six ports can be written as:

$$\begin{bmatrix} \widehat{\mathbf{H}}_{\mathrm{AB}} \\ \widehat{\mathbf{H}}_{\mathrm{CD}} \\ \widehat{\mathbf{H}}_{\mathrm{EF}} \end{bmatrix} = \begin{bmatrix} \widehat{y}_{\mathrm{AB}}^{\mathrm{AB}} & \widehat{y}_{\mathrm{AB}}^{\mathrm{CD}} & \widehat{y}_{\mathrm{AB}}^{\mathrm{EF}} \\ \widehat{y}_{\mathrm{CD}}^{\mathrm{AB}} & \widehat{y}_{\mathrm{CD}}^{\mathrm{CD}} & \widehat{y}_{\mathrm{CD}}^{\mathrm{EF}} \\ \widehat{y}_{\mathrm{EF}}^{\mathrm{AB}} & \widehat{y}_{\mathrm{EF}}^{\mathrm{CD}} & \widehat{y}_{\mathrm{EF}}^{\mathrm{EF}} \end{bmatrix} \begin{bmatrix} \widehat{\mathbf{E}}_{\mathrm{AB}} \\ \widehat{\mathbf{E}}_{\mathrm{CD}} \\ \widehat{\mathbf{E}}_{\mathrm{EF}} \end{bmatrix} \tag{3.140}$$

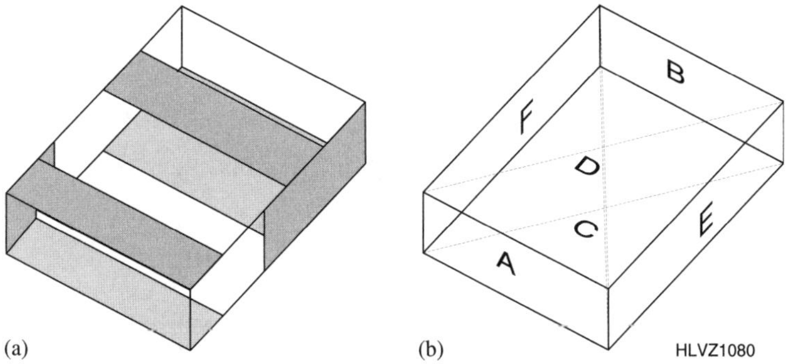

Fig. 3.43 The reduced junction (a) of the 3D waveguide junction in Fig. 3.42 (zoomed), and the associated inner junction (b) (Reproduced by permission of Copernicus Gesellschaft mbH)

$\widehat{\mathbf{F}}_{\mathrm{UV}}$ ($\mathbf{F}$ is equivalent to $\mathbf{E}$ or $\mathbf{H}$) is a supervector with the fields on opposite ports:

$$\widehat{\mathbf{E}}_{\mathrm{UV}} = [\mathbf{E}_{\mathrm{U}}^{\mathrm{t}}, \mathbf{E}_{\mathrm{V}}^{\mathrm{t}}]^{\mathrm{t}} \qquad \widehat{\mathbf{H}}_{\mathrm{UV}} = [\mathbf{H}_{\mathrm{U}}^{\mathrm{t}}, -\mathbf{H}_{\mathrm{V}}^{\mathrm{t}}]^{\mathrm{t}} \tag{3.141}$$

All submatrices are determined by short circuiting the ports alternately. By short circuiting the ports C, D, E and F the submatrices

$$\widehat{y}_{\mathrm{AB}}^{\mathrm{AB}} \quad \widehat{y}_{\mathrm{CD}}^{\mathrm{AB}} \quad \widehat{y}_{\mathrm{EF}}^{\mathrm{AB}}$$

can be calculated. The other submatrices can be computed by an analogous procedure.

3.4.2 Basic equations

We determine all field components from the two tangential electric or magnetic field components in the waveguide. As basic components we can also use E_x together with H_y, or E_y in combination with H_x. From the first two equations of Ampère's law and the law of electromagnetic induction (both in vectorial form) we obtain:

$$\frac{\partial}{\partial \overline{z}} \begin{bmatrix} -\widetilde{H}_x \\ \widetilde{H}_y \end{bmatrix} = -\mathrm{j}[R_{\mathrm{E}}] \begin{bmatrix} E_y \\ E_x \end{bmatrix} \qquad [R_{\mathrm{E}}] = \begin{bmatrix} D_{\overline{x}} D_{\overline{x}} + \varepsilon_r & -D_{\overline{x}} D_{\overline{y}} \\ -D_{\overline{y}} D_{\overline{x}} & D_{\overline{y}} D_{\overline{y}} + \varepsilon_r \end{bmatrix} \tag{3.142}$$

$$\frac{\partial}{\partial \overline{z}} \begin{bmatrix} E_y \\ E_x \end{bmatrix} = -\mathrm{j}[R_{\mathrm{H}}] \begin{bmatrix} -\widetilde{H}_x \\ \widetilde{H}_y \end{bmatrix} \qquad [R_{\mathrm{H}}] = \frac{1}{\varepsilon_r} \begin{bmatrix} D_{\overline{y}} D_{\overline{y}} + \varepsilon_r & D_{\overline{y}} D_{\overline{x}} \\ D_{\overline{x}} D_{\overline{y}} & D_{\overline{x}} D_{\overline{x}} + \varepsilon_r \end{bmatrix} \tag{3.143}$$

From the third equation of the law of induction and Faraday's law we obtain the field components $\widetilde{H}_z$ and E_z, which we had introduced into the first ones.

$$\widetilde{H}_z = -\mathrm{j}[-D_{\overline{x}} \; D_{\overline{y}}] \begin{bmatrix} E_y \\ E_x \end{bmatrix} \qquad E_z = -\mathrm{j}\varepsilon_r^{-1}[D_{\overline{y}} \; D_{\overline{x}}] \begin{bmatrix} -\widetilde{H}_x \\ \widetilde{H}_y \end{bmatrix} \tag{3.144}$$

By introducing eq. (3.142) into eq. (3.143) we obtain:

$$\frac{\partial^2}{\partial \overline{z}^2} \begin{bmatrix} E_y \\ E_x \end{bmatrix} - \begin{bmatrix} Q_{11}^{\mathrm{E}} & 0 \\ 0 & Q_{22}^{\mathrm{E}} \end{bmatrix} \begin{bmatrix} E_y \\ E_x \end{bmatrix} = \begin{bmatrix} 0 \\ 0 \end{bmatrix} \tag{3.145}$$

where $[Q^{\mathrm{E}}] = -[R_{\mathrm{H}}][R_{\mathrm{E}}]$. Alternatively, eq. (3.143) can be introduced into eq. (3.142), which results in a wave equation for $[-\widetilde{H}_x, \widetilde{H}_y]^{\mathrm{t}}$.

3.4.3 Discretisation scheme for propagation between A and B

The first step in the analysis consists of finding a general description of the field relation between the two opposite ports of a straight rectangular waveguide, e.g. ports A and B. The discretisation is performed in such a way that the columns of the discretised values in y-direction are stacked on each other (see Fig. 3.44). The column on the left side is stacked on the highest position and the right column is stacked on the lowest position. With the supervectors of the discretised field components:

$$\widehat{\mathbf{E}} = [\widehat{\mathbf{E}}_y^{\mathrm{t}}, \widehat{\mathbf{E}}_x^{\mathrm{t}}]^{\mathrm{t}} \qquad \widehat{\mathbf{H}} = [-\widehat{\mathbf{H}}_x^{\mathrm{t}}, \widehat{\mathbf{H}}_y^{\mathrm{t}}]^{\mathrm{t}} \tag{3.146}$$

the discretised forms of eqs. (3.142) and (3.143) are:

$$\frac{\mathrm{d}}{\mathrm{d}\overline{z}} \widehat{\mathbf{H}} = -\mathrm{j}\widehat{\boldsymbol{R}}_{\mathrm{E}} \widehat{\mathbf{E}} \qquad \frac{\mathrm{d}}{\mathrm{d}\overline{z}} \widehat{\mathbf{E}} = -\mathrm{j}\widehat{\boldsymbol{R}}_{\mathrm{H}} \widehat{\mathbf{H}} \tag{3.147}$$

with:

$$\widehat{\boldsymbol{R}}_{\mathrm{E}} = \begin{bmatrix} \epsilon_r \widehat{\boldsymbol{I}}_\bullet - \widehat{\boldsymbol{D}}_{\overline{x}}^{\bullet\mathrm{t}} \widehat{\boldsymbol{D}}_{\overline{x}}^{\bullet} & \widehat{\boldsymbol{D}}_{\overline{x}}^{\bullet\mathrm{t}} \widehat{\boldsymbol{D}}_{\overline{y}}^{\circ} \\ \widehat{\boldsymbol{D}}_{\overline{y}}^{\circ\mathrm{t}} \widehat{\boldsymbol{D}}_{\overline{x}}^{\bullet} & \epsilon_r \widehat{\boldsymbol{I}}_\circ - \widehat{\boldsymbol{D}}_{\overline{y}}^{\circ\mathrm{t}} \widehat{\boldsymbol{D}}_{\overline{y}}^{\circ} \end{bmatrix} \qquad \widehat{\boldsymbol{D}}_{\overline{x}}^{\circ,\bullet} = \boldsymbol{D}_{\overline{x}}^{\circ,\bullet} \otimes \boldsymbol{I}_y^{\circ,\bullet} \tag{3.148}$$

$$\widehat{\boldsymbol{R}}_{\mathrm{H}} = \epsilon_r^{-1} \begin{bmatrix} \epsilon_r \widehat{\boldsymbol{I}}_\bullet - \widehat{\boldsymbol{D}}_{\overline{y}}^{\bullet\mathrm{t}} \widehat{\boldsymbol{D}}_{\overline{y}}^{\bullet} & -\widehat{\boldsymbol{D}}_{\overline{y}}^{\bullet\mathrm{t}} \widehat{\boldsymbol{D}}_{\overline{x}}^{\circ} \\ -\widehat{\boldsymbol{D}}_{\overline{x}}^{\circ\mathrm{t}} \widehat{\boldsymbol{D}}_{\overline{y}}^{\bullet} & \epsilon_r \widehat{\boldsymbol{I}}_\circ - \widehat{\boldsymbol{D}}_{\overline{x}}^{\circ\mathrm{t}} \widehat{\boldsymbol{D}}_{\overline{x}}^{\circ} \end{bmatrix} \qquad \widehat{\boldsymbol{D}}_{\overline{y}}^{\circ,\bullet} = \boldsymbol{I}_x^{\circ,\bullet} \otimes \boldsymbol{D}_{\overline{y}}^{\circ,\bullet} \tag{3.149}$$

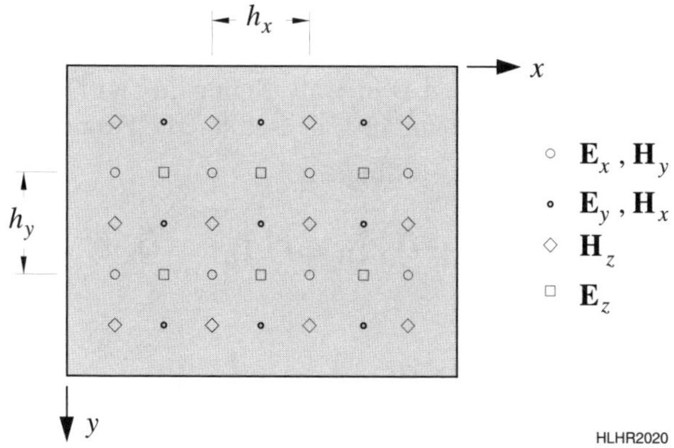

Fig. 3.44 2D discretisation scheme in the cross-section (Reproduced by permission of Copernicus Gesellschaft mbH)

Eqs. (3.147) are completely analogous to the well-known ones for coupled multi-conductor transmission lines. Therefore, we will show how the solutions can be obtained in an analogous manner. Combining eqs. (3.147), we obtain for the wave equations:

$$\frac{\mathrm{d}^2}{\mathrm{d}\overline{z}^2}\widehat{\mathbf{H}} - \widehat{\boldsymbol{Q}}^{\mathrm{H}}\widehat{\mathbf{H}} = \mathbf{0} \qquad \frac{\mathrm{d}^2}{\mathrm{d}\overline{z}^2}\widehat{\mathbf{E}} - \widehat{\boldsymbol{Q}}^{\mathrm{E}}\widehat{\mathbf{E}} = \mathbf{0} \tag{3.150}$$

where:

$$\widehat{\boldsymbol{Q}}^{\mathrm{H}} = -\widehat{\boldsymbol{R}}_{\mathrm{E}}\widehat{\boldsymbol{R}}_{\mathrm{H}} \qquad \widehat{\boldsymbol{Q}}^{\mathrm{E}} = -\widehat{\boldsymbol{R}}_{\mathrm{H}}\widehat{\boldsymbol{R}}_{\mathrm{E}} \tag{3.151}$$

The matrices $\widehat{\boldsymbol{R}}$ are symmetric and the relation between the two $\widehat{\boldsymbol{Q}}$ matrices is given by:

$$\widehat{\boldsymbol{R}}_{\mathrm{E,H}} = \widehat{\boldsymbol{R}}_{\mathrm{E,H}}^{\mathrm{t}} \qquad \widehat{\boldsymbol{Q}}_{\mathrm{H}} = \widehat{\boldsymbol{Q}}_{\mathrm{E}}^{\mathrm{t}} = \widehat{\boldsymbol{Q}}_{\mathrm{E}} \tag{3.152}$$

If we divide $\widehat{\boldsymbol{Q}}^{\mathrm{E,H}}$ into blocks, we obtain for the submatrices:

$$\begin{aligned} \boldsymbol{Q}_{11}^{\mathrm{E,H}} &= \widehat{\boldsymbol{D}}_{\overline{x}}^{\bullet\mathrm{t}}\widehat{\boldsymbol{D}}_{\overline{x}}^{\bullet} + \widehat{\boldsymbol{D}}_{\overline{y}}^{\bullet\mathrm{t}}\widehat{\boldsymbol{D}}_{\overline{y}}^{\bullet} - \epsilon_r\widehat{\boldsymbol{I}} \qquad \boldsymbol{Q}_{12}^{\mathrm{E,H}} = \boldsymbol{0} \\ \boldsymbol{Q}_{22}^{\mathrm{E,H}} &= \widehat{\boldsymbol{D}}_{\overline{x}}^{\circ\mathrm{t}}\widehat{\boldsymbol{D}}_{\overline{x}}^{\circ} + \widehat{\boldsymbol{D}}_{\overline{y}}^{\circ\mathrm{t}}\widehat{\boldsymbol{D}}_{\overline{y}}^{\circ} - \varepsilon_r\widehat{\boldsymbol{I}} \qquad \boldsymbol{Q}_{21}^{\mathrm{E,H}} = \boldsymbol{0} \end{aligned} \tag{3.153}$$

where we have taken into account the conditions $-\boldsymbol{D}_{\overline{y}}^{\bullet\mathrm{t}} = \boldsymbol{D}_{\overline{y}}^{\circ}$ and $-\boldsymbol{D}_{\overline{x}}^{\circ\mathrm{t}} = \boldsymbol{D}_{\overline{x}}^{\bullet}$. The wave equations for the x- and y-field components are decoupled. In detailed form, we may write:[3]

$$\boldsymbol{Q}_{11}^{\mathrm{E,H}} = (\boldsymbol{I}_{x\mathrm{D}} \otimes \overline{\boldsymbol{D}}_{\overline{y}\mathrm{N}}^{\mathrm{t}}\overline{\boldsymbol{D}}_{\overline{y}\mathrm{N}}) + (\overline{\boldsymbol{D}}_{\overline{x}\mathrm{D}}^{\mathrm{t}}\overline{\boldsymbol{D}}_{\overline{x}\mathrm{D}} \otimes \boldsymbol{I}_{y\mathrm{N}}) - \varepsilon_r\widehat{\boldsymbol{I}}_{\bullet} \tag{3.154}$$

$$\boldsymbol{Q}_{22}^{\mathrm{E,H}} = (\boldsymbol{I}_{x\mathrm{N}} \otimes \overline{\boldsymbol{D}}_{\overline{y}\mathrm{D}}^{\mathrm{t}}\overline{\boldsymbol{D}}_{\overline{y}\mathrm{D}}) + (\overline{\boldsymbol{D}}_{\overline{x}\mathrm{N}}^{\mathrm{t}}\overline{\boldsymbol{D}}_{\overline{x}\mathrm{N}} \otimes \boldsymbol{I}_{y\mathrm{D}}) - \varepsilon_r\widehat{\boldsymbol{I}}_{\circ} \tag{3.155}$$

[3]Instead of the symbols • and ∘ we use the subscripts N (for Neumann BC) and D (for Dirichlet BC). So we have e.g. $\boldsymbol{D}_{\overline{x}}^{\bullet} = \boldsymbol{D}_{\overline{x}\mathrm{D}}$. For the adequate BCs, see Fig. 3.44.

$\boldsymbol{I}_x(\boldsymbol{I}_y)$ is an identity matrix, whose size is the number of discretisation points in x- (y)-direction. $\boldsymbol{D}_x(\boldsymbol{D}_y)$ is the difference operator matrix in x- (y)-direction. The subscripts N and D identify Neumann and Dirichlet boundary conditions, respectively. The symbol $\otimes$ is used for the Kronecker product. By transformation of $\widehat{\widehat{\mathbf{H}}}$ and $\widehat{\mathbf{E}}$ according to:

$$\widehat{\mathbf{E}} = \widehat{\boldsymbol{T}}_{\mathrm{E}}\widehat{\widehat{\mathbf{E}}} \qquad \widehat{\widehat{\mathbf{H}}} = \widehat{\boldsymbol{T}}_{\mathrm{H}}\widehat{\widehat{\mathbf{H}}} \qquad \widehat{\boldsymbol{Q}}_{\mathrm{H}}\widehat{\boldsymbol{T}}_{\mathrm{H}} = \widehat{\boldsymbol{\Gamma}}^2\widehat{\boldsymbol{T}}_{\mathrm{H}} \qquad \widehat{\boldsymbol{Q}}_{\mathrm{E}}\widehat{\boldsymbol{T}}_{\mathrm{E}} = \widehat{\boldsymbol{\Gamma}}^2\widehat{\boldsymbol{T}}_{\mathrm{E}} \tag{3.156}$$

with:

$$\widehat{\boldsymbol{T}}_{\mathrm{H}} = \begin{bmatrix} \boldsymbol{T}_e & 0 \\ 0 & \boldsymbol{T}_h \end{bmatrix} = \begin{bmatrix} \boldsymbol{T}_{x\mathrm{D}} \otimes \boldsymbol{T}_{y\mathrm{N}} & 0 \\ 0 & \boldsymbol{T}_{x\mathrm{N}} \otimes \boldsymbol{T}_{y\mathrm{D}} \end{bmatrix} \qquad \widehat{\boldsymbol{T}}_{\mathrm{E}} = \widehat{\boldsymbol{R}}_{\mathrm{H}}\widehat{\boldsymbol{T}}_{\mathrm{H}}\widehat{\boldsymbol{\beta}}^{-1} \tag{3.157}$$

we obtain ($\widehat{\widehat{\mathbf{F}}} = \widehat{\widehat{\mathbf{E}}}$ or $\widehat{\widehat{\mathbf{H}}}$):

$$\frac{\mathrm{d}^2}{\mathrm{d}\overline{z}^2}\widehat{\widehat{\mathbf{F}}} - \widehat{\boldsymbol{\Gamma}}^2\widehat{\widehat{\mathbf{F}}} = 0 \qquad \widehat{\boldsymbol{\Gamma}}^2 = \mathrm{Diag}(\widehat{\boldsymbol{\Gamma}}^2_\bullet, \widehat{\boldsymbol{\Gamma}}^2_\circ) \qquad \widehat{\boldsymbol{\Gamma}}^2_{\circ,\bullet} = \widehat{\overline{\boldsymbol{\lambda}}}^2_{y\circ,\bullet} + \widehat{\overline{\boldsymbol{\lambda}}}^2_{x\circ,\bullet} - \varepsilon_r\widehat{\boldsymbol{I}}_{\circ,\bullet} \tag{3.158}$$

where:

$$\begin{aligned} \widehat{\overline{\boldsymbol{\lambda}}}^2_{x\circ} &= \overline{\boldsymbol{\lambda}}^2_{x\mathrm{N}} \otimes \boldsymbol{I}_{y\mathrm{D}} & \widehat{\overline{\boldsymbol{\lambda}}}^2_{x\bullet} &= \overline{\boldsymbol{\lambda}}^2_{x\mathrm{D}} \otimes \boldsymbol{I}_{y\mathrm{N}} \\ \widehat{\overline{\boldsymbol{\lambda}}}^2_{y\circ} &= \boldsymbol{I}_{x\mathrm{N}} \otimes \overline{\boldsymbol{\lambda}}^2_{y\mathrm{D}} & \widehat{\overline{\boldsymbol{\lambda}}}^2_{y\bullet} &= \boldsymbol{I}_{x\mathrm{D}} \otimes \overline{\boldsymbol{\lambda}}^2_{y\mathrm{N}} \\ \widehat{\boldsymbol{I}}_\circ &= \boldsymbol{I}_{x\mathrm{N}} \otimes \boldsymbol{I}_{y\mathrm{D}} & \widehat{\boldsymbol{I}}_\bullet &= \boldsymbol{I}_{x\mathrm{D}} \otimes \boldsymbol{I}_{y\mathrm{N}} \end{aligned} \tag{3.159}$$

We determine the transformation matrix $\widehat{\boldsymbol{T}}_{\mathrm{E}}$ from $\widehat{\boldsymbol{T}}_{\mathrm{H}}$ according to:

$$\widehat{\boldsymbol{T}}_{\mathrm{E}} = \widehat{\boldsymbol{R}}_{\mathrm{H}}\widehat{\boldsymbol{T}}_{\mathrm{H}}\widehat{\boldsymbol{\beta}}^{-1} = \widehat{\boldsymbol{T}}_{\mathrm{H}}\begin{bmatrix} \varepsilon_r\widehat{\boldsymbol{I}}_\bullet - \boldsymbol{I}_{x\mathrm{D}} \otimes \overline{\boldsymbol{\lambda}}^2_{y\mathrm{N}} & \overline{\boldsymbol{\delta}}_{x\mathrm{N}} \otimes \overline{\boldsymbol{\delta}}_{y\mathrm{D}} \\ \overline{\boldsymbol{\delta}}_{x\mathrm{D}} \otimes \overline{\boldsymbol{\delta}}_{y\mathrm{N}} & \varepsilon_r\widehat{\boldsymbol{I}}_\circ - \overline{\boldsymbol{\lambda}}^2_{x\mathrm{N}} \otimes \boldsymbol{I}_{y\mathrm{D}} \end{bmatrix}\widehat{\boldsymbol{\beta}}^{-1}\varepsilon_r^{-1} \tag{3.160}$$

with $\widehat{\boldsymbol{\Gamma}}^2 = -\widehat{\boldsymbol{\beta}}^2$ and e.g. $\overline{\boldsymbol{\delta}}_{x\mathrm{D}} = \boldsymbol{T}^{\mathrm{t}}_{x\mathrm{N}}\overline{\boldsymbol{D}}_{\overline{x}\mathrm{D}}\boldsymbol{T}_{x\mathrm{D}}$. Further, we use $\overline{\boldsymbol{\delta}}_{\mathrm{N}} = -\overline{\boldsymbol{\delta}}^{\mathrm{t}}_{\mathrm{D}}$. The multiplication with $\boldsymbol{\beta}^{-1}$ is introduced for normalisation.

The above wave equation has the following special solution for a waveguide section with ports A and B of distance d_{AB}:

$$\frac{\mathrm{d}}{\mathrm{d}\overline{z}}\begin{bmatrix} \widehat{\widehat{\mathbf{F}}}_{\mathrm{A}} \\ \widehat{\widehat{\mathbf{F}}}_{\mathrm{B}} \end{bmatrix} = \begin{bmatrix} -\widehat{\boldsymbol{\gamma}} & \widehat{\boldsymbol{\alpha}} \\ -\widehat{\boldsymbol{\alpha}} & \widehat{\boldsymbol{\gamma}} \end{bmatrix}\begin{bmatrix} \widehat{\widehat{\mathbf{F}}}_{\mathrm{A}} \\ \widehat{\widehat{\mathbf{F}}}_{\mathrm{B}} \end{bmatrix} \tag{3.161}$$

with:

$$\widehat{\boldsymbol{\alpha}} = \widehat{\boldsymbol{\Gamma}}/\sinh(\widehat{\boldsymbol{\Gamma}}\overline{d}_{\mathrm{AB}}) \qquad \widehat{\boldsymbol{\gamma}} = \widehat{\boldsymbol{\Gamma}}/\tanh(\widehat{\boldsymbol{\Gamma}}\overline{d}_{\mathrm{AB}}) \tag{3.162}$$

By using the different transformation matrices $\widehat{\boldsymbol{T}}_{\mathrm{E}}$ and $\widehat{\boldsymbol{T}}_{\mathrm{H}}$, we obtain for the GTL equations in transformed domain:

$$\frac{\mathrm{d}}{\mathrm{d}\overline{y}}\widehat{\widehat{\mathbf{E}}} = -\mathrm{j}\widehat{\boldsymbol{\beta}}\widehat{\widehat{\mathbf{H}}} \quad \frac{\mathrm{d}}{\mathrm{d}\overline{y}}\widehat{\widehat{\mathbf{H}}} = -\mathrm{j}\widehat{\boldsymbol{\beta}}\widehat{\widehat{\mathbf{E}}} \tag{3.163}$$

By introducing these GTL equations in transformed domain into the special solution in eq. (3.161), we obtain:

$$\begin{bmatrix} \widehat{\overline{\mathbf{H}}}_{\mathrm{A}} \\ -\widehat{\overline{\mathbf{H}}}_{\mathrm{B}} \end{bmatrix} = \begin{bmatrix} \widehat{\overline{\boldsymbol{y}}}_1 & \widehat{\overline{\boldsymbol{y}}}_2 \\ \widehat{\overline{\boldsymbol{y}}}_2 & \widehat{\overline{\boldsymbol{y}}}_1 \end{bmatrix} \begin{bmatrix} \widehat{\overline{\mathbf{E}}}_{\mathrm{A}} \\ \widehat{\overline{\mathbf{E}}}_{\mathrm{B}} \end{bmatrix} \quad \text{or} \quad \widehat{\overline{\mathbf{H}}}_{\mathrm{AB}} = \widehat{\overline{\boldsymbol{y}}}^{\mathrm{AB}}_{\mathrm{AB}} \widehat{\overline{\mathbf{E}}}_{\mathrm{AB}} \tag{3.164}$$

where:

$$\widehat{\overline{\boldsymbol{y}}}_1 = \boldsymbol{I}/\tanh(\widehat{\boldsymbol{\Gamma}}\overline{d}_{\mathrm{AB}}) = \mathrm{Diag}(\widehat{\overline{\boldsymbol{y}}}^{\bullet}_1, \widehat{\overline{\boldsymbol{y}}}^{\circ}_1) \quad \widehat{\overline{\boldsymbol{y}}}_2 = -\boldsymbol{I}/\sinh(\widehat{\boldsymbol{\Gamma}}\overline{d}_{\mathrm{AB}}) = \mathrm{Diag}(\widehat{\overline{\boldsymbol{y}}}^{\bullet}_2, \widehat{\overline{\boldsymbol{y}}}^{\circ}_2) \tag{3.165}$$

Inverting eq. (3.164) results in:

$$\begin{bmatrix} \widehat{\overline{\mathbf{E}}}_{\mathrm{A}} \\ \widehat{\overline{\mathbf{E}}}_{\mathrm{B}} \end{bmatrix} = \begin{bmatrix} \widehat{\overline{\boldsymbol{z}}}_1 & \widehat{\overline{\boldsymbol{z}}}_2 \\ \widehat{\overline{\boldsymbol{z}}}_2 & \widehat{\overline{\boldsymbol{z}}}_1 \end{bmatrix} \begin{bmatrix} \widehat{\overline{\mathbf{H}}}_{\mathrm{A}} \\ -\widehat{\overline{\mathbf{H}}}_{\mathrm{B}} \end{bmatrix} \quad \text{or} \quad \widehat{\overline{\mathbf{E}}}_{\mathrm{AB}} = \widehat{\overline{\boldsymbol{z}}}^{\mathrm{AB}}_{\mathrm{AB}} \widehat{\overline{\mathbf{H}}}_{\mathrm{AB}} \tag{3.166}$$

with:

$$\widehat{\overline{\boldsymbol{z}}}_1 = \boldsymbol{I}/\tanh(\widehat{\boldsymbol{\Gamma}}\overline{d}_{\mathrm{AB}}) = \mathrm{Diag}(\widehat{\overline{\boldsymbol{z}}}^{\bullet}_1, \widehat{\overline{\boldsymbol{z}}}^{\circ}_1) \quad \widehat{\overline{\boldsymbol{z}}}_2 = \boldsymbol{I}/\sinh(\widehat{\boldsymbol{\Gamma}}\overline{d}_{\mathrm{AB}}) = \mathrm{Diag}(\widehat{\overline{\boldsymbol{y}}}^{\bullet}_2, \widehat{\overline{\boldsymbol{y}}}^{\circ}_2) \tag{3.167}$$

The field components $\widehat{\mathbf{E}}_z$ and $\widehat{\mathbf{H}}_z$ are obtained by:

$$\mathrm{j}\varepsilon_r \widehat{\mathbf{E}}_z = [\boldsymbol{I}_{x\mathrm{D}} \otimes \overline{\boldsymbol{D}}_{\overline{y}\mathrm{N}} \quad \overline{\boldsymbol{D}}_{\overline{x}\mathrm{N}} \otimes \boldsymbol{I}_{y\mathrm{D}}]\widehat{\mathbf{H}} \qquad \mathrm{j}\widehat{\mathbf{H}}_z = [-\overline{\boldsymbol{D}}_{\overline{x}\mathrm{D}} \otimes \boldsymbol{I}_{y\mathrm{N}} \quad \boldsymbol{I}_{x\mathrm{N}} \otimes \overline{\boldsymbol{D}}_{\overline{y}\mathrm{D}}]\widehat{\mathbf{E}} \tag{3.168}$$

The transformation of $\widehat{\mathbf{E}}_z$ and $\widehat{\mathbf{H}}_z$ has to be performed according to:

$$\widehat{\mathbf{E}}_z = (\boldsymbol{T}_{x\mathrm{D}} \otimes \boldsymbol{T}_{y\mathrm{D}})\widehat{\overline{\mathbf{E}}}_z \quad \widehat{\mathbf{H}}_z = (\boldsymbol{T}_{x\mathrm{N}} \otimes \boldsymbol{T}_{y\mathrm{N}})\widehat{\overline{\mathbf{H}}}_z \tag{3.169}$$

For eq. (3.168), we obtain in transformed domain:

$$\begin{aligned} \mathrm{j}\varepsilon_r \widehat{\overline{\mathbf{E}}}_z &= [\boldsymbol{I}_{x\mathrm{D}} \otimes \boldsymbol{T}^{\mathrm{t}}_{y\mathrm{D}}\overline{\boldsymbol{D}}_{\overline{y}\mathrm{N}}\boldsymbol{T}_{y\mathrm{N}} \quad \boldsymbol{T}^{\mathrm{t}}_{x\mathrm{D}}\overline{\boldsymbol{D}}_{\overline{x}\mathrm{N}}\boldsymbol{T}_{x\mathrm{N}} \otimes \boldsymbol{I}_{y\mathrm{D}}]\widehat{\overline{\mathbf{H}}} \\ \mathrm{j}\widehat{\overline{\mathbf{H}}}_z &= [-\boldsymbol{T}^{\mathrm{t}}_{x\mathrm{N}}\overline{\boldsymbol{D}}_{\overline{x}\mathrm{D}}\boldsymbol{T}_{x\mathrm{D}} \otimes \boldsymbol{I}_{y\mathrm{N}} \quad \boldsymbol{I}_{x\mathrm{N}} \otimes \boldsymbol{T}^{\mathrm{t}}_{y\mathrm{N}}\overline{\boldsymbol{D}}_{\overline{y}\mathrm{D}}\boldsymbol{T}_{y\mathrm{D}}]\widehat{\overline{\mathbf{E}}} \end{aligned} \tag{3.170}$$

or:

$$\mathrm{j}\varepsilon_r \widehat{\overline{\mathbf{E}}}_z = [\boldsymbol{I}_{x\mathrm{D}} \otimes \overline{\boldsymbol{\delta}}_{y\mathrm{N}} \quad \overline{\boldsymbol{\delta}}_{x\mathrm{N}} \otimes \boldsymbol{I}_{y\mathrm{D}}]\widehat{\overline{\mathbf{H}}} \tag{3.171}$$

$$\mathrm{j}\widehat{\overline{\mathbf{H}}}_z = [-\overline{\boldsymbol{\delta}}_{x\mathrm{D}} \otimes \boldsymbol{I}_{y\mathrm{N}} \quad \boldsymbol{I}_{x\mathrm{N}} \otimes \overline{\boldsymbol{\delta}}_{y\mathrm{D}}]\boldsymbol{\beta}^{-1}\widehat{\overline{\mathbf{E}}} = \mathrm{j}[\widehat{\overline{\boldsymbol{y}}}^{\bullet}_3, \widehat{\overline{\boldsymbol{y}}}^{\circ}_3]\widehat{\overline{\mathbf{E}}} \tag{3.172}$$

where we have $\widehat{\overline{\boldsymbol{y}}}^{\bullet}_3 = -\overline{\boldsymbol{\delta}}_{x\mathrm{D}} \otimes \boldsymbol{I}_{y\mathrm{N}}\boldsymbol{\Gamma}^{-1}_{\bullet}$ and $\widehat{\overline{\boldsymbol{y}}}^{\circ}_3 = \boldsymbol{I}_{x\mathrm{N}} \otimes \overline{\boldsymbol{\delta}}_{y\mathrm{D}}\boldsymbol{\Gamma}^{-1}_{\circ}$. For the impedance/admittance transformation through waveguide sections, see Chapter 2.5.

3.4.4 Discontinuities

The impedance/admittance transformation through waveguide discontinuities is described in Chapter 2.5 and will not be repeated here.

3.4.5 Coupling to other ports

To obtain the coupling matrices in eq. (3.140), we short circuit (to obtain y matrices) some of the ports. e.g. to compute the elements in the left column, we short circuit ports C, D, E and F. We will now show how these terms can be determined. The elements in the other columns can be computed in an analogous way. The field between the two ports A and B at an arbitrary location z can be calculated from the field at ports A and B by:

$$\widehat{\widehat{\mathbf{F}}}(z) = \boldsymbol{\Lambda}_{\mathrm{A}}^{\mathrm{d}}\widehat{\widehat{\mathbf{F}}}_{\mathrm{A}} + \boldsymbol{\Lambda}_{\mathrm{B}}^{\mathrm{d}}\widehat{\widehat{\mathbf{F}}}_{\mathrm{B}} \tag{3.173}$$

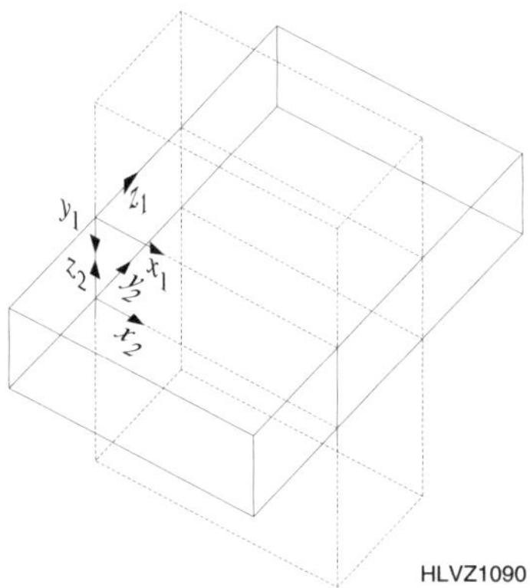

Fig. 3.45 Coupling from ports A and B to E-plane ports C and D

$\widehat{\widehat{\mathbf{F}}}$ is the vector with the relevant field components. Only magnetic field components are responsible for the coupling from ports A and B to the other ports. The electric field causes only tangential field components, which are zero at the other two ports. The diagonal matrices $\boldsymbol{\Lambda}^{\mathrm{d}}$ are given by:

$$\boldsymbol{\Lambda}_{\mathrm{A}}^{\mathrm{d}} = \frac{\sinh(\boldsymbol{\Gamma}_z \overline{z}_-)}{\sinh(\boldsymbol{\Gamma}_z \overline{d}_{\mathrm{AB}})} \qquad \boldsymbol{\Lambda}_{\mathrm{B}}^{\mathrm{d}} = \frac{\sinh(\boldsymbol{\Gamma}_z \overline{z})}{\sinh(\boldsymbol{\Gamma}_z \overline{d}_{\mathrm{AB}})} \tag{3.174}$$

where $\overline{z}_- = \overline{d}_{\mathrm{AB}} - \overline{z}$ and $\overline{d}_{\mathrm{AB}} = k_0 d_{\mathrm{AB}}$. d_{AB} is the distance between ports A and B. The diagonal matrix $\boldsymbol{\Gamma}_z$ is equal to $\boldsymbol{\Gamma}_\bullet(\boldsymbol{\Gamma}_\circ)$ for $\widehat{\widehat{\mathbf{F}}} = \widehat{\widehat{\mathbf{H}}}_x(\widehat{\widehat{\mathbf{H}}}_y)$. For $\widehat{\widehat{\mathbf{F}}} = \widehat{\widehat{\mathbf{H}}}_z$, we obtain ($\boldsymbol{\Gamma}_z^2 = \boldsymbol{\Gamma}_\diamond^2$):

$$\boldsymbol{\Gamma}_\diamond^2 = \boldsymbol{I}_{x\mathrm{N}} \otimes \overline{\boldsymbol{\lambda}}_{y\mathrm{N}}^2 + \overline{\boldsymbol{\lambda}}_{x\mathrm{N}}^2 \otimes \boldsymbol{I}_{y\mathrm{N}} - \varepsilon_r \boldsymbol{I}_{x\mathrm{N}} \otimes \boldsymbol{I}_{y\mathrm{N}} \tag{3.175}$$

3.4.5.1 Coupling from ports A and B to ports C and D (Fig. 3.45)

At ports $R \equiv$ C and $R \equiv$ D, the field $\mathbf{H}_x$ and $\mathbf{H}_z$ are determined by:

$$\mathbf{H}_{x\mathrm{R}}(z) = \boldsymbol{T}_{x\mathrm{D}} \otimes \boldsymbol{T}_{y\mathrm{N}}^{\Delta\mathrm{R}}(\boldsymbol{\Lambda}_{\bullet\mathrm{A}}^{d}\widehat{\widehat{\mathbf{H}}}_{x\mathrm{A}} + \boldsymbol{\Lambda}_{\bullet\mathrm{B}}^{d}\widehat{\widehat{\mathbf{H}}}_{x\mathrm{B}}) \tag{3.176}$$

$$\mathbf{H}_{z\mathrm{R}}(z) = \boldsymbol{T}_{x\mathrm{N}} \otimes \boldsymbol{T}_{y\mathrm{N}}^{\Delta\mathrm{R}}(\boldsymbol{\Lambda}_{\diamond\mathrm{A}}^{d}\widehat{\widehat{\mathbf{H}}}_{z\mathrm{A}} + \boldsymbol{\Lambda}_{\diamond\mathrm{B}}^{d}\widehat{\widehat{\mathbf{H}}}_{z\mathrm{B}}) \tag{3.177}$$

Note that all terms $\boldsymbol{\Lambda}^d_{\mathrm{A,B}}$ in these equations are different, because of the different parameter $\boldsymbol{\Gamma}_z$ in eq. (3.174). The row vectors $\boldsymbol{T}^{\Delta\mathrm{R}}_{y\mathrm{N}}$ are given by:

$$\boldsymbol{T}^{\Delta\mathrm{C}}_{y\mathrm{N}} = \frac{1}{8}(9\boldsymbol{T}^{(\mathrm{N}_y)}_{y\mathrm{N}} - \boldsymbol{T}^{(\mathrm{N}_y-1)}_{y\mathrm{N}}) \quad \boldsymbol{T}^{\Delta\mathrm{D}}_{y\mathrm{N}} = \frac{1}{8}(9\boldsymbol{T}^{(1)}_{y\mathrm{N}} - \boldsymbol{T}^{(2)}_{y\mathrm{N}}) \tag{3.178}$$

The superscripts 1, 2, $\mathrm{N}_y - 1, \mathrm{N}_y$ identify the row of $\boldsymbol{T}_{y\mathrm{N}}$. The relevant fields at ports $\mathrm{R} \equiv \mathrm{E}$ or $\mathrm{R} \equiv \mathrm{F}$ are expressed by:

$$\mathbf{H}_{y\mathrm{R}}(z) = \boldsymbol{T}^{\Delta\mathrm{R}}_{x\mathrm{N}} \otimes \boldsymbol{T}_{y\mathrm{D}}(\boldsymbol{\Lambda}^{\mathrm{d}}_{\circ\mathrm{A}}\hat{\hat{\mathbf{H}}}_{y\mathrm{A}} + \boldsymbol{\Lambda}^{\mathrm{d}}_{\circ\mathrm{B}}\hat{\hat{\mathbf{H}}}_{y\mathrm{B}}) \tag{3.179}$$

$$\mathbf{H}_{z\mathrm{R}}(z) = \boldsymbol{T}^{\Delta\mathrm{R}}_{x\mathrm{N}} \otimes \boldsymbol{T}_{y\mathrm{N}}(\boldsymbol{\Lambda}^{\mathrm{d}}_{\diamond\mathrm{A}}\hat{\hat{\mathbf{H}}}_{z\mathrm{A}} + \boldsymbol{\Lambda}^{\mathrm{d}}_{\diamond\mathrm{B}}\hat{\hat{\mathbf{H}}}_{z\mathrm{B}}) \tag{3.180}$$

The row vectors $\boldsymbol{T}^{\Delta\mathrm{R}}_{x\mathrm{N}}$ can be calculated in a similar way to $\boldsymbol{T}^{\Delta\mathrm{R}}_{y\mathrm{N}}$.

The eqs. (3.176)–(3.180) must now be evaluated at the discrete points z_i of the ports. These points may be different at each port. Let us take a look at eq. (3.180). We change the order of the product between $\boldsymbol{T}_{x\mathrm{D}} \otimes \boldsymbol{T}^{\Delta\mathrm{R}}_{y\mathrm{N}}$ and $\boldsymbol{\Lambda}^{\mathrm{d}}_{\bullet}$. The term $\boldsymbol{\Lambda}^{\mathrm{d}}_{\bullet}(z_i)$ should now be the ith row of the matrix $\boldsymbol{\Lambda}_{\bullet}$. The part $\mathbf{H}^{\mathrm{A}}_{x\mathrm{R}}$ of $\mathbf{H}_{x\mathrm{R}}$ (related to the field $\hat{\hat{\mathbf{H}}}_{x\mathrm{A}}$) can be expressed as:

$$\hat{\mathbf{H}}_{x\mathrm{R}} = \begin{bmatrix} \boldsymbol{\Lambda}_{\bullet A}(\boldsymbol{T}^{1}_{x\mathrm{D}} \otimes \boldsymbol{T}^{\Delta\mathrm{R}}_{y\mathrm{N}})^{\mathrm{d}} \\ \vdots \\ \boldsymbol{\Lambda}_{\bullet A}(\boldsymbol{T}^{\mathrm{N}x}_{x\mathrm{D}} \otimes \boldsymbol{T}^{\Delta\mathrm{R}}_{y\mathrm{N}})^{\mathrm{d}} \end{bmatrix} \hat{\hat{\mathbf{H}}}_{x\mathrm{A}} = \mathbf{V}^{\mathrm{A}}_{x\mathrm{R}}\hat{\hat{\mathbf{H}}}_{x\mathrm{A}} \tag{3.181}$$

$(\boldsymbol{T}^{\mathrm{k}}_{x\mathrm{D}} \otimes \boldsymbol{T}^{\Delta\mathrm{R}}_{y\mathrm{N}})^d$ is a diagonal matrix constructed from the Kronecker product between the kth row of $\boldsymbol{T}_{x\mathrm{D}}$ and $\boldsymbol{T}^{\Delta\mathrm{R}}_{y\mathrm{N}}$. By defining $\mathbf{V}^{\mathrm{A}}_{x\mathrm{R}}$ as:

$$\hat{\hat{\mathbf{H}}}^{\mathrm{A}}_{x\mathrm{R}} = \overline{\mathbf{V}}^{\mathrm{A}}_{x\mathrm{R}}\hat{\hat{\mathbf{H}}}_{x\mathrm{A}} \tag{3.182}$$

we obtain from eq. (3.181) after transformation:

$$\overline{\mathbf{V}}^{\mathrm{A}}_{\mathrm{R}} = (\boldsymbol{T}^{\mathrm{t}}_{x\mathrm{D}} \otimes \boldsymbol{T}^{\mathrm{t}}_{y_2\mathrm{N}})\mathbf{V}^{\mathrm{A}}_{x\mathrm{R}} \tag{3.183}$$

$\boldsymbol{T}_{y_2\mathrm{N}}$ is the transformation matrix for z-direction in port R. $\overline{\mathbf{V}}^{\mathrm{B}}_{x\mathrm{R}}$ and $\overline{\mathbf{V}}^{\mathrm{A,B}}_{z\mathrm{R}}$ can be calculated in an analogous manner. To summarise, the results may be written as follows:

$$\hat{\hat{\mathbf{H}}}_{\mathrm{CD}} = \begin{bmatrix} -\hat{\hat{\mathbf{H}}}_{x\mathrm{C}} \\ \hat{\hat{\mathbf{H}}}_{z\mathrm{C}} \\ \hat{\hat{\mathbf{H}}}_{x\mathrm{D}} \\ -\hat{\hat{\mathbf{H}}}_{z\mathrm{D}} \end{bmatrix} = \begin{bmatrix} \overline{\mathbf{V}}^{\mathrm{A}}_{x\mathrm{C}} & & -\overline{\mathbf{V}}^{\mathrm{B}}_{x\mathrm{C}} & \\ & \overline{\mathbf{V}}^{\mathrm{A}}_{z\mathrm{C}} & & -\overline{\mathbf{V}}^{\mathrm{B}}_{z\mathrm{C}} \\ -\overline{\mathbf{V}}^{\mathrm{A}}_{x\mathrm{D}} & & \overline{\mathbf{V}}^{\mathrm{B}}_{x\mathrm{D}} & \\ & -\overline{\mathbf{V}}^{\mathrm{A}}_{z\mathrm{D}} & & \overline{\mathbf{V}}^{\mathrm{B}}_{z\mathrm{D}} \end{bmatrix} \begin{bmatrix} -\hat{\hat{\mathbf{H}}}_{x\mathrm{A}} \\ \hat{\hat{\mathbf{H}}}_{z\mathrm{A}} \\ \hat{\hat{\mathbf{H}}}_{x\mathrm{B}} \\ -\hat{\hat{\mathbf{H}}}_{z\mathrm{B}} \end{bmatrix}$$

$$= \hat{\overline{\mathbf{V}}}^{\mathrm{AB}}_{\mathrm{CD}}\hat{\hat{\mathbf{H}}}_{\mathrm{AB}} \tag{3.184}$$

The field quantities on the right side must be expressed in terms of an electric field, with the help of eqs. (3.172) and (3.164):

$$\begin{bmatrix} -\widehat{\overline{\mathbf{H}}}_{x\mathrm{A}} \\ \widehat{\overline{\mathbf{H}}}_{z\mathrm{A}} \\ \widehat{\overline{\mathbf{H}}}_{x\mathrm{B}} \\ -\widehat{\overline{\mathbf{H}}}_{z\mathrm{B}} \end{bmatrix} = \begin{bmatrix} -\widehat{\overline{\boldsymbol{y}}}_1^{\bullet} & & -\widehat{\overline{\boldsymbol{y}}}_2^{\bullet} & \\ \widehat{\overline{\boldsymbol{y}}}_3^{\bullet} & \widehat{\overline{\boldsymbol{y}}}_3^{\circ} & & \\ \widehat{\overline{\boldsymbol{y}}}_2^{\bullet} & & \widehat{\overline{\boldsymbol{y}}}_1^{\bullet} & \\ & & -\widehat{\overline{\boldsymbol{y}}}_3^{\bullet} & -\widehat{\overline{\boldsymbol{y}}}_3^{\circ} \end{bmatrix} \begin{bmatrix} \widehat{\overline{\mathbf{E}}}_{y\mathrm{A}} \\ \widehat{\overline{\mathbf{E}}}_{x\mathrm{A}} \\ \widehat{\overline{\mathbf{E}}}_{y\mathrm{B}} \\ \widehat{\overline{\mathbf{E}}}_{x\mathrm{B}} \end{bmatrix} \tag{3.185}$$

or:

$$\widehat{\overline{\mathbf{H}}}_{\mathrm{AB}} = \widehat{\overline{\mathbf{Y}}}_{\mathrm{AB}}^{\mathrm{AB}} \widehat{\overline{\mathbf{E}}}_{\mathrm{AB}} \tag{3.186}$$

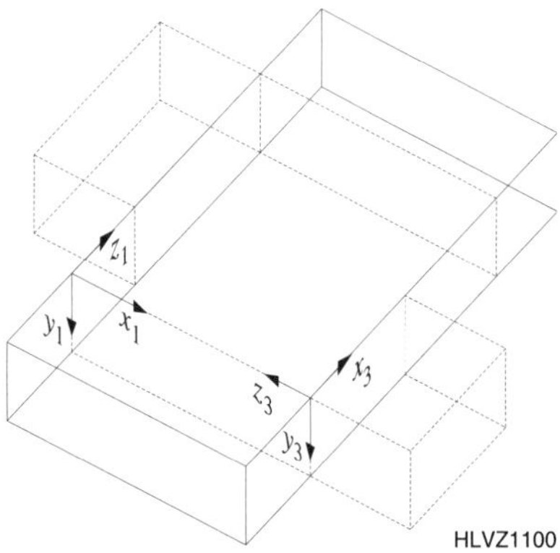

Fig. 3.46 Coupling from ports A and B to H-plane ports E and F

Now the short-circuit parameter matrix $\overline{\boldsymbol{y}}_{\mathrm{CD}}^{\mathrm{AB}}$ is given by:

$$\overline{\boldsymbol{y}}_{\mathrm{CD}}^{\mathrm{AB}} = \widehat{\overline{\mathbf{V}}}_{\mathrm{CD}}^{\mathrm{AB}} \widehat{\overline{\mathbf{Y}}}_{\mathrm{AB}}^{\mathrm{AB}} \tag{3.187}$$

3.4.5.2 Coupling from ports A and B to ports E and F (Fig. 3.46)

The field part $\mathbf{H}_{y\mathrm{R}}^{\mathrm{A}}$ of $\mathbf{H}_{y\mathrm{R}}$ related to $\widehat{\overline{\mathbf{H}}}_{y\mathrm{A}}$ can be expressed as:

$$\widehat{\mathbf{H}}_{y\mathrm{R}}^{\mathrm{A}} = \begin{bmatrix} (\boldsymbol{T}_{x\mathrm{N}}^{\Delta\mathrm{R}} \otimes \boldsymbol{T}_{y\mathrm{D}}) \boldsymbol{\Lambda}_{\circ\mathrm{A}}^{\mathrm{d}}(z_{\mathrm{N}_1}) \\ \vdots \\ (\boldsymbol{T}_{x\mathrm{N}}^{\Delta\mathrm{R}} \otimes \boldsymbol{T}_{y\mathrm{D}}) \boldsymbol{\Lambda}_{\circ\mathrm{A}}^{\mathrm{d}}(z_{\mathrm{N}_z}) \end{bmatrix} \widehat{\overline{\mathbf{H}}}_{y\mathrm{A}} = \mathbf{V}_{y\mathrm{R}}^{\mathrm{A}} \widehat{\overline{\mathbf{H}}}_{y\mathrm{A}} \tag{3.188}$$

By defining $\overline{\mathbf{V}}_{y\mathrm{R}}^{\mathrm{A}}$ as:

$$\widehat{\overline{\mathbf{H}}}_{y\mathrm{R}}^{\mathrm{A}} = \overline{\mathbf{V}}_{y\mathrm{R}}^{\mathrm{A}} \widehat{\overline{\mathbf{H}}}_{y\mathrm{A}} \tag{3.189}$$

we obtain from eq. (3.188) after transformation:

$$\overline{\mathbf{V}}_{y\mathrm{R}}^{\mathrm{A}} = (\boldsymbol{T}_{x_3\mathrm{N}} \otimes \boldsymbol{T}_{y\mathrm{D}}) \mathbf{V}_{y\mathrm{R}}^{\mathrm{A}} \tag{3.190}$$

Analogously, $\overline{\mathbf{V}}_{y\mathrm{B}}^{\mathrm{B}}$ and $\overline{\mathbf{V}}_{x_3\mathrm{R}}^{\mathrm{A,B}}$ can be determined. To summarise, the results may be written as follows:

$$\hat{\hat{\mathbf{H}}}_{\mathrm{EF}} = \begin{bmatrix} -\hat{\hat{\mathbf{H}}}_{x_3\mathrm{E}} \\ \hat{\hat{\mathbf{H}}}_{y_3\mathrm{E}} \\ \hat{\hat{\mathbf{H}}}_{x_3\mathrm{F}} \\ -\hat{\hat{\mathbf{H}}}_{y_3\mathrm{F}} \end{bmatrix} = \begin{bmatrix} -\overline{\mathbf{V}}_{x_3\mathrm{E}}^{\mathrm{A}} & & -\overline{\mathbf{V}}_{x_3\mathrm{E}}^{\mathrm{B}} & \\ & \overline{\mathbf{V}}_{y_3\mathrm{E}}^{\mathrm{A}} & & -\overline{\mathbf{V}}_{y_3\mathrm{E}}^{\mathrm{B}} \\ -\overline{\mathbf{V}}_{x_3\mathrm{F}}^{\mathrm{A}} & & \overline{\mathbf{V}}_{x_3\mathrm{F}}^{\mathrm{B}} & \\ & \overline{\mathbf{V}}_{y_3\mathrm{F}}^{\mathrm{A}} & & -\overline{\mathbf{V}}_{y_3\mathrm{F}}^{\mathrm{B}} \end{bmatrix} \begin{bmatrix} -\hat{\hat{\mathbf{H}}}_{x\mathrm{A}} \\ \hat{\hat{\mathbf{H}}}_{z\mathrm{A}} \\ \hat{\hat{\mathbf{H}}}_{x\mathrm{B}} \\ -\hat{\hat{\mathbf{H}}}_{z\mathrm{B}} \end{bmatrix}$$

$$= \hat{\overline{\mathbf{V}}}_{\mathrm{EF}}^{\mathrm{AB}} \hat{\hat{\mathbf{H}}}_{\mathrm{AB}} \tag{3.191}$$

With the help of eq. (3.186), we obtain for the parameter $\overline{\boldsymbol{y}}_{\mathrm{EF}}^{\mathrm{AB}}$:

$$\overline{\boldsymbol{y}}_{\mathrm{EF}}^{\mathrm{AB}} = \hat{\overline{\mathbf{V}}}_{\mathrm{EF}}^{\mathrm{AB}} \mathbf{Y}_{\mathrm{AB}}^{\mathrm{AB}} \tag{3.192}$$

All the other transformed submatrices in eq. (3.140) can be computed analogously.

3.4.6 Impedance/admittance transformation

In this section we derive the formulas for the impedance/admittance transformation from one port to another. Let us combine the impedance/admittance matrices of two opposite ports according to:

$$\hat{\hat{\mathbf{Z}}}_{\mathrm{UV}} = \mathrm{Diag}\,(\hat{\hat{\mathbf{Z}}}_{\mathrm{U}}, -\hat{\hat{\mathbf{Z}}}_{\mathrm{V}}) \qquad \widehat{\widehat{\mathbf{Y}}}_{\mathrm{UV}} = \mathrm{Diag}\,(\widehat{\widehat{\mathbf{Y}}}_{\mathrm{U}}, -\widehat{\widehat{\mathbf{Y}}}_{\mathrm{V}}) \tag{3.193}$$

The impedances $\hat{\hat{\mathbf{Z}}}_{\mathrm{UV}}$ (admittances $\widehat{\widehat{\mathbf{Y}}}_{\mathrm{UV}}$) are obtained from the input impedances (admittances) of the connecting waveguide by a transformation that is analogous to eq. (3.25). By introducing the above relation for ports CD and EF into the transformed eq. (3.140), we obtain:

$$\begin{bmatrix} \hat{\hat{\mathbf{H}}}_{\mathrm{AB}} \\ \hat{\hat{\mathbf{H}}}_{\mathrm{CD}} \\ \hat{\hat{\mathbf{H}}}_{\mathrm{EF}} \end{bmatrix} = \begin{bmatrix} \hat{\overline{\boldsymbol{y}}}_{\mathrm{AB}}^{\mathrm{AB}} & \hat{\overline{\boldsymbol{y}}}_{\mathrm{AB}}^{\mathrm{CD}} & \hat{\overline{\boldsymbol{y}}}_{\mathrm{AB}}^{\mathrm{EF}} \\ \hat{\overline{\boldsymbol{y}}}_{\mathrm{CD}}^{\mathrm{AB}} & \hat{\overline{\boldsymbol{y}}}_{\mathrm{CD}}^{\mathrm{CD}} & \hat{\overline{\boldsymbol{y}}}_{\mathrm{CD}}^{\mathrm{EF}} \\ \hat{\overline{\boldsymbol{y}}}_{\mathrm{EF}}^{\mathrm{AB}} & \hat{\overline{\boldsymbol{y}}}_{\mathrm{EF}}^{\mathrm{CD}} & \hat{\overline{\boldsymbol{y}}}_{\mathrm{EF}}^{\mathrm{EF}} \end{bmatrix} \begin{bmatrix} \hat{\hat{\mathbf{E}}}_{\mathrm{AB}} \\ \hat{\hat{\mathbf{Z}}}_{\mathrm{CD}}\hat{\hat{\mathbf{H}}}_{\mathrm{CD}} \\ \hat{\hat{\mathbf{Z}}}_{\mathrm{EF}}\hat{\hat{\mathbf{H}}}_{\mathrm{EF}} \end{bmatrix} \tag{3.194}$$

Now, by defining $\hat{\overline{\mathbf{V}}}_{\mathrm{EF}}^{\mathrm{CD}}$ as:

$$\hat{\overline{\mathbf{V}}}_{\mathrm{EF}}^{\mathrm{CD}} = \begin{bmatrix} \hat{\overline{\boldsymbol{y}}}_{\mathrm{CD}}^{\mathrm{CD}}\hat{\hat{\mathbf{Z}}}_{\mathrm{CD}} - \boldsymbol{I} & \hat{\overline{\boldsymbol{y}}}_{\mathrm{CD}}^{\mathrm{EF}}\hat{\hat{\mathbf{Z}}}_{\mathrm{EF}} \\ \hat{\overline{\boldsymbol{y}}}_{\mathrm{EF}}^{\mathrm{CD}}\hat{\hat{\mathbf{Z}}}_{\mathrm{CD}} & \hat{\overline{\boldsymbol{y}}}_{\mathrm{EF}}^{\mathrm{EF}}\hat{\hat{\mathbf{Z}}}_{\mathrm{EF}} - \boldsymbol{I} \end{bmatrix} \tag{3.195}$$

we obtain:

$$\hat{\hat{\mathbf{H}}}_{\mathrm{AB}} = \widehat{\widehat{\mathbf{Y}}}_{\mathrm{AB}}^{\mathrm{AB}}\hat{\hat{\mathbf{E}}}_{\mathrm{AB}}$$

$$\widehat{\widehat{\mathbf{Y}}}_{\mathrm{AB}}^{\mathrm{AB}} = \hat{\overline{\boldsymbol{y}}}_{\mathrm{AB}}^{\mathrm{AB}} - \left[\hat{\overline{\boldsymbol{y}}}_{\mathrm{AB}}^{\mathrm{CD}}\hat{\hat{\mathbf{Z}}}_{\mathrm{CD}} \quad \hat{\overline{\boldsymbol{y}}}_{\mathrm{AB}}^{\mathrm{EF}}\hat{\hat{\mathbf{Z}}}_{\mathrm{EF}}\right] \hat{\overline{\mathbf{V}}}_{\mathrm{EF}}^{\mathrm{CD}-1} \begin{bmatrix} \hat{\overline{\boldsymbol{y}}}_{\mathrm{CD}}^{\mathrm{AB}} \\ \hat{\overline{\boldsymbol{y}}}_{\mathrm{EF}}^{\mathrm{AB}} \end{bmatrix} \tag{3.196}$$

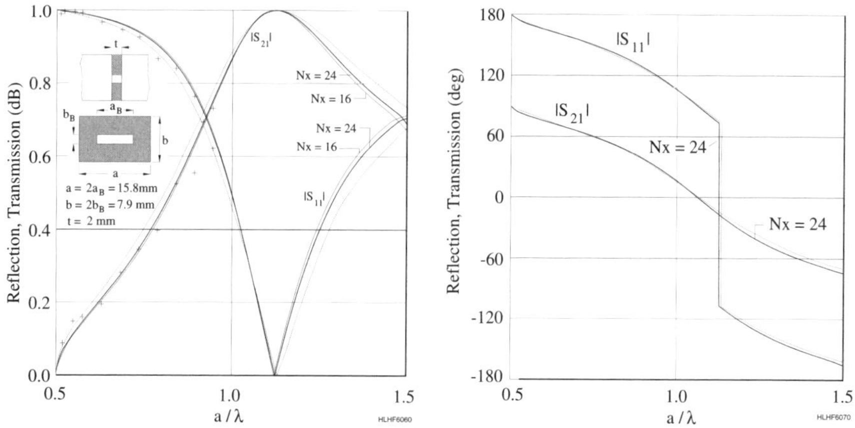

Fig. 3.47 Finite thickness iris in a rectangular waveguide. Magnitude and phase of the scattering coefficients as function of frequency (results from [20] and [21]) (Reproduced by permission of Copernicus Gesellschaft mbH)

We obtain the following input impedance at port A:

$$\widehat{\overline{\overline{\mathbf{Y}}}}_{\mathrm{A}} = \widehat{\overline{\overline{\mathbf{Y}}}}^{\mathrm{AB}}_{\mathrm{AB11}} - \widehat{\overline{\overline{\mathbf{Y}}}}^{\mathrm{AB}}_{\mathrm{AB12}}\left(\widehat{\overline{\overline{\mathbf{Y}}}}^{\mathrm{AB}}_{\mathrm{AB22}} + \widehat{\overline{\overline{\mathbf{Y}}}}_{\mathrm{B}}\right)^{-1}\widehat{\overline{\overline{\mathbf{Y}}}}^{\mathrm{AB}}_{\mathrm{AB21}} \tag{3.197}$$

where $\widehat{\overline{\overline{\mathbf{Y}}}}^{\mathrm{AB}}_{\mathrm{AB}ik}$ $i, k = 1, 2$ are the submatrices of $\widehat{\overline{\overline{\mathbf{Y}}}}^{\mathrm{AB}}_{\mathrm{AB}}$.

3.4.7 Numerical results

The proposed algorithm was used to examine an iris with finite thickness (see Fig. 3.47). The results of the determined scattering parameters were compared with other numerical methods [20], [21], and showed a very good agreement.

References

[1] R. Pregla, 'The Method of Lines for the Analysis of Dielectric Waveguides Bends', *J. Lightwave Technol.*, vol. 14, no. 4, pp. 634–639, Apr. 1996.

[2] M. Thorburn, A. Biswas and V. K. Tripathi, 'Application of the Method of Lines to Cylindrical Inhomogeneous Propagation Structures', *Electron. Lett.*, vol. 26, no. 3, pp. 170–171, Feb. 1990.

[3] W. Pascher and R. Pregla, 'Analysis of Rectangular Waveguide Junctions by the Method of Lines', *IEE Trans. on Microwave Theory Tech.*, vol. 43, pp. 2649–2653, 1995.

[4] U. Rogge and R. Pregla, 'Method of Lines for the Analysis of Dielectric Waveguides', *J. Lightwave Technol.*, vol. 11, no. 12, pp. 2015–2020, 1993.

[5] R. Pregla, 'The Method of Lines as Generalized Transmission Line Technique for the Analysis of Multilayered Structures', *AEÜ*, vol. 50, no. 5, pp. 293–300, Sep. 1996.

[6] A. Weisshaar, S. M. Goodnick and V. J. Tripathi, 'A Rigorous and Efficient Method of Moments Solution for Curved Waveguide Bends', *IEEE Trans. Microwave Theory Tech.*, vol. MTT-40, no. 12, pp. 2200–2206, 1992.

[7] B. Gimeno and M. Guglielmi, 'Multimode network representation for H- and E-plane uniform bends in rectangular waveguide', in *IEEE MTT-S Int. Symp. Dig.*, 1995, pp. 241–244.

[8] R. Pregla, 'Concatenations of Waveguide Sections', *IEE Proc. -H, Microwave Antennas Propagation*, vol. 144, no. 2, pp. 119–125, Apr. 1997.

[9] M. Mongiardo, A. Morini and T. Rozzi, 'Analysis and Design of Full-Band Matched Waveguide Bends', *IEEE Trans. Microwave Theory Tech.*, vol. MTT-43, no. 12, pp. 2965–2971, 1992.

[10] A. Klaassen, H. Barth and W. Menzel, 'Effektiver Entwurf und Aufbau von mm-Wellen-Hohlleiter-Bandpaßfiltern', in *Proc. Conf Microw. and Optronics – MIOP'90*, Stuttgart, Germany, 1990, pp. 534–536.

[11] J. Machać, 'Analysis of Discoutinuities in Waveguiding Structures by MAB Method', *IEE Proc. -H, Microwave Antennas Propagation*, vol. 139, pp. 351–357, Apr. 1992.

[12] L. Q. Bui, D. Ball and I. Itoh, 'Broadband Millimeter-Wave E-Plane Bandpass Filters', *IEEE Trans. Microwave Theory Tech.*, vol. MTT-32, pp. 1655–1658, Dec. 1984.

[13] U. Schulz, 'Evaluation of the Equivalent Circuit Parameters of Posts and Diaphragms in Waveguides by the Method of Lines', *AEÜ*, vol. 39, no. 3, pp. 203–207, 1985.

[14] R. Pregla, 'Analysis of a Bend Discontinuity by the Method of Lines', *Frequenz*, vol. 45, pp. 213–216, 1991.

[15] N. Marcuvitz, *Waveguide Handbook*, P. Peregrinus Ltd., London, Great Britain, 1986.

[16] E. Kühn, 'A Mode-Matching Method for Solving Field Problems in Waveguide and Resonator Circuits', *AEÜ*, vol. 27, no. 12, pp. 511–518, 1973.

[17] F. Arndt, I. Ahrens, U. Papziner, U. Wiechmann and R. Wilkeit, 'Optimized *E*-Plane T-Junction Series Power Dividers', *IEEE Trans. Microwave Theory Tech.*, vol. MTT-35, no. 11, pp. 1052–1059, 1987.

[18] Z. Ma and E. Yamashita, 'Port Reflection Coefficient Method for Solving Multi-Port Microwave Network Problems', *IEEE Trans. Microwave Theory Tech.*, vol. MTT-43, no. 2, pp. 331–337, Feb. 1995.

[19] R. Pregla, 'MoL-Analysis of Rectangular Waveguide Junctions by an Impedance/Admittance Transfer Concept and Crossed Discretisation Lines', *AEÜ*, vol. 53, pp. 83–91, 1999.

[20] H. Patzelt, *Berechnung von Querschnittssprüngen im Rechteckhohlleiter, inhomogenen Rechteckhohlleitertransformatoren, Blenden endlicher Dicke im Rechteckhohlleiter und längsgekoppelten Rechteckhohlleiterfiltern mit der Methode der Orthogonal-Reihen-Entwicklung*, PhD thesis, Universität Bremen, 1979.

[21] H. Patzelt and F. Arndt, 'Double-Plane Steps in Rectangular Waveguides and their Application for Transformers, Irises, and Filters', *IEEE Trans. Microwave Theory Tech.*, vol. MTT–30, no. 5, pp. 771–776, 1982.

Further Reading

[22] L. Vietzorreck and R. Pregla, 'Hybrid Analysis of 3-D MMIC Elements by the Method of Lines', *IEEE Trans. Microwave Theory Tech.*, vol. MTT-44, no. 12, pp. 2580–2586, Dec. 1996.

[23] R. Pregla and L. Vietzorreck, 'Analysis of Planar Circuits by the Method of Lines using Crossed Discretization Lines', in *Progress in Electromagnetics Research Symp. (PIERS)*, Boston, USA, July 1997, p. 766.

[24] L. Vietzorreck and R. Pregla, 'Analysis of MMIC Junctions and Multiports by the Method of Lines', in *IEEE MTT-S Int. Symp. Dig.*, Baltimore, USA, June 1998, vol. 3, pp. 1547–1550.

[25] N. Kaneda, Y. Qian and T. Itoh, 'A Broad-Band Microstrip-to-Waveguide Transition Using Quasi-Yagi Antenna', *IEEE Trans. Microwave Theory Tech.*, vol. MTT-47, no. 12, pp. 2562–2567, 1999.

[26] R. Pregla, 'Novel Algorithms for the Analysis of Optical Fiber Structures with Anisotropic Materials', in *Int. Conf. on Transparent Optical Networks*, Kielce, Polen, June 1999, pp. 49–52.

[27] R. Pregla, 'The Method of Lines for Modeling of Integrated Optics Structures', in *Latsis Symp. on Computational Electromagnetics*, H. Baggenstos, (Ed.), ETH Zürich, 1995, pp. 216–229.

CHAPTER 4

ANALYSIS OF WAVEGUIDE STRUCTURES IN CYLINDRICAL COORDINATES

4.1 INTRODUCTION

There are various structures with cylindrical geometry, such as circular waveguides and circuits of circular waveguides, which also have dielectric inserts, circuits of general microstrip lines on cylindrical bodies, circular antennas, graded index fibres, fibre gratings, curved integrated microwave and optical waveguide bends, passive and active resonators, even vertical cavity lasers. Obviously, these should be analysed in suitable coordinates, i.e. cylindrical coordinates. These cylindrical coordinates are a particular example of general orthogonal coordinates (see Chapter 8 and [1]).

Let us begin with an overview of papers dealing with the MoL in cylindrical coordinates. The first paper that used the Method of Lines to solve wave equations in cylindrical coordinates was [2]. A treatment of microstrip lines with arbitrary cross-section, accomplished with cylindrical functions, appeared in [3]. In [4] the analysis of antennas composed of microstrip and microslot resonators using cylindrical bodies was explained. Dipoles have been analysed in [5] and general axially symmetric antennas in [6]. General formulas for eigenmodes of conformal waveguides on cylindrical bodies are given in [7], which describe the relation of the fields from one boundary of a cylindrical layer to another. The first results from discontinuities in cylindrical waveguides were reported in [8]. Bends in optical waveguides were analysed in [9], [10] and [11]. Fibres gratings where analysed in [1], [12] and [13].

The formulation of the field problem in this chapter is given in such a way that generalised transmission line equations are obtained, providing a straight-forward solution of relevant coupled second-order partial differential equations. The permittivity can be a function of all coordinates. The waveguides may have discontinuities in the direction of propagation. The analytical solution should be obtained in this direction. There are three main propagation directions, which coincide with the three coordinates. Therefore, we have to develop separate equations for each of these three directions. A generalised description of the transformation of fields from one cylindrical boundary surface to another in the form of z and y matrices is provided. These field transformations result in impedance/admittance transformation formulas for layers, sections and interfaces that are numerically stable.

With such general formulae, computer programming is made very easy.

4.2 GENERALISED TRANSMISSION LINE (GTL) EQUATIONS

4.2.1 Material parameters in a cylindrical coordinate system

In this subsection we would like to describe the material parameters for anisotropic materials in cylindrical coordinates. We assume that one of the crystal axes coincides with the z coordinate. The relation between the field components F_u in Cartesian ($u = x, y, z$) and the field components F_v in cylindrical ($v = r, \phi, z$) ones is given by: (See. Fig. 4.1.)

$$\begin{bmatrix} F_x \\ F_y \\ F_z \end{bmatrix} = \begin{bmatrix} \cos\phi & -\sin\phi & 0 \\ \sin\phi & \cos\phi & 0 \\ 0 & 0 & 1 \end{bmatrix} \begin{bmatrix} F_r \\ F_\phi \\ F_z \end{bmatrix} \tag{4.1}$$

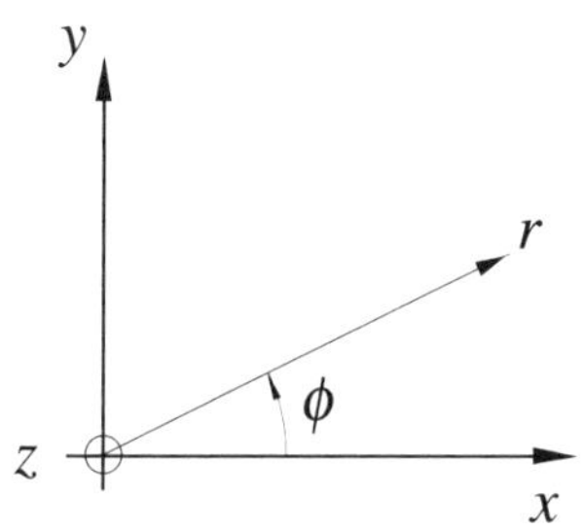

Fig. 4.1 Relation between the coordinate axes

We have a material that has anisotropic behaviour described in Cartesian coordinates with the following tensor:

$$\overleftrightarrow{\nu}_r = \begin{bmatrix} \nu_{xx} & \nu_{xy} & 0 \\ \nu_{yx} & \nu_{yy} & 0 \\ 0 & 0 & \nu_{zz} \end{bmatrix} = \begin{bmatrix} \overleftrightarrow{\nu}_{rc} & 0 \\ 0 & \nu_{zz} \end{bmatrix} \qquad \nu = \varepsilon \text{ or } \mu \tag{4.2}$$

$\overleftrightarrow{\nu}_{rc}$ (the subscript c means 'cross-section') has to be changed to $\overleftrightarrow{\nu}_{rc}^{\mathrm{c}}$ (the superscript c means 'cylindrical'), which results from:

$$\overleftrightarrow{\nu}_{rc}^{\mathrm{c}} = \begin{bmatrix} \cos\phi & \sin\phi \\ -\sin\phi & \cos\phi \end{bmatrix} \begin{bmatrix} \nu_{xx} & \nu_{xy} \\ \nu_{yx} & \nu_{yy} \end{bmatrix} \begin{bmatrix} \cos\phi & -\sin\phi \\ \sin\phi & \cos\phi \end{bmatrix} = \begin{bmatrix} \nu_{rr} & \nu_{r\phi} \\ \nu_{\phi r} & \nu_{\phi\phi} \end{bmatrix} \tag{4.3}$$

where:

$$\begin{aligned}
\nu_{rr} &= \nu_{xx}\cos^2\phi + \nu_{yy}\sin^2\phi + (\nu_{xy} + \nu_{yx})\sin\phi\cos\phi \\
\nu_{\phi\phi} &= \nu_{yy}\cos^2\phi + \nu_{xx}\sin^2\phi - (\nu_{xy} + \nu_{yx})\sin\phi\cos\phi \\
\nu_{r\phi} &= \nu_{xy}\cos^2\phi - \nu_{yx}\sin^2\phi + (\nu_{yy} - \nu_{xx})\sin\phi\cos\phi \\
\nu_{\phi r} &= \nu_{yx}\cos^2\phi - \nu_{xy}\sin^2\phi + (\nu_{yy} - \nu_{xx})\sin\phi\cos\phi
\end{aligned} \tag{4.4}$$

For gyrotropic media we have $\nu_{xy} = -\mathrm{j}\kappa$ and $\nu_{yx} = \mathrm{j}\kappa$. The expressions in eq. (4.4) reduce to:

$$\begin{aligned} \nu_{rr} &= \nu_{xx}\cos^2\phi + \nu_{yy}\sin^2\phi \quad & \nu_{r\phi} &= -\mathrm{j}\kappa + (\nu_{yy} - \nu_{xx})\sin\phi\cos\phi \\ \nu_{\phi\phi} &= \nu_{yy}\cos^2\phi + \nu_{xx}\sin^2\phi \quad & \nu_{\phi r} &= \mathrm{j}\kappa + (\nu_{yy} - \nu_{xx})\sin\phi\cos\phi \end{aligned} \tag{4.5}$$

For $\nu_{xx} = \nu_{yy}$ the equations simplify further. We obtain:

$$\nu_{rr} = \nu_{\phi\phi} = \nu_{xx} \qquad \nu_{r\phi} = -\mathrm{j}\kappa \qquad \nu_{\phi r} = \mathrm{j}\kappa \tag{4.6}$$

In the next subsection, in which we develop the field equations, we will introduce the material parameter tensors according to eq. (4.3). For the sake of brevity, as in Cartesian coordinates, the coordinates and the dimensions are normalised with the free space wave number k_o according to $\overline{z} = k_o z$, $\overline{r} = k_o r$ or generally $\overline{u} = k_o u$.

4.2.2 GTL equations for z-direction

In this subsection we will develop full vectorial generalised transmission line equations for the z-direction. Therefore, the material parameters in the section under consideration must be functions of r and ϕ only. One example is the conformal waveguide antenna structure sketched in Fig. 4.2. The cross-sections can be more complex. An example is shown in Fig. 4.3a, where one of the layers is inhomogeneous in the azimuthal direction. A further example is the fibre cross-section in Fig. 4.3b. From Ampère's law (second and first equation) and from the law of induction (first and second equation), we obtain:

$$-\mathrm{j}\begin{bmatrix} \varepsilon_{\phi\phi} & \varepsilon_{\phi r} \\ \varepsilon_{r\phi} & \varepsilon_{rr} \end{bmatrix} \begin{bmatrix} E_\phi \\ E_r \end{bmatrix} = \frac{\partial}{\partial \overline{z}} \begin{bmatrix} -\widetilde{H}_r \\ \widetilde{H}_\phi \end{bmatrix} + \begin{bmatrix} D_{\overline{r}} \\ -\overline{r}^{-1} D_\phi \end{bmatrix} \widetilde{H}_z \tag{4.7}$$

$$-\mathrm{j}\begin{bmatrix} \mu_{rr} & -\mu_{r\phi} \\ -\mu_{\phi r} & \mu_{\phi\phi} \end{bmatrix} \begin{bmatrix} -\widetilde{H}_r \\ \widetilde{H}_\phi \end{bmatrix} = \frac{\partial}{\partial \overline{z}} \begin{bmatrix} E_\phi \\ E_r \end{bmatrix} - \begin{bmatrix} \overline{r}^{-1} D_\phi \\ D_{\overline{r}} \end{bmatrix} E_z \tag{4.8}$$

We have ordered the transverse field components in such a way that the inner product of the field vectors with the transverse field components is proportional to the Poynting vector in z-direction. We calculate the longitudinal field components $\widetilde{H}_z$ and E_z from the third equations of the law of electromagnetic induction and Ampère's law, respectively, and obtain:

$$\widetilde{H}_z = \mathrm{j}\mu_{zz}^{-1}\overline{r}^{-1}[D_{\overline{r}} - D_\phi][E^z] \qquad E_z = -\mathrm{j}\varepsilon_{zz}^{-1}\overline{r}^{-1}[D_\phi D_{\overline{r}}][H^z] \tag{4.9}$$

where we use the notations:

$$[E^z] = \begin{bmatrix} \overline{r}E_\phi \\ E_r \end{bmatrix} \qquad [H^z] = \begin{bmatrix} -\widetilde{H}_r \\ \overline{r}\widetilde{H}_\phi \end{bmatrix} \tag{4.10}$$

Fig. 4.2 Conformal waveguide antenna structure with propagation in z-direction (R. Pregla, 'Efficient Analysis of Conformal Antennas with Anisotropic Material (0682)', in *AP 2000 Millennium Conference on Antennas and Propagation.* © 2000 European Space Agency (ESA))

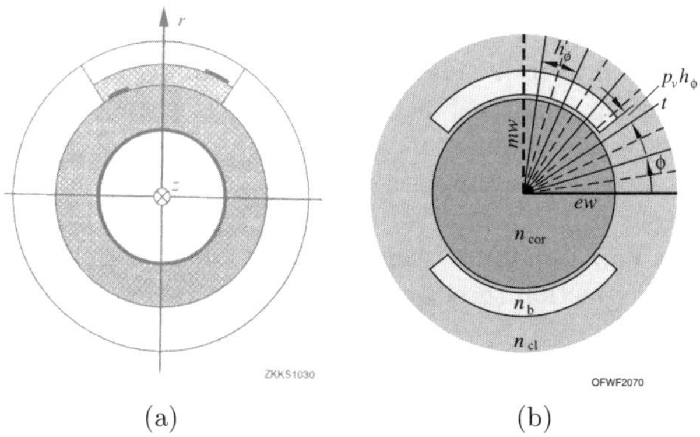

Fig. 4.3 Radial layered cross-sections with azimuthal inhomogeneous layers: (a) coupled microstrip waveguides on cylindrical body (b) cross-section of a bow-tie-type fibre (Reproduced by permission of Springer Netherlands)

By introducing $\widetilde{H}_z$ and E_z from eq. (4.9) into eqs. (4.7) and (4.8), we can derive relations between the transverse field components in the following way:

$$\frac{\partial}{\partial \overline{z}}[H^z] = -\mathrm{j}[R_{\mathrm{E}}^z][E^z] \qquad \frac{\partial}{\partial \overline{z}}[E^z] = -\mathrm{j}[R_{\mathrm{H}}^z][H^z] \tag{4.11}$$

$$[R_{\mathrm{E}}^z] = \begin{bmatrix} D_{\overline{r}}\mu_{zz}^{-1}\overline{r}^{-1}D_{\overline{r}} + \varepsilon_{\phi\phi}\overline{r}^{-1} & -D_{\overline{r}}\mu_{zz}^{-1}\overline{r}^{-1}D_{\phi} + \varepsilon_{\phi r} \\ -D_{\phi}\mu_{zz}^{-1}\overline{r}^{-1}D_{\overline{r}} + \varepsilon_{r\phi} & D_{\phi}\mu_{zz}^{-1}\overline{r}^{-1}D_{\phi} + \varepsilon_{rr}\overline{r} \end{bmatrix} \tag{4.12}$$

$$[R_{\mathrm{H}}^z] = \begin{bmatrix} D_{\phi}\varepsilon_{zz}^{-1}\overline{r}^{-1}D_{\phi} + \mu_{rr}\overline{r} & D_{\phi}\varepsilon_{zz}^{-1}\overline{r}^{-1}D_{\overline{r}} - \mu_{r\phi} \\ D_{\overline{r}}\varepsilon_{zz}^{-1}\overline{r}^{-1}D_{\phi} - \mu_{\phi r} & D_{\overline{r}}\varepsilon_{zz}^{-1}\overline{r}^{-1}D_{\overline{r}} + \mu_{\phi\phi}\overline{r}^{-1} \end{bmatrix} \tag{4.13}$$

By combining the two first-order differential equation systems given in eq. (4.11), we obtain the second-order differential equations for the transverse electric or magnetic fields:

$$\frac{\partial^2}{\partial \overline{z}^2}[E^z][Q^z_{\mathrm{E}}][E^z] = \begin{bmatrix}0\\0\end{bmatrix} \qquad \frac{\partial^2}{\partial \overline{z}^2}[H^z] - [Q^z_{\mathrm{H}}][H^z] = \begin{bmatrix}0\\0\end{bmatrix} \tag{4.14}$$

where:

$$-[R^z_{\mathrm{E}}][R^z_{\mathrm{H}}] = [Q^z_{\mathrm{H}}] = \begin{bmatrix}Q^z_{\mathrm{H11}} & Q^z_{\mathrm{H12}}\\ Q^z_{\mathrm{H21}} & Q^z_{\mathrm{H22}}\end{bmatrix} \qquad -[R^z_{\mathrm{H}}][R^z_{\mathrm{E}}] = [Q^z_{\mathrm{E}}] = \begin{bmatrix}Q^z_{\mathrm{E11}} & Q^z_{\mathrm{E12}}\\ Q^z_{\mathrm{E21}} & Q^z_{\mathrm{E22}}\end{bmatrix} \tag{4.15}$$

Now we split the matrices $[R^z_{\mathrm{E,H}}]$ into two parts:

$$[R^z_{\mathrm{E}}] = [R^{zc}_{\mathrm{E}}] + [\varepsilon^{zc}_{rt}] \qquad [R^z_{\mathrm{H}}] = [R^{zc}_{\mathrm{H}}] + [\mu^{zc}_{rt}] \tag{4.16}$$

Using the abbreviations $\overline{\mu}_{zz} = \mu_{zz}\overline{r}$ and $\overline{\varepsilon}_{zz} = \varepsilon_{zz}\overline{r}$, we may write:

$$[R^{zc}_{\mathrm{E}}] = \begin{bmatrix} D_{\overline{r}}\overline{\mu}^{-1}_{zz}D_{\overline{r}} & -D_{\overline{r}}\overline{\mu}^{-1}_{zz}D_{\phi}\\ -D_{\phi}\overline{\mu}^{-1}_{zz}D_{\overline{r}} & D_{\phi}\overline{\mu}^{-1}_{zz}D_{\phi}\end{bmatrix} \qquad [\varepsilon^{zc}_{rt}] = \begin{bmatrix}\varepsilon_{\phi\phi}\overline{r}^{-1} & \varepsilon_{\phi r}\\ \varepsilon_{r\phi} & \varepsilon_{rr}\overline{r}\end{bmatrix} \tag{4.17}$$

$$[R^{zc}_{\mathrm{H}}] = \begin{bmatrix} D_{\phi}\overline{\varepsilon}^{-1}_{zz}D_{\phi} & D_{\phi}\overline{\varepsilon}^{-1}_{zz}D_{\overline{r}}\\ D_{\overline{r}}\overline{\varepsilon}^{-1}_{zz}D_{\phi} & D_{\overline{r}}\overline{\varepsilon}^{-1}_{zz}D_{\overline{r}}\end{bmatrix} \qquad [\mu^{zc}_{rt}] = \begin{bmatrix}\mu_{rr}\overline{r} & -\mu_{r\phi}\\ -\mu_{\phi r} & \mu_{\phi\phi}\overline{r}^{-1}\end{bmatrix} \tag{4.18}$$

Because of $D_{\phi}D_{\overline{r}} = D_{\overline{r}}D_{\phi}$, the products of the matrices $[R^{zc}_{\mathrm{E}}]$ and $[R^{zc}_{\mathrm{H}}]$ are always equal to zero.Therefore, we obtain for the matrices $[Q^z_{\mathrm{H}}]$ and $[Q^z_{\mathrm{E}}]$:

$$-[Q^z_{\mathrm{E}}] = [R^{zc}_{\mathrm{H}}][\varepsilon^{zc}_{rt}] + [\mu^{zc}_{rt}][R^{zc}_{\mathrm{E}}] + [\mu^{zc}_{rt}][\varepsilon^{zc}_{rt}] \tag{4.19}$$

$$-[Q^z_{\mathrm{H}}] = [R^{zc}_{\mathrm{E}}][\mu^{zc}_{rt}] + [\varepsilon^{zc}_{rt}][R^{\phi c}_{\mathrm{H}}] + [\varepsilon^{zc}_{rt}][\mu^{zc}_{rt}] \tag{4.20}$$

The matrices $[Q^z_{\mathrm{H}}]$ and $[Q^z_{\mathrm{E}}]$ simplify for the case $\varepsilon_{r\phi} = \varepsilon_{\phi r} = 0$, $\mu_{r\phi} = \mu_{\phi r} = 0$ because $[\varepsilon^{zc}_{rt}]$ and $[\mu^{zc}_{rt}]$ are diagonal matrices.

We saw in Section 4.2.1 that for this case we must have $\nu_{rr} = \nu_{\phi\phi}$. Nevertheless, we will distinguish between these two values, because they are discretised on different positions.

4.2.2.1 Special case: ϕ-independent material parameters

In the case of ϕ-independent material parameters ($\nu_{xy} = \nu_{yx} = 0$, $\nu_{xx} = \nu_{yy} \neq f(\phi)$) and circular cross-sections, we can write for the ϕ-dependence of the field components:

$$E_r, H_{\phi} \sim \cos(m\phi) \qquad E_{\phi}, H_r \sim \sin(m\phi) \tag{4.21}$$

In this case, we obtain for the matrices $[R^z_{\mathrm{E}}]$ and $[R^z_{\mathrm{H}}]$:

$$[R^z_{\mathrm{E}}] = \begin{bmatrix} D_{\overline{r}}\mu^{-1}_{zz}\overline{r}^{-1}D_{\overline{r}} + \varepsilon_{\phi\phi}\overline{r}^{-1} & mD_{\overline{r}}\mu^{-1}_{zz}\overline{r}^{-1}\\ -m\mu^{-1}_{zz}\overline{r}^{-1}D_{\overline{r}} & \overline{r}\varepsilon_{rr} - m^2\mu^{-1}_{zz}\overline{r}^{-1}\end{bmatrix} \tag{4.22}$$

$$[R^z_{\mathrm{H}}] = \begin{bmatrix} \overline{r}\mu_{rr} - m^2\varepsilon^{-1}_{zz}\overline{r}^{-1} & -m\varepsilon^{-1}_{zz}\overline{r}^{-1}D_{\overline{r}}\\ mD_{\overline{r}}\varepsilon^{-1}_{zz}\overline{r}^{-1} & D_{\overline{r}}\varepsilon^{-1}_{zz}\overline{r}^{-1}D_{\overline{r}} + \mu_{\phi\phi}\overline{r}^{-1}\end{bmatrix} \tag{4.23}$$

For ϕ-independent fields ($m = 0$) we get:

$$[R_{\mathrm{E}}^{z}] = \begin{bmatrix} D_{\overline{r}}\mu_{zz}^{-1}\overline{r}^{-1}D_{\overline{r}} + \varepsilon_{\phi\phi}\overline{r}^{-1} & 0 \\ 0 & \overline{r}\varepsilon_{rr} \end{bmatrix} \tag{4.24}$$

$$[R_{\mathrm{H}}^{z}] = \begin{bmatrix} \overline{r}\mu_{rr} & 0 \\ 0 & D_{\overline{r}}\varepsilon_{zz}^{-1}\overline{r}^{-1}D_{\overline{r}} + \mu_{\phi\phi}\overline{r}^{-1} \end{bmatrix} \tag{4.25}$$

and:

$$-[Q_{\mathrm{E}}^{z}] = \begin{bmatrix} \overline{r}\mu_{rr}D_{\overline{r}}\mu_{zz}^{-1}\overline{r}^{-1}D_{\overline{r}} + \varepsilon_{\phi\phi}\mu_{rr} & 0 \\ 0 & D_{\overline{r}}\varepsilon_{zz}^{-1}\overline{r}^{-1}D_{\overline{r}}\overline{r}\varepsilon_{rr} + \varepsilon_{rr}\mu_{\phi\phi} \end{bmatrix} \tag{4.26}$$

$$-[Q_{\mathrm{H}}^{z}] = \begin{bmatrix} D_{\overline{r}}\mu_{zz}^{-1}\overline{r}^{-1}D_{\overline{r}}\overline{r}\mu_{rr} + \varepsilon_{\phi\phi}\mu_{rr} & 0 \\ 0 & \overline{r}\varepsilon_{rr}D_{\overline{r}}\varepsilon_{zz}^{-1}\overline{r}^{-1}D_{\overline{r}} + \varepsilon_{rr}\mu_{\phi\phi} \end{bmatrix} \tag{4.27}$$

As we can see, we obtain decoupled TE_z and TM_z modes in this case.

4.2.2.2 Special case: homogeneous isotropic material parameters

In the case of homogeneous isotropic materials according to:

$$\varepsilon_{rr} = \varepsilon_{\phi\phi} = \varepsilon_{zz} = \varepsilon_r, \mu_{rr} = \mu_{\phi\phi} = \mu_{zz} = \mu_r$$

we obtain from eqs. (4.12) and (4.13):

$$[R_{\mathrm{E}}^{z}] = \frac{1}{\mu_r}\begin{bmatrix} D_{\overline{r}}\overline{r}^{-1}D_{\overline{r}} + \varepsilon_r\mu_r\overline{r}^{-1} & -D_{\overline{r}}\overline{r}^{-1}D_{\phi} \\ -D_{\phi}\overline{r}^{-1}D_{\overline{r}} & D_{\phi}\overline{r}^{-1}D_{\phi} + \varepsilon_r\mu_r\overline{r} \end{bmatrix} \tag{4.28}$$

$$[R_{\mathrm{H}}^{z}] = \frac{1}{\varepsilon_r}\begin{bmatrix} D_{\phi}\overline{r}^{-1}D_{\phi} + \varepsilon_r\mu_r\overline{r} & D_{\phi}\overline{r}^{-1}D_{\overline{r}} \\ D_{\overline{r}}\overline{r}^{-1}D_{\phi} & D_{\overline{r}}\overline{r}^{-1}D_{\overline{r}} + \varepsilon_r\mu_r\overline{r}^{-1} \end{bmatrix} \tag{4.29}$$

which reduce to eqs. (4.24) and (4.25) for ϕ-independent fields ($D_\phi = 0$).

4.2.2.3 Special case: gyrotropic material with respect to the z-direction

For the analysis of the Faraday rotation in circular hollow waveguides, we assume a ϕ-dependence of the field components according to $e^{\mathrm{j}m\phi}$ and introduce the following expressions into eqs. (4.12) and (4.13):

$$D_\phi = \mathrm{j}m \qquad \varepsilon_{r\phi} = -\mathrm{j}\kappa_e \qquad \varepsilon_{\phi r} = \mathrm{j}\kappa_e \qquad \mu_{r\phi} = -\mathrm{j}\kappa_m \qquad \mu_{\phi r} = \mathrm{j}\kappa_m \tag{4.30}$$

We redefine the vectors in eq. (4.10) according to:

$$[E^z] = \begin{bmatrix} \mathrm{j}\overline{r}E_\phi \\ E_r \end{bmatrix} \qquad [H^z] = \begin{bmatrix} -\mathrm{j}\widetilde{H}_r \\ \overline{r}\widetilde{H}_\phi \end{bmatrix} \tag{4.31}$$

and obtain instead of the terms in eqs. (4.17) and (4.18):

$$[R_{\mathrm{E}}^{zc}] = \begin{bmatrix} D_{\overline{r}}\overline{\mu}_{zz}^{-1}D_{\overline{r}} & mD_{\overline{r}}\overline{\mu}_{zz}^{-1} \\ -m\overline{\mu}_{zz}^{-1}D_{\overline{r}} & -m^2\overline{\mu}_{zz}^{-1} \end{bmatrix} \qquad [\varepsilon_{rt}^{zc}] = \begin{bmatrix} \varepsilon_{\phi\phi}\overline{r}^{-1} & -\kappa_e \\ -\kappa_e & \varepsilon_{rr}\overline{r} \end{bmatrix} \tag{4.32}$$

$$[R_{\mathrm{H}}^{zc}] = \begin{bmatrix} -m^2\overline{\varepsilon}_{zz}^{-1} & -m\overline{\varepsilon}_{zz}^{-1}D_{\overline{r}} \\ mD_{\overline{r}}\overline{\varepsilon}_{zz}^{-1} & D_{\overline{r}}\overline{\varepsilon}_{zz}^{-1}D_{\overline{r}} \end{bmatrix} \qquad [\mu_{rt}^{zc}] = \begin{bmatrix} \mu_{rr}\overline{r} & -\kappa_m \\ -\kappa_m & \mu_{\phi\phi}\overline{r}^{-1} \end{bmatrix} \tag{4.33}$$

We obtain the longitudinal components from eq. (4.9) according to:

$$\widetilde{H}_z = (\mu_{zz}\overline{r})^{-1}[D_{\overline{r}}m][E^z] \qquad E_z = -\mathrm{j}(\varepsilon_{zz}\overline{r})^{-1}[mD_{\overline{r}}][H^z] \tag{4.34}$$

The discretisation and solution will be described in Section 4.4.4.1.

4.2.3 GTL equations for ϕ-direction

Let us now examine the case where we would like to have an analytical solution in ϕ-direction. An example is the conformal waveguide antenna structure sketched in Fig. 4.4.

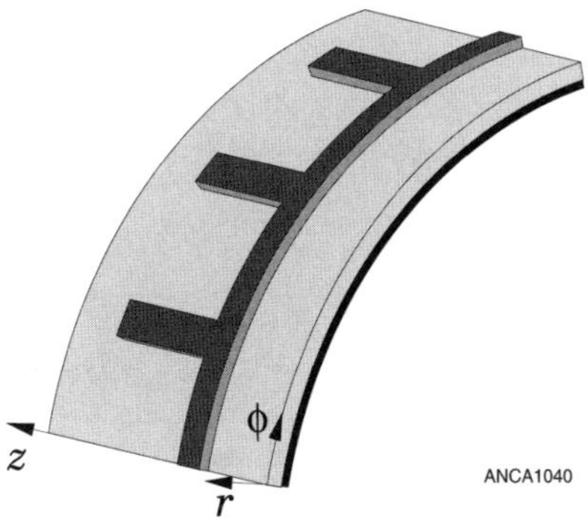

Fig. 4.4 Conformal waveguide antenna structure with propagation in ϕ-direction (R. Pregla, 'Efficient Analysis of Conformal Antennas with Anisotropic Material (0682)', in *AP 2000 Millennium Conference on Antennas and Propagation.* © 2000 European Space Agency (ESA))

To develop suitable equations we assume that the material parameters ν_{rr}, $\nu_{\phi\phi}$ and ν_{zz} and the off-diagonal elements ν_{rz} and ν_{zr} are functions of r and z only. The other off-diagonal elements in the material tensors are zero. (The case with arbitrary material parameters is described in Chapter 8.) As before, we obtain from Ampère's law (third and first equation):

$$-\mathrm{j}\begin{bmatrix}\varepsilon_{rr} & \varepsilon_{rz}\\ \varepsilon_{zr} & \varepsilon_{zz}\end{bmatrix}\begin{bmatrix}E_r\\ E_z\end{bmatrix} = \overline{r}^{-1}\frac{\partial}{\partial\phi}\begin{bmatrix}-\widetilde{H}_z\\ \widetilde{H}_r\end{bmatrix} + \begin{bmatrix}D_{\overline{z}}\\ -\overline{r}^{-1}D_{\overline{r}}\overline{r}\end{bmatrix}\widetilde{H}_\phi \tag{4.35}$$

and from the law of induction (third and first equation):

$$-\mathrm{j}\begin{bmatrix}\mu_{zz} & -\mu_{zr}\\ -\mu_{rz} & \mu_{rr}\end{bmatrix}\begin{bmatrix}-\widetilde{H}_z\\ \widetilde{H}_r\end{bmatrix} = \overline{r}^{-1}\frac{\partial}{\partial\phi}\begin{bmatrix}E_r\\ E_z\end{bmatrix} - \begin{bmatrix}\overline{r}^{-1}D_{\overline{r}}\overline{r}\\ D_{\overline{z}}\end{bmatrix}E_\phi \tag{4.36}$$

The azimuthal field components $\widetilde{H}_\phi$ and E_ϕ are obtained from the second equations of the law of induction and of Ampère's law, respectively:

$$\widetilde{H}_\phi = \mathrm{j}\mu_{\phi\phi}^{-1}[D_{\overline{z}} - D_{\overline{r}}]\begin{bmatrix}E_r\\ E_z\end{bmatrix} \qquad E_\phi = -\mathrm{j}\varepsilon_{\phi\phi}^{-1}[D_{\overline{r}}\ D_{\overline{z}}]\begin{bmatrix}-\widetilde{H}_z\\ \widetilde{H}_r\end{bmatrix} \tag{4.37}$$

Now we use the notations:

$$[E^\phi] = \begin{bmatrix} E_r \\ E_z \end{bmatrix} \qquad [H^\phi] = \begin{bmatrix} -\widetilde{H}_z \\ \widetilde{H}_r \end{bmatrix} \tag{4.38}$$

By introducing $\widetilde{H}_\phi$ and E_ϕ from eq. (4.37) into eqs. (4.35) and (4.36), we are able to give a system of coupled differential equations for the transverse components:

$$\frac{\partial}{\partial\phi}[H^\phi] = -\mathrm{j}[R_\mathrm{E}^\phi][E^\phi] \qquad \frac{\partial}{\partial\phi}[E^\phi] = -\mathrm{j}[R_\mathrm{H}^\phi][H^\phi] \tag{4.39}$$

with:

$$[R_\mathrm{E}^\phi] = \begin{bmatrix} D_{\overline{z}}\mu_{\phi\phi}^{-1}\overline{r}D_{\overline{z}} + \varepsilon_{rr}\overline{r} & -D_{\overline{z}}\mu_{\phi\phi}^{-1}\overline{r}D_{\overline{r}} + \varepsilon_{rz}\overline{r} \\ -D_{\overline{r}}\mu_{\phi\phi}^{-1}\overline{r}D_{\overline{z}} + \varepsilon_{zr}\overline{r} & D_{\overline{r}}\mu_{\phi\phi}^{-1}\overline{r}D_{\overline{r}} + \varepsilon_{zz}\overline{r} \end{bmatrix} \tag{4.40}$$

$$[R_\mathrm{H}^\phi] = \begin{bmatrix} D_{\overline{r}}\varepsilon_{\phi\phi}^{-1}\overline{r}D_{\overline{r}} + \mu_{zz}\overline{r} & D_{\overline{r}}\varepsilon_{\phi\phi}^{-1}\overline{r}D_{\overline{z}} - \mu_{zr}\overline{r} \\ D_{\overline{z}}\varepsilon_{\phi\phi}^{-1}\overline{r}D_{\overline{r}} - \mu_{rz}\overline{r} & D_{\overline{z}}\varepsilon_{\phi\phi}^{-1}\overline{r}D_{\overline{z}} + \mu_{rr}\overline{r} \end{bmatrix} \tag{4.41}$$

By combining the two first-order differential equation systems (4.39), we obtain:

$$\frac{\partial^2}{\partial\phi^2}[E^\phi] - [Q_\mathrm{E}^\phi][E^\phi] = \begin{bmatrix} 0 \\ 0 \end{bmatrix} \qquad -[Q_\mathrm{E}^\phi] = [R_\mathrm{H}^\phi][R_\mathrm{E}^\phi] \tag{4.42}$$

$$\frac{\partial^2}{\partial\phi^2}[H^\phi] - [Q_\mathrm{H}^\phi][H^\phi] = \begin{bmatrix} 0 \\ 0 \end{bmatrix} \qquad -[Q_\mathrm{H}^\phi] = [R_\mathrm{E}^\phi][R_\mathrm{H}^\phi] \tag{4.43}$$

where:

$$[Q_\mathrm{E}^\phi] = \begin{bmatrix} Q_\mathrm{E11}^\phi & Q_\mathrm{E12}^\phi \\ Q_\mathrm{E21}^\phi & Q_\mathrm{E22}^\phi \end{bmatrix} \qquad [Q_\mathrm{H}^\phi] = \begin{bmatrix} Q_\mathrm{H11}^\phi & Q_\mathrm{H12}^\phi \\ Q_\mathrm{H21}^\phi & Q_\mathrm{H22}^\phi \end{bmatrix} \tag{4.44}$$

Now we split the matrices $[R_\mathrm{E,H}^\phi]$ into two parts:

$$[R_\mathrm{E}^\phi] = [R_\mathrm{E}^{\phi\mathrm{C}}] + \overline{r}[\varepsilon_{rt}^{\phi\mathrm{C}}] \qquad [R_\mathrm{H}^\phi] = [R_\mathrm{H}^{\phi\mathrm{C}}] + \overline{r}[\mu_{rt}^{\phi\mathrm{C}}] \tag{4.45}$$

By using the abbreviations $\overline{\mu}_{\phi\phi} = \mu_{\phi\phi}/\overline{r}$ and $\overline{\varepsilon}_{\phi\phi} = \varepsilon_{\phi\phi}/\overline{r}$, we may write for these matrices:

$$\left[R_\mathrm{E}^{\phi\mathrm{C}}\right] = \begin{bmatrix} D_{\overline{z}}\overline{\mu}_{\phi\phi}^{-1}D_{\overline{z}} & -D_{\overline{z}}\overline{\mu}_{\phi\phi}^{-1}D_{\overline{r}} \\ -D_{\overline{r}}\overline{\mu}_{\phi\phi}^{-1}D_{\overline{z}} & D_{\overline{r}}\overline{\mu}_{\phi\phi}^{-1}D_{\overline{r}} \end{bmatrix} \qquad [\varepsilon_{rt}^{\phi\mathrm{C}}] = \begin{bmatrix} \varepsilon_{rr} & \varepsilon_{rz} \\ \varepsilon_{zr} & \varepsilon_{zz} \end{bmatrix} \tag{4.46}$$

$$\left[R_\mathrm{H}^{\phi\mathrm{C}}\right] = \begin{bmatrix} D_{\overline{r}}\overline{\varepsilon}_{\phi\phi}^{-1}D_{\overline{r}} & D_{\overline{r}}\overline{\varepsilon}_{\phi\phi}^{-1}D_{\overline{z}} \\ D_{\overline{z}}\overline{\varepsilon}_{\phi\phi}^{-1}D_{\overline{r}} & D_{\overline{z}}\overline{\varepsilon}_{\phi\phi}^{-1}D_{\overline{z}} \end{bmatrix} \qquad [\mu_{rt}^{\phi\mathrm{C}}] = \begin{bmatrix} \mu_{zz} & -\mu_{zr} \\ -\mu_{rz} & \mu_{rr} \end{bmatrix} \tag{4.47}$$

Since we may switch the order of the derivatives, i.e. $D_{\overline{r}}D_{\overline{z}} = D_{\overline{z}}D_{\overline{r}}$, the products of the matrices $[R_\mathrm{E}^{\phi\mathrm{C}}]$ and $[R_\mathrm{H}^{\phi\mathrm{C}}]$ are always equal to zero. Therefore, we may write for $[Q^\phi_\mathrm{H}]$ and $[Q^\phi_\mathrm{E}]$:

$$-[Q^\phi_\mathrm{E}] = [R^{\phi\mathrm{C}}_\mathrm{H}]\overline{r}[\varepsilon^{\phi\mathrm{C}}_{rt}] + \overline{r}[\mu^{\phi\mathrm{C}}_{rt}][R^{\phi\mathrm{C}}_\mathrm{E}] + \overline{r}^2[\mu^{\phi\mathrm{C}}_{rt}][\varepsilon^{\phi\mathrm{C}}_{rt}] \tag{4.48}$$

$$-[Q^\phi_\mathrm{H}] = [R^{\phi\mathrm{C}}_\mathrm{E}]\overline{r}[\mu^{\phi\mathrm{C}}_{rt}] + \overline{r}[\varepsilon^{\phi\mathrm{C}}_{rt}][R^{\phi\mathrm{C}}_\mathrm{H}] + \overline{r}^2[\varepsilon^{\phi\mathrm{C}}_{rt}][\mu^{\phi\mathrm{C}}_{rt}] \tag{4.49}$$

The matrices $[Q^\phi_\mathrm{H}]$ and $[Q^\phi_\mathrm{E}]$ can then easily be determined.

4.2.3.1 Special case: modes in rectangular waveguide bends

In this subsection, the basic formulas will be given for the analysis of circular rectangular waveguide bends, as shown in Fig. 4.5. The complete analysis was performed in Section 3.2. If the material parameters do not depend on z, we can describe the z dependence of the field components in the following form:

$$E_z \sim \cos(k_{\overline{z}}\overline{z}) \quad \widetilde{H}_z \sim \sin(k_{\overline{z}}\overline{z}) \tag{4.50}$$

$$\widetilde{H}_r \sim \cos(k_{\overline{z}}\overline{z}) \quad \widetilde{E}_r \sim \sin(k_{\overline{z}}\overline{z}) \tag{4.51}$$

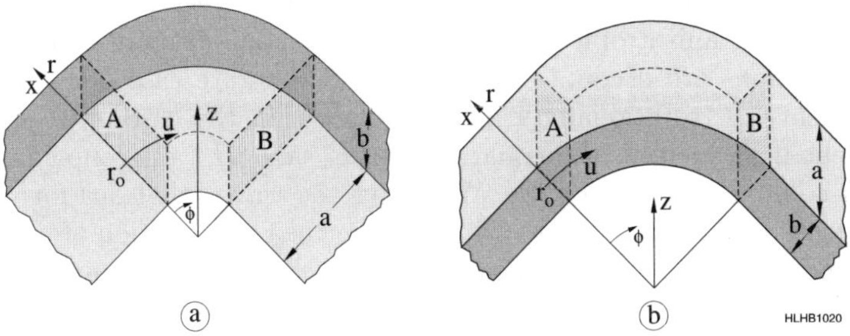

Fig. 4.5 H–plane (a) and E-plane (b) bend of a rectangular waveguide

In the case of $\mu_{rr} = \mu_{\phi\phi} = \mu_{zz} = 1$ and an isotropic dielectric, the matrices $[R^\phi_\mathrm{E}]$ and $[R^\phi_\mathrm{H}]$ in eqs. (4.40) and (4.41) are given by:

$$[R^\phi_\mathrm{E}] = \begin{bmatrix} \overline{r}\varepsilon_q & k_{\overline{z}}\overline{r}D_{\overline{r}} \\ -k_{\overline{z}}D_{\overline{r}}\overline{r} & D_{\overline{r}}\overline{r}D_{\overline{r}} + \overline{r}\varepsilon_r \end{bmatrix} \tag{4.52}$$

$$[R^\phi_\mathrm{H}] = \begin{bmatrix} D_{\overline{r}}\overline{r}\varepsilon_r^{-1}D_{\overline{r}} + \overline{r} & -k_{\overline{z}}D_{\overline{r}}\overline{r}\varepsilon_r^{-1} \\ k_{\overline{z}}\overline{r}\varepsilon_r^{-1}D_{\overline{r}} & \overline{r}\varepsilon_r^{-1}\varepsilon_q \end{bmatrix} \tag{4.53}$$

where $\varepsilon_q = \varepsilon_r - k^2_{\overline{z}}$. The matrices Q^ϕ are:

$$-[Q^\phi_\mathrm{E}] = \begin{bmatrix} D_{\overline{r}}\overline{r}\varepsilon_r^{-1}D_{\overline{r}}\overline{r}\varepsilon_r + \overline{r}^2\varepsilon_q & -k_{\overline{z}}(D_{\overline{r}}\overline{r}^2 - \overline{r}^2D_{\overline{r}}) \\ k_{\overline{z}}(\overline{r}\varepsilon_r^{-1}D_{\overline{r}}\overline{r}\varepsilon_r - \overline{r}D_{\overline{r}}\overline{r}) & \overline{r}D_{\overline{r}}\overline{r}D_{\overline{r}} + \overline{r}^2\varepsilon_q \end{bmatrix} \tag{4.54}$$

$$-[Q^\phi_\mathrm{H}] = \begin{bmatrix} \overline{r}\varepsilon_r D_{\overline{r}}\overline{r}\varepsilon_r^{-1}D_{\overline{r}} + \overline{r}^2\varepsilon_q & k_{\overline{z}}(\overline{r}D_{\overline{r}}\overline{r} - \overline{r}\varepsilon_r D_{\overline{r}}\overline{r}\varepsilon_r^{-1}) \\ -k_{\overline{z}}(D_{\overline{r}}\overline{r}^2 - \overline{r}^2D_{\overline{r}}) & D_{\overline{r}}\overline{r}D_{\overline{r}}\overline{r} + \overline{r}^2\varepsilon_q \end{bmatrix} \tag{4.55}$$

In the case of homogeneous dielectric, it follows that Q^{ϕ}_{E21} (i.e. the lower-left block of Q^{ϕ}_{E}) and Q^{ϕ}_{H12} (the upper-right block of Q^{ϕ}_{H}) are equal to zero, and we can write two independent wave equations for $F_z = E_z$ and $F_z = \widetilde{H}_z$:

$$\frac{\partial}{\overline{r}^2 \partial \phi^2} F_z + \overline{r}^{-1} D_{\overline{r}} \overline{r} D_{\overline{r}} F_z + \epsilon_q F_z = 0 \tag{4.56}$$

The remaining required components can be determined with the help of the GTL equations (see Section 3.2).

4.2.4 Analysis of circular (coaxial) waveguides with azimuthally-magnetised ferrites and azimuthally-magnetised solid plasma

In this subsection we would like to develop a system of partial differential equations for longitudinal components of electric and magnetic vectors in the azimuthally-magnetised solid-plasma coaxial waveguide and cylindrical waveguide. The anisotropic ferrite is magnetised azimuthally to remanence. These equations are decoupled only in the degenerate case of azimuthally-independent (i.e. rotationally symmetric) fields. In the case of azimuthally-dependent field vectors, the equations for E_z and H_z are coupled, leading to a six-component hybrid wave. We derive the wave equations for the field components E_z and H_z from eq. (4.39). From the first equation of (4.35) and the second equation of (4.36), we obtain the r components as functions of E_z and $\widetilde{H}_z$, by:

$$-\mathrm{j}(\varepsilon_{rr} + D_{\overline{z}} \mu^{-1}_{\phi\phi} D_{\overline{z}})\ E_r = \mathrm{j}\ (\varepsilon_{rz} - D_{\overline{z}} \mu^{-1}_{\phi\phi} D_{\overline{r}})\ E_z - \overline{r}^{-1} D_{\phi} \widetilde{H}_z \tag{4.57}$$

$$-\mathrm{j}(\mu_{rr} - D_{\overline{z}} \varepsilon^{-1}_{\phi\phi} D_{\overline{z}})\ \widetilde{H}_r = \mathrm{j}\ (\mu_{rz} - D_{\overline{z}} \varepsilon^{-1}_{\phi\phi} D_{\overline{r}})\ \widetilde{H}_z + \overline{r}^{-1} D_{\overline{\phi}} E_z \tag{4.58}$$

Introducing these equations into the second equation in system (4.42) and into the first equation in system (4.43), we obtain for propagation in z-direction according to $\exp(-\mathrm{j}\sqrt{\varepsilon_{re}}\overline{z})$ and $D_z = -\mathrm{j}\sqrt{\varepsilon_{re}}$, and using $\varepsilon_{rz} = -j\kappa_e$, $\varepsilon_{zr} = j\kappa_e$, $\mu_{rz} = -j\kappa_m$ and $\mu_{zr} = j\kappa_m$, the following coupled second-order differential equations for E_z and H_z:

$$\begin{aligned}
&\frac{\varepsilon_{rr}\mu_{\phi\phi} - \varepsilon_{re}}{\mu_{rr}\epsilon_{\phi\phi} - \varepsilon_{re}} \frac{\varepsilon_{\phi\phi}}{\varepsilon_{rr}} \frac{1}{\overline{r}^2} \frac{\partial^2 E_z}{\partial \phi^2} + \frac{1}{\overline{r}} D_{\overline{r}} \overline{r} D_{\overline{r}} E_z + \left(\mu_{\phi\phi}\varepsilon_{\perp} - \frac{\varepsilon_{zz}}{\varepsilon_{rr}} \varepsilon_{re} - \frac{\kappa_e}{\varepsilon_{rr}} \frac{\sqrt{\varepsilon_{re}}}{\overline{r}} \right) E_z \\
&+ \frac{\sqrt{\varepsilon_{re}}}{\varepsilon_{rr}} \frac{\mu_{rr}\varepsilon_{\phi\phi} - \varepsilon_{rr}\mu_{\phi\phi}}{\mu_{rr}\varepsilon_{\phi\phi} - \varepsilon_{re}} \frac{1}{\overline{r}} D_{\overline{r}} D_{\phi} \widetilde{H}_z \\
&+ \left(\kappa_m \frac{\varepsilon_{\phi\phi}}{\varepsilon_{rr}} \frac{\varepsilon_{rr}\mu_{\phi\phi} - \varepsilon_{re}}{\mu_{rr}\varepsilon_{\phi\phi} - \varepsilon_{re}} + \kappa_e \frac{\mu_{\phi\phi}}{\varepsilon_{rr}} \right) \frac{1}{\overline{r}} D_{\phi} \widetilde{H}_z = 0
\end{aligned} \tag{4.59}$$

$$\frac{\mu_{rr}\varepsilon_{\phi\phi}-\varepsilon_{re}}{\varepsilon_{rr}\mu_{\phi\phi}-\varepsilon_{re}}\frac{\mu_{\phi\phi}}{\mu_{rr}}\frac{1}{\overline{r}^2}\frac{\partial^2\widetilde{H}_z}{\partial\phi^2}+\frac{1}{\overline{r}}D_{\overline{r}}\overline{r}D_{\overline{r}}\widetilde{H}_z$$
$$+\left(\varepsilon_{\phi\phi}\mu_\perp-\frac{\mu_{zz}}{\mu_{rr}}\varepsilon_{re}-\frac{\kappa_m}{\mu_{rr}}\frac{\sqrt{\varepsilon_{re}}}{\overline{r}}\right)\widetilde{H}_z+\frac{\sqrt{\varepsilon_{re}}}{\mu_{rr}}\frac{\mu_{rr}\varepsilon_{\phi\phi}-\varepsilon_{rr}\mu_{\phi\phi}}{\varepsilon_{rr}\mu_{\phi\phi}-\varepsilon_{re}}\frac{1}{\overline{r}}D_{\overline{r}}D_\phi E_z$$
$$-\left(\kappa_e\frac{\mu_{\phi\phi}}{\mu_{rr}}\frac{\mu_{rr}\varepsilon_{\phi\phi}-\varepsilon_{re}}{\varepsilon_{rr}\mu_{\phi\phi}-\varepsilon_{re}}-\kappa_m\frac{\varepsilon_{\phi\phi}}{\mu_{rr}}\right)\frac{1}{\overline{r}}D_\phi E_z=0 \tag{4.60}$$

where:

$$\varepsilon_\perp=\varepsilon_{zz}\left(1-\frac{\kappa_e^2}{\varepsilon_{rr}\varepsilon_{zz}}\right)\qquad \mu_\perp=\mu_{zz}\left(1-\frac{\kappa_m^2}{\mu_{rr}\mu_{zz}}\right) \tag{4.61}$$

The r components of the field are now given by:

$$\begin{bmatrix}(\varepsilon_{rr}\mu_{\phi\phi}-\varepsilon_{re})E_r\\(\mu_{rr}\varepsilon_{\phi\phi}-\varepsilon_{re})\widetilde{H}_r\end{bmatrix}=\mathrm{j}\begin{bmatrix}\kappa_e\mu_{\phi\phi}-\sqrt{\varepsilon_{re}}D_{\overline{r}} & -\overline{r}^{-1}\mu_{\phi\phi}D_\phi\\ \overline{r}^{-1}\varepsilon_{\phi\phi}D_\phi & \kappa_m\varepsilon_{\phi\phi}-\sqrt{\varepsilon_{re}}D_{\overline{r}}\end{bmatrix}\begin{bmatrix}E_z\\\widetilde{H}_z\end{bmatrix} \tag{4.62}$$

and the ϕ components by:

$$\begin{bmatrix}(\mu_{rr}\varepsilon_{\phi\phi}-\varepsilon_{re})E_\phi\\(\varepsilon_{rr}\mu_{\phi\phi}-\varepsilon_{re})\widetilde{H}_\phi\end{bmatrix}=\mathrm{j}\begin{bmatrix}-\sqrt{\varepsilon_{re}}\overline{r}^{-1}D_\phi & -\sqrt{\varepsilon_{re}}\kappa_m+\mu_{rr}D_{\overline{r}}\\ \kappa_e\sqrt{\varepsilon_{re}}-\varepsilon_{rr}D_{\overline{r}} & -\sqrt{\varepsilon_{re}}\overline{r}^{-1}D_\phi\end{bmatrix}\begin{bmatrix}E_z\\\widetilde{H}_z\end{bmatrix} \tag{4.63}$$

In the next subsections we will specialise these equations for circular waveguides with ferrites magnetised azimuthally to remanence and to azimuthally-magnetised solid-plasma coaxial waveguides.

4.2.4.1 Circular waveguides with azimuthally-magnetised ferrites

We will first apply the equations in Section 4.2.4 to the case of circular waveguides with azimuthally-magnetised ferrites. To study the propagation problem, we have to solve the following second-order coupled differential equations for the longitudinal components that are obtained from eqs. (4.59) and (4.60) by setting $\kappa_e=0, \varepsilon_{rr}=\varepsilon_{\phi\phi}=\varepsilon_{zz}=\varepsilon_r, \mu_{rr}=\mu_{zz}=\mu_1, \mu_{\phi\phi}=\mu_2$ in the ferrite region:

$$\frac{\varepsilon_r\mu_2-\varepsilon_{re}}{\varepsilon_r\mu_1-\varepsilon_{re}}\frac{1}{\overline{r}^2}\frac{\partial^2E_z}{\partial\phi^2}+\frac{1}{\overline{r}}D_{\overline{r}}\overline{r}D_{\overline{r}}E_z+(\varepsilon_r\mu_2-\varepsilon_{re})E_z$$
$$+\frac{\mu_1-\mu_2}{\mu_1\varepsilon_r-\varepsilon_{re}}\frac{\sqrt{\varepsilon_{re}}}{\overline{r}}D_{\overline{r}}D_\phi\widetilde{H}_z+\kappa_m\frac{\varepsilon_r\mu_2-\varepsilon_{re}}{\varepsilon_r\mu_1-\varepsilon_{re}}\frac{1}{\overline{r}}D_\phi\widetilde{H}_z=0 \tag{4.64}$$

$$\frac{\varepsilon_r\mu_1-\varepsilon_{re}}{\varepsilon_r\mu_2-\varepsilon_{re}}\frac{\mu_2}{\mu_1}\frac{1}{\overline{r}^2}\frac{\partial^2\widetilde{H}_z}{\partial\phi^2}+\frac{1}{\overline{r}}D_{\overline{r}}\overline{r}D_{\overline{r}}\widetilde{H}_z+\left(\varepsilon_r\mu_\perp-\varepsilon_{re}-\frac{\kappa_m}{\mu_1}\frac{\sqrt{\varepsilon_{re}}}{\overline{r}}\right)\widetilde{H}_z$$
$$+\frac{\varepsilon_r}{\mu_1}\frac{\mu_1-\mu_2}{\varepsilon_r\mu_2-\varepsilon_{re}}\frac{\sqrt{\varepsilon_{re}}}{\overline{r}}D_{\overline{r}}D_\phi E_z+\kappa_m\frac{\varepsilon_r}{\mu_1}\frac{1}{\overline{r}}D_\phi E_z=0 \tag{4.65}$$

Usually we have $\mu_1 = \mu_2 = 1$, which results in a further simplification:

$$\frac{1}{\overline{r}} D_{\overline{r}} \overline{r} D_{\overline{r}} E_z + \frac{1}{\overline{r}^2} \frac{\partial^2 E_z}{\partial \phi^2} + (\varepsilon_r - \varepsilon_{re}) E_z + \frac{\kappa_m}{\overline{r}} \frac{\partial \widetilde{H}_z}{\partial \phi} = 0 \tag{4.66}$$

$$\frac{1}{\overline{r}} D_{\overline{r}} \overline{r} D_{\overline{r}} \widetilde{H}_z + \frac{1}{\overline{r}^2} \frac{\partial^2 \widetilde{H}_z}{\partial \phi^2} + \left(\varepsilon_r \mu_\perp - \varepsilon_{re} - \frac{\kappa_m}{\overline{r}} \sqrt{\varepsilon_{re}} \right) \widetilde{H}_z$$
$$+ \varepsilon_r \frac{\kappa_m}{\overline{r}} \frac{\partial E_z}{\partial \phi} = 0 \tag{4.67}$$

where $\mu_\perp = 1 - \kappa_m^2$ denotes the effective magnetic permeability of the ferrite. As we can see, the electric and magnetic field components E_z and $\widetilde{H}_z$ are coupled by the off-diagonal element κ_m of permeability tensor, resulting in different solutions for E_z and $\widetilde{H}_z$ for opposite signs of κ_m (M_r). In the case of rotationally symmetric fields, the two equations are decoupled, yielding again solutions for $\widetilde{H}_z$ that depend only on the remanent magnetisation M_r. Worth noting is that when $\kappa_m = 0$, the ferrite reduces to a dielectric of permittivity ε_r; the equations for E_z and $\widetilde{H}_z$ are of the same kind and the method of separation of the variables results in two Bessel equations for the radial dependence of z-directed components.

The azimuthal dependence of the fields is now assumed to be:

$$E_r, E_z \sim \sin(m\phi) \qquad \widetilde{H}_r, \widetilde{H}_z \sim \cos(m\phi) \tag{4.68}$$

The component $\widetilde{H}_\phi$ has the same azimuthal dependence as E_z and the component E_ϕ the same azimuthal dependence as $\widetilde{H}_z$. Now we multiply these equations with $\overline{r}_o^2 = (k_o r_o)^2$, where r_0 is the outer guide radius. By using the abbreviations $r_n = \overline{r}/\overline{r}_o = r/r_o, \varepsilon_{ren} = \varepsilon_{re}/\varepsilon_r, f_n = \overline{r}_o\sqrt{\varepsilon_r}, E_{un} = \sqrt{\varepsilon_r} E_u$ and $u = r, \varphi, z$, we obtain the following second-order partial differential equations for E_{zn} and $\widetilde{H}_z$:

$$\frac{1}{r_n} D_{r_n} r_n D_{r_n} E_{zn} - \frac{m^2}{r_n^2} E_{zn} + f_n^2 (1 - \varepsilon_{ren}) E_{zn} - \kappa_m m \frac{f_n}{r_n} \widetilde{H}_z = 0 \tag{4.69}$$

$$\frac{1}{r_n} D_{r_n} r_n D_{r_n} E_{zn} \widetilde{H}_z - \frac{m^2}{r_n^2} \widetilde{H}_z + f_n^2 \left(\mu_\perp - \varepsilon_{ren} - \kappa_m \frac{\sqrt{\varepsilon_{ren}}}{r_n f_n} \right) \widetilde{H}_z$$
$$+ \kappa_m m \frac{f_n}{r_n} E_{zn} = 0 \tag{4.70}$$

$D_{r_n} = \partial/\partial r_n$ denotes the differential operator with respect to normalised r_n. The normalised propagation constant ε_{ren} can be computed as a function of the normalised frequency f_n with the off-diagonal component of permeability tensor κ_m as a parameter.

The transverse electric and magnetic field components are expressed in terms of E_{zn}, $\widetilde{H}_z$ using the following matrix form:

$$(1-\varepsilon_{ren})f_n\begin{bmatrix}E_{rn}\\E_{\phi n}\end{bmatrix}=\mathrm{j}\begin{bmatrix}-\sqrt{\varepsilon_{ren}}D_{\overline{r}} & mr_n^{-1}\\-\sqrt{\varepsilon_{ren}}mr_n^{-1} & -\kappa_m\sqrt{\varepsilon_{ren}}f_n+D_{\overline{r}}\end{bmatrix}\begin{bmatrix}E_{zn}\\\widetilde{H}_z\end{bmatrix} \tag{4.71}$$

$$(1-\varepsilon_{ren})f_n\begin{bmatrix}\widetilde{H}_\phi\\\widetilde{H}_r\end{bmatrix}=\mathrm{j}\begin{bmatrix}-D_{\overline{r}} & m\sqrt{\varepsilon_{ren}}r_n^{-1}\\mr_n^{-1} & \kappa_m f_n-\sqrt{\varepsilon_{ren}}D_{\overline{r}}\end{bmatrix}\begin{bmatrix}E_{zn}\\\widetilde{H}_z\end{bmatrix} \tag{4.72}$$

4.2.4.2 Coaxial waveguides with azimuthally magnetised plasma

In this subsection we will apply the equations in Section 4.2.4 to the case of circular waveguides with azimuthally-magnetised plasma. To study the propagation problem we have to solve a second-order coupled differential equation system for the longitudinal components. This is obtained from eqs. (4.59) and (4.60) by setting $\kappa_m=0, \varepsilon_{rr}=\varepsilon_{zz}=\varepsilon_1, \varepsilon_{\phi\phi}=\varepsilon_2, \mu_{rr}=\mu_{\phi\phi}=\mu_{zz}=\mu_1=1$ in the magnetised plasma region:

$$\frac{1}{\overline{r}}\frac{\partial}{\partial\overline{r}}\overline{r}\frac{\partial E_z}{\partial\overline{r}}+\frac{\varepsilon_1-\varepsilon_{re}}{\varepsilon_2-\varepsilon_{re}}\frac{\varepsilon_2}{\varepsilon_1}\frac{1}{\overline{r}^2}\frac{\partial^2E_z}{\partial\phi^2}+\left(\varepsilon_\perp-\varepsilon_{re}-\frac{\kappa_e}{\varepsilon_1}\frac{\sqrt{\varepsilon_{re}}}{\overline{r}}\right)E_z+\frac{\kappa_e}{\varepsilon_1}\frac{1}{\overline{r}}\frac{\partial\widetilde{H}_z}{\partial\phi}$$
$$+\frac{\sqrt{\varepsilon_{re}}}{\varepsilon_1}\frac{\varepsilon_2-\varepsilon_1}{\varepsilon_2-\varepsilon_{re}}\frac{1}{\overline{r}}\frac{\partial}{\partial\overline{r}}\frac{\partial\widetilde{H}_z}{\partial\phi}=0 \tag{4.73}$$

$$\frac{1}{\overline{r}}\frac{\partial}{\partial\overline{r}}\overline{r}\frac{\partial\widetilde{H}_z}{\partial\overline{r}}+\frac{\varepsilon_2-\varepsilon_{re}}{\varepsilon_1-\varepsilon_{re}}\frac{1}{\overline{r}^2}\frac{\partial^2\widetilde{H}_z}{\partial\phi^2}+(\varepsilon_2-\varepsilon_{re})\widetilde{H}_z-\frac{\varepsilon_2-\varepsilon_{re}}{\varepsilon_1-\varepsilon_{re}}\frac{\kappa_e}{\overline{r}}\frac{\partial E_z}{\partial\phi}$$
$$-\frac{\varepsilon_1-\varepsilon_2}{\varepsilon_1-\varepsilon_{re}}\frac{\sqrt{\varepsilon_{re}}}{\overline{r}}\frac{\partial}{\partial\overline{r}}\frac{\partial E_z}{\partial\phi}=0 \tag{4.74}$$

The transverse field components E_r, E_ϕ and H_r, H_ϕ are expressed in terms of E_z and H_z in matrix notation:

$$\begin{bmatrix}(\varepsilon_1-\varepsilon_{re})E_r\\(\varepsilon_2-\varepsilon_{re})E_\phi\end{bmatrix}=\mathrm{j}\begin{bmatrix}\kappa_e-\sqrt{\varepsilon_{re}}D_{\overline{r}} & -\overline{r}^{-1}D_\phi\\-\sqrt{\varepsilon_{re}}\overline{r}^{-1}D_\phi & D_{\overline{r}}\end{bmatrix}\begin{bmatrix}E_z\\\widetilde{H}_z\end{bmatrix} \tag{4.75}$$

$$\begin{bmatrix}(\varepsilon_2-\varepsilon_{re})\widetilde{H}_r\\(\varepsilon_1-\varepsilon_{re})\widetilde{H}_\phi\end{bmatrix}=\mathrm{j}\begin{bmatrix}\varepsilon_2\overline{r}^{-1}D_\phi & -\sqrt{\varepsilon_{re}}D_{\overline{r}}\\\kappa_e\sqrt{\varepsilon_{re}}-\varepsilon_1D_{\overline{r}} & -\sqrt{\varepsilon_{re}}\overline{r}^{-1}D_\phi\end{bmatrix}\begin{bmatrix}E_z\\\widetilde{H}_z\end{bmatrix} \tag{4.76}$$

Generally, the field vectors depend on the azimuthal coordinate ϕ and the differential operator $D_\phi=\partial/\partial\phi\neq0$. The azimuthal dependence of the fields is now assumed to have the form given in eq. (4.68).

We multiply eqs. (4.73) and (4.74) by $\overline{r}_o^2=(k_or_o)^2$, where r_0 is the outer guide radius. By using the abbreviations $r_n=\overline{r}/\overline{r}_o=r/r_o$ and $f_n=\overline{r}_o$, we obtain the following second-order ordinary differential equations for E_z

and $\widetilde{H}_z$:

$$\frac{1}{r_n} D_{r_n} r_n D_{r_n} E_z + \left(-\frac{\varepsilon_1 - \varepsilon_{re}}{\varepsilon_2 - \varepsilon_{re}} \frac{\varepsilon_2}{\varepsilon_1} \frac{m^2}{r_n^2} + (\varepsilon_\perp - \varepsilon_{re}) f_n^2 - \sqrt{\varepsilon_{re}} \frac{\kappa_e}{\varepsilon_1} \frac{f_n}{r_n} \right) E_z$$
$$- \left(\frac{\kappa_e}{\varepsilon_1} \frac{f_n}{r_n} m \right) \widetilde{H}_z - \frac{\sqrt{\varepsilon_{re}}}{\varepsilon_1} \frac{\varepsilon_2 - \varepsilon_1}{\varepsilon_2 - \varepsilon_{re}} \frac{m}{r_n} D_{r_n} \widetilde{H}_z = 0 \tag{4.77}$$

$$\frac{1}{r_n} D_{r_n} r_n D_{r_n} \widetilde{H}_z + \left(-\frac{\varepsilon_2 - \varepsilon_{re}}{\varepsilon_1 - \varepsilon_{re}} \frac{m^2}{r_n^2} + (\varepsilon_2 - \varepsilon_{re}) f_n^2 \right) \widetilde{H}_z$$
$$- \frac{\varepsilon_2 - \varepsilon_{re}}{\varepsilon_1 - \varepsilon_{re}} \frac{m \kappa_e f_n}{r_n} E_z - \sqrt{\varepsilon_{re}} \frac{\varepsilon_1 - \varepsilon_2}{\varepsilon_1 - \varepsilon_{re}} \frac{m}{r_n} D_{r_n} E_z = 0 \tag{4.78}$$

The normalised propagation constant ε_{re} can be computed as a function of the normalised frequency f_n, with the ratio of off-diagonal and diagonal components of permittivity tensor κ_e/ε_1 as a parameter.

The transverse electric and magnetic field components are expressed in terms of E_{zn}, $\widetilde{H}_z$, using the following matrix form:

$$\begin{bmatrix} (\varepsilon_1 - \varepsilon_{re}) f_n E_r \\ (\varepsilon_2 - \varepsilon_{re}) f_n E_\phi \end{bmatrix} = \mathrm{j} \begin{bmatrix} \kappa_e f_n - \sqrt{\varepsilon_{re}} D_{r_n} & \dfrac{m}{r_n} \\ -\sqrt{\varepsilon_{re}} \dfrac{m}{r_n} & D_{r_n} \end{bmatrix} \begin{bmatrix} E_z \\ \widetilde{H}_z \end{bmatrix} \tag{4.79}$$

$$\begin{bmatrix} (\varepsilon_1 - \varepsilon_{re}) f_n \widetilde{H}_\phi \\ (\varepsilon_2 - \varepsilon_{re}) f_n \widetilde{H}_r \end{bmatrix} = \mathrm{j} \begin{bmatrix} \kappa_e f_n \sqrt{\varepsilon_{re}} - \varepsilon_1 D_{r_n} & \sqrt{\varepsilon_{re}} \dfrac{m}{r_n} \\ \varepsilon_2 \dfrac{m}{r_n} & -\sqrt{\varepsilon_{re}} D_{r_n} \end{bmatrix} \begin{bmatrix} E_z \\ \widetilde{H}_z \end{bmatrix} \tag{4.80}$$

$D_{r_n} = \partial/\partial r_n$ denotes the differential operator with respect to the normalised radius r_n.

4.2.5 GTL equations for r-direction

Let us now derive GTL equations for the r-direction. These equation can be used e.g. for the analysis of radial waveguides as presented in Figs. 4.6 and 4.7.

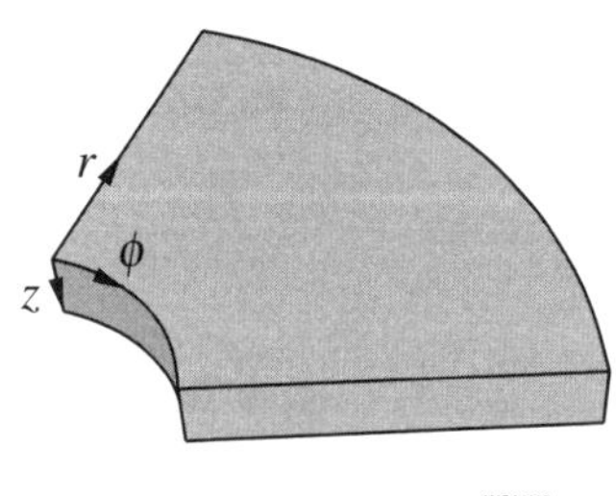

Fig. 4.6 Hollow waveguide with wave propagation in r-direction

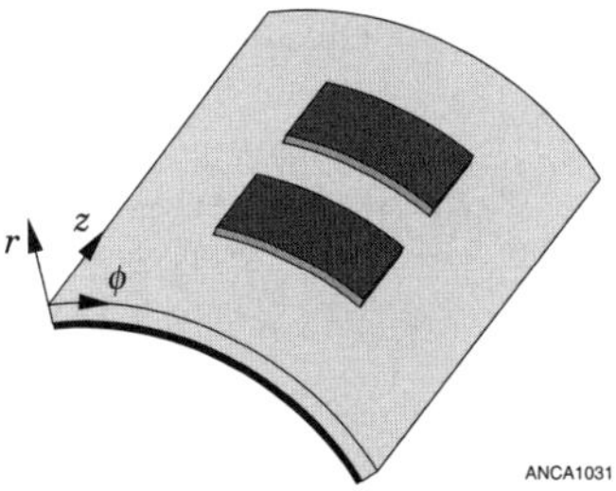

Fig. 4.7 Patches on cylinder

Formally, we introduce the off-diagonal parameters $\nu_{z\phi}, \nu_{\phi z}$. The material parameters may be functions of z and ϕ only. The other off-diagonal elements in the material tensors are assumed to be zero. (The general case with full tensors is described in Chapter 8.) Later on we will also assume $\nu_{rr} = \nu_{\phi\phi} = \nu_{rt}$. First we write the equations quite generally. From the second and third equations of Ampère's law and the law of induction, respectively, we obtain:

$$-\mathrm{j}\begin{bmatrix}\varepsilon_{\phi\phi} & \overline{r}\varepsilon_{\phi z}\\ \overline{r}\varepsilon_{z\phi} & \overline{r}^2\varepsilon_{zz}\end{bmatrix}[\widehat{E}^r] = \overline{r}\frac{\partial}{\partial\overline{r}}[\widehat{H}^r] + \begin{bmatrix}-D_{\overline{z}}\\ D_\phi\end{bmatrix}\overline{r}\widetilde{H}_r \qquad [\widehat{H}^r]=\begin{bmatrix}\widetilde{H}_z\\ -\overline{r}\widetilde{H}_\phi\end{bmatrix} \tag{4.81}$$

$$-\mathrm{j}\begin{bmatrix}\overline{r}^2\mu_{zz} & -\overline{r}\mu_{z\phi}\\ -\overline{r}\mu_{\phi z} & \mu_{\phi\phi}\end{bmatrix}[\widehat{H}^r] = \overline{r}\frac{\partial}{\partial\overline{r}}[\widehat{E}^r] - \begin{bmatrix}D_\phi\\ D_{\overline{z}}\end{bmatrix}\overline{r}E_r \qquad [\widehat{E}^r]=\begin{bmatrix}\overline{r}E_\phi\\ E_z\end{bmatrix} \tag{4.82}$$

We determine the radial field components $\widetilde{H}_r$ and E_r from the first equations of the law of induction and from Ampère's law, respectively, and obtain:

$$\overline{r}\widetilde{H}_r = \mathrm{j}\mu_{rr}^{-1}[-D_{\overline{z}} \quad D_\phi][\widehat{E}^r] \qquad \overline{r}E_r = -\mathrm{j}\varepsilon_{rr}^{-1}[D_\phi \quad D_{\overline{z}}][\widehat{H}^r] \tag{4.83}$$

By introducing eq. (4.83) into eqs. (4.81) and (4.82), the generalised transmission line equations for propagation in r-direction take the form:

$$\overline{r}\frac{\partial}{\partial\overline{r}}[\widehat{H}^r] = -\mathrm{j}([R_{\mathrm{E}}^{rc\mathrm{D}}] + [\widehat{\varepsilon}_{tr}^{rc}])[\widehat{E}^r] \qquad \overline{r}\frac{\partial}{\partial\overline{r}}[\widehat{E}^r] = -\mathrm{j}([R_{\mathrm{H}}^{rc\mathrm{D}}] + [\widehat{\mu}_{tr}^{rc}])[\widehat{H}^r] \tag{4.84}$$

or, if we assume $\nu_{z\phi} = \nu_{\phi z} = 0$:

$$\overline{r}\frac{\partial}{\partial\overline{r}}[\widehat{H}^r] = -\mathrm{j}([R_{\mathrm{E}}^{rc}] + \overline{r}^2[\widehat{\varepsilon}_{zz}])[\widehat{E}^r] \qquad \overline{r}\frac{\partial}{\partial\overline{r}}[\widehat{E}^r] = -\mathrm{j}([R_{\mathrm{H}}^{rc}] + \overline{r}^2[\widehat{\mu}_{zz}])[\widehat{H}^r] \tag{4.85}$$

We define the matrices $R_{\mathrm{E,H}}$ in the following way:

$$[R_{\mathrm{E}}^{rc}] = [R_{\mathrm{E}}^{rc\mathrm{D}}] + [\widehat{\varepsilon}_{\phi\phi}] \qquad [R_{\mathrm{E}}^{rc\mathrm{D}}] = \begin{bmatrix} D_{\overline{z}}\mu_{rr}^{-1}D_{\overline{z}} & -D_{\overline{z}}\mu_{rr}^{-1}D_{\phi} \\ -D_{\phi}\mu_{rr}^{-1}D_{\overline{z}} & D_{\phi}\mu_{rr}^{-1}D_{\phi} \end{bmatrix} \tag{4.86}$$

$$[R_{\mathrm{H}}^{rc}] = [R_{\mathrm{H}}^{rc\mathrm{D}}] + [\widehat{\mu}_{\phi\phi}] \qquad [R_{\mathrm{H}}^{rc\mathrm{D}}] = \begin{bmatrix} D_{\phi}\varepsilon_{rr}^{-1}D_{\phi} & D_{\phi}\varepsilon_{rr}^{-1}D_{\overline{z}} \\ D_{\overline{z}}\varepsilon_{rr}^{-1}D_{\phi} & D_{\overline{z}}\varepsilon_{rr}^{-1}D_{\overline{z}} \end{bmatrix} \tag{4.87}$$

The matrices with the material parameters are given as:

$$\begin{aligned} \left[\widehat{\varepsilon}_{tr}^{rc}\right] &= \begin{bmatrix} \varepsilon_{\phi\phi} & \overline{r}\varepsilon_{\phi z} \\ \overline{r}\varepsilon_{z\phi} & \overline{r}^2\varepsilon_{zz} \end{bmatrix} & \left[\widehat{\mu}_{tr}^{rc}\right] &= \begin{bmatrix} \overline{r}^2\mu_{zz} & -\overline{r}\mu_{z\phi} \\ -\overline{r}\mu_{\phi z} & \mu_{\phi\phi} \end{bmatrix} \\ \left[\widehat{\varepsilon}_{\phi\phi}\right] &= \begin{bmatrix} \varepsilon_{\phi\phi} & 0 \\ 0 & 0 \end{bmatrix} & \left[\widehat{\varepsilon}_{zz}\right] &= \begin{bmatrix} 0 & 0 \\ 0 & \varepsilon_{zz} \end{bmatrix} \\ \left[\widehat{\mu}_{\phi\phi}\right] &= \begin{bmatrix} 0 & 0 \\ 0 & \mu_{\phi\phi} \end{bmatrix} & \left[\widehat{\mu}_{zz}\right] &= \begin{bmatrix} \mu_{zz} & 0 \\ 0 & 0 \end{bmatrix} \end{aligned} \tag{4.88}$$

The first-order differential eqs. (4.84) can be combined to a first-order differential equation system and then be solved by finite differences. By combining the two first-order differential equation systems (4.85) to second-order differential equations, we obtain:

$$\overline{r}\frac{\partial}{\partial\overline{r}}\left(\overline{r}\frac{\partial}{\partial\overline{r}}[\widehat{E}^r]\right) + [Q_{\mathrm{E}}^{rc}][\widehat{E}^r] + \overline{r}^2[S_{\mathrm{E}}^{rc}][\widehat{E}^r] + \mathrm{j}2\overline{r}^2[\widehat{\mu}_{zz}][\widehat{H}^r] = \begin{bmatrix} 0 \\ 0 \end{bmatrix} \tag{4.89}$$

$$\overline{r}\frac{\partial}{\partial\overline{r}}\left(\overline{r}\frac{\partial}{\partial\overline{r}}[\widehat{H}^r]\right) + [Q_{\mathrm{H}}^{rc}][\widehat{H}^r] + \overline{r}^2[S_{\mathrm{H}}^{rc}][\widehat{H}^r] + \mathrm{j}2\overline{r}^2[\widehat{\varepsilon}_{zz}][\widehat{E}^r] = \begin{bmatrix} 0 \\ 0 \end{bmatrix} \tag{4.90}$$

The matrices $[Q_{\mathrm{E,H}}^{rc}]$ are given by:

$$\begin{aligned} [Q_{\mathrm{E}}^{rc}] &= [R_{\mathrm{H}}^{rc}][R_{\mathrm{E}}^{rc}] = [R_{\mathrm{H}}^{rc\mathrm{D}}][\widehat{\varepsilon}_{\phi\phi}] + [\widehat{\mu}_{\phi\phi}][R_{\mathrm{E}}^{rc\mathrm{D}}] \\ &= \begin{bmatrix} D_{\phi}\varepsilon_{rr}^{-1}D_{\phi}\varepsilon_{\phi\phi} & 0 \\ D_{\overline{z}}\varepsilon_{rr}^{-1}D_{\phi}\varepsilon_{\phi\phi} - \mu_{\phi\phi}D_{\phi}\mu_{rr}^{-1}D_{\overline{z}} & \mu_{\phi\phi}D_{\phi}\mu_{rr}^{-1}D_{\phi} \end{bmatrix} \end{aligned} \tag{4.91}$$

$$\begin{aligned} [Q_{\mathrm{H}}^{rc}] &= [R_{\mathrm{E}}^{rc}][R_{\mathrm{H}}^{rc}] = [R_{\mathrm{E}}^{rc\mathrm{D}}][\widehat{\mu}_{\phi\phi}] + [\widehat{\varepsilon}_{\phi\phi}][R_{\mathrm{H}}^{rc\mathrm{D}}] \\ &= \begin{bmatrix} \varepsilon_{\phi\phi}D_{\phi}\varepsilon_{rr}^{-1}D_{\phi} & \varepsilon_{\phi\phi}D_{\phi}\varepsilon_{rr}^{-1}D_{\overline{z}} - D_{\overline{z}}\mu_{rr}^{-1}D_{\phi}\mu_{\phi\phi} \\ 0 & D_{\phi}\mu_{rr}^{-1}D_{\phi}\mu_{\phi\phi} \end{bmatrix} \end{aligned} \tag{4.92}$$

and the matrices $[S_{\mathrm{E,H}}^{rc}]$ by:

$$\begin{aligned}[S_{\mathrm{E}}^{rc}] &= [R_{\mathrm{H}}^{rc}][\widehat{\varepsilon}_{zz}] + [\widehat{\mu}_{zz}][R_{\mathrm{E}}^{rc}] \\ &= [R_{\mathrm{H}}^{rc\mathrm{D}}][\widehat{\varepsilon}_{zz}] + [\widehat{\mu}_{zz}][R_{\mathrm{E}}^{rc\mathrm{D}}] + [\widehat{\varepsilon}_{\phi\phi}][\widehat{\mu}_{zz}] + [\widehat{\varepsilon}_{zz}][\widehat{\mu}_{\phi\phi}] \\ &= \begin{bmatrix} \mu_{zz} D_{\overline{z}} \mu_{rr}^{-1} D_{\overline{z}} + \varepsilon_{\phi\phi}\mu_{zz} & D_{\phi}\varepsilon_{rr}^{-1} D_{\overline{z}}\varepsilon_{zz} - \mu_{zz} D_{\overline{z}}\mu_{rr}^{-1} D_{\phi} \\ 0 & D_{\overline{z}}\varepsilon_{rr}^{-1} D_{\overline{z}}\varepsilon_{zz} + \varepsilon_{zz}\mu_{\phi\phi} \end{bmatrix}\end{aligned} \tag{4.93}$$

$$\begin{aligned}[S_{\mathrm{H}}^{rc}] &= [R_{\mathrm{E}}^{rc}][\widehat{\mu}_{zz}] + [\widehat{\varepsilon}_{zz}][R_{\mathrm{H}}^{rc}] \\ &= [R_{\mathrm{E}}^{rc\mathrm{D}}][\widehat{\mu}_{zz}] + [\widehat{\varepsilon}_{zz}][R_{\mathrm{H}}^{rc\mathrm{D}}] + [\widehat{\varepsilon}_{\phi\phi}][\widehat{\mu}_{zz}] + [\widehat{\varepsilon}_{zz}][\widehat{\mu}_{\phi\phi}] \\ &= \begin{bmatrix} \varepsilon_{\phi\phi}\mu_{zz} + D_{\overline{z}}\mu_{rr}^{-1} D_{\overline{z}}\mu_{zz} & 0 \\ \varepsilon_{zz} D_{\overline{z}}\varepsilon_{rr}^{-1} D_{\phi} - D_{\phi}\mu_{rr}^{-1} D_{\overline{z}}\mu_{zz} & \varepsilon_{zz}\mu_{\phi\phi} + \varepsilon_{zz} D_{\overline{z}}\varepsilon_{rr}^{-1} D_{\overline{z}} \end{bmatrix}\end{aligned} \tag{4.94}$$

where $D_{\overline{z}} D_{\phi} = D_{\phi} D_{\overline{z}}$ was used. The two eqs. (4.89) and (4.90) may be combined to make a single equation in the form:

$$\overline{r}\frac{\partial}{\partial \overline{r}}\left(\overline{r}\frac{\partial}{\partial \overline{r}}[\widehat{F}^{r}]\right) + [\widehat{Q}_{\mathrm{F}}^{rc}][\widehat{F}^{r}] + \overline{r}^{2}[\widehat{S}_{\mathrm{F}}^{rc}][\widehat{F}^{r}] = [\widehat{0}] \qquad [\widehat{F}^{r}] = \begin{bmatrix} \widehat{E}^{r} \\ \widehat{H}^{r} \end{bmatrix} \tag{4.95}$$

where:

$$[\widehat{Q}_{\mathrm{F}}^{rc}] = \begin{bmatrix} [Q_{\mathrm{E}}^{rc}] & [0] \\ [0] & [Q_{\mathrm{H}}^{rc}] \end{bmatrix} \qquad [\widehat{S}_{\mathrm{F}}^{rc}] = \begin{bmatrix} [S_{\mathrm{E}}^{rc}] & \mathrm{j}2[\widehat{\mu}_{zz}] \\ \mathrm{j}2[\widehat{\varepsilon}_{zz}] & [S_{\mathrm{H}}^{rc}] \end{bmatrix} \tag{4.96}$$

Eqs. (4.89) and (4.90) have the shape of a Bessel differential equation. However, in their general form they cannot be solved analytically. Therefore, we will now continue with some important special cases.

4.2.5.1 Layered structures on cylinders

For multilayered structures on cylinders, e.g. patches (see Fig. 4.7), we can determine the fields from the components E_z and $\widetilde{H}_z$. We assume homogeneous layers with uniaxial anisotropic material parameters according to $\mu_{rr} = \mu_{\phi\phi} = \mu_{rt}$, μ_{zz} and $\varepsilon_{rr} = \varepsilon_{\phi\phi} = \varepsilon_{rt}$, ε_{zz}. In this case, we find that the second equation of system (4.89) and the first equation of system (4.90) contain only the components E_z and $\widetilde{H}_z$. Therefore, we can write the following wave equations for each layer:

$$\overline{r}\frac{\partial}{\partial \overline{r}}\left(\overline{r}\frac{\partial}{\partial \overline{r}}\widetilde{H}_z\right) + D_{\phi}D_{\phi}\widetilde{H}_z + \mu_{zz}\mu_{rt}^{-1}(\varepsilon_{rt}\mu_{rt} + D_{\overline{z}}D_{\overline{z}})\overline{r}^{2}\widetilde{H}_z = 0 \tag{4.97}$$

$$\overline{r}\frac{\partial}{\partial \overline{r}}\left(\overline{r}\frac{\partial}{\partial \overline{r}}E_z\right) + D_{\phi}D_{\phi}E_z + \varepsilon_{zz}\varepsilon_{rt}^{-1}(\mu_{rt}\varepsilon_{rt} + D_{\overline{z}}D_{\overline{z}})\overline{r}^{2}E_z = 0 \tag{4.98}$$

The ϕ components are obtained from eq. (4.85):

$$\begin{bmatrix}(D_{\overline{z}}D_{\overline{z}}+\varepsilon_{rt}\mu_{rt})\overline{r}E_\phi\\(D_{\overline{z}}D_{\overline{z}}+\varepsilon_{rt}\mu_{rt})\overline{r}\widetilde{H}_\phi\end{bmatrix}=\begin{bmatrix}D_{\overline{z}}D_\phi & \mathrm{j}\mu_{rt}\overline{r}\dfrac{\partial}{\partial\overline{r}}\\-\mathrm{j}\varepsilon_{rt}\overline{r}\dfrac{\partial}{\partial\overline{r}} & D_{\overline{z}}D_\phi\end{bmatrix}\begin{bmatrix}E_z\\\widetilde{H}_z\end{bmatrix} \tag{4.99}$$

4.2.5.2 Eigenmodes in z-direction

Homogeneous layers in ϕ-direction:

We assume wave propagation in z-direction according to $D_{\overline{z}}=-\mathrm{j}\sqrt{\varepsilon_{re}}$ and homogeneous circular layers where the material parameters are independent of ϕ (see e.g. Fig. 4.8).

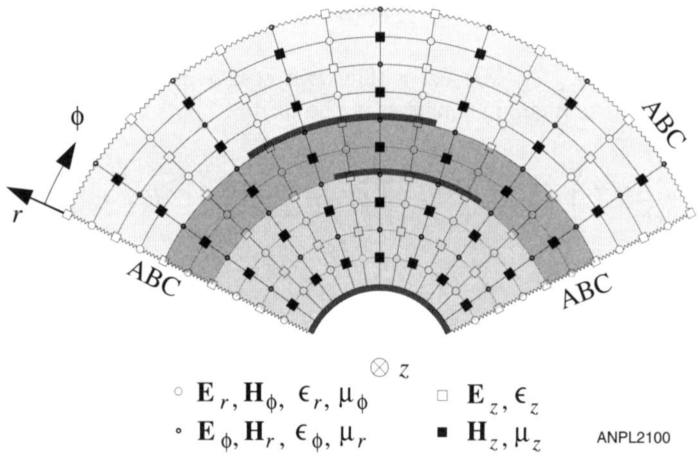

Fig. 4.8 Cross-section of a cylindrical waveguide with metallic strips between the layers (R. Pregla, 'Efficient Analysis of Conformal Antennas with Anisotropic Material (0682)', in *AP 2000 Millennium Conference on Antennas and Propagation.* © 2000 European Space Agency (ESA))

With an anisotropic material according to $\mu_{rr}=\mu_{\phi\phi}=\mu_{rt}$ and $\varepsilon_{rr}=\varepsilon_{\phi\phi}=\varepsilon_{rt}$, we obtain the following wave equations for E_z and $\widetilde{H}_z$ from eqs. (4.97) and (4.98):

$$\overline{r}\frac{\partial}{\partial\overline{r}}\left(\overline{r}\frac{\partial}{\partial\overline{r}}\widetilde{H}_z\right)+D_\phi D_\phi\widetilde{H}_z+\mu_{zz}\mu_{rt}^{-1}(\varepsilon_{rt}\mu_{rt}-\varepsilon_{re})\overline{r}^2\widetilde{H}_z=0 \tag{4.100}$$

$$\overline{r}\frac{\partial}{\partial\overline{r}}\left(\overline{r}\frac{\partial}{\partial\overline{r}}E_z\right)+D_\phi D_\phi E_z+\varepsilon_{zz}\varepsilon_{rt}^{-1}(\varepsilon_{rt}\mu_{rt}-\varepsilon_{re})\overline{r}^2E_z=0 \tag{4.101}$$

The other field components are obtained from eq. (4.84). The further procedure is described in Section 4.3.2.

Inhomogeneous layers in ϕ-direction:
For the case of inhomogeneous circular layers (see e.g. the circular layer between the strips of the cross-section in Fig. 4.3a) we rewrite eq. (4.84) into the following form:

$$\overline{r}\frac{\partial}{\partial \overline{r}}[\widehat{H}^{re}] = -[R_{\mathrm{E}}^{re}][\widehat{E}^{re}] \qquad \overline{r}\frac{\partial}{\partial \overline{r}}[\widehat{E}^{re}] = -[R_{\mathrm{H}}^{re}][H^{re}] \tag{4.102}$$

The superscript e symbolises eigenmodes. We have further:

$$[R_{\mathrm{E}}^{re}] = \begin{bmatrix} \varepsilon_{re}\mu_{rr}^{-1} - \varepsilon_{\phi\phi} & \sqrt{\varepsilon_{re}}\mu_{rr}^{-1}D_\phi \\ \sqrt{\varepsilon_{re}}D_\phi\mu_{rr}^{-1} & D_\phi\mu_{rr}^{-1}D_\phi + \overline{r}^2\varepsilon_{zz} \end{bmatrix} \qquad [E^{re}] = \begin{bmatrix} \overline{r}E_\phi \\ -\mathrm{j}E_z \end{bmatrix} \tag{4.103}$$

$$[R_{\mathrm{H}}^{re}] = \begin{bmatrix} D_\phi\varepsilon_{rr}^{-1}D_\phi + \overline{r}^2\mu_{zz} & -\sqrt{\varepsilon_{re}}D_\phi\varepsilon_{rr}^{-1} \\ -\sqrt{\varepsilon_{re}}\varepsilon_{rr}^{-1}D_\phi & \varepsilon_{re}\varepsilon_{rr}^{-1} - \mu_{\phi\phi} \end{bmatrix} \qquad [H^{re}] = \begin{bmatrix} \mathrm{j}\widetilde{H}_z \\ \overline{r}\widetilde{H}_\phi \end{bmatrix} \tag{4.104}$$

For the components in r-direction we obtain:

$$\overline{r}\widetilde{H}_r = -\mu_{rr}^{-1}[\sqrt{\varepsilon_{re}}D_\phi][E^{re}] \qquad \overline{r}E_r = \varepsilon_{rr}^{-1}[-D_\phi \ \ \sqrt{\varepsilon_{re}}][H^{re}] \tag{4.105}$$

We can solve eq. (4.102) e.g. with finite differences (see Section 4.4.3.7).

4.2.5.3 z-independent fields
In the case of z-independent fields ($\partial/\partial z = 0$), we obtain from eqs. (4.86) and (4.87):

$$[R_{\mathrm{E}}^{rc}] = \begin{bmatrix} \varepsilon_{\phi\phi} & 0 \\ 0 & D_\phi\mu_{rr}^{-1}D_\phi \end{bmatrix} \qquad [R_{\mathrm{H}}^{rc}] = \begin{bmatrix} D_\phi\varepsilon_{rr}^{-1}D_\phi & 0 \\ 0 & \mu_{\phi\phi} \end{bmatrix} \tag{4.106}$$

Because of the zero off-diagonal elements, the equations are decoupled. In explicit form we can write eq. (4.85) as:

$$-\mathrm{j}\varepsilon_{\phi\phi}E_\phi = \frac{\partial}{\partial \overline{r}}\widetilde{H}_z \tag{4.107}$$

$$\mathrm{j}\mu_{\phi\phi}\widetilde{H}_\phi = \frac{\partial}{\partial \overline{r}}E_z \tag{4.108}$$

$$-\overline{r}\frac{\partial}{\partial \overline{r}}(\overline{r}\widetilde{H}_\phi) = -\mathrm{j}\left(\frac{\partial}{\partial\phi}\left(\frac{1}{\mu_{rr}}\frac{\partial}{\partial\phi}\right) + \varepsilon_{zz}\overline{r}^2\right)E_z \tag{4.109}$$

$$\overline{r}\frac{\partial}{\partial \overline{r}}(\overline{r}E_\phi) = -\mathrm{j}\left(\frac{\partial}{\partial\phi}\left(\frac{1}{\varepsilon_{rr}}\frac{\partial}{\partial\phi}\right) + \mu_{zz}\overline{r}^2\right)H_z \tag{4.110}$$

By introducing eqs. (4.107) and (4.108) into (4.110) and (4.109), respectively, we obtain the following wave equations:

$$\overline{r}\frac{\partial}{\partial \overline{r}}\left(\frac{\overline{r}}{\mu_{\phi\phi}}\frac{\partial}{\partial \overline{r}}E_z\right) + \frac{\partial}{\partial\phi}\left(\frac{1}{\mu_{rr}}\frac{\partial}{\partial\phi}E_z\right) + \varepsilon_{zz}\overline{r}^2E_z = 0 \tag{4.111}$$

$$\overline{r}\frac{\partial}{\partial \overline{r}}\left(\frac{\overline{r}}{\varepsilon_{\phi\phi}}\frac{\partial}{\partial \overline{r}}\widetilde{H}_z\right) + \frac{\partial}{\partial\phi}\left(\frac{1}{\varepsilon_{rr}}\frac{\partial}{\partial\phi}\widetilde{H}_z\right) + \mu_{zz}\overline{r}^2\widetilde{H}_z = 0 \tag{4.112}$$

The components E_ϕ and H_ϕ are obtained from eqs. (4.107) and (4.108); the components E_r and H_r from eq. (4.83) according to:

$$\bar{r}E_r = -\mathrm{j}\varepsilon_{rr}^{-1}\frac{\partial}{\partial\phi}\widetilde{H}_z \qquad \bar{r}\widetilde{H}_r = \mathrm{j}\mu_{rr}^{-1}\frac{\partial}{\partial\phi}E_z \tag{4.113}$$

These equations can be used for the analysis of wave propagation, e.g. in horns, as shown in Fig. 4.4 (E_z modes).

4.3 DISCRETISATION OF THE FIELDS AND SOLUTIONS

In this section we show the discretisation of the fields in the cross-sections and explain the principles of the solution.

4.3.1 Equations for propagation in z-direction

The discretisation scheme for this case is shown in Fig. 4.9. In the following, we mark 2D discretised quantities with a hat ($\widehat{\ }$). This is true for the fields, as well as for the finite difference operators.

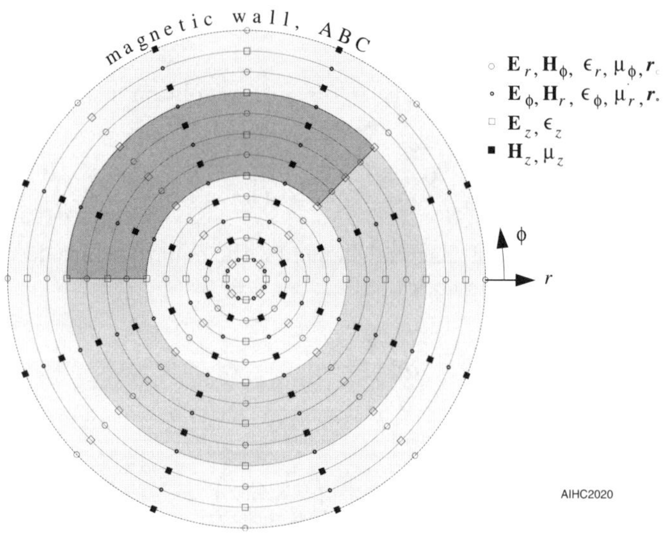

Fig. 4.9 Cross-section of cylindrical waveguide with inhomogeneous layers

The discretised field components are collected in vectors. To order the elements, we start at the waveguide centre and proceed in the radial direction. The values of two neighbouring points in the radial direction are placed below each other. The material parameters and the discretised radii are collected in diagonal matrices. Then the differential operators are replaced by central differences. All central differences in the radial or azimuthal direction are put together in the difference operators (matrices) $\boldsymbol{D}_{\bar{r}}^{\circ,\bullet}$ and $\boldsymbol{D}_{\phi}^{\circ,\bullet}$, respectively.

To obtain expressions for the whole structure we can describe the difference operators (inhomogeneous cross-sections without metallic subsections) by Kronecker products of the difference operators for the radial ($\boldsymbol{D}_{\overline{r}}^{\circ,\bullet}$) or azimuthal direction ($\boldsymbol{D}_{\phi}^{\circ,\bullet}$), and by unit matrices of the order $\mathrm{N}_{r}^{\circ,\bullet}$ and $\mathrm{N}_{\phi}^{\circ,\bullet}$, the number of the radial or azimuthal discretisation points in a radial or azimuthal circle, respectively. So we have (analogous to the Cartesian coordinates in Section 2.4.1) the following expressions:

$$\widehat{\boldsymbol{D}}_{\overline{r}}^{\circ,\bullet} = \boldsymbol{I}_{\phi}^{\circ,\bullet} \otimes \boldsymbol{D}_{\overline{r}}^{\circ,\bullet} \quad \widehat{\boldsymbol{D}}_{\phi}^{\circ,\bullet} = \boldsymbol{D}_{\phi}^{\circ,\bullet} \otimes \boldsymbol{I}_{r}^{\circ,\bullet} \tag{4.114}$$

$\boldsymbol{I}_{r}^{\circ,\bullet}$ and $\boldsymbol{I}_{\phi}^{\circ,\bullet}$ are unit matrices of the order $\mathrm{N}_{r}^{\circ,\bullet}$ and $\mathrm{N}_{\phi}^{\circ,\bullet}$, respectively. The difference operators $\boldsymbol{D}_{\overline{r}}^{\circ,\bullet}$ have to fulfil Dirichlet or Neumann boundary conditions in the case of magnetic walls. If we discretise completely in the azimuthal direction, we must introduce periodic boundary conditions into $\boldsymbol{D}_{\phi}^{\circ,\bullet}$. The conditions have to be exchanged in the case of electric walls. We must choose different places for the electric and magnetic walls. If we need absorbing boundary conditions (ABC), they must replace the Dirichlet conditions. In the case of metallic inserts (see Fig. 4.8), a reduction of the vectors and matrices analogous to the Cartesian case (see Section 4.4.1) has to be performed.

To proceed, we define the supervectors:

$$\widehat{\mathbf{H}} = [-\widetilde{\mathbf{H}}_{r}^{\mathrm{t}}, \overline{\boldsymbol{r}}_{\circ}\widetilde{\mathbf{H}}_{\phi}^{\mathrm{t}}]^{\mathrm{t}} \qquad \widehat{\mathbf{E}} = [\overline{\boldsymbol{r}}_{\bullet}\mathbf{E}_{\phi}^{\mathrm{t}}, \mathbf{E}_{r}^{\mathrm{t}}]^{\mathrm{t}} \tag{4.115}$$

and use the following abbreviations:

$$\begin{aligned} &\widehat{\overline{\boldsymbol{\epsilon}}}_{r} \equiv \widehat{\boldsymbol{\epsilon}}_{rr}\widehat{\overline{\boldsymbol{r}}}_{\circ} \quad &&\widehat{\overline{\boldsymbol{\epsilon}}}_{\phi} \equiv \widehat{\boldsymbol{\epsilon}}_{\phi\phi}\widehat{\overline{\boldsymbol{r}}}_{\bullet}^{-1} \quad &&\widehat{\overline{\boldsymbol{\epsilon}}}_{z} \equiv \widehat{\boldsymbol{\epsilon}}_{zz}\widehat{\overline{\boldsymbol{r}}}_{\square} \\ &\widehat{\overline{\boldsymbol{\mu}}}_{r} \equiv \widehat{\boldsymbol{\mu}}_{rr}\widehat{\overline{\boldsymbol{r}}}_{\bullet} \quad &&\widehat{\overline{\boldsymbol{\mu}}}_{\phi} \equiv \widehat{\boldsymbol{\mu}}_{\phi\phi}\widehat{\overline{\boldsymbol{r}}}_{\circ}^{-1} \quad &&\widehat{\overline{\boldsymbol{\mu}}}_{z} \equiv \widehat{\boldsymbol{\mu}}_{zz}\widehat{\overline{\boldsymbol{r}}}_{\blacksquare} \end{aligned} \tag{4.116}$$

The discretisation of eqs. (4.11) and (4.14) results in:

$$\frac{\mathrm{d}}{\mathrm{d}\overline{z}}\widehat{\mathbf{H}} = -\mathrm{j}\widehat{\boldsymbol{R}}_{\mathrm{E}}^{z}\widehat{\mathbf{E}} \;\; \frac{\mathrm{d}}{\mathrm{d}\overline{z}}\widehat{\mathbf{E}} = -\mathrm{j}\widehat{\boldsymbol{R}}_{\mathrm{H}}^{z}\widehat{\mathbf{H}} \tag{4.117}$$

$$\frac{\mathrm{d}^2}{\mathrm{d}\overline{z}^2}\widehat{\mathbf{H}} - \widehat{\boldsymbol{Q}}_{\mathrm{H}}^{z}\widehat{\mathbf{H}} = \widehat{\mathbf{0}} \;\; \frac{\mathrm{d}^2}{\mathrm{d}\overline{z}^2}\widehat{\mathbf{E}} - \widehat{\boldsymbol{Q}}_{\mathrm{E}}^{z}\widehat{\mathbf{E}} = \widehat{\mathbf{0}} \tag{4.118}$$

with: (See eqs. (4.12) and (4.13).)

$$\widehat{\boldsymbol{R}}_{\mathrm{E}}^{z} = \begin{bmatrix} \widehat{\overline{\boldsymbol{\epsilon}}}_{\phi} - \widehat{\overline{\boldsymbol{D}}}_{\overline{r}}^{\bullet\mathrm{t}}\widehat{\overline{\boldsymbol{\mu}}}_{z}^{-1}\widehat{\overline{\boldsymbol{D}}}_{\overline{r}}^{\bullet} & \widehat{\mathcal{E}}_{\phi r} + \widehat{\overline{\boldsymbol{D}}}_{\overline{r}}^{\bullet\mathrm{t}}\widehat{\overline{\boldsymbol{\mu}}}_{z}^{-1}\widehat{\overline{\boldsymbol{D}}}_{\phi}^{\circ} \\ \widehat{\overline{\boldsymbol{D}}}_{\phi}^{\circ\mathrm{t}}\widehat{\overline{\boldsymbol{\mu}}}_{z}^{-1}\widehat{\overline{\boldsymbol{D}}}_{\mathrm{r}}^{\bullet} + \widehat{\mathcal{E}}_{r\phi} & \widehat{\boldsymbol{\epsilon}}_{r} - \widehat{\overline{\boldsymbol{D}}}_{\phi}^{\circ\mathrm{t}}\widehat{\overline{\boldsymbol{\mu}}}_{z}^{-1}\widehat{\overline{\boldsymbol{D}}}_{\phi}^{\circ} \end{bmatrix} \tag{4.119}$$

$$\widehat{\boldsymbol{R}}_{\mathrm{H}}^{z} = \begin{bmatrix} \widehat{\overline{\boldsymbol{\mu}}}_{r} - \widehat{\overline{\boldsymbol{D}}}_{\phi}^{\bullet\mathrm{t}}\widehat{\overline{\boldsymbol{\epsilon}}}_{z}^{-1}\widehat{\overline{\boldsymbol{D}}}_{\phi}^{\bullet} & \widehat{\mathcal{M}}_{r\phi} - \widehat{\overline{\boldsymbol{D}}}_{\phi}^{\bullet\mathrm{t}}\widehat{\overline{\boldsymbol{\epsilon}}}_{z}^{-1}\widehat{\overline{\boldsymbol{D}}}_{\overline{r}}^{\circ} \\ \widehat{\mathcal{M}}_{\phi r} - \widehat{\overline{\boldsymbol{D}}}_{\overline{r}}^{\circ\mathrm{t}}\widehat{\overline{\boldsymbol{\epsilon}}}_{z}^{-1}\widehat{\overline{\boldsymbol{D}}}_{\phi}^{\bullet} & \widehat{\overline{\boldsymbol{\mu}}}_{\phi} - \widehat{\overline{\boldsymbol{D}}}_{\overline{r}}^{\circ\mathrm{t}}\widehat{\overline{\boldsymbol{\epsilon}}}_{z}^{-1}\widehat{\overline{\boldsymbol{D}}}_{\overline{r}}^{\circ} \end{bmatrix} \tag{4.120}$$

where e.g.:

$$\begin{array}{rclcrcl}
\widehat{\mathcal{E}}_{\phi r} & = & \widehat{\boldsymbol{M}}_{\phi}^{\square}\widehat{\boldsymbol{M}}_{r}^{\circ}\widehat{\varepsilon}_{\phi r} & \text{or} & \widehat{\mathcal{E}}_{\phi r} & = & \widehat{\boldsymbol{M}}_{r}^{\blacksquare}\widehat{\boldsymbol{M}}_{\phi}^{\circ}\widehat{\varepsilon}_{\phi r} \\
\widehat{\mathcal{E}}_{r\phi} & = & \widehat{\boldsymbol{M}}_{\phi}^{\blacksquare}\widehat{\boldsymbol{M}}_{r}^{\bullet}\widehat{\varepsilon}_{r\phi} & \text{or} & \widehat{\mathcal{E}}_{r\phi} & = & \widehat{\boldsymbol{M}}_{r}^{\square}\widehat{\boldsymbol{M}}_{\phi}^{\bullet}\widehat{\varepsilon}_{r\phi} \\
\widehat{\mathcal{M}}_{\phi r} & = & \widehat{\boldsymbol{M}}_{\phi}^{\blacksquare}\widehat{\boldsymbol{M}}_{r}^{\bullet}\widehat{\mu}_{\phi r} & \text{or} & \widehat{\mathcal{M}}_{\phi r} & = & \widehat{\boldsymbol{M}}_{r}^{\square}\widehat{\boldsymbol{M}}_{\phi}^{\bullet}\widehat{\mu}_{\phi r} \\
\widehat{\mathcal{M}}_{r\phi} & = & \widehat{\boldsymbol{M}}_{\phi}^{\square}\widehat{\boldsymbol{M}}_{r}^{\circ}\widehat{\mu}_{r\phi} & \text{or} & \widehat{\mathcal{M}}_{r\phi} & = & \widehat{\boldsymbol{M}}_{r}^{\blacksquare}\widehat{\boldsymbol{M}}_{\phi}^{\circ}\widehat{\mu}_{r\phi}
\end{array} \tag{4.121}$$

The matrices $\widehat{\boldsymbol{M}}$ are introduced for interpolation between the discretisation schemes (see Section 4.4.3). Their subscript indicates the direction of interpolation and their superscript marks the discretisation scheme that needs to be interpolated.

The submatrices of $\widehat{\boldsymbol{Q}}_{\mathrm{E}}^{z} = -\widehat{\boldsymbol{R}}_{\mathrm{H}}^{z}\widehat{\boldsymbol{R}}_{\mathrm{E}}^{z}$, in the case of diagonal material tensors, can be computed very easily by splitting $\widehat{\boldsymbol{R}}_{\mathrm{E,H}}^{z}$ in the following way:

$$\widehat{\boldsymbol{R}}_{\mathrm{E,H}}^{z} = \widehat{\boldsymbol{R}}_{\mathrm{E,H}}^{zc} + \left\{ \begin{array}{c} \widehat{\boldsymbol{\epsilon}} \\ \widehat{\boldsymbol{\mu}} \end{array} \right\} \qquad \begin{array}{c} \widehat{\boldsymbol{R}}_{\mathrm{E}}^{z}\widehat{\boldsymbol{R}}_{\mathrm{H}}^{z} = \widehat{\boldsymbol{R}}_{\mathrm{E}}^{zc}\widehat{\boldsymbol{\mu}} + \widehat{\boldsymbol{\epsilon}}\widehat{\boldsymbol{R}}_{\mathrm{H}}^{zc} + \widehat{\boldsymbol{\epsilon}}\widehat{\boldsymbol{\mu}} \\ \widehat{\boldsymbol{R}}_{\mathrm{H}}^{z}\widehat{\boldsymbol{R}}_{\mathrm{E}}^{z} = \widehat{\boldsymbol{R}}_{\mathrm{H}}^{zc}\widehat{\boldsymbol{\epsilon}} + \widehat{\boldsymbol{\mu}}\widehat{\boldsymbol{R}}_{\mathrm{E}}^{zc} + \widehat{\boldsymbol{\mu}}\widehat{\boldsymbol{\epsilon}} \end{array} \tag{4.122}$$

4.3.1.1 Absorbing boundary conditions (ABC)

The difference operators for absorbing boundary conditions replace the operators with Dirichlet conditions (see Fig. 4.8). Details can be found in Section A.2.

$$\widehat{\boldsymbol{R}}_{\mathrm{E}}^{az} = \begin{bmatrix} \widehat{\overline{\boldsymbol{\epsilon}}}_{\phi} + \widehat{\overline{\boldsymbol{D}}}_{\overline{r}}^{\blacksquare}\widehat{\overline{\boldsymbol{\mu}}}_{z}^{-1}\widehat{\overline{\boldsymbol{D}}}_{\overline{r}}^{\bullet a} & -\widehat{\overline{\boldsymbol{D}}}_{\overline{r}}^{\blacksquare}\widehat{\overline{\boldsymbol{\mu}}}_{z}^{-1}\widehat{\overline{\boldsymbol{D}}}_{\phi}^{\circ a} \\ -\widehat{\overline{\boldsymbol{D}}}_{\phi}^{\blacksquare}\widehat{\overline{\boldsymbol{\mu}}}_{z}^{-1}\widehat{\overline{\boldsymbol{D}}}_{\overline{r}}^{\bullet a} & \widehat{\boldsymbol{\epsilon}}_{r} + \widehat{\overline{\boldsymbol{D}}}_{\phi}^{\blacksquare}\widehat{\overline{\boldsymbol{\mu}}}_{z}^{-1}\widehat{\overline{\boldsymbol{D}}}_{\phi}^{\circ a} \end{bmatrix} \tag{4.123}$$

$$\widehat{\boldsymbol{R}}_{\mathrm{H}}^{az} = \begin{bmatrix} \widehat{\overline{\boldsymbol{\mu}}}_{r} + \widehat{\overline{\boldsymbol{D}}}_{\phi}^{\square a}\widehat{\overline{\boldsymbol{\epsilon}}}_{z}^{-1}\widehat{\overline{\boldsymbol{D}}}_{\phi}^{\bullet} & \widehat{\overline{\boldsymbol{D}}}_{\phi}^{\square a}\widehat{\overline{\boldsymbol{\epsilon}}}_{z}^{-1}\widehat{\overline{\boldsymbol{D}}}_{\overline{r}}^{\circ} \\ \widehat{\overline{\boldsymbol{D}}}_{\overline{r}}^{\square a}\widehat{\overline{\boldsymbol{\epsilon}}}_{z}^{-1}\widehat{\overline{\boldsymbol{D}}}_{\phi}^{\bullet} & \widehat{\overline{\boldsymbol{\mu}}}_{\phi} + \widehat{\overline{\boldsymbol{D}}}_{\overline{r}}^{\square a}\widehat{\overline{\boldsymbol{\epsilon}}}_{z}^{-1}\widehat{\overline{\boldsymbol{D}}}_{\overline{r}}^{\circ} \end{bmatrix} \tag{4.124}$$

We may also move the ABCs half a discretisation distance at each border to the inner side in Fig. 4.8. Then we may write for the matrices:

$$\widehat{\boldsymbol{R}}_{\mathrm{E}}^{az} = \begin{bmatrix} \widehat{\overline{\boldsymbol{\epsilon}}}_{\phi} + \widehat{\overline{\boldsymbol{D}}}_{\overline{r}}^{\blacksquare a}\widehat{\overline{\boldsymbol{\mu}}}_{z}^{-1}\widehat{\overline{\boldsymbol{D}}}_{\overline{r}}^{\bullet} & -\widehat{\overline{\boldsymbol{D}}}_{\overline{r}}^{\blacksquare a}\widehat{\overline{\boldsymbol{\mu}}}_{z}^{-1}\widehat{\overline{\boldsymbol{D}}}_{\phi}^{\circ} \\ -\widehat{\overline{\boldsymbol{D}}}_{\phi}^{\blacksquare a}\widehat{\overline{\boldsymbol{\mu}}}_{z}^{-1}\widehat{\overline{\boldsymbol{D}}}_{\overline{r}}^{\bullet} & \widehat{\boldsymbol{\epsilon}}_{r} + \widehat{\overline{\boldsymbol{D}}}_{\phi}^{\blacksquare a}\widehat{\overline{\boldsymbol{\mu}}}_{z}^{-1}\widehat{\overline{\boldsymbol{D}}}_{\phi}^{\circ} \end{bmatrix} \tag{4.125}$$

$$\widehat{\boldsymbol{R}}_{\mathrm{H}}^{az} = \begin{bmatrix} \widehat{\overline{\boldsymbol{\mu}}}_{r} + \widehat{\overline{\boldsymbol{D}}}_{\phi}^{\square}\widehat{\overline{\boldsymbol{\epsilon}}}_{z}^{-1}\widehat{\overline{\boldsymbol{D}}}_{\phi}^{\bullet a} & \widehat{\overline{\boldsymbol{D}}}_{\phi}^{\square}\widehat{\overline{\boldsymbol{\epsilon}}}_{z}^{-1}\widehat{\overline{\boldsymbol{D}}}_{\overline{r}}^{\circ a} \\ \widehat{\overline{\boldsymbol{D}}}_{\overline{r}}^{\square}\widehat{\overline{\boldsymbol{\epsilon}}}_{z}^{-1}\widehat{\overline{\boldsymbol{D}}}_{\phi}^{\bullet a} & \widehat{\overline{\boldsymbol{\mu}}}_{\phi} + \widehat{\overline{\boldsymbol{D}}}_{\overline{r}}^{\square}\widehat{\overline{\boldsymbol{\epsilon}}}_{z}^{-1}\widehat{\overline{\boldsymbol{D}}}_{\overline{r}}^{\circ a} \end{bmatrix} \tag{4.126}$$

4.3.1.2 ϕ-independent material parameters

In case of ϕ-independent material parameters and:

$$E_r, H_\phi \sim \cos(m\phi) \; E_\phi, H_r \sim \sin(m\phi) \tag{4.127}$$

the matrices simplify to:

$$\boldsymbol{R}_{\mathrm{E}}^{z} = \begin{bmatrix} \overline{\boldsymbol{\epsilon}}_{\phi} - \overline{\boldsymbol{D}}_{\overline{r}}^{\bullet\mathrm{t}} \overline{\boldsymbol{\mu}}_{z}^{-1} \overline{\boldsymbol{D}}_{\overline{r}}^{\bullet} & m\overline{\boldsymbol{D}}_{\overline{r}}^{\bullet\mathrm{t}} \overline{\boldsymbol{\mu}}_{z}^{-1} \\ m\overline{\boldsymbol{\mu}}_{z}^{-1} \overline{\boldsymbol{D}}_{\overline{r}}^{\bullet} & \overline{\boldsymbol{\epsilon}}_{r} + m^{2}\overline{\boldsymbol{\mu}}_{z}^{-1} \end{bmatrix} \tag{4.128}$$

$$\boldsymbol{R}_{\mathrm{H}}^{z} = \begin{bmatrix} \overline{\boldsymbol{\mu}}_{r} + m^{2}\overline{\boldsymbol{\epsilon}}_{z}^{-1} & m\overline{\boldsymbol{\epsilon}}_{z}^{-1} \overline{\boldsymbol{D}}_{\overline{r}}^{\circ} \\ m\overline{\boldsymbol{D}}_{\overline{r}}^{\circ\mathrm{t}} \overline{\boldsymbol{\epsilon}}_{z}^{-1} & \overline{\boldsymbol{\mu}}_{\phi} - \overline{\boldsymbol{D}}_{\overline{r}}^{\circ\mathrm{t}} \overline{\boldsymbol{\epsilon}}_{\mathrm{z}}^{-1} \overline{\boldsymbol{D}}_{\overline{r}}^{\circ} \end{bmatrix} \tag{4.129}$$

Only one-dimensional discretisation in r-direction is required. The matrices $\widehat{\boldsymbol{R}}_{\mathrm{E,H}}^{z}$ are symmetric:

$$\widehat{\boldsymbol{R}}_{\mathrm{E,H}}^{z} = \widehat{\boldsymbol{R}}_{\mathrm{E,H}}^{z\mathrm{t}} \tag{4.130}$$

However, they are not Hermitian in case of complex elements.

4.3.2 Equations for propagation in ϕ-direction

Here, we discretise according to Fig. 4.11. To order the discretised fields, we start on the left, put each column (z-direction) in a vector and proceed in radial direction. We define the supervectors of the discretised field components as:

$$\widehat{\mathbf{H}} = [-\widetilde{\mathbf{H}}_{z}^{\mathrm{t}}, \widetilde{\mathbf{H}}_{r}^{\mathrm{t}}]^{\mathrm{t}} \quad \widehat{\mathbf{E}} = [\mathbf{E}_{r}^{\mathrm{t}}, \mathbf{E}_{z}^{\mathrm{t}}]^{\mathrm{t}} \tag{4.131}$$

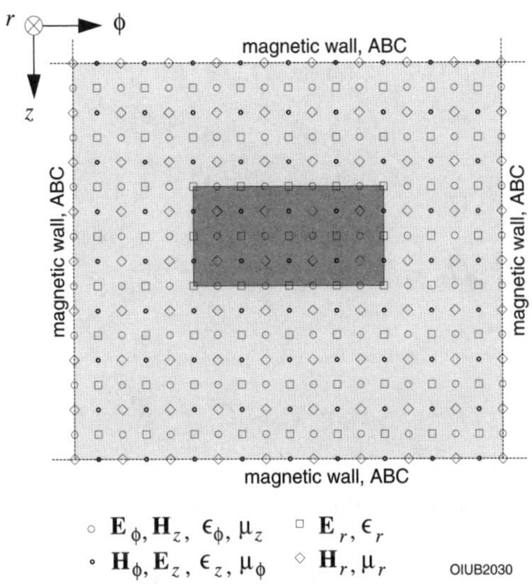

Fig. 4.10 Discretisation for analytical solution in r-direction

After discretisation, eq. (4.39) takes the form

$$\frac{\mathrm{d}}{\overline{r}_0 \mathrm{d}\phi}\widehat{\mathbf{H}} = -\mathrm{j}(\widehat{\boldsymbol{R}}_{\mathrm{E}}^{\phi\mathrm{D}} + [\widehat{\epsilon}])\widehat{\mathbf{E}} \qquad \frac{\mathrm{d}}{\overline{r}_0 \mathrm{d}\phi}\widehat{\mathbf{E}} = -\mathrm{j}(\widehat{\boldsymbol{R}}_{\mathrm{H}}^{\phi\mathrm{D}} + [\widehat{\boldsymbol{\mu}}])\widehat{\mathbf{H}} \tag{4.132}$$

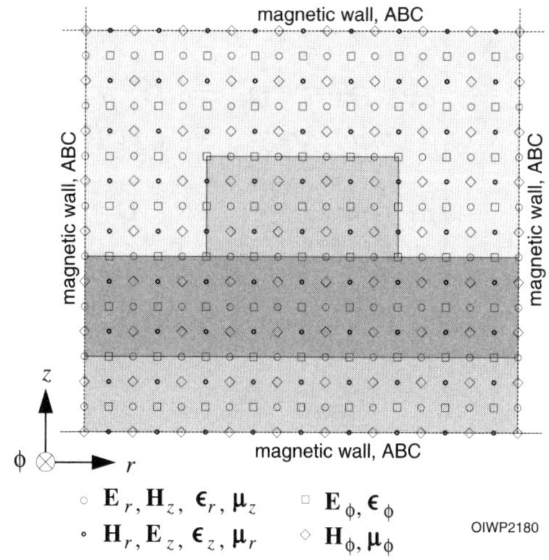

Fig. 4.11 Discretisation for analytical solution in ϕ-direction

where for diagonal material tensors

$$\widehat{\boldsymbol{R}}_{\mathrm{E}}^{\phi\mathrm{D}}=\begin{bmatrix} -\widehat{\overline{\boldsymbol{D}}}_{\overline{z}}^{\circ\mathrm{t}}\,\widehat{\boldsymbol{\mu}}_{\phi}^{-1}\widehat{\overline{\mathbf{r}}}_{\mathrm{n}}^{\diamond}\widehat{\overline{\boldsymbol{D}}}_{\overline{z}}^{\circ} & \widehat{\overline{\boldsymbol{D}}}_{\overline{z}}^{\circ\mathrm{t}}\,\widehat{\boldsymbol{\mu}}_{\phi}^{-1}\widehat{\overline{\mathbf{r}}}_{\mathrm{n}}^{\diamond}\widehat{\overline{\boldsymbol{D}}}_{\overline{r}}^{\bullet} \\ \widehat{\overline{\boldsymbol{D}}}_{\overline{r}}^{\bullet\mathrm{t}}\,\widehat{\boldsymbol{\mu}}_{\phi}^{-1}\widehat{\overline{\mathbf{r}}}_{\mathrm{n}}^{\diamond}\widehat{\overline{\boldsymbol{D}}}_{\overline{z}}^{\circ} & -\widehat{\overline{\boldsymbol{D}}}_{\overline{r}}^{\bullet\mathrm{t}}\,\widehat{\boldsymbol{\mu}}_{\phi}^{-1}\widehat{\overline{\mathbf{r}}}_{\mathrm{n}}^{\diamond}\widehat{\overline{\boldsymbol{D}}}_{\overline{r}}^{\bullet} \end{bmatrix} \qquad [\widehat{\boldsymbol{\epsilon}}]=\begin{bmatrix} \widehat{\boldsymbol{\epsilon}}_{\mathrm{r}}\widehat{\overline{\boldsymbol{r}}}_{\mathrm{n}}^{\circ} & \\ & \widehat{\boldsymbol{\epsilon}}_{\mathrm{z}}\widehat{\overline{\boldsymbol{r}}}_{\mathrm{n}}^{\bullet} \end{bmatrix} \tag{4.133}$$

$$\widehat{\boldsymbol{R}}_{\mathrm{H}}^{\phi\mathrm{D}}=\begin{bmatrix} -\widehat{\overline{\boldsymbol{D}}}_{\overline{r}}^{\circ\mathrm{t}}\,\widehat{\boldsymbol{\epsilon}}_{\phi}^{-1}\widehat{\overline{\mathbf{r}}}_{\mathrm{n}}^{\square}\widehat{\overline{\boldsymbol{D}}}_{\overline{r}}^{\circ} & -\widehat{\overline{\boldsymbol{D}}}_{\overline{r}}^{\circ\mathrm{t}}\,\widehat{\boldsymbol{\epsilon}}_{\phi}^{-1}\widehat{\overline{\mathbf{r}}}_{\mathrm{n}}^{\square}\widehat{\overline{\boldsymbol{D}}}_{\overline{z}}^{\bullet} \\ -\widehat{\overline{\boldsymbol{D}}}_{\overline{z}}^{\bullet\mathrm{t}}\,\widehat{\boldsymbol{\epsilon}}_{\phi}^{-1}\widehat{\overline{\mathbf{r}}}_{\mathrm{n}}^{\square}\widehat{\overline{\boldsymbol{D}}}_{\overline{r}}^{\circ} & -\widehat{\overline{\boldsymbol{D}}}_{\overline{z}}^{\bullet\mathrm{t}}\,\widehat{\boldsymbol{\epsilon}}_{\phi}^{-1}\widehat{\overline{\mathbf{r}}}_{\mathrm{n}}^{\square}\widehat{\overline{\boldsymbol{D}}}_{\overline{z}}^{\bullet} \end{bmatrix} \qquad [\widehat{\boldsymbol{\mu}}]=\begin{bmatrix} \widehat{\boldsymbol{\mu}}_{\mathrm{z}}\widehat{\overline{\boldsymbol{r}}}_{\mathrm{n}}^{\circ} & \\ & \widehat{\boldsymbol{\mu}}_{\mathrm{r}}\widehat{\overline{\boldsymbol{r}}}_{\mathrm{n}}^{\bullet} \end{bmatrix} \tag{4.134}$$

and where:

$$\overline{\mathbf{r}}_{\mathrm{n}}=\overline{\mathbf{r}}/\overline{r}_0 \qquad \widehat{\overline{\mathbf{r}}}_{\mathrm{n}}^{\circ}=\overline{\mathbf{r}}_{\mathrm{n}}^{\circ}\otimes\boldsymbol{I}_{\mathrm{z}}^{\circ} \qquad \widehat{\overline{\mathbf{r}}}_{\mathrm{n}}^{\bullet}=\overline{\mathbf{r}}_{\mathrm{n}}^{\bullet}\otimes\boldsymbol{I}_{\mathrm{z}}^{\bullet}$$

Again, the matrices $\widehat{\boldsymbol{R}}^{\phi}$ are symmetric but not Hermitian (in the case of complex elements). Combining eqs. (4.132) we obtain:

$$\frac{\mathrm{d}^2}{\overline{r}_{\circ}^2\mathrm{d}\phi^2}\widehat{\mathbf{H}}-\widehat{\boldsymbol{Q}}_{\mathrm{H}}^{\phi}\widehat{\mathbf{H}}=\mathbf{0} \qquad \frac{\mathrm{d}^2}{\overline{r}_{\circ}^2\mathrm{d}\phi^2}\widehat{\mathbf{E}}-\widehat{\boldsymbol{Q}}_{\mathrm{E}}^{\phi}\widehat{\mathbf{E}}=\mathbf{0} \tag{4.135}$$

For the matrices $\widehat{\boldsymbol{Q}}_{\mathrm{E}}^{\phi}$ and $\widehat{\boldsymbol{Q}}_{\mathrm{H}}^{\phi}$ we obtain:

$$-\boldsymbol{Q}_{\mathrm{H}}^{\phi}=\widehat{\boldsymbol{R}}_{\mathrm{E}}^{\phi}\widehat{\boldsymbol{R}}_{\mathrm{H}}^{\phi}=\widehat{\boldsymbol{R}}_{\mathrm{E}}^{\phi\mathrm{D}}[\widehat{\boldsymbol{\mu}}]+[\widehat{\boldsymbol{\epsilon}}]\widehat{\boldsymbol{R}}_{\mathrm{H}}^{\phi\mathrm{D}}+[\widehat{\boldsymbol{\epsilon}}][\widehat{\boldsymbol{\mu}}] \tag{4.136}$$

$$-\boldsymbol{Q}_{\mathrm{E}}^{\phi}=\widehat{\boldsymbol{R}}_{\mathrm{H}}^{\phi}\widehat{\boldsymbol{R}}_{\mathrm{E}}^{\phi}=\widehat{\boldsymbol{R}}_{\mathrm{H}}^{\phi\mathrm{D}}[\widehat{\boldsymbol{\epsilon}}]+[\widehat{\boldsymbol{\mu}}]\widehat{\boldsymbol{R}}_{\mathrm{E}}^{\phi\mathrm{D}}+[\widehat{\boldsymbol{\mu}}][\widehat{\boldsymbol{\epsilon}}] \tag{4.137}$$

where we use $\widehat{\boldsymbol{R}}_{\mathrm{E}}^{\phi\mathrm{D}}\widehat{\boldsymbol{R}}_{\mathrm{H}}^{\phi\mathrm{D}} = \widehat{\boldsymbol{R}}_{\mathrm{H}}^{\phi\mathrm{D}}\widehat{\boldsymbol{R}}_{\mathrm{E}}^{\phi\mathrm{D}} = \boldsymbol{0}$ because of the relation $\widehat{\boldsymbol{D}}_{\overline{r}}^{\bullet}\widehat{\boldsymbol{D}}_{\overline{z}}^{\bullet\mathrm{t}} = \widehat{\boldsymbol{D}}_{\overline{z}}^{\circ}\widehat{\boldsymbol{D}}_{\overline{r}}^{\circ\mathrm{t}}$.

4.3.3 Solution of the wave equations in z- and ϕ-direction

Next, we have to solve the discretised wave equations for the z- or ϕ-direction (eqs. (4.118) and (4.135)). For this purpose we introduce a new variable:

$$\overline{u} = \overline{r}_0\phi$$

With this new variable the equations for z and u in cylindrical coordinates are identical to the ones in Cartesian coordinates (eq. (2.31)). Therefore, the further steps, like eigenmode determination, occur in the same way as described in Section 4.2.5 and are not repeated here.

4.3.4 Equations for propagation in r-direction

For GTL-expressions in the radial direction, we define the following supervectors:

$$\widehat{\mathbf{H}} = [\widetilde{\mathbf{H}}_z^{\mathrm{t}}, -\overline{r}\widetilde{\mathbf{H}}_\phi^{\mathrm{t}}]^{\mathrm{t}} \quad \widehat{\mathbf{E}} = [\overline{r}\mathbf{E}_\phi^{\mathrm{t}}, \mathbf{E}_z^{\mathrm{t}}]^{\mathrm{t}} \tag{4.138}$$

In discretised form, the eqs. (4.84)–(4.87) are:

$$\overline{r}\frac{\mathrm{d}}{\mathrm{d}\overline{r}}\widehat{\mathbf{H}} = -\mathrm{j}\widehat{\boldsymbol{R}}_{\mathrm{E}}^{r}\widehat{\mathbf{E}} \qquad \overline{r}\frac{\mathrm{d}}{\mathrm{d}\overline{r}}\widehat{\mathbf{E}} = -\mathrm{j}\widehat{\boldsymbol{R}}_{\mathrm{H}}^{r}\widehat{\mathbf{H}} \tag{4.139}$$

where we have for the diagonal material tensors and the combined matrices:

$$\widehat{\boldsymbol{R}}_{\mathrm{E}}^{r} = \begin{bmatrix} -\widehat{\overline{\boldsymbol{D}}}_{\overline{z}}^{\circ\mathrm{t}}\widehat{\mu}_r^{-1}\widehat{\overline{\boldsymbol{D}}}_{\overline{z}}^{\circ} + \widehat{\epsilon}_\phi & \widehat{\overline{\boldsymbol{D}}}_{\overline{z}}^{\circ\mathrm{t}}\widehat{\mu}_r^{-1}\widehat{\overline{\boldsymbol{D}}}_\phi^{\bullet} \\ \widehat{\overline{\boldsymbol{D}}}_\phi^{\bullet\mathrm{t}}\widehat{\mu}_r^{-1}\widehat{\overline{\boldsymbol{D}}}_{\overline{z}}^{\circ} & -\widehat{\overline{\boldsymbol{D}}}_\phi^{\bullet\mathrm{t}}\widehat{\mu}_r^{-1}\widehat{\overline{\boldsymbol{D}}}_\phi^{\bullet} + \overline{r}^2\widehat{\epsilon}_z \end{bmatrix} \tag{4.140}$$

$$\widehat{\boldsymbol{R}}_{\mathrm{H}}^{r} = \begin{bmatrix} -\widehat{\overline{\boldsymbol{D}}}_\phi^{\circ\mathrm{t}}\widehat{\epsilon}_r^{-1}\widehat{\overline{\boldsymbol{D}}}_\phi^{\circ} + \overline{r}^2\widehat{\mu}_z & -\widehat{\overline{\boldsymbol{D}}}_\phi^{\circ\mathrm{t}}\widehat{\epsilon}_r^{-1}\widehat{\overline{\boldsymbol{D}}}_{\overline{z}}^{\bullet} \\ -\widehat{\overline{\boldsymbol{D}}}_{\overline{z}}^{\bullet\mathrm{t}}\widehat{\epsilon}_r^{-1}\widehat{\overline{\boldsymbol{D}}}_\phi^{\circ} & -\widehat{\overline{\boldsymbol{D}}}_{\overline{z}}^{\bullet\mathrm{t}}\widehat{\epsilon}_r^{-1}\widehat{\overline{\boldsymbol{D}}}_{\overline{z}}^{\bullet} + \widehat{\mu}_\phi \end{bmatrix} \tag{4.141}$$

For arbitrary problems we cannot solve the equations analytically. In these cases we can use finite differences, as described in Section 4.5.3

4.4 SOLUTION IN RADIAL DIRECTION

4.4.1 Discretisation in z-direction – circular dielectric resonators

4.4.1.1 Introduction

Dielectric resonators are widely used in microwave and millimetre wave circuits. They stabilise the oscillators or serve as filter elements. Fig. 4.12 shows longitudinal section of various forms of cylindrical resonators with circular cross-section. We will analyse these resonators by discretisation in z-direction and analytical solution in r-direction. The devices are open in z-direction. Therefore, ABCs are used for the modelling of these structures.

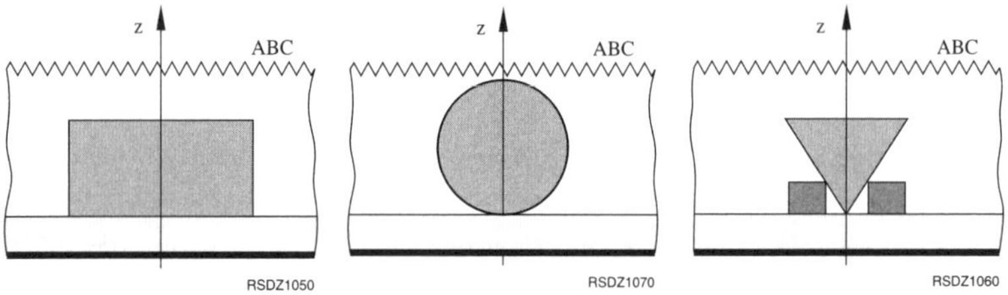

Fig. 4.12 Various types of circular dielectric resonators (D. Kremer and R. Pregla, 'The Method of Lines for the Hybrid Analysis of Multilayered Dielectric Resonators', in *IEEE MTT-S Int. Symp. Dig.* pp. 491–494. © 1995 Institute of Electrical and Electronics Engineers (IEEE))

4.4.1.2 Basic analysis equations

Fig. 4.13 shows the radial segmentation. In each segment the material parameters do not change in r-direction. We assume $\varepsilon_{rr} = \varepsilon_{\phi\phi} = \varepsilon_{rt}$, $\mu_{rr} = \mu_{\phi\phi} = \mu_{rt}$. Furthermore, these parameters should not be functions of ϕ. Therefore, the azimuthal field dependence may be written in the following form:

$$H_z, H_r, E_\phi \sim \sin(m\phi) \quad E_z, E_r, H_\phi \sim \cos(m\phi) \tag{4.142}$$

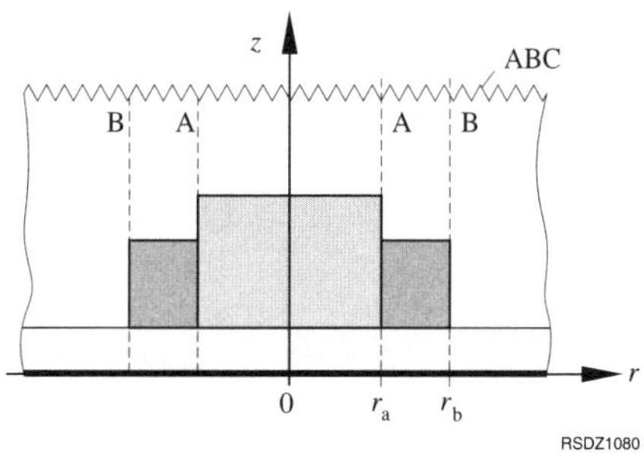

Fig. 4.13 Radial segmentation of circular dielectric resonator (D. Kremer and R. Pregla, 'The Method of Lines for the Hybrid Analysis of Multilayered Dielectric Resonators', in *IEEE MTT-S Int. Symp. Dig.* pp. 491–494. © 1995 Institute of Electrical and Electronics Engineers (IEEE))

The matrices $[R_{\mathrm{E}}^{rc}]$ and $[R_{\mathrm{H}}^{rc}]$ in eqs. (4.86) and (4.87) are given as:

$$[R_{\mathrm{E}}^{rc}] = \begin{bmatrix} D_{\overline{z}}\mu_{rt}^{-1}D_{\overline{z}} + \varepsilon_{rt} & mD_{\overline{z}}\mu_{rt}^{-1} \\ -m\mu_{rt}^{-1}D_{\overline{z}} & -m^2\mu_{rt}^{-1} \end{bmatrix} \tag{4.143}$$

$$[R_{\mathrm{H}}^{rc}] = \begin{bmatrix} -m^2\varepsilon_{rt}^{-1} & -m\varepsilon_{rt}^{-1}D_{\overline{z}} \\ mD_{\overline{z}}\varepsilon_{rt}^{-1} & D_{\overline{z}}\varepsilon_{rt}^{-1}D_{\overline{z}} + \mu_{rt} \end{bmatrix} \tag{4.144}$$

The matrices $[Q_{\rm E}^{rc}]$, $[Q_{\rm H}^{rc}]$ in eqs. (4.91) and (4.92) become:

$$[Q_{\rm E,H}^{rc}] = -\begin{bmatrix} m^2 & 0 \\ 0 & m^2 \end{bmatrix} \tag{4.145}$$

and the matrices $[S_{\rm E}^{rc}]$ and $[S_{\rm H}^{rc}]$ in eqs. (4.93) and (4.94) are:

$$[S_{\rm E}^{rc}] = \begin{bmatrix} \mu_{zz} D_{\overline{z}} \mu_{rt}^{-1} D_{\overline{z}} + \varepsilon_{rt}\mu_{zz} & m(\mu_{zz} D_{\overline{z}} \mu_{rt}^{-1} - \varepsilon_{rt}^{-1} D_{\overline{z}} \varepsilon_{zz}) \\ 0 & \varepsilon_{zz}\mu_{rt} + D_{\overline{z}}\varepsilon_{rt}^{-1} D_{\overline{z}}\varepsilon_{zz} \end{bmatrix} \tag{4.146}$$

$$[S_{\rm H}^{rc}] = \begin{bmatrix} D_{\overline{z}} \mu_{rt}^{-1} D_{\overline{z}} \mu_{zz} + \varepsilon_{rt}\mu_{zz} & 0 \\ m(\varepsilon_{zz} D_{\overline{z}} \varepsilon_{rt}^{-1} - \mu_{rt}^{-1} D_{\overline{z}} \mu_{zz}) & \varepsilon_{zz}\mu_{rt} + \varepsilon_{zz} D_{\overline{z}}\varepsilon_{rt}^{-1} D_{\overline{z}} \end{bmatrix} \tag{4.147}$$

Because of the zero elements in $[S_{\rm E,H}^{rc}]$, we obtain from the second equation in (4.89) and from the first equation in (4.90) the following differential equations for E_z and $\widetilde{H}_z$:

$$\frac{1}{\overline{r}}\frac{\partial}{\partial \overline{r}}\left(\overline{r}\frac{\partial E_z}{\partial \overline{r}}\right) - \frac{m^2}{\overline{r}^2}E_z + D_{\overline{z}}\varepsilon_{rt}^{-1} D_{\overline{z}}\varepsilon_{zz} E_z + \mu_{rt}\varepsilon_{zz} E_z = 0 \tag{4.148}$$

$$\frac{1}{\overline{r}}\frac{\partial}{\partial \overline{r}}\left(\overline{r}\frac{\partial \widetilde{H}_z}{\partial \overline{r}}\right) - \frac{m^2}{\overline{r}^2}\widetilde{H}_z + D_{\overline{z}}\mu_{rt}^{-1} D_{\overline{z}}\mu_{zz} \widetilde{H}_z + \varepsilon_{rt}\mu_{zz} \widetilde{H}_z = 0 \tag{4.149}$$

Note: in contrast to Section 4.2.5.1, the structure may vary in z-direction here. Therefore, we cannot simply introduce $-m^2$ for the second derivative with respect to ϕ in eqs. (4.97) and (4.98).

We obtain from eq. (4.85):

$$\overline{r}\frac{\partial}{\partial \overline{r}}E_z = -{\rm j}\left[mD_{\overline{z}}\varepsilon_{rt}^{-1} \quad D_{\overline{z}}\varepsilon_{rt}^{-1}D_{\overline{z}} + \mu_{rt}\right]\begin{bmatrix} \widetilde{H}_z \\ -\overline{r}\widetilde{H}_\phi \end{bmatrix}$$

$$\overline{r}\frac{\partial}{\partial \overline{r}}\widetilde{H}_z = -{\rm j}\left[D_{\overline{z}}\mu_{rt}^{-1}D_{\overline{z}} + \varepsilon_{rt} \quad mD_{\overline{z}}\mu_{rt}^{-1}\right]\begin{bmatrix} \overline{r}\widetilde{E}_\phi \\ \widetilde{E}_z \end{bmatrix}$$

or, after rearranging:

$$\begin{bmatrix} (\mu_{zz} D_{\overline{z}}\mu_{rt}^{-1} D_{\overline{z}} + \varepsilon_{rt}\mu_{zz})\overline{r}E_\phi \\ (\varepsilon_{zz} D_{\overline{z}}\varepsilon_{rt}^{-1} D_{\overline{z}} + \varepsilon_{zz}\mu_{rt})\overline{r}\widetilde{H}_\phi \end{bmatrix}$$

$$= \begin{bmatrix} -m\mu_{zz} D_{\overline{z}}\mu_{rt}^{-1}\varepsilon_{zz}^{-1} & {\rm j}\overline{r}\dfrac{\partial}{\partial \overline{r}} \\ -{\rm j}\overline{r}\dfrac{\partial}{\partial \overline{r}} & m\varepsilon_{zz} D_{\overline{z}}\varepsilon_{rt}^{-1}\mu_{zz}^{-1} \end{bmatrix}\begin{bmatrix} \varepsilon_{zz}E_z \\ \mu_{zz}\widetilde{H}_z \end{bmatrix} \tag{4.150}$$

We will use this equation in discretised form to determine E_ϕ and $\widetilde{H}_\phi$ from E_z and $\widetilde{H}_z$.

4.4.1.3 Discretisation

For the solution of eqs. (4.148), and (4.149), we discretise the fields in z-direction according to Fig. 4.14. Only a few discretisation lines are introduced. We use the variables $\varepsilon_{zz}E_z$ and $\mu_{zz}H_z$ instead of E_z and $\widetilde{H}_z$, because these quantities are continuous in z-direction. Due to the discretisation we obtain:

$$\begin{aligned}
\varepsilon_{zz}, \varepsilon_{rt} &\longrightarrow \boldsymbol{\epsilon}_e, \boldsymbol{\epsilon}_h \\
\mu_{zz}, \mu_{rt} &\longrightarrow \boldsymbol{\mu}_h, \boldsymbol{\mu}_e \\
\varepsilon_{zz} D_{\overline{z}} \varepsilon_{rt}^{-1} D_{\overline{z}}(\varepsilon_{zz}E_z) &\longrightarrow -\boldsymbol{\epsilon}_e \overline{\boldsymbol{D}}_e^{\,t} \boldsymbol{\epsilon}_h^{-1} \overline{\boldsymbol{D}}_e(\boldsymbol{\epsilon}_e \mathbf{E}_z) = -\overline{\boldsymbol{P}}_e^{\epsilon}(\boldsymbol{\epsilon}_e \mathbf{E}_z) \\
\mu_{zz} D_{\overline{z}} \mu_{rt}^{-1} D_{\overline{z}}(\mu_{zz}\widetilde{H}_z) &\longrightarrow -\boldsymbol{\mu}_h \overline{\boldsymbol{D}}_h^{\,t} \boldsymbol{\mu}_e^{-1} \overline{\boldsymbol{D}}_h(\boldsymbol{\mu}_h \widetilde{\mathbf{H}}_z) = -\overline{\boldsymbol{P}}_h^{\mu}(\boldsymbol{\mu}_h \widetilde{\mathbf{H}}_z)
\end{aligned} \tag{4.151}$$

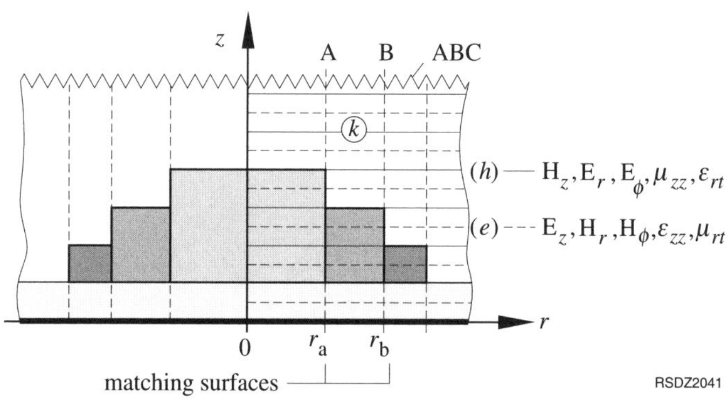

Fig. 4.14 Field transformation between surfaces A and B

The subscripts e and h are related to E_z and H_z, respectively. Without ABCs the difference operators $\overline{\boldsymbol{D}}_e$ and $\overline{\boldsymbol{D}}_h$ according to Fig. 4.14 have to be constructed for Neumann and Dirichlet boundary conditions, respectively. When using ABCs (as shown in Fig. 4.14) we have to replace the difference operators $\overline{\boldsymbol{D}}_h$ and $-\overline{\boldsymbol{D}}_e^{\,t}$ with $\overline{\boldsymbol{D}}_h^{\,a}$. The remaining difference operators $\overline{\boldsymbol{D}}_e$ and $\overline{\boldsymbol{D}}_h^{\,t}$ have the same form as before. We transform the fields according to:

$$\boldsymbol{\epsilon}_e \mathbf{E}_z = \boldsymbol{T}_e^{\varepsilon} \overline{\mathbf{E}}_z \qquad \boldsymbol{\mu}_h \widetilde{\mathbf{H}}_z = \boldsymbol{T}_h^{\mu} \overline{\mathbf{H}}_z \tag{4.152}$$

and:

$$(\boldsymbol{T}_{e,h}^{\alpha})^{-1}\left(\boldsymbol{\epsilon}_{e,h}\boldsymbol{\mu}_{e,h} - \overline{\boldsymbol{P}}_{e,h}^{\alpha}\right)\boldsymbol{T}_{e,h}^{\alpha} = \boldsymbol{k}_{re,h}^2 \qquad \alpha = \varepsilon, \mu \tag{4.153}$$

$\boldsymbol{k}_{re,h}^2$ is a diagonal matrix. We obtain the discretised and transformed wave equations:

$$\overline{r}\frac{\mathrm{d}}{\mathrm{d}\overline{r}}\left(\overline{r}\frac{\mathrm{d}}{\mathrm{d}\overline{r}}\overline{\mathbf{F}}\right) + ((\boldsymbol{k}_{re,h}\overline{r})^2 - m^2\boldsymbol{I})\overline{\mathbf{F}} = \mathbf{0} \qquad \overline{\mathbf{F}} = \overline{\mathbf{E}}_z, \overline{\mathbf{H}}_z \tag{4.154}$$

The general solution is given by:

$$\overline{\mathbf{F}} = J_m(\boldsymbol{k}_{re,h}\overline{r})\mathbf{A} + Y_m(\boldsymbol{k}_{re,h}\overline{r})\mathbf{B} \tag{4.155}$$

where J_m and Y_m are Bessel and Neumann functions, respectively, of the mth order. Since the terms $\boldsymbol{k}^2_{re,h}$ are diagonal matrices, the arguments of the cylinder functions and therefore J_m and Y_m are diagonal matrices.

4.4.1.4 Admittance matrix equations

From the general solution we obtain the following expression for the radii $\overline{r} = \overline{r}_a$ and $\overline{r} = \overline{r}_b$ (see Fig. 4.14):

$$\begin{bmatrix} \overline{\mathbf{F}}_{\mathrm{A}} \\ \overline{\mathbf{F}}_{\mathrm{B}} \end{bmatrix} = \begin{bmatrix} J_m(\boldsymbol{k}_r\overline{r}_a) & Y_m(\boldsymbol{k}_r\overline{r}_a) \\ J_m(\boldsymbol{k}_r\overline{r}_b) & Y_m(\boldsymbol{k}_r\overline{r}_b) \end{bmatrix} \begin{bmatrix} \mathbf{A} \\ \mathbf{B} \end{bmatrix} \tag{4.156}$$

Therefore, we obtain for the derivatives:

$$\overline{r}\frac{\partial}{\partial \overline{r}} \begin{bmatrix} \overline{\mathbf{F}}_{\mathrm{A}} \\ \overline{\mathbf{F}}_{\mathrm{B}} \end{bmatrix} = \widehat{\boldsymbol{p}}_m^{-1} \begin{bmatrix} \overline{\boldsymbol{r}}_m & \frac{2}{\pi}\boldsymbol{I} \\ -\frac{2}{\pi}\boldsymbol{I} & \overline{\boldsymbol{q}}_m \end{bmatrix} \begin{bmatrix} \overline{\mathbf{F}}_{\mathrm{A}} \\ \overline{\mathbf{F}}_{\mathrm{B}} \end{bmatrix} \tag{4.157}$$

where $\overline{r}$ is equal to $\overline{r}_a$ for the upper line (matching surface A) and equal to $\overline{r}_b$ for the lower line (matching surface B). $\boldsymbol{p}_m$, $\overline{\boldsymbol{r}}_m$ and $\overline{\boldsymbol{q}}_m$ are normalised cross-products (cf. formula 9.1.32 in [14]):

$$\begin{aligned}
\boldsymbol{p}_m &= J_m(\boldsymbol{t}_{\mathrm{A}})Y_m(\boldsymbol{t}_{\mathrm{B}}) - J_m(\boldsymbol{t}_{\mathrm{B}})Y_m(\boldsymbol{t}_{\mathrm{A}}) & \widehat{\boldsymbol{p}}_m &= \mathrm{Diag}(\boldsymbol{p}_m, \boldsymbol{p}_m) \\
\boldsymbol{r}_m &= J'_m(\boldsymbol{t}_{\mathrm{A}})Y_m(\boldsymbol{t}_{\mathrm{B}}) - Y'_m(\boldsymbol{t}_{\mathrm{A}})J_m(\boldsymbol{t}_{\mathrm{B}}) & \overline{\boldsymbol{r}}_m &= \boldsymbol{t}_{\mathrm{A}}\boldsymbol{r}_m \\
\boldsymbol{q}_m &= J_m(\boldsymbol{t}_{\mathrm{A}})Y'_m(\boldsymbol{t}_{\mathrm{B}}) - J'_m(\boldsymbol{t}_{\mathrm{B}})Y_m(\boldsymbol{t}_{\mathrm{A}}) & \overline{\boldsymbol{q}}_m &= \boldsymbol{t}_{\mathrm{B}}\boldsymbol{q}_m \\
\boldsymbol{s}_m &= J'_m(\boldsymbol{t}_{\mathrm{A}})Y'_m(\boldsymbol{t}_{\mathrm{B}}) - J'_m(\boldsymbol{t}_{\mathrm{B}})Y'_m(\boldsymbol{t}_{\mathrm{A}}) & \overline{\boldsymbol{s}}_m &= \boldsymbol{t}_{\mathrm{A}}\boldsymbol{t}_{\mathrm{B}}\boldsymbol{s}_m \\
\boldsymbol{W}_{\mathrm{P}} &= J_m(\boldsymbol{t}_{\mathrm{P}})Y'_m(\boldsymbol{t}_{\mathrm{P}}) - J'_m(\boldsymbol{t}_{\mathrm{P}})Y_m(\boldsymbol{t}_{\mathrm{P}}) = \tfrac{2}{\pi}\boldsymbol{t}_{\mathrm{P}}^{-1} & \overline{\boldsymbol{W}}_{\mathrm{P}} &= \boldsymbol{t}_{\mathrm{P}}\boldsymbol{W}_{\mathrm{P}} = \tfrac{2}{\pi}\boldsymbol{I}
\end{aligned} \tag{4.158}$$

$R'_m = \mathrm{d}R_m(t)/\mathrm{d}t$ is the derivative of the Bessel functions with respect to the argument t. In our case the arguments $\boldsymbol{t}_{\mathrm{A,B}}$ are diagonal matrices according to:

$$\boldsymbol{t}_{\mathrm{A}} = \boldsymbol{k}_{\overline{r}}\overline{r}_a \qquad \boldsymbol{t}_{\mathrm{B}} = \boldsymbol{k}_{\overline{r}}\overline{r}_b \tag{4.159}$$

The arguments $\boldsymbol{t}_{\mathrm{A,B}}$ must have the subscript e or h (the same as $\boldsymbol{k}_r$). (Please distinguish between the value for the cross-product $\boldsymbol{r}_m$ (bold) and the coordinate r (not bold).) From eq. (4.150) we obtain with eq. (4.152) and $\mathbf{E}_\phi = \boldsymbol{T}_h^\mu\overline{\mathbf{E}}_\phi$ and $\mathbf{H}_\phi = \boldsymbol{T}_e^\epsilon\overline{\mathbf{H}}_\phi$:

$$\begin{bmatrix} \overline{r}\overline{\mathbf{E}}_\phi \\ -\overline{r}\overline{\mathbf{H}}_\phi \end{bmatrix} = \widehat{\boldsymbol{k}}_r^{-2} \begin{bmatrix} -\widetilde{\boldsymbol{\delta}}_e & \mathrm{j}\overline{r}\dfrac{\partial}{\partial \overline{r}}\boldsymbol{I}_h \\ \mathrm{j}\overline{r}\dfrac{\partial}{\partial \overline{r}}\boldsymbol{I}_e & -\widetilde{\boldsymbol{\delta}}_h \end{bmatrix} \begin{bmatrix} \overline{\mathbf{E}}_z \\ \overline{\mathbf{H}}_z \end{bmatrix} \tag{4.160}$$

The following abbreviations were used:

$$\begin{aligned}
\widehat{\boldsymbol{k}}_r^2 &= \mathrm{Diag}(\widehat{\boldsymbol{k}}_{rh}^2, \widehat{\boldsymbol{k}}_{re}^2) \quad \widehat{\boldsymbol{k}}_{re,h}^2 = \mathrm{Diag}\,(\boldsymbol{k}_{re,h}^2, \boldsymbol{k}_{re,h}^2)\\
\widetilde{\boldsymbol{\delta}}_e &= (\boldsymbol{T}_h^{\mu})^{-1}(m\boldsymbol{\mu}_h\overline{\boldsymbol{D}}_e\boldsymbol{\mu}_e^{-1}\boldsymbol{\epsilon}_e^{-1})\boldsymbol{T}_e^{\epsilon}\\
\widetilde{\boldsymbol{\delta}}_h &= (\boldsymbol{T}_e^{\varepsilon})^{-1}(m\boldsymbol{\epsilon}_e\overline{\boldsymbol{D}}_h\boldsymbol{\epsilon}_h^{-1}\boldsymbol{\mu}_h^{-1})\boldsymbol{T}_h^{\mu}\\
\widehat{\widehat{\boldsymbol{k}}}_r^2 &= \mathrm{Diag}\,(\widehat{\boldsymbol{k}}_{rh}^2, \widehat{\boldsymbol{k}}_{re}^2) \quad \widehat{\widehat{\boldsymbol{\delta}}}_{e,h} = \mathrm{Diag}\,(\widetilde{\boldsymbol{\delta}}_{e,h}, \widetilde{\boldsymbol{\delta}}_{e,h})
\end{aligned} \tag{4.161}$$

Here, $\widetilde{\boldsymbol{\delta}}_e$ and $\widetilde{\boldsymbol{\delta}}_h$ are diagonal matrices only for special cases. The ϕ-components at the surfaces A and B can be determined by:

$$\begin{bmatrix} \overline{r}_{\mathrm{A}}\overline{\mathbf{E}}_{\phi\mathrm{A}} \\ \overline{r}_{\mathrm{B}}\overline{\mathbf{E}}_{\phi\mathrm{B}} \\ \mathrm{j}\overline{r}_{\mathrm{A}}\overline{\mathbf{H}}_{\phi\mathrm{A}} \\ \mathrm{j}\overline{r}_{\mathrm{B}}\overline{\mathbf{H}}_{\phi\mathrm{B}} \end{bmatrix} = \widehat{\widehat{\boldsymbol{k}}}_r^{-2} \begin{bmatrix} \begin{bmatrix} -\widetilde{\boldsymbol{\delta}}_e & \boldsymbol{0} \\ \boldsymbol{0} & -\widetilde{\boldsymbol{\delta}}_e \end{bmatrix} & \widehat{\boldsymbol{p}}_{mh}^{-1} \begin{bmatrix} \overline{\boldsymbol{r}}_{mh} & \frac{2}{\pi}\boldsymbol{I}_h \\ -\frac{2}{\pi}\boldsymbol{I}_h & \overline{\boldsymbol{q}}_{mh} \end{bmatrix} \\ \widehat{\boldsymbol{p}}_{me}^{-1} \begin{bmatrix} \overline{\boldsymbol{r}}_{me} & \frac{2}{\pi}\boldsymbol{I}_e \\ -\frac{2}{\pi}\boldsymbol{I}_e & \overline{\boldsymbol{q}}_{me} \end{bmatrix} & \begin{bmatrix} \widetilde{\boldsymbol{\delta}}_h & \boldsymbol{0} \\ \boldsymbol{0} & \widetilde{\boldsymbol{\delta}}_h \end{bmatrix} \end{bmatrix} \begin{bmatrix} \overline{\mathbf{E}}_{z\mathrm{A}} \\ \overline{\mathbf{E}}_{z\mathrm{B}} \\ \mathrm{j}\overline{\mathbf{H}}_{z\mathrm{A}} \\ \mathrm{j}\overline{\mathbf{H}}_{z\mathrm{B}} \end{bmatrix} \tag{4.162}$$

After rearrangement we obtain:

$$\begin{bmatrix} \mathrm{j}\overline{\mathbf{H}}_{z\mathrm{A}} \\ \mathrm{j}\overline{\mathbf{H}}_{z\mathrm{B}} \\ \mathrm{j}\overline{r}_{\mathrm{A}}\overline{\mathbf{H}}_{\phi\mathrm{A}} \\ \mathrm{j}\overline{r}_{\mathrm{B}}\overline{\mathbf{H}}_{\phi\mathrm{B}} \end{bmatrix} = \begin{bmatrix} \widehat{\overline{\boldsymbol{s}}}_{mh}^{-1} \begin{bmatrix} \overline{\boldsymbol{q}}_{mh} & -\frac{2}{\pi}\boldsymbol{I}_h \\ \frac{2}{\pi}\boldsymbol{I}_h & \overline{\boldsymbol{r}}_{mh} \end{bmatrix} \widehat{\boldsymbol{k}}_{rh}^2 & \widehat{\overline{\boldsymbol{s}}}_{mh}^{-1} \begin{bmatrix} \overline{\boldsymbol{q}}_{mh} & -\frac{2}{\pi}\boldsymbol{I}_h \\ \frac{2}{\pi}\boldsymbol{I}_h & \overline{\boldsymbol{r}}_{mh} \end{bmatrix} \widehat{\widehat{\boldsymbol{\delta}}}_e \\ \widehat{\boldsymbol{k}}_{re}^{-2}\widehat{\widehat{\boldsymbol{\delta}}}_h\widehat{\overline{\boldsymbol{s}}}_{mh}^{-1} \begin{bmatrix} \overline{\boldsymbol{q}}_{mh} & -\frac{2}{\pi}\boldsymbol{I}_h \\ \frac{2}{\pi}\boldsymbol{I}_h & \overline{\boldsymbol{r}}_{mh} \end{bmatrix} \widehat{\boldsymbol{k}}_{rh}^2 & \begin{bmatrix} \gamma_{\mathrm{E}_1} & -\boldsymbol{\alpha}_{\mathrm{E}} \\ \boldsymbol{\alpha}_{\mathrm{E}} & \gamma_{\mathrm{E}_2} \end{bmatrix} \end{bmatrix}$$

$$\times \begin{bmatrix} \overline{r}_{\mathrm{A}}\overline{\mathbf{E}}_{\phi\mathrm{A}} \\ \overline{r}_{\mathrm{B}}\overline{\mathbf{E}}_{\phi\mathrm{B}} \\ \overline{\mathbf{E}}_{z\mathrm{A}} \\ \overline{\mathbf{E}}_{z\mathrm{B}} \end{bmatrix} \tag{4.163}$$

with:

$$\begin{bmatrix} \gamma_{\mathrm{E}_1} & -\boldsymbol{\alpha}_{\mathrm{E}} \\ \boldsymbol{\alpha}_{\mathrm{E}} & \gamma_{\mathrm{E}_2} \end{bmatrix} = \widehat{\boldsymbol{k}}_{re}^{-2}\widehat{\boldsymbol{p}}_{me}^{-1} \begin{bmatrix} \overline{\boldsymbol{r}}_{me} & \frac{2}{\pi}\boldsymbol{I}_e \\ -\frac{2}{\pi}\boldsymbol{I}_e & \overline{\boldsymbol{q}}_{me} \end{bmatrix} + \widehat{\boldsymbol{k}}_{re}^{-2}\widehat{\widehat{\boldsymbol{\delta}}}_h\widehat{\overline{\boldsymbol{s}}}_{mh}^{-1} \begin{bmatrix} \overline{\boldsymbol{q}}_{mh} & -\frac{2}{\pi}\boldsymbol{I}_h \\ \frac{2}{\pi}\boldsymbol{I}_h & \overline{\boldsymbol{r}}_{mh} \end{bmatrix} \widehat{\widehat{\boldsymbol{\delta}}}_e \tag{4.164}$$

where we used $\boldsymbol{p}_m\boldsymbol{s}_m - \boldsymbol{q}_m\boldsymbol{r}_m = 4(\pi^2\boldsymbol{t}_a\boldsymbol{t}_b)^{-1}$ (cf. eq. 9.1.34 in [14]) and:

$$\left(\widehat{\boldsymbol{p}}_m^{-1} \begin{bmatrix} \overline{\boldsymbol{r}}_m & \frac{2}{\pi}\boldsymbol{I} \\ -\frac{2}{\pi}\boldsymbol{I} & \overline{\boldsymbol{q}}_m \end{bmatrix}\right)^{-1} = \widehat{\overline{\boldsymbol{s}}}_m^{-1} \begin{bmatrix} \overline{\boldsymbol{q}}_m & -\frac{2}{\pi}\boldsymbol{I} \\ \frac{2}{\pi}\boldsymbol{I} & \overline{\boldsymbol{r}}_m \end{bmatrix} \tag{4.165}$$

A further rearrangement results in:

$$\begin{bmatrix} \overline{\mathbf{H}}_{zA} \\ \overline{r}_A\overline{\mathbf{H}}_{\phi A} \\ -\overline{\mathbf{H}}_{zB} \\ -\overline{r}_B\overline{\mathbf{H}}_{\phi B} \end{bmatrix} = \begin{bmatrix} -\widetilde{\boldsymbol{q}}_{mh}\boldsymbol{k}_{rh}^2 & -\widetilde{\boldsymbol{q}}_{mh}\widetilde{\boldsymbol{\delta}}_e & \widetilde{\mathbf{w}}_{mh}\boldsymbol{k}_{rh}^2 & \widetilde{\mathbf{w}}_{mh}\widetilde{\boldsymbol{\delta}}_e \\ -\widetilde{\boldsymbol{\delta}}_{hk}\widetilde{\boldsymbol{q}}_{mh} & -\boldsymbol{\gamma}_{E_1} & \widetilde{\boldsymbol{\delta}}_{hk}\widetilde{\mathbf{w}}_{mh} & \boldsymbol{\alpha}_E \\ \widetilde{\mathbf{w}}_{mh}\boldsymbol{k}_{rh}^2 & \widetilde{\mathbf{w}}_{mh}\widetilde{\boldsymbol{\delta}}_e & \widetilde{\mathbf{r}}_{mh}\boldsymbol{k}_{rh}^2 & \widetilde{\mathbf{r}}_{mh}\widetilde{\boldsymbol{\delta}}_e \\ \widetilde{\boldsymbol{\delta}}_{hk}\widetilde{\mathbf{w}}_{mh} & \boldsymbol{\alpha}_E & \widetilde{\boldsymbol{\delta}}_{hk}\widetilde{\mathbf{r}}_{mh} & \widetilde{\boldsymbol{\gamma}}_{E_2} \end{bmatrix} \begin{bmatrix} \mathrm{j}\overline{r}_A\overline{\mathbf{E}}_{\phi A} \\ \mathrm{j}\overline{\mathbf{E}}_{zA} \\ \mathrm{j}\overline{r}_B\overline{\mathbf{E}}_{\phi B} \\ \mathrm{j}\overline{\mathbf{E}}_{zB} \end{bmatrix} \tag{4.166}$$

with:

$$\widetilde{\boldsymbol{\delta}}_{hk} = \boldsymbol{k}_{re}^{-2}\widetilde{\boldsymbol{\delta}}_h\boldsymbol{k}_{rh}^2 \quad \widetilde{\boldsymbol{q}}_{mh} = \overline{\mathbf{s}}_{mh}^{-1}\overline{\boldsymbol{q}}_{mh} \quad \widetilde{\mathbf{r}}_{mh} = \overline{\mathbf{s}}_{mh}^{-1}\overline{\mathbf{r}}_{mh} \quad \widetilde{\mathbf{w}}_{mh} = \frac{2}{\pi}\overline{\mathbf{s}}_{mh}^{-1} \tag{4.167}$$

By using the abbreviations:

$$\overline{\mathbf{H}}_{A,B} = \begin{bmatrix} \overline{\mathbf{H}}_{zA,B} \\ \overline{r}_{A,B}\overline{\mathbf{H}}_{\phi A,B} \end{bmatrix} \qquad \overline{\mathbf{E}}_{A,B} = \begin{bmatrix} \mathrm{j}\overline{r}_{A,B}\overline{\mathbf{E}}_{\phi A,B} \\ \mathrm{j}\overline{\mathbf{E}}_{zA,B} \end{bmatrix} \tag{4.168}$$

the relation for the transverse fields of a waveguide section k may be written in the following form:

$$\begin{bmatrix} \overline{\mathbf{H}}_A \\ -\overline{\mathbf{H}}_B \end{bmatrix} = \begin{bmatrix} \overline{\boldsymbol{y}}_{1A} & \overline{\boldsymbol{y}}_2 \\ \overline{\boldsymbol{y}}_2 & \overline{\boldsymbol{y}}_{1B} \end{bmatrix} \begin{bmatrix} \overline{\mathbf{E}}_A \\ \overline{\mathbf{E}}_B \end{bmatrix} \tag{4.169}$$

The submatrices $\overline{\boldsymbol{y}}_{1A,B}$ and $\overline{\boldsymbol{y}}_2$ can immediately be obtained from eq. (4.166). Now the admittance transformation can be performed analogously to Section 2.5. For the transfer matrices according to:

$$\begin{bmatrix} \overline{\mathbf{E}}_B \\ \overline{\mathbf{H}}_B \end{bmatrix} = \begin{bmatrix} \overline{\mathbf{V}}_1 & \overline{\mathbf{Z}} \\ \overline{\mathbf{Y}} & \overline{\mathbf{V}}_2 \end{bmatrix} \begin{bmatrix} \overline{\mathbf{E}}_A \\ \overline{\mathbf{H}}_A \end{bmatrix} \tag{4.170}$$

we obtain:

$$\begin{aligned} \overline{\mathbf{V}}_1 &= \overline{\boldsymbol{y}}_2^{-1}\overline{\boldsymbol{y}}_{1A} & \qquad \overline{\mathbf{Z}} &= -\overline{\boldsymbol{y}}_2^{-1} \\ \overline{\mathbf{V}}_2 &= \overline{\boldsymbol{y}}_{1B}\overline{\boldsymbol{y}}_2^{-1} & \qquad \overline{\mathbf{Y}} &= \overline{\boldsymbol{y}}_2 - \overline{\boldsymbol{y}}_{1B}\overline{\boldsymbol{y}}_2^{-1}\overline{\boldsymbol{y}}_{1A} \end{aligned} \tag{4.171}$$

All these formulas are valid inside section k. For the matching at the interface between e.g. sections k and $k+1$ we have to determine the fields in the original domain, which leads to a matching condition in the transformed domain.

4.4.1.5 Centre and outer regions

For the centre and infinite outer regions we derive separate equations. In the inner region only the first term of eq. (4.155) can be used. The second term is infinite in the centre (i.e. for $r = 0$). In the infinite outer region we have to use the modified Bessel function K_m. Instead of eq. (4.157) we obtain:

$$\overline{r}\frac{\mathrm{d}}{\mathrm{d}\overline{r}}\overline{\mathbf{F}} = \boldsymbol{t}R'_m(\boldsymbol{t})R_m^{-1}(\boldsymbol{t})\overline{\mathbf{F}} \tag{4.172}$$

where R_m is equal to $J_m(K_m)$ in the inner (infinite outer) region. Again, R'_m is the derivative with respect to the argument. By using the abbreviation:

$$\boldsymbol{\Lambda}_{e,h}^{\mathrm{C}} = \boldsymbol{t}_{\mathrm{C}}^{e,h}R'_m(\boldsymbol{t}_{\mathrm{C}}^{e,h})R_m^{-1}(\boldsymbol{t}_{\mathrm{C}}^{e,h}) \tag{4.173}$$

we obtain instead of eq. (4.169):

$$\widehat{\boldsymbol{k}}_r^2\begin{bmatrix}\overline{r}_{\mathrm{C}}\overline{\mathbf{E}}_{\phi\mathrm{C}}\\ -\overline{r}_{\mathrm{C}}\overline{\mathbf{H}}_{\phi\mathrm{C}}\end{bmatrix} = \begin{bmatrix}-\widetilde{\boldsymbol{\delta}}_e & \mathrm{j}\boldsymbol{\Lambda}_h^{\mathrm{C}}\\ \mathrm{j}\boldsymbol{\Lambda}_e^{\mathrm{C}} & -\widetilde{\boldsymbol{\delta}}_h\end{bmatrix}\begin{bmatrix}\overline{\mathbf{E}}_{z\mathrm{C}}\\ \overline{\mathbf{H}}_{z\mathrm{C}}\end{bmatrix} \tag{4.174}$$

with $\mathrm{C} = \mathrm{A}, \mathrm{B}$. We have to use A(B) for the outer (inner) interface of the inner (outer) region. After rearranging, and by using the definitions in eq. (4.173), we obtain:

$$\overline{\mathbf{H}}_{\mathrm{C}} = \overline{\mathbf{Y}}_{\mathrm{C}}\overline{\mathbf{E}}_{\mathrm{C}} \qquad \overline{\mathbf{Y}}_{\mathrm{C}} = \begin{bmatrix}-\boldsymbol{\Lambda}_{h\mathrm{C}}^{-1}\boldsymbol{k}_{rh}^2 & -\boldsymbol{\Lambda}_{h\mathrm{C}}^{-1}\widetilde{\boldsymbol{\delta}}_e\\ -\boldsymbol{k}_{re}^{-2}\widetilde{\boldsymbol{\delta}}_h\boldsymbol{\Lambda}_{h\mathrm{C}}^{-1}\boldsymbol{k}_{rh}^2 & -\boldsymbol{k}_{re}^{-2}(\boldsymbol{\Lambda}_{e\mathrm{C}} + \widetilde{\boldsymbol{\delta}}_h\boldsymbol{\Lambda}_{h\mathrm{C}}^{-1}\widetilde{\boldsymbol{\delta}}_e)\end{bmatrix} \tag{4.175}$$

4.4.1.6 Numerical results

A conical dielectric resonator was analysed with the algorithm demonstrated here and the results were compared with those presented in [15]. In radial direction the shape of the cone was approximated by a set of single coaxial rings, which consisted of inhomogeneous permittivities in z-direction. In Fig. 4.15 the resonant frequency of the TE_{011} mode is presented versus the height L. A comparison of our results with the numerical results in [15] shows a good agreement.

4.4.2 Discretisation in z-direction – propagation in ϕ-direction

4.4.2.1 Introduction

Waveguide bends are important components in integrated circuits. They connect different parts of the circuit or change the propagation direction e.g. in directional couplers separating the two coupled waveguides. In this subsection we study the propagation in such circular waveguide bends, e.g. the rib waveguide circular bend shown in Fig. 4.16 or the microstrip circular bend in Fig. 4.17. The analysis in this subsection is very similar to that presented in the section in which we examined circular resonators. Nevertheless, some of the equations are different. In Section 4.4.1 the azimuthal dependence of the fields was given by sine and cosine functions. Here, we assume a propagation according to $e^{-\mathrm{j}\nu\phi} = e^{-\mathrm{j}n_{\mathrm{eff}}\overline{u}}$. Therefore, we obtain different signs of the

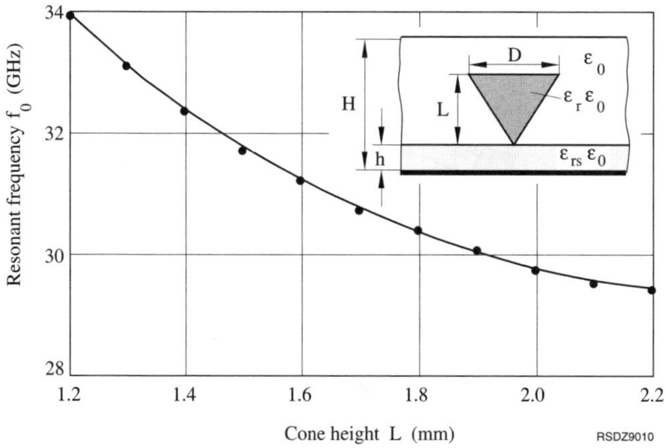

Fig. 4.15 Resonant frequency of TE_{011} mode in a conical resonator versus cone height L. —— method in [15] • • • MoL. $\varepsilon_r = 29.57$, $h = 250$ μm, $\varepsilon_{rS} = 10$, $H = 3.0$ mm, $D = 2.5$ mm (D. Kremer and R. Pregla, 'The Method of Lines for the Hybrid Analysis of Multilayered Cylindrical Resonator Structures', *IEEE Trans. Microwave Theory Tech.*, vol. MTT-45, no. 12, pp. 2152–2155. © 1997 Institute of Electrical and Electronics Engineers (IEEE))

derivative with respect to ϕ (in comparison to Section 4.4.1). The effective index is defined as $n_{\text{eff}} = \sqrt{\varepsilon_{re}}$, $\overline{u} = \phi\overline{r}_o$, $\nu = n_{\text{eff}}\overline{r}_o$ and $\overline{r}_o = r_o k_0$. For r_o we choose a suitable radius, e.g. the distance between the origin of the coordinate system and the middle of the rib in Fig. 4.16 (labelled R). In order to consider the radiation loss, the effective refractive index n_{eff} of the eigenmodes of the bent waveguide must be complex. As in the previous case, we assume $\varepsilon_{rr} = \varepsilon_{\phi\phi} = \varepsilon_{rt}$, $\mu_{rr} = \mu_{\phi\phi} = \mu_{rt}$. Furthermore, these parameters should not be functions of ϕ.

In r-direction we again use the segmentation shown in Fig. 4.14. In each circular segment the material parameters do not depend on r.

4.4.2.2 Basic analysis equations

By using $D_\phi = -\mathrm{j}\nu = -\mathrm{j}n_{\text{eff}}\overline{r}_o$, the matrices $[R_{\mathrm{E}}^{rc}]$ and $[R_{\mathrm{H}}^{rc}]$ in eqs. (4.86) and (4.87) are given by:

$$[R_{\mathrm{E}}^{rc}] = \begin{bmatrix} D_{\overline{z}}\mu_{rt}^{-1}D_{\overline{z}} + \varepsilon_{rt} & \mathrm{j}\nu D_{\overline{z}}\mu_{rt}^{-1} \\ \mathrm{j}\nu\mu_{rt}^{-1}D_{\overline{z}} & -\nu^2\mu_{rt}^{-1} \end{bmatrix} \tag{4.176}$$

$$[R_{\mathrm{H}}^{rc}] = \begin{bmatrix} -\nu^2\varepsilon_{rt}^{-1} & -\mathrm{j}\nu\varepsilon_{rt}^{-1}D_{\overline{z}} \\ -\mathrm{j}\nu D_{\overline{z}}\varepsilon_{rt}^{-1} & D_{\overline{z}}\varepsilon_{rt}^{-1}D_{\overline{z}} + \mu_{rt} \end{bmatrix} \tag{4.177}$$

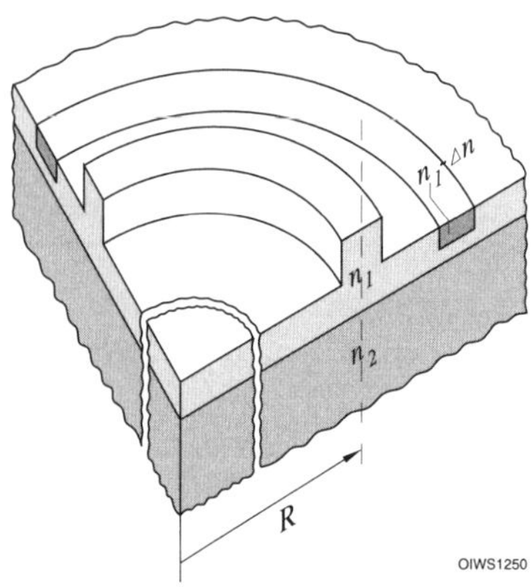

Fig. 4.16 Rib waveguide circular bend (W. Pascher and R. Pregla, 'Vectorial Analysis of Bends in Optical Strip Waveguides by the Method of Lines', *Radio Sci.*, vol. 28, pp. 1229–1233. © 1993 American Geophysical Union (AGU))

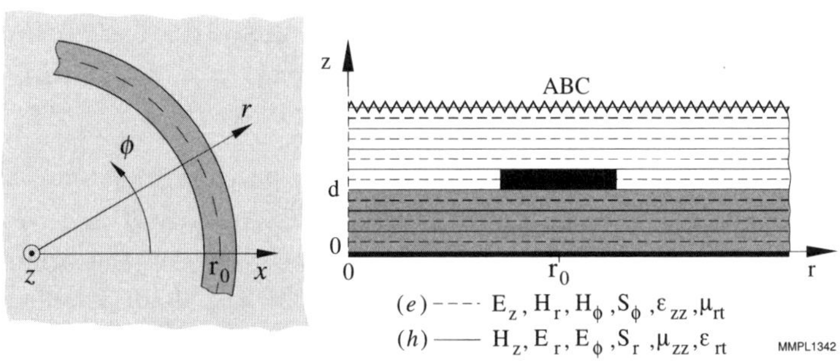

Fig. 4.17 Microstrip waveguide circular bend (a) top view (b) with horizontal discretisation lines

For the matrices $[Q_{\mathrm{E}}^{rc}]$, $[Q_{\mathrm{H}}^{rc}]$ in eqs. (4.91) and (4.92) we have:

$$[Q_{\mathrm{E,H}}^{rc}] = -\begin{bmatrix} \nu^2 & 0 \\ 0 & \nu^2 \end{bmatrix} \tag{4.178}$$

The operators $[S_{\mathrm{E}}^{rc}]$ and $[S_{\mathrm{H}}^{rc}]$ in eqs. (4.93) and (4.94) are:

$$[S_{\mathrm{E}}^{rc}] = \begin{bmatrix} \mu_{zz} D_{\overline{z}} \mu_{rt}^{-1} D_{\overline{z}} + \varepsilon_{rt}\mu_{zz} & \mathrm{j}\nu(\mu_{zz} D_{\overline{z}} \mu_{rt}^{-1} - \varepsilon_{rt}^{-1} D_{\overline{z}} \varepsilon_{zz}) \\ 0 & \varepsilon_{zz}\mu_{rt} + D_{\overline{z}} \varepsilon_{rt}^{-1} D_{\overline{z}} \varepsilon_{zz} \end{bmatrix} \tag{4.179}$$

$$[S_{\mathrm{H}}^{rc}] = \begin{bmatrix} D_{\overline{z}} \mu_{rt}^{-1} D_{\overline{z}} \mu_{zz} + \varepsilon_{rt}\mu_{zz} & 0 \\ \mathrm{j}\nu(\mu_{rt}^{-1} D_{\overline{z}} \mu_{zz} - \varepsilon_{zz} D_{\overline{z}} \varepsilon_{rt}^{-1}) & \varepsilon_{zz}\mu_{rt} + \varepsilon_{zz} D_{\overline{z}} \varepsilon_{rt}^{-1} D_{\overline{z}} \end{bmatrix} \tag{4.180}$$

Because of the zero elements in $[S^{rc}_{\mathrm{E,H}}]$, we obtain from the second equation in (4.89) and from the first equation in (4.90), respectively, the following differential equations for E_z and $\widetilde{H}_z$:

$$\overline{r}\frac{\partial}{\partial \overline{r}}\left(\overline{r}\frac{\partial E_z}{\partial \overline{r}}\right) - \nu^2 E_z + \overline{r}^2(D_{\overline{z}}\varepsilon_{rt}^{-1}D_{\overline{z}}\varepsilon_{zz} + \mu_{rt}\varepsilon_{zz})E_z = 0 \qquad (4.181)$$

$$\overline{r}\frac{\partial}{\partial \overline{r}}\left(\overline{r}\frac{\partial \widetilde{H}_z}{\partial \overline{r}}\right) - \nu^2 \widetilde{H}_z + \overline{r}^2(D_{\overline{z}}\mu_{rt}^{-1}D_{\overline{z}}\mu_{zz} + \varepsilon_{rt}\mu_{zz})\widetilde{H}_z = 0 \qquad (4.182)$$

As in the previous section, we cannot derive these equations simply from eqs. (4.97) and (4.98), because the material parameters may depend on z here.

From eq. (4.84) (rows with E_z and $\widetilde{H}_z$ on the left side) we obtain:

$$\overline{r}\frac{\partial}{\partial \overline{r}}E_z = \begin{bmatrix} -\nu D_{\overline{z}}\varepsilon_{rt}^{-1} & D_{\overline{z}}\varepsilon_{rt}^{-1}D_{\overline{z}} + \mu_{rt}\end{bmatrix}\begin{bmatrix}\widetilde{H}_z \\ \mathrm{j}\overline{r}\widetilde{H}_\phi\end{bmatrix}$$

$$\overline{r}\frac{\partial}{\partial \overline{r}}\widetilde{H}_z = \begin{bmatrix} D_{\overline{z}}\mu_{rt}^{-1}D_{\overline{z}} + \varepsilon_{rt} & \nu D_{\overline{z}}\mu_{rt}^{-1}\end{bmatrix}\begin{bmatrix}-\mathrm{j}\overline{r}\widetilde{E}_\phi \\ \widetilde{E}_z\end{bmatrix}$$

or after rearranging:

$$\begin{bmatrix}(\mu_{zz}D_{\overline{z}}\mu_{rt}^{-1}D_{\overline{z}} + \varepsilon_{rt}\mu_{zz})\mathrm{j}\overline{r}E_\phi \\ (\varepsilon_{zz}D_{\overline{z}}\varepsilon_{rt}^{-1}D_{\overline{z}} + \varepsilon_{zz}\mu_{rt})\mathrm{j}\overline{r}\widetilde{H}_\phi\end{bmatrix} = \begin{bmatrix}\nu\mu_{zz}D_{\overline{z}}\mu_{rt}^{-1}\varepsilon_{zz}^{-1} & -\overline{r}\frac{\partial}{\partial \overline{r}} \\ \overline{r}\frac{\partial}{\partial \overline{r}} & \nu\varepsilon_{zz}D_{\overline{z}}\varepsilon_{rt}^{-1}\mu_{zz}^{-1}\end{bmatrix}\begin{bmatrix}\varepsilon_{zz}E_z \\ \mu_{zz}\widetilde{H}_z\end{bmatrix} \qquad (4.183)$$

Eq. (4.183) gives a relation between the ϕ and z components of the electric and magnetic fields. In the following, we will use this equation in discretised form to compute E_ϕ and $\widetilde{H}_\phi$ from E_z and $\widetilde{H}_z$.

4.4.2.3 Discretisation

To solve eqs. (4.181) and (4.182) we discretise the fields in z-direction according to Fig. 4.18. However, instead of E_z and $\widetilde{H}_z$ we use $\varepsilon_{zz}E_z$ and $\mu_{zz}H_z$ as variables, because these quantities are continuous in z-direction. The discretisation of the fields and of eqs. (4.181)–(4.183) is done completely analogously to that for eqs. (4.148)–(4.150). Therefore the details can be found in eq. (4.151). We transform the discretised quantities according to:

$$\boldsymbol{\epsilon}_e\mathbf{E}_z = \boldsymbol{T}_e^{\varepsilon}\overline{\mathbf{E}}_z \qquad \boldsymbol{\mu}_h\widetilde{\mathbf{H}}_z = \boldsymbol{T}_h^{\mu}\overline{\mathbf{H}}_z \qquad (4.184)$$

and have to solve the following eigenvalue problem:

$$(\boldsymbol{T}^{\alpha}_{e,h})^{-1}(\boldsymbol{\epsilon}_{e,h}\boldsymbol{\mu}_{e,h} - \overline{\boldsymbol{P}}^{\alpha}_{e,h})\boldsymbol{T}^{\alpha}_{e,h} = \boldsymbol{k}^2_{re,h} \qquad \alpha = \varepsilon, \mu \qquad (4.185)$$

$\boldsymbol{k}^2_{re,h}$ is a diagonal matrix of eigenvalues. The discretised and transformed wave equations are:

$$\overline{r}\frac{\mathrm{d}}{\mathrm{d}\overline{r}}\left(\overline{r}\frac{\mathrm{d}}{\mathrm{d}\overline{r}}\overline{\mathbf{F}}\right) + ((\boldsymbol{k}_{re,h}\overline{r})^2 - \nu^2\boldsymbol{I})\overline{\mathbf{F}} = \mathbf{0} \qquad \overline{\mathbf{F}} = \overline{\mathbf{E}}_z, \overline{\mathbf{H}}_z \qquad (4.186)$$

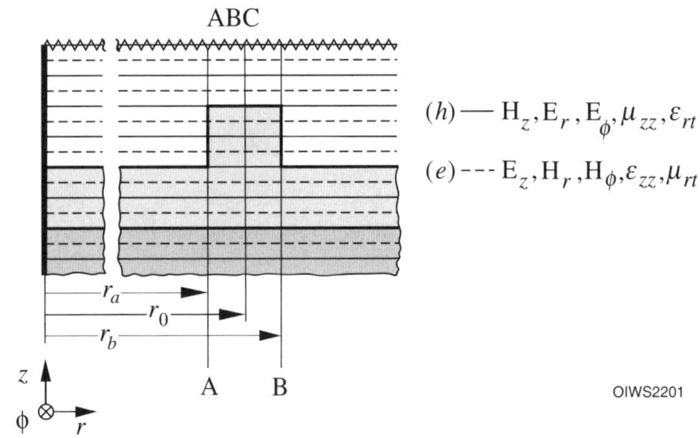

Fig. 4.18 Cross-section of rib waveguide bend with discretisation lines in r-direction

The general solution is given by:

$$\overline{\mathbf{F}} = J_\nu(\boldsymbol{k}_{re,h}\overline{r})\mathbf{A} + Y_\nu(\boldsymbol{k}_{re,h}\overline{r})\mathbf{B} \tag{4.187}$$

J_ν and Y_ν are Bessel and Neumann functions of order ν. Usually, this order is complex. Because of $\boldsymbol{k}^2_{re,h}$ the arguments of the Bessel functions are diagonal matrices. Therefore, J_ν and Y_ν are also diagonal matrices.

4.4.2.4 Admittance matrix equations

From the general solution we determine the field at the radii $\overline{r} = \overline{r}_a$ and $\overline{r} = \overline{r}_b$ (see Fig. 4.18):

$$\begin{bmatrix} \overline{\mathbf{F}}_{\mathrm{A}} \\ \overline{\mathbf{F}}_{\mathrm{B}} \end{bmatrix} = \begin{bmatrix} J_\nu(\boldsymbol{k}_r\overline{r}_a) & Y_\nu(\boldsymbol{k}_r\overline{r}_a) \\ J_\nu(\boldsymbol{k}_r\overline{r}_b) & Y_\nu(\boldsymbol{k}_r\overline{r}_b) \end{bmatrix} \begin{bmatrix} \mathbf{A} \\ \mathbf{B} \end{bmatrix} \tag{4.188}$$

We obtain for the derivatives at these places:

$$\overline{r}_{a,b}\frac{\partial}{\partial \overline{r}} \begin{bmatrix} \overline{\mathbf{F}}_{\mathrm{A}} \\ \overline{\mathbf{F}}_{\mathrm{B}} \end{bmatrix} = \widehat{\boldsymbol{p}}_\nu^{-1} \begin{bmatrix} \overline{\boldsymbol{r}}_\nu & \frac{2}{\pi}\boldsymbol{I} \\ -\frac{2}{\pi}\boldsymbol{I} & \overline{\boldsymbol{q}}_\nu \end{bmatrix} \begin{bmatrix} \overline{\mathbf{F}}_{\mathrm{A}} \\ \overline{\mathbf{F}}_{\mathrm{B}} \end{bmatrix} \tag{4.189}$$

where $\overline{r}_{a,b}$ is equal to $\overline{r}_a$ for the upper line (matching surface A) and equal to $\overline{r}_b$ for the lower line (matching surface B). $\boldsymbol{p}_\nu$, $\overline{\boldsymbol{r}}_\nu$ and $\overline{\boldsymbol{q}}_\nu$ are normalised cross-products (see formula 9.1.32 in [14]):

$$\begin{aligned}
\boldsymbol{p}_\nu &= J_\nu(\boldsymbol{t}_\mathrm{A})Y_\nu(\boldsymbol{t}_\mathrm{B}) - J_\nu(\boldsymbol{t}_\mathrm{B})Y_\nu(\boldsymbol{t}_\mathrm{A}) & \widehat{\boldsymbol{p}}_\nu &= \mathrm{Diag}(\boldsymbol{p}_\nu, \boldsymbol{p}_\nu)\\
\boldsymbol{r}_\nu &= J'_\nu(\boldsymbol{t}_\mathrm{A})Y_\nu(\boldsymbol{t}_\mathrm{B}) - Y'_\nu(\boldsymbol{t}_\mathrm{A})J_\nu(\boldsymbol{t}_\mathrm{B}) & \overline{\boldsymbol{r}}_\nu &= \boldsymbol{t}_\mathrm{A}\boldsymbol{r}_\nu\\
\boldsymbol{q}_\nu &= J_\nu(\boldsymbol{t}_\mathrm{A})Y'_\nu(\boldsymbol{t}_\mathrm{B}) - J'_\nu(\boldsymbol{t}_\mathrm{B})Y_\nu(\boldsymbol{t}_\mathrm{A}) & \overline{\boldsymbol{q}}_\nu &= \boldsymbol{t}_\mathrm{B}\boldsymbol{q}_\nu\\
\boldsymbol{s}_\nu &= J'_\nu(\boldsymbol{t}_\mathrm{A})Y'_\nu(\boldsymbol{t}_\mathrm{B}) - J'_\nu(\boldsymbol{t}_\mathrm{B})Y'_\nu(\boldsymbol{t}_\mathrm{A}) & \overline{\boldsymbol{s}}_\nu &= \boldsymbol{t}_\mathrm{A}\boldsymbol{t}_\mathrm{B}\boldsymbol{s}_\nu\\
\boldsymbol{W}_\mathrm{P} &= J_\nu(\boldsymbol{t}_\mathrm{P})Y'_\nu(\boldsymbol{t}_\mathrm{P}) - J'_\nu(\boldsymbol{t}_\mathrm{P})Y_\nu(\boldsymbol{t}_\mathrm{P}) = \frac{2}{\pi}\boldsymbol{t}_\mathrm{P}^{-1} & \overline{\boldsymbol{W}} &= \boldsymbol{t}_\mathrm{P}\boldsymbol{W}_\mathrm{P} = \frac{2}{\pi}\boldsymbol{I}
\end{aligned} \tag{4.190}$$

$R'_\nu = \mathrm{d}R_\nu(t)/\mathrm{d}t$ is the derivative of the Bessel functions R_ν with respect to the argument t. In our case, the arguments $\boldsymbol{t}_{\mathrm{A,B}}$ are diagonal matrices according to:

$$\boldsymbol{t}_\mathrm{A} = \boldsymbol{k}_{\overline{r}}\overline{r}_a \qquad \boldsymbol{t}_\mathrm{B} = \boldsymbol{k}_{\overline{r}}\overline{r}_b \tag{4.191}$$

Note: we must distinguish between the value of the cross-product $\boldsymbol{r}_\nu$ (bold) and the coordinate r (not bold). We obtain from eq. (4.183):

$$\begin{bmatrix} \mathrm{j}\overline{r}\overline{\mathbf{E}}_\phi \\ \mathrm{j}\overline{r}\overline{\mathbf{H}}_\phi \end{bmatrix} = \widehat{\boldsymbol{k}}_r^{-2} \begin{bmatrix} \widetilde{\boldsymbol{\delta}}_e & -\overline{r}\dfrac{\partial}{\partial\overline{r}}\boldsymbol{I}_h \\ \overline{r}\dfrac{\partial}{\partial\overline{r}}\boldsymbol{I}_e & \widetilde{\boldsymbol{\delta}}_h \end{bmatrix} \begin{bmatrix} \overline{\mathbf{E}}_z \\ \overline{\mathbf{H}}_z \end{bmatrix} \tag{4.192}$$

The following abbreviations were used:

$$\begin{aligned}
&\widehat{\boldsymbol{k}}_r^2 = \mathrm{Diag}(\boldsymbol{k}_{rh}^2, \boldsymbol{k}_{re}^2) \quad \widehat{\boldsymbol{k}}_{re,h}^2 = \mathrm{Diag}\left(\boldsymbol{k}_{re,h}^2, \boldsymbol{k}_{re,h}^2\right)\\
&\widetilde{\boldsymbol{\delta}}_e = (\boldsymbol{T}_h^\mu)^{-1}(\nu\boldsymbol{\mu}_h\overline{\boldsymbol{D}}_e\boldsymbol{\mu}_e^{-1}\boldsymbol{\epsilon}_e^{-1})\boldsymbol{T}_e^\epsilon\\
&\widetilde{\boldsymbol{\delta}}_h = (\boldsymbol{T}_e^\varepsilon)^{-1}(\nu\boldsymbol{\epsilon}_e\overline{\boldsymbol{D}}_h\boldsymbol{\epsilon}_h^{-1}\boldsymbol{\mu}_h^{-1})\boldsymbol{T}_h^\mu\\
&\widehat{\widehat{\boldsymbol{k}}}_r^2 = \mathrm{Diag}\left(\widehat{\boldsymbol{k}}_{rh}^2, \widehat{\boldsymbol{k}}_{re}^2\right) \quad \widehat{\widetilde{\boldsymbol{\delta}}}_{e,h} = \mathrm{Diag}\left(\widetilde{\boldsymbol{\delta}}_{e,h}, \widetilde{\boldsymbol{\delta}}_{e,h}\right)
\end{aligned}$$

Now the ϕ components at surfaces A and B can be determined from eq. (4.192) by using eq. (4.189). We obtain:

$$\begin{bmatrix} \mathrm{j}\overline{r}_\mathrm{A}\overline{\mathbf{E}}_{\phi\mathrm{A}} \\ \mathrm{j}\overline{r}_\mathrm{B}\overline{\mathbf{E}}_{\phi\mathrm{B}} \\ \mathrm{j}\overline{r}_\mathrm{A}\overline{\mathbf{H}}_{\phi\mathrm{A}} \\ \mathrm{j}\overline{r}_\mathrm{B}\overline{\mathbf{H}}_{\phi\mathrm{B}} \end{bmatrix} = \widehat{\widehat{\boldsymbol{k}}}_r^{-2} \begin{bmatrix} \begin{bmatrix} \widetilde{\boldsymbol{\delta}}_e & \boldsymbol{0} \\ \boldsymbol{0} & \widetilde{\boldsymbol{\delta}}_e \end{bmatrix} & -\widehat{\boldsymbol{p}}_{\nu h}^{-1}\begin{bmatrix} \overline{\boldsymbol{r}}_{\nu h} & \frac{2}{\pi}\boldsymbol{I}_h \\ -\frac{2}{\pi}\boldsymbol{I}_h & \overline{\boldsymbol{q}}_{\nu h} \end{bmatrix} \\ \widehat{\boldsymbol{p}}_{\nu e}^{-1}\begin{bmatrix} \overline{\boldsymbol{r}}_{\nu e} & \frac{2}{\pi}\boldsymbol{I}_e \\ -\frac{2}{\pi}\boldsymbol{I}_e & \overline{\boldsymbol{q}}_{\nu e} \end{bmatrix} & \begin{bmatrix} \widetilde{\boldsymbol{\delta}}_h & \boldsymbol{0} \\ \boldsymbol{0} & \widetilde{\boldsymbol{\delta}}_h \end{bmatrix} \end{bmatrix} \begin{bmatrix} \overline{\mathbf{E}}_{z\mathrm{A}} \\ \overline{\mathbf{E}}_{z\mathrm{B}} \\ \overline{\mathbf{H}}_{z\mathrm{A}} \\ \overline{\mathbf{H}}_{z\mathrm{B}} \end{bmatrix} \tag{4.193}$$

After rearrangement we obtain:

$$\begin{bmatrix} \overline{\mathbf{H}}_{z\mathrm{A}} \\ \overline{\mathbf{H}}_{z\mathrm{B}} \\ \mathrm{j}\overline{r}_{\mathrm{A}}\overline{\mathbf{H}}_{\phi\mathrm{A}} \\ \mathrm{j}\overline{r}_{\mathrm{B}}\overline{\mathbf{H}}_{\phi\mathrm{B}} \end{bmatrix} = \begin{bmatrix} -\widehat{\overline{\mathbf{s}}}_{\nu h}^{-1} \begin{bmatrix} \overline{\boldsymbol{q}}_{\nu h} & -\frac{2}{\pi}\boldsymbol{I}_h \\ \frac{2}{\pi}\boldsymbol{I}_h & \overline{\boldsymbol{r}}_{\nu h} \end{bmatrix} \widehat{\boldsymbol{k}}_{rh}^2 & \widehat{\overline{\mathbf{s}}}_{\nu h}^{-1} \begin{bmatrix} \overline{\boldsymbol{q}}_{\nu h} & -\frac{2}{\pi}\boldsymbol{I}_h \\ \frac{2}{\pi}\boldsymbol{I}_h & \overline{\boldsymbol{r}}_{\nu h} \end{bmatrix} \widehat{\widehat{\boldsymbol{\delta}}}_e \\ -\widehat{\boldsymbol{k}}_{re}^{-2}\widehat{\widehat{\boldsymbol{\delta}}}_h\widehat{\overline{\mathbf{s}}}_{\nu h}^{-1} \begin{bmatrix} \overline{\boldsymbol{q}}_{\nu h} & -\frac{2}{\pi}\boldsymbol{I}_h \\ \frac{2}{\pi}\boldsymbol{I}_h & \overline{\boldsymbol{r}}_{\nu h} \end{bmatrix} \widehat{\boldsymbol{k}}_{rh}^2 & \begin{bmatrix} \boldsymbol{\gamma}_{\mathrm{E}_1} & -\boldsymbol{\alpha}_{\mathrm{E}} \\ \boldsymbol{\alpha}_{\mathrm{E}} & \boldsymbol{\gamma}_{\mathrm{E}_2} \end{bmatrix} \end{bmatrix} \times \begin{bmatrix} \mathrm{j}\overline{r}_{\mathrm{A}}\overline{\mathbf{E}}_{\phi\mathrm{A}} \\ \mathrm{j}\overline{r}_{\mathrm{B}}\overline{\mathbf{E}}_{\phi\mathrm{B}} \\ \overline{\mathbf{E}}_{z\mathrm{A}} \\ \overline{\mathbf{E}}_{z\mathrm{B}} \end{bmatrix} \tag{4.194}$$

with:

$$\begin{bmatrix} \boldsymbol{\gamma}_{\mathrm{E}_1} & -\boldsymbol{\alpha}_{\mathrm{E}} \\ \boldsymbol{\alpha}_{\mathrm{E}} & \boldsymbol{\gamma}_{\mathrm{E}_2} \end{bmatrix} = \widehat{\boldsymbol{k}}_{re}^{-2}\widehat{\boldsymbol{p}}_{\nu e}^{-1} \begin{bmatrix} \overline{\boldsymbol{r}}_{\nu e} & \frac{2}{\pi}\boldsymbol{I}_e \\ -\frac{2}{\pi}\boldsymbol{I}_e & \overline{\boldsymbol{q}}_{\nu e} \end{bmatrix} + \widehat{\boldsymbol{k}}_{re}^{-2}\widehat{\widehat{\boldsymbol{\delta}}}_h\widehat{\overline{\mathbf{s}}}_{\nu h}^{-1} \begin{bmatrix} \overline{\boldsymbol{q}}_{\nu h} & -\frac{2}{\pi}\boldsymbol{I}_h \\ \frac{2}{\pi}\boldsymbol{I}_h & \overline{\boldsymbol{r}}_{\nu h} \end{bmatrix} \widehat{\widehat{\boldsymbol{\delta}}}_e \tag{4.195}$$

where we use $\boldsymbol{p}_\nu \boldsymbol{s}_\nu - \boldsymbol{q}_\nu \boldsymbol{r}_\nu = 4(\pi^2 \boldsymbol{t}_a \boldsymbol{t}_b)^{-1}$ (see eq. (9.1.34) in [14]) and:

$$\left(\widehat{\boldsymbol{p}}_\nu^{-1} \begin{bmatrix} \overline{\boldsymbol{r}}_\nu & \frac{2}{\pi}\boldsymbol{I} \\ -\frac{2}{\pi}\boldsymbol{I} & \overline{\boldsymbol{q}}_\nu \end{bmatrix}\right)^{-1} = \widehat{\overline{\mathbf{s}}}_\nu^{-1} \begin{bmatrix} \overline{\boldsymbol{q}}_\nu & -\frac{2}{\pi}\boldsymbol{I} \\ \frac{2}{\pi}\boldsymbol{I} & \overline{\boldsymbol{r}}_\nu \end{bmatrix} \tag{4.196}$$

A further rearrangement results in:

$$\begin{bmatrix} \mathbf{H}_{z\mathrm{A}} \\ \mathrm{j}r_{\mathrm{A}}\mathbf{H}_{\phi\mathrm{A}} \\ -\mathbf{H}_{z\mathrm{B}} \\ -\mathrm{j}r_{\mathrm{B}}\mathbf{H}_{\phi\mathrm{B}} \end{bmatrix} = \begin{bmatrix} -\widetilde{\boldsymbol{q}}_{\nu h}\boldsymbol{k}_{rh}^2 & \widetilde{\boldsymbol{q}}_{\nu h}\widetilde{\boldsymbol{\delta}}_e & \widetilde{\mathbf{w}}_{\nu h}\boldsymbol{k}_{rh}^2 & -\widetilde{\mathbf{w}}_{\nu h}\widetilde{\boldsymbol{\delta}}_e \\ -\widetilde{\boldsymbol{\delta}}_{hk}\widetilde{\boldsymbol{q}}_{\nu h} & \boldsymbol{\gamma}_{\mathrm{E}_1} & \widetilde{\boldsymbol{\delta}}_{hk}\widetilde{\mathbf{w}}_{\nu h} & -\boldsymbol{\alpha}_{\mathrm{E}} \\ \widetilde{\mathbf{w}}_{\nu h}\boldsymbol{k}_{rh}^2 & -\widetilde{\mathbf{w}}_{\nu h}\widetilde{\boldsymbol{\delta}}_e & \widetilde{\boldsymbol{r}}_{\nu h}\boldsymbol{k}_{rh}^2 & -\widetilde{\boldsymbol{r}}_{\nu h}\widetilde{\boldsymbol{\delta}}_e \\ \widetilde{\boldsymbol{\delta}}_{hk}\widetilde{\mathbf{w}}_{\nu h} & -\boldsymbol{\alpha}_{\mathrm{E}} & \widetilde{\boldsymbol{\delta}}_{hk}\widetilde{\boldsymbol{r}}_{\nu h} & -\widetilde{\boldsymbol{\gamma}}_{\mathrm{E}_2} \end{bmatrix} \begin{bmatrix} \mathrm{j}r_{\mathrm{A}}\mathbf{E}_{\phi\mathrm{A}} \\ \mathbf{E}_{z\mathrm{A}} \\ \mathrm{j}r_{\mathrm{B}}\mathbf{E}_{\phi\mathrm{B}} \\ \mathbf{E}_{z\mathrm{B}} \end{bmatrix} \tag{4.197}$$

with:

$$\widetilde{\boldsymbol{\delta}}_{hk} = \boldsymbol{k}_{re}^{-2}\widetilde{\boldsymbol{\delta}}_h\boldsymbol{k}_{rh}^2 \qquad \widetilde{\boldsymbol{q}}_{\nu h} = \overline{\mathbf{s}}_{\nu h}^{-1}\overline{\boldsymbol{q}}_{\nu h} \qquad \widetilde{\boldsymbol{r}}_{\nu h} = \overline{\mathbf{s}}_{\nu h}^{-1}\overline{\boldsymbol{r}}_{\nu h} \qquad \widetilde{\mathbf{w}}_{\nu h} = \frac{2}{\pi}\overline{\mathbf{s}}_{\nu h}^{-1} \tag{4.198}$$

By introducing the abbreviations:

$$\overline{\mathbf{H}}_{\mathrm{A,B}} = \begin{bmatrix} \overline{\mathbf{H}}_{z\mathrm{A,B}} \\ \mathrm{j}\overline{r}_{\mathrm{A,B}}\overline{\mathbf{H}}_{\phi\mathrm{A,B}} \end{bmatrix} \qquad \overline{\mathbf{E}}_{\mathrm{A,B}} = \begin{bmatrix} \mathrm{j}\overline{r}_{\mathrm{A,B}}\overline{\mathbf{E}}_{\phi\mathrm{A,B}} \\ \overline{\mathbf{E}}_{z\mathrm{A,B}} \end{bmatrix} \tag{4.199}$$

we may write the relation for the transverse fields of a waveguide section k in the following form:

$$\begin{bmatrix} \overline{\mathbf{H}}_{\mathrm{A}} \\ -\overline{\mathbf{H}}_{\mathrm{B}} \end{bmatrix} = \begin{bmatrix} \overline{\boldsymbol{y}}_{1\mathrm{A}} & \overline{\boldsymbol{y}}_2 \\ \overline{\boldsymbol{y}}_2 & \overline{\boldsymbol{y}}_{1\mathrm{B}} \end{bmatrix} \begin{bmatrix} \overline{\mathbf{E}}_{\mathrm{A}} \\ \overline{\mathbf{E}}_{\mathrm{B}} \end{bmatrix} \tag{4.200}$$

The matrices $\overline{\mathbf{y}}_{1\mathrm{A,B}}$ and $\overline{\mathbf{y}}_2$ can easily be obtained from eq. (4.197). The admittance transformation can be performed analogously to Section 2.5. For the transfer matrices according to:

$$\begin{bmatrix} \overline{\mathbf{E}}_\mathrm{B} \\ \overline{\mathbf{H}}_\mathrm{B} \end{bmatrix} = \begin{bmatrix} \overline{\mathbf{V}}_1 & \overline{\mathbf{Z}} \\ \overline{\mathbf{Y}} & \overline{\mathbf{V}}_2 \end{bmatrix} \begin{bmatrix} \overline{\mathbf{E}}_\mathrm{A} \\ \overline{\mathbf{H}}_\mathrm{A} \end{bmatrix} \tag{4.201}$$

we obtain:

$$\begin{array}{ll} \overline{\mathbf{V}}_1 = \overline{\mathbf{y}}_2^{-1}\overline{\mathbf{y}}_{1\mathrm{A}} & \overline{\mathbf{V}}_2 = \overline{\mathbf{y}}_{1\mathrm{B}}\overline{\mathbf{y}}_2^{-1} \\ \overline{\mathbf{Z}} = -\overline{\mathbf{y}}_2^{-1} & \overline{\mathbf{Y}} = \overline{\mathbf{y}}_2 - \overline{\mathbf{y}}_{1\mathrm{B}}\overline{\mathbf{y}}_2^{-1}\overline{\mathbf{y}}_{1\mathrm{A}} \end{array} \tag{4.202}$$

These formulas are for the homogeneous (with respect to r) section k. At the interfaces e.g. between sections k and $k+1$ we must determine the fields in the original domain leading to a matching condition for the transformed fields and impedances/admittances.

4.4.2.5 Center and outer regions

For the inner and infinite outer regions we must again derive separate equations. In the inner region only the first term of eq. (4.187) can be used because the second term (with the Neumann function) is infinite in the centre. In the infinite outer region we have to use the modified Bessel function K_ν. Instead of eq. (4.189) we obtain:

$$\overline{r}\frac{\mathrm{d}}{\mathrm{d}\overline{r}}\overline{\mathbf{F}} = \boldsymbol{t}R'_\nu(\boldsymbol{t})R_\nu^{-1}(\boldsymbol{t})\overline{\mathbf{F}} \tag{4.203}$$

where R_ν is equal to J_ν (K_ν) in the inner (infinite outer) region. Again, R'_ν is the derivative with respect to the argument. By using the abbreviation:

$$\boldsymbol{\Lambda}_{e,h}^\mathrm{C} = \boldsymbol{t}_\mathrm{C}^{e,h}R'_\nu(\boldsymbol{t}_\mathrm{C}^{e,h})R_\nu^{-1}(\boldsymbol{t}_\mathrm{C}^{e,h}) \tag{4.204}$$

we obtain instead of eq. (4.193):

$$\widehat{\boldsymbol{k}}_r^2 \begin{bmatrix} \mathrm{j}\overline{r}_\mathrm{C}\overline{\mathbf{E}}_{\phi\mathrm{C}} \\ \mathrm{j}\overline{r}_\mathrm{C}\overline{\mathbf{H}}_{\phi\mathrm{C}} \end{bmatrix} = \begin{bmatrix} \widetilde{\boldsymbol{\delta}}_e & -\boldsymbol{\Lambda}_h^\mathrm{C} \\ \boldsymbol{\Lambda}_e^\mathrm{C} & \widetilde{\boldsymbol{\delta}}_h \end{bmatrix} \begin{bmatrix} \overline{\mathbf{E}}_{z\mathrm{C}} \\ \overline{\mathbf{H}}_{z\mathrm{C}} \end{bmatrix} \tag{4.205}$$

with C = A,B. A (B) stands for the outer (inner) interface of the inner (outer) region. After rearranging, and by using the definitions in eq. (4.204), we obtain:

$$\overline{\mathbf{H}}_\mathrm{C} = \overline{\mathbf{Y}}_\mathrm{C}\overline{\mathbf{E}}_\mathrm{C} \qquad \overline{\mathbf{Y}}_\mathrm{C} = \begin{bmatrix} -\boldsymbol{\Lambda}_{h\mathrm{C}}^{-1}\boldsymbol{k}_{rh}^2 & \boldsymbol{\Lambda}_{h\mathrm{C}}^{-1}\widetilde{\boldsymbol{\delta}}_e \\ -\boldsymbol{k}_{re}^{-2}\widetilde{\boldsymbol{\delta}}_h\boldsymbol{\Lambda}_{h\mathrm{C}}^{-1}\boldsymbol{k}_{rh}^2 & \boldsymbol{k}_{re}^{-2}\left(\boldsymbol{\Lambda}_{e\mathrm{C}} + \widetilde{\boldsymbol{\delta}}_h\boldsymbol{\Lambda}_{h\mathrm{C}}^{-1}\widetilde{\boldsymbol{\delta}}_e\right) \end{bmatrix} \tag{4.206}$$

4.4.2.6 Numerical results

The fundamental and the first higher-order mode of a 90° bend in a rib waveguide with $R = 10$ mm (Fig. 4.19a) were determined. Due to the very high ratio $R/\lambda \approx 10^4$, the complex orders of the Bessel functions were very large, with a comparatively small imaginary part ($\nu \approx 10^5 - 5\text{j}$). Moreover, the argument for the first eigenvalue is nearly equal to the argument. For these values of the parameters, the Bessel functions should be computed by uniform asymptotic series (see formula 9.3.35ff in [14]). Alternatively, for general sections the cross-products can be computed by means of the multiplication theorem (see formula 9.3.74 in [14]). For more details on this problem see also [9] and [10]. The radiation loss is given by:

$$L = -8.68 \cdot \frac{\pi}{2} \text{Im}(n_{\text{eff}})\overline{R} \text{ dB} \tag{4.207}$$

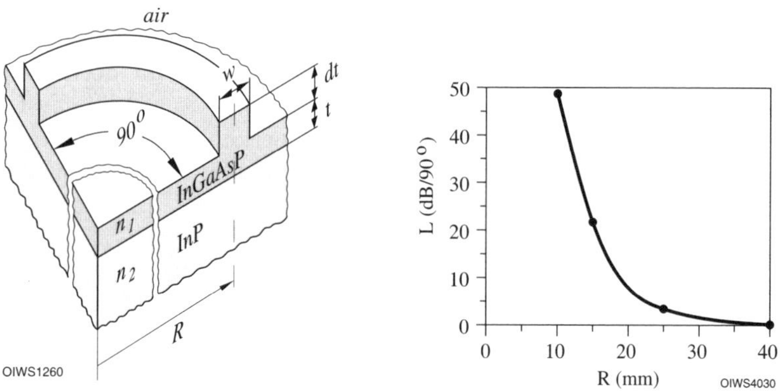

(a) (b)

Fig. 4.19 Circular rib waveguide (a) geometry with refractive indices: $n_1 = 3.23001$, $n_2 = 3.20528$ and dimensions: $w = 2$ μm, $t = 1.0$ μm, $dt = 0.3$ μm, $\lambda = 1.286$ μm (b) radiation loss L vs. the bending radius R for HE_{00} mode ((a) and (b) W. Pascher and R. Pregla, 'Vectorial Analysis of Bends in Optical Strip Waveguides by the Method of Lines', *Radio Sci.*, vol. 28, pp. 1229–1233. © 1993 American Geophysical Union (AGU))

For the HE_{00} mode we obtain a loss of 48.7 dB/90° and a mode shift of 0.15 μm, and for the EH_{00} mode we find 35.9 dB/90° and 0.14 μm. The dependence of the loss L on the bending radius R is given in Fig. 4.19b for HE_{00} mode. Fig. 4.20 shows the intensity distribution in the cross-section of the rib waveguide for the HE_{00} and EH_{00} modes, respectively. The loss can be reduced by introducing a zone with decreased refractive index on the outer side of the bend, as shown in Fig. 4.16.

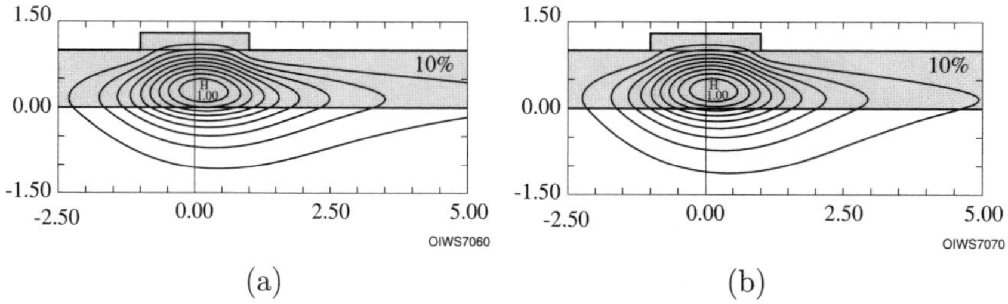

Fig. 4.20 Field intensity distribution of HE_{00} mode (a) and EH_{00} mode in a circular rib waveguide (b). Parameters as in Fig. 4.19a ((a) and (b) W. Pascher and R. Pregla, 'Vectorial Analysis of Bends in Optical Strip Waveguides by the Method of Lines', *Radio Sci.*, vol. 28, pp. 1229–1233. © 1993 American Geophysical Union (AGU))

4.4.3 Discretisation in ϕ-direction – eigenmodes in circular multilayered waveguides

4.4.3.1 Introduction

In this section the determination of the eigenmodes in structures such as those shown in Fig. 4.21 will be described. An important feature is that the materials are homogeneous in the azimuthal direction. The number of circular layers in these structures may be arbitrary. An arbitrary number of metallic strips or cylinders can be placed between the layers of the structure, and the layers can begin at $r = 0$ and extend to infinity. The permittivities and permeabilities in the layers can be uniaxial-anisotropic and complex. Structures with an r-dependent permittivity (e.g. graded index fibres) can be modelled by a sufficient number of distinct layers. A special example is the Bragg fibre sketched in Fig. 4.22.

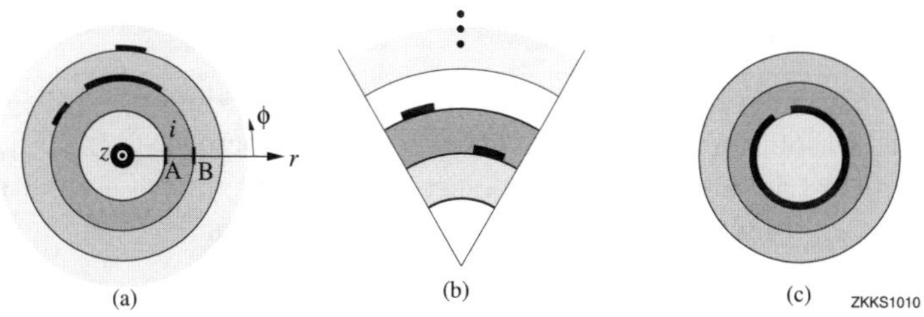

Fig. 4.21 Cross-sections of cylindrical multilayer structures with metallisations (a) general cross-section (b) sectorial cross-section (c) slot structure

In contrast to the procedures described in the sections above, the discretisation is performed in ϕ-direction. The analytical solution, however, is found again in r-direction. The discretisation in ϕ-direction is similar to the discretisation in Cartesian coordinates. Therefore, the procedure is analogous to the determination of eigenmodes in planar waveguide structures. However, for the analytical solutions in r-direction we again obtain Bessel

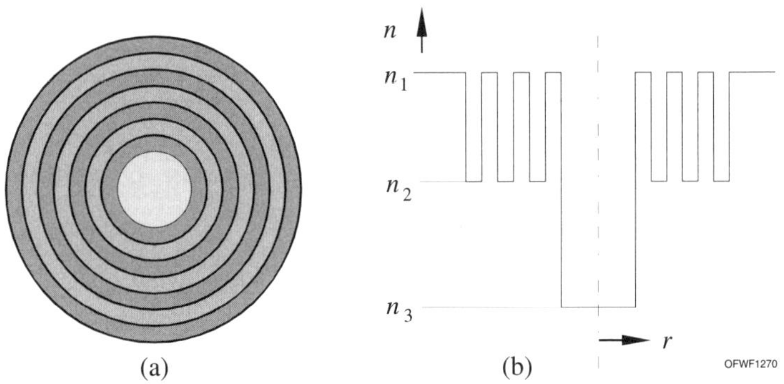

Fig. 4.22 Cross-section (a) and refractive index profile (b) of a Bragg fibre

functions. The goal of our derivations is to obtain general transformation formulas for fields between two boundary layers, i.e. from a surface A to a surface B in the homogeneous layers. From these equations we derive the admittance/impedance transformation formulas.

4.4.3.2 Basic analysis equation

The whole field may be obtained from the components in z-direction, E_z and H_z. These are the only Cartesian components in the cylindrical coordinate system. For these components, we obtain from eqs. (4.97) and (4.98) the following wave equations:

$$\begin{aligned} \overline{r}\frac{\partial}{\partial\overline{r}}\left(\overline{r}\frac{\partial\widetilde{H}_z}{\partial\overline{r}}\right) + D_\phi D_\phi \widetilde{H}_z + \mu_{zz}\mu_{rt}^{-1}(\varepsilon_{rt}\mu_{rt} - \varepsilon_{re})\overline{r}^2\widetilde{H}_z = 0 \\ \overline{r}\frac{\partial}{\partial\overline{r}}\left(\overline{r}\frac{\partial E_z}{\partial\overline{r}}\right) + D_\phi D_\phi E_z + \varepsilon_{zz}\varepsilon_{rt}^{-1}(\varepsilon_{rt}\mu_{rt} - \varepsilon_{re})\overline{r}^2 E_z = 0 \end{aligned} \tag{4.208}$$

Here we assume a propagation in z-direction according to $\exp(-j\sqrt{\varepsilon_{re}}\overline{z})$. Therefore, we obtain for the derivative $\partial/\partial\overline{z}$ the factor $-j\sqrt{\varepsilon_{re}}$. ε_{re} is the effective dielectric constant. The transverse components are computed from eqs. (4.83) and (4.99). We obtain for the ϕ components:

$$\varepsilon_{dt}\begin{bmatrix}\overline{r}E_\phi \\ \overline{r}\widetilde{H}_\phi\end{bmatrix} = -\begin{bmatrix}\sqrt{\varepsilon_{re}}D_\phi & -\mu_{rt}\overline{r}\dfrac{\partial}{\partial\overline{r}} \\ \varepsilon_{rt}\overline{r}\dfrac{\partial}{\partial\overline{r}} & \sqrt{\varepsilon_{re}}D_\phi\end{bmatrix}\begin{bmatrix}\mathrm{j}E_z \\ \mathrm{j}\widetilde{H}_z\end{bmatrix} \tag{4.209}$$

and for the r components:

$$\varepsilon_{dt}\begin{bmatrix}\overline{r}E_r \\ \overline{r}\widetilde{H}_r\end{bmatrix} = -\begin{bmatrix}\sqrt{\varepsilon_{re}}\overline{r}\dfrac{\partial}{\partial\overline{r}} & \mu_{rt}D_\phi \\ -\varepsilon_{rt}D_\phi & \sqrt{\varepsilon_{re}}\overline{r}\dfrac{\partial}{\partial\overline{r}}\end{bmatrix}\begin{bmatrix}\mathrm{j}E_z \\ \mathrm{j}\widetilde{H}_z\end{bmatrix} \tag{4.210}$$

with the abbreviation $\varepsilon_{dt} = \varepsilon_{rt}\mu_{rt} - \varepsilon_{re}$.

4.4.3.3 Discretisation, transformation, basic solution

For the solution of the wave equations (4.208) and for the determination of the field components, a discretisation in ϕ-direction is performed. In principle, the analysis is the same as that in Cartesian coordinates. Therefore, all that is known for the discretisation in Cartesian coordinates can be used here for the ϕ-direction. In Fig. 4.23 three different possibilities are shown. To save memory and computational effort, absorbing boundary conditions (ABC) are suitable (a). If the whole cross-section is of interest, periodic boundary conditions have to be used, described in the most general case in the appendices. In the case of symmetry, or if the structure is only a part of the whole circle, Dirichlet and Neumann boundary conditions are suitable. The following descriptions and definitions are used for the discretisation:

$$\begin{array}{rclrcl} E_z & \rightarrow & \mathbf{E}_z & H_z & \rightarrow & \mathbf{H}_z \\ h\dfrac{\partial E_z}{\partial \phi} & \rightarrow & \boldsymbol{D}_e\mathbf{E}_z & h\dfrac{\partial H_z}{\partial \phi} & \rightarrow & \boldsymbol{D}_h\mathbf{H}_z \\ h^2\dfrac{\partial^2 E_z}{\partial \phi^2} & \rightarrow & \boldsymbol{D}_h\boldsymbol{D}_e\mathbf{E}_z & h^2\dfrac{\partial^2 H_z}{\partial \phi^2} & \rightarrow & \boldsymbol{D}_e\boldsymbol{D}_h\mathbf{H}_z \\ \boldsymbol{T}_h^{-1}\boldsymbol{D}_e\boldsymbol{T}_e & = & \boldsymbol{\delta}_e & \boldsymbol{T}_e^{-1}\boldsymbol{D}_h\boldsymbol{T}_h & = & \boldsymbol{\delta}_h \\ \boldsymbol{T}_e^{-1}(\boldsymbol{D}_h\boldsymbol{D}_e)\boldsymbol{T}_e & = & -\boldsymbol{\lambda}_e^2 & \boldsymbol{T}_h^{-1}(\boldsymbol{D}_e\boldsymbol{D}_h)\boldsymbol{T}_h & = & -\boldsymbol{\lambda}_h^2 \end{array} \tag{4.211}$$

h is the angular discretisation distance in ϕ-direction (in radian). $\boldsymbol{D}_e$ and $\boldsymbol{D}_h$ are the difference operators for the first order differential quotients. After suitable normalisation of the eigenvectors, we obtain, in the case of Neumann and Dirichlet conditions (cf. the appendices):

$$\boldsymbol{D}_h = -\boldsymbol{D}_e^t \qquad \boldsymbol{T}_e^{-1} = \boldsymbol{T}_e^t \qquad \boldsymbol{T}_h^{-1} = \boldsymbol{T}_h^t \tag{4.212}$$

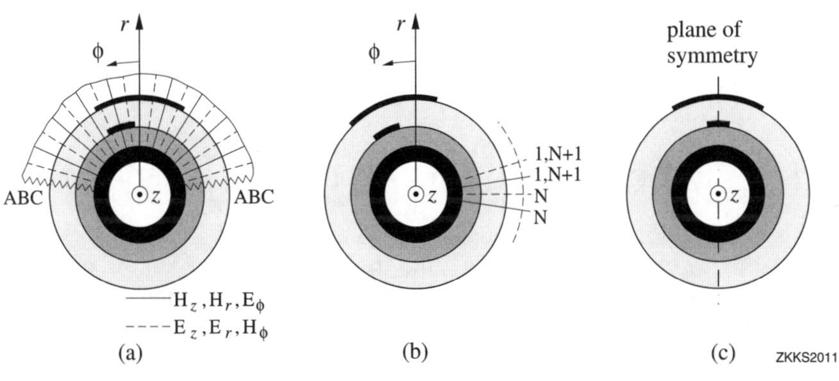

Fig. 4.23 Cross-section with discretisation lines (Reproduced by permission of Springer Netherlands)

In the case of periodic boundary conditions (cf. the appendices) we may write:

$$\boldsymbol{D}_h = -\boldsymbol{D}_e^{*t} \qquad \boldsymbol{T}_e^{-1} = \boldsymbol{T}_e^{*t} \qquad \boldsymbol{T}_h^{-1} = \boldsymbol{T}_h^{*l} \tag{4.213}$$

Finally, for the case of absorbing boundary conditions (cf. the appendices):

$$\boldsymbol{D}_e = \boldsymbol{D}_a \qquad \boldsymbol{D}_h = -\boldsymbol{D}^t \qquad \text{or} \qquad \boldsymbol{D}_h = \boldsymbol{D}_a \qquad \boldsymbol{D}_e = -\boldsymbol{D}^t \tag{4.214}$$

$\boldsymbol{D}$ corresponds to $\boldsymbol{D}_a$. We get this operator from $\boldsymbol{D}_a$ by setting the parameters $a = 1$ and $b = c = 0$. $\boldsymbol{D}$ fulfills the Dirichlet conditions. Furthermore, we have: (See eq. (4.211).)

$$\boldsymbol{\delta}_e\boldsymbol{\delta}_h = -\boldsymbol{\lambda}_h^2 \qquad\qquad \boldsymbol{\delta}_h\boldsymbol{\delta}_e = -\boldsymbol{\lambda}_e^2 \tag{4.215}$$

$$\overline{\boldsymbol{\lambda}}^2 = h^{-2}\boldsymbol{\lambda}^2 \qquad\qquad \widetilde{\boldsymbol{\delta}} = h^{-1}\sqrt{\varepsilon_{re}}\boldsymbol{\delta} = \sqrt{\varepsilon_{re}}\overline{\boldsymbol{\delta}} \tag{4.216}$$

We should remember that we have homogeneous layers in this section. Therefore we do not have to discretise the material parameters. Consequently, eqs. (4.208)–(4.210) are only valid for such homogeneous layers. The discretisation of eqs. (4.209) and (4.210) gives:

$$\varepsilon_{dt}\begin{bmatrix}\overline{r}\overline{\mathbf{E}}_\phi \\ \overline{r}\overline{\mathbf{H}}_\phi\end{bmatrix} = -\begin{bmatrix}\widetilde{\boldsymbol{\delta}}_e & -\mu_{rt}\overline{r}\dfrac{\partial}{\partial\overline{r}}\boldsymbol{I}_h \\ \varepsilon_{rt}\overline{r}\dfrac{\partial}{\partial\overline{r}}\boldsymbol{I}_e & \widetilde{\boldsymbol{\delta}}_h\end{bmatrix}\begin{bmatrix}\mathrm{j}\overline{\mathbf{E}}_z \\ \mathrm{j}\overline{\mathbf{H}}_z\end{bmatrix} \tag{4.217}$$

$$\varepsilon_{dt}\begin{bmatrix}\overline{r}\overline{\mathbf{E}}_r \\ \overline{r}\overline{\mathbf{H}}_r\end{bmatrix} = -\begin{bmatrix}\sqrt{\varepsilon_{re}}\overline{r}\dfrac{\partial}{\partial\overline{r}}\boldsymbol{I}_e & \mu_{rt}\overline{\boldsymbol{\delta}}_h \\ -\varepsilon_{rt}\overline{\boldsymbol{\delta}}_e & \sqrt{\varepsilon_{re}}\overline{r}\dfrac{\partial}{\partial\overline{r}}\boldsymbol{I}_h\end{bmatrix}\begin{bmatrix}\mathrm{j}\overline{\mathbf{E}}_z \\ \mathrm{j}\overline{\mathbf{H}}_z\end{bmatrix} \tag{4.218}$$

The bars above the field components indicate that the quantities are transformed according to:

$$\mathbf{E}_z = \boldsymbol{T}_e\overline{\mathbf{E}}_z \qquad \mathbf{H}_z = \boldsymbol{T}_h\overline{\mathbf{H}}_z \tag{4.219}$$

The quantities $\mathbf{E}_r$, $\widetilde{\mathbf{H}}_\phi$ are transformed with $\boldsymbol{T}_e$ and $\widetilde{\mathbf{H}}_r$, $\mathbf{E}_\phi$ with $\boldsymbol{T}_h$. The discretisation and transformation of the wave eq. (4.208) ($\mathbf{F}_z = \mathbf{E}_z$ or $\widetilde{\mathbf{H}}_z$) results in:

$$\overline{r}\frac{\mathrm{d}}{\mathrm{d}\overline{r}}\left(\overline{r}\frac{\mathrm{d}\overline{\mathbf{F}}_z}{\mathrm{d}\overline{r}}\right) - \overline{\boldsymbol{\lambda}}^2\overline{\mathbf{F}}_z + \varepsilon_{de,dh}\overline{r}^2\overline{\mathbf{F}}_z = \mathbf{0} \quad \rightarrow \quad t\frac{\mathrm{d}}{\mathrm{d}t}\left(t\frac{\mathrm{d}\overline{\mathbf{F}}_z}{\mathrm{d}t}\right) + (t^2\boldsymbol{I} - \overline{\boldsymbol{\lambda}}^2)\overline{\mathbf{F}}_z = \mathbf{0} \tag{4.220}$$

with $\varepsilon_{de} = \varepsilon_{dt}\varepsilon_{zz}/\varepsilon_{rt}$, $\varepsilon_{dh} = \varepsilon_{dt}\mu_{zz}/\mu_{rt}$, $t_{e,h} = \sqrt{\varepsilon_{de,dh}}\ \overline{r}$. For $\overline{\boldsymbol{\lambda}}^2$ and t we use the subscript e in the case of $\overline{\mathbf{F}}_z = \overline{\mathbf{E}}_z$, and h for $\overline{\mathbf{F}}_z = \overline{\mathbf{H}}_z$. The general solution for the kth component is:

$$\overline{F}_{zk} = J_{\overline{\lambda}_k}(t)A_k + Y_{\overline{\lambda}_k}(t)B_k \tag{4.221}$$

In the following we use $\boldsymbol{\nu} = \overline{\boldsymbol{\lambda}}$ instead of $\overline{\boldsymbol{\lambda}}$.

The boldface subscript $\boldsymbol{\nu}$ on both the Bessel functions, e.g. $\mathcal{C}_{\boldsymbol{\nu}}$, indicates that we have a diagonal matrix with Bessel functions of the same kind and all with the same argument but with different orders ($\mathcal{C}_{\nu_k}$ gives the order). The same is true for the functions that are constructed from them, e.g. the cross-products. This is the difference from the situation in Sections 4.4.1 and 4.4.2 where we had diagonal matrices composed of Bessel functions with constant order but different arguments (cf. eqs. (4.155) and (4.187)).

4.4.3.4 Fields on the interfaces A and B

From the general solution in eq. (4.221) we can give the fields at the interfaces A and B of a layer (cf. Fig. 4.24) in matrix form:

$$\begin{bmatrix} \overline{\mathbf{F}}_{z\mathrm{A}} \\ \overline{\mathbf{F}}_{z\mathrm{B}} \end{bmatrix} = \begin{bmatrix} J_{\boldsymbol{\nu}}(t_{\mathrm{A}}) & Y_{\boldsymbol{\nu}}(t_{\mathrm{A}}) \\ J_{\boldsymbol{\nu}}(t_{\mathrm{B}}) & Y_{\boldsymbol{\nu}}(t_{\mathrm{B}}) \end{bmatrix} \begin{bmatrix} \mathbf{A} \\ \mathbf{B} \end{bmatrix} \tag{4.222}$$

or:

$$\begin{bmatrix} \mathbf{A} \\ \mathbf{B} \end{bmatrix} = \widehat{p_{\boldsymbol{\nu}}}^{-1} \begin{bmatrix} Y_{\boldsymbol{\nu}}(t_{\mathrm{B}}) & -Y_{\boldsymbol{\nu}}(t_{\mathrm{A}}) \\ -J_{\boldsymbol{\nu}}(t_{\mathrm{B}}) & J_{\boldsymbol{\nu}}(t_{\mathrm{A}}) \end{bmatrix} \begin{bmatrix} \overline{\mathbf{F}}_{z\mathrm{A}} \\ \overline{\mathbf{F}}_{z\mathrm{B}} \end{bmatrix} \tag{4.223}$$

where:

$$p_{\boldsymbol{\nu}} = J_{\boldsymbol{\nu}}(t_{\mathrm{A}})Y_{\boldsymbol{\nu}}(t_{\mathrm{B}}) - J_{\boldsymbol{\nu}}(t_{\mathrm{B}})Y_{\boldsymbol{\nu}}(t_{\mathrm{A}}) \quad \text{and} \quad \widehat{p_{\boldsymbol{\nu}}} = \mathrm{Diag}(p_{\boldsymbol{\nu}}, p_{\boldsymbol{\nu}}). \tag{4.224}$$

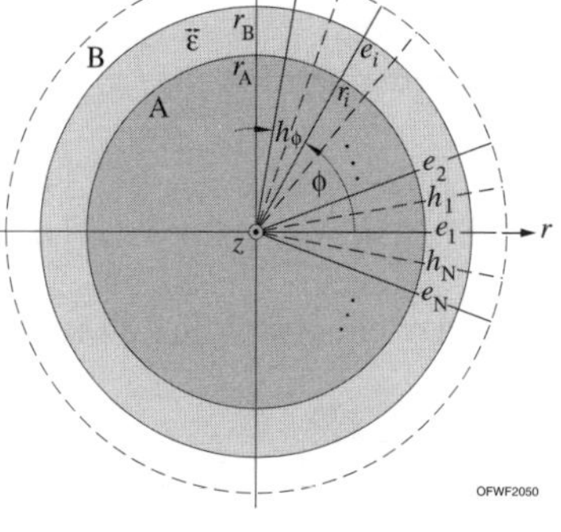

Fig. 4.24 Circular fibre cross-section with discretisation lines in radial direction

Here, the hat (^) above a quantity indicates that we have a block diagonal matrix composed from two matrices without hats.

By differentiating eq. (4.221) with respect to the argument and introducing eq. (4.223), we obtain:

$$\frac{\mathrm{d}}{\mathrm{d}t} \begin{bmatrix} \overline{\mathbf{F}}_{z\mathrm{A}} \\ \overline{\mathbf{F}}_{z\mathrm{B}} \end{bmatrix} = \widehat{p_{\boldsymbol{\nu}}}^{-1} \begin{bmatrix} r_{\boldsymbol{\nu}} & \frac{2}{\pi t_{\mathrm{A}}} \boldsymbol{I} \\ -\frac{2}{\pi t_{\mathrm{B}}} \boldsymbol{I} & q_{\boldsymbol{\nu}} \end{bmatrix} \begin{bmatrix} \overline{\mathbf{F}}_{z\mathrm{A}} \\ \overline{\mathbf{F}}_{z\mathrm{B}} \end{bmatrix} \tag{4.225}$$

Here, as in eq. (4.189), we used cross-products according to definitions 9.1.32 in [14]:

$$\begin{aligned} r_\nu &= J'_\nu(t_\mathrm{A})Y_\nu(t_\mathrm{B}) - Y'_\nu(t_\mathrm{A})J_\nu(t_\mathrm{B}) \\ q_\nu &= J_\nu(t_\mathrm{A})Y'_\nu(t_\mathrm{B}) - J'_\nu(t_\mathrm{B})Y_\nu(t_\mathrm{A}) \\ s_\nu &= J'_\nu(t_\mathrm{A})Y'_\nu(t_\mathrm{B}) - J'_\nu(t_\mathrm{B})Y'_\nu(t_\mathrm{A}) \\ \boldsymbol{W}_P &= J_\nu(t_\mathrm{P})Y'_\nu(t_\mathrm{P}) - J'_\nu(t_\mathrm{P})Y_\nu(t_\mathrm{P}) = \frac{2}{\pi}t_\mathrm{P}^{-1}\boldsymbol{I} \end{aligned} \tag{4.226}$$

which we normalise according to:

$$\overline{r}_\nu = t_\mathrm{A} r_\nu \qquad \overline{q}_\nu = t_\mathrm{B} q_\nu \qquad \overline{s}_\nu = t_\mathrm{A} t_\mathrm{B} s_\nu \tag{4.227}$$

All these equations are very similar to those in the previous section. As we mentioned before, the order of the Bessel functions is different here, while we have a constant argument. This was the other way around before. The normalisation is always connected with one of the derivatives of the Bessel functions. Instead of eq. (4.220) we can now write:

$$\varepsilon_{dt}\begin{bmatrix} \overline{r}_\mathrm{A}\overline{\mathbf{E}}_{\phi\mathrm{A}} \\ \overline{r}_\mathrm{B}\overline{\mathbf{E}}_{\phi\mathrm{B}} \\ \overline{r}_\mathrm{A}\overline{\mathbf{H}}_{\phi\mathrm{A}} \\ \overline{r}_\mathrm{B}\overline{\mathbf{H}}_{\phi\mathrm{B}} \end{bmatrix} = \begin{bmatrix} \widehat{\widetilde{\boldsymbol{\delta}}}_e & \mu_{rt}\boldsymbol{\Lambda}_h \\ \varepsilon_{rt}\boldsymbol{\Lambda}_e & -\widehat{\widetilde{\boldsymbol{\delta}}}_h \end{bmatrix} \begin{bmatrix} -\mathrm{j}\overline{\mathbf{E}}_{z\mathrm{A}} \\ -\mathrm{j}\overline{\mathbf{E}}_{z\mathrm{B}} \\ \mathrm{j}\overline{\mathbf{H}}_{z\mathrm{A}} \\ \mathrm{j}\overline{\mathbf{H}}_{z\mathrm{B}} \end{bmatrix} \tag{4.228}$$

where the $\boldsymbol{\Lambda}$ and $\widehat{\widetilde{\boldsymbol{\delta}}}$ matrices are given by:

$$\boldsymbol{\Lambda} = \widehat{p}_\nu^{-1}\begin{bmatrix} \overline{r}_\nu & \frac{2}{\pi}\boldsymbol{I} \\ -\frac{2}{\pi}\boldsymbol{I} & \overline{q}_\nu \end{bmatrix} \qquad \widehat{\widetilde{\boldsymbol{\delta}}} = \begin{bmatrix} \widetilde{\boldsymbol{\delta}} & \boldsymbol{0} \\ \boldsymbol{0} & \widetilde{\boldsymbol{\delta}} \end{bmatrix} \tag{4.229}$$

The inversion of matrix $\boldsymbol{\Lambda}$ yields, with 9.1.34 in [14]:

$$\left(\widehat{p}_\nu^{-1}\begin{bmatrix} \overline{r}_\nu & \frac{2}{\pi}\boldsymbol{I} \\ -\frac{2}{\pi}\boldsymbol{I} & \overline{q}_\nu \end{bmatrix}\right)^{-1} = \widehat{\overline{s}}_\nu^{-1}\begin{bmatrix} \overline{q}_\nu & -\frac{2}{\pi}\boldsymbol{I} \\ \frac{2}{\pi}\boldsymbol{I} & \overline{r}_\nu \end{bmatrix} \tag{4.230}$$

Therefore, eq. (4.228) takes the following form:

$$\begin{bmatrix} \mathrm{j}\overline{\mathbf{H}}_{z\mathrm{A}} \\ \mathrm{j}\overline{\mathbf{H}}_{z\mathrm{B}} \\ \overline{r}_\mathrm{A}\overline{\mathbf{H}}_{\phi\mathrm{A}} \\ \overline{r}_\mathrm{B}\overline{\mathbf{H}}_{\phi\mathrm{B}} \end{bmatrix} = \begin{bmatrix} \varepsilon_{dt}\widehat{\overline{s}}_{\nu_h}^{-1}\begin{bmatrix} \overline{q}_{\nu_h} & -\frac{2}{\pi}\boldsymbol{I}_h \\ \frac{2}{\pi}\boldsymbol{I}_h & \overline{r}_{\nu_h} \end{bmatrix} & -\widehat{\overline{s}}_{\nu_h}^{-1}\begin{bmatrix} \overline{q}_{\nu_h}\widetilde{\boldsymbol{\delta}}_e & -\frac{2}{\pi}\widetilde{\boldsymbol{\delta}}_e \\ \frac{2}{\pi}\widetilde{\boldsymbol{\delta}}_e & \overline{r}_{\nu_h}\widetilde{\boldsymbol{\delta}}_e \end{bmatrix} \\ -\widehat{\overline{s}}_{\nu_e}^{-1}\begin{bmatrix} \widetilde{\boldsymbol{\delta}}_h\overline{q}_{\nu_h} & -\frac{2}{\pi}\widetilde{\boldsymbol{\delta}}_h \\ \frac{2}{\pi}\widetilde{\boldsymbol{\delta}}_h & \widetilde{\boldsymbol{\delta}}_h\overline{r}_{\nu_h} \end{bmatrix} & \begin{bmatrix} \boldsymbol{\gamma}_{\mathrm{E}_1} & \boldsymbol{\alpha}_\mathrm{E} \\ -\boldsymbol{\alpha}_\mathrm{E} & \boldsymbol{\gamma}_{\mathrm{E}_2} \end{bmatrix} \end{bmatrix} \begin{bmatrix} \overline{r}_\mathrm{A}\overline{\mathbf{E}}_{\phi\mathrm{A}} \\ \overline{r}_\mathrm{B}\overline{\mathbf{E}}_{\phi\mathrm{B}} \\ -\mathrm{j}\overline{\mathbf{E}}_{z\mathrm{A}} \\ -\mathrm{j}\overline{\mathbf{E}}_{z\mathrm{B}} \end{bmatrix} \tag{4.231}$$

with $\widetilde{s}_\nu = \mu_{rt}\overline{s}_\nu$ and:

$$\varepsilon_{dt}\begin{bmatrix} \gamma_{\mathrm{E}_1} & \boldsymbol{\alpha}_\mathrm{E} \\ -\boldsymbol{\alpha}_\mathrm{E} & \gamma_{\mathrm{E}_2} \end{bmatrix} = \varepsilon_{rt}\widehat{p}_{\boldsymbol{\nu}_e}^{-1}\begin{bmatrix} \overline{r}_{\boldsymbol{\nu}_e} & \frac{2}{\pi}\boldsymbol{I}_e \\ -\frac{2}{\pi}\boldsymbol{I}_e & \overline{q}_{\boldsymbol{\nu}_e} \end{bmatrix} - \varepsilon_{re}\widehat{\widetilde{s}}_{\boldsymbol{\nu}_e}^{-1}\begin{bmatrix} \overline{q}_{\boldsymbol{\nu}_e}\overline{\boldsymbol{\lambda}}_e^2 & -\frac{2}{\pi}\overline{\boldsymbol{\lambda}}_e^2 \\ \frac{2}{\pi}\overline{\boldsymbol{\lambda}}_e^2 & \overline{r}_{\boldsymbol{\nu}_e}\overline{\boldsymbol{\lambda}}_e^2 \end{bmatrix} \tag{4.232}$$

or:

$$\begin{aligned} \gamma_{\mathrm{E}_1} &= \varepsilon_{dt}^{-1}(\varepsilon_{rt}\overline{r}_{\boldsymbol{\nu}_e}p_{\boldsymbol{\nu}_e}^{-1} - \varepsilon_{re}\overline{q}_{\boldsymbol{\nu}_e}\widetilde{s}_{\boldsymbol{\nu}_e}^{-1}\overline{\boldsymbol{\lambda}}_e^2) \\ \gamma_{\mathrm{E}_2} &= \varepsilon_{dt}^{-1}(\varepsilon_{rt}\overline{q}_{\boldsymbol{\nu}_e}p_{\boldsymbol{\nu}_e}^{-1} - \varepsilon_{re}\overline{r}_{\boldsymbol{\nu}_e}\widetilde{s}_{\boldsymbol{\nu}_e}^{-1}\overline{\boldsymbol{\lambda}}_e^2) \\ \boldsymbol{\alpha}_\mathrm{E} &= \frac{2}{\pi}\varepsilon_{dt}^{-1}(\varepsilon_{rt}p_{\boldsymbol{\nu}_e}^{-1} + \varepsilon_{re}\widetilde{s}_{\boldsymbol{\nu}_e}^{-1}\overline{\boldsymbol{\lambda}}_e^2) \end{aligned} \tag{4.233}$$

For this formulation we used the relations $\boldsymbol{\delta}_e\boldsymbol{u}_e = \boldsymbol{u}_h\boldsymbol{\delta}_e$ and $\boldsymbol{\delta}_h\boldsymbol{u}_h = \boldsymbol{u}_e\boldsymbol{\delta}_h$ [17]. In these relations, $\boldsymbol{u}_{e,h}$ is one of the diagonal matrices, e.g. $r_{\boldsymbol{\nu}}, p_{\boldsymbol{\nu}}, q_{\boldsymbol{\nu}}, s_{\boldsymbol{\nu}}$.

4.4.3.5 Transformation in homogeneous layers

The last step in our derivation is the rewriting of eq. (4.231). With the definitions:

$$\overline{\mathbf{H}}_{\mathrm{A,B}} = \begin{bmatrix} \mathrm{j}\overline{\mathbf{H}}_{z\mathrm{A,B}} \\ \overline{r}_{\mathrm{A,B}}\overline{\mathbf{H}}_{\phi\mathrm{A,B}} \end{bmatrix} \qquad \overline{\mathbf{E}}_{\mathrm{A,B}} = \begin{bmatrix} \overline{r}_{\mathrm{A,B}}\overline{\mathbf{E}}_{\phi\mathrm{A,B}} \\ -\mathrm{j}\overline{\mathbf{E}}_{z\mathrm{A,B}} \end{bmatrix} \tag{4.234}$$

we obtain the following shortened form:

$$\begin{bmatrix} \overline{\mathbf{H}}_\mathrm{A} \\ -\overline{\mathbf{H}}_\mathrm{B} \end{bmatrix} = \begin{bmatrix} \overline{\boldsymbol{y}}_{1\mathrm{A}} & \overline{\boldsymbol{y}}_2 \\ \overline{\boldsymbol{y}}_2 & \overline{\boldsymbol{y}}_{1\mathrm{B}} \end{bmatrix}\begin{bmatrix} \overline{\mathbf{E}}_\mathrm{A} \\ \overline{\mathbf{E}}_\mathrm{B} \end{bmatrix} \tag{4.235}$$

where ($\widetilde{s}_{\boldsymbol{\nu}} = \mu_{rt}\overline{s}_{\boldsymbol{\nu}}$):

$$\overline{\boldsymbol{y}}_{1\mathrm{A}} = \begin{bmatrix} \varepsilon_{dt}\overline{q}_{\boldsymbol{\nu}_h}\widetilde{s}_{\boldsymbol{\nu}_h}^{-1} & -\overline{q}_{\boldsymbol{\nu}_h}\widetilde{s}_{\boldsymbol{\nu}_h}^{-1}\widetilde{\boldsymbol{\delta}}_e \\ -\widetilde{\boldsymbol{\delta}}_h\overline{q}_{\boldsymbol{\nu}_h}\widetilde{s}_{\boldsymbol{\nu}_h}^{-1} & \gamma_{\mathrm{E}_1} \end{bmatrix} \qquad \overline{\boldsymbol{y}}_2 = \begin{bmatrix} -\frac{2}{\pi}\varepsilon_{dt}\widetilde{s}_{\boldsymbol{\nu}_h}^{-1} & \frac{2}{\pi}\widetilde{s}_{\boldsymbol{\nu}_h}^{-1}\widetilde{\boldsymbol{\delta}}_e \\ \frac{2}{\pi}\widetilde{\boldsymbol{\delta}}_h\widetilde{s}_{\boldsymbol{\nu}_h}^{-1} & \boldsymbol{\alpha}_\mathrm{E} \end{bmatrix}$$

$$\overline{\boldsymbol{y}}_{1\mathrm{B}} = \begin{bmatrix} -\varepsilon_{dt}\overline{r}_{\boldsymbol{\nu}_h}\widetilde{s}_{\boldsymbol{\nu}_h}^{-1} & \overline{r}_{\boldsymbol{\nu}_h}\widetilde{s}_{\boldsymbol{\nu}_h}^{-1}\widetilde{\boldsymbol{\delta}}_e \\ \widetilde{\boldsymbol{\delta}}_h\overline{r}_{\boldsymbol{\nu}_h}\widetilde{s}_{\boldsymbol{\nu}_h}^{-1} & -\gamma_{\mathrm{E}_2} \end{bmatrix} \tag{4.236}$$

With $\overline{\mathbf{H}}_\mathrm{A} = \overline{\mathbf{Y}}_\mathrm{A}\overline{\mathbf{E}}_\mathrm{A}$ and $\overline{\mathbf{H}}_\mathrm{B} = \overline{\mathbf{Y}}_\mathrm{B}\overline{\mathbf{E}}_\mathrm{B}$, the following formulas for the transformation of admittances are obtained:

$$\overline{\mathbf{Y}}_\mathrm{A} = \overline{\boldsymbol{y}}_{1\mathrm{A}} - \overline{\boldsymbol{y}}_2(\overline{\boldsymbol{y}}_{1\mathrm{B}} + \overline{\mathbf{Y}}_\mathrm{B})^{-1}\overline{\boldsymbol{y}}_2 \tag{4.237}$$

$$-\overline{\mathbf{Y}}_\mathrm{B} = \overline{\boldsymbol{y}}_{1\mathrm{B}} - \overline{\boldsymbol{y}}_2(\overline{\boldsymbol{y}}_{1\mathrm{A}} + (-\overline{\mathbf{Y}}_\mathrm{A}))^{-1}\overline{\boldsymbol{y}}_2 \tag{4.238}$$

From eq. (4.235) we can also derive transfer matrix expressions:

$$\begin{bmatrix}\overline{\mathbf{E}}_{\mathrm{B}}\\ \overline{\mathbf{H}}_{\mathrm{B}}\end{bmatrix} = \begin{bmatrix}\overline{\boldsymbol{V}}_{1\mathrm{A}} & \overline{\mathbf{Z}}_{0\mathrm{A}}\\ \overline{\mathbf{Y}}_{0\mathrm{A}} & \overline{\boldsymbol{V}}_{2\mathrm{A}}\end{bmatrix}\begin{bmatrix}\overline{\mathbf{E}}_{\mathrm{A}}\\ \overline{\mathbf{H}}_{\mathrm{A}}\end{bmatrix} = \begin{bmatrix}-\boldsymbol{y}_2^{-1}\boldsymbol{y}_{1\mathrm{A}} & \boldsymbol{y}_2^{-1}\\ \boldsymbol{y}_{1\mathrm{B}}\boldsymbol{y}_2^{-1}\boldsymbol{y}_{1\mathrm{A}} - \boldsymbol{y}_2 & -\boldsymbol{y}_{1\mathrm{B}}\boldsymbol{y}_2^{-1}\end{bmatrix}\begin{bmatrix}\overline{\mathbf{E}}_{\mathrm{A}}\\ \overline{\mathbf{H}}_{\mathrm{A}}\end{bmatrix} \tag{4.239}$$

$$\begin{bmatrix}\overline{\mathbf{E}}_{\mathrm{A}}\\ \overline{\mathbf{H}}_{\mathrm{A}}\end{bmatrix} = \begin{bmatrix}\overline{\boldsymbol{V}}_{1\mathrm{B}} & \overline{\mathbf{Z}}_{0\mathrm{B}}\\ \overline{\mathbf{Y}}_{0\mathrm{B}} & \overline{\boldsymbol{V}}_{2\mathrm{B}}\end{bmatrix}\begin{bmatrix}\overline{\mathbf{E}}_{\mathrm{B}}\\ \overline{\mathbf{H}}_{\mathrm{B}}\end{bmatrix} = \begin{bmatrix}-\boldsymbol{y}_2^{-1}\boldsymbol{y}_{1\mathrm{B}} & -\boldsymbol{y}_2^{-1}\\ \boldsymbol{y}_2 - \boldsymbol{y}_{1\mathrm{A}}\boldsymbol{y}_2^{-1}\boldsymbol{y}_{1\mathrm{B}} & -\boldsymbol{y}_{1\mathrm{A}}\boldsymbol{y}_2^{-1}\end{bmatrix}\begin{bmatrix}\overline{\mathbf{E}}_{\mathrm{B}}\\ \overline{\mathbf{H}}_{\mathrm{B}}\end{bmatrix} \tag{4.240}$$

The submatrices in eq. (4.239) are determined as:

$$\begin{aligned}
&\overline{\boldsymbol{V}}_{1\mathrm{A}} = \frac{\pi}{2}\begin{bmatrix}\overline{q}_{\boldsymbol{\nu}_h} & -\varepsilon_{dt}^{-1}(\overline{r}_{\boldsymbol{\nu}_h} + \overline{q}_{\boldsymbol{\nu}_h})\widetilde{\boldsymbol{\delta}}_e\\ \boldsymbol{0} & -\overline{r}_{\boldsymbol{\nu}_e}\end{bmatrix} \quad \overline{\boldsymbol{V}}_{2\mathrm{A}} = \frac{\pi}{2}\begin{bmatrix}-\overline{r}_{\boldsymbol{\nu}_h} & \boldsymbol{0}\\ \varepsilon_{dt}^{-1}(\overline{q}_{\boldsymbol{\nu}_e} + \overline{r}_{\boldsymbol{\nu}_e})\widetilde{\boldsymbol{\delta}}_h & \overline{q}_{\boldsymbol{\nu}_e}\end{bmatrix}\\
&\overline{\mathbf{Z}}_{0\mathrm{A}} = \frac{\pi}{2\varepsilon_{rt}}\begin{bmatrix}-\varepsilon_{dt}^{-1}(\mu_{rt}\varepsilon_{rt}\overline{s}_{\boldsymbol{\nu}_h} + \varepsilon_{re}p_{\boldsymbol{\nu}_h}\overline{\boldsymbol{\lambda}}_h^2) & p_{\boldsymbol{\nu}_h}\widetilde{\boldsymbol{\delta}}_e\\ p_{\boldsymbol{\nu}_e}\widetilde{\boldsymbol{\delta}}_h & \varepsilon_{dt}p_{\boldsymbol{\nu}_e}\end{bmatrix}\\
&\overline{\mathbf{Y}}_{0\mathrm{A}} = \frac{\pi}{2\mu_{rt}}\begin{bmatrix}\varepsilon_{dt}p_{\boldsymbol{\nu}_h} & -p_{\boldsymbol{\nu}_h}\widetilde{\boldsymbol{\delta}}_e\\ -p_{\boldsymbol{\nu}_e}\widetilde{\boldsymbol{\delta}}_h & -\varepsilon_{dt}^{-1}(\varepsilon_{rt}\mu_{rt}\overline{s}_{\boldsymbol{\nu}_e} + \varepsilon_{re}p_{\boldsymbol{\nu}_e}\overline{\boldsymbol{\lambda}}_e^2)\end{bmatrix}
\end{aligned} \tag{4.241}$$

and for eq. (4.240):

$$\begin{aligned}
&\overline{\boldsymbol{V}}_{1\mathrm{B}} = \frac{\pi}{2}\begin{bmatrix}-\overline{r}_{\boldsymbol{\nu}_h} & \varepsilon_{dt}^{-1}(\overline{q}_{\boldsymbol{\nu}_h} + r_{\boldsymbol{\nu}_h})\widetilde{\boldsymbol{\delta}}_e\\ \boldsymbol{0} & \overline{q}_{\boldsymbol{\nu}_e}\end{bmatrix} \quad \overline{\boldsymbol{V}}_{2\mathrm{B}} = \frac{\pi}{2}\begin{bmatrix}\overline{q}_{\boldsymbol{\nu}_h} & \boldsymbol{0}\\ -\varepsilon_{dt}^{-1}(\overline{q}_{\boldsymbol{\nu}_e} + \overline{r}_{\boldsymbol{\nu}_e})\widetilde{\boldsymbol{\delta}}_h & -\overline{r}_{\boldsymbol{\nu}_e}\end{bmatrix}\\
&\overline{\mathbf{Z}}_{0\mathrm{B}} = \frac{\pi}{2\varepsilon_{rt}}\begin{bmatrix}\varepsilon_{dt}^{-1}(\mu_{rt}\varepsilon_{rt}\overline{s}_{\boldsymbol{\nu}_h} + \varepsilon_{re}p_{\boldsymbol{\nu}_h}\overline{\boldsymbol{\lambda}}_h^2) & -p_{\boldsymbol{\nu}_h}\widetilde{\boldsymbol{\delta}}_e\\ -p_{\boldsymbol{\nu}e}\widetilde{\boldsymbol{\delta}}_h & -\varepsilon_{dt}p_{\boldsymbol{\nu}e}\end{bmatrix}\\
&\overline{\mathbf{Y}}_{0\mathrm{B}} = \frac{\pi}{2\mu_{rt}}\begin{bmatrix}-\varepsilon_{dt}p_{\boldsymbol{\nu}_h} & p_{\boldsymbol{\nu}_h}\widetilde{\boldsymbol{\delta}}_e\\ p_{\boldsymbol{\nu}_e}\widetilde{\boldsymbol{\delta}}_h & \varepsilon_{dt}^{-1}(\varepsilon_{rt}\mu_{rt}\overline{s}_{\boldsymbol{\nu}_e} + \varepsilon_{re}p_{\boldsymbol{\nu}_e}\overline{\boldsymbol{\lambda}}_e^2)\end{bmatrix}
\end{aligned} \tag{4.242}$$

The radial field components can be obtained from eq. (4.218).

4.4.3.6 Centre and outer regions

For the centre and outer regions we derive separate equations. In the centre region only the first term of eq. (4.221) can be used, because the second one approaches infinity for $r \to 0$. In the outer region we must use the modified Bessel function $K_{\boldsymbol{\nu}}$ to assure the correct decrease of the fields. Instead of eq. (4.225) we obtain:

$$\frac{\mathrm{d}}{\mathrm{d}t}\overline{\mathbf{F}}_{z\mathrm{A,B}} = R'_{\boldsymbol{\nu}}R_{\boldsymbol{\nu}}^{-1}\overline{\mathbf{F}}_{z\mathrm{A,B}} \tag{4.243}$$

where R_ν is equal to J_ν (K_ν) in the inner (outer) region. Instead of eq. (4.228) we obtain now, with $\boldsymbol{\Lambda}^{\mathrm{C}}_{e,h} = t_{\mathrm{C}} R'_{\boldsymbol{\nu}_{e,h}}(t_{\mathrm{C}}) R^{-1}_{\boldsymbol{\nu}_{e,h}}(t_{\mathrm{C}})(\mathrm{C} = \mathrm{A}, \mathrm{B})$:

$$\varepsilon_{dt} \begin{bmatrix} \overline{r}_{\mathrm{C}} \overline{\mathbf{E}}_{\phi \mathrm{C}} \\ \overline{r}_{\mathrm{C}} \overline{\mathbf{H}}_{\phi \mathrm{C}} \end{bmatrix} = \begin{bmatrix} \widetilde{\boldsymbol{\delta}}_e & \mu_{rt} \boldsymbol{\Lambda}^{\mathrm{C}}_h \\ \varepsilon_{rt} \boldsymbol{\Lambda}^{\mathrm{C}}_e & -\widetilde{\boldsymbol{\delta}}_h \end{bmatrix} \begin{bmatrix} -\mathrm{j} \overline{\mathbf{E}}_{z\mathrm{C}} \\ \mathrm{j} \overline{\mathbf{H}}_{z\mathrm{C}} \end{bmatrix} \tag{4.244}$$

This is transformed to:

$$\begin{bmatrix} \mathrm{j} \overline{\mathbf{H}}_{z\mathrm{C}} \\ \overline{r}_{\mathrm{C}} \overline{\mathbf{H}}_{\phi \mathrm{C}} \end{bmatrix} = \begin{bmatrix} \varepsilon_{dt} \mu_{rt}^{-1} \boldsymbol{\Lambda}^{-1}_{h\mathrm{C}} & -\mu_{rt}^{-1} \boldsymbol{\Lambda}^{-1}_{h\mathrm{C}} \widetilde{\boldsymbol{\delta}}_e \\ -\mu_{rt}^{-1} \widetilde{\boldsymbol{\delta}}_h \boldsymbol{\Lambda}^{-1}_{h\mathrm{C}} & \varepsilon_{dt}^{-1} (\varepsilon_{rt} \boldsymbol{\Lambda}_{e\mathrm{C}} - \varepsilon_{re} \mu_{rt}^{-1} \overline{\boldsymbol{\lambda}}^2_e \boldsymbol{\Lambda}^{-1}_{e\mathrm{C}}) \end{bmatrix} \begin{bmatrix} \overline{r}_{\mathrm{C}} \overline{\mathbf{E}}_{\phi \mathrm{C}} \\ -\mathrm{j} \overline{\mathbf{E}}_{z\mathrm{C}} \end{bmatrix} \tag{4.245}$$

With the definitions in eq. (4.234) we can now write:

$$\overline{\mathbf{H}}_{\mathrm{C}} = \overline{\mathbf{Y}}_{\mathrm{C}} \overline{\mathbf{E}}_{\mathrm{C}} \tag{4.246}$$

where:

$$\overline{\mathbf{Y}}_{\mathrm{C}} = \begin{bmatrix} \varepsilon_{dt} \mu_{rt}^{-1} \boldsymbol{\Lambda}^{-1}_{h\mathrm{C}} & -\mu_{rt}^{-1} \boldsymbol{\Lambda}^{-1}_{h\mathrm{C}} \widetilde{\boldsymbol{\delta}}_e \\ -\mu_{rt}^{-1} \widetilde{\boldsymbol{\delta}}_h \boldsymbol{\Lambda}^{-1}_{h\mathrm{C}} & \varepsilon_{dt}^{-1} (\varepsilon_{rt} \boldsymbol{\Lambda}_{e\mathrm{C}} - \varepsilon_{re} \mu_{rt}^{-1} \overline{\boldsymbol{\lambda}}^2_e \boldsymbol{\Lambda}^{-1}_{e\mathrm{C}}) \end{bmatrix} \tag{4.247}$$

For the radial field components we obtain from eq. (4.218):

$$\varepsilon_{dt} \begin{bmatrix} \overline{r} \overline{\mathbf{E}}_r \\ \overline{r} \widetilde{\mathbf{H}}_r \end{bmatrix} = - \begin{bmatrix} -\sqrt{\varepsilon_{re}}\, t \dfrac{\partial}{\partial t} & \mu_{rt} \overline{\boldsymbol{\delta}}_h \\ \varepsilon_{rt} \overline{\boldsymbol{\delta}}_e & \sqrt{\varepsilon_{re}}\, t \dfrac{\partial}{\partial t} \end{bmatrix} \begin{bmatrix} -\mathrm{j} \mathbf{E}_z \\ \mathrm{j} \widetilde{\mathbf{H}}_z \end{bmatrix} \tag{4.248}$$

At the radius $r = r_{\mathrm{C}}$ ($\rightarrow\ t = t_{\mathrm{C}}$) we obtain:

$$\varepsilon_{dt} \begin{bmatrix} \overline{r}_{\mathrm{C}} \mathbf{E}_{r\mathrm{C}} \\ \overline{r}_{\mathrm{C}} \widetilde{\mathbf{H}}_{r\mathrm{C}} \end{bmatrix} = - \begin{bmatrix} -\sqrt{\varepsilon_{re}} \boldsymbol{\Lambda}^{\mathrm{C}}_e & \mu_{rt} \overline{\boldsymbol{\delta}}_h \\ \varepsilon_{rt} \overline{\boldsymbol{\delta}}_e & \sqrt{\varepsilon_{re}} \boldsymbol{\Lambda}^{\mathrm{C}}_h \end{bmatrix} \begin{bmatrix} -\mathrm{j} \mathbf{E}_{z\mathrm{C}} \\ \mathrm{j} \widetilde{\mathbf{H}}_{z\mathrm{C}} \end{bmatrix} \tag{4.249}$$

4.4.3.7 Azimuthal inhomogeneous layers

For the case of inhomogeneous circular layers as in Fig. 4.3 we use eqs. (4.102)–(4.104) of Section 4.2.5.2. The discretisation in ϕ-direction is performed as shown in Fig. 4.25. As we know, ABCs are put on Dirichlet boundaries. Therefore, with the position of the lines as shown in Fig. 4.25 we could replace the ABCs with magnetic walls. To model metallic walls, we shift the boundaries to the dashed lines.

Then we obtain:

$$\overline{r} \frac{\mathrm{d}}{\mathrm{d}\overline{r}} [\widehat{\mathbf{H}}^{re}] = -[\boldsymbol{R}^{re}_{\mathrm{E}}][\widehat{\mathbf{E}}^{re}] \quad \overline{r} \frac{\mathrm{d}}{\mathrm{d}\overline{r}} [\widehat{\mathbf{E}}^{re}] = -[\boldsymbol{R}^{re}_{\mathrm{H}}][\widehat{\mathbf{H}}^{re}] \tag{4.250}$$

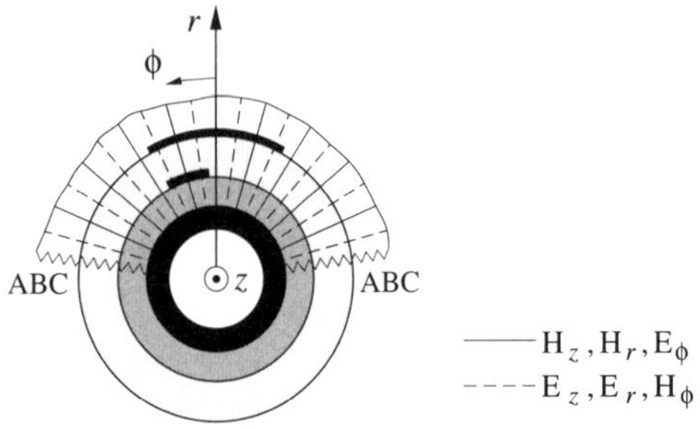

Fig. 4.25 One-dimensional in ϕ-direction discretised circular cross-sections for a microstrip on cylindrical body: — h-lines, – – – e-lines (Reproduced by permission of Springer Netherlands)

where the superscript e symbolise eigenmodes and where:

$$[\boldsymbol{R}_{\mathrm{E}}^{re}] = \begin{bmatrix} \varepsilon_{re}\boldsymbol{\mu}_{rr}^{-1} - \boldsymbol{\epsilon}_{\phi\phi} & \sqrt{\varepsilon_{re}}\boldsymbol{\mu}_{rr}^{-1}\overline{\boldsymbol{D}}_{\phi}^{e} \\ \sqrt{\varepsilon_{re}}\overline{\boldsymbol{D}}_{\phi}^{h}\boldsymbol{\mu}_{rr}^{-1} & \overline{\boldsymbol{D}}_{\phi}^{h}\boldsymbol{\mu}_{rr}^{-1}\overline{\boldsymbol{D}}_{\phi}^{e} + \overline{r}^{2}\boldsymbol{\epsilon}_{zz} \end{bmatrix} \quad [\widehat{\mathbf{E}}^{re}] = \begin{bmatrix} \overline{r}\mathbf{E}_{\phi} \\ -\mathrm{j}\mathbf{E}_{z} \end{bmatrix} \tag{4.251}$$

$$[\boldsymbol{R}_{\mathrm{H}}^{re}] = \begin{bmatrix} \overline{\boldsymbol{D}}_{\phi}^{e}\boldsymbol{\epsilon}_{rr}^{-1}\overline{\boldsymbol{D}}_{\phi}^{h} + \overline{r}^{2}\boldsymbol{\mu}_{zz} & -\sqrt{\varepsilon_{re}}\overline{\boldsymbol{D}}_{\phi}^{e}\boldsymbol{\epsilon}_{rr}^{-1} \\ -\sqrt{\varepsilon_{re}}\boldsymbol{\epsilon}_{rr}^{-1}\overline{\boldsymbol{D}}_{\phi}^{h} & \varepsilon_{re}\boldsymbol{\epsilon}_{rr}^{-1} - \boldsymbol{\mu}_{\phi\phi} \end{bmatrix} \quad [\widehat{\mathbf{H}}^{re}] = \begin{bmatrix} \mathrm{j}\widetilde{\mathbf{H}}_{z} \\ \overline{r}\widetilde{\mathbf{H}}_{\phi} \end{bmatrix} \tag{4.252}$$

For the components in r-direction we obtain:

$$\overline{r}\widetilde{\mathbf{H}}_{r} = -\boldsymbol{\mu}_{rr}^{-1}[\sqrt{\varepsilon_{re}}\boldsymbol{I}\ \overline{\boldsymbol{D}}_{\phi}^{e}][\widehat{\mathbf{E}}^{re}] \quad \overline{r}\mathbf{E}_{r} = \boldsymbol{\epsilon}_{rr}^{-1}[-\overline{\boldsymbol{D}}_{\phi}^{h}\ \sqrt{\varepsilon_{re}}\boldsymbol{I}][\widehat{\mathbf{H}}^{re}] \tag{4.253}$$

To solve the coupled differential equations in eq. (4.250) we can apply the impedance/admittance transformation with finite differences as presented in Section 2.5.3.

4.4.3.8 Modes with azimuthal independent fields

If we examine modes with azimuthal independent fields ($D_\phi = 0$) we do not need to discretise. The fields are determined by using the previous results and by introducing $\lambda = 0$ or $\nu = 0$ and $\delta = 0$. We can now summarise the results in the following tables:

TM$_z$ mode: E_z	TE$_z$ mode: $\widetilde{H}_z$
$\overline{r}\widetilde{H}_\phi = -\varepsilon_{rt}\varepsilon_{dt}^{-1}\overline{r}\frac{\partial}{\partial\overline{r}}(\mathrm{j}E_z)$	$\overline{r}E_\phi = \mu_{rt}\varepsilon_{dt}^{-1}\overline{r}\frac{\partial}{\partial\overline{r}}(\mathrm{j}\widetilde{H}_z)$
$\overline{r}E_r = -\sqrt{\varepsilon_{re}}\varepsilon_{dt}^{-1}\overline{r}\frac{\partial}{\partial\overline{r}}(\mathrm{j}E_z)$	$\overline{r}\widetilde{H}_r = -\sqrt{\varepsilon_{re}}\varepsilon_{dt}^{-1}\overline{r}\frac{\partial}{\partial\overline{r}}(\mathrm{j}\widetilde{H}_z)$

(4.254)

E_z and $\widetilde{H}_z$ have to solve the wave equations (4.208) with $D_\phi = 0$ (or eq. (4.220) with $\boldsymbol{\lambda} = 0$).

The principle solution is given in eq. (4.221). In this case the order of the Bessel functions is zero. The relation of the fields between the two sides of a layer is given as:

$$\varepsilon_{dt}\begin{bmatrix}\overline{r}_{\mathrm{A}}\widetilde{H}_{\phi\mathrm{A}}\\ \overline{r}_{\mathrm{B}}\widetilde{H}_{\phi\mathrm{B}}\end{bmatrix}=\varepsilon_{rt}[\Lambda]\begin{bmatrix}-\mathrm{j}E_{z\mathrm{A}}\\ -\mathrm{j}E_{z\mathrm{B}}\end{bmatrix}\quad\Bigg|\quad\varepsilon_{dt}\begin{bmatrix}\overline{r}_{\mathrm{A}}\widetilde{E}_{\phi\mathrm{A}}\\ \overline{r}_{\mathrm{B}}\widetilde{E}_{\phi\mathrm{B}}\end{bmatrix}=\mu_{rt}[\Lambda]\begin{bmatrix}\mathrm{j}\widetilde{H}_{z\mathrm{A}}\\ \mathrm{j}\widetilde{H}_{z\mathrm{B}}\end{bmatrix}\tag{4.255}$$

where:

$$[\Lambda]=\widehat{p}_0^{\,-1}\begin{bmatrix}\overline{r}_0 & \frac{2}{\pi}\\ -\frac{2}{\pi} & \overline{q}_0\end{bmatrix}\qquad[\Lambda]^{-1}=\widehat{\overline{s}}_0^{\,-1}\begin{bmatrix}\overline{q}_0 & -\frac{2}{\pi}\\ \frac{2}{\pi} & \overline{r}_0\end{bmatrix}\tag{4.256}$$

For the cross-products we use eqs. (4.226) and (4.227). Finally, using the following definitions (C = A or B):

$$E_{\mathrm{C}}=-\mathrm{j}E_{z\mathrm{C}}\quad\widetilde{H}_{\mathrm{C}}=\overline{r}_{\mathrm{C}}\widetilde{H}_{\phi\mathrm{C}}\quad\Big|\quad E_{\mathrm{C}}=\overline{r}_{\mathrm{C}}E_{\phi\mathrm{C}}\quad\widetilde{H}_{\mathrm{C}}=\mathrm{j}\widetilde{H}_{z\mathrm{C}}\tag{4.257}$$

the field relations are described by:

$$\begin{bmatrix}\widetilde{H}_{\mathrm{A}}\\ -\widetilde{H}_{\mathrm{B}}\end{bmatrix}=\begin{bmatrix}y_{1\mathrm{A}} & y_2\\ y_2 & y_{1\mathrm{B}}\end{bmatrix}\begin{bmatrix}E_{\mathrm{A}}\\ E_{\mathrm{B}}\end{bmatrix}\quad\Bigg|\quad\begin{bmatrix}E_{\mathrm{A}}\\ E_{\mathrm{B}}\end{bmatrix}=\begin{bmatrix}z_{1\mathrm{A}} & z_2\\ z_2 & z_{1\mathrm{B}}\end{bmatrix}\begin{bmatrix}\widetilde{H}_{\mathrm{A}}\\ -\widetilde{H}_{\mathrm{B}}\end{bmatrix}\tag{4.258}$$

For the matrix parameters we have:

$$\begin{array}{rcl|rcl}
y_{1\mathrm{A}} &=& \varepsilon_{rt}\varepsilon_{dt}^{-1}\overline{r}_0 p_0^{-1} & z_{1\mathrm{A}} &=& \mu_{rt}\varepsilon_{dt}^{-1}\overline{r}_0 p_0^{-1}\\
y_{1\mathrm{B}} &=& -\varepsilon_{rt}\varepsilon_{dt}^{-1}\overline{q}_0 p_0^{-1} & z_{1\mathrm{B}} &=& -\mu_{rt}\varepsilon_{dt}^{-1}\overline{q}_0 p_0^{-1}\\
y_2 &=& \frac{2}{\pi}\varepsilon_{rt}\varepsilon_{dt}^{-1}p_0^{-1} & z_2 &=& -\frac{2}{\pi}\mu_{rt}\varepsilon_{dt}^{-1}p_0^{-1}\\
z_{1\mathrm{A}} &=& \varepsilon_{dt}\varepsilon_{rt}^{-1}\overline{q}_0\overline{s}_0^{-1} & y_{1\mathrm{A}} &=& \varepsilon_{dt}\mu_{rt}^{-1}\overline{q}_0\overline{s}_0^{-1}\\
z_{1\mathrm{B}} &=& -\varepsilon_{dt}\varepsilon_{rt}^{-1}\overline{r}_0\overline{s}_0^{-1} & y_{1\mathrm{B}} &=& -\varepsilon_{dt}\mu_{rt}^{-1}\overline{r}_0\overline{s}_0^{-1}\\
z_2 &=& \frac{2}{\pi}\varepsilon_{dt}\varepsilon_{rt}^{-1}\overline{s}_0^{-1} & y_2 &=& -\frac{2}{\pi}\varepsilon_{dt}\mu_{rt}^{-1}\overline{s}_0^{-1}
\end{array}\tag{4.259}$$

For the centre and outer regions the following equations are valid:

$$\begin{array}{rl|rl}
\mathrm{j}E_z &= J_0(t)A & \mathrm{j}\widetilde{H}_z &= J_0(t)A\\
\overline{r}\frac{\partial}{\partial\overline{r}}(\mathrm{j}E_z) &= t\frac{\partial}{\partial t}(\mathrm{j}E_z) & \overline{r}\frac{\partial}{\partial\overline{r}}(\mathrm{j}\widetilde{H}_z) &= t\frac{\partial}{\partial t}(\mathrm{j}\widetilde{H}_z)\\
&= tJ_0'(t)/J_0(t)(\mathrm{j}E_z) & &= tJ_0'(t)/J_0(t)(\mathrm{j}\widetilde{H}_z)\\
\overline{r}_{\mathrm{C}}\widetilde{H}_{\phi\mathrm{C}} &= \underbrace{\frac{\varepsilon_{rt}}{\varepsilon_{dt}}t_{\mathrm{C}}\frac{J_0'(t_{\mathrm{C}})}{J_0(t_{\mathrm{C}})}}_{Y_{\mathrm{C}}}(-\mathrm{j}E_{z\mathrm{C}}) & \overline{r}_{\mathrm{C}}E_{\phi\mathrm{C}} &= \underbrace{\frac{\mu_{rt}}{\varepsilon_{dt}}t_{\mathrm{C}}\frac{J_0'(t_{\mathrm{C}})}{J_0(t_{\mathrm{C}})}}_{Z_{\mathrm{C}}}(\mathrm{j}\widetilde{H}_{z\mathrm{C}})\\
\widetilde{H}_{\mathrm{C}} &= Y_{\mathrm{C}}E_{\mathrm{C}} & E_{\mathrm{C}} &= Z_{\mathrm{C}}\widetilde{H}_{\mathrm{C}}
\end{array}\tag{4.260}$$

In the outer region, J_0 must be replaced by K_0.

4.4.3.9 Numerical results

Microstrip on cylindrical body:

First we would like to show the results for a microstrip on cylindrical body (see Fig. 4.26).

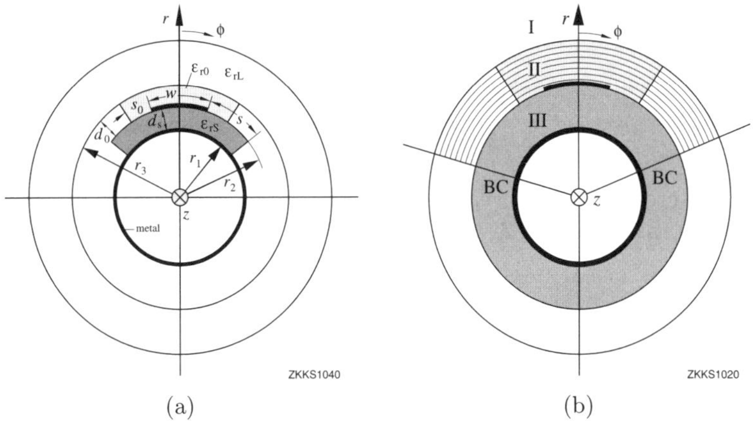

Fig. 4.26 Microstrip on cylindrical body with finite substrate and superstrates (L. A. Greda and R. Pregla, 'Modeling of planar periodic antennas using Floqut's theorem by the Method of Lines', in *15th Int. Conf. Micr. Radar Wir. Comm. MIKON.* © 2004 Institute of Electrical and Electronics Engineers (IEEE))

In Fig. 4.26a the substrate and the superstrate have finite width, which means that those layers are inhomogeneous. The principle of the analysis is sketched in Fig. 4.26b. Here, only the superstrate layer is of finite width. For the homogeneous layer III we use the formulas derived in Section 4.4.3.5 and for the outer region I we use the expressions of Section 4.4.3.6. Layer II is inhomogeneous. Therefore, we divide this layer into a suitable number of sublayers and use the formulas of Section 4.4.3.7 and the FD procedure described in Section 2.5.3. Fig. 4.27 shows the effective permittivity of the microstrip waveguide on a cylindrical body. The substrate is of finite widths. The dashed curves show where a superstrate with the width of the microstrip is included ($s_o = 0.0$). The thickness of the substrate and of the superstrate is 1 mm.

For the next example we analysed circularly homogeneous fibres, as shown in Fig. 4.24. Fig. 4.28 is a plot of the dispersion curve for the fundamental (HE_{11}) and higher-order modes in a fibre with refractive index $n_1 = 1.6$ in the core and $n_2 = 1.5$ in the cladding. We used the normalised frequency V and the normalised propagation constant B, defined by the equations:

$$V = \overline{a}\sqrt{\varepsilon_1 - \varepsilon_2} \quad \text{and} \quad B = \frac{\varepsilon_{re} - \varepsilon_2}{\varepsilon_1 - \varepsilon_2} \tag{4.261}$$

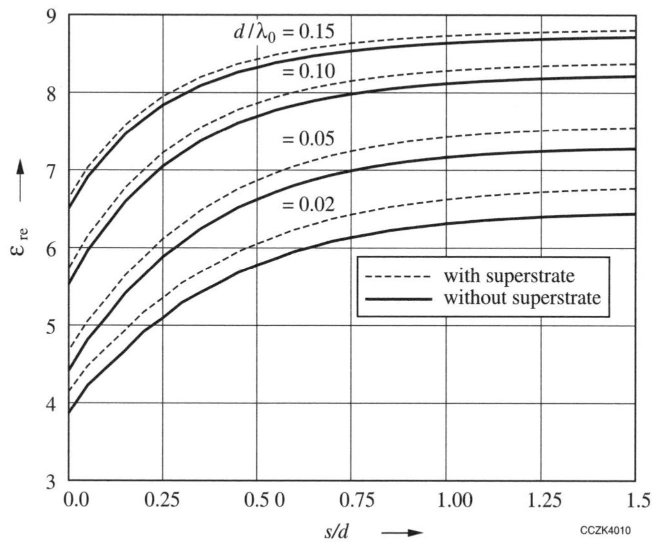

Fig. 4.27 Dispersion diagrams of the fundamental microstrip mode on a cylindrical body and finite substrate; $R = 10$ mm, $\varepsilon_r = 9.7$ (L. A. Greda and R. Pregla, 'Modeling of planar periodic antennas using Floqut's theorem by the Method of Lines', in *15th Int. Conf. Micr. Radar Wir. Comm. MIKON.* © 2004 Institute of Electrical and Electronics Engineers (IEEE))

a is the fibre radius, $\overline{a} = k_0 a$ and $\varepsilon = n^2$. The results obtained by the algorithm described and those of Stuwe [18] agree very well.

For cross-sections without circular symmetry (see e.g. Fig. 4.29), the FD impedance/admittance transformation algorithm described in Section 2.5.3 for azimuthal inhomogeneous layers was very efficient. For example, we wanted to determine results for the horizontal and vertical polarised fundamental modes HE_{11}. The discretisation is shown in Fig. 4.30. Because of the symmetries, we used only a quarter of the cross-section. Normally an e-discretisation line should lie on the interface between two dielectrics. In the case of the problem in Fig. 4.30 this was not possible. Therefore, the values of the permittivities of the line t were chosen according to the interpolation formula:

$$\varepsilon_{rt} = \tfrac{1}{2}(\varepsilon_{r1} + \varepsilon_{r2}) + p(\varepsilon_{r1} - \varepsilon_{r2}) \qquad -0.5 \leq p \leq 0.5 \tag{4.262}$$

In our case, ε_{r1} and ε_{r2} are equal to $\varepsilon_{r1} = n_{cl}^2$ and $\varepsilon_{r2} = n_b^2$, respectively. For p we have to introduce $p = -p_\mathrm{h}$ and $p = p_\mathrm{v}$ for horizontal and vertical polarisation, respectively. Results for the dispersion curves of the fundamental mode in the fibre with the cross-section in Fig. 4.29 are given in Fig. 4.31. The highest curve is identical to the corresponding one in Fig. 4.28. For this

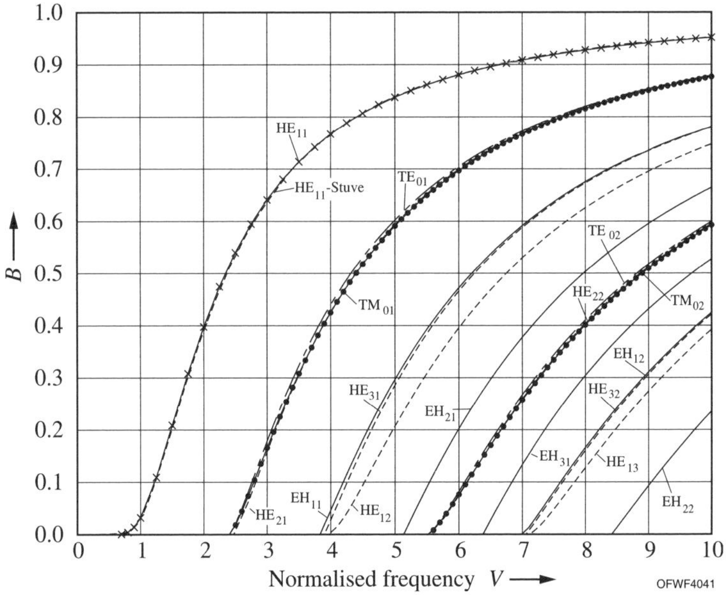

Fig. 4.28 Dispersion diagrams of the fundamental mode HE_{11} and higher-order modes in a fibre with refractive index $n_1 = 1.6$ in the core and $n_2 = 1.5$ in the cladding

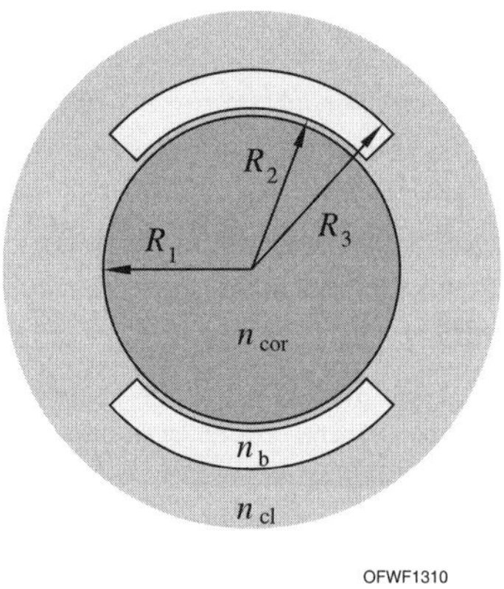

Fig. 4.29 Cross-section of a circular inhomogeneous fibre (Reproduced by permission of Springer Netherlands)

case, the accuracy of the algorithm was checked by introducing $n_b = n_{cl}$. The algorithm is able to incorporate anisotropic material into the analysis.

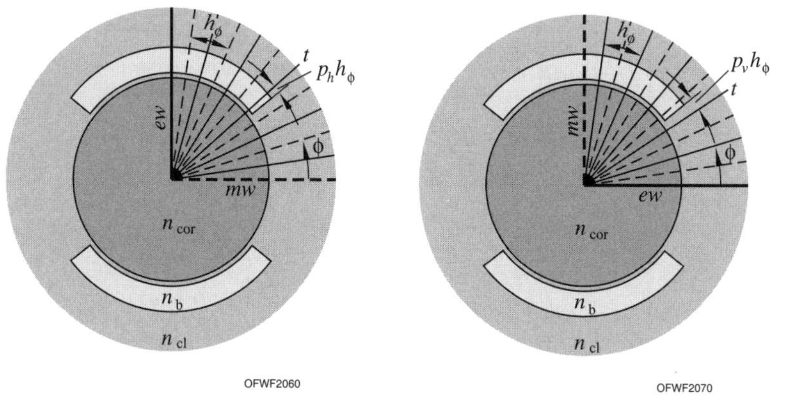

(a) (b)

Fig. 4.30 Cross-section of a circular inhomogeneous fibre with discretisation lines in radial direction. (a) Horizontal and (b) vertical polarisation e-line: — E_z, E_r, H_ϕ, ε_z, ε_r h-line: - - - H_z, H_r, E_ϕ, ε_ϕ

In particular, the layers may be inhomogeneous and an arbitrary anisotropy can be examined with the FD. Results for the fibre sketched in Fig. 4.29 are shown in Fig. 4.32. Here we introduced anisotropic material. In the dashed part of the curve (4) we obtain complex results. The real part of B and the imaginary part ε_{reI} of ε_{re} are given in the insert of Fig. 4.32. The imaginary part is obtained because of loss caused by radiation.

4.4.3.10 Metallic losses

Losses due to metallic strips can be approximately considered as they are shown in [5]. On the surface of metallisation it is assumed that for the tangential fields (subscript t) the following expression is valid:

$$\eta_m \vec{H}_t = -\vec{e}_n \times \vec{E}_t \qquad \eta_m = \frac{1+\mathrm{j}}{\sqrt{2}}\sqrt{\frac{k_o \eta_o}{\kappa}} \tag{4.263}$$

where $\vec{e}_n$ is the normal unit vector, κ is the conductivity and η_m is the surface resistance. This assumption indicates that the penetration depth of the field is minimal compared to the thickness of the strip (high-frequency approximation).

Using analogous definitions to those in eq. (4.234), we can write $\vec{e}_n = \pm\vec{e}_r$ for the outward or inward direction of the surface:

$$\overline{\mathbf{H}}_{\mathrm{M}}^{\circ,i} = \pm\mathbf{Y}_t^s \overline{\mathbf{E}}_{\mathrm{M}}^{\circ,i} \quad \text{with} \quad \mathbf{Y}_t^s = -\mathrm{j}\frac{\eta_o}{\eta_m}\mathrm{Diag}\,(\overline{\rho}^{-1}\boldsymbol{I}, \overline{\rho}\boldsymbol{I}) \tag{4.264}$$

The expressions in eq. (4.264) have to be used in the matching process for the strips in the layer interfaces, instead of the ideal condition $\mathbf{E}_{\mathrm{M}}^i = \mathbf{E}_{\mathrm{M}}^\circ = \mathbf{0}$. The procedure is analogous to that given for planar structures.

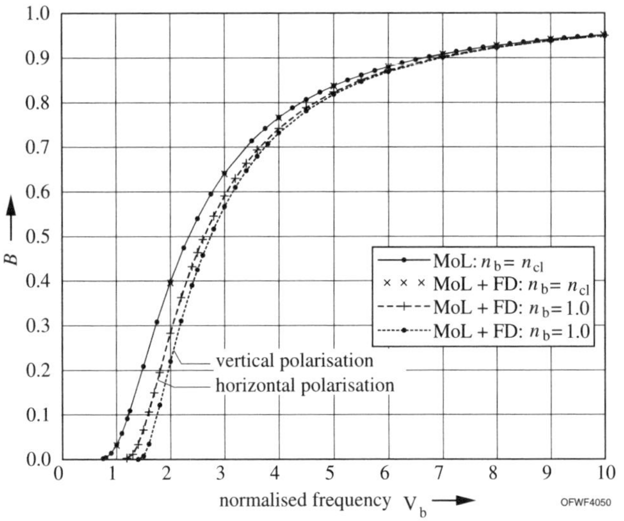

Fig. 4.31 Dispersion diagrams of the fundamental mode HE_{11} in a fibre with cross-section in Fig. 4.29. Refractive indices: $n_{cor} = 1.6$, $n_{cl} = 1.5$ and $n_b = 1.0$. $R_2 = R_1$ (Reproduced by permission of Springer Netherlands)

4.4.4 Eigenmodes of circular waveguides with magnetised ferrite or plasma – discretisation in r-direction

4.4.4.1 Longitudinal magnetised ferrite

Introduction:

In this subsection we would like to determine the eigenmodes in circular hollow waveguides. The geometry of the problem is shown in Fig. 4.33. It is an infinitely long, uniform hollow waveguide of radius a with a perfectly conducting wall. In the centre it is filled with a rod of longitudinally-magnetised ferrite material of radius b where monochromatic waves can propagate.

A cylindrical coordinate system (r, ϕ, z) with z-axis along the geometric axis of the structure is the obvious one to use. So the perfectly-conducting surface of the guide wall is a coordinate surface, described simply by $r = a$.

Discretisation:

For the modes H_{11} and E_{11} we assume $\kappa_e = 0$ and discretise the fields and equations in Section 4.2.2.3 according to Fig. 4.34. We obtain the following equations for the supervectors in eq. (4.31):

$$[\widehat{\mathbf{E}}^z] = \begin{bmatrix} \mathrm{j}\overline{\mathbf{r}}^{\bullet}\mathbf{E}_{\phi} \\ \mathbf{E}_r \end{bmatrix} \qquad [\widehat{\mathbf{H}}^z] = \begin{bmatrix} -\mathrm{j}\widetilde{\mathbf{H}}_r \\ \overline{\mathbf{r}}^{\circ}\widetilde{\mathbf{H}}_{\phi} \end{bmatrix} \tag{4.265}$$

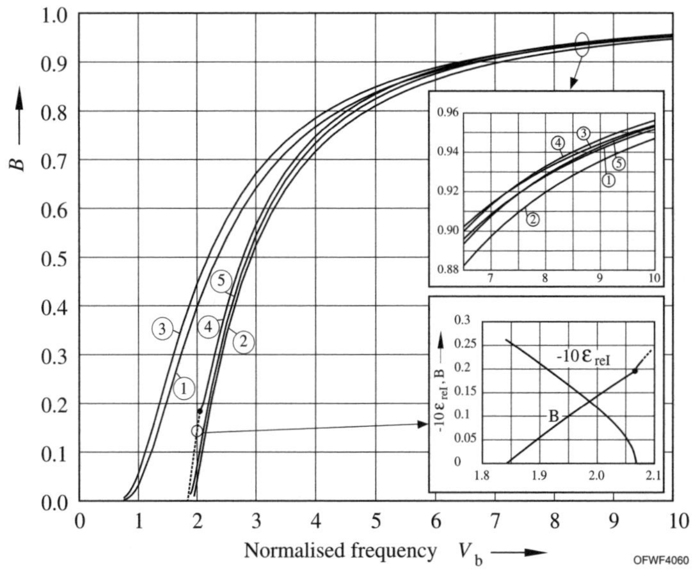

Fig. 4.32 Dispersion diagrams of the fundamental mode HE_{11} (hp = horizontal and vp = vertical polarisation) in a fibre of dimensions as in Fig. 4.29b with refractive indices:
(1) $n_{cor} = 1.6$, $n_{cl} = 1.5$, $n_b = 1.5$
(2) $n_{cor} = 1.6$, $n_{cor_z} = 1.55$, $n_{cl} = 1.5$, $n_b = 1.5$
(3) $n_{cor} = 1.6$, $n_{cor_z} = 1.55$, $n_{cl} = 1.5$, $n_b = 1.55$
(4) $n_{cor} = 1.6$, $n_{cor_z} = 1.55$, $n_{cl} = 1.5$, $n_b = 1.55$, $n_{b_z} = 1.55$ - vp
(5) $n_{cor} = 1.6$, $n_{cor_z} = 1.55$, $n_{cl} = 1.5$, $n_b = 1.55$, $n_{b_z} = 1.55$ - hp
The subscript z always means the value for the z-direction
(Reproduced by permission of Springer Netherlands)

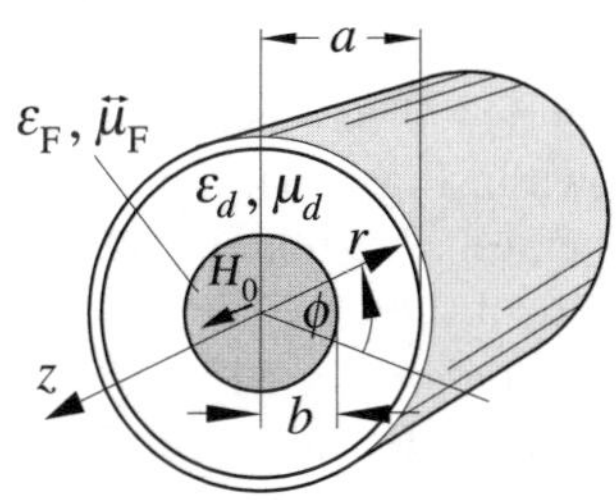

Fig. 4.33 Circular waveguide with longitudinally-magnetised ferrite

The matrices in eqs. (4.32) and (4.33) are:

$$[\boldsymbol{R}_{\mathrm{E}}^{zc}] = \begin{bmatrix} \overline{\boldsymbol{D}}_{\overline{r}}^{\circ}\overline{\boldsymbol{\mu}}_{zz}^{-1}\overline{\boldsymbol{D}}_{\overline{r}}^{\bullet} & m\overline{\boldsymbol{D}}_{\overline{r}}^{\circ}\overline{\boldsymbol{\mu}}_{zz}^{-1} \\ -m\overline{\boldsymbol{\mu}}_{zz}^{-1}\overline{\boldsymbol{D}}_{\overline{r}}^{\bullet} & -m^2\overline{\boldsymbol{\mu}}_{zz}^{-1} \end{bmatrix} \qquad [\widehat{\boldsymbol{\varepsilon}}_{rt}^{zc}] = \begin{bmatrix} \boldsymbol{\epsilon}_{\phi\phi}\overline{\mathbf{r}}^{\bullet-1} & \\ & \boldsymbol{\epsilon}_{rr}\overline{\mathbf{r}}^{\circ} \end{bmatrix} \tag{4.266}$$

$$[\boldsymbol{R}_{\mathrm{H}}^{zc}] = \begin{bmatrix} -m^2\overline{\boldsymbol{\epsilon}}_{zz}^{-1} & -m\overline{\boldsymbol{\epsilon}}_{zz}^{-1}\overline{\boldsymbol{D}}_{\overline{r}}^{\circ} \\ m\overline{\boldsymbol{D}}_{\overline{r}}^{\bullet}\overline{\boldsymbol{\epsilon}}_{zz}^{-1} & \boldsymbol{D}_{\overline{r}}^{\bullet}\overline{\boldsymbol{\epsilon}}_{zz}^{-1}\overline{\boldsymbol{D}}_{\overline{r}}^{\circ} \end{bmatrix} \qquad [\widehat{\boldsymbol{\mu}}_{rt}^{zc}] = \begin{bmatrix} \boldsymbol{\mu}_{rr}\overline{\mathbf{r}}^{\bullet} & -\boldsymbol{M}^{\circ}\boldsymbol{\kappa}_m^{\circ} \\ -\boldsymbol{M}^{\bullet}\boldsymbol{\kappa}_m^{\bullet} & \boldsymbol{\mu}_{\phi\phi}\overline{\mathbf{r}}^{\circ-1} \end{bmatrix} \tag{4.267}$$

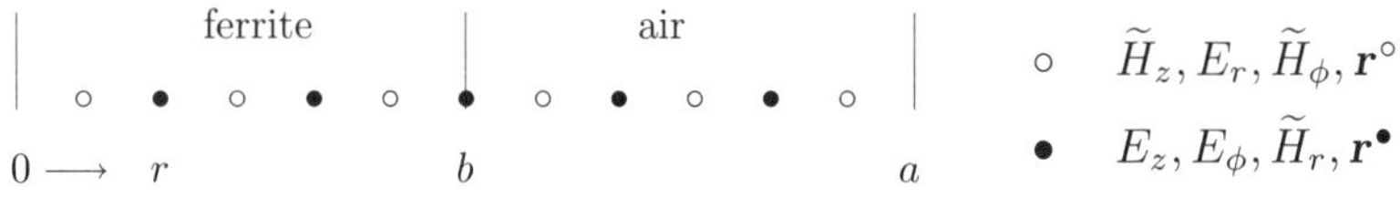

Fig. 4.34 Discretization scheme in case of H_{11} and E_{11} modes

The material parameters (permittivities and permeabilities) are discretised at the same points as the associated field components and are combined in diagonal matrices. The values at interfaces are chosen as arithmetic mean values. The matrices $\boldsymbol{M}^{\circ,\bullet}$ interpolate the values on the $\circ$ points to obtain values on the $\bullet$ points and vice versa. In principle they are given by $\boldsymbol{M}^{\circ,\bullet} = 0.5\boldsymbol{D}^{\circ,\bullet}$ for linear interpolation. Because $\widehat{H}_\phi$ (even with Dirichlet boundary conditions for $\boldsymbol{D}^\bullet-$, see Fig. 4.34) is not equal to zero at $r = 0$, we have to choose $\boldsymbol{M}^\bullet(1,1) = 1$. We have to take special care on the outer side if we have Dirichlet boundary conditions and completely ferrite-filled waveguide (see Section 4.4.3). The longitudinal components are obtained from eq. (4.34):

$$\widetilde{\mathbf{H}}_z = \overline{\boldsymbol{\mu}}_{zz}^{-1}[\overline{\boldsymbol{D}}_{\overline{r}}^{\bullet}\ \ m\boldsymbol{I}^{\circ}][\widehat{\mathbf{E}}^z] \qquad \mathbf{E}_z = -\mathrm{j}\overline{\boldsymbol{\epsilon}}_{zz}^{-1}[m\boldsymbol{I}^{\bullet}\ \ \overline{\boldsymbol{D}}_{\overline{r}}^{\circ}][\widehat{\mathbf{H}}^z] \tag{4.268}$$

where $\overline{\boldsymbol{\epsilon}}_{zz} = \boldsymbol{\epsilon}_{zz}\mathbf{r}^\bullet$ and $\overline{\boldsymbol{\mu}}_{zz} = \boldsymbol{\mu}_{zz}\mathbf{r}^\circ$.

We obtain the modes H_{01} and E_{01} with azimuthal-independent fields with $m = 0$. The matrices are given by:

$$[\boldsymbol{R}_{\mathrm{E}}^z] = \begin{bmatrix} \overline{\boldsymbol{D}}_{\overline{r}}^{\circ}\overline{\boldsymbol{\mu}}_{zz}^{-1}\overline{\boldsymbol{D}}_{\overline{r}}^{\bullet} + \boldsymbol{\epsilon}_{\phi\phi}\overline{\boldsymbol{r}}_{\mathrm{H}}^{\bullet\,-1} & \boldsymbol{0} \\ \boldsymbol{0} & \boldsymbol{\epsilon}_{rr}\overline{\boldsymbol{r}}_{\mathrm{E}}^{\circ} \end{bmatrix} \tag{4.269}$$

$$[\boldsymbol{R}_{\mathrm{H}}^z] = \begin{bmatrix} \boldsymbol{\mu}_{rr}\overline{\boldsymbol{r}}_{\mathrm{H}}^{\bullet} & -\boldsymbol{M}^{\circ}\boldsymbol{\kappa}_m^{\circ} \\ -\boldsymbol{M}^{\bullet}\boldsymbol{\kappa}_m^{\bullet} & \overline{\boldsymbol{D}}_{\overline{r}}^{\bullet}\overline{\boldsymbol{\epsilon}}_{zz}^{-1}\overline{\boldsymbol{D}}_{\overline{r}}^{\circ} + \boldsymbol{\mu}_{\phi\phi}\overline{\boldsymbol{r}}_{\mathrm{E}}^{\circ\,-1} \end{bmatrix} \tag{4.270}$$

In the case of $\kappa_m = 0$, the two modes are completely decoupled. For the E_{01} mode we cannot use the discretisation scheme in Fig. 4.34 any longer. The correct discretisation scheme is shown in Fig. 4.35. Thus, the modes must be calculated with two different discretisation schemes. The scheme for H_{01} is the same as the one shown in Fig. 4.34. For $\kappa \neq 0$ a special interpolation is necessary between the schemes. The matrix $\boldsymbol{M}^\circ$ should interpolate between the $\widetilde{H}_\phi$ values on the $\circ$ points of the E_{01} scheme to obtain the E_ϕ values on the $\bullet$ points of the H_{01} scheme. The matrix $\boldsymbol{M}^\bullet$ interpolates the $\widetilde{H}_r$ values on the $\bullet$ points of the H_{01} scheme to the E_r values on the $\circ$ points of the E_{01} scheme. We assume that the number of $\circ$ points is the same in both schemes. Fig. 4.36 explains the relation between the two relevant systems of the schemes. To obtain the matrix $\boldsymbol{M}^\circ$, we calculate the vectors of distances $\mathbf{d}_l^\circ$ and $\mathbf{d}_r^\circ$ from the $\bullet$ points in Fig. 4.36 to the next $\circ$ points on the left and right sides of the $\bullet$ points, respectively. This can easily be done with the help of the matrices $\mathbf{r}_{\mathrm{E,H}}^{\circ\bullet}$. Using the MATLAB notation, we obtain:

$$\mathbf{d}_l^\circ = \mathbf{r}_{\mathrm{H}}^\bullet - \mathbf{r}_{\mathrm{E}}^\circ(1 : N^\circ - 1) \qquad \mathbf{d}_r^\circ = \mathbf{r}_{\mathrm{E}}^\circ(2 : N^\circ) - \mathbf{r}_{\mathrm{H}}^\bullet \tag{4.271}$$

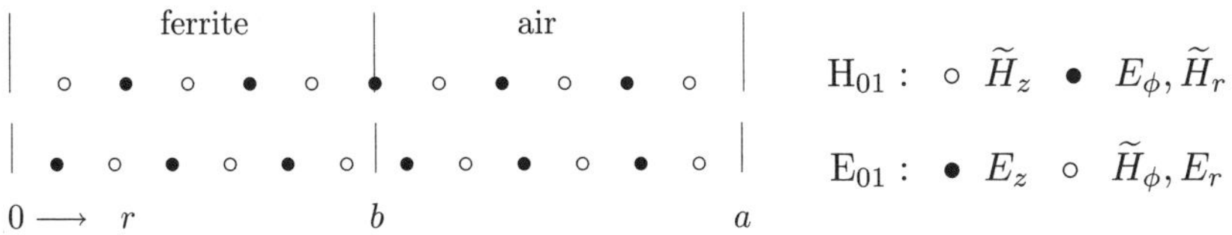

Fig. 4.35 Discretisation schemes in the case of E_{01} and H_{01} modes. The number of ∘ points should be the same in both schemes

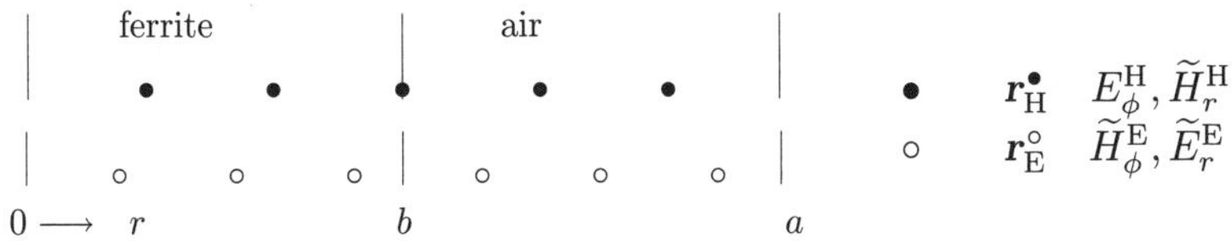

Fig. 4.36 Interpolation between the H_{01} and the E_{01} discretisation schemes

To obtain the matrix $\boldsymbol{M}^\bullet$, we calculate the vectors of distances $\mathbf{d}_l^\bullet$ and $\mathbf{d}_r^\bullet$ from the ∘ points in Fig. 4.36 to the next • points on the left and right sides of the ∘ points, respectively:

$$\mathbf{d}_l^\bullet = [\mathbf{r}_\mathrm{E}^\circ(1); \mathbf{r}_\mathrm{E}^\circ(2 : N^\circ) - \mathbf{r}_\mathrm{H}^\bullet] \qquad \mathbf{d}_r^\bullet = [\mathbf{r}_\mathrm{H}^\bullet - \mathbf{r}_\mathrm{E}^\circ(1 : N^\circ - 1); a - \mathbf{r}_\mathrm{E}^\circ(N^\circ)] \tag{4.272}$$

Note that the summation of the two vectors in eqs. (4.269) and (4.270) results in a vector of discretisation distances in the systems of ∘ and • points. The value f_i for linear interpolation between the values f_l and f_r and the distances d_l and d_r to left and right from f_i to f_l and f_r is given by $f_i = (f_l d_r + f_r d_l)/(d_l + d_r)$. Therefore, we may write for the matrix $\boldsymbol{M}^{\circ\bullet}$:

$$\boldsymbol{M}^\circ = \frac{1}{h^\circ}\begin{bmatrix} \ddots & \ddots & & \\ & \mathbf{d}_r^\circ & \mathbf{d}_l^\circ & \\ & & \ddots & \ddots \end{bmatrix} \qquad \boldsymbol{M}^\bullet = \frac{1}{h^\bullet}\left[\begin{bmatrix} \ddots & & \\ \ddots & \mathbf{d}_l^\bullet & \\ & \mathbf{d}_r^\bullet & \ddots \\ & & \ddots \end{bmatrix}\right] \tag{4.273}$$

where h° and $h^\bullet$ are the discretisation distances. For $\boldsymbol{M}^\bullet$, the two outer columns have to be cancelled. But for $\kappa \neq 0$ and $H_r \neq 0$ at the outer boundary in the case of the completely ferrite-filled waveguide, the value $d_N = \mathbf{r}_\mathrm{E}^\circ(N^\circ) - \mathbf{r}_\mathrm{H}^\bullet(N)$ in the outer column is required for extrapolation. For linear extrapolation, the value at the outer boundary is given as $H_r(a) = 2H_N - H_{N-1}$.

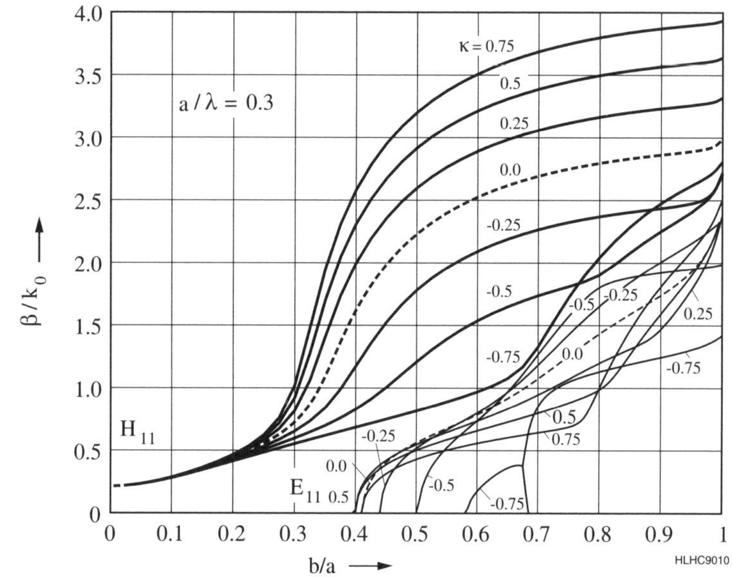

Fig. 4.37 Normalised propagation constant for the H_{11} and E_{11} modes versus b/a in ferrite-filled circular waveguides: $a/\lambda = 0.3$, $\epsilon_r = 10.0$

Numerical results:

Figs. 4.37 and 4.38 show the computed curves for the propagation constants of the fundamental and the first higher-order modes. The propagation constants are obtained from the eigenvalue matrices $\boldsymbol{\Gamma}^2$ of the matrices $\boldsymbol{R}_{\mathrm{H}}^{z}\boldsymbol{R}_{\mathrm{E}}^{z}$ or $\boldsymbol{R}_{\mathrm{E}}^{z}\boldsymbol{R}_{\mathrm{H}}^{z}$. It is interesting to see that in the case of $\kappa = 0$, the curves for the H_{01} and E_{01} modes cross, which is not the case for $\kappa \neq 0$, where the modes are coupled. For comparison see [19].

4.4.4.2 Circular waveguide with azimuthal magnetised ferrite

Introduction:

This subsection aims to present the use of the Method of Lines in numerical analysis of cylindrical wave guiding structures containing remanent ferrite magnetised azimuthally to remanence. The geometry of the problem is shown in Fig. 4.39. We have a waveguide that is completely filled with ferrite. This ferrite has a square hysteresis loop, and is magnetised azimuthally to remanence. A monochromatic wave is propagating. The basic analysis equations are given in Section 4.2.4.1.

Discretisation:

The coupled second-order differential equations for longitudinal components in eq. (4.70) cannot be solved analytically. Even the case of azimuthal

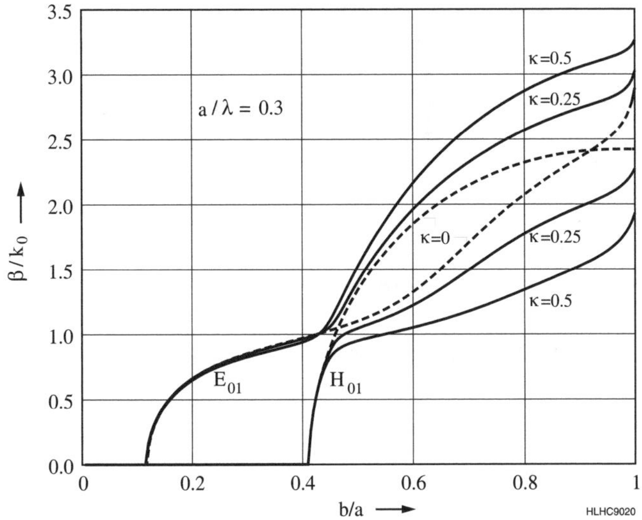

Fig. 4.38 Normalised propagation constant for the H_{01} and E_{01} modes versus b/a in ferrite-filled circular waveguides: $a/\lambda = 0.3$, $\epsilon_r = 10.0$

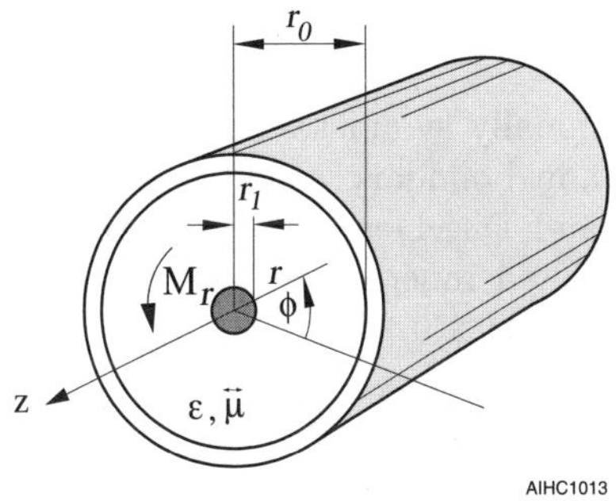

Fig. 4.39 Coaxial waveguide completely filled with magnetised ferrite

independent fields can only be described by higher transcendental functions. Therefore the solution is performed using discretisation in r-direction on two different line systems according to Fig. 4.40.

We use two different discretisation line systems. The system of full lines is marked e. This is because the longitudinal component of the electric field (E_z) is discretised on these lines. The system of dashed lines is marked h. On these lines we determine the longitudinal magnetic component (H_z). The definition of the discretised quantities is as follows:

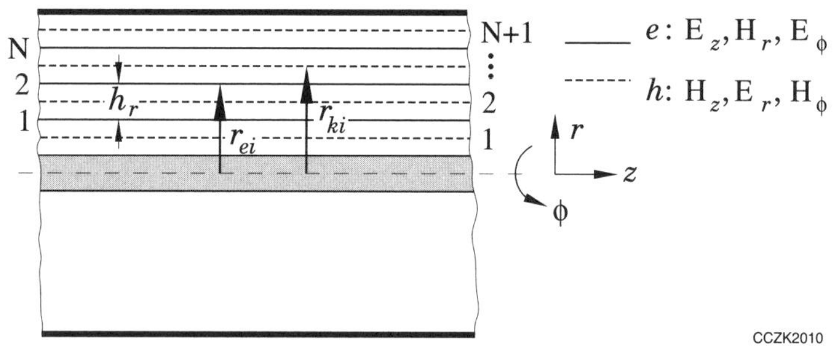

Fig. 4.40 Longitudinal section with discretisation lines

$$\begin{array}{llll} \text{e-line system} & & \text{h-line system} & \\ E_{zn} & \to \mathbf{E}_{zn} & \widetilde{H}_z & \to \widetilde{\mathbf{H}}_z \\ r_n & \to \boldsymbol{r}_e & r_n & \to \boldsymbol{r}_h \\ D_{\overline{r}} & \to \overline{h}_r^{-1}\boldsymbol{D}_e = \overline{\boldsymbol{D}}_e & D_{\overline{r}} & \to \overline{h}_r^{-1}\boldsymbol{D}_h = \overline{\boldsymbol{D}}_h \end{array} \tag{4.274}$$

$\mathbf{E}_{zn}$ and $\widetilde{\mathbf{H}}_z$ are one-dimensional vectors, $\boldsymbol{r}_{e,h}$ are diagonal matrices of the discretised radii $r_{e,hi}$ and the difference operators $\boldsymbol{D}_{e,h}$ are matrices with bandwidth 2. The difference operators should fulfil the following boundary conditions:

(a) If $m = 0$ (rotationally symmetric fields) and TM_z modes, $\boldsymbol{D}_e$ should fulfil the Neumann boundary conditions at the centre of the circular guide ($r = 0$) and Dirichlet boundary conditions at the inner guide surface ($r = r_1$) and inner guide wall ($r = r_0$) of the coaxial guide. ($H_\phi(0) \equiv 0$ and $E_z(a) \equiv 0$.)

(b) If $m = 0$ and TE_z modes, $\boldsymbol{D}_h$ should satisfy Neumann boundary conditions:

$$\frac{\partial H_z}{\partial \overline{r}} = \kappa_m \mu_{rr}^{-1} \sqrt{\varepsilon_{re}} H_z \longrightarrow \frac{\partial H_z}{\partial r_n} = \kappa_m f_n \sqrt{\varepsilon_{ren}} H_z \tag{4.275}$$

at guide axis ($r = r_1$) and at inner guide wall ($r = r_0$). ($E_\phi(0) = 0$.)

(c) If $m \neq 0$ (i.e. azimuthally-dependent fields), the difference operator $\boldsymbol{D}_e$ should fulfil Dirichlet boundary conditions at $r = r_1$ and $r = r_0$. This is because the field component E_z must be zero there. The component H_z must also be zero at the guide axis ($r = 0$). Nevertheless, the difference operator $\boldsymbol{D}_h$ should be constructed there for Neumann boundary conditions. Since E_ϕ must be zero at $r = r_1$ and $r = r_0$, the difference operator $\boldsymbol{D}_h$ for the $\mathbf{H}_z$ component has to fulfil Neumann conditions there in view of eq. (4.275). Further details on the construction of $\boldsymbol{D}_h$ are outlined below.

The matrix form of discretised equations for longitudinal field components is:

$$\begin{bmatrix} \overline{\boldsymbol{P}}_e^r + f_n^2(1-\varepsilon_{ren})\boldsymbol{I}_e & -m\kappa_m f_n \mathbf{r}_e^{-1}\boldsymbol{M}_h \\ m\kappa_m f_n \mathbf{r}_h^{-1}\boldsymbol{M}_e & \overline{\boldsymbol{P}}_h^r + f_n^2(\mu_\perp - \varepsilon_{ren})\boldsymbol{I}_h - \kappa_m f_n \sqrt{\varepsilon_{ren}}\,\overline{\mathbf{r}}_h^{-1} \end{bmatrix} \begin{bmatrix} \mathbf{E}_{zn} \\ \widetilde{\mathbf{H}}_z \end{bmatrix} = \mathbf{0} \tag{4.276}$$

with:

$$\begin{aligned} \overline{\boldsymbol{P}}_e^r &= -\mathbf{r}_e^{-1}\overline{\boldsymbol{D}}_e^{\mathrm{t}}\boldsymbol{r}_h\overline{\boldsymbol{D}}_e - m^2\mathbf{r}_e^{-2} \\ \overline{\boldsymbol{P}}_h^r &= -\mathbf{r}_h^{-1}\overline{\boldsymbol{D}}_h^{\circ\mathrm{t}}\mathbf{r}_e^{*}\overline{\boldsymbol{D}}_h^{*} - m^2\mathbf{r}_h^{-2} \end{aligned} \tag{4.277}$$

The matrices $\boldsymbol{M}_e$ and $\boldsymbol{M}_h$ are introduced to facilitate interpolation between the two discretisation line systems. For linear interpolation they are given by $\boldsymbol{M}_{e,h} = 0.5|\boldsymbol{D}_{e,h}|$, where the difference operators $\boldsymbol{D}_{e,h}$ are the known operators for the discretisation of the first derivatives. The superscript t denotes 'transposed matrices'.

The matrix form given by eq. (4.276) describes an indirect eigenvalue system. Its determinant must be zero, resulting in a relation between the normalised propagation constant ϵ_{ren} and normalised frequency f_n, with normalised guide radius and off-diagonal component of permeability tensor κ_m as parameters.

Construction of the difference operator $\boldsymbol{D}_h$:

If eq. (4.275) is to be fulfilled the difference operator $\boldsymbol{D}_h$ must have a special form. This special form is given by:

$$\boldsymbol{D}_h^* = \begin{bmatrix} p_l & & & \\ \cdots & \cdots & \cdots & \cdots \\ -1 & 1 & & \\ & \ddots & \ddots & \\ & & -1 & 1 \\ \cdots & \cdots & \cdots & \cdots \\ & & & p_r \end{bmatrix} \quad \text{and} \quad \boldsymbol{D}_h^{\circ} = \begin{bmatrix} 1 & & & \\ \cdots & \cdots & \cdots & \cdots \\ -1 & 1 & & \\ & \ddots & \ddots & \\ & & -1 & 1 \\ \cdots & \cdots & \cdots & \cdots \\ & & & 1 \end{bmatrix} \tag{4.278}$$

where p_l and p_r are obtained by introducing linear approximations into eq. (4.275):

$$p_l = \frac{H_z(r_i + h/2) - H_z(r_i - h/2)}{H_z(r_i + h/2)} = \frac{2g}{1+g} \tag{4.279}$$

$$p_r = \frac{H_z(r_a + h/2) - H_z(r_a - h/2)}{H_z(r_a - h/2)} = \frac{2g}{1-g} \tag{4.280}$$

with $g = \frac{1}{2}\kappa_m\, f_n\, h_n \sqrt{\varepsilon_{ren}}$ and $h_n = h/r_0$. This special form of $\boldsymbol{D}_h^*$ also results in values at the metallic walls for the first derivative. Therefore, the matrix $\mathbf{r}_e^*$ must also include the radii at the inner and outer walls.

Other field components:

The discretised equations for transverse field components are given in the following matrix form ($\varepsilon_{rd} = 1 - \varepsilon_{ren}$):

$$\varepsilon_{rd} f_n \begin{bmatrix} \mathbf{E}_{rn} \\ \mathbf{E}_{\phi n} \end{bmatrix} = \begin{bmatrix} -\sqrt{\varepsilon_{ren}}\overline{\boldsymbol{D}}_e & m\mathbf{r}_h^{-1} \\ -m\sqrt{\varepsilon_{ren}}\mathbf{r}_e^{-1} & -\kappa_m f_n \sqrt{\varepsilon_{ren}}\mathbf{M}_h + \overline{\boldsymbol{D}}_h \end{bmatrix} \begin{bmatrix} \mathrm{j}\mathbf{E}_{zn} \\ \mathrm{j}\widetilde{\mathbf{H}}_z \end{bmatrix} \tag{4.281}$$

$$\varepsilon_{rd} f_n \begin{bmatrix} \widetilde{\mathbf{H}}_\phi \\ \widetilde{\mathbf{H}}_r \end{bmatrix} = \begin{bmatrix} -\overline{\boldsymbol{D}}_e & m\sqrt{\varepsilon_{ren}}\mathbf{r}_h^{-1} \\ m\mathbf{r}_e^{-1} & \kappa_m f_n \boldsymbol{M}_h - \sqrt{\varepsilon_{ren}}\overline{\boldsymbol{D}}_h \end{bmatrix} \begin{bmatrix} \mathrm{j}\mathbf{E}_{zn} \\ \mathrm{j}\widetilde{\mathbf{H}}_z \end{bmatrix} \tag{4.282}$$

In these equations the usual form of $\overline{\boldsymbol{D}}_h$ can be used. If it is obtained from $\boldsymbol{D}_h^*$, the Matrix $\boldsymbol{M}_h$ must also be converted to $\boldsymbol{M}_h^*$. This matrix is obtained from $\boldsymbol{D}_h^*$ by replacing p_l and p_r with q_l, and q_r with $q_l = p_l/g$ and $q_r = p_r/g$.

The matrix $\mathbf{r}_e$ must also be replaced with $\mathbf{r}_e^*$. As we can see, if $\mathbf{E}_z$ and $\mathbf{H}_z$ are assumed to be imaginary (or $\mathrm{j}\mathbf{E}_z$ and $\mathrm{j}\mathbf{H}_z$ to be real), the other components are real. This allows the computation of the energy flow in z-direction. The z component of the Poynting vector is:

$$\boldsymbol{S}_z = \eta_0^{-1}(\mathbf{E}_r\widetilde{\mathbf{H}}_\phi - \mathbf{E}_\phi\widetilde{\mathbf{H}}_r) \tag{4.283}$$

In the case of rotationally-symmetric waves, i.e. $m = 0$, only the second term differs from zero. The radial and longitudinal components of the magnetic flux density are given by:

$$\mathbf{B}_r = \mu_0(\mathbf{H}_r - \kappa_m(\mathrm{j}\mathbf{H}_z)) \quad \mathbf{B}_z = \mathrm{j}\mu_0(\kappa_m\mathbf{H}_r - (\mathrm{j}\mathbf{H}_z)) \tag{4.284}$$

The azimuthal component is given by $\mathbf{B}_\phi = \mu_0\mathbf{H}_\phi$.

Numerical results for field components:

Solving the indirect eigenvalue equations for longitudinal components of electric and magnetic fields in eq. (4.276) yields the normalised propagation constant ε_{ren} as a function of the normalised frequency f_n and the values of all electromagnetic fields in the gyrotropic structure.

As an illustrative example, we show in Fig. 4.41 the distribution of components E_ϕ and H_r of the rotationally-symmetric TE_{01} mode propagating in a coaxial guide vs. reduced distance r/r_0 from the guide axis.

We have the ratio between inner and outer radius $\rho = r_1/r_0 = 0{,}4$. The guide is filled with azimuthally-magnetised ferrite with $\kappa_m = -0{,}8$.

As we can see, both H_r and H_z differ from zero at the inner and outer surfaces of the guide as a result of anisotropic medium – a feature not observed in the case of isotropic filling. For the radial component of a magnetic field we have $H_r = 0$ at those intermediate surfaces whose position depends on guide geometry and ferrite parameters.

The transverse electric field fulfils $E_\phi = 0$ at both surfaces of a coaxial guide (the electric field must meet the ideal conductor perpendicularly).

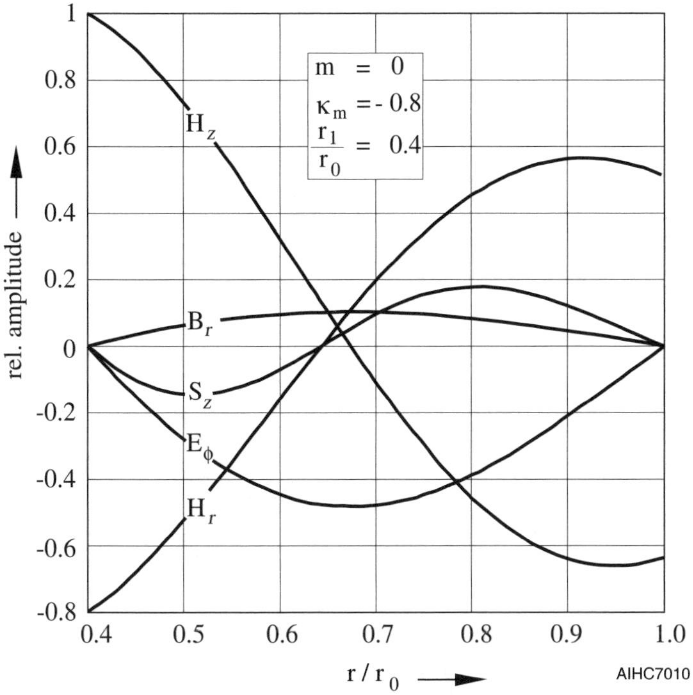

Fig. 4.41 Field distribution in an azimuthally-magnetized ferrite-filled coaxial guide ('Analysis of Circular Waveguides with Azimuthally Magnetized Ferrite', in *27th European Microwave Conference.* © 1997 European Microwave Association (EuMA))

It reaches a maximum (highest density of electric field lines) at a distance from the guide axis that depends on the sign and magnitude of κ_m.

The curve labelled S_z shows the radial distribution of the Poynting vector, which in the case of TE_{01} mode has only z component, i.e. $S_z = -\frac{1}{2}E_\phi H_r^*$, $\rightarrow \vec{S} \equiv (0, 0, S_z)$. Of special interest is the existence of a region near the coaxial inner conductor, in which we have $S_z < 0$. This suggests that the power is transmitted in a direction opposite to the propagation of the phase front, i.e. the integral of S_z over this region is negative. However, the total power, defined by the integral of Poynting vector S over the whole guide cross-section, propagates in $+z$-direction.

Numerical results for normalised phase characteristics:
Using the discretised form of the coupled partial differential equations for longitudinal components E_z and H_z, the normalised phase constant $\overline{\beta} = \sqrt{\varepsilon_{ren}}$ is computed as a function of normalised frequency f_n.

A representative set of theoretically calculated $\sqrt{\varepsilon_{ren}}$ versus normalised frequency f_n characteristics of gyrotropic coaxial guide for TE_{01} mode is plotted in Fig. 4.42. We have examined positive and negative remanent magnetisation (thick and thin lines, respectively) parameterised with κ_m for discrete values of inner to outer guide radius ratios.

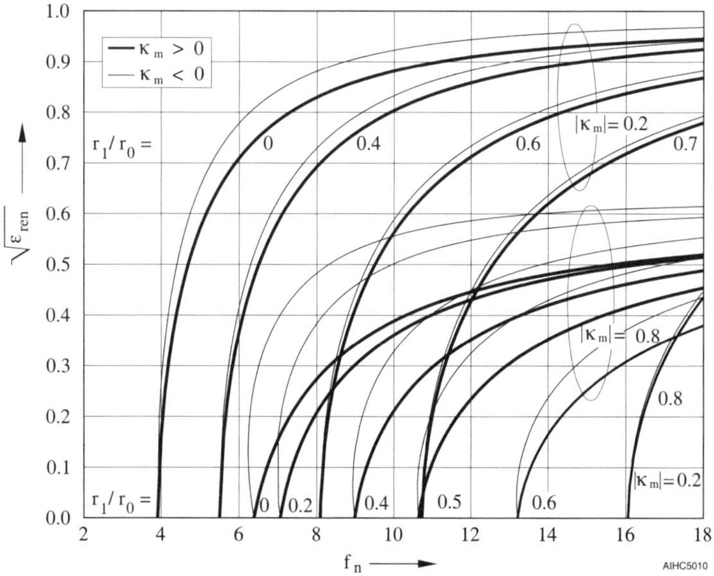

Fig. 4.42 Normalised phase constant as a function of normalised frequency f_n of coaxial gyrotropic waveguide with r_1/r_0 as a parameter ('Analysis of Circular Waveguides with Azimuthally Magnetized Ferrite', in *27th European Microwave Conference.* © 1997 European Microwave Association (EuMA))

The graphs show that the structure exhibits substantially different normalised phase constants $\overline{\beta}_+$ and $\overline{\beta}_-$, corresponding to the remanent magnetisation ($+M_r$ or $-M_r$). A reversal of M_r produces a normalised phase shift of propagating TE_{01} mode $\Delta\overline{\beta} = \overline{\beta}_- - \overline{\beta}_+$, which can easily be computed from the curves. It can also be seen that $\Delta\overline{\beta}$ shows a decreasing tendency with r_1/r_0. Worth noting is that the phase characteristics for one and the same κ_m possess a common asymptote. A section of coaxial guide with ferrite magnetised azimuthally to remanence, provided with mode purity filters and quarter wave transformers at either end appears to be a promising device configuration for single-bit nonreciprocal phases. Therefore it should be applicable to a wide range of specifications at microwave frequencies. In principle, any arbitrarily large differential phase shift can be obtained by adjusting the length of ferrite toroid (tube). The graphs allow us to quantify the dispersion effects of the structure and make knowledgeable tradeoffs, to obtain best operation dynamics of the device, i.e. substantial phase increment and zero phase slope over a given frequency band. Fig. 4.43 is a plot of normalised phase characteristics vs. normalised radius of a circular guide of radius r_0, completely filled with a ferrite, magnetised azimuthally to remanence with κ_m as a parameter for both signs of M_r. Latching the ferrite between the two stable states of M_r provides a discrete differential phase shift. In general, the observed effect of increased κ_m (strong anisotropy of the ferrite filling) is a substantial increase of normalised differential phase shift.

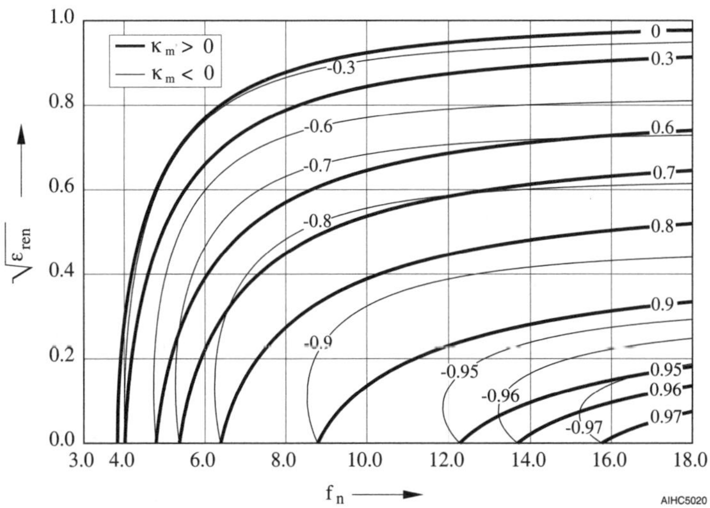

Fig. 4.43 Normalised phase constant as a function of normalised frequency f_n of circular gyrotropic waveguide ('Analysis of Circular Waveguides with Azimuthally Magnetized Ferrite', in *27th European Microwave Conference.* © 1997 European Microwave Association (EuMA))

4.4.4.3 Coaxial waveguide filled with azimuthally-magnetised plasma

Introduction:

This section deals with the analysis of the transmission characteristics of an azimuthally-magnetised solid-plasma circular guide. The capability of the structure to support TM_{01} backward wave propagation for one direction of magnetostatic field is explored and interpreted. The geometry of the problem is shown in Fig. 4.44. We have an infinitely long, perfectly conducting uniform coaxial waveguide with outer and inner radii r_0 and r_1. It is loaded with one-carrier semiconductor that is subjected to an external *DC* magnetic field B_0 applied in the azimuthal direction. A monochromatic wave can propagate. The basic equations for analysis are given in Section 4.2.4.2.

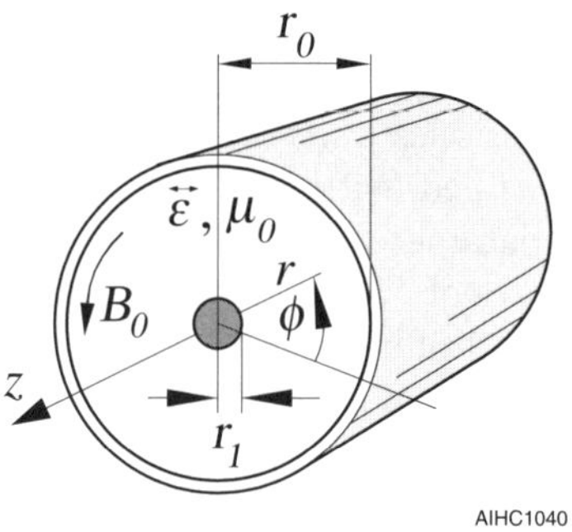

Fig. 4.44 Circular waveguide with magnetised plasma

Discretisation:

The general case of the coupled differential equations in eq. (4.77) and (4.78) cannot be solved analytically. Even the case of azimuthal independent fields can only be described by higher transcendental functions. Therefore, a numerical procedure will be described in this subsection. For this purpose, the field will be discretised in r-direction. This means we calculate the field not continually but only on the discretisation lines shown in Fig. 4.40. We use two different discretisation line systems. The system of full lines is marked by e. On these lines we discretise the longitudinal component of the electric field (E_z). The system of dashed lines is marked by h. Here, we discretise H_z. We have introduced two different discretisation line systems because E_z and H_z have to fulfil two different boundary conditions. E_z, as a tangential component on a perfectly-conducting surface, must be zero at $r = r_1$ and $r = r_0$, i.e. it must fulfil Dirichlet boundary conditions. $\widehat{H}_z$ is not zero but its normal derivative. With $\mathrm{d}\widehat{H}_z/\mathrm{d}r_n = 0$ at $r = r_1$ and $r = r_0$ (Neumann boundary condition), $\widehat{E}_\phi$ becomes zero at these places, too (see eq. (4.79)). This is also true for the normal field component $\widehat{H}_r$. We number the discretisation lines as shown in Fig. 4.40, from inner radius in outwards direction, and present all field quantities on discretisation lines $\widehat{E}_{zi}$ and $\widehat{H}_{zi}$ as vectors $\mathbf{E}_z$ and $\mathbf{H}_z$.

$\mathbf{E}_z$ has N components and $\mathbf{H}_z$ has N+1 components. We also discretise the radius. We collect the values of $r_{e,i}$ $(i = 1 \ldots \mathrm{N})$ and $r_{h,i}$ $(i = 1 \ldots \mathrm{N}+1)$ in diagonal matrices $\boldsymbol{r}_e$ and $\boldsymbol{r}_h$, respectively.

We can summarise our results:

$$
\begin{array}{ll}
\text{e-line system} & \text{h-line system} \\
E_z \rightarrow \mathbf{E}_z & \widetilde{H}_z \rightarrow \widetilde{\mathbf{H}}_z \\
r_n \rightarrow \boldsymbol{r}_e & r_n \rightarrow \boldsymbol{r}_h \\
\dfrac{\mathrm{d}}{\mathrm{d}r_n} \rightarrow h_n^{-1}\boldsymbol{D}_e = \overline{\boldsymbol{D}}_e & \dfrac{\mathrm{d}}{\mathrm{d}r_n} \rightarrow h_n^{-1}\boldsymbol{D}_h = \overline{\boldsymbol{D}}_h
\end{array}
\tag{4.285}
$$

$\widehat{\mathbf{E}}_z$ and $\widehat{\mathbf{H}}_z$ are one-dimensional vectors, $\boldsymbol{r}_{e,h}$ are diagonal matrices of the discretised radii and the difference operators $\boldsymbol{D}_{e,h}$ are matrices with bandwidth 2. The difference operators should fulfil the following boundary conditions:

(a) If $m = 0$ (rotationally symmetric fields) and TE$_z$ modes, $\boldsymbol{D}_h$ should fulfil Neumann conditions at the centre of the circular guide ($r = 0$), at the inner guide surface ($r = r_1$) and at the inner guide wall ($r = r_0$) of the coaxial guide because we have to fulfil $E_\phi(0) \equiv 0$.

(b) If $m = 0$ and TM$_z$ modes, $\boldsymbol{D}_e$ must satisfy the Dirichlet conditions at the inner guide surface ($r = r_1$) and inner guide wall ($r = r_0$) of the coaxial guide ($E_z = 0$). For $r_i = 0$, the condition:

$$\frac{\mathrm{d}E_z}{\mathrm{d}r_n} = \kappa_e \overline{r}_0 \sqrt{\varepsilon_{re}}/\varepsilon_1 E_z \tag{4.286}$$

must be fulfilled at the guide axis ($r = 0$) because H_ϕ must be zero there ($H_\phi(0) \equiv 0$).

(c) For azimuthally-dependent fields ($m \neq 0$), the difference operator $\boldsymbol{D}_e$ should fulfil the Dirichlet conditions at $r = r_1$ and $r = r_0$ since the field components E_z and E_ϕ must be zero at the surface of a perfect conductor. The component E_z must also be zero at the guide axis ($r = 0$). Nevertheless, the difference operator $\boldsymbol{D}_e$ should be constructed there for Neumann boundary conditions according to eq. (4.286), since H_ϕ must also be zero there. Since E_ϕ must be zero at $r = r_1$ and $r = r_0$, the difference operator $\boldsymbol{D}_h$ for $\mathbf{H}_z$ component has to fulfil the Neumann boundary condition there. At the guide axis, $\boldsymbol{D}_h$ has to fulfil the Dirichlet condition. Further details on the construction of $\boldsymbol{D}_e$ are outlined below.

The matrix form of discretised equations for longitudinal field components (see eqs. (4.77) and (4.78)) is:

$$\begin{bmatrix} \overline{\boldsymbol{P}}^r_{ee} & \overline{\boldsymbol{P}}^r_{eh} \\ \overline{\boldsymbol{P}}^r_{he} & \overline{\boldsymbol{P}}^r_{hh} \end{bmatrix} \begin{bmatrix} \mathbf{E}_z \\ \widetilde{\mathbf{H}}_z \end{bmatrix} = 0 \tag{4.287}$$

with:

$$\begin{aligned}
\overline{\boldsymbol{P}}^r_{ee} &= -\boldsymbol{r}_e^{-1}\overline{\boldsymbol{D}}_e^t \boldsymbol{r}_h \overline{\boldsymbol{D}}_e - m^2 \frac{\varepsilon_1 - \varepsilon_{re}}{\varepsilon_2 - \varepsilon_{re}} \frac{\varepsilon_2}{\varepsilon_1} \boldsymbol{r}_e^{-2} + (\varepsilon_\perp - \varepsilon_{re}) f_n^2 \boldsymbol{I}_e - \frac{\sqrt{\varepsilon_{re}}\kappa_e f_n}{\varepsilon_1} \boldsymbol{r}_e^{-1} \\
\overline{\boldsymbol{P}}^r_{eh} &= m \frac{\sqrt{\varepsilon_{re}}}{\varepsilon_1} \frac{\varepsilon_1 - \varepsilon_2}{\varepsilon_2 - \varepsilon_{re}} \boldsymbol{r}_e^{-1} \overline{\boldsymbol{D}}_h - \frac{m\kappa_e f_n}{\varepsilon_1} \boldsymbol{r}_e^{-1} \boldsymbol{M}_h \\
\overline{\boldsymbol{P}}^r_{he} &= -m\sqrt{\varepsilon_{re}} \frac{\varepsilon_1 - \varepsilon_2}{\varepsilon_1 - \varepsilon_{re}} \boldsymbol{r}_e^{-1} \overline{\boldsymbol{D}}_e - m f_n \kappa_e \frac{\varepsilon_2 - \varepsilon_{re}}{\varepsilon_1 - \varepsilon_{re}} \boldsymbol{r}_e^{-1} \boldsymbol{M}_e \\
\overline{\boldsymbol{P}}^r_{hh} &= -\boldsymbol{r}_h^{-1} \overline{\boldsymbol{D}}_h^t \boldsymbol{r}_e \overline{\boldsymbol{D}}_h - m^2 \frac{\varepsilon_2 - \varepsilon_{re}}{\varepsilon_1 - \varepsilon_{re}} \boldsymbol{r}_h^{-2} + (\varepsilon_2 - \varepsilon_{re}) f_n^2 \boldsymbol{I}_h
\end{aligned} \tag{4.288}$$

The matrices $\boldsymbol{M}_e$ and $\boldsymbol{M}_h$ are introduced to facilitate interpolation between the two discretisation line systems. For linear interpolation they are given by $\boldsymbol{M}_{e,h} = 0.5|\boldsymbol{D}_{e,h}|$, where the difference operators $\boldsymbol{D}_{e,h}$ are the known operators.

Eq. (4.287) is an indirect eigenvalue system. Its determinant must be zero, resulting in a relation between the normalised propagation constant ϵ_e and normalised frequency f_n, where the normalised guide radius and $\varepsilon_1, \varepsilon_2$ and κ_e are used as parameters.

The discretised equations for transverse field components are given in the following matrix form:

$$\begin{bmatrix} (\varepsilon_1 - \varepsilon_{re}) f_n \mathbf{E}_r \\ (\varepsilon_2 - \varepsilon_{re}) f_n \mathbf{E}_\phi \end{bmatrix} = \begin{bmatrix} \kappa_e f_n \boldsymbol{M}_e - \sqrt{\varepsilon_{re}} \overline{\boldsymbol{D}}_e & m \mathbf{r}_h^{-1} \\ -m\sqrt{\varepsilon_{re}} \mathbf{r}_e^{-1} & \overline{\boldsymbol{D}}_h \end{bmatrix} \begin{bmatrix} \mathrm{j}\mathbf{E}_z \\ \mathrm{j}\widetilde{\mathbf{H}}_z \end{bmatrix} \tag{4.289}$$

$$\begin{bmatrix} (\varepsilon_1 - \varepsilon_{re}) f_n \widetilde{\mathbf{H}}_\phi \\ (\varepsilon_2 - \varepsilon_{re}) f_n \widetilde{\mathbf{H}}_r \end{bmatrix} = \begin{bmatrix} \kappa_e f_n \sqrt{\varepsilon_{re}} \boldsymbol{M}_e - \varepsilon_1 \overline{\boldsymbol{D}}_e & m\sqrt{\varepsilon_{re}} \mathbf{r}_h^{-1} \\ m \varepsilon_2 \overline{\mathbf{r}}_e^{-1} & -\sqrt{\varepsilon_{re}} \overline{\boldsymbol{D}}_h \end{bmatrix} \begin{bmatrix} \mathrm{j}\mathbf{E}_z \\ \mathrm{j}\widetilde{\mathbf{H}}_z \end{bmatrix} \tag{4.290}$$

As we can see, if $\mathbf{E}_z$ and $\mathbf{H}_z$ are imaginary (or $\mathrm{j}\mathbf{E}_z$ and $\mathrm{j}\mathbf{H}_z$ are real) the other components are real for real permittivities.

This allows the computation of the flow of energy in z-direction. The z component of the Poynting vector is:

$$\boldsymbol{S}_z = \eta_0^{-1} (\mathbf{E}_r \widetilde{\mathbf{H}}_\phi - \mathbf{E}_\phi \widetilde{\mathbf{H}}_r) \tag{4.291}$$

In the case of rotationally symmetric waves, i.e. $m = 0$, only the second term differs from zero.

Construction of the difference operator $\boldsymbol{D}_e$:

To fulfil eq. (4.286) the difference operator $\boldsymbol{D}_e$ should have the following form:

$$\boldsymbol{D}_e = \begin{bmatrix} p & & & & \\ -1 & 1 & & & \\ & \ddots & \ddots & & \\ & & -1 & 1 & \\ & & & 1 & \end{bmatrix} \tag{4.292}$$

The parameter p is obtained by introducing the following linear approximations into eq. (4.286):

$$p = \frac{E_z(r_i + h/2) - E_z(r_i - h/2)}{E_z(r_i + h/2)} = \frac{2g}{1+g} \tag{4.293}$$

with $g = \frac{1}{2}\kappa_e \overline{r}_0 h_n \sqrt{\varepsilon_e}/\varepsilon_1$; $h_n = h/r_0$. To construct $\boldsymbol{D}_e^{\mathrm{t}}$ we introduce $p = 1$.

Numerical results: normalised phase characteristics:

A representative set of the theoretically-calculated dispersion characteristics $\sqrt{\varepsilon_{re}\kappa_L}$ as a function of $f_n\sqrt{\kappa_L}$ of solid-plasma coaxial waveguide for TM_{01} mode is plotted in Figs. 4.45 and 4.46 for $+B_0$ and $-B_0$ (solid and dotted lines, respectively). The curves are parameterised with κ_e/κ_L. We assume an inner to outer guide radius ratio of $\rho = r_1/r_0 = 0.0$ and $\rho = r_1/r_0 = 0.3$, respectively. κ_L is the atomic lattice permittivity. We assume lossless plasma with $\varepsilon_1 = \varepsilon_2 = \kappa_L$ [20–22]. The graphs show that the structure exhibits substantially different normalised phase constants $\overline{\beta}_+$ and $\overline{\beta}_-$, corresponding to $+B_0$ and $-B_0$. A reversal of magnetostatic field from $-B_0$ to $+B_0$ produces a normalised phase shift of propagating TM_{01} mode, which can easily be computed from the curves.

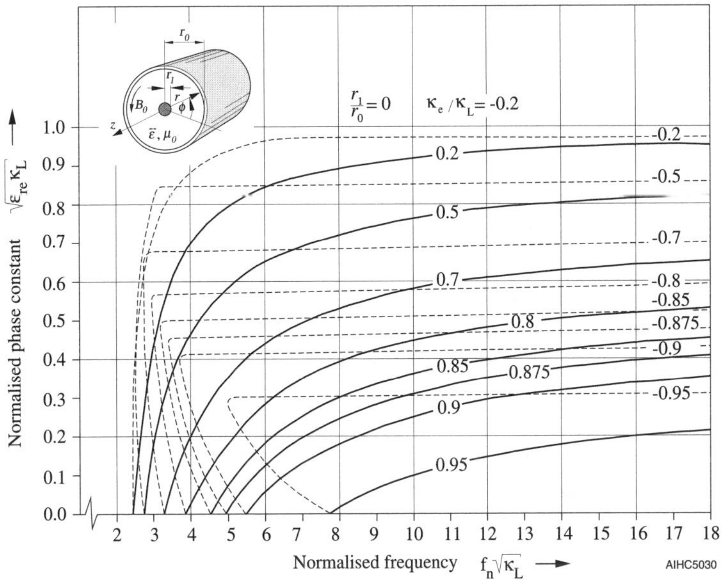

Fig. 4.45 Normalised phase characteristics of azimuthally-magnetised millimetre-wave solid-plasma circular guide for $\rho = r_1/r_2 = 0.0$

Thus, a section of azimuthally-magnetised solid-plasma coaxial guide, provided with matching transformers at either end, appears to be a promising device configuration for single-bit nonreciprocal phaser. It should be applicable to a wide range of specifications at millimetre frequencies. The graphs allow us to quantify the dispersion effects of the structure. By making a knowledgeable trade-off, we obtain best operation dynamics of the device, i.e. a substantial phase increase and zero phase slope over a given frequency band.

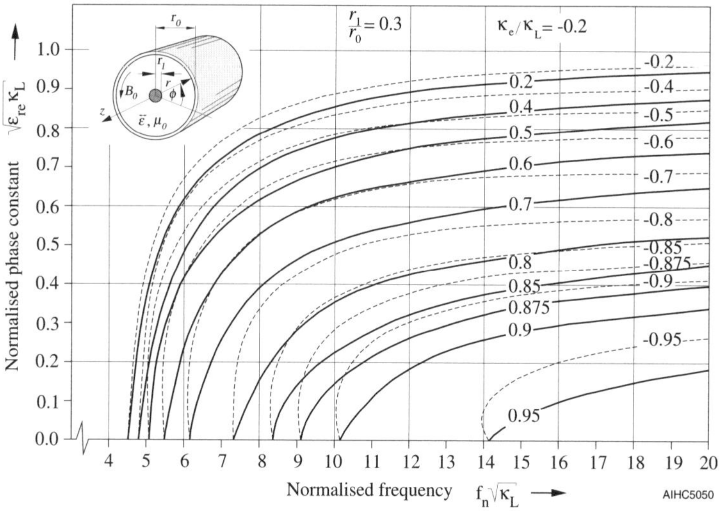

Fig. 4.46 Normalised phase characteristics of azimuthally-magnetised millimetre-wave solid-plasma circular guide for $\rho = r_1/r_2 = 0.3$

4.4.5 Waveguide bends – discretisation in r-direction

4.4.5.1 Introduction

In this subsection an alternative method for the analysis of waveguide bends will be described. In the analysis described in Section 4.4.2 the fields are expressed in terms of Bessel and Hankel functions of large complex orders and complex arguments. This in turn complicates the numerical calculations, demanding special computer codes. In this subsection an algorithm for the full wave analysis of dielectric bends is suggested, eliminating the use of cylindrical functions. The analysis is valid not only for the simple cross-sections shown in Figs. 4.47 and 4.48, but also for the cross-section of the general model of the waveguide in Fig. 4.49 – a multilayered structure with an arbitrary number of layers. The permittivity in each layer may depend on the radius r, can be complex and/or can change abruptly. This makes it possible not only to analyse lossy dielectric media but to take into account lossy metallisations [23], [24]. In $\pm z$-direction the structure may be either infinite or closed by a metallic or magnetic wall. At the inner radius of the calculation window Dirichlet or Neumann boundary conditions may be introduced. At the outer radius of the calculation window, radiation can be modelled by introducing absorbing boundary conditions (ABC) (cf. the appendices). In Section 4.4.2 the structures are discretised in the Cartesian z-direction. A radial discretisation was used in the MoL, e.g. in [25], [26], [27] and [28]. In these publications the structures consist of homogenous layers. In this subsection we introduce radial discretisation for the analysis of structures with inhomogeneous layers. Furthermore, even in the case of homogenous layers the discretisation procedure differs substantially from [27]. To construct the difference operators for the derivatives in the radial direction in each of the wave equations we use two different diagonal matrices of the discretised radii. Only in this way can the field components be given accurately. This is equivalent to the modelling of dielectric steps.

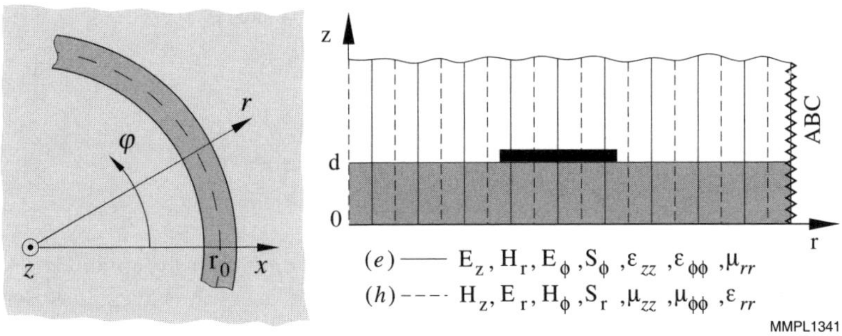

Fig. 4.47 Microstrip waveguide bend (left) and cross-section with discretisation lines (right)

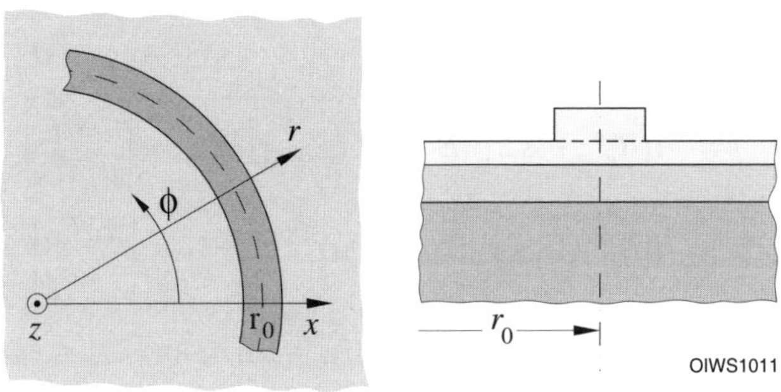

Fig. 4.48 Circular optical waveguide bend (left), with rib waveguide cross-section (right) as an example

4.4.5.2 Basic equations for eigenmodes in planar circular bends

We normalise the radii with r_o, where r_o is the mean radius of the bend (see e.g. Fig. 4.48): $\overline{r}_n = \overline{r}/\overline{r}_o = r/r_o = r_n$, $\overline{r}_o = k_o r_o$. The uniaxial anisotropic material parameters ($\varepsilon_{rr} = \varepsilon_{\phi\phi}$, ε_{zz} and $\mu_{rr} = \mu_{\phi\phi}$, μ_{zz}) of each layer may be functions of radial coordinate r only. We distinguish between e.g. ε_{rr} and $\varepsilon_{\phi\phi}$ because both parameters will be discretised on different discretisation lines. We assume a wave propagation in $+\phi$-direction according to $\exp(-\mathrm{j}k_\phi \phi) = \exp(-\mathrm{j}\sqrt{\varepsilon_{re}}\,\overline{u})$ with $\overline{u} = \phi r_o k_o$. Therefore, we have $D_\phi =$

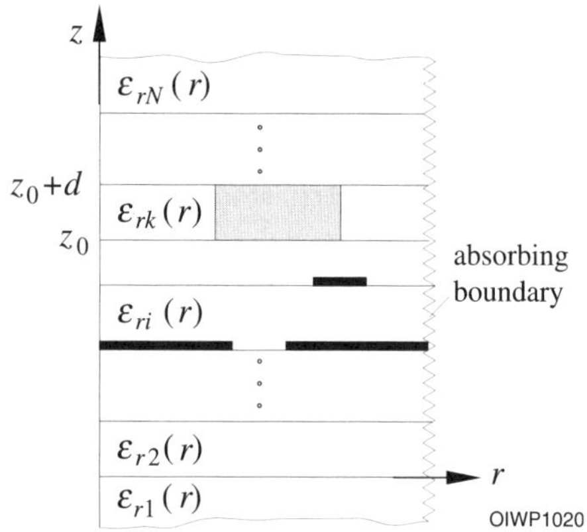

Fig. 4.49 General analysis model for multilayered circular waveguide structures (R. Pregla, 'The Method of Lines for the Analysis of Dielectric Waveguides Bends', *J. Lightwave Technol.*, vol. 14, no. 4, pp. 634–639. © 1996 Institute of Electrical and Electronics Engineers (IEEE))

$\partial/\partial\phi = -\mathrm{j}k_\phi = -\mathrm{j}\sqrt{\varepsilon_{re}}\overline{r}_o$. The problem is to solve the two coupled generalised transmission line equations in (4.11) for the z-direction for each layer specified. As in eq. (4.17) we split the matrices $[R^c_{\mathrm{E,H}}]$ into two parts ($[R^z_{\mathrm{E}}] = [R^{zc}_{\mathrm{E}}]+[\varepsilon^{zc}_{rt}]$, $[R^z_{\mathrm{H}}] = [R^{zc}_{\mathrm{H}}] + [\mu^{zc}_{rt}]$) and obtain:

$$\frac{\partial}{\partial\overline{z}}[\widehat{H}^{ze}] = -([R^{zc}_{\mathrm{E}}] + [\varepsilon^{zc}_{rt}])[\widehat{E}^{ze}] \qquad [\widehat{H}^{ze}] = \begin{bmatrix} -\widetilde{H}_r \\ \mathrm{j}r_n\widetilde{H}_\phi \end{bmatrix} \tag{4.294}$$

$$\frac{\partial}{\partial\overline{z}}[\widehat{E}^{ze}] = -([R^{zc}_{\mathrm{H}}] + [\mu^{zc}_{rt}])[\widehat{H}^{ze}right] \qquad [\widehat{E}^{ze}] = \begin{bmatrix} \mathrm{j}r_nE_\phi \\ E_r \end{bmatrix} \tag{4.295}$$

By using the abbreviations $\overline{\mu}_{zz} = \mu_{zz}\overline{r}_n = \mu_{zz}r_n$ and $\overline{\varepsilon}_{zz} = \varepsilon_{zz}\overline{r}_n = \varepsilon_{zz}r_n$, we may write for these matrices:

$$[R^{zc}_{\mathrm{E}}] = \begin{bmatrix} D_{\overline{r}}\overline{\mu}^{-1}_{zz}D_{\overline{r}} & -\sqrt{\varepsilon_{re}}D_{\overline{r}}\overline{\mu}^{-1}_{zz} \\ -\sqrt{\varepsilon_{re}}\overline{\mu}^{-1}_{zz}D_{\overline{r}} & \varepsilon_{re}\overline{\mu}^{-1}_{zz} \end{bmatrix} \qquad [\widehat{\varepsilon}^{zc}_{rt}] = \begin{bmatrix} \varepsilon_{\phi\phi}r_n^{-1} & \mathrm{j}\varepsilon_{\phi r} \\ \mathrm{j}\varepsilon_{r\phi} & -\varepsilon_{rr}r_n \end{bmatrix} \tag{4.296}$$

$$[R^{zc}_{\mathrm{H}}] = \begin{bmatrix} \varepsilon_{re}\overline{\varepsilon}^{-1}_{zz} & \sqrt{\varepsilon_{re}}\overline{\varepsilon}^{-1}_{zz}D_{\overline{r}} \\ \sqrt{\varepsilon_{re}}D_{\overline{r}}\overline{\varepsilon}^{-1}_{zz} & D_{\overline{r}}\overline{\varepsilon}^{-1}_{zz}D_{\overline{r}} \end{bmatrix} \qquad [\widehat{\mu}^{zc}_{rt}] = \begin{bmatrix} -\mu_{rr}r_n & -\mathrm{j}\mu_{r\phi} \\ -\mathrm{j}\mu_{\phi r} & \mu_{\phi\phi}r_n^{-1} \end{bmatrix} \tag{4.297}$$

The combination of eqs. (4.294) and (4.295) yields:

$$\frac{\partial^2}{d\overline{z}^2}[\widehat{E}^{ze}] - [Q^z_{\mathrm{E}}][\widehat{E}^{ze}] = 0 \qquad \frac{\partial^2}{d\overline{z}^2}[\widehat{H}^{ze}] - [Q^z_{\mathrm{H}}][\widehat{H}^{ze}] = 0 \tag{4.298}$$

The product of the two matrices R_{H}^{zc} and R_{E}^{zc} is equal to zero. Therefore, we obtain for the two matrices Q_{H}^{z} and Q_{E}^{z}:

$$\begin{aligned}
[Q_{\mathrm{E}}^{z}] &= [R_{\mathrm{H}}^{z}][R_{\mathrm{E}}^{z}] = [R_{\mathrm{H}}^{zc}][\widehat{\varepsilon}_{rt}^{zc}] + [\widehat{\mu}_{rt}^{zc}][R_{\mathrm{E}}^{zc}] + [\widehat{\mu}_{rt}^{zc}][\widehat{\varepsilon}_{rt}^{zc}] \\
[Q_{\mathrm{H}}^{z}] &= [R_{\mathrm{E}}^{z}][R_{\mathrm{H}}^{z}] = [R_{\mathrm{E}}^{zc}][\widehat{\mu}_{rt}^{zc}] + [\widehat{\varepsilon}_{rt}^{zc}][R_{\mathrm{H}}^{\phi c}] + [\widehat{\varepsilon}_{rt}^{zc}][\widehat{\mu}_{rt}^{zc}]
\end{aligned} \tag{4.299}$$

For the z components we obtain from eq. (4.9):

$$\widetilde{H}_z = \overline{\mu}_{zz}^{-1}[D_{\overline{r}} \ - \sqrt{\varepsilon_{re}}][\widehat{E}^{ze}] \qquad E_z = -\overline{\varepsilon}_{zz}^{-1}[\sqrt{\varepsilon_{re}} \ \ D_{\overline{r}}][\widehat{H}^{ze}] \tag{4.300}$$

4.4.5.3 Discretisation

The partial differential eqs. (4.294), (4.295) and (4.298) are discretised with respect to r. We obtain ordinary differential equations that can be solved analytically. In this case, not only the field components but also the permittivities and the radial coordinate r have to be discretised. We now assume that the off-diagonal material tensor components are equal to zero. If necessary, these can be introduced analogously to the sections before. The discretisation (see Fig. 4.50) is done on two different line systems, yielding:

$\circ$ h-discretisation lines $\qquad\qquad$ $\bullet$ e-discretisation lines

$$\begin{array}{rcl@{\qquad}rcl}
E_r, \widetilde{H}_\phi, \widetilde{H}_z & \longrightarrow & \mathbf{E}_r, \mathbf{H}_\phi, \mathbf{H}_z & \widetilde{H}_r, E_\phi, E_z & \longrightarrow & \mathbf{H}_r, \mathbf{E}_\phi, \mathbf{E}_z \\
\varepsilon_{rr}, \mu_{\phi\phi}, \mu_{zz} & \longrightarrow & \boldsymbol{\epsilon}_r, \boldsymbol{\mu}_\phi, \boldsymbol{\mu}_z & \mu_{rr}, \varepsilon_{\phi\phi}, \varepsilon_{zz} & \longrightarrow & \boldsymbol{\mu}_r, \boldsymbol{\epsilon}_\phi, \boldsymbol{\epsilon}_z \\
\overline{r}_n = r_n & \longrightarrow & \overline{\mathbf{r}}_\circ & \overline{r}_n = r_n & \longrightarrow & \overline{\mathbf{r}}_\bullet \\
D_{\overline{r}} = \frac{\partial}{\partial \overline{r}} & \longrightarrow & \overline{h}^{-1}\mathbf{D}_{\overline{r}}^{\circ} = \overline{\mathbf{D}}_{\overline{r}}^{\circ} & D_{\overline{r}} = \frac{\partial}{\partial \overline{r}} & \longrightarrow & \overline{h}^{-1}\mathbf{D}_{\overline{r}}^{\bullet} = \overline{\mathbf{D}}_{\overline{r}}^{\bullet}
\end{array} \tag{4.301}$$

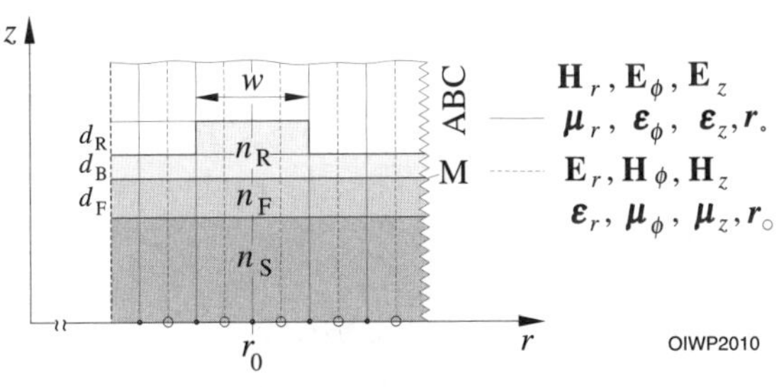

Fig. 4.50 Cross-section of the rib waveguide bend with the two systems of discretisation lines (R. Pregla, 'The Method of Lines for the Analysis of Dielectric Waveguides Bends', *J. Lightwave Technol.*, vol. 14, no. 4, pp. 634–639. © 1996 Institute of Electrical and Electronics Engineers (IEEE))

$\overline{h} = k_o h$ is the normalised discretisation distance. $\varepsilon_u, \mu_u (u = r, \phi, z)$, $\overline{\mathbf{r}}_\circ$ and $\overline{\mathbf{r}}_\bullet$ are diagonal matrices. The symbols $\circ$ and $\bullet$ indicate the discretisation line system to which the quantities belong. The superscripts on the difference operators $\overline{\boldsymbol{D}}$ indicate for which line system the difference operator has to be used. In view of what has been said before, the discretisation of the wave equations runs as follows: we introduce ABCs into the difference operator $\overline{\boldsymbol{D}}_{\overline{r}}^{\bullet}$ and write $\overline{\boldsymbol{D}}_{\overline{r}}^{\bullet a}$ instead. Then we also have to use $-\overline{\boldsymbol{D}}_{\overline{r}}^{\bullet a}$ instead of $\overline{\boldsymbol{D}}_{\overline{r}}^{\circ t}$ and $-\overline{\boldsymbol{D}}_{\overline{r}}^{\circ}$ instead of $\overline{\boldsymbol{D}}_{\overline{r}}^{\bullet t}$. For $\overline{\boldsymbol{D}}_{\overline{r}}^{\circ}$, Neumann boundary conditions have to be used at the outer side of the bend. At the inner side of the cross-section in Fig. 4.50 we use Dirichlet (for $\overline{\boldsymbol{D}}_{\overline{r}}^{\circ}$) and Neumann (for $\overline{\boldsymbol{D}}_{\overline{r}}^{\bullet}$) conditions. Using the definitions of supervectors for the fields in eqs. (4.294), and (4.295) we obtain in discretised form:

$$\widehat{\mathbf{H}}^{ze} = \begin{bmatrix} -\mathbf{H}_r \\ \mathrm{j}\overline{\mathbf{r}}_{\circ}\mathbf{H}_{\phi} \end{bmatrix} \qquad \widehat{\mathbf{E}}^{ze} = \begin{bmatrix} \mathrm{j}\overline{\mathbf{r}}_{\bullet}\mathbf{E}_{\phi} \\ \mathbf{E}_r \end{bmatrix} \tag{4.302}$$

$$\frac{\mathrm{d}}{\mathrm{d}\overline{z}}\widehat{\mathbf{H}}^{ze} = -(\boldsymbol{R}_{\mathrm{E}}^{zc} + \widehat{\boldsymbol{\epsilon}}_{rt}^{zc})\widehat{\mathbf{E}}_{\phi r}^{e} \qquad \frac{\mathrm{d}}{\mathrm{d}\overline{z}}\widehat{\mathbf{E}}^{ze} = -(\boldsymbol{R}_{\mathrm{H}}^{zc} + \widehat{\boldsymbol{\mu}}_{rt}^{zc})\widehat{\mathbf{H}}^{ze} \tag{4.303}$$

By using the abbreviations $\overline{\boldsymbol{\mu}}_z^i = (\boldsymbol{\mu}_{zz}\overline{\mathbf{r}}_{\circ})^{-1}$ and $\overline{\boldsymbol{\epsilon}}_z^i = (\boldsymbol{\epsilon}_{zz}\overline{\mathbf{r}}_{\bullet})^{-1}$, we write for the operator matrices:

$$\boldsymbol{R}_{\mathrm{E}}^{zc} = \begin{bmatrix} \overline{\boldsymbol{D}}_{\overline{r}}^{\circ}\overline{\boldsymbol{\mu}}_z^i\overline{\boldsymbol{D}}_{\overline{r}}^{\bullet a} & -\sqrt{\varepsilon_{re}}\overline{\boldsymbol{D}}_{\overline{r}}^{\circ}\overline{\boldsymbol{\mu}}_z^i \\ -\sqrt{\varepsilon_{re}}\overline{\boldsymbol{\mu}}_z^i\overline{\boldsymbol{D}}_{\overline{r}}^{\bullet a} & \varepsilon_{re}\overline{\boldsymbol{\mu}}_z^i \end{bmatrix} \qquad [\widehat{\boldsymbol{\epsilon}}_{\phi}] = \begin{bmatrix} \boldsymbol{\epsilon}_{\phi\phi}\overline{\mathbf{r}}_{\bullet}^{-1} & \mathrm{j}\boldsymbol{M}^{\circ}\boldsymbol{\varepsilon}_{r\phi} \\ \mathrm{j}\boldsymbol{M}^{\bullet}\boldsymbol{\varepsilon}_{r\phi} & -\overline{\mathbf{r}}_{\circ}\boldsymbol{\epsilon}_{rr} \end{bmatrix} \tag{4.304}$$

$$\boldsymbol{R}_{\mathrm{H}}^{zc} = \begin{bmatrix} \varepsilon_{re}\overline{\boldsymbol{\epsilon}}_z^i & \sqrt{\varepsilon_{re}}\overline{\boldsymbol{\epsilon}}_z^i\overline{\boldsymbol{D}}_{\overline{r}}^{\circ} \\ \sqrt{\varepsilon_{re}}\overline{\boldsymbol{D}}_{\overline{r}}^{\bullet a}\overline{\boldsymbol{\epsilon}}_z^i & \overline{\boldsymbol{D}}_{\overline{r}}^{\bullet a}\overline{\boldsymbol{\epsilon}}_z^i\overline{\boldsymbol{D}}_{\overline{r}}^{\circ} \end{bmatrix} \qquad [\widehat{\boldsymbol{\mu}}_{\phi}] = \begin{bmatrix} -\overline{\mathbf{r}}_{\bullet}\boldsymbol{\mu}_{rr} & -\mathrm{j}\boldsymbol{M}^{\circ}\boldsymbol{\mu}_{r\phi} \\ -\mathrm{j}\boldsymbol{M}^{\bullet}\boldsymbol{\mu}_{\phi r} & \boldsymbol{\mu}_{\phi\phi}\overline{\mathbf{r}}_{\circ}^{-1} \end{bmatrix} \tag{4.305}$$

It should be mentioned that ε_{rr} and $\varepsilon_{\phi\phi}$ must have the same value ($\varepsilon_{rr} = \varepsilon_{\phi\phi}$) for the algorithm described in this section. However, because of discretisation on different places the matrices $\boldsymbol{\epsilon}_{rr}$ and $\boldsymbol{\epsilon}_{\phi\phi}$ for inhomogeneous layers are different. The matrices $\boldsymbol{M}^{\circ,\bullet}$ are interpolation matrices between the two discretisation line systems. The matrices $\boldsymbol{Q}_{\mathrm{E,H}}^{z}$ should be determined numerically by eq. (4.299). With the matrices $\boldsymbol{R}_{\mathrm{E,H}}^{z}$ we can calculate the eigenvector and eigenvalue matrices $\boldsymbol{T}_{\mathrm{E,H}}^{z}$ $\boldsymbol{\Gamma}_{\mathrm{E,H}}^{z}$ analogously to Section 2.3.2 according to $-\widehat{\boldsymbol{R}}_{\mathrm{E}}^{z}\widehat{\boldsymbol{R}}_{\mathrm{H}}^{z}\boldsymbol{T}_{\mathrm{H}}^{z} = \boldsymbol{T}_{\mathrm{H}}^{z}\boldsymbol{\Gamma}_{\mathrm{H}}^{z2}$ and $-\widehat{\boldsymbol{R}}_{\mathrm{H}}\widehat{\boldsymbol{R}}_{\mathrm{E}}\boldsymbol{T}_{\mathrm{E}} = \boldsymbol{T}_{\mathrm{E}}\boldsymbol{\Gamma}_{\mathrm{E}}^{z2}$.

4.4.5.4 Field and impedance transformation between layer interfaces

The relation between the electric and magnetic field in the transformed domain at lower and upper planes A and B of a layer can be described analogously to Section 2.3.2 by:

$$\begin{bmatrix} \widehat{\mathbf{H}}_{\mathrm{A}}^{ze} \\ -\widehat{\mathbf{H}}_{\mathrm{B}}^{ze} \end{bmatrix} = \begin{bmatrix} \overline{\boldsymbol{y}}_1 & \overline{\boldsymbol{y}}_2 \\ \overline{\boldsymbol{y}}_2 & \overline{\boldsymbol{y}}_1 \end{bmatrix} \begin{bmatrix} \widehat{\mathbf{E}}_{\mathrm{A}}^{ze} \\ \widehat{\mathbf{E}}_{\mathrm{B}}^{ze} \end{bmatrix} \qquad \begin{bmatrix} \widehat{\mathbf{E}}_{\mathrm{A}}^{ze} \\ \widehat{\mathbf{E}}_{\mathrm{B}}^{ze} \end{bmatrix} = \begin{bmatrix} \overline{\mathbf{z}}_1 & \overline{\mathbf{z}}_2 \\ \overline{\mathbf{z}}_2 & \overline{\mathbf{z}}_1 \end{bmatrix} \begin{bmatrix} \widehat{\mathbf{H}}_{\mathrm{A}}^{ze} \\ -\widehat{\mathbf{H}}_{\mathrm{B}}^{ze} \end{bmatrix} \tag{4.306}$$

with:

$$\begin{aligned} \overline{\boldsymbol{y}}_1 &= \overline{\mathbf{Y}}_0/\tanh(\boldsymbol{\Gamma}^z\overline{d}) \qquad & \overline{\mathbf{z}}_1 &= \overline{\mathbf{Z}}_0/\tanh(\boldsymbol{\Gamma}^z\overline{d}) \\ \overline{\boldsymbol{y}}_2 &= -\overline{\mathbf{Y}}_0/\sinh(\boldsymbol{\Gamma}^z\overline{d}) \qquad & \overline{\mathbf{z}}_2 &= \overline{\mathbf{Z}}_0/\sinh(\boldsymbol{\Gamma}^z\overline{d}) \end{aligned} \tag{4.307}$$

$\overline{\boldsymbol{Y}}_0$ and $\overline{\boldsymbol{Z}}_0$ are equal to $\boldsymbol{I}$ (the unit matrix). Now we can use the algorithm developed in Section 2.3.2 for the impedance/admittance transformation from one side of a layer to the other and from one layer to the other. For the transformation through the layers we use eq. (2.89). For transformation in the other direction (A $\rightarrow$ B) the formulas must be rewritten. For the transformation from one layer to another (from layer 2 to layer 1) without metallisation, the formula $\widehat{\overline{\boldsymbol{Z}}}_1 = \widehat{\boldsymbol{T}}_{\mathrm{E1}}^{-1}\widehat{\boldsymbol{T}}_{\mathrm{E2}}\widehat{\overline{\boldsymbol{Z}}}_2\widehat{\boldsymbol{T}}_{\mathrm{H2}}^{-1}\widehat{\boldsymbol{T}}_{\mathrm{H1}}$ (analogous to eq. (2.86)) or the formula $\widehat{\overline{\boldsymbol{Y}}}_1 = \widehat{\boldsymbol{T}}_{\mathrm{H1}}^{-1}\widehat{\boldsymbol{T}}_{\mathrm{H2}}\widehat{\overline{\boldsymbol{Y}}}_2\widehat{\boldsymbol{T}}_{\mathrm{E2}}^{-1}\widehat{\boldsymbol{T}}_{\mathrm{E1}}$ must be used. If we have metallisation, between the layers, an expression similar to eq. (2.164)) must be used.

4.4.5.5 Formulas for a rib waveguide bend

For the structure shown in Fig. 4.48 the further analysis runs as follows. In plane M (see Fig. 4.50) we obtain by the admittance transformations described (see also eqs. (2.74) and (2.75):

$$\widehat{\mathbf{H}}_M^{\mathrm{I}} = -\mathbf{Y}_t^{\mathrm{I}}\widehat{\mathbf{E}}_M^{\mathrm{I}} \qquad \widehat{\mathbf{H}}_M^{\mathrm{II}} = \mathbf{Y}_t^{\mathrm{II}}\widehat{\mathbf{E}}_M^{\mathrm{II}} \tag{4.308}$$

We obtain $\mathbf{Y}_t^{\mathrm{I}}$ from the lower part and $\mathbf{Y}_t^{\mathrm{II}}$ from the upper part. In the lower part we start with the substrate and $\mathbf{Y}_t^{(\mathrm{substrate})} = \boldsymbol{y}_1^{(\mathrm{substrate})}$, where the substrate thickness is infinite. In the upper part we start the transformation in the air region with $\mathbf{Y}_t^{(\mathrm{air})} = \boldsymbol{y}_1^{(\mathrm{air})}$. Again the thickness of this layer is infinite. Subtracting eq. (4.308a) from (4.308b) yields:

$$\widehat{\mathbf{H}}_M^{\mathrm{I}} - \widehat{\mathbf{H}}_M^{\mathrm{II}} = (\widehat{\mathbf{Y}}_t^{\mathrm{I}} + \widehat{\mathbf{Y}}_t^{\mathrm{II}})\widehat{\mathbf{E}}_M = \mathbf{0} \tag{4.309}$$

We assume that the electric and magnetic fields are equal in plane M. Eq. (4.309) is an indirect eigenvalue problem for the effective propagation constant ε_{re}, which is complex because of the radiation. The radiation loss is obtained from the imaginary part of $\sqrt{\varepsilon_{re}}$. The effect of dielectric loss may be considered by a second calculation, using complex permittivities. This does not complicate the numerical algorithm as the computation already includes complex quantities because of the consideration of radiation loss. If the field does not change much compared to the first calculation, the dielectric loss may be determined from the difference of the loss values.

4.4.5.6 Interface conditions and non-equidistant discretisation

The consideration of the interface conditions at permittivity steps in radial direction r leads to a monotonic convergence behaviour. This can be achieved by choosing suitable values for the permittivities in the vicinity of the steps (cf. the appendices). However, in certain cases the steps cannot be placed on suitable lines. In the ideal case shown in Fig. 4.50, the step is placed on a • line ($p = 0$). For the permittivity on this place we choose:

$$\varepsilon_{rt} = (\varepsilon_{r1} + \varepsilon_{r2})/2 + p\,(\varepsilon_{r1} - \varepsilon_{r2}); \qquad -0.5 \le p \le 0.5 \tag{4.310}$$

where ε_{r1} (ε_{r2}) is the permittivity on the left (right) side of the steps. For the right rib step of the structure shown in Fig. 4.50, ε_{r1} and ε_{r2} are equal to $\varepsilon_{r1} = n_R^2$ and $\varepsilon_{r2} = 1$, respectively, n being the refractive index. The term with p allows a shift of the step from the • line by ph. It should be mentioned that a non-equidistant discretisation can also be introduced. The principles are analogous to the discretisation in Cartesian coordinate direction (cf. the appendices).

4.4.5.7 Numerical results

To validate the algorithm a 90° bend of a rib waveguide was analysed. Fig. 4.51 shows the field component E_r in the middle of the film as a function of the position. Because of the leftward bending, the field is shifted to the right. At the inner side of the bend the field decreases very rapidly, whereas at the outer side it has a finite value even at a large distance from the rib. The rib does not have a great influence on the field in the film. Therefore, an effective index analysis of rib waveguide bends would give incorrect results. The field distribution obtained with the algorithm in Section 4.4.2 is also included in Fig. 4.51. As we can see, the difference is very small. The reason for the (small) difference is the finite height of the cross-section in the algorithm of Section 4.4.2. Here, on the other hand, we used an infinite air and substrate thickness. The radiation loss for a 90° circular bend is plotted as a function of the bend radius r_o (parameterised with rib height d_R) for TE and TM modes in Fig. 4.52. Noteworthy is the rapid decrease of the loss with r_o. For comparison, the numerical results from [29] (obtained by a scalar MoL procedure using Bessel and Hankel functions) and measured results from [30] are included. The diagram shows that the loss is very sensitive to the rib height. Unlike other algorithms, no higher transcendental functions (e.g. Bessel and Hankel functions) are required to compute the losses. By choosing large values for r_0, the algorithm automatically gives results for straight guides.

4.4.6 Uniaxial anisotropic fibres with circular and noncircular cross-section – discretisation in ϕ-direction

4.4.6.1 Introduction

Optical fibres only have a circular cross-section in the ideal case. Therefore, structures with noncircular cross-sections are of particular interest, especially elliptical and star-like cross-sections (cf. Figs 4.53).

An algorithm for the analysis of fibres with nearly arbitrary cross-sections will be presented in this section. This algorithm is based on the last section, and we use discretisation lines of varying length. Impedance/admittance transformation formulas between the interfaces of the layers are developed. The materials are assumed to have uniaxial anisotropy.

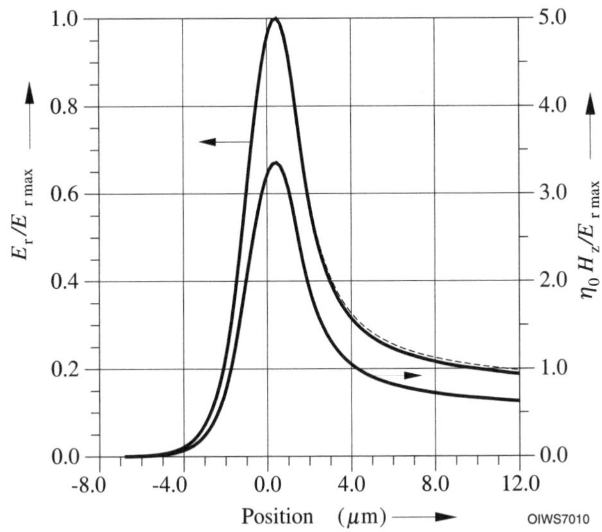

Fig. 4.51 Distribution of the field component E_r in the middle of the film of the structure in Fig. 4.50: $w = 2.9$ μm, $d_R = 0.32$ μm, $d_B = 0$, $d_F = 1.12$ μm, $n_S = 3.3042$, $n_F = 3.3735$, $n_R - 3.3735$, $n_o = 1$ — this algorithm, --- algorithm in Section 4.2 (R. Pregla, 'The Method of Lines for the Analysis of Dielectric Waveguides Bends', *J. Lightwave Technol.*, vol. 14, no. 4, pp. 634–639. © 1996 Institute of Electrical and Electronics Engineers (IEEE))

4.4.6.2 Matching conditions for the interface E

In this subsection we would like to analyse layers with varying thicknesses (see Fig. 4.54) by using the algorithm previously developed and discretisation lines of different lengths. Matching the tangential fields at a noncircular interface (the layers are labelled I and II) results in:

$$\begin{bmatrix} \overline{\mathbf{r}}_h(\mathbf{E}^{\mathrm{I}}_\phi + \tan\boldsymbol{\alpha}_h \boldsymbol{M}_e \mathbf{E}^{\mathrm{I}}_r) \\ -\mathrm{j}\mathbf{E}^{\mathrm{I}}_z \end{bmatrix} = \begin{bmatrix} \overline{\mathbf{r}}_h(\mathbf{E}^{\mathrm{II}}_\phi + \tan\boldsymbol{\alpha}_h \boldsymbol{M}_e \mathbf{E}^{\mathrm{II}}_r) \\ -\mathrm{j}\mathbf{E}^{\mathrm{II}}_z \end{bmatrix} \tag{4.311}$$

$$\begin{bmatrix} \mathrm{j}\mathbf{H}^{\mathrm{I}}_z \\ \overline{\mathbf{r}}_e(\widetilde{\mathbf{H}}^{\mathrm{I}}_\phi + \tan\boldsymbol{\alpha}_e \boldsymbol{M}_h \widetilde{\mathbf{H}}^{\mathrm{I}}_r) \end{bmatrix} = \begin{bmatrix} \mathrm{j}\mathbf{H}^{\mathrm{II}}_z \\ \overline{\mathbf{r}}_e(\widetilde{\mathbf{H}}^{\mathrm{II}}_\phi + \tan\boldsymbol{\alpha}_e \boldsymbol{M}_h \widetilde{\mathbf{H}}^{\mathrm{II}}_r) \end{bmatrix} \tag{4.312}$$

The matching of the longitudinal component $\mathbf{E}_z$ is done as before. For the transverse tangential component we must also take into account the r components, because they are parts of the tangential interface components, as can be seen in Fig. 4.54. We divide the equations that contain ϕ components by $\cos\boldsymbol{\alpha}_{e,h}$. If the radius $\overline{r}_i(\phi)$ of the interface is known in analytical form, we may derive $\tan\boldsymbol{\alpha}$ from $\tan\alpha = \overline{r}_i^{-1}\mathrm{d}\overline{r}_i/\mathrm{d}\phi = \overline{r}_i^{-1}\overline{r}_i'(\phi)$ and by discretisation for the two line systems. The values have to be collected in diagonal matrices. Because we generally need r_i in discretised form for the two discretisation line systems as diagonal matrices $\overline{\mathbf{r}}_e$ and $\overline{\mathbf{r}}_h$, respectively, we can also construct special

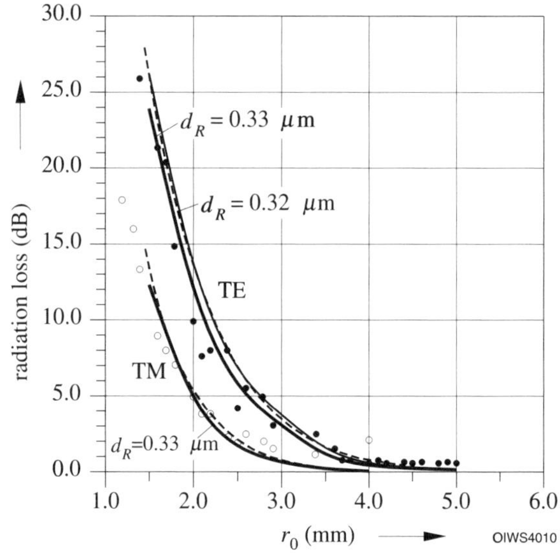

Fig. 4.52 Radiation loss of a 90° bend $a = 8.68 \cdot \frac{\pi}{2}\overline{r}_o \mathrm{Im}(\sqrt{\varepsilon_{re}})$[dB] as function of bend radius r_o. Dimensions as in Fig. 4.51. —— this algorithm, - - - results from [29] for $d_\mathrm{R} = 0.33\mu$m, ○ and • measured results [30] for TM and TE modes, respectively (R. Pregla, 'The Method of Lines for the Analysis of Dielectric Waveguides Bends', *J. Lightwave Technol.*, vol. 14, no. 4, pp. 634–639. © 1996 Institute of Electrical and Electronics Engineers (IEEE))

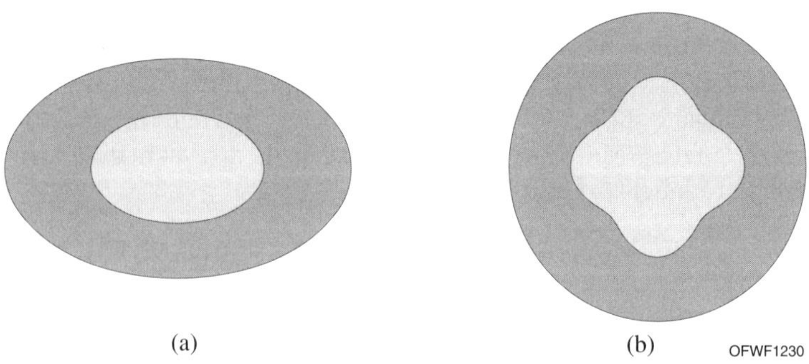

Fig. 4.53 (a) Dielectric elliptical waveguide (b) waveguide with a star-formed cross-section (R. Pregla and O. Conradi, 'Modeling of uniaxial anisotropic fibers with noncircular cross-section by the Method of Lines', *J. Lightwave Technol.*, vol. 21, pp. 1294–1299. © 2003 Institute of Electrical and Electronics Engineers (IEEE))

column-vector matrices (symbolised by the superscript e) $\overline{\mathbf{r}}_e^\mathrm{e} = \mathrm{diag}\,(\overline{\mathbf{r}}_e^\mathrm{e})$ and $\overline{\mathbf{r}}_h^\mathrm{e} = \mathrm{diag}\,(\overline{\mathbf{r}}_h^\mathrm{e})$. A small extension is required in the case of non-periodic boundary conditions. Namely, we have to take into account the value of $\overline{r}_i$ at the Dirichlet boundary. Therefore, for constructing $\tan \boldsymbol{\alpha}_{e,h}$ (only for this

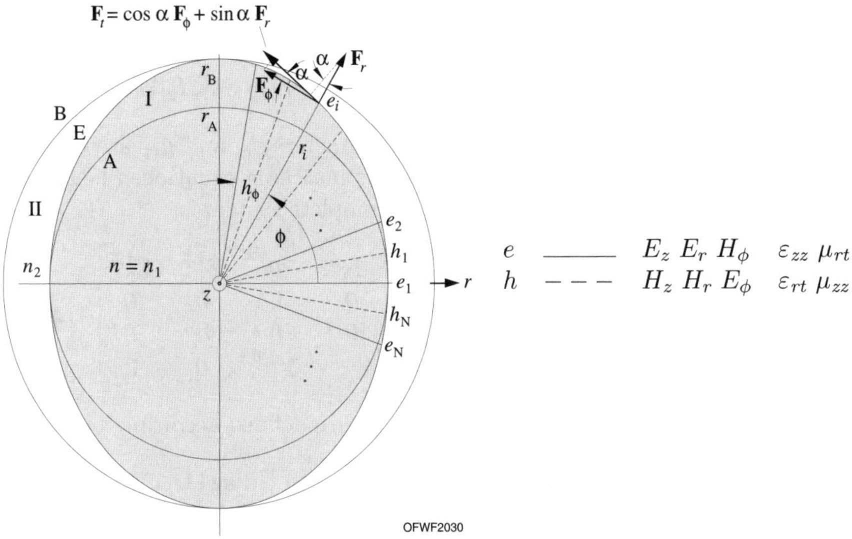

Fig. 4.54 Noncircular cross-section with discretisation lines and interface conditions (R. Pregla and O. Conradi, 'Modeling of uniaxial anisotropic fibers with noncircular cross-section by the Method of Lines', *J. Lightwave Technol.*, vol. 21, pp. 1294–1299. © 2003 Institute of Electrical and Electronics Engineers (IEEE))

task), the difference operators $\overline{\boldsymbol{D}}_e$ and $\overline{\boldsymbol{D}}_h$ must also be changed to $\overline{\boldsymbol{D}}_e^{\text{e}}$ and $\overline{\boldsymbol{D}}_h^{\text{e}}$. The Dirichlet condition must be changed to a Neumann condition. We may now write for $\tan \boldsymbol{\alpha}_{e,h}$ in discretised form:

$$\tan \boldsymbol{\alpha}_e = \overline{\mathbf{r}}_e^{-1} \mathrm{diag}\,(\overline{\boldsymbol{D}}_h^{\text{e}} \overline{\mathbf{r}}_h^{\text{e}}) \qquad \tan \boldsymbol{\alpha}_h = \overline{\mathbf{r}}_h^{-1} \mathrm{diag}\,(\overline{\boldsymbol{D}}_e^{\text{e}} \overline{\mathbf{r}}_e^{\text{e}}) \tag{4.313}$$

We introduce interpolation matrices $\boldsymbol{M}_{e,h}$ in eqs. (4.311) and (4.312) between the discretisation line systems as the r components are not discretised on the same line system as the ϕ components. The interpolation matrices are defined by linear and quadratic interpolation, respectively, in the case of periodic boundary conditions, as:

$$\begin{aligned} \boldsymbol{M}_e &= \frac{1}{2}\begin{bmatrix} 1 & 1 & & & \\ & 1 & 1 & & \\ & & \ddots & \ddots & \\ 1 & & & & 1 \end{bmatrix} = \boldsymbol{M}_h^{\text{t}} \\ \boldsymbol{M}_e &= \frac{1}{8}\begin{bmatrix} 3 & 6 & -1 & & \\ & 3 & 6 & -1 & \\ & & \ddots & \ddots & \\ -1 & & & & \\ 6 & -1 & & & 3 \end{bmatrix} = \boldsymbol{M}_h^{\text{t}} \end{aligned} \tag{4.314}$$

For $\mathbf{E}_r$ and $\mathbf{H}_r$, we have (see eqs. (4.253) and (4.234)):

$$\overline{\mathbf{r}}\widetilde{\mathbf{H}}_r = -\mu_{rt}^{-1}[\sqrt{\varepsilon_{re}}\boldsymbol{I}_h \;\; \overline{\boldsymbol{D}}_e]\widehat{\mathbf{E}} \qquad \overline{\mathbf{r}}\mathbf{E}_r = \varepsilon_{rt}^{-1}[-\overline{\boldsymbol{D}}_h \;\; \sqrt{\varepsilon_{re}}\boldsymbol{I}_e]\widehat{\widetilde{\mathbf{H}}} \tag{4.315}$$

The layers are labelled I and II. The parameters ε_{rt} and μ_{rt} are different in both layers. Due to the r components, the matching equations (4.311) and (4.312) are coupled with each other. The complete system is:

$$\widehat{\boldsymbol{P}}_{\mathrm{E}}^{\mathrm{I}}\begin{bmatrix}\widehat{\mathbf{E}}^{\mathrm{I}}\\ \widehat{\mathbf{H}}^{\mathrm{I}}\end{bmatrix} = \widehat{\boldsymbol{P}}_{\mathrm{E}}^{\mathrm{II}}\begin{bmatrix}\widehat{\mathbf{E}}^{\mathrm{II}}\\ \widehat{\mathbf{H}}^{\mathrm{II}}\end{bmatrix} \quad \text{where} \quad \widehat{\boldsymbol{P}}_{\mathrm{E}}^{\mathrm{I,II}} = \begin{bmatrix}\boldsymbol{I}_h & \boldsymbol{0} & \boldsymbol{Z}_1^{\mathrm{I,II}} & \boldsymbol{Z}_2^{\mathrm{I,II}}\\ \boldsymbol{0} & \boldsymbol{I}_e & \boldsymbol{0} & \boldsymbol{0}\\ \boldsymbol{0} & \boldsymbol{0} & \boldsymbol{I}_h & \boldsymbol{0}\\ \mathbf{Y}_2^{\mathrm{I,II}} & \mathbf{Y}_1^{\mathrm{I,II}} & \boldsymbol{0} & \boldsymbol{I}_e\end{bmatrix} \tag{4.316}$$

The impedance and admittance matrices are defined according to:

$$\begin{aligned}\widehat{\boldsymbol{Z}}_1^{\mathrm{I,II}} &= -\varepsilon_{rt}^{-1}\mathrm{diag}\,(\overline{\boldsymbol{D}}_e^{\mathrm{e}}\overline{\mathbf{r}}_e^{\mathrm{e}})\boldsymbol{M}_e\overline{\boldsymbol{r}}_e^{-1}\overline{\boldsymbol{D}}_h \quad & \widehat{\boldsymbol{Z}}_2^{\mathrm{I,II}} &= \varepsilon_{rt}^{-1}\sqrt{\varepsilon_{re}}\mathrm{diag}\,(\overline{\boldsymbol{D}}_e^{\mathrm{e}}\overline{\mathbf{r}}_e^{\mathrm{e}})\boldsymbol{M}_e\overline{\boldsymbol{r}}_e^{-1}\\ \widehat{\mathbf{Y}}_1^{\mathrm{I,II}} &= -\mu_{rt}^{-1}\mathrm{diag}\,(\overline{\boldsymbol{D}}_h^{\mathrm{e}}\overline{\mathbf{r}}_h^{\mathrm{e}})\boldsymbol{M}_h\overline{\boldsymbol{r}}_h^{-1}\overline{\boldsymbol{D}}_e \quad & \widehat{\mathbf{Y}}_2^{\mathrm{I,II}} &= -\mu_{rt}^{-1}\sqrt{\varepsilon_{re}}\mathrm{diag}\,(\overline{\boldsymbol{D}}_h^{\mathrm{e}}\overline{\mathbf{r}}_h^{\mathrm{e}})\boldsymbol{M}_h\overline{\boldsymbol{r}}_h^{-1}\end{aligned} \tag{4.317}$$

The parameters ε_{rt} and μ_{rt} must be marked with I or II. The transfer matrix relation between both sides of interface E is now given by:

$$\begin{bmatrix}\widehat{\mathbf{E}}_{\mathrm{E}}^{\mathrm{I}}\\ \widehat{\mathbf{H}}_{\mathrm{E}}^{\mathrm{I}}\end{bmatrix} = (\widehat{\boldsymbol{P}}_{\mathrm{E}}^{\mathrm{I}})^{-1}\widehat{\boldsymbol{P}}_{\mathrm{E}}^{\mathrm{II}}\begin{bmatrix}\widehat{\mathbf{E}}_{\mathrm{E}}^{\mathrm{II}}\\ \widehat{\mathbf{H}}_{\mathrm{E}}^{\mathrm{II}}\end{bmatrix} = \begin{bmatrix}\widehat{\boldsymbol{V}}_{\mathrm{E}}^{\mathrm{E}} & \widehat{\boldsymbol{Z}}_{\mathrm{E}}^{\mathrm{E}}\\ \widehat{\mathbf{Y}}_{\mathrm{H}}^{\mathrm{E}} & \widehat{\boldsymbol{V}}_{\mathrm{H}}^{\mathrm{E}}\end{bmatrix}\begin{bmatrix}\widehat{\mathbf{E}}_{\mathrm{E}}^{\mathrm{II}}\\ \widehat{\mathbf{H}}_{\mathrm{E}}^{\mathrm{II}}\end{bmatrix} \tag{4.318}$$

Defining impedances at both sides of the interface E by $\widehat{\mathbf{E}}_{\mathrm{E}}^{\mathrm{I,II}} = \widehat{\boldsymbol{Z}}_{\mathrm{E}}^{\mathrm{I,II}}\widehat{\mathbf{H}}_{\mathrm{E}}^{\mathrm{I,II}}$, the impedance transformation from side II to side I yields:

$$\widehat{\boldsymbol{Z}}_{\mathrm{E}}^{\mathrm{I}} = \left(\widehat{\boldsymbol{V}}_{\mathrm{E}}^{\mathrm{E}}\widehat{\boldsymbol{Z}}_{\mathrm{E}}^{\mathrm{II}} + \widehat{\boldsymbol{Z}}_{\mathrm{E}}^{\mathrm{E}}\right)\left(\widehat{\mathbf{Y}}_{\mathrm{H}}^{\mathrm{E}}\widehat{\boldsymbol{Z}}_{\mathrm{E}}^{\mathrm{II}} + \widehat{\boldsymbol{V}}_{\mathrm{H}}^{\mathrm{E}}\right)^{-1} \tag{4.319}$$

With the help of the general transfer matrix relations, e.g. in eq. (4.239) and (4.240), we can write the original fields in plane E as functions of the transformed fields in plane B and A, respectively:

$$\begin{bmatrix}\widehat{\mathbf{E}}_{\mathrm{E}}^{\mathrm{II}}\\ \widehat{\mathbf{H}}_{\mathrm{E}}^{\mathrm{II}}\end{bmatrix} = \begin{bmatrix}\widehat{\boldsymbol{V}}_{1\mathrm{B}} & \widehat{\boldsymbol{Z}}_{\mathrm{B}}\\ \widehat{\mathbf{Y}}_{\mathrm{B}} & \widehat{\boldsymbol{V}}_{2\mathrm{B}}\end{bmatrix}\begin{bmatrix}\widehat{\overline{\mathbf{E}}}_{\mathrm{B}}\\ \widehat{\overline{\mathbf{H}}}_{\mathrm{B}}\end{bmatrix} \qquad \begin{bmatrix}\widehat{\mathbf{E}}_{\mathrm{E}}^{\mathrm{I}}\\ \widehat{\mathbf{H}}_{\mathrm{E}}^{\mathrm{I}}\end{bmatrix} = \begin{bmatrix}\widehat{\boldsymbol{V}}_{1\mathrm{A}} & \widehat{\boldsymbol{Z}}_{\mathrm{A}}\\ \widehat{\mathbf{Y}}_{\mathrm{A}} & \widehat{\boldsymbol{V}}_{2\mathrm{A}}\end{bmatrix}\begin{bmatrix}\widehat{\overline{\mathbf{E}}}_{\mathrm{A}}\\ \widehat{\overline{\mathbf{H}}}_{\mathrm{A}}\end{bmatrix} \tag{4.320}$$

Because of different line lengths, we obtain from eq. (4.241), with $\widehat{\boldsymbol{T}} = \mathrm{Diag}(\boldsymbol{T_h}, \boldsymbol{T_e})$ e.g.:

$$\boldsymbol{V}_{1\mathrm{A}} = \widehat{\boldsymbol{T}} \bullet \frac{\pi}{2}\begin{bmatrix}\overline{q}_{\boldsymbol{\nu}_h}(\mathbf{t}_{\mathrm{E}}, t_{\mathrm{A}}) & -\varepsilon_{dt}^{-1}\left(\overline{r}_{\boldsymbol{\nu}_h}(\mathbf{t}_{\mathrm{E}}, t_{\mathrm{A}}) + \overline{q}_{\boldsymbol{\nu}_h}(\mathbf{t}_{\mathrm{E}}, t_{\mathrm{A}})\right)\widetilde{\boldsymbol{\delta}}_e\\ \boldsymbol{0} & -\overline{\mathbf{r}}_{\boldsymbol{\nu}_e}(\mathbf{t}_{\mathrm{E}}, t_{\mathrm{A}})\end{bmatrix} \tag{4.321}$$

The multiplication sign '•' means that the matrices have to be multiplied element by element (array multiplication). The different line lengths of the normalised radii (see eq. (4.220)) at interface E are collected in the column vector $\mathbf{t}_\mathrm{E}$. The cross-products of the Bessel functions have to be written as row vectors determined by the row vector $\boldsymbol{\nu}$. The arguments in the cross-products are the column vectors $\mathbf{t}_\mathrm{E}$. Therefore, the cross-products are full matrices. The other matrices in eq. (4.241) must be constructed analogously. To obtain impedance transformation formulas, we determine the z-matrix parameters by:

$$
\begin{aligned}
\tilde{\mathbf{z}}^{\mathrm{II}}_{11} &= \mathbf{V}_{1\mathrm{B}}\mathbf{Y}_\mathrm{B}^{-1} & \tilde{\mathbf{z}}^{\mathrm{I}}_{11} &= -\mathbf{Y}_\mathrm{A}^{-1}\mathbf{V}_{2\mathrm{A}} \\
\tilde{\mathbf{z}}^{\mathrm{II}}_{12} &= \mathbf{V}_{1\mathrm{B}}\mathbf{Y}_\mathrm{B}^{-1}\mathbf{V}_{2\mathrm{B}} - \mathbf{Z}_\mathrm{B} & \tilde{\mathbf{z}}^{\mathrm{I}}_{12} &= -\mathbf{Y}_\mathrm{A}^{-1} \\
\tilde{\mathbf{z}}^{\mathrm{II}}_{22} &= \mathbf{Y}_\mathrm{B}^{-1}\mathbf{V}_{2\mathrm{B}} & \tilde{\mathbf{z}}^{\mathrm{I}}_{22} &= -\mathbf{V}_{1\mathrm{A}}\mathbf{Y}_\mathrm{A}^{-1} \\
\tilde{\mathbf{z}}^{\mathrm{II}}_{21} &= \mathbf{Y}_\mathrm{B}^{-1} & \tilde{\mathbf{z}}^{\mathrm{I}}_{21} &= -\mathbf{V}_{1\mathrm{A}}\mathbf{Y}_\mathrm{A}^{-1}\mathbf{V}_{2\mathrm{A}} + \mathbf{Z}_\mathrm{A}
\end{aligned}
\tag{4.322}
$$

The tilde in these parameters indicates that they connect fields in the transformed domain on one side and fields in the original domain on the other side of the section. With these matrices, the input impedances in plane E^{II} of section II and plane A^{I} of section I can be calculated by:

$$
\widehat{\mathbf{Z}}^{\mathrm{II}}_\mathrm{E} = \tilde{\mathbf{z}}^{\mathrm{II}}_{11} - \tilde{\mathbf{z}}^{\mathrm{II}}_{12}\left(\tilde{\mathbf{z}}^{\mathrm{II}}_{22} + \overline{\mathbf{Z}}^{\mathrm{II}}_\mathrm{B}\right)^{-1}\tilde{\mathbf{z}}^{\mathrm{II}}_{21} \qquad \overline{\mathbf{Z}}^{\mathrm{I}}_\mathrm{A} = \tilde{\mathbf{z}}^{\mathrm{I}}_{11} - \tilde{\mathbf{z}}^{\mathrm{I}}_{12}\left(\tilde{\mathbf{z}}^{\mathrm{I}}_{22} + \widehat{\mathbf{Z}}^{\mathrm{I}}_\mathrm{E}\right)^{-1}\tilde{\mathbf{z}}^{\mathrm{I}}_{21} \tag{4.323}
$$

In eq. (4.323) the impedance of the outer region $\overline{\mathbf{Z}}^{\mathrm{II}}_\mathrm{B}$ is transformed to the elliptic region $\widehat{\mathbf{Z}}^{\mathrm{II}}_\mathrm{E}$. Using (4.319) the tangential field can be matched between regions E^{II} and E^{I}. To find a solution the impedance $\overline{\mathbf{Z}}^{\mathrm{I}}_\mathrm{A}$ must be matched at both sides of the inner region A.

4.4.6.3 Numerical results

To validate the developed algorithm with discretisation lines of different lengths, we calculate the cut-off wavelength of elliptic hollow waveguides. Exact values can be determined from the non-vanishing roots $e\chi'_i$ and $o\chi'_i$ of the derivatives of the radial Mathieu functions Re'_m and Ro'_m [31]:

$$
Re'_m(\xi_\mathrm{b}, e\chi'_i) = 0 \qquad Ro'_m(\xi_\mathrm{b}, o\chi'_i) = 0 \quad \text{with} \quad \cosh\xi_\mathrm{b} = 1/e \tag{4.324}
$$

In our case, we use $m = 1$ for the azimuthal dependence of the H_{11} modes. If eccentricity $e = 0$, the waveguide is circular and we obtain from the adequate Bessel function $\lambda_c/s = 1.0/1.84118 = 0.54318$. In Fig. 4.55 the normalised cut-off wavelengths are plotted against the eccentricity e. We see a very good agreement up $e \approx 0.7$ or $b/a \approx 0.7$.

Next we examine elliptic dielectric waveguide. The electric fields of the fundamental modes are shown in Fig. 4.56 for an axis ratio $r_2/r_1 = 0.7$.

The determined effective index is shown in Fig. 4.57. Here we have the refractive index of the elliptical core $n_1 = 1.539$, with air as the outer region. The axis ratio of the ellipse (half axes a and b) is $b/a = 0.5$.

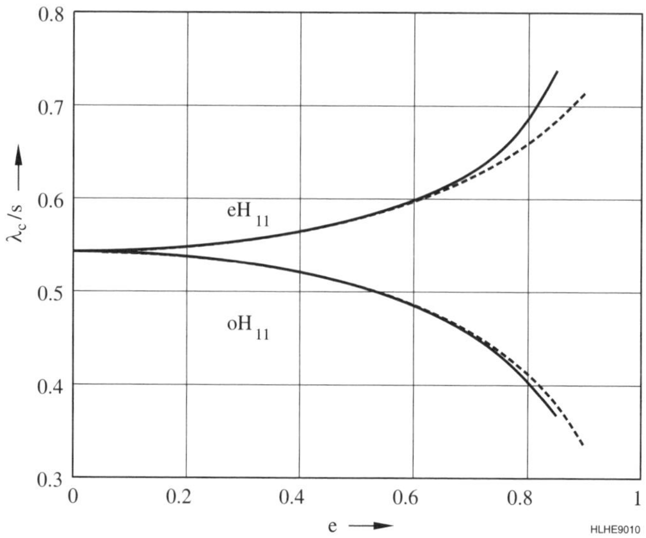

Fig. 4.55 Normalised cutoff wavelength of the H_{11} mode in an elliptical hollow waveguide as a function of the eccentricity $e = \sqrt{1-(b/a)^2}$, s is the circumference of the boundary ellipse: $s = 4aE(e)$ with the elliptic integral $E(e)$. - - - result from the algorithm described in Section 8.7

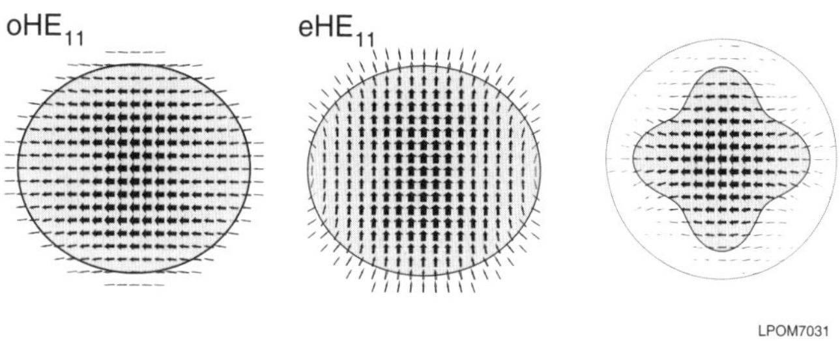

Fig. 4.56 Electric field vector for fibres with elliptical cross-section and for a star fibre with an axis ratio $r_2/r_1 = 0.7$

The results obtained by the presented algorithm are compared with numerical results obtained by the MoL and a Fourier expansion in the azimuthal direction. We also show results from [32].

In the case of an elliptic waveguide, the modes are divided into different polarisations with different propagation, constants. Therefore we show two curves in Fig. 4.57 (and for the fields in Fig. 4.56). The effective index is plotted against normalised frequency $V_b = 2\pi\, b/\lambda\, \sqrt{n_1^2 - n_2^2}$ for the proposed elliptic waveguide. We can see that the results differ slightly. The reason for this might be the small axis ratio $b/a = 0.5$.

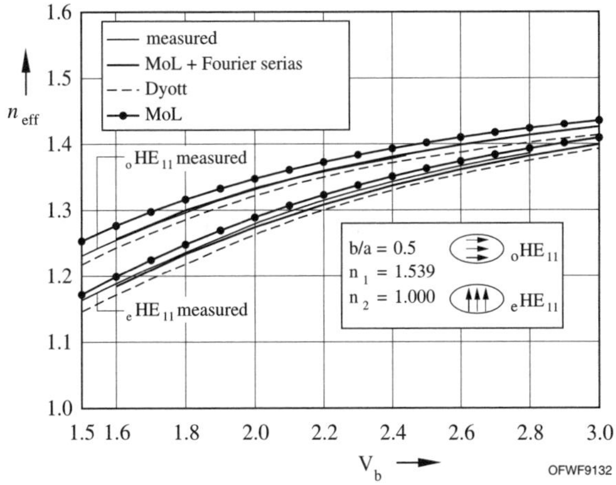

Fig. 4.57 Effective index of fibres with elliptical cross-section versus normalised frequency V_b. MoL + Fourier: the azimuthal field dependence is described by Fourier series (R. Pregla and O. Conradi, 'Modeling of uniaxial anisotropic fibers with noncircular cross-section by the Method of Lines', *J. Lightwave Technol.*, vol. 21, pp. 1294–1299. © 2003 Institute of Electrical and Electronics Engineers (IEEE))

Additionally, a waveguide with a star-like cross-section (see the insert in Fig. 4.58) has been analysed. This kind of waveguide leads to decoupled fundamental modes and therefore to a defined polarisation of the electric field vectors [34]. The electric field is shown in Fig. 4.56.

In Fig. 4.58 we show the effective index versus waveguide axis ratio r_2/r_1. For decreasing ratios r_2/r_1, the effective index also decreases because of the smaller cross-section of the core. The numerical results in Fig. 4.58 have been compared with the previously mentioned Fourier expansion and the well-known 2D discretisations in cylindrical and elliptical coordinates. All those results are based on the MoL. We can see that the results agree very well between $0.7 < r_2/r_1 < 1.0$. As in the case of elliptic cross-section, the reason for the differences between the curves might be the difference in the length of discretisation lines used. However, for ratios $r_2/r_1 \geq 0.7$ the results agree very well. For higher differences in the lengths of the discretisation lines the results are slightly different.

It should be mentioned that the fibre waveguides with noncircular cross-section can also be analysed by using the algorithm of impedance/admittance transformation by finite differences (see Section 2.5.3). Fig. 4.59 shows a more general cross-section. We have three main regions (or layers), I, II and III. In the homogeneous regions I (inside the circle C_1) and III (outside the circle C_2) the field in r-direction is determined analytically. The inhomogeneous layer II is partitioned in sublayers. In this layer the FD-algorithm (see Section 2.5.3) can be used to describe the radial field dependence.

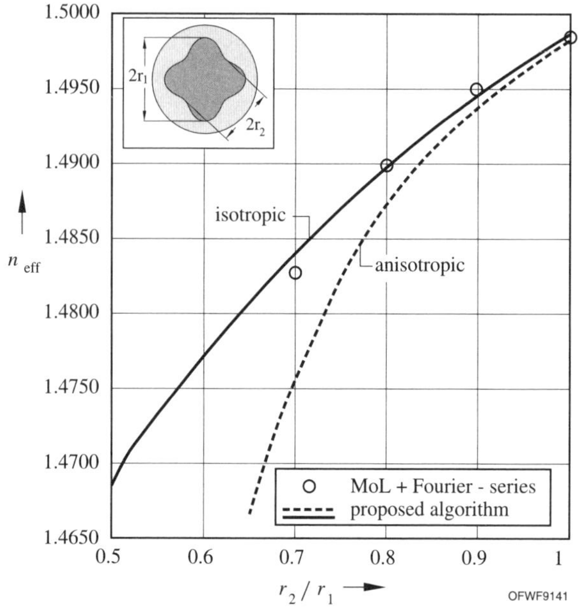

Fig. 4.58 Effective index of fibres with cross-section of star form versus the axis ratio r_2/r_1. $r_1/\lambda = 1$, $n_1 = 1.539$, $n_2 = 1.0$, anisotropic case: $n_{1t} = n_1$, $n_{1zz} = 1.535$ MoL + Fourier: the azimuthal field dependence is described by Fourier series (R. Pregla and O. Conradi, 'Modeling of uniaxial anisotropic fibers with noncircular cross-section by the Method of Lines', *J. Lightwave Technol.*, vol. 21, pp. 1294–1299. © 2003 Institute of Electrical and Electronics Engineers (IEEE))

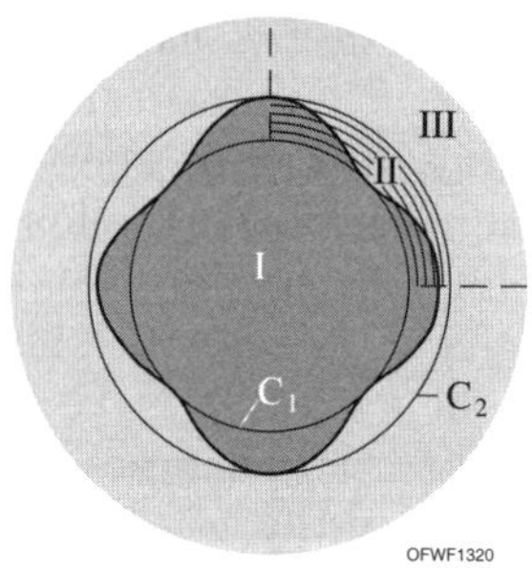

Fig. 4.59 Cross-section of a noncircular fibre with partition of the inhomogeneous layer

4.5 DISCONTINUITIES IN CIRCULAR WAVEGUIDES – ONE-DIMENSIONAL DISCRETISATION IN RADIAL DIRECTION

4.5.1 Introduction

In this section we will present an analysis of waveguides as shown in Fig. 4.60. We begin with waveguides that have rotational symmetry.

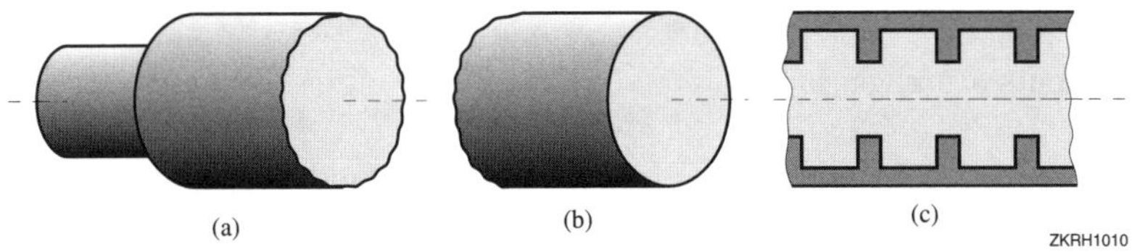

Fig. 4.60 Circular waveguide discontinuities (a) diameter step (b) open end (c) filter

If the fields have rotational symmetry as well, we can introduce $\partial/\partial\phi = 0$ ($m = 0$). We see from eqs. (4.22) and (4.23) that the equations for TE_z and TM_z modes decouple. The coaxial line and its fundamental TEM mode are also rotationally symmetric. If we only have discontinuities in z-direction, the fields remain independent of ϕ. The equations which will be derived in the next subsections for TM_z modes can be used for the analysis of a variety of discontinuities in coaxial lines. Some typical examples are shown in Fig. 4.61. As a special example we will analyse the diameter step depicted in Fig. 4.61b. The material parameters should be given by $\varepsilon_{rr} = \varepsilon_{\phi\phi} = \varepsilon_{zz} = \varepsilon_r$ and $\mu_{rr} = \mu_{\phi\phi} = \mu_{zz} = \mu_r$, which may be functions of the radius; also steps in r-direction are possible. We would like to derive equations for discontinuities and open ends. The basic equations are (4.11) and (4.14), together with eqs. (4.22) and (4.23).

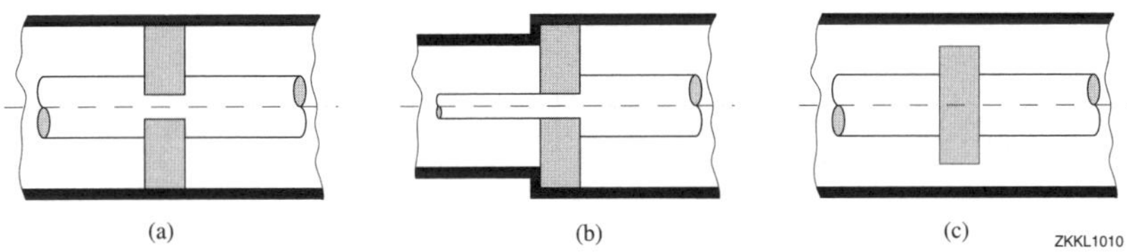

Fig. 4.61 Examples of various discontinuities in coaxial line (a) conductor support (b) diameter step (c) filter structure (Reproduced by permission of IGTE (Institut für Grundlagen und Theorie der Elektrotechnik) Graz)

4.5.2 Basic equations for rotational symmetry

In the case of rotational symmetry of the field, we can see from eqs. (4.22) and (4.23) that the equations decouple for TE_z and TM_z modes. To obtain symmetric matrices in the equations, all the field components are normalised with $\sqrt{r}$ according to:

$$F_{un} = \sqrt{r}F_u, \quad u = r, \phi, z, \quad F = E, H \tag{4.325}$$

Using the abbreviations $D_{r\mathrm{n}} = \frac{1}{\sqrt{\overline{r}}}\frac{\partial}{\partial \overline{r}}\sqrt{\overline{r}}$ and $D_{r\mathrm{n}}^{\mathrm{t}} = \sqrt{\overline{r}}\frac{\partial}{\partial \overline{r}}\frac{1}{\sqrt{\overline{r}}}$, the field components are given by:

TE$_z$ modes | TM$_z$ modes

$$
\begin{array}{ll|ll}
\frac{\partial E_{\phi\mathrm{n}}}{\partial \overline{z}} = -\mathrm{j}R_{\mathrm{H}}^{z}(-H_{r\mathrm{n}}) & R_{\mathrm{H}}^{z} = \mu_r & \frac{\partial}{\partial \overline{z}}\widetilde{H}_{\phi\mathrm{n}} = -\mathrm{j}R_{\mathrm{E}}^{z}E_{r\mathrm{n}} & R_{\mathrm{E}}^{z} = \varepsilon_r \\
\frac{\partial(-H_{r\mathrm{n}})}{\partial \overline{z}} = -\mathrm{j}R_{\mathrm{E}}^{z}E_{\phi\mathrm{n}} & & \frac{\partial}{\partial \overline{z}}E_{r\mathrm{n}} = -\mathrm{j}R_{\mathrm{H}}^{z}\widetilde{H}_{\phi\mathrm{n}} & \\
R_{\mathrm{E}}^{z} = \varepsilon_r + D_{\overline{r}\mathrm{n}}^{\mathrm{t}}\mu_r^{-1}D_{\overline{r}\mathrm{n}} & & R_{\mathrm{H}}^{z} = \mu_r + D_{\overline{r}\mathrm{n}}^{\mathrm{t}}\varepsilon_r^{-1}D_{\overline{r}\mathrm{n}} & \\
\mu_r H_{z\mathrm{n}} = \mathrm{j}D_{\overline{r}\mathrm{n}}E_{\phi\mathrm{n}} & & \varepsilon_r E_{z\mathrm{n}} = -\mathrm{j}D_{\overline{r}\mathrm{n}}\widetilde{H}_{\phi\mathrm{n}} &
\end{array}
\tag{4.326}
$$

4.5.3 Solution of the equations for rotational symmetry

Next, we discretise the fields in r-direction to obtain ordinary differential equations. Fig. 4.62 shows the discretisation lines for the two typical modes of the circular waveguides. In Fig. 4.63 we see the details of the discretisation in the case of a diameter step structure in a coaxial line. The figures also show which boundary conditions have to be fulfilled by the components. The discretisation is given by:

$$
\begin{array}{rl|rl}
\sqrt{\overline{r}}E_{\phi} & \rightarrow \mathbf{E}_{\phi\mathrm{n}} & \sqrt{\overline{r}}\widetilde{H}_{\phi} & \rightarrow \mathbf{H}_{\phi\mathrm{n}} \\
\sqrt{\overline{r}}H_{r} & \rightarrow \mathbf{H}_{r\mathrm{n}} & \sqrt{\overline{r}}\widetilde{E}_{r} & \rightarrow \mathbf{E}_{r\mathrm{n}} \\
\varepsilon_r & \rightarrow \boldsymbol{\varepsilon_r}^{\circ}, \boldsymbol{\varepsilon_r}^{\bullet} & \varepsilon_r & \rightarrow \boldsymbol{\varepsilon_r}^{\circ}, \boldsymbol{\varepsilon_r}^{\bullet} \\
\mu_r & \rightarrow \boldsymbol{\mu_r}^{\circ}, \boldsymbol{\mu_r}^{\bullet} & \mu_r & \rightarrow \boldsymbol{\mu_r}^{\circ}, \boldsymbol{\mu_r}^{\bullet} \\
\overline{r} & \rightarrow \mathbf{r}_{\circ}, \mathbf{r}_{\bullet} & \overline{r} & \rightarrow \mathbf{r}_{\circ}, \mathbf{r}_{\bullet} \\
\sqrt{\overline{r}^{-1}}D_{\overline{r}}\sqrt{\overline{r}} & \rightarrow \sqrt{\mathbf{r}_{\circ}^{-1}}\overline{\mathbf{D}}_{\overline{r}}^{\bullet}\sqrt{\mathbf{r}_{\bullet}} = \overline{\mathbf{D}}_{\overline{r}\mathrm{n}}^{\bullet} & \sqrt{\overline{r}^{-1}}D_{\overline{r}}\sqrt{\overline{r}} & \rightarrow \sqrt{\mathbf{r}_{\bullet}^{-1}}\overline{\mathbf{D}}_{\overline{r}}^{\circ}\sqrt{\mathbf{r}_{\circ}} = \overline{\mathbf{D}}_{\overline{r}\mathrm{n}}^{\circ}
\end{array}
\tag{4.327}
$$

The modal matrices are obtained by transformation to principle axes, i.e.:

$$
\begin{array}{rl|rl}
-\boldsymbol{R}_{\mathrm{H}}^{z}\boldsymbol{R}_{\mathrm{E}}^{z}\boldsymbol{T}_{\mathrm{E}} & = \boldsymbol{T}_{\mathrm{E}}\boldsymbol{\Gamma}_{\bullet}^{2} & -\boldsymbol{R}_{\mathrm{E}}^{z}\boldsymbol{R}_{\mathrm{H}}^{z}\boldsymbol{T}_{\mathrm{H}} & = \boldsymbol{T}_{\mathrm{H}}\boldsymbol{\Gamma}_{\circ}^{2} \\
\boldsymbol{T}_{\mathrm{H}} & = \boldsymbol{R}_{\mathrm{E}}^{z}\boldsymbol{T}_{\mathrm{E}}\boldsymbol{\beta}_{\bullet}^{-1} & \boldsymbol{T}_{\mathrm{E}} & = \boldsymbol{R}_{\mathrm{H}}^{z}\boldsymbol{T}_{\mathrm{H}}\boldsymbol{\beta}_{\circ}^{-1}
\end{array}
\tag{4.328}
$$

where $\boldsymbol{\Gamma}_{\bullet,\circ}^{2} = -\boldsymbol{\beta}_{\bullet,\circ}^{2}$. A subscript indicating the examined case has to be added. As before, we obtain the following relation for the fields in cross-sections A ($z = z_{\mathrm{A}}$) and B ($z = z_{\mathrm{B}} = z_{\mathrm{A}} + d$) of a homogeneous waveguide section:

$$
\begin{bmatrix} (-\overline{\mathbf{H}}_{r\mathrm{nA}}) \\ -(-\overline{\mathbf{H}}_{r\mathrm{nB}}) \end{bmatrix} = \begin{bmatrix} \overline{\mathbf{y}}_1 & \overline{\mathbf{y}}_2 \\ \overline{\mathbf{y}}_2 & \overline{\mathbf{y}}_1 \end{bmatrix} \begin{bmatrix} \overline{\mathbf{E}}_{\phi\mathrm{nA}} \\ \overline{\mathbf{E}}_{\phi\mathrm{nB}} \end{bmatrix} \qquad \begin{bmatrix} \overline{\mathbf{E}}_{\mathrm{rnA}} \\ \overline{\mathbf{E}}_{\mathrm{rnB}} \end{bmatrix} = \begin{bmatrix} \overline{\mathbf{z}}_1 & \overline{\mathbf{z}}_2 \\ \overline{\mathbf{z}}_2 & \overline{\mathbf{z}}_1 \end{bmatrix} \begin{bmatrix} \overline{\mathbf{H}}_{\phi\mathrm{nA}} \\ -\overline{\mathbf{H}}_{\phi\mathrm{nB}} \end{bmatrix}
\tag{4.329}
$$

$$
\begin{array}{rl|rl}
\overline{\mathbf{y}}_1 = & \overline{\boldsymbol{I}}_{\bullet}/\tanh\left(\boldsymbol{\Gamma}_{\bullet}\overline{d}\right) & \overline{\mathbf{z}}_1 = & \overline{\boldsymbol{I}}_{\circ}/\tanh(\boldsymbol{\Gamma}_{\circ}\overline{d}) \\
\overline{\mathbf{y}}_2 = & -\overline{\boldsymbol{I}}_{\bullet}/\sinh\left(\boldsymbol{\Gamma}_{\bullet}\overline{d}\right) & \overline{\mathbf{z}}_2 = & \overline{\boldsymbol{I}}_{\circ}/\sinh(\boldsymbol{\Gamma}_{\circ}\overline{d})
\end{array}
\tag{4.330}
$$

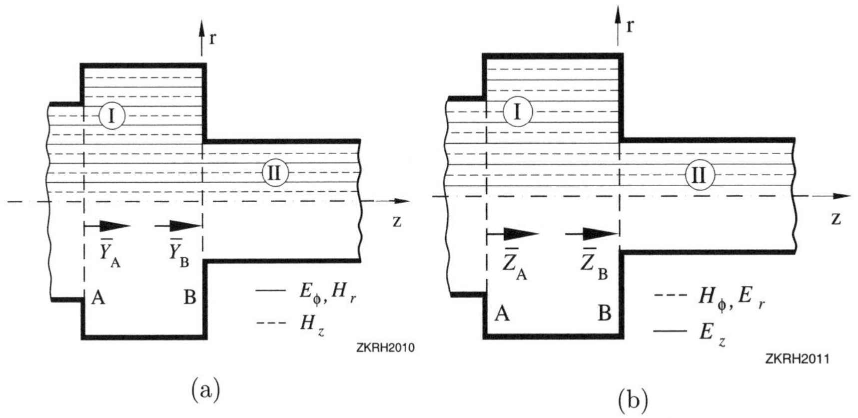

• ——— $E_\phi, H_r, r_\bullet, \mu_r^\bullet, \varepsilon_r^\bullet$
○ – – – $H_z, r_\circ, \mu_r^\circ$

• ——— $E_z, r_\bullet, \varepsilon_r^\bullet$
○ – – – $H_\phi, E_r, r_\circ, \varepsilon_r^\circ, \mu_r^\circ$

Fig. 4.62 Longitudinal section of two connected circular waveguides with discretisation lines for TE_z modes (a) and for TM_z modes (b)

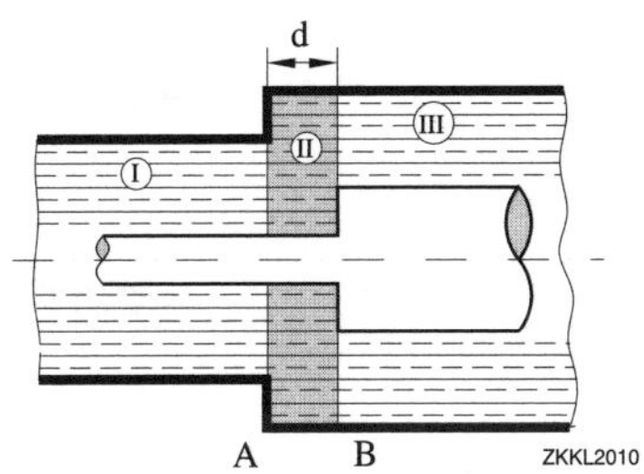

• ——— $E_z, \mathbf{r}_\bullet$
○ – – – $E_r, H_\phi, \mathbf{r}_\circ$

Fig. 4.63 Diameter step in a coaxial line with discretisation lines and partition in three regions (Reproduced by permission of IGTE (Institut für Grundlagen und Theorie der Elektrotechnik) Graz)

4.5.4 Admittance and impedance transformation

The admittance/impedance transformation between cross-sections A and B of a section has to be performed according to Section 2.5. The admittance/impedance transformation at a concatenation of two different waveguides, as in Fig. 4.64, is also given there. It should be remembered that the field components are normalised with $\sqrt{r}$. Therefore, the impedances/admittances are also normalised with the radius. However, the denormalisation can be done very easily.

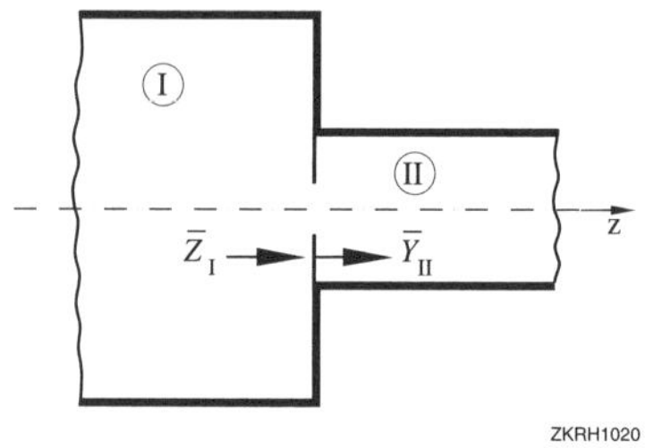

Fig. 4.64 Concatenation of two circular waveguides of different diameters and an iris

4.5.5 Open ending circular waveguide

To model the radiating field, absorbing boundary conditions (ABC) are introduced into the difference operators. The discretisation schemes are sketched in Fig. 4.65.

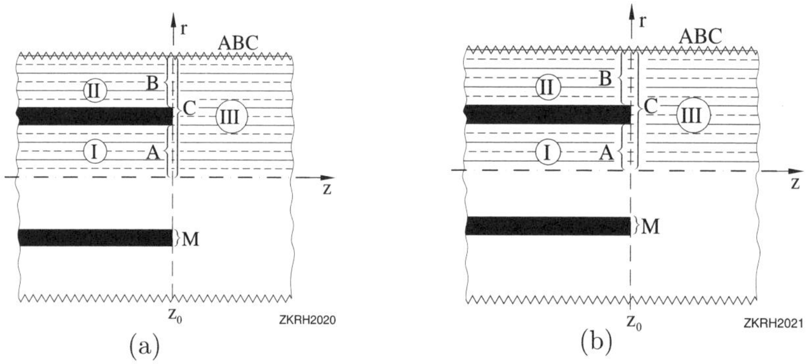

• —— $E_\phi, H_r, r_\bullet$ • —— $E_z, r_\bullet$

∘ – – – $H_z, r_\circ$ ∘ – – – $H_\phi, E_r, r_\circ$

Fig. 4.65 Open end of a circular waveguide with discretisation lines (Reproduced by permission of IGTE (Institut für Grundlagen und Theorie der Elektrotechnik) Graz)

At the open end, the relation between the electric and magnetic fields in regions I, II and III can be described with the help of impedance matrices according to the definitions $\overline{\mathbf{H}}_{\mathrm{A,B}} = \overline{\mathbf{Y}}_{\mathrm{A,B}}\overline{\mathbf{E}}_{\mathrm{A,B}}$, $\overline{\mathbf{E}}_{\mathrm{A,B}} = \overline{\mathbf{Z}}_{\mathrm{A,B}}\overline{\mathbf{H}}_{\mathrm{A,B}}$:

$$\begin{array}{rl|rl} \overline{\mathbf{E}}_{\phi\mathrm{A}} &= \overline{\mathbf{Z}}_{\mathrm{A}}(-\overline{\mathbf{H}}_{r\mathrm{A}}) & \overline{\mathbf{E}}_{r\mathrm{A}} &= \overline{\mathbf{Z}}_{\mathrm{A}}\overline{\mathbf{H}}_{\phi\mathrm{A}} \\ \overline{\mathbf{E}}_{\phi\mathrm{C}} &= \overline{\mathbf{Z}}_{\mathrm{C}}(-\overline{\mathbf{H}}_{r\mathrm{C}}) & \overline{\mathbf{E}}_{r\mathrm{C}} &= \overline{\mathbf{Z}}_{\mathrm{C}}\overline{\mathbf{H}}_{\phi\mathrm{C}} \\ \overline{\mathbf{E}}_{\phi\mathrm{B}} &= \overline{\mathbf{Z}}_{\mathrm{B}}(-\overline{\mathbf{H}}_{r\mathrm{B}}) & \overline{\mathbf{E}}_{r\mathrm{B}} &= \overline{\mathbf{Z}}_{\mathrm{B}}\overline{\mathbf{H}}_{\phi\mathrm{B}} \end{array} \tag{4.331}$$

where $\overline{\boldsymbol{Z}}_{\mathrm{C}} = \boldsymbol{I}^{\mathrm{III}}_{\bullet,\circ} = \overline{\boldsymbol{Y}}_{\mathrm{C}}^{-1}$. In waveguide section II the waves are propagating in $-z$-direction, which results in a minus sign or $\overline{\boldsymbol{Z}}_{\mathrm{B}} = -\boldsymbol{I}^{\mathrm{II}}_{\bullet,\circ}$. In the matching process we can work with normalised quantities because we use identical normalisations on both sides of the discontinuities. The impedance $\overline{\boldsymbol{Z}}_{\mathrm{B}}$ is obtained from the matching process at $z = z_0$.

The electric field is divided according to: ($u = \phi$ for TE and $u = r$ for TM modes.)

$$\mathbf{E}_{u\mathrm{C}} = \begin{bmatrix} \mathbf{E}_{u\mathrm{A}} \\ \mathbf{E}_{u\mathrm{M}} \\ \mathbf{E}_{u\mathrm{B}} \end{bmatrix} \tag{4.332}$$

where the parts $\mathbf{E}_{u\mathrm{A,B}}$ stand for the fields in region A, B. The subvector $\mathbf{E}_{u\mathrm{M}}$ represents the electric field on the non-ideal surface of the metallic ring between regions B and C. For ideal metal, $\mathbf{E}_{u\mathrm{M}}$ is a vector whose components are zero. For the vectors in transformed domain we obtain:

$$\boldsymbol{T}_{\mathrm{E}}^{\mathrm{III}}\overline{\mathbf{E}}_{u\mathrm{C}} = \begin{bmatrix} \boldsymbol{T}_{\mathrm{EA}}^{\mathrm{III}} \\ \boldsymbol{T}_{\mathrm{EM}}^{\mathrm{III}} \\ \boldsymbol{T}_{\mathrm{EB}}^{\mathrm{III}} \end{bmatrix} \overline{\mathbf{E}}_{u\mathrm{C}} = \begin{bmatrix} \boldsymbol{T}_{\mathrm{E}}^{\mathrm{I}}\overline{\mathbf{E}}_{u\mathrm{A}} \\ \mathbf{E}_{u\mathrm{M}} \\ \boldsymbol{T}_{\mathrm{E}}^{\mathrm{II}}\overline{\mathbf{E}}_{u\mathrm{B}} \end{bmatrix} \tag{4.333}$$

$\boldsymbol{T}_{\mathrm{EA}}^{\mathrm{III}}$, $\boldsymbol{T}_{\mathrm{EB}}^{\mathrm{III}}$ and $\boldsymbol{T}_{\mathrm{EM}}^{\mathrm{III}}$ are parts of $\boldsymbol{T}_{\mathrm{E}}^{\mathrm{III}}$ associated with the regions A, B and metallisation M, respectively. They were obtained by matrix partition technique, as proposed in Section 2.3.7.

The vector $\overline{\mathbf{E}}_{u\mathrm{C}}$ as a function of $\overline{\mathbf{E}}_{u\mathrm{A}}$, $\overline{\mathbf{E}}_{u\mathrm{M}}$ and $\overline{\mathbf{E}}_{u\mathrm{B}}$ is therefore given as:

$$\overline{\mathbf{E}}_{u\mathrm{C}} = \boldsymbol{T}_{\mathrm{E}}^{\mathrm{III}^{-1}} \begin{bmatrix} \boldsymbol{T}_{\mathrm{E}}^{\mathrm{I}}\overline{\mathbf{E}}_{u\mathrm{A}} \\ \mathbf{E}_{u\mathrm{M}} \\ \boldsymbol{T}_{\mathrm{E}}^{\mathrm{II}}\overline{\mathbf{E}}_{u\mathrm{B}} \end{bmatrix} \tag{4.334}$$

This can be written in the following form (using the partition parts $\boldsymbol{T}_{\mathrm{E\,A}}^{\mathrm{III}^{-1}}$, $\boldsymbol{T}_{\mathrm{E\,B}}^{\mathrm{III}^{-1}}$ and $\boldsymbol{T}_{\mathrm{E\,M}}^{\mathrm{III}^{-1}}$ of $\boldsymbol{T}_{\mathrm{E}}^{\mathrm{III}^{-1}}$):

$$\overline{\mathbf{E}}_{u\mathrm{C}} = \boldsymbol{T}_{\mathrm{E\,A}}^{\mathrm{III}^{-1}}\boldsymbol{T}_{\mathrm{E}}^{\mathrm{I}}\overline{\mathbf{E}}_{u\mathrm{A}} + \boldsymbol{T}_{\mathrm{E\,B}}^{\mathrm{III}^{-1}}\boldsymbol{T}_{\mathrm{E}}^{\mathrm{II}}\overline{\mathbf{E}}_{u\mathrm{B}} + \boldsymbol{T}_{\mathrm{E\,M}}^{\mathrm{III}^{-1}}\mathbf{E}_{u\mathrm{M}} \tag{4.335}$$

As with the electric field, we divide the magnetic one according to: ($v = r$ for TE and $v = \phi$ for TM modes.)

$$\mathbf{H}_{v\mathrm{C}} = \begin{bmatrix} \mathbf{H}_{v\mathrm{A}} \\ \mathbf{J}_{\mathrm{M}} \\ \mathbf{H}_{v\mathrm{B}} \end{bmatrix} \tag{4.336}$$

$\mathbf{J}_{\mathrm{M}}$ is the current density in the metallic ring between regions A and B, i.e. on the same places as $\mathbf{E}_{u\mathrm{M}}$. With the transformed quantities the equation takes the form:

$$\boldsymbol{T}_{\mathrm{H}}^{\mathrm{III}}\overline{\mathbf{H}}_{v\mathrm{C}} = \begin{bmatrix} \boldsymbol{T}_{\mathrm{HA}}^{\mathrm{III}} \\ \boldsymbol{T}_{\mathrm{HM}}^{\mathrm{III}} \\ \boldsymbol{T}_{\mathrm{HB}}^{\mathrm{III}} \end{bmatrix} \overline{\mathbf{H}}_{v\mathrm{C}} = \begin{bmatrix} \boldsymbol{T}_{\mathrm{H}}^{\mathrm{I}}\overline{\mathbf{H}}_{v\mathrm{A}} \\ \mathbf{J}_{\mathrm{M}} \\ \boldsymbol{T}_{\mathrm{H}}^{\mathrm{II}}\overline{\mathbf{H}}_{v\mathrm{B}} \end{bmatrix} \tag{4.337}$$

The following three equations result from this array expression:

$$\boldsymbol{T}_{\mathrm{H}}^{\mathrm{I}}\overline{\mathbf{H}}_{v\mathrm{A}} = \boldsymbol{T}_{\mathrm{HA}}^{\mathrm{III}}\overline{\mathbf{H}}_{v\mathrm{C}} \tag{4.338}$$

$$\boldsymbol{T}_{\mathrm{H}}^{\mathrm{II}}\overline{\mathbf{H}}_{v\mathrm{B}} = \boldsymbol{T}_{\mathrm{HB}}^{\mathrm{III}}\overline{\mathbf{H}}_{v\mathrm{C}} \tag{4.339}$$

$$\mathbf{J}_{\mathrm{M}} = \boldsymbol{T}_{\mathrm{HM}}^{\mathrm{III}}\overline{\mathbf{H}}_{v\mathrm{C}} \tag{4.340}$$

Next, we replace $\overline{\mathbf{H}}_{v\mathrm{C}}$ in eqs. (4.338)–(4.340) with $\overline{\boldsymbol{Y}}_{\mathrm{C}}\overline{\mathbf{E}}_{u\mathrm{C}}$ or $-\overline{\boldsymbol{Y}}_{\mathrm{C}}\overline{\mathbf{E}}_{u\mathrm{C}}$ for TM_z or TE_z modes, respectively. Then, by introducing eq. (4.331) into eqs. (4.338) and (4.339), we obtain the associated equations for tangential electric field vectors $\overline{\mathbf{E}}_{u\mathrm{A}}$ and $\overline{\mathbf{E}}_{u\mathrm{B}}$:

$$\overline{\mathbf{E}}_{u\mathrm{A}} = \overline{\boldsymbol{Z}}_{\mathrm{A}}\boldsymbol{T}_{\mathrm{H}}^{\mathrm{I}\;-1}\boldsymbol{T}_{\mathrm{HA}}^{\mathrm{III}}\overline{\boldsymbol{Y}}_{\mathrm{C}}\overline{\mathbf{E}}_{u\mathrm{C}} \tag{4.341}$$

$$\overline{\mathbf{E}}_{u\mathrm{B}} = \overline{\boldsymbol{Z}}_{\mathrm{B}}\boldsymbol{T}_{\mathrm{H}}^{\mathrm{II}\;-1}\boldsymbol{T}_{\mathrm{HB}}^{\mathrm{III}}\overline{\boldsymbol{Y}}_{\mathrm{C}}\overline{\mathbf{E}}_{u\mathrm{C}} \tag{4.342}$$

By introducing eq. (4.335) into the above equations we obtain:

$$\begin{bmatrix} (-)\overline{\mathbf{H}}_{v\mathrm{A}} \\ (-)\overline{\mathbf{H}}_{v\mathrm{B}} \\ (-)\mathbf{J}_{\mathrm{M}} \end{bmatrix} = \begin{bmatrix} \overline{\boldsymbol{y}}_{\mathrm{AA}} & \overline{\boldsymbol{y}}_{\mathrm{AB}} & \overline{\boldsymbol{y}}_{\mathrm{AM}} \\ \overline{\boldsymbol{y}}_{\mathrm{BA}} & \overline{\boldsymbol{y}}_{\mathrm{BB}} & \overline{\boldsymbol{y}}_{\mathrm{BM}} \\ \overline{\boldsymbol{y}}_{\mathrm{MA}} & \overline{\boldsymbol{y}}_{\mathrm{MB}} & \overline{\boldsymbol{y}}_{\mathrm{MM}} \end{bmatrix} \begin{bmatrix} \overline{\mathbf{E}}_{u\mathrm{A}} \\ \overline{\mathbf{E}}_{u\mathrm{B}} \\ \mathbf{E}_{u\mathrm{M}} \end{bmatrix} \tag{4.343}$$

where the (−) sign is valid for the TE_z modes. The submatrices of the admittance are:

$$\begin{aligned}
\overline{\boldsymbol{y}}_{\mathrm{AA}} &= \boldsymbol{T}_{\mathrm{H}}^{\mathrm{I}\;-1}\boldsymbol{T}_{\mathrm{HA}}^{\mathrm{III}}\overline{\boldsymbol{Y}}_{\mathrm{C}}\boldsymbol{T}_{\mathrm{E\,A}}^{\mathrm{III}\;-1}\boldsymbol{T}_{\mathrm{E}}^{\mathrm{I}} & \overline{\boldsymbol{y}}_{\mathrm{AB}} &= \boldsymbol{T}_{\mathrm{H}}^{\mathrm{I}\;-1}\boldsymbol{T}_{\mathrm{HA}}^{\mathrm{III}}\overline{\boldsymbol{Y}}_{\mathrm{C}}\boldsymbol{T}_{\mathrm{E\,B}}^{\mathrm{III}\;-1}\boldsymbol{T}_{\mathrm{E}}^{\mathrm{II}} \\
\overline{\boldsymbol{y}}_{\mathrm{BA}} &= \boldsymbol{T}_{\mathrm{H}}^{\mathrm{II}\;-1}\boldsymbol{T}_{\mathrm{HB}}^{\mathrm{III}}\overline{\boldsymbol{Y}}_{\mathrm{C}}\boldsymbol{T}_{\mathrm{E\,A}}^{\mathrm{III}\;-1}\boldsymbol{T}_{\mathrm{E}}^{\mathrm{I}} & \overline{\boldsymbol{y}}_{\mathrm{BB}} &= \boldsymbol{T}_{\mathrm{H}}^{\mathrm{II}\;-1}\boldsymbol{T}_{\mathrm{HB}}^{\mathrm{III}}\overline{\boldsymbol{Y}}_{\mathrm{C}}\boldsymbol{T}_{\mathrm{E\,B}}^{\mathrm{III}\;-1}\boldsymbol{T}_{\mathrm{E}}^{\mathrm{II}} \\
\overline{\boldsymbol{y}}_{\mathrm{MA}} &= \boldsymbol{T}_{\mathrm{HM}}^{\mathrm{III}}\overline{\boldsymbol{Y}}_{\mathrm{C}}\boldsymbol{T}_{\mathrm{EA}}^{\mathrm{III}\;-1}\boldsymbol{T}_{\mathrm{E}}^{\mathrm{I}} & \overline{\boldsymbol{y}}_{\mathrm{MB}} &= \boldsymbol{T}_{\mathrm{HM}}^{\mathrm{III}}\overline{\boldsymbol{Y}}_{\mathrm{C}}\boldsymbol{T}_{\mathrm{E\,B}}^{\mathrm{III}\;-1}\boldsymbol{T}_{\mathrm{E}}^{\mathrm{II}} \\
\overline{\boldsymbol{y}}_{\mathrm{AM}} &= \boldsymbol{T}_{\mathrm{H}}^{\mathrm{I}\;-1}\boldsymbol{T}_{\mathrm{HA}}^{\mathrm{III}}\overline{\boldsymbol{Y}}_{\mathrm{C}}\boldsymbol{T}_{\mathrm{E\,M}}^{\mathrm{III}\;-1} \\
\overline{\boldsymbol{y}}_{\mathrm{BM}} &= \boldsymbol{T}_{\mathrm{H}}^{\mathrm{II}\;-1}\boldsymbol{T}_{\mathrm{HB}}^{\mathrm{III}}\overline{\boldsymbol{Y}}_{\mathrm{C}}\boldsymbol{T}_{\mathrm{E\,M}}^{\mathrm{III}\;-1} \\
\overline{\boldsymbol{y}}_{\mathrm{MM}} &= \boldsymbol{T}_{\mathrm{HM}}^{\mathrm{III}}\overline{\boldsymbol{Y}}_{\mathrm{C}}\boldsymbol{T}_{\mathrm{E\,M}}^{\mathrm{III}\;-1}
\end{aligned} \tag{4.344}$$

First, we consider ideal metal with $\mathbf{E}_{u\mathrm{M}} = \mathbf{0}$. In this case the last equation in system (4.343) decouples from the first ones. Now, taking into account eq. (4.331) and $\overline{\boldsymbol{Y}}_{\mathrm{B}} = \overline{\boldsymbol{Z}}_{\mathrm{B}}^{-1}$, we get from the second equation in (4.343):

$$\overline{\mathbf{E}}_{u\mathrm{B}} = -(\overline{\boldsymbol{y}}_{\mathrm{BB}} - \overline{\boldsymbol{Y}}_{\mathrm{B}})^{-1}\overline{\boldsymbol{y}}_{\mathrm{BA}}\overline{\mathbf{E}}_{u\mathrm{A}} \tag{4.345}$$

This expression may be introduced into the first equation in (4.343), resulting in:

$$(-)\overline{\mathbf{H}}_{v\mathrm{A}} = (\overline{\boldsymbol{y}}_{\mathrm{AA}} - \overline{\boldsymbol{y}}_{\mathrm{AB}}(\overline{\boldsymbol{y}}_{\mathrm{BB}} - \overline{\boldsymbol{Y}}_{\mathrm{B}})^{-1}\overline{\boldsymbol{y}}_{\mathrm{BA}})\overline{\mathbf{E}}_{\mathrm{uA}} = \overline{\boldsymbol{Y}}_{\mathrm{A}}\overline{\mathbf{E}}_{u\mathrm{A}} \tag{4.346}$$

The admittance $\overline{\boldsymbol{Y}}_{\mathrm{A}}$ is therefore given by:

$$\overline{\boldsymbol{Y}}_{\mathrm{A}} = \overline{\boldsymbol{y}}_{\mathrm{AA}} - \overline{\boldsymbol{y}}_{\mathrm{AB}}(\overline{\boldsymbol{y}}_{\mathrm{BB}} - \overline{\boldsymbol{Y}}_{\mathrm{B}})^{-1}\overline{\boldsymbol{y}}_{\mathrm{BA}} \tag{4.347}$$

which has the well-known form. (Remember that $\overline{\boldsymbol{Y}}_{\mathrm{B}}$ relates to the negative z-direction.)

In the case of non-ideal metal wall approximate boundary conditions for the tangential fields on the metallic surface:

$$\eta_m \vec{H}_t = \vec{e}_r \times \vec{E}_t$$

can be used. The further procedure is analogous to that in Section 4.6.2.6 concerning antennas.

Let us assume that the waveguide is fed by the TE_{01} or TM_{01} mode, respectively. The field propagating in section I towards A is known. The vector of the forward propagating field for this mode may be written as $\overline{\mathbf{E}}_{u\mathrm{A}}^{f} = [1, 0, 0 \ldots 0]^{\mathrm{t}}$. (We sorted the eigenmodes in such a way that the above-mentioned modes are on the first place.) With the input impedance $\overline{\boldsymbol{Z}}_{\mathrm{A}} = \overline{\boldsymbol{Y}}_{\mathrm{A}}^{-1}$ and the characteristic impedance being a unit matrix $\boldsymbol{I}_{\bullet,\circ}^{\mathrm{I}}$, we obtain:

$$\overline{\boldsymbol{H}}_{v\mathrm{A}} = 2(\overline{\boldsymbol{Z}}_{\mathrm{A}} + \boldsymbol{I}_{\bullet,\circ}^{\mathrm{I}})^{-1} \overline{\boldsymbol{E}}_{u\mathrm{A}}^{f} \qquad \overline{\boldsymbol{E}}_{u\mathrm{A}} = \overline{\boldsymbol{Z}}_{\mathrm{A}} \overline{\boldsymbol{H}}_{v\mathrm{A}} \tag{4.348}$$

Using this result, we can calculate the field $\overline{\mathbf{E}}_{u\mathrm{B}}$ with eq. (4.345) and then the other quantities, including the surface current density $\mathbf{J}_{\mathrm{M}}$, by eq. (4.343).

4.5.6 Numerical results for discontinuities in circular waveguides

Fig. 4.66 shows the reflection coefficient R for the E_{01} mode as a function of frequency in the circular waveguide with an open end. The diameter of the waveguide is 30 mm and we have the cut-off frequency $f_c = 7.654$ GHz. The reflection coefficient decreases starting from 1 for $f = f_c$ monotonically with increasing frequency. The phase of the reflection coefficient also decreases from 180° at $f = f_c$ with increasing frequency. Fig. 4.67 shows the distribution of the field component E_r. At the centre of the waveguide this component must vanish and at the metallic edge we obtain a singularity. The phase difference at the metal (i.e. between the inside and the outside of the waveguide) is 180°. The remaining two field components are sketched in Fig. 4.68.

4.5.7 Numerical results for coaxial line discontinuities and coaxial filter devices

As our first example, we give numerical results for the diameter step depicted in Fig. 4.63. We assume that region III is of infinite length. Therefore, the input impedance of that region is equal to its characteristic impedance $\overline{\boldsymbol{Z}}^{\mathrm{III}} = \overline{\boldsymbol{Z}}_0^{\mathrm{III}}$ $(= \boldsymbol{I}_{\circ}^{\mathrm{III}})$. Using the impedance transformation equations in Section 2.5, we can transform $\overline{\boldsymbol{Z}}_0^{\mathrm{III}}$ through interface III/II, section II and interface II/I and obtain the load impedance of section I. From this impedance the reflection coefficient S_{11} can be calculated, followed by the fields in each of the other cross-sections. Results for a diameter step in a coaxial line are given in Fig. 4.69. The reflection coefficient R and the transmission coefficient T are presented.

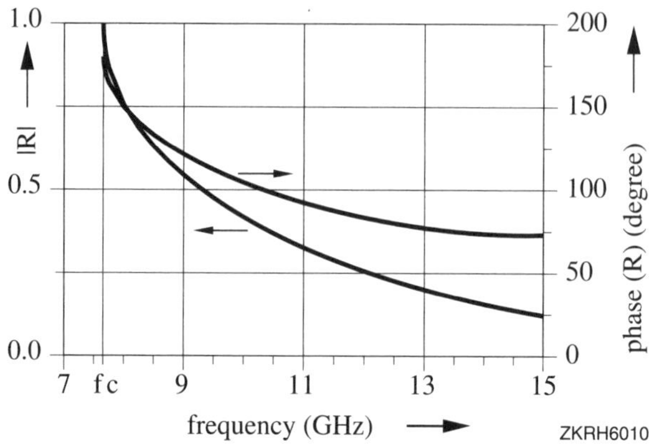

Fig. 4.66 Reflection coefficient of the E_{01} mode in the circular waveguide with open end (Reproduced by permission of IGTE (Institut für Grundlagen und Theorie der Elektrotechnik) Graz)

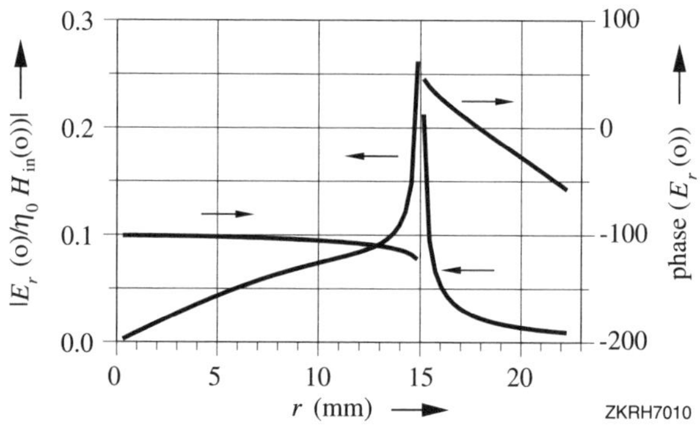

Fig. 4.67 Field component E_r at the open end of a circular waveguide

As we can see, the reflection coefficient increases with the frequency. However, even for $f = 10$ GHz the absolute value is less than 0.08. This is due to the fact that the distance d was optimised.

The other discontinuities in Fig. 4.61 can be analysed in an analogous manner. Next, we show the results of a more complex structure. We can see a coaxial line filter (reported in [35]) in Fig. 4.70. All dimensions given in this figure are in millimetres. The structure is symmetric to the plane S. For the analysis, we examined the structure between the cross-sections A and B. If the sections outside this region are ideal, they do not change the filter behaviour. The analysis of this structure is straightforward. We again start with the characteristic impedance matrix in cross-section B and transform this impedance through all the interfaces and sections to the input

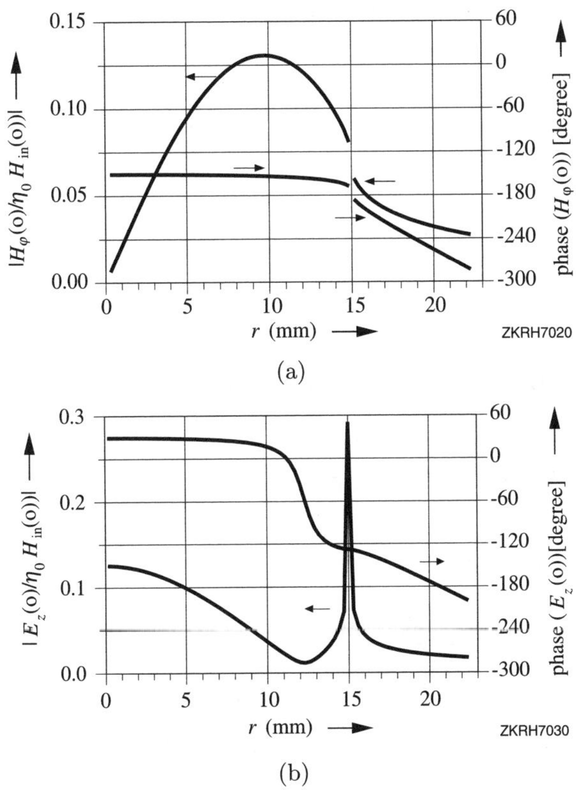

Fig. 4.68 Distribution of the field components H_ϕ and E_z of the E_{01} mode at the open end of a circular waveguide (Reproduced by permission of IGTE (Institut für Grundlagen und Theorie der Elektrotechnik) Graz)

(cross-section A) by using the equations in Section 2.5. From this impedance we compute the scattering parameter S_{11} for the fundamental mode and the fields in the cross-sections, and eventually the scattering parameter S_{21}. Results for the coaxial line filter in Fig. 4.70 are shown in Fig. 4.71. A comparison with the measured ones of [35] shows a good agreement.

Fig. 4.72 shows a periodic structure in the coaxial line. Such a structure can be analysed with the Floquet-mode concept described in Chapter 5.

4.5.8 Non-rotational modes in circular waveguides

This section is of particular importance for the analysis of circular dielectric and fibre waveguides. The reason is that the fields of the fundamental modes do not have a rotational symmetry. Some typical structures are shown in Fig. 4.73a, b, c.

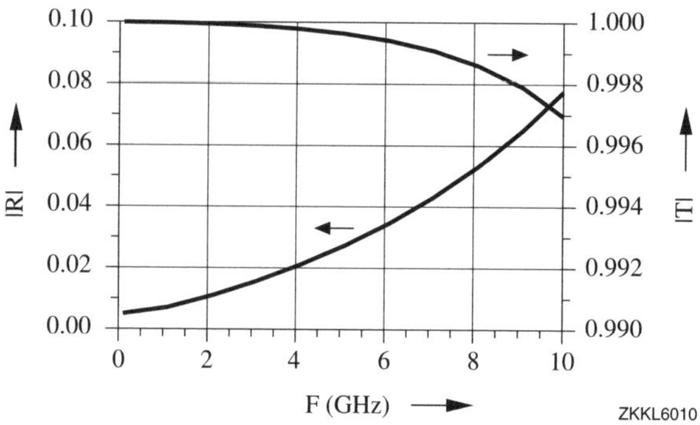

Fig. 4.69 Reflection and transmission determined for a diameter step in a coaxial line (see Fig. 4.63). The inner diameters are 4 mm and 7 mm, the outer diameters are 12 mm and 21 mm, respectively. $d = 0.5$ mm (Reproduced by permission of IGTE (Institut für Grundlagen und Theorie der Elektrotechnik) Graz)

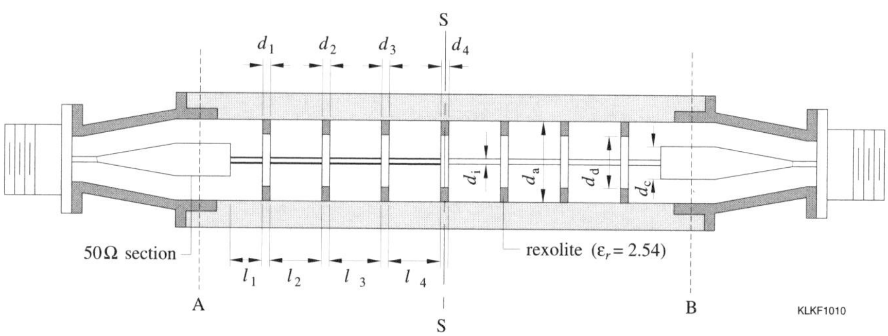

Fig. 4.70 Coaxial line filter (see page 371 in [35]). Dimensions – symmetric to plane S – in mm: $d_\mathrm{i} = 1.8542, d_\mathrm{a} = 22.7838, d_\mathrm{c} = 9.8806, d_\mathrm{d} = 17.4752$ $d_1 = 2.3368, d_2 = d_3 = d_4 = 2.9972$ $l_1 = 10.8712, l_2 = 18.3642, l_3 = l_4 = 18.9738$

Not only longitudinally homogeneous but also inhomogeneous ones like fibre gratings can be studied. Fibre grating structures are analysed by calculating Floquet modes for one period following the description in Section 4.5. If $m > 0$ (azimuthal dependency), we have to start with the operators $[R_\mathrm{E}^z]$ and $[R_\mathrm{H}^z]$ given in eqs. (4.22) and (4.23).

For the discretisation we use the scheme shown in Fig. 4.74 (only one-dimensional discretisation in r-direction is required). Then, the discretised

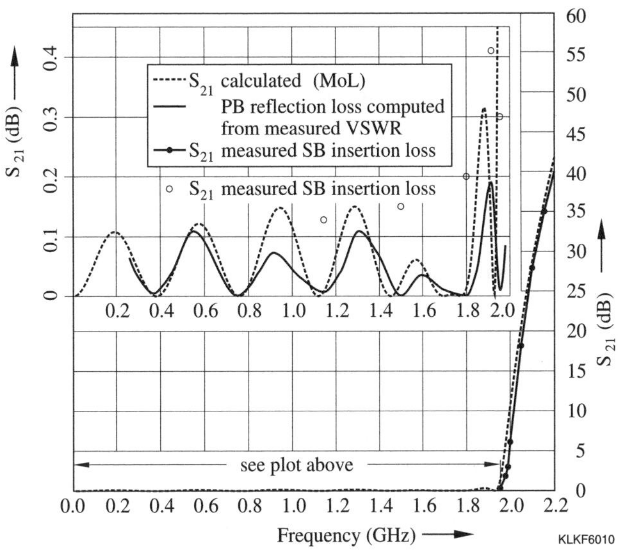

Fig. 4.71 Scattering parameters as functions of the frequency for the filter in Fig. 4.70 (measured results see p. 372 in [35])

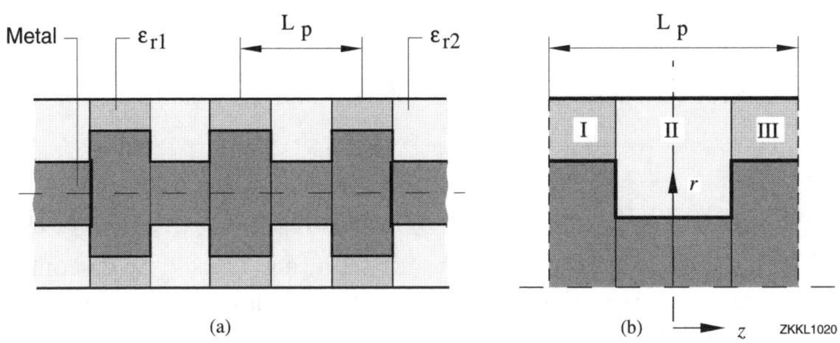

Fig. 4.72 Periodic coaxial filter structure: (a) longitudinal section (b) one period

operators read as: ($\mu_r = 1.$)

$$\boldsymbol{R}_{\mathrm{E}} = \begin{bmatrix} -\overline{\boldsymbol{D}}^{\bullet t}_{\overline{r}}\,\overline{\mathbf{r}}_{\circ}^{-1}\overline{\boldsymbol{D}}^{\bullet}_{\overline{r}} + \boldsymbol{\epsilon}_{\bullet}\overline{\mathbf{r}}_{\bullet}^{-1} & -m\overline{\boldsymbol{D}}^{\bullet t}_{\overline{r}}\,\overline{\mathbf{r}}_{\circ}^{-1} \\ -m\overline{\mathbf{r}}_{\circ}^{-1}\overline{\boldsymbol{D}}^{\bullet}_{\overline{r}} & \boldsymbol{\epsilon}_{\circ}\overline{\mathbf{r}}_{\circ} - m^2\overline{\mathbf{r}}_{\circ}^{-1} \end{bmatrix} \tag{4.349}$$

$$\boldsymbol{R}_{\mathrm{H}} = \begin{bmatrix} \overline{\mathbf{r}}_{\bullet} - m^2\overline{\mathbf{r}}_{\bullet}^{-1}\boldsymbol{\epsilon}_{\bullet z}^{-1} & -m\overline{\mathbf{r}}_{\bullet}^{-1}\boldsymbol{\epsilon}_{\bullet z}^{-1}\overline{\boldsymbol{D}}^{\circ}_{\overline{r}} \\ -m\overline{\boldsymbol{D}}^{\circ t}_{\overline{r}}\,\overline{\mathbf{r}}_{\bullet}^{-1}\boldsymbol{\epsilon}_{\bullet z}^{-1} & -\overline{\boldsymbol{D}}^{\circ t}_{\overline{r}}\,\overline{\mathbf{r}}_{\bullet}^{-1}\boldsymbol{\epsilon}_{\bullet z}^{-1}\overline{\boldsymbol{D}}^{\circ}_{\overline{r}} + \overline{\mathbf{r}}_{\circ}^{-1} \end{bmatrix} \tag{4.350}$$

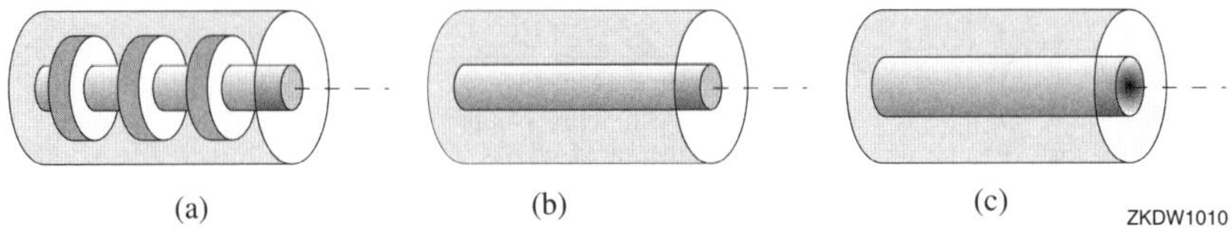

Fig. 4.73 Dielectric waveguides and fibres (a) dielectric insert in a hollow guide (b) monomode fibre (c) gradient fibre

For the products we obtain:

$$
\boldsymbol{R}_{\mathrm{E}}\boldsymbol{R}_{\mathrm{H}} = (\boldsymbol{R}_{\mathrm{H}}\boldsymbol{R}_{\mathrm{E}})^{\mathrm{t}} =
\begin{bmatrix}
\boldsymbol{\epsilon}_\bullet - m^2\overline{\mathbf{r}}_\bullet^{-2}\boldsymbol{\epsilon}_\bullet\boldsymbol{\epsilon}_{\bullet z}^{-1} - \overline{\boldsymbol{D}}_{\overline{r}}^{\bullet\mathrm{t}}\overline{\mathbf{r}}_\circ^{-1}\overline{\boldsymbol{D}}_{\overline{r}}^{\bullet}\overline{\mathbf{r}}_\bullet & -m(\overline{\mathbf{r}}_\bullet^{-2}\boldsymbol{\epsilon}_\bullet\boldsymbol{\epsilon}_{\bullet z}^{-1}\overline{\boldsymbol{D}}_{\overline{r}}^{\circ} + \overline{\boldsymbol{D}}_{\overline{r}}^{\bullet\mathrm{t}}\overline{\mathbf{r}}_\circ^{-2}) \\
-m(\overline{\mathbf{r}}_\circ^{-1}\overline{\boldsymbol{D}}_{\overline{r}}^{\bullet}\overline{\mathbf{r}}_\bullet + \overline{\mathbf{r}}_\circ\boldsymbol{\epsilon}_\circ\overline{\boldsymbol{D}}_{\overline{r}}^{\circ\mathrm{t}}\overline{\mathbf{r}}_\bullet^{-1}\boldsymbol{\epsilon}_{\bullet z}^{-1}) & \boldsymbol{\epsilon}_\circ - m^2\overline{\mathbf{r}}_\circ^{-2} - \boldsymbol{\epsilon}_\circ\overline{\mathbf{r}}_\circ\overline{\boldsymbol{D}}_{\overline{r}}^{\circ\mathrm{t}}\overline{\mathbf{r}}_\bullet^{-1}\boldsymbol{\epsilon}_\bullet^{-1}\overline{\boldsymbol{D}}_{\overline{r}}^{\circ}
\end{bmatrix}
\tag{4.351}
$$

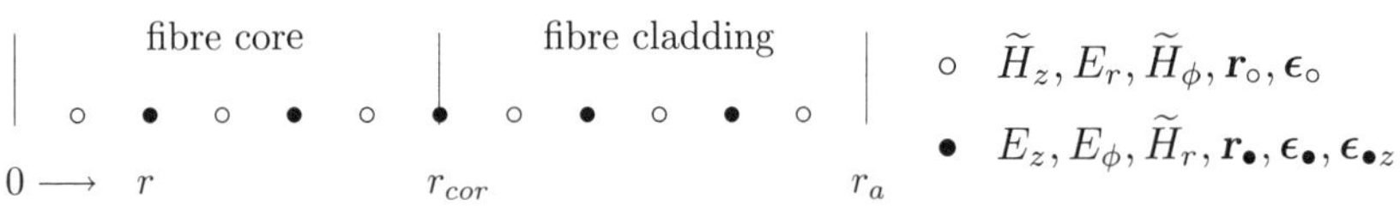

Fig. 4.74 Discretisation scheme in case of H_{11} and E_{11} modes: $\epsilon_\circ = \epsilon_{rr}, \epsilon_\bullet = \epsilon_{\phi\phi}, \epsilon_{\bullet z} = \epsilon_{zz}$

We combine the fields in supervectors according to: $\widehat{\mathbf{H}} = [-\widetilde{\mathbf{H}}_r^{\mathrm{t}}, \overline{\mathbf{r}}_\circ\widetilde{\mathbf{H}}_\phi^{\mathrm{t}}]^{\mathrm{t}}$ and $\widehat{\mathbf{E}} = [\overline{\mathbf{r}}_\bullet\mathbf{E}_\phi^{\mathrm{t}}, \mathbf{E}_r^{\mathrm{t}}]^{\mathrm{t}}$. To take into account radiation (e.g. caused by discontinuities at concatenations), we replace the Dirichlet boundary conditions at $r = r_a$ in the operators with absorbing boundary conditions (ABC). Then $\overline{\boldsymbol{D}}_{\overline{r}}^{\bullet}$ and $-\overline{\boldsymbol{D}}_{\overline{r}}^{\circ\mathrm{t}}$ are replaced by $\overline{\boldsymbol{D}}_{\overline{r}}^{\bullet a}$ and $\overline{\boldsymbol{D}}_{\overline{r}}^{\bullet\mathrm{t}}$ has to be replaced by $-\overline{\boldsymbol{D}}_{\overline{r}}^{\circ}$ in the eqs. (4.349) to (4.351). $\overline{\boldsymbol{D}}_{\overline{r}}^{\circ}$ remains the same as before.

4.5.9 Numerical results and discussion

Results for periodic structure are presented in Chapter 5. Here we only show the principle distribution of the field components (amplitudes) of the fundamental HE_{11} mode as functions of the radius in Fig. 4.75. We used Neumann boundary conditions for the components on the circles and Dirichlet boundary conditions for the components on the bullets (see Fig. 4.74).

The curves for $\mathbf{E}_r$ and $\mathbf{E}_\phi$ resp. for $\mathbf{H}_r$ and $\mathbf{H}_\phi$ are practically lying on each other. At the centre ($r = 0$) the values must be identical.

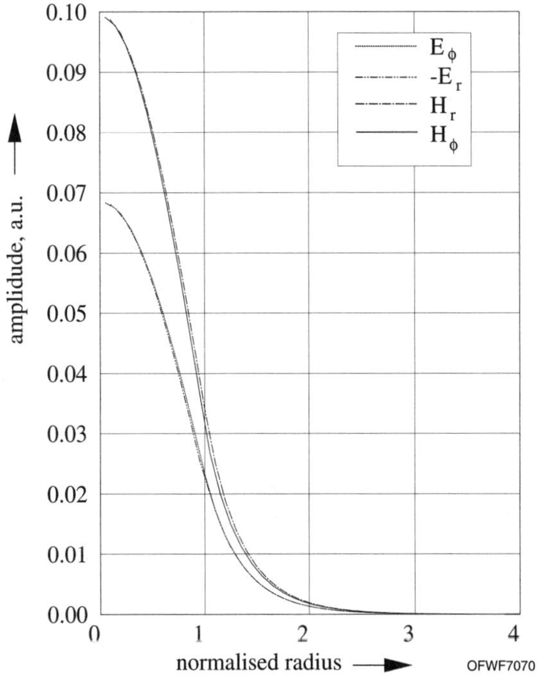

Fig. 4.75 Principle radial dependence of the transverse field components (R. Pregla, 'Modeling of Optical Waveguide Structures with General Anisotropy in Arbitrary Orthogonal Coordinate Systems', *IEEE J. of Sel. Topics in Quantum Electronics*, vol. 8, pp. 1217–1224. © 2002 Institute of Electrical and Electronics Engineers (IEEE))

4.6 ANALYSIS OF GENERAL AXIALLY SYMMETRIC ANTENNAS WITH COAXIAL FEED LINES

4.6.1 Introduction

Planar antennas consist of a metallised (ground side) substrate and metallic patches on the back side. The patches are fed either by striplines on the side of the patches, or by coaxial lines from the ground side. The patches may have rectangular, circular or any other form.

In this section we will show how the Method of Lines (MoL) can be used to analyse circular patch antennas with coaxial feed lines (Fig. 4.76) and how various monopole antennas (Fig. 4.77) can be examined. The analysis of the coaxial feeding lines was shown in the last section.

In the substrate and above the patch the discretisation is carried out in the vertical direction. In the radial direction the field is described analytically by Bessel functions.

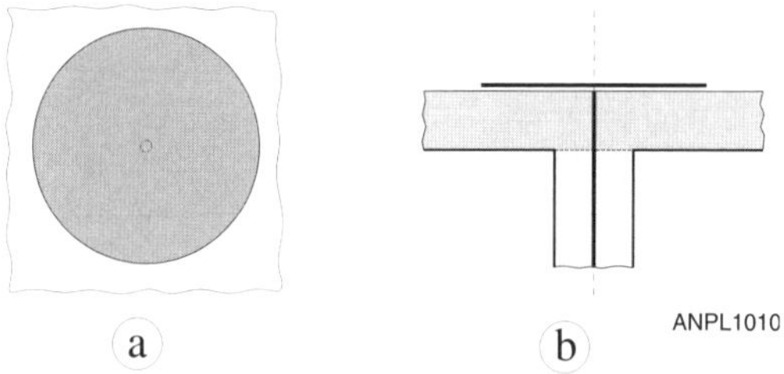

Fig. 4.76 Circular patch antenna (a) top view (b) longitudinal section

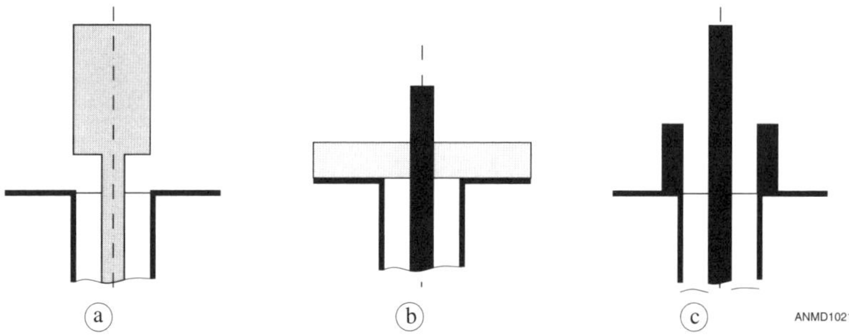

Fig. 4.77 Different types of monopole antennas (a) conventional type (b) monopole antenna partially burned in a grounded dielectric substrate (c) master sleeve antenna

In the transition region from coaxial line to parallel-plate radial waveguide, the field is decomposed into two parts. These parts can be treated as a continuation of the fields in the coaxial line and parallel-plate radial waveguide, respectively. Therefore, these parts are also discretised on the continuation of the discretisation lines of neighbouring regions that cross each other. In this way, the wave transition from the vertical (coaxial line) to the radial direction can be described very easily.

An impedance/admittance transformation procedure is given for all of the regions, which allows us to calculate the input impedance in the coaxial line through a successive transformation from the aperture through the different sections of the considered structure. By introducing a source wave in the coaxial line, the field in the whole structure can be computed. To simulate an open structure at a suitable distance above the patch, absorbing boundary conditions (ABC) are introduced. Results are presented for the input impedance and for the field distribution.

4.6.2 Theory

The general structure to be considered is sketched in Fig. 4.78. All the antenna forms presented in Figs. 4.76 and 4.77 are special cases of this general one.

If the planes P_1 and P_2 are moved towards each other so that they coincide, the structure reduces to a normal patch antenna.

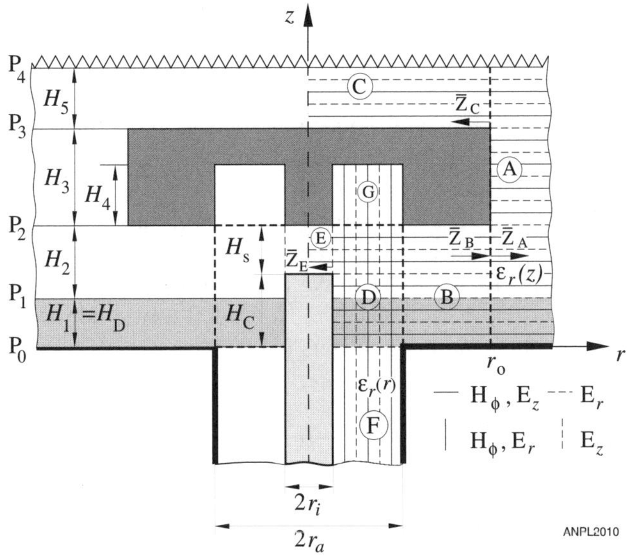

Fig. 4.78 General structure of a circular planar antenna (R. Pregla, 'The analysis of general axially symmetric antennas with a coaxial feed line by the method of lines', *IEEE Trans. Antennas. Propagation*, vol. 46, no. 10, pp. 1433–1443. © 1998 Institute of Electrical and Electronics Engineers (IEEE))

The patch may be of finite thickness but may also be infinitely thin. If we increase H_3 (Fig. 4.78) in such a way that it reaches the order of a quarter wavelength, a normal coaxial-fed monopole (Fig 4.77a) is obtained. In the case of the monopole shown in Fig. 4.77b, plane P_2 is replaced by ABCs. Generally, ABCs are introduced in plane P_4. Because the efficiency of ABCs depends on the angle of incidence (see Section 4.2), they must be chosen carefully. If we assume that the radiation into region C occurs towards the ABCs from the upper metallic edge, the height H_5 should be greater than r_0. The reason for this is that we can very easily achieve high absorption for incidence angles up to 45°. In region A the reflected radiation field at the ABCs does not return to the antenna. Therefore, the problem is not so sensitive in this region.

When we partition the structure in subregions, we must assure that the field in each of these subregions can be described completely after the discretisation. Yet there is some degree of freedom. For example, regions B and D in Fig. 4.78 could be combined. However, as we will see later, the field impedance (admittance) transformation from one region to another is very simple and highly accurate. Therefore, a higher number of subregions is advantageous. We could also e.g. divide region A into three subregions;

two are the continuations of subregions B and C with crossed discretisation lines, while the third is the region between the other two. We would use vertical discretisation lines here. In this subsection we will not make use of this possibility, however the accuracy could be improved with this division.

4.6.2.1 Basic equations

Here we derive the fundamental equations for the above mentioned general structure. The structure has circular symmetry and the field components depend on the azimuthal angle ϕ. Hence, we have $D_\phi = \partial/\partial\phi = 0$. The analysis of the coaxial lines has already been described (Section 4.5). The permittivity ε_r is assumed to be a function of r: $\varepsilon_r = \varepsilon_r(r)$. The only non-zero components are H_ϕ, E_r and E_z. We need suitable GTL equations for H_ϕ and E_z for propagation in r-direction in the radial waveguide sections. These are obtained from the general equations (4.83) with $D_\phi = 0$:

$$\frac{1}{\overline{r}}\frac{\partial}{\partial\overline{r}}(-\overline{r}\widetilde{H}_\phi) = -\mathrm{j}R^r_{\mathrm{E}}E_z \qquad \frac{\partial}{\partial\overline{r}}E_z = -\mathrm{j}R^r_{\mathrm{H}}(-\widetilde{H}_\phi) \tag{4.352}$$

with:

$$R^r_{\mathrm{E}} = \varepsilon_r \qquad R^r_{\mathrm{H}} = 1 + D_{\overline{z}}\varepsilon_r^{-1}D_{\overline{z}} \tag{4.353}$$

The wave equations are obtained very easily and read:

$$\begin{aligned} \frac{\partial}{\partial\overline{r}}\left(\frac{1}{\overline{r}}\frac{\partial}{\partial\overline{r}}(\overline{r}\widetilde{H}_\phi)\right) + Q^r_{\mathrm{H}}\widetilde{H}_\phi = 0 \qquad & Q^r_{\mathrm{H}} = R^r_{\mathrm{E}}R^r_{\mathrm{H}} \\ \frac{1}{\overline{r}}\frac{\partial}{\partial\overline{r}}\left(\overline{r}\frac{\partial}{\partial\overline{r}}E_z\right) + Q^r_{\mathrm{E}}E_z = 0 \qquad & Q^r_{\mathrm{E}} = R^r_{\mathrm{H}}R^r_{\mathrm{E}} \end{aligned} \tag{4.354}$$

The remaining electric field component E_r is given by:

$$\mathrm{j}\varepsilon_r E_r = -\frac{\partial\widetilde{H}_\phi}{\partial\overline{z}} \tag{4.355}$$

4.6.2.2 Discretisation, transformation and general solution

We use different directions of discretisation depending on the various regions of the structure (Fig. 4.78). In the planar regions (A, B, C and E) the discretisation is done in z-direction. In the transition region D (coaxial line–parallel-plate radial waveguide), crossed discretisation line systems are used.

4.6.2.3 Coaxial line sections

Generally, the coaxial line part consists of concatenations of sections with different radii (see Fig. 4.79a). The discretisation lines are shown in Fig. 4.78, and in more detail in Fig. 4.79a.

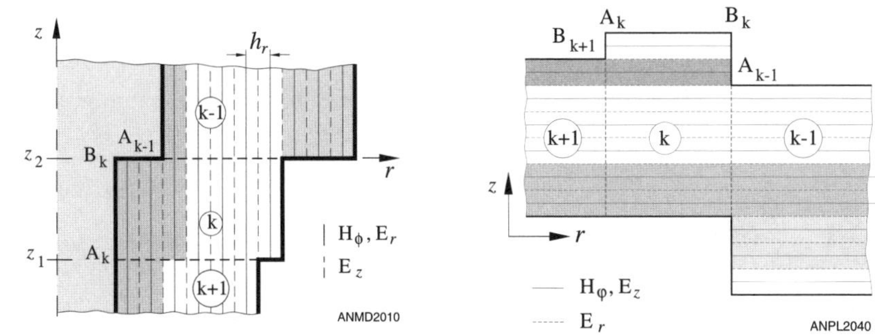

(a) (b)

Fig. 4.79 Concatenations of the waveguide sections with discretisation lines: (a) coaxial sections: the permittivities of the sections are functions of r (b) radial sections: the permittivities of the sections are functions of z (R. Pregla, 'The analysis of general axially symmetric antennas with a coaxial feed line by the method of lines', *IEEE Trans. Antennas. Propagation*, vol. 46, no. 10, pp. 1433–1443. © 1998 Institute of Electrical and Electronics Engineers (IEEE))

4.6.2.4 Radial waveguide sections

Generally, the radial waveguide may consist of concatenations of sections with different heights. We obtain the solution for the fields in each of these sections by discretising the equations in z-direction. The discretisation lines are shown in Fig. 4.78, and in more detail in Fig. 4.79b. The discretisation for the kth section yields:

$$\widetilde{H}_\phi \longrightarrow \widetilde{\mathbf{H}}_\phi \qquad \varepsilon_r \longrightarrow \boldsymbol{\varepsilon}_h, \boldsymbol{\varepsilon}_e \qquad \frac{\partial}{\partial \bar{z}} \longrightarrow \bar{h}_z^{-1} \boldsymbol{D}_{\bar{z}} = \overline{\boldsymbol{D}}_{\bar{z}} \tag{4.356}$$

$$E_z \longrightarrow \mathbf{E}_z \qquad R^r_{\mathrm{E}} \longrightarrow \boldsymbol{\varepsilon}_e \qquad R^r_{\mathrm{H}} \longrightarrow \boldsymbol{I}_e - \overline{\boldsymbol{D}}^{t}_{\bar{z}} \boldsymbol{\varepsilon}_h^{-1} \overline{\boldsymbol{D}}_{\bar{z}} \tag{4.357}$$

H_ϕ and E_z are discretised on the same (full) lines (see Figs. 4.78 and 4.79b). $\boldsymbol{\varepsilon}_e$ denotes the diagonal matrix of the permittivities on these lines, and $\boldsymbol{\varepsilon}_h$ is a diagonal matrix with the permittivities on the dashed lines. The difference operator matrix $\boldsymbol{D}_z$ operating for the full lines has to fulfil Neumann boundary conditions. To avoid confusion, we omit the label k at the discretised fields. Again, we perform a transformation with the matrices $\boldsymbol{T}^r_{\mathrm{E,H}}$ to diagonalise the second-order system matrices $\boldsymbol{Q}^r_{\mathrm{E,H}}$ according to:

$$\widetilde{\mathbf{H}}_\phi = \boldsymbol{T}^r_{\mathrm{H}} \overline{\mathbf{H}}_\phi \qquad \mathbf{E}_z = \boldsymbol{T}^r_{\mathrm{E}} \overline{\mathbf{E}}_z \qquad \boldsymbol{T}^{r\,-1}_{\mathrm{H}} \boldsymbol{Q}^r_{\mathrm{H}} \boldsymbol{T}^r_{\mathrm{H}} = \boldsymbol{T}^{r\,-1}_{\mathrm{E}} \boldsymbol{Q}^r_{\mathrm{E}} \boldsymbol{T}^r_{\mathrm{E}} = \boldsymbol{k}_r^2 \tag{4.358}$$

The diagonal matrix $\boldsymbol{k}_r^2$ contains all the eigenvalues, and $\boldsymbol{T}^r_{\mathrm{E,H}}$ the corresponding eigenvectors, of the matrices $\boldsymbol{Q}^r_{\mathrm{E,H}}$.

By using the relation between the eigenvector matrices given by:

$$\boldsymbol{T}^r_{\mathrm{E}} = \boldsymbol{R}^r_{\mathrm{H}} \boldsymbol{T}^r_{\mathrm{H}} \boldsymbol{k}_r^{-1} \qquad \boldsymbol{T}^r_{\mathrm{H}} = \boldsymbol{R}^r_{\mathrm{E}} \boldsymbol{T}^r_{\mathrm{E}} \boldsymbol{k}_r^{-1} \tag{4.359}$$

we obtain discretised GTL equations in the transformed domain, reading:

$$\frac{1}{\overline{r}}\frac{\partial}{\partial \overline{r}}(-\overline{r}\overline{\mathbf{H}}_\phi) = -\mathrm{j}\boldsymbol{k}_r\overline{\mathbf{E}}_z \qquad \frac{\partial}{\partial \overline{r}}\overline{\mathbf{E}}_z = -\mathrm{j}\boldsymbol{k}_r(-\overline{\mathbf{H}}_\phi) \tag{4.360}$$

The wave equations reduce to Bessel's equations:

$$\frac{\partial}{\partial \overline{r}}\left(\frac{1}{\overline{r}}\frac{\partial}{\partial \overline{r}}(\overline{r}\overline{\mathbf{H}}_\phi)\right) + \boldsymbol{k}_r^2\overline{\mathbf{H}}_\phi = \mathbf{0} \qquad \frac{1}{\overline{r}}\frac{\partial}{\partial \overline{r}}\left(\overline{r}\frac{\partial}{\partial \overline{r}}\overline{\mathbf{E}}_z\right) + \boldsymbol{k}_r^2\overline{\mathbf{E}}_z = \mathbf{0} \tag{4.361}$$

Their general solution is expressed in terms of first- and zeroth-order Bessel and Neumann functions J_1, Y_1 and J_0, Y_0, respectively:

$$\overline{\mathbf{H}}_\phi = \mathrm{J}_1(\boldsymbol{k}_r\overline{r})\mathbf{A} + \mathrm{Y}_1(\boldsymbol{k}_r\overline{r})\mathbf{B} \qquad \overline{\mathbf{E}}_z = \mathrm{J}_0(\boldsymbol{k}_r\overline{r})\mathbf{C} + \mathrm{Y}_0(\boldsymbol{k}_r\overline{r})\mathbf{D} \tag{4.362}$$

The arguments contain the diagonal matrix $\boldsymbol{k}_r$. Therefore, $\mathcal{C}_1(\boldsymbol{k}_r\overline{r})$ ($\mathcal{C}$ is the general notation of cylindrical functions, J or Y) is also a diagonal matrix. It will be abbreviated in the following by $\boldsymbol{\mathcal{C}}_1$ or $\boldsymbol{\mathcal{C}}_{1\mathrm{A}}$ (if $\overline{r} = \overline{r}_\mathrm{A}$). For the fields on the cylinders A ($\overline{r} = \overline{r}_\mathrm{A}$) and B ($\overline{r} = \overline{r}_\mathrm{B}$) we may therefore write:

$$\begin{bmatrix}\overline{\mathbf{H}}_{\phi\mathrm{A}}\\ \overline{\mathbf{H}}_{\phi\mathrm{B}}\end{bmatrix} = \begin{bmatrix}J_1(\boldsymbol{k}_r\overline{r}_\mathrm{A}) & Y_1(\boldsymbol{k}_r\overline{r}_\mathrm{A})\\ J_1(\boldsymbol{k}_r\overline{r}_\mathrm{B}) & Y_1(\boldsymbol{k}_r\overline{r}_\mathrm{B})\end{bmatrix}\begin{bmatrix}\mathbf{A}\\ \mathbf{B}\end{bmatrix} \tag{4.363}$$

$$\mathrm{j}\begin{bmatrix}\overline{\mathbf{E}}_{z\mathrm{A}}\\ \overline{\mathbf{E}}_{z\mathrm{B}}\end{bmatrix} = \begin{bmatrix}J_0(\boldsymbol{k}_r\overline{r}_\mathrm{A}) & Y_0(\boldsymbol{k}_r\overline{r}_\mathrm{A})\\ J_0(\boldsymbol{k}_r\overline{r}_\mathrm{B}) & Y_0(\boldsymbol{k}_r\overline{r}_\mathrm{B})\end{bmatrix}\begin{bmatrix}\mathbf{A}\\ \mathbf{B}\end{bmatrix} \tag{4.364}$$

Formula 9.1.28 of [14] has been used to obtain the relation between $\overline{\mathbf{E}}_{z\mathrm{A,B}}$ and $\overline{\mathbf{H}}_{\phi\mathrm{A,B}}$. Using Hankel functions instead of Bessel and Neumann functions in eq. (4.362), we can define the characteristic impedance matrix (normalised with respect to the free-space wave impedance η_0) for the wave propagating in forward direction at a position r:

$$\overline{\mathbf{E}}_{z\mathrm{f}} = \overline{\boldsymbol{Z}}_{r0}(-\overline{\mathbf{H}}_{\phi\mathrm{f}}) \quad \overline{\boldsymbol{Z}}_{r0} = jH_0^{(2)}(\boldsymbol{k}_r\overline{r})(H_1^{(2)}(\boldsymbol{k}_r\overline{r}))^{-1} \tag{4.365}$$

For $r \to \infty$, the characteristic impedance matrix approaches the unit matrix $\overline{\boldsymbol{Z}}_{r0} = \boldsymbol{I}$.

The relation between the electric and magnetic field components at the two boundaries A and B of a section is therefore given by:

$$\begin{bmatrix}\overline{\mathbf{E}}_{z\mathrm{A}}\\ \overline{\mathbf{E}}_{z\mathrm{B}}\end{bmatrix} = -\mathrm{j}\widehat{\boldsymbol{p}}_1^{-1}\begin{bmatrix}\boldsymbol{J}_{0\mathrm{A}} & \mathbf{Y}_{0\mathrm{A}}\\ \boldsymbol{J}_{0\mathrm{B}} & \mathbf{Y}_{0\mathrm{B}}\end{bmatrix}\begin{bmatrix}\mathbf{Y}_{1\mathrm{B}} & -\mathbf{Y}_{1\mathrm{A}}\\ -\boldsymbol{J}_{1\mathrm{B}} & \boldsymbol{J}_{1\mathrm{A}}\end{bmatrix}\begin{bmatrix}\overline{\mathbf{H}}_{\phi\mathrm{A}}\\ \overline{\mathbf{H}}_{\phi\mathrm{B}}\end{bmatrix} \tag{4.366}$$

$\boldsymbol{J}_{0\mathrm{A,B}}$, $\boldsymbol{J}_{1\mathrm{A,B}}$, $\mathbf{Y}_{0\mathrm{A,B}}$, and $\mathbf{Y}_{1\mathrm{A,B}}$ denote diagonal matrices of the corresponding functions in eqs. (4.363) and (4.364). Furthermore, we have:

$$\boldsymbol{p}_1 = \boldsymbol{J}_{1\mathrm{A}}\mathbf{Y}_{1\mathrm{B}} - \boldsymbol{J}_{1\mathrm{B}}\mathbf{Y}_{1\mathrm{A}} \quad \widehat{\boldsymbol{p}}_1 = \mathrm{Diag}(\boldsymbol{p}_1, \boldsymbol{p}_1) \tag{4.367}$$

$\boldsymbol{p}_1$ is the generalised cross-product of Bessel functions p_1 (see formula 9.1.32 in [14]). In our case, $\boldsymbol{p}_1$ is a diagonal matrix that implicitly contains $\boldsymbol{k}_r$.

Therefore, the relation between the electric and magnetic field components at the two section boundaries A ($r = r_A$) and B ($r = R_B$) is given by:

$$\begin{bmatrix} \overline{\mathbf{E}}_{zA} \\ \overline{\mathbf{E}}_{zB} \end{bmatrix} = \begin{bmatrix} \overline{\mathbf{z}}_{AA} & \overline{\mathbf{z}}_{AB} \\ \overline{\mathbf{z}}_{BA} & \overline{\mathbf{z}}_{BB} \end{bmatrix} \begin{bmatrix} (-\overline{\mathbf{H}}_{\phi A}) \\ -(-\overline{\mathbf{H}}_{\phi B}) \end{bmatrix} \tag{4.368}$$

with:

$$\begin{aligned} \overline{\mathbf{z}}_{AA} &= -\mathrm{j}\boldsymbol{p}_1^{-1}\boldsymbol{q}_0 & \overline{\mathbf{z}}_{AB} &= \tfrac{2}{\pi}(\mathrm{j}\overline{r}_A\boldsymbol{k}_r\boldsymbol{p}_1)^{-1} \\ \overline{\mathbf{z}}_{BB} &= \mathrm{j}\boldsymbol{p}_1^{-1}\boldsymbol{r}_0 & \overline{\mathbf{z}}_{BA} &= \tfrac{2}{\pi}(\mathrm{j}\overline{r}_B\boldsymbol{k}_r\boldsymbol{p}_1)^{-1} \end{aligned} \tag{4.369}$$

The further generalised cross-products of Bessel functions $\boldsymbol{q}_0$ and $\boldsymbol{r}_0$ also correspond to the notations in formula 9.1.32 in [14], except that they are diagonal matrices. In addition, the following relation for the general Bessel function $\mathcal{C}$ was used: $\mathcal{C}_0'(x) = -\mathcal{C}_1(x)$, where in our case $x = \boldsymbol{k}_r\overline{r}$.

Instead of eq. (4.368) we may also write:

$$\begin{bmatrix} (-\overline{\mathbf{H}}_{\phi A}) \\ -(-\overline{\mathbf{H}}_{\phi B}) \end{bmatrix} = \begin{bmatrix} \overline{\mathbf{y}}_{AA} & \overline{\mathbf{y}}_{AB} \\ \overline{\mathbf{y}}_{BA} & \overline{\mathbf{y}}_{BB} \end{bmatrix} \begin{bmatrix} \overline{\mathbf{E}}_{zA} \\ \overline{\mathbf{E}}_{zB} \end{bmatrix} \tag{4.370}$$

where:

$$\begin{aligned} \overline{\boldsymbol{y}}_{AA} &= \mathrm{j}\boldsymbol{p}_0^{-1}\boldsymbol{r}_0 & \overline{\boldsymbol{y}}_{AB} &= \mathrm{j}\tfrac{2}{\pi}(\boldsymbol{k}_r\overline{r}_A\boldsymbol{p}_0)^{-1} \\ \overline{\boldsymbol{y}}_{BB} &= -\mathrm{j}\boldsymbol{p}_0^{-1}\boldsymbol{q}_0 & \overline{\boldsymbol{y}}_{BA} &= \mathrm{j}\tfrac{2}{\pi}(\boldsymbol{k}_r\overline{r}_B\boldsymbol{p}_0)^{-1} \end{aligned} \tag{4.371}$$

The transformation of impedances at concatenations of radial sections with offsets (e.g. between sections k and $k-1$ in Fig. 4.79b) occurs analogously to that in the previous section.

4.6.2.5 Study of sections A, C and E

The solution for region A in Fig. 4.78 must describe an outgoing wave. Therefore, the fields must be described by Hankel functions of the second kind. Regions C and E include the origin of the radial coordinate r. The Neumann functions Y_0 and Y_1 in solution (4.362) contain a singularity at the origin, but physically the fields must be finite. Therefore, we may only use the first terms, i.e. the Bessel functions J_0 and J_1, in these regions. The special solutions for regions A, C and E in Fig. 4.78 are written as follows:

$$\begin{aligned} \text{region A:}\quad & \overline{\mathbf{H}}_\phi = \mathrm{H}_1^{(2)}(\boldsymbol{k}_{rA}\overline{r})\mathrm{H}_1^{(2)-1}(\boldsymbol{k}_{rA}\overline{r}_0)\mathbf{A}_A \\ & \overline{\mathbf{E}}_z = -\mathrm{j}\mathrm{H}_0^{(2)}(\boldsymbol{k}_{rA}\overline{r})\mathrm{H}_1^{(2)-1}(\boldsymbol{k}_{rA}\overline{r}_0)\mathbf{A}_A \\ \text{region C:}\quad & \overline{\mathbf{H}}_\phi = \mathrm{J}_1(\boldsymbol{k}_{rC}\overline{r})\mathrm{J}_1^{-1}(\boldsymbol{k}_{rC}\overline{r}_0)\mathbf{A}_C \\ & \overline{\mathbf{E}}_z = -\mathrm{j}\mathrm{J}_0(\boldsymbol{k}_{rC}\overline{r})\mathrm{J}_1^{-1}(\boldsymbol{k}_{rC}\overline{r}_0)\mathbf{A}_C \\ \text{region E:}\quad & \overline{\mathbf{H}}_\phi = \mathrm{J}_1(\boldsymbol{k}_{rE}\overline{r})\mathrm{J}_1^{-1}(\boldsymbol{k}_{rE}\overline{r}_i)\mathbf{A}_E \\ & \overline{\mathbf{E}}_z = -\mathrm{j}\mathrm{J}_0(\boldsymbol{k}_{rE}\overline{r})\mathrm{J}_1^{-1}(\boldsymbol{k}_{rE}\overline{r}_i)\mathbf{A}_E \end{aligned} \tag{4.372}$$

where $\mathrm{H}_1^{(2)}$ and $\mathrm{H}_0^{(2)}$ are first- and zeroth-order Hankel functions of the second kind, respectively. These solutions are normalised so that the magnetic field components at $r = r_0$ or $r = r_i$ are given by $\mathbf{A}_\mathrm{A}$, $\mathbf{A}_\mathrm{C}$ and $\mathbf{A}_\mathrm{E}$, respectively. For open structures in z-direction (as in regions A and C of Fig. 4.78), the operator $\overline{\boldsymbol{D}}_z$ in eq. (4.370) has to be replaced by the operator $\overline{\boldsymbol{D}}_z^a$. This in turn requires introduction of ABCs.

The electric field components $\mathbf{E}_r$ in the spatial domain are given by:

$$\mathbf{E}_{r\mathrm{A}} = \mathrm{j}\varepsilon_{h\mathrm{A}}^{-1}\overline{\boldsymbol{D}}_{\bar{z}\mathrm{A}}^{a}\boldsymbol{T}_{\mathrm{H_A}}^{r}\overline{\mathbf{H}}_{\phi\mathrm{A}} \tag{4.373}$$
$$\mathbf{E}_{r\mathrm{C}} = \mathrm{j}\varepsilon_{h\mathrm{C}}^{-1}\overline{\boldsymbol{D}}_{\bar{z}\mathrm{C}}^{a}\boldsymbol{T}_{\mathrm{H_C}}^{r}\overline{\mathbf{H}}_{\phi\mathrm{C}} \tag{4.374}$$
$$\mathbf{E}_{r\mathrm{E}} = \mathrm{j}\varepsilon_{h\mathrm{E}}^{-1}\overline{\boldsymbol{D}}_{\bar{z}\mathrm{E}}\boldsymbol{T}_{\mathrm{H_E}}^{r}\overline{\mathbf{H}}_{\phi\mathrm{E}} \tag{4.375}$$

We define the impedance matrices $\overline{\mathbf{Z}}_\mathrm{A}$, $\overline{\mathbf{Z}}_\mathrm{C}$ and $\overline{\mathbf{Z}}_\mathrm{E}$ as:

$$\overline{\mathbf{E}}_{z\mathrm{A}}(\bar{r}_0) = \overline{\mathbf{Z}}_\mathrm{A}(\bar{r}_0)(-\overline{\mathbf{H}}_{\phi\mathrm{A}}(\bar{r}_0)) \tag{4.376}$$
$$\overline{\mathbf{E}}_{z\mathrm{B}}(\bar{r}_0) = \overline{\mathbf{Z}}_\mathrm{B}(\bar{r}_0)(-\overline{\mathbf{H}}_{\phi\mathrm{B}}(\bar{r}_0)) \tag{4.377}$$
$$\overline{\mathbf{E}}_{z\mathrm{C}}(\bar{r}_0) = -\overline{\mathbf{Z}}_\mathrm{C}(\bar{r}_0)(-\overline{\mathbf{H}}_{\phi\mathrm{C}}(\bar{r}_0)) \tag{4.378}$$
$$\overline{\mathbf{E}}_{z\mathrm{E}}(\bar{r}_i) = -\overline{\mathbf{Z}}_\mathrm{E}(\bar{r}_i)(-\overline{\mathbf{H}}_{\phi\mathrm{E}}(\bar{r}_i)) \tag{4.379}$$

The signs are chosen according to the Poynting vector. (In regions C and E we are looking in $-r$-direction.)

From eq. (4.372) the following expressions are obtained for the impedance matrices $\overline{\mathbf{Z}}_\mathrm{A}$, $\overline{\mathbf{Z}}_\mathrm{C}$ and $\overline{\mathbf{Z}}_\mathrm{E}$:

$$\overline{\mathbf{Z}}_\mathrm{A}(\bar{r}_0) = \mathrm{j}\mathrm{H}_0^{(2)}(\boldsymbol{k}_{r\mathrm{A}}\bar{r}_0)\mathrm{H}_1^{(2)-1}(\boldsymbol{k}_{r\mathrm{A}}\bar{r}_0) \tag{4.380}$$
$$\overline{\mathbf{Z}}_\mathrm{C}(\bar{r}_0) = -\mathrm{j}\mathrm{J}_0(\boldsymbol{k}_{r\mathrm{C}}\bar{r}_0)\mathrm{J}_1^{-1}(\boldsymbol{k}_{r\mathrm{C}}\bar{r}_0) \tag{4.381}$$
$$\overline{\mathbf{Z}}_\mathrm{E}(\bar{r}_i) = -\mathrm{j}\mathrm{J}_0(\boldsymbol{k}_{r\mathrm{E}}\bar{r}_i)\mathrm{J}_1^{-1}(\boldsymbol{k}_{r\mathrm{E}}\bar{r}_i) \tag{4.382}$$

4.6.2.6 Impedance in aperture B

The impedance $\overline{\mathbf{Z}}_\mathrm{B}$ (see Fig. 4.78) as defined in eq. (4.377) is obtained from the matching process at $\bar{r} = \bar{r}_0$. The electric field of aperture A is divided into three parts according to:

$$\mathbf{E}_{z\mathrm{A}} = \begin{bmatrix} \mathbf{E}_{z\mathrm{B}} \\ \mathbf{E}_{z\mathrm{M}} \\ \mathbf{E}_{z\mathrm{C}} \end{bmatrix} \tag{4.383}$$

The sub-vector $\mathbf{E}_{z\mathrm{M}}$ represents the electric field on the non-ideal surface of the metallic cylinder between regions B and C at position $r = r_0$. For ideal metal, $\mathbf{E}_{z\mathrm{M}}$ must be a vector whose components are zero. For the vectors in transformed domain we obtain:

$$\boldsymbol{T}_{\mathrm{E_A}}^{r}\overline{\mathbf{E}}_{z\mathrm{A}} = \begin{bmatrix} \boldsymbol{T}_{\mathrm{E_{AB}}}^{r} \\ \boldsymbol{T}_{\mathrm{E_{AM}}}^{r} \\ \boldsymbol{T}_{\mathrm{E_{AC}}}^{r} \end{bmatrix}\overline{\mathbf{E}}_{z\mathrm{A}} = \begin{bmatrix} \boldsymbol{T}_{\mathrm{E_B}}^{r}\overline{\mathbf{E}}_{z\mathrm{B}} \\ \mathbf{E}_{z\mathrm{M}} \\ \boldsymbol{T}_{\mathrm{E_C}}^{r}\overline{\mathbf{E}}_{z\mathrm{C}} \end{bmatrix} \tag{4.384}$$

$\boldsymbol{T}^r_{\mathrm{E}_{AB}}$, $\boldsymbol{T}^r_{\mathrm{E_{AC}}}$ and $\boldsymbol{T}^r_{\mathrm{E_{AM}}}$ are parts of $\boldsymbol{T}^r_{\mathrm{E_A}}$ associated with regions B, C and metallisation, respectively. They were obtained by the matrix partition technique proposed in Section 2.5 (see also [36]).

As shown in Section 2.5, we can also partition the inverse of $\boldsymbol{T}^r_{\mathrm{E_A}}$. By taking the relevant parts of it, we can write the vector $\overline{\mathbf{E}}_{z\mathrm{A}}$ as a function of $\overline{\mathbf{E}}_{z\mathrm{B}}$, $\overline{\mathbf{E}}_{z\mathrm{M}}$ and $\overline{\mathbf{E}}_{z\mathrm{C}}$. We obtain:

$$\overline{\mathbf{E}}_{z\mathrm{A}} = (\boldsymbol{T}^r_{\mathrm{E_{AB}}})^{-1}\boldsymbol{T}^r_{\mathrm{E_B}}\overline{\mathbf{E}}_{z\mathrm{B}} + (\boldsymbol{T}^r_{\mathrm{E_{AC}}})^{-1}\boldsymbol{T}^r_{\mathrm{E_C}}\overline{\mathbf{E}}_{z\mathrm{C}} + (\boldsymbol{T}^r_{\mathrm{E_{AM}}})^{-1}\mathbf{E}_{z\mathrm{M}} \tag{4.385}$$

The magnetic field at the interface is divided in a similar way to the electric one:

$$\mathbf{H}_{\phi\mathrm{A}} = \begin{bmatrix} \mathbf{H}_{\phi\mathrm{B}} \\ \mathbf{J}_{\mathrm{M}} \\ \mathbf{H}_{\phi\mathrm{C}} \end{bmatrix} \tag{4.386}$$

$\mathbf{J}_{\mathrm{M}}$ is the current density on the metallic front at $r = r_0$ between regions B and C, i.e. on the same places as $\mathbf{E}_{z\mathrm{M}}$. We obtain for the transformed quantities:

$$\boldsymbol{T}^r_{\mathrm{H_A}}\overline{\mathbf{H}}_{\phi\mathrm{A}} = \begin{bmatrix} \boldsymbol{T}^r_{\mathrm{H_{AB}}} \\ \boldsymbol{T}^r_{\mathrm{H_{AM}}} \\ \boldsymbol{T}^r_{\mathrm{H_{AC}}} \end{bmatrix} \overline{\mathbf{H}}_{\phi\mathrm{A}} = \begin{bmatrix} \boldsymbol{T}^r_{\mathrm{H_B}}\overline{\mathbf{H}}_{\phi\mathrm{B}} \\ \mathbf{J}_{\mathrm{M}} \\ \boldsymbol{T}^r_{\mathrm{H_C}}\overline{\mathbf{H}}_{\phi\mathrm{C}} \end{bmatrix} \tag{4.387}$$

$\boldsymbol{T}^r_{\mathrm{H_{AB}}}$, $\boldsymbol{T}^r_{\mathrm{H_{AC}}}$ and $\boldsymbol{T}^r_{\mathrm{H_{AM}}}$ are the parts of $\boldsymbol{T}^r_{\mathrm{H_A}}$ associated with regions B, C and the metallisation, respectively. Again, they were obtained by matrix partition technique. The following three equations result from this array expression:

$$\boldsymbol{T}^r_{\mathrm{H_B}}\overline{\mathbf{H}}_{\phi\mathrm{B}} = \boldsymbol{T}^r_{\mathrm{H_{AB}}}\overline{\mathbf{H}}_{\phi\mathrm{A}} \tag{4.388}$$

$$\boldsymbol{T}^r_{\mathrm{H_C}}\overline{\mathbf{H}}_{\phi\mathrm{C}} = \boldsymbol{T}^r_{\mathrm{H_{AC}}}\overline{\mathbf{H}}_{\phi\mathrm{A}} \tag{4.389}$$

$$\mathbf{J}_{\mathrm{M}} = \boldsymbol{T}^r_{\mathrm{H_{AM}}}\overline{\mathbf{H}}_{\phi\mathrm{A}} \tag{4.390}$$

The associated equations for $\overline{\mathbf{E}}_{z\mathrm{B}}$ and $\overline{\mathbf{E}}_{z\mathrm{C}}$ are obtained using eqs. (4.376)–(4.378)

$$\overline{\mathbf{E}}_{z\mathrm{B}} = \overline{\boldsymbol{Z}}_{\mathrm{B}}(\boldsymbol{T}^r_{\mathrm{H_B}})^{-1}\boldsymbol{T}^r_{\mathrm{H_{AB}}}\overline{\boldsymbol{Y}}_{\mathrm{A}}\overline{\mathbf{E}}_{z\mathrm{A}} \tag{4.391}$$

$$\overline{\mathbf{E}}_{z\mathrm{C}} = -\overline{\boldsymbol{Z}}_{\mathrm{C}}(\boldsymbol{T}^r_{\mathrm{H_C}})^{-1}\boldsymbol{T}^r_{\mathrm{H_{AC}}}\overline{\boldsymbol{Y}}_{\mathrm{A}}\overline{\mathbf{E}}_{z\mathrm{A}} \tag{4.392}$$

By replacing $\overline{\mathbf{H}}_{\phi\mathrm{A}}$ in eqs. (4.388)–(4.390) with $-\overline{\boldsymbol{Y}}_{\mathrm{A}}\overline{\mathbf{E}}_{z\mathrm{A}}$ and introducing eq. (4.385) into these equations, we obtain:

$$\begin{bmatrix} -\overline{\mathbf{H}}_{\phi\mathrm{B}} \\ -\overline{\mathbf{H}}_{\phi\mathrm{C}} \\ -\mathbf{J}_{\mathrm{M}} \end{bmatrix} = \begin{bmatrix} \overline{\boldsymbol{y}}_{\mathrm{BB}} & \overline{\boldsymbol{y}}_{\mathrm{BC}} & \overline{\boldsymbol{y}}_{\mathrm{BM}} \\ \overline{\boldsymbol{y}}_{\mathrm{CB}} & \overline{\boldsymbol{y}}_{\mathrm{CC}} & \overline{\boldsymbol{y}}_{\mathrm{CM}} \\ \overline{\boldsymbol{y}}_{\mathrm{MB}} & \overline{\boldsymbol{y}}_{\mathrm{MC}} & \overline{\boldsymbol{y}}_{\mathrm{MM}} \end{bmatrix} \begin{bmatrix} \overline{\mathbf{E}}_{z\mathrm{B}} \\ \overline{\mathbf{E}}_{z\mathrm{C}} \\ \mathbf{E}_{z\mathrm{M}} \end{bmatrix} \tag{4.393}$$

with:

$$
\begin{aligned}
\overline{\boldsymbol{y}}_{\mathrm{BB}} &= (\boldsymbol{T}^r_{\mathrm{H_B}})^{-1}\boldsymbol{T}^r_{\mathrm{H_AB}}\overline{\mathbf{Y}}_{\mathrm{A}}(\boldsymbol{T}^r_{\mathrm{E_{AB}}})^{-1}\boldsymbol{T}^r_{\mathrm{E_B}} \\
\overline{\boldsymbol{y}}_{\mathrm{CB}} &= (\boldsymbol{T}^r_{\mathrm{H_C}})^{-1}\boldsymbol{T}^r_{\mathrm{H_{AC}}}\overline{\mathbf{Y}}_{\mathrm{A}}(\boldsymbol{T}^r_{\mathrm{E_{AB}}})^{-1}\boldsymbol{T}^r_{\mathrm{E_B}} \\
\overline{\boldsymbol{y}}_{\mathrm{MB}} &= \boldsymbol{T}^r_{\mathrm{H_{AM}}}\overline{\mathbf{Y}}_{\mathrm{A}}(\boldsymbol{T}^r_{\mathrm{E_{AB}}})^{-1}\boldsymbol{T}^r_{\mathrm{E_B}} \\
\overline{\boldsymbol{y}}_{\mathrm{BC}} &= (\boldsymbol{T}^r_{\mathrm{H_B}})^{-1}\boldsymbol{T}^r_{\mathrm{H_{AB}}}\overline{\mathbf{Y}}_{\mathrm{A}}(\boldsymbol{T}^r_{\mathrm{E_{AC}}})^{-1}\boldsymbol{T}^r_{\mathrm{E_C}} \\
\overline{\boldsymbol{y}}_{\mathrm{CC}} &= (\boldsymbol{T}^r_{\mathrm{H_C}})^{-1}\boldsymbol{T}^r_{\mathrm{H_{AC}}}\overline{\mathbf{Y}}_{\mathrm{A}}(\boldsymbol{T}^r_{\mathrm{E_{AC}}})^{-1}\boldsymbol{T}^r_{\mathrm{E_C}} \\
\overline{\boldsymbol{y}}_{\mathrm{MC}} &= \boldsymbol{T}^r_{\mathrm{H_{AM}}}\overline{\mathbf{Y}}_{\mathrm{A}}(\boldsymbol{T}^r_{\mathrm{E_{AC}}})^{-1}\boldsymbol{T}^r_{\mathrm{E_C}} \\
\overline{\boldsymbol{y}}_{\mathrm{BM}} &= (\boldsymbol{T}^r_{\mathrm{H_B}})^{-1}\boldsymbol{T}^r_{\mathrm{H_{AB}}}\overline{\mathbf{Y}}_{\mathrm{A}}(\boldsymbol{T}^r_{\mathrm{E_{AM}}})^{-1} \\
\overline{\boldsymbol{y}}_{\mathrm{CM}} &= (\boldsymbol{T}^r_{\mathrm{H_C}})^{-1}\boldsymbol{T}^r_{\mathrm{H_{AC}}}\overline{\mathbf{Y}}_{\mathrm{A}}(\boldsymbol{T}^r_{\mathrm{E_{AM}}})^{-1} \\
\overline{\boldsymbol{y}}_{\mathrm{MM}} &= \boldsymbol{T}^r_{\mathrm{H_{AM}}}\overline{\mathbf{Y}}_{\mathrm{A}}(\boldsymbol{T}^r_{\mathrm{E_{AM}}})^{-1}
\end{aligned}
\tag{4.394}
$$

Let us consider the case of an ideal metal, i.e. $\mathbf{E}_{z\mathrm{M}} = \mathbf{0}$. The last equation in system (4.393) decouples from the first ones. Now, by considering eq. (4.378) and using $\overline{\mathbf{Y}}_{\mathrm{C}} = \overline{\boldsymbol{Z}}_{\mathrm{C}}^{-1}$, we obtain from the second expression in eq. (4.393):

$$
\overline{\mathbf{E}}_{z\mathrm{C}} = -(\overline{\mathbf{Y}}_{\mathrm{C}} + \overline{\boldsymbol{y}}_{\mathrm{CC}})^{-1}\overline{\boldsymbol{y}}_{\mathrm{CB}}\overline{\mathbf{E}}_{z\mathrm{B}} \tag{4.395}
$$

This may be introduced into the first expression in (4.393):

$$
-\overline{\mathbf{H}}_{\phi\mathrm{B}} = (\overline{\boldsymbol{y}}_{\mathrm{BB}} - \overline{\boldsymbol{y}}_{\mathrm{BC}}(\overline{\mathbf{Y}}_{\mathrm{C}} + \overline{\boldsymbol{y}}_{\mathrm{CC}})^{-1}\overline{\boldsymbol{y}}_{\mathrm{CB}})\overline{\mathbf{E}}_{\mathrm{zB}} \tag{4.396}
$$

Therefore, the admittance $\overline{\mathbf{Y}}_{\mathrm{B}}$ is given by:

$$
\overline{\mathbf{Y}}_{\mathrm{B}} = \overline{\boldsymbol{y}}_{\mathrm{BB}} - \overline{\boldsymbol{y}}_{\mathrm{BC}}(\overline{\mathbf{Y}}_{\mathrm{C}} + \overline{\boldsymbol{y}}_{\mathrm{CC}})^{-1}\overline{\boldsymbol{y}}_{\mathrm{CB}} \tag{4.397}
$$

and has the same form as that in previous cases.

In the case of non-ideal metal walls (which occur particularly for the analysis of monopoles), we use approximate boundary conditions for the tangential fields on the metallic surface $\eta_m \vec{H}_t = \vec{e}_r \times \vec{E}_t$, which result in:

$$
\mathbf{J}_{z\mathrm{M}} = \frac{\eta_0}{\eta_m}\mathbf{E}_{z\mathrm{M}} = \mathbf{Y}_{\mathrm{M}}\mathbf{E}_{z\mathrm{M}} \tag{4.398}
$$

$\boldsymbol{Y}_{\mathrm{M}}$ is a diagonal matrix with the elements η_0/η_m. η_m is the wave impedance in the lossy material with conductivity κ, and is determined as:

$$
\eta_m = \sqrt{\frac{jk_0\eta_0}{\kappa}} \tag{4.399}
$$

$\boldsymbol{E}_{z\mathrm{M}}$ in the first two expressions of eq. (4.393) is substituted as:

$$
\mathbf{E}_{z\mathrm{M}} = -(\mathbf{Y}_{\mathrm{M}} + \overline{\boldsymbol{y}}_{\mathrm{MM}})^{-1}(\overline{\boldsymbol{y}}_{\mathrm{MB}}\overline{\mathbf{E}}_{z\mathrm{B}} + \overline{\boldsymbol{y}}_{\mathrm{MC}}\overline{\mathbf{E}}_{z\mathrm{C}}) \tag{4.400}
$$

This relation was obtained from the last expression in eq. (4.393) and by considering eq. (4.398). Now we obtain new parameters $\overline{\boldsymbol{y}}^{\mathrm{n}}_{\mathrm{BB}}, \overline{\boldsymbol{y}}^{\mathrm{n}}_{\mathrm{BC}}, \overline{\boldsymbol{y}}^{\mathrm{n}}_{\mathrm{CB}}$ and

$\overline{\mathbf{y}}^{\mathrm{n}}_{\mathrm{CC}}$, which have to be used in eqs. (4.395)–(4.397) instead of $\overline{\mathbf{y}}_{\mathrm{BB}}, \overline{\mathbf{y}}_{\mathrm{BC}}, \overline{\mathbf{y}}_{\mathrm{CB}}$ and $\overline{\mathbf{y}}_{\mathrm{CC}}$:

$$\begin{aligned}
\overline{\mathbf{y}}^{\mathrm{n}}_{\mathrm{BB}} &= \overline{\mathbf{y}}_{\mathrm{BB}} - \overline{\mathbf{y}}_{\mathrm{BM}}(\mathbf{Y}_{\mathrm{M}} + \overline{\mathbf{y}}_{\mathrm{MM}})^{-1}\overline{\mathbf{y}}_{\mathrm{MB}} \\
\overline{\mathbf{y}}^{\mathrm{n}}_{\mathrm{BC}} &= \overline{\mathbf{y}}_{\mathrm{BC}} - \overline{\mathbf{y}}_{\mathrm{BM}}(\mathbf{Y}_{\mathrm{M}} + \overline{\mathbf{y}}_{\mathrm{MM}})^{-1}\overline{\mathbf{y}}_{\mathrm{MC}} \\
\overline{\mathbf{y}}^{\mathrm{n}}_{\mathrm{CB}} &= \overline{\mathbf{y}}_{\mathrm{CB}} - \overline{\mathbf{y}}_{\mathrm{CM}}(\mathbf{Y}_{\mathrm{M}} + \overline{\mathbf{y}}_{\mathrm{MM}})^{-1}\overline{\mathbf{y}}_{\mathrm{MB}} \\
\overline{\mathbf{y}}^{\mathrm{n}}_{\mathrm{CC}} &= \overline{\mathbf{y}}_{\mathrm{CC}} - \overline{\mathbf{y}}_{\mathrm{CM}}(\mathbf{Y}_{\mathrm{M}} + \overline{\mathbf{y}}_{\mathrm{MM}})^{-1}\overline{\mathbf{y}}_{\mathrm{MC}}
\end{aligned} \tag{4.401}$$

Let us assume that the antenna is fed by the fundamental mode of the radial waveguide. In that case the field $\overline{\mathbf{E}}_{\mathrm{B}}$ is known. With this knowledge and the admittance $\overline{\mathbf{Y}}_{\mathrm{B}}$ we can calculate the fields $\overline{\mathbf{E}}_{z\mathrm{C}}$, $\overline{\mathbf{H}}_{\phi\mathrm{B}}$ and $\overline{\mathbf{E}}_{z\mathrm{M}}$ by using eqs. (4.395), (4.396) and (4.400), respectively. The other quantities, particularly the surface current density $\mathbf{J}_{\mathrm{M}}$, are then determined from eq. (4.393).

4.6.3 Regions with crossed lines

As can be seen in Fig. 4.78, region D has interfaces with regions of radial (F and G) and vertical (B and E) discretisation direction. By prolonging the lines into region D, we obtain a crossed discretisation line system in this area. A similar problem occured in Section 3.3 (see particularly Figs. 3.25 and 3.26), where we were dealing with junctions of rectangular waveguides.

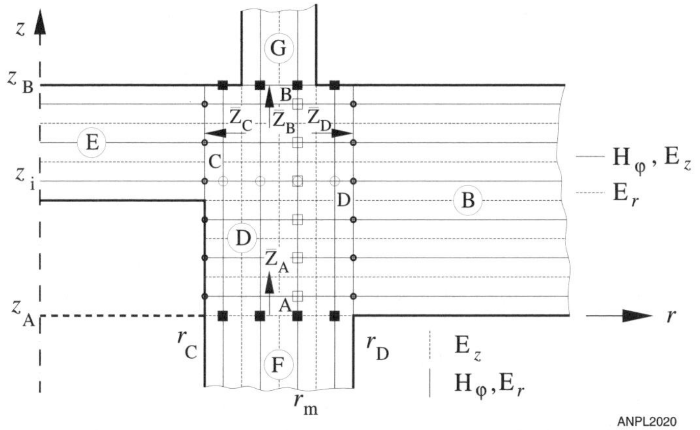

Fig. 4.80 General region with crossed discretisation lines (R. Pregla, 'The analysis of general axially symmetric antennas with a coaxial feed line by the method of lines', *IEEE Trans. Antennas. Propagation*, vol. 46, no. 10, pp. 1433–1443. © 1998 Institute of Electrical and Electronics Engineers (IEEE))

The general region with crossed lines is sketched in Fig. 4.80. In the vertical direction this region is bounded by planes A and B at z_{A} and z_{B}, and in the radial direction by surfaces C and D at r_{C} and r_{D}, respectively. The four surfaces of region D can be seen as generalised ports. Our goal is to obtain a general relation between the fields at these four ports. The side walls of the connecting waveguides are metallic. Therefore, we describe the relation of the fields at the inner side of the four generalised ports A–D (Fig. 4.80) with short-circuit matrix parameters:

$$\begin{bmatrix} \hat{\overline{\mathbf{H}}}_{\mathrm{A}} \\ -\hat{\overline{\mathbf{H}}}_{\mathrm{B}} \\ \hat{\overline{\mathbf{H}}}_{\mathrm{C}} \\ -\hat{\overline{\mathbf{H}}}_{\mathrm{D}} \end{bmatrix} = \begin{bmatrix} \hat{\overline{\mathbf{y}}}^{\mathrm{AB}}_{\mathrm{AB11}} & \hat{\overline{\mathbf{y}}}^{\mathrm{AB}}_{\mathrm{AB12}} & \hat{\overline{\mathbf{y}}}^{\mathrm{CD}}_{\mathrm{AB11}} & \hat{\overline{\mathbf{y}}}^{\mathrm{CD}}_{\mathrm{AB12}} \\ \hat{\overline{\mathbf{y}}}^{\mathrm{AB}}_{\mathrm{AB21}} & \hat{\overline{\mathbf{y}}}^{\mathrm{AB}}_{\mathrm{AB22}} & \hat{\overline{\mathbf{y}}}^{\mathrm{CD}}_{\mathrm{AB21}} & \hat{\overline{\mathbf{y}}}^{\mathrm{CD}}_{\mathrm{AB22}} \\ \hat{\overline{\mathbf{y}}}^{\mathrm{AB}}_{\mathrm{CD11}} & \hat{\overline{\mathbf{y}}}^{\mathrm{AB}}_{\mathrm{CD12}} & \hat{\overline{\mathbf{y}}}^{\mathrm{CD}}_{\mathrm{CD11}} & \hat{\overline{\mathbf{y}}}^{\mathrm{CD}}_{\mathrm{CD12}} \\ \hat{\overline{\mathbf{y}}}^{\mathrm{AB}}_{\mathrm{CD21}} & \hat{\overline{\mathbf{y}}}^{\mathrm{AB}}_{\mathrm{CD22}} & \hat{\overline{\mathbf{y}}}^{\mathrm{CD}}_{\mathrm{CD21}} & \hat{\overline{\mathbf{y}}}^{\mathrm{CD}}_{\mathrm{CD22}} \end{bmatrix} \begin{bmatrix} \hat{\overline{\mathbf{E}}}_{\mathrm{A}} \\ \hat{\overline{\mathbf{E}}}_{\mathrm{B}} \\ \hat{\overline{\mathbf{E}}}_{\mathrm{C}} \\ \hat{\overline{\mathbf{E}}}_{\mathrm{D}} \end{bmatrix} \tag{4.402}$$

In this section we assume that the field components (especially those on discretisation lines in z-direction) are not normalised with the coordinate r. (The denormalisation of the previous results is very easy.) By combining the fields of opposite ports in supervectors according to:

$$\hat{\overline{\mathbf{E}}}_{\mathrm{AB}} = [\hat{\overline{\mathbf{E}}}^{\mathrm{t}}_{\mathrm{A}}, \hat{\overline{\mathbf{E}}}^{\mathrm{t}}_{\mathrm{B}}]^{\mathrm{t}} \quad \hat{\overline{\mathbf{H}}}_{\mathrm{AB}} = [\hat{\overline{\mathbf{H}}}^{\mathrm{t}}_{\mathrm{A}}, -\hat{\overline{\mathbf{H}}}^{\mathrm{t}}_{\mathrm{B}}]^{\mathrm{t}} \tag{4.403}$$

we can write eq. (4.402) in a more compact form:

$$\begin{bmatrix} \hat{\overline{\mathbf{H}}}_{\mathrm{AB}} \\ \hat{\overline{\mathbf{H}}}_{\mathrm{CD}} \end{bmatrix} = \begin{bmatrix} \hat{\overline{\mathbf{y}}}^{\mathrm{AB}}_{\mathrm{AB}} & \hat{\overline{\mathbf{y}}}^{\mathrm{CD}}_{\mathrm{AB}} \\ \hat{\overline{\mathbf{y}}}^{\mathrm{AB}}_{\mathrm{CD}} & \hat{\overline{\mathbf{y}}}^{\mathrm{CD}}_{\mathrm{CD}} \end{bmatrix} \begin{bmatrix} \hat{\overline{\mathbf{E}}}_{\mathrm{AB}} \\ \hat{\overline{\mathbf{E}}}_{\mathrm{CD}} \end{bmatrix} \tag{4.404}$$

The four matrices in this equation are obtained by short-circuiting the ports. If ports C and D are short-circuited, matrices $\hat{\overline{\mathbf{y}}}^{\mathrm{AB}}_{\mathrm{AB}}$ and $\hat{\overline{\mathbf{y}}}^{\mathrm{AB}}_{\mathrm{CD}}$ are obtained using the vertical discretisation line system. Analogously, by short-circuiting ports A and B, matrices $\hat{\overline{\mathbf{y}}}^{\mathrm{CD}}_{\mathrm{CD}}$ and $\hat{\overline{\mathbf{y}}}^{\mathrm{CD}}_{\mathrm{AB}}$ can be computed using the horizontal discretisation line system. Thus, this procedure leads automatically to crossed discretisation lines (crossing each other at 90 degrees), as shown in Fig. 4.80. This also means that all field components are linear superpositions of two parts, e.g.:

$$H^{\mathrm{D}}_{\phi} = H^{\mathrm{D}}_{\phi r} + H^{\mathrm{D}}_{\phi z} \tag{4.405}$$

The second subscript denotes the direction of discretisation. We have to determine the fields on surfaces C and D on the discretisation lines oriented in the horizontal direction. The same holds for planes A and B, i.e. we need the field on the vertical discretisation lines. For both cases, we must express the fields through a part caused by the examined discretisation lines system and through a second contribution caused by the vertical line system. We would particularly like to obtain the relation when we short-circuit ports C and D. With $\hat{\overline{\mathbf{E}}}_{\mathrm{CD}} = \mathbf{0}$ we obtain from eq. (4.404):

$$\hat{\overline{\mathbf{H}}}_{\mathrm{AB}} = \hat{\overline{\mathbf{y}}}^{\mathrm{AB}}_{\mathrm{AB}} \hat{\overline{\mathbf{E}}}_{\mathrm{AB}} \qquad \hat{\overline{\mathbf{H}}}_{\mathrm{CD}} = \hat{\overline{\mathbf{y}}}^{\mathrm{AB}}_{\mathrm{CD}} \hat{\overline{\mathbf{E}}}_{\mathrm{AB}} \tag{4.406}$$

Analogously, we obtain from eq. (4.404) when short-circuiting ports A and B (i.e. $\widehat{\overline{\mathbf{E}}}_{\mathrm{AB}} = \mathbf{0}$):

$$\widehat{\overline{\mathbf{H}}}_{\mathrm{AB}} = \widehat{\overline{\boldsymbol{y}}}^{\mathrm{CD}}_{\mathrm{AB}} \widehat{\overline{\mathbf{E}}}_{\mathrm{CD}} \qquad \widehat{\overline{\mathbf{H}}}_{\mathrm{CD}} = \widehat{\overline{\boldsymbol{y}}}^{\mathrm{CD}}_{\mathrm{CD}} \widehat{\overline{\mathbf{E}}}_{\mathrm{CD}} \tag{4.407}$$

These equations show us that the main diagonal matrices in eq. (4.404) are equal to our results in eqs. (4.329b) (inverted version) and (4.370) (replace A and B with C and D, respectively). Therefore, we only have to determine the off-diagonal submatrices. For discretisation lines in z-direction, we can write:

$$\widehat{\overline{\mathbf{H}}}_{\mathrm{CD}} = \widehat{\overline{\boldsymbol{V}}}^{\mathrm{AB}}_{\mathrm{CD}} \widehat{\overline{\mathbf{H}}}_{\mathrm{AB}} = \begin{bmatrix} \overline{\boldsymbol{V}}_{\mathrm{CA}} & \overline{\boldsymbol{V}}_{\mathrm{CB}} \\ \overline{\boldsymbol{V}}_{\mathrm{DA}} & \overline{\boldsymbol{V}}_{\mathrm{DB}} \end{bmatrix} \widehat{\overline{\mathbf{H}}}_{\mathrm{AB}} \tag{4.408}$$

and for discretisation lines in r-direction:

$$\widehat{\overline{\mathbf{H}}}_{\mathrm{AB}} = \widehat{\overline{\boldsymbol{V}}}^{\mathrm{CD}}_{\mathrm{AB}} \widehat{\overline{\mathbf{H}}}_{\mathrm{CD}} = \begin{bmatrix} \overline{\boldsymbol{V}}_{\mathrm{AC}} & \overline{\boldsymbol{V}}_{\mathrm{AD}} \\ \overline{\boldsymbol{V}}_{\mathrm{BC}} & \overline{\boldsymbol{V}}_{\mathrm{BD}} \end{bmatrix} \widehat{\overline{\mathbf{H}}}_{\mathrm{CD}} \tag{4.409}$$

Instead of eq. (4.404) we may write:

$$\begin{bmatrix} \widehat{\overline{\mathbf{H}}}_{\mathrm{AB}} \\ \widehat{\overline{\mathbf{H}}}_{\mathrm{CD}} \end{bmatrix} = \begin{bmatrix} \widehat{\overline{\boldsymbol{I}}}^{\mathrm{AB}}_{\mathrm{AB}} & \widehat{\overline{\boldsymbol{V}}}^{\mathrm{CD}}_{\mathrm{AB}} \\ \widehat{\overline{\boldsymbol{V}}}^{\mathrm{AB}}_{\mathrm{CD}} & \widehat{\overline{\boldsymbol{I}}}^{\mathrm{CD}}_{\mathrm{CD}} \end{bmatrix} \begin{bmatrix} \widehat{\overline{\boldsymbol{y}}}^{\mathrm{AB}}_{\mathrm{AB}} & \widehat{\overline{\boldsymbol{0}}} \\ \widehat{\overline{\boldsymbol{0}}} & \widehat{\overline{\boldsymbol{y}}}^{\mathrm{CD}}_{\mathrm{CD}} \end{bmatrix} \begin{bmatrix} \widehat{\overline{\mathbf{E}}}_{\mathrm{AB}} \\ \widehat{\overline{\mathbf{E}}}_{\mathrm{CD}} \end{bmatrix} \tag{4.410}$$

4.6.3.1 Coupling from ports A and B to ports C and D

In this case we introduce electric walls at ports C and D. Therefore, we have only $\overline{\mathbf{H}}_\phi$ as tangential components here. We may write for the z-dependence of the magnetic field from the discretisation lines in z-direction (discretisation in r-direction):

$$\overline{\mathbf{H}}_{\phi r}(z) = \frac{\sinh(\boldsymbol{\Gamma}(\overline{z}_{AB} - \overline{z}))}{\sinh(\boldsymbol{\Gamma}\overline{\zeta})} \overline{\mathbf{H}}_{\phi r\mathrm{A}} + \frac{\sinh(\boldsymbol{\Gamma}\overline{z})}{\sinh(\boldsymbol{\Gamma}\overline{z}_{AB})} \overline{\mathbf{H}}_{\phi r\mathrm{B}} \tag{4.411}$$

where $\overline{z}_{AB} = k_0(z_{\mathrm{B}} - z_{\mathrm{A}})$. Because of the diagonal matrices, we may write such a fraction. This equation must now be evaluated at the particular z-positions of the discretisation lines in r-direction. At such a position z_i (see Fig. 4.80) we may write:

$$\overline{\mathbf{H}}_{\phi r}(z_i) = \boldsymbol{\Lambda}^{\mathrm{d}}_{\mathrm{A}i} \overline{\mathbf{H}}_{\phi r\mathrm{A}} + \boldsymbol{\Lambda}^{\mathrm{d}}_{\mathrm{B}i} \overline{\mathbf{H}}_{\phi r\mathrm{B}} \tag{4.412}$$

with diagonal matrices $\boldsymbol{\Lambda}^{\mathrm{d}}_i$ given by the expressions:

$$\boldsymbol{\Lambda}^{\mathrm{d}}_{\mathrm{A}i} = \sinh(\boldsymbol{\Gamma}(\overline{z}_{AB} - \overline{z}_i))(\sinh(\boldsymbol{\Gamma}\overline{z}_{AB}))^{-1} \tag{4.413}$$

$$\boldsymbol{\Lambda}^{\mathrm{d}}_{\mathrm{B}i} = \sinh(\boldsymbol{\Gamma}\overline{z}_i)(\sinh(\boldsymbol{\Gamma}\overline{z}_{AB}))^{-1} \tag{4.414}$$

The vector $\mathbf{H}_{\phi r}(z_i) = \boldsymbol{T}^z_{\mathrm{H}} \overline{\mathbf{H}}_{\phi r}(z_i)$ gives the values $H_{\phi r}$ at the points marked by circles ○ in Fig. 4.80. For the matching procedure, we need the values of the field at the boundaries $r = r_\mathrm{C}$ and $r = r_\mathrm{D}$. Because of the Neumann conditions, we must extrapolate to obtain these values. For this purpose, we use the transformation matrix $\boldsymbol{T}_r$ (see the appendices). We obtain at the position z_i on the left ($\mathrm{R} \equiv \mathrm{C}$, $r = r_\mathrm{C}$) or right ($\mathrm{R} \equiv \mathrm{D}$, $r = r_\mathrm{D}$) boundary (marked by • at the surfaces C and D):

$$H_{\phi r\mathrm{R}}(z_i) = \mathbf{T}^z_{\mathrm{H}_{\Delta\mathrm{R}}} \overline{\mathbf{H}}_{\phi r}(z_i) = \mathbf{T}^z_{\mathrm{H}_{\Delta\mathrm{R}}} (\boldsymbol{\Lambda}^\mathrm{d}_{\mathrm{A}i} \overline{\mathbf{H}}_{\phi r\mathrm{A}} + \boldsymbol{\Lambda}^\mathrm{d}_{\mathrm{B}i} \overline{\mathbf{H}}_{\phi r\mathrm{B}}) \tag{4.415}$$

$\mathrm{R} \equiv \mathrm{C(D)}$ denotes the left (right) surface of region D.

$\mathbf{T}^z_{\mathrm{H}_{\Delta\mathrm{R}}}$ is a row vector constructed from the weighted first (last) rows of matrix $\boldsymbol{T}^z_\mathrm{H}$. This vector is constructed as:

$$\mathbf{T}^z_{\mathrm{H}_{\Delta\mathrm{C}}} = \tfrac{1}{8}(15\mathbf{T}^z_{\mathrm{H}1} - 10\mathbf{T}^z_{\mathrm{H}2} + 3\mathbf{T}^z_{\mathrm{H}3}) \tag{4.416}$$

$$\mathbf{T}^z_{\mathrm{H}_{\Delta\mathrm{D}}} = \tfrac{1}{8}(15\mathbf{T}^z_{\mathrm{H,N}} - 10\mathbf{T}^z_{\mathrm{H,N-1}} + 3\mathbf{T}^z_{\mathrm{H,N-2}}) \tag{4.417}$$

$\mathbf{T}^z_{\mathrm{H}1}$, $\mathbf{T}^z_{\mathrm{H}2}$ and $\mathbf{T}^z_{\mathrm{H}3}$ are the first three and $\mathbf{T}^z_{\mathrm{HN}}$, $\mathbf{T}^z_{\mathrm{H,N-1}}$ and $\mathbf{T}^z_{\mathrm{H,N-2}}$ the last three row vectors of the transformation matrix $\boldsymbol{T}^z_\mathrm{H}$. We want to change the order of the product between $\mathbf{T}^z_{\mathrm{H}_{\Delta\mathrm{R}}}$ and $\boldsymbol{\Lambda}^\mathrm{d}_i$. Therefore, we must reshape these quantities in a suitable matrix form. $\mathbf{T}^z_{\mathrm{H}_{\Delta\mathrm{R}}}$ should be reformed to a diagonal matrix $\boldsymbol{T}^{z\mathrm{d}}_{\mathrm{H}_{\Delta\mathrm{R}}}$, and the diagonal matrix $\boldsymbol{\Lambda}^\mathrm{d}_i$ becomes a row vector $\mathbf{\Lambda}_i$. After the reordering, the result for all values z_i can be written simultaneously:

$$\mathbf{H}_{\phi r\mathrm{R}} = \boldsymbol{\Lambda}_\mathrm{A} \boldsymbol{T}^{z\mathrm{d}}_{\mathrm{H}_{\Delta\mathrm{R}}} \overline{\mathbf{H}}_{\phi r\mathrm{A}} + \boldsymbol{\Lambda}_\mathrm{B} \boldsymbol{T}^{z\mathrm{d}}_{\mathrm{H}_{\Delta\mathrm{R}}} \overline{\mathbf{H}}_{\phi r\mathrm{B}} \tag{4.418}$$

The full matrices $\boldsymbol{\Lambda}_\mathrm{A}$ and $\boldsymbol{\Lambda}_\mathrm{B}$ are given by:

$$(\boldsymbol{\Lambda}_\mathrm{A})_{ik} = \sinh(\boldsymbol{\Gamma}_k(\overline{z}_{AB} - \overline{z}_i))(\sinh(\boldsymbol{\Gamma}_k \overline{z}_{AB}))^{-1} \tag{4.419}$$

$$(\boldsymbol{\Lambda}_\mathrm{B})_{ik} = \sinh(\boldsymbol{\Gamma}_k \overline{z}_i)(\sinh(\boldsymbol{\Gamma}_k \overline{z}_{AB}))^{-1} \tag{4.420}$$

Therefore, we have in transformed domain:

$$\overline{\mathbf{H}}_{\phi r\mathrm{R}} = \boldsymbol{T}^{r\,-1}_\mathrm{H} \boldsymbol{\Lambda}_\mathrm{A} \boldsymbol{T}^{z\mathrm{d}}_{\mathrm{H}_{\Delta\mathrm{R}}} \overline{\mathbf{H}}_{\phi r\mathrm{A}} + \boldsymbol{T}^{r\,-1}_\mathrm{H} \boldsymbol{\Lambda}_\mathrm{B} \boldsymbol{T}^{z\mathrm{d}}_{\mathrm{H}_{\Delta\mathrm{R}}} \overline{\mathbf{H}}_{\phi r\mathrm{B}} \tag{4.421}$$

The relations between the magnetic fields discretised in r-direction at the four ports A,B,C and D can now be expressed in the following matrix form:

$$\begin{bmatrix} (-\overline{\mathbf{H}}_{\phi r\mathrm{C}}) \\ -(-\overline{\mathbf{H}}_{\phi r\mathrm{D}}) \end{bmatrix} = \begin{bmatrix} \overline{\boldsymbol{V}}_\mathrm{CA} & \overline{\boldsymbol{V}}_\mathrm{CB} \\ \overline{\boldsymbol{V}}_\mathrm{DA} & \overline{\boldsymbol{V}}_\mathrm{DB} \end{bmatrix} \begin{bmatrix} \overline{\mathbf{H}}_{\phi r\mathrm{A}} \\ -\overline{\mathbf{H}}_{\phi r\mathrm{B}} \end{bmatrix} = \widehat{\overline{\mathbf{V}}}{}^\mathrm{CD}_\mathrm{AB} \begin{bmatrix} \overline{\mathbf{H}}_{\phi r\mathrm{A}} \\ -\overline{\mathbf{H}}_{\phi r\mathrm{B}} \end{bmatrix} \tag{4.422}$$

where:

$$\begin{aligned} \overline{\boldsymbol{V}}_\mathrm{CA} &= -\boldsymbol{T}^{r\,-1}_\mathrm{H} \boldsymbol{\Lambda}_\mathrm{A} \boldsymbol{T}^{z\mathrm{d}}_{\mathrm{H}_{\Delta\mathrm{C}}} & \overline{\boldsymbol{V}}_\mathrm{CB} &= \boldsymbol{T}^{r\,-1}_\mathrm{H} \boldsymbol{\Lambda}_\mathrm{B} \boldsymbol{T}^{z\mathrm{d}}_{\mathrm{H}_{\Delta\mathrm{C}}} \\ \overline{\boldsymbol{V}}_\mathrm{DA} &= \boldsymbol{T}^{r\,-1}_\mathrm{H} \boldsymbol{\Lambda}_\mathrm{A} \boldsymbol{T}^{z\mathrm{d}}_{\mathrm{H}_{\Delta\mathrm{D}}} & \overline{\boldsymbol{V}}_\mathrm{DB} &= -\boldsymbol{T}^{r\,-1}_\mathrm{H} \boldsymbol{\Lambda}_\mathrm{B} \boldsymbol{T}^{z\mathrm{d}}_{\mathrm{H}_{\Delta\mathrm{D}}} \end{aligned} \tag{4.423}$$

The part $\overline{\mathbf{H}}_{\phi r}$ does not cause an E_z component on the side walls of region D due to the Dirichlet boundary conditions (electric walls) for this component.

4.6.3.2 Coupling from ports C and D to ports A and B

Next, we want to determine the fields at ports A and B from those at ports C and D (see Fig. 4.80). We take the solutions for discretisation in z-direction and use the expressions described in Section 4.6.2.2.

In this case, we have electric walls at ports A and B. Therefore, we have $\overline{\mathbf{H}}_\phi$ as the only tangential field component. For the r-dependence of the magnetic field from the discretisation lines we can adapt the general solution in eq.4.362, which we write in the following form:

$$\overline{\mathbf{H}}_{\phi z}(r) = \begin{bmatrix} J_1(\boldsymbol{k}_r\overline{r}) & Y_1(\boldsymbol{k}_r\overline{r}) \end{bmatrix} \begin{bmatrix} \mathbf{A} \\ \mathbf{B} \end{bmatrix} \tag{4.424}$$

For surfaces C and D we obtain:

$$\begin{bmatrix} \overline{\mathbf{H}}_{\phi z\mathrm{C}} \\ \overline{\mathbf{H}}_{\phi z\mathrm{D}} \end{bmatrix} = \begin{bmatrix} J_1(\boldsymbol{k}_r\overline{r}_\mathrm{C}) & Y_1(\boldsymbol{k}_r\overline{r}_\mathrm{C}) \\ J_1(\boldsymbol{k}_r\overline{r}_\mathrm{D}) & Y_1(\boldsymbol{k}_r\overline{r}_\mathrm{D}) \end{bmatrix} \begin{bmatrix} \mathbf{A} \\ \mathbf{B} \end{bmatrix} \tag{4.425}$$

The inversion yields:

$$\begin{bmatrix} \mathbf{A} \\ \mathbf{B} \end{bmatrix} = \widehat{\boldsymbol{p}}_1^{-1} \begin{bmatrix} Y_1(\boldsymbol{k}_r\overline{r}_\mathrm{D}) & -Y_1(\boldsymbol{k}_r\overline{r}_\mathrm{C}) \\ -J_1(\boldsymbol{k}_r\overline{r}_\mathrm{D}) & J_1(\boldsymbol{k}_r\overline{r}_\mathrm{C}) \end{bmatrix} \begin{bmatrix} \overline{\mathbf{H}}_{\phi z\mathrm{C}} \\ \overline{\mathbf{H}}_{\phi z\mathrm{D}} \end{bmatrix} \tag{4.426}$$

in view of: (Formula 9.1.32 [14].)

$$\boldsymbol{p}_1^{\mathrm{CD}} = J_1(\boldsymbol{k}_r\overline{r}_\mathrm{C})Y_1(\boldsymbol{k}_r\overline{r}_\mathrm{D}) - J_1(\boldsymbol{k}_r\overline{r}_\mathrm{D})Y_1(\boldsymbol{k}_r\overline{r}_\mathrm{C}) \tag{4.427}$$

The superscripts (C and D) at $\boldsymbol{p}_1$ here and in the following denote the radii that have to be introduced into the arguments of the Bessel functions. The order is correlated to the order of the radii in the functions.

Therefore, by introducing eq. (4.426) into eq. (4.424) we obtain for position $r = r_m$:

$$\overline{\mathbf{H}}_{\phi z}(r_m) = \frac{\boldsymbol{p}_1^{m\mathrm{D}}}{\boldsymbol{p}_1^{\mathrm{CD}}}\overline{\mathbf{H}}_{\phi z\mathrm{C}} + \frac{\boldsymbol{p}_1^{\mathrm{C}m}}{\boldsymbol{p}_1^{\mathrm{CD}}}\overline{\mathbf{H}}_{\phi z\mathrm{D}} \tag{4.428}$$

(the superscript m at p_1 indicates that $\overline{r}_\mathrm{D}$ or $\overline{r}_m$ has to be introduced into eq. (4.427) instead of $\overline{r}_\mathrm{C}$) or similarly to eq. (4.412):

$$\overline{\mathbf{H}}_{\phi z}(r_m) = \boldsymbol{\Lambda}_{\mathrm{C}m}^{\mathrm{d}}\overline{\mathbf{H}}_{\phi z\mathrm{C}} + \boldsymbol{\Lambda}_{\mathrm{D}m}^{\mathrm{d}}\overline{\mathbf{H}}_{\phi z\mathrm{D}} \tag{4.429}$$

with:

$$\boldsymbol{\Lambda}_{\mathrm{C}m}^{\mathrm{d}} = \boldsymbol{p}_1^{m\mathrm{D}}(\boldsymbol{p}_1^{\mathrm{CD}})^{-1} \qquad \boldsymbol{\Lambda}_{\mathrm{D}m}^{\mathrm{d}} = \boldsymbol{p}_1^{\mathrm{C}m}(\boldsymbol{p}_1^{\mathrm{CD}})^{-1} \tag{4.430}$$

The vector $\mathbf{H}_{\phi z}(\overline{r}_m) = \boldsymbol{T}_\mathrm{H}^z\overline{\mathbf{H}}_{\phi z}(\overline{r}_m)$ gives the values $H_{\phi z}$ at the points marked by empty squares □ in Fig. 4.80. The values at the transitions points (marked by ▪) in planes A (R ≡ A) and B (R ≡ B) are obtained through analogous process, by means of $\boldsymbol{T}_z$. The following equation holds:

$$\mathbf{H}_{\phi z\mathrm{R}} = \boldsymbol{\Lambda}_\mathrm{C}\boldsymbol{T}_{\mathrm{H}\Delta\mathrm{R}}^{rd}\overline{\mathbf{H}}_{\phi z\mathrm{C}} + \boldsymbol{\Lambda}_\mathrm{D}\boldsymbol{T}_{\mathrm{H}\Delta\mathrm{R}}^{rd}\overline{\mathbf{H}}_{\phi z\mathrm{D}} \tag{4.431}$$

where $\boldsymbol{\Lambda}_{\rm C}$ and $\boldsymbol{\Lambda}_{\rm D}$ are full matrices. They are constructed like $\boldsymbol{\Lambda}_{\rm A,B}$ in eq. (4.419) and (4.420):

$$(\boldsymbol{\Lambda}_{\rm C})_{mn} = p_1^{m\rm D}/p_1^{\rm CD} \qquad (\boldsymbol{\Lambda}_{\rm D})_{mn} = p_1^{{\rm C}m}/p_1^{\rm CD} \tag{4.432}$$

The nth component of $\boldsymbol{k}_r$ has to be introduced into the argument of the Bessel functions.

The general solution is:

$$\begin{bmatrix} \overline{\mathbf{H}}_{\phi z\rm A} \\ -\overline{\mathbf{H}}_{\phi z\rm B} \end{bmatrix} \begin{bmatrix} \overline{\mathbf{V}}_{\rm AC} & \overline{\mathbf{V}}_{\rm AD} \\ \overline{\mathbf{V}}_{\rm BC} & \overline{\mathbf{V}}_{\rm BD} \end{bmatrix} \begin{bmatrix} (-\overline{\mathbf{H}}_{\phi z\rm C}) \\ -(-\overline{\mathbf{H}}_{\phi z\rm D}) \end{bmatrix} = \widehat{\overline{\mathbf{V}}}{}^{\rm AB}_{\rm CD} \begin{bmatrix} (-\overline{\mathbf{H}}_{\phi z\rm C}) \\ -(-\overline{\mathbf{H}}_{\phi z\rm D}) \end{bmatrix} \tag{4.433}$$

with:

$$\begin{aligned} \overline{\mathbf{V}}_{\rm AC} &= -(\boldsymbol{T}_{\rm H}^{z})^{-1}\boldsymbol{\Lambda}_{\rm C}\boldsymbol{T}_{\rm H\Delta A}^{r\rm d} & \overline{\mathbf{V}}_{\rm AD} &= (\boldsymbol{T}_{\rm H}^{z})^{-1}\boldsymbol{\Lambda}_{\rm D}\boldsymbol{T}_{\rm H\Delta A}^{r\rm d} \\ \overline{\mathbf{V}}_{\rm BC} &= (\boldsymbol{T}_{\rm H}^{z})^{-1}\boldsymbol{\Lambda}_{\rm C}\boldsymbol{T}_{\rm H\Delta B}^{r\rm d} & \overline{\mathbf{V}}_{\rm BD} &= -(\boldsymbol{T}_{\rm H}^{z})^{-1}\boldsymbol{\Lambda}_{\rm D}\boldsymbol{T}_{\rm H\Delta B}^{r\rm d} \end{aligned} \tag{4.434}$$

The extrapolation of the field to planes A and B is performed with the diagonal matrices $\boldsymbol{T}_{\rm H\Delta A}^{r\rm d}$ and $\boldsymbol{T}_{\rm H\Delta B}^{r\rm d}$. These matrices are constructed from the row vectors, analogously to eqs. (4.422) and (4.423).

4.6.4 Two special cases

4.6.4.1 I: Region D with short-circuited ports C and B

In this subsection we assume that ports B and C are short-circuited. Now region D transforms the waves from the coaxial line directly to the radial parallel-plate transmission line. This transition region is shown in detail in Fig. 4.81.

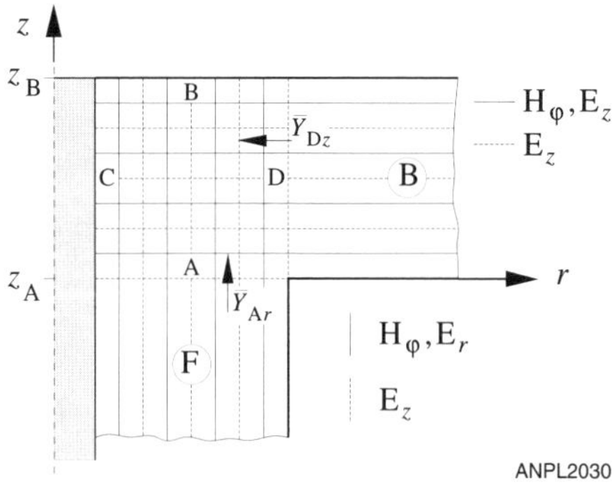

Fig. 4.81 Transition section with crossed lines (R. Pregla, 'The analysis of general axially symmetric antennas with a coaxial feed line by the method of lines', *IEEE Trans. Antennas. Propagation*, vol. 46, no. 10, pp. 1433–1443. © 1998 Institute of Electrical and Electronics Engineers (IEEE))

By introducing $\overline{\mathbf{E}}_{\mathrm{B,C}} = \mathbf{0}$, we obtain directly from eq. (4.402) the following expression:

$$\begin{bmatrix} \overline{\mathbf{H}}_{\mathrm{A}} \\ -\overline{\mathbf{H}}_{\mathrm{D}} \end{bmatrix} = \begin{bmatrix} \overline{\boldsymbol{y}}^{\mathrm{AB}}_{\mathrm{AB11}} & \overline{\boldsymbol{y}}^{\mathrm{CD}}_{\mathrm{AB12}} \\ \overline{\boldsymbol{y}}^{\mathrm{AB}}_{\mathrm{CD21}} & \overline{\boldsymbol{y}}^{\mathrm{CD}}_{\mathrm{CD22}} \end{bmatrix} \begin{bmatrix} \overline{\mathbf{E}}_{\mathrm{A}} \\ \overline{\mathbf{E}}_{\mathrm{D}} \end{bmatrix} \tag{4.435}$$

4.6.4.2 II: Region D with layered dielectric

The situation with two dielectric layers is shown in Fig. 4.82. The upper side (port B) is connected with the open space. Therefore, ABCs have to be introduced on this interface (for the discretisation lines in r-direction). For the lines in z-direction, the region has to be subdivided into two subregions D_1 and D_2. The magnetic field $\mathbf{H}_{\phi r}$ at an arbitrary position z_i must be expressed by $\mathbf{H}_{\phi r\mathrm{A}}$. Taking into account the ABCs at plane B or, which is identical in this case, by assuming an infinite height of region D_1, the following results are valid:

$$\overline{\mathbf{Z}}_{\mathrm{B}r} = \overline{\mathbf{Z}}_{\mathrm{Fr}_1} = \overline{\mathbf{Z}}_{01} = \boldsymbol{I} \tag{4.436}$$

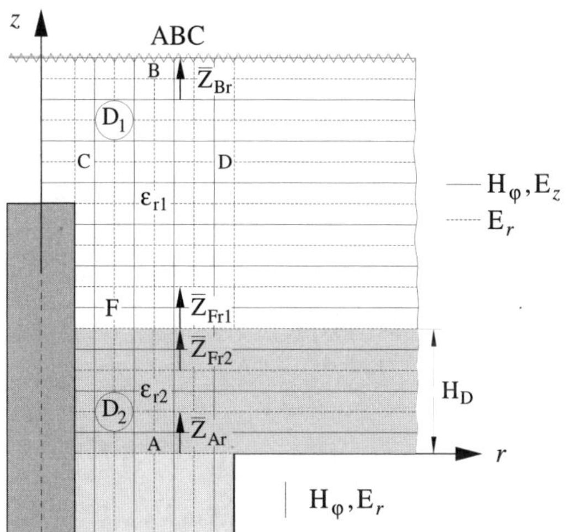

Fig. 4.82 Region D with multilayered dielectric (R. Pregla, 'The analysis of general axially symmetric antennas with a coaxial feed line by the method of lines', *IEEE Trans. Antennas. Propagation*, vol. 46, no. 10, pp. 1433–1443. © 1998 Institute of Electrical and Electronics Engineers (IEEE))

The magnetic field depends on z_i in the following way:

$$\overline{\mathbf{H}}_{\phi r} = e^{-\boldsymbol{\Gamma}_1(\overline{z}_i - \overline{z}_{\mathrm{F}})}\overline{\mathbf{H}}_{\phi r\mathrm{F}} \tag{4.437}$$

In region D_2 we adapt solution (4.329):

$$\begin{bmatrix} \overline{\mathbf{E}}_{rr\mathrm{A}} \\ \overline{\mathbf{E}}_{rr\mathrm{F}} \end{bmatrix} = \begin{bmatrix} \overline{\boldsymbol{z}}_1 & \overline{\boldsymbol{z}}_2 \\ \overline{\boldsymbol{z}}_2 & \overline{\boldsymbol{z}}_1 \end{bmatrix}_{|\mathrm{D}_2} \begin{bmatrix} \overline{\mathbf{H}}_{\phi r\mathrm{A}} \\ -\overline{\mathbf{H}}_{\phi r\mathrm{F}} \end{bmatrix} \tag{4.438}$$

By introducing $\overline{\mathbf{E}}_{rr\mathrm{F}} = \overline{\boldsymbol{Z}}_{\mathrm{Fr2}}\overline{\mathbf{H}}_{\phi r\mathrm{F}}$ with $\overline{\boldsymbol{Z}}_{\mathrm{Fr2}} = (\boldsymbol{T}^z_{\mathrm{E2}})^{-1}\boldsymbol{T}^z_{\mathrm{E1}}\overline{\boldsymbol{I}}(\boldsymbol{T}^z_{\mathrm{H1}})^{-1}\boldsymbol{T}^z_{\mathrm{H2}}$ into the second equation of system (4.438) we obtain:

$$\overline{\mathbf{H}}_{\phi r\mathrm{F}} = \overline{\boldsymbol{V}}_{\mathrm{FA}}\overline{\mathbf{H}}_{\phi r\mathrm{A}} \tag{4.439}$$

$$\overline{\boldsymbol{V}}_{\mathrm{FA}} = (\cosh(\boldsymbol{\Gamma}_2\overline{H}_{\mathrm{D}}) + \overline{\boldsymbol{Z}}_{\mathrm{Fr2}}\sinh(\boldsymbol{\Gamma}_2\overline{H}_{\mathrm{D}}))^{-1} \tag{4.440}$$

The first equation in system (4.438) now yields:

$$\overline{\boldsymbol{Z}}_{\mathrm{A}r} = \overline{\boldsymbol{z}}_1 - \overline{\boldsymbol{z}}_2\overline{\boldsymbol{V}}_{\mathrm{FA}} \qquad \overline{\boldsymbol{Y}}_{\mathrm{A}r} = \overline{\boldsymbol{Z}}_{\mathrm{A}r}^{-1} \tag{4.441}$$

Now we can write instead of eq. (4.418):

$$\mathbf{H}_{\phi r\mathrm{R}} = \begin{bmatrix} (\boldsymbol{\Lambda}_{\mathrm{A}} + \boldsymbol{\Lambda}_{\mathrm{F2}}\overline{\boldsymbol{V}}_{\mathrm{FA}})\overline{\boldsymbol{T}}^{zd}_{\mathrm{H}^{\mathrm{II}}_{\Delta\mathrm{R}}} \\ \boldsymbol{\Lambda}_{\mathrm{F1}}\overline{\boldsymbol{V}}_{\mathrm{FA}}\overline{\boldsymbol{T}}^{zd}_{\mathrm{H}^{\mathrm{I}}_{\Delta\mathrm{R}}} \end{bmatrix} \overline{\mathbf{H}}_{\phi r\mathrm{A}} \tag{4.442}$$

with:

$$(\boldsymbol{\Lambda}_{\mathrm{A}})_{ik} = \sinh(\boldsymbol{\Gamma}_{2k}(\overline{H}_{\mathrm{D}} - \overline{z}_i))(\sinh(\boldsymbol{\Gamma}_{2k}\overline{H}_{\mathrm{D}}))^{-1} \tag{4.443}$$

$$(\boldsymbol{\Lambda}_{\mathrm{F}_2})_{ik} = \sinh(\boldsymbol{\Gamma}_{2k}\overline{z}_i)(\sinh(\boldsymbol{\Gamma}_{2k}\overline{H}_{\mathrm{D}}))^{-1} \tag{4.444}$$

$$(\boldsymbol{\Lambda}_{\mathrm{F}_1})_{ik} = e^{-\boldsymbol{\Gamma}_{1k}(\overline{z}_i - \overline{H}_{\mathrm{D}})} \tag{4.445}$$

The index i describes the discrete positions $\overline{\mathbf{z}}_i$ in region D_2 (eqs. (4.443) and (4.444)) and the position $\overline{\mathbf{z}}_i$ in D_1 (eq. (4.445)). Equations analogous to eq. (4.422) are:

$$\begin{bmatrix} (-\overline{\mathbf{H}}_{\phi r\mathrm{C}}) \\ -(-\overline{\mathbf{H}}_{\phi r\mathrm{D}}) \end{bmatrix} = \begin{bmatrix} \overline{\boldsymbol{V}}_{\mathrm{CA}} \\ \overline{\boldsymbol{V}}_{\mathrm{DA}} \end{bmatrix} \overline{\mathbf{H}}_{\phi r\mathrm{A}} \tag{4.446}$$

with:

$$\overline{\boldsymbol{V}}_{\mathrm{CA}} = -\boldsymbol{T}^{r\,-1}_{\mathrm{H}} \begin{bmatrix} (\boldsymbol{\Lambda}_{\mathrm{A}} + \boldsymbol{\Lambda}_{\mathrm{F2}}\overline{\boldsymbol{V}}_{\mathrm{FA}})\boldsymbol{T}^{zd}_{\mathrm{H}^{\mathrm{II}}_{\Delta\mathrm{C}}} \\ \boldsymbol{\Lambda}_{\mathrm{F1}}\overline{\boldsymbol{V}}_{\mathrm{FA}}\boldsymbol{T}^{zd}_{\mathrm{H}^{\mathrm{I}}_{\Delta\mathrm{C}}} \end{bmatrix} \tag{4.447}$$

$$\overline{\boldsymbol{V}}_{\mathrm{DA}} = \boldsymbol{T}^{r\,-1}_{\mathrm{H}} \begin{bmatrix} (\boldsymbol{\Lambda}_{\mathrm{A}} + \boldsymbol{\Lambda}_{\mathrm{F2}}\overline{\boldsymbol{V}}_{\mathrm{FA}})\boldsymbol{T}^{zd}_{\mathrm{H}^{\mathrm{II}}_{\Delta\mathrm{D}}} \\ \boldsymbol{\Lambda}_{\mathrm{F1}}\overline{\boldsymbol{V}}_{\mathrm{FA}}\boldsymbol{T}^{zd}_{\mathrm{H}^{\mathrm{I}}_{\Delta\mathrm{D}}} \end{bmatrix} \tag{4.448}$$

We now may summarise and write:

$$\begin{bmatrix} \widehat{\overline{\mathbf{H}}}_{\mathrm{A}} \\ \widehat{\overline{\mathbf{H}}}_{\mathrm{CD}} \end{bmatrix} = \begin{bmatrix} \overline{\boldsymbol{I}}^{\mathrm{A}}_{\mathrm{A}} & [\overline{\boldsymbol{V}}_{\mathrm{AC}}\overline{\boldsymbol{V}}_{\mathrm{AD}}] \\ \begin{bmatrix} \overline{\boldsymbol{V}}_{\mathrm{CA}} \\ \overline{\boldsymbol{V}}_{\mathrm{DA}} \end{bmatrix} & \widehat{\overline{\boldsymbol{I}}}^{\mathrm{CD}}_{\mathrm{CD}} \end{bmatrix} \begin{bmatrix} \overline{\boldsymbol{Y}}_{\mathrm{Ar}} & \widehat{\overline{\boldsymbol{0}}} \\ \widehat{\overline{\boldsymbol{0}}} & \widehat{\overline{\boldsymbol{y}}}^{\mathrm{CD}}_{\mathrm{CD}} \end{bmatrix} \begin{bmatrix} \overline{\mathbf{E}}_{\mathrm{A}} \\ \widehat{\overline{\mathbf{E}}}_{\mathrm{CD}} \end{bmatrix} \tag{4.449}$$

The fields at port B can be determined separately.

4.6.5 Port relations of section D

We now know the general relation between the fields at the four ports A, B, C and D of section D in Fig. 4.80. We can now e.g. calculate the input impedance/admittance of one port if the load impedances of the other ports are known. The inversion of eq. (4.410) may be written in the following form:

$$\begin{bmatrix} \overline{\mathbf{E}}_{\mathrm{AB}} \\ \overline{\mathbf{E}}_{\mathrm{CD}} \end{bmatrix} = \begin{bmatrix} \hat{\overline{\mathbf{z}}}^{\mathrm{AB}}_{\mathrm{AB}} & \hat{\overline{\mathbf{z}}}^{\mathrm{AB}}_{\mathrm{CD}} \\ \hat{\overline{\mathbf{z}}}^{\mathrm{CD}}_{\mathrm{AB}} & \hat{\overline{\mathbf{z}}}^{\mathrm{CD}}_{\mathrm{CD}} \end{bmatrix} \begin{bmatrix} \overline{\mathbf{H}}_{\mathrm{AB}} \\ -\overline{\mathbf{H}}_{\mathrm{CD}} \end{bmatrix} \tag{4.450}$$

Knowing the sources and the loads, all the fields at the ports can be calculated. Let us assume that port A is the port connected with the waveguide from the source. If so, the impedance matrices $\overline{\mathbf{Z}}_{\mathrm{B}}$, $\overline{\mathbf{Z}}_{\mathrm{C}}$, $\overline{\mathbf{Z}}_{\mathrm{D}}$ defined by:

$$\begin{aligned} \overline{\mathbf{E}}_{z\mathrm{D}} &= \overline{\mathbf{Z}}_{\mathrm{D}}(-\overline{\mathbf{H}}_{\phi\mathrm{D}}) \qquad & \overline{\mathbf{E}}_{z\mathrm{C}} &= \overline{\mathbf{Z}}_{\mathrm{C}}\overline{\mathbf{H}}_{\phi\mathrm{C}} \\ \overline{\mathbf{E}}_{r\mathrm{B}} &= \overline{\mathbf{Z}}_{\mathrm{B}}\overline{\mathbf{H}}_{\phi\mathrm{B}} \qquad & \overline{\mathbf{E}}_{r\mathrm{A}} &= \overline{\mathbf{Z}}_{\mathrm{A}}\overline{\mathbf{H}}_{\phi\mathrm{A}} \end{aligned} \tag{4.451}$$

are the load impedance matrices. They are defined inside region D. We obtain them from the admittances of the neighbouring regions B, E and G by transformation according to the equations in section 2.5. E.g. $\overline{\mathbf{Y}}_{\mathrm{C}}$ is computed from $\overline{\mathbf{Z}}_{\mathrm{E}}$ in eq. (4.382). The load admittance matrix of the coaxial feed line has to be determined by transformation according to the equations in Section 2.5 for connecting waveguide sections. Then we can determine the input admittance matrix $\overline{\mathbf{Y}}_{\mathrm{A}}$ with the help of eq. (4.449).

In some cases, an admittance matrix does not exist. Therefore, we use the port relation in eq. (4.450).

We first combine the port impedance matrices in the following way:

$$\hat{\overline{\mathbf{Z}}}_{\mathrm{AB}} = \mathrm{Diag}(\overline{\mathbf{Z}}_{\mathrm{A}}, -\overline{\mathbf{Z}}_{\mathrm{B}}) \qquad \hat{\overline{\mathbf{Z}}}_{\mathrm{CD}} = \mathrm{Diag}(\overline{\mathbf{Z}}_{\mathrm{C}}, \overline{\mathbf{Z}}_{\mathrm{D}}) \tag{4.452}$$

Instead of eq. (4.450), we can therefore write:

$$\mathbf{0} = \begin{bmatrix} \hat{\overline{\mathbf{z}}}^{\mathrm{AB}}_{\mathrm{AB}} - \hat{\overline{\mathbf{Z}}}_{\mathrm{AB}} & \hat{\overline{\mathbf{z}}}^{\mathrm{AB}}_{\mathrm{CD}} \\ \hat{\overline{\mathbf{z}}}^{\mathrm{CD}}_{\mathrm{AB}} & \hat{\overline{\mathbf{z}}}^{\mathrm{CD}}_{\mathrm{CD}} + \hat{\overline{\mathbf{Z}}}_{\mathrm{CD}} \end{bmatrix} \begin{bmatrix} \hat{\overline{\mathbf{H}}}_{\mathrm{AB}} \\ -\hat{\overline{\mathbf{H}}}_{\mathrm{CD}} \end{bmatrix} \tag{4.453}$$

Introducing the second equation of this system into the first one results in:

$$\hat{\overline{\mathbf{Z}}}_{\mathrm{AB}}\hat{\overline{\mathbf{H}}}_{\mathrm{AB}} = \left(\hat{\overline{\mathbf{z}}}^{\mathrm{AB}}_{\mathrm{AB}} - \hat{\overline{\mathbf{z}}}^{\mathrm{AB}}_{\mathrm{CD}}\left(\hat{\overline{\mathbf{z}}}^{\mathrm{CD}}_{\mathrm{CD}} + \hat{\overline{\mathbf{Z}}}_{\mathrm{CD}}\right)^{-1}\hat{\overline{\mathbf{z}}}^{\mathrm{CD}}_{\mathrm{AB}}\right)\hat{\overline{\mathbf{H}}}_{\mathrm{AB}} \tag{4.454}$$

Now we replace the bracket on the right side with an equivalent matrix:

$$\begin{bmatrix} \overline{\mathbf{Z}}_{\mathrm{A}}\overline{\mathbf{H}}_{\mathrm{A}} \\ \overline{\mathbf{Z}}_{\mathrm{B}}\overline{\mathbf{H}}_{\mathrm{B}} \end{bmatrix} = \begin{bmatrix} \overline{\mathbf{z}}^{\mathrm{n}}_{\mathrm{AA}} & \overline{\mathbf{z}}^{\mathrm{n}}_{\mathrm{AB}} \\ \overline{\mathbf{z}}^{\mathrm{n}}_{\mathrm{BA}} & \overline{\mathbf{z}}^{\mathrm{n}}_{\mathrm{BA}} \end{bmatrix} \begin{bmatrix} \overline{\mathbf{H}}_{\mathrm{A}} \\ -\overline{\mathbf{H}}_{\mathrm{B}} \end{bmatrix} \tag{4.455}$$

We obtain for the input impedance $\overline{\mathbf{Z}}_{\mathrm{A}}$:

$$\overline{\mathbf{Z}}_{\mathrm{A}} = \overline{\mathbf{z}}^{\mathrm{n}}_{\mathrm{AA}} - \overline{\mathbf{z}}^{\mathrm{n}}_{\mathrm{AB}}(\overline{\mathbf{z}}^{\mathrm{n}}_{\mathrm{BB}} + \overline{\mathbf{Z}}_{\mathrm{B}})^{-1}\overline{\mathbf{z}}^{\mathrm{n}}_{\mathrm{BA}} \tag{4.456}$$

The last equation again has the known form.

The formulas for the special subsections I and II can be obtained in an analogous manner.

4.6.6 Numerical results

In what follows, illustrative numerical results are given for different structures. In the first example we present results for a dipole antenna. Figs. 4.83 and 4.84 show current distributions (see also [5]). The finite radius of the antenna is taken into account in the results obtained with the MoL. Therefore, the values at the end are not zero. This is also the case in the measured results of Mack, taken from Collin–Zucker [37]. The results of R. W. King (also taken from [37]) were obtained by an integral equation method. There, the radius of the dipole stubs was assumed to be approximately zero.

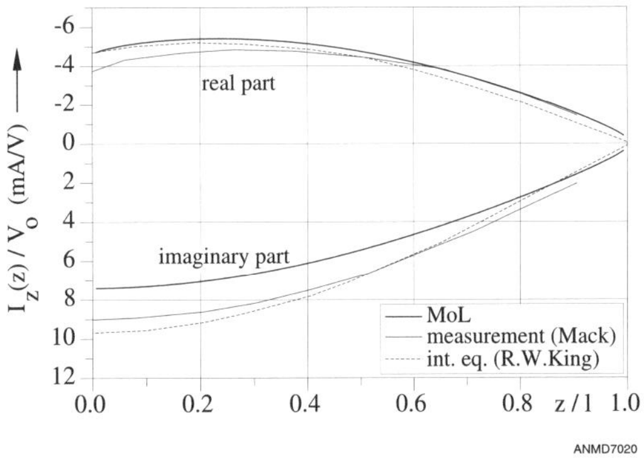

Fig. 4.83 Current distribution on one stub of a $\lambda/2$-dipole antenna. The results for comparison were taken from Collin–Zucker [37] (R. Pregla, 'The analysis of general axially symmetric antennas with a coaxial feed line by the method of lines', *IEEE Trans. Antennas. Propagation*, vol. 46, no. 10, pp. 1433–1443. © 1998 Institute of Electrical and Electronics Engineers (IEEE))

The input impedance of the dipole is presented in Fig. 4.85. It was determined from the field in the gap between the stubs and the input current. Curves a, b and c show increasing values for three different gaps. The impedance in the series resonance is about 71Ω in all three cases. For the dashed curve we took into account losses. We assumed a dipole of copper with conductivity $\kappa = 56\Omega^{-1}$ m/mm^2.

The loss resistance has the value 0.023 Ω in the series-type equivalent circuit for the $2H_P = \lambda/2$ dipole. In the parallel-type equivalent circuit ($2H_P = \lambda$ dipole) we have 39 MΩ, 9.6 MΩ and 6.2 MΩ, respectively.

Next, results for a coaxial-line-feed monopole antenna are plotted in Fig. 4.86. The antenna input impedance for two different positions is plotted in the complex plane. Position (1) is in the outer-gap side of the monopole and the ground, position (2) is on the end of the coaxial feed line. The diagram exhibits the well-known spiral behaviour. The first real input impedance has a value of about 36.6 Ohms.

The input impedances of a monopole, partly buried in a grounded dielectric substrate, are plotted in the Smith chart presented in Fig. 4.87.

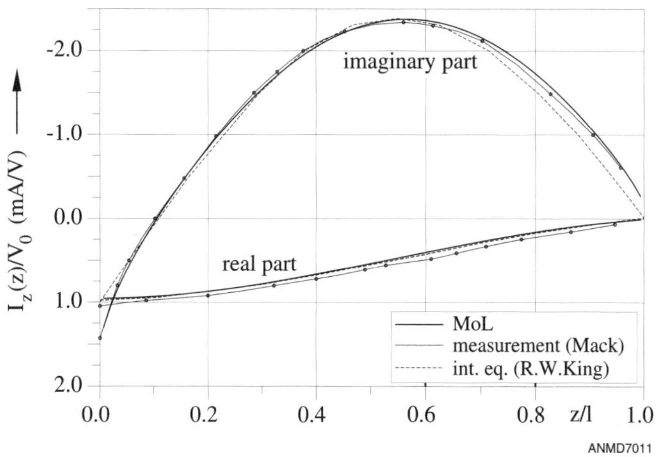

Fig. 4.84 Current distribution on one stub of a λ-dipole antenna. The results for comparison were taken from Collin–Zucker [37] (R. Pregla, 'New Approach for the Analysis of Cylindrical Antennas by the Method of Lines', *Electron. Lett.*, vol. 30, no. 8, pp. 614–615. © 1994 The Institution of Engineering and Technology (IET))

Numerical results obtained by the modal expansion method [38] are shown for comparison. Measured values reported by Alexopoulus [39] are also introduced. All results are in very good agreement. In Fig. 4.88 the input impedance of a centred microstrip disk antenna is plotted in a Smith chart with a centre impedance of 50Ω. The parameter is k_0R_p. For comparison, results from [40] obtained by moment-method solution are given. Up to $k_0R_p \approx 6.5$, there is quite a good agreement.

Besides the discretisation, no further approximations were introduced in the MoL.

Fig. 4.89b is a plot (polar coordinates) of the input reflection coefficient for the circular-patch antenna sketched in Fig. 4.89a. The antenna is fed by a coaxial line. Further elements can be introduced in the feed line, e.g. to match the input impedance.

4.6.7 Further structures and remarks

The presented MoL algorithm can also be used to study various transitions between coaxial and parallel-plate lines (see Fig. 4.90). Fig. 4.91 shows other forms of circular planar antennas and Fig. 4.92 further forms of monopoles and dipoles. A dipole with a filter element in the lower part to prevent a wave propagating down the mast is sketched in Fig. 4.93. The adequate discretisation lines are also shown. The analysis occurs as described earlier.

A special case in this chapter is again the concatenation of waveguide sections of different direction. Here, crossed lines (this should not be confused with 2D discretisation) were introduced.

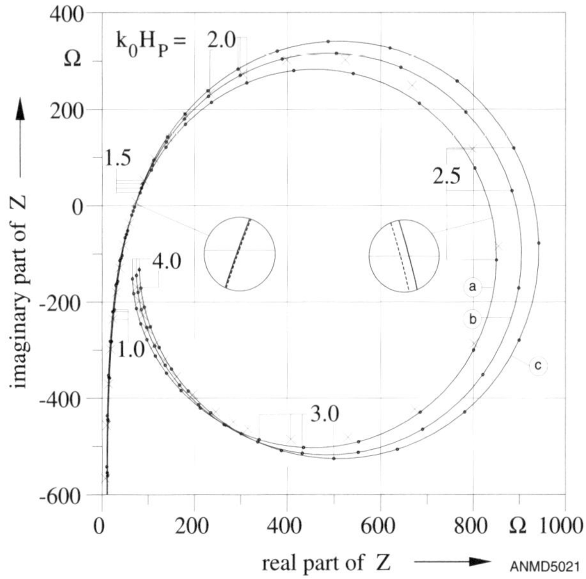

Fig. 4.85 Input impedance loci of a dipole antenna parameterised with $k_0 H_p$ (k_0 = free space wave number) — lossless - - - with loss (R. Pregla, 'The analysis of general axially symmetric antennas with a coaxial feed line by the method of lines', *IEEE Trans. Antennas. Propagation*, vol. 46, no. 10, pp. 1433–1443. © 1998 Institute of Electrical and Electronics Engineers (IEEE))

The determination of the field should be done in reverse direction. Starting at the position of the source and using the impedances/admittances calculated earlier, the fields can be successively calculated from section to section. The far field of the antenna can be obtained from the tangential fields in its surrounding area by usual near-field/far-field transformation, from which the radiation pattern can be extracted. This transformation is well developed. To improve the accuracy, a further partition, especially of region A (see, Fig. 4.78), can be introduced. In principle, all the necessary formulas are given for this case. For radial discretisation, ABCs have to be introduced on the outer side.

4.7 DEVICES IN CYLINDRICAL COORDINATES – TWO-DIMENSIONAL DISCRETISATION

4.7.1 Discretisation in r- and ϕ-direction

In this section we would like to verify the proposed algorithm in Section 4.3.1. Therefore, patch antennas on a cylindrical body (see Fig. 4.94) are analysed. In Section 6.6.3 we will show similar structures. There we deal with conformal

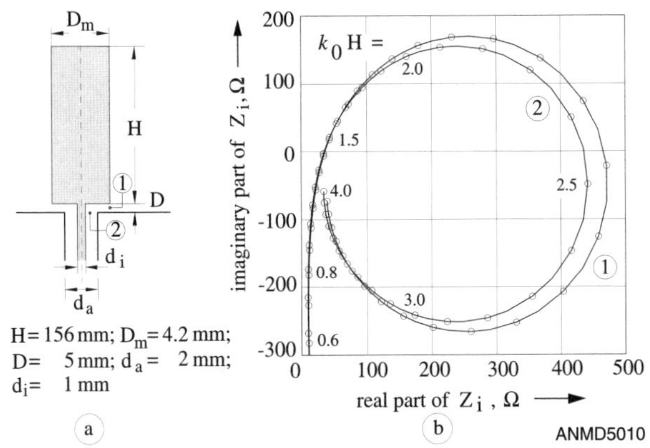

Fig. 4.86 Input impedance of a coaxial-line-fed monopole antenna at two different positions (a) sketch of the antenna with dimensions (b) input impedance in the complex plane (R. Pregla, 'The analysis of general axially symmetric antennas with a coaxial feed line by the method of lines', *IEEE Trans. Antennas. Propagation*, vol. 46, no. 10, pp. 1433–1443. © 1998 Institute of Electrical and Electronics Engineers (IEEE))

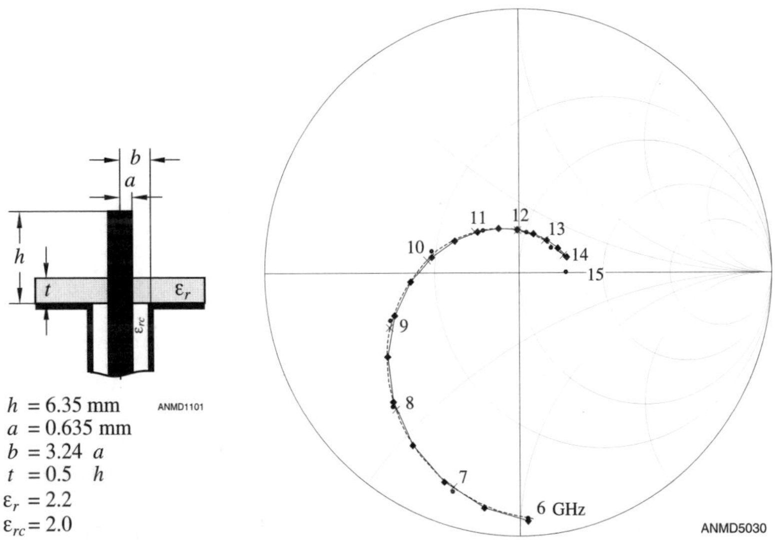

Fig. 4.87 Input impedance of the monopole partly buried in a dielectric substrate in a Smith chart. o——o MoL x - - - x numerical results of [41] ••• experimental results of [39] (R. Pregla, 'The analysis of general axially symmetric antennas with a coaxial feed line by the method of lines', *IEEE Trans. Antennas. Propagation*, vol. 46, no. 10, pp. 1433–1443. © 1998 Institute of Electrical and Electronics Engineers (IEEE))

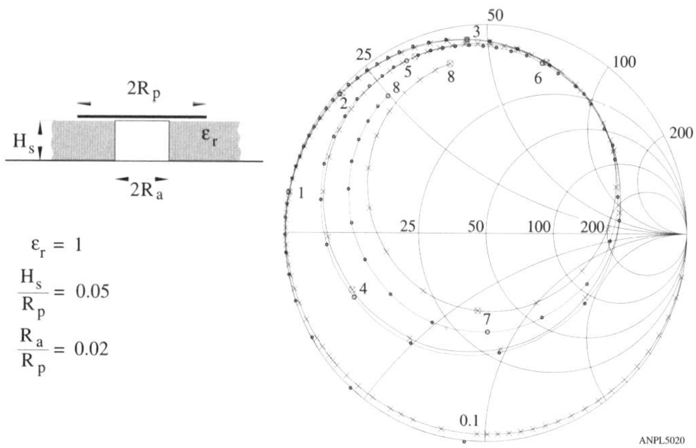

Fig. 4.88 Input impedance of a planar circular antenna at $r = R_a$ in a Smith chart. Parameter k_0R_p (k_0 = free space wave number). • – – – • Results from [40] (R. Pregla, 'The analysis of general axially symmetric antennas with a coaxial feed line by the method of lines', *IEEE Trans. Antennas. Propagation*, vol. 46, no. 10, pp. 1433–1443. © 1998 Institute of Electrical and Electronics Engineers (IEEE))

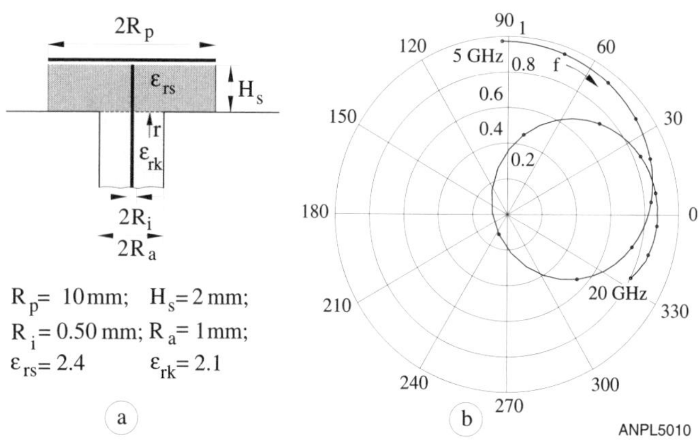

Fig. 4.89 Reflection coefficient r of a coaxial-line-fed circular-patch antenna on the marked position. The frequency step between two marked points is 1 GHz. (a) sketch of the antenna with dimensions (b) reflection coefficient in the complex plane (R. Pregla, 'The analysis of general axially symmetric antennas with a coaxial feed line by the method of lines', *IEEE Trans. Antennas. Propagation*, vol. 46, no. 10, pp. 1433–1443. © 1998 Institute of Electrical and Electronics Engineers (IEEE))

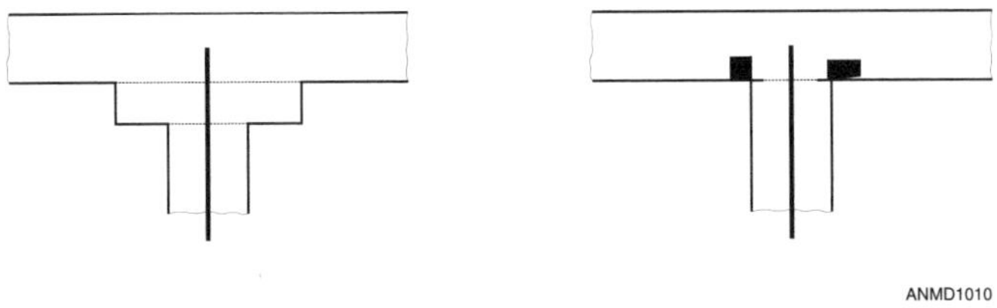

Fig. 4.90 Transition from coaxial-line to parallel-plate guide

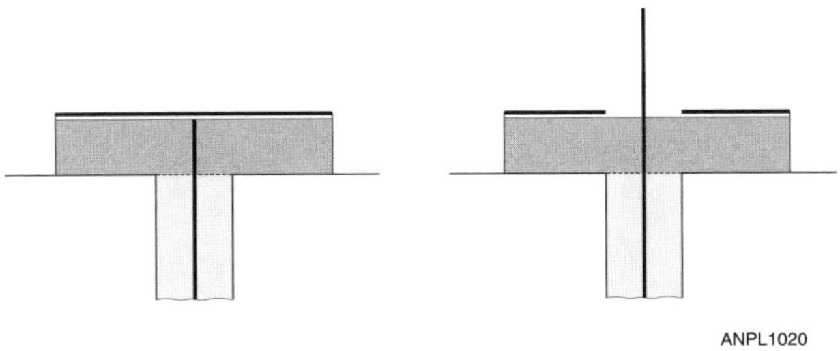

Fig. 4.91 Other quasi-planar antenna structures

antennas on a sphere. The antennas are fed by microstrip lines. The discretisation scheme is shown in Fig. 4.8.

4.7.2 Numerical results

Fig. 4.95a shows a diagram of the scattering parameters as a function of frequency. Devices with one and two patches were analysed. Fig. 4.95b shows the radiated power as a function of frequency. The devices were not optimised. Also, the near fields and the current distributions on the metallisations can be obtained with the algorithm. The radiation characteristic can be obtained in a further step from the current distribution on the metallic strips.

4.7.3 Discretisation in r- and z-direction

In this section we would like to show an example of a complex wave-guiding structure (Fig. 4.96). The four inner sections are curved and we use cylindrical coordinate systems with propagation in ϕ-direction for their analysis. The discretisation has to be performed in r- and z-directions. The direction of the z coordinate changes if the direction of the r coordinate changes. The impedance matching must be done analogous to the S bends in rectangular waveguides. The discretisation scheme is shown in Fig. 4.11. The waveguides on the left and right sides are straight waveguides. These parts are described by Cartesian coordinates. Numerical results can be found in [42], where the connection of straight waveguides to a curved region is described.

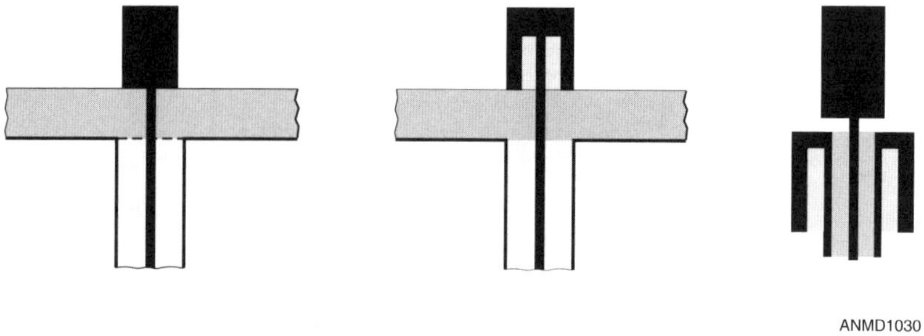

Fig. 4.92 Monopoles and dipoles

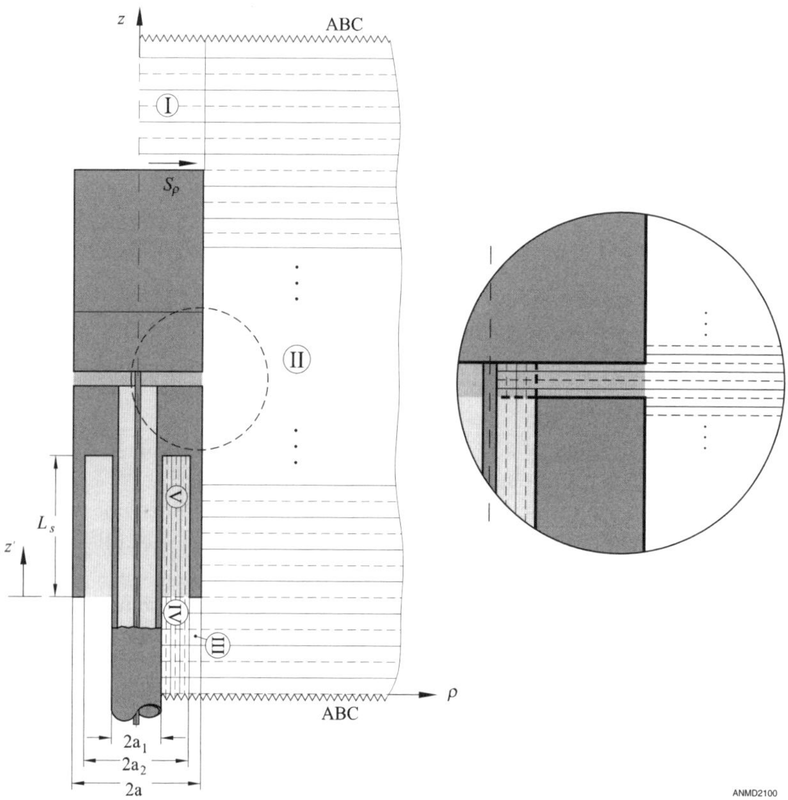

Fig. 4.93 Dipole with filter element and discretisation-line scheme

4.7.4 Discretisation in ϕ- and z-direction

We would like to describe the analysis procedure for structures such as the circular resonator in Fig. 4.97. It is periodic in azimuthal direction. Therefore,

Fig. 4.94 Conformal waveguide antenna structure with propagation in z-direction (R. Pregla, 'Efficient Analysis of Conformal Antennas with Anisotropic Material (0682)', in *AP 2000 Millennium Conference on Antennas and Propagation.* © 2000 European Space Agency (ESA))

we only need to describe one period. The phase variation in the azimuthal direction must fulfil the condition for resonance of the whole structure. We use discretisation lines in the radial direction.

4.7.5 GTL equations for r-direction

For the analysis, we start with the wave equations for the electric or magnetic field (eqs. (4.89) and (4.90)):

$$\overline{r}\frac{\partial}{\partial\overline{r}}\left(\overline{r}\frac{\partial}{\partial\overline{r}}[\widehat{E}^r]\right) + [U_{\mathrm{E}}^{rc}][\widehat{E}^r] + \overline{r}^2[V_{\mathrm{E}}^{rc}][\widehat{E}^r] + \mathrm{j}2\overline{r}^2[\widehat{\mu}_{zz}][\widehat{H}^r] = \begin{bmatrix}0\\0\end{bmatrix} \qquad (4.457)$$

$$\overline{r}\frac{\partial}{\partial\overline{r}}\left(\overline{r}\frac{\partial}{\partial\overline{r}}[\widehat{H}^r]\right) + [U_{\mathrm{H}}^{rc}][\widehat{H}^r] + \overline{r}^2[V_{\mathrm{H}}^{rc}][\widehat{H}^r] + \mathrm{j}2\overline{r}^2[\widehat{\varepsilon}_{zz}][\widehat{E}^r] = \begin{bmatrix}0\\0\end{bmatrix} \qquad (4.458)$$

where, by using $D_{\overline{z}}D_\phi = D_\phi D_{\overline{z}}$, the matrices $[U_{\mathrm{E,H}}^{rc}]$ are given as:

$$\begin{aligned}[U_{\mathrm{E}}^{rc}] &= [R_{\mathrm{H}}^{rc}][R_{\mathrm{E}}^{rc}] = [R_{\mathrm{H}}^{rc\mathrm{D}}][\widehat{\varepsilon}_{\phi\phi}] + [\widehat{\mu}_{\phi\phi}][R_{\mathrm{E}}^{rc\mathrm{D}}] \\ &= \begin{bmatrix} D_\phi\varepsilon_{rr}^{-1}D_\phi\varepsilon_{\phi\phi} & 0 \\ D_{\overline{z}}\varepsilon_{rr}^{-1}D_\phi\varepsilon_{\phi\phi} - \mu_{\phi\phi}D_\phi\mu_{rr}^{-1}D_{\overline{z}} & \mu_{\phi\phi}D_\phi\mu_{rr}^{-1}D_\phi\end{bmatrix}\end{aligned} \qquad (4.459)$$

$$\begin{aligned}[U_{\mathrm{H}}^{rc}] &= [R_{\mathrm{E}}^{rc}][R_{\mathrm{H}}^{rc}] = [R_{\mathrm{E}}^{rc\mathrm{D}}][\widehat{\mu}_{\phi\phi}] + [\widehat{\varepsilon}_{\phi\phi}][R_{\mathrm{H}}^{rc\mathrm{D}}] \\ &= \begin{bmatrix} \varepsilon_{\phi\phi}D_\phi\varepsilon_{rr}^{-1}D_\phi & \varepsilon_{\phi\phi}D_\phi\varepsilon_{rr}^{-1}D_{\overline{z}} - D_{\overline{z}}\mu_{rr}^{-1}D_\phi\mu_{\phi\phi} \\ 0 & D_\phi\mu_{rr}^{-1}D_\phi\mu_{\phi\phi}\end{bmatrix}\end{aligned} \qquad (4.460)$$

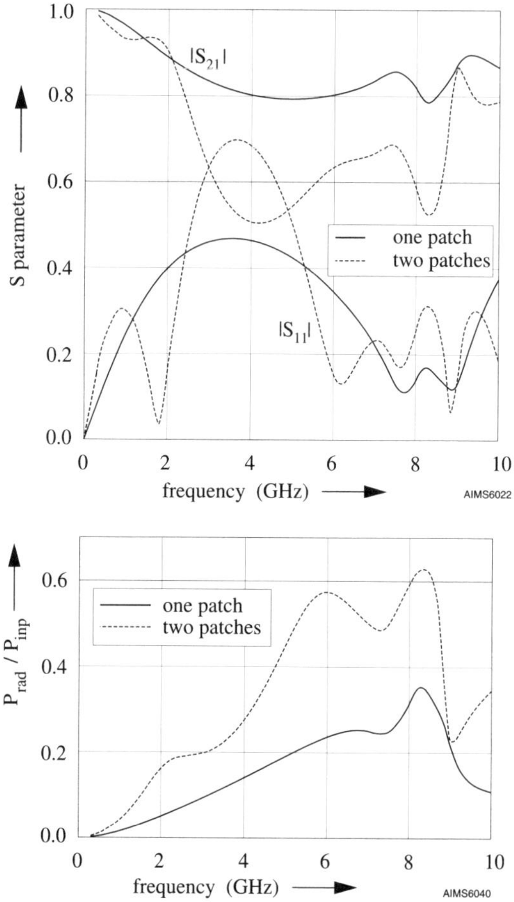

Fig. 4.95 Patch antenna on a cylindrical body: Upper diagram: scattering parameters as a function of frequency Lower diagram: radiated power in relation to the input power (Reproduced by permission of IGTE (Institut für Grundlagen und Theorie der Elektrotechnik) Graz)

and the matrices $[V^{rc}_{\mathrm{E,H}}]$ as:

$$[V^{rc}_{\mathrm{E}}] = [R^{rc}_{\mathrm{H}}][\widehat{\varepsilon}_{zz}] + [\widehat{\mu}_{zz}][R^{rc}_{\mathrm{E}}] =$$

$$= \begin{bmatrix} \mu_{zz} D_{\overline{z}} \mu_{rr}^{-1} D_{\overline{z}} + \varepsilon_{\phi\phi}\mu_{zz} & D_{\phi}\varepsilon_{rr}^{-1} D_{\overline{z}}\varepsilon_{zz} - \mu_{zz} D_{\overline{z}}\mu_{rr}^{-1} D_{\phi} \\ 0 & D_{\overline{z}}\varepsilon_{rr}^{-1} D_{\overline{z}}\varepsilon_{zz} + \varepsilon_{zz}\mu_{\phi\phi} \end{bmatrix} \tag{4.461}$$

$$[V^{rc}_{\mathrm{H}}] = [R^{rc}_{\mathrm{E}}][\widehat{\mu}_{zz}] + [\widehat{\varepsilon}_{zz}][R^{rc}_{\mathrm{H}}]$$

$$= \begin{bmatrix} \varepsilon_{\phi\phi}\mu_{zz} + D_{\overline{z}}\mu_{rr}^{-1} D_{\overline{z}}\mu_{zz} & 0 \\ \varepsilon_{zz} D_{\overline{z}}\varepsilon_{rr}^{-1} D_{\phi} - D_{\phi}\mu_{rr}^{-1} D_{\overline{z}}\mu_{zz} & \varepsilon_{zz}\mu_{\phi\phi} + \varepsilon_{zz} D_{\overline{z}}\varepsilon_{rr}^{-1} D_{\overline{z}} \end{bmatrix} \tag{4.462}$$

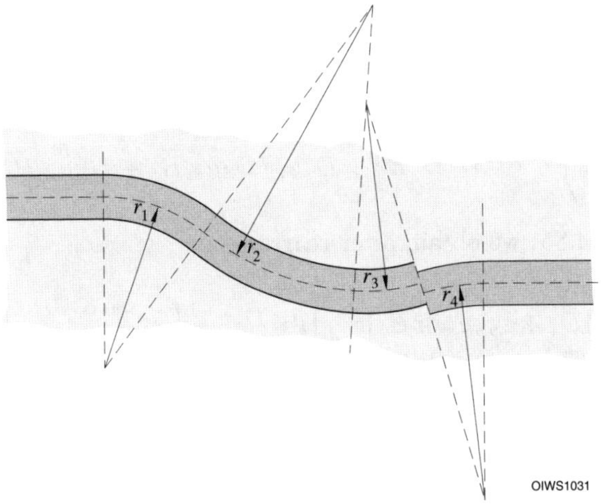

Fig. 4.96 Rib waveguide S bend with propagation in different directions

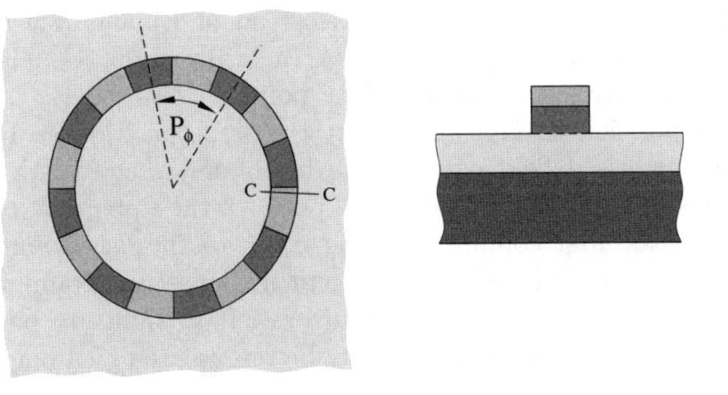

Fig. 4.97 Top view and cross-sections of a circular rib waveguide resonator. The structure is periodic in azimuthal direction (R. Pregla, 'Analysis of complex photonic structures', *ICTON Conf.* © 2007 Institute of Electrical and Electronics Engineers (IEEE))

We assume ϕ-independent material parameters (as we find in the inner and outer regions of our resonator) and introduce $\varepsilon_{rr} = \varepsilon_{\phi\phi}$ and $\mu_{rr} = \mu_{\phi\phi}$. The matrices U will become diagonal.

Because of the zero elements in $[V^{rc}_{\mathrm{E,H}}]$, we obtain from the second equation in (4.457) and from the first equation in (4.458), respectively, the following differential equations for E_z and $\widetilde{H}_z$:

$$\frac{1}{\overline{r}}\frac{\partial}{\partial \overline{r}}\left(\overline{r}\frac{\partial E_z}{\partial \overline{r}}\right) + \overline{r}^{-2} D_\phi D_\phi E_z + D_{\overline{z}} \varepsilon_{rr}^{-1} D_{\overline{z}} \varepsilon_{zz} E_z + \mu_{\phi\phi}\varepsilon_{zz} E_z = 0 \quad (4.463)$$

$$\frac{1}{\overline{r}}\frac{\partial}{\partial \overline{r}}\left(\overline{r}\frac{\partial \widetilde{H}_z}{\partial \overline{r}}\right) + \overline{r}^{-2} D_\phi D_\phi \widetilde{H}_z + D_{\overline{z}} \mu_{rr}^{-1} D_{\overline{z}} \mu_{zz} \widetilde{H}_z + \varepsilon_{\phi\phi}\mu_{zz} \widetilde{H}_z = 0 \quad (4.464)$$

From eq. (4.85) we obtain after rearranging:

$$\begin{bmatrix} (\mu_{zz} D_{\overline{z}} \mu_{rr}^{-1} D_{\overline{z}} + \varepsilon_{\phi\phi}\mu_{zz})\overline{r} E_\phi \\ (\varepsilon_{zz} D_{\overline{z}} \varepsilon_{rr}^{-1} D_{\overline{z}} + \varepsilon_{zz}\mu_{\phi\phi})\overline{r} \widetilde{H}_\phi \end{bmatrix} = \begin{bmatrix} \mu_{zz} D_{\overline{z}} \mu_{rr}^{-1} D_\phi \varepsilon_{zz}^{-1} & \mathrm{j}\overline{r}\frac{\partial}{\partial \overline{r}} \\ -\mathrm{j}\overline{r}\frac{\partial}{\partial \overline{r}} & \varepsilon_{zz} D_{\overline{z}} \varepsilon_{rr}^{-1} D_\phi \mu_{zz}^{-1} \end{bmatrix} \times \begin{bmatrix} \varepsilon_{zz} E_z \\ \mu_{zz} \widetilde{H}_z \end{bmatrix} \quad (4.465)$$

Hence, E_ϕ and $\widetilde{H}_\phi$ can be derived in terms of E_z and $\widetilde{H}_z$.

4.7.5.1 Discretisation

For the solution of eqs. (4.463) and (4.464) we discretise the fields in z-direction according to Fig. 4.98. Only a few discretisation lines are introduced. We use $\varepsilon_{zz} E_z$ and $\mu_{zz} H_z$ (instead of E_z and $\widetilde{H}_z$) as variables, because these quantities are continuous in z-direction. Usually, the field components are discretised on different points. However, $\mathbf{E}_\phi$ is discretised at the same position as $\mathbf{H}_z$, and $\mathbf{E}_z$ at the same position as $\mathbf{H}_\phi$. Now we collect the discretised field components in column vectors. We order these components starting with the adequate point in the upper-left point and go downwards in the first column. Coming down to the last point in the first column we continue with the highest point in the second column and so on. In this way, the values in the columns from left to right are put below one another. Each of the column vectors may be represented by a boldface capital letter of the field component and by a subscript i for the ith column. The collection of all these vectors in the total column vector is marked by a hat ($\hat{\ }$).

If we have N_ϕ° columns of $\circ$ discretisation points and N_z° points in a column, the total number of points is $\mathrm{N}_\phi^\circ \mathrm{N}_z^\circ$. The number of $\bullet$ columns and rows should be $\mathrm{N}_\phi^\bullet$ and $\mathrm{N}_z^\bullet$, respectively. With periodic boundary conditions we have $\mathrm{N}_\phi^\circ = \mathrm{N}_\phi^\bullet$. The values of the permittivities and permeabilities are collected in the same order as the field components, not in column vectors but in the main diagonal of a diagonal matrix. In general, each of the components of the tensor is discretised on a different permittivity or permeability point. Therefore, we have three different permittivity and permeability matrices each. Even in the case of isotropic materials, we need to have three different matrices of each of the parameters. For the solution of eqs. (4.463) and (4.464), the multilayered cross-section consists of homogeneous layers. Furthermore, we introduce periodic boundary conditions on the left and right sides. Therefore, we need

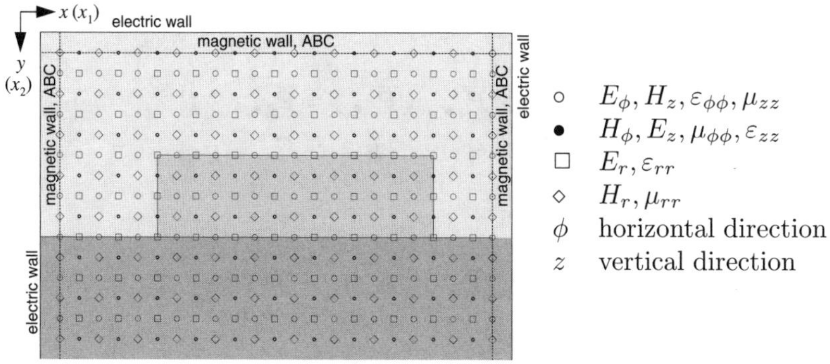

Fig. 4.98 Cross-section (ϕ- z-plane) of a period in the resonator of Fig. 4.97 with discretisation points, periodic boundary conditions on the left and right sides, magnetic walls or ABCs on upper and lower sides. (R. Pregla, 'Analysis of complex photonic structures', *ICTON Conf.* © 2007 Institute of Electrical and Electronics Engineers (IEEE))

only two different permittivity and permeability matrices. The differential operators are replaced by central differences. All the central differences in one row or one column are collected to the difference operators (matrices) $\boldsymbol{D}_{\overline{z}}^{\circ,\bullet}$ and $\boldsymbol{D}_{\phi}^{\circ,\bullet}$, respectively. The collection of all rows and all columns is marked by a hat $(\widehat{\ })$.

We obtain by discretisation in one column (vertical or z-direction):

$$\begin{aligned}
\varepsilon_{zz}, \varepsilon_{rr} = \varepsilon_{\phi\phi} &\longrightarrow \boldsymbol{\epsilon}_\bullet, \boldsymbol{\epsilon}_\circ\\
\mu_{zz}, \mu_{rr} = \mu_{\phi\phi} &\longrightarrow \boldsymbol{\mu}_\circ, \boldsymbol{\mu}_\bullet\\
\varepsilon_{zz} D_{\overline{z}} \varepsilon_{rr}^{-1} D_{\overline{z}}(\varepsilon_{zz} E_z) &\longrightarrow -\boldsymbol{\epsilon}_\bullet \overline{\boldsymbol{D}}_{\overline{z}\bullet}^{\mathrm{t}} \boldsymbol{\epsilon}_\circ^{-1} \overline{\boldsymbol{D}}_{\overline{z}\bullet}(\boldsymbol{\epsilon}_\bullet \mathbf{E}_z) = -\overline{\boldsymbol{P}}_{z\bullet}^{\epsilon}(\boldsymbol{\epsilon}_\bullet \mathbf{E}_z)\\
\mu_{zz} D_{\overline{z}} \mu_{rr}^{-1} D_{\overline{z}}(\mu_{zz} \widetilde{H}_z) &\longrightarrow -\boldsymbol{\mu}_\circ \overline{\boldsymbol{D}}_{\overline{z}\circ}^{\mathrm{t}} \boldsymbol{\mu}_\bullet^{-1} \overline{\boldsymbol{D}}_{\overline{z}\circ}(\boldsymbol{\mu}_\circ \widetilde{\mathbf{H}}_z) = -\overline{\boldsymbol{P}}_{z\circ}^{\mu}(\boldsymbol{\mu}_\circ \widetilde{\mathbf{H}}_z)
\end{aligned}$$

The subscripts $\bullet, \circ$ are related to E_z and H_z, respectively. Without ABCs, the difference operators $\overline{\boldsymbol{D}}_{\overline{z}\bullet}$ and $\overline{\boldsymbol{D}}_{\overline{z}\circ}$ according to Fig. 4.98 have to be constructed for Dirichlet and Neumann boundary conditions, respectively. Using ABCs as shown in Fig. 4.98, we have to replace the difference operators $\overline{\boldsymbol{D}}_{z\bullet}$ and $-\overline{\boldsymbol{D}}_{\overline{z}\circ}^{\mathrm{t}}$ with $\overline{\boldsymbol{D}}_{\overline{z}\bullet}^{\mathrm{a}}$. The remaining difference operators $\overline{\boldsymbol{D}}_{\overline{z}\circ}$ and $\overline{\boldsymbol{D}}_{\overline{z}\bullet}^{\mathrm{t}}$ have the same form as before.

We obtain by discretisation in one row (horizontal or ϕ-direction):

$$D_\phi D_\phi \longrightarrow -\overline{\boldsymbol{D}}_{\phi\circ}^{\mathrm{t}} \overline{\boldsymbol{D}}_{\phi\circ} = -\overline{\boldsymbol{P}}_{\phi\circ} \qquad D_\phi D_\phi \longrightarrow -\overline{\boldsymbol{D}}_{\phi\bullet}^{\mathrm{t}} \overline{\boldsymbol{D}}_{\phi\bullet} = -\overline{\boldsymbol{P}}_{\phi\bullet} \quad (4.466)$$

Periodic boundary conditions [17] have to be introduced into the $\overline{\boldsymbol{D}}_{\phi\circ,\bullet}$ operators. To do this correctly, the left- and right-side boundaries have to be shifted together by a small (infinitesimal) amount to the right or left. The complete operators are now given by Kronecker products with unit matrices

$\boldsymbol{I}_\phi^{\bullet,\circ}$ and $\boldsymbol{I}_z^{\bullet,\circ}$ $(\alpha = \varepsilon, \mu)$:

$$\widehat{\overline{\boldsymbol{P}}}^{\alpha}_{z\bullet,\circ} = \boldsymbol{I}_\phi^{\bullet,\circ} \otimes \overline{\boldsymbol{P}}^{\alpha}_{z\bullet,\circ} \qquad \widehat{\overline{\boldsymbol{P}}}_{\phi\bullet,\circ} = \overline{\boldsymbol{P}}_{\phi\bullet,\circ} \otimes \boldsymbol{I}_z^{\bullet,\circ}$$
$$\widehat{\boldsymbol{\epsilon}}_{\bullet,\circ} = \boldsymbol{I}_\phi^{\bullet,\circ} \otimes \boldsymbol{\epsilon}_{\bullet,\circ} \qquad \widehat{\boldsymbol{\mu}}_{\bullet,\circ} = \boldsymbol{I}_\phi^{\bullet,\circ} \otimes \boldsymbol{\mu}_{\bullet,\circ}$$

We transform according to:

$$\widehat{\boldsymbol{\epsilon}}_{\bullet}\widehat{\mathbf{E}}_z = \widehat{\boldsymbol{T}}^{\varepsilon}_{\bullet}\widehat{\widehat{\mathbf{E}}}_z = (\boldsymbol{T}_{\phi\bullet} \otimes \boldsymbol{T}^{\varepsilon}_{z\bullet})\widehat{\widehat{\mathbf{E}}}_z \qquad \widehat{\boldsymbol{\mu}}_{\circ}\widehat{\widehat{\mathbf{H}}}_z = \widehat{\boldsymbol{T}}^{\mu}_{\circ}\widehat{\widehat{\mathbf{H}}}_z = (\boldsymbol{T}_{\phi\circ} \otimes \boldsymbol{T}^{\mu}_{z\circ})\widehat{\widehat{\mathbf{H}}}_z \tag{4.467}$$

and:

$$(\boldsymbol{T}^{\alpha}_{z\bullet,\circ})^{-1}(\boldsymbol{\epsilon}_{\bullet,\circ}\boldsymbol{\mu}_{\bullet,\circ} - \overline{\boldsymbol{P}}^{\alpha}_{z\bullet,\circ})\boldsymbol{T}^{\alpha}_{z\bullet,\circ} = \boldsymbol{k}^2_{r\bullet,\circ} \qquad (\boldsymbol{T}_{\phi\bullet,\circ})^{-1}\overline{\boldsymbol{P}}_{\phi\bullet,\circ}\boldsymbol{T}_{\phi\bullet,\circ} = \boldsymbol{\nu}^2_{\bullet,\circ} \tag{4.468}$$

where $\boldsymbol{k}^2_{r\bullet,\circ}$ and $\boldsymbol{\nu}^2_{\bullet,\circ}$ are diagonal matrices. We obtain as discretised and transformed wave equations (4.463) and (4.464):

$$\overline{r}\frac{\mathrm{d}}{\mathrm{d}\overline{r}}\left(\overline{r}\frac{\mathrm{d}}{\mathrm{d}\overline{r}}\widehat{\widehat{\mathbf{F}}}\right) + ((\widehat{\boldsymbol{k}}_{r\bullet,\circ}\overline{r})^2 - \widehat{\boldsymbol{\nu}}^2_{\bullet,\circ})\widehat{\widehat{\mathbf{F}}} = \mathbf{0} \qquad \widehat{\widehat{\mathbf{F}}} = \widehat{\widehat{\mathbf{E}}}_z, \widehat{\widehat{\mathbf{H}}}_z \tag{4.469}$$
$$\widehat{\boldsymbol{k}}^2_{r\bullet,\circ} = \boldsymbol{I}_\phi^{\bullet,\circ} \otimes \boldsymbol{k}^2_{r\bullet,\circ} \qquad \widehat{\boldsymbol{\nu}}^2_{\bullet,\circ} = \boldsymbol{\nu}^2_{\bullet,\circ} \otimes \boldsymbol{I}_z^{\bullet,\circ}$$

The general solution is given by:

$$\widehat{\widehat{\mathbf{F}}} = J_{\widehat{\boldsymbol{\nu}}\bullet,\circ}(\widehat{\boldsymbol{t}}_{\bullet,\circ})\mathbf{A} + Y_{\widehat{\boldsymbol{\nu}}\bullet,\circ}(\widehat{\boldsymbol{t}}_{\bullet,\circ})\mathbf{B} \qquad \widehat{\boldsymbol{t}}_{\bullet,\circ} = \widehat{\boldsymbol{k}}_{r\bullet,\circ}\overline{r} \tag{4.470}$$

where $J_{\widehat{\boldsymbol{\nu}}_{\bullet,\circ}}$ and $Y_{\widehat{\boldsymbol{\nu}}_{\bullet,\circ}}$ are Bessel and Neumann functions, respectively. Their order $\widehat{\boldsymbol{\nu}}\bullet,\circ$ is a diagonal matrix. The arguments are diagonal matrices, as $\widehat{\boldsymbol{k}}_{r\bullet,\circ}$. Place i in the argument matrix and i in the order matrix correspond to each other. Therefore, $J_{\widehat{\boldsymbol{\nu}}_{\bullet,\circ}}$ and $Y_{\widehat{\boldsymbol{\nu}}_{\bullet,\circ}}$ are also diagonal matrices. From the solution in eq. (4.470) impedance/admittance transformation algorithms can be developed. Besides the fact that eqs. (4.463) and (4.464) can only be written for homogeneous layers, the solution algorithm shows that the diagonalisation is also only possible for homogeneous layers. So we cannot use this solution for the rib-ring region in our resonator. In this region we use the transformation with finite differences [43].

4.7.5.2 Centre and outer regions

For the centre and infinite outer regions we must derive separate equations. In the inner region only the first term of eq. (4.470) can be used, because the second one is infinite in the centre. In the infinite outer region we have to use the modified Hankel function K_m. To determine the quantities in eq. (4.465) we need the derivatives of the Bessel's function:

$$\overline{r}\frac{\mathrm{d}}{\mathrm{d}\overline{r}}\overline{\mathbf{F}} = \widehat{\boldsymbol{t}}_{\bullet,\circ}R'_{\widehat{\boldsymbol{\nu}}_{\bullet,\circ}}(\widehat{\boldsymbol{t}}_{\bullet,\circ})R^{-1}_{\widehat{\boldsymbol{\nu}}_{\bullet,\circ}}(\widehat{\boldsymbol{t}}_{\bullet,\circ})\widehat{\widehat{\mathbf{F}}} \tag{4.471}$$

where $R_{\widehat{\nu}_{\bullet,\circ}}$ is equal to $J_{\widehat{\nu}_{\bullet,\circ}}(K_{\widehat{\nu}_{\bullet,\circ}})$ in the inner (infinite outer) region. $R'_{\widehat{\nu}_{\bullet,\circ}}$ is the derivative with respect to the argument. Now we obtain from eq. (4.465):

$$\begin{bmatrix} \widehat{\mathbf{k}}^2_{r\circ}(\overline{r}_{\mathrm{C}}\widehat{\widehat{\mathbf{E}}}_{\phi\mathrm{C}}) \\ -\widehat{\mathbf{k}}^2_{r\bullet}(\overline{r}_{\mathrm{C}}\widehat{\widehat{\mathbf{H}}}_{\phi\mathrm{C}}) \end{bmatrix} = \begin{bmatrix} \widehat{\boldsymbol{\Delta}}_{\bullet} & \mathrm{j}\widehat{\boldsymbol{\Lambda}}^{\mathrm{C}}_{\circ} \\ \mathrm{j}\widehat{\boldsymbol{\Lambda}}^{\mathrm{C}}_{\bullet} & -\widehat{\boldsymbol{\Delta}}_{\circ} \end{bmatrix} \begin{bmatrix} \widehat{\widehat{\mathbf{E}}}_{z\mathrm{C}} \\ \widehat{\widehat{\mathbf{H}}}_{z\mathrm{C}} \end{bmatrix} \tag{4.472}$$

$$\begin{aligned} \widehat{\boldsymbol{\Lambda}}^{\mathrm{C}}_{\bullet,\circ} &= \widehat{\boldsymbol{t}}^{\bullet,\circ}_{\mathrm{C}} R'_{\widehat{\nu}_{\bullet,\circ}}(\widehat{\boldsymbol{t}}^{\bullet,\circ}_{\mathrm{C}}) R^{-1}_{\widehat{\nu}_{\bullet,\circ}}(\widehat{\boldsymbol{t}}^{\bullet,\circ}_{\mathrm{C}}) \\ \widehat{\boldsymbol{\Delta}}_{\bullet} &= \boldsymbol{T}^{-1}_{\phi\circ}\boldsymbol{D}_{\phi\bullet}\boldsymbol{T}_{\phi\bullet} \otimes (\boldsymbol{T}^{\mu}_{z\circ})^{-1}\boldsymbol{\mu}_{\circ}\boldsymbol{D}_{\overline{z}\bullet}(\boldsymbol{\varepsilon}_{\bullet}\boldsymbol{\mu}_{\bullet})^{-1}\boldsymbol{T}^{\varepsilon}_{z\bullet} \\ \widehat{\boldsymbol{\Delta}}_{\circ} &= \boldsymbol{T}^{-1}_{\phi\bullet}\boldsymbol{D}_{\phi\circ}\boldsymbol{T}_{\phi\circ} \otimes (\boldsymbol{T}^{\varepsilon}_{z\bullet})^{-1}\boldsymbol{\varepsilon}_{\bullet}\boldsymbol{D}_{\overline{z}\circ}(\boldsymbol{\varepsilon}_{\circ}\boldsymbol{\mu}_{\circ})^{-1}\boldsymbol{T}^{\mu}_{z\circ} \end{aligned}$$

We have C = A, B, where A (B) is the outer (inner) interface of the inner (outer) region. After rearrangement, the admittances at interfaces A and B can be calculated and must be matched with the impedances at the inner and outer side of the circular rib waveguide. This results in an indirect eigenvalue problem for the resonant frequency.

References

[1] R. Pregla, 'Modeling of Optical Waveguide Structures with General Anisotropy in Arbitrary Orthogonal Coordinate Systems', *IEEE J. of Sel. Topics in Quantum Electronics*, vol. 8, pp. 1217–1224, Dec. 2002.

[2] Y. Xu, 'Application of the Method of Lines to Solve Problems in Cylindrical Coordinates', *Microwave Optical Technology Lett.*, vol. 1, no. 5, pp. 173–175, July 1988.

[3] Y. J. He and S. F. Li, 'Analysis of Arbitrary Cross-Sections Using the Method of Lines', *IEEE Trans. Microwave Theory Tech.*, vol. MTT–42, no. 1, pp. 162–164, Jan. 1994.

[4] R. Pregla and L. Vietzorreck, 'Calculation of Input Impedances of Planar Antennas with the Method of Lines', in *Progress in Electromagnetics Research Symp. (PIERS)*, Noordwijk, The Netherlands, 1994, pp. CD–ROM.

[5] R. Pregla, 'New Approach for the Analysis of Cylindrical Antennas by the Method of Lines', *Electron. Lett.*, vol. 30, no. 8, pp. 614–615, 1994.

[6] R. Pregla, 'The Analysis of General Axially Symmetric Antennas with a Coaxial Feed Line by the Method of Lines', *IEEE Trans. Antennas. Propagation*, vol. 46, no. 10, pp. 1433–1443, Oct. 1998.

[7] R. Pregla, 'General Formulas for the Method of Lines in Cylindrical Coordinates', *IEEE Trans. Microwave Theory Tech.*, vol. MTT-43, no. 7, pp. 1617–1620, 1995.

[8] R. Pregla, 'The Method of Lines for the Analysis of Discontinuities in Cylindrical Waveguides', in 6th *Int. IGTE Symposium*, Graz, Austria, Sep. 1994, pp. 174–179.

[9] W. Pascher and R. Pregla, 'Analysis of Curved Optical Waveguides by the Vectorial Method of Lines', in *Int. Conf. on Integrated Optics and Optical Fibre Communications*, Paris, France, 1991, pp. 237–240.

[10] W. Pascher and R. Pregla, 'Vectorial Analysis of Bends in Optical Strip Waveguides by the Method of Lines', *Radio Sci.*, vol. 28, pp. 1229–1233, 1993.

[11] R. Pregla, 'The Method of Lines for the Analysis of Dielectric Waveguides Bends', *J. Lightwave Technol.*, vol. 14, no. 4, pp. 634–639, Apr. 1996.

[12] I. A. Goncharenko, S. F. Helfert and R. Pregla, 'General Analysis of Fibre Grating Structures', *J. of Optics A: Pure and Appl. Optics*, vol. 1, pp. 25–31, 1999.

[13] I. A. Goncharenko, S. F. Helfert and R. Pregla, 'Analysis of Nonlinear Properties of Fibre Grating Structures', *AEÜ*, vol. 53, pp. 25–31, 1999.

[14] M. Abramowitz and I. A. Stegun, *Handbook of Mathematical Functions*, Chapter 9, Dover Publ., New York, 1965.

[15] W. K. Hui and I. Wolff, 'A Multicomposite, Multilayered Cylindrical Dielectric Resonator for Application in MMIC's', in *IEEE MTT-S Int. Symp. Dig.*, Albuquerque, USA, 1992, pp. 929–932.

[16] D. Kremer and R. Pregla, 'The Method of Lines for the Hybrid Analysis of Multilayered Cylindrical Resonator Structures', *IEEE Trans. Microwave Theory Tech.*, vol. MTT-45, no. 12, pp. 2152–2155, Dec. 1997.

[17] R. Pregla and W. Pascher, 'The Method of Lines', in *Numerical Techniques for Microwave and Millimeter Wave Passive Structures*, T. Itoh (Ed.), pp. 381–446. J. Wiley Publ., New York, USA, 1989.

[18] P. Stuwe, 'Properties of Optical Waveguides', in *Optical Sensors*, W. Goepel (Ed.), Sensors, vol. 6, pp. 143–172, VCH-Verlag, Weinheim, Germany, 1992.

[19] R. A. Waldron, *Ferrites–An Introduction for Microwave Engineers*, D. Van Nostrand Company LTD, London, UK, 1961.

[20] V. I. Miteva and K. P. Ivanov, 'Some Fundamental Properties of Azimuthally Magnetized Solid-Plasma Circular Guide', in *European Microwave Conference*, London, UK, 1989, pp. 522–527.

[21] K. P. Ivanov, V. I. Miteva, R. Pregla and W. Pascher, 'Analysis of Solid-Plasma Coaxial Waveguide by the Method of Lines', in *Bianisotropics'97*, Glasgow, June 1997, pp. 157–160.

[22] C. M. Krowne, 'Waveguiding Structures Employing the Solid-State Magnetoplasma Effect for Microwave and Milimetre-Wave Propagation', *IEE Proc. -H, Microwave Antennas Propagation*, vol. 140, no. 3, pp. 147–164, 1993.

[23] F. J. Schmückle and R. Pregla, 'The Method of Lines for the Analysis of Planar Waveguides with Finite Metallization Thickness', *IEEE Trans. Microwave Theory Tech.*, vol. MTT-39, pp. 101–107, 1991.

[24] F. J. Schmückle and R. Pregla, 'The Method of Lines for the Analysis of Lossy Planar Waveguide Structures', *IEEE Trans. Microwave Theory Tech.*, vol. MTT-38, pp. 1473–1479, 1990.

[25] H. Diestel, 'A Uniform Analysis of Straight and Circular Planar Multiconductor Systems', *AEÜ*, vol. 42, pp. 54–58, 1988.

[26] H. Diestel, 'A Quasi-TEM Analysis for Curved and Straight Planar Multiconductor Systems', *IEEE Trans. Microwave Theory Tech.*, vol. MTT-37, no. 4, pp. 748–753, 1989.

[27] M. Thorburn, A. Biswas and V. K. Tripathi, 'Application of the Method of Lines to Cylindrical Inhomogeneous Propagation Structures', *Electron. Lett.*, vol. 26, no. 3, pp. 170–171, Feb. 1990.

[28] M. A. Thornburn, A. Biswas and V. K. Tripathi, 'Application of the Method of Lines to Planar Transmission Lines in Waveguides with composite Cross-sectional Geometries', *IEE Proc. -H, Microwave Antennas Propagation*, vol. 139, no. 6, pp. 542–544, 1992. (Special Issue On Gyroelectric Waveguides And Their Circuit Application).

[29] J.-S. Gu, *Numerical Analysis of Directionally Varying Optical Waveguides*, PhD thesis, ETH Zürich, 1991, chap. 6.2.

[30] R. J. Deri and R. J. Hawkins, 'Polarization, Scattering and Coherent Effects in Semiconductor Rib Waveguide Bends', *Electron. Lett.*, vol. 13, no. 10, pp. 922–924, 1988.

[31] N. Marcuvitz, *Waveguide Handbook*, P. Peregrinus Ltd., London, Great Britain, 1986.

[32] C. Yeh, 'Elliptical Dielectric Waveguides', *J. Appl. Phys*, vol. 33, no. 11, pp. 3235–3243, 1962.

[33] R. Pregla and O. Conradi, 'Modeling of Uniaxial Anisotropic Fibers with Noncircular Cross-Section by the Method of Lines', *J. Lightwave Technol.*, vol. 21, pp. 1294–1299, 2003.

[34] B. Widenberg and A. Karlsson, 'Investigation of Dielectric Waveguides with Arbitrary Cross-Section Using the T-Matrix-Method', in *U.R.S.I intern. Symp. Electromagn. Theo.*, Victoria, Canada, 2001, pp. 419–421.

[35] G. Matthaei, L. Young and E. M. T. Jones, *Microwave Filters, Impedance-Matching Networks, and Coupling Structures*, Artech House, Dedham, MA, 1980.

[36] R. Pregla, 'The Method of Lines as Generalized Transmission Line Technique for the Analysis of Multilayered Structures', *AEÜ*, vol. 50, no. 5, pp. 293–300, Sep. 1996.

[37] R. E. Collin and F. J. Zucker (Eds.), *Antenna Theory*, McGraw-Hill, New York, USA, 1969.

[38] Z. Shen and R. H. MacPhie, 'Modeling of a Monopole Partially Buried in a Grounded Dielectric Substrate by the Modal Expansion Method', *IEEE Trans. Antennas. Propagation*, vol. 44, no. 11, pp. 1535–1536, Nov. 1996.

[39] C. L. Chi and N. G. Alexopoulos, 'Radiation by a Probe Through a Substrate', *IEEE Trans. Antennas. Propagation*, vol. 34, pp. 1080–1091, 1986.

[40] S. Pinhas, S. Shtrikman and D. Treves, 'Moment-Method Solution of the Center-Fed Microstrip Disk Antenna Invoking Feed and Edge Current Singularities', *IEEE Trans. Antennas. Propagation*, vol. 37, no. 12, pp. 1516–1522, Dec. 1989.

[41] Z. Shen and R. H. MacPhie, 'Rigorous Evaluation of the Input Impedance of a Sleeve Monopole by Modal-Expansion Method', *IEEE Trans. Antennas. Propagation*, vol. 44, no. 12, pp. 1584–1591, Dec. 1996.

[42] S. Helfert, 'Analysis of Curved Bends in Arbitrary Optical Devices using Cylindrical Coordinates', *Opt. Quantum Electron.*, vol. 30, pp. 359–368, 1998.

[43] R. Pregla, 'Modeling of Optical Waveguides and Devices by Combination of the Method of Lines and Finite Differences of Second Order Accuracy', *Opt. Quantum Electron.*, vol. 38, no. 1–3, pp. 3–17, 2006, Special Issue on Optical Waveguide Theory and Numerical Modelling.

Further Reading

[44] K. P. Ivanov, 'Propagation Along Azimuthally Magnetized Ferrite-Load Circular Guide', *Radio Sci.*, vol. 19, pp. 1305–1310, 1984.

[45] R. Pregla, 'The Impedance/Admittance Transformation – An Efficient Concept for the Analysis of Optical Waveguide Structures', in *OSA Integr. Photo. Resear. Tech. Dig.*, Santa Barbara, USA, July 1999, pp. 40–42.

[46] R. Pregla, 'Novel Algorithms for the Analysis of Optical Fiber Structures with Anisotropic Materials', in *Int. Conf. on Transparent Optical Networks*, Kielce, Polen, June 1999, pp. 49–52.

[47] R. Pregla, 'Analysis of Planar Microwave and Millimeterwave Circuits with Anisotropic Layers Based on Generalized Transmission Line Equations and on the Method of Lines', in *IEEE MTT-S Int. Symp. Dig.*, Boston, USA, June 2000, vol. 1, pp. 125–128.

[48] R. Pregla, J. Gerdes, E. Ahlers and S. Helfert, 'MoL-BPM Algorithms for Waveguide Bends and Vectorial Fields', in *OSA Integr. Photo. Resear. Tech. Dig.*, New Orleans, USA, 1992, vol. 10, pp. 32–33.

[49] R. Pregla, 'Higher Order Approximation for the Difference Operators in the Method of Lines', *IEEE Microwave Guided Wave Lett.*, vol. 5, no. 2, pp. 53–55, 1995.

[50] A. Weisshaar and V. K.Tripathi, 'Frequency-Dependent Transmission Characteristics of Curved Microstrip Bends', *Electron. Lett.*, vol. 25, no. 17, pp. 1138–1139, 1989.

[51] R. Pregla and E. Ahlers, 'New Vector-BPM in Cylindrical Coordinates Based on the Method of Lines', in *OSA Integr. Photo. Resear. Tech. Dig.*, Dana Point, USA, Feb. 1995, vol. 7, pp. 24–26.

[52] I. A. Goncharenko, S. F. Helfert and R. Pregla, 'Radiation Loss and Mode Field Distribution in Curved Holey Fibers', *AEÜ*, vol. 59, pp. 185–191, 2005.

[53] R. Pregla, 'MoL-BPM Method of Lines Based Beam Propagation Method', in *Methods for Modeling and Simulation of Guided-Wave Optoelectronic Devices (PIER 11)*, W. P. Huang (Ed.), Progress in Electromagnetic Research, pp. 51–102. EMW Publishing, Cambridge, Massachusetts, USA, 1995.

[54] E. Ahlers and R. Pregla, '3D-Modeling of Concatenations of Straight and Curved Waveguides', *Opt. Quantum Electron.*, vol. 29, pp. 151–156, Feb. 1997.

[55] R. Pregla and E. Ahlers, 'Method of Lines for Analysis of Arbitrarily Curved Waveguide Bends', *Electron. Lett.*, vol. 30, no. 18, pp. 1478–1479, Sep. 1994.

[56] R. Pregla and E. Ahlers, 'The Method of Lines for Analysis of Discontinuities in Optical Waveguides', *Electron. Lett.*, vol. 29, no. 21, pp. 1845–1846, Oct. 1993.

CHAPTER 5

ANALYSIS OF PERIODIC STRUCTURES

5.1 INTRODUCTION

Periodic structures play an important role in microwaves and optics, e.g. gratings are used to design special filters, phase or delay elements or equalisers. Let us first look on some examples of such periodic structures.

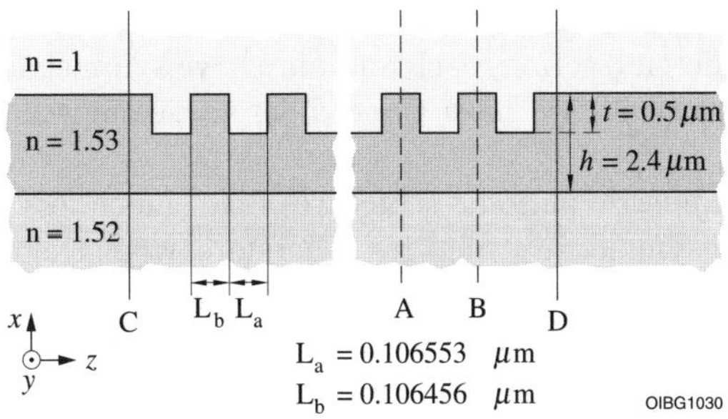

Fig. 5.1 Bragg grating (COST 240) (Reproduced by permission of Elsevier)

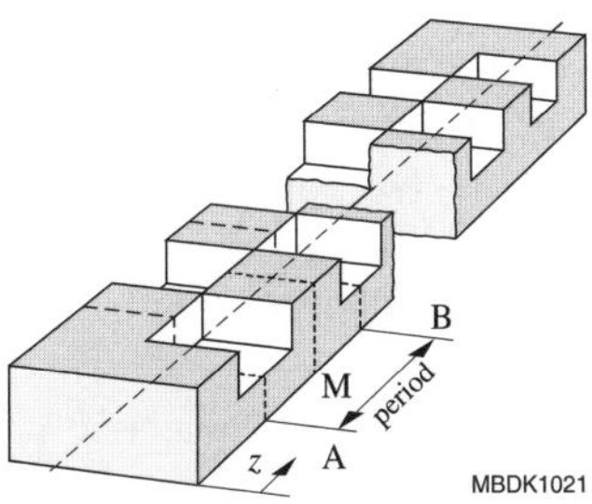

Fig. 5.2 Polarisation converter (Reproduced by permission of Elsevier)

First, a Bragg grating is shown in Fig. 5.1. The next example is a polarisation converter, shown in Fig. 5.2. Here we have additional different sections at the input and output of the structure. The meander line shown in Fig. 5.3a is a typical example of a periodic structure at microwave frequencies. One period of this periodic structure is sketched in Fig. 5.3b.

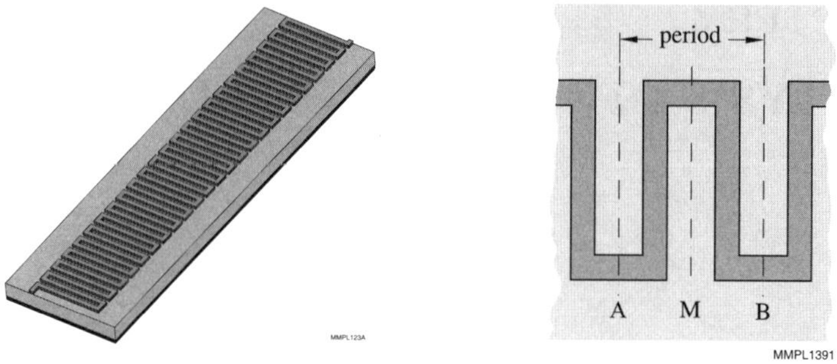

(a) (b)

Fig. 5.3 Microstrip meander line (a) and one period of this periodic structure (b) (Reproduced by permission of Elsevier)

Different types of fibre Bragg grating are shown in Fig. 5.4. The first two consist of two different sections each and may have symmetric periods. In the third example, however, we have three different sections and unsymmetric periods.

Modelling symmetric periodic structures takes much less numerical effort than analysing unsymmetric structures. Therefore, when examining a periodic structure we should try to consider symmetric periods. E.g. to achieve this, we could take a period defined by the two vertical dashed lines for the Bragg grating shown in Fig. 5.1. Generally, we should take a single period that starts and ends in the centre of a homogeneous section.

Fig. 5.5a shows a period of the structure shown in Fig. 5.4 as an example. The periodic part may of course be more complicated. In Fig. 5.5b we have a Sine grating, which is divided in the presented example into four homogeneous sections. Yet, it still looks the same from the left as from the right. The grating presented in Fig. 5.4c, however, cannot be divided into such symmetric periods and a more complicated algorithm will be required for its analysis.

The above examples all show structures with a one-dimensional periodicity. Two- and three-dimensional periodic structures are a current topic of research. These structures are called photonic crystals (PC) or bandgap structures.

In Fig. 5.6 a structure consisting of dielectric rods in air is sketched as an example of such a PC. PCs prohibit the propagation of electromagnetic waves (in our case in the x–y-plane) of some frequencies. By introducing defects (e.g. by removing some of the rods), various circuits may be designed. Fig. 5.7 shows such a defect waveguide as an example. One of the problems of interest is the determination of the bandgaps in such structures. We will show how the band diagram can be determined with the MoL.

In the following we will first describe the analysis of general one-dimensional periodic structures. The results are the background for the algorithms that are developed for 2D and 3D structures. After considering

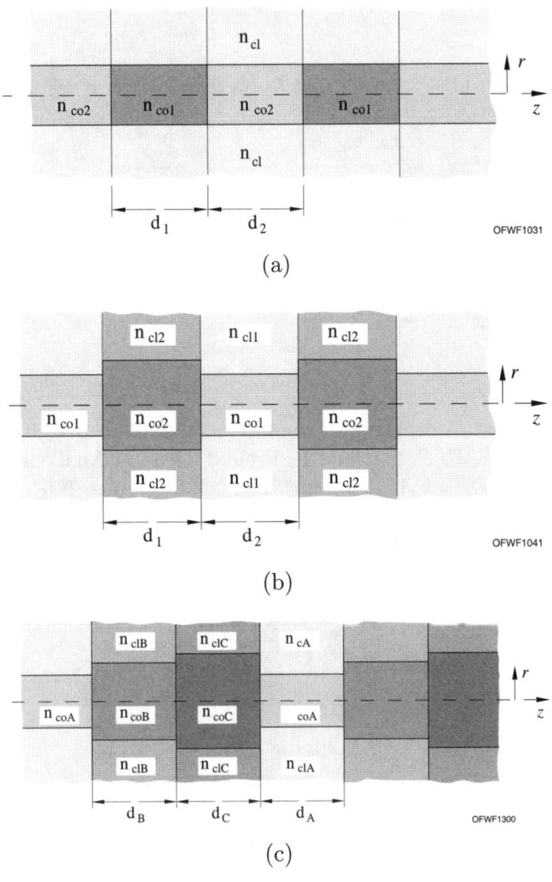

Fig. 5.4 Different types of period for fibre Bragg gratings (a) symmetric period with equal diameters of sections (b) symmetric period with unequal diameters of sections (c) unsymmetric period with unequal diameters of sections (I. A. Goncharenko, S. F. Helfert and R. Pregla, 'General Analysis of Fibre Grating Structures', in *J. of Optics A: Pure and Appl. Optics*, 1999. IOP-publishing)

the 1D case we will show how symmetric periodic structures can be analysed very efficiently [1], and give a description of unsysmmetric periodic structures. We will then give numerical results for several of the examples above. We will finish by describing in detail the examination of photonic crystal structures and of circuits consisting of PCs.

5.2 PRINCIPLE BEHAVIOUR OF PERIODIC STRUCTURES

Before we describe the analysis procedures for arbitrary periodic structures, let us begin with their general behaviour. For this purpose we shall examine

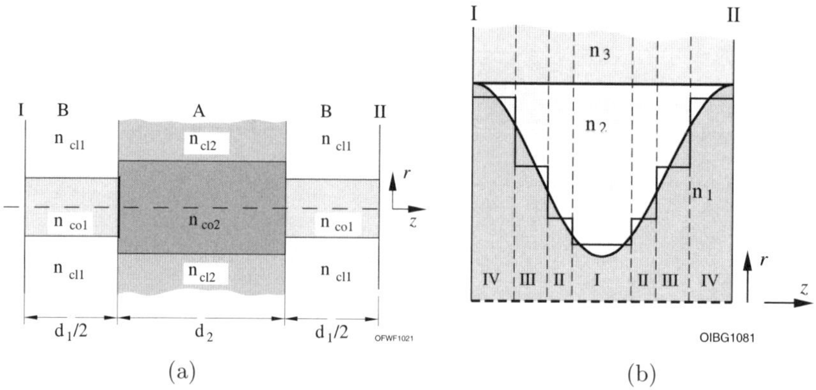

Fig. 5.5 One segment of (a) a periodical fibre structure and (b) a sine fibre grating (I. A. Goncharenko, S. F. Helfert and R. Pregla, 'General Analysis of Fibre Grating Structures', in *J. of Optics A: Pure and Appl. Optics*, 1999. IOP-publishing)

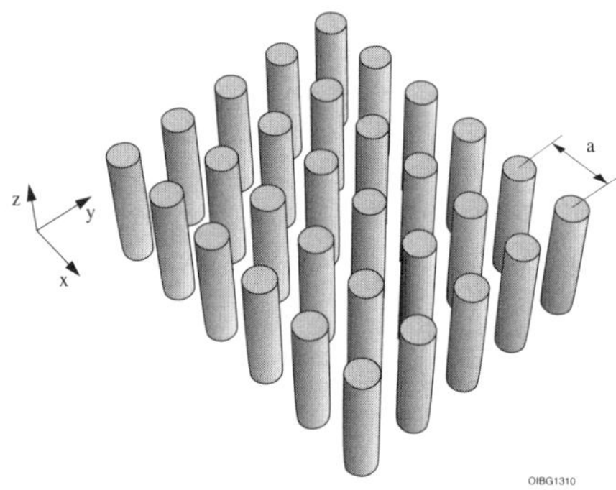

Fig. 5.6 2D photonic crystal of dielectric rods in air (S. F. Helfert, 'The method of lines for the calculation of band structures in photonic crystals', in *ICTON Conf.* © 2003 Institute of Electrical and Electronics Engineers (IEEE)

the parallel-plate waveguide with magnetic side walls in Fig. 5.8. This structure could e.g. be a model for Bragg gratings.

The electric and magnetic fields of the fundamental TEM mode are constant in each cross-section. The characteristic quantities are:

$$\beta = nk_o \qquad Z_{\mathrm{F}} = \eta_o/n \qquad \eta_o = \sqrt{\mu_o/\varepsilon_o} = 120\pi\ \Omega \tag{5.1}$$

where β is the propagation constant, k_o the free space wave number, Z_{F} the characteristic field impedance and n the effective refractive index of the dielectric. With $\theta = \beta l$, we describe the relation of the fields between ports i

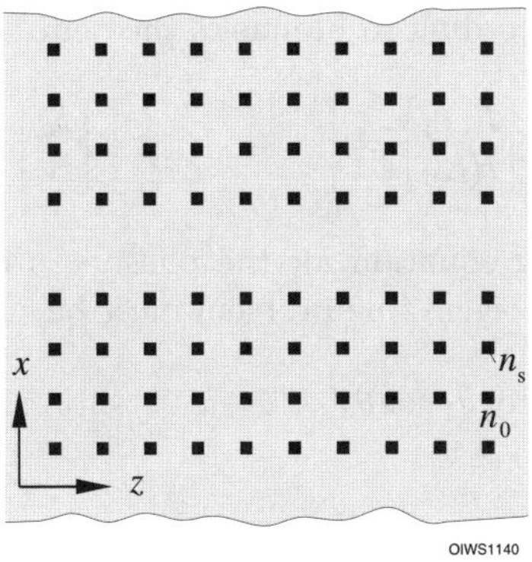

Fig. 5.7 Defect waveguide in a photonic crystal (A. Barcz, S. Helfert and R. Pregla, 'Modeling of 2D photonic crystals by using the Method of Lines', in *ICTON Conf.* © 2002 Institute of Electrical and Electronics Engineers (IEEE))

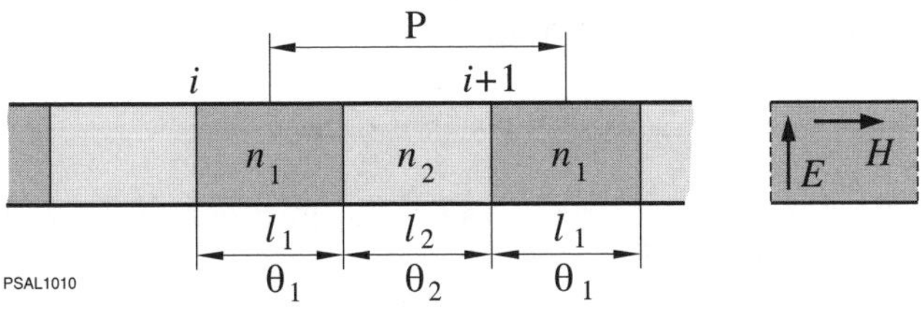

Fig. 5.8 Periodic parallel-plate structure

and $i+1$ by:

$$\begin{aligned} \begin{bmatrix} E_i \\ H_i \end{bmatrix} &= [M] \begin{bmatrix} E_{i+1} \\ H_{i+1} \end{bmatrix} \\ [M] &= \begin{bmatrix} \cos\theta_1 & jZ_{\mathrm{F1}}\sin\theta_1 \\ jZ_{\mathrm{F1}}^{-1}\sin\theta_1 & \cos\theta_1 \end{bmatrix} \begin{bmatrix} \cos\theta_2 & jZ_{\mathrm{F2}}\sin\theta_2 \\ jZ_{\mathrm{F2}}^{-1}\sin\theta_2 & \cos\theta_2 \end{bmatrix} \end{aligned} \tag{5.2}$$

The two matrices in this product in $[M]$ are nothing other than the well-known transfer matrices of the TEM transmission line sections. With $\zeta = Z_{\mathrm{F1}}/Z_{\mathrm{F2}}$, we obtain for this product:

$$\begin{aligned} &[M] \\ &= \begin{bmatrix} \cos\theta_1\cos\theta_2 - \zeta\sin\theta_1\sin\theta_2 & jZ_{\mathrm{F1}}(\sin\theta_1\cos\theta_2 + \zeta^{-1}\sin\theta_2\cos\theta_1) \\ jZ_{\mathrm{F1}}^{-1}(\sin\theta_1\cos\theta_2 + \zeta\sin\theta_2\cos\theta_1) & \cos\theta_1\cos\theta_2 - \zeta^{-1}\sin\theta_1\sin\theta_2 \end{bmatrix} \end{aligned} \tag{5.3}$$

With the ansatz according to Floquet's theorem:

$$\begin{bmatrix} E_{i+1} \\ H_{i+1} \end{bmatrix} = \begin{bmatrix} e^{-j\psi} & \\ & e^{-j\psi} \end{bmatrix} \begin{bmatrix} E_i \\ H_i \end{bmatrix} \tag{5.4}$$

we obtain the following equation for the phase ψ: (The determinant of the matrix in the resulting eigenvalue problem must be zero.)

$$e^{j2\psi} - e^{j\psi}(2\cos\theta_1\cos\theta_2 - (\zeta + \zeta^{-1})\sin\theta_1\sin\theta_2) + 1 = 0$$

or:

$$\cos\psi = \cos\theta_1\cos\theta_2 - a\sin\theta_1\sin\theta_2 \tag{5.5}$$

with:

$$a = \frac{1}{2}(\zeta + \zeta^{-1}) = \frac{1}{2}\left(\frac{Z_1}{Z_2} + \frac{Z_2}{Z_1}\right) = \frac{1}{2}\left(\frac{n_2}{n_1} + \frac{n_1}{n_2}\right) \geq 1$$

We now choose sections of equal electrical length (i.e. $l_1 n_1 = l_2 n_2$), leading to $\theta_1 = \theta_2 = \theta$. Therefore, we obtain:

$$\cos\psi = \cos^2\theta - a\sin^2\theta \tag{5.6}$$

In Fig. 5.9 the curves of $\cos\psi$ and ψ as functions of θ are drawn. The periodic structure shows filter behaviour. We have a transmission band (TB) for $|\cos\psi| \leq 1$ ist. In the stop band (SB) (or band gap) the phase is constant and equal to π. To show the change from the homogeneous line to the periodic one, the phase curve $\psi = 2\theta$ of the homogeneous line is introduced into the diagram, too.

The field decay or 'transmission loss' in the stop band is obtained from:

$$\begin{aligned} \cos(\pi - j\alpha\Lambda) &= -\cosh\alpha\Lambda = \cos^2\theta - a\sin^2\theta \\ \cosh(\alpha\Lambda) &= a\sin^2\theta - \cos^2\theta \end{aligned} \tag{5.7}$$

with $\Lambda = l_1 + l_2$. The centre of the stop band is determined as $\theta_o = \pi/2$. The related frequency f_o ($\theta = k_o nl$) is given by $f_o = c/(4nl)$. The maximum of the transmission loss (at the centre) is equal to:

$$\Lambda\alpha_{\max} = \ln(a + \sqrt{a^2 - 1}) = \left|\ln\frac{n_1}{n_2}\right| \approx \frac{\Delta n}{n}$$

where $n = (n_1 + n_2)/2$ and $|n_1 - n_2| = \Delta n$. The approximate value for the example in Fig. 5.9 is given by $\Lambda\alpha_{\max} = 0.5$, which is in good agreement with the exact value of 0.5108. The band edges of the stop band are determined from the condition $\cos\psi = -1$. Therefore, we obtain:

$$(a + 1)\sin^2\theta_c = 2 \tag{5.8}$$

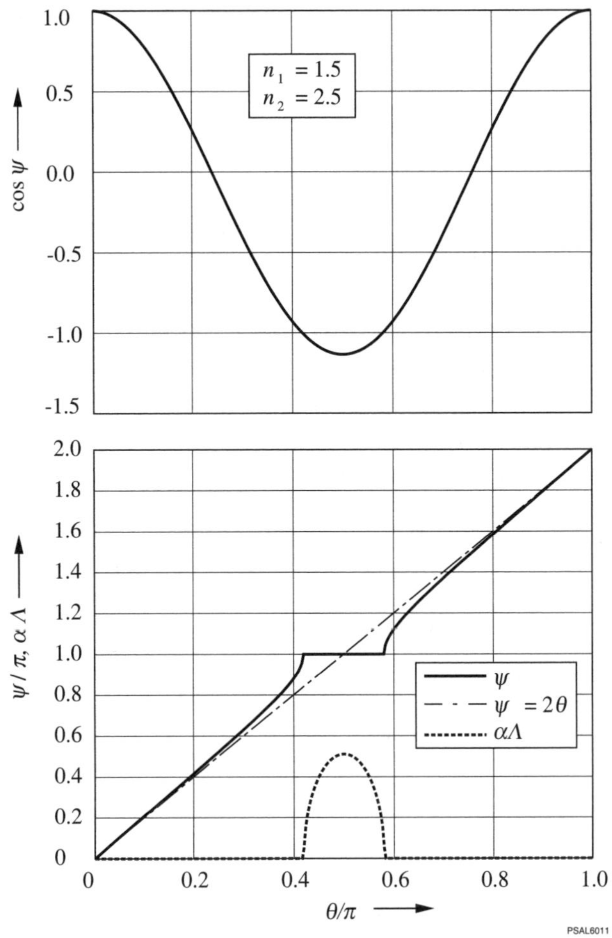

Fig. 5.9 Curves of $\cos\psi$ and ψ as a function of θ

With $\theta = \dfrac{\pi}{2}\dfrac{f}{f_o}$ and $\theta_{c1,2} = \dfrac{\pi}{2}\dfrac{f_o \pm \Delta f}{f_o}$ we obtain:

$$(a+1)\cos^2\left(\frac{\pi}{2}\frac{\Delta f}{f_o}\right) = 2$$

or:

$$\frac{2\Delta f}{f_o} = \frac{4}{\pi}\arcsin\sqrt{\frac{a-1}{a+1}} = \frac{4}{\pi}\arcsin\left|\frac{n_1 - n_2}{n_1 + n_2}\right| \tag{5.9}$$

For small band stop filters we have $|n_1 - n_2| \ll n_1 + n_2$. Therefore, the following approximation may be introduced for the relative bandwidth:

$$\frac{2\Delta f}{f_o} \approx \frac{2}{\pi}\frac{\Delta n}{n} \tag{5.10}$$

The exact value of the relative bandwidth is 0.315 in our example and the approximate one is 0.3183. Thus the approximations are very good even for relatively large values of $\Delta n/n$.

5.3 GENERAL THEORY OF PERIODIC STRUCTURES

5.3.1 Port relations for general two ports

As is well known, we can describe a two port with open-circuit or short-circuit matrix parameters (see Section 2.5 and [3]). Therefore, we can also describe a periodic structure in such a way. Hence, we may write for one period:

$$\begin{bmatrix} \mathbf{E}_\mathrm{A} \\ \mathbf{E}_\mathrm{B} \end{bmatrix} = \begin{bmatrix} \mathbf{z}_{11} & \mathbf{z}_{12} \\ \mathbf{z}_{21} & \mathbf{z}_{22} \end{bmatrix} \begin{bmatrix} \mathbf{H}_\mathrm{A} \\ -\mathbf{H}_\mathrm{B} \end{bmatrix} \qquad \begin{bmatrix} \mathbf{H}_\mathrm{A} \\ -\mathbf{H}_\mathrm{B} \end{bmatrix} = \begin{bmatrix} \mathbf{y}_{11} & \mathbf{y}_{12} \\ \mathbf{y}_{21} & \mathbf{y}_{22} \end{bmatrix} \begin{bmatrix} \mathbf{E}_\mathrm{A} \\ \mathbf{E}_\mathrm{B} \end{bmatrix} \tag{5.11}$$

From eq. (5.11) we easily obtain the following transfer matrix relations:

$$\begin{bmatrix} \mathbf{E}_\mathrm{B} \\ \mathbf{H}_\mathrm{B} \end{bmatrix} = \begin{bmatrix} \mathbf{z}_{22}\mathbf{z}_{12}^{-1} & \mathbf{z}_{21} - \mathbf{z}_{22}\mathbf{z}_{12}^{-1}\mathbf{z}_{11} \\ -\mathbf{z}_{12}^{-1} & \mathbf{z}_{12}^{-1}\mathbf{z}_{11} \end{bmatrix} \begin{bmatrix} \mathbf{E}_\mathrm{A} \\ \mathbf{H}_\mathrm{A} \end{bmatrix} \tag{5.12}$$

$$\begin{bmatrix} \mathbf{E}_\mathrm{A} \\ \mathbf{H}_\mathrm{A} \end{bmatrix} = \begin{bmatrix} \mathbf{z}_{11}\mathbf{z}_{21}^{-1} & \mathbf{z}_{11}\mathbf{z}_{21}^{-1}\mathbf{z}_{22} - \mathbf{z}_{12} \\ \mathbf{z}_{21}^{-1} & \mathbf{z}_{21}^{-1}\mathbf{z}_{22} \end{bmatrix} \begin{bmatrix} \mathbf{E}_\mathrm{B} \\ \mathbf{H}_\mathrm{B} \end{bmatrix} \tag{5.13}$$

$$\begin{bmatrix} \mathbf{E}_\mathrm{A} \\ \mathbf{H}_\mathrm{A} \end{bmatrix} = \begin{bmatrix} -\mathbf{y}_{21}^{-1}\mathbf{y}_{22} & -\mathbf{y}_{21}^{-1} \\ \mathbf{y}_{12} - \mathbf{y}_{11}\mathbf{y}_{21}^{-1}\mathbf{y}_{22} & -\mathbf{y}_{11}\mathbf{y}_{21}^{-1} \end{bmatrix} \begin{bmatrix} \mathbf{E}_\mathrm{B} \\ \mathbf{H}_\mathrm{B} \end{bmatrix} \tag{5.14}$$

$$\begin{bmatrix} \mathbf{E}_\mathrm{B} \\ \mathbf{H}_\mathrm{B} \end{bmatrix} = \begin{bmatrix} -\mathbf{y}_{12}^{-1}\mathbf{y}_{11} & \mathbf{y}_{12}^{-1} \\ \mathbf{y}_{22}\mathbf{y}_{12}^{-1}\mathbf{y}_{11} - \mathbf{y}_{21} & -\mathbf{y}_{22}\mathbf{y}_{12}^{-1} \end{bmatrix} \begin{bmatrix} \mathbf{E}_\mathrm{A} \\ \mathbf{H}_\mathrm{A} \end{bmatrix} \tag{5.15}$$

The vectors $\mathbf{E}_{\mathrm{A,B}}$ and $\mathbf{H}_{\mathrm{A,B}}$ contain the transversal field components at ports A and B (see e.g. Fig. 5.1). Instead of the fields in the original domain, we can use the ones in the transformed domain.

5.3.2 Floquet modes for symmetric periods

The analysis of periodic structures can be simplified for symmetric (with respect to the periodicity) periods. In this case we may introduce $\mathbf{z}_{22} = \mathbf{z}_{11}, \mathbf{z}_{12} = \mathbf{z}_{21}$ ($\mathbf{y}_{22} = \mathbf{y}_{11}, \mathbf{y}_{12} = \mathbf{y}_{21}$). Now, by using the Floquet modal matrices $\boldsymbol{S}_\mathrm{E}$ and $\boldsymbol{S}_\mathrm{H}$, we perform a transformation to Floquet modes according to:

$$\mathbf{E}_{\mathrm{A,B}} = \boldsymbol{S}_\mathrm{E}\widetilde{\mathbf{E}}_{\mathrm{A,B}} \qquad \mathbf{H}_{\mathrm{A,B}} = \boldsymbol{S}_\mathrm{H}\widetilde{\mathbf{H}}_{\mathrm{A,B}} \tag{5.16}$$

5.3.2.1 Floquet modes from transfer matrix equations

By using these Floquet modal matrices $\boldsymbol{S}_{\mathrm{E}}$ and $\boldsymbol{S}_{\mathrm{H}}$ we obtain from eqs. (5.12) and (5.15):

$$\begin{bmatrix}\widetilde{\mathbf{E}}_{\mathrm{B}}\\ \widetilde{\mathbf{H}}_{\mathrm{B}}\end{bmatrix}=\begin{bmatrix}\boldsymbol{S}_{\mathrm{E}}^{-1}\mathbf{z}_{11}\mathbf{z}_{12}^{-1}\boldsymbol{S}_{\mathrm{E}} & -\boldsymbol{S}_{\mathrm{E}}^{-1}(\mathbf{z}_{11}\mathbf{z}_{12}^{-1}\mathbf{z}_{11}-\mathbf{z}_{12})\boldsymbol{S}_{\mathrm{H}}\\ -\boldsymbol{S}_{\mathrm{H}}^{-1}\mathbf{z}_{12}^{-1}\boldsymbol{S}_{\mathrm{E}} & \boldsymbol{S}_{\mathrm{H}}^{-1}\mathbf{z}_{12}^{-1}\mathbf{z}_{11}\boldsymbol{S}_{\mathrm{H}}\end{bmatrix}\begin{bmatrix}\widetilde{\mathbf{E}}_{\mathrm{A}}\\ \widetilde{\mathbf{H}}_{\mathrm{A}}\end{bmatrix} \tag{5.17}$$

and:

$$\begin{bmatrix}\widetilde{\mathbf{E}}_{\mathrm{B}}\\ \widetilde{\mathbf{H}}_{\mathrm{B}}\end{bmatrix}=\begin{bmatrix}-\boldsymbol{S}_{\mathrm{E}}^{-1}\mathbf{y}_{12}^{-1}\mathbf{y}_{11}\boldsymbol{S}_{\mathrm{E}} & \boldsymbol{S}_{\mathrm{E}}^{-1}\mathbf{y}_{12}^{-1}\boldsymbol{S}_{\mathrm{H}}\\ \boldsymbol{S}_{\mathrm{H}}^{-1}(\mathbf{y}_{11}\mathbf{y}_{12}^{-1}\mathbf{y}_{11}-\mathbf{y}_{12})\boldsymbol{S}_{\mathrm{E}} & -\boldsymbol{S}_{\mathrm{H}}^{-1}\mathbf{y}_{11}\mathbf{y}_{12}^{-1}\boldsymbol{S}_{\mathrm{H}}\end{bmatrix}\begin{bmatrix}\widetilde{\mathbf{E}}_{\mathrm{A}}\\ \widetilde{\mathbf{H}}_{\mathrm{A}}\end{bmatrix} \tag{5.18}$$

By using Floquet's theorem, we transform the period structure into an equivalent transmission line. Therefore, eqs. (5.17) and (5.18) should be equivalent to the following transmission line relation for the Floquet modes:

$$\begin{bmatrix}\widetilde{\mathbf{E}}_{\mathrm{B}}\\ \widetilde{\mathbf{H}}_{\mathrm{B}}\end{bmatrix}=\begin{bmatrix}\cosh\boldsymbol{\Gamma}_{\mathrm{F}} & -\widetilde{\boldsymbol{Z}}_0\sinh\boldsymbol{\Gamma}_{\mathrm{F}}\\ -\widetilde{\boldsymbol{Y}_0}\sinh\boldsymbol{\Gamma}_{\mathrm{F}} & \cosh\boldsymbol{\Gamma}_{\mathrm{F}}\end{bmatrix}\begin{bmatrix}\widetilde{\mathbf{E}}_{\mathrm{A}}\\ \widetilde{\mathbf{H}}_{\mathrm{A}}\end{bmatrix} \tag{5.19}$$

$\boldsymbol{\Gamma}_{\mathrm{F}}$ is the matrix of phase constants and $\widetilde{\boldsymbol{Z}}_0$ ($\widetilde{\boldsymbol{Y}_0}^{-1}=\widetilde{\boldsymbol{Z}}_0^{-1}$) is the matrix of characteristic impedances. To induce this equivalence we have to solve the following eigenvalue problems:

$$\begin{aligned}-\mathbf{y}_{12}^{-1}\mathbf{y}_{11}\boldsymbol{S}_{\mathrm{E}} &= \mathbf{z}_{11}\mathbf{z}_{12}^{-1}\boldsymbol{S}_{\mathrm{E}} = \boldsymbol{S}_{\mathrm{E}}\boldsymbol{\lambda}_{\mathrm{E}}\\ -\mathbf{y}_{11}\mathbf{y}_{12}^{-1}\boldsymbol{S}_{\mathrm{H}} &= \mathbf{z}_{12}^{-1}\mathbf{z}_{11}\boldsymbol{S}_{\mathrm{H}} = \boldsymbol{S}_{\mathrm{H}}\boldsymbol{\lambda}_{\mathrm{H}}\end{aligned} \tag{5.20}$$

which result in:

$$\boldsymbol{\lambda}_{\mathrm{E}}=\boldsymbol{\lambda}_{\mathrm{H}}=\boldsymbol{\lambda}_{\mathrm{C}}=\cosh\boldsymbol{\Gamma}_{\mathrm{F}} \tag{5.21}$$

By using the addition theorems for the hyperbolic functions we can also write:

$$\sinh\boldsymbol{\Gamma}_{\mathrm{F}}=\sqrt{\boldsymbol{\lambda}_{\mathrm{C}}^2-\boldsymbol{I}}$$

The modal matrices $\boldsymbol{S}_{\mathrm{E}}$ and $\boldsymbol{S}_{\mathrm{H}}$ are not independent of each other. Because they are determined from the product of $\mathbf{z}_{11}$ and $\mathbf{z}_{12}^{-1}$, just the order of these matrices is exchanged (see eq. (5.20)). Therefore, the following relations hold:

$$\boldsymbol{S}_{\mathrm{E}}=\mathbf{z}_{11}\boldsymbol{S}_{\mathrm{H}}\boldsymbol{\delta}_1 \qquad \boldsymbol{S}_{\mathrm{H}}=\mathbf{z}_{12}^{-1}\boldsymbol{S}_{\mathrm{E}}\boldsymbol{\delta}_2 \tag{5.22}$$

We introduce the diagonal matrices $\boldsymbol{\delta}_{1,2}$ because the amplitudes of eigenvectors can be chosen arbitrarily. The selection of these values $\boldsymbol{\delta}_{1,2}$ is shown in the next section. For self-consistency, they have to fulfil the condition:

$$\boldsymbol{\delta}_1\boldsymbol{\delta}_2=\boldsymbol{\lambda}_{\mathrm{C}}^{-1} \tag{5.23}$$

We will now introduce eq. (5.22) into the off-diagonal submatrices in eq. (5.17). We start with the lower-left submatrix. We obtain:

$$\begin{aligned} \boldsymbol{S}_{\mathrm{H}}^{-1}\boldsymbol{z}_{12}^{-1}\boldsymbol{S}_{\mathrm{E}} &= \boldsymbol{S}_{\mathrm{H}}^{-1}\boldsymbol{z}_{12}^{-1}\boldsymbol{z}_{11}\boldsymbol{S}_{\mathrm{H}}\boldsymbol{\delta}_1 &= \boldsymbol{\lambda}_{\mathrm{C}}\boldsymbol{\delta}_1 \\ &= \boldsymbol{\delta}_2^{-1}\boldsymbol{S}_{\mathrm{E}}^{-1}\boldsymbol{z}_{12}\boldsymbol{z}_{12}^{-1}\boldsymbol{S}_{\mathrm{E}} &= \boldsymbol{\delta}_2^{-1} \end{aligned} \tag{5.24}$$

We substitute $\boldsymbol{S}_{\mathrm{E}}$ with the left eq. (5.22) (first line) or $\boldsymbol{S}_{\mathrm{H}}$ (second line) with the right equation (5.22). Since both expressions on the right side must be identical, we have again $\boldsymbol{\delta}_1\boldsymbol{\delta}_2 = \boldsymbol{\lambda}_{\mathrm{C}}^{-1}$. We proceed analogously for the right upper matrix in eq. (5.17) and obtain:

$$\begin{aligned} \boldsymbol{S}_{\mathrm{E}}^{-1}(\boldsymbol{z}_{11}\boldsymbol{z}_{12}^{-1}\boldsymbol{z}_{11} - \boldsymbol{z}_{12})\boldsymbol{S}_{\mathrm{H}} &= \boldsymbol{S}_{\mathrm{E}}^{-1}((\boldsymbol{z}_{11}\boldsymbol{z}_{12}^{-1})^2 - \boldsymbol{I})\boldsymbol{S}_{\mathrm{E}}\boldsymbol{\delta}_2 = (\boldsymbol{\lambda}_{\mathrm{C}}^2 - \boldsymbol{I})\boldsymbol{\delta}_2 \\ &= \boldsymbol{\delta}_1^{-1}\boldsymbol{S}_{\mathrm{H}}^{-1}(\boldsymbol{z}_{12}^{-1}\boldsymbol{z}_{11} - \boldsymbol{z}_{11}^{-1}\boldsymbol{z}_{12})\boldsymbol{S}_{\mathrm{H}} \\ &= \boldsymbol{\delta}_1^{-1}(\boldsymbol{\lambda}_{\mathrm{C}} - \boldsymbol{\lambda}_{\mathrm{C}}^{-1}) \end{aligned} \tag{5.25}$$

Since the expressions on the right side (first and third line) must be identical, we obtain again:

$$\boldsymbol{\delta}_1\boldsymbol{\delta}_2 = \boldsymbol{\lambda}_{\mathrm{C}}^{-1}$$

Remember, λ_{C} and $\boldsymbol{\delta}_{1,2}$ are diagonal matrices, therefore we may exchange the order of the products. $\boldsymbol{S}_{\mathrm{E}}$ and $\boldsymbol{S}_{\mathrm{E}}$ were replaced through use of eq. (5.22). Furthermore, we take into account:

$$\boldsymbol{S}_{\mathrm{H}}^{-1}\boldsymbol{z}_{11}^{-1}\boldsymbol{z}_{12}\boldsymbol{S}_{\mathrm{H}} = (\boldsymbol{S}_{\mathrm{H}}^{-1}\boldsymbol{z}_{12}^{-1}\boldsymbol{z}_{11}\boldsymbol{S}_{\mathrm{H}})^{-1} = \boldsymbol{\lambda}_{\mathrm{C}}^{-1} \tag{5.26}$$

As we know, the off-diagonals of the matrix in eq. (5.17) should be equal to $\widetilde{\boldsymbol{Y}}_0 \cdot \sinh \boldsymbol{\Gamma}_{\mathrm{F}}$ and $\widetilde{\boldsymbol{Z}}_0 \cdot \sinh \boldsymbol{\Gamma}_{\mathrm{F}}$, respectively. By using eqs. (5.24) and (5.25), we obtain:

$$\boldsymbol{\lambda}_{\mathrm{C}}\boldsymbol{\delta}_1 = \widetilde{\boldsymbol{Y}}_0\sqrt{\boldsymbol{\lambda}_{\mathrm{C}}^2 - \boldsymbol{I}} \quad \text{and} \quad (\boldsymbol{\lambda}_{\mathrm{C}}^2 - \boldsymbol{I})\boldsymbol{\delta}_2 = \widetilde{\boldsymbol{Z}}_0\sqrt{\boldsymbol{\lambda}_{\mathrm{C}}^2 - \boldsymbol{I}}$$

Now, we have various possibilities when choosing the quantities $\boldsymbol{\delta}_{1,2}$ in conjunction with $\widetilde{\boldsymbol{Z}}_0$ and $\widetilde{\boldsymbol{Y}}_0$. Two important ones will be described here.

1. By choosing $\widetilde{\boldsymbol{Z}}_0 = \widetilde{\boldsymbol{Y}}_0 = \boldsymbol{I}$ we obtain:

$$\boldsymbol{\delta}_1 = \sqrt{\boldsymbol{\lambda}_{\mathrm{C}}^2 - \boldsymbol{I}}/\boldsymbol{\lambda}_{\mathrm{C}} \qquad \boldsymbol{\delta}_2 = \boldsymbol{I}/\sqrt{\boldsymbol{\lambda}_{\mathrm{C}}^2 - \boldsymbol{I}} \tag{5.27}$$

2. We can also choose $\boldsymbol{\delta}_1 = \boldsymbol{\delta}_2 = \boldsymbol{\delta}$ and obtain:

$$\boldsymbol{\delta} = \boldsymbol{\lambda}_{\mathrm{C}}^{-\frac{1}{2}} \qquad \widetilde{\boldsymbol{Z}}_0 = ((\boldsymbol{\lambda}_{\mathrm{C}}^2 - \boldsymbol{I})/\boldsymbol{\lambda}_{\mathrm{C}})^{\frac{1}{2}} = \widetilde{\boldsymbol{Y}}_0^{-1} \tag{5.28}$$

The problem now is to calculate the eigenvalues $\boldsymbol{\lambda}_{\mathrm{C}}$ and the eigenvectors $\boldsymbol{S}_{\mathrm{E,H}}$ in an efficient way.

5.3.2.2 Floquet modes from impedance/admittance equations

The last section shows how Floquet modes can be introduced as eigenmodes of periodic structures and how they are related to modes in homogeneous sections. However, for a computational determination of these Floquet modes with eq. (5.20) we must invert the transfer impedance (or admittance) $\mathbf{z}_{12}$ (or $\mathbf{y}_{12}$). This may cause problems if we have long sections and/or high losses.

We will now show how these problems can be avoided. We will use the open-circuit and short-circuit parameter formulation in eq. (5.11) to determine the Floquet-modes. Then, using the transformation according to eq. (5.16) instead of eq. (5.11), we can write for the equivalent homogeneous lines:

$$\begin{bmatrix} \widetilde{\mathbf{E}}_{\mathrm{A}} \\ \widetilde{\mathbf{E}}_{\mathrm{B}} \end{bmatrix} = \widetilde{\mathbf{Z}}_0 \begin{bmatrix} (\tanh \boldsymbol{\Gamma}_{\mathrm{F}})^{-1} & (\sinh \boldsymbol{\Gamma}_{\mathrm{F}})^{-1} \\ (\sinh \boldsymbol{\Gamma}_{\mathrm{F}})^{-1} & (\tanh \boldsymbol{\Gamma}_{\mathrm{F}})^{-1} \end{bmatrix} \begin{bmatrix} \widetilde{\mathbf{H}}_{\mathrm{A}} \\ -\widetilde{\mathbf{H}}_{\mathrm{B}} \end{bmatrix} \tag{5.29}$$

$$\begin{bmatrix} \widetilde{\mathbf{H}}_{\mathrm{A}} \\ -\widetilde{\mathbf{H}}_{\mathrm{B}} \end{bmatrix} = \widetilde{\mathbf{Y}}_0 \begin{bmatrix} (\tanh \boldsymbol{\Gamma}_{\mathrm{F}})^{-1} & -(\sinh \boldsymbol{\Gamma}_{\mathrm{F}})^{-1} \\ -(\sinh \boldsymbol{\Gamma}_{\mathrm{F}})^{-1} & (\tanh \boldsymbol{\Gamma}_{\mathrm{F}})^{-1} \end{bmatrix} \begin{bmatrix} \widetilde{\mathbf{E}}_{\mathrm{A}} \\ \widetilde{\mathbf{E}}_{\mathrm{B}} \end{bmatrix} \tag{5.30}$$

$\boldsymbol{\Gamma}_{\mathrm{F}}$ and $\widetilde{\mathbf{Z}}_0 = \widetilde{\mathbf{Y}}_0^{-1}$ are the Floquet modes' phase and characteristic impedance matrix, respectively. Comparing the submatrices in eqs. (5.29) and (5.30) with the transformed submatrices in eq. (5.11), we see that the following relations hold:

$$\mathbf{S}_{\mathrm{E}}^{-1} \mathbf{z}_{11} \mathbf{S}_{\mathrm{H}} = \widetilde{\mathbf{Z}}_0 (\tanh \boldsymbol{\Gamma}_{\mathrm{F}})^{-1} \qquad \mathbf{S}_{\mathrm{H}}^{-1} \mathbf{y}_{11} \mathbf{S}_{\mathrm{E}} = \widetilde{\mathbf{Y}}_0 (\tanh \boldsymbol{\Gamma}_{\mathrm{F}})^{-1} \tag{5.31}$$

From these two equations we can determine $\boldsymbol{\Gamma}_{\mathrm{F}}$ and $\widetilde{\mathbf{Z}}_0 = \widetilde{\mathbf{Y}}_0^{-1}$. The multiplication of these two equations results in:

$$\tanh \boldsymbol{\Gamma}_{\mathrm{F}} = \boldsymbol{I}/\sqrt{\mathbf{S}_{\mathrm{E}}^{-1} \mathbf{z}_{11} \mathbf{y}_{11} \mathbf{S}_{\mathrm{E}}} = \boldsymbol{I}/\sqrt{\mathbf{S}_{\mathrm{H}}^{-1} \mathbf{y}_{11} \mathbf{z}_{11} \mathbf{S}_{\mathrm{H}}} = \boldsymbol{I}/\sqrt{\boldsymbol{\lambda}_{\mathrm{os}}} \tag{5.32}$$

where $\boldsymbol{\lambda}_{\mathrm{os}}$ is the eigenvalue matrix of $\mathbf{z}_{11}\mathbf{y}_{11}$ or $\mathbf{y}_{11}\mathbf{z}_{11}$. $\boldsymbol{\lambda}_{\mathrm{os}}$ is related to $\boldsymbol{\lambda}_{\mathrm{C}}$ by:

$$\boldsymbol{\lambda}_{\mathrm{os}} = \boldsymbol{\lambda}_{\mathrm{C}}^2 / (\boldsymbol{\lambda}_{\mathrm{C}}^2 - \boldsymbol{I}) \qquad \boldsymbol{\lambda}_{\mathrm{C}} = \sqrt{\boldsymbol{\lambda}_{\mathrm{os}} / (\boldsymbol{\lambda}_{\mathrm{os}} - \boldsymbol{I})} \tag{5.33}$$

The following relations hold for the new eigenvector matrices $\mathbf{S}_{\mathrm{E}}$ and $\mathbf{S}_{\mathrm{H}}$:

$$\mathbf{S}_{\mathrm{E}} = \mathbf{z}_{11} \mathbf{S}_{\mathrm{H}} \boldsymbol{\delta}_{\mathrm{a}} \qquad \mathbf{S}_{\mathrm{H}} = \mathbf{y}_{11} \mathbf{S}_{\mathrm{E}} \boldsymbol{\delta}_{\mathrm{b}} \tag{5.34}$$

As before, we introduce the diagonal matrices $\boldsymbol{\delta}_{\mathrm{a,b}}$ because the eigenvectors are unique up to multiplication by a non-zero constant. For self-consistency, the condition $\boldsymbol{\delta}_{\mathrm{a}}\boldsymbol{\delta}_{\mathrm{b}} = \boldsymbol{\delta}_{\mathrm{b}}\boldsymbol{\delta}_{\mathrm{a}} = \boldsymbol{I}/\boldsymbol{\lambda}_{\mathrm{os}}$ has to be fulfilled. By dividing the left expression in eq. (5.31) by the right one, we obtain:

$$\begin{aligned} \widetilde{\mathbf{Z}}_0 &= \sqrt{\mathbf{S}_{\mathrm{E}}^{-1} \mathbf{z}_{11} \mathbf{S}_{\mathrm{H}} \mathbf{S}_{\mathrm{E}}^{-1} \mathbf{y}_{11}^{-1} \mathbf{S}_{\mathrm{H}}} \\ &= \sqrt{(\mathbf{S}_{\mathrm{E}}^{-1} \mathbf{z}_{11} \mathbf{y}_{11} \mathbf{S}_{\mathrm{E}}) \boldsymbol{\delta}_{\mathrm{b}} \boldsymbol{\delta}_{\mathrm{a}}^{-1} (\mathbf{S}_{\mathrm{H}}^{-1} \mathbf{y}_{11} \mathbf{z}_{11} \mathbf{S}_{\mathrm{H}})^{-1}} = \sqrt{\boldsymbol{\delta}_{\mathrm{b}} \boldsymbol{\delta}_{\mathrm{a}}^{-1}} \end{aligned} \tag{5.35}$$

The selection $\boldsymbol{\delta}_{\mathrm{a}} = \boldsymbol{\delta}_{\mathrm{b}}$ results in $\widetilde{\boldsymbol{Z}}_0 = \boldsymbol{I}$.

Because $\mathbf{z}_{11}$ and $\boldsymbol{y}_{11}$ are the open-circuit input impedance and short-circuit input admittance, respectively, they can be computed in a numerically stable way. $\boldsymbol{\Gamma}_{\mathrm{F}}$ and $\boldsymbol{S}_{\mathrm{E},H}$ are also obtained numerically stable. We could determine $\mathbf{z}_{11}$ and $\boldsymbol{y}_{11}$ by performing an impedance and an admittance transformation along the whole period (i.e. two transformations). To reduce the numerical effort, we will show in the next sections how $\boldsymbol{\Gamma}_{\mathrm{F}}$ and $\boldsymbol{S}_{\mathrm{E},H}$ can be computed from half of the periods, which roughly cuts the numerical effort in half.

5.3.2.3 Symmetric and anti-symmetric excitation

As mentioned in the last section, we can use half of the periods to determine the Floquet modes, in the case of symmetric periods. We obtain from eq. (5.11) for the even and odd excitation case:

1. Even case (magnetic wall in symmetry plane M): $\mathbf{E}_{\mathrm{A}} = \mathbf{E}_{\mathrm{B}}, \mathbf{H}_{\mathrm{A}} = -\mathbf{H}_{\mathrm{B}}$

$$\mathbf{E}_{\mathrm{A}} = (\mathbf{z}_{11} + \mathbf{z}_{12})\mathbf{H}_{\mathrm{A}} \longrightarrow \mathbf{Z}_{\mathrm{even}} = \mathbf{z}_{11} + \mathbf{z}_{12} = \mathbf{z}_{11\mathrm{h}} \tag{5.36}$$

$$\mathbf{H}_{\mathrm{A}} = (\boldsymbol{y}_{11} + \boldsymbol{y}_{12})\mathbf{E}_{\mathrm{A}} \longrightarrow \mathbf{Y}_{\mathrm{even}} = \boldsymbol{y}_{11} + \boldsymbol{y}_{12} = \mathbf{z}_{11\mathrm{h}}^{-1} \tag{5.37}$$

2. Odd case (electric wall in symmetry plane M): $\mathbf{E}_{\mathrm{A}} = -\mathbf{E}_{\mathrm{B}}, \mathbf{H}_{\mathrm{A}} = \mathbf{H}_{\mathrm{B}}$

$$\mathbf{E}_{\mathrm{A}} = (\mathbf{z}_{11} - \mathbf{z}_{12})\mathbf{H}_{\mathrm{A}} \longrightarrow \mathbf{Z}_{\mathrm{odd}} = \mathbf{z}_{11} - \mathbf{z}_{12} = \boldsymbol{y}_{11\mathrm{h}}^{-1} \tag{5.38}$$

$$\mathbf{H}_{\mathrm{A}} = (\boldsymbol{y}_{11} - \boldsymbol{y}_{12})\mathbf{E}_{\mathrm{A}} \longrightarrow \mathbf{Y}_{\mathrm{odd}} = \boldsymbol{y}_{11} - \boldsymbol{y}_{12} = \boldsymbol{y}_{11\mathrm{h}} \tag{5.39}$$

The subscript h symbolises the half of the period. The matrices $\mathbf{z}_{11}$ $(\boldsymbol{y}_{11})$ and $\mathbf{z}_{12}$ $(\boldsymbol{y}_{12})$ can now be obtained from $\mathbf{z}_{11\mathrm{h}}$ and $\boldsymbol{y}_{11\mathrm{h}}$:

$$\mathbf{z}_{11} = \tfrac{1}{2}(\mathbf{z}_{11\mathrm{h}} + \boldsymbol{y}_{11\mathrm{h}}^{-1}) \qquad \mathbf{z}_{12} = \tfrac{1}{2}(\mathbf{z}_{11\mathrm{h}} - \boldsymbol{y}_{11\mathrm{h}}^{-1}) \tag{5.40}$$

$$\boldsymbol{y}_{11} = \tfrac{1}{2}(\mathbf{z}_{11\mathrm{h}}^{-1} + \boldsymbol{y}_{11\mathrm{h}}) \qquad \boldsymbol{y}_{12} = \tfrac{1}{2}(\mathbf{z}_{11\mathrm{h}}^{-1} - \boldsymbol{y}_{11\mathrm{h}}) \tag{5.41}$$

5.3.2.4 Determination of Floquet modes from half of the periods

The required quantities for Floquet modes can be calculated very efficiently by using the open- and short-circuit matrix parameters of half of the periods. Using the relations in eqs. (5.40) and (5.41) we obtain e.g.:

$$\begin{aligned}\mathbf{z}_{11}\mathbf{z}_{12}^{-1} &= (\mathbf{z}_{11\mathrm{h}} + \boldsymbol{y}_{11\mathrm{h}}^{-1})(\mathbf{z}_{11\mathrm{h}} - \boldsymbol{y}_{11\mathrm{h}}^{-1})^{-1}\\ &= (\mathbf{z}_{11\mathrm{h}}\boldsymbol{y}_{11\mathrm{h}} + \boldsymbol{I})(\mathbf{z}_{11\mathrm{h}}\boldsymbol{y}_{11\mathrm{h}} - \boldsymbol{I})^{-1}\end{aligned} \tag{5.42}$$

$$\begin{aligned}-\boldsymbol{y}_{12}^{-1}\boldsymbol{y}_{11} &= (\boldsymbol{y}_{11\mathrm{h}} - \mathbf{z}_{11\mathrm{h}}^{-1})^{-1}(\boldsymbol{y}_{11\mathrm{h}} + \mathbf{z}_{11\mathrm{h}}^{-1})\\ &= (\mathbf{z}_{11\mathrm{h}}\boldsymbol{y}_{11\mathrm{h}} - \boldsymbol{I})^{-1}(\mathbf{z}_{11\mathrm{h}}\boldsymbol{y}_{11\mathrm{h}} + \boldsymbol{I})\end{aligned} \tag{5.43}$$

$$\begin{aligned}\mathbf{z}_{12}^{-1}\mathbf{z}_{11} &= (\mathbf{z}_{11\mathrm{h}} - \boldsymbol{y}_{11\mathrm{h}}^{-1})^{-1}(\mathbf{z}_{11\mathrm{h}} + \boldsymbol{y}_{11\mathrm{h}}^{-1})\\ &= (\boldsymbol{y}_{11\mathrm{h}}\mathbf{z}_{11\mathrm{h}} - \boldsymbol{I})^{-1}(\boldsymbol{y}_{11\mathrm{h}}\mathbf{z}_{11\mathrm{h}} + \boldsymbol{I})\end{aligned} \tag{5.44}$$

$$\begin{aligned}-\boldsymbol{y}_{11}\boldsymbol{y}_{12}^{-1} &= (\boldsymbol{y}_{11\mathrm{h}} + \mathbf{z}_{11\mathrm{h}}^{-1})(\boldsymbol{y}_{11\mathrm{h}} - \mathbf{z}_{11\mathrm{h}}^{-1})^{-1}\\ &= (\boldsymbol{y}_{11\mathrm{h}}\mathbf{z}_{11\mathrm{h}} + \boldsymbol{I})(\boldsymbol{y}_{11\mathrm{h}}\mathbf{z}_{11\mathrm{h}} - \boldsymbol{I})^{-1}\end{aligned} \tag{5.45}$$

$\boldsymbol{I}$ is an adequate unit matrix. We obtain the eigenvalues/eigenvectors by introducing eqs. (5.42) and (5.44) into eq. (5.20):

$$\begin{aligned}\boldsymbol{S}_{\mathrm{E}}^{-1}\boldsymbol{z}_{11}\boldsymbol{z}_{12}^{-1}\boldsymbol{S}_{\mathrm{E}} = \boldsymbol{\lambda}_{\mathrm{C}} &= \boldsymbol{S}_{\mathrm{E}}^{-1}(\boldsymbol{z}_{11\mathrm{h}}\boldsymbol{y}_{11\mathrm{h}} + \boldsymbol{I})\boldsymbol{S}_{\mathrm{E}}\boldsymbol{S}_{\mathrm{E}}^{-1}(\boldsymbol{z}_{11\mathrm{h}}\boldsymbol{y}_{11\mathrm{h}} - \boldsymbol{I})^{-1}\boldsymbol{S}_{\mathrm{E}} \\ &= (\boldsymbol{S}_{\mathrm{E}}^{-1}\boldsymbol{z}_{11\mathrm{h}}\boldsymbol{y}_{11\mathrm{h}}\boldsymbol{S}_{\mathrm{E}} + \boldsymbol{I})(\boldsymbol{S}_{\mathrm{E}}^{-1}\boldsymbol{z}_{11\mathrm{h}}\boldsymbol{y}_{11\mathrm{h}}\boldsymbol{S}_{\mathrm{E}} - \boldsymbol{I})^{-1}\end{aligned} \tag{5.46}$$

or:

$$\begin{aligned}\boldsymbol{S}_{\mathrm{H}}^{-1}\boldsymbol{z}_{12}^{-1}\boldsymbol{z}_{11}\boldsymbol{S}_{\mathrm{H}} = \boldsymbol{\lambda}_{\mathrm{C}} &= \boldsymbol{S}_{\mathrm{H}}^{-1}(\boldsymbol{y}_{11\mathrm{h}}\boldsymbol{z}_{11\mathrm{h}} - \boldsymbol{I})^{-1}\boldsymbol{S}_{\mathrm{H}}\boldsymbol{S}_{\mathrm{H}}^{-1}(\boldsymbol{y}_{11\mathrm{h}}\boldsymbol{z}_{11\mathrm{h}} + \boldsymbol{I})\boldsymbol{S}_{\mathrm{H}} \\ &= (\boldsymbol{S}_{\mathrm{H}}^{-1}\boldsymbol{y}_{11\mathrm{h}}\boldsymbol{z}_{11\mathrm{h}}\boldsymbol{S}_{\mathrm{H}} - \boldsymbol{I})^{-1}(\boldsymbol{S}_{\mathrm{H}}^{-1}\boldsymbol{y}_{11\mathrm{h}}\boldsymbol{z}_{11\mathrm{h}}\boldsymbol{S}_{\mathrm{H}} + \boldsymbol{I})\end{aligned} \tag{5.47}$$

So we have:

$$\boldsymbol{\lambda}_{\mathrm{C}} = (\boldsymbol{\lambda}_{\mathrm{h}} + \boldsymbol{I})(\boldsymbol{\lambda}_{\mathrm{h}} - \boldsymbol{I})^{-1} = (\boldsymbol{\lambda}_{\mathrm{h}} - \boldsymbol{I})^{-1}(\boldsymbol{\lambda}_{\mathrm{h}} + \boldsymbol{I}) \tag{5.48}$$

where we defined:

$$\boldsymbol{\lambda}_{\mathrm{h}} = \boldsymbol{S}_{\mathrm{E}}^{-1}\boldsymbol{z}_{11\mathrm{h}}\boldsymbol{y}_{11\mathrm{h}}\boldsymbol{S}_{\mathrm{E}} = \boldsymbol{S}_{\mathrm{H}}^{-1}\boldsymbol{y}_{11\mathrm{h}}\boldsymbol{z}_{11\mathrm{h}}\boldsymbol{S}_{\mathrm{H}} \tag{5.49}$$

This is a fundamental equation for determining the Floquet modes with symmetric periods. As with eq. (5.32) and in contrast to eq. (5.20), we determine the Floquet modes from the input impedance and admittance without the need for an inversion. Therefore, these expressions are numerically stable.

Equivalent equations can be obtained by using eqs. (5.43) and (5.45). The parameters $\boldsymbol{\delta}_{1,2}$ in eq. (5.27) (with $\widetilde{\boldsymbol{Z}}_0 = \widetilde{\boldsymbol{Y}}_0 = \boldsymbol{I}$): are now given by:

$$\boldsymbol{\delta}_1 = 2\sqrt{\boldsymbol{\lambda}_{\mathrm{h}}}/(\boldsymbol{\lambda}_{\mathrm{h}} + \boldsymbol{I}) \qquad \boldsymbol{\delta}_2 = \tfrac{1}{2}(\boldsymbol{\lambda}_{\mathrm{h}} - \boldsymbol{I})/\sqrt{\boldsymbol{\lambda}_{\mathrm{h}}} \tag{5.50}$$

Introducing these values into eq. (5.22) by using the relations (5.40) and (5.49), we obtain the following relations for the eigenvector matrices $\boldsymbol{S}_{\mathrm{E}}$ and $\boldsymbol{S}_{\mathrm{H}}$:

$$\boldsymbol{S}_{\mathrm{E}} = \boldsymbol{z}_{11\mathrm{h}}\boldsymbol{S}_{\mathrm{H}}/\sqrt{\boldsymbol{\lambda}_{\mathrm{h}}} \qquad \boldsymbol{S}_{\mathrm{H}} = \boldsymbol{y}_{11\mathrm{h}}\boldsymbol{S}_{\mathrm{E}}/\sqrt{\boldsymbol{\lambda}_{\mathrm{h}}} \tag{5.51}$$

These formulas are analogous to those in eq. (5.34), with $\boldsymbol{\delta}_a = \boldsymbol{\delta}_b = \boldsymbol{I}/\sqrt{\boldsymbol{\lambda}_{\mathrm{os}}}$. Due to the relation $\cosh x = (1 + \tanh^2 \frac{x}{2})/(1 - \tanh^2 \frac{x}{2})$, we have the following relation between $\boldsymbol{\lambda}_{\mathrm{h}}$ and $\boldsymbol{\Gamma}_{\mathrm{F}}$:

$$\tanh\left(\frac{1}{2}\boldsymbol{\Gamma}_{\mathrm{F}}\right) = \boldsymbol{\lambda}_{\mathrm{h}}^{-\frac{1}{2}} = \boldsymbol{I}/\sqrt{\boldsymbol{S}_{\mathrm{E}}^{-1}\boldsymbol{z}_{11\mathrm{h}}\boldsymbol{y}_{11\mathrm{h}}\boldsymbol{S}_{\mathrm{E}}} = \boldsymbol{I}/\sqrt{\boldsymbol{S}_{\mathrm{H}}^{-1}\boldsymbol{y}_{11\mathrm{h}}\boldsymbol{z}_{11\mathrm{h}}\boldsymbol{S}_{\mathrm{H}}} \tag{5.52}$$

This expression is similar to that given in eq. (5.32) for $\tanh \boldsymbol{\Gamma}_{\mathrm{F}}$. We immediately obtain the result in eq. (5.32) from that in eq. (5.52) if we examine the half of two concatenated periods. The result can also be obtained from $\tanh \boldsymbol{\Gamma}_{\mathrm{F}} = \sinh \boldsymbol{\Gamma}_{\mathrm{F}}/\cosh \boldsymbol{\Gamma}_{\mathrm{F}}$ by introducing the impedance/admittance matrices in eq. (5.11).

By introducing the result of eq. (5.48) into eq. (5.28) (with $\boldsymbol{\delta}_1 = \boldsymbol{\delta}_2$), we obtain for the wave impedance:

$$\widetilde{\boldsymbol{Z}}_0 = \sqrt{(\boldsymbol{\lambda}_{\mathrm{C}}^2 - \boldsymbol{I})/\boldsymbol{\lambda}_{\mathrm{C}}} = 2\sqrt{\boldsymbol{\lambda}_{\mathrm{h}}/(\boldsymbol{\lambda}_{\mathrm{h}}^2 - \boldsymbol{I})}$$

5.3.3 Concatenation of N symmetric periods

In this subsection we would like to give the relevant formulas for N concatenated periodic sections (see Fig. 5.1). Each periodic section was described like a homogeneous waveguide. Therefore, if we concatenate N periods we need only replace $\boldsymbol{\Gamma}_\mathrm{F}$ with $N\boldsymbol{\Gamma}_\mathrm{F}$. Eq. (5.19) now reads:

$$\begin{bmatrix} \widetilde{\mathbf{E}}_\mathrm{D} \\ \widetilde{\mathbf{H}}_\mathrm{D} \end{bmatrix} = \begin{bmatrix} \cosh(N\boldsymbol{\Gamma}_\mathrm{F}) & -\widetilde{\boldsymbol{Z}}_0 \sinh(N\boldsymbol{\Gamma}_\mathrm{F}) \\ -\widetilde{\boldsymbol{Y}}_0 \sinh(N\boldsymbol{\Gamma}_\mathrm{F}) & \cosh(N\boldsymbol{\Gamma}_\mathrm{F}) \end{bmatrix} \begin{bmatrix} \widetilde{\mathbf{E}}_\mathrm{C} \\ \widetilde{\mathbf{H}}_\mathrm{C} \end{bmatrix} \tag{5.53}$$

Ports C and D are the input and output ports of the whole structure. Furthermore, we can again write the field relations between the generalised two ports C and D with open-circuit impedance or short-circuit admittance parameter ($\boldsymbol{z}$ and $\boldsymbol{y}$)-matrices:

$$\begin{bmatrix} \widetilde{\mathbf{E}}_\mathrm{C} \\ \widetilde{\mathbf{E}}_\mathrm{D} \end{bmatrix} = \begin{bmatrix} \widetilde{\boldsymbol{z}}_1 & \widetilde{\boldsymbol{z}}_2 \\ \widetilde{\boldsymbol{z}}_2 & \widetilde{\boldsymbol{z}}_1 \end{bmatrix} \begin{bmatrix} \widetilde{\mathbf{H}}_\mathrm{C} \\ -\widetilde{\mathbf{H}}_\mathrm{D} \end{bmatrix} \qquad \begin{bmatrix} \widetilde{\mathbf{H}}_\mathrm{C} \\ -\widetilde{\mathbf{H}}_\mathrm{D} \end{bmatrix} = \begin{bmatrix} \widetilde{\boldsymbol{y}}_1 & \widetilde{\boldsymbol{y}}_2 \\ \widetilde{\boldsymbol{y}}_2 & \widetilde{\boldsymbol{y}}_1 \end{bmatrix} \begin{bmatrix} \widetilde{\mathbf{E}}_\mathrm{C} \\ \widetilde{\mathbf{E}}_\mathrm{D} \end{bmatrix} \tag{5.54}$$

The parameters are defined as:

$$\begin{aligned} \widetilde{\boldsymbol{z}}_1 &= \widetilde{\boldsymbol{Z}}_0/\tanh(N\widetilde{\boldsymbol{\Gamma}}_\mathrm{F}) & \widetilde{\boldsymbol{z}}_2 &= \widetilde{\boldsymbol{Z}}_0/\sinh(N\widetilde{\boldsymbol{\Gamma}}_\mathrm{F}) \\ \widetilde{\boldsymbol{y}}_1 &= \widetilde{\boldsymbol{Y}}_0/\tanh(N\widetilde{\boldsymbol{\Gamma}}_\mathrm{F}) & \widetilde{\boldsymbol{y}}_2 &= -\widetilde{\boldsymbol{Y}}_0/\sinh(N\widetilde{\boldsymbol{\Gamma}}_\mathrm{F}) \end{aligned} \tag{5.55}$$

By using $\widetilde{\mathbf{E}}_\mathrm{C,D} = \widetilde{\boldsymbol{Z}}_\mathrm{C,D}\widetilde{\mathbf{H}}_\mathrm{C,D}$ and $\widetilde{\mathbf{H}}_\mathrm{C,D} = \widetilde{\boldsymbol{Y}}_\mathrm{C,D}\widetilde{\mathbf{E}}_\mathrm{C,D}$, the impedance/admittance transformation for the whole periodic structure is performed by:

$$\widetilde{\boldsymbol{Z}}_\mathrm{C} = \widetilde{\boldsymbol{z}}_1 - \widetilde{\boldsymbol{z}}_2(\widetilde{\boldsymbol{z}}_1 + \widetilde{\boldsymbol{Z}}_\mathrm{D})^{-1}\widetilde{\boldsymbol{z}}_2 \qquad \widetilde{\boldsymbol{Y}}_\mathrm{C} = \widetilde{\boldsymbol{y}}_1 - \widetilde{\boldsymbol{y}}_2(\widetilde{\boldsymbol{y}}_1 + \widetilde{\boldsymbol{Y}}_\mathrm{D})^{-1}\widetilde{\boldsymbol{y}}_2 \tag{5.56}$$

The relations between the impedances/admittances with tilde (~) (Floquet impedances) and bar (–) (mode impedances) are given by:

$$\widetilde{\boldsymbol{Z}}_\mathrm{D} = \boldsymbol{S}_\mathrm{E}^{-1}\boldsymbol{T}_\mathrm{E}\overline{\boldsymbol{Z}}_\mathrm{Di}\boldsymbol{T}_\mathrm{H}^{-1}\boldsymbol{S}_\mathrm{H} = \widetilde{\boldsymbol{Y}}_\mathrm{D}^{-1} \qquad \overline{\boldsymbol{Z}}_\mathrm{Ci} = \boldsymbol{T}_\mathrm{E}^{-1}\boldsymbol{S}_\mathrm{E}\widetilde{\boldsymbol{Z}}_\mathrm{C}\boldsymbol{S}_\mathrm{H}^{-1}\boldsymbol{T}_\mathrm{H} = \overline{\boldsymbol{Y}}_\mathrm{Ci}^{-1} \tag{5.57}$$

$\overline{\boldsymbol{Z}}_\mathrm{Di}$ and $\overline{\boldsymbol{Z}}_\mathrm{Ci}$ are the values at the inner side of ports D and C in the transformed domain. This expression is very similar to that for transforming impedances/admittances at the interface between two homogeneous sections (eqs. (2.152) and (2.153)). Since the connecting waveguides at the outer side (o) of ports C and D can be completely different from the sections on the inner side (i), a further transformation may be required. This can be done by the known transformation between two different waveguides (see Section 2.5). We have e.g. at port D:

$$\overline{\boldsymbol{Z}}_\mathrm{Di} = \boldsymbol{T}_\mathrm{Ei}^{-1}\boldsymbol{J}_\mathrm{i}^\mathrm{ct}(\boldsymbol{J}_\mathrm{o}^\mathrm{c}\boldsymbol{T}_\mathrm{Ho}\overline{\boldsymbol{Y}}_\mathrm{Do}\boldsymbol{T}_\mathrm{Eo}^{-1}\boldsymbol{J}_\mathrm{o}^\mathrm{ct})^{-1}\boldsymbol{J}_\mathrm{i}^\mathrm{c}\boldsymbol{T}_\mathrm{Hi} = \overline{\boldsymbol{Y}}_\mathrm{Di}^{-1} \tag{5.58}$$

5.3.4 Floquet modes for unsymmetric periods

Floquet modes can also be calculated for unsymmetric circuits consisting of e.g. concatenated various sections (e.g. see Fig. 5.10). In the general case we must calculate the Floquet modes from the combination of the E and H fields. The procedure is different from that for symmetric periods. We define the supervectors:

$$\widehat{\mathbf{F}}_{\mathrm{A,B}} = [\mathbf{E}^{\mathrm{t}}_{\mathrm{A,B}}, \mathbf{H}^{\mathrm{t}}_{\mathrm{A,B}}]^{\mathrm{t}} \tag{5.59}$$

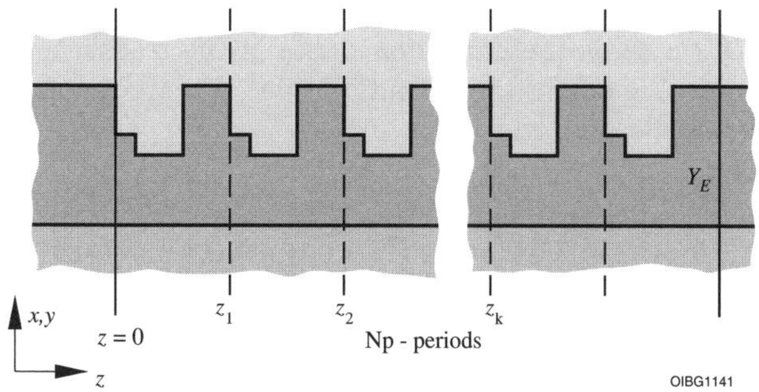

Fig. 5.10 Bragg-grating with unsymmetrical periods

These supervectors are now transformed into Floquet mode supervectors $\widehat{\widehat{\mathbf{F}}}_{\mathrm{A,B}}$ according to:

$$\widehat{\mathbf{F}}_{\mathrm{A,B}} = \widehat{\mathbf{S}}_{\mathrm{EH}}\widehat{\widehat{\mathbf{F}}}_{\mathrm{A,B}} \qquad \widehat{\widehat{\mathbf{F}}}_{\mathrm{A,B}} = \left[\widetilde{\mathbf{F}}^{\mathrm{t}}_{\mathrm{A,B,f}}, \widetilde{\mathbf{F}}^{\mathrm{t}}_{\mathrm{A,B,b}}\right]^{\mathrm{t}} \qquad \widehat{\mathbf{S}}_{\mathrm{EH}} = \begin{bmatrix} \mathbf{S}_{\mathrm{Ef}} & \mathbf{S}_{\mathrm{Ef}} \\ \mathbf{S}_{\mathrm{Hb}} & \mathbf{S}_{\mathrm{Hb}} \end{bmatrix} \tag{5.60}$$

The subscripts 'f' and 'b' indicate 'forward' and 'backward', respectively. Let us now introduce Floquet's theorem into eq. (5.13) according to:

$$\widehat{\widehat{\mathbf{F}}}_{\mathrm{A}} = e^{\widehat{\boldsymbol{\Gamma}}_{\mathrm{F}}}\widehat{\widehat{\mathbf{F}}}_{\mathrm{B}} \tag{5.61}$$

The diagonal matrix $e^{\widehat{\boldsymbol{\Gamma}}_{\mathrm{F}}}$ is obtained from the eigenvalue problem:

$$\widehat{\mathbf{S}}^{-1}_{\mathrm{EH}} \begin{bmatrix} \mathbf{z}_{11}\mathbf{z}_{21}^{-1} & \mathbf{z}_{11}\mathbf{z}_{21}^{-1}\mathbf{z}_{22} - \mathbf{z}_{12} \\ \mathbf{z}_{21}^{-1} & \mathbf{z}_{21}^{-1}\mathbf{z}_{22} \end{bmatrix} \widehat{\mathbf{S}}_{\mathrm{EH}} = e^{\widehat{\boldsymbol{\Gamma}}_{\mathrm{F}}} \tag{5.62}$$

Instead of eq. (5.13) we may use eq. (5.12). The examination of the transfer matrix in eq. (5.62) shows that its determinant can be computed as det(transfer matrix) = det $(\mathbf{z}_{12}^{-1})$ det$(\mathbf{z}_{21}) = 1$. To obtain this relation, we used the Schur complement. The matrices $\mathbf{z}_{12}$ and $\mathbf{z}_{21}$ have identical eigenvalues in the isotropic case. The value of 1 was found numerically. This means that the product of all eigenvalues $\exp(\Gamma_{\mathrm{F}_i})$ is also equal to 1 and the sum of $(\Gamma_{\mathrm{F}})_i$ results in 0. Now we order $\widehat{\boldsymbol{\Gamma}}_{\mathrm{F}}$ according to $\widehat{\boldsymbol{\Gamma}}_{\mathrm{F}} = \mathrm{diag}(\boldsymbol{\Gamma}_{\mathrm{F1}}, -\boldsymbol{\Gamma}_{\mathrm{F2}})$, where the real parts of $\boldsymbol{\Gamma}_{\rm F1}$ and $\boldsymbol{\Gamma}_{\rm F2}$ are positive. Usually, we introduce ABCs at the side walls, which results in complex values for $\boldsymbol{\Gamma}_{\rm F1,2}$. Let us now assume N periods between the input and the output ports C and D of the finite periodic device (e.g. see Fig. 5.1). The relation between the fields at these ports in Floquet mode domain can be written as:

$$\begin{bmatrix} \widetilde{\mathbf{F}}_{\rm fC} \\ \widetilde{\mathbf{F}}_{\rm bC} \end{bmatrix} = \begin{bmatrix} e^{N\boldsymbol{\Gamma}_{\rm F1}} & \\ & e^{-N\boldsymbol{\Gamma}_{\rm F2}} \end{bmatrix} \begin{bmatrix} \widetilde{\mathbf{F}}_{\rm fD} \\ \widetilde{\mathbf{F}}_{\rm bD} \end{bmatrix} \tag{5.63}$$

The relation between the fields at port D is given by $\mathbf{E}_{\rm D} = \boldsymbol{Z}_{\rm D}\mathbf{H}_{\rm D}$, where $\boldsymbol{Z}_{\rm D} = \boldsymbol{T}_{\rm E}\overline{\boldsymbol{Z}}_{\rm D}\boldsymbol{T}_{\rm H}^{-1}$. $\overline{\boldsymbol{Z}}_{\rm D}$ is the input impedance in the transformed domain of the connecting waveguide at the output. $\boldsymbol{T}_{\rm E}$ and $\boldsymbol{T}_{\rm H}$ are the modal matrices of this waveguide. So we have:

$$\begin{bmatrix} \widetilde{\mathbf{F}}_{\rm fD} \\ \widetilde{\mathbf{F}}_{\rm bD} \end{bmatrix} = \widehat{\mathbf{S}}_{\rm EH}^{-1} \begin{bmatrix} \boldsymbol{Z}_{\rm D}\mathbf{H}_{\rm D} \\ \mathbf{H}_{\rm D} \end{bmatrix} = \begin{bmatrix} \boldsymbol{S}_{\rm Ef}^{\rm i} & \boldsymbol{S}_{\rm Eb}^{\rm i} \\ \boldsymbol{S}_{\rm Hf}^{\rm i} & \boldsymbol{S}_{\rm Hb}^{\rm i} \end{bmatrix} \begin{bmatrix} \boldsymbol{Z}_{\rm D}\mathbf{H}_{\rm D} \\ \mathbf{H}_{\rm D} \end{bmatrix} \tag{5.64}$$

where the superscript $(^{\rm i})$ indicates the submatrices of the inverted matrix $\widehat{\mathbf{S}}_{\rm EH}$. It should however be clarified that the determination of these submatrices requires the inversion of the whole matrix $\widehat{\mathbf{S}}_{\rm EH}$, i.e. we cannot simply invert all the blocks of this matrix. Therefore, $\boldsymbol{S}_{\rm Ef}^{\rm i}$ is not the inverse of $\boldsymbol{S}_{\rm Ef}$!

Introducing this relation into eq. (5.63) results in:

$$\begin{bmatrix} \widetilde{\mathbf{F}}_{\rm fC} \\ \widetilde{\mathbf{F}}_{\rm bC} \end{bmatrix} = \begin{bmatrix} e^{N\boldsymbol{\Gamma}_{\rm F1}} & \\ & e^{-N\boldsymbol{\Gamma}_{\rm F2}} \end{bmatrix} \begin{bmatrix} \boldsymbol{S}_{\rm EE}^{\rm i} & \boldsymbol{S}_{\rm EH}^{\rm i} \\ \boldsymbol{S}_{\rm HE}^{\rm i} & \boldsymbol{S}_{\rm HH}^{\rm i} \end{bmatrix} \begin{bmatrix} \boldsymbol{Z}_{\rm D}\mathbf{H}_{\rm D} \\ \mathbf{H}_{\rm D} \end{bmatrix} \tag{5.65}$$

From the last equation we obtain:

$$\widetilde{\mathbf{F}}_{\rm bC} = \widetilde{\boldsymbol{R}}_{\rm C}\widetilde{\mathbf{F}}_{\rm fC} \qquad \widetilde{\mathbf{R}}_{\rm C} = e^{-N\boldsymbol{\Gamma}_{\rm F2}}(\boldsymbol{S}_{\rm Hf}^{\rm i}\boldsymbol{Z}_{\rm D} + \boldsymbol{S}_{\rm Hb}^{\rm i})(\boldsymbol{S}_{\rm Ef}^{\rm i}\boldsymbol{Z}_{\rm D} + \boldsymbol{S}_{\rm Eb}^{\rm i})^{-1}e^{-N\boldsymbol{\Gamma}_{\rm F1}} \tag{5.66}$$

These expressions are similar to the formulas with reflection coefficients given in [4]. At port C we obtain the real fields with:

$$\begin{bmatrix} \mathbf{E}_{\rm C} \\ \mathbf{H}_{\rm C} \end{bmatrix} = \begin{bmatrix} \boldsymbol{S}_{\rm Ef} & \boldsymbol{S}_{\rm Eb} \\ \boldsymbol{S}_{\rm Hf} & \boldsymbol{S}_{\rm Hb} \end{bmatrix} \begin{bmatrix} \widetilde{\mathbf{F}}_{\rm fC} \\ \widetilde{\boldsymbol{R}}_{\rm C}\widetilde{\mathbf{F}}_{\rm fC} \end{bmatrix} \tag{5.67}$$

The load impedance of the input waveguide $\boldsymbol{Z}_{\rm C}$ at port C is now given by:

$$\mathbf{E}_{\rm C} = \boldsymbol{Z}_{\rm C}\mathbf{H}_{\rm C} \qquad \boldsymbol{Z}_{\rm C} = (\boldsymbol{S}_{\rm Ef} + \boldsymbol{S}_{\rm Eb}\widetilde{\boldsymbol{R}}_{\rm C})(\boldsymbol{S}_{\rm Hf} + \boldsymbol{S}_{\rm Hb}\widetilde{\boldsymbol{R}}_{\rm C})^{-1} \tag{5.68}$$

By using the modal matrices of the input waveguide (they may be different from those of the output waveguide), we can also calculate the impedance matrix $\overline{\boldsymbol{Z}}_{\rm C}$ in the transformed domain (mode domain).

The transfer matrix for ports C and D can easily be determined as:

$$\begin{bmatrix} \mathbf{E}_{\rm C} \\ \mathbf{H}_{\rm C} \end{bmatrix} = \begin{bmatrix} \boldsymbol{S}_{\rm Ef} & \boldsymbol{S}_{\rm Eb} \\ \boldsymbol{S}_{\rm Hf} & \boldsymbol{S}_{\rm Hb} \end{bmatrix} \begin{bmatrix} e^{N\boldsymbol{\Gamma}_{\rm F1}} & \\ & e^{-N\boldsymbol{\Gamma}_{\rm F2}} \end{bmatrix} \begin{bmatrix} \boldsymbol{S}_{\rm Ef}^{\rm i} & \boldsymbol{S}_{\rm Eb}^{\rm i} \\ \boldsymbol{S}_{\rm Hf}^{\rm i} & \boldsymbol{S}_{\rm Hb}^{\rm i} \end{bmatrix} \begin{bmatrix} \mathbf{E}_{\rm D} \\ \mathbf{H}_{\rm D} \end{bmatrix} \tag{5.69}$$

or:

$$\begin{bmatrix} \mathbf{E}_{\mathrm{C}} \\ \mathbf{H}_{\mathrm{C}} \end{bmatrix} = \begin{bmatrix} \mathbf{V}_{\mathrm{E}} & \mathbf{Z}_{\mathrm{F}} \\ \mathbf{Y}_{\mathrm{F}} & \mathbf{V}_{\mathrm{H}} \end{bmatrix} \begin{bmatrix} \mathbf{E}_{\mathrm{D}} \\ \mathbf{H}_{\mathrm{D}} \end{bmatrix} \tag{5.70}$$

with:

$$\begin{aligned} \mathbf{V}_{\mathrm{E}} &= \mathbf{S}_{\mathrm{Ef}} e^{N\boldsymbol{\Gamma}_{\mathrm{F1}}} \mathbf{S}_{\mathrm{Ef}}^{\mathrm{i}} + \mathbf{S}_{\mathrm{Eb}} e^{-N\boldsymbol{\Gamma}_{\mathrm{F2}}} \mathbf{S}_{\mathrm{Hf}}^{\mathrm{i}} \\ \mathbf{V}_{\mathrm{H}} &= \mathbf{S}_{\mathrm{Hf}} e^{N\boldsymbol{\Gamma}_{\mathrm{F1}}} \mathbf{S}_{\mathrm{Eb}}^{\mathrm{i}} + \mathbf{S}_{\mathrm{Hb}} e^{-N\boldsymbol{\Gamma}_{\mathrm{F2}}} \mathbf{S}_{\mathrm{Hb}}^{\mathrm{i}} \\ \mathbf{Z}_{\mathrm{F}} &= \mathbf{S}_{\mathrm{Ef}} e^{N\boldsymbol{\Gamma}_{\mathrm{F1}}} \mathbf{S}_{\mathrm{Eb}}^{\mathrm{i}} + \mathbf{S}_{\mathrm{Eb}} e^{-N\boldsymbol{\Gamma}_{\mathrm{F2}}} \mathbf{S}_{\mathrm{Hb}}^{\mathrm{i}} \\ \mathbf{Y}_{\mathrm{F}} &= \mathbf{S}_{\mathrm{Hf}} e^{N\boldsymbol{\Gamma}_{\mathrm{F1}}} \mathbf{S}_{\mathrm{Ef}}^{\mathrm{i}} + \mathbf{S}_{\mathrm{Hb}} e^{-N\boldsymbol{\Gamma}_{\mathrm{F2}}} \mathbf{S}_{\mathrm{Hf}}^{\mathrm{i}} \end{aligned} \tag{5.71}$$

By rearranging the expressions in eq. (5.71), we can determine the general $\mathbf{z}$-matrix and $\mathbf{y}$-matrix parameters for ports C and D (see eq. (5.11)):

$$\begin{aligned} \begin{bmatrix} \mathbf{z}_{11} & \mathbf{z}_{12} \\ \mathbf{z}_{21} & \mathbf{z}_{22} \end{bmatrix} &= \begin{bmatrix} \mathbf{V}_{\mathrm{E}} \mathbf{Y}_{\mathrm{F}}^{-1} & \mathbf{V}_{\mathrm{E}} \mathbf{Y}_{\mathrm{F}}^{-1} \mathbf{V}_{\mathrm{H}} - \mathbf{Z}_{\mathrm{F}} \\ \mathbf{Y}_{\mathrm{F}}^{-1} & \mathbf{Y}_{\mathrm{F}}^{-1} \mathbf{V}_{\mathrm{H}} \end{bmatrix} \\ \begin{bmatrix} \mathbf{y}_{11} & \mathbf{y}_{12} \\ \mathbf{y}_{21} & \mathbf{y}_{22} \end{bmatrix} &= \begin{bmatrix} \mathbf{V}_{\mathrm{H}} \mathbf{Z}_{\mathrm{F}}^{-1} & \mathbf{Y}_{\mathrm{F}} - \mathbf{V}_{\mathrm{H}} \mathbf{Z}_{\mathrm{F}}^{-1} \mathbf{V}_{\mathrm{E}} \\ -\mathbf{Z}_{\mathrm{F}}^{-1} & \mathbf{Z}_{\mathrm{F}}^{-1} \mathbf{V}_{\mathrm{E}} \end{bmatrix} \end{aligned} \tag{5.72}$$

Special care is required when the exponential terms are numerically determined in the products.

5.3.5 Some further general relations in periodic structures

We will now provide some further relations in periodic structures that may be of interest and helpful.

5.3.5.1 Characteristic impedances

We first define wave impedances at the ports. At the output port (B) we choose $\mathbf{Z}_{02}$ as load impedance, i.e. $\mathbf{E}_{\mathrm{B}} = \mathbf{Z}_{02}\mathbf{H}_{\mathrm{B}}$. The input impedance (at port A) is $\mathbf{Z}_{01}$, with $\mathbf{E}_{\mathrm{A}} = \mathbf{Z}_{01}\mathbf{H}_{\mathrm{B}}$. The impedances are transformed according to:

$$\mathbf{Z}_{01} = \mathbf{z}_{11} - \mathbf{z}_{12}(\mathbf{z}_{22} + \mathbf{Z}_{02})^{-1}\mathbf{z}_{21} \qquad \mathbf{Z}_{02} = -\mathbf{z}_{22} + \mathbf{z}_{21}(\mathbf{z}_{11} - \mathbf{Z}_{01})^{-1}\mathbf{z}_{12} \tag{5.73}$$

Analogously, we use port A as output. As load impedance we choose $\mathbf{Z}_{01}$ ($\mathbf{E}_{\mathrm{A}} = -\mathbf{Z}_{01}\mathbf{H}_{\mathrm{A}}$). We should now obtain at port B the input impedance $\mathbf{Z}_{02}$, with $\mathbf{E}_{\mathrm{B}} = -\mathbf{Z}_{02}\mathbf{H}_{\mathrm{B}}$. Here, the impedance transformation formulas are:

$$\mathbf{Z}_{01} = -\mathbf{z}_{11} + \mathbf{z}_{12}(\mathbf{z}_{22} - \mathbf{Z}_{02})^{-1}\mathbf{z}_{21} \qquad \mathbf{Z}_{02} = \mathbf{z}_{22} - \mathbf{z}_{21}(\mathbf{z}_{11} + \mathbf{Z}_{01})^{-1}\mathbf{z}_{12} \tag{5.74}$$

After some lengthy but not difficult computations we can determine $\mathbf{Z}_{01}$ and $\mathbf{Z}_{02}$:

$$\mathbf{Z}_{01} = \sqrt{\mathbf{y}_{11}^{-1}\mathbf{z}_{11}} \qquad \mathbf{Z}_{02} = \sqrt{\mathbf{y}_{22}^{-1}\mathbf{z}_{22}} \tag{5.75}$$

In deriving eq. (5.75), we took into account that the open-circuit parameter (z) matrix is the inverse of the short-circuit parameter (y) matrix (see eq. (5.11)).

5.3.5.2 *Difference equations*

In principle we could use eq. (5.62) to determine the Floquet modes in unsymmetric periodic structures. However, to construct this transfer matrix we would need to multiply with exponentially increasing terms (the solution of backward propagating modes). Therefore, as before, numerical problems can occur, particularly for long sections and high losses. So we present a method to determine the Floquet modes in a much more stable way here. For this purpose, we start with the equations for two periods of an unsymmetric periodic structure (see Fig. 5.11). From eq. (5.11) we obtain the following general difference equations:

$$\mathbf{z}_{12}\mathbf{H}_{n+1} + \mathbf{z}_{21}\mathbf{H}_{n-1} = (\mathbf{z}_{11} + \mathbf{z}_{22})\mathbf{H}_n \tag{5.76}$$

$$-\mathbf{y}_{12}\mathbf{E}_{n+1} - \mathbf{y}_{21}\mathbf{E}_{n-1} = (\mathbf{y}_{11} + \mathbf{y}_{22})\mathbf{E}_n \tag{5.77}$$

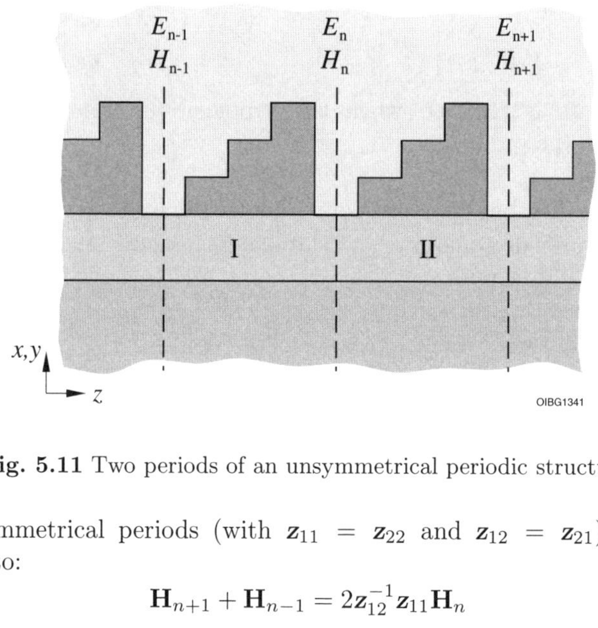

Fig. 5.11 Two periods of an unsymmetrical periodic structure

For symmetrical periods (with $\mathbf{z}_{11} = \mathbf{z}_{22}$ and $\mathbf{z}_{12} = \mathbf{z}_{21}$), eq. (5.77) simplifies to:

$$\mathbf{H}_{n+1} + \mathbf{H}_{n-1} = 2\mathbf{z}_{12}^{-1}\mathbf{z}_{11}\mathbf{H}_n \tag{5.78}$$

Therefore, by introducing:

$$\mathbf{H}_{\mathrm{n}} = \boldsymbol{S}_{\mathrm{H}}\widetilde{\mathbf{H}}_{\mathrm{n}} \qquad \widetilde{\mathbf{H}}_{\mathrm{n}} = e^{-\boldsymbol{\Gamma}_{\mathrm{F}}}\widetilde{\mathbf{H}}_{\mathrm{n}-1} \tag{5.79}$$

we obtain again from eq. (5.78):

$$\cosh \boldsymbol{\Gamma}_{\mathrm{F}} = \boldsymbol{S}_{\mathrm{H}}^{-1}\mathbf{z}_{12}^{-1}\mathbf{z}_{11}\boldsymbol{S}_{\mathrm{H}} \tag{5.80}$$

We can determine an analogous equation for the electric field.

For the unsymmetric case, we rewrite eq. (5.77) for the two different ports n and $n+1$:

$$\begin{aligned}\mathbf{z}_{21}\mathbf{H}_{n-1} + \mathbf{z}_{12}\mathbf{H}_{n+1} - (\mathbf{z}_{11} + \mathbf{z}_{22})\mathbf{H}_n &= 0\\ \mathbf{z}_{21}\mathbf{H}_{n} + \mathbf{z}_{12}\mathbf{H}_{n+2} - (\mathbf{z}_{11} + \mathbf{z}_{22})\mathbf{H}_{n+1} &= 0\end{aligned} \tag{5.81}$$

Now we introduce the wave ansatz according to:

$$\mathbf{H}_{n+1} = s_\mathrm{F}\mathbf{H}_n \quad \text{with} \quad s_\mathrm{F} = e^{-\Gamma_\mathrm{F}} \tag{5.82}$$

and obtain:

$$\begin{aligned}\mathbf{z}_{21}\mathbf{H}_{n} - s_\mathrm{F}(\mathbf{z}_{11} + \mathbf{z}_{22})\mathbf{H}_n + s_\mathrm{F}\mathbf{z}_{12}\mathbf{H}_{n+1} &= \mathbf{0}\\ \mathbf{z}_{21}\mathbf{H}_{n} - (\mathbf{z}_{11} + \mathbf{z}_{22})\mathbf{H}_{n+1} + s_\mathrm{F}\mathbf{z}_{12}\mathbf{H}_{n+1} &= \mathbf{0}\end{aligned} \tag{5.83}$$

or:

$$\left(\begin{bmatrix}\mathbf{z}_{21} & \boldsymbol{0}\\ \mathbf{z}_{21} & -(\mathbf{z}_{11}+\mathbf{z}_{22})\end{bmatrix} - s_\mathrm{F}\begin{bmatrix}\mathbf{z}_{11}+\mathbf{z}_{22} & -\mathbf{z}_{12}\\ \boldsymbol{0} & -\mathbf{z}_{12}\end{bmatrix}\right)\begin{bmatrix}\mathbf{H}_n\\ \mathbf{H}_{n+1}\end{bmatrix} = \mathbf{0} \tag{5.84}$$

Instead of using both expressions from eq. (5.81), we can use eq. (5.82). The equivalent problem reads in this case:

$$\left(\begin{bmatrix}\boldsymbol{0} & \boldsymbol{I}\\ \mathbf{z}_{21} & -(\mathbf{z}_{11}+\mathbf{z}_{22})\end{bmatrix} - s_\mathrm{F}\begin{bmatrix}\boldsymbol{I} & \boldsymbol{0}\\ \boldsymbol{0} & -\mathbf{z}_{12}\end{bmatrix}\right)\begin{bmatrix}\mathbf{H}_n\\ \mathbf{H}_{n+1}\end{bmatrix} = \mathbf{0} \tag{5.85}$$

This is a general eigenvalue problem and is analogous to that described by Helfert in [6]. After a suitable ordering of the eigenvalues we may write:

$$\mathbf{s}_\mathrm{F} = \mathrm{diag}(\mathbf{s}_\mathrm{F}^\mathrm{f}, \mathbf{s}_\mathrm{F}^\mathrm{b}) = \mathrm{diag}(\exp(-\boldsymbol{\Gamma}_\mathrm{F}^\mathrm{f}), \exp(\boldsymbol{\Gamma}_\mathrm{F}^\mathrm{b})) \tag{5.86}$$

With:

$$\begin{bmatrix}\mathbf{H}_n\\ \mathbf{H}_{n+1}\end{bmatrix} = \widehat{\mathbf{S}}_\mathrm{H}\begin{bmatrix}\widetilde{\mathbf{F}}_\mathrm{f}\\ \widetilde{\mathbf{F}}_\mathrm{b}\end{bmatrix} \tag{5.87}$$

we may write for the corresponding eigenvector matrix $\widehat{\mathbf{S}}_\mathrm{H}$:

$$\widehat{\mathbf{S}}_\mathrm{H} = \begin{bmatrix}\mathbf{S}_\mathrm{H}^\mathrm{f} & \mathbf{S}_\mathrm{H}^\mathrm{b}\\ \mathbf{S}_\mathrm{H}^\mathrm{f}\mathbf{s}_\mathrm{F}^\mathrm{f} & \mathbf{S}_\mathrm{H}^\mathrm{b}\mathbf{s}_\mathrm{F}^\mathrm{b}\end{bmatrix} \tag{5.88}$$

Numerically, it was found that the matrices $\boldsymbol{\Gamma}_\mathrm{F}^\mathrm{f}$ and $\boldsymbol{\Gamma}_\mathrm{F}^\mathrm{b}$ are equal to each other with isotropic material. The matrix $\widehat{\mathbf{S}}_\mathrm{E}$ can be determined from eq. (5.11). Note: a difference equation can also be developed from the transfer matrix expressions given in eqs. (5.12)–(5.15).

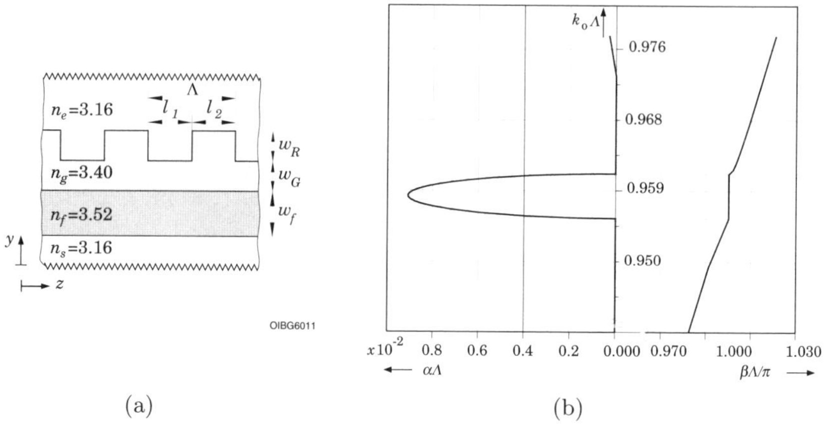

Fig. 5.12 Complex phase of one period of the Bragg grating in Fig. 5.1 as a function of the frequency. The frequency is plotted on the ordinate

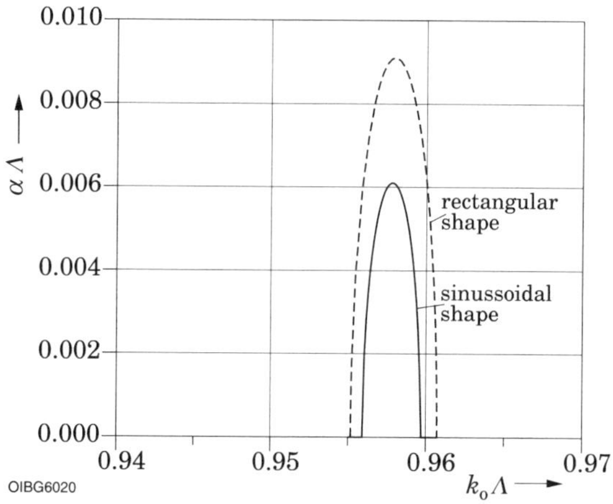

Fig. 5.13 Damping of one period of the Bragg grating in Fig. 5.1 with rectangular and sinusoidal shape as a function of the frequency

5.4 NUMERICAL RESULTS FOR PERIODIC STRUCTURES IN ONE DIRECTION

The complex phase of one period of the Bragg grating (Fig. 5.1) as a function of the frequency is shown in Fig. 5.12. Note: here, the frequency is plotted on the ordinate.

Fig. 5.13 shows the damping of the phase for different shapes.

The basic equations for the analysis of 2D structures are given in Section 5.3.1. Results for the reflectivity ($|S_{11}|^2$) of the fundamental mode of the Bragg grating in Fig. 5.1 are shown in Fig. 5.14 for different numbers of periods N. It is an interesting phenomenon that the curves are not symmetrical. The results are identical to those published in [7] and [8] – analysed with different algorithms. The curves are calculated with absorbing boundary conditions. Electric and magnetic walls can also be used. In the latter cases, the curves are superimposed by very small oscillation, as can be seen for the case $N = 2000$ m (metallic walls). For a planar Bragg grating, the CPU times were measured in [7]. It was found that the CPU time for analysing a grating with 1000 periods could be reduced from 2.5 h to 2 min if the Floquet approach was used instead of the general impedance transformation. For fibre gratings with constant material parameters in the azimuthal direction (like the ones for which we will show numerical results in this subsection), only a one-dimensional discretisation of the cross-section is required, because the φ-dependence of the fields can be given analytically. Therefore, the numerical effort is similar to the planar case and a comparable time saving is expected. However, since the analysis *without* 'Floquet' requires a lot of CPU time, this comparison has not been made for fibre structures. The algorithm is a combination of the numerically stable impedance transformation and Floquet's theorem. Therefore, the numerical effort does not depend on the number of periods and the numerical problems can be minimised. One more advantage is that in this method (as usual in the MoL) the mode coupling due to inhomogenities is automatically taken into account.

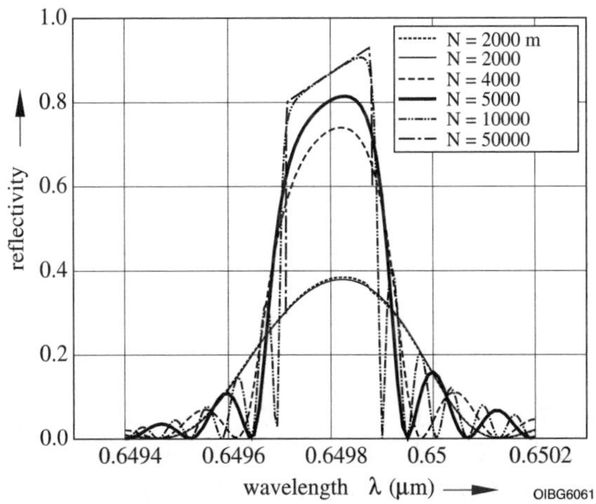

Fig. 5.14 Reflectivity of the Bragg grating in Fig. 5.1 with different numbers of periods as a function of the wavelength (Reproduced by permission of Elsevier)

Results for the reflectivity ($|S_{11}|^2$) of the fibre Bragg grating in Fig. 5.4 are shown in Fig. 5.15a for two different numbers of periods M. The numerical examinations were done for a device with constant cladding index that consisted of two different homogeneous regions. Curves for the reflectivity and transmission as a function of the wavelength and for 2000 and 5000 periods were calculated. The refractive index in the core of the first homogeneous region was $n_{co1} = 1.45$. For the difference between core and cladding index we chose $\Delta = 0.005$. It is an interesting phenomenon that the curves are in principle not symmetric. This is because the radiation of optical waves in the grating is larger for smaller wavelengths. Similar results are presented in [9]. In Fig. 5.15b numerical results are shown for a fibre Bragg grating with unsymmetric periods. The curve in the middle is identical to that shown in Fig. 5.15a. For the other two curves, section A was split into two subsections A and C each of half the length the former section. Then the core index of section C was chosen as $n_C = 0.9995 n_A$ for the left curve and $n_C = 1.0005 n_A$ for the right one. The dashed line is for anisotropic material ($n_{cz} = 1.5$). At the interfaces between the different homogeneous sections, guided modes as well as radiation modes are excited. These radiation modes were taken into account with the introduction of absorbing boundary conditions into the finite difference scheme (see Chapter 2). Anisotropic material was introduced in Figs. 5.16 and 5.17. In this case, the component of the tensor in the direction of propagation is different to the transverse components. The isotropic case was already examined in [2]. The curve with the anisotropic materials is shifted slightly to the right. The gratings have $N = 5000$ and $N = 8000$ periods. The size of each period $\Lambda = d_1 + d_2$ is 0.528 μm and the relative difference between core refractive indices in the two sections $\Delta n_{co} = (n_{co2} - n_{co1})/n_{co1}$ has been chosen to be 0.001. As mentioned above, the method permits us to analyse a more general periodic fibre structure with varying core and cladding index. In our examinations we kept the core radius constant in both sections. An improvement of the reflectivity can be achieved when the condition:

$$n_1^{\mathrm{e}} \overline{d}_1 = n_2^{\mathrm{e}} \overline{d}_2 = \pi/2 \tag{5.89}$$

is fulfilled for each interface between the layers in periodic structures. Here $n_{1,2}^{\mathrm{e}}$ are effective indices in two different sections of the grating. The analysis of apodised and chirped fibre gratings, as well as of the non-linear properties of gratings, can easily be performed.

As a further example of a cylindrical periodic structure, the magnetron resonator [10] in Fig. 5.18a is shown. This resonator is designed to operate in the fundamental $\pi/2$ mode around 38 GHz, with $N = 16$ slots of depth 1.385 mm, an anode radius of 2.25 mm and a cathode radius of 1.3 mm. Fig. 5.18b shows the resonant frequencies of the TE_{l10} and TE_{l20} modes ($l = 0, 1, 2, \ldots, N/2$) with axial open-boundary conditions. The results obtained by [10] and the MoL are in very good agreement.

A typical periodic microwave structure is the microstrip meander line. A meander line on RT/duroid microwave laminate with 23 periods (see Fig. 5.3a) has been fabricated, measured and compared with the calculated results. To match the input and output impedances of the meander line to the 50 Ω connecting line of width w_1, and to minimise insertion loss, the first and last

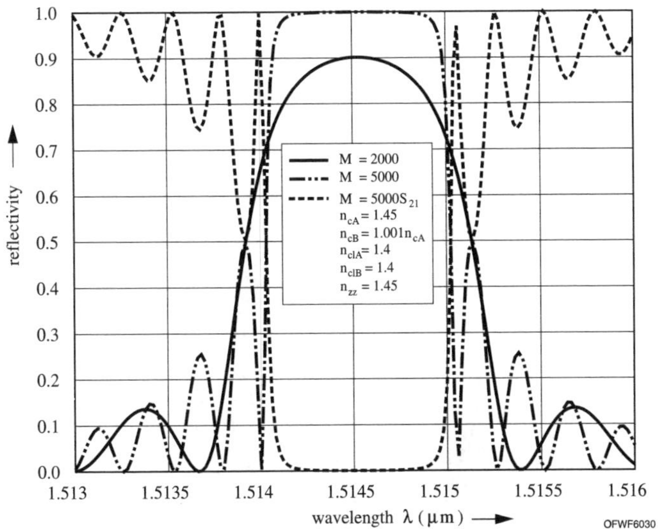

(a)

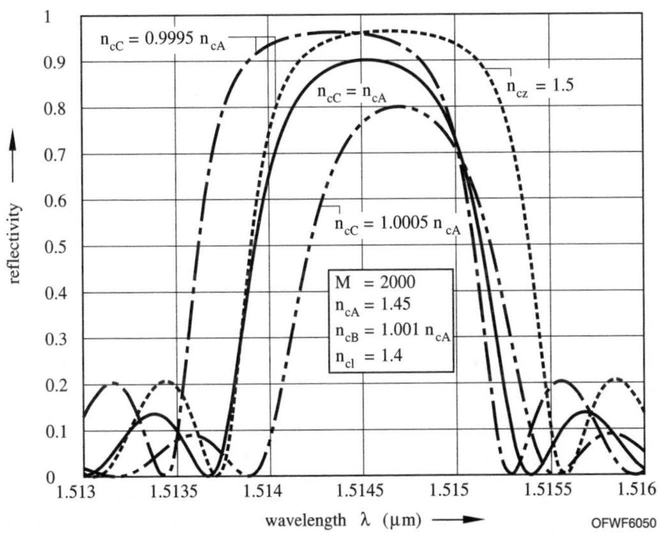

(b)

Fig. 5.15 Reflectivity and transmitivity of the fibre Bragg gratings in Fig. 5.4 (a) with symmetric periods and with two different numbers of periods (b) with unsymmetric periods as a function of the wavelength (R. Pregla, 'Analysis of gratings with symmetrical and unsymmetrical periods', in *ICTON Conf.* © 2004 Institute of Electrical and Electronics Engineers (IEEE))

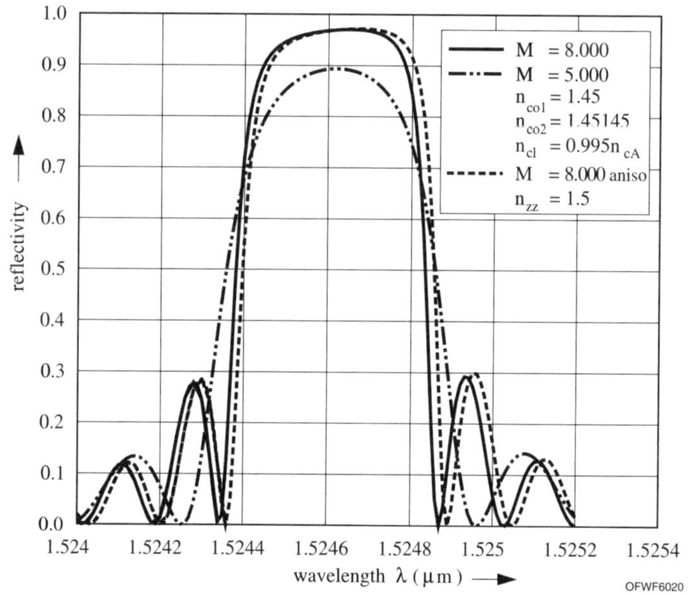

Fig. 5.16 Reflectivity as a function of wavelength of the fibre grating with anisotropic material (R. Pregla, 'Modeling of Optical Waveguide Structures with General Anisotropy in Arbitrary Orthogonal Coordinate Systems', in *IEEE J. of Sel. Topics in Quantum Electronics.* © 2002 Institute of Electrical and Electronics Engineers (IEEE))

strips (or the outer half periods) of the meander lines were used as impedance transformators. Therefore, their width w_2 and the distance d_1 to the next strips were changed (see Fig. 5.19).

Fig. 5.20a shows the measured values of the scattering parameters in dB (with Hewlett Packard 8720D network analyser) against the numerical results. As we can see, the measured and the calculated results are in very good agreement. Because a lossless structure was assumed in the analysis, the transmission coefficient $|S_{21}|$ can be computed from the relation $|S_{11}|^2 + |S_{21}|^2 = 1$. However, to check the accuracy of the algorithm, we also calculated S_{21} by transforming the fields from the input to the output. The difference between the theoretical (see above formula) and numerical results is approximately 10^{-14}, which shows the high accuracy of the proposed approach. The measured and the calculated phase of the whole periodic structure are also in very good agreement (Fig. 5.20b).

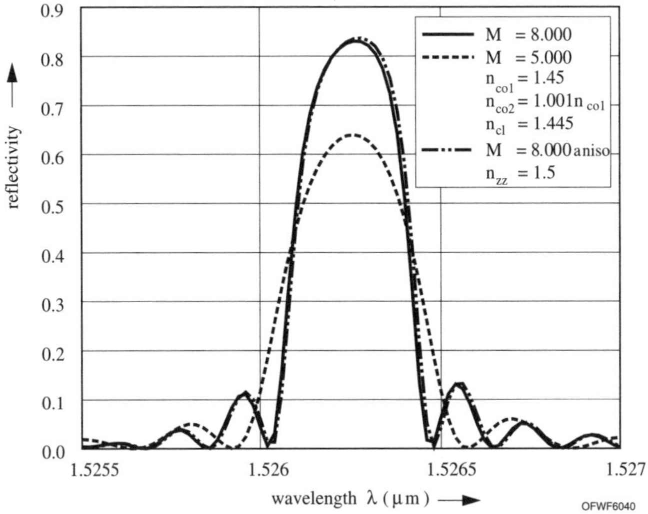

Fig. 5.17 Reflectivity as a function of a wavelength of a fibre grating with anisotropic material

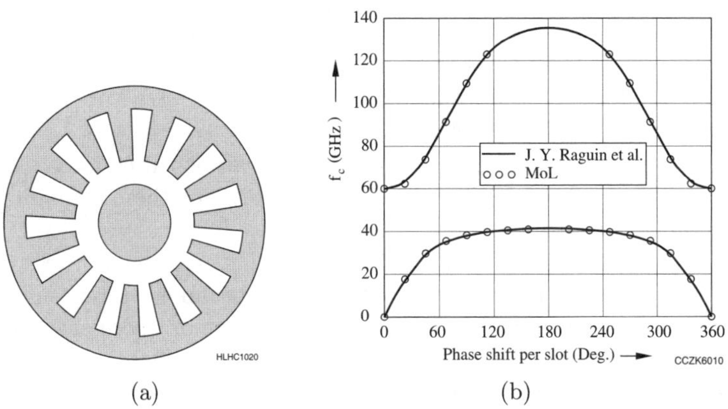

Fig. 5.18 (a) magnetron resonator (b) dispersion diagram associated with TE_{l10} and TE_{l20} modes (L. Greda and R. Pregla, 'MoL – Analysis of Periodic Structures', in *IEEE MTT-S Int. Symp. Dig.* © 2003 Institute of Electrical and Electronics Engineers (IEEE))

5.5 ANALYSIS OF PHOTONIC CRYSTALS

5.5.1 Determination of band diagrams

Waveguide devices can be made of photonic crystals for frequency bands in which the dispersion diagram shows a stop band. For the simple example in Fig. 5.8 we obtain a stop band between the values θ_{c1} and θ_{c2} or between the

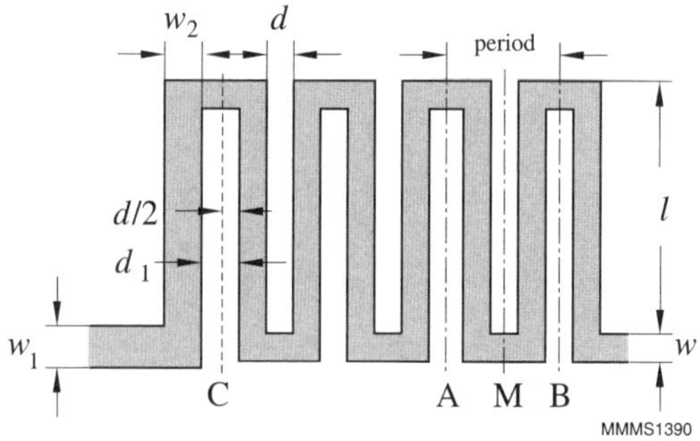

Fig. 5.19 Input part of the analysed meander line with dimensions: $w_1 = 0.6$ mm, $w_2 = 0.4$ mm, $w = 0.4$ mm, $d_1 = 0.6$ mm, $d = 0.4$ mm, $l = 6.4$ mm. Substrate: $\varepsilon_r = 10.2$, thickness: $h = 0.635$ mm (L. Greda and R. Pregla, 'MoL – Analysis of Periodic Structures', in *IEEE MTT-S Int. Symp. Dig.* © 2003 Institute of Electrical and Electronics Engineers (IEEE))

frequencies $f_{c1} = f_o - \Delta f$ and $f_{c2} = f_o + \Delta f$. In the diagram shown in Fig. 5.9 we showed only one solution of eq. (5.6). However, due to the symmetry of the trigonometric functions, we obtain the same value ψ_1 not only for θ_1 but for $\pi - \theta_1$. Therefore, the curve in Fig. 5.9 has been extended symmetrically to $\theta = \pi/2$. (Note that we show θ as a function of ψ in band diagrams.) The band diagram for the simple one-dimensional periodic structure in Fig. 5.8 is given in Fig. 5.21. With $\Theta = k_o\Lambda/2$, we may write $\Theta/\pi = \Lambda/\lambda$.

In photonic crystals we have a periodicity in various directions (e.g. the x- and y-direction in Fig. 5.6). Waves can propagate along these main directions, but also under arbitrary angles.

To identify the band gaps we must in principle examine all these possible angles. However, due to the periodicity and the symmetry, it is sufficient to examine only the so-called irreducible Brillouin zone or, more accurately, the edges of this irreducible Brillouin zone (for more details see e.g. [11] or the first chapters of a book about solid state physics like [12]). As an example, Fig. 5.22 shows the irreducible Brillouin zone of the square lattice presented in Fig. 5.6.

Let us examine the direction of propagation on the edge of the triangle. Between points Γ and X we have $k_{x_2} = 0$. This describes a wave propagating in x_1-direction with changing propagation constant k_{x_1}. In the X point we have $k_{x_1}a = \pi$. On the line between X- and M-direction we have propagation in x_2-direction with a varying wave number k_{x_2}. For k_{x_1} we always have π/a. At point M we have $k_{x_2} = k_{x_1} = \pi/a$. Finally, for the connection between Γ and M we have $k_{x_1} = k_{x_2}$. Therefore, this describes a propagation in diagonal direction. As mentioned before, at the M point we have $k_{x_1} = k_{x_2} = \pi/a$.

To compute the band diagrams in Γ–X-, resp. X–M-direction we use elementary cells of size $a \times a$. Two examples are shown on the left side of

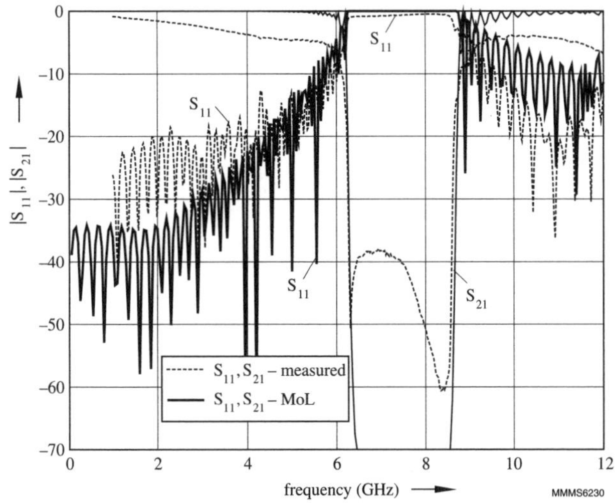

(a)

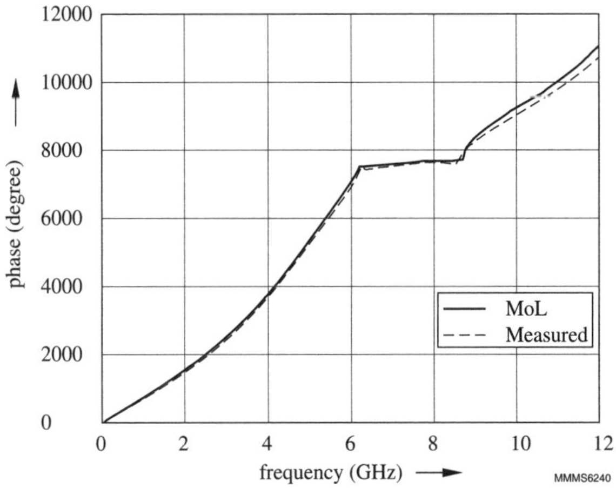

(b)

Fig. 5.20 Meander line in Fig. 5.19 with 23 periods: (a) absolute values of the scattering parameters (b) total phase in degree (L. Greda and R. Pregla, 'MoL – Analysis of Periodic Structures', in *IEEE MTT-S Int. Symp. Dig.* © 2003 Institute of Electrical and Electronics Engineers (IEEE))

Fig. 5.23. For the Γ–X part of the diagram we discretise in vertical direction lines (lines in x_1-direction). On the top and bottom of this cell, periodic boundary conditions (PBC) are introduced into the difference operators (see Section A.2.9). Since the phase factor β_x (which is equal to k_{x_2}) is equal to zero, fields on the top and bottom are identical. We vary the frequency and determine the Floquet modes (as described in Section 5.3.2.4) and their phase

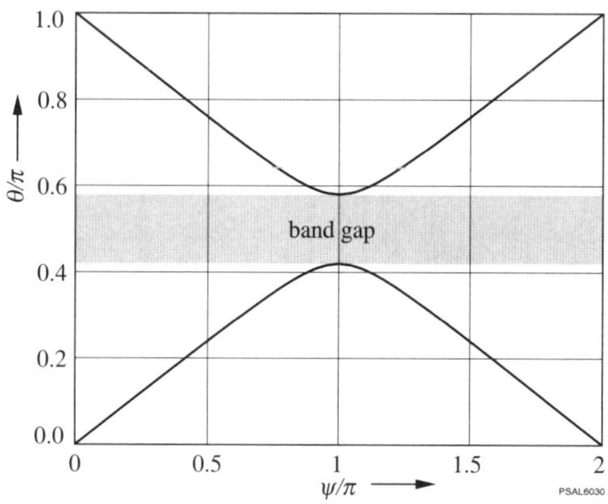

Fig. 5.21 Band diagram of the one-dimensional periodic structure in Fig. 5.8

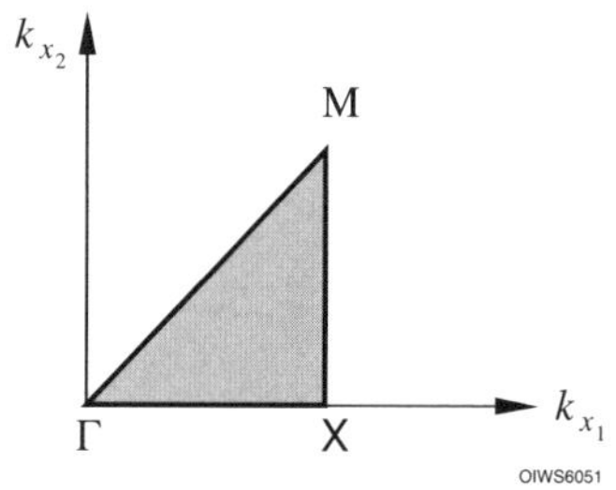

Fig. 5.22 Irreducible Brillouin zone of the square lattice in Fig. 5.6

factor Γ_F. This is different to other methods for computing the band diagrams, where the frequency is determined as a function of both propagation constants k_{x_1} and k_{x_2}.

When determining the X–M part of the diagram we proceed in a similar way. We discretise in the horizontal direction and use periodic boundary conditions on the left and on the right. We have to introduce the condition $k_{x_1}a = \pi$. Therefore, the fields on the left and right are the negative of each other. The further steps are identical to those for the Γ–X-direction. Due to the symmetry of the elementary cell, these two calculations only differ in the choice of the periodic boundaries.

To compute the bands for the Γ–X-direction, we could use the same cell as before. However, an iteration is required to enforce the condition $k_{x_1} = k_{x_2}$ in this case. Therefore, we choose the elementary cell shown on the right side of Fig. 5.23. The further steps occur analogously to those for the Γ–X band. We discretise in ξ_2-direction and introduce the condition $k_{\xi_2} = 0$ into

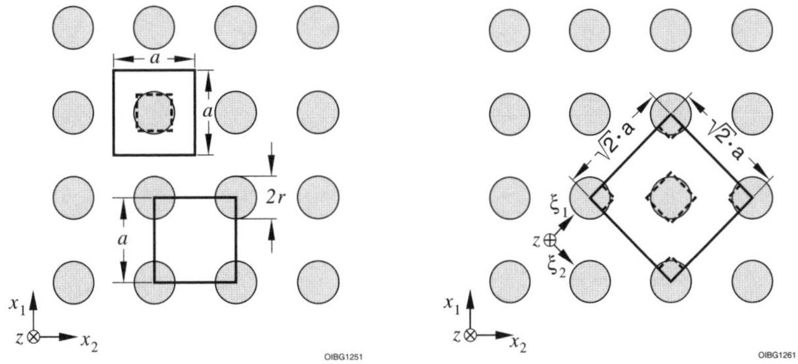

Fig. 5.23 Cells for determining the dispersion diagrams for the different conditions of the irreducible Brillouin zone

the finite difference operator. Then we determine the Floquet modes for the ξ_1-direction.

The band diagram for the 2D square lattice and the TM modes is given in Fig. 5.24. We follow the labelling used by Joannopoulos [11] here. In this case, the electric field is parallel to the rods. We have approximated the round dielectric rods with squares of the same area. Nevertheless, our results agree very well with those in [11]. Only at the M point is there a small step in the highest curve. This is because the squares as approximations of the circles are rotated by 45 degrees (see Fig. 5.23). More accurate approximations of the round rods can be obtained by staircase approximations (simply by using a cross) or by discretisation lines of varying lengths [5].

Fig. 5.25 shows examples of PCs with hexagonal lattice. In Fig. 5.25a we again have dielectric rods in air, whereas Fig. 5.25b shows air holes in a dielectric substrate. The first structure has band gaps for the TM modes, whereas the band gaps appear in the second one for TE modes [11]. The irreducible Brillouin zone of the hexagonal lattice is presented in Fig. 5.26. To compute the band diagram, we use rectangular cells as shown in Fig. 5.25. To compute the Γ–M band we discretise in x_2-direction and determine the Floquet modes in x_1-direction, and for the M–K band we just do this the opposite way around. For the PBCs we must introduce $k_{x_2} = 0$ (Γ–M) and $\beta a\sqrt{3} = 2\pi$ (M–K). Both conditions require that the fields on the boundaries perpendicular to the direction of propagation are identical.

For the Γ–K-direction we have to use the right rectangular, where we discretise in ζ_2-direction and determine the Floquet modes for the ξ_1-direction. Here, the fields at the boundaries in ξ_2-direction must be equal. A closer inspection shows that we therefore have the same conditions as for the M–K band and both parts are obtained in one program run.

Fig. 5.27 shows the band diagram for the hexagonal lattice shown in Fig. 5.25a for the TM modes. The dielectric rods have the diameter $2r = 0.4a$,

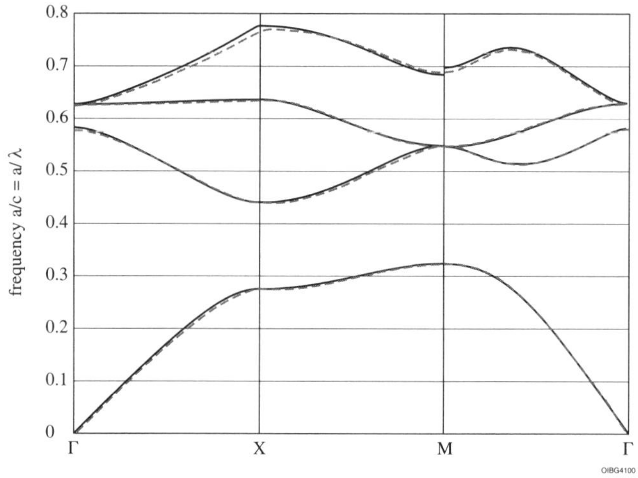

Fig. 5.24 Band diagram for the square lattice of Fig. 5.6 and TM modes: dielectric rods with $2r = 0.4a$ and $\varepsilon_r = 8.9$. - - - results from [11]

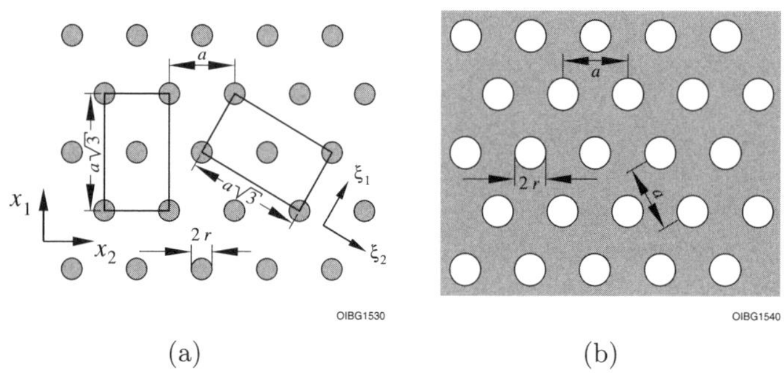

Fig. 5.25 Hexagonal lattices: (a) dielectric rods in air (b) air holes in dielectric substrate

and the permittivity is given by $\varepsilon_r = 12.0$. The results were compared with those in [13]. Again, an excellent agreement was found even though we used a quadratic approximation of the rods.

Fig. 5.28 shows the band diagram for the hexagonal lattice of Fig. 5.25b and TE modes. The air holes have the diameter $2r = 0.6a$ and the permittivity of the substrate is given by $\varepsilon_r = 12.0$. A comparison with the results of [13] shows a larger deviation than in the previous cases. This is caused by the quadratic approximation that we used for the rods. The corners produce a singularity for the E-field, whereas circular structures do not have such singularities. To improve the results we could use a finer staircase

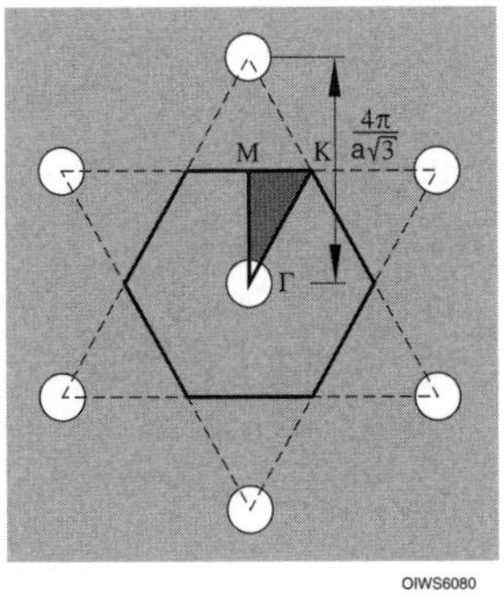

Fig. 5.26 Irreducible Brillouin zone of the hexagonal lattices sketched in the reciprocal lattice space [11]

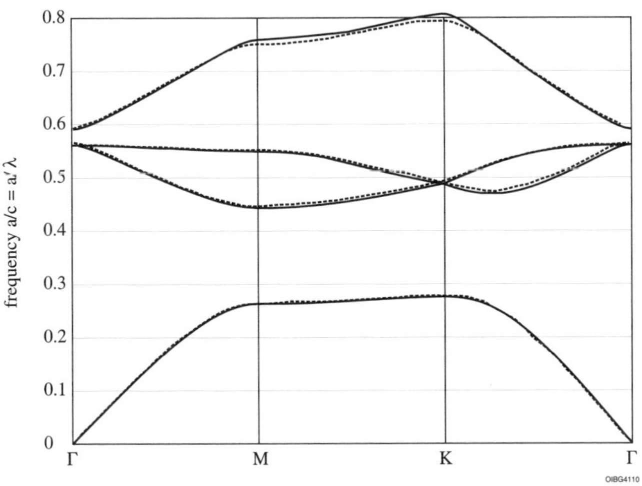

Fig. 5.27 Band diagram for the hexagonal lattice of Fig. 5.25a and TM modes: rods with $2r = 0.4a$ and $\varepsilon_r = 12.0$. - - - results from [13]

approximation or use discretisation lines of varying lengths (see Section 6.1 and [5]).

5.5.2 Waveguide circuits in photonic crystals

By introducing defects into PCs, we can design waveguides and other devices. Examples are given in Fig. 5.29. In the first row we have a straight waveguide and a 60 degree sharp bend. The second row shows a 90 degree sharp bend and a general junction. Such sharp bends cannot be realised with e.g. rib or channel waveguides. Therefore, PCs present new possibilities in miniaturising optical circuits.

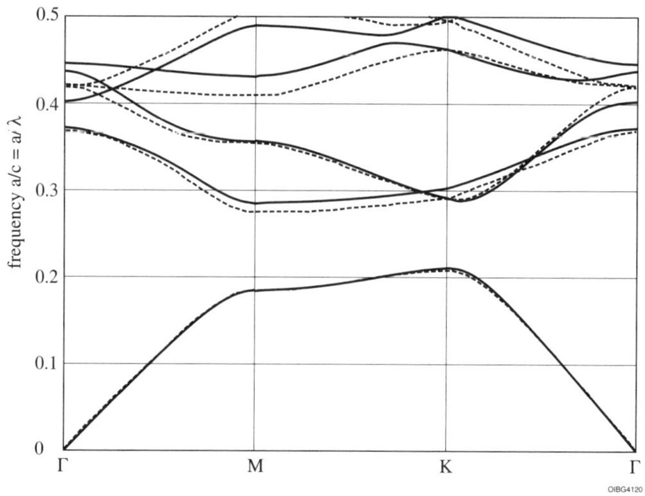

Fig. 5.28 Band diagram for the hexagonal lattice of Fig. 5.25b and TE modes: air holes with $2r = 0.6a$. Permittivity $\varepsilon_r = 12.0$. - - - results from [13]

To analyse the wave propagation in the two-dimensional structures only one-dimensional discretisation is necessary. The field decays very fast in the transverse direction. Therefore, ABCs should be introduced at the side walls of the computation window and only a very few rods on each side are required. For the three-dimensional structures, the discretisation is done as shown in the example presented in Fig. 5.30. The analysis of the wave propagation in the periodic waveguides (or waveguides in photonic crystals) is now straight forward. In the z-direction we should choose (if possible, symmetric) periods and use the algorithms described before.

General waveguide devices consist of concatenations of waveguide sections and junctions. The analysis is in principle analogous to that of planar structures (see Section 2.4.1). To analyse the junction in Fig. 5.29d, we should take into account the fact that the fields are propagating in perpendicular directions, usually in the direction of the discretisation lines. Therefore, we must use crossed discretisation lines in the junction region for the analysis.

The details are demonstrated by the example given in Fig. 5.31a. The junction region is bounded by four ports (A to D). In general, all of these four ports have connecting waveguides, in our case the waveguides $\mathrm{W_A}$ to $\mathrm{W_D}$. These connecting waveguides may be different to the waveguides inside the junction. For sharp bends (which may be considered as a special case of junction) we have only two ports with connecting waveguides (e.g. the ports A and D with $\mathrm{W_A}$ and $\mathrm{W_D}$). To derive relations for the ports, consider Fig. 5.31b, where only the junction region is sketched. Going from port A to port B, we assume K concatenated waveguide sections. This number is generally

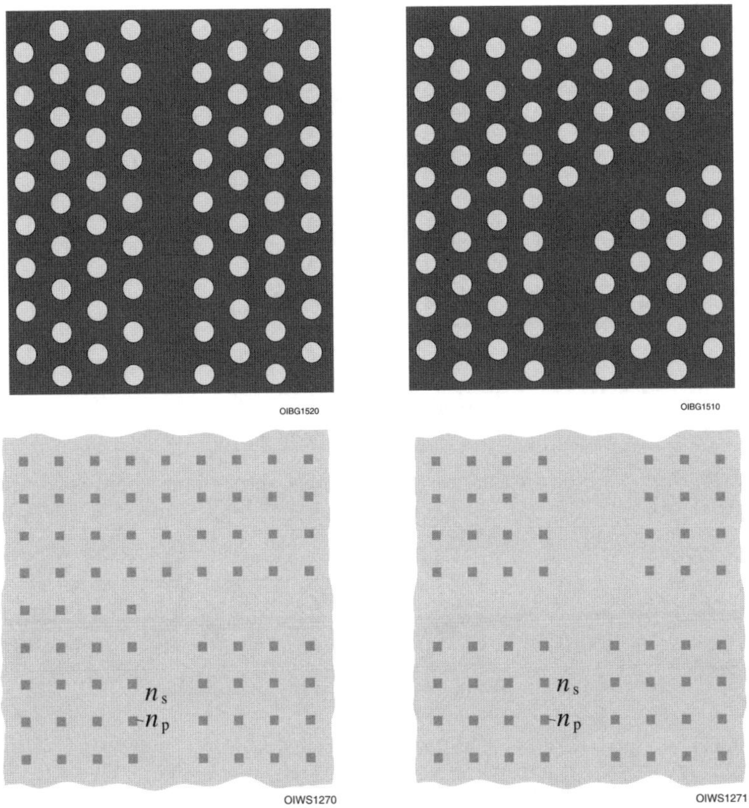

Fig. 5.29 Waveguide circuits in photonic crystals (Top: Reproduced by permission of CRC Press; bottom: R. Pregla, 'Analysis of waveguide junctions and sharp bends with general anisotropic material by using orthogonal propagating waves', in *ICTON Conf.* © 2003 Institute of Electrical and Electronics Engineers (IEEE))

much greater than that in 'normal junctions' (see e.g. Section 5.4). The two ports of the section numbered k are labelled A^k and B^k. At the side walls of the connecting waveguides we assume absorbing boundary conditions (ABC). These are placed where we would otherwise introduce Dirichlet walls. Under these conditions, we can use the formulas described in Section 6.4 or in Section 8.4 in the case of arbitrary anisotropy.

5.5.3 Numerical results for photonic crystal circuits

5.5.3.1 Waveguides

Fig. 5.32 shows the dispersion diagram of the TM mode in the waveguide of Fig. 5.33. For $d/a = 0.25$ $a = 0.6$ μm and $n_s = 3.4$ $n_0 = 1.0$ the band gap is between $\lambda = 1.26$ μm and $\lambda = 1.7$ μm.

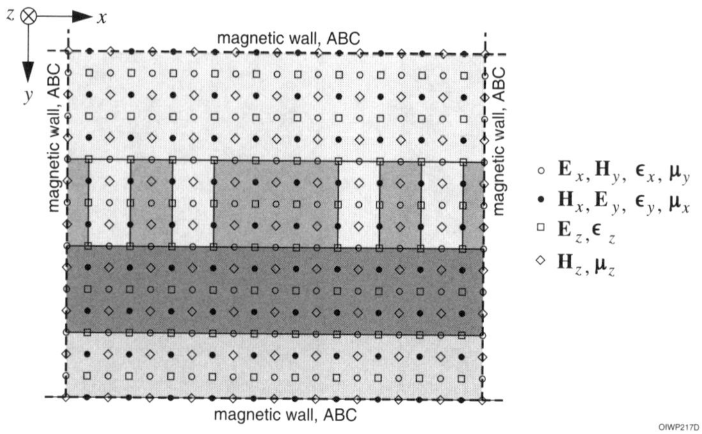

Fig. 5.30 Discretisation in the cross-section of a three-dimensional waveguide in photonic crystals

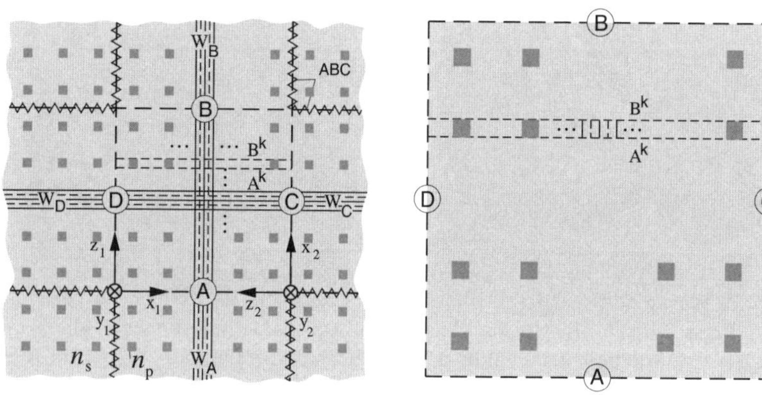

(a) (b)

Fig. 5.31 Junction of waveguides in photonic crystals (a) and inner junction (enlarged) (b) (R. Pregla, 'Analysis of waveguide junctions and sharp bends with general anisotropic material by using orthogonal propagating waves', in *ICTON Conf.* © 2003 Institute of Electrical and Electronics Engineers (IEEE))

Fig. 5.33 shows the lateral distribution of the field component E_y, determined for an infinite number of lateral periods, compared with a structure that was cut after seven rods. We can clearly see how the field decreases (alternating) to zero. As mentioned before, only very few periods are required. The waveguides were designed for the frequency region in which we have a band gap for the lateral direction. Therefore, for the phase in lateral direction, we have $\Gamma d = a + \mathrm{j}\pi$ (see Fig. 5.12). For this reason, we can introduce

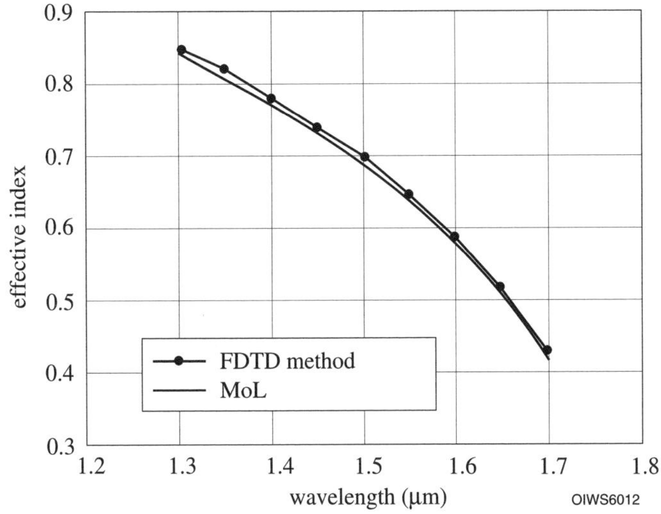

Fig. 5.32 Dispersion diagram of the TM mode in a two-dimensional photonic crystal such as in Fig. 5.7; comparison between MoL and FDTD [14] (Reproduced by permission of CRC Press)

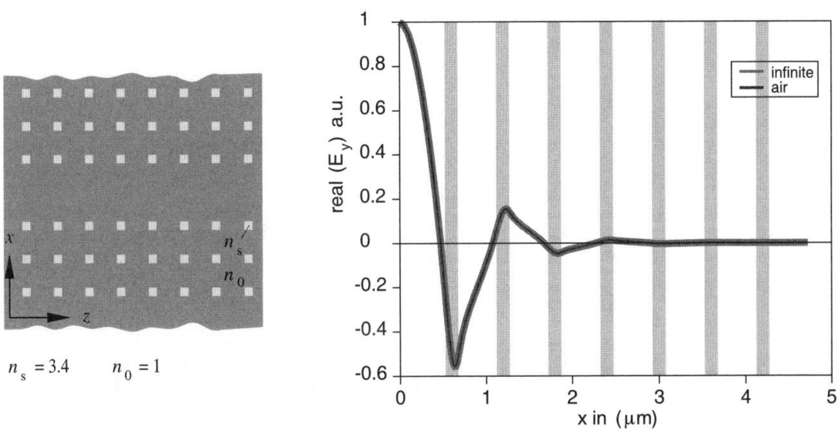

Fig. 5.33 Lateral field distribution for the TM mode in a two-dimensional photonic crystal of Fig. 5.7 with $n_0 = 1.0$ and $n_S = 3.4$

periodic conditions as shown in Fig. 5.34 into the discretisation scheme. The relation between the fields in a distance of one period can be given as:

$$F_{N+P} = F_N e^{-\Gamma} \quad F_{N+i} = F_{N-P+i} e^{-\Gamma} \quad 0 \le i \le P \quad s = e^{-\Gamma} \tag{5.90}$$

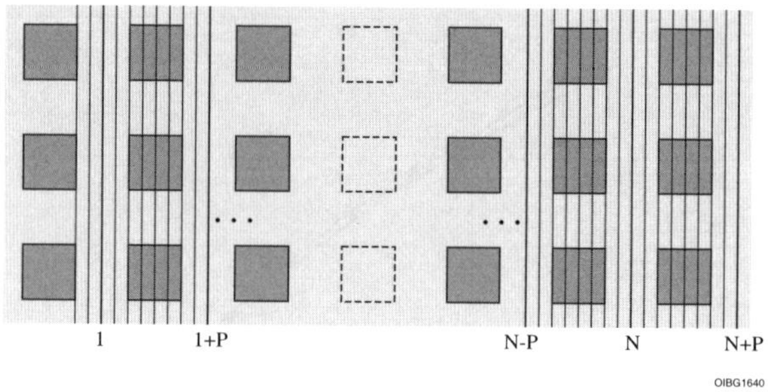

Fig. 5.34 Discretisation with periodic boundary conditions

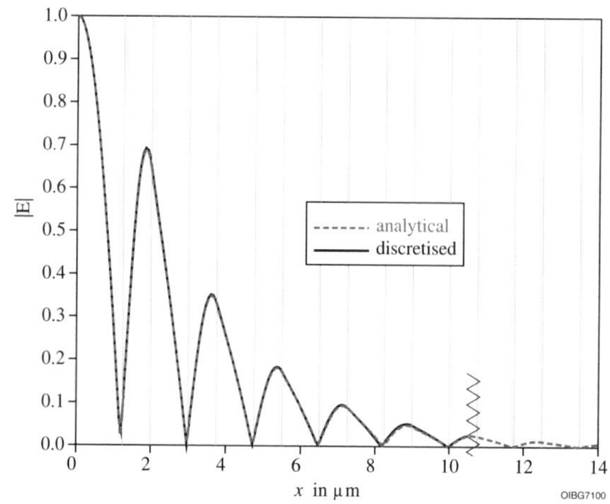

Fig. 5.35 Electric field distribution of the fundamental Floquet mode in a photonic crystal; comparison of the analytical determined field with that by discretisation and ABCs for periodic structures

This s can be introduced into the difference operator in the following way.

$$\boldsymbol{D} = \begin{matrix} -1 \\ \\ \\ \\ \\ \\ \\ \end{matrix}\begin{bmatrix} 1 & & & & \\ -1 & 1 & & & \\ & \ddots & \ddots & & \\ \cdots & \cdots & \cdots & \cdots & \cdots \\ & & \ddots & \ddots & \\ & & & -1 & 1 \\ & & & & -1 \end{bmatrix}_1 \longrightarrow \boldsymbol{D} = \begin{bmatrix} 1 & & -s & & \\ -1 & 1 & & & \\ & \ddots & \ddots & & \\ \cdots & \cdots & \cdots & \cdots & \cdots \\ & & \ddots & \ddots & \\ & & & -1 & 1 \\ & & s & & -1 \end{bmatrix} \tag{5.91}$$

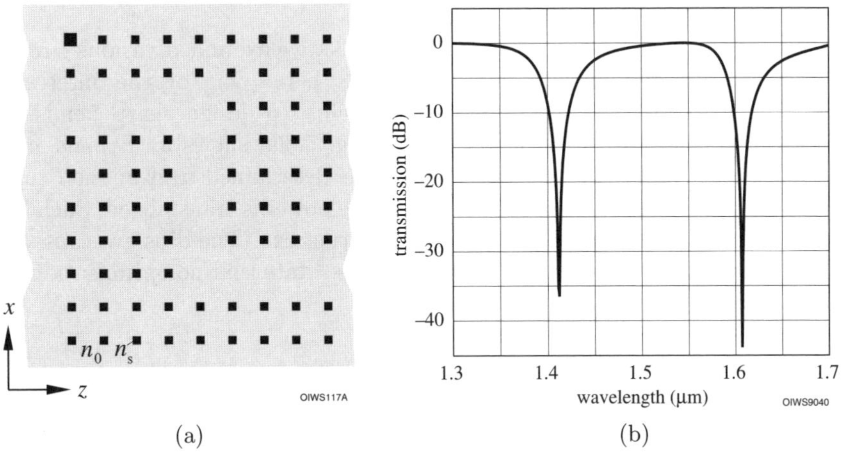

Fig. 5.36 S bend: (a) structure (b) numerical results for the transmission (Reproduced by permission of CRC Press)

We applied the derived formulas to a photonic crystal waveguide. Fig. 5.35 shows the electric field distribution of the fundamental Floquet mode. The curve labelled 'discretised' was computed with the formulas derived above with ABC. In the curve 'analytical', we used analytic expressions in the horizontal direction (see Fig. 5.34). We can recognise a good agreement between these two curves.

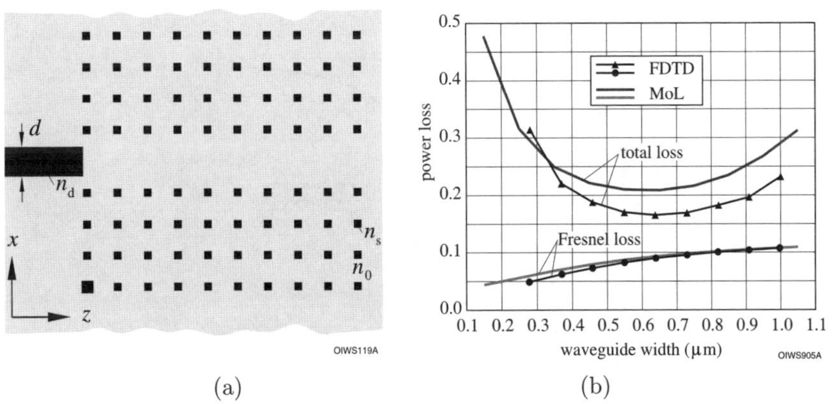

Fig. 5.37 Connection of a slab waveguide with a PC waveguide (a); loss as function of waveguide width d, comparison of the MoL with the FDTD [14], wavelength $\lambda = 1.5\ \mu m$ (b) (Reproduced by permission of CRC Press)

5.5.3.2 Numerical results for PC circuits

Based on the waveguide presented in Fig. 5.33, we examined various circuits, for which we will show numerical results here. One of the interesting applications of PCs is the potential possibility to design sharp bends and concatenations of them with low losses. Fig. 5.36a shows an S bend made of photonic crystals. In Fig. 5.36b we see the determined transmission curve. For some wavelength regions, nearly all of the power is transmitted. Such high values would not be possible with 'usual' structures. The dips are caused by reflection and not by radiation. We should also state that no optimisation was done at all.

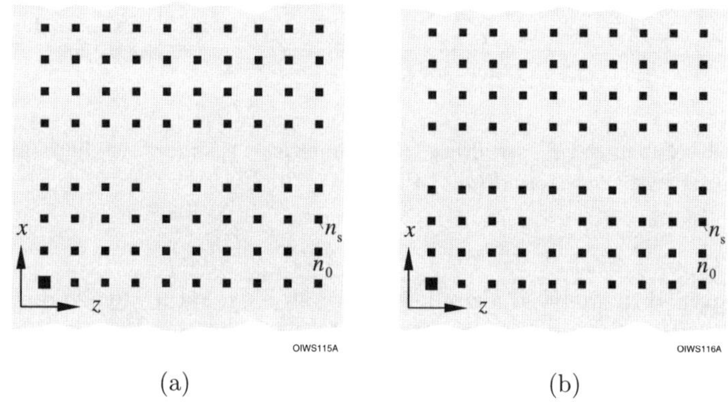

Fig. 5.38 Resonator structures with photonic crystals

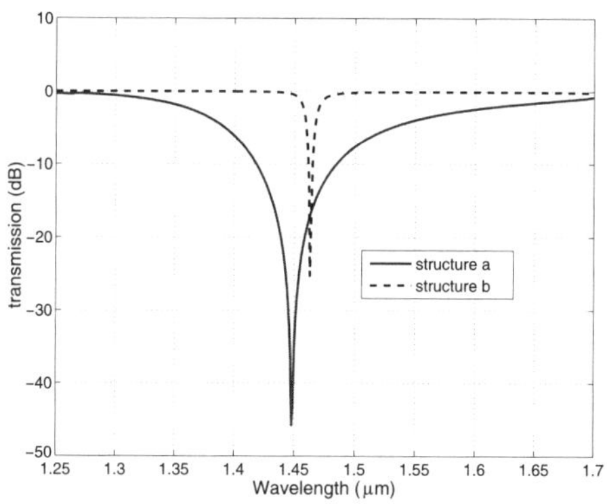

Fig. 5.39 Stop-band characteristic of the filter structure shon in Fig. 5.38

In integrated circuits we will need to connect PC waveguides to regular ones. This coupling should occur with a high efficiency. An example is shown in Fig. 5.37a. We computed the power loss as a function of the waveguide

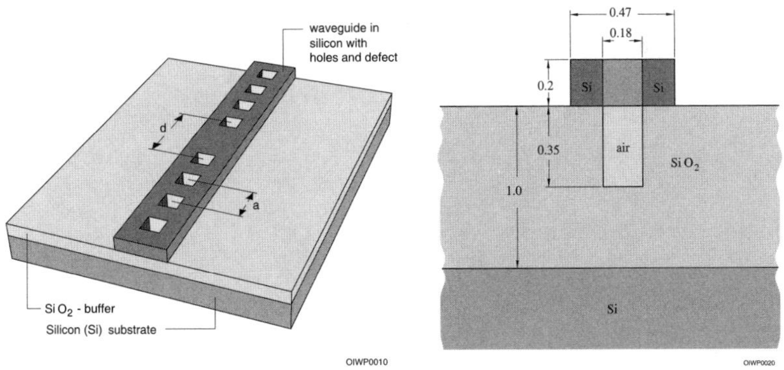

(a) (b)

Fig. 5.40 Waveguide filter with holes (a) and cross-section of the waveguide (b) (A. Barcz, S. F. Helfert and R. Pregla, 'The method of lines applied to numerical simulation of 2D and 3D bandgap structures', in *ICTON Conf.* © 2003 Institute of Electrical and Electronics Engineers (IEEE))

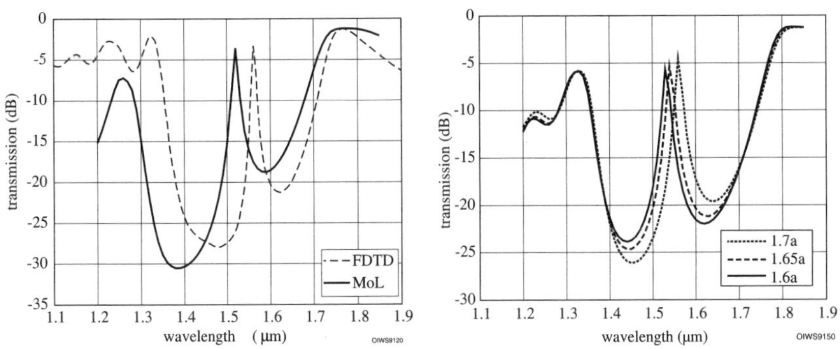

(a) (b)

Fig. 5.41 Transmission characteristic of the filter in Fig. 5.40 (a) $d = 0.45$ μm and (b) three different values of d (A. Barcz, S. F. Helfert and R. Pregla, 'The method of lines applied to numerical simulation of 2D and 3D bandgap structures', in *ICTON Conf.* © 2003 Institute of Electrical and Electronics Engineers (IEEE))

width d in Fig. 5.37b. A minimum was found for $d \approx= 0.65$ μm (wavelength $\lambda = 1.5$ μm). The same optimum was found by the FDTD [14]. However, the total loss obtained with the latter method was slightly lower.

5.5.3.3 Numerical results for filters

Resonator structures (filters) can be made very easily from PCs, e.g. simply by removing one of the rods. Two examples are shown in Fig. 5.38a and b.

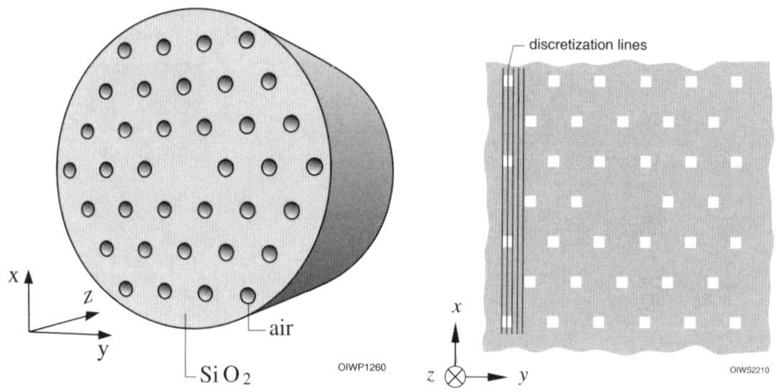

(a) (b)

Fig. 5.42 Holey fibre (a) and model with discretisation lines (b) (A. Barcz, S. F. Helfert and R. Pregla, 'The method of lines applied to numerical simulation of 2D and 3D bandgap structures', in *ICTON Conf.* © 2003 Institute of Electrical and Electronics Engineers (IEEE))

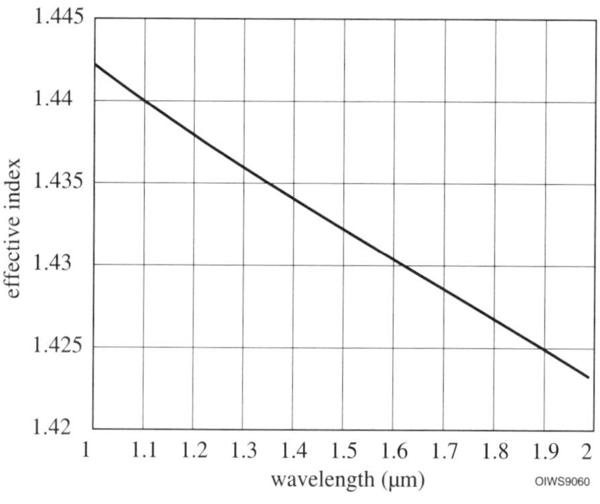

Fig. 5.43 Holey fibre – determined effective index (A. Barcz, S. F. Helfert and R. Pregla, 'The method of lines applied to numerical simulation of 2D and 3D bandgap structures', in *ICTON Conf.* © 2003 Institute of Electrical and Electronics Engineers (IEEE))

A comparison of the determined filter characteristics is presented in Fig. 5.39. As we can see, we have a sharper resonance for configuration a.

The structures we have shown so far are two-dimensional. The filter sketched in Fig. 5.40 is an example of a 3D device. The transmission characteristic computed with the MoL was compared with that of the

FDTD [15] in Fig. 5.41a. The round holes were replaced by square ones for the MoL computations, whereas a rounder structure was examined with the FDTD. This might explain the slight shift of the determined curves. Fig. 5.41b shows the influence of the distance between the holes (in the middle of the structure) on the transmission. We see a shift of the curve to the left when we increase the distance.

5.5.3.4 Numerical results for holey fibres

Another PC structure is the holey fibre plotted in Fig. 5.42a. Here the waves propagate in the direction of the holes (and not perpendicular to them as in the previous examples). For the analysis, we used the discretisation scheme shown in Fig. 5.42b. We determined the propagation constant (effective index) of the fundamental mode. The results are shown in Fig. 5.43.

References

[1] R. Pregla, 'Efficient Modeling of Periodic Structures', *AEÜ*, vol. 57, pp. 185–189, 2003.

[2] I. A. Goncharenko, S. F. Helfert and R. Pregla, 'General Analysis of Fibre Grating Structures', *J. of Optics A: Pure and Appl. Optics*, vol. 1, pp. 25–31, 1999.

[3] R. Pregla, 'Efficient and Accurate Modeling of Planar Anisotropic Microwave Structures by the Method of Lines', *IEEE Trans. Microwave Theory Tech.*, vol. 50, pp. 1469–1479, June 2002.

[4] S. F. Helfert, 'Efficient Analysis of Non Symmetric Periodic Optical Devices', in *OSA Integr. Photo. Resear. Tech. Dig.*, Victoria, Canada, Mar. 1998, vol. 4, pp. 372–374.

[5] S. F. Helfert, 'Modeling of Structures with Curved Boundaries: Applied to the Determination of Band Structures in Photonic Crystals', in *ICTON Conf.*, Wraclow, Poland, 2004, vol. 6, pp. 118–121.

[6] S. F. Helfert, 'Determination of Floquet-modes in Asymmetric Periodic Structures', *Opt. Quantum Electron.*, vol. 37, pp. 185–197, 2005, Special Issue on Optical Waveguide Theory and Numerical Modelling.

[7] S. F. Helfert and R. Pregla, 'Efficient Analysis of Periodic Structures', *J. Lightwave Technol.*, vol. 16, no. 9, pp. 1694–1702, Sep. 1998.

[8] J. Čtyroký, S. Helfert and R. Pregla, 'Analysis of a Deep Waveguide Bragg Grating', *Opt. Quantum Electron.*, vol. 30, pp. 343–358, 1998.

[9] J. L. Archambault and S. G. Grubb, 'Fiber Gratings in Lasers and Amplifiers', *J. Lightwave Technol.*, vol. 15, no. 8, pp. 1378–1390, 1997.

[10] J.-Y. Raguin, H.-G. Unger and D. M. Vavriv, 'Recent Advances in the Analysis of Space-Harmonic Millimeter-Wave Magnetrons with Secondary-Emission Cathode', in *Int. Symp. on Recent Advances in Microwave Technology*, Malaga, Spain, Dec. 1999, pp. 63–66.

[11] J. D. Joannopoulus, R. D. Meade and J. N. Winn, *Photonic Crystals–Molding the Flow of Light*, Princeton University Press, 1995.

[12] C. Kittel, *Introduction to Solid State Physics*, Wiley, New York, 6 edition, 1986.

[13] R. K. Chakrabarty 'Photonic Crystals and their Applications', Physics and Photonics, Physics 461/661. http://physics.unr.edu/faculty/bruch/Phys461/5.pdf.

[14] R. Stoffer, H. J. W. M. Hoekstra, R. M. de Ridder, E. van Groesen and F. P. H. van Beckum, 'Numerical Studies of 2D Photonic Crystals: Waveguides, Coupling Between Waveguides and Filters', *Opt. Quantum Electron.*, vol. 32, pp. 947–961, 2000, Special Issue on Optical Waveguide Theory and Numerical Modelling.

[15] B. Kuhlow and Przyrembel, Heinrich Hertz Institut Berlin, private communication.

Further Reading

[16] S. F. Helfert, 'The Method of Lines for the Calculation of Band Structures in Photonic Crystals', in *ICTON Conf.*, Warsaw, Poland, 2003, vol. 5, pp. 122–125.

[17] S. F. Helfert, 'Numerical Stable Determination of Floquet-Modes and the Application to the Computation of Band Structures', *Opt. Quantum Electron.*, vol. 36, pp. 87–107, 2004, Special Issue on Optical Waveguide Theory and Numerical Modelling.

[18] A. Barcz, S. Helfert and R. Pregla, 'Modeling of 2D Photonic Crystals by Using the Method of Lines', in *ICTON Conf.*, Warsaw, Poland, 2002, vol. 4, pp. 45–48.

[19] A. Barcz, S.F. Helfert and R. Pregla, 'The Method of Lines Applied to Numerical Simulation of 2D and 3D Bandgap Structures', in *ICTON Conf.*, Warsaw, Poland, 2003, vol. 5, pp. 126–129.

[20] A. Barcz, S.F. Helfert and R. Pregla, 'Numerical Analysis of Couplers and Novel Filters with the Method of Lines', in *ICTON Conf.*, Wraclow, Poland, 2004, vol. 6, pp. 122–125.

[21] R. Pregla and S. F. Helfert, 'The Method of Lines for the Analysis of Photonic Bandgap Structures', in *Electromagnetic Theory and Applications for Photonic Crystals*, Kiyotoshi Yasumoto (Ed.), pp. 295–350. CRC Press, Boca Raton Fl, London, 2006.

[22] R. Pregla, 'Analysis of Waveguide Junctions and Sharp Bends with General Anisotropic Material by Using Orthogonal Propagating Waves', in *ICTON Conf.*, Warsaw, Poland, 2003, vol. 5, pp. 116–121.

[23] I. A. Goncharenko, S. F. Helfert and R. Pregla, 'Radiation Loss and Mode Field Distribution in Curved Holey Fibers', *AEÜ*, vol. 59, pp. 185–191, 2005.

[24] I. A. Goncharenko, S. F. Helfert and R. Pregla, 'Analysis of Nonlinear Properties of Fibre Grating Structures', *AEÜ*, vol. 53, pp. 25–31, 1999.

CHAPTER 6

ANALYSIS OF COMPLEX STRUCTURES

In this chapter we will give an analysis of structures that are more complex than the ones presented in Chapter 2. In particular, we will examine structures in which the thickness of the layers is variable, as well as sharp bends and junctions. The first two problems can be solved with lines of varying length. For the junction problem we start with structures that are analysed with lines in one direction and continue with devices for which we use crossed discretisation lines.

6.1 LAYERS OF VARIABLE THICKNESS

6.1.1 Introduction

In an ideal case, the layers in multilayered waveguide structures for microwave and millimetre waves and for optical devices would have a constant thickness. However, in planar waveguides for optical frequencies especially, the rib may have a varying thickness. Examples are shown in Fig. 6.1. Fig. 6.2a shows a quasi-planar waveguide in which the central microstrip is on a dielectric layer of varying thickness above the main substrate with the side metallisations. In this section we will show an algorithm for analysing multilayered waveguides with variable layer thicknesses, as in Fig. 6.18b.

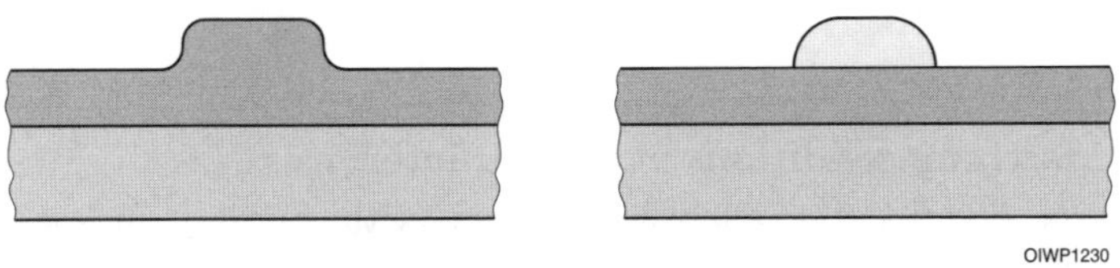

Fig. 6.1 Cross-sections of a multilayered rib guide and quasi-planar waveguide (Reproduced by permission of Springer Netherlands)

In general, the substrates may be multilayered and the materials may have biaxial or specific anisotropic properties. Based on Fig. 6.3, the following principles are important:

- Each curved interface between layers has to be modelled by two layers.
- Each of this layers must have one straight boundary (parallel to the boundaries of the 'normal' layers).

Analysis of Electromagnetic Fields and Waves R. Pregla

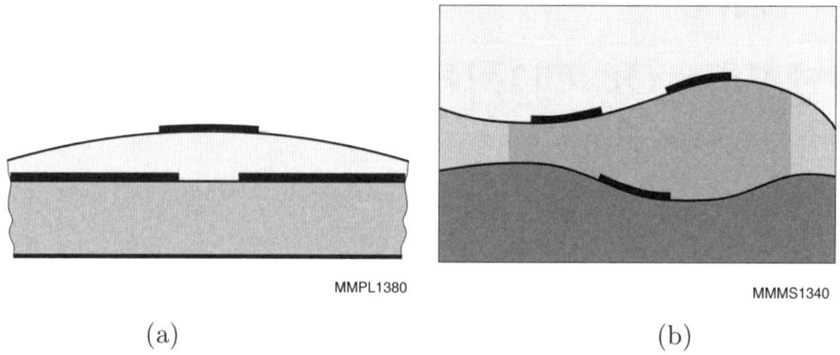

Fig. 6.2 Cross-sections of multilayered quasi-planar waveguides with one (a) or more (b) layers of variable thickness

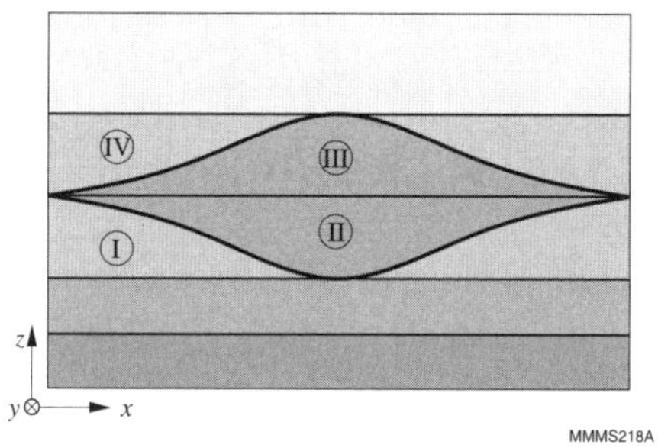

Fig. 6.3 Cross-sections with a layer of variable thickness

- To obtain a numerically stable algorithm, these layers should be as thin as possible.

6.1.2 Matching conditions at curved interfaces

Let us now describe the analysis of layers with varying thicknesses (see Fig. 6.4). We will use the algorithms developed earlier and discretisation lines of different lengths.

At each interface, the tangential fields have to be continuous. Therefore, matching the fields at a curved interface (the layers are labelled I and II) results in the following expression:

$$\begin{bmatrix} -\mathrm{j}\mathbf{E}_y^{\mathrm{I}} \\ \mathbf{E}_x^{\mathrm{I}} + \boldsymbol{d}_h' \boldsymbol{M}_e \mathbf{E}_z^{\mathrm{I}} \end{bmatrix} = \begin{bmatrix} -\mathrm{j}\mathbf{E}_y^{\mathrm{II}} \\ \mathbf{E}_x^{\mathrm{II}} + \boldsymbol{d}_h' \boldsymbol{M}_e \mathbf{E}_z^{\mathrm{II}} \end{bmatrix} \tag{6.1}$$

$$\begin{bmatrix} \mathbf{H}_x^{\mathrm{I}} + \boldsymbol{d}_e' \boldsymbol{M}_h \mathbf{H}_z^{\mathrm{I}} \\ -\mathrm{j}\mathbf{H}_y^{\mathrm{I}} \end{bmatrix} = \begin{bmatrix} \mathbf{H}_x^{\mathrm{II}} + \boldsymbol{d}_e' \boldsymbol{M}_h \mathbf{H}_z^{\mathrm{II}} \\ -\mathrm{j}\mathbf{H}_y^{\mathrm{II}} \end{bmatrix} \tag{6.2}$$

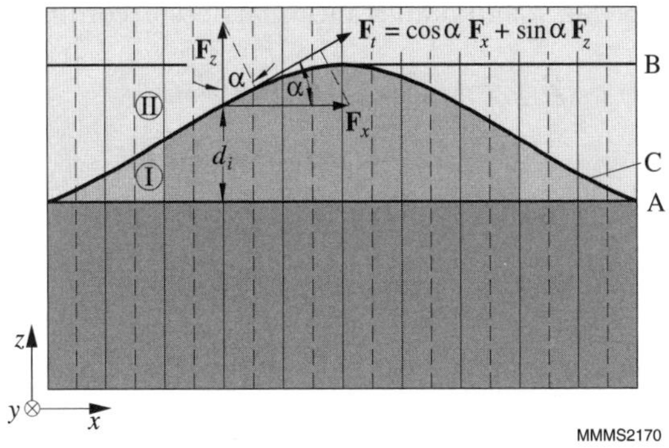

e ——— $E_y\ E_z\ H_x$ $\varepsilon_{yy}\ \varepsilon_{xy}\ \varepsilon_{zz}\ \mu_{xx}\ \mu_{yx}$

h – – – $H_y\ H_z\ E_x$ $\varepsilon_{xx}\ \varepsilon_{yx}\ \mu_{xy}\ \mu_{yy}\ \mu_{zz}$

Fig. 6.4 Interface conditions at curved interfaces (Reproduced by permission of Springer Netherlands)

The transverse components consist of the y component and a second one made of both the x and z components (F_t see Fig. 6.4). We divide the equations containing the x components by $\cos \boldsymbol{\alpha}_{e,h}$ and then use the abbreviation $\tan(\boldsymbol{\alpha}_{e,h})$ by $\mathbf{d}'_{e,h}$. If the variable thickness d_i is known in analytical form, we may derive $\tan \boldsymbol{\alpha}_{e,h}$ from $\tan \alpha = \mathrm{d}d_i/\mathrm{d}x$ and discretise this function on the two-line systems. The values have to be collected in diagonal matrices. However, we can also obtain the differential quotient numerically. The values d_i are determined at the discretisation points x_i and we can obtain an approximation of the derivative with the help of the difference matrices $\overline{\boldsymbol{D}}_{e,h}$. Therefore, we may also write for $\mathbf{d}'_e$ and $\mathbf{d}'_h$:

$$\mathbf{d}'_e = \mathrm{diag}(\overline{\boldsymbol{D}}_h \overline{\mathbf{d}}_h) \qquad \mathbf{d}'_h = \mathrm{diag}(\overline{\boldsymbol{D}}_e \overline{\mathbf{d}}_e) \tag{6.3}$$

Furthermore, we introduce interpolation matrices $\boldsymbol{M}_{e,h}$ between the discretisation line systems, because the z components are not discretised on the same line system as the x components. By introducing the z components obtained from eq. (2.57), we obtain the following matching relation:

$$\widehat{\boldsymbol{P}}_{\mathrm{C}}^{\mathrm{I}} \begin{bmatrix} \widehat{\mathbf{E}}_{\mathrm{C}}^{\mathrm{I}} \\ \widehat{\mathbf{H}}_{\mathrm{C}}^{\mathrm{I}} \end{bmatrix} = \widehat{\boldsymbol{P}}_{\mathrm{C}}^{\mathrm{II}} \begin{bmatrix} \widehat{\mathbf{E}}_{\mathrm{C}}^{\mathrm{II}} \\ \widehat{\mathbf{H}}_{\mathrm{C}}^{\mathrm{II}} \end{bmatrix} \tag{6.4}$$

with:

$$\widehat{\boldsymbol{P}}_{\mathrm{C}}^{\mathrm{I,II}} = \begin{bmatrix} \boldsymbol{I}_e & \boldsymbol{0} & \boldsymbol{0} & \boldsymbol{0} \\ \boldsymbol{0} & \boldsymbol{I}_h & \mathbf{Z}_{xx} & \mathbf{Z}_{xy} \\ -\mathbf{Y}_{xy} & -\mathbf{Y}_{xx} & \boldsymbol{I}_e & \boldsymbol{0} \\ \boldsymbol{0} & \boldsymbol{0} & \boldsymbol{0} & \boldsymbol{I}_h \end{bmatrix} \tag{6.5}$$

We define impedance and admittance matrices according to:

$$\begin{aligned}
\widehat{\boldsymbol{Z}}_{xx}^{\mathrm{I,II}} &= \sqrt{\varepsilon_{re}}\,\mathrm{diag}(\overline{\boldsymbol{D}}_e\overline{\boldsymbol{d}}_e)\boldsymbol{M}_e\boldsymbol{\epsilon}_{zz}^{\mathrm{I,II}-1}\\
\widehat{\boldsymbol{Z}}_{xy}^{\mathrm{I,II}} &= \mathrm{diag}(\overline{\boldsymbol{D}}_e\overline{\boldsymbol{d}}_e)\boldsymbol{M}_e\boldsymbol{\epsilon}_{zz}^{\mathrm{I,II}-1}\overline{\boldsymbol{D}}_h\\
\widehat{\mathbf{Y}}_{xy}^{\mathrm{I,II}} &= \mathrm{diag}(\overline{\boldsymbol{D}}_h\overline{\boldsymbol{d}}_h)\boldsymbol{M}_h\boldsymbol{\mu}_{zz}^{\mathrm{I,II}-1}\overline{\boldsymbol{D}}_e\\
\widehat{\mathbf{Y}}_{xx}^{\mathrm{I,II}} &= \sqrt{\varepsilon_{re}}\,\mathrm{diag}(\overline{\boldsymbol{D}}_h\overline{\boldsymbol{d}}_h)\boldsymbol{M}_h\boldsymbol{\mu}_{zz}^{\mathrm{I,II}-1}
\end{aligned} \tag{6.6}$$

Therefore, the relation between both sides of interface C (see Fig. 6.4) is given by:

$$\begin{bmatrix}\widehat{\mathbf{E}}_{\mathrm{C}}^{\mathrm{I}}\\ \widehat{\mathbf{H}}_{\mathrm{C}}^{\mathrm{I}}\end{bmatrix} = (\widehat{\boldsymbol{P}}_{\mathrm{C}}^{\mathrm{I}})^{-1}\widehat{\boldsymbol{P}}_{\mathrm{C}}^{\mathrm{II}}\begin{bmatrix}\widehat{\mathbf{E}}_{\mathrm{C}}^{\mathrm{II}}\\ \widehat{\mathbf{H}}_{\mathrm{C}}^{\mathrm{II}}\end{bmatrix} = \begin{bmatrix}\widehat{\mathbf{V}}_{\mathrm{E}}^{\mathrm{C}} & \widehat{\mathbf{Z}}_{\mathrm{E}}^{\mathrm{C}}\\ \widehat{\mathbf{Y}}_{\mathrm{H}}^{\mathrm{C}} & \widehat{\mathbf{V}}_{\mathrm{H}}^{\mathrm{C}}\end{bmatrix}\begin{bmatrix}\widehat{\mathbf{E}}_{\mathrm{C}}^{\mathrm{II}}\\ \widehat{\mathbf{H}}_{\mathrm{C}}^{\mathrm{II}}\end{bmatrix} \tag{6.7}$$

By defining impedances on both sides of interface C by $\widehat{\mathbf{E}}_{\mathrm{C}}^{\mathrm{I,II}} = \widehat{\mathbf{Z}}_{\mathrm{C}}^{\mathrm{I,II}}\widehat{\mathbf{H}}_{\mathrm{C}}^{\mathrm{I,II}}$, the impedance transformation from side II to side I yields:

$$\widehat{\mathbf{Z}}_{\mathrm{C}}^{\mathrm{I}} = (\widehat{\mathbf{V}}_{\mathrm{E}}^{\mathrm{C}}\widehat{\mathbf{Z}}_{\mathrm{C}}^{\mathrm{II}})(\widehat{\mathbf{Y}}_{\mathrm{H}}^{\mathrm{C}}\widehat{\mathbf{Z}}_{\mathrm{C}}^{\mathrm{II}} + \widehat{\mathbf{V}}_{\mathrm{H}}^{\mathrm{C}})^{-1} \tag{6.8}$$

With the help of the transfer matrix relations in e.g. eq. (2.141), we can write the original fields in interface C (see Fig. 6.4) as functions of the transformed fields in either plane $\mathrm{B} \equiv \mathrm{B}^{\mathrm{I}}$ or $\mathrm{A} \equiv \mathrm{A}^{\mathrm{I}}$:

$$\begin{bmatrix}\widehat{\mathbf{E}}_{\mathrm{C}}^{\mathrm{II}}\\ \widehat{\mathbf{H}}_{\mathrm{C}}^{\mathrm{II}}\end{bmatrix} = \begin{bmatrix}\widehat{\mathbf{V}}_{\mathrm{E}} & \widehat{\mathbf{Z}}_{\mathrm{E}}\\ \widehat{\mathbf{Y}}_{\mathrm{H}} & \widehat{\mathbf{V}}_{\mathrm{H}}\end{bmatrix}\begin{bmatrix}\widehat{\overline{\mathbf{E}}}_{\mathrm{B}}\\ \widehat{\overline{\mathbf{H}}}_{\mathrm{B}}\end{bmatrix} \qquad \begin{bmatrix}\widehat{\mathbf{E}}_{\mathrm{C}}^{\mathrm{I}}\\ \widehat{\mathbf{H}}_{\mathrm{C}}^{\mathrm{I}}\end{bmatrix} = \begin{bmatrix}\widehat{\mathbf{V}}_{\mathrm{E}} & -\widehat{\mathbf{Z}}_{\mathrm{E}}\\ -\widehat{\mathbf{Y}}_{\mathrm{H}} & \widehat{\mathbf{V}}_{\mathrm{H}}\end{bmatrix}\begin{bmatrix}\widehat{\overline{\mathbf{E}}}_{\mathrm{A}}\\ \widehat{\overline{\mathbf{H}}}_{\mathrm{A}}\end{bmatrix} \tag{6.9}$$

Due to the different line lengths, we have:

$$\begin{aligned}
\widehat{\mathbf{V}}_{\mathrm{E,H}} &= \widehat{\boldsymbol{T}}_{\mathrm{E,H}} \bullet \cosh(\widehat{\overline{\mathbf{d}}}\boldsymbol{\Gamma}^{\mathrm{r}})\\
\widehat{\mathbf{Z}}_{\mathrm{E}} &= \widehat{\boldsymbol{T}}_{\mathrm{E}} \bullet \sinh(\widehat{\overline{\mathbf{d}}}\boldsymbol{\Gamma}^{\mathrm{r}})\\
\widehat{\mathbf{Y}}_{\mathrm{H}} &= \widehat{\boldsymbol{T}}_{\mathrm{H}} \bullet \sinh(\widehat{\overline{\mathbf{d}}}\boldsymbol{\Gamma}^{\mathrm{r}})\\
\widehat{\overline{\mathbf{d}}} &= [\overline{\boldsymbol{d}}_e^{\mathrm{t}}, \overline{\boldsymbol{d}}_h^{\mathrm{t}}]^{\mathrm{t}}
\end{aligned} \tag{6.10}$$

The multiplication sign '•' means that the matrices have to be multiplied element by element (array multiplication). All $\overline{d}_i$ (see Fig. 6.4) are now collected in the column vector $\overline{\mathbf{d}}$. $\boldsymbol{\Gamma}^{\mathrm{r}}$ is a row vector containing the diagonal elements of $\boldsymbol{\Gamma}$. Therefore, the product $\widehat{\overline{\mathbf{d}}}\boldsymbol{\Gamma}^{\mathrm{r}}$ results in a full matrix. To obtain numerically more stable impedance transformation formulas, we determine the z-matrix parameters by:

$$\begin{aligned}
\widetilde{\mathbf{z}}_{11}^{\mathrm{II}} &= \widetilde{\mathbf{z}}_{22}^{\mathrm{I}} = \widehat{\mathbf{V}}_{\mathrm{E}}\widehat{\mathbf{Y}}_{\mathrm{H}}^{-1}\\
\widetilde{\mathbf{z}}_{12}^{\mathrm{II}} &= \widetilde{\mathbf{z}}_{21}^{\mathrm{I}} = \widehat{\mathbf{V}}_{\mathrm{E}}\widehat{\mathbf{Y}}_{\mathrm{H}}^{-1}\widehat{\mathbf{V}}_{\mathrm{H}} - \mathbf{Z}_{\mathrm{E}}\\
\widetilde{\mathbf{z}}_{22}^{\mathrm{II}} &= \widetilde{\mathbf{z}}_{11}^{\mathrm{I}} = \widehat{\mathbf{Y}}_{\mathrm{H}}^{-1}\widehat{\mathbf{V}}_{\mathrm{H}}\\
\widetilde{\mathbf{z}}_{21}^{\mathrm{II}} &= \widetilde{\mathbf{z}}_{12}^{\mathrm{I}} = \widehat{\mathbf{Y}}_{\mathrm{H}}^{-1}
\end{aligned} \tag{6.11}$$

The tilde in these parameters indicates that they connect fields in the transformed domain on one side of the section and fields in the original domain on the other. With these matrices, the input impedances in interface C^{II} of section II and plane A of section I can be calculated by:

$$\widehat{\mathbf{Z}}_{\mathrm{C}}^{\mathrm{II}} = \widetilde{\mathbf{z}}_{11}^{\mathrm{II}} - \widetilde{\mathbf{z}}_{12}^{\mathrm{II}}(\widetilde{\mathbf{z}}_{22}^{\mathrm{II}} + \overline{\mathbf{Z}}_{\mathrm{B}}^{\mathrm{II}})^{-1}\widetilde{\mathbf{z}}_{21}^{\mathrm{II}} \tag{6.12}$$

$$\widehat{\mathbf{Z}}_{\mathrm{A}}^{\mathrm{I}} = \widetilde{\mathbf{z}}_{11}^{\mathrm{I}} - \widetilde{\mathbf{z}}_{12}^{\mathrm{I}}(\widetilde{\mathbf{z}}_{22}^{\mathrm{I}} + \widehat{\mathbf{Z}}_{\mathrm{C}}^{\mathrm{II}})^{-1}\widetilde{\mathbf{z}}_{21}^{\mathrm{I}} \tag{6.13}$$

In eq. (6.13) we introduced the impedance $\overline{\mathbf{Z}}_{\mathrm{B}}^{\mathrm{II}}$ on the upper side of layer II as the load impedance at port B. With impedance $\widehat{\mathbf{Z}}_{\mathrm{A}}^{\mathrm{I}}$, the impedance transformation through the layer below the plane A can be performed.

6.2 MICROSTRIP SHARP BEND

In this section we will analyse sharp microstrip bends (see Fig. 6.5). Again we will use discretisation lines of different lengths. Since the propagation direction changes from z to z', we will not consider the anisotropy given in eq. (2.5) but will restrict ourselves to the case in which the components in the off-diagonal are zero and in which the relation $\nu_{xx} = \nu_{zz}$ holds. However, ν_{yy} can be different from these values.

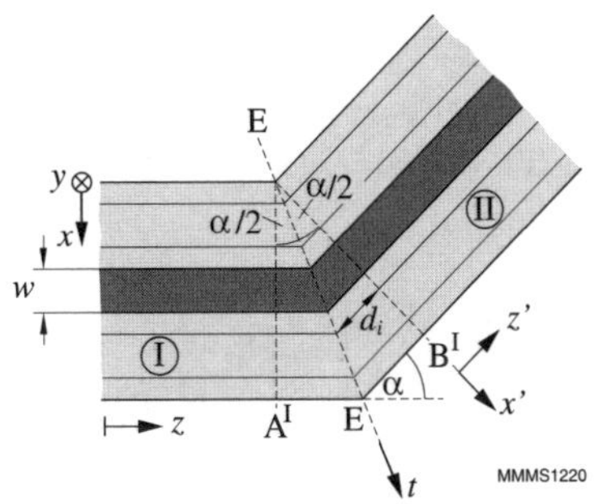

Fig. 6.5 Sharp microstrip bend (R. Pregla, 'Efficient and Accurate Modeling of Planar Anisotropic Microwave Structures by the Method of Lines', *IEEE Trans. Microwave Theory Tech.*, vol. 50, pp. 1469–1479. © 2002 Institute of Electrical and Electronics Engineers (IEEE))

As before, we must match the transverse fields at interfaces. In the tilted plane E, they are given by:

$$\widehat{\mathbf{E}}_t^{\mathrm{I,II}} = \begin{bmatrix} \widehat{\mathbf{E}}_y^{\mathrm{I,II}} \\ \widehat{\mathbf{E}}_x^{\mathrm{I,II}} \cos\frac{\alpha}{2} \pm \widehat{\mathbf{E}}_z^{\mathrm{I,II}} \sin\frac{\alpha}{2} \end{bmatrix} \quad \widehat{\mathbf{H}}_t^{\mathrm{I,II}} = \begin{bmatrix} \widehat{\mathbf{H}}_x^{\mathrm{I,II}} \cos\frac{\alpha}{2} \pm \widehat{\mathbf{H}}_z^{\mathrm{I,II}} \sin\frac{\alpha}{2} \\ \widehat{\mathbf{H}}_y^{\mathrm{I,II}} \end{bmatrix} \tag{6.14}$$

The waveguide parts of the bend are labelled I and II. Therefore, we also use the expressions x and z for the components in waveguide II (instead of x'

and z'). The components in the directions of propagation are obtained from eqs. (2.8) and (2.12):

$$\widehat{\mathbf{E}}_z^{\mathrm{I,II}} = -j\varepsilon_{zz}^{-1}\begin{bmatrix}\widehat{\boldsymbol{D}}_{\overline{y}}^{\bullet} & \widehat{\boldsymbol{D}}_{\overline{x}}^{\circ}\end{bmatrix}\begin{bmatrix}-\widehat{\mathbf{H}}_x^{\mathrm{I,II}} \\ \widehat{\mathbf{H}}_y^{\mathrm{I,II}}\end{bmatrix} \quad \widehat{\mathbf{H}}_z^{\mathrm{I,II}} = \mathrm{j}\mu_{zz}^{-1}\begin{bmatrix}\widehat{\boldsymbol{D}}_{\overline{x}}^{\bullet} & -\widehat{\boldsymbol{D}}_{\overline{y}}^{\circ}\end{bmatrix}\begin{bmatrix}\widehat{\mathbf{E}}_y^{\mathrm{I,II}} \\ \widehat{\mathbf{E}}_x^{\mathrm{I,II}}\end{bmatrix} \tag{6.15}$$

By matching the tangential electric and magnetic fields in plane E, we obtain ($t_\alpha = \tan(\alpha/2)$):

$$\begin{bmatrix}\widehat{\mathbf{E}}_y^{\mathrm{I}} \\ \widehat{\mathbf{E}}_x^{\mathrm{I}}\end{bmatrix} - \begin{bmatrix}\widehat{\mathbf{E}}_y^{\mathrm{II}} \\ \widehat{\mathbf{E}}_x^{\mathrm{II}}\end{bmatrix} = \begin{bmatrix}\mathbf{0} \\ -t_\alpha(\widehat{\mathbf{E}}_z^{\mathrm{I}} + \widehat{\mathbf{E}}_z^{\mathrm{II}})\end{bmatrix} = \begin{bmatrix}\boldsymbol{0} & \boldsymbol{0} \\ \widehat{\boldsymbol{Z}}_{xx} & \widehat{\boldsymbol{Z}}_{xy}\end{bmatrix}\begin{bmatrix}-(\widehat{\mathbf{H}}_x^{\mathrm{I}} + \widehat{\mathbf{H}}_x^{\mathrm{II}}) \\ \widehat{\mathbf{H}}_y^{\mathrm{I}} + \widehat{\mathbf{H}}_y^{\mathrm{II}}\end{bmatrix} \tag{6.16}$$

$$\begin{bmatrix}-\widehat{\mathbf{H}}_x^{\mathrm{I}} \\ \widehat{\mathbf{H}}_y^{\mathrm{I}}\end{bmatrix} - \begin{bmatrix}-\widehat{\mathbf{H}}_x^{\mathrm{II}} \\ \widehat{\mathbf{H}}_y^{\mathrm{II}}\end{bmatrix} = \begin{bmatrix}t_\alpha(\widehat{\mathbf{H}}_z^{\mathrm{I}} + \widehat{\mathbf{H}}_z^{\mathrm{II}}) \\ \mathbf{0}\end{bmatrix} = \begin{bmatrix}\widehat{\boldsymbol{Y}}_{xy} & \widehat{\boldsymbol{Y}}_{xx} \\ \boldsymbol{0} & \boldsymbol{0}\end{bmatrix}\begin{bmatrix}\widehat{\mathbf{E}}_y^{\mathrm{I}} + \widehat{\mathbf{E}}_y^{\mathrm{II}} \\ \widehat{\mathbf{E}}_x^{\mathrm{I}} + \widehat{\mathbf{E}}_x^{\mathrm{II}}\end{bmatrix} \tag{6.17}$$

We define impedance and admittance matrices according to:

$$\begin{aligned}\widehat{\boldsymbol{Z}}_{xx} &= \mathrm{j}t_\alpha\widehat{\boldsymbol{M}}_x^{\square}\boldsymbol{\epsilon}_{zz}^{-1}\widehat{\boldsymbol{D}}_{\overline{y}}^{\bullet} \quad & \widehat{\boldsymbol{Z}}_{xy} &= \mathrm{j}t_\alpha\widehat{\boldsymbol{M}}_x^{\square}\boldsymbol{\epsilon}_{zz}^{-1}\widehat{\boldsymbol{D}}_{\overline{x}}^{\circ} \\ \widehat{\boldsymbol{Y}}_{xy} &= \mathrm{j}t_\alpha\widehat{\boldsymbol{M}}_x^{\diamond}\boldsymbol{\mu}_{zz}^{-1}\widehat{\boldsymbol{D}}_{\overline{x}}^{\bullet} \quad & \widehat{\boldsymbol{Y}}_{xx} &= -\mathrm{j}t_\alpha\widehat{\boldsymbol{M}}_x^{\diamond}\boldsymbol{\mu}_{zz}^{-1}\widehat{\boldsymbol{D}}_{\overline{y}}^{\circ}\end{aligned} \tag{6.18}$$

Suitable interpolation matrices $\widehat{\boldsymbol{M}}_x^{\square,\diamond}$ are introduced between the discretisation line systems. The combination of the above equations yields the following matching relation:

$$\widehat{\boldsymbol{P}}_{\mathrm{E}}^{\mathrm{I}}\begin{bmatrix}\widehat{\mathbf{E}}_{\mathrm{E}}^{\mathrm{I}} \\ \widehat{\mathbf{H}}_{\mathrm{E}}^{\mathrm{I}}\end{bmatrix} = \widehat{\boldsymbol{P}}_{\mathrm{E}}^{\mathrm{II}}\begin{bmatrix}\widehat{\mathbf{E}}_{\mathrm{E}}^{\mathrm{II}} \\ \widehat{\mathbf{H}}_{\mathrm{E}}^{\mathrm{II}}\end{bmatrix} \tag{6.19}$$

with:

$$\widehat{\boldsymbol{P}}_{\mathrm{E}}^{\mathrm{I,II}} = \begin{bmatrix}\widehat{\boldsymbol{I}}^{\bullet} & & & \\ & \widehat{\boldsymbol{I}}^{\circ} & \mp\widehat{\boldsymbol{Z}}_{xx} & \mp\widehat{\boldsymbol{Z}}_{xy} \\ \mp\widehat{\boldsymbol{Y}}_{xy} & \mp\widehat{\boldsymbol{Y}}_{xx} & \widehat{\boldsymbol{I}}^{\bullet} & \\ & & & \widehat{\boldsymbol{I}}^{\circ}\end{bmatrix} \tag{6.20}$$

The transfer matrix relation between both sides of plane E is therefore given by:

$$\begin{bmatrix}\widehat{\mathbf{E}}_{\mathrm{E}}^{\mathrm{I}} \\ \widehat{\mathbf{H}}_{\mathrm{E}}^{\mathrm{I}}\end{bmatrix} = (\widehat{\boldsymbol{P}}_{\mathrm{E}}^{\mathrm{I}})^{-1}\widehat{\boldsymbol{P}}_{\mathrm{E}}^{\mathrm{II}}\begin{bmatrix}\widehat{\mathbf{E}}_{\mathrm{E}}^{\mathrm{II}} \\ \widehat{\mathbf{H}}_{\mathrm{E}}^{\mathrm{II}}\end{bmatrix} = \begin{bmatrix}\widehat{\boldsymbol{V}}_{\mathrm{E}}^{\mathrm{E}} & \widehat{\boldsymbol{Z}}_{\mathrm{E}}^{\mathrm{E}} \\ \widehat{\boldsymbol{Y}}_{\mathrm{H}}^{\mathrm{E}} & \widehat{\boldsymbol{V}}_{\mathrm{H}}^{\mathrm{E}}\end{bmatrix}\begin{bmatrix}\widehat{\mathbf{E}}_{\mathrm{E}}^{\mathrm{II}} \\ \widehat{\mathbf{H}}_{\mathrm{E}}^{\mathrm{II}}\end{bmatrix} \tag{6.21}$$

By defining impedances on both sides of plane E by $\widehat{\mathbf{E}}_{\mathrm{E}}^{\mathrm{I,II}} = \widehat{\mathbf{Z}}_{\mathrm{E}}^{\mathrm{I,II}}\widehat{\mathbf{H}}_{\mathrm{E}}^{\mathrm{I,II}}$, the impedance transformation from side II to side I is given by:

$$\widehat{\boldsymbol{Z}}_{\mathrm{E}}^{\mathrm{I}} = (\widehat{\boldsymbol{V}}_{\mathrm{E}}^{\mathrm{E}}\widehat{\boldsymbol{Z}}_{\mathrm{E}}^{\mathrm{II}} + \widehat{\boldsymbol{Z}}_{\mathrm{E}}^{\mathrm{E}})(\widehat{\boldsymbol{Y}}_{\mathrm{H}}^{\mathrm{E}}\widehat{\boldsymbol{Z}}_{\mathrm{E}}^{\mathrm{II}} + \widehat{\boldsymbol{V}}_{\mathrm{H}}^{\mathrm{E}})^{-1} \tag{6.22}$$

With the help of the transfer matrix relations in eq. (2.48) we can write the original fields in plane E as functions of the transformed fields in either plane B $\equiv$ B^{I} or A $\equiv$ A^{I}:

$$\begin{bmatrix} \widehat{\mathbf{E}}_{\mathrm{E}}^{\mathrm{II}} \\ \widehat{\mathbf{H}}_{\mathrm{E}}^{\mathrm{II}} \end{bmatrix} = \begin{bmatrix} \widehat{\boldsymbol{V}}_{\mathrm{E}} & \widehat{\boldsymbol{Z}}_{\mathrm{E}} \\ \widehat{\mathbf{Y}}_{\mathrm{H}} & \widehat{\boldsymbol{V}}_{\mathrm{H}} \end{bmatrix} \begin{bmatrix} \widehat{\widehat{\mathbf{E}}}_{\mathrm{B}} \\ \widehat{\widehat{\mathbf{H}}}_{\mathrm{B}} \end{bmatrix} = \widehat{\boldsymbol{M}}_{\mathrm{T}}^{\mathrm{II}} \begin{bmatrix} \widehat{\widehat{\mathbf{E}}}_{\mathrm{B}} \\ \widehat{\widehat{\mathbf{H}}}_{\mathrm{B}} \end{bmatrix} \tag{6.23}$$

$$\begin{bmatrix} \widehat{\mathbf{E}}_{\mathrm{E}}^{\mathrm{I}} \\ \widehat{\mathbf{H}}_{\mathrm{E}}^{\mathrm{I}} \end{bmatrix} = \begin{bmatrix} \widehat{\boldsymbol{V}}_{\mathrm{E}} & -\widehat{\boldsymbol{Z}}_{\mathrm{E}} \\ -\widehat{\mathbf{Y}}_{\mathrm{H}} & \widehat{\boldsymbol{V}}_{\mathrm{H}} \end{bmatrix} \begin{bmatrix} \widehat{\widehat{\mathbf{E}}}_{\mathrm{A}} \\ \widehat{\widehat{\mathbf{H}}}_{\mathrm{A}} \end{bmatrix} = \widehat{\boldsymbol{M}}_{\mathrm{T}}^{\mathrm{I}} \begin{bmatrix} \widehat{\widehat{\mathbf{E}}}_{\mathrm{A}} \\ \widehat{\widehat{\mathbf{H}}}_{\mathrm{A}} \end{bmatrix} \tag{6.24}$$

Because of the lines with different lengths, we have:

$$\begin{aligned} \widehat{\boldsymbol{V}}_{\mathrm{E,H}} &= \widehat{\boldsymbol{T}}_{\mathrm{E,H}} \bullet \cosh(\widehat{\widehat{\mathbf{d}}}\mathbf{\Gamma}^{\mathrm{r}}) \\ \widehat{\boldsymbol{Z}}_{\mathrm{E}} &= \widehat{\boldsymbol{T}}_{\mathrm{E}} \bullet \sinh(\widehat{\widehat{\mathbf{d}}}\mathbf{\Gamma}^{\mathrm{r}})\widehat{\widehat{\boldsymbol{Z}}}_{0} \\ \widehat{\mathbf{Y}}_{\mathrm{H}} &= \widehat{\boldsymbol{T}}_{\mathrm{H}} \bullet \sinh(\widehat{\widehat{\mathbf{d}}}\mathbf{\Gamma}^{\mathrm{r}})\widehat{\widehat{\mathbf{Y}}}_{0} \\ \widehat{\widehat{\mathbf{d}}} &= [(\boldsymbol{J}_{\bullet}^{\mathrm{t}}(\overline{\mathbf{d}}^{\bullet} \otimes \mathbf{I}_{y}^{\bullet}))^{\mathrm{t}}, (\boldsymbol{J}_{\circ}^{\mathrm{t}}(\overline{\mathbf{d}}^{\circ} \otimes \mathbf{I}_{y}^{\circ}))^{\mathrm{t}}]^{\mathrm{t}} \end{aligned} \tag{6.25}$$

The multiplication sign '$\bullet$' means that the matrices have to be multiplied element by element (array multiplication). As before, we collect all $\overline{d}_i$ (see Fig. 6.5) in a column vector $\overline{\mathbf{d}}$. $\mathbf{I}_y^{\circ,\bullet}$ is a column vector of the order $N_y^{\circ,\bullet}$ whose elements are ones. The matrices $\boldsymbol{J}_{\bullet,\circ}$ are defined as in eq. (2.127). $\mathbf{\Gamma}^{\mathrm{r}}$ is a row vector containing the diagonal elements of $\boldsymbol{\Gamma}$. Therefore, the product $\widehat{\widehat{\mathbf{d}}}\mathbf{\Gamma}^{\mathrm{r}}$ results in a full matrix. To obtain more numerically stable impedance transformation formulas, we determine the z-matrix parameters by:

$$\begin{aligned} \widetilde{\mathbf{z}}_{11}^{\mathrm{II}} &= \widetilde{\mathbf{z}}_{22}^{\mathrm{I}} = \widehat{\boldsymbol{V}}_{\mathrm{E}}\widehat{\mathbf{Y}}_{\mathrm{H}}^{-1} \\ \widetilde{\mathbf{z}}_{12}^{\mathrm{II}} &= \widetilde{\mathbf{z}}_{21}^{\mathrm{I}} = \widehat{\boldsymbol{V}}_{\mathrm{E}}\widehat{\mathbf{Y}}_{\mathrm{H}}^{-1}\widehat{\boldsymbol{V}}_{\mathrm{H}} - \widehat{\boldsymbol{Z}}_{\mathrm{E}} \\ \widetilde{\mathbf{z}}_{22}^{\mathrm{II}} &= \widetilde{\mathbf{z}}_{11}^{\mathrm{I}} = \widehat{\mathbf{Y}}_{\mathrm{H}}^{-1}\widehat{\boldsymbol{V}}_{\mathrm{H}} \\ \widetilde{\mathbf{z}}_{21}^{\mathrm{II}} &= \widetilde{\mathbf{z}}_{12}^{\mathrm{I}} = \widehat{\mathbf{Y}}_{\mathrm{H}}^{-1} \end{aligned} \tag{6.26}$$

The tilde on these parameters indicates that they connect the fields in the transformed domain on one side of the section and the fields in the original domain on the other side. With these matrices, the input impedances in plane E^{II} of section II and plane A^{I} of section I can be calculated by:

$$\widehat{\boldsymbol{Z}}_{\mathrm{E}}^{\mathrm{II}} = \widetilde{\mathbf{z}}_{11}^{\mathrm{II}} - \widetilde{\mathbf{z}}_{12}^{\mathrm{II}}(\widetilde{\mathbf{z}}_{22}^{\mathrm{II}} + \overline{\boldsymbol{Z}}_{\mathrm{o}}^{\mathrm{II}})^{-1}\widetilde{\mathbf{z}}_{21}^{\mathrm{II}} \tag{6.27}$$

$$\widehat{\boldsymbol{Z}}_{\mathrm{A}}^{\mathrm{I}} = \widetilde{\mathbf{z}}_{11}^{\mathrm{I}} - \widetilde{\mathbf{z}}_{12}^{\mathrm{I}}(\widetilde{\mathbf{z}}_{22}^{\mathrm{I}} + \widehat{\boldsymbol{Z}}_{\mathrm{E}}^{\mathrm{II}})^{-1}\widetilde{\mathbf{z}}_{21}^{\mathrm{I}} \tag{6.28}$$

We introduce the characteristic impedance of the infinitely long waveguide II as the load impedance at port B^{I} in eq. (6.27). From the input impedance $\widehat{\boldsymbol{Z}}_{\mathrm{A}}^{\mathrm{I}}$ and the source mode, the reflected mode and the fields can be calculated. Transforming the fields in the opposite direction results in the fields at port B^{I}. From that the scattering parameter, S_{21} can be determined.

6.3 IMPEDANCE TRANSFORMATION AT DISCONTINUITIES

In Section 2.5 we described the general impedance transformation through sections and at concatenations of waveguide sections. Here, we show the transformation at special concatenations.

6.3.1 Impedance transformation at concatenated junctions

At the concatenations of waveguides (as at $z = z_0$ in Fig. 6.6), the relation between the fields in regions I, II and III can be described by the impedance matrices according to:

$$\widehat{\overline{\mathbf{E}}}_{\mathrm{A}}^{\mathrm{I}} = \widehat{\overline{\mathbf{Z}}}_{\mathrm{A}}^{\mathrm{I}} \widehat{\overline{\mathbf{H}}}_{\mathrm{A}}^{\mathrm{I}} \qquad \widehat{\overline{\mathbf{E}}}_{\mathrm{B}}^{\mathrm{II}} = -\widehat{\overline{\mathbf{Z}}}_{\mathrm{B}}^{\mathrm{II}} \widehat{\overline{\mathbf{H}}}_{\mathrm{B}}^{\mathrm{II}} \qquad \widehat{\overline{\mathbf{E}}}_{\mathrm{C}}^{\mathrm{III}} = \widehat{\overline{\mathbf{Z}}}_{\mathrm{C}}^{\mathrm{III}} \widehat{\overline{\mathbf{H}}}_{\mathrm{C}}^{\mathrm{III}} \tag{6.29}$$

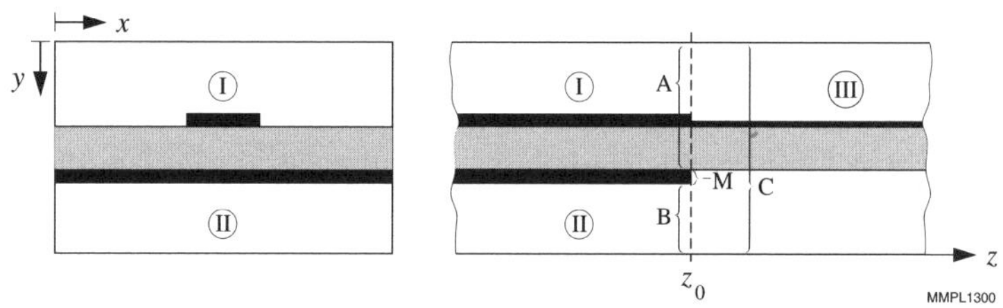

Fig. 6.6 Concatenation of waveguides (R. Pregla, 'Efficient and Accurate Modeling of Planar Anisotropic Microwave Structures by the Method of Lines', *IEEE Trans. Microwave Theory Tech.*, vol. 50, pp. 1469–1479. © 2002 Institute of Electrical and Electronics Engineers (IEEE))

For the matching procedure in waveguide III, we split the tangential electric and magnetic field vectors according to the different parts of the cross-section (A, B and M for the structure shown in Fig. 6.6). We obtain:

$$\widehat{\overline{\mathbf{E}}}_{\mathrm{C}}^{\mathrm{III}} = \begin{bmatrix} \widehat{\overline{\mathbf{E}}}_{\mathrm{A}}^{\mathrm{III}} \\ \widehat{\overline{\mathbf{E}}}_{\mathrm{M}}^{\mathrm{III}} \\ \widehat{\overline{\mathbf{E}}}_{\mathrm{B}}^{\mathrm{III}} \end{bmatrix} = \begin{bmatrix} \boldsymbol{J}_{\mathrm{III A}} \widehat{\overline{\mathbf{E}}}_{\mathrm{C}}^{\mathrm{III}} \\ \boldsymbol{J}_{\mathrm{III M}} \widehat{\overline{\mathbf{E}}}_{\mathrm{C}}^{\mathrm{III}} \\ \boldsymbol{J}_{\mathrm{III B}} \widehat{\overline{\mathbf{E}}}_{\mathrm{C}}^{\mathrm{III}} \end{bmatrix} \qquad \widehat{\overline{\mathbf{H}}}_{\mathrm{C}}^{\mathrm{III}} = \begin{bmatrix} \widehat{\overline{\mathbf{H}}}_{\mathrm{A}}^{\mathrm{III}} \\ \widehat{\overline{\mathbf{H}}}_{\mathrm{M}}^{\mathrm{III}} \\ \widehat{\overline{\mathbf{H}}}_{\mathrm{B}}^{\mathrm{III}} \end{bmatrix} = \begin{bmatrix} \boldsymbol{J}_{\mathrm{III A}} \widehat{\overline{\mathbf{H}}}_{\mathrm{C}}^{\mathrm{III}} \\ \boldsymbol{J}_{\mathrm{III M}} \widehat{\overline{\mathbf{H}}}_{\mathrm{C}}^{\mathrm{III}} \\ \boldsymbol{J}_{\mathrm{III B}} \widehat{\overline{\mathbf{H}}}_{\mathrm{C}}^{\mathrm{III}} \end{bmatrix} \tag{6.30}$$

The subvector $\mathbf{E}_{\mathrm{M}}$ represents the electric field on the non-ideal metallic end between regions A and B. $\mathbf{E}_{\mathrm{M}}$ must be a vector whose components are zero for an ideal metal. The matrices $\boldsymbol{J}$ are quasi-unit matrices. We use them to select the field components on special lines and relate the fields to the different regions. E.g. $\boldsymbol{J}_{\mathrm{III M}}$ selects the lines in region III which end at the metal M. The combination of these matrices $\boldsymbol{J}$ for a waveguide cross-section (here $\boldsymbol{J}_{\mathrm{III A,B,M}}$) gives a unit matrix. For details see Section 2.5.

By matching the tangential electric field, using the vectors in the transformed domain, we obtain:

$$\widehat{\boldsymbol{T}}_{\mathrm{E III}}\widehat{\widehat{\mathbf{E}}}_{\mathrm{C}}^{\mathrm{III}} = \begin{bmatrix} \boldsymbol{J}_{\mathrm{IIIA}}\widehat{\boldsymbol{T}}_{\mathrm{EIII}} \\ \boldsymbol{J}_{\mathrm{IIIM}}\widehat{\boldsymbol{T}}_{\mathrm{EIII}} \\ \boldsymbol{J}_{\mathrm{IIIB}}\widehat{\boldsymbol{T}}_{\mathrm{EIII}} \end{bmatrix} \widehat{\widehat{\mathbf{E}}}_{\mathrm{C}}^{\mathrm{III}} = \begin{bmatrix} \widehat{\boldsymbol{T}}_{\mathrm{EI}}\widehat{\widehat{\mathbf{E}}}_{\mathrm{A}} \\ \widehat{\mathbf{E}}_{\mathrm{M}} \\ \widehat{\boldsymbol{T}}_{\mathrm{EII}}\widehat{\widehat{\mathbf{E}}}_{\mathrm{B}} \end{bmatrix} \tag{6.31}$$

Therefore, we can give the vector $\widehat{\widehat{\mathbf{E}}}_{\mathrm{C}}$ as a function of $\widehat{\widehat{\mathbf{E}}}_{\mathrm{A}}$, $\widehat{\mathbf{E}}_{\mathrm{M}}$ and $\widehat{\widehat{\mathbf{E}}}_{\mathrm{B}}$:

$$\widehat{\widehat{\mathbf{E}}}_{\mathrm{C}}^{\mathrm{III}} = \widehat{\boldsymbol{T}}_{\mathrm{EIII}}^{-1} \begin{bmatrix} \boldsymbol{J}_{\mathrm{IIIA}}^{\mathrm{t}}\widehat{\boldsymbol{T}}_{\mathrm{EI}}\widehat{\widehat{\mathbf{E}}}_{\mathrm{A}} \\ \boldsymbol{J}_{\mathrm{IIIM}}^{\mathrm{t}}\widehat{\mathbf{E}}_{\mathrm{M}} \\ \boldsymbol{J}_{\mathrm{IIIB}}^{\mathrm{t}}\widehat{\boldsymbol{T}}_{\mathrm{EII}}\widehat{\widehat{\mathbf{E}}}_{\mathrm{B}} \end{bmatrix} \tag{6.32}$$

This expression can be written in the following form:

$$\widehat{\widehat{\mathbf{E}}}_{\mathrm{C}} = \widehat{\boldsymbol{T}}_{\mathrm{EIII}}^{-1}\boldsymbol{J}_{\mathrm{IIIA}}^{\mathrm{t}}\widehat{\boldsymbol{T}}_{\mathrm{EI}}\widehat{\widehat{\mathbf{E}}}_{\mathrm{A}} + \widehat{\boldsymbol{T}}_{\mathrm{EIII}}^{-1}\boldsymbol{J}_{\mathrm{IIIB}}^{\mathrm{t}}\widehat{\boldsymbol{T}}_{\mathrm{EII}}\widehat{\widehat{\mathbf{E}}}_{\mathrm{B}} + \widehat{\boldsymbol{T}}_{\mathrm{EIII}}^{-1}\boldsymbol{J}_{\mathrm{IIIM}}^{\mathrm{t}}\widehat{\mathbf{E}}_{\mathrm{M}} \tag{6.33}$$

Analogously, we can match the tangential magnetic field and obtain in transformed domain:

$$\widehat{\boldsymbol{T}}_{\mathrm{HIII}}\widehat{\widehat{\mathbf{H}}}_{\mathrm{C}}^{\mathrm{III}} = \begin{bmatrix} \boldsymbol{J}_{\mathrm{IIIA}}\widehat{\boldsymbol{T}}_{\mathrm{HIII}} \\ \boldsymbol{J}_{\mathrm{IIIM}}\widehat{\boldsymbol{T}}_{\mathrm{HIII}} \\ \boldsymbol{J}_{\mathrm{IIIB}}\widehat{\boldsymbol{T}}_{\mathrm{HIII}} \end{bmatrix} \widehat{\widehat{\mathbf{H}}}_{\mathrm{C}}^{\mathrm{III}} = \begin{bmatrix} \widehat{\mathbf{H}}_{\mathrm{A}} \\ \widehat{\mathbf{S}}_{\mathrm{M}} \\ \widehat{\mathbf{H}}_{\mathrm{B}} \end{bmatrix} = \begin{bmatrix} \widehat{\boldsymbol{T}}_{\mathrm{HI}}\widehat{\widehat{\mathbf{H}}}_{\mathrm{A}}^{\mathrm{I}} \\ \widehat{\mathbf{S}}_{\mathrm{M}} \\ \widehat{\boldsymbol{T}}_{\mathrm{HII}}\widehat{\widehat{\mathbf{H}}}_{\mathrm{B}}^{\mathrm{II}} \end{bmatrix} \tag{6.34}$$

$\widehat{\mathbf{S}}_{\mathrm{M}}$ is the current density on the metallic end between regions A and B and is therefore determined on the same positions as $\widehat{\mathbf{E}}_{\mathrm{M}}$. The following three equations result from this array:

$$\widehat{\boldsymbol{T}}_{\mathrm{HI}}\widehat{\widehat{\mathbf{H}}}_{\mathrm{A}}^{\mathrm{I}} = \boldsymbol{J}_{\mathrm{IIIA}}\widehat{\boldsymbol{T}}_{\mathrm{HIII}}\widehat{\widehat{\mathbf{H}}}_{\mathrm{C}}^{\mathrm{III}} \tag{6.35}$$

$$\widehat{\boldsymbol{T}}_{\mathrm{HII}}\widehat{\widehat{\mathbf{H}}}_{\mathrm{B}}^{\mathrm{II}} = \boldsymbol{J}_{\mathrm{IIIB}}\widehat{\boldsymbol{T}}_{\mathrm{HIII}}\widehat{\widehat{\mathbf{H}}}_{\mathrm{C}}^{\mathrm{III}} \tag{6.36}$$

$$\widehat{\mathbf{S}}_{\mathrm{M}} = \boldsymbol{J}_{\mathrm{IIIM}}\widehat{\boldsymbol{T}}_{\mathrm{H\,III}}\widehat{\widehat{\mathbf{H}}}_{\mathrm{C}}^{\mathrm{III}} \tag{6.37}$$

The associated equations for the tangential electric field vectors $\widehat{\widehat{\mathbf{E}}}_{\mathrm{A}}$ and $\widehat{\widehat{\mathbf{E}}}_{\mathrm{B}}$ are obtained by using eq. (6.29):

$$\widehat{\widehat{\mathbf{E}}}_{\mathrm{A}}^{\mathrm{I}} = \overline{\boldsymbol{Z}}_{\mathrm{A}}\widehat{\boldsymbol{T}}_{\mathrm{HI}}^{-1}\boldsymbol{J}_{\mathrm{IIIA}}\widehat{\boldsymbol{T}}_{\mathrm{HIII}}\overline{\boldsymbol{Y}}_{\mathrm{C}}\widehat{\widehat{\mathbf{E}}}_{\mathrm{C}}^{\mathrm{III}} \tag{6.38}$$

$$\widehat{\widehat{\mathbf{E}}}_{\mathrm{B}}^{\mathrm{II}} = -\overline{\boldsymbol{Z}}_{\mathrm{B}}\widehat{\boldsymbol{T}}_{\mathrm{HII}}^{-1}\boldsymbol{J}_{\mathrm{IIIB}}\widehat{\boldsymbol{T}}_{\mathrm{HIII}}\widehat{\overline{\boldsymbol{Y}}}_{\mathrm{C}}\widehat{\widehat{\mathbf{E}}}_{\mathrm{C}}^{\mathrm{III}} \tag{6.39}$$

Now we replace $\widehat{\widehat{\mathbf{H}}}_{\mathrm{C}}$ in eqs. (6.35)–(6.37) with $\overline{\boldsymbol{Y}}_{\mathrm{C}}\widehat{\widehat{\mathbf{E}}}_{\mathrm{C}}$. By introducing eq. (6.33) into eqs. (6.38) and (6.39), we obtain:

$$\begin{bmatrix} \widehat{\widehat{\mathbf{H}}}_{\mathrm{A}} \\ -\widehat{\widehat{\mathbf{H}}}_{\mathrm{B}} \\ \widehat{\mathbf{S}}_{\mathrm{M}} \end{bmatrix} = \begin{bmatrix} \widehat{\widehat{\boldsymbol{y}}}_{\mathrm{AA}} & \widehat{\widehat{\boldsymbol{y}}}_{\mathrm{AB}} & \widehat{\widehat{\boldsymbol{y}}}_{\mathrm{AM}} \\ \widehat{\widehat{\boldsymbol{y}}}_{\mathrm{BA}} & \widehat{\widehat{\boldsymbol{y}}}_{\mathrm{BB}} & \widehat{\widehat{\boldsymbol{y}}}_{\mathrm{BM}} \\ \widehat{\widehat{\boldsymbol{y}}}_{\mathrm{MA}} & \widehat{\widehat{\boldsymbol{y}}}_{\mathrm{MB}} & \widehat{\boldsymbol{y}}_{\mathrm{MM}} \end{bmatrix} \begin{bmatrix} \widehat{\widehat{\mathbf{E}}}_{\mathrm{A}} \\ \widehat{\widehat{\mathbf{E}}}_{\mathrm{B}} \\ \widehat{\mathbf{E}}_{\mathrm{M}} \end{bmatrix} \tag{6.40}$$

with:

$$
\begin{aligned}
\widehat{\widehat{\boldsymbol{y}}}_{\mathrm{AA}} &= \widehat{\boldsymbol{T}}_{\mathrm{HI}}^{-1}\boldsymbol{J}_{\mathrm{IIIA}}\widehat{\boldsymbol{T}}_{\mathrm{HIII}}\widehat{\widehat{\overline{\boldsymbol{Y}}}}_{\mathrm{C}}\widehat{\boldsymbol{T}}_{\mathrm{EIII}}^{-1}\boldsymbol{J}_{\mathrm{IIIA}}^{\mathrm{t}}\widehat{\boldsymbol{T}}_{\mathrm{EI}} \\
\widehat{\widehat{\boldsymbol{y}}}_{\mathrm{BA}} &= -\widehat{\boldsymbol{T}}_{\mathrm{HII}}^{-1}\boldsymbol{J}_{\mathrm{IIIB}}\widehat{\boldsymbol{T}}_{\mathrm{HIII}}\widehat{\widehat{\overline{\boldsymbol{Y}}}}_{\mathrm{C}}\widehat{\boldsymbol{T}}_{\mathrm{EIII}}^{-1}\boldsymbol{J}_{\mathrm{IIIA}}^{\mathrm{t}}\widehat{\boldsymbol{T}}_{\mathrm{EI}} \\
\widehat{\widehat{\boldsymbol{y}}}_{\mathrm{MA}} &= \boldsymbol{J}_{\mathrm{IIIM}}\widehat{\boldsymbol{T}}_{\mathrm{HIII}}\widehat{\widehat{\overline{\boldsymbol{Y}}}}_{\mathrm{C}}\widehat{\boldsymbol{T}}_{\mathrm{EIII}}^{-1}\boldsymbol{J}_{\mathrm{IIIA}}^{\mathrm{t}}\widehat{\boldsymbol{T}}_{\mathrm{EI}} \\
\widehat{\widehat{\boldsymbol{y}}}_{\mathrm{AB}} &= \widehat{\boldsymbol{T}}_{\mathrm{HI}}^{-1}\boldsymbol{J}_{\mathrm{IIIA}}\widehat{\boldsymbol{T}}_{\mathrm{HIII}}\widehat{\widehat{\overline{\boldsymbol{Y}}}}_{\mathrm{C}}\widehat{\boldsymbol{T}}_{\mathrm{EIII}}^{-1}\boldsymbol{J}_{\mathrm{IIIB}}^{\mathrm{t}}\widehat{\boldsymbol{T}}_{\mathrm{EII}} \\
\widehat{\widehat{\boldsymbol{y}}}_{\mathrm{BB}} &= -\widehat{\boldsymbol{T}}_{\mathrm{HII}}^{-1}\boldsymbol{J}_{\mathrm{IIIB}}\widehat{\boldsymbol{T}}_{\mathrm{HIII}}\widehat{\widehat{\overline{\boldsymbol{Y}}}}_{\mathrm{C}}\widehat{\boldsymbol{T}}_{\mathrm{EIII}}^{-1}\boldsymbol{J}_{\mathrm{IIIB}}^{\mathrm{t}}\widehat{\boldsymbol{T}}_{\mathrm{EII}} \\
\widehat{\widehat{\boldsymbol{y}}}_{\mathrm{MB}} &= \boldsymbol{J}_{\mathrm{IIIM}}\widehat{\boldsymbol{T}}_{\mathrm{HIII}}\widehat{\widehat{\overline{\boldsymbol{Y}}}}_{\mathrm{C}}\widehat{\boldsymbol{T}}_{\mathrm{EIII}}^{-1}\boldsymbol{J}_{\mathrm{IIIB}}^{\mathrm{t}}\widehat{\boldsymbol{T}}_{\mathrm{EII}} \\
\widehat{\widehat{\boldsymbol{y}}}_{\mathrm{AM}} &= \widehat{\boldsymbol{T}}_{\mathrm{HI}}^{-1}\boldsymbol{J}_{\mathrm{IIIA}}\widehat{\boldsymbol{T}}_{\mathrm{HIII}}\widehat{\widehat{\overline{\boldsymbol{Y}}}}_{\mathrm{C}}\widehat{\boldsymbol{T}}_{\mathrm{EIII}}^{-1}\boldsymbol{J}_{\mathrm{IIIM}}^{\mathrm{t}} \\
\widehat{\widehat{\boldsymbol{y}}}_{\mathrm{BM}} &= -\widehat{\boldsymbol{T}}_{\mathrm{HII}}^{-1}\boldsymbol{J}_{\mathrm{IIIB}}\widehat{\boldsymbol{T}}_{\mathrm{HIII}}\widehat{\widehat{\overline{\boldsymbol{Y}}}}_{\mathrm{C}}\widehat{\boldsymbol{T}}_{\mathrm{EIII}}^{-1}\boldsymbol{J}_{\mathrm{IIIM}}^{\mathrm{t}} \\
\widehat{\widehat{\boldsymbol{y}}}_{\mathrm{MM}} &= \boldsymbol{J}_{\mathrm{IIIM}}\widehat{\boldsymbol{T}}_{\mathrm{HIII}}\widehat{\widehat{\overline{\boldsymbol{Y}}}}_{\mathrm{C}}\widehat{\boldsymbol{T}}_{\mathrm{EIII}}^{-1}\boldsymbol{J}_{\mathrm{IIIM}}^{\mathrm{t}}
\end{aligned}
\tag{6.41}
$$

First we consider the case of ideal material with $\widehat{\mathbf{E}}_{\mathrm{M}} = \mathbf{0}$. In this case, the last equation in system (6.40) decouples from the others. The fields $\widehat{\widehat{\mathbf{H}}}_{\mathrm{A}}$, $\widehat{\widehat{\mathbf{H}}}_{\mathrm{B}}$, $\widehat{\widehat{\mathbf{E}}}_{\mathrm{A}}$ and $\widehat{\widehat{\mathbf{E}}}_{\mathrm{B}}$ are now related in the same way as described in Section 2.5. The impedance/admittance transformation can be performed as described there. In the case of a non-ideal metal wall, the approximate boundary condition for the tangential fields on the metallic surface

$$\eta_m \vec{H}_t = \vec{e}_r \times \vec{E}_t$$

can be used. The further procedure is analogous to that in Section 4.6, where we dealt with antennas. Let us assume that the waveguide is fed by the fundamental mode in waveguide I. In that case, field $\widehat{\widehat{\mathbf{E}}}_{\mathrm{A}}$ is known. With that knowledge and the admittance $\widehat{\widehat{\overline{\mathbf{Y}}}}_{\mathrm{A}}$ we can calculate fields $\widehat{\widehat{\mathbf{H}}}_{\mathrm{A}}$ and all the other quantities, including the surface current density $\widehat{\mathbf{S}}_{\mathrm{M}}$.

6.4 ANALYSIS OF PLANAR WAVEGUIDE JUNCTIONS

To analyse waveguide junctions in planar technology we will use crossed discretisation lines. The details of the procedure are demonstrated by the example given in Fig. 6.7. It shows a planar waveguide junction in the form of a microstrip crossing. The junction region J is bounded by the four ports A to D. In general, we have connecting waveguides at all of these four ports; in our case the waveguides $\mathrm{W_A}$ to $\mathrm{W_D}$. The connecting waveguides may also differ from the waveguides inside the junction (see e.g. waveguide $\mathrm{W_D}$). Only the junction region is sketched in Fig. 6.7b. Between ports A and B we have three concatenated waveguide sections I, II and III. The ports of this section are denoted $\mathrm{A^I}$, $\mathrm{B^I}$ to $\mathrm{A^{III}}$, $\mathrm{B^{III}}$. We choose magnetic boundaries at the side walls of the connecting waveguides. Therefore, we describe the relation of the fields at the inner side of the four generalised ports A, B, C and D with open-circuit matrix parameters in the form:

$$\begin{bmatrix} \hat{\hat{\mathbf{E}}}_{\mathrm{A}} \\ \hat{\hat{\mathbf{E}}}_{\mathrm{B}} \\ \hat{\hat{\mathbf{E}}}_{\mathrm{C}} \\ \hat{\hat{\mathbf{E}}}_{\mathrm{D}} \end{bmatrix} = \begin{bmatrix} \hat{\hat{\mathbf{z}}}^{\mathrm{AB}}_{\mathrm{AB11}} & \hat{\hat{\mathbf{z}}}^{\mathrm{AB}}_{\mathrm{AB12}} & \hat{\hat{\mathbf{z}}}^{\mathrm{CD}}_{\mathrm{AB11}} & \hat{\hat{\mathbf{z}}}^{\mathrm{CD}}_{\mathrm{AB12}} \\ \hat{\hat{\mathbf{z}}}^{\mathrm{AB}}_{\mathrm{AB21}} & \hat{\hat{\mathbf{z}}}^{\mathrm{AB}}_{\mathrm{AB22}} & \hat{\hat{\mathbf{z}}}^{\mathrm{CD}}_{\mathrm{AB21}} & \hat{\hat{\mathbf{z}}}^{\mathrm{CD}}_{\mathrm{AB22}} \\ \hat{\hat{\mathbf{z}}}^{\mathrm{AB}}_{\mathrm{CD11}} & \hat{\hat{\mathbf{z}}}^{\mathrm{AB}}_{\mathrm{CD12}} & \hat{\hat{\mathbf{z}}}^{\mathrm{CD}}_{\mathrm{CD11}} & \hat{\hat{\mathbf{z}}}^{\mathrm{CD}}_{\mathrm{CD12}} \\ \hat{\hat{\mathbf{z}}}^{\mathrm{AB}}_{\mathrm{CD21}} & \hat{\hat{\mathbf{z}}}^{\mathrm{AB}}_{\mathrm{CD22}} & \hat{\hat{\mathbf{z}}}^{\mathrm{CD}}_{\mathrm{CD21}} & \hat{\hat{\mathbf{z}}}^{\mathrm{CD}}_{\mathrm{CD22}} \end{bmatrix} \begin{bmatrix} \hat{\hat{\mathbf{H}}}_{\mathrm{A}} \\ -\hat{\hat{\mathbf{H}}}_{\mathrm{B}} \\ \hat{\hat{\mathbf{H}}}_{\mathrm{C}} \\ -\hat{\hat{\mathbf{H}}}_{\mathrm{D}} \end{bmatrix} \tag{6.42}$$

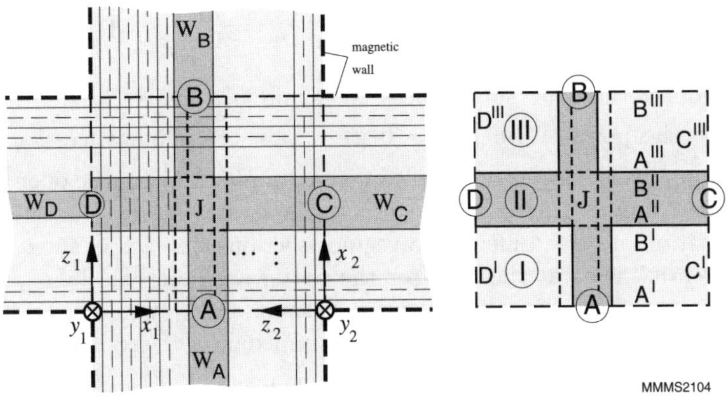

(a) (b)

Fig. 6.7 Junction of planar waveguides (a) and inner junctions (b) (R. Pregla, 'Efficient and Accurate Modeling of Planar Anisotropic Microwave Structures by the Method of Lines', *IEEE Trans. Microwave Theory Tech.*, vol. 50, pp. 1469–1479. © 2002 Institute of Electrical and Electronics Engineers (IEEE))

We should keep in mind that we have two components of the tangential field in each of the column vectors at the ports. By combining the fields of opposite ports into supervectors:

$$\hat{\hat{\mathbf{E}}}_{\mathrm{AB}} = \left[\hat{\hat{\mathbf{E}}}^{t}_{\mathrm{A}}, \hat{\hat{\mathbf{E}}}^{t}_{\mathrm{B}}\right]^{t} \quad \hat{\hat{\mathbf{H}}}_{\mathrm{AB}} = \left[\hat{\hat{\mathbf{H}}}^{t}_{\mathrm{A}}, -\hat{\hat{\mathbf{H}}}^{t}_{\mathrm{B}}\right]^{t} \tag{6.43}$$

we can write eq. (6.42) in a more compact form:

$$\begin{bmatrix} \hat{\hat{\mathbf{E}}}_{\mathrm{AB}} \\ \hat{\hat{\mathbf{E}}}_{\mathrm{CD}} \end{bmatrix} = \begin{bmatrix} \hat{\hat{\mathbf{z}}}^{\mathrm{AB}}_{\mathrm{AB}} & \hat{\hat{\mathbf{z}}}^{\mathrm{CD}}_{\mathrm{AB}} \\ \hat{\hat{\mathbf{z}}}^{\mathrm{AB}}_{\mathrm{CD}} & \hat{\hat{\mathbf{z}}}^{\mathrm{CD}}_{\mathrm{CD}} \end{bmatrix} \begin{bmatrix} \hat{\hat{\mathbf{H}}}_{\mathrm{AB}} \\ \hat{\hat{\mathbf{H}}}_{\mathrm{CD}} \end{bmatrix} \tag{6.44}$$

The four matrices in this equation are obtained by open-circuiting the ports. If ports C and D are open-circuited, matrices $\hat{\hat{\mathbf{z}}}^{\mathrm{AB}}_{\mathrm{AB}}$ and $\hat{\hat{\mathbf{z}}}^{\mathrm{AB}}_{\mathrm{CD}}$ are determined. By open-circuiting ports A and B, matrices $\widehat{\overline{\overline{\mathbf{z}}}}{}^{\mathrm{CD}}_{\mathrm{CD}}$ and $\widehat{\overline{\overline{\mathbf{z}}}}{}^{\mathrm{CD}}_{\mathrm{AB}}$ can be calculated. We would like first to demonstrate the determination of the two main diagonal submatrices in eq. (6.44).

6.4.1 Main diagonal submatrices

Let us start with $\widehat{\overline{\overline{\mathbf{z}}}}{}^{\mathrm{AB}}_{\mathrm{AB}}$. We assume that ports C and D are open-circuited, which means $\widehat{\overline{\overline{\mathbf{H}}}}_{\mathrm{CD}} = \mathbf{0}$. In that case, the field relation between ports A and B is described by:

$$\widehat{\overline{\overline{\mathbf{E}}}}_{\mathrm{AB}} = \widehat{\overline{\overline{\mathbf{z}}}}{}^{\mathrm{AB}}_{\mathrm{AB}} \widehat{\overline{\overline{\mathbf{H}}}}_{\mathrm{AB}} \quad \text{or} \quad \begin{bmatrix} \widehat{\overline{\overline{\mathbf{E}}}}_{\mathrm{A}} \\ \widehat{\overline{\overline{\mathbf{E}}}}_{\mathrm{B}} \end{bmatrix} = \begin{bmatrix} \widehat{\overline{\overline{\mathbf{z}}}}{}^{\mathrm{AB}}_{\mathrm{AB11}} & \widehat{\overline{\overline{\mathbf{z}}}}{}^{\mathrm{AB}}_{\mathrm{AB12}} \\ \widehat{\overline{\overline{\mathbf{z}}}}{}^{\mathrm{AB}}_{\mathrm{AB21}} & \widehat{\overline{\overline{\mathbf{z}}}}{}^{\mathrm{AB}}_{\mathrm{AB22}} \end{bmatrix} \begin{bmatrix} \widehat{\overline{\overline{\mathbf{H}}}}_{\mathrm{A}} \\ -\widehat{\overline{\overline{\mathbf{H}}}}_{\mathrm{B}} \end{bmatrix} \tag{6.45}$$

To obtain the four submatrices we again use the technique of open-circuiting the ports. With $\widehat{\overline{\overline{\mathbf{H}}}}_{\mathrm{B}} = 0$ we obtain the submatrices $\widehat{\overline{\overline{\mathbf{z}}}}{}^{\mathrm{AB}}_{\mathrm{AB11}}$ and $\widehat{\overline{\overline{\mathbf{z}}}}{}^{\mathrm{AB}}_{\mathrm{AB21}}$. $\widehat{\overline{\overline{\mathbf{z}}}}{}^{\mathrm{AB}}_{\mathrm{AB11}}$ is the input impedance matrix in plane A^{I} for the open port B.

In the example shown in Fig. 6.7, ports A and B are connected via a concatenation of three different waveguide sections. In each of these sections the tangential fields at the ends (or the ports) are described by expressions similar to eq. (2.47). For the transition between the waveguide sections, the formulas in Section 2.5 hold. The input impedance of section III in plane $\mathrm{A}^{\mathrm{III}}$ is determined as $\widehat{\overline{\overline{\mathbf{z}}}}{}^{\mathrm{III}}_{1}$ from the formula in eq. (2.47) (see. eq. (6.45)). From $\widehat{\overline{\overline{\boldsymbol{Z}}}}{}^{\mathrm{III}}_{A} = \widehat{\overline{\overline{\mathbf{z}}}}{}^{\mathrm{III}}_{1}$ we obtain $\widehat{\overline{\overline{\boldsymbol{Z}}}}{}^{\mathrm{II}}_{\mathrm{B}}$ with the help of the impedance transformation formula at concatenations in Section 2.5, by:

$$\widehat{\overline{\overline{\boldsymbol{Z}}}}{}^{\mathrm{II}}_{\mathrm{B}} = \widehat{\boldsymbol{T}}^{-1}_{\mathrm{EII\,c}}(\widehat{\boldsymbol{T}}_{\mathrm{HIII\,c}}(\widehat{\overline{\overline{\boldsymbol{Z}}}}{}^{\mathrm{III}}_{A})^{-1}\widehat{\boldsymbol{T}}^{-1}_{\mathrm{EIII\,c}})^{-1}\widehat{\boldsymbol{T}}_{\mathrm{HII\,c}} \tag{6.46}$$

$\widehat{\overline{\overline{\boldsymbol{Z}}}}{}^{\mathrm{II}}_{\mathrm{B}}$ has to be transformed by using the adequate transformation formula for waveguide section II to obtain $\widehat{\overline{\overline{\boldsymbol{Z}}}}{}^{\mathrm{II}}_{\mathrm{A}}$. This procedure must be repeated for section I. So the impedance matrix $\widehat{\overline{\overline{\mathbf{z}}}}{}^{\mathrm{AB}}_{\mathrm{AB11}}$ is determined as input impedance matrix of section I (plane A^{I}) by $\widehat{\overline{\overline{\mathbf{z}}}}{}^{\mathrm{AB}}_{\mathrm{AB11}} = \widehat{\overline{\overline{\boldsymbol{Z}}}}{}^{\mathrm{I}}_{\mathrm{A}}$. It should be mentioned that this impedance matrix is a full matrix.

To calculate matrix $\widehat{\overline{\overline{\mathbf{z}}}}{}^{\mathrm{AB}}_{\mathrm{AB21}}$ we must proceed in the opposite direction. With $\widehat{\overline{\overline{\boldsymbol{Z}}}}{}^{\mathrm{I}}_{\mathrm{A}}$ and eq. (2.47) we obtain:[1]

$$\widehat{\overline{\overline{\mathbf{H}}}}{}^{\mathrm{I}}_{\mathrm{B}} = (\widehat{\overline{\overline{\mathbf{z}}}}{}^{\mathrm{I}}_{2})^{-1}(\widehat{\overline{\overline{\mathbf{z}}}}{}^{\mathrm{I}}_{1} - \widehat{\overline{\overline{\boldsymbol{Z}}}}{}^{\mathrm{I}}_{\mathrm{A}})\widehat{\overline{\overline{\mathbf{H}}}}{}^{\mathrm{I}}_{\mathrm{A}} \tag{6.47}$$

$$\widehat{\overline{\overline{\mathbf{E}}}}{}^{\mathrm{I}}_{\mathrm{B}} = (\widehat{\overline{\overline{\mathbf{z}}}}{}^{\mathrm{I}}_{2} - \widehat{\overline{\overline{\mathbf{z}}}}{}^{\mathrm{I}}_{1}(\widehat{\overline{\overline{\mathbf{z}}}}{}^{\mathrm{I}}_{2})^{-1}(\widehat{\overline{\overline{\mathbf{z}}}}{}^{\mathrm{I}}_{1} - \widehat{\overline{\overline{\boldsymbol{Z}}}}{}^{\mathrm{I}}_{\mathrm{A}}))\widehat{\overline{\overline{\mathbf{H}}}}{}^{\mathrm{I}}_{\mathrm{A}} = \widehat{\overline{\overline{\boldsymbol{Z}}}}{}^{\mathrm{I}}_{\mathrm{B}}\widehat{\overline{\overline{\mathbf{H}}}}{}^{\mathrm{I}}_{\mathrm{B}} \tag{6.48}$$

[1] For numerically stable calculations, use the algorithms described in Section 2.5.4

The next step is the transformation of $\widehat{\overline{\overline{\mathbf{H}}}}_{\mathrm{B}}^{\mathrm{I}}$ and $\widehat{\overline{\overline{\mathbf{E}}}}_{\mathrm{B}}^{\mathrm{I}}$ into the values $\widehat{\overline{\overline{\mathbf{E}}}}_{\mathrm{A}}^{\mathrm{II}}$ and $\widehat{\overline{\overline{\mathbf{H}}}}_{\mathrm{A}}^{\mathrm{II}}$. We obtain (cf. eq. (2.158) in chapter 2.5):

$$\widehat{\overline{\overline{\mathbf{E}}}}_{\mathrm{A}}^{\mathrm{II}} = \widehat{\boldsymbol{T}}_{\mathrm{EII\,c}}^{-1}\widehat{\boldsymbol{T}}_{\mathrm{EI\,c}}\widehat{\overline{\overline{\mathbf{E}}}}_{\mathrm{B}}^{\mathrm{I}} \qquad \widehat{\overline{\overline{\mathbf{H}}}}_{\mathrm{A}}^{\mathrm{II}} = (\widehat{\overline{\overline{\mathbf{Z}}}}_{\mathrm{A}}^{\mathrm{II}})^{-1}\widehat{\overline{\overline{\mathbf{E}}}}_{\mathrm{A}}^{\mathrm{II}} \tag{6.49}$$

This procedure must be repeated for the other sections. To generalise it, we may replace the superscript I and II in eqs. (6.47)–(6.49) with $k-1$ and k, respectively. Then, from the equation for $\widehat{\overline{\overline{\mathbf{E}}}}_{\mathrm{B}}^{\mathrm{k}}$ with $k=3$, in which all formerly calculated quantities have been introduced and which is proportionate to $\widehat{\overline{\overline{\mathbf{H}}}}_{\mathrm{A}}$, the transmittance matrix $\widehat{\overline{\overline{\mathbf{z}}}}_{\mathrm{AB21}}^{\mathrm{AB}}$ is obtained. The analogous procedure holds for the other two submatrices in eq. (6.45). Port A must be open-circuited and the input impedance at port B looking towards A must be calculated. This input impedance is identical to the submatrix $\widehat{\overline{\overline{\mathbf{z}}}}_{\mathrm{AB22}}^{\mathrm{AB}}$. For the calculation of $\widehat{\overline{\overline{\mathbf{z}}}}_{\mathrm{AB12}}^{\mathrm{AB}}$, we must transform the magnetic field column vector $\widehat{\overline{\overline{\mathbf{H}}}}_{\mathrm{B}}$ from port B through the different waveguide sections to port A, taking into account the impedances computed earlier. $\widehat{\overline{\overline{\mathbf{z}}}}_{\mathrm{AB12}}^{\mathrm{AB}}$ is the matrix that connects $\widehat{\overline{\overline{\mathbf{E}}}}_{\mathrm{A}}$ with $-\widehat{\overline{\overline{\mathbf{H}}}}_{\mathrm{B}}$. Now $\widehat{\overline{\overline{\mathbf{z}}}}_{\mathrm{AB}}^{\mathrm{AB}}$ is completely determined.

In an analogous way, we can compute the submatrix in $\widehat{\overline{\overline{\mathbf{z}}}}_{\mathrm{CD}}^{\mathrm{CD}}$ in eq. (6.44). In this case, ports A and B must be open-circuited, with $\widehat{\overline{\overline{\mathbf{H}}}}_{\mathrm{AB}} = \mathbf{0}$. We now have to calculate the off-diagonal submatrix $\widehat{\overline{\overline{\mathbf{z}}}}_{\mathrm{CD}}^{\mathrm{AB}}$.

6.4.2 Off-diagonal submatrices – coupling to perpendicular ports

The off-diagonal submatrix $\widehat{\overline{\overline{\mathbf{z}}}}_{\mathrm{CD}}^{\mathrm{AB}}$ is defined by the equation:

$$\widehat{\overline{\overline{\mathbf{E}}}}_{\mathrm{CD}} = \widehat{\overline{\overline{\mathbf{z}}}}_{\mathrm{CD}}^{\mathrm{AB}}\widehat{\overline{\overline{\mathbf{H}}}}_{\mathrm{AB}} \text{or} \begin{bmatrix} \widehat{\overline{\overline{\mathbf{E}}}}_{\mathrm{C}} \\ \widehat{\overline{\overline{\mathbf{E}}}}_{\mathrm{D}} \end{bmatrix} = \begin{bmatrix} \widehat{\overline{\overline{\mathbf{z}}}}_{\mathrm{CD11}}^{\mathrm{AB}} & \widehat{\overline{\overline{\mathbf{z}}}}_{\mathrm{CD12}}^{\mathrm{AB}} \\ \widehat{\overline{\overline{\mathbf{z}}}}_{\mathrm{CD21}}^{\mathrm{AB}} & \widehat{\overline{\overline{\mathbf{z}}}}_{\mathrm{CD22}}^{\mathrm{AB}} \end{bmatrix} \begin{bmatrix} \widehat{\overline{\overline{\mathbf{H}}}}_{\mathrm{A}} \\ -\widehat{\overline{\overline{\mathbf{H}}}}_{\mathrm{B}} \end{bmatrix} \tag{6.50}$$

which is obtained from eq. (6.42) with the condition $\widehat{\overline{\overline{\mathbf{H}}}}_{\mathrm{CD}} = \mathbf{0}$. The two submatrices on the left side are obtained by introducing the condition $\widehat{\overline{\overline{\mathbf{H}}}}_{\mathrm{B}} = \mathbf{0}$. For the two submatrices on the right side we set $\widehat{\overline{\overline{\mathbf{H}}}}_{\mathrm{A}} = \mathbf{0}$. We will now demonstrate the procedure for the two submatrices on the left side. The procedure for those on the right side is analogous. The tangential electric field components at ports C and D (at the magnetic walls) are given by (see Fig. 6.8):

$$\widehat{\mathbf{E}}_{\mathrm{C,D}} = \begin{bmatrix} \widehat{\mathbf{E}}_{y2}^{\mathrm{C,D}} \\ \widehat{\mathbf{E}}_{x2}^{\mathrm{C,D}} \end{bmatrix} = \begin{bmatrix} \widehat{\mathbf{E}}_{y1}^{\mathrm{C,D}} \\ \widehat{\mathbf{E}}_{z1}^{\mathrm{C,D}} \end{bmatrix} = \begin{bmatrix} \widehat{\mathbf{E}}_{y}^{\mathrm{C,D}} \\ \widehat{\mathbf{E}}_{z}^{\mathrm{C,D}} \end{bmatrix} \tag{6.51}$$

We have only these electric field components at ports C and D. The tangential magnetic fields are zero because of the magnetic walls. Hence

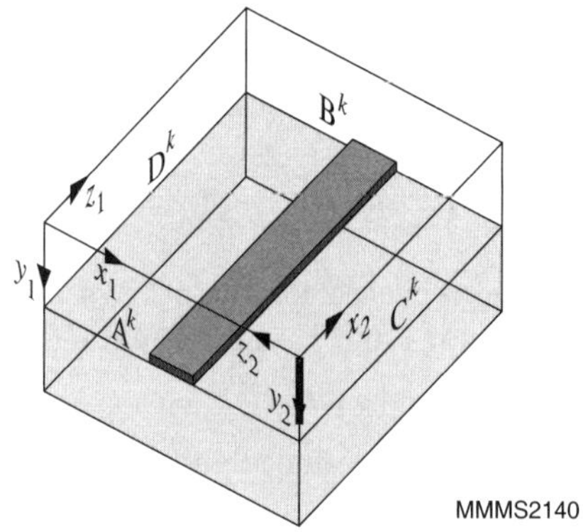

Fig. 6.8 Coupling from ports A^k and B^k to ports C^k and D^k (Reproduced by permission of Union Radio-Scientifique Internationale–International Union of Radio Science (URSI))

we have to determine $\widehat{\mathbf{E}}_y^{\mathrm{C,D}}$ and $\widehat{\mathbf{E}}_z^{\mathrm{C,D}}$, caused by $\widehat{\widehat{\mathbf{H}}}_{\mathrm{A}}$. These calculations must be divided into various parts. This is because ports C and D must be partitioned into subports (in our example, three) C^k and D^k according to the side wall areas of the waveguide sections in z_1-direction. Let us now calculate the tangential electric fields at subports C^k and D^k. Once we have obtained the tangential electric fields at all subports, we have only to arrange these field components in a column vector in the order of the subport numbers.

The field in a plane at an arbitrary position $z \equiv z_1$ between the two subports A^k and B^k of a section k ($k = \mathrm{I, II}$ or III in the example of Fig. 6.7b) can be calculated from the fields at ports A^k and B^k by:

$$\widehat{\widehat{\mathbf{F}}}(z_i^k) = \boldsymbol{\Lambda}_{\mathrm{A^k}}^{\mathrm{d}} \widehat{\widehat{\mathbf{F}}}_{\mathrm{A^k}} + \boldsymbol{\Lambda}_{\mathrm{B^k}}^{\mathrm{d}} \widehat{\widehat{\mathbf{F}}}_{\mathrm{B^k}} \tag{6.52}$$

$\widehat{\widehat{\mathbf{F}}}$ is the supervector of the transverse electric or magnetic field components. The fields $\widehat{\widehat{\mathbf{F}}}_{\mathrm{A^k}}$ and $\widehat{\widehat{\mathbf{F}}}_{\mathrm{B^k}}$ both have to be calculated from the fields at port A (!). Even with $\mathbf{H}_{\mathrm{B}} = \mathbf{0}$, we have magnetic fields at both subports A^k and B^k in subsection k, which both have to be calculated from $\mathbf{H}_{\mathrm{A}}$ (see above). Only if $\mathbf{B}^k$ is identical to $\mathbf{B}$ do we have $\mathbf{H}_{\mathrm{B}^k} = \mathbf{0}$. (We assume that section k does not have a metallic interface at ports C and D.) The diagonal matrices $\boldsymbol{\Lambda}_{\mathrm{A}^k}^{\mathrm{d}}$ and $\boldsymbol{\Lambda}_{\mathrm{B}^k}^{\mathrm{d}}$ in eq. (6.52) are given by:

$$\boldsymbol{\Lambda}_{\mathrm{A^k}}^{\mathrm{d}} = \frac{\sinh(\boldsymbol{\Gamma}_z^k \overline{z}_-^k)}{\sinh(\boldsymbol{\Gamma}_z^k \overline{d}_{\mathrm{AB^k}})} \qquad \boldsymbol{\Lambda}_{\mathrm{B^k}}^{\mathrm{d}} = \frac{\sinh(\boldsymbol{\Gamma}_z^k \overline{z}^k)}{\sinh(\boldsymbol{\Gamma}_z^k \overline{d}_{\mathrm{AB^k}})} \tag{6.53}$$

with $\overline{z}_-^k = \overline{d}_k - \overline{z}^k$ and $\overline{d}_k = k_0 d_k$. $z^k = 0$ in plane of port A^k. d_k is the distance between ports A^k and B^k. The diagonal matrix $\boldsymbol{\Gamma}_z^k$ is obtained as in eq. (2.34). We describe the determination of field $\widehat{\mathbf{E}}_{y\mathrm{R}^k}$ at ports $\mathrm{R}^k \equiv \mathrm{C}^k$ and $\mathrm{R}^k \equiv \mathrm{D}^k$ below.

Field $\mathbf{E}_{yn}^k$ in the nth column of the cross-section at z^k is given by:

$$\mathbf{E}_{yn}^k(z^k) = \boldsymbol{T}_{\mathrm{E}_{yn}}(\boldsymbol{\Lambda}_{\mathrm{A^k}}^{\mathrm{d}}\hat{\hat{\mathbf{E}}}_{\mathrm{A^k}} + \boldsymbol{\Lambda}_{\mathrm{B^k}}^{\mathrm{d}}\hat{\hat{\mathbf{E}}}_{\mathrm{B^k}}) \tag{6.54}$$

where $\boldsymbol{T}_{\mathrm{E}_{yn}}$ is the nth block of matrix $\boldsymbol{T}_{\mathrm{E}_y}$ for waveguide section k (we will not introduce a sub or superscript k for this identification). We define the matrices $\boldsymbol{T}_{\mathrm{E}_y}^{\mathrm{C^k}}$ and $\boldsymbol{T}_{\mathrm{E}_y}^{\mathrm{D^k}}$ by:

$$\boldsymbol{T}_{\mathrm{E}_y}^{\mathrm{C}^k} = \tfrac{1}{8}(9\boldsymbol{T}_{\mathrm{E}_{y\mathrm{N}}} - \boldsymbol{T}_{\mathrm{E}_{y\mathrm{N}_-}}) \quad \boldsymbol{T}_{\mathrm{E}y}^{\mathrm{D}^k} = \tfrac{1}{8}(9\boldsymbol{T}_{\mathrm{E}_{y1}} - \boldsymbol{T}_{\mathrm{E}_{y2}}) \tag{6.55}$$

where $\boldsymbol{T}_{\mathrm{E}_{y\mathrm{N}}}$ ($\boldsymbol{T}_{\mathrm{E}_{y1}}$) and $\boldsymbol{T}_{\mathrm{E}_{y\mathrm{N}_-}}$ ($\boldsymbol{T}_{\mathrm{E}_{y2}}$) are the last (first) and last but one (second) blocks of matrix $\boldsymbol{T}_{\mathrm{E}_y}$, respectively. $\boldsymbol{T}_{\mathrm{E}_y}^{\mathrm{R}^k}$ are rectangular matrices. The relevant field components $\mathbf{E}_{y\mathrm{R}}$ at ports $\mathrm{R}^k \equiv \mathrm{C^k}$ or $\mathrm{R}^k \equiv \mathrm{D}^k$ are now expressed by:

$$\mathbf{E}_{y\mathrm{R}}(z^k) = \boldsymbol{T}_{\mathrm{E}_y}^{\mathrm{R}}(\boldsymbol{\Lambda}_{\mathrm{A^k}}^{\mathrm{d}}\hat{\hat{\mathbf{E}}}_{\mathrm{A^k}} + \boldsymbol{\Lambda}_{\mathrm{B^k}}^{\mathrm{d}}\hat{\hat{\mathbf{E}}}_{\mathrm{B^k}}) \tag{6.56}$$

Eq. (6.56) must now be discretised at the discretisation points z_i^k of the ports. The field part $\mathbf{E}_{y\mathrm{R}^k}^{\mathrm{A}^k}$, related to $\hat{\hat{\mathbf{E}}}_{\mathrm{A}^k}$, can be written as:

$$\widehat{\mathbf{E}}_{y\mathrm{R}^k}^{\mathrm{A}^k} = \begin{bmatrix} \boldsymbol{T}_{\mathrm{E}_y}^{\mathrm{R^k}}\boldsymbol{\Lambda}_{\mathrm{A^k}}^{\mathrm{d}}(z_1^k) \\ \vdots \\ \boldsymbol{T}_{\mathrm{E}_y}^{\mathrm{R^k}}\boldsymbol{\Lambda}_{\mathrm{A}^k}^{\mathrm{d}}(z_{\mathrm{N}_z}^k) \end{bmatrix} \hat{\hat{\mathbf{E}}}_{\mathrm{A^k}} = \widehat{\mathbf{V}}_{y\mathrm{R}}^{\mathrm{A}^k}\hat{\hat{\mathbf{E}}}_{\mathrm{A^k}} \tag{6.57}$$

$z_{\mathrm{N}_z}^k$ is the column at subport C^k or D^k in z_1-direction. The field vectors $\hat{\hat{\mathbf{E}}}_{\mathrm{A}^k}$ and $\hat{\hat{\mathbf{E}}}_{\mathrm{B}^k}$ are both determined from $\hat{\hat{\mathbf{H}}}_{\mathrm{A}}$ by the procedure described in Section 6.4.1. These relations may be expressed as:

$$\hat{\hat{\mathbf{E}}}_{\mathrm{A}^k} = \hat{\hat{\mathbf{Z}}}_{\mathrm{AA}^k}\hat{\hat{\mathbf{H}}}_{\mathrm{A}} \qquad \hat{\hat{\mathbf{E}}}_{\mathrm{B}^k} = \hat{\hat{\mathbf{Z}}}_{\mathrm{AB}^k}\hat{\hat{\mathbf{H}}}_{\mathrm{A}} \tag{6.58}$$

The fields at all the subports C^k and D^k (subports k of ports C and D) are now collected in the common column vectors to obtain the total column vectors for $\widehat{\mathbf{E}}_y$ at ports C and D, respectively:

$$\widehat{\mathbf{E}}_{y\mathrm{R}} = \begin{bmatrix} \widehat{\mathbf{E}}_{y\mathrm{R}^1}^{\mathrm{A}^1} + \widehat{\mathbf{E}}_{y\mathrm{R}^1}^{\mathrm{B}^1} \\ \vdots \\ \widehat{\mathbf{E}}_{y\mathrm{R}^k}^{\mathrm{A}^k} + \widehat{\mathbf{E}}_{y\mathrm{R}^k}^{\mathrm{B}^k} \\ \vdots \end{bmatrix} = \begin{bmatrix} \widehat{\mathbf{V}}_{y\mathrm{R}^1}^{\mathrm{A}^1}\hat{\hat{\mathbf{Z}}}_{\mathrm{AA}^1} + \widehat{\mathbf{V}}_{y\mathrm{R}^1}^{\mathrm{B}^1}\hat{\hat{\mathbf{Z}}}_{\mathrm{AB}^1} \\ \vdots \\ \widehat{\mathbf{V}}_{y\mathrm{R}^k}^{\mathrm{A}^k}\hat{\hat{\mathbf{Z}}}_{\mathrm{AA}^k} + \widehat{\mathbf{V}}_{y\mathrm{R}^k}^{\mathrm{B}^k}\hat{\hat{\mathbf{Z}}}_{\mathrm{AB}^k} \\ \vdots \end{bmatrix} \overline{\mathbf{H}}_{\mathrm{A}} = \mathbf{Z}_{y\mathrm{R}}^{\mathrm{A}}\hat{\hat{\mathbf{H}}}_{\mathrm{A}} \tag{6.59}$$

This is the upper part in the vector of eq. (6.51). We obtain the tangential electric field component $\widehat{\mathbf{E}}_z^{\mathrm{C,D}}$ from the magnetic field $\hat{\hat{\mathbf{H}}}(z_i)$ according to eq. (2.12) (or in discretised form):

$$\widehat{\mathbf{E}}_z(z^k) = -\mathrm{j}\epsilon_{zz}^{-1}[\widehat{\boldsymbol{D}}_{\overline{y}}^{\bullet} \quad \widehat{\boldsymbol{D}}_{\overline{x}}^{\circ}]\widehat{\mathbf{H}}(z^k) = -\mathrm{j}\epsilon_{zz}^{-1}[\widehat{\boldsymbol{D}}_{\overline{y}}^{\bullet} \quad \widehat{\boldsymbol{D}}_{\overline{x}}^{\circ}]\widehat{\mathbf{T}}_{\mathrm{H}}\hat{\hat{\mathbf{H}}}(z^k) \tag{6.60}$$

To calculate $\mathbf{E}_z$ at ports C^k and D^k by an extrapolation process, we need the values of $\mathbf{E}_z$ that belong to the last two and the first two columns of $\widehat{\mathbf{E}}_z$ in the cross-section at z^k. The nth column can be calculated by:

$$\mathbf{E}^k_{zn}(z^k) = -\mathrm{j}(\boldsymbol{\epsilon}^k_{zzn})^{-1}[\overline{\boldsymbol{D}}^{\bullet}_{\overline{y}}\boldsymbol{T}^k_{\mathrm{H}_{xn}} \quad \overline{\boldsymbol{D}}^{\circ}_{\overline{x}n}\boldsymbol{T}^k_{\mathrm{H}_{yn}}]\widehat{\widehat{\mathbf{H}}}(z^k) \tag{6.61}$$

where $\boldsymbol{\epsilon}^k_{zzn}$ is the diagonal matrix of the ε_{rzz} values for that column. $\widehat{\boldsymbol{T}}^k_{\mathrm{H}}$ is divided into two parts and these two parts are divided further into two subparts each, as described in Section 6.4.2. The upper part $\boldsymbol{T}^k_{\mathrm{H}_x}$ belongs to $\mathbf{H}^k_x$ and the lower part $\boldsymbol{T}^k_{\mathrm{H}_y}$ to $\mathbf{H}^k_y$, with the same number of rows each. $\boldsymbol{T}^k_{\mathrm{H}_{xn}}$ is the matrix of the nth block of rows related to the nth column. The extrapolation of the columns in the first part of eq. (6.61) to the ports with the matrices $\boldsymbol{T}^{\mathrm{R^k}}_{xn}$ is done by a method analogous to eq. (6.55). The difference matrix $\boldsymbol{D}^{\bullet}_{\overline{y}}$ is used at the port as usual. The extrapolation of the second part is different because the difference operator $\boldsymbol{D}^{\circ}_{\overline{x}}$ must be applied to the columns. Therefore, $\boldsymbol{D}^{\circ}_{\overline{x}n}\boldsymbol{T}^k_{\mathrm{H}_{yn}}$ has to be understood in the following way: $\boldsymbol{D}^{\circ}_{\overline{x}n}$ must be applied to two neighbouring vertical columns of the $\mathbf{H}^k_y$ quantities or the $\mathbf{H}^k_{yn}$ columns (at $\circ$ discretisation points). Then the columns of the derivatives can be extrapolated in a way analogous to eq. (6.55), because the derivatives have to fulfil Neumann boundary conditions. The columns of the derivatives can be expressed by the central differences of the neighbouring columns. The field component $\mathbf{H}^k_{yn}$ itself has to fulfil Dirichlet conditions. By using the central differences for this case, we obtain:

$$\boldsymbol{T}^{\mathrm{C}^k}_{\mathrm{H}_y} = \tfrac{1}{8}(10\boldsymbol{T}^k_{\mathrm{H}_{y\mathrm{N}}} - \boldsymbol{T}^k_{\mathrm{H}_{y\mathrm{N}_-}}) \qquad \boldsymbol{T}^{\mathrm{D}^k}_{\mathrm{H}_y} = \tfrac{1}{8}(10\boldsymbol{T}^k_{\mathrm{H}_{y1}} - \boldsymbol{T}^k_{\mathrm{H}_{y2}}) \tag{6.62}$$

Eq. (6.62) can easily be checked. If the behaviour of $\mathbf{H}^k_y(x)$ at the boundary is linear, we have e.g. $\boldsymbol{T}^k_{\mathrm{H}_{y2}} = 2\boldsymbol{T}^k_{\mathrm{H}_{y1}}$ and therefore $\boldsymbol{T}^{\mathrm{D}^k}_{\mathrm{H}_y} = \boldsymbol{T}^k_{\mathrm{H}_{y1}}$. The values at subports C^k and D^k are given by:

$$\begin{aligned}\mathbf{E}^k_{z\mathrm{D}} &= \boldsymbol{T}^{\mathrm{D}}_{\mathrm{H}_z}\widehat{\widehat{\mathbf{H}}}(z^k) \qquad & \boldsymbol{T}^{\mathrm{D}}_{\mathrm{H}_z} &= -\mathrm{j}(\boldsymbol{\epsilon}^k_{zzn})^{-1}[\overline{\boldsymbol{D}}^{\bullet}_{\overline{y}}\boldsymbol{T}^{\mathrm{D}^k}_{\mathrm{H}_x} \quad \overline{h}_x^{-1}\boldsymbol{T}^{\mathrm{D}^k}_{\mathrm{H}_y}]\\ \mathbf{E}^k_{z\mathrm{C}} &= \boldsymbol{T}^{\mathrm{C}}_{\mathrm{H}_z}\widehat{\widehat{\mathbf{H}}}(z^k) \qquad & \boldsymbol{T}^{\mathrm{C}}_{\mathrm{H}_z} &= -\mathrm{j}(\boldsymbol{\epsilon}^k_{zzn})^{-1}[\overline{\boldsymbol{D}}^{\bullet}_{\overline{y}}\boldsymbol{T}^{\mathrm{C}^k}_{\mathrm{H}_x} \quad \overline{h}_x^{-1}\boldsymbol{T}^{\mathrm{C}^k}_{\mathrm{H}_y}]\end{aligned} \tag{6.63}$$

$\overline{h}_x$ in eq. (6.63) is the normalised discretisation distance in x-direction. The vector of the magnetic field components $\widehat{\widehat{\mathbf{H}}}$ in eq. (6.63) has to be replaced by eq. (6.52). $\widehat{\widehat{\mathbf{H}}}_{\mathrm{A^k}}$ and $\widehat{\widehat{\mathbf{H}}}_{\mathrm{B^k}}$ are obtained with eqs. (6.47)–(6.49). The field component $\mathbf{E}_{z\mathrm{R}}$ at ports $\mathrm{R}^k \equiv \mathrm{C}^k$ or $\mathrm{R}^k \equiv \mathrm{D}^k$ is now expressed by:

$$\mathbf{E}^k_{z\mathrm{R}}(z^k) = \boldsymbol{T}^{\mathrm{R}}_{\mathrm{H}_z}(\boldsymbol{\Lambda}^{\mathrm{d}}_{\mathrm{A^k}}\widehat{\widehat{\mathbf{H}}}_{\mathrm{A^k}} + \boldsymbol{\Lambda}^{\mathrm{d}}_{\mathrm{B^k}}\widehat{\widehat{\mathbf{H}}}_{\mathrm{B^k}}) \tag{6.64}$$

Eq. (6.64) must be discretised at the discretisation points z_i^k of the ports. The field part $\mathbf{E}_{z\mathrm{R}^k}^{\mathrm{A}^k}$, related to $\hat{\overline{\mathbf{H}}}_{\mathrm{A}^k}$, can be written as:

$$\widehat{\mathbf{E}}_{z\mathrm{R}^k}^{\mathrm{A}^k} = \begin{bmatrix} \boldsymbol{T}_{\mathrm{H}_z}^{\mathrm{R}} \boldsymbol{\Lambda}_{\mathrm{A}^k}^{\mathrm{d}}(z_1^k) \\ \vdots \\ \boldsymbol{T}_{\mathrm{H}_z}^{\mathrm{R}} \boldsymbol{\Lambda}_{\mathrm{A}^k}^{\mathrm{d}}(z_{\mathrm{N}_z}^k) \end{bmatrix} \hat{\overline{\mathbf{H}}}_{\mathrm{A}^k} = \widehat{\boldsymbol{Z}}_{z\mathrm{R}}^{\mathrm{A}^k} \hat{\overline{\mathbf{H}}}_{\mathrm{A}^k} \tag{6.65}$$

Both field vectors ($\hat{\overline{\mathbf{H}}}_{\mathrm{A}^k}$ and $\hat{\overline{\mathbf{H}}}_{\mathrm{B}^k}$) are determined from $\overline{\mathbf{H}}_{\mathrm{A}}$ by the procedure described in Section 6.4.1. The relation may be given as:

$$\hat{\overline{\mathbf{H}}}_{\mathrm{A}^k} = \hat{\overline{\mathbf{V}}}_{\mathrm{AA}^k} \hat{\overline{\mathbf{H}}}_{\mathrm{A}} \qquad \hat{\overline{\mathbf{H}}}_{\mathrm{B}^k} = \hat{\overline{\mathbf{V}}}_{\mathrm{AB}^k} \hat{\overline{\mathbf{H}}}_{\mathrm{A}} \tag{6.66}$$

The fields of all the subports C^k and D^k (subports k of ports C and D) are now collected in common column vectors to obtain the total column vectors for $\widehat{\mathbf{E}}_z$ at ports C and D:

$$\widehat{\mathbf{E}}_{z\mathrm{R}} = \begin{bmatrix} \widehat{\mathbf{E}}_{z\mathrm{R}^1}^{\mathrm{A}^1} + \widehat{\mathbf{E}}_{z\mathrm{R}^1}^{\mathrm{B}^1} \\ \vdots \\ \widehat{\mathbf{E}}_{z\mathrm{R}^k}^{\mathrm{A}^k} + \widehat{\mathbf{E}}_{z\mathrm{R}^k}^{\mathrm{B}^k} \\ \vdots \end{bmatrix} = \begin{bmatrix} \widehat{\boldsymbol{Z}}_{z\mathrm{R}^1}^{\mathrm{A}^1} \hat{\overline{\mathbf{V}}}_{\mathrm{AA}^1} + \widehat{\boldsymbol{Z}}_{z\mathrm{R}^1}^{\mathrm{B}^1} \hat{\overline{\mathbf{V}}}_{\mathrm{AB}^1} \\ \vdots \\ \widehat{\boldsymbol{Z}}_{z\mathrm{R}^k}^{\mathrm{A}^k} \hat{\overline{\mathbf{V}}}_{\mathrm{AA}^k} + \widehat{\boldsymbol{Z}}_{z\mathrm{R}^k}^{\mathrm{B}^k} \hat{\overline{\mathbf{V}}}_{\mathrm{AB}^k} \\ \vdots \end{bmatrix} \hat{\overline{\mathbf{H}}}_{\mathrm{A}} = \widehat{\boldsymbol{Z}}_{z\mathrm{R}}^{\mathrm{A}} \hat{\overline{\mathbf{H}}}_{\mathrm{A}} \tag{6.67}$$

In an analogous manner, $\widehat{\boldsymbol{Z}}_{y\mathrm{R}}^{\mathrm{B}}$ and $\widehat{\boldsymbol{Z}}_{z\mathrm{R}}^{\mathrm{B}}$ (under the condition $\hat{\overline{\mathbf{H}}}_{\mathrm{A}} = \mathbf{0}$) can be calculated. The results may be summarised as follows:

$$\hat{\overline{\mathbf{E}}}_{\mathrm{CD}} = \begin{bmatrix} \hat{\overline{\mathbf{E}}}_{y_2}^{\mathrm{C}} \\ \hat{\overline{\mathbf{E}}}_{x_2}^{\mathrm{C}} \\ \hat{\overline{\mathbf{E}}}_{y_2}^{\mathrm{D}} \\ \hat{\overline{\mathbf{E}}}_{x_2}^{\mathrm{D}} \end{bmatrix} = \begin{bmatrix} \widehat{\boldsymbol{T}}_{\mathrm{C}}^{-1} \widehat{\boldsymbol{Z}}_{y\mathrm{C}}^{\mathrm{A}} & \widehat{\boldsymbol{T}}_{\mathrm{C}}^{-1} \widehat{\boldsymbol{Z}}_{y\mathrm{C}}^{\mathrm{B}} \\ \widehat{\boldsymbol{T}}_{\mathrm{C}}^{-1} \widehat{\boldsymbol{Z}}_{z\mathrm{C}}^{\mathrm{A}} & \widehat{\boldsymbol{T}}_{\mathrm{C}}^{-1} \widehat{\boldsymbol{Z}}_{z\mathrm{C}}^{\mathrm{B}} \\ \widehat{\boldsymbol{T}}_{\mathrm{D}}^{-1} \widehat{\boldsymbol{Z}}_{y\mathrm{D}}^{\mathrm{A}} & \widehat{\boldsymbol{T}}_{\mathrm{D}}^{-1} \widehat{\boldsymbol{Z}}_{y\mathrm{D}}^{\mathrm{B}} \\ \widehat{\boldsymbol{T}}_{\mathrm{D}}^{-1} \widehat{\boldsymbol{Z}}_{z\mathrm{D}}^{\mathrm{A}} & \widehat{\boldsymbol{T}}_{\mathrm{D}}^{-1} \widehat{\boldsymbol{Z}}_{z\mathrm{D}}^{\mathrm{B}} \end{bmatrix} \begin{bmatrix} -\hat{\overline{\mathbf{H}}}_{x}^{\mathrm{A}} \\ \hat{\overline{\mathbf{H}}}_{y}^{\mathrm{A}} \\ \hat{\overline{\mathbf{H}}}_{x}^{\mathrm{B}} \\ -\hat{\overline{\mathbf{H}}}_{y}^{\mathrm{B}} \end{bmatrix} = \hat{\overline{\mathbf{Z}}}_{\mathrm{CD}}^{\mathrm{AB}} \hat{\overline{\mathbf{H}}}_{\mathrm{AB}} \tag{6.68}$$

We transform the field column vectors on the left side with the transformation matrices $\widehat{\boldsymbol{T}}_{\mathrm{CD}}$ and $\widehat{\boldsymbol{T}}_{\mathrm{CD}}$ for ports C and D, respectively. The other matrix parameters in eq. (6.42) are obtained in an analogous way. Now, to analyse the behaviour of a special device, the load impedances at the ports must be calculated.

6.5 NUMERICAL RESULTS

Let us now show some numerical results obtained by the presented algorithm. When comparing our results with other methods, we usually found very good agreement.

6.5.1 Discontinuities in microstrips

The diagrams in Fig. 6.10 show the magnitude and phase of the scattering parameters for a concatenation of two microstrips consisting of different widths (parameters see Fig. 6.9). The results of Koster and Jansen [1] were obtained by a spectral domain approach. A mode-matching technique was used by Schmidt [2]. The results of the algorithm presented in this chapter are in particularly good agreement with Worm's results, obtained by discretisation in x,z-plane (lines to the strips) [3]. The fundamental mode couples into higher-order modes at frequencies above ~6 GHz. By taking into account these higher-order modes, the power balance is fulfilled within a deviation of the order 10^{-12}.

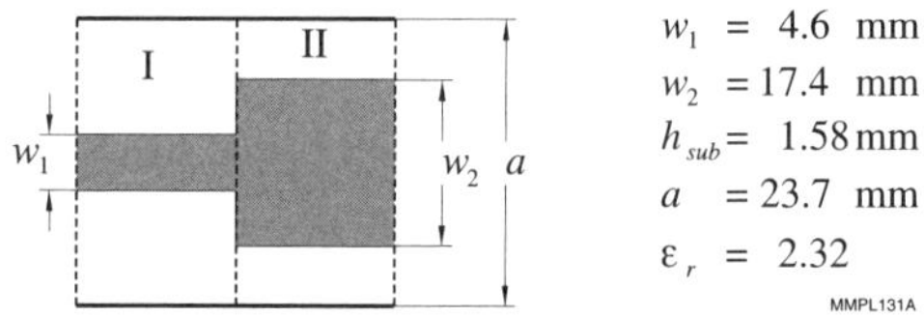

Fig. 6.9 Concatenation of two microstrips of different widths (impedance step) (R. Pregla, 'Analysis of Planar Microwave and Millimeterwave Circuits with Anisotropic Layers Based on Generalized Transmission Line Equations and on the Method of Lines', in *IEEE MTT-S Int. Symp. Dig.*, vol. 1, pp. 125–128. © 2000 Institute of Electrical and Electronics Engineers (IEEE))

Fig. 6.11 is a plot of dispersion curves for a microstrip on an anisotropic substrate. We used a 2D discretisation according to Fig. 2.27. The curves in the middle were obtained by an algorithm with one-dimensional discretisation [4], [5]. These curves are the most accurate and can be used as references. The thin lines are from [6], with discretisation lines vertical to the strips. The results of the thick lines were obtained by using eq. (2.34).

Fig. 6.12 shows the results for an impedance step of a microstrip on an anisotropic substrate. We compare them with those given in [6]. In [6], microstrips with finite length were examined. The diagram also contains results for two steps. The numerical results for the scattering parameters of a transition from microstrip to a suspended substrate line in a rectangular waveguide WR51 are given in Fig. 6.13. In these calculations we used the formulas of Section 6.3.1. As can be seen, the reflections can be reduced by using a larger strip width in waveguide II.

In Fig. 6.14, numerical results for the reflection coefficient as a function of frequency of a microstrip sharp bend are given with the bend angle as parameter. The results were obtained using the algorithm described in Section 6.2. The dashed lines correspond to a substrate with anisotropic properties. The results marked by a ∘ were obtained by a magnetic wall (mw) model and one-dimensional discretisation [7]. In Fig. 6.15 a 90° bend was analysed, constructed by a concatenation of sharp bends of smaller bend

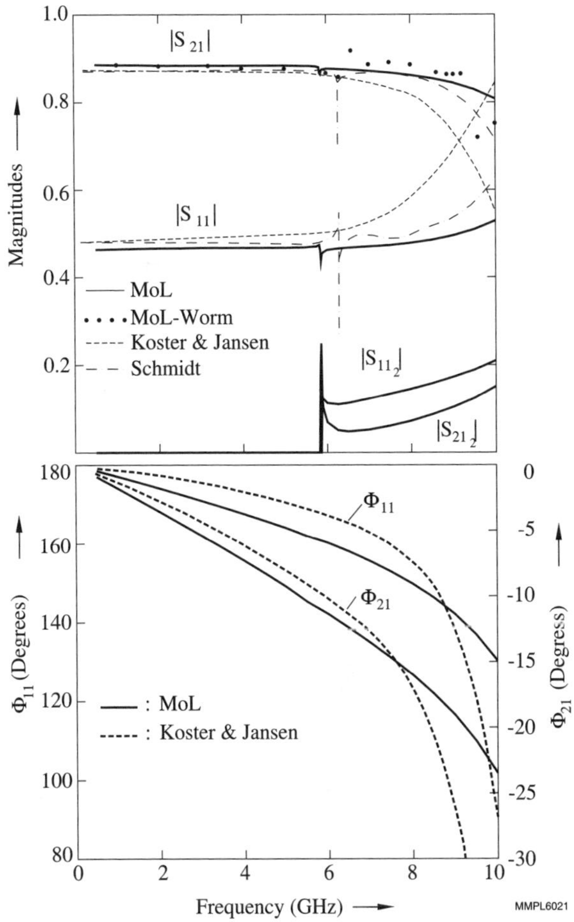

Fig. 6.10 Scattering parameters of the microstrip impedance step in Fig. 6.9 (see Fig. 10 in [1]) (R. Pregla, 'Analysis of Planar Microwave and Millimeterwave Circuits with Anisotropic Layers Based on Generalized Transmission Line Equations and on the Method of Lines', in *IEEE MTT-S Int. Symp. Dig.*, vol. 1, pp. 125–128. © 2000 Institute of Electrical and Electronics Engineers (IEEE))

angles. As can be seen, we can construct a 90° in this way with a total reflection coefficient smaller than 0.05 in the whole frequency range.

6.5.1.1 Microstrip meander lines

A microstrip meander line with discretisation lines scheme is shown in Fig. 6.16.

For the analysis of microstrip meander lines, periodic boundary conditions according to Floquet's theorem were introduced into the difference operators

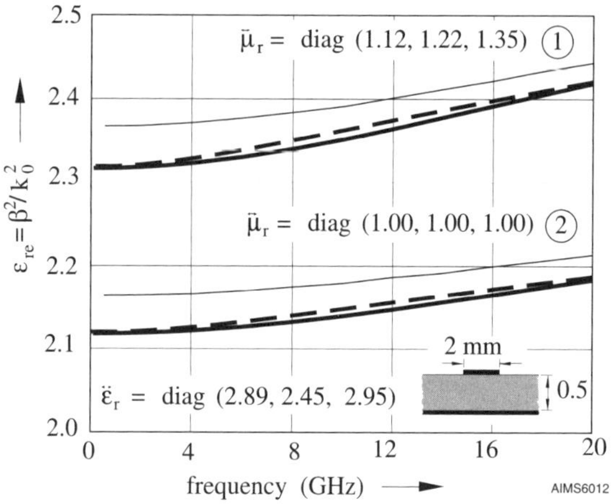

Fig. 6.11 Dispersion curves for a microstrip on an anisotropic substrate (R. Pregla, 'Efficient and Accurate Modeling of Planar Anisotropic Microwave Structures by the Method of Lines', *IEEE Trans. Microwave Theory Tech.*, vol. 50, pp. 1469–1479. © 2002 Institute of Electrical and Electronics Engineers (IEEE))

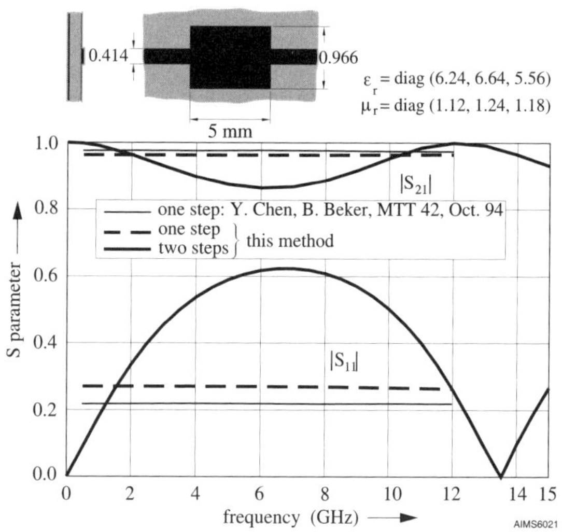

Fig. 6.12 Microstrip impedance steps on an anisotropic substrate (R. Pregla, 'Analysis of Planar Microwave and Millimeterwave Circuits with Anisotropic Layers Based on Generalized Transmission Line Equations and on the Method of Lines', in *IEEE MTT-S Int. Symp. Dig.*, vol. 1, pp. 125–128. © 2000 Institute of Electrical and Electronics Engineers (IEEE))

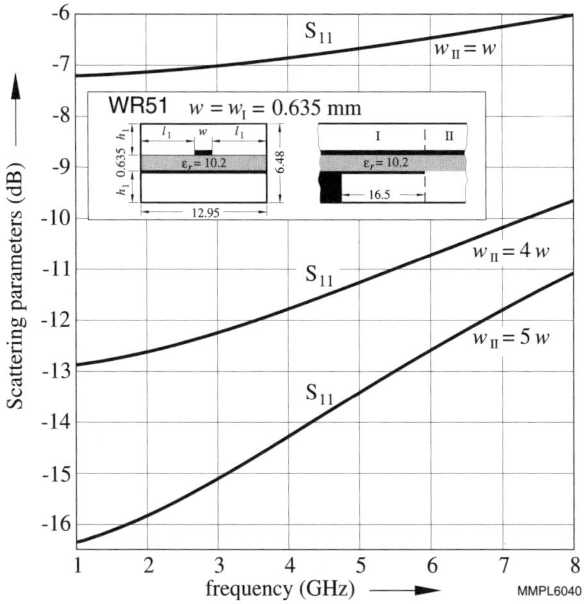

Fig. 6.13 Scattering parameters in a waveguide transition (R. Pregla, 'Efficient and Accurate Modeling of Planar Anisotropic Microwave Structures by the Method of Lines', in *IEEE Trans. Microwave Theory Tech.*, vol. 50, pp. 1469–1479. © 2002 Institute of Electrical and Electronics Engineers (IEEE))

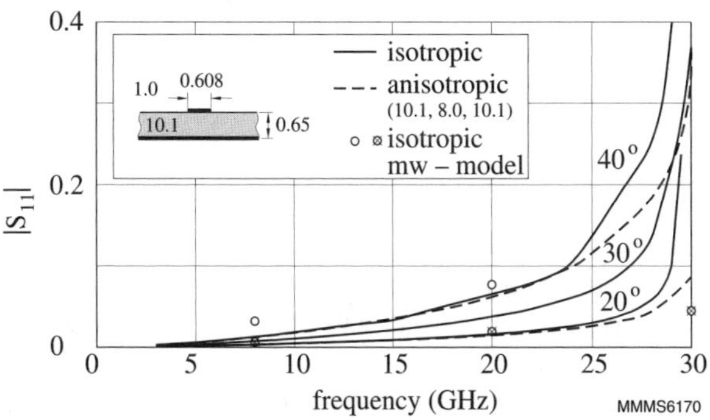

Fig. 6.14 Scattering parameter $|S_{11}|$ for a microstrip sharp bend. Substrate thickness = 0.65 mm, w = 0.608 mm, $\varepsilon_r = 10.1$ - - - $\overleftrightarrow{\epsilon_r} = \mathrm{diag}(10.1, 8.0, 10.1)$, ∘ from [7] (R. Pregla, 'Efficient and Accurate Modeling of Planar Anisotropic Microwave Structures by the Method of Lines', in *IEEE Trans. Microwave Theory Tech.*, vol. 50, pp. 1469–1479. © 2002 Institute of Electrical and Electronics Engineers (IEEE))

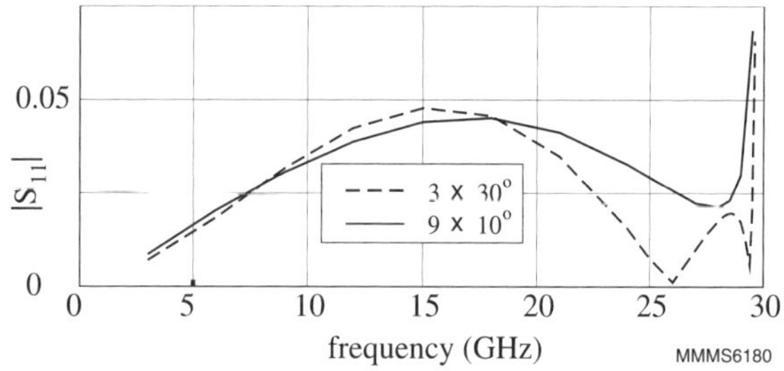

Fig. 6.15 Scattering parameter $|S_{11}|$ for concatenated sharp microstrip bends to form a 90° bend. Dimensions of the microstrip are as in Fig. 6.17 (R. Pregla, 'Efficient and Accurate Modeling of Planar Anisotropic Microwave Structures by the Method of Lines', *IEEE Trans. Microwave Theory Tech.*, vol. 50, pp. 1469–1479. © 2002 Institute of Electrical and Electronics Engineers (IEEE))

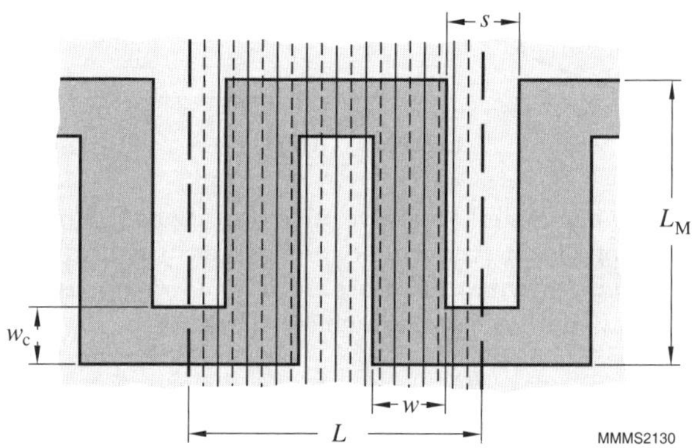

Fig. 6.16 One period of a microstrip meander line with discretisation lines

(see Section A.2.9). This is different to Chapter 5, where the periodicity was mainly treated in the analytic part. The dispersion curves for a microstrip meander line are given in Fig. 6.17. The dashed curve is copied from [8, 9], where the discretisation lines are perpendicular to the strip. The full thick line was obtained by using 2D discretisation. The impedance transformation from both ends (upper and lower ends) results in an indirect eigenvalue problem, e.g. in the middle of length L_M. The thin line is for an anisotropic substrate. The arithmetic mean value of the permittivities is equal to that of the isotropic case. A microstrip meander line with stubs is shown in Fig. 6.18. Dispersion curves for the case of end stubs are shown in Fig. 6.19.

6.5.1.2 Microstrip filter structure

Fig. 6.20 shows the determined scattering parameters for a filter as shown in Fig. 2.1a. In the upper diagram the filter contains only one resonator. The

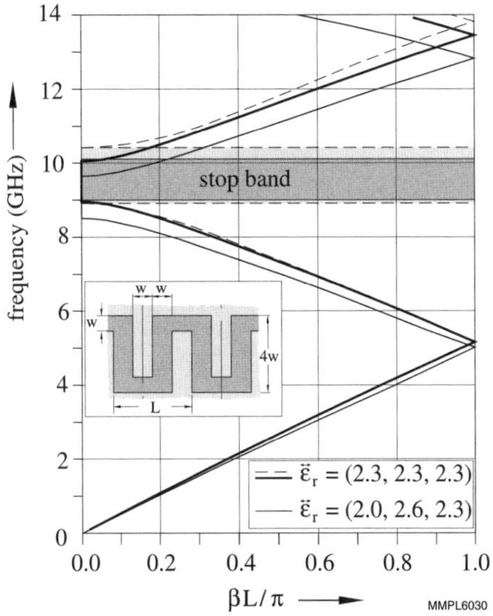

Fig. 6.17 Dispersion curves for a microstrip meander line. substrate thickness = 0.79 mm, w = 2.37 mm, ε_r = 2.3 - - - results of Worm [8, 9] with lines perpendicular to the structure (R. Pregla, 'Efficient and Accurate Modeling of Planar Anisotropic Microwave Structures by the Method of Lines', *IEEE Trans. Microwave Theory Tech.*, vol. 50, pp. 1469–1479. © 2002 Institute of Electrical and Electronics Engineers (IEEE))

results for an isotropic substrate are in good agreement with other results. The filter for the lower diagram consists of two resonators. Curves for isotropic and anisotropic substrates are shown.

The scattering parameters of the next two microwave filters were computed using the non-equidistant difference operators described in Section A.4. This was necessary because both filters feature big differences between the widths of microstrips and the intermediate gaps. The first analysed filter is a printed microstrip filter composed of T-shaped port elements and loop resonators (Fig. 6.21). The magnitude of scattering parameters (Fig. 6.22 [15]) is in very good agreement with measured and calculated results reported in [10].

The second analysed filter is a printed microstrip filter composed of two open-loop resonators (Fig. 6.23). The obtained results for this filter (Fig. 6.24) are also in good agreement with the results reported by Chang [10] and Bozzi [11].

6.5.2 Waveguide junctions

6.5.2.1 Coplanar T-junctions

T-junctions play an important role in a wide number of micro and millimetre-wave circuits, e.g. in couplers, shifters, power dividers, filters and

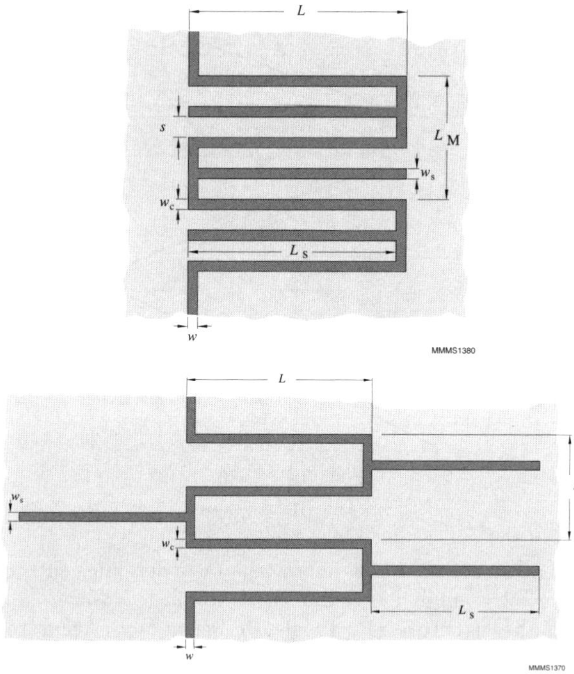

Fig. 6.18 Microstrip meander line with stubs

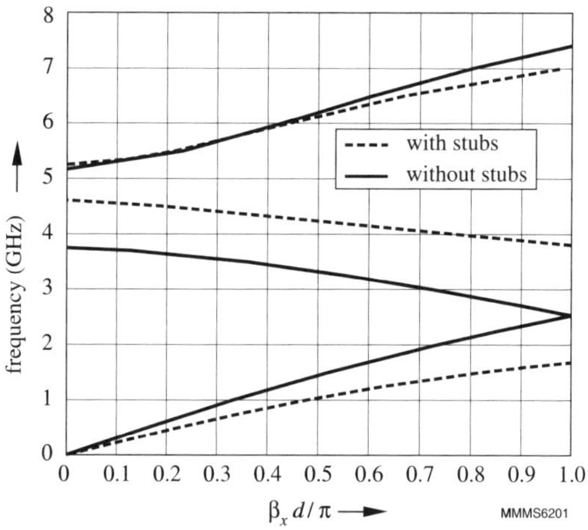

Fig. 6.19 Dispersion curves for a microstrip meander line with end stubs. Substrate thickness = 0.79 mm, $w = 2.37$ mm, $\varepsilon_r = 2.3$

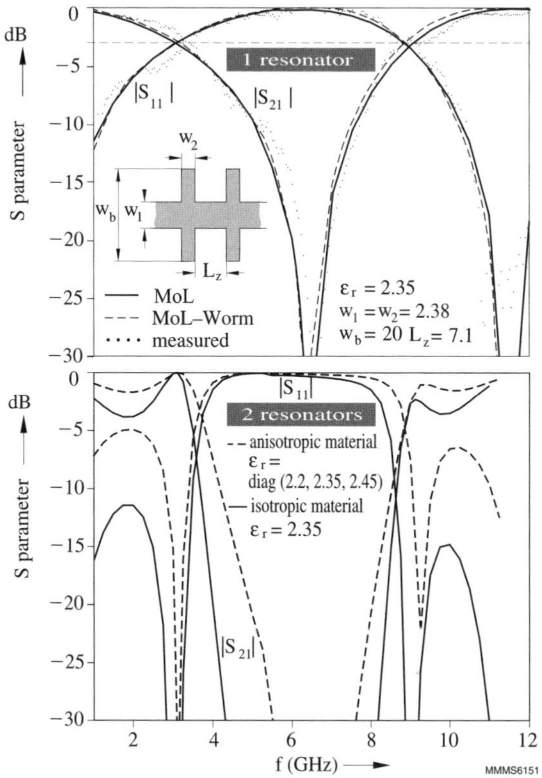

Fig. 6.20 Scattering parameters for a microstrip filter (R. Pregla, 'Analysis of Planar Microwave and Millimeterwave Circuits with Anisotropic Layers Based on Generalized Transmission Line Equations and on the Method of Lines', in *IEEE MTT-S Int. Symp. Dig.*, vol. 1, pp. 125–128. © 2000 Institute of Electrical and Electronics Engineers (IEEE))

multiplexers [12]. Coplanar waveguides (CPW) are used in monolithic microwave integrated circuits (MMICs). Their principal advantage is the location of the signal grounds on the same substrate surface as the signal line. This eliminates the need for via holes and simplifies the fabrication process. The other advantages of CPW lines are reduced parasitic inductance, low dispersion and low loss. A disadvantage of the CPW is the existence of two modes: the coplanar mode and the parasitic slot-line mode. The latter mode is removed by short-circuiting it with air bridges.

In modelling T-junctions with accurate results, the full 3D analysis must be used. This analysis requires considerable numerical effort. But the use of two crossed 2D line systems for modelling the central junction region instead of a full 3D discretisation allows us to significantly reduce the numerical effort. Only the cross-section need be discretised; in the direction of propagation

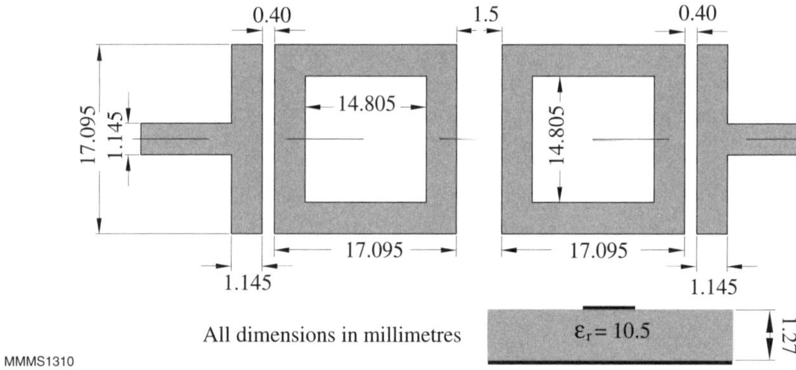

Fig. 6.21 Printed microstrip filter composed of T-shaped port elements and loop resonators (L. Greda and R. Pregla, 'Hybrid Analysis of Three-Dimensional Structures by the Method of Lines Using Novel Nonequidistant Discretization', in *IEEE MTT-S Int. Symp. Dig.*, pp. 1877–1880. © 2002 Institute of Electrical and Electronics Engineers (IEEE))

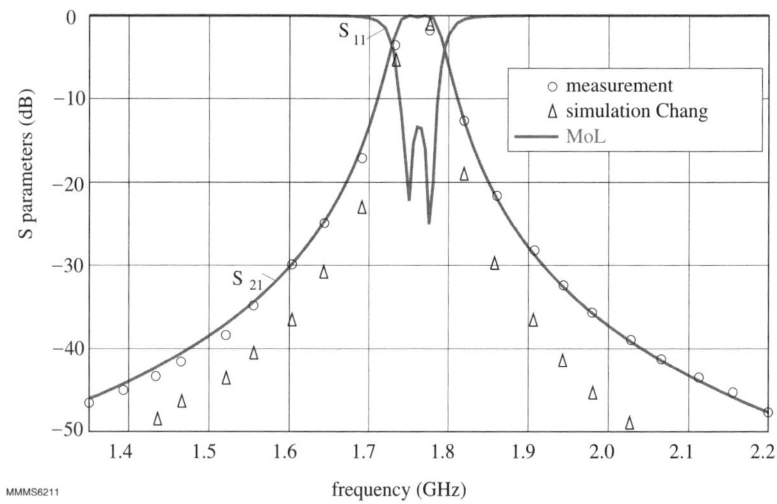

Fig. 6.22 Scattering parameters for the filter shown in Fig 6.21 (L. Greda and R. Pregla, 'Hybrid Analysis of Three-Dimensional Structures by the Method of Lines Using Novel Nonequidistant Discretization', in *IEEE MTT-S Int. Symp. Dig.*, pp. 1877–1880. © 2002 Institute of Electrical and Electronics Engineers (IEEE))

an analytical solution is obtained. For this reason, the size of the discretisation window depends only on the cross-section of the analysed structure. This property – once again – makes the Method of Lines particularly suitable for long structures. The impedance transformation concept used

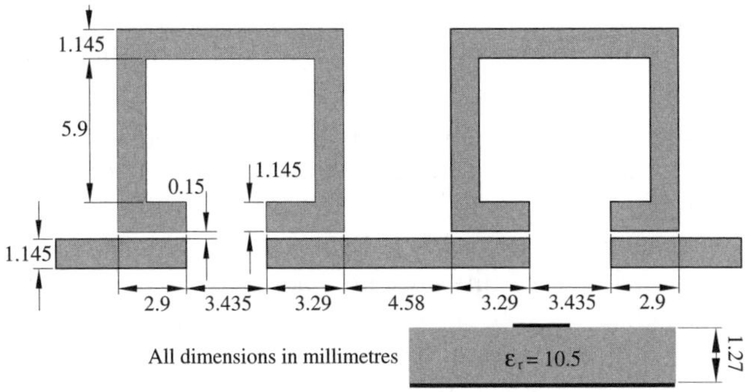

Fig. 6.23 Printed microstrip filter composed of two open-loop resonators (R. Pregla, S. F. Helfert, L. Greda and A. Barcz, 'The impedance/admittance transformation concept in engineering electrodynamics', in *Fields, Networks, Computational Methods, and Systems in Modern Electrodynamics – A Tribute to Leopold B. Felsen*, M. Mongiardo P. Russer (Ed.), pp. 179–186. © 2004 Springer Germany)

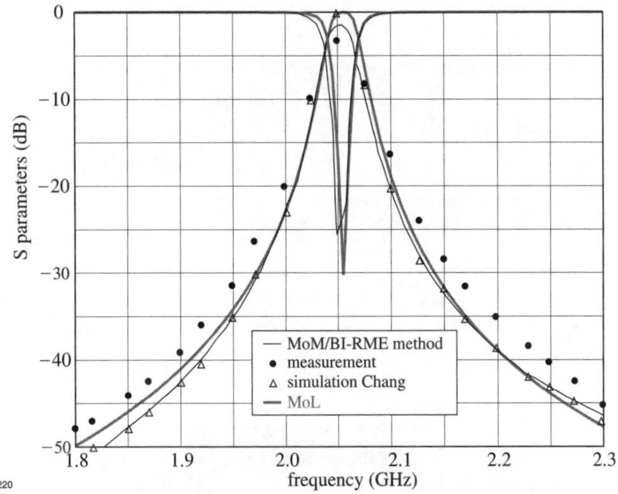

Fig. 6.24 Scattering parameters for the filter shown in Fig. 6.23 (L. Greda and R. Pregla, 'Modeling of Planar Microwave Filters', in *European Microwave Week*. © 2003 European Microwave Association (EuMA))

in this approach is numerically stable and gives correct results even for very long sections. The metal and dielectric loss, as well as anisotropic material parameters, can easily be incorporated. In the case of structures with big differences in the dimensions of the elements of the cross-section, non-equidistant discretisation can be used. As an example of a coplanar T-junction, the structure shown in Fig. 6.25 is analysed. The central junction region is bounded by three ports. In general, waveguides of any dimension to all these ports can be connected. The side walls of connecting waveguides can be chosen to be either magnetic or electric. The field relations at the inner side of each port can be described with the help of open-circuit (magnetic wall) or short-circuit (electric wall) matrix parameters. These parameters are calculated by open- or short-circuiting the ports. The scattering parameters of a coplanar transmission line with a shorted stub and air-bridges (Fig. 6.25) have been calculated. The results presented here were obtained for two lengths of the shorted stub: $l_1 = 100$ μm and $l_1 = 500$ μm. An ideal, infinitely thin metal and lossless dielectric were assumed. The magnitude of the scattering parameters shown in Fig. 6.26 is in very good agreement with results obtained by the FDTD method and Sonnet Software [16]. The electric field distribution of two propagating modes in this structure, obtained from the eigenvector matrix $\widehat{\boldsymbol{T}}_{\mathrm{E}}$, is presented in Fig. 6.27.

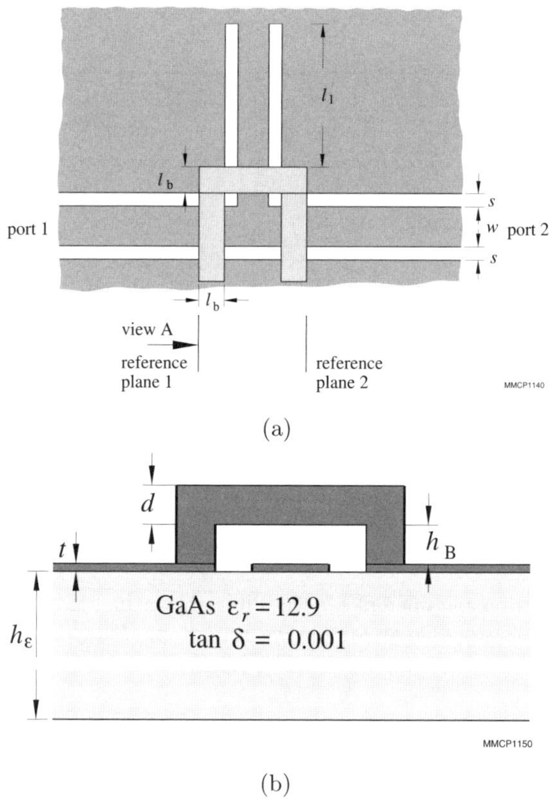

Fig. 6.25 Coplanar T-junction: (a) top view (b) cross-section (Reproduced by permission of Elsevier)

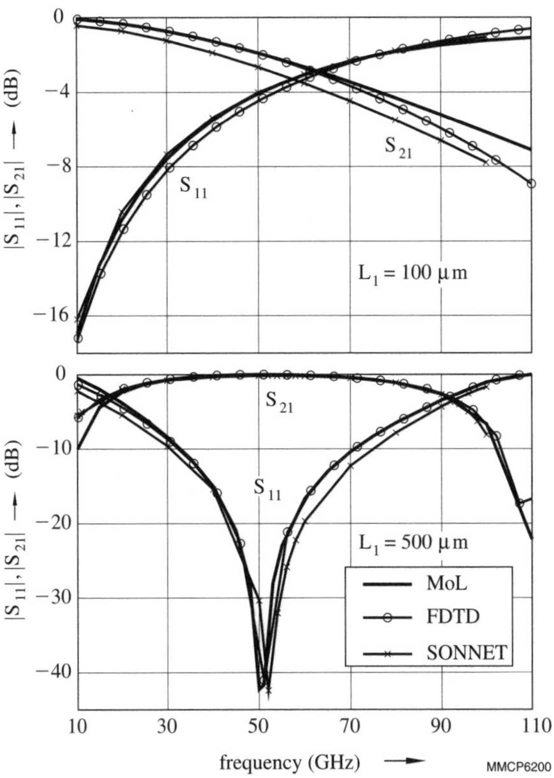

Fig. 6.26 Scattering parameters of a coplanar T-junction. The dimensions of the structure (see Fig. 6.25) are: $w = 20\ \mu\text{m}$, $s = 15\ \mu\text{m}$, $l_b = 20\ \mu\text{m}$, $h_\varepsilon = 100\ \mu\text{m}$, $t \to 0$, $h_B = 3\ \mu\text{m}$, $d = 5\ \mu\text{m}$ (Reproduced by permission of Elsevier)

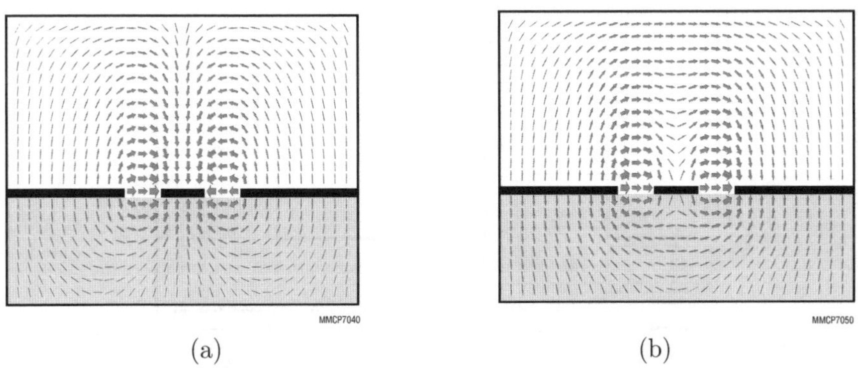

Fig. 6.27 Electric field distribution (a) of the coplanar mode and (b) of the parasitic slot-line mode (Reproduced by permission of Elsevier)

6.5.2.2 Microstrip junction on anisotropic substrate

To demonstrate the validity of the algorithm for junctions in the case of anisotropic materials, Fig. 6.28 shows the results of scattering parameters for a symmetrical microstrip junction [18]. The substrate has a thickness $h = 0.635$ mm and the microstrip has a width $w = 0.6$ mm.

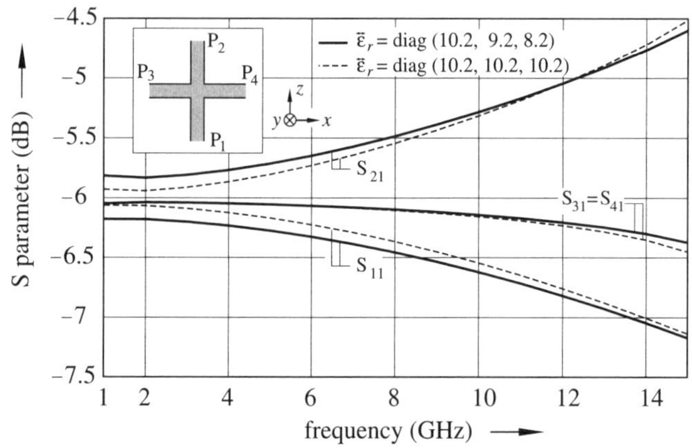

Fig. 6.28 Scattering parameters for a microstrip junction on isotropic and anisotropic substrate (Reproduced by permission of Union Radio-Scientifique Internationale–International Union of Radio Science (URSI))

6.5.2.3 Junctions of planar and rectangular waveguides

In modified form, the algorithm for the analysis of planar waveguide junctions described in Section 6.4 can also be used for junctions of planar and rectangular waveguides. An example is shown in Fig. 6.29, where we have a transition from a microstrip to a rectangular waveguide [13]. In this case, the discretisation lines of the rectangular waveguide cross the microstrip line in the vertical direction. Waveguide-to-microstrip and waveguide-to-coplanar line transitions are essential parts of microwave circuits. There are several

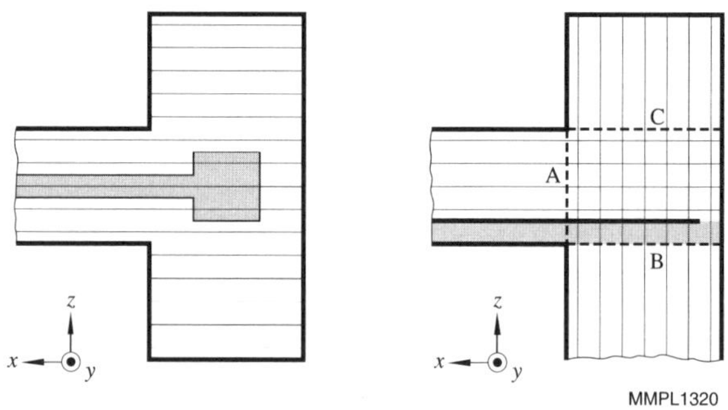

Fig. 6.29 Rectangular waveguide-to-microstrip transition [13]

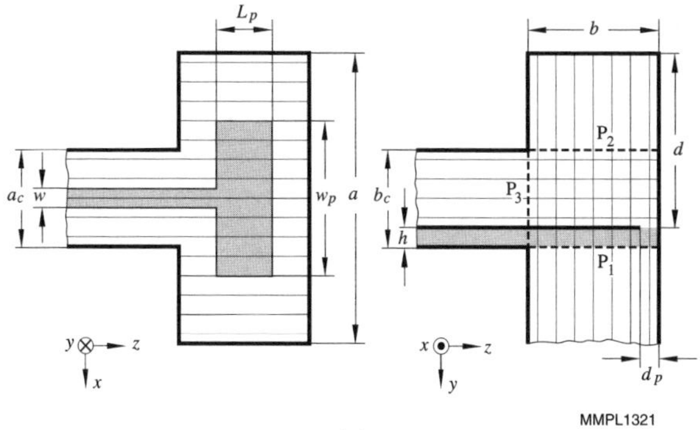

(a)

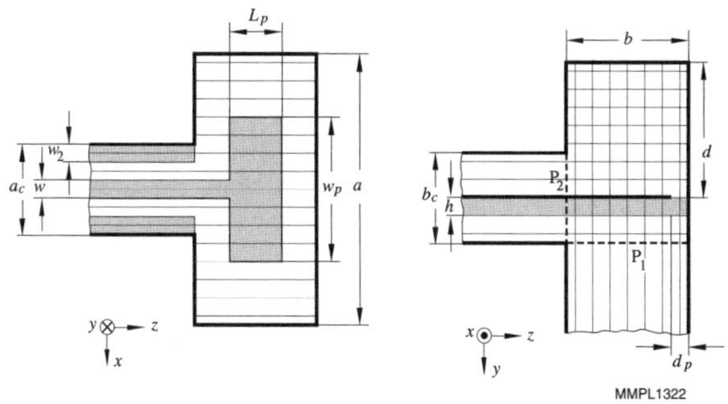

(b)

Fig. 6.30 Rectangular waveguide-to-microstrip (a) and waveguide-to-coplanar line (b) transitions with two possible discretisation schemes (L. Greda and R. Pregla, 'Efficient analysis of waveguide-to-microstrip and waveguide-to-coplanar line transitions', in *IEEE MTT-S Int. Symp. Dig.*, pp. 1241–1244. © 2001 Institute of Electrical and Electronics Engineers (IEEE))

types of such transitions. The most well-known include a ridged-waveguide taper, a finline taper and an E-plane probe. A new transition type with a rectangular patch instead of a strip probe, proposed by Machac *et al.* [14], has significantly broader bandwidth. This new transition type can occur in two versions: microstrip and coplanar line (Fig. 6.30). The analysis procedure will be demonstrated using the example of the waveguide-to-microstrip transition given in Fig. 6.30a. The procedure for analysis of waveguide-to-coplanar line transitions is analogous.

The use of two crossed two-dimensional line systems for modelling the central region, instead of a full three-dimensional discretisation, allows us to significantly reduce the numerical effort. Additionally, to reduce the total number of lines needed for modelling the central region, a non-equidistant discretisation can be used. The discretisation scheme for one of the

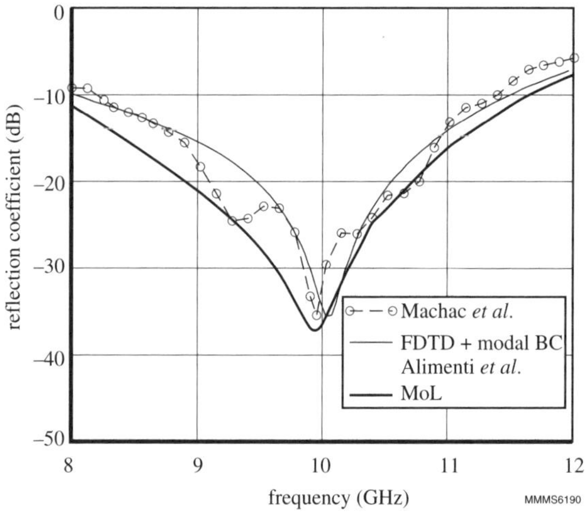

Fig. 6.31 Magnitude of the reflection coefficient of waveguide-to-microstrip transition. The dimensions of the structure (description in Fig. 6.30a) are: a = 22.86 mm, b = 10.16 mm, h = 0.794 mm, w = 2.3 mm, L_p = 6 mm, w_p = 12 mm, d_p = 3.16 mm, d = 5.3 mm, a_c = 8.0 mm, b_c = 5.0 mm (L. Greda and R. Pregla, 'Efficient analysis of waveguide-to-microstrip and waveguide-to-coplanar line transitions', in *IEEE MTT-S Int. Symp. Dig.*, pp. 1241–1244. © 2001 Institute of Electrical and Electronics Engineers (IEEE))

discretisation line systems (discretisation lines in z-direction) in the transition region is shown in principle (actually, the discretisation is 2D) in Fig. 6.29. The discretisation scheme for the other discretisation line system (in y-direction) is completely analogous.

For the structures depicted in Fig. 6.30, there are two possibilities of analysis: with the subdivision on two- or three-port junctions. In the case of the two-port analysis, more lines are needed to model the central region, but there are less impedance/admittance transformations performed and lower programming effort is required. The choice of subdivision method depends on the distance between port P2 (Fig. 6.30a) and the back short. In cases where this distance is small, the two-port analysis is more efficient, otherwise it is worth using the three-port analysis.

The reflection coefficient of the structure shown in Fig. 6.30a is presented in Fig. 6.31 [17]. It has been calculated and compared with experimental [14] and theoretical results using FDTD + modal BC [13].

The waveguide type R100 (WR90) and a substrate with permittivity ϵ_r = 2.35 were used. This result is in good agreement with experimental and theoretical results. A slightly lower reflection coefficient is obtained than the measured one, because of the assumption of an ideal metal and substrate.

References

[1] N. H. Koster and R. H. Jansen, 'The Microstrip Step Discontinuity: A Revised Description', *IEEE Trans. Microwave Theory Tech.*, vol. MTT-34, pp. 213–223, 1986.

[2] L. P. Schmidt, 'Rigorous Computation of the Frequency Dependent Properties of Filters and Coupled Resonators Composed from Transverse Microstrip Discontinuities', in *10th Eur. Microwave Conf.*, Warsaw, Poland, 1980, pp. 436–440.

[3] S. B. Worm, 'Full-Wave Analysis of Discontinuities in Planar Waveguides by the Method of lines Using a Source Approach', *IEEE Trans. Microwave Theory Tech.*, vol. MTT-38, pp. 1510–1514, 1990.

[4] R. Pregla, 'The Analysis of Wave Propagation in General Anisotropic Multilayered Waveguides by the Method of Lines', in *U.R.S.I intern. Symp. Electromagn. Theo.*, Thessaloniki, Greece, May 1998, pp. 51–53.

[5] R. Pregla, 'A Generalized Algorithm for Analysis of Planar Multilayered Anisotropic Waveguide Structures by the Method of Lines', *AEÜ*, vol. 52, no. 2, pp. 94–98, 1998.

[6] Y. Chen and B. Beker, 'Study of Microstrip Step Discontinuities on Bianisotropic Substrates Using the Method of Lines and Transverse Resonance Technique', *IEEE Trans. Microwave Theory Tech.*, vol. 42, no. 10, pp. 1945–1950, Oct. 1994.

[7] R. Pregla, 'Analysis of a Bend Discontinuity by the Method of Lines', *Frequenz*, vol. 45, pp. 213–216, 1991.

[8] S. B. Worm, *Analysis of Planar Microwave Structures with Arbitrary Contour (in German)*, PhD thesis, FernUniversität – Hagen, 1983.

[9] S. B. Worm and R. Pregla, 'Hybrid Mode Analysis of Arbitrarily Shaped Planar Microwave Structures by the Method of Lines', *IEEE Trans. Microwave Theory Tech.*, vol. MTT-32, pp. 191–196, 1984.

[10] C.-C. Yu and K. Chang, 'Novel Compact Elliptic-Function Narrow-Band Bandpass Filters Using Microstrip Open-Loop Resonators with Coupled and Crossing Lines', *IEEE Trans. Microwave Theory Tech.*, vol. MTT-46, no. 7, pp. 952–958, 1998.

[11] M. Bozzi, L. Perregrini, A. A. Melcón, M. Gugliemi and G. Gonciauro, 'MoM/BI–RME Analysis of Boxed MMICs with Arbitrarily Shaped Metallizations', *IEEE Trans. Microwave Theory Tech.*, vol. MTT-49, no. 12, pp. 2227–2234, 2001.

[12] J. M. Rebollar, J. Esteban and J. E. Page, 'Fullwave Analysis of Three and Four-Port Rectangular Waveguide Junctions', *IEEE Trans. Microwave Theory Tech.*, vol. MTT-42, no. 2, pp. 256–263, Feb. 1994.

[13] F. Alimenti, P. Mezzanotte, L. Roselli and R. Sorrentino, 'A Revised Formulation of Modal Absorbing and Matched Modal Source Boundary Conditions for the Efficient FDTD Analysis of Waveguide Structures', *IEEE Trans. Microwave Theory Tech.*, vol. MTT-48, no. 1, pp. 50–59, 2000.

[14] J. Machac and W. Menzel, 'On the Design of Waveguide-to-Microstrip and Waveguide-to-Coplanar Line Transitions', in *European Microwave Conference*, Madrid , Spain, 1993, pp. 615–616.

[15] Ł. Gręda and R. Pregla, 'Hybrid Analysis of Three-Dimensional Structures by the Method of Lines Using Novel Nonequidistant Discretization', in *IEEE MTT-S Int. Symp. Dig.*, Seattle, USA, June 2002, pp. 1877–1880.

[16] Ł. Gręda and R. Pregla, 'Analysis of Coplanar T-junctions by the Method of Lines', *AEÜ*, vol. 55, pp. 313–318, 2001.

[17] Ł. Gręda and R. Pregla, 'Efficient Analysis of Waveguide-to-Microstrip and Waveguide-to-Coplanar Line Transitions', in *IEEE MTT-S Int. Symp. Dig.*, Phoenix, USA, May 2001, pp. 1241–1244.

[18] R. Pregla and L. A. Greda, 'Modeling of Waveguide Junctions with General Anisotropic Material by Using Orthogonal Propagating Waves', in *URSI-B-Symposium*, Pisa/Italy, May 2004.

Further Reading

[19] A. Barcz, 'Analysis of the Wave Propagation in Planar Structures with Substrates of Magnetized Ferrites', PhD thesis, FernUniversität in Hagen, 2006.

CHAPTER 7

PRECISE RESOLUTION WITH AN ENHANCED AND GENERALISED LINE ALGORITHM

7.1 INTRODUCTION

Modern integrated circuits for microwave and millimetre frequencies consist of various composite layers with embedded metallic strips. The trend towards miniaturisation results in closer positioning of all elements. Usually, the dimensions of the waveguide cross-sections are small compared to the lengths of sections in the circuits.

Figs. 7.1–7.5 give an example of a complex cross-section. Such a waveguide structure is e.g. important for Lange couplers in millimetre wave technology. In this case the metallic strips may have equal dimensions. Generally, the dimensions in the cross-section are different. The slot widths between the strips and their heights may be small compared to the widths of the metallic strips and the thicknesses of the layers. To analyse such waveguide structures precisely, the modelling algorithms should take these factors into account. Here we will show a special discretisation scheme with crossed discretisation lines. For eigenmode problems, this means that we use two one-dimensional discretisation schemes that cross each other orthogonally. So we do not use 2D-discretisation! This type of discretisation is analogous to the special type of Mode Matching Technique (MMT) described by Kühn [1]. The right part of Fig. 7.1 shows how this discretisation [2] should be performed for the example structure. We have some regions that are bounded by metallisations on two sides. In these regions we use discretisation lines parallel to the metallisation surface. In regions H_i $(i = 1, 2, 3, 4)$ we have horizontal, and in region V_1 vertical discretisation lines. Regions C_1 and C_2 connect the regions with horizontal and vertical discretisation lines. Therefore, in these regions the lines from the neighbouring regions should have a continuation. This in turn requires the use of crossed lines in regions C_1 and C_2. The numbers of discretisation lines in horizontal and vertical directions can be chosen separately. Therefore, the number of lines e.g. between the strips should be large enough to obtain a high accuracy without increasing the numerical effort too much. In the regions with unidirectional discretisation lines, the field description is as shown in Section 7.2 in the relevant other chapters.

The principle of crossed discretisation lines can also be used in waveguides that are described in non-Cartesian coordinates. In Fig. 7.6 two different

Analysis of Electromagnetic Fields and Waves R. Pregla

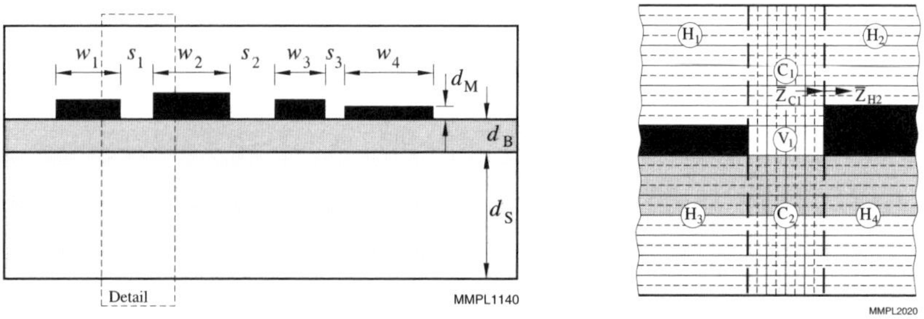

Fig. 7.1 Cross-section of a millimetre-wave circuit with details for alternative discretisation (R. Pregla, 'MoL-Mode Analysis with Precise Resolution by an Enhanced and Generalized Line Algorithm', in *IEEE MTT-S Int. Symp. Dig.*, vol. 3, pp. 1543–1546. © 1998 Institute of Electrical and Electronics Engineers (IEEE))

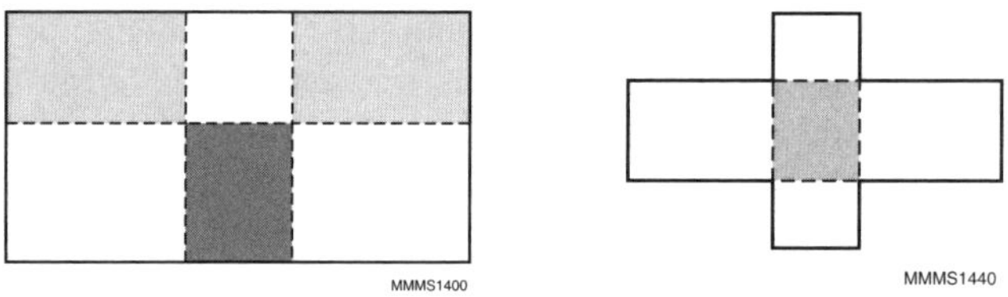

(a) (b)

Fig. 7.2 Cross-sections of rectangular waveguides: ridge waveguide (a) and cross waveguide (b)

examples are shown in cylindrical coordinates. The first one is a coaxial ridge guide used in mode transformers [3]. The second is the cross-section of a magnetron resonator [4].

In this chapter, the concept of crossed discretisation lines in cylindrical coordinates is extended to more complex structures. The cross-sections of the waveguides in Fig. 7.6 are impressive examples of this extension.

7.2 CROSSED DISCRETISATION LINES AND CARTESIAN COORDINATES

7.2.1 Theoretical background

The field description in regions with crossed lines can be given in the following way: an arbitrary region R (e.g. region C_1 in Fig. 7.1), with four ports and the (tangential) fields there, is shown in Fig. 7.8. If we know these tangential fields, we can determine the Huygen's sources. Because of the law of unambiguity of electromagnetic field theory, the field outside region R is then exactly determined. Inside region R the material is linear and without sources. Further, the material here is isotropic and homogeneous.

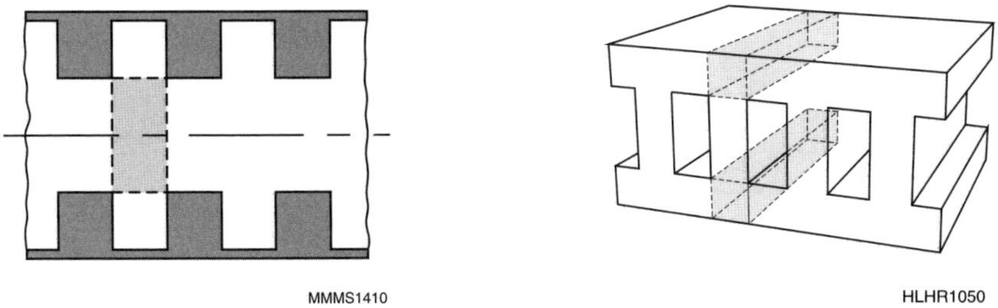

(a) (b)

Fig. 7.3 (a) Corrugated waveguide (rectangular or circular) (b) branch waveguide coupler

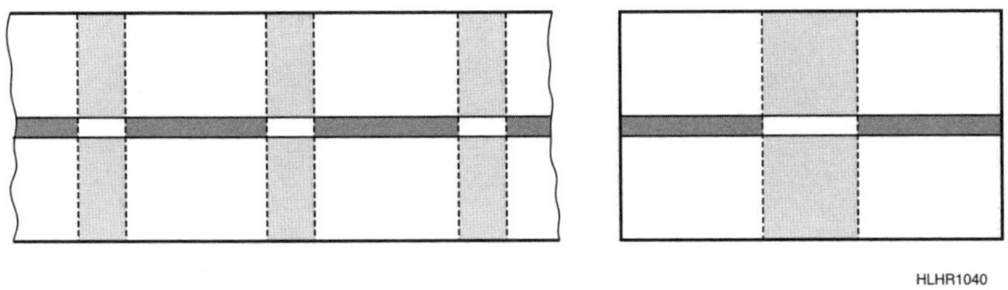

(a) (b)

Fig. 7.4 Waveguides coupled by irises: (a) longitudinal section (b) cross-section

We define supervectors of the fields at opposite ports according to:

$$\begin{aligned} \widehat{\mathbf{H}}_{\mathrm{AB}} &= [\widehat{\mathbf{H}}_{\mathrm{A}}^{\mathrm{t}}, -\widehat{\mathbf{H}}_{\mathrm{B}}^{\mathrm{t}}]^{\mathrm{t}} \qquad & \widehat{\mathbf{E}}_{\mathrm{AB}} &= [\widehat{\mathbf{E}}_{\mathrm{A}}^{\mathrm{t}}, \widehat{\mathbf{E}}_{\mathrm{B}}^{\mathrm{t}}]^{\mathrm{t}} \\ \widehat{\mathbf{H}}_{\mathrm{CD}} &= [\widehat{\mathbf{H}}_{\mathrm{C}}^{\mathrm{t}}, -\widehat{\mathbf{H}}_{\mathrm{D}}^{\mathrm{t}}]^{\mathrm{t}} \qquad & \widehat{\mathbf{E}}_{\mathrm{CD}} &= [\widehat{\mathbf{E}}_{\mathrm{C}}^{\mathrm{t}}, \widehat{\mathbf{E}}_{\mathrm{D}}^{\mathrm{t}}]^{\mathrm{t}} \end{aligned} \tag{7.1}$$

$\widehat{\mathbf{E}}_{\mathrm{U}}$ and $\widehat{\mathbf{H}}_{\mathrm{U}}$ (U = A, B, C, D) are supervectors of the discretised tangential fields at port U. Each supervector contains two discretised tangential components (see below). The discretisation is performed according to the rules described in Chapter 2.

The fields inside region R are completely determined if the tangential fields at the boundaries are known. Due to the linearity of the materials and of Maxwell's equations, the following relation for the tangential fields at ports A to D of the general region R holds (for the discretisation shown in Fig. 7.9):

$$\begin{bmatrix} \widehat{\mathbf{H}}_{\mathrm{A}} \\ -\widehat{\mathbf{H}}_{\mathrm{B}} \\ \widehat{\mathbf{H}}_{\mathrm{C}} \\ -\widehat{\mathbf{H}}_{\mathrm{D}} \end{bmatrix} = \begin{bmatrix} \widehat{y}_{\mathrm{AA}} & \cdots & \widehat{y}_{\mathrm{AD}} \\ \vdots & & \vdots \\ \widehat{y}_{\mathrm{DA}} & \cdots & \widehat{y}_{\mathrm{DD}} \end{bmatrix} \begin{bmatrix} \widehat{\mathbf{E}}_{\mathrm{A}} \\ \widehat{\mathbf{E}}_{\mathrm{B}} \\ \widehat{\mathbf{E}}_{\mathrm{C}} \\ \widehat{\mathbf{E}}_{\mathrm{D}} \end{bmatrix} \tag{7.2}$$

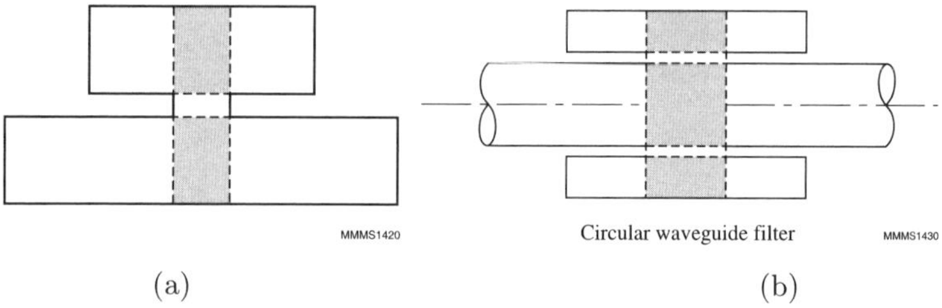

Fig. 7.5 Filters by coupled resonators: (a) rectangular waveguides (b) circular waveguides

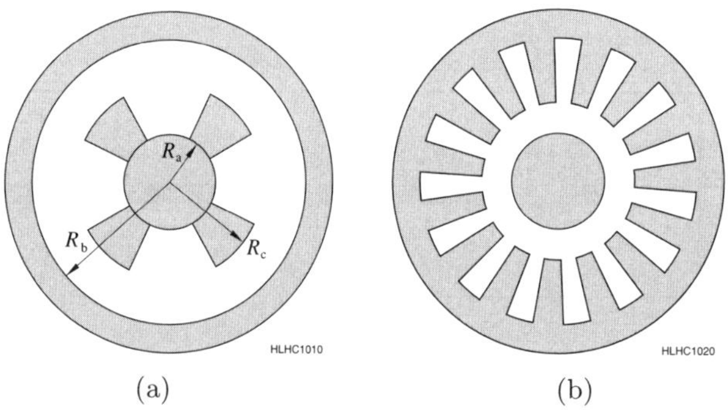

Fig. 7.6 Cross-section of a (a) coaxial ridge guide and (b) magnetron resonator (Reproduced by permission of IGTE (Institut für Grundlagen und Theorie der Elektrotechnik) Graz)

or (for the discretisation shown in Fig. 7.10):

$$\begin{bmatrix} \widehat{\mathbf{E}}_{\mathrm{A}} \\ \widehat{\mathbf{E}}_{\mathrm{B}} \\ \widehat{\mathbf{E}}_{\mathrm{C}} \\ \widehat{\mathbf{E}}_{\mathrm{D}} \end{bmatrix} = \begin{bmatrix} \widehat{\mathbf{z}}_{\mathrm{AA}} & \cdots & \widehat{\mathbf{z}}_{\mathrm{AD}} \\ \vdots & & \vdots \\ \widehat{\mathbf{z}}_{\mathrm{DA}} & \cdots & \widehat{\mathbf{z}}_{\mathrm{DD}} \end{bmatrix} \begin{bmatrix} \widehat{\widehat{\mathbf{H}}}_{\mathrm{A}} \\ -\widehat{\widehat{\mathbf{H}}}_{\mathrm{B}} \\ \widehat{\widehat{\mathbf{H}}}_{\mathrm{C}} \\ -\widehat{\widehat{\mathbf{H}}}_{\mathrm{D}} \end{bmatrix} \tag{7.3}$$

In a more compact form, we can write:

$$\begin{bmatrix} \widehat{\mathbf{H}}_{\mathrm{AB}} \\ \widehat{\mathbf{H}}_{\mathrm{CD}} \end{bmatrix} = \begin{bmatrix} \widehat{\mathbf{y}}_{\mathrm{AB}}^{\mathrm{AB}} & \widehat{\mathbf{y}}_{\mathrm{AB}}^{\mathrm{CD}} \\ \widehat{\mathbf{y}}_{\mathrm{CD}}^{\mathrm{AB}} & \widehat{\mathbf{y}}_{\mathrm{CD}}^{\mathrm{CD}} \end{bmatrix} \begin{bmatrix} \widehat{\mathbf{E}}_{\mathrm{AB}} \\ \widehat{\mathbf{E}}_{\mathrm{CD}} \end{bmatrix} \quad \begin{bmatrix} \widehat{\mathbf{E}}_{\mathrm{AB}} \\ \widehat{\mathbf{E}}_{\mathrm{CD}} \end{bmatrix} = \begin{bmatrix} \widehat{\mathbf{z}}_{\mathrm{AB}}^{\mathrm{AB}} & \widehat{\mathbf{z}}_{\mathrm{AB}}^{\mathrm{CD}} \\ \widehat{\mathbf{z}}_{\mathrm{CD}}^{\mathrm{AB}} & \widehat{\mathbf{z}}_{\mathrm{CD}}^{\mathrm{CD}} \end{bmatrix} \begin{bmatrix} \widehat{\mathbf{H}}_{\mathrm{AB}} \\ \widehat{\mathbf{H}}_{\mathrm{CD}} \end{bmatrix} \tag{7.4}$$

Here, we combined two field vectors (including signs) to a supervector and four submatrices to super-submatrices. The admittance matrices can be calculated in the following way: short-circuiting ports C and D (by using metallic walls) results in $\widehat{\mathbf{E}}_{\mathrm{CD}} = \mathbf{0}$. The matrices $\widehat{\mathbf{y}}_{\mathrm{AB}}^{\mathrm{AB}}$ and $\widehat{\mathbf{y}}_{\mathrm{CD}}^{\mathrm{AB}}$ are determined

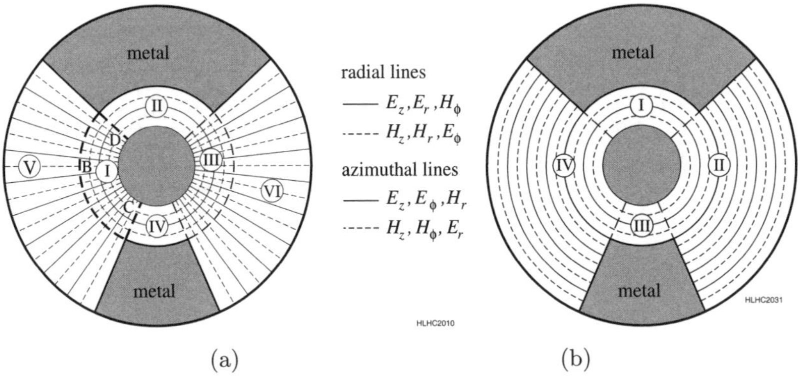

Fig. 7.7 Cross-section of a general circular ridge guide with (a) crossed (Reproduced by permission of IGTE (Institut für Grundlagen und Theorie der Elektrotechnik) Graz) and (b) conventional discretisation lines (L. Greda and R. Pregla, 'MoL – Analysis of Periodic Structures', in *IEEE MTT-S Int. Symp. Dig.*, pp. 1967–1970. © 2003 Institute of Electrical and Electronics Engineers (IEEE))

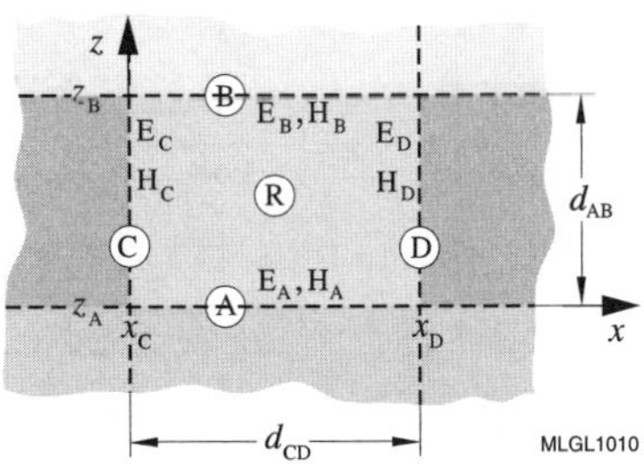

Fig. 7.8 General region R with port and field definitions

by vertical discretisation lines from the magnetic fields at ports A and B and from the magnetic fields at ports C and D, which are caused by the electric field $\widehat{\mathbf{E}}_{\mathrm{AB}}$. Similarly, short-circuiting ports A and B (i.e. introducing metallic walls there) results in $\widehat{\mathbf{E}}_{\mathrm{AB}} = \mathbf{0}$. The submatrices $\widehat{\mathbf{y}}_{\mathrm{CD}}^{\mathrm{CD}}$ and $\widehat{\mathbf{y}}_{\mathrm{AB}}^{\mathrm{CD}}$ are determined from the magnetic fields at ports C and D and from the magnetic fields at ports A and B, respectively, caused by the electric field $\mathbf{E}_{\mathrm{CD}}$. Here we use horizontal discretisation lines. The corresponding equations are as follows:

$$\begin{bmatrix} \widehat{\mathbf{H}}_{\mathrm{AB}}^{\mathrm{V}} \\ \widehat{\mathbf{H}}_{\mathrm{CD}}^{\mathrm{V}} \end{bmatrix} = \begin{bmatrix} \widehat{\mathbf{y}}_{\mathrm{AB}}^{\mathrm{AB}} \\ \widehat{\mathbf{y}}_{\mathrm{CD}}^{\mathrm{AB}} \end{bmatrix} \widehat{\mathbf{E}}_{\mathrm{AB}} \qquad \begin{bmatrix} \widehat{\mathbf{H}}_{\mathrm{AB}}^{\mathrm{H}} \\ \widehat{\mathbf{H}}_{\mathrm{CD}}^{\mathrm{H}} \end{bmatrix} = \begin{bmatrix} \widehat{\mathbf{y}}_{\mathrm{AB}}^{\mathrm{CD}} \\ \widehat{\mathbf{y}}_{\mathrm{CD}}^{\mathrm{CD}} \end{bmatrix} \widehat{\mathbf{E}}_{\mathrm{CD}} \tag{7.5}$$

The impedance matrices are calculated similarly. Now we open-circuit ports C and D (i.e. introduce magnetic walls there), leading to $\widehat{\mathbf{H}}_{\mathrm{CD}} = \mathbf{0}$. In this case, the electric fields at ports A to D are caused by the magnetic

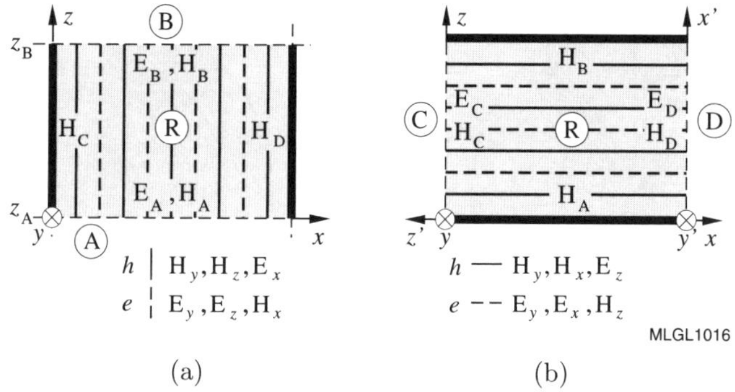

Fig. 7.9 Field calculation in a region R using crossed lines and y matrices: (a) ports C and D short-circuited: analysis with vertical lines (b) ports A and B short-circuited: analysis with horizontal lines (R. Pregla, 'MoL-Mode Analysis with Precise Resolution by an Enhanced and Generalized Line Algorithm", in *IEEE MTT-S Int. Symp. Dig.*, vol. 3, pp. 1543–1546. © 1998 Institute of Electrical and Electronics Engineers (IEEE))

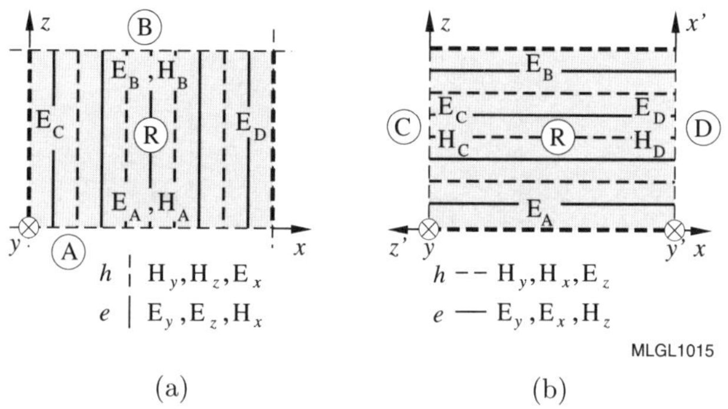

Fig. 7.10 Field calculation in a region R using crossed lines and z matrices: (a) ports C and D open-circuited: analysis with vertical lines (b) ports A and B open-circuited: analysis with horizontal lines

field $\widehat{\mathbf{H}}_{\mathrm{AB}}$. From these electric fields we determine the impedance matrices $\widehat{\mathbf{z}}_{\mathrm{AB}}^{\mathrm{AB}}$ and $\widehat{\mathbf{z}}_{\mathrm{CD}}^{\mathrm{AB}}$. Here, we again use vertical discretisation lines (parallel to the magnetic side walls).

Analogously, by open-circuiting ports A and B and using horizontal discretisation lines, we obtain the submatrices $\widehat{\mathbf{z}}_{\mathrm{AB}}^{\mathrm{CD}}$ and $\widehat{\mathbf{z}}_{\mathrm{CD}}^{\mathrm{CD}}$. The field equations for these two cases are:

$$\begin{bmatrix} \widehat{\mathbf{E}}_{\mathrm{AB}}^{\mathrm{V}} \\ \widehat{\mathbf{E}}_{\mathrm{CD}}^{\mathrm{V}} \end{bmatrix} = \begin{bmatrix} \widehat{\mathbf{z}}_{\mathrm{AB}}^{\mathrm{AB}} \\ \widehat{\mathbf{z}}_{\mathrm{CD}}^{\mathrm{AB}} \end{bmatrix} \widehat{\mathbf{H}}_{\mathrm{AB}} \qquad \begin{bmatrix} \widehat{\mathbf{E}}_{\mathrm{AB}}^{\mathrm{H}} \\ \widehat{\mathbf{E}}_{\mathrm{CD}}^{\mathrm{H}} \end{bmatrix} = \begin{bmatrix} \widehat{\mathbf{z}}_{\mathrm{AB}}^{\mathrm{CD}} \\ \widehat{\mathbf{z}}_{\mathrm{CD}}^{\mathrm{CD}} \end{bmatrix} \widehat{\mathbf{H}}_{\mathrm{CD}} \tag{7.6}$$

Analogous equations can be written for regions with unidirectional discretisation lines.

In the following we assume wave propagation in y-direction according to $\exp(-j\sqrt{\varepsilon_{re}}\overline{y})$ and use the abbreviation $\varepsilon_d = \varepsilon_r - \varepsilon_{re}$.

7.2.2 Lines in vertical direction

7.2.2.1 Basic equations

The general formulas are taken from Section 2.3.2. We assume that region R is homogeneous and isotropic. The quantities for this case should be marked with a 'V' as sub or superscript. However, this has not been done explicitly everywhere. We define the following supervectors with the field components determined on the discretisation lines given in Figs. 7.9a and 7.10a:

$$\widehat{\mathbf{H}}^{\mathrm{V}} = \begin{bmatrix} \widetilde{\mathbf{H}}_{\mathrm{x}} \\ -\mathrm{j}\widetilde{\mathbf{H}}_{\mathrm{y}} \end{bmatrix} \qquad \widehat{\mathbf{E}}^{\mathrm{V}} = \begin{bmatrix} -\mathrm{j}\mathbf{E}_{\mathrm{y}} \\ \mathbf{E}_{\mathrm{x}} \end{bmatrix} \tag{7.7}$$

We obtain the following GTL equations in discretised form:

$$\frac{\mathrm{d}}{\mathrm{d}\overline{z}}\widehat{\widetilde{\mathbf{H}}}^{\mathrm{V}} = -\widehat{\boldsymbol{R}}_{\mathrm{E}}^{\mathrm{V}}\widehat{\mathbf{E}}^{\mathrm{V}} \qquad \widehat{\boldsymbol{R}}_{\mathrm{E}}^{\mathrm{V}} = \begin{bmatrix} \varepsilon_r \boldsymbol{I}_e - \overline{\boldsymbol{D}}_e^{\mathrm{t}}\overline{\boldsymbol{D}}_e & \sqrt{\varepsilon_{re}}\overline{\boldsymbol{D}}_h \\ -\sqrt{\varepsilon_{re}}\overline{\boldsymbol{D}}_e & \varepsilon_d \boldsymbol{I}_h \end{bmatrix} \tag{7.8}$$

$$\frac{\mathrm{d}}{\mathrm{d}\overline{z}}\widehat{\mathbf{E}}^{\mathrm{V}} = -\widehat{\boldsymbol{R}}_{\mathrm{H}}^{\mathrm{V}}\widehat{\mathbf{H}}^{\mathrm{V}} \qquad \widehat{\boldsymbol{R}}_{\mathrm{H}}^{\mathrm{V}} = \begin{bmatrix} -\varepsilon_d \boldsymbol{I}_e & \sqrt{\varepsilon_{re}}\overline{\boldsymbol{D}}_h \\ -\sqrt{\varepsilon_{re}}\overline{\boldsymbol{D}}_e & \overline{\boldsymbol{D}}_h^{\mathrm{t}}\overline{\boldsymbol{D}}_h - \varepsilon_r \boldsymbol{I}_h \end{bmatrix} \varepsilon_r^{-1} \tag{7.9}$$

The discretisation in x-direction is performed according to the description given in Chapter 2.

The vectors of discretised field components E_z and H_z are given by:

$$\mathbf{E}_z^{\mathrm{V}} = \varepsilon_r^{-1}[\sqrt{\varepsilon_{re}}\boldsymbol{I}_e \quad \overline{\boldsymbol{D}}_h]\widehat{\widetilde{\mathbf{H}}}^{\mathrm{V}} \qquad \widetilde{\mathbf{H}}_z^{\mathrm{V}} = -[\overline{\boldsymbol{D}}_e \quad \sqrt{\varepsilon_{re}}\boldsymbol{I}_h]\widehat{\mathbf{E}}^{\mathrm{V}} \tag{7.10}$$

As usual, $\overline{\boldsymbol{D}}_{e,h}$ (the subscripts e and h correspond to E_y and H_y, respectively) are the difference operators related to the first-order differential operators in x-direction for the components on e or h discretisation lines, respectively. They are normalised with $k_o h$ and we have $\boldsymbol{D}_h = -\boldsymbol{D}_e^{\mathrm{t}}$.

By combining the GTL equations, the discretised wave equations have the form:

$$\frac{\mathrm{d}^2}{\mathrm{d}\overline{z}^2}\widehat{\mathbf{E}}^{\mathrm{V}} - \widehat{\boldsymbol{Q}}_{\mathrm{E}}^{\mathrm{V}}\widehat{\mathbf{E}}^{\mathrm{V}} = \mathbf{0} \qquad \widehat{\boldsymbol{Q}}_{\mathrm{E}}^{\mathrm{V}} = \widehat{\boldsymbol{R}}_{\mathrm{H}}^{\mathrm{V}}\widehat{\boldsymbol{R}}_{\mathrm{E}}^{\mathrm{V}} \tag{7.11}$$

$$\frac{\mathrm{d}^2}{\mathrm{d}\overline{z}^2}\widehat{\mathbf{H}}^{\mathrm{V}} - \widehat{\boldsymbol{Q}}_{\mathrm{H}}\widehat{\widetilde{\mathbf{H}}}^{\mathrm{V}} = \mathbf{0} \qquad \widehat{\boldsymbol{Q}}_{\mathrm{H}}^{\mathrm{V}} = \widehat{\boldsymbol{R}}_{\mathrm{E}}^{\mathrm{V}}\widehat{\boldsymbol{R}}_{\mathrm{H}}^{\mathrm{V}} \tag{7.12}$$

with:

$$\widehat{\boldsymbol{Q}}_{\mathrm{E}}^{\mathrm{V}} = \widehat{\boldsymbol{Q}}_{\mathrm{H}}^{\mathrm{V}} = \begin{bmatrix} \overline{\boldsymbol{D}}_e^{\mathrm{t}}\overline{\boldsymbol{D}}_e - \varepsilon_d \boldsymbol{I}_e & \boldsymbol{0} \\ \boldsymbol{0} & \overline{\boldsymbol{D}}_h^{\mathrm{t}}\overline{\boldsymbol{D}}_h - \varepsilon_d \boldsymbol{I}_h \end{bmatrix} = \begin{bmatrix} \boldsymbol{Q}_e^{\mathrm{V}} & \boldsymbol{0} \\ \boldsymbol{0} & \boldsymbol{Q}_h^{\mathrm{V}} \end{bmatrix} \tag{7.13}$$

Now we transform to the main axis (diagonalisation) according to:

$$\widehat{\mathbf{E}}^{\mathrm{V}} = \widehat{\boldsymbol{T}}_{\mathrm{E}}^{\mathrm{V}}\widehat{\widehat{\mathbf{E}}}^{\mathrm{V}} \qquad \widehat{\mathbf{H}}^{\mathrm{V}} = \widehat{\boldsymbol{T}}_{\mathrm{H}}^{\mathrm{V}}\widehat{\widehat{\mathbf{H}}}^{\mathrm{V}} \qquad \widehat{\boldsymbol{T}}_{\mathrm{E}}^{-1}\widehat{\boldsymbol{Q}}_{\mathrm{E}}^{\mathrm{V}}\widehat{\boldsymbol{T}}_{\mathrm{E}} = \widehat{\boldsymbol{T}}_{\mathrm{H}}^{-1}\widehat{\boldsymbol{Q}}_{\mathrm{H}}^{\mathrm{V}}\widehat{\boldsymbol{T}}_{\mathrm{H}} = \widehat{\boldsymbol{\Gamma}}_{\mathrm{V}}^{2} \quad (7.14)$$

with:

$$\widehat{\boldsymbol{T}}_{\mathrm{H}}^{\mathrm{V}} = \widehat{\boldsymbol{T}}_{\mathrm{E}}^{\mathrm{V}} = \begin{bmatrix} \boldsymbol{T}_{e}^{\mathrm{V}} & \boldsymbol{0} \\ \boldsymbol{0} & \boldsymbol{T}_{h}^{\mathrm{V}} \end{bmatrix} \qquad \widehat{\boldsymbol{\Gamma}}_{\mathrm{V}}^{2} = \begin{bmatrix} \boldsymbol{\Gamma}_{\mathrm{V}e}^{2} & \boldsymbol{0} \\ \boldsymbol{0} & \boldsymbol{\Gamma}_{\mathrm{V}h}^{2} \end{bmatrix} \quad \boldsymbol{\Gamma}_{\mathrm{V}e,h}^{2} = \overline{\boldsymbol{\lambda}}_{e,h}^{2} - \varepsilon_{d}\boldsymbol{I}_{e,h} \quad (7.15)$$

$\overline{\boldsymbol{\lambda}}_{e,h}^{2}$ are eigenvalues of $\overline{\boldsymbol{D}}_{e}^{\mathrm{t}}\overline{\boldsymbol{D}}_{e}$ and $\overline{\boldsymbol{D}}_{h}^{\mathrm{t}}\overline{\boldsymbol{D}}_{h}$, respectively. $\boldsymbol{\Gamma}_{h}^{2}$, $\boldsymbol{T}_{h}$ and $\boldsymbol{\Gamma}_{e}^{2}$, $\boldsymbol{T}_{e}$ are eigensolutions of $\boldsymbol{Q}_{h}$ and $\boldsymbol{Q}_{e}$. We obtain decoupled equations:

$$\frac{\mathrm{d}^{2}}{\mathrm{d}\overline{z}^{2}}\widehat{\widehat{\mathbf{E}}}^{\mathrm{V}} - \widehat{\boldsymbol{\Gamma}}_{\mathrm{V}}^{2}\widehat{\widehat{\mathbf{E}}}^{\mathrm{V}} = \mathbf{0} \qquad \frac{\mathrm{d}}{\mathrm{d}\overline{z}^{2}}\widehat{\widehat{\mathbf{H}}}^{\mathrm{V}} - \widehat{\boldsymbol{\Gamma}}_{\mathrm{V}}^{2}\widehat{\widehat{\mathbf{H}}}^{\mathrm{V}} = \mathbf{0} \quad (7.16)$$

Note: to increase the accuracy of $\overline{\boldsymbol{\lambda}}_{e,h}^{2}$ and $\boldsymbol{\Gamma}_{\mathrm{V}e,h}^{2}$, the recurrence procedure in Section A.3 can be used. Due to eq. (7.13) (i.e. $\widehat{\boldsymbol{Q}}_{\mathrm{E}}^{\mathrm{V}} = \widehat{\boldsymbol{Q}}_{\mathrm{H}}^{\mathrm{V}}$), we could choose for matrices $\widehat{\boldsymbol{T}}_{\mathrm{E}}^{\mathrm{V}}$ and $\widehat{\boldsymbol{T}}_{\mathrm{H}}^{\mathrm{V}}$ to be equal to each other. In fact, we will use different matrices (!), as will be shown in what follows, leading to a simple relation between the transverse electric and magnetic fields in planes A and B. To obtain this relation we use the special solution for the equations in (7.16), i.e. for the relation between the fields $\overline{\mathbf{F}} = \overline{\mathbf{E}}(\overline{\mathbf{H}})$ and the derivatives in the planes $\mathrm{A}(z = z_{\mathrm{A}})$ and $\mathrm{B}(z = z_{\mathrm{B}} = z_{\mathrm{A}} + d_{\mathrm{AB}})$:

$$\frac{\mathrm{d}}{\mathrm{d}\overline{z}}\begin{bmatrix} \widehat{\overline{\mathbf{F}}}_{\mathrm{A}} \\ \widehat{\overline{\mathbf{F}}}_{\mathrm{B}} \end{bmatrix} = \begin{bmatrix} -\widehat{\boldsymbol{\gamma}} & \widehat{\boldsymbol{\alpha}} \\ -\widehat{\boldsymbol{\alpha}} & \widehat{\boldsymbol{\gamma}} \end{bmatrix}\begin{bmatrix} \widehat{\overline{\mathbf{F}}}_{\mathrm{A}} \\ \widehat{\overline{\mathbf{F}}}_{\mathrm{B}} \end{bmatrix} \qquad \begin{aligned} \widehat{\boldsymbol{\gamma}} &= \widehat{\boldsymbol{\Gamma}}_{\mathrm{V}}/\tanh(\widehat{\boldsymbol{\Gamma}}_{\mathrm{V}}\overline{d}_{\mathrm{AB}}) \\ \widehat{\boldsymbol{\alpha}} &= \widehat{\boldsymbol{\Gamma}}_{\mathrm{V}}/\sinh(\widehat{\boldsymbol{\Gamma}}_{\mathrm{V}}\overline{d}_{\mathrm{AB}}) \end{aligned} \quad (7.17)$$

As we can see from eqs. (7.8) and (7.9), the derivative of the transverse $\widehat{\overline{\mathbf{E}}}$ field is proportionate to the transverse $\widehat{\overline{\mathbf{H}}}$ field and vice versa. Therefore, we only have to introduce the GTL equations into the special solution (7.17) on the left side to obtain the desired result. To do this adequately, we need only determine one of the transformation matrices ($\widehat{\boldsymbol{T}}_{\mathrm{E}}^{\mathrm{V}}$ or $\widehat{\boldsymbol{T}}_{\mathrm{H}}^{\mathrm{V}}$) with the procedure described above. Because the $\widehat{\boldsymbol{Q}}_{\mathrm{E,H}}^{\mathrm{V}}$ matrices are products of the $\boldsymbol{R}_{\mathrm{E,H}}^{\mathrm{V}}$ matrices, the second transformation matrix ($\widehat{\boldsymbol{T}}_{\mathrm{E}}^{\mathrm{V}}$ or $\widehat{\boldsymbol{T}}_{\mathrm{H}}^{\mathrm{V}}$) is obtained in the following way: (we distinguish between the calculation of $\boldsymbol{z}$ and of $\boldsymbol{y}$ matrices) for the calculation of $\boldsymbol{z}$ matrices we determine the $\widehat{\boldsymbol{T}}_{\mathrm{E}}^{\mathrm{V}}$ and $\widehat{\boldsymbol{T}}_{\mathrm{H}}^{\mathrm{V}}$ matrices according to:

$$\widehat{\boldsymbol{T}}_{\mathrm{E}}^{\mathrm{V}} = \begin{bmatrix} \boldsymbol{T}_{e}^{\mathrm{V}} & \boldsymbol{0} \\ \boldsymbol{0} & \boldsymbol{T}_{h}^{\mathrm{V}} \end{bmatrix} \qquad \widehat{\boldsymbol{T}}_{\mathrm{H}}^{\mathrm{V}} = \widehat{\boldsymbol{R}}_{\mathrm{E}}^{\mathrm{V}}\widehat{\boldsymbol{T}}_{\mathrm{E}}^{\mathrm{V}}\widehat{\boldsymbol{\Gamma}}_{\mathrm{V}}^{-1} = \widehat{\boldsymbol{T}}_{\mathrm{E}}^{\mathrm{V}}\begin{bmatrix} \varepsilon_{r}\boldsymbol{I}_{e} - \overline{\boldsymbol{\lambda}}_{e}^{2} & \widetilde{\boldsymbol{\delta}}_{h}^{\mathrm{V}} \\ \widetilde{\boldsymbol{\delta}}_{h}^{\mathrm{t}} & \varepsilon_{d}\boldsymbol{I}_{h} \end{bmatrix}\widehat{\boldsymbol{\Gamma}}_{\mathrm{V}}^{-1} \quad (7.18)$$

with $\widetilde{\boldsymbol{\delta}}_{h}^{\mathrm{V}} = \sqrt{\varepsilon_{re}}\,\boldsymbol{T}_{e}^{\mathrm{V}\mathrm{t}}\overline{\boldsymbol{D}}_{h}\boldsymbol{T}_{h}^{\mathrm{V}}$.

For the calculation of $\boldsymbol{y}$ matrices we determine $\widehat{\boldsymbol{T}}_{\mathrm{H}}^{\mathrm{V}}$ and $\widehat{\boldsymbol{T}}_{\mathrm{E}}^{\mathrm{V}}$ according to:

$$\widehat{\boldsymbol{T}}_{\mathrm{H}}^{\mathrm{V}} = \begin{bmatrix} \boldsymbol{T}_e^{\mathrm{V}} & \boldsymbol{0} \\ \boldsymbol{0} & \boldsymbol{T}_h^{\mathrm{V}} \end{bmatrix} \quad \widehat{\boldsymbol{T}}_{\mathrm{E}}^{\mathrm{V}} = \widehat{\boldsymbol{R}}_{\mathrm{H}}^{\mathrm{V}} \widehat{\boldsymbol{T}}_{\mathrm{H}}^{\mathrm{V}} \widehat{\boldsymbol{\Gamma}}_{\mathrm{V}}^{-1} = \widehat{\boldsymbol{T}}_{\mathrm{H}}^{\mathrm{V}} \begin{bmatrix} -\varepsilon_d \boldsymbol{I}_e & \widetilde{\boldsymbol{\delta}}_h \\ \widetilde{\boldsymbol{\delta}}_h^{\mathrm{t}} & \overline{\boldsymbol{\lambda}}_h^{2} - \varepsilon_r \boldsymbol{I}_h \end{bmatrix} \widehat{\boldsymbol{\Gamma}}_{\mathrm{V}}^{-1} \varepsilon_r^{-1} \tag{7.19}$$

The multiplication by $\widehat{\boldsymbol{\Gamma}}_{\mathrm{V}}^{-1}$ is introduced for normalisation. By using these different transformation matrices $\widehat{\boldsymbol{T}}_{\mathrm{E}}^{\mathrm{V}}$ and $\widehat{\boldsymbol{T}}_{\mathrm{H}}^{\mathrm{V}}$, we obtain for the GTL equations in the transformed domain:

$$\frac{\mathrm{d}}{\mathrm{d}\overline{y}} \widehat{\overline{\mathbf{E}}}^{\mathrm{V}} = -\widehat{\boldsymbol{\Gamma}}_{\mathrm{V}} \widehat{\overline{\mathbf{H}}}^{\mathrm{V}} \qquad \frac{\mathrm{d}}{\mathrm{d}\overline{y}} \widehat{\overline{\mathbf{H}}}^{\mathrm{V}} = -\widehat{\boldsymbol{\Gamma}}_{\mathrm{V}} \widehat{\overline{\mathbf{E}}}^{\mathrm{V}} \tag{7.20}$$

7.2.2.2 Field transformation equations between planes A and B

By introducing eqs. (7.20) into the special solution (7.17), we obtain for the field transformation between planes A and B:

$$\begin{bmatrix} \widehat{\overline{\mathbf{H}}}_{\mathrm{A}}^{\mathrm{V}} \\ -\widehat{\overline{\mathbf{H}}}_{\mathrm{B}}^{\mathrm{V}} \end{bmatrix} = \begin{bmatrix} \widehat{\overline{\boldsymbol{y}}}_1^{\mathrm{V}} & \widehat{\overline{\boldsymbol{y}}}_2^{\mathrm{V}} \\ \widehat{\overline{\boldsymbol{y}}}_2^{\mathrm{V}} & \widehat{\overline{\boldsymbol{y}}}_1^{\mathrm{V}} \end{bmatrix} \begin{bmatrix} \widehat{\overline{\mathbf{E}}}_{\mathrm{A}}^{\mathrm{V}} \\ \widehat{\overline{\mathbf{E}}}_{\mathrm{B}}^{\mathrm{V}} \end{bmatrix} \qquad \widehat{\overline{\mathbf{H}}}_{\mathrm{AB}}^{\mathrm{V}} = \widehat{\overline{\boldsymbol{y}}}_{\mathrm{AB}}^{\mathrm{AB}} \widehat{\overline{\mathbf{E}}}_{\mathrm{AB}}^{\mathrm{V}} \tag{7.21}$$

$$\begin{bmatrix} \widehat{\overline{\mathbf{E}}}_{\mathrm{A}}^{\mathrm{V}} \\ \widehat{\overline{\mathbf{E}}}_{\mathrm{B}}^{\mathrm{V}} \end{bmatrix} = \begin{bmatrix} \widehat{\overline{\boldsymbol{z}}}_1^{\mathrm{V}} & \widehat{\overline{\boldsymbol{z}}}_2^{\mathrm{V}} \\ \widehat{\overline{\boldsymbol{z}}}_2^{\mathrm{V}} & \widehat{\overline{\boldsymbol{z}}}_1^{\mathrm{V}} \end{bmatrix} \begin{bmatrix} \widehat{\overline{\mathbf{H}}}_{\mathrm{A}}^{\mathrm{V}} \\ -\widehat{\overline{\mathbf{H}}}_{\mathrm{B}}^{\mathrm{V}} \end{bmatrix} \qquad \widehat{\overline{\mathbf{E}}}_{\mathrm{AB}}^{\mathrm{V}} = \widehat{\overline{\boldsymbol{z}}}_{\mathrm{AB}}^{\mathrm{AB}} \widehat{\overline{\mathbf{H}}}_{\mathrm{AB}}^{\mathrm{V}} \tag{7.22}$$

with:

$$\begin{aligned} \widehat{\overline{\boldsymbol{y}}}_1^{\mathrm{V}} &= 1/\tanh(\widehat{\boldsymbol{\Gamma}}_{\mathrm{V}} \overline{d}_{\mathrm{AB}}) = \mathrm{diag}(\overline{\boldsymbol{y}}_{1e}^{\mathrm{V}}, \overline{\boldsymbol{y}}_{1h}^{\mathrm{V}}) \\ \widehat{\overline{\boldsymbol{y}}}_2^{\mathrm{V}} &= -1/\sinh(\widehat{\boldsymbol{\Gamma}}_{\mathrm{V}} \overline{d}_{\mathrm{AB}}) = \mathrm{diag}(\overline{\boldsymbol{y}}_{2e}^{\mathrm{V}}, \overline{\boldsymbol{y}}_{2h}^{\mathrm{V}}) \\ \widehat{\overline{\boldsymbol{z}}}_1^{\mathrm{V}} &= 1/\tanh(\widehat{\boldsymbol{\Gamma}}_{\mathrm{V}} \overline{d}_{\mathrm{AB}}) = \mathrm{diag}(\overline{\boldsymbol{z}}_{1e}^{\mathrm{V}}, \overline{\boldsymbol{z}}_{1h}^{\mathrm{V}}) \\ \widehat{\overline{\boldsymbol{z}}}_2^{\mathrm{V}} &= 1/\sinh(\widehat{\boldsymbol{\Gamma}}_{\mathrm{V}} \overline{d}_{\mathrm{AB}}) = \mathrm{diag}(\overline{\boldsymbol{z}}_{2e}^{\mathrm{V}}, \overline{\boldsymbol{z}}_{2h}^{\mathrm{V}}) \end{aligned} \tag{7.23}$$

and $\overline{d}_{\mathrm{AB}} = k_0 d_{\mathrm{AB}} = k_0(z_{\mathrm{B}} - z_{\mathrm{A}})$. These equations are valid only when transformation matrices $\widehat{\boldsymbol{T}}_{\mathrm{E}}$ and $\widehat{\boldsymbol{T}}_{\mathrm{H}}$ are used according to eqs. (7.18) and (7.19). It should be mentioned that both systems of equations (7.21) and (7.22) can be used, e.g. for the case shown in Fig. 7.9a. They are inverse to each other. The fields $\overline{\mathbf{H}}_z^{\mathrm{V}}$ and $\overline{\mathbf{E}}_z^{\mathrm{V}}$ can be calculated as functions of $\widehat{\overline{\mathbf{E}}}^{\mathrm{V}}$ and $\widehat{\overline{\mathbf{H}}}^{\mathrm{V}}$, respectively. With the help of eq. (7.10) the field components E_z and H_z are given by:

$$\varepsilon_r \mathbf{E}_z^{\mathrm{V}} = \left[\sqrt{\varepsilon_{re}} \boldsymbol{I}_e \quad \overline{\boldsymbol{D}}_h\right] \widehat{\boldsymbol{T}}_{\mathrm{H}}^{\mathrm{V}} \widehat{\overline{\mathbf{H}}}^{\mathrm{V}} \qquad \widetilde{\mathbf{H}}_z^{\mathrm{V}} = -\left[\overline{\boldsymbol{D}}_e \quad \sqrt{\varepsilon_{re}} \boldsymbol{I}_h\right] \widehat{\boldsymbol{T}}_{\mathrm{E}}^{\mathrm{V}} \widehat{\overline{\mathbf{E}}}^{\mathrm{V}} \tag{7.24}$$

By using the transformation equations according to (7.18) for z matrices, we obtain:

$$\overline{\mathbf{E}}_z^{\mathrm{V}} = \boldsymbol{T}_e^{\mathrm{Vt}} \mathbf{E}_z^{\mathrm{V}} = \left[\sqrt{\varepsilon_{re}} \boldsymbol{I}_e \quad \overline{\boldsymbol{\delta}}_h\right] \widehat{\boldsymbol{\Gamma}}_V^{-1} \widehat{\overline{\mathbf{H}}}^{\mathrm{V}} \qquad \overline{\mathbf{H}}_z^{\mathrm{V}} = \boldsymbol{T}_h^{\mathrm{Vt}} \widetilde{\mathbf{H}}_z^{\mathrm{V}} = -\left[\overline{\boldsymbol{\delta}}_e \quad \sqrt{\varepsilon_{re}} \boldsymbol{I}_h\right] \widehat{\overline{\mathbf{E}}} \tag{7.25}$$

Likewise, the use of transformation equations according to (7.19) for y matrices gives:

$$\overline{\mathbf{E}}_z^{\mathrm{V}} = \boldsymbol{T}_e^{\mathrm{Vt}}\mathbf{E}_z^{\mathrm{V}} = \varepsilon_r^{-1}\left[\sqrt{\varepsilon_{re}}\boldsymbol{I}_e \quad \overline{\boldsymbol{\delta}}_h\right]\widehat{\overline{\mathbf{H}}}^{\mathrm{V}} \qquad \overline{\mathbf{H}}_z^{\mathrm{V}} = \boldsymbol{T}_h^{\mathrm{Vt}}\widetilde{\mathbf{H}}_z^{\mathrm{V}} = \left[\overline{\boldsymbol{\delta}}_e \quad \boldsymbol{I}_h\right]\widehat{\boldsymbol{\Gamma}}_{\mathrm{V}}^{-1}\widehat{\overline{\mathbf{E}}}^{\mathrm{V}} \tag{7.26}$$

These equations are particularly valid at ports A and B.

7.2.2.3 Field transformation equation from planes A/B to planes C/D

In order to determine $\overline{\boldsymbol{y}}_{\mathrm{CD}}^{\mathrm{AB}}(\overline{\boldsymbol{z}}_{\mathrm{CD}}^{\mathrm{AB}})$ we have to calculate the tangential fields at ports C and D from the ones at ports A and B. The fields described by the above equations do not have a tangential electric (magnetic) field at ports C and D, because of the electric, resp., magnetic walls. The tangential magnetic (electric) field can be obtained as follows: for the z-dependence of field component $\overline{\mathbf{F}}_u$ we may write:

$$\overline{\mathbf{F}}_u(z) = \frac{\sinh(\boldsymbol{\Gamma}_f^{\mathrm{V}}(\overline{d}_{\mathrm{AB}} - \overline{z}))}{\sinh(\boldsymbol{\Gamma}_f^{\mathrm{V}}\overline{d}_{\mathrm{AB}})}\overline{\mathbf{F}}_{u\mathrm{A}} + \frac{\sinh(\boldsymbol{\Gamma}_f^{\mathrm{V}}\overline{z})}{\sinh(\boldsymbol{\Gamma}_f^{\mathrm{V}}\overline{d}_{\mathrm{AB}})}\overline{\mathbf{F}}_{u\mathrm{B}} \tag{7.27}$$

where $\mathbf{F}$ stands for $\mathbf{H}$ ($\mathbf{E}$) in case of $\overline{\mathbf{y}}$-matrices ($\overline{\mathbf{z}}$-matrices). u is equal to y and z. f has to be replaced by $h(e)$ for $\mathbf{F} = \mathbf{H}$ ($\mathbf{F} = \mathbf{E}$), which means at the position of the Neumann boundary conditions. As before, the division by the diagonal matrix $\sinh(\boldsymbol{\Gamma}_f^{\mathrm{V}}\overline{d}_{\mathrm{AB}})$ stands for the multiplication with its inverse. This equation must now be evaluated at the discrete z-positions where we have horizontal discretisation lines. At such a position z_i we may write the equation:

$$\overline{\mathbf{F}}_u(z_i) = \boldsymbol{\Lambda}_{\mathrm{A}i}^{\mathrm{d}}\overline{\mathbf{F}}_{u\mathrm{A}} + \boldsymbol{\Lambda}_{\mathrm{B}i}^{\mathrm{d}}\overline{\mathbf{F}}_{u\mathrm{B}} \tag{7.28}$$

with diagonal matrices $\boldsymbol{\Lambda}_i^{\mathrm{d}}$ given by the expressions:

$$\boldsymbol{\Lambda}_{\mathrm{A}i}^{\mathrm{d}} = \sinh(\boldsymbol{\Gamma}_f^{\mathrm{V}}(\overline{d}_{\mathrm{AB}} - \overline{z}_i))(\sinh(\boldsymbol{\Gamma}_f^{\mathrm{V}}\overline{d}_{\mathrm{AB}}))^{-1} \tag{7.29}$$

$$\boldsymbol{\Lambda}_{\mathrm{B}i}^{\mathrm{d}} = \sinh(\boldsymbol{\Gamma}_f^{\mathrm{V}}\overline{z}_i)(\sinh(\boldsymbol{\Gamma}_f^{\mathrm{V}}\overline{d}_{\mathrm{AB}}))^{-1} \tag{7.30}$$

The vector $\mathbf{F}_u(z_i) = \boldsymbol{T}_f^{\mathrm{V}}\overline{\mathbf{F}}_u(z_i)$ gives the values F_u at the crossing points of vertical and horizontal lines at $z = z_i$. $\boldsymbol{T}_f^{\mathrm{V}}$ must be the matrix for Neumann boundary conditions. Thus, f should be equal to $h(e)$ when $\mathbf{F} = \mathbf{H}$ ($\mathbf{F} = \mathbf{E}$). We need the values of the field $\mathbf{F}_u$ at boundaries C and D with $x = x_{\mathrm{C}}$ and $x = x_{\mathrm{D}}$, respectively. Because of the Neumann conditions, we must extrapolate from points near the boundary to the boundary. This can easily be done with the transformation matrix $\boldsymbol{T}_f^{\mathrm{V}}$. We obtain at the position z_i on the left (R $\equiv$ C, $x = x_{\mathrm{C}}$) or right (R $\equiv$ D, $x = x_{\mathrm{D}}$) boundary:

$$\mathbf{F}_{u\mathrm{R}}(z_i) = \mathbf{T}_{f\mathrm{R}}^{\mathrm{V}}\overline{\mathbf{F}}_u(z_i) = \mathbf{T}_{f\mathrm{R}}^{\mathrm{V}}(\boldsymbol{\Lambda}_{\mathrm{A}i}^{\mathrm{d}}\overline{\mathbf{F}}_{u\mathrm{A}} + \boldsymbol{\Lambda}_{\mathrm{B}i}^{\mathrm{d}}\overline{\mathbf{F}}_{u\mathrm{B}}) \tag{7.31}$$

$\mathbf{T}_{f\mathrm{R}}$ is a row vector constructed from weighted first (last) rows of the matrix $\boldsymbol{T}_f$. We have:

$$\mathbf{T}_{f\mathrm{C}}^{\mathrm{V}} = \tfrac{1}{8}(9\mathbf{T}_{f1} - \mathbf{T}_{f2}) \qquad \mathbf{T}_{f\mathrm{D}}^{\mathrm{V}} = \tfrac{1}{8}(9\mathbf{T}_{f,\mathrm{N}} - \mathbf{T}_{f,\mathrm{N}-1}) \tag{7.32}$$

$\mathbf{T}_{f1}$ and $\mathbf{T}_{f2}$ are the first two and $\mathbf{T}_{f\mathrm{N}}$ and $\mathbf{T}_{f,\mathrm{N}-1}$ the last two row vectors of the transformation matrix $\boldsymbol{T}_f^{\mathrm{v}}$. If we want to change the order of the product between $\mathbf{T}_{f\mathrm{R}}^{\mathrm{v}}$ and $\boldsymbol{\Lambda}_i^{\mathrm{d}}$ we must also modify these quantities in a suitable way. $\mathbf{T}_{f\mathrm{R}}^{\mathrm{v}}$ becomes a diagonal matrix $\boldsymbol{T}_{f\mathrm{R}}^{\mathrm{d}}$ and $\boldsymbol{\Lambda}_i^{\mathrm{d}}$ is modified to a row vector $\boldsymbol{\Lambda}_i$. In this way we have the same result for their products, and we can write expressions for all values of z_i, $i = 1, 2, \ldots, \mathrm{N}_z^{\mathrm{H}}$ (N_z^{H} is the number of horizontal discretisation lines) with one formula:

$$\mathbf{F}_{u\mathrm{R}} = \boldsymbol{\Lambda}_{\mathrm{A}}^{\mathrm{v}} \boldsymbol{T}_{f\mathrm{R}}^{\mathrm{d}} \overline{\mathbf{F}}_{u\mathrm{A}} + \boldsymbol{\Lambda}_{\mathrm{B}}^{\mathrm{v}} \boldsymbol{T}_{f\mathrm{R}}^{\mathrm{d}} \overline{\mathbf{F}}_{u\mathrm{B}} \tag{7.33}$$

The full matrices $\boldsymbol{\Lambda}_{\mathrm{A}}^{\mathrm{v}}$ and $\boldsymbol{\Lambda}_{\mathrm{B}}^{\mathrm{v}}$ consist of the following components:

$$(\boldsymbol{\Lambda}_{\mathrm{A}}^{\mathrm{v}})_{ik} = \sinh(\boldsymbol{\Gamma}_{fk}^{\mathrm{v}}(\overline{d}_{\mathrm{AB}} - \overline{z}_i))(\sinh(\boldsymbol{\Gamma}_{fk}^{\mathrm{v}}\overline{d}_{\mathrm{AB}}))^{-1} \tag{7.34}$$

$$(\boldsymbol{\Lambda}_{\mathrm{B}}^{\mathrm{v}})_{ik} = \sinh(\boldsymbol{\Gamma}_{fk}^{\mathrm{v}}\overline{z}_i)(\sinh(\boldsymbol{\Gamma}_{fk}^{\mathrm{v}}\overline{d}_{\mathrm{AB}}))^{-1} \tag{7.35}$$

$\boldsymbol{\Gamma}_{fk}^{\mathrm{v}}$ is the kth component of the diagonal propagation matrix $\boldsymbol{\Gamma}_f^{\mathrm{v}}$ in z-direction. In the transformed domain we therefore have:

$$\overline{\mathbf{F}}_{u\mathrm{R}} = \boldsymbol{T}_{\mathrm{H}g}^{-1} \boldsymbol{\Lambda}_{\mathrm{A}} \boldsymbol{T}_{f\mathrm{R}}^{\mathrm{d}} \overline{\mathbf{F}}_{u\mathrm{A}} + \boldsymbol{T}_{\mathrm{H}g}^{-1} \boldsymbol{\Lambda}_{\mathrm{B}} \boldsymbol{T}_{f\mathrm{R}}^{\mathrm{d}} \overline{\mathbf{F}}_{u\mathrm{B}} \tag{7.36}$$

where $\boldsymbol{T}_{\mathrm{H}g}$ is the adequate transformation matrix for horizontal discretisation lines. g is either e or h, depending on which line system (e or h) the component $\mathbf{F}_{u\mathrm{R}}$ is calculated from.

The relation between the field components at ports C and D depends on the field components at ports A and B. They are associated with the discretisation lines in the horizontal direction and can now be expressed in the following matrix form:

$$\begin{bmatrix} \overline{\mathbf{F}}_{u\mathrm{C}} \\ \overline{\mathbf{F}}_{u\mathrm{D}} \end{bmatrix} = \begin{bmatrix} \overline{\boldsymbol{V}}_{\mathrm{CA}} & \overline{\boldsymbol{V}}_{\mathrm{CB}} \\ \overline{\boldsymbol{V}}_{\mathrm{DA}} & \overline{\boldsymbol{V}}_{\mathrm{DB}} \end{bmatrix} \begin{bmatrix} \overline{\mathbf{F}}_{u\mathrm{A}} \\ \overline{\mathbf{F}}_{u\mathrm{B}} \end{bmatrix} \tag{7.37}$$

where:

$$\begin{aligned} \overline{\boldsymbol{V}}_{\mathrm{CA}} &= \boldsymbol{T}_{\mathrm{H}g}^{-1} \boldsymbol{\Lambda}_{\mathrm{A}} \boldsymbol{T}_{f\mathrm{C}}^{\mathrm{d}} \quad & \overline{\boldsymbol{V}}_{\mathrm{CB}} &= \boldsymbol{T}_{\mathrm{H}g}^{-1} \boldsymbol{\Lambda}_{\mathrm{B}} \boldsymbol{T}_{f\mathrm{C}}^{\mathrm{d}} \\ \overline{\boldsymbol{V}}_{\mathrm{DA}} &= \boldsymbol{T}_{\mathrm{H}g}^{-1} \boldsymbol{\Lambda}_{\mathrm{A}} \boldsymbol{T}_{f\mathrm{D}}^{\mathrm{d}} \quad & \overline{\boldsymbol{V}}_{\mathrm{DB}} &= \boldsymbol{T}_{\mathrm{H}g}^{-1} \boldsymbol{\Lambda}_{\mathrm{B}} \boldsymbol{T}_{f\mathrm{D}}^{\mathrm{d}} \end{aligned} \tag{7.38}$$

We can write the dependence of $\overline{\mathbf{H}}_{z\mathrm{C}}^{\mathrm{v}}$, $\overline{\mathbf{H}}_{y\mathrm{C}}^{\mathrm{v}}$, $\overline{\mathbf{H}}_{z\mathrm{D}}^{\mathrm{v}}$ and $\overline{\mathbf{H}}_{y\mathrm{D}}^{\mathrm{v}}$ from the tangential magnetic fields at ports A and B in the following way:

$$\begin{bmatrix} \overline{\mathbf{H}}_{z\mathrm{C}}^{\mathrm{v}} \\ -\mathrm{j}\overline{\mathbf{H}}_{y\mathrm{C}}^{\mathrm{v}} \\ -\overline{\mathbf{H}}_{z\mathrm{D}}^{\mathrm{v}} \\ \mathrm{j}\overline{\mathbf{H}}_{y\mathrm{D}}^{\mathrm{v}} \end{bmatrix} = \begin{bmatrix} \overline{\boldsymbol{V}}_{\mathrm{CA}}^{\mathrm{D}} & & -\overline{\boldsymbol{V}}_{\mathrm{CB}}^{\mathrm{D}} & \\ & \overline{\boldsymbol{V}}_{\mathrm{CA}}^{\mathrm{N}} & & -\overline{\boldsymbol{V}}_{\mathrm{CB}}^{\mathrm{N}} \\ -\overline{\boldsymbol{V}}_{\mathrm{DA}}^{\mathrm{D}} & & \overline{\boldsymbol{V}}_{\mathrm{DB}}^{\mathrm{D}} & \\ & -\overline{\boldsymbol{V}}_{\mathrm{DA}}^{\mathrm{N}} & & \overline{\boldsymbol{V}}_{\mathrm{DB}}^{\mathrm{N}} \end{bmatrix} \begin{bmatrix} \overline{\mathbf{H}}_{z\mathrm{A}}^{\mathrm{v}} \\ -\mathrm{j}\overline{\mathbf{H}}_{y\mathrm{A}}^{\mathrm{v}} \\ -\overline{\mathbf{H}}_{z\mathrm{B}}^{\mathrm{v}} \\ \mathrm{j}\overline{\mathbf{H}}_{y\mathrm{B}}^{\mathrm{v}} \end{bmatrix} \tag{7.39}$$

or:

$$\widehat{\overline{\mathbf{H}}}_{\mathrm{CD}}^{\mathrm{v}} = \widehat{\overline{\boldsymbol{V}}}_{\mathrm{H}}^{\mathrm{v}} \widehat{\overline{\mathbf{H}}}_{\mathrm{AB}}^{\mathrm{vT}} \tag{7.40}$$

The superscript N (D) indicates that the quantities have to be calculated for horizontal discretisation lines with Neumann (Dirichlet) boundary conditions. For the superscript N (D) we use $h(e)$ as the corresponding subscript for g in eq. (7.38).

Analogously, we write $\overline{\mathbf{E}}_{y\mathrm{C}}^{\mathrm{V}}$, $\overline{\mathbf{E}}_{z\mathrm{C}}^{\mathrm{V}}$, $\overline{\mathbf{E}}_{y\mathrm{D}}^{\mathrm{V}}$ and $\overline{\mathbf{E}}_{z\mathrm{D}}^{\mathrm{V}}$ as functions of the tangential electric fields at ports A and B:

$$\begin{bmatrix} -\mathrm{j}\overline{\mathbf{E}}_{y\mathrm{C}}^{\mathrm{V}} \\ \overline{\mathbf{E}}_{z\mathrm{C}}^{\mathrm{V}} \\ -\mathrm{j}\overline{\mathbf{E}}_{y\mathrm{D}}^{\mathrm{V}} \\ \overline{\mathbf{E}}_{z\mathrm{D}}^{\mathrm{V}} \end{bmatrix} = \begin{bmatrix} \overline{\boldsymbol{V}}_{\mathrm{CA}}^{\mathrm{N}} & & \overline{\boldsymbol{V}}_{\mathrm{CB}}^{\mathrm{N}} & \\ & \overline{\boldsymbol{V}}_{\mathrm{CA}}^{\mathrm{D}} & & \overline{\boldsymbol{V}}_{\mathrm{CB}}^{\mathrm{D}} \\ \overline{\boldsymbol{V}}_{\mathrm{DA}}^{\mathrm{N}} & & \overline{\boldsymbol{V}}_{\mathrm{DB}}^{\mathrm{N}} & \\ & \overline{\boldsymbol{V}}_{\mathrm{DA}}^{\mathrm{D}} & & \overline{\boldsymbol{V}}_{\mathrm{DB}}^{\mathrm{D}} \end{bmatrix} \begin{bmatrix} -\mathrm{j}\overline{\mathbf{E}}_{y\mathrm{A}}^{\mathrm{V}} \\ \overline{\mathbf{E}}_{z\mathrm{A}}^{\mathrm{V}} \\ -\mathrm{j}\overline{\mathbf{E}}_{y\mathrm{B}}^{\mathrm{V}} \\ \overline{\mathbf{E}}_{z\mathrm{B}}^{\mathrm{V}} \end{bmatrix} \tag{7.41}$$

or:

$$\widehat{\overline{\overline{\mathbf{E}}}}_{\mathrm{CD}}^{\mathrm{V}} = \widehat{\overline{\boldsymbol{V}}}_{\mathrm{E}}^{\mathrm{V}} \widehat{\overline{\overline{\mathbf{E}}}}_{\mathrm{AB}}^{\mathrm{VT}} \tag{7.42}$$

We use the subscript $e(h)$ in eq. (7.38) for g when we have the superscript N (D). The magnetic (electric) fields on the right side in eqs. (7.39) and (7.41) should be replaced by the electric (magnetic) fields at ports A and B. For y-matrices we combine eq. (7.21) with the right expression in eq. (7.24) and obtain:

$$\begin{bmatrix} \overline{\mathbf{H}}_{z\mathrm{A}}^{\mathrm{V}} \\ -\mathrm{j}\overline{\mathbf{H}}_{y\mathrm{A}}^{\mathrm{V}} \\ -\overline{\mathbf{H}}_{z\mathrm{B}}^{\mathrm{V}} \\ \mathrm{j}\overline{\mathbf{H}}_{y\mathrm{B}}^{\mathrm{V}} \end{bmatrix} = \begin{bmatrix} \overline{\boldsymbol{\delta}}_{e}^{\mathrm{V}} \boldsymbol{\Gamma}_{\mathrm{V}e}^{-1} & \boldsymbol{\Gamma}_{\mathrm{V}h}^{-1} & \boldsymbol{0} & \boldsymbol{0} \\ \boldsymbol{0} & \overline{\boldsymbol{y}}_{1h}^{\mathrm{V}} & \boldsymbol{0} & \overline{\boldsymbol{y}}_{2h}^{\mathrm{V}} \\ \boldsymbol{0} & \boldsymbol{0} & -\overline{\boldsymbol{\delta}}_{e}^{\mathrm{V}} \boldsymbol{\Gamma}_{\mathrm{V}e}^{-1} & -\boldsymbol{\Gamma}_{\mathrm{V}h}^{-1} \\ \boldsymbol{0} & \overline{\boldsymbol{y}}_{2h}^{\mathrm{V}} & \boldsymbol{0} & \overline{\boldsymbol{y}}_{1h}^{\mathrm{V}} \end{bmatrix} \begin{bmatrix} \widehat{\overline{\overline{\mathbf{E}}}}_{\mathrm{A}}^{\mathrm{V}} \\ \widehat{\overline{\overline{\mathbf{E}}}}_{\mathrm{B}}^{\mathrm{V}} \end{bmatrix} \tag{7.43}$$

or:

$$\widehat{\overline{\overline{\mathbf{H}}}}_{\mathrm{AB}}^{\mathrm{VT}} = \widehat{\overline{\boldsymbol{y}}}_{\mathrm{AB}}^{\mathrm{V}} \widehat{\overline{\overline{\mathbf{E}}}}_{\mathrm{AB}}^{\mathrm{V}} \tag{7.44}$$

We proceed analogously, for z-matrices. We combine eq. (7.21) and the left expression in eq. (7.24) and obtain:

$$\begin{bmatrix} -\mathrm{j}\overline{\mathbf{E}}_{y\mathrm{A}}^{\mathrm{V}} \\ \overline{\mathbf{E}}_{z\mathrm{A}}^{\mathrm{V}} \\ -\mathrm{j}\overline{\mathbf{E}}_{y\mathrm{B}}^{\mathrm{V}} \\ \overline{\mathbf{E}}_{z\mathrm{B}}^{\mathrm{V}} \end{bmatrix} = \begin{bmatrix} \overline{\mathbf{z}}_{1e}^{\mathrm{V}} & \boldsymbol{0} & \overline{\mathbf{z}}_{2e}^{\mathrm{V}} & \boldsymbol{0} \\ \sqrt{\varepsilon_{re}}\boldsymbol{\Gamma}_{\mathrm{V}e}^{-1} & \overline{\boldsymbol{\delta}}_{h}^{\mathrm{V}} \boldsymbol{\Gamma}_{\mathrm{V}h}^{-1} & \boldsymbol{0} & \boldsymbol{0} \\ \overline{\mathbf{z}}_{2e}^{\mathrm{V}} & \boldsymbol{0} & \overline{\mathbf{z}}_{1e}^{\mathrm{V}} & \boldsymbol{0} \\ \boldsymbol{0} & \boldsymbol{0} & -\sqrt{\varepsilon_{re}}\boldsymbol{\Gamma}_{\mathrm{V}e}^{-1} & -\overline{\boldsymbol{\delta}}_{h}^{\mathrm{V}} \boldsymbol{\Gamma}_{\mathrm{V}h}^{-1} \end{bmatrix} \begin{bmatrix} \widehat{\overline{\overline{\mathbf{H}}}}_{\mathrm{A}}^{\mathrm{V}} \\ -\widehat{\overline{\overline{\mathbf{H}}}}_{\mathrm{B}}^{\mathrm{V}} \end{bmatrix} \tag{7.45}$$

or:

$$\widehat{\overline{\overline{\mathbf{E}}}}_{\mathrm{AB}}^{\mathrm{VT}} = \widehat{\overline{\mathbf{z}}}_{\mathrm{AB}}^{\mathrm{V}} \widehat{\overline{\overline{\mathbf{H}}}}_{\mathrm{AB}}^{\mathrm{V}} \tag{7.46}$$

The off-diagonal submatrices in the left side of eq. (7.4) are given by:

$$\widehat{\boldsymbol{y}}_{\mathrm{CD}}^{\mathrm{AB}} = \widehat{\overline{\boldsymbol{V}}}_{\mathrm{H}}^{\mathrm{V}} \widehat{\overline{\boldsymbol{y}}}_{\mathrm{AB}}^{\mathrm{V}} \qquad \widehat{\mathbf{z}}_{\mathrm{CD}}^{\mathrm{AB}} = \widehat{\overline{\boldsymbol{V}}}_{\mathrm{E}}^{\mathrm{V}} \widehat{\overline{\mathbf{z}}}_{\mathrm{AB}}^{\mathrm{V}} \tag{7.47}$$

Hence, we obtain:

$$\widehat{\overline{\boldsymbol{y}}}_{\mathrm{CD}}^{\mathrm{AB}} = \begin{bmatrix} \overline{\boldsymbol{V}}_{\mathrm{CA}}^{\mathrm{D}}\overline{\boldsymbol{\delta}}_e^{\mathrm{V}}\boldsymbol{\Gamma}_{\mathrm{V}e}^{-1} & \overline{\boldsymbol{V}}_{\mathrm{CA}}^{\mathrm{D}}\boldsymbol{\Gamma}_{\mathrm{V}h}^{-1} & \overline{\boldsymbol{V}}_{\mathrm{CB}}^{\mathrm{D}}\overline{\boldsymbol{\delta}}_e^{\mathrm{V}}\boldsymbol{\Gamma}_{\mathrm{V}e}^{-1} & \overline{\boldsymbol{V}}_{\mathrm{CB}}^{\mathrm{D}}\boldsymbol{\Gamma}_{\mathrm{V}h}^{-1} \\ \boldsymbol{0} & \overline{\boldsymbol{V}}_{\mathrm{CA}}^{\mathrm{N}}\overline{\boldsymbol{y}}_{1h}^{\mathrm{V}} - \overline{\boldsymbol{V}}_{\mathrm{CB}}^{\mathrm{N}}\overline{\boldsymbol{y}}_{2h}^{\mathrm{V}} & \boldsymbol{0} & \overline{\boldsymbol{V}}_{\mathrm{CA}}^{\mathrm{N}}\overline{\boldsymbol{y}}_{2h}^{\mathrm{V}} - \overline{\boldsymbol{V}}_{\mathrm{CB}}^{\mathrm{N}}\overline{\boldsymbol{y}}_{1h}^{\mathrm{V}} \\ -\overline{\boldsymbol{V}}_{\mathrm{DA}}^{\mathrm{D}}\overline{\boldsymbol{\delta}}_e^{\mathrm{V}}\boldsymbol{\Gamma}_{\mathrm{V}e}^{-1} & -\overline{\boldsymbol{V}}_{\mathrm{DA}}^{\mathrm{D}}\boldsymbol{\Gamma}_{\mathrm{V}h}^{-1} & -\overline{\boldsymbol{V}}_{\mathrm{DB}}^{\mathrm{D}}\overline{\boldsymbol{\delta}}_e^{\mathrm{V}}\boldsymbol{\Gamma}_{\mathrm{V}e}^{-1} & -\overline{\boldsymbol{V}}_{\mathrm{DB}}^{\mathrm{D}}\boldsymbol{\Gamma}_{\mathrm{V}h}^{-1} \\ \boldsymbol{0} & \overline{\boldsymbol{V}}_{\mathrm{DB}}^{\mathrm{N}}\overline{\boldsymbol{y}}_{2h}^{\mathrm{V}} - \overline{\boldsymbol{V}}_{\mathrm{DA}}^{\mathrm{N}}\overline{\boldsymbol{y}}_{1h}^{\mathrm{V}} & \boldsymbol{0} & \overline{\boldsymbol{V}}_{\mathrm{DB}}^{\mathrm{N}}\overline{\boldsymbol{y}}_{1h}^{\mathrm{V}} - \overline{\boldsymbol{V}}_{\mathrm{DA}}^{\mathrm{N}}\overline{\boldsymbol{y}}_{2h}^{\mathrm{V}} \end{bmatrix} \tag{7.48}$$

and:

$$\widehat{\overline{\boldsymbol{z}}}_{\mathrm{CD}}^{\mathrm{AB}} = \begin{bmatrix} \overline{\boldsymbol{V}}_{\mathrm{CA}}^{\mathrm{N}}\overline{\boldsymbol{z}}_{1e}^{\mathrm{V}} + \overline{\boldsymbol{V}}_{\mathrm{CB}}^{\mathrm{N}}\overline{\boldsymbol{z}}_{2e}^{\mathrm{V}} & \boldsymbol{0} & \overline{\boldsymbol{V}}_{\mathrm{CA}}^{\mathrm{N}}\overline{\boldsymbol{z}}_{2e}^{\mathrm{V}} + \overline{\boldsymbol{V}}_{\mathrm{CB}}^{\mathrm{N}}\overline{\boldsymbol{z}}_{1e}^{\mathrm{V}} & \boldsymbol{0} \\ \sqrt{\varepsilon_{re}}\,\overline{\boldsymbol{V}}_{\mathrm{CA}}^{\mathrm{D}}\boldsymbol{\Gamma}_{\mathrm{V}e}^{-1} & \overline{\boldsymbol{V}}_{\mathrm{CA}}^{\mathrm{D}}\overline{\boldsymbol{\delta}}_h^{\mathrm{V}}\boldsymbol{\Gamma}_{\mathrm{V}h}^{-1} & -\sqrt{\varepsilon_{re}}\,\overline{\boldsymbol{V}}_{\mathrm{CB}}^{\mathrm{D}}\boldsymbol{\Gamma}_{\mathrm{V}e}^{-1} & -\overline{\boldsymbol{V}}_{\mathrm{CB}}^{\mathrm{D}}\overline{\boldsymbol{\delta}}_h^{\mathrm{V}}\boldsymbol{\Gamma}_{\mathrm{V}h}^{-1} \\ \overline{\boldsymbol{V}}_{\mathrm{DA}}^{\mathrm{N}}\overline{\boldsymbol{z}}_{1e}^{\mathrm{V}} + \overline{\boldsymbol{V}}_{\mathrm{DB}}^{\mathrm{N}}\overline{\boldsymbol{z}}_{2e}^{\mathrm{V}} & \boldsymbol{0} & \overline{\boldsymbol{V}}_{\mathrm{DA}}^{\mathrm{N}}\overline{\boldsymbol{z}}_{2e}^{\mathrm{V}} + \overline{\boldsymbol{V}}_{\mathrm{DB}}^{\mathrm{N}}\overline{\boldsymbol{z}}_{1e}^{\mathrm{V}} & \boldsymbol{0} \\ \sqrt{\varepsilon_{re}}\,\overline{\boldsymbol{V}}_{\mathrm{DA}}^{\mathrm{D}}\boldsymbol{\Gamma}_{\mathrm{V}e}^{-1} & \overline{\boldsymbol{V}}_{\mathrm{DA}}^{\mathrm{D}}\overline{\boldsymbol{\delta}}_h^{\mathrm{V}}\boldsymbol{\Gamma}_{\mathrm{V}h}^{-1} & -\sqrt{\varepsilon_{re}}\,\overline{\boldsymbol{V}}_{\mathrm{DB}}^{\mathrm{D}}\boldsymbol{\Gamma}_{\mathrm{V}e}^{-1} & -\overline{\boldsymbol{V}}_{\mathrm{DB}}^{\mathrm{D}}\overline{\boldsymbol{\delta}}_h^{\mathrm{V}}\boldsymbol{\Gamma}_{\mathrm{V}h}^{-1} \end{bmatrix} \tag{7.49}$$

7.2.3 Lines in horizontal direction

7.2.3.1 Field equations

The equations for discretisation lines in horizontal directions can be obtained from those for vertical lines. Therefore, we first denote the coordinates in eqs. (7.7)–(7.10), x', y', z'. Then $-x, y, z$ are substituted by z', y', x', respectively. Therefore, we also have to change $\mathrm{d}\overline{z}$ into $-\mathrm{d}\overline{x}$, as well as $\widetilde{\mathbf{H}}_{\mathrm{z}}, \mathbf{E}_{\mathrm{z}}$ into $-\widetilde{\mathbf{H}}_{\mathrm{x}}, -\mathbf{E}_{\mathrm{x}}$, respectively. The quantities for this case are now marked with an H (for 'horizontal') as sub or superscript. Accordingly, we define the supervectors:

$$\widehat{\mathbf{H}}^{\mathrm{H}} = \begin{bmatrix} \widetilde{\mathbf{H}}_{\mathrm{z}} \\ -\mathrm{j}\mathbf{H}_{\mathrm{y}} \end{bmatrix} \quad \widehat{\mathbf{E}}^{\mathrm{H}} = \begin{bmatrix} -\mathrm{j}\mathbf{E}_{\mathrm{y}} \\ \mathbf{E}_{\mathrm{z}} \end{bmatrix} \tag{7.50}$$

and obtain GTL equations in discretised form. (The subscripts e and h at D correspond to E_y and H_y, respectively; the discretisation is done in z-direction.) We have:

$$\frac{\mathrm{d}}{\mathrm{d}\overline{x}}\widehat{\mathbf{H}}^{\mathrm{H}} = -\widehat{\boldsymbol{R}}_{\mathrm{E}}^{\mathrm{H}}\widehat{\mathbf{E}}^{\mathrm{H}} \qquad \widehat{\boldsymbol{R}}_{\mathrm{E}}^{\mathrm{H}} = -\begin{bmatrix} \varepsilon_r\boldsymbol{I}_e - \overline{\boldsymbol{D}}_e^{\mathrm{t}}\overline{\boldsymbol{D}}_e & \sqrt{\varepsilon_{re}}\overline{\boldsymbol{D}}_h \\ -\sqrt{\varepsilon_{re}}\overline{\boldsymbol{D}}_e & \varepsilon_d\boldsymbol{I}_h \end{bmatrix} \tag{7.51}$$

$$\frac{\mathrm{d}}{\mathrm{d}\overline{x}}\widehat{\mathbf{E}}^{\mathrm{H}} = -\widehat{\boldsymbol{R}}_{\mathrm{H}}^{\mathrm{H}}\widehat{\mathbf{H}}^{\mathrm{H}} \qquad \widehat{\boldsymbol{R}}_{\mathrm{H}}^{\mathrm{H}} = \frac{1}{\varepsilon_r}\begin{bmatrix} \varepsilon_d\boldsymbol{I}_e & -\sqrt{\varepsilon_{re}}\overline{\boldsymbol{D}}_h \\ \sqrt{\varepsilon_{re}}\overline{\boldsymbol{D}}_e & \varepsilon_r\boldsymbol{I}_h - \overline{\boldsymbol{D}}_h^{\mathrm{t}}\overline{\boldsymbol{D}}_h \end{bmatrix} \tag{7.52}$$

The components E_x and H_x are given by:

$$\mathbf{E}_x^{\mathrm{H}} = -\varepsilon_r^{-1}\left[\sqrt{\varepsilon_{re}}\boldsymbol{I}_e \quad \overline{\boldsymbol{D}}_h\right]\widehat{\mathbf{H}}^{\mathrm{H}} \qquad \mathbf{H}_x^{\mathrm{H}} = \left[\overline{\boldsymbol{D}}_e \quad \sqrt{\varepsilon_{re}}\boldsymbol{I}_h\right]\widehat{\mathbf{E}}^{\mathrm{H}} \tag{7.53}$$

$\overline{\boldsymbol{D}}_{e,h}$ is the difference operator related to the first-order differential operator in z-direction for the components on e or h discretisation lines, respectively. It is normalised with $k_o h$, where $\boldsymbol{D}_\mathrm{h} = -\boldsymbol{D}_\mathrm{e}^\mathrm{t}$. By combining the GTL equations, we obtain in this case the following discretised wave equations:

$$\frac{\mathrm{d}^2}{\mathrm{d}\overline{y}^2}\widehat{\mathbf{E}}^\mathrm{H} - \widehat{\boldsymbol{Q}}_\mathrm{E}^\mathrm{H} = \mathbf{0} \qquad \frac{d^2}{d\overline{y}^2}\widehat{\mathbf{H}}^\mathrm{H} - \widehat{\boldsymbol{Q}}_\mathrm{H}^\mathrm{H} = \mathbf{0} \qquad \widehat{\boldsymbol{Q}}_\mathrm{E}^\mathrm{H} = \widehat{\boldsymbol{R}}_\mathrm{H}^\mathrm{H}\widehat{\boldsymbol{R}}_\mathrm{E}^\mathrm{H} \qquad \widehat{\boldsymbol{Q}}_\mathrm{H}^\mathrm{H} = \widehat{\boldsymbol{R}}_\mathrm{E}^\mathrm{H}\widehat{\boldsymbol{R}}_\mathrm{H}^\mathrm{H} \tag{7.54}$$

with:

$$\widehat{\boldsymbol{Q}}_\mathrm{E}^\mathrm{H} = \widehat{\boldsymbol{Q}}_\mathrm{H}^\mathrm{H} = \begin{bmatrix} \overline{\boldsymbol{D}}_e^\mathrm{t}\overline{\boldsymbol{D}}_e - \varepsilon_d \boldsymbol{I}_e & \boldsymbol{0} \\ \boldsymbol{0} & \overline{\boldsymbol{D}}_h^t\overline{\boldsymbol{D}}_h - \varepsilon_d \boldsymbol{I}_h \end{bmatrix} = \begin{bmatrix} \boldsymbol{Q}_e^\mathrm{H} & \boldsymbol{0} \\ \boldsymbol{0} & \boldsymbol{Q}_h^\mathrm{H} \end{bmatrix} \tag{7.55}$$

The hat ($\widehat{\ }$) on the quantities indicates supervectors or supermatrices. We transform to the main axis (diagonalisation) according to:

$$\widehat{\mathbf{E}}^\mathrm{H} = \widehat{\boldsymbol{T}}_\mathrm{E}\widehat{\overline{\mathbf{E}}}^\mathrm{H} \qquad \widehat{\mathbf{H}}^\mathrm{H} = \widehat{\boldsymbol{T}}_\mathrm{H}\widehat{\overline{\mathbf{H}}}^\mathrm{H} \qquad \widehat{\boldsymbol{T}}_\mathrm{E}^{-1}\widehat{\boldsymbol{Q}}_\mathrm{E}\widehat{\boldsymbol{T}}_\mathrm{E} = \widehat{\boldsymbol{T}}_\mathrm{H}^{-1}\widehat{\boldsymbol{Q}}_\mathrm{H}\widehat{\boldsymbol{T}}_\mathrm{H} = \widehat{\boldsymbol{\Gamma}}_\mathrm{H}^2 \tag{7.56}$$

$$\widehat{\boldsymbol{T}}_\mathrm{H}^\mathrm{H} = \widehat{\boldsymbol{T}}_\mathrm{E}^\mathrm{H} = \begin{bmatrix} \boldsymbol{T}_e^\mathrm{H} & \boldsymbol{0} \\ \boldsymbol{0} & \boldsymbol{T}_h^\mathrm{H} \end{bmatrix} \qquad \widehat{\boldsymbol{\Gamma}}_\mathrm{H}^2 = \begin{bmatrix} \boldsymbol{\Gamma}_{\mathrm{H}e}^2 & \boldsymbol{0} \\ \boldsymbol{0} & \boldsymbol{\Gamma}_{\mathrm{H}h}^2 \end{bmatrix} \qquad \boldsymbol{\Gamma}_{\mathrm{H}e,h}^2 = \overline{\boldsymbol{\lambda}}_{e,h}^2 - \varepsilon_d \boldsymbol{I}_{e,h} \tag{7.57}$$

$\overline{\boldsymbol{\lambda}}_{e,h}^2$ are eigenvalues of $\overline{\boldsymbol{D}}_e^\mathrm{t}\overline{\boldsymbol{D}}_e$ and $\overline{\boldsymbol{D}}_h^\mathrm{t}\overline{\boldsymbol{D}}_h$, respectively. $\boldsymbol{\Gamma}_h^2$, $\boldsymbol{T}_h^\mathrm{H}$ and $\boldsymbol{\Gamma}_e^2$, $\boldsymbol{T}_e^\mathrm{H}$ are eigensolutions of $\boldsymbol{Q}_h^\mathrm{H}$ and $\boldsymbol{Q}_e^\mathrm{H}$, respectively. We obtain decoupled equations:

$$\frac{\mathrm{d}^2}{\mathrm{d}\overline{z}^2}\widehat{\overline{\mathbf{E}}}^\mathrm{H} - \widehat{\boldsymbol{\Gamma}}_\mathrm{H}^2\widehat{\overline{\mathbf{E}}}^\mathrm{H} = \mathbf{0} \qquad \frac{\mathrm{d}}{\mathrm{d}\overline{z}^2}\widehat{\overline{\mathbf{H}}}^\mathrm{H} - \widehat{\boldsymbol{\Gamma}}_\mathrm{H}^2\widehat{\overline{\mathbf{H}}}^\mathrm{H} = \mathbf{0} \tag{7.58}$$

To increase the accuracy of $\overline{\boldsymbol{\lambda}}_{e,h}^2$ and $\boldsymbol{\Gamma}_{\mathrm{V}e,h}^2$ the recurrence procedure in Section A.3 can be used. As we can see in eq. (7.55), we have $\widehat{\boldsymbol{Q}}_\mathrm{E}^\mathrm{H} = \widehat{\boldsymbol{Q}}_\mathrm{H}^\mathrm{H}$. Therefore, the matrices $\widehat{\boldsymbol{T}}_\mathrm{E}^\mathrm{H}$ and $\widehat{\boldsymbol{T}}_\mathrm{H}^\mathrm{H}$ can be chosen to be equal to each other. However, we will use different matrices again in this case, as will be shown later. We distinguish between the calculations of $\boldsymbol{z}$ and $\boldsymbol{y}$ matrices.

For the calculation of $\boldsymbol{z}$ matrices we determine the matrices $\widehat{\boldsymbol{T}}_\mathrm{E}^\mathrm{H}$ and $\widehat{\boldsymbol{T}}_\mathrm{H}^\mathrm{H}$ according to:

$$\widehat{\boldsymbol{T}}_\mathrm{E}^\mathrm{H} = \begin{bmatrix} \boldsymbol{T}_e^\mathrm{H} & \boldsymbol{0} \\ \boldsymbol{0} & \boldsymbol{T}_h^\mathrm{H} \end{bmatrix} \qquad \widehat{\boldsymbol{T}}_\mathrm{H}^\mathrm{H} = \widehat{\boldsymbol{R}}_\mathrm{E}^\mathrm{H}\widehat{\boldsymbol{T}}_\mathrm{E}^\mathrm{H}\widehat{\boldsymbol{\Gamma}}_\mathrm{H}^{-1} = -\widehat{\boldsymbol{T}}_\mathrm{E}^\mathrm{H}\begin{bmatrix} \varepsilon_r \boldsymbol{I}_e - \overline{\boldsymbol{\lambda}}_e^2 & \widetilde{\boldsymbol{\delta}}_h \\ \widetilde{\boldsymbol{\delta}}_h^\mathrm{t} & \varepsilon_d \boldsymbol{I}_h \end{bmatrix}\widehat{\boldsymbol{\Gamma}}_\mathrm{H}^{-1} \tag{7.59}$$

where $\widetilde{\boldsymbol{\delta}}_h^\mathrm{H} = \sqrt{\varepsilon_{re}}\,\boldsymbol{T}_e^{\mathrm{H}\,\mathrm{t}}\overline{\boldsymbol{D}}_h\boldsymbol{T}_h^\mathrm{H}$. For the calculation of $\boldsymbol{y}$ matrices we compute the matrices $\widehat{\boldsymbol{T}}_\mathrm{H}^\mathrm{H}$ and $\widehat{\boldsymbol{T}}_\mathrm{E}^\mathrm{H}$ as:

$$\widehat{\boldsymbol{T}}_\mathrm{H}^\mathrm{H} = \begin{bmatrix} \boldsymbol{T}_e^\mathrm{H} & \boldsymbol{0} \\ \boldsymbol{0} & \boldsymbol{T}_h^\mathrm{H} \end{bmatrix} \qquad \widehat{\boldsymbol{T}}_\mathrm{E}^\mathrm{H} = \boldsymbol{R}_\mathrm{H}^\mathrm{H}\widehat{\boldsymbol{T}}_\mathrm{H}^\mathrm{H}\boldsymbol{\Gamma}_\mathrm{H}^{-1} = -\widehat{\boldsymbol{T}}_\mathrm{H}^\mathrm{H}\begin{bmatrix} -\varepsilon_d \boldsymbol{I}_e & \widetilde{\boldsymbol{\delta}}_h \\ \widetilde{\boldsymbol{\delta}}_h^\mathrm{t} & \overline{\boldsymbol{\lambda}}_h^2 - \varepsilon_r \boldsymbol{I}_h \end{bmatrix}\widehat{\boldsymbol{\Gamma}}_\mathrm{H}^{-1}\varepsilon_r^{-1} \tag{7.60}$$

The multiplication by $\boldsymbol{\Gamma}_{\mathrm{H}}^{-1}$ is introduced for normalisation. By using these transformation matrices $\widehat{\boldsymbol{T}}_{\mathrm{E}}$ and $\widehat{\boldsymbol{T}}_{\mathrm{H}}$, we obtain for the GTL equations in the transformed domain:

$$\frac{\mathrm{d}}{\mathrm{d}\overline{y}}\widehat{\overline{\mathbf{E}}}^{\mathrm{H}} = -\widehat{\boldsymbol{\Gamma}}_{\mathrm{H}}\widehat{\overline{\mathbf{H}}}^{\mathrm{H}} \qquad \frac{\mathrm{d}}{\mathrm{d}\overline{y}}\widehat{\overline{\mathbf{H}}}^{\mathrm{H}} = -\widehat{\boldsymbol{\Gamma}}_{\mathrm{H}}\widehat{\overline{\mathbf{E}}}^{\mathrm{H}} \tag{7.61}$$

7.2.3.2 Field transformation between planes C and D

For the field transformation between planes C and D we obtain:

$$\begin{bmatrix} \widehat{\overline{\mathbf{H}}}_{\mathrm{C}}^{\mathrm{H}} \\ -\widehat{\overline{\mathbf{H}}}_{\mathrm{D}}^{\mathrm{H}} \end{bmatrix} = \begin{bmatrix} \widehat{\overline{\boldsymbol{y}}}_1^{\mathrm{H}} & \widehat{\overline{\boldsymbol{y}}}_2^{\mathrm{H}} \\ \widehat{\overline{\boldsymbol{y}}}_2^{\mathrm{H}} & \widehat{\overline{\boldsymbol{y}}}_1^{\mathrm{H}} \end{bmatrix} \begin{bmatrix} \widehat{\overline{\mathbf{E}}}_{\mathrm{C}}^{\mathrm{H}} \\ \widehat{\overline{\mathbf{E}}}_{\mathrm{D}}^{\mathrm{H}} \end{bmatrix} \qquad \widehat{\overline{\mathbf{H}}}_{\mathrm{CD}}^{\mathrm{H}} = \widehat{\overline{\boldsymbol{y}}}_{\mathrm{CD}}^{\mathrm{CD}}\widehat{\overline{\mathbf{E}}}_{\mathrm{CD}}^{\mathrm{H}} \tag{7.62}$$

$$\begin{bmatrix} \widehat{\overline{\mathbf{E}}}_{\mathrm{C}}^{\mathrm{H}} \\ \widehat{\overline{\mathbf{E}}}_{\mathrm{D}}^{\mathrm{H}} \end{bmatrix} = \begin{bmatrix} \widehat{\overline{\boldsymbol{z}}}_1^{\mathrm{H}} & \widehat{\overline{\boldsymbol{z}}}_2^{\mathrm{H}} \\ \widehat{\overline{\boldsymbol{z}}}_2^{\mathrm{H}} & \widehat{\overline{\boldsymbol{z}}}_1^{\mathrm{H}} \end{bmatrix} \begin{bmatrix} \widehat{\overline{\mathbf{H}}}_{\mathrm{C}}^{\mathrm{H}} \\ -\widehat{\overline{\mathbf{H}}}_{\mathrm{D}}^{\mathrm{H}} \end{bmatrix} \qquad \widehat{\overline{\mathbf{E}}}_{\mathrm{CD}}^{\mathrm{H}} = \widehat{\overline{\boldsymbol{z}}}_{\mathrm{CD}}^{\mathrm{CD}}\widehat{\overline{\mathbf{H}}}_{\mathrm{CD}}^{\mathrm{H}} \tag{7.63}$$

with:

$$\begin{aligned} \widehat{\overline{\boldsymbol{y}}}_1^{\mathrm{H}} &= 1/\tanh(\widehat{\boldsymbol{\Gamma}}_{\mathrm{H}}\overline{d}_{\mathrm{CD}}) = \mathrm{Diag}(\overline{\boldsymbol{y}}_{1e}^{\mathrm{H}}, \overline{\boldsymbol{y}}_{1h}^{\mathrm{H}}) \\ \widehat{\overline{\boldsymbol{y}}}_2^{\mathrm{H}} &= -1/\sinh(\widehat{\boldsymbol{\Gamma}}_{\mathrm{H}}\overline{d}_{\mathrm{CD}}) = \mathrm{Diag}(\overline{\boldsymbol{y}}_{2e}^{\mathrm{H}}, \overline{\boldsymbol{y}}_{2h}^{\mathrm{H}}) \\ \widehat{\overline{\boldsymbol{z}}}_1^{\mathrm{H}} &= 1/\tanh(\widehat{\boldsymbol{\Gamma}}_{\mathrm{H}}\overline{d}_{\mathrm{CD}}) = \mathrm{Diag}(\overline{\boldsymbol{z}}_{1e}^{\mathrm{H}}, \overline{\boldsymbol{z}}_{1h}^{\mathrm{H}}) \\ \widehat{\overline{\boldsymbol{z}}}_2^{\mathrm{H}} &= 1/\sinh(\widehat{\boldsymbol{\Gamma}}_{\mathrm{H}}\overline{d}_{\mathrm{CD}}) = \mathrm{Diag}(\overline{\boldsymbol{z}}_{2e}^{\mathrm{H}}, \overline{\boldsymbol{z}}_{2h}^{\mathrm{H}}) \end{aligned} \tag{7.64}$$

The fields $\overline{\mathbf{H}}_x^{\mathrm{H}}$ and $\overline{\mathbf{E}}_x^{\mathrm{H}}$ can now be calculated as functions of $\widehat{\overline{\mathbf{E}}}_{\mathrm{C,D}}^{\mathrm{H}}$ and $\widehat{\overline{\mathbf{H}}}_{\mathrm{C,D}}^{\mathrm{H}}$, respectively. From eq. (7.53) we obtain:

$$\mathbf{E}_x^{\mathrm{H}} = -\varepsilon_r^{-1}\left[\sqrt{\varepsilon_{re}}\boldsymbol{I}_e^{\mathrm{H}} \quad \overline{\boldsymbol{D}}_{\mathrm{H}h}\right]\widehat{\boldsymbol{T}}_{\mathrm{H}}^{\mathrm{H}}\widehat{\overline{\mathbf{H}}}^{\mathrm{H}} \qquad \mathbf{H}_x^{\mathrm{H}} = \left[\overline{\boldsymbol{D}}_{\mathrm{H}e} \quad \sqrt{\varepsilon_{re}}\boldsymbol{I}_{\mathrm{h}}^{\mathrm{H}}\right]\widehat{\boldsymbol{T}}_{\mathrm{E}}^{\mathrm{H}}\widehat{\overline{\mathbf{E}}}^{\mathrm{H}} \tag{7.65}$$

By using the transformation matrices according to eq. (7.59) for z-matrices, we get:

$$\overline{\mathbf{E}}_x^{\mathrm{H}} = \mathbf{T}_e^{\mathrm{t}}\mathbf{E}_x^{\mathrm{H}} = \left[\sqrt{\varepsilon_{re}}\boldsymbol{I}_e \quad \overline{\boldsymbol{\delta}}_h\right]\widehat{\boldsymbol{\Gamma}}_{\mathrm{H}}^{-1}\widehat{\overline{\mathbf{H}}}^{\mathrm{H}} \qquad \overline{\mathbf{H}}_x^{\mathrm{H}} = \boldsymbol{T}_h^{\mathrm{t}}\widetilde{\mathbf{H}}_x^{\mathrm{H}} = \left[\overline{\boldsymbol{\delta}}_e \quad \sqrt{\varepsilon_{re}}\boldsymbol{I}_h\right]\widehat{\overline{\mathbf{E}}}^{\mathrm{H}} \tag{7.66}$$

Applying the transformation matrices according to eq. (7.60) for y-matrices, we have:

$$\overline{\mathbf{E}}_x^{\mathrm{H}} = \mathbf{T}_e^{\mathrm{t}}\mathbf{E}_x^{\mathrm{H}} = -\varepsilon_r^{-1}\left[\sqrt{\varepsilon_{re}}\boldsymbol{I}_e \quad \overline{\boldsymbol{\delta}}_h\right]\widehat{\overline{\mathbf{H}}}^{\mathrm{H}} \qquad \overline{\mathbf{H}}_x^{\mathrm{H}} = \boldsymbol{T}_h^{\mathrm{t}}\widetilde{\mathbf{H}}_x^{\mathrm{H}} = \left[\overline{\boldsymbol{\delta}}_e \quad \boldsymbol{I}_h\right]\widehat{\boldsymbol{\Gamma}}_{\mathrm{H}}^{-1}\widehat{\overline{\mathbf{E}}}^{\mathrm{H}} \tag{7.67}$$

These equations are also valid at ports C and D.

7.2.3.3 Field transformation from planes C/D to planes A/B

Now we need to write the fields at ports A and B as functions of the fields at ports C and D. Here we have discretisation lines in the horizontal direction and we obtain similar expressions as those from discretisation lines in the vertical direction:

$$\begin{bmatrix} \mathbf{F}_{u\mathrm{A}} \\ \mathbf{F}_{u\mathrm{B}} \end{bmatrix} = \begin{bmatrix} \overline{\boldsymbol{V}}_{\mathrm{AC}} & \overline{\boldsymbol{V}}_{\mathrm{AD}} \\ \overline{\boldsymbol{V}}_{\mathrm{BC}} & \overline{\boldsymbol{V}}_{\mathrm{BD}} \end{bmatrix} \begin{bmatrix} \mathbf{F}_{u\mathrm{C}} \\ \mathbf{F}_{u\mathrm{D}} \end{bmatrix} \tag{7.68}$$

with:

$$\begin{aligned} \overline{\boldsymbol{V}}_{\mathrm{AC}} &= \boldsymbol{T}_{\mathrm{V}g}^{-1}\boldsymbol{\Lambda}_{\mathrm{C}}\boldsymbol{T}_{f\mathrm{A}}^{\mathrm{d}} & \overline{\boldsymbol{V}}_{\mathrm{AD}} &= \boldsymbol{T}_{\mathrm{V}g}^{-1}\boldsymbol{\Lambda}_{\mathrm{D}}\boldsymbol{T}_{f\mathrm{A}}^{\mathrm{d}} \\ \overline{\boldsymbol{V}}_{\mathrm{BC}} &= \boldsymbol{T}_{\mathrm{V}g}^{-1}\boldsymbol{\Lambda}_{\mathrm{C}}\boldsymbol{T}_{f\mathrm{B}}^{\mathrm{d}} & \overline{\boldsymbol{V}}_{\mathrm{BD}} &= \boldsymbol{T}_{\mathrm{V}g}^{-1}\boldsymbol{\Lambda}_{\mathrm{D}}\boldsymbol{T}_{f\mathrm{B}}^{\mathrm{d}} \end{aligned} \tag{7.69}$$

$\boldsymbol{T}_{\mathrm{V}g}$ is the adequate transformation matrix for horizontal discretisation lines. g is either e or h depending on the line system where the component $\mathbf{F}_{u\mathrm{R}}$ is calculated:

$$\begin{bmatrix} \overline{\mathbf{H}}_{\mathrm{yA}}^{\mathrm{H}} \\ -\mathrm{j}\overline{\mathbf{H}}_{\mathrm{yA}}^{\mathrm{H}} \\ -\overline{\mathbf{H}}_{\mathrm{xB}}^{\mathrm{H}} \\ \mathrm{j}\overline{\mathbf{H}}_{\mathrm{yB}}^{\mathrm{H}} \end{bmatrix} = \begin{bmatrix} \overline{\boldsymbol{V}}_{\mathrm{AC}}^{\mathrm{D}} & & -\overline{\boldsymbol{V}}_{\mathrm{AD}}^{\mathrm{D}} & \\ & \overline{\boldsymbol{V}}_{\mathrm{AC}}^{\mathrm{N}} & & -\overline{\boldsymbol{V}}_{\mathrm{AD}}^{\mathrm{N}} \\ -\overline{\boldsymbol{V}}_{\mathrm{BC}}^{\mathrm{D}} & & \overline{\boldsymbol{V}}_{\mathrm{BD}}^{\mathrm{D}} & \\ & -\overline{\boldsymbol{V}}_{\mathrm{BC}}^{\mathrm{N}} & & \overline{\boldsymbol{V}}_{\mathrm{BD}}^{\mathrm{N}} \end{bmatrix} \begin{bmatrix} \overline{\mathbf{H}}_{\mathrm{xC}}^{\mathrm{H}} \\ -\mathrm{j}\overline{\mathbf{H}}_{\mathrm{yC}}^{\mathrm{H}} \\ -\overline{\mathbf{H}}_{\mathrm{xD}}^{\mathrm{H}} \\ \mathrm{j}\overline{\mathbf{H}}_{\mathrm{yD}}^{\mathrm{H}} \end{bmatrix} \tag{7.70}$$

or:

$$\widehat{\overline{\mathbf{H}}}_{\mathrm{AB}}^{\mathrm{H}} = \widehat{\overline{\boldsymbol{V}}}_{\mathrm{H}}^{\mathrm{H}}\widehat{\overline{\mathbf{H}}}_{\mathrm{CD}}^{\mathrm{H}T} \tag{7.71}$$

For the superscript N (D), the subscript g in eq. (7.69) is $h(e)$. Likewise, we can write the tangential electric field components at ports A and B as functions of the adequate electric field components at ports C and D:

$$\begin{bmatrix} -\mathrm{j}\overline{\mathbf{E}}_{\mathrm{yA}}^{\mathrm{H}} \\ \overline{\mathbf{E}}_{\mathrm{xA}}^{\mathrm{H}} \\ -\mathrm{j}\overline{\mathbf{E}}_{\mathrm{yB}}^{\mathrm{H}} \\ \overline{\mathbf{E}}_{\mathrm{xB}}^{\mathrm{H}} \end{bmatrix} = \begin{bmatrix} \overline{\boldsymbol{V}}_{\mathrm{AC}}^{\mathrm{N}} & & \overline{\boldsymbol{V}}_{\mathrm{AD}}^{\mathrm{N}} & \\ & \overline{\boldsymbol{V}}_{\mathrm{AC}}^{\mathrm{D}} & & \overline{\boldsymbol{V}}_{\mathrm{AD}}^{\mathrm{D}} \\ \overline{\boldsymbol{V}}_{\mathrm{BC}}^{\mathrm{N}} & & \overline{\boldsymbol{V}}_{\mathrm{BD}}^{\mathrm{N}} & \\ & \overline{\boldsymbol{V}}_{\mathrm{BC}}^{\mathrm{D}} & & \overline{\boldsymbol{V}}_{\mathrm{BD}}^{\mathrm{D}} \end{bmatrix} \begin{bmatrix} -\mathrm{j}\overline{\mathbf{E}}_{\mathrm{yC}}^{\mathrm{H}} \\ \overline{\mathbf{E}}_{\mathrm{xC}}^{\mathrm{H}} \\ -\mathrm{j}\overline{\mathbf{E}}_{\mathrm{yD}}^{\mathrm{H}} \\ \overline{\mathbf{E}}_{\mathrm{xD}}^{\mathrm{H}} \end{bmatrix} \tag{7.72}$$

or:

$$\widehat{\overline{\mathbf{E}}}_{\mathrm{AB}}^{\mathrm{H}} = \widehat{\overline{\boldsymbol{V}}}_{\mathrm{E}}^{\mathrm{H}}\widehat{\overline{\mathbf{E}}}_{\mathrm{CD}}^{\mathrm{H}T} \tag{7.73}$$

For the superscript N (D), the subscript g in eq. (7.69) should be $e(h)$. Combining eqs. (7.62) and (7.67) results in the following expression:

$$\begin{bmatrix} \overline{\mathbf{H}}_{\mathrm{xC}}^{\mathrm{H}} \\ -\mathrm{j}\overline{\mathbf{H}}_{\mathrm{yC}}^{\mathrm{H}} \\ -\overline{\mathbf{H}}_{\mathrm{xD}}^{\mathrm{H}} \\ \mathrm{j}\overline{\mathbf{H}}_{\mathrm{yD}}^{\mathrm{H}} \end{bmatrix} = \begin{bmatrix} \overline{\boldsymbol{\delta}}_{e}^{\mathrm{H}}\boldsymbol{\Gamma}_{\mathrm{H}e}^{-1} & \boldsymbol{\Gamma}_{\mathrm{H}h}^{-1} & \boldsymbol{0} & \boldsymbol{0} \\ \boldsymbol{0} & \overline{\boldsymbol{y}}_{1h}^{\mathrm{H}} & \boldsymbol{0} & \overline{\boldsymbol{y}}_{2h}^{\mathrm{H}} \\ \boldsymbol{0} & \boldsymbol{0} & -\overline{\boldsymbol{\delta}}_{e}^{\mathrm{H}}\boldsymbol{\Gamma}_{\mathrm{H}e}^{-1} & -\boldsymbol{\Gamma}_{\mathrm{H}h}^{-1} \\ \boldsymbol{0} & \overline{\boldsymbol{y}}_{2h}^{\mathrm{H}} & \boldsymbol{0} & \overline{\boldsymbol{y}}_{1h}^{\mathrm{H}} \end{bmatrix} \begin{bmatrix} \widehat{\overline{\mathbf{E}}}_{\mathrm{C}}^{\mathrm{H}} \\ \widehat{\overline{\mathbf{E}}}_{\mathrm{D}}^{\mathrm{H}} \end{bmatrix} \tag{7.74}$$

or:

$$\widehat{\overline{\mathbf{H}}}_{\mathrm{CD}}^{\mathrm{H}T} = \widehat{\overline{\boldsymbol{y}}}_{\mathrm{CD}}^{\mathrm{H}}\widehat{\overline{\mathbf{E}}}_{\mathrm{CD}}^{\mathrm{H}} \tag{7.75}$$

For z-matrices, we combine eqs. (7.63) and (7.66) and obtain:

$$\begin{bmatrix} -\mathrm{j}\overline{\mathbf{E}}^{\mathrm{H}}_{\mathrm{yC}} \\ \overline{\mathbf{E}}^{\mathrm{H}}_{\mathrm{xC}} \\ -\mathrm{j}\overline{\mathbf{E}}^{\mathrm{H}}_{\mathrm{yD}} \\ \overline{\mathbf{E}}^{\mathrm{H}}_{\mathrm{xD}} \end{bmatrix} = \begin{bmatrix} \overline{\boldsymbol{z}}^{\mathrm{H}}_{1e} & \boldsymbol{0} & \overline{\boldsymbol{z}}^{\mathrm{H}}_{2e} & \boldsymbol{0} \\ \sqrt{\varepsilon_{re}}\boldsymbol{\Gamma}^{-1}_{\mathrm{H}e} & \overline{\boldsymbol{\delta}}^{\mathrm{H}}_{h}\boldsymbol{\Gamma}^{-1}_{\mathrm{H}h} & \boldsymbol{0} & \boldsymbol{0} \\ \overline{\boldsymbol{z}}^{\mathrm{H}}_{2e} & \boldsymbol{0} & \overline{\boldsymbol{z}}^{\mathrm{H}}_{1e} & \boldsymbol{0} \\ \boldsymbol{0} & \boldsymbol{0} & -\sqrt{\varepsilon_{re}}\boldsymbol{\Gamma}^{-1}_{\mathrm{H}e} & -\overline{\boldsymbol{\delta}}^{\mathrm{H}}_{e}\boldsymbol{\Gamma}^{-1}_{\mathrm{H}h} \end{bmatrix} \begin{bmatrix} \widehat{\overline{\mathbf{H}}}^{\mathrm{H}}_{\mathrm{C}} \\ -\widehat{\overline{\mathbf{H}}}^{\mathrm{H}}_{\mathrm{D}} \end{bmatrix} \tag{7.76}$$

or:

$$\widehat{\overline{\mathbf{E}}}^{\mathrm{HT}}_{\mathrm{CD}} = \widehat{\overline{\boldsymbol{z}}}^{\mathrm{H}}_{\mathrm{CD}}\widehat{\overline{\mathbf{H}}}^{\mathrm{H}}_{\mathrm{CD}} \tag{7.77}$$

The off-diagonal submatrices on the right-hand side of eq. (7.4) are given by:

$$\widehat{\overline{\boldsymbol{y}}}^{\mathrm{CD}}_{\mathrm{AB}} = \widehat{\overline{\boldsymbol{V}}}^{\mathrm{H}}_{\mathrm{H}}\widehat{\overline{\boldsymbol{y}}}^{\mathrm{H}}_{\mathrm{CD}} \qquad \widehat{\overline{\boldsymbol{z}}}^{\mathrm{CD}}_{\mathrm{AB}} = \widehat{\overline{\boldsymbol{V}}}^{\mathrm{H}}_{\mathrm{E}}\widehat{\overline{\boldsymbol{z}}}^{\mathrm{H}}_{\mathrm{CD}} \tag{7.78}$$

Hence we obtain:

$$\widehat{\overline{\boldsymbol{y}}}^{\mathrm{CD}}_{\mathrm{AB}} = \begin{bmatrix} \overline{\boldsymbol{V}}^{\mathrm{D}}_{\mathrm{AC}}\overline{\boldsymbol{\delta}}^{\mathrm{H}}_{e}\boldsymbol{\Gamma}^{-1}_{\mathrm{H}e} & \overline{\boldsymbol{V}}^{\mathrm{D}}_{\mathrm{AC}}\boldsymbol{\Gamma}^{-1}_{\mathrm{H}h} & \overline{\boldsymbol{V}}^{\mathrm{D}}_{\mathrm{AD}}\overline{\boldsymbol{\delta}}^{\mathrm{H}}_{e}\boldsymbol{\Gamma}^{-1}_{\mathrm{H}e} & \overline{\boldsymbol{V}}^{\mathrm{D}}_{\mathrm{AD}}\boldsymbol{\Gamma}^{-1}_{\mathrm{H}h} \\ \boldsymbol{0} & \overline{\boldsymbol{V}}^{\mathrm{N}}_{\mathrm{AC}}\overline{\boldsymbol{y}}^{\mathrm{H}}_{1h} - \overline{\boldsymbol{V}}^{\mathrm{N}}_{\mathrm{AD}}\overline{\boldsymbol{y}}^{\mathrm{H}}_{2h} & \boldsymbol{0} & \overline{\boldsymbol{V}}^{\mathrm{N}}_{\mathrm{AC}}\overline{\boldsymbol{y}}^{\mathrm{H}}_{2h} - \overline{\boldsymbol{V}}^{\mathrm{N}}_{\mathrm{AD}}\overline{\boldsymbol{y}}^{\mathrm{H}}_{1h} \\ -\overline{\boldsymbol{V}}^{\mathrm{D}}_{\mathrm{BC}}\overline{\boldsymbol{\delta}}^{\mathrm{H}}_{e}\boldsymbol{\Gamma}^{-1}_{\mathrm{H}e} & -\overline{\boldsymbol{V}}^{\mathrm{D}}_{\mathrm{BC}}\boldsymbol{\Gamma}^{-1}_{\mathrm{H}h} & -\overline{\boldsymbol{V}}^{\mathrm{D}}_{\mathrm{BD}}\overline{\boldsymbol{\delta}}^{\mathrm{H}}_{e}\boldsymbol{\Gamma}^{-1}_{\mathrm{H}c} & -\overline{\boldsymbol{V}}^{\mathrm{D}}_{\mathrm{BC}}\boldsymbol{\Gamma}^{-1}_{\mathrm{N}h} \\ \boldsymbol{0} & \overline{\boldsymbol{V}}^{\mathrm{N}}_{\mathrm{BD}}\overline{\boldsymbol{y}}^{\mathrm{H}}_{2h} - \overline{\boldsymbol{V}}^{\mathrm{N}}_{\mathrm{BC}}\overline{\boldsymbol{y}}^{\mathrm{H}}_{1h} & \boldsymbol{0} & \overline{\boldsymbol{V}}^{\mathrm{H}}_{\mathrm{BD}}\overline{\boldsymbol{y}}^{\mathrm{H}}_{1h} - \overline{\boldsymbol{V}}^{\mathrm{N}}_{\mathrm{BC}}\overline{\boldsymbol{y}}^{\mathrm{H}}_{2h} \end{bmatrix} \tag{7.79}$$

and:

$$\widehat{\overline{\boldsymbol{z}}}^{\mathrm{CD}}_{\mathrm{AB}} = \begin{bmatrix} \overline{\boldsymbol{V}}^{\mathrm{N}}_{\mathrm{AC}}\overline{\boldsymbol{z}}^{\mathrm{H}}_{1e} + \overline{\boldsymbol{V}}^{\mathrm{N}}_{\mathrm{AD}}\overline{\boldsymbol{z}}^{\mathrm{H}}_{2e} & \boldsymbol{0} & \overline{\boldsymbol{V}}^{\mathrm{N}}_{\mathrm{AC}}\overline{\boldsymbol{z}}^{\mathrm{H}}_{2e} + \overline{\boldsymbol{V}}^{\mathrm{N}}_{\mathrm{AD}}\overline{\boldsymbol{z}}^{\mathrm{H}}_{1e} & \boldsymbol{0} \\ \sqrt{\varepsilon_{re}}\overline{\boldsymbol{V}}^{\mathrm{D}}_{\mathrm{AC}}\boldsymbol{\Gamma}^{-1}_{\mathrm{H}e} & \overline{\boldsymbol{V}}^{\mathrm{D}}_{\mathrm{AD}}\overline{\boldsymbol{\delta}}^{\mathrm{H}}_{h}\boldsymbol{\Gamma}^{-1}_{\mathrm{H}h} & -\sqrt{\varepsilon_{re}}\overline{\boldsymbol{V}}^{\mathrm{D}}_{\mathrm{AC}}\boldsymbol{\Gamma}^{-1}_{\mathrm{H}e} & -\overline{\boldsymbol{V}}^{\mathrm{D}}_{\mathrm{AD}}\overline{\boldsymbol{\delta}}^{\mathrm{H}}_{e}\boldsymbol{\Gamma}^{-1}_{\mathrm{H}h} \\ \overline{\boldsymbol{V}}^{\mathrm{N}}_{\mathrm{BC}}\overline{\boldsymbol{z}}^{\mathrm{H}}_{1e} + \overline{\boldsymbol{V}}^{\mathrm{N}}_{\mathrm{BD}}\overline{\boldsymbol{z}}^{\mathrm{H}}_{2e} & \boldsymbol{0} & \overline{\boldsymbol{V}}^{\mathrm{N}}_{\mathrm{BC}}\overline{\boldsymbol{z}}^{\mathrm{H}}_{2e} + \overline{\boldsymbol{V}}^{\mathrm{N}}_{\mathrm{BD}}\overline{\boldsymbol{z}}^{\mathrm{H}}_{1e} & \boldsymbol{0} \\ \sqrt{\varepsilon_{re}}\overline{\boldsymbol{V}}^{\mathrm{D}}_{\mathrm{BC}}\boldsymbol{\Gamma}^{-1}_{\mathrm{H}e} & \overline{\boldsymbol{V}}^{\mathrm{D}}_{\mathrm{BC}}\overline{\boldsymbol{\delta}}^{\mathrm{H}}_{h}\boldsymbol{\Gamma}^{-1}_{\mathrm{H}h} & -\sqrt{\varepsilon_{re}}\overline{\boldsymbol{V}}^{\mathrm{D}}_{\mathrm{BD}}\boldsymbol{\Gamma}^{-1}_{\mathrm{H}e} & -\overline{\boldsymbol{V}}^{\mathrm{D}}_{\mathrm{BD}}\overline{\boldsymbol{\delta}}^{\mathrm{H}}_{e}\boldsymbol{\Gamma}^{-1}_{\mathrm{H}h} \end{bmatrix} \tag{7.80}$$

7.3 SPECIAL STRUCTURES IN CARTESIAN COORDINATES

In Section 3.3 we dealt with junctions of rectangular waveguides. In the current section we will describe further particular examples for which the algorithm with crossed discretisation lines was successfully used.

7.3.1 Groove guide

The first example is the groove guide shown in Fig. 7.11a. We would like to obtain the cut-off frequencies for the fundamental TE_y-modes. Therefore, ε_{re} has to be introduced into the equations. Because of symmetry, only a quarter of the cross-section need be examined (cf. Fig. 7.11b). TE_y modes do not have E_y components. We have a mixed problem with metallic and magnetic walls. It is not necessary to calculate all matrix parameters. Ports B and C of region E are open-circuited and short-circuited, respectively. Thus, we have only two important ports A and D where the fields are of interest. Therefore, the general equation derived in the previous sections simplifies to:

$$\begin{bmatrix} \overline{\mathbf{H}}_{\mathrm{A}} \\ -\overline{\mathbf{H}}_{\mathrm{D}} \end{bmatrix} = \begin{bmatrix} \overline{\boldsymbol{y}}_{\mathrm{A}}^{\mathrm{A}} & \overline{\boldsymbol{y}}_{\mathrm{A}}^{\mathrm{D}} \\ \overline{\boldsymbol{y}}_{\mathrm{D}}^{\mathrm{A}} & \overline{\boldsymbol{y}}_{\mathrm{D}}^{\mathrm{D}} \end{bmatrix} \begin{bmatrix} \overline{\mathbf{E}}_{\mathrm{A}} \\ \overline{\mathbf{E}}_{\mathrm{D}} \end{bmatrix} \tag{7.81}$$

with:

$$\overline{\mathbf{E}}_{\mathrm{A}} = \overline{\mathbf{E}}_{x\mathrm{A}} \qquad \overline{\mathbf{H}}_{\mathrm{A}} = -\mathrm{j}\overline{\mathbf{H}}_{y\mathrm{A}} \qquad \overline{\mathbf{E}}_{\mathrm{D}} = \overline{\mathbf{E}}_{z\mathrm{D}} \qquad \overline{\mathbf{H}}_{\mathrm{D}} = -\mathrm{j}\overline{\mathbf{H}}_{y\mathrm{D}} \tag{7.82}$$

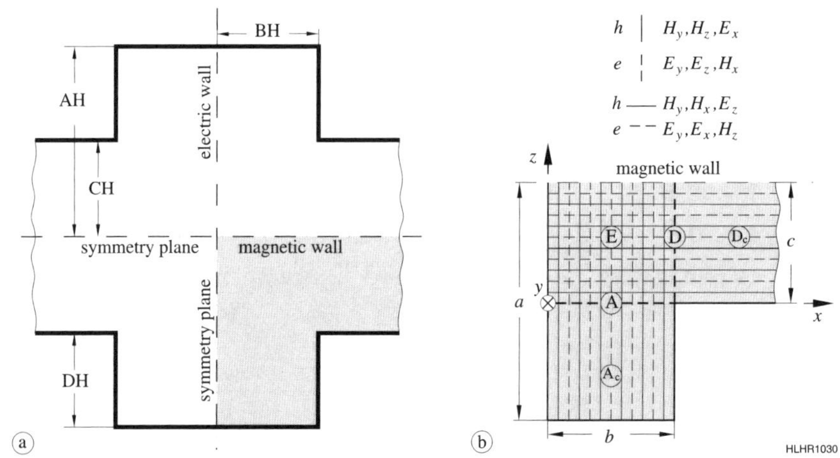

Fig. 7.11 Cross-section of a groove guide (a) and a quarter (b) of the cross-section

By short-circuiting port D we calculate the y submatrices on the left. $\overline{\boldsymbol{y}}_{\mathrm{A}}^{\mathrm{A}}$ is equal to the input admittance at port A of region E. The upper port is open-circuited. Therefore, we obtain from eqs. (7.22) and (7.23):

$$\overline{\mathbf{E}}_{\mathrm{A}} = \overline{\mathbf{z}}_{1h}^{\mathrm{V}}\overline{\mathbf{H}}_{\mathrm{A}} \qquad \overline{\boldsymbol{y}}_{\mathrm{A}}^{\mathrm{A}} = (\overline{\mathbf{z}}_{1h}^{\mathrm{V}})^{-1} = \tanh(\boldsymbol{\Gamma}_{h}^{\mathrm{V}}\overline{c}) \tag{7.83}$$

Matrix $\overline{\boldsymbol{y}}_{\mathrm{D}}^{\mathrm{A}}$ is obtained from eq. (7.39) and eq. (7.83). We obtain:

$$\overline{\mathbf{H}}_{y\mathrm{D}} = \overline{\boldsymbol{V}}_{\mathrm{DA}}^{\mathrm{N}}\overline{\mathbf{H}}_{y\mathrm{A}} \longrightarrow -\overline{\mathbf{H}}_{\mathrm{D}} = -\overline{\boldsymbol{V}}_{\mathrm{DA}}^{\mathrm{N}}\overline{\mathbf{H}}_{\mathrm{A}} = -\overline{\boldsymbol{V}}_{\mathrm{DA}}^{\mathrm{N}}\overline{\boldsymbol{y}}_{\mathrm{A}}^{\mathrm{A}}\overline{\mathbf{E}}_{\mathrm{A}} \tag{7.84}$$

Therefore, we have:

$$\overline{\boldsymbol{y}}_{\mathrm{D}}^{\mathrm{A}} = -\overline{\boldsymbol{V}}_{\mathrm{DA}}^{\mathrm{N}}\overline{\boldsymbol{y}}_{\mathrm{A}}^{\mathrm{A}} \tag{7.85}$$

with:

$$\overline{\boldsymbol{V}}_{\mathrm{DA}} = \boldsymbol{T}_{\mathrm{H}}^{\mathrm{t}}\boldsymbol{\Lambda}_{\mathrm{A}}\boldsymbol{T}_{\mathrm{VD}}^{\mathrm{d}} \qquad (\boldsymbol{\Lambda}_{\mathrm{A}})_{ik} = \frac{\sinh(\boldsymbol{\Gamma}_{hk}^{\mathrm{V}}(\overline{c} - \overline{z}_i))}{\sinh(\boldsymbol{\Gamma}_{hk}^{\mathrm{V}}\overline{c})} \tag{7.86}$$

and:

$$\boldsymbol{T}_{\mathrm{H}} = \boldsymbol{T}_{h}^{\mathrm{H}} \qquad \boldsymbol{T}_{\mathrm{V}} = \boldsymbol{T}_{h}^{\mathrm{V}} \tag{7.87}$$

By short-circuiting port A we calculate the y submatrices on the right. $\overline{\boldsymbol{y}}_{\mathrm{D}}^{\mathrm{D}}$ is equal to the input admittance at port D of region E. The left side port (C) is short-circuited. Therefore, we obtain from eqs. (7.62) and (7.64), considering $\overline{\mathbf{E}}_{\mathrm{C}} = \mathbf{0}$:

$$-\overline{\mathbf{H}}_{\mathrm{D}} = \overline{\boldsymbol{y}}_{1h}^{\mathrm{H}} \overline{\mathbf{E}}_{\mathrm{D}} \qquad \overline{\boldsymbol{y}}_{\mathrm{D}}^{\mathrm{D}} = \overline{\boldsymbol{y}}_{1h}^{\mathrm{H}} = 1/\tanh(\boldsymbol{\Gamma}_h^{\mathrm{H}} \overline{b}) \tag{7.88}$$

Matrix $\overline{\boldsymbol{y}}_{\mathrm{A}}^{\mathrm{D}}$ is obtained from eq. (7.70) and eq. (7.88). We have:

$$\overline{\mathbf{H}}_{y\mathrm{A}} = \overline{\boldsymbol{V}}_{\mathrm{AD}}^{\mathrm{N}} \overline{\mathbf{H}}_{y\mathrm{D}} \longrightarrow \overline{\mathbf{H}}_{\mathrm{A}} = \overline{\boldsymbol{V}}_{\mathrm{AD}}^{\mathrm{N}} \overline{\mathbf{H}}_{\mathrm{D}} = -\overline{\boldsymbol{V}}_{\mathrm{AD}}^{\mathrm{N}} \overline{\boldsymbol{y}}_{\mathrm{D}}^{\mathrm{D}} \overline{\mathbf{E}}_{\mathrm{D}} \tag{7.89}$$

Therefore, we obtain:

$$\overline{\boldsymbol{y}}_{\mathrm{A}}^{\mathrm{D}} = -\overline{\boldsymbol{V}}_{\mathrm{AD}}^{\mathrm{N}} \overline{\boldsymbol{y}}_{\mathrm{D}}^{\mathrm{D}} \tag{7.90}$$

with:

$$\overline{\boldsymbol{V}}_{\mathrm{AD}}^{\mathrm{N}} = \boldsymbol{T}_{\mathrm{V}}^{\mathrm{t}} \boldsymbol{\Lambda}_{\mathrm{D}} \boldsymbol{T}_{\mathrm{HA}}^{\mathrm{d}} \qquad (\boldsymbol{\Lambda}_{\mathrm{D}})_{ik} = \frac{\cosh(\boldsymbol{\Gamma}_h^{\mathrm{H}} \overline{x}_i)}{\cosh(\boldsymbol{\Gamma}_h^{\mathrm{H}} \overline{b})} \tag{7.91}$$

For the connecting waveguide (sections) at ports A and D we have the relation:

$$\overline{\mathbf{H}}_{y\mathrm{D_C}} = \overline{\boldsymbol{Y}}_{o\mathrm{H}} \overline{\mathbf{E}}_{x\mathrm{D_C}} \quad \overline{\mathbf{H}}_{\mathrm{A_C}} = -\overline{\boldsymbol{Y}}_{\mathrm{A_C}} \overline{\mathbf{E}}_{\mathrm{A_C}} \quad \overline{\boldsymbol{Y}}_{\mathrm{A_C}} = 1/\tanh(\boldsymbol{\Gamma}_h^{\mathrm{V}} (\overline{a} - \overline{c}) \tag{7.92}$$

This is because port A is loaded by a short-circuited waveguide. The waveguide at port D is infinitely long. The load matrix is therefore equal to the wave admittance matrix $\overline{\boldsymbol{Y}}_{o\mathrm{H}} = \boldsymbol{I}_h$. The introduction of all the obtained relations into eq. (7.81) yields the final result:

$$\begin{bmatrix} \overline{\boldsymbol{y}}_{\mathrm{A}}^{\mathrm{A}} + \overline{\boldsymbol{Y}}_{\mathrm{A_C}} & -\overline{\boldsymbol{V}}_{\mathrm{AD}}^{\mathrm{N}} \overline{\boldsymbol{y}}_{\mathrm{D}}^{\mathrm{D}} \\ -\overline{\boldsymbol{V}}_{\mathrm{DA}}^{\mathrm{N}} \overline{\boldsymbol{y}}_{\mathrm{A}}^{\mathrm{A}} & \overline{\boldsymbol{y}}_{\mathrm{D}}^{\mathrm{D}} + \overline{\boldsymbol{Y}}_{o\mathrm{H}} \end{bmatrix} \begin{bmatrix} \overline{\mathbf{E}}_{y\mathrm{A}} \\ \overline{\mathbf{E}}_{z\mathrm{D}} \end{bmatrix} = 0 \tag{7.93}$$

The determinant of the system matrix in this equation must be zero, which yields the cut-off frequencies.

7.3.2 Coplanar waveguide

In this subsection we show the analysis of the coplanar waveguide presented in Fig. 7.12. Due to the symmetry, we only need analyse the right half of the cross-section. For the fundamental mode we have to introduce a magnetic wall at the symmetry plane. We start with regions H_1 and H_2. (The fields are labelled according to the number of the section.) In these regions we have only horizontal discretisation lines. For the field description we use eqs. (7.62) and (7.63). At port C_1 in region H_1 the vector $\overline{\mathbf{H}}_{\mathrm{C}^1}$ is zero. In region H_2 we assume at port D_2 (which can also be moved to infinity) an electric wall with $\overline{\mathbf{E}}_{\mathrm{D}^2}$ being equal to zero. Thus, we have:

$$\begin{aligned} &\mathrm{H}_1 : \widehat{\overline{\mathbf{E}}}_{\mathrm{CD}^1} = \widehat{\overline{\mathbf{z}}}_{\mathrm{CD}^1}^{\mathrm{CD}} \widehat{\overline{\mathbf{H}}}_{\mathrm{CD}^1} \quad \overline{\mathbf{H}}_{\mathrm{C}^1} = \mathbf{0} \rightarrow \overline{\mathbf{E}}_{\mathrm{D}^1} = -\overline{\mathbf{z}}_{1^1} \overline{\mathbf{H}}_{\mathrm{D}^1} \\ &\mathrm{H}_2 : \widehat{\overline{\mathbf{H}}}_{\mathrm{CD}^2} = \widehat{\overline{\boldsymbol{y}}}_{\mathrm{CD}^2}^{\mathrm{CD}} \widehat{\overline{\mathbf{E}}}_{\mathrm{CD}^2} \quad \overline{\mathbf{E}}_{\mathrm{D}^2} = \mathbf{0} \rightarrow \overline{\mathbf{H}}_{\mathrm{C}^2} = \overline{\boldsymbol{y}}_{1^2} \overline{\mathbf{E}}_{\mathrm{C}^2} \end{aligned} \tag{7.94}$$

with:

$$\hat{\bar{\mathbf{y}}}_{1^2} = 1/\tanh(\hat{\boldsymbol{\Gamma}}_{\mathrm{H}_2}\bar{c}) \quad \hat{\bar{\mathbf{z}}}_{1^1} = 1/\tanh(\hat{\boldsymbol{\Gamma}}_{\mathrm{H}_1}\bar{w}/2) \quad \hat{\bar{\mathbf{z}}}_{1^1}^{-1} = \tanh(\hat{\boldsymbol{\Gamma}}_{\mathrm{H}_1}\bar{w}/2) \tag{7.95}$$

If there is a step from region H_1 to region R_I we have to transform $\hat{\bar{\mathbf{z}}}_{1^1}^{-1}$ to the inner side of port C of region R_I according to:

$$\hat{\bar{\boldsymbol{Z}}}_{\mathrm{C}^\mathrm{I}} = \hat{\boldsymbol{T}}_{\mathrm{E}^\mathrm{I}_\mathrm{c}}^{-1}\hat{\boldsymbol{T}}_{\mathrm{E}^1}\hat{\bar{\mathbf{z}}}_{1^1}\hat{\boldsymbol{T}}_{\mathrm{H}^1}^{\mathrm{t}}\hat{\boldsymbol{T}}_{\mathrm{H}^\mathrm{I}\mathrm{c}} \tag{7.96}$$

The subscript c at $\boldsymbol{T}$ means that the matrices have to be reduced to the common part. In general, the transformation from region H_2 to region R_I is done analogously to eq. (7.96). In Fig. 7.12 we do not have a step and the permittivities are equal. Therefore, such a transformation is not necessary. We combine the impedances at ports C_I and D_I according to:

$$\hat{\bar{\mathbf{E}}}_{\mathrm{CD}^\mathrm{I}} = \hat{\bar{\boldsymbol{Z}}}_{\mathrm{CD}^\mathrm{I}}\hat{\bar{\mathbf{H}}}_{\mathrm{CD}^\mathrm{I}} \qquad \hat{\bar{\boldsymbol{Z}}}_{\mathrm{CD}^\mathrm{I}} = \mathrm{Diag}(-\hat{\bar{\boldsymbol{Z}}}_{\mathrm{C}^\mathrm{I}}, -\hat{\bar{\mathbf{y}}}_{1^2}^{-1}) \tag{7.97}$$

which represents the field matching at ports C_I and D_I of region I. For region R_I we obtain from eq. (7.4a) with $\hat{\bar{\mathbf{E}}}_{\mathrm{B}^\mathrm{I}} = \mathbf{0}$:

$$\begin{bmatrix} \hat{\bar{\mathbf{H}}}_{\mathrm{A}^\mathrm{I}} \\ \hat{\bar{\mathbf{H}}}_{\mathrm{CD}^\mathrm{I}} \end{bmatrix} = \begin{bmatrix} \bar{\mathbf{y}}_{\mathrm{A}^\mathrm{I}}^{\mathrm{A}} & \hat{\bar{\mathbf{y}}}_{\mathrm{A}^\mathrm{I}}^{\mathrm{CD}} \\ \hat{\bar{\mathbf{y}}}_{\mathrm{CD}^\mathrm{I}}^{\mathrm{A}} & \hat{\bar{\mathbf{y}}}_{\mathrm{CD}^\mathrm{I}}^{\mathrm{CD}} \end{bmatrix} \begin{bmatrix} \hat{\bar{\mathbf{E}}}_{\mathrm{A}^\mathrm{I}} \\ \hat{\bar{\mathbf{E}}}_{\mathrm{CD}^\mathrm{I}} \end{bmatrix} \tag{7.98}$$

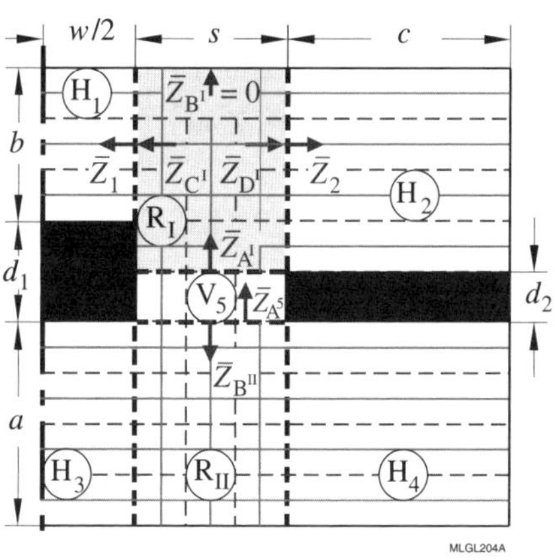

Fig. 7.12 Cross-section of a coplanar waveguide (CPW). Only half of the structure is shown because of the symmetry

We have withdrawn lines and rows belonging to port B. By introducing eq. (7.97) we obtain from the second expression in eq. (7.98):

$$\hat{\bar{\mathbf{H}}}_{\mathrm{CD}^\mathrm{I}} = -(\hat{\bar{\mathbf{y}}}_{\mathrm{CD}^\mathrm{I}}^{\mathrm{CD}}\hat{\bar{\boldsymbol{Z}}}_{\mathrm{CD}^\mathrm{I}} - \hat{\boldsymbol{I}})^{-1}\hat{\bar{\mathbf{y}}}_{\mathrm{CD}^\mathrm{I}}^{\mathrm{A}}\mathbf{E}_{\mathrm{A}^\mathrm{I}} \tag{7.99}$$

which may be introduced into the first expression, yielding $\widehat{\widehat{\mathbf{H}}}_{\mathrm{A^I}} = \widehat{\widehat{\mathbf{Y}}}_{\mathrm{A^I}}\widehat{\mathbf{E}}_{\mathrm{A^I}}$ with:

$$\widehat{\widehat{\mathbf{Y}}}_{\mathrm{A^I}} = \widehat{\widehat{\boldsymbol{y}}}_{1^{\mathrm{I}}}^{\mathrm{V}} - \widehat{\boldsymbol{y}}_{\mathrm{A^I}}^{\mathrm{CD}}\widehat{\widehat{\mathbf{Z}}}_{\mathrm{CD^I}}(\widehat{\widehat{\boldsymbol{y}}}_{\mathrm{CD^I}}^{\mathrm{CD}}\widehat{\widehat{\mathbf{Z}}}_{\mathrm{CD^I}} - \widehat{\boldsymbol{I}})^{-1}\widehat{\widehat{\boldsymbol{y}}}_{\mathrm{CD^I}}^{\mathrm{A}} \tag{7.100}$$

or:

$$\widehat{\widehat{\mathbf{Y}}}_{\mathrm{A^I}} = \widehat{\widehat{\boldsymbol{y}}}_{1^{\mathrm{I}}}^{\mathrm{V}} - \widehat{\boldsymbol{y}}_{\mathrm{A^I}}^{\mathrm{CD}}(\widehat{\boldsymbol{y}}_{\mathrm{CD^I}}^{\mathrm{CD}} - \widehat{\widehat{\mathbf{Z}}}_{\mathrm{CD^I}}^{-1})^{-1}\widehat{\widehat{\boldsymbol{y}}}_{\mathrm{CD^I}}^{\mathrm{A}} \tag{7.101}$$

If there is no step between regions $\mathrm{H_1}$ and $\mathrm{R_I}$ we obtain (instead of eq. (7.97)) as result of the matching:

$$\widehat{\widehat{\mathbf{H}}}_{\mathrm{CD^I}} = \widehat{\widehat{\mathbf{Y}}}_{\mathrm{CD^I}}\widehat{\widehat{\mathbf{E}}}_{\mathrm{CD^I}} \qquad \widehat{\widehat{\mathbf{Y}}}_{\mathrm{CD^I}} = \mathrm{Diag}(-\widehat{\widehat{\mathbf{z}}}_{1^1}^{-1}, -\widehat{\widehat{\boldsymbol{y}}}_{1^2}) \tag{7.102}$$

By introducing this equation into eq. (7.98) we obtain for the admittance:

$$\widehat{\widehat{\mathbf{Y}}}_{\mathrm{A^I}} = \widehat{\widehat{\boldsymbol{y}}}_{\mathrm{A^I}}^{\mathrm{A}} - \widehat{\boldsymbol{y}}_{\mathrm{A^I}}^{\mathrm{CD}}(\widehat{\boldsymbol{y}}_{\mathrm{CD^I}}^{\mathrm{CD}} - \widehat{\widehat{\mathbf{Y}}}_{\mathrm{CD^I}})^{-1}\widehat{\widehat{\boldsymbol{y}}}_{\mathrm{CD^I}}^{\mathrm{A}} \tag{7.103}$$

In the lower region we have:

$$\begin{aligned} &\mathrm{H_3}: \widehat{\widehat{\mathbf{E}}}_{\mathrm{CD^3}} = \widehat{\mathbf{z}}_{\mathrm{CD^3}}^{\mathrm{CD}}\widehat{\widehat{\mathbf{H}}}_{\mathrm{CD^3}} \quad \overline{\mathbf{H}}_{\mathrm{C^3}} = \mathbf{0} \rightarrow \overline{\mathbf{E}}_{\mathrm{D^3}} = -\widehat{\widehat{\mathbf{z}}}_{1^3}\overline{\mathbf{H}}_{\mathrm{D^3}} \\ &\mathrm{H_4}: \widehat{\widehat{\mathbf{H}}}_{\mathrm{CD^4}} = \widehat{\widehat{\boldsymbol{y}}}_{\mathrm{CD^4}}^{\mathrm{CD}}\widehat{\widehat{\mathbf{E}}}_{\mathrm{CD^4}} \quad \overline{\mathbf{E}}_{\mathrm{D^4}} = \mathbf{0} \rightarrow \overline{\mathbf{H}}_{\mathrm{C^4}} = \widehat{\widehat{\boldsymbol{y}}}_{1^4}\overline{\mathbf{E}}_{\mathrm{C^4}} \end{aligned} \tag{7.104}$$

with:

$$\widehat{\widehat{\boldsymbol{y}}}_{1^4} = 1/\tanh(\widehat{\boldsymbol{\Gamma}}_{\mathrm{H_4}}\overline{c}) \quad \widehat{\widehat{\mathbf{z}}}_{1^3} = 1/\tanh(\widehat{\boldsymbol{\Gamma}}_{\mathrm{H_3}}\overline{w}/2) \quad \widehat{\widehat{\mathbf{z}}}_{1^3}^{-1} = \tanh(\widehat{\boldsymbol{\Gamma}}_{\mathrm{H_3}}\overline{w}/2) \tag{7.105}$$

Ports $\mathrm{D_3}$ and $\mathrm{C_{II}}$ and $\mathrm{C_4}$ and $\mathrm{D_{II}}$ are identical. Therefore, we can write as matching equations:

$$\widehat{\widehat{\mathbf{H}}}_{\mathrm{CD^{II}}} = \widehat{\widehat{\mathbf{Y}}}_{\mathrm{CD^{II}}}\widehat{\widehat{\mathbf{E}}}_{\mathrm{CD^{II}}} \qquad \widehat{\widehat{\mathbf{Y}}}_{\mathrm{CD^{II}}} = \mathrm{Diag}(-\widehat{\widehat{\mathbf{z}}}_{1^3}^{-1}, -\widehat{\widehat{\boldsymbol{y}}}_{1^4}) \tag{7.106}$$

For region $\mathrm{R_{II}}$ we have:

$$\begin{bmatrix} -\widehat{\widehat{\mathbf{H}}}_{\mathrm{B^{II}}} \\ \widehat{\widehat{\mathbf{H}}}_{\mathrm{CD^{II}}} \end{bmatrix} = \begin{bmatrix} \widehat{\widehat{\boldsymbol{y}}}_{\mathrm{B^{II}}}^{\mathrm{B}} & \widehat{\boldsymbol{y}}_{\mathrm{B^{II}}}^{\mathrm{CD}} \\ \widehat{\widehat{\boldsymbol{y}}}_{\mathrm{CD^{II}}}^{\mathrm{B}} & \widehat{\boldsymbol{y}}_{\mathrm{CD^{II}}}^{\mathrm{CD}} \end{bmatrix} \begin{bmatrix} \widehat{\widehat{\mathbf{E}}}_{\mathrm{B^{II}}} \\ \widehat{\widehat{\mathbf{E}}}_{\mathrm{CD^{II}}} \end{bmatrix} \tag{7.107}$$

Therefore, we obtain with eq. (7.106):

$$-\widehat{\widehat{\mathbf{H}}}_{\mathrm{B^{II}}} = \widehat{\widehat{\mathbf{Y}}}_{\mathrm{B^{II}}}\widehat{\widehat{\mathbf{E}}}_{\mathrm{B^{II}}} \qquad \widehat{\widehat{\mathbf{Y}}}_{\mathrm{B^{II}}} = \widehat{\widehat{\boldsymbol{y}}}_{\mathrm{B^{II}}}^{\mathrm{B}} - \widehat{\boldsymbol{y}}_{\mathrm{B^{II}}}^{\mathrm{CD}}(\widehat{\boldsymbol{y}}_{\mathrm{CD^{II}}}^{\mathrm{CD}} - \widehat{\widehat{\mathbf{Y}}}_{\mathrm{CD^{II}}})^{-1}\widehat{\widehat{\boldsymbol{y}}}_{\mathrm{CD^{II}}}^{\mathrm{B}} \tag{7.108}$$

The admittances $\widehat{\widehat{\mathbf{Y}}}_{\mathrm{A^I}}$ and $\widehat{\widehat{\mathbf{Y}}}_{\mathrm{B^{II}}}$ are connected by region $\boldsymbol{V}_5$ with discretisation lines in the vertical direction. They are the load admittances of ports $\mathrm{B_5}$ and $\mathrm{A_5}$, respectively. By using eq. (7.21) with $\widehat{\widehat{\mathbf{H}}}_{\mathrm{B^5}} = \widehat{\widehat{\mathbf{Y}}}_{\mathrm{A^I}}\widehat{\widehat{\mathbf{E}}}_{\mathrm{B^5}}$ we obtain:

$$\widehat{\widehat{\mathbf{H}}}_{\mathrm{A^5}} = \widehat{\widehat{\mathbf{Y}}}_{\mathrm{A^5}}\widehat{\widehat{\mathbf{E}}}_{\mathrm{A^5}} \quad \text{with} \quad \widehat{\widehat{\mathbf{Y}}}_{\mathrm{A^5}} = \widehat{\widehat{\boldsymbol{y}}}_{1^5} - \widehat{\widehat{\boldsymbol{y}}}_{2^5}(\widehat{\widehat{\boldsymbol{y}}}_{1^5} + \widehat{\widehat{\mathbf{Y}}}_{\mathrm{A^I}})^{-1}\widehat{\widehat{\boldsymbol{y}}}_{2^5} \tag{7.109}$$

Since the ports A_5 and B_{II} are identical, the fields $\widehat{\mathbf{E}}_{B^{II}}, \widehat{\mathbf{H}}_{B^{II}}$ and $\widehat{\mathbf{E}}_{A^5}, \widehat{\mathbf{H}}_{A^5}$, respectively, must also be equal. Therefore, we obtain the following eigenvalue equation:

$$(\widehat{\mathbf{Y}}_{B^{II}} + \widehat{\mathbf{Y}}_{A^5})\widehat{\mathbf{E}}_{A^5} = \widehat{\mathbf{0}} \tag{7.110}$$

From the condition that the determinant of the system matrix must vanish, we obtain ε_{re}.

7.4 CROSSED DISCRETISATION LINES AND CYLINDRICAL COORDINATES

7.4.1 Principle of analysis

In this section we will extend the principle of crossed discretisation lines that we described before in Cartesian coordinates to problems which can be described with cylindrical coordinates. Fig. 7.13 shows an arbitrary region R with crossed lines (see e.g. regions I and III in Fig. 7.7) and the definitions of the ports and of the fields. The description of the fields can again be given with eqs. (7.2)–(7.4). The admittance matrices can be calculated in the following way: short-circuiting the ports C and D by using electric walls results in $\widehat{\mathbf{E}}_{CD} = \mathbf{0}$. To determine the matrices $\widehat{\boldsymbol{y}}_{AB}^{AB}$ and $\widehat{\boldsymbol{y}}_{CD}^{AB}$ we use radial discretisation lines. The electric field $\widehat{\mathbf{E}}_{AB}$ at ports A and B causes magnetic fields at all ports (i.e. A to D). From these magnetic fields we obtain the impedances. Similarly, short-circuiting ports A and B by using electric walls results in $\widehat{\mathbf{E}}_{AB} = \mathbf{0}$. Here, we have an electric field at Ports C and D that causes magnetic fields at all ports. From these we determine the submatrices $\widehat{\boldsymbol{y}}_{CD}^{CD}$ and $\widehat{\boldsymbol{y}}_{AB}^{CD}$. Now we use azimuthal discretisation lines. The corresponding equations are as follows:

$$\begin{bmatrix} \widehat{\mathbf{H}}_{AB}^{R} \\ \widehat{\mathbf{H}}_{CD}^{R} \end{bmatrix} = \begin{bmatrix} \widehat{\boldsymbol{y}}_{AB}^{AB} \\ \widehat{\boldsymbol{y}}_{CD}^{AB} \end{bmatrix} \widehat{\mathbf{E}}_{AB} \qquad \begin{bmatrix} \widehat{\mathbf{H}}_{AB}^{A} \\ \widehat{\mathbf{H}}_{CD}^{A} \end{bmatrix} = \begin{bmatrix} \widehat{\boldsymbol{y}}_{AB}^{CD} \\ \widehat{\boldsymbol{y}}_{CD}^{CD} \end{bmatrix} \widehat{\mathbf{E}}_{CD} \tag{7.111}$$

The superscripts R and A at the magnetic field vectors indicate the radial and azimuthal direction of discretisation lines. For the regions with unidirectional discretisation lines (e.g. II, IV, V and VI in Fig. 7.7a) only the main diagonal matrices in eq. (7.4) have to be determined.

7.4.2 General formulas for eigenmode calculation

The modes in the ridge waveguide with homogeneous dielectric with the material parameters ε_r and $\mu_r = 1$ can be classified into TE_z and TM_z modes. They can be determined from H_z and E_z, respectively (see Section 4.4.3). $\varepsilon_{de,dh}$ should be replaced by $\varepsilon_d = \varepsilon_r - \varepsilon_{re}$. First we will determine the main diagonal matrices $\widehat{\boldsymbol{y}}_{AB}^{AB}$ and $\widehat{\boldsymbol{y}}_{CD}^{CD}$ in (7.111). To obtain the first one, we use discretisation lines in the radial direction; for the second we take discretisation lines in the azimuthal direction.

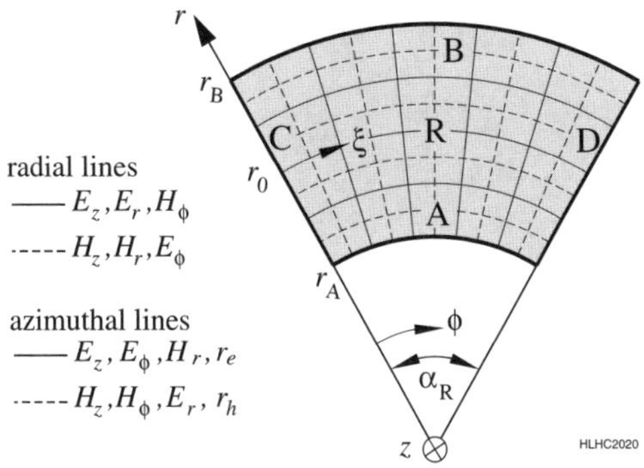

Fig. 7.13 Region in cylindrical coordinates with crossed discretisation lines (Reproduced by permission of IGTE (Institut für Grundlagen und Theorie der Elektrotechnik) Graz)

7.4.3 Discretisation lines in radial direction

In this subsection we determine the field dependence in the radial direction. We discretise in ϕ-direction and obtain the matrix $\widehat{\mathbf{y}}_{\mathrm{AB}}^{\mathrm{AB}}$ in eq. (7.111).

The results are also valid for regions V and VI in Fig. 7.7. In principle, the analysis is the same as in Cartesian coordinates. Therefore, all that we know about discretisation in Cartesian coordinates can be used for the ϕ-direction as well. If the region R is a whole circular layer, periodic boundary conditions have to be used, which are described in the appendices. For the structure in Fig. 7.7 (or in regions V and VI), Dirichlet and Neumann boundary conditions are relevant. All required formulas are found in Section 4.4.3. (For cross-products of the Bessel functions see [5].) Here we will introduce specialisations. We distinguish now between the TM and the TE case. We obtain ($\varepsilon_{dt} = \varepsilon_d, \varepsilon_{rt} = \varepsilon_r, \mu_{rt} = 1$) for:

TM$_z$ modes:

$$\begin{bmatrix} \bar{r}_{\mathrm{A}}\overline{\mathbf{H}}_{\phi}^{\mathrm{A}} \\ -\bar{r}_{\mathrm{B}}\overline{\mathbf{H}}_{\phi}^{\mathrm{B}} \end{bmatrix} = \widehat{\overline{\mathbf{y}}}_{\mathrm{ABe}}^{\mathrm{AB}} \begin{bmatrix} -\mathrm{j}\overline{\mathbf{E}}_{z}^{\mathrm{A}} \\ -\mathrm{j}\overline{\mathbf{E}}_{z}^{\mathrm{B}} \end{bmatrix} \qquad \widehat{\overline{\mathbf{y}}}_{\mathrm{ABe}}^{\mathrm{AB}} = \begin{bmatrix} \overline{\mathbf{y}}_{1\mathrm{A}}^{r} & \overline{\mathbf{y}}_{2}^{r} \\ \overline{\mathbf{y}}_{2}^{r} & \overline{\mathbf{y}}_{1\mathrm{B}}^{r} \end{bmatrix}_{|e} = \varepsilon_r \varepsilon_d^{-1} \widehat{\overline{\mathbf{\Lambda}}}_e \tag{7.112}$$

TE$_z$ modes:

$$\begin{bmatrix} \bar{r}_{\mathrm{A}}\overline{\mathbf{E}}_{\phi}^{\mathrm{A}} \\ \bar{r}_{\mathrm{B}}\overline{\mathbf{E}}_{\phi}^{\mathrm{B}} \end{bmatrix} = \widehat{\overline{\mathbf{z}}}_{\mathrm{ABh}}^{\mathrm{AB}} \begin{bmatrix} \mathrm{j}\overline{\mathbf{H}}_{z}^{\mathrm{A}} \\ -\mathrm{j}\overline{\mathbf{H}}_{z}^{\mathrm{B}} \end{bmatrix} \qquad \widehat{\overline{\mathbf{z}}}_{\mathrm{ABh}}^{\mathrm{AB}} = \begin{bmatrix} \overline{\mathbf{z}}_{1\mathrm{A}}^{r} & \overline{\mathbf{z}}_{2}^{r} \\ \overline{\mathbf{z}}_{2}^{r} & \overline{\mathbf{z}}_{1\mathrm{B}}^{r} \end{bmatrix}_{|h} = \varepsilon_d^{-1} \widehat{\overline{\mathbf{\Lambda}}}_h \tag{7.113}$$

where $\widehat{\boldsymbol{\Lambda}}_{e,h}$ and $\widehat{\boldsymbol{\Lambda}}_{e,h}^{-1}$ are given by:

$$\widehat{\boldsymbol{\Lambda}}_e = \widehat{p}_{\boldsymbol{\nu}_e}^{-1} \begin{bmatrix} \overline{r}_{\boldsymbol{\nu}_e} & \frac{2}{\pi}\boldsymbol{I}_e \\ \frac{2}{\pi}\boldsymbol{I}_e & -\overline{q}_{\boldsymbol{\nu}_e} \end{bmatrix} \qquad \widehat{\boldsymbol{\Lambda}}_e^{-1} = \widehat{s}_{\boldsymbol{\nu}_e}^{-1} \begin{bmatrix} \overline{q}_{\boldsymbol{\nu}_e} & \frac{2}{\pi}\boldsymbol{I}_e \\ \frac{2}{\pi}\boldsymbol{I}_e & -\overline{r}_{\boldsymbol{\nu}_e} \end{bmatrix} \tag{7.114}$$

$$\widehat{\boldsymbol{\Lambda}}_h = \widehat{p}_{\boldsymbol{\nu}_h}^{-1} \begin{bmatrix} \overline{r}_{\boldsymbol{\nu}_h} & -\frac{2}{\pi}\boldsymbol{I}_h \\ -\frac{2}{\pi}\boldsymbol{I}_h & -\overline{q}_{\boldsymbol{\nu}_h} \end{bmatrix} \qquad \widehat{\boldsymbol{\Lambda}}_h^{-1} = \widehat{s}_{\boldsymbol{\nu}_h}^{-1} \begin{bmatrix} \overline{q}_{\boldsymbol{\nu}_h} & -\frac{2}{\pi}\boldsymbol{I}_h \\ -\frac{2}{\pi}\boldsymbol{I}_h & -\overline{r}_{\boldsymbol{\nu}_h} \end{bmatrix} \tag{7.115}$$

For the cross-products, see the referenced section. Eqs. (7.112) and (7.113) can easily be inverted. Hence, we obtain e.g.:

$$\widehat{\boldsymbol{y}}_{\mathrm{ABh}}^{\mathrm{AB}} = \varepsilon_d \widehat{\boldsymbol{\Lambda}}_h^{-1} \tag{7.116}$$

The impedance/admittance transformation is performed in the usual way.

7.4.4 Discretisation lines in azimuthal direction

In this subsection we determine the field dependence in the azimuthal direction to obtain the matrix $\widehat{\boldsymbol{y}}_{\mathrm{CD}}^{\mathrm{CD}}$ in eq. (7.111). The results are also valid in regions II and IV in Fig. 7.7. We will adopt the previously obtained results (see Section 3.2.1). For the TM$_z$ modes we have to introduce $k_z = \mathrm{j}\sqrt{\varepsilon_{re}}$ and for the TE$_z$ modes, $k_z = -\mathrm{j}\sqrt{\varepsilon_{re}}$. The following relation holds between the main tangential field components at ports C and D of region R of the waveguide (see Fig. 7.13):

TM$_z$ modes:

$$\begin{bmatrix} \overline{\mathbf{E}}_{zn}^{\mathrm{C}} \\ \overline{\mathbf{E}}_{zn}^{\mathrm{D}} \end{bmatrix} = \begin{bmatrix} \overline{\mathbf{z}}_1^{\phi} & \overline{\mathbf{z}}_2^{\phi} \\ \overline{\mathbf{z}}_2^{\phi} & \overline{\mathbf{z}}_1^{\phi} \end{bmatrix}_{|e} \begin{bmatrix} \overline{\mathbf{H}}_{rn}^{\mathrm{C}} \\ -\overline{\mathbf{H}}_{rn}^{\mathrm{D}} \end{bmatrix} \tag{7.117}$$

TE$_z$ modes:

$$\begin{bmatrix} \overline{\mathbf{E}}_{rn}^{\mathrm{C}} \\ \overline{\mathbf{E}}_{rn}^{\mathrm{D}} \end{bmatrix} = \begin{bmatrix} \overline{\mathbf{z}}_1^{\phi} & \overline{\mathbf{z}}_2^{\phi} \\ \overline{\mathbf{z}}_2^{\phi} & \overline{\mathbf{z}}_1^{\phi} \end{bmatrix}_{|h} \begin{bmatrix} (-\overline{\mathbf{H}}_{zn}^{\mathrm{C}}) \\ -(-\overline{\mathbf{H}}_{zn}^{\mathrm{D}}) \end{bmatrix} \tag{7.118}$$

with:

$$\overline{\mathbf{z}}_1 = \overline{\boldsymbol{I}}_f / \tanh(\boldsymbol{\Gamma}_{\xi}^{f}\overline{a}_{\mathrm{R}}) \qquad \overline{\mathbf{z}}_2 = \overline{\boldsymbol{I}}_f / \sinh(\boldsymbol{\Gamma}_{\xi}^{f}\overline{a}_{\mathrm{R}}) \qquad \overline{a}_{\mathrm{R}} = \alpha_{\mathrm{R}}\overline{r}_0 \tag{7.119}$$

The superscripts C and D label the ports (see Fig. 7.13). The characteristic impedances of the modes are given in this case by identity matrices. Note that the fields in the TM$_z$ (TE$_z$) case are calculated from E_z (H_z). Therefore, the parameters are marked with the subscript $e(h)$. Sometimes these subscripts are omitted for brevity. The inversion of the above relations yields:

TM$_z$ modes:

$$\begin{bmatrix} \overline{\mathbf{H}}_{rn}^{\mathrm{C}} \\ -\overline{\mathbf{H}}_{rn}^{\mathrm{D}} \end{bmatrix} = \begin{bmatrix} \overline{\boldsymbol{y}}_1^{\phi} & \overline{\boldsymbol{y}}_2^{\phi} \\ \overline{\boldsymbol{y}}_2^{\phi} & \overline{\boldsymbol{y}}_1^{\phi} \end{bmatrix}_{|e} \begin{bmatrix} \overline{\mathbf{E}}_{zn}^{\mathrm{C}} \\ \overline{\mathbf{E}}_{zn}^{\mathrm{D}} \end{bmatrix} = \widehat{\boldsymbol{y}}_{\mathrm{CDe}}^{\mathrm{CD}} \begin{bmatrix} \overline{\mathbf{E}}_{zn}^{\mathrm{C}} \\ \overline{\mathbf{E}}_{zn}^{\mathrm{D}} \end{bmatrix} \tag{7.120}$$

TE_z modes:

$$\begin{bmatrix} (-\overline{\mathbf{H}}_{zn}^{\mathrm{C}}) \\ -(-\overline{\mathbf{H}}_{zn}^{\mathrm{D}}) \end{bmatrix} = \begin{bmatrix} \overline{\boldsymbol{y}}_1^{\phi} & \overline{\boldsymbol{y}}_2^{\phi} \\ \overline{\boldsymbol{y}}_2^{\phi} & \overline{\boldsymbol{y}}_1^{\phi} \end{bmatrix}_{|h} \begin{bmatrix} \overline{\mathbf{E}}_{rn}^{\mathrm{C}} \\ \overline{\mathbf{E}}_{rn}^{\mathrm{D}} \end{bmatrix} = \widehat{\overline{\boldsymbol{y}}}_{\mathrm{CD}h}^{\mathrm{CD}} \begin{bmatrix} \overline{\mathbf{E}}_{rn}^{\mathrm{C}} \\ \overline{\mathbf{E}}_{rn}^{\mathrm{D}} \end{bmatrix} \tag{7.121}$$

where:

$$\overline{\boldsymbol{y}}_1 = \overline{\boldsymbol{I}}_f / \tanh(\boldsymbol{\Gamma}_{\xi}^f \overline{a}_{\mathrm{R}}) \overline{\boldsymbol{y}}_2 = -\overline{\boldsymbol{I}}_f / \sinh(\boldsymbol{\Gamma}_{\xi}^f \overline{a}_{\mathrm{R}}) \tag{7.122}$$

7.4.5 Coupling to neighbouring ports

Now we would like to determine the off-diagonal matrices $\overline{\boldsymbol{y}}_{\mathrm{AB}}^{\mathrm{CD}}$ and $\overline{\boldsymbol{y}}_{\mathrm{CD}}^{\mathrm{AB}}$ in eq. (7.111).

7.4.5.1 Field coupling from ports C/D to ports A/B

In order to determine $\overline{\boldsymbol{y}}_{\mathrm{AB}}^{\mathrm{CD}}$, we have to compute the tangential fields at ports A and B from those at ports C and D. The fields described by the above equations do not have a tangential electric field at ports A and B. In the TM_z case we have to calculate the tangential magnetic field component $\overline{\mathbf{H}}_{\phi}$ and in the TE_z case we must determine the tangential magnetic field component $\overline{\mathbf{H}}_z$. In both cases we must start with the z components. For the ξ dependence of field component $\overline{\mathbf{F}}_z$ we may write:

$$\overline{\mathbf{F}}_{zn}(\xi) = \frac{\sinh(\boldsymbol{\Gamma}_{\xi}^f (\overline{a}_{\mathrm{R}} - \overline{\xi}))}{\sinh(\boldsymbol{\Gamma}_{\xi}^f \overline{a}_{\mathrm{R}})} \overline{\mathbf{F}}_{zn}^{\mathrm{C}} + \frac{\sinh(\boldsymbol{\Gamma}_{\xi}^f \overline{\xi})}{\sinh(\boldsymbol{\Gamma}_{\xi}^f \overline{a}_{\mathrm{R}})} \overline{\mathbf{F}}_{zn}^{\mathrm{D}} \tag{7.123}$$

where $\overline{\mathbf{F}}_{zn}$ is identical to $\overline{\mathbf{E}}_{zn}$ in the TM_z and to $\overline{\mathbf{H}}_{zn}$ in the TE_z case, f has to be replaced by e and h, respectively. In the first case (TM_z) Dirichlet and in the second case (TE_z) Neumann boundary conditions are imposed at ports A and B. The division by the diagonal matrix $\sinh(\boldsymbol{\Gamma}_{\xi f} \overline{d}_{\mathrm{R}})$ must be understood as multiplication with its inverse. This equation must now be evaluated at the particular discretisation points ξ where we have discretisation lines in radial direction. At such a position ξ_i we may write:

$$\overline{\mathbf{F}}_{zn}(\xi_i) = \boldsymbol{\Lambda}_{\mathrm{C}i}^{\mathrm{d}} \overline{\mathbf{F}}_{zn}^{\mathrm{C}} + \boldsymbol{\Lambda}_{\mathrm{D}i}^{\mathrm{d}} \overline{\mathbf{F}}_{zn}^{\mathrm{C}} \tag{7.124}$$

with diagonal matrices $\boldsymbol{\Lambda}_i^{\mathrm{d}}$ given by the expressions:

$$\boldsymbol{\Lambda}_{\mathrm{C}i}^{\mathrm{d}} = \sinh(\boldsymbol{\Gamma}_{\xi}^f (\overline{a}_{\mathrm{R}} - \overline{\xi}_i)) / \sinh(\boldsymbol{\Gamma}_{\xi}^f \overline{a}_{\mathrm{R}}) \tag{7.125}$$

$$\boldsymbol{\Lambda}_{\mathrm{D}i}^{\mathrm{d}} = \sinh(\boldsymbol{\Gamma}_{\xi}^f \overline{\xi}_i) / \sinh(\boldsymbol{\Gamma}_{\xi}^f \overline{a}_{\mathrm{R}}) \tag{7.126}$$

We first proceed with the TE_z case. The vector $\mathbf{H}_{zn}(\xi_i) = \boldsymbol{T}_{\mathrm{H}}^{\xi h} \overline{\mathbf{H}}_{zn}(\xi_i)$ gives the values H_{zn} in the crossing points of the azimuthal lines with the radial line at $\xi = \xi_i$. $\boldsymbol{T}_{\mathrm{H}}^{\xi h}$ is the matrix for the Neumann boundary condition. We need the values of the field $\mathbf{H}_{zn}$ at the boundaries or ports A and B with

$r = r_{\mathrm{A}}$ and $r = r_{\mathrm{B}}$, respectively. Because of the Neumann conditions we must extrapolate from points near the port boundary to the boundary. This can be done easily with the expressions given in Appendix A.6 and by using the transformation matrix $\boldsymbol{T}_{\mathrm{H}}^{\xi h}$. We obtain at position ξ_i at the port A (R $\equiv$ A, $r = r_{\mathrm{A}}$) or port B (R $\equiv$ B, $r = r_{\mathrm{B}}$) boundary:

$$\mathbf{H}_{zn}^{\mathrm{R}}(\xi_i) = \mathbf{T}_{\mathrm{HR}}^{\xi h}\overline{\mathbf{H}}_{zn}(\xi_i) = \mathbf{T}_{\mathrm{HR}}^{\xi h}(\boldsymbol{\Lambda}_{\mathrm{C}i}^{\mathrm{d}}\overline{\mathbf{H}}_{zn}^{\mathrm{C}} + \boldsymbol{\Lambda}_{\mathrm{D}i}^{\mathrm{d}}\overline{\mathbf{H}}_{zn}^{\mathrm{D}}) \tag{7.127}$$

R $\equiv$ A (B) denotes port A (port B) of region R. $\mathbf{T}_{\mathrm{HR}}^{\xi h}$ is a row vector constructed from the weighted first (last) rows of the matrix $\boldsymbol{T}_{\mathrm{H}}^{\xi h}$. These vectors are constructed as:

$$\mathbf{T}_{\mathrm{HA}}^{\xi h} = \alpha_{\mathrm{A}}\mathbf{T}_{\mathrm{H1}}^{\xi h} - \beta_{\mathrm{A}}\mathbf{T}_{\mathrm{H2}}^{\xi h} \qquad \mathbf{T}_{\mathrm{HB}}^{\xi h} = \alpha_{\mathrm{B}}\mathbf{T}_{\mathrm{HN}}^{\xi h} - \beta_{\mathrm{B}}\mathbf{T}_{\mathrm{HN-1}}^{\xi h} \tag{7.128}$$

$\mathbf{T}_{\mathrm{H1}}^{\xi h}, \mathbf{T}_{\mathrm{H2}}^{\xi h}$ the first two and $\mathbf{T}_{\mathrm{HN}}^{\xi h}, \mathbf{T}_{\mathrm{HN-1}}^{\xi h}$ the last two row vectors of the transformation matrix $\boldsymbol{T}_{\mathrm{H}}^{\xi h}$. For $p_{\mathrm{A,B}} = 1/2$ (see Appendix A.6) we have $\alpha_{\mathrm{A,B}} = 9/8$ and $\beta_{\mathrm{A,B}} = 1/8$. If we want to give expressions for all values of ξ_i at the same time, we need to change the order of $\mathbf{T}_{\mathrm{HR}}^{\xi h}$ and $\boldsymbol{\Lambda}_i^{\mathrm{d}}$. For this purpose, we transform the vector $\mathbf{T}_{\mathrm{HR}}^{\xi h}$ to a diagonal matrix $\boldsymbol{T}_{\mathrm{HR}}^{\xi h\mathrm{d}}$ and $\boldsymbol{\Lambda}_i^{\mathrm{d}}$ becomes a row vector $\boldsymbol{\Lambda}_i$. Then we can write for all values of ξ_i, $i = 1, 2 \ldots \mathrm{N}_r^{\mathrm{R}}$ (N_r^{R} is the number of radial discretisation lines):

$$\mathbf{H}_{zn}^{\mathrm{R}} = \boldsymbol{\Lambda}_{\mathrm{C}}\boldsymbol{T}_{\mathrm{HR}}^{\xi h\mathrm{d}}\overline{\mathbf{H}}_{zn}^{\mathrm{C}} + \boldsymbol{\Lambda}_{\mathrm{D}}\boldsymbol{T}_{\mathrm{HR}}^{\xi h\mathrm{d}}\overline{\mathbf{H}}_{zn}^{\mathrm{D}} \tag{7.129}$$

The (full) matrices $\boldsymbol{\Lambda}_{\mathrm{C}}$ and $\boldsymbol{\Lambda}_{\mathrm{D}}$ have the following components:

$$(\boldsymbol{\Lambda}_{\mathrm{C}})_{ik} = \sinh(\boldsymbol{\Gamma}_{\xi k}^{h}(\overline{a}_{\mathrm{R}} - \overline{\xi}_i))/\sinh(\boldsymbol{\Gamma}_{\xi k}^{h}\overline{a}_{\mathrm{R}}) \tag{7.130}$$

$$(\boldsymbol{\Lambda}_{\mathrm{D}})_{ik} = \sinh(\boldsymbol{\Gamma}_{\xi k}^{h}\overline{\xi}_i)/\sinh(\boldsymbol{\Gamma}_{k}^{h}\overline{a}_{\mathrm{R}}) \tag{7.131}$$

$\boldsymbol{\Gamma}_{\xi hk}$ is the kth component of the diagonal matrix $\boldsymbol{\Gamma}_{\xi h}$, which describes the propagation in ξ-direction. In transformed domain we therefore have:

$$\overline{\mathbf{H}}_{zn}^{\mathrm{R}} = \boldsymbol{T}_{rh}^{-1}\boldsymbol{\Lambda}_{\mathrm{C}}\boldsymbol{T}_{\xi h\mathrm{R}}^{\mathrm{d}}\overline{\mathbf{H}}_{zn}^{\mathrm{C}} + \boldsymbol{T}_{rh}^{-1}\boldsymbol{\Lambda}_{\mathrm{D}}\boldsymbol{T}_{\xi h\mathrm{R}}^{\mathrm{d}}\overline{\mathbf{H}}_{zn}^{\mathrm{D}} \tag{7.132}$$

where $\boldsymbol{T}_{rh}$ is the adequate transformation matrix for radial discretisation lines. Now we can give the fields at the ports C and D as functions of those at ports A and B (after denormalisation), where we have discretisation lines in radial direction. We obtain the following expression:

$$\begin{bmatrix} \mathrm{j}\overline{\mathbf{H}}_z^{\mathrm{A}} \\ -\mathrm{j}\overline{\mathbf{H}}_z^{\mathrm{B}} \end{bmatrix} = \begin{bmatrix} -\mathrm{j}\overline{\mathbf{V}}_{\mathrm{AC}}^{\mathrm{N}} & \mathrm{j}\overline{\mathbf{V}}_{\mathrm{AD}}^{\mathrm{N}} \\ \mathrm{j}\overline{\mathbf{V}}_{\mathrm{BC}}^{\mathrm{N}} & -\mathrm{j}\overline{\mathbf{V}}_{\mathrm{BD}}^{\mathrm{N}} \end{bmatrix} \begin{bmatrix} (-\overline{\mathbf{H}}_{zn}^{\mathrm{C}}) \\ -(-\overline{\mathbf{H}}_{zn}^{\mathrm{D}}) \end{bmatrix} = \widehat{\mathbf{V}}_{\mathrm{CD}}^{\mathrm{AB}}\widehat{\mathbf{H}}_{\mathrm{CD}} \tag{7.133}$$

where:

$$\begin{aligned} \overline{\mathbf{V}}_{\mathrm{AC}}^{\mathrm{N}} &= \sqrt{\overline{r}_{\mathrm{A}n}}\boldsymbol{T}_{rh}^{-1}\boldsymbol{\Lambda}_{\mathrm{C}}\boldsymbol{T}_{\xi\mathrm{A}}^{\mathrm{d}} & \overline{\mathbf{V}}_{\mathrm{AD}}^{\mathrm{N}} &= \sqrt{\overline{r}_{\mathrm{A}n}}\boldsymbol{T}_{rh}^{-1}\boldsymbol{\Lambda}_{\mathrm{D}}\boldsymbol{T}_{\xi\mathrm{A}}^{\mathrm{d}} \\ \overline{\mathbf{V}}_{\mathrm{BC}}^{\mathrm{N}} &= \sqrt{\overline{r}_{\mathrm{A}n}}\boldsymbol{T}_{rh}^{-1}\boldsymbol{\Lambda}_{\mathrm{C}}\boldsymbol{T}_{\xi\mathrm{B}}^{\mathrm{d}} & \overline{\mathbf{V}}_{\mathrm{BD}}^{\mathrm{N}} &= \sqrt{\overline{r}_{\mathrm{A}n}}\boldsymbol{T}_{rh}^{-1}\boldsymbol{\Lambda}_{\mathrm{D}}\boldsymbol{T}_{\xi\mathrm{B}}^{\mathrm{d}} \end{aligned} \tag{7.134}$$

The superscript N means that the quantities have to be calculated for radial discretisation lines with Neumann boundary conditions. To calculate the matrix $\overline{\boldsymbol{y}}^{\mathrm{CD}}_{\mathrm{ABh}}$ we have to replace $\overline{\mathbf{H}}_{zn\mathrm{C,D}}$ with $\overline{\mathbf{E}}_r^{\mathrm{C,D}}$ using the relation (7.121):

$$\overline{\boldsymbol{y}}^{\mathrm{CD}}_{\mathrm{ABh}} = \widehat{\overline{\boldsymbol{V}}}^{\mathrm{AB}}_{\mathrm{CD}} \overline{\boldsymbol{y}}^{\mathrm{CD}}_{\mathrm{CDh}} \tag{7.135}$$

In the TM$_z$ case we use eq. (7.124) for $\overline{\mathbf{E}}_{zn}$. The vector $\mathbf{E}_{zn}(\xi_i) = \boldsymbol{T}_{\mathrm{E}}^{\xi e}\overline{\mathbf{E}}_{zn}(\xi_i)$ gives the values E_{zn} in the crossing points of the azimuthal lines with the radial line at $\xi = \xi_i$. $\boldsymbol{T}_{\mathrm{E}}^{\xi e}$ is the matrix for Dirichlet boundary conditions. From these values we must calculate the values of $H_{\phi n}$ at boundaries A and B. Using the equation $\mathrm{j}\varepsilon_d/\varepsilon_r \mathbf{H}_{\phi n} = \overline{\boldsymbol{D}}_{en}\boldsymbol{T}_{\mathrm{E}}^{e}\overline{\mathbf{E}}_{zn}$ we can write, analogously to eq. (7.127):

$$\mathrm{j}\varepsilon_d/\varepsilon_r \mathbf{H}^{\mathrm{R}}_{\phi n}(\xi_i) = \mathbf{T}^{\xi e}_{\mathrm{ER}}\overline{\mathbf{E}}_{zn}(\xi_i) = \mathbf{T}^{\xi e}_{\mathrm{ER}}(\boldsymbol{\Lambda}^{\mathrm{d}}_{\mathrm{C}i}\overline{\mathbf{E}}^{\mathrm{C}}_{zn} + \boldsymbol{\Lambda}^{\mathrm{d}}_{\mathrm{D}i}\overline{\mathbf{E}}^{\mathrm{D}}_{zn}) \tag{7.136}$$

In the matrices $\boldsymbol{\Lambda}^{\mathrm{d}}_{\mathrm{C,D}i}$ we must now introduce $f = e$. The vector $\mathbf{T}^{\xi e}_{\mathrm{ER}}$ is here given by:

$$\mathbf{T}^{\xi e}_{\mathrm{EA}} = c_1\mathbf{T}^{\xi e}_{\mathrm{E1}} - c_2\mathbf{T}^{\xi e}_{\mathrm{E2}} \qquad \mathbf{T}^{\xi e}_{\mathrm{EB}} = c_{\mathrm{N}}\mathbf{T}^{\xi e}_{en} - c_{\mathrm{N}-1}\mathbf{T}^{\xi e}_{en-1} \tag{7.137}$$

with e.g.:

$$c_1 = \tfrac{9}{8}d_{11} + \tfrac{1}{8}d_{21} \qquad c_2 = \tfrac{1}{8}d_{22} \tag{7.138}$$

We assume that the first components in the matrix $\overline{\boldsymbol{D}}_{en}$ (the normalised difference matrix for r-direction) are given by d_{11}, $-d_{21}$ and d_{22}. Therefore, the first two values for the derivatives of $\overline{\mathbf{E}}_{zn}(\xi_i)$ on the h-lines with Neumann boundary conditions are given by:

$$\mathbf{E}'_{zn1}(\xi_i) = d_{11}\mathbf{T}^{\xi e}_{\mathrm{E1}}\overline{\mathbf{E}}_{zn}(\xi_i) \qquad \mathbf{E}'_{zn2}(\xi_i) = (-d_{21}\mathbf{T}^{\xi e}_{\mathrm{E1}} + d_{22}\mathbf{T}^{\xi e}_{\mathrm{E2}})\overline{\mathbf{E}}_{zn}(\xi_i)$$

By extrapolation to boundary A (see Appendix A.6), we obtain the result in eqs. (7.136) and (7.137). The procedure for boundary B is analogous. The method for obtaining $\overline{\mathbf{H}}^{\mathrm{AB}}_{\phi n}$ as a function of $\overline{\mathbf{E}}^{\mathrm{CD}}_{zn}$ (and therefore the matrix $\overline{\boldsymbol{y}}^{\mathrm{CD}}_{\mathrm{ABe}}$) is now straightforward.

7.4.5.2 Field coupling from ports A/B to ports C/D

For the determination of $\overline{\boldsymbol{y}}^{\mathrm{AB}}_{\mathrm{CD}}$, we have to calculate the tangential fields at ports C and D from the fields at ports A and B. This can in principle be done as shown in the previous section. Therefore, we will give only a few equations.

Again, we do not have tangential electric fields at boundaries C and D. $\overline{\mathbf{F}}_z = \overline{\mathbf{H}}_z$ ($\overline{\mathbf{E}}_z$) as a function of r or t for the TE$_z$ (TM$_z$) case is given by: (Analogously to eq. (7.123).)

$$\overline{\mathbf{F}}_z(t) = p_{\boldsymbol{\nu}}^{-1}(p^{\mathrm{B}}_{\boldsymbol{\nu}}(t)\overline{\mathbf{F}}^{\mathrm{A}}_z + p^{\mathrm{A}}_{\boldsymbol{\nu}}(t)\overline{\mathbf{F}}^{\mathrm{A}}_z) \tag{7.139}$$

where:

$$p_{\nu}^{\mathrm{A}}(t) = J_{\nu}(t_{\mathrm{A}})Y_{\nu}(t) - J_{\nu}(t)Y_{\nu}(t_{\mathrm{A}}) \tag{7.140}$$

$$p_{\nu}^{\mathrm{B}}(t) = J_{\nu}(t)Y_{\nu}(t_{\mathrm{B}}) - J_{\nu}(t_{\mathrm{B}})Y_{\nu}(t) \tag{7.141}$$

The evaluation of this relation at the positions r_i or t_i (for the azimuthal discretisation lines) occurs analogously to that in the previous subsection. The transformation of H_z in the TE$_z$ case to boundaries C and D is analogous to the previous one. The relation between $\overline{\mathbf{H}}_z^{\mathrm{A,B}}$ and $\overline{\mathbf{E}}_{\phi}^{\mathrm{A,B}}$ is given according to eq. (7.113).

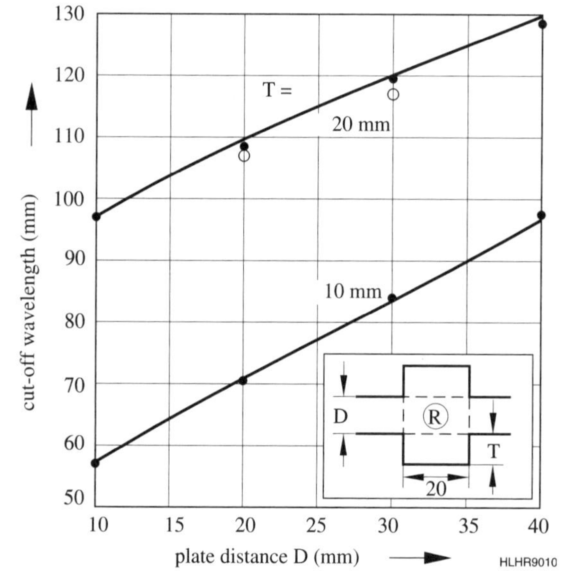

Fig. 7.14 Cut-off wavelength λ_c vs. plate distance of a groove guide. — MoL • • • MMT [6] ○ ○ ○ experiment (R. Pregla, 'MoL-Mode Analysis with Precise Resolution by an Enhanced and Generalized Line Algorithm', in *IEEE MTT-S Int. Symp. Dig.*, vol. 3, pp. 1543–1546. © 1998 Institute of Electrical and Electronics Engineers (IEEE))

The relation between $\overline{\mathbf{H}}_r^{\mathrm{A,B}}$ and $\overline{\mathbf{E}}_z^{\mathrm{A,B}}$ for the TM$_z$ case is:

$$\overline{\mathbf{H}}_r = j\frac{\varepsilon_r}{\varepsilon_d}\boldsymbol{D}_{\phi}^{e}\boldsymbol{T}_e\overline{\mathbf{E}}_z \tag{7.142}$$

The first components of the matrix $\boldsymbol{D}_{\phi}^{e}$ are $d_{11} = -d_{21} = d_{21} = 1$. Therefore, we here have:

$$c_1 = c_{\mathrm{N}} = \tfrac{10}{8} \qquad c_2 = c_{\mathrm{N}-1} = \tfrac{1}{8} \tag{7.143}$$

7.4.6 Steps of the analysis procedure

With the procedure described in the previous section, all matrices in eq. (7.111) can be obtained. Eq. (7.111) is also valid for other regions in the cross-section. In other regions, some of the ports may be short-circuited. The algorithm then works as follows: let us first assume a symmetric cross-section. Then we can start at the two ports in the symmetry plane and transform the impedance/admittance in the opposite direction to a common port. There we can formulate an eigenvalue problem. In azimuthal periodic structures, Floquet's theorem can be introduced. For asymmetric cross-sections a transfer-matrix formulation is used to combine the field transformation formulas and to obtain an eigenvalue problem. The solution gives all desired quantities.

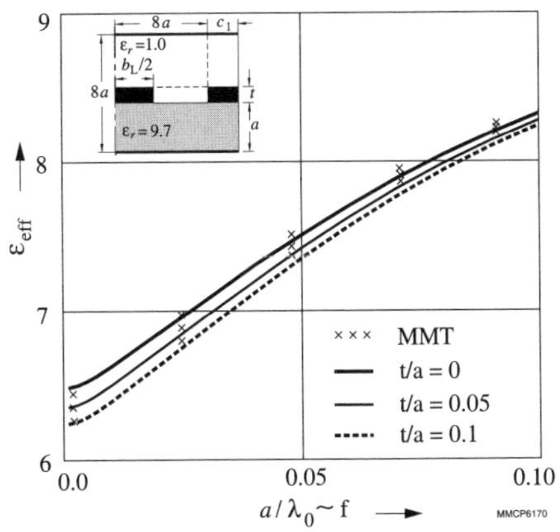

Fig. 7.15 Effective dielectric constant of a microstrip line with finite conductor thickness as a function of normalised frequency with $a = 1$ mm, $b_L/a = 1$

7.5 NUMERICAL RESULTS

The proposed algorithm has now been applied to a variety of open and shielded dielectric waveguides, including the insulated image guide. The latter provides the most useful canonical configuration for modelling the open dielectric waveguiding geometries used in millimetre through optical frequency ranges.

As our first example we determine the cut-off frequencies of the groove guide in Fig. 7.11. Fig. 7.14 shows a diagram of the cut-off wavelength as a function of the plate distance D, with the groove deepness T as parameter. For comparison, the results of M. Sachidananda [6] obtained by Mode Matching Technique (MMT) are included. Measured results obtained by T. Nakahara and N. Kurauchi and taken form [6] are also shown. Only a quarter of the structure was used for the analysis.

As our second example we examine a microstrip line with finite conductor thickness. In fact, the structure is a coplanar line. However, since the slot is wide enough, the metallisation does not influence the propagation constant. Fig. 7.15 shows the effective dielectric constant ε_{re} as a function of the normalised frequency, parameterised with the normalised strip thickness. The results are compared with results using Mode Matching Technique [7].

Results for a magnetron resonator have already been shown in Fig. 5.18.

References

[1] E. Kühn, 'A Mode-Matching Method for Solving Field Problems in Waveguide and Resonator Circuits', *AEÜ*, vol. 27, no. 12, pp. 511–518, 1973.

[2] R. Pregla, 'MoL-Mode Analysis with Precise Resolution by an Enhanced and Generalized Line Algorithm', in *IEEE MTT-S Int. Symp. Dig.*, Baltimore, USA, June 1998, vol. 3, pp. 1543–1546.

[3] H. Z. Zhang, 'A Wideband Orthogonal-Mode Junction Using a Junction of a Quad-Ridged Coaxial Waveguide and Four Ridged Sectoral Waveguides', *IEEE Microw. Wireless Compon. Lett.*, vol. 5, pp. 172–174, 2002.

[4] J.-Y. Raguin, H.-G. Unger and D. M. Vavriv, 'Recent Advances in the Analysis of Space-Harmonic Millimeter-Wave Magnetrons with Secondary-Emission Cathode', in *Int. Symp. on Recent Advances in Microwave Technology*, Malaga, Spain, Dec. 1999, pp. 63–66.

[5] M. Abramowitz and I. A. Stegun, *Handbook of Mathematical Functions*, chapter 9, Dover Publ., New York, 1965.

[6] M. Sachidananda, 'Rigorous Analysis of a Groove Guide', *IEE Proc. -H, Microwave Antennas Propagation*, vol. 139, pp. 449–452, May 1992.

[7] G. Kowalski and R. Pregla, 'Dispersion Characteristics of Shielded Microstrips with Finite Thickness', *AEÜ*, vol. 25, pp. 193–196, 1971.

Further Reading

[8] R. Pregla, 'The Analysis of General Axially Symmetric Antennas with a Coaxial Feed Line by the Method of Lines', *IEEE Trans. Antennas. Propagation*, vol. 46, no. 10, pp. 1433–1443, Oct. 1998.

[9] W. Pascher and R. Pregla, 'Analysis of Rectangular Waveguide Junctions by the Method of Lines', *IEEE Trans. Microwave Theory Tech.*, vol. MTT-43, no. 12, pp. 2649–2653, Dec. 1995.

[10] R. Pregla, 'MoL Analysis of Rectangular Waveguide Junctions by an Impedance/Admittance Transfer Concept and Crossed Discretisation Lines', *Int. J. of Electronics and Communications (AEÜ)*, vol. 53, no. 2, 1999, pp. 83–91.

[11] R. Pregla, 'Analysis of Planar Microwave and Millimeterwave Circuits with Anisotropic Layers on Generalized Transmission Line Equations and on the Method of Lines', *IEEE MTT-S Int. Microwave Symp. Dig.*, Boston, USA, 2000, pap. TU2E-5.

[12] R. Pregla, 'Efficient and Accurate Modeling of Planar Anisotropic Microwave Structures by the Method of Lines', *IEEE Trans. on Microwave Theory and Tech.*, vol. 50, no. 6, 2002, pp. 1469–1479.

[13] L. Vietzorreck and R. Pregla, 'Analysis of MMIC Junctions and Multiports by the Method of Lines', *IEEE MTT-S Int. Microwave Symp. Dig.*, Baltimore, USA, 1998, pp. 1547–1550.

[14] R. Pregla, 'Efficient and Accurate Modeling of Planar Microwave Structures with Anisotropic Layers by the Method of Lines (MoL)', *ISRAMT'99*, Malaga, Spain, Dec. 1999, pp. 699–703.

[15] R. Pregla, 'Efficient Modeling of Conformal Antennas', *Millennium Conf. on Antennas and Propagation*, Davos, Switzerland, Apr. 2000, paper 0682.

[16] R. Pregla, 'General Formulas for the Method of Lines in Cylindrical Coordinates' *IEEE Trans. on Microwave Theory and Tech.*, vol. 43, no. 7, 1995, pp. 1617–1620.

CHAPTER 8

WAVEGUIDE STRUCTURES WITH MATERIALS OF GENERAL ANISOTROPY IN ARBITRARY ORTHOGONAL COORDINATE SYSTEMS

8.1 GENERALISED TRANSMISSION LINE EQUATIONS

In this section we derive the generalised transmission line (GTL) equations that are analogous to the well-known equations for coupled multi-conductor transmission lines. We have shown the latter in the previous sections. These transmission line equations are solved by determining modal matrices [1].

8.1.1 Material properties

For the formulation of GTL equations in an arbitrary coordinate system, the material parameters (permeability and permittivity tensors) are assumed to have the following general form:

$$\overleftrightarrow{\nu}_r = \begin{bmatrix} \nu_{11} & \nu_{12} & \nu_{13} \\ \nu_{21} & \nu_{22} & \nu_{23} \\ \nu_{31} & \nu_{32} & \nu_{33} \end{bmatrix} \quad \text{with} \quad \overleftrightarrow{\nu}_r = \overleftrightarrow{\varepsilon}_r \quad \text{or} \quad \overleftrightarrow{\mu}_r \tag{8.1}$$

Even if the tensor elements are complex, no symmetry is assumed. We would like to develop the GTL expressions for the x_3-direction. The device under study is divided into homogeneous sections in this direction. Hence the material parameters in each cross-section are functions of x_1 and x_2 only.

8.1.2 Maxwell's equations in matrix notation

For the determination of GTL equations it is very efficient to write Maxwell's equations in matrix notation (we only need the curl equations).[1] The curl operator in general orthogonal coordinates (see Fig. 8.1) may be written in following form:

$$\overline{\nabla} \times \vec{A} = \begin{bmatrix} 0 & -(g_2g_3)^{-1}D_{\overline{x}_3} & (g_2g_3)^{-1}D_{\overline{x}_2} \\ (g_1g_3)^{-1}D_{\overline{x}_3} & 0 & -(g_1g_3)^{-1}D_{\overline{x}_1} \\ -(g_1g_2)^{-1}D_{\overline{x}_2} & (g_1g_2)^{-1}D_{\overline{x}_1} & 0 \end{bmatrix} \begin{bmatrix} g_1A_1 \\ g_2A_2 \\ g_3A_3 \end{bmatrix} \tag{8.2}$$

[1]The author would like to thank Dr. Kamen P. Ivanov for pointing to this formulation.

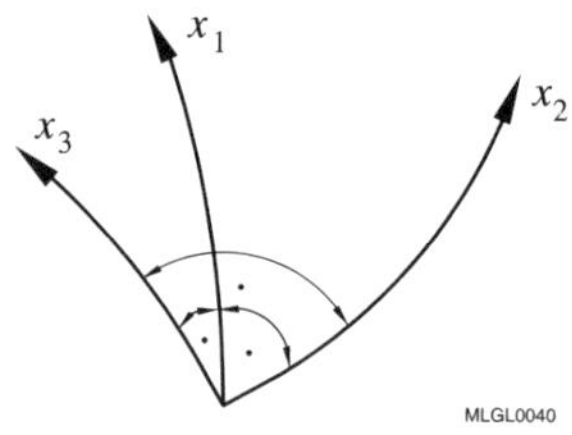

Fig. 8.1 General orthogonal coordinate system (R. Pregla, 'Modeling of Optical Waveguide Structures with General Anisotropy in Arbitrary Orthogonal Coordinate Systems', *IEEE J. of Sel. Topics in Quantum Electronics*, vol. 8, pp. 1217–1224. © 2002 Institute of Electrical and Electronics Engineers (IEEE))

A_i and g_i $(i = 1, 2, 3)$ are the field components and the metric factors, respectively. The metric factors are given by:

$$g_{\mathrm{i}}^2 = \left(\frac{\partial x}{\partial x_{\mathrm{i}}}\right)^2 + \left(\frac{\partial y}{\partial x_{\mathrm{i}}}\right)^2 + \left(\frac{\partial z}{\partial x_{\mathrm{i}}}\right)^2 \tag{8.3}$$

Additionally, we use the abbreviation $D_{\overline{x}_i} = \partial/\partial\overline{x}_i$. The coordinates x_i are normalised with the free space wave number k_0 (i.e. $\overline{x}_i = k_0 x_i$). In shorter form, we may write eq. (8.2):

$$\overline{\nabla} \times \vec{A} = [G_2]^{-1}[\widehat{D}_{\mathrm{c}}][A_n] \qquad [A_n] = [G_1][A] \tag{8.4}$$

with:

$$\left[\widehat{D}_{\mathrm{c}}\right] = \begin{bmatrix} 0 & -D_{\overline{x}_3} & D_{\overline{x}_2} \\ D_{\overline{x}_3} & 0 & -D_{\overline{x}_1} \\ -D_{\overline{x}_2} & D_{\overline{x}_1} & 0 \end{bmatrix} \tag{8.5}$$

$$\begin{aligned} [A] &= [A_1, A_2, A_3]^{\mathrm{t}} \\ [G_1] &= \mathrm{diag}(g_1, g_2, g_3) \\ [G_2] &= g_1 g_2 g_3 \, [G_1]^{-1} \end{aligned} \tag{8.6}$$

The Maxwell's curl equations in an arbitrary orthogonal coordinate system can now be written in the following matrix form: (The magnetic field components are normalised with the free space wave impedance $\eta_0 = \sqrt{\mu_0/\varepsilon_0}$, i.e. $\widetilde{H} = \eta_0 H$.)

$$\mathrm{j}[\overline{\varepsilon}_r][E_n] = [\widehat{D}_{\mathrm{c}}][\widetilde{H}_n] \qquad -\mathrm{j}[\overline{\mu}_r][\widetilde{H}_n] = [\widehat{D}_{\mathrm{c}}][E_n] \tag{8.7}$$

with:

$$[\overline{\nu}_r] = [G_2]\overleftrightarrow{\nu}_r[G_1]^{-1} \qquad \overleftrightarrow{\nu}_r = \overleftrightarrow{\varepsilon}_r, \overleftrightarrow{\mu}_r \tag{8.8}$$

or:

$$[\overline{\nu}_r] = \begin{bmatrix} g_2 g_3 g_1^{-1} \nu_{11} & g_3 \nu_{12} & g_2 \nu_{13} \\ g_3 \nu_{21} & g_1 g_3 g_2^{-1} \nu_{22} & g_1 \nu_{23} \\ g_2 \nu_{31} & g_1 \nu_{32} & g_1 g_2 g_3^{-1} \nu_{33} \end{bmatrix} \tag{8.9}$$

By using the definitions:

$$[\widehat{D}_\mathrm{d}] = [D_{\overline{x}_1}, D_{\overline{x}_2}, D_{\overline{x}_3}] \quad \text{and} \quad \overline{\rho} = g_1 g_2 g_3 \rho \tag{8.10}$$

we obtain for the Maxwell's divergence relations:

$$[\widehat{D}_\mathrm{d}][\overline{\varepsilon}_r][E_n] = \overline{\rho}/\varepsilon_0 \qquad [\widehat{D}_\mathrm{d}][\overline{\mu}_r][\widetilde{H}_n] = 0 \tag{8.11}$$

For the gradient operator in matrix notation we may write:

$$\overline{\nabla}\phi = [G_1]^{-1}[\widehat{D}_\mathrm{d}]^\mathrm{t}\phi \tag{8.12}$$

Eqs. (8.7) and (8.11) have the exact same form as in Cartesian coordinates. Here we only have to use normalised field components and normalised permeability and permittivity tensors. Therefore, we can directly transform the GTL equation from Cartesian coordinates [2] into arbitrary orthogonal coordinate systems. Note that we have $[\widehat{D}_\mathrm{d}][\widehat{D}_\mathrm{c}] = [0, 0, 0]$. This relation is still valid if the differential operators are replaced by difference operators. Therefore the divergence relations are also fulfilled by the curl relations in discretised form in charge-free regions.

8.1.3 Generalised transmission line equations in Cartesian coordinates for general anisotropic material

We will first write the GTL equations for general anisotropic material in the Cartesian coordinate case. GTL equations (but not in matrix notation) are also described in [3]. The adequate cross-section of the waveguide is shown in Fig. 8.2.

Here we define supervectors according to:

$$[\widetilde{\mathrm{H}}] = [-\widetilde{H}_x, \widetilde{H}_y]^\mathrm{t}, \qquad [\mathrm{E}] = [E_y, E_x]^\mathrm{t} \tag{8.13}$$

The first two equations of Ampère's law and of law of can be written as:

$$\frac{\partial}{\partial \overline{z}}[\widetilde{\mathrm{H}}] = -\mathrm{j}\begin{bmatrix} \varepsilon_{yy} & M_y^\square M_x^\circ \varepsilon_{yx} \\ M_x^\square M_y^\bullet \varepsilon_{xy} & \varepsilon_{xx} \end{bmatrix}[\mathrm{E}] - \begin{bmatrix} M_y^\square \varepsilon_{yz} \\ M_x^\square \varepsilon_{xz} \end{bmatrix}\mathrm{j}E_z - \mathrm{j}\begin{bmatrix} -D_{\overline{x}}^\diamond \\ D_{\overline{y}}^\diamond \end{bmatrix}\mathrm{j}\widetilde{H}_z \tag{8.14}$$

$$\frac{\partial}{\partial \overline{z}}[\mathrm{E}] = -\mathrm{j}\begin{bmatrix} \mu_{xx} & -M_x^\diamond M_y^\circ \mu_{xy} \\ -M_y^\diamond M_x^\bullet \mu_{yx} & \mu_{yy} \end{bmatrix}[\widetilde{\mathrm{H}}] - \begin{bmatrix} -M_x^\diamond \mu_{xz} \\ M_y^\diamond \mu_{yz} \end{bmatrix}\mathrm{j}\widetilde{H}_z - \mathrm{j}\begin{bmatrix} D_{\overline{y}}^\square \\ D_{\overline{x}}^\square \end{bmatrix}\mathrm{j}E_z \tag{8.15}$$

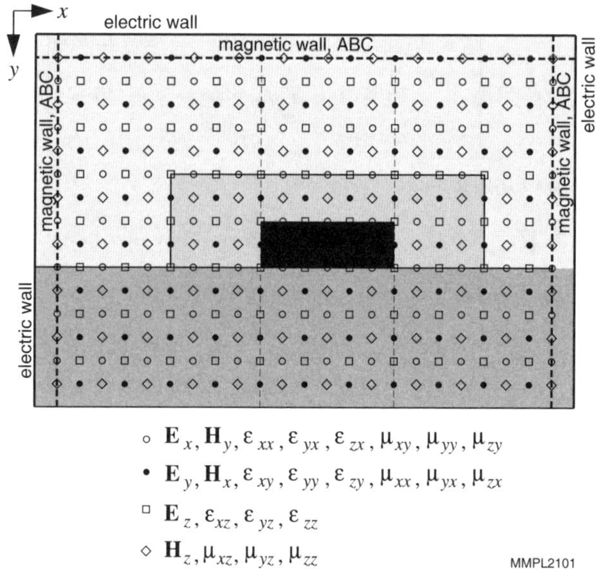

Fig. 8.2 Inhomogeneous anisotropic cross-section with discretisation points (R. Pregla, 'Modeling of Optical Waveguide Structures with General Anisotropy in Arbitrary Orthogonal Coordinate Systems', *IEEE J. of Sel. Topics in Quantum Electronics*, vol. 8, pp. 1217–1224. © 2002 Institute of Electrical and Electronics Engineers (IEEE))

Again, we normalise the coordinates $u = x, y, z$ with the free space wave number k_0 (i.e. to $\overline{u} = k_0 u$) and the magnetic field components H with the free space wave impedance $\eta_0 = \sqrt{\mu_0/\varepsilon_0}$ (i.e. $\widetilde{H} = \eta_0 H$). Further, we use the abbreviation $D_{\overline{u}} = \partial/\partial\overline{u}$ with $\overline{u} = \overline{x}, \overline{y}, \overline{z}$.

The components E_z, H_z are given by:

$$\mathrm{j}\varepsilon_{zz} E_z = [D_{\overline{y}}^{\bullet} \quad D_{\overline{x}}^{\circ}][\widetilde{\mathrm{H}}] - \mathrm{j}[M_y^{\bullet}\varepsilon_{zy} \quad M_x^{\circ}\varepsilon_{zx}]\widehat{\mathrm{E}} \tag{8.16}$$

$$\mathrm{j}\mu_{zz}\widetilde{H}_z = [-D_{\overline{x}}^{\bullet} \quad D_{\overline{y}}^{\circ}]\widehat{\mathrm{E}} - \mathrm{j}[-M_x^{\bullet}\mu_{zx} \quad M_y^{\circ}\mu_{zy}][\widetilde{\mathrm{H}}] \tag{8.17}$$

By inserting eqs. (8.16) and (8.17) into eqs. (8.14) and (8.15) we may write:

$$\frac{\partial}{\partial\overline{z}}[\widetilde{\mathrm{H}}] = -\mathrm{j}[R_{\mathrm{E}}][\mathrm{E}] - [S_{\mathrm{H}}][\widetilde{\mathrm{H}}] \quad \frac{\partial}{\partial\overline{z}}[\mathrm{E}] = -[S_{\mathrm{E}}][\mathrm{E}] - \mathrm{j}[R_{\mathrm{H}}][\widetilde{\mathrm{H}}] \tag{8.18}$$

We obtain for the matrices $[R_{\mathrm{E,H}}^z]$:

$$[R_{\mathrm{E}}^z] = \begin{bmatrix} \varepsilon_{yy} + D_{\overline{x}}^{\diamond}\mu_{zz}^{-1}D_{\overline{x}}^{\bullet} & M_y^{\square}M_x^{\circ}\varepsilon_{yx} - D_{\overline{x}}^{\diamond}\mu_{zz}^{-1}D_{\overline{y}}^{\circ} \\ M_x^{\square}M_y^{\bullet}\varepsilon_{xy} - D_{\overline{y}}^{\diamond}\mu_{zz}^{-1}D_{\overline{x}}^{\bullet} & \varepsilon_{xx} + D_{\overline{y}}^{\diamond}\mu_{zz}^{-1}D_{\overline{y}}^{\circ} \end{bmatrix} - \begin{bmatrix} M_y^{\square}\varepsilon_{yz}\varepsilon_{zz}^{-1}M_y^{\bullet}\varepsilon_{zy} & M_y^{\square}\varepsilon_{yz}\varepsilon_{zz}^{-1}M_x^{\circ}\varepsilon_{zx} \\ M_x^{\square}\varepsilon_{xz}\varepsilon_{zz}^{-1}M_y^{\bullet}\varepsilon_{zy} & M_x^{\square}\varepsilon_{xz}\varepsilon_{zz}^{-1}M_x^{\circ}\varepsilon_{zx} \end{bmatrix} \tag{8.19}$$

$$
\begin{aligned}
[R_{\mathrm{H}}^{z}] = & \begin{bmatrix} D_{\bar{y}}^{\square}\varepsilon_{zz}^{-1}D_{\bar{y}}^{\bullet} + \mu_{xx} & D_{\bar{y}}^{\square}\varepsilon_{zz}^{-1}D_{\bar{x}}^{\circ} - M_{x}^{\diamond}M_{y}^{\circ}\mu_{xy} \\ D_{\bar{x}}^{\square}\varepsilon_{zz}^{-1}D_{\bar{y}}^{\bullet} - M_{y}^{\diamond}M_{x}^{\bullet}\mu_{yx} & D_{\bar{x}}^{\square}\varepsilon_{zz}^{-1}D_{\bar{x}}^{\circ} + \mu_{yy} \end{bmatrix} \\
& - \begin{bmatrix} M_{x}^{\diamond}\mu_{xz}\mu_{zz}^{-1}M_{x}^{\bullet}\mu_{zx} & -M_{x}^{\diamond}\mu_{xz}\mu_{zz}^{-1}M_{y}^{\circ}\mu_{zy} \\ -M_{y}^{\diamond}\mu_{yz}\mu_{zz}^{-1}M_{x}^{\bullet}\mu_{zx} & M_{y}^{\diamond}\mu_{yz}\mu_{zz}^{-1}M_{y}^{\circ}\mu_{zy} \end{bmatrix}
\end{aligned} \tag{8.20}
$$

and for the matrices $[S_{\mathrm{E,H}}^{z}]$:

$$
[S_{\mathrm{H}}^{z}] = \begin{bmatrix} M_{y}^{\square}\varepsilon_{yz}\varepsilon_{zz}^{-1} \\ M_{x}^{\square}\varepsilon_{xz}\varepsilon_{zz}^{-1} \end{bmatrix} [D_{\bar{y}}^{\bullet} \quad D_{\bar{x}}^{\circ}] + \begin{bmatrix} -D_{\bar{x}}^{\diamond}\mu_{zz}^{-1} \\ D_{\bar{y}}^{\diamond}\mu_{zz}^{-1} \end{bmatrix} [-M_{x}^{\bullet}\mu_{zx} \quad M_{y}^{\circ}\mu_{zy}] \tag{8.21}
$$

$$
[S_{\mathrm{E}}^{z}] = \begin{bmatrix} -M_{x}^{\diamond}\mu_{xz}\mu_{zz}^{-1} \\ M_{y}^{\diamond}\mu_{yz}\mu_{zz}^{-1} \end{bmatrix} [-D_{\bar{x}}^{\bullet} \quad D_{\bar{y}}^{\circ}] + \begin{bmatrix} D_{\bar{y}}^{\square}\varepsilon_{zz}^{-1} \\ D_{\bar{x}}^{\square}\varepsilon_{zz}^{-1} \end{bmatrix} [M_{y}^{\bullet}\varepsilon_{zy} \quad M_{x}^{\circ}\varepsilon_{zx}] \tag{8.22}
$$

In the process of discretisation, later on we have to interpolate some of the field components. This will be done with interpolation matrices $\boldsymbol{M}$. Therefore, we have already introduced interpolation operators $M_{x,y}$ into the above equations. The superscript and the subscript indicate the direction of interpolation. We need these M-matrices if two components from different discretisation points need to be combined [4], [5]. The subscript indicates the interpolation direction and the superscript marks the discretisation system from which the components have to be interpolated. It should be mentioned that there are other possible choices for these matrices. We could e.g. choose $M_{x}^{\diamond}M_{y}^{\circ}$ in eq. (8.14) instead of $M_{y}^{\square}M_{x}^{\circ}$. More details are given in Section 8.1.6.

8.1.4 Generalised transmission line equations for general anisotropic material in arbitrary orthogonal coordinates

Let us now derive GTL equations for arbitrary orthogonal coordinate systems [6]. By using the abbreviations:

$$
[\mathrm{E}] = [\mathrm{E}_{\mathrm{xn}_1}, \mathrm{E}_{\mathrm{xn}_2}]^{\mathrm{t}} \qquad [\widetilde{\mathrm{H}}] = [\widetilde{\mathrm{H}}_{\mathrm{xn}_2}, -\widetilde{\mathrm{H}}_{\mathrm{xn}_1}]^{\mathrm{t}} \tag{8.23}
$$

we obtain from Maxwell's curl equations for the x_3-direction:

$$
\frac{\partial}{\partial \overline{x}_3}[\mathrm{E}] = -\mathrm{j}[\mathrm{R}_{\mathrm{H}}^{\mathrm{x}_3}][\widetilde{\mathrm{H}}] - [\mathrm{S}_{\mathrm{E}}^{\mathrm{x}_3}][\mathrm{E}] \quad \frac{\partial}{\partial \overline{\mathrm{x}}_3}[\widetilde{\mathrm{H}}] = -\mathrm{j}[\mathrm{R}_{\mathrm{E}}^{\mathrm{x}_3}][\mathrm{E}] - [\mathrm{S}_{\mathrm{H}}^{\mathrm{x}_3}][\widetilde{\mathrm{H}}] \tag{8.24}
$$

On the right side of each equation we have an additional term – the matrices $S_{\mathrm{E,H}}$ – proportional to the vector on the left side (see also [2]). These matrices $S_{\mathrm{E,H}}$ vanish if there are gyrotropic material tensors with respect to x_3 ($\nu_{13} = \nu_{31} = \nu_{23} = \nu_{32} = 0$). Again, we obtain these equations from the first two scalar equations in (8.7), into which the third one was introduced.

The components E_{xn_3}, H_{xn_3} are given by:

$$
E_{xn_3} = -\overline{\varepsilon}_{33}^{-1}[\overline{\varepsilon}_{31} \quad \overline{\varepsilon}_{32}][\mathrm{E}] - \mathrm{j}\overline{\varepsilon}_{33}^{-1}[D_{\overline{x}_1} \quad D_{\overline{x}_2}][\widetilde{\mathrm{H}}] \tag{8.25}
$$

$$
\widetilde{H}_{xn_3} = -\overline{\mu}_{33}^{-1}[\overline{\mu}_{32} \quad -\overline{\mu}_{31}][\widetilde{\mathrm{H}}] + \mathrm{j}\overline{\mu}_{33}^{-1}[-D_{\overline{x}_2} \quad D_{\overline{x}_1}][\mathrm{E}] \tag{8.26}
$$

The matrices $[R^{x_3}_{\mathrm{E,H}}]$ and $[S^{x_3}_{\mathrm{E,H}}]$ contain derivatives with respect to x_1 and x_2 as well as the material parameters.

$$[R^{x_3}_{\mathrm{H}}] = \begin{bmatrix} D_{\overline{x}_1}\overline{\varepsilon}_{33}^{-1}D_{\overline{x}_1} + \overline{\mu}_{r22} & D_{\overline{x}_1}\overline{\varepsilon}_{33}^{-1}D_{\overline{x}_2} - \overline{\mu}_{r21} \\ D_{\overline{x}_2}\overline{\varepsilon}_{33}^{-1}D_{\overline{x}_1} - \overline{\mu}_{r12} & D_{\overline{x}_2}\overline{\varepsilon}_{33}^{-1}D_{\overline{x}_2} + \overline{\mu}_{r11} \end{bmatrix} - \begin{bmatrix} \overline{\mu}_{23}\overline{\mu}_{33}^{-1}\overline{\mu}_{32} & -\overline{\mu}_{23}\overline{\mu}_{33}^{-1}\overline{\mu}_{31} \\ -\overline{\mu}_{13}\overline{\mu}_{33}^{-1}\overline{\mu}_{32} & \overline{\mu}_{13}\overline{\mu}_{33}^{-1}\overline{\mu}_{31} \end{bmatrix} \tag{8.27}$$

$$[R^{x_3}_{\mathrm{E}}] = \begin{bmatrix} D_{\overline{x}_2}\overline{\mu}_{33}^{-1}D_{\overline{x}_2} + \overline{\varepsilon}_{r11} & \overline{\varepsilon}_{r12} - D_{\overline{x}_2}\overline{\mu}_{33}^{-1}D_{\overline{x}_1} \\ \overline{\varepsilon}_{r21} - D_{\overline{x}_1}\overline{\mu}_{33}^{-1}D_{\overline{x}_2} & D_{\overline{x}_1}\overline{\mu}_{33}^{-1}D_{\overline{x}_1} + \overline{\varepsilon}_{r22} \end{bmatrix} - \begin{bmatrix} \overline{\varepsilon}_{13}\overline{\varepsilon}_{33}^{-1}\overline{\varepsilon}_{31} & \overline{\varepsilon}_{13}\overline{\varepsilon}_{33}^{-1}\overline{\varepsilon}_{32} \\ \overline{\varepsilon}_{23}\overline{\varepsilon}_{33}^{-1}\overline{\varepsilon}_{31} & \overline{\varepsilon}_{23}\overline{\varepsilon}_{33}^{-1}\overline{\varepsilon}_{32} \end{bmatrix} \tag{8.28}$$

$$[S^{x_3}_{\mathrm{H}}] = \begin{bmatrix} \overline{\varepsilon}_{13} \\ \overline{\varepsilon}_{23} \end{bmatrix} [\overline{\varepsilon}_{33}^{-1}D_{\overline{x}_1} \quad \overline{\varepsilon}_{33}^{-1}D_{\overline{x}_2}] + \begin{bmatrix} D_{\overline{x}_2} \\ -D_{\overline{x}_1} \end{bmatrix} [\overline{\mu}_{33}^{-1}\mu_{32} \quad -\overline{\mu}_{33}^{-1}\overline{\mu}_{31}] \tag{8.29}$$

$$[S^{x_3}_{\mathrm{E}}] = \begin{bmatrix} -\overline{\mu}_{23} \\ \overline{\mu}_{13} \end{bmatrix} [-\overline{\mu}_{33}^{-1}D_{\overline{x}_2} \quad \overline{\mu}_{33}^{-1}D_{\overline{x}_1}] + \begin{bmatrix} D_{\overline{x}_1} \\ D_{\overline{x}_2} \end{bmatrix} [\overline{\varepsilon}_{33}^{-1}\overline{\varepsilon}_{31} \quad \overline{\varepsilon}_{33}^{-1}\overline{\varepsilon}_{32}] \tag{8.30}$$

Matrices M must be introduced analogously to the previous section. The combination of the eqs. (8.24) yields:

$$\frac{\mathrm{d}}{\mathrm{d}\overline{x}_3}\begin{bmatrix} \mathrm{E} \\ \widetilde{\mathrm{H}} \end{bmatrix} + \begin{bmatrix} S^{x_3}_{\mathrm{E}} & \mathrm{j}R^{x_3}_{\mathrm{H}} \\ \mathrm{j}R^{x_3}_{\mathrm{E}} & S^{x_3}_{\mathrm{H}} \end{bmatrix}\begin{bmatrix} \mathrm{E} \\ \widetilde{\mathrm{H}} \end{bmatrix} = \begin{bmatrix} 0 \\ 0 \end{bmatrix} \tag{8.31}$$

We can also combine the equations in the following form:

$$\left([S^{x_3}_{\mathrm{E}}] + \frac{\mathrm{d}}{\mathrm{d}\overline{x}_3}\right)[R^{x_3}_{\mathrm{E}}]^{-1}\left([S^{x_3}_{\mathrm{H}}] + \frac{\mathrm{d}}{\mathrm{d}\overline{x}_3}\right)[\widetilde{\mathrm{H}}] + [R^{x_3}_{\mathrm{H}}][\widetilde{\mathrm{H}}] = [0] \tag{8.32}$$

$$\left([S^{x_3}_{\mathrm{H}}] + \frac{\mathrm{d}}{\mathrm{d}\overline{x}_3}\right)[R^{x_3}_{\mathrm{H}}]^{-1}\left([S^{x_3}_{\mathrm{E}}] + \frac{\mathrm{d}}{\mathrm{d}\overline{x}_3}\right)[\mathrm{E}] + [R^{x_3}_{\mathrm{E}}][\mathrm{E}] = [0] \tag{8.33}$$

Multiplying with $[R^{x_3}_{\mathrm{E}}]$ and $[R^{x_3}_{\mathrm{H}}]$ (they must be independent of $\overline{x}_3$), respectively, we obtain:

$$\frac{\mathrm{d}}{\mathrm{d}\overline{x}_3^2}[\widetilde{\mathrm{H}}] + ([S^{x_3}_{\mathrm{H}}] + [R^{x_3}_{\mathrm{E}}][S^{x_3}_{\mathrm{E}}][R^{x_3}_{\mathrm{E}}]^{-1})\frac{\mathrm{d}}{\mathrm{d}\overline{z}}[\widetilde{\mathrm{H}}] + ([R^{x_3}_{\mathrm{E}}][S^{x_3}_{\mathrm{E}}][R^{x_3}_{\mathrm{E}}]^{-1}[S^{x_3}_{\mathrm{H}}] + [R^{x_3}_{\mathrm{E}}][R^{x_3}_{\mathrm{H}}])\widetilde{\mathrm{H}} = [0] \tag{8.34}$$

$$\frac{\mathrm{d}}{\mathrm{d}\overline{x}_3^2}[\mathrm{E}] + ([\mathrm{S}^{\mathrm{x}_3}_{\mathrm{E}}] + [\mathrm{R}^{\mathrm{x}_3}_{\mathrm{H}}][\mathrm{S}^{\mathrm{x}_3}_{\mathrm{H}}][\mathrm{R}^{\mathrm{x}_3}_{\mathrm{H}}]^{-1})\frac{\mathrm{d}}{\mathrm{d}\overline{\mathrm{x}}_3}[\mathrm{E}] + ([R^{x_3}_{\mathrm{H}}][S^{x_3}_{\mathrm{H}}][R^{x_3}_{\mathrm{H}}]^{-1}[S^{x_3}_{\mathrm{E}}] + [R^{x_3}_{\mathrm{H}}][R^{x_3}_{\mathrm{E}}])[\mathrm{E}] = [0] \tag{8.35}$$

If $S^{x_3}_{\mathrm{E,H}} = 0$ and if the metric factors do not depend on $\overline{x}_3$ we immediately obtain:

$$\frac{\mathrm{d}^2}{\mathrm{d}\overline{x}_3^2}\begin{bmatrix} g_1E_{x1} \\ g_2E_{x2} \end{bmatrix} + [R^{x_3}_{\mathrm{H}}][R^{x_3}_{\mathrm{E}}]\begin{bmatrix} g_1E_{x1} \\ g_2E_{x2} \end{bmatrix} = \begin{bmatrix} 0 \\ 0 \end{bmatrix} \tag{8.36}$$

$$\frac{\mathrm{d}^2}{\mathrm{d}\overline{x}_3^2}\begin{bmatrix} g_2 H_{x2} \\ -g_1 H_{x1}\end{bmatrix} + [R_{\mathrm{E}}^{x_3}][R_{\mathrm{H}}^{x_3}]\begin{bmatrix} g_2 H_{x2} \\ -g_1 H_{x1}\end{bmatrix} = \begin{bmatrix}0\\0\end{bmatrix} \tag{8.37}$$

The matrices $R_{\mathrm{E,H}}^{x_3}$ in these equations consist only of the first parts of eqs. (8.27) and (8.28). To calculate the products, we again split the two matrices into two parts each according to:

$$[R_{\mathrm{E}}^{x_3}] = [R_{\mathrm{E}}^{x_3\mathrm{D}}] + [\widehat{\varepsilon}_{rt}] \qquad [R_{\mathrm{E}}^{x_3\mathrm{D}}] = \begin{bmatrix} D_{\overline{x}_2}\overline{\mu}_{33}^{-1}D_{\overline{x}_2} & -D_{\overline{x}_2}\overline{\mu}_{33}^{-1}D_{\overline{x}_1} \\ -D_{\overline{x}_1}\overline{\mu}_{33}^{-1}D_{\overline{x}_2} & D_{\overline{x}_1}\overline{\mu}_{33}^{-1}D_{\overline{x}_1}\end{bmatrix} \tag{8.38}$$

$$[R_{\mathrm{H}}^{x_3}] = [R_{\mathrm{H}}^{x_3\mathrm{D}}] + [\widehat{\mu}_{rt}] \qquad [R_{\mathrm{H}}^{x_3\mathrm{D}}] = \begin{bmatrix} D_{\overline{x}_1}\overline{\varepsilon}_{33}^{-1}D_{\overline{x}_1} & D_{\overline{x}_1}\overline{\varepsilon}_{33}^{-1}D_{\overline{x}_2} \\ D_{\overline{x}_2}\overline{\varepsilon}_{33}^{-1}D_{\overline{x}_1} & D_{\overline{x}_2}\overline{\varepsilon}_{33}^{-1}D_{\overline{x}_2}\end{bmatrix} \tag{8.39}$$

The matrices with the material parameters are:

$$[\widehat{\varepsilon}_{rt}] = \begin{bmatrix}\overline{\varepsilon}_{r11} & \overline{\varepsilon}_{r12} \\ \overline{\varepsilon}_{r21} & \overline{\varepsilon}_{r22}\end{bmatrix} \qquad [\widehat{\mu}_{rt}] = \begin{bmatrix}\overline{\mu}_{r22} & -\overline{\mu}_{r21} \\ -\overline{\mu}_{r12} & \overline{\mu}_{r11}\end{bmatrix} \tag{8.40}$$

With these definitions, and due to $[R_{\mathrm{H}}^{x_3\mathrm{D}}][R_{\mathrm{E}}^{x_3\mathrm{D}}] = [R_{\mathrm{E}}^{x_3\mathrm{D}}][R_{\mathrm{H}}^{x_3\mathrm{D}}] = 0$, the products can be written as:

$$[Q_{\mathrm{E}}^{x_3}] = [R_{\mathrm{H}}^{x_3}][R_{\mathrm{E}}^{x_3}] = [R_{\mathrm{H}}^{x_3\mathrm{D}}][\widehat{\varepsilon}_{rt}] + [\widehat{\mu}_{rt}][R_{\mathrm{E}}^{x_3\mathrm{D}}] + [\widehat{\mu}_{rt}][\widehat{\varepsilon}_{rt}] \tag{8.41}$$

$$[Q_{\mathrm{H}}^{x_3}] = [R_{\mathrm{E}}^{x_3}][R_{\mathrm{H}}^{x_3}] = [R_{\mathrm{E}}^{x_3\mathrm{D}}][\widehat{\mu}_{rt}] + [\widehat{\varepsilon}_{rt}][R_{\mathrm{H}}^{x_3\mathrm{D}}] + [\widehat{\varepsilon}_{rt}][\widehat{\mu}_{rt}] \tag{8.42}$$

The submatrices of $[Q_{\mathrm{E,H}}^{x_3}]$ can easily be determined.

8.1.5 Boundary conditions

It is well known that the fulfilment of the boundary conditions for the electromagnetic fields at electric or magnetic walls is not a trivial problem in the case of anisotropic materials. One possible way to circumvent this problem is to use periodic boundary conditions [4]. We will study the problem in more detail in this section and will show how the conditions can be fulfilled in the case of discretised fields. The results are not only valid for the MoL but may also be used in arbitrary finite difference methods. Because we have shown that the GTL equations have the same form in Cartesian as in arbitrary orthogonal coordinate systems, we can explain the problem for Cartesian coordinates. For other coordinates we only have to replace x, y, z with x_1, x_2, x_3 and the associated field components and the material tensor elements with the normalized ones, respectively. Here we will assume that the waveguide sections are surrounded by metallic walls. As an example, a cross-section is shown in Fig. 8.2. The case of magnetic walls is dual and the treatment of absorbing boundary conditions is described in the appendices.

At metallic walls (see Fig. 8.2) the condition $E_z = 0$ must be fulfilled. From eq. (8.16) we obtain the following expression:

$$\mathrm{j}\varepsilon_{zz}E_z = -D_{\overline{y}}^{\bullet}\widetilde{H}_x + D_{\overline{x}}^{\circ}\widetilde{H}_y - \mathrm{j}(M_y^{\bullet}\varepsilon_{zy}E_y + M_x^{\circ}\varepsilon_{zx}E_x) = 0 \tag{8.43}$$

Furthermore, we have to fulfil the following conditions at electric walls:

- $x = \text{const.}$: $E_y = 0$ and

$$B_x = \mu_{xx}\widetilde{H}_x + M_x^{\diamond} M_y^{\circ} \mu_{xy}\widetilde{H}_y + M_x^{\diamond}\mu_{xz}\widetilde{H}_z = 0 \tag{8.44}$$

- $y = \text{const.}$: $E_x = 0$ and

$$B_y = M_y^{\diamond} M_x^{\bullet} \mu_{yx}\widetilde{H}_x + \mu_{yy}\widetilde{H}_y + M_y^{\diamond}\mu_{yz}\widetilde{H}_z = 0 \tag{8.45}$$

First we fulfil the following conditions at metallic walls: $E_x = 0$ for $y =$ const. and $E_y = 0$ for $x =$ const. Hence, in the GTL equation system for [E] the conditions for the transversal electric field components E_x and E_y must be fulfilled. Therefore, $D_{\overline{x}}^{\bullet}$ and $D_{\overline{y}}^{\diamond}$ in $[R_{\mathrm{E}}^{z}]$ have to fulfil Dirichlet boundary conditions. $D_{\overline{x}}^{\diamond}$ and $D_{\overline{y}}^{\diamond}$ have to be constructed for Neumann boundary conditions (also in the matrix $[S_{\mathrm{H}}^{z}]$). $D_{\overline{y}}^{\bullet}$ and $D_{\overline{x}}^{\circ}$ must also have the Neumann form.

Because we have fulfilled $E_x = 0$ or $E_y = 0$ at the corresponding electric walls, the conditions $\partial E_y/\partial\overline{z} = 0$ and $\partial E_x/\partial\overline{z} = 0$ are also fulfilled there. We rewrite these equations for [E] in the following form:

$$\begin{aligned}\frac{\partial E_y}{\partial \overline{z}} &= -\mathrm{j}D_{\overline{y}}^{\square}\varepsilon_{zz}^{-1}(-D_{\overline{y}}^{\bullet}\widetilde{H}_x + D_{\overline{x}}^{\circ}\widetilde{H}_y - \mathrm{j}(M_y^{\bullet}\varepsilon_{zy}E_y + M_x^{\circ}\varepsilon_{zx}E_x)) \\ &\quad + \mathrm{j}M_x^{\diamond}\mu_{xz}\mu_{zz}^{-1}(-D_{\overline{x}}^{\bullet}E_y + D_{\overline{y}}^{\circ}E_x - \mathrm{j}(M_x^{\bullet}\mu_{zx}\widetilde{H}_x + M_y^{\circ}\mu_{zy}\widetilde{H}_y)) \\ &\quad - (\mu_{xx}\widetilde{H}_x + M_x^{\diamond}M_y^{\circ}\mu_{xy}\widetilde{H}_y) = D_{\overline{y}}^{\square}E_z + \mathrm{j}\widetilde{B}_x\end{aligned} \tag{8.46}$$

$$\begin{aligned}\frac{\partial E_x}{\partial \overline{z}} &= -\mathrm{j}D_{\overline{x}}^{\square}\varepsilon_{zz}^{-1}(-D_{\overline{y}}^{\bullet}\widetilde{H}_x + D_{\overline{x}}^{\circ}\widetilde{H}_y - \mathrm{j}(M_y^{\bullet}\varepsilon_{zy}E_y + M_x^{\circ}\varepsilon_{zx}E_x)) \\ &\quad - \mathrm{j}M_y^{\diamond}\mu_{yz}\mu_{zz}^{-1}(-D_{\overline{x}}^{\bullet}E_y + D_{\overline{y}}^{\circ}E_x - \mathrm{j}(M_x^{\bullet}\mu_{zx}\widetilde{H}_x + M_y^{\circ}\mu_{zy}\widetilde{H}_y)) \\ &\quad + (M_y^{\diamond}M_x^{\bullet}\mu_{yx}\widetilde{H}_x + \mu_{yy}\widetilde{H}_y) = D_{\overline{x}}^{\square}E_z - \mathrm{j}\widetilde{B}_y\end{aligned} \tag{8.47}$$

The terms inside the parentheses in the first lines of these equations are proportional to E_z (see eq. (8.43)), which has to be zero at metallic walls. Therefore, the differential (difference) operators on the left sides of the brackets ($D_{\overline{y}}^{\square}$ and $D_{\overline{x}}^{\square}$, respectively) must be constructed for Dirichlet conditions. The terms in parentheses in the second lines are equal to $\mathrm{j}\mu_{zz}\widetilde{H}_z$. Therefore, the second lines result in $-M_x^{\diamond}\mu_{xz}\widetilde{H}_z$ and $M_y^{\diamond}\mu_{yz}\widetilde{H}_z$, respectively. By adding these values to the corresponding last lines we obtain the terms of eqs. (8.44) and (8.45), respectively. Because the other terms are zero at the corresponding boundaries, these terms must also be zero. Hence the conditions for B_x and B_y according to eqs. (8.44) and (8.45), respectively, are also fulfilled.

8.1.6 Interpolation matrices

In this subsection we refer to the GTL equations in Section 8.1.3. We assume electric walls at the boundaries. Some of the matrices M for interpolation are formed for Neumann and others for Dirichlet boundary conditions. We use linear or quadratic interpolation, respectively. Details about interpolation matrices can be found in Section A.4.3. Here we give special $\boldsymbol{M}$ matrices:

$$\boldsymbol{M}_x^{\circ} = \frac{1}{2}\begin{bmatrix} 1 & 1 & & \\ & 1 & 1 & \\ & & \ddots & \ddots \end{bmatrix} \qquad \boldsymbol{M}_x^{\circ} = \frac{1}{8}\begin{bmatrix} 3 & 6 & -1 & & \\ & 3 & 6 & -1 & \\ & & \ddots & \ddots & \ddots \end{bmatrix}$$
$$\boldsymbol{M}_x^{\square} = \frac{1}{2}\begin{bmatrix} 1 & & & \\ 1 & 1 & & \\ & \ddots & \ddots & \end{bmatrix} \qquad \boldsymbol{M}_x^{\square} = \frac{1}{8}\begin{bmatrix} 6 & -1 & & \\ 3 & 6 & -1 & \\ & \ddots & \ddots & \ddots \end{bmatrix} \tag{8.48}$$

In summary, the following $\boldsymbol{M}$ matrices are used in the GTL equations in Section 8.1.3 (for alternative choices this table is not valid):

- Neumann form: $\boldsymbol{M}_x^{\circ}$, $\boldsymbol{M}_y^{\bullet}$, $\boldsymbol{M}_x^{\diamond}$, $\boldsymbol{M}_y^{\diamond}$
- Dirichlet form: $\boldsymbol{M}_x^{\square}$, $\boldsymbol{M}_y^{\square}$
- Special Dirichlet form: $\boldsymbol{M}_x^{\bullet}$, $\boldsymbol{M}_y^{\circ}$

A special Dirichlet form is required when the field component that is discretised for Dirichlet boundary conditions is not zero at the boundary. In that case, we obtain the field quantity (e.g. H_0) at the boundary of the calculation window by extrapolation from the neighbouring discretisation points 1, 2 and 3 (with values H_1, H_2, H_3) by:

$$\widetilde{H}_0 = a_1\widetilde{H}_1 + a_2\widetilde{H}_2 + a_3\widetilde{H}_3 \tag{8.49}$$

We have $a_0 = a_1 + 1$, $a_1 = 3, a_2 = -3, a_3 = 1$ in the case of quadratic and $a_1 = 2, a_2 = -1, a_3 = 0$ in the case of linear extrapolation. We obtain therefore e.g. for $\boldsymbol{M}_x^{\bullet}$ in the case of linear or quadratic interpolation, respectively:

$$\boldsymbol{M}_x^{\bullet} = \frac{1}{2}\begin{bmatrix} a_0 & a_2 & a_3 & \\ 1 & 1 & & \\ & \ddots & \ddots & \end{bmatrix} \qquad \boldsymbol{M}_x^{\bullet} = \frac{1}{8}\begin{bmatrix} 15 & -10 & 3 & \\ 3 & 6 & -1 & \\ & \ddots & \ddots & \ddots \end{bmatrix} \tag{8.50}$$

$\boldsymbol{M}_y^{\circ}$ must be formed in the analogous way.

8.2 DISCRETISATION

8.2.1 Two-dimensional discretisation

We will use the discretisation procedure described in Section 2.4.1. The equations will be given only for the Cartesian case. The discretisation scheme is shown in Fig. 8.2. In general, the metric factors also have to be discretised. This is done analogously to the material parameters in the main diagonal of the tensors.

Supervectors of the discretised field components are defined as:

$$\widehat{\mathbf{E}} = [\mathbf{E}_y^{\mathrm{t}}, \mathbf{E}_x^{\mathrm{t}}]^{\mathrm{t}} \qquad \widehat{\mathbf{H}} = [-\mathbf{H}_x^{\mathrm{t}}, \mathbf{H}_y^{\mathrm{t}}]^{\mathrm{t}} \tag{8.51}$$

The discretisation of the GTL equations (8.18)–(8.22) or (8.24)–(8.30) (with $\overline{x}_3 = \overline{z}$ according to Fig. 8.2) yields the following expressions:

$$\frac{\mathrm{d}}{\mathrm{d}\overline{z}}\widehat{\mathbf{H}} = -\mathrm{j}\boldsymbol{R}_{\mathrm{E}}^{z}\widehat{\mathbf{E}} - \mathbf{S}_{\mathrm{H}}^{z}\widehat{\mathbf{H}} \quad \frac{\mathrm{d}}{\mathrm{d}\overline{z}}\widehat{\mathbf{E}} = -\mathbf{S}_{\mathrm{E}}^{z}\widehat{\mathbf{E}} - \mathrm{j}\boldsymbol{R}_{\mathrm{H}}^{z}\widehat{\mathbf{H}} \tag{8.52}$$

For the matrices $\boldsymbol{R}_{\mathrm{E}}^{z}$ and $\boldsymbol{R}_{\mathrm{H}}^{z}$ we obtain:

$$\begin{aligned}\boldsymbol{R}_{\mathrm{E}}^{z} = &\begin{bmatrix} \boldsymbol{\epsilon}_{yy} + \overline{\boldsymbol{D}}_{\overline{x}}^{\diamond}\boldsymbol{\mu}_{zz}^{-1}\overline{\boldsymbol{D}}_{\overline{x}}^{\bullet} & \boldsymbol{M}_{yx}^{\square\circ}\boldsymbol{\epsilon}_{yx} - \overline{\boldsymbol{D}}_{\overline{x}}^{\diamond}\boldsymbol{\mu}_{zz}^{-1}\overline{\boldsymbol{D}}_{\overline{y}}^{\circ} \\ \boldsymbol{M}_{xy}^{\square\bullet}\boldsymbol{\epsilon}_{xy} - \overline{\boldsymbol{D}}_{\overline{y}}^{\diamond}\boldsymbol{\mu}_{zz}^{-1}\overline{\boldsymbol{D}}_{\overline{x}}^{\bullet} & \boldsymbol{\epsilon}_{xx} + \overline{\boldsymbol{D}}_{\overline{y}}^{\diamond}\boldsymbol{\mu}_{zz}^{-1}\overline{\boldsymbol{D}}_{\overline{y}}^{\circ} \end{bmatrix} \\ &- \begin{bmatrix} \boldsymbol{M}_{y}^{\square}\boldsymbol{\epsilon}_{yz}\boldsymbol{\epsilon}_{zz}^{-1}\boldsymbol{M}_{y}^{\bullet}\boldsymbol{\epsilon}_{zy} & \boldsymbol{M}_{y}^{\square}\boldsymbol{\epsilon}_{yz}\boldsymbol{\epsilon}_{zz}^{-1}\boldsymbol{M}_{x}^{\circ}\boldsymbol{\epsilon}_{zx} \\ \boldsymbol{M}_{x}^{\square}\boldsymbol{\epsilon}_{xz}\boldsymbol{\epsilon}_{zz}^{-1}\boldsymbol{M}_{y}^{\bullet}\boldsymbol{\epsilon}_{zy} & \boldsymbol{M}_{x}^{\square}\boldsymbol{\epsilon}_{xz}\boldsymbol{\epsilon}_{zz}^{-1}\boldsymbol{M}_{x}^{\circ}\boldsymbol{\epsilon}_{zx} \end{bmatrix}\end{aligned} \tag{8.53}$$

$$\begin{aligned}\boldsymbol{R}_{\mathrm{H}}^{z} = &\begin{bmatrix} \overline{\boldsymbol{D}}_{\overline{y}}^{\square}\boldsymbol{\epsilon}_{zz}^{-1}\overline{\boldsymbol{D}}_{\overline{y}}^{\bullet} + \boldsymbol{\mu}_{xx} & \overline{\boldsymbol{D}}_{\overline{y}}^{\square}\boldsymbol{\epsilon}_{zz}^{-1}\overline{\boldsymbol{D}}_{\overline{x}}^{\circ} - \boldsymbol{M}_{xy}^{\diamond\circ}\boldsymbol{\mu}_{xy} \\ \overline{\boldsymbol{D}}_{\overline{x}}^{\square}\boldsymbol{\epsilon}_{zz}^{-1}\overline{\boldsymbol{D}}_{\overline{y}}^{\bullet} - \boldsymbol{M}_{yx}^{\diamond\bullet}\boldsymbol{\mu}_{yx} & \overline{\boldsymbol{D}}_{\overline{x}}^{\square}\boldsymbol{\epsilon}_{zz}^{-1}\overline{\boldsymbol{D}}_{\overline{x}}^{\circ} + \boldsymbol{\mu}_{yy} \end{bmatrix} \\ &- \begin{bmatrix} \boldsymbol{M}_{x}^{\diamond}\boldsymbol{\mu}_{xz}\boldsymbol{\mu}_{zz}^{-1}\boldsymbol{M}_{x}^{\bullet}\boldsymbol{\mu}_{zx} & -\boldsymbol{M}_{x}^{\diamond}\boldsymbol{\mu}_{xz}\boldsymbol{\mu}_{zz}^{-1}\boldsymbol{M}_{y}^{\circ}\boldsymbol{\mu}_{zy} \\ -\boldsymbol{M}_{y}^{\diamond}\boldsymbol{\mu}_{yz}\boldsymbol{\mu}_{zz}^{-1}\boldsymbol{M}_{x}^{\bullet}\boldsymbol{\mu}_{zx} & \boldsymbol{M}_{y}^{\diamond}\boldsymbol{\mu}_{yz}\boldsymbol{\mu}_{zz}^{-1}\boldsymbol{M}_{y}^{\circ}\boldsymbol{\mu}_{zy} \end{bmatrix}\end{aligned} \tag{8.54}$$

with:

$$\begin{aligned} \boldsymbol{M}_{yx}^{\square\circ} &= \boldsymbol{M}_{y}^{\square}\boldsymbol{M}_{x}^{\circ} & \boldsymbol{M}_{xy}^{\square\bullet} &= \boldsymbol{M}_{x}^{\square}\boldsymbol{M}_{y}^{\bullet} \\ \boldsymbol{M}_{xy}^{\diamond\circ} &= \boldsymbol{M}_{x}^{\diamond}\boldsymbol{M}_{y}^{\circ} & \boldsymbol{M}_{yx}^{\diamond\bullet} &= \boldsymbol{M}_{y}^{\diamond}\boldsymbol{M}_{x}^{\bullet} \end{aligned}$$

The matrices $\boldsymbol{S}_{\mathrm{E}}^{z}$ and $\boldsymbol{S}_{\mathrm{H}}^{z}$ are:

$$\boldsymbol{S}_{\mathrm{H}}^{z} = \begin{bmatrix} \boldsymbol{M}_{y}^{\square}\boldsymbol{\epsilon}_{yz}\boldsymbol{\epsilon}_{zz}^{-1} \\ \boldsymbol{M}_{x}^{\square}\boldsymbol{\epsilon}_{xz}\boldsymbol{\epsilon}_{zz}^{-1} \end{bmatrix} \begin{bmatrix} \overline{\boldsymbol{D}}_{\overline{y}}^{\bullet} & \overline{\boldsymbol{D}}_{\overline{x}}^{\circ} \end{bmatrix} + \begin{bmatrix} -\overline{\boldsymbol{D}}_{\overline{x}}^{\diamond}\boldsymbol{\mu}_{zz}^{-1} \\ \overline{\boldsymbol{D}}_{\overline{y}}^{\diamond}\boldsymbol{\mu}_{zz}^{-1} \end{bmatrix} \begin{bmatrix} -\boldsymbol{M}_{x}^{\bullet}\boldsymbol{\mu}_{zx} & \boldsymbol{M}_{y}^{\circ}\boldsymbol{\mu}_{zy} \end{bmatrix} \tag{8.55}$$

$$\boldsymbol{S}_{\mathrm{E}}^{z} = \begin{bmatrix} -\boldsymbol{M}_{x}^{\diamond}\boldsymbol{\mu}_{xz}\boldsymbol{\mu}_{zz}^{-1} \\ \boldsymbol{M}_{y}^{\diamond}\boldsymbol{\mu}_{yz}\boldsymbol{\mu}_{zz}^{-1} \end{bmatrix} \begin{bmatrix} -\overline{\boldsymbol{D}}_{\overline{x}}^{\bullet} & \overline{\boldsymbol{D}}_{\overline{y}}^{\circ} \end{bmatrix} + \begin{bmatrix} \overline{\boldsymbol{D}}_{\overline{y}}^{\square}\boldsymbol{\epsilon}_{zz}^{-1} \\ \overline{\boldsymbol{D}}_{\overline{x}}^{\square}\boldsymbol{\epsilon}_{zz}^{-1} \end{bmatrix} \begin{bmatrix} \boldsymbol{M}_{y}^{\bullet}\boldsymbol{\epsilon}_{zy} & \boldsymbol{M}_{x}^{\circ}\boldsymbol{\epsilon}_{zx} \end{bmatrix} \tag{8.56}$$

8.2.2 One-dimensional discretisation

For eigenmode problems with arbitrary layers (see Fig. 8.3) we assume propagation in y-direction. The z-direction (for analytical solution) is now one of the directions in the cross-section. To obtain the equations needed we replace $\boldsymbol{D}_{\overline{y}}$ by $\overline{D}_{\overline{y}}\boldsymbol{I} = -\mathrm{j}\sqrt{\varepsilon_{\mathrm{re}}}\boldsymbol{I}$. $\boldsymbol{I}$ is the unit matrix and $\varepsilon_{\mathrm{re}}$ is the effective dielectric constant for propagation in y-direction. Furthermore, all matrices $\boldsymbol{M}_y$ are unit matrices and can be withdrawn. The matrices $\overline{\boldsymbol{D}}_{\overline{x}}^{\bullet\square}$ and $\boldsymbol{M}_{x}^{\bullet\square}$ should be replaced by $\overline{\boldsymbol{D}}_{\overline{x}}^{e}$ and $\boldsymbol{M}_{x}^{e}$, respectively. They are used for the quantities on full lines. The matrices $\overline{\boldsymbol{D}}_{\overline{x}}^{\circ\diamond}$ and $\boldsymbol{M}_{x}^{\circ\diamond}$ should be replaced by $\overline{\boldsymbol{D}}_{\overline{x}}^{h}$ and $\boldsymbol{M}_{x}^{h}$, respectively. They are used for the quantities on dashed lines.

The discretisation of eq. (8.18) according to Fig. 8.3 yields the following GTL equations:

$$\frac{\mathrm{d}}{\mathrm{d}\overline{z}}\widetilde{\mathbf{H}} = -\mathrm{j}\boldsymbol{R}_{\mathrm{E}}\mathbf{E} - \boldsymbol{S}_{\mathrm{H}}\widetilde{\mathbf{H}} \quad \frac{\mathrm{d}}{\mathrm{d}\overline{z}}\mathbf{E} = -\boldsymbol{S}_{\mathrm{E}}\mathbf{E} - \mathrm{j}\boldsymbol{R}_{\mathrm{H}}\widetilde{\mathbf{H}} \tag{8.57}$$

with:

$$\boldsymbol{R}_{\mathrm{E}} = \begin{bmatrix} \boldsymbol{\epsilon}_{yy} + \overline{\boldsymbol{D}}^{h}_{\overline{x}}\boldsymbol{\mu}_{zz}^{-1}\overline{\boldsymbol{D}}^{e}_{\overline{x}} & \boldsymbol{M}^{h}_{x}\boldsymbol{\epsilon}_{yx} - \overline{\boldsymbol{D}}^{h}_{\overline{x}}\boldsymbol{\mu}_{zz}^{-1}\overline{D}_{\overline{y}} \\ \boldsymbol{\epsilon}_{xy}\boldsymbol{M}^{e}_{x} - \overline{D}_{\overline{y}}\boldsymbol{\mu}_{zz}^{-1}\overline{\boldsymbol{D}}^{e}_{\overline{x}} & \boldsymbol{\epsilon}_{xx} + \overline{D}_{\overline{y}}\boldsymbol{\mu}_{zz}^{-1}\overline{D}_{\overline{y}} \end{bmatrix} - \begin{bmatrix} \boldsymbol{\epsilon}_{yz}\boldsymbol{\epsilon}_{zz}^{-1}\boldsymbol{\epsilon}_{zy} & \boldsymbol{\epsilon}_{yz}\boldsymbol{\epsilon}_{zz}^{-1}\boldsymbol{M}^{h}_{x}\boldsymbol{\epsilon}_{zx} \\ \boldsymbol{\epsilon}_{xz}\boldsymbol{M}^{e}_{x}\boldsymbol{\epsilon}_{zz}^{-1}\boldsymbol{\epsilon}_{zy} & \boldsymbol{\epsilon}_{xz}\boldsymbol{M}^{e}_{x}\boldsymbol{\epsilon}_{zz}^{-1}\boldsymbol{M}^{h}_{x}\boldsymbol{\epsilon}_{zx} \end{bmatrix} \tag{8.58}$$

$$\boldsymbol{R}_{\mathrm{H}} = \begin{bmatrix} \boldsymbol{\mu}_{xx} + \overline{D}_{\overline{y}}\boldsymbol{\epsilon}_{zz}^{-1}\overline{D}_{\overline{y}} & \overline{D}_{\overline{y}}\boldsymbol{\epsilon}_{zz}^{-1}\overline{\boldsymbol{D}}^{h}_{\overline{x}} - \boldsymbol{M}^{h}_{x}\boldsymbol{\mu}_{xy} \\ \overline{\boldsymbol{D}}^{e}_{\overline{x}}\boldsymbol{\epsilon}_{zz}^{-1}\overline{D}_{\overline{y}} - \boldsymbol{M}^{e}\boldsymbol{\mu}_{yx} & \boldsymbol{\mu}_{yy} + \overline{\boldsymbol{D}}^{e}_{\overline{x}}\boldsymbol{\epsilon}_{zz}^{-1}\overline{\boldsymbol{D}}^{h}_{\overline{x}} \end{bmatrix} - \begin{bmatrix} \boldsymbol{M}^{h}_{x}\boldsymbol{\mu}_{xz}\boldsymbol{\mu}_{zz}^{-1}\boldsymbol{M}^{e}_{x}\boldsymbol{\mu}_{zx} & -\boldsymbol{M}^{h}_{x}\boldsymbol{\mu}_{xz}\boldsymbol{\mu}_{zz}^{-1}\boldsymbol{\mu}_{zy} \\ -\boldsymbol{\mu}_{yz}\boldsymbol{\mu}_{zz}^{-1}\boldsymbol{M}^{e}\boldsymbol{\mu}_{zx} & \boldsymbol{\mu}_{yz}\boldsymbol{\mu}_{zz}^{-1}\boldsymbol{\mu}_{zy} \end{bmatrix} \tag{8.59}$$

$$\boldsymbol{S}_{\mathrm{H}} = \begin{bmatrix} -\overline{\boldsymbol{D}}^{h}_{\overline{x}}\boldsymbol{\mu}_{zz}^{-1} \\ \overline{D}_{\overline{y}}\boldsymbol{\mu}_{zz}^{-1} \end{bmatrix} \begin{bmatrix} -\boldsymbol{M}^{e}_{x}\boldsymbol{\mu}_{zx} & \boldsymbol{\mu}_{zy} \end{bmatrix} + \begin{bmatrix} \boldsymbol{\epsilon}_{yz}\boldsymbol{\epsilon}_{zz}^{-1} \\ \boldsymbol{M}^{e}_{x}\boldsymbol{\epsilon}_{xz}\boldsymbol{\epsilon}_{zz}^{-1} \end{bmatrix} \begin{bmatrix} \overline{D}_{\overline{y}}\boldsymbol{I} & \overline{\boldsymbol{D}}^{h}_{\overline{x}} \end{bmatrix} \tag{8.60}$$

$$\boldsymbol{S}_{\mathrm{E}} = \begin{bmatrix} \overline{D}_{\overline{y}}\boldsymbol{\epsilon}_{zz}^{-1} \\ \overline{\boldsymbol{D}}^{e}_{\overline{x}}\boldsymbol{\epsilon}_{zz}^{-1} \end{bmatrix} \begin{bmatrix} \boldsymbol{\epsilon}_{zy} & \boldsymbol{M}^{h}_{x}\boldsymbol{\epsilon}_{zx} \end{bmatrix} + \begin{bmatrix} -\boldsymbol{M}^{h}_{x}\boldsymbol{\mu}_{xz}\boldsymbol{\mu}_{zz}^{-1} \\ \boldsymbol{\mu}_{yz}\boldsymbol{\mu}_{zz}^{-1} \end{bmatrix} \begin{bmatrix} -\overline{\boldsymbol{D}}^{e}_{\overline{x}} & \overline{D}_{\overline{y}}\boldsymbol{I} \end{bmatrix} \tag{8.61}$$

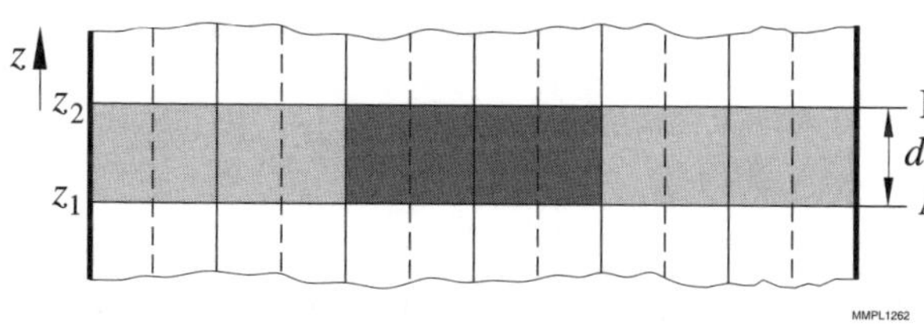

– – – $h \circ \diamond$ H_y H_z E_x ε_{xx} ε_{yx} ε_{zx} μ_{xy} μ_{yy} μ_{zy} μ_{xz} μ_{yz} μ_{zz}

——— $e \bullet \square$ E_y E_z H_x ε_{xy} ε_{yy} ε_{zy} ε_{xz} ε_{yz} ε_{zz} μ_{xx} μ_{yx} μ_{zx}

Fig. 8.3 General inhomogeneous anisotropic layer with discretisation lines (R. Pregla, 'Modeling of Optical Waveguide Structures with General Anisotropy in Arbitrary Orthogonal Coordinate Systems', *IEEE J. of Sel. Topics in Quantum Electronics*, vol. 8, pp. 1217–1224. © 2002 Institute of Electrical and Electronics Engineers (IEEE))

The matrices $\boldsymbol{M}^{e}_{x}$, $\boldsymbol{M}^{h}_{x}$ are interpolation matrices used to determine the discretised fields on one line system from those on the other. For details, see Section 8.1.6.

8.3 SOLUTION OF THE DIFFERENTIAL EQUATIONS

8.3.1 General solution

In this section we discuss the solution of the general GTL equations in the Cartesian case. The other orthogonal coordinate systems can be treated in an analogous way. The discretised equation (8.31) (with $\overline{x}_3 = \overline{z}$) can now be written in the form:

$$\frac{\mathrm{d}}{\mathrm{d}\overline{z}}\widehat{\mathbf{F}} + \widehat{\boldsymbol{Q}}\widehat{\mathbf{F}} = \mathbf{0} \tag{8.62}$$

with:

$$\widehat{\mathbf{F}} = \begin{bmatrix} \widehat{\mathbf{E}} \\ \widehat{\mathbf{H}} \end{bmatrix} \quad \text{and} \quad \widehat{\boldsymbol{Q}} = \begin{bmatrix} \boldsymbol{S}_{\mathrm{E}} & \mathrm{j}\boldsymbol{R}_{\mathrm{H}} \\ \mathrm{j}\boldsymbol{R}_{\mathrm{E}} & \boldsymbol{S}_{\mathrm{H}} \end{bmatrix} \tag{8.63}$$

To solve the differential equation (8.62) we transform to principle axes:

$$\widehat{\mathbf{F}} = \widehat{\boldsymbol{T}}_{\mathrm{EH}}\widehat{\widehat{\mathbf{F}}} \quad \text{where} \quad \widehat{\boldsymbol{T}}_{\mathrm{EH}}^{-1}\widehat{\boldsymbol{Q}}\widehat{\boldsymbol{T}}_{\mathrm{EH}} = \widehat{\boldsymbol{\Gamma}} \tag{8.64}$$

The matrix of eigenvalues is divided into two parts by using the following ordering principle: $\mathrm{Re}(\boldsymbol{\Gamma}_1, \boldsymbol{\Gamma}_2) \geq 0$:

$$\widehat{\boldsymbol{\Gamma}} = \begin{bmatrix} \boldsymbol{\Gamma}_1 & \boldsymbol{0} \\ \boldsymbol{0} & -\boldsymbol{\Gamma}_2 \end{bmatrix} \tag{8.65}$$

The relation $\boldsymbol{\Gamma}_2 = \boldsymbol{\Gamma}_1^*$ is found numerically for symmetrical tensors. The transformed wave equation and its solution is given by:

$$\frac{\mathrm{d}}{\mathrm{d}\overline{z}}\widehat{\widehat{\mathbf{F}}} + \widehat{\boldsymbol{\Gamma}}\widehat{\widehat{\mathbf{F}}} = \mathbf{0} \qquad \widehat{\widehat{\mathbf{F}}} = \begin{bmatrix} e^{-\boldsymbol{\Gamma}_1\overline{z}} & \boldsymbol{0} \\ \boldsymbol{0} & e^{-\boldsymbol{\Gamma}_2(\overline{d}-\overline{z})} \end{bmatrix}\widehat{\widehat{\mathbf{F}}}_{\mathrm{AB}} \qquad \widehat{\widehat{\mathbf{F}}}_{\mathrm{AB}} = \begin{bmatrix} \overline{\mathbf{F}}_{\mathrm{A}} \\ \overline{\mathbf{F}}_{\mathrm{B}} \end{bmatrix} \tag{8.66}$$

$\widehat{\widehat{\mathbf{F}}}_{\mathrm{AB}}$ describes the field at $z = 0$ where the part $\overline{\mathbf{F}}_{\mathrm{B}}$ is normalised with $e^{\boldsymbol{\Gamma}_2\overline{d}}$. The transformation matrix is partitioned into four parts with identical sizes, according to:

$$\widehat{\boldsymbol{T}}_{\mathrm{EH}} = \begin{bmatrix} \boldsymbol{T}_{11} & \boldsymbol{T}_{12} \\ \boldsymbol{T}_{21} & \boldsymbol{T}_{22} \end{bmatrix} \tag{8.67}$$

For the fields at interfaces A and B of a section or a layer we obtain:

$$\begin{bmatrix} \mathbf{E}_{\mathrm{A}} \\ \mathbf{E}_{\mathrm{B}} \end{bmatrix} = \begin{bmatrix} \boldsymbol{T}_{11} & \boldsymbol{T}_{12}e^{-\boldsymbol{\Gamma}_2\overline{d}} \\ \boldsymbol{T}_{11}e^{-\boldsymbol{\Gamma}_1\overline{d}} & \boldsymbol{T}_{12} \end{bmatrix}\widehat{\widehat{\mathbf{F}}}_{\mathrm{AB}} = \widehat{\boldsymbol{M}}_{\mathrm{E}}\widehat{\widehat{\mathbf{F}}}_{\mathrm{AB}} \tag{8.68}$$

$$\begin{bmatrix} \mathbf{H}_{\mathrm{A}} \\ -\mathbf{H}_{\mathrm{B}} \end{bmatrix} = \begin{bmatrix} \boldsymbol{T}_{21} & \boldsymbol{T}_{22}e^{-\boldsymbol{\Gamma}_2\overline{d}} \\ -\boldsymbol{T}_{21}e^{-\boldsymbol{\Gamma}_1\overline{d}} & -\boldsymbol{T}_{22} \end{bmatrix}\widehat{\widehat{\mathbf{F}}}_{\mathrm{AB}} = \widehat{\boldsymbol{M}}_{\mathrm{H}}\widehat{\widehat{\mathbf{F}}}_{\mathrm{AB}} \tag{8.69}$$

8.3.2 Field relation between interfaces A and B

A combination of these two equations results in the following relation of the fields between interfaces A and B:

$$\begin{bmatrix} \mathbf{E}_{\mathrm{A}} \\ \mathbf{E}_{\mathrm{B}} \end{bmatrix} = \begin{bmatrix} \mathbf{z}_{11} & \mathbf{z}_{12} \\ \mathbf{z}_{21} & \mathbf{z}_{22} \end{bmatrix} \begin{bmatrix} \mathbf{H}_{\mathrm{A}} \\ -\mathbf{H}_{\mathrm{B}} \end{bmatrix} = \widehat{\mathbf{z}}_{\mathrm{AB}} \begin{bmatrix} \mathbf{H}_{\mathrm{A}} \\ -\mathbf{H}_{\mathrm{B}} \end{bmatrix} \tag{8.70}$$

$$\begin{bmatrix} \mathbf{H}_{\mathrm{A}} \\ -\mathbf{H}_{\mathrm{B}} \end{bmatrix} = \begin{bmatrix} \mathbf{y}_{11} & \mathbf{y}_{12} \\ \mathbf{y}_{21} & \mathbf{y}_{22} \end{bmatrix} \begin{bmatrix} \mathbf{E}_{\mathrm{A}} \\ \mathbf{E}_{\mathrm{B}} \end{bmatrix} = \widehat{\mathbf{y}}_{\mathrm{AB}} \begin{bmatrix} \mathbf{E}_{\mathrm{A}} \\ \mathbf{E}_{\mathrm{B}} \end{bmatrix} \tag{8.71}$$

where:

$$\widehat{\mathbf{z}}_{\mathrm{AB}} = \widehat{\boldsymbol{M}}_{\mathrm{E}} \widehat{\boldsymbol{M}}_{\mathrm{H}}^{-1} \tag{8.72}$$

$$\widehat{\mathbf{y}}_{\mathrm{AB}} = \widehat{\boldsymbol{M}}_{\mathrm{H}} \widehat{\boldsymbol{M}}_{\mathrm{E}}^{-1} \tag{8.73}$$

The impedance/admittance transformation is as described in Section 2.5. If we consider arbitrary orthogonal coordinate systems, it is not always possible to solve eq. (8.62) analytically. In such cases the FD impedance/admittance transformation algorithms described in Section 2.5.3 can be used.

8.4 ANALYSIS OF WAVEGUIDE JUNCTIONS AND SHARP BENDS WITH GENERAL ANISOTROPIC MATERIAL BY USING ORTHOGONAL PROPAGATING WAVES

8.4.1 Introduction

Complex waveguide devices consist of concatenations of waveguide sections and junctions. Fig. 8.4 shows examples of microstrip devices and photonic crystals. Waveguide devices realised with photonic crystals allow us to design waveguide junctions and sharp bends with low losses. In these circuit elements the incoming and outgoing waves propagate orthogonally to each other. The analysis should take this fact into account in order to model the wave propagation in an efficient and accurate way. This section shows how this can be done [7], [5], [8]. The fields are discretised on lines and the waves propagate in the direction of these lines. Therefore, crossed discretisation lines are used for the analysis. We will develop the equations for arbitrary anisotropic materials.

8.4.2 Theory

8.4.2.1 Port relations

The analysis of waveguide junctions is demonstrated for the example given in Fig. 8.5a. The figure shows a microstrip waveguide junction. The junction region is bounded by the four ports A to D. In general, all of these four ports have connecting waveguides, in this case the waveguides $\mathrm{W_A}$ to $\mathrm{W_D}$. The connecting waveguides may be different to the waveguides inside the junction. If there are sharp bends we only have two ports with connecting waveguides, i.e. ports A and D with $\mathrm{W_A}$ and $\mathrm{W_D}$, respectively. Only the junction region is

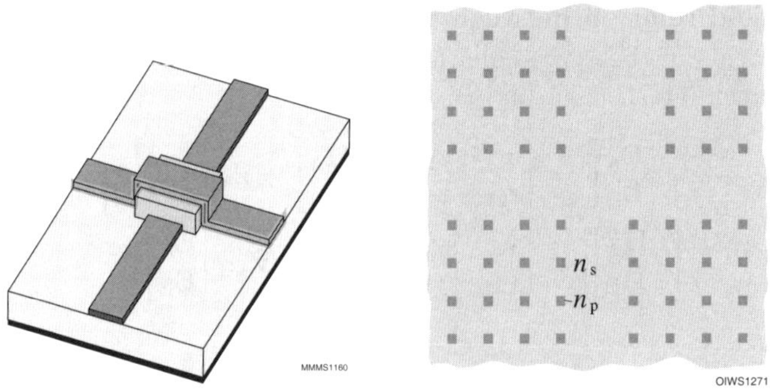

(a) (b)

Fig. 8.4 Waveguide junctions: (a) microstrip cross over (Reproduced by permission of Union Radio-Scientifique Internationale–International Union of Radio Science (URSI)), (b) photonic crystal waveguides (R. Pregla, 'Analysis of waveguide junctions and sharp bends with general anisotropic material by using orthogonal propagating waves', in *ICTON Conf.*, vol. 5, pp. 116–121. © 2003 Institute of Electrical and Electronics Engineers (IEEE))

sketched in Fig. 8.5b. The number of concatenated waveguide sections between ports A and B is K. The ports (subports) of section k are labelled A^k and B^k (see Fig. 8.5c). At the side walls of the connecting waveguides we assume absorbing boundary conditions (ABC). These are positioned at the places where we would otherwise introduce Dirichlet conditions. According to the uniqueness theorem, the field outside the inner region is determined if we know the tangential field on the outer boundary. The fields at the outer and inner sides of the boundary are clearly related (see Section 2.5.2.2). We describe the relation of the tangential fields at the inner side of the four generalised ports A, B, C and D using open-circuit matrix parameters in the form [5]:

$$\begin{bmatrix} \widehat{\mathbf{E}}_{\mathrm{A}} \\ \widehat{\mathbf{E}}_{\mathrm{B}} \\ \widehat{\mathbf{E}}_{\mathrm{C}} \\ \widehat{\mathbf{E}}_{\mathrm{D}} \end{bmatrix} = \begin{bmatrix} \widehat{\mathbf{z}}^{\mathrm{AB}}_{\mathrm{AB11}} & \widehat{\mathbf{z}}^{\mathrm{AB}}_{\mathrm{AB12}} & \widehat{\mathbf{z}}^{\mathrm{CD}}_{\mathrm{AB11}} & \widehat{\mathbf{z}}^{\mathrm{CD}}_{\mathrm{AB12}} \\ \widehat{\mathbf{z}}^{\mathrm{AB}}_{\mathrm{AB21}} & \widehat{\mathbf{z}}^{\mathrm{AB}}_{\mathrm{AB22}} & \widehat{\mathbf{z}}^{\mathrm{CD}}_{\mathrm{AB21}} & \widehat{\mathbf{z}}^{\mathrm{CD}}_{\mathrm{AB22}} \\ \widehat{\mathbf{z}}^{\mathrm{AB}}_{\mathrm{CD11}} & \widehat{\mathbf{z}}^{\mathrm{AB}}_{\mathrm{CD12}} & \widehat{\mathbf{z}}^{\mathrm{CD}}_{\mathrm{CD11}} & \widehat{\mathbf{z}}^{\mathrm{CD}}_{\mathrm{CD12}} \\ \widehat{\mathbf{z}}^{\mathrm{AB}}_{\mathrm{CD21}} & \widehat{\mathbf{z}}^{\mathrm{AB}}_{\mathrm{CD22}} & \widehat{\mathbf{z}}^{\mathrm{CD}}_{\mathrm{CD21}} & \widehat{\mathbf{z}}^{\mathrm{CD}}_{\mathrm{CD22}} \end{bmatrix} \begin{bmatrix} \widehat{\mathbf{H}}_{\mathrm{A}} \\ -\widehat{\mathbf{H}}_{\mathrm{B}} \\ \widehat{\mathbf{H}}_{\mathrm{C}} \\ -\widehat{\mathbf{H}}_{\mathrm{D}} \end{bmatrix} \tag{8.74}$$

We have collected the discretised field components in column vectors, as usual for the MoL [6], [5]. We should keep in mind that each of the field column vectors contains the two tangential field components at the ports, i.e. $\widehat{\mathbf{E}}_{\mathrm{A}} = [\mathbf{E}^{\mathrm{t}}_y, \mathbf{E}^{\mathrm{t}}_x]^{\mathrm{t}}, \widehat{\mathbf{H}}_{\mathrm{A}} = [-\mathbf{H}^{\mathrm{t}}_x, \mathbf{H}^{\mathrm{t}}_y]^{\mathrm{t}}$. By combining the fields on opposite ports in supervectors, we can write eq. (8.74) in a more compact form:

$$\begin{bmatrix} \widehat{\mathbf{E}}_{\mathrm{AB}} \\ \widehat{\mathbf{E}}_{\mathrm{CD}} \end{bmatrix} = \begin{bmatrix} \widehat{\mathbf{z}}^{\mathrm{AB}}_{\mathrm{AB}} & \widehat{\mathbf{z}}^{\mathrm{CD}}_{\mathrm{AB}} \\ \widehat{\mathbf{z}}^{\mathrm{AB}}_{\mathrm{CD}} & \widehat{\mathbf{z}}^{\mathrm{CD}}_{\mathrm{CD}} \end{bmatrix} \begin{bmatrix} \widehat{\mathbf{H}}_{\mathrm{AB}} \\ \widehat{\mathbf{H}}_{\mathrm{CD}} \end{bmatrix} \qquad \begin{aligned} \widehat{\mathbf{E}}_{\mathrm{UV}} &= [\widehat{\mathbf{E}}^{\mathrm{t}}_{\mathrm{U}}, \widehat{\mathbf{E}}^{\mathrm{t}}_{\mathrm{V}}]^{\mathrm{t}} \\ \widehat{\mathbf{H}}_{\mathrm{UV}} &= [\widehat{\mathbf{H}}^{\mathrm{t}}_{\mathrm{U}}, -\widehat{\mathbf{H}}^{\mathrm{t}}_{\mathrm{V}}]^{\mathrm{t}} \end{aligned} \tag{8.75}$$

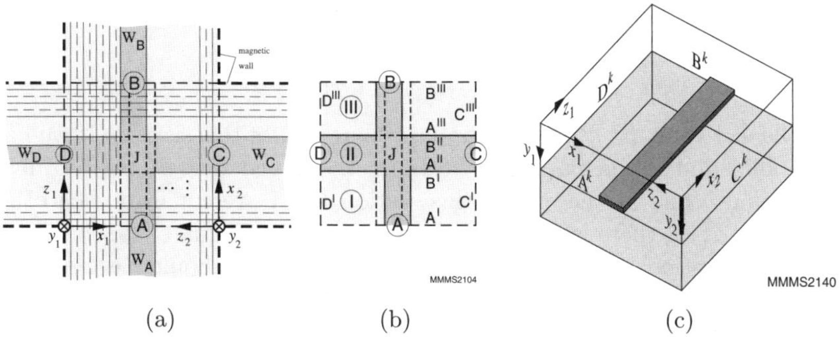

Fig. 8.5 Junction of microstrip waveguides (a) (R. Pregla, 'Efficient and Accurate Modeling of Planar Anisotropic Microwave Structures by the Method of Lines', *IEEE Trans. Microwave Theory Tech.*, vol. 50, pp. 1469–1479. © 2002 Institute of Electrical and Electronics Engineers (IEEE)), inner junction (b) (Reproduced by permission of Copernicus Gesellschaft mbH), subsection k ($k = \mathrm{I}$ or III in (b)) (c)

U is identical to A or C and V to B or D, respectively. The four submatrices in this equation are obtained by open-circuiting the ports and placing magnetic walls there. If ports C and D are open-circuited, the matrices $\widehat{\mathbf{z}}_{\mathrm{AB}}^{\mathrm{AB}}$ and $\widehat{\mathbf{z}}_{\mathrm{CD}}^{\mathrm{AB}}$ are obtained by using discretisation lines from port A to port B in the vertical direction. By open-circuiting ports A and B, the matrices $\widehat{\mathbf{z}}_{\mathrm{CD}}^{\mathrm{CD}}$ and $\widehat{\mathbf{z}}_{\mathrm{AB}}^{\mathrm{CD}}$ can be calculated by using discretisation lines from port C to port D in the horizontal direction. Both procedures are independent of each other. Next we determine the two main diagonal submatrices in eq. (8.75).

8.4.3 Main diagonal submatrices

Let us start with $\widehat{\mathbf{z}}_{\mathrm{AB}}^{\mathrm{AB}}$. We assume that ports C and D are open-circuited, which means $\widehat{\mathbf{H}}_{\mathrm{CD}} = \mathbf{0}$. The relation of the fields between ports A and B is described by:

$$\widehat{\mathbf{E}}_{\mathrm{AB}} = \widehat{\mathbf{z}}_{\mathrm{AB}}^{\mathrm{AB}} \widehat{\mathbf{H}}_{\mathrm{AB}} \quad \text{or} \quad \begin{bmatrix} \widehat{\mathbf{E}}_{\mathrm{A}} \\ \widehat{\mathbf{E}}_{\mathrm{B}} \end{bmatrix} = \begin{bmatrix} \widehat{\mathbf{z}}_{\mathrm{AB11}}^{\mathrm{AB}} & \widehat{\mathbf{z}}_{\mathrm{AB12}}^{\mathrm{AB}} \\ \widehat{\mathbf{z}}_{\mathrm{AB21}}^{\mathrm{AB}} & \widehat{\mathbf{z}}_{\mathrm{AB22}}^{\mathrm{AB}} \end{bmatrix} \begin{bmatrix} \widehat{\mathbf{H}}_{\mathrm{A}} \\ -\widehat{\mathbf{H}}_{\mathrm{B}} \end{bmatrix} \tag{8.76}$$

To obtain the four submatrices we again use the technique of open-circuiting the ports. With $\widehat{\mathbf{H}}_{\mathrm{B}} = 0$ we obtain the submatrices $\widehat{\mathbf{z}}_{\mathrm{AB11}}^{\mathrm{AB}}$ and $\widehat{\mathbf{z}}_{\mathrm{AB21}}^{\mathrm{AB}}$. $\widehat{\mathbf{z}}_{\mathrm{AB11}}^{\mathrm{AB}}$ is the input impedance matrix in plane (or port) A for the open port B. In this case, ports A and B are connected via a concatenation of K different waveguide sections. For each of these sections the tangential fields at the ends (or the ports) are described by eqs. (8.70)–(8.73). We assume a section k with the open-circuit matrices $\mathbf{z}^k$ and with impedances at the ports A^k and B^k according to:

$$\widehat{\mathbf{E}}_{\mathrm{A,B}}^{k} = \widehat{\mathbf{Z}}_{\mathrm{A,B}}^{k} \widehat{\mathbf{H}}_{\mathrm{A,B}}^{k} \tag{8.77}$$

In that case, $\widehat{\mathbf{Z}}_{\mathrm{B}}^{k}$ has to be transformed according to:

$$\widehat{\mathbf{Z}}_{\mathrm{A}}^{k} = \widehat{\mathbf{z}}_{11}^{k} - \widehat{\mathbf{z}}_{12}^{k}(\widehat{\mathbf{z}}_{22}^{k} + \widehat{\mathbf{Z}}_{\mathrm{B}}^{k})^{-1}\widehat{\mathbf{z}}_{21}^{k} \tag{8.78}$$

from the end (port B^k) to the beginning (port A^k) of this section. If we have no metallisation in the waveguide sections, the impedances at the beginning of a section k and at the end of this section $k-1$ are equal to each other, or $\widehat{\mathbf{Z}}_{\mathrm{B}}^{k-1} = \widehat{\mathbf{Z}}_{\mathrm{A}}^{k}$. If there is metal, we have to use the adequate transformation formula in Section 5.5.2.4 (see also eq. (63) of [6]). It should be mentioned that this impedance matrix is a full matrix. Starting at the end or in section K with B^K=B and $\mathbf{H}_{\mathrm{B}}^{K} = \mathbf{H}_{\mathrm{B}} = \mathbf{0}$ (or $\widehat{\mathbf{Z}}_{\mathrm{B}}^{K} = \infty$), we obtain the impedance $\widehat{\mathbf{Z}}_{\mathrm{A}}^{K} = \widehat{\mathbf{z}}_{11}^{K}$ at A^K. With $\widehat{\mathbf{Z}}_{\mathrm{B}}^{K-1} = \widehat{\mathbf{Z}}_{\mathrm{A}}^{K}$ and $k = K-1$ we can calculate the input impedance of the next section. By repeating this process until we reach the first section, we finally obtain the input impedance $\widehat{\mathbf{Z}}_{\mathrm{A}}^{1}$, which is equal to $\widehat{\mathbf{z}}_{\mathrm{AB}11}^{\mathrm{AB}}$.

To calculate the matrix $\widehat{\mathbf{z}}_{\mathrm{AB}21}^{\mathrm{AB}}$ we must go in opposite direction. The transformation in a section k is performed by:

$$\widehat{\mathbf{H}}_{\mathrm{B}}^{k} = (\widehat{\mathbf{z}}_{12}^{k})^{-1}(\widehat{\mathbf{z}}_{11}^{k} - \widehat{\mathbf{Z}}_{\mathrm{A}}^{k})\widehat{\mathbf{H}}_{\mathrm{A}}^{k} \tag{8.79}$$

$$\widehat{\mathbf{E}}_{\mathrm{B}}^{k} = (\widehat{\mathbf{z}}_{21}^{k} - \widehat{\mathbf{z}}_{22}^{k}(\widehat{\mathbf{z}}_{12}^{k})^{-1}(\widehat{\mathbf{z}}_{11}^{k} - \widehat{\mathbf{Z}}_{\mathrm{A}}^{k}))\widehat{\mathbf{H}}_{\mathrm{A}} = \widehat{\mathbf{Z}}_{\mathrm{B}}^{k}\widehat{\mathbf{H}}_{\mathrm{B}}^{k} \tag{8.80}$$

The next step is the transformation of the field at the concatenations. We have:

$$\widehat{\mathbf{H}}_{\mathrm{B}}^{k} = \widehat{\mathbf{H}}_{\mathrm{A}}^{k+1} \qquad \widehat{\mathbf{E}}_{\mathrm{B}}^{k} = \widehat{\mathbf{E}}_{\mathrm{A}}^{k+1} \tag{8.81}$$

if we assume no metallisation. Otherwise we again refer to Section 2.5.2.2. We start with $k = 1$ and $\widehat{\mathbf{Z}}_{\mathrm{A}}^{1} = \widehat{\mathbf{z}}_{\mathrm{AB}11}^{\mathrm{AB}}$. This procedure must be repeated for all other sections. From the equation for $\widehat{\mathbf{E}}_{\mathrm{B}}^{\mathrm{k}}$ with $k = K$ (in which all the former calculated quantities may have been introduced and which is then proportional to $\widehat{\mathbf{H}}_{\mathrm{A}}$), the transmittance impedance matrix $\widehat{\mathbf{z}}_{\mathrm{AB}21}^{\mathrm{AB}}$ is obtained. An analogous procedure is performed for the two other submatrices in eq. (8.76). Now port A must be open-circuited and the input impedance at port B looking towards A must be calculated. This input impedance is identical to the submatrix $\widehat{\mathbf{z}}_{\mathrm{AB}22}^{\mathrm{AB}}$. For the calculation of $\widehat{\mathbf{z}}_{\mathrm{AB}12}^{\mathrm{AB}}$ we must transform the magnetic field column vector $\widehat{\mathbf{H}}_{\mathrm{B}}$ from port B through the different waveguide sections to port A, taking into account the impedances determined before. $\widehat{\mathbf{z}}_{\mathrm{AB}12}^{\mathrm{AB}}$ is the matrix that connects $\widehat{\mathbf{E}}_{\mathrm{A}}$ with $-\widehat{\mathbf{H}}_{\mathrm{B}}$. Now $\widehat{\mathbf{z}}_{\mathrm{AB}}^{\mathrm{AB}}$ is completely determined.

In an analogous way to $\widehat{\mathbf{z}}_{\mathrm{AB}}^{\mathrm{AB}}$, the submatrix in $\widehat{\mathbf{z}}_{\mathrm{CD}}^{\mathrm{CD}}$ in eq. (8.75) can be determined. However, in this case ports A and B must be open-circuited (i.e. we introduce magnetic walls with $\widehat{\mathbf{H}}_{\mathrm{AB}} = \mathbf{0}$). Now we have L different waveguide sections concatenated between ports C and D (not extra marked) and proceed as before.

In the next subsection we will show how the off-diagonal submatrix $\widehat{\mathbf{z}}_{\mathrm{CD}}^{\mathrm{AB}}$ can be calculated.

8.4.4 Off-diagonal submatrices – coupling to other ports

The off-diagonal submatrix $\widehat{\mathbf{z}}_{\mathrm{CD}}^{\mathrm{AB}}$ in eq. (8.75) is defined as:

$$\widehat{\mathbf{E}}_{\mathrm{CD}} = \widehat{\mathbf{z}}_{\mathrm{CD}}^{\mathrm{AB}} \widehat{\mathbf{H}}_{\mathrm{AB}} \quad \text{or} \quad \begin{bmatrix} \widehat{\mathbf{E}}_{\mathrm{C}} \\ \widehat{\mathbf{E}}_{\mathrm{D}} \end{bmatrix} = \begin{bmatrix} \widehat{\mathbf{z}}_{\mathrm{CD11}}^{\mathrm{AB}} & \widehat{\mathbf{z}}_{\mathrm{CD12}}^{\mathrm{AB}} \\ \widehat{\mathbf{z}}_{\mathrm{CD21}}^{\mathrm{AB}} & \widehat{\mathbf{z}}_{\mathrm{CD22}}^{\mathrm{AB}} \end{bmatrix} \begin{bmatrix} \widehat{\mathbf{H}}_{\mathrm{A}} \\ -\widehat{\mathbf{H}}_{\mathrm{B}} \end{bmatrix} \tag{8.82}$$

for the condition $\widehat{\mathbf{H}}_{\mathrm{CD}} = \mathbf{0}$ (magnetic walls have to be introduced there). We use discretisation lines from port A to port B and must calculate the tangential electric field components on the magnetic walls at ports C and D. The two left-side submatrices are obtained by setting $\widehat{\mathbf{H}}_{\mathrm{B}} = \mathbf{0}$ and the two right-side ones by setting $\widehat{\mathbf{H}}_{\mathrm{A}} = \mathbf{0}$. We will now demonstrate the procedure for the two left-side submatrices. The procedure for the right-side ones is analogous. The tangential electric field components at ports C and D (at the magnetic walls there) are given by (see Fig. 8.6):

$$\widehat{\mathbf{E}}_{\mathrm{C,D}} = \begin{bmatrix} \widehat{\mathbf{E}}_{y2}^{\mathrm{C,D}} \\ \widehat{\mathbf{E}}_{x2}^{\mathrm{C,D}} \end{bmatrix} = \begin{bmatrix} \widehat{\mathbf{E}}_{y1}^{\mathrm{C,D}} \\ \widehat{\mathbf{E}}_{z1}^{\mathrm{C,D}} \end{bmatrix} = \begin{bmatrix} \widehat{\mathbf{E}}_{y}^{\mathrm{C,D}} \\ \widehat{\mathbf{E}}_{z}^{\mathrm{C,D}} \end{bmatrix} \tag{8.83}$$

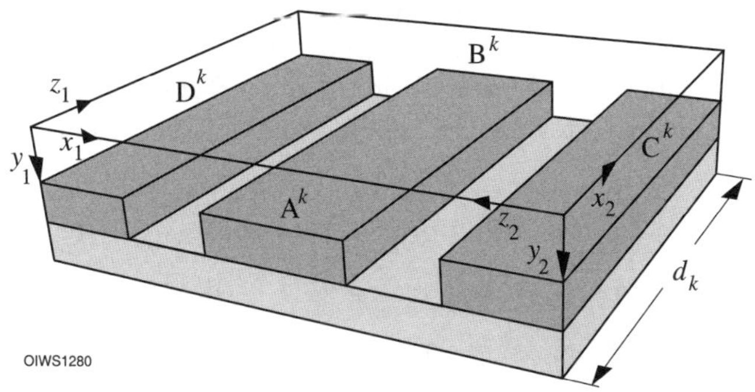

Fig. 8.6 Subsection of the inner junction, to define the coordinates for the coupling from ports A^{k} and B^{k} to ports C^{k} and D^{k} (R. Pregla, 'Analysis of waveguide junctions and sharp bends with general anisotropic material by using orthogonal propagating waves', in *ICTON Conf.*, vol. 5, pp. 116–121. © 2003 Institute of Electrical and Electronics Engineers (IEEE))

Only these electric field components occur for the coupling from ports A and B to ports C and D. The tangential magnetic field components at the boundaries C and D are zero (because of the magnetic walls). Therefore, we have to determine $\widehat{\mathbf{E}}_{y}^{\mathrm{C,D}}$ and $\widehat{\mathbf{E}}_{z}^{\mathrm{C,D}}$ caused by $\widehat{\mathbf{H}}_{\mathrm{A}}$. The calculations must be done in parts. This is because ports C and D must each be partitioned in K subsections with subports C^k and D^k depending on the sidewall areas of the waveguide sections in z_1-direction. Let us now calculate the tangential electric fields at subports C^k and D^k. Once we have obtained the tangential electric fields at all subports we have only to arrange each of these field components in a column vector in the order of the subports, numbers.

The field components in a plane at an arbitrary position $z_k \equiv z_{k1}$ between the two subports A^k and B^k of a section $k\,(k = 1, 2, \ldots, K)$ in the example of Fig. 8.6 can be calculated from the fields at ports A^k and B^k. For the tangential electric and magnetic fields at an arbitrary position z_k, obtained from the general solution (see Section 8.3), we can write:

$$\begin{aligned}\widehat{\mathbf{E}}^k(z_k) &= [\boldsymbol{T}_{11}^k \quad \boldsymbol{T}_{12}^k]\begin{bmatrix} e^{-\boldsymbol{\Gamma}_1^k \bar{z}_k} & \boldsymbol{0} \\ \boldsymbol{0} & e^{-\boldsymbol{\Gamma}_2^k(\bar{d}_k - \bar{z}_k)}\end{bmatrix}(\widehat{\boldsymbol{M}}_{\mathrm{H}}^k)^{-1}\widehat{\mathbf{H}}_{\mathrm{AB}}^k \\ &= [\boldsymbol{T}_{11}^k e^{-\boldsymbol{\Gamma}_1^k \bar{z}_k} \quad \boldsymbol{T}_{12}^k e^{-\boldsymbol{\Gamma}_2^k(\bar{d}_k - \bar{z}_k)}](\widehat{\boldsymbol{M}}_{\mathrm{H}}^k)^{-1}\widehat{\mathbf{H}}_{\mathrm{AB}}^k \end{aligned} \tag{8.84}$$

$$\begin{aligned}\widehat{\mathbf{H}}^k(z_k) &= [\boldsymbol{T}_{21}^k \quad \boldsymbol{T}_{22}^k]\begin{bmatrix} e^{-\boldsymbol{\Gamma}_1^k \bar{z}_k} & \boldsymbol{0} \\ \boldsymbol{0} & e^{-\boldsymbol{\Gamma}_2^k(\bar{d}_k - \bar{z}_k)}\end{bmatrix}(\widehat{\boldsymbol{M}}_{\mathrm{H}}^k)^{-1}\widehat{\mathbf{H}}_{\mathrm{AB}}^k \\ &= [\boldsymbol{T}_{21}^k e^{-\boldsymbol{\Gamma}_1^k \bar{z}_k} \quad \boldsymbol{T}_{22}^k e^{-\boldsymbol{\Gamma}_2^k(\bar{d}_k - \bar{z}_k)}](\widehat{\boldsymbol{M}}_{\mathrm{H}}^k)^{-1}\widehat{\mathbf{H}}_{\mathrm{AB}}^k \end{aligned} \tag{8.85}$$

where we define $z^k = 0$ in the plane of port A^k. Furthermore, we introduce $\bar{d}_k = k_0 d_k$, where d_k is the distance between ports A^k and B^k. Even with $\mathbf{H}_{\mathrm{B}} = \mathbf{0}$ we have magnetic fields at both subports A^k and B^k in subsection k. However, both fields $\widehat{\mathbf{H}}_{\mathrm{A}^k}$ and $\widehat{\mathbf{H}}_{\mathrm{B}^k}$ have to be calculated from the fields at port A (!) (see above). Only if $\mathbf{B}^k$ is identical to $\mathbf{B}$ do we also have $\mathbf{H}_{\mathrm{B}^k} = \mathbf{0}$. The nth column vectors of the components of $\widehat{\mathbf{E}}(\bar{z}_k)$ and $\widehat{\mathbf{H}}(\bar{z}_k)$ can be determined as:

$$\begin{aligned} \mathbf{E}_{yn}^k(z_k) &= [\boldsymbol{T}_{11_{yn}}^k e^{-\boldsymbol{\Gamma}_1^k \bar{z}_k} \quad \boldsymbol{T}_{12_{yn}}^k e^{-\boldsymbol{\Gamma}_2^k(\bar{d}_k - \bar{z}_k)}](\widehat{\boldsymbol{M}}_{\mathrm{H}}^k)^{-1}\widehat{\mathbf{H}}_{\mathrm{AB}}^k \\ \mathbf{E}_{xn}^k(z_k) &= [\boldsymbol{T}_{11_{xn}}^k e^{-\boldsymbol{\Gamma}_1^k \bar{z}_k} \quad \boldsymbol{T}_{12_{xn}}^k e^{-\boldsymbol{\Gamma}_2^k(\bar{d}_k - \bar{z}_k)}](\widehat{\boldsymbol{M}}_{\mathrm{H}}^k)^{-1}\widehat{\mathbf{H}}_{\mathrm{AB}}^k \\ -\mathbf{H}_{xn}^k(z_k) &= [\boldsymbol{T}_{21_{xn}}^k e^{-\boldsymbol{\Gamma}_1^k \bar{z}_k} \quad \boldsymbol{T}_{22_{xn}}^k e^{-\boldsymbol{\Gamma}_2^k(\bar{d}_k - \bar{z}_k)}](\widehat{\boldsymbol{M}}_{\mathrm{H}}^k)^{-1}\widehat{\mathbf{H}}_{\mathrm{AB}}^k \\ \mathbf{H}_{yn}^k(z_k) &= [\boldsymbol{T}_{21_{yn}}^k e^{-\boldsymbol{\Gamma}_1^k \bar{z}_k} \quad \boldsymbol{T}_{22_{yn}}^k e^{-\boldsymbol{\Gamma}_2^k(\bar{d}_k - \bar{z}_k)}](\widehat{\boldsymbol{M}}_{\mathrm{H}}^k)^{-1}\widehat{\mathbf{H}}_{\mathrm{AB}}^k \end{aligned} \tag{8.86}$$

where e.g. $\boldsymbol{T}_{11_{yn}}^k$ $(\boldsymbol{T}_{12_{yn}}^k)$ in the first equation is the nth block of the matrix $\boldsymbol{T}_{11_y}^k$ $(\boldsymbol{T}_{12_y}^k)$. $\boldsymbol{T}_{11_y}^k$ $(\boldsymbol{T}_{12_y}^k)$ is the upper part of $\boldsymbol{T}_{11}^k$ $(\boldsymbol{T}_{12}^k)$. The matrices in the other equations are obtained in an analogous way. The field $\widehat{\mathbf{E}}_{y\mathrm{R}}^k$ at ports $\mathrm{R}^k \equiv \mathrm{C}^k$ and $\mathrm{R}^k \equiv \mathrm{D}^k$ can be determined as follows: we define the matrices $\boldsymbol{T}_{11,12_y}^{\mathrm{Ck}}$ and $\boldsymbol{T}_{11,12_y}^{\mathrm{Dk}}$ by:

$$\boldsymbol{T}_{11,12_y}^{\mathrm{C}^k} = \tfrac{1}{8}(9\boldsymbol{T}_{11,12_{y\mathrm{N}}}^k - \boldsymbol{T}_{11,12_{y\mathrm{N}_-}}^k) \tag{8.87}$$

$$\boldsymbol{T}_{11,12_y}^{\mathrm{D}^k} = \tfrac{1}{8}(9\boldsymbol{T}_{11,12_{y1}}^k - \boldsymbol{T}_{11,12_{y2}}^k) \tag{8.88}$$

Here, $\boldsymbol{T}_{11,12_{y\mathrm{N}}}^k$ and $\boldsymbol{T}_{11,12_{y\mathrm{N}_-}}^k$, $(\mathrm{N}_- = N - 1)$ are the last and next to last blocks of the matrix $\boldsymbol{T}_{11,12_y}^k$. The relevant field component $\mathbf{E}_{y\mathrm{R}}^k$ at ports

$\mathrm{R}^k \equiv \mathrm{C}^k$ or $\mathrm{R}^k \equiv \mathrm{D}^k$ is now expressed as:

$$\mathbf{E}^k_{y\mathrm{R}}(z_k) = [\boldsymbol{T}^{\mathrm{R}^k}_{11_y} e^{-\boldsymbol{\Gamma}^k_1 \overline{z}_k} \quad \boldsymbol{T}^{\mathrm{R}^k}_{12_y} e^{-\boldsymbol{\Gamma}^k_2 (\overline{d}_k - \overline{z}_k)}](\widehat{\boldsymbol{M}}^k_{\mathrm{H}})^{-1}\widehat{\mathbf{H}}^k_{\mathrm{AB}} = \widehat{\mathbf{Z}}^{\mathrm{AB}^k}_{y\mathrm{R}^k}(\overline{z}_k)\widehat{\mathbf{H}}^k_{\mathrm{AB}} \tag{8.89}$$

By splitting the matrix $\widehat{\mathbf{Z}}^{\mathrm{AB}k}_{y\mathrm{R}^k}(\overline{z}_k)$ into a left and right part $\mathbf{Z}^{\mathrm{A}k}_{y\mathrm{R}^k}(\overline{z}_k)$ and $\mathbf{Z}^{\mathrm{B}k}_{y\mathrm{R}^k}(\overline{z}_k)$ (according to the subvectors $\widehat{\mathbf{H}}^k_{\mathrm{A}}$ and $-\widehat{\mathbf{H}}^k_{\mathrm{B}}$ of $\widehat{\mathbf{H}}^k_{\mathrm{AB}}$), we can write instead of eq. (8.89):

$$\mathbf{E}^k_{y\mathrm{R}}(z_k) = \mathbf{Z}^{\mathrm{A}^k}_{y\mathrm{R}^k}(\overline{z}_k)\widehat{\mathbf{H}}^k_{\mathrm{A}} + \mathbf{Z}^{\mathrm{B}^k}_{y\mathrm{R}^k}(\overline{z}_k)(-\widehat{\mathbf{H}}^k_{\mathrm{B}}) \tag{8.90}$$

Eq. (8.90) must now be discretised at points $\overline{z}^k_i = (\overline{z}_k)_i$ of ports C^k and D^k. The field parts $\mathbf{E}^{\mathrm{A}^k}_{y\mathrm{R}^k}$ and $\mathbf{E}^{\mathrm{B}^k}_{y\mathrm{R}^k}$ at the subports related to $\widehat{\mathbf{H}}^k_{\mathrm{A}}$ and $-\widehat{\mathbf{H}}^k_{\mathrm{B}}$, respectively, are collected in column vectors and can be expressed as:

$$\begin{aligned}
\mathbf{E}^{\mathrm{A}^k}_{y\mathrm{R}^k} &= \begin{bmatrix} \mathbf{Z}^{\mathrm{A}k}_{y\mathrm{R}^k}(\overline{z}^k_1) \\ \vdots \\ \mathbf{Z}^{\mathrm{A}^k}_{y\mathrm{R}^k}(\overline{z}^k_{\mathrm{N}^k_z}) \end{bmatrix} \widehat{\mathbf{H}}^k_{\mathrm{A}} = \widehat{\mathbf{Z}}^{\mathrm{A}^k}_{y\mathrm{R}}\widehat{\mathbf{H}}^k_{\mathrm{A}} \\
\mathbf{E}^{\mathrm{R}^k}_{y\mathrm{R}^k} &= \begin{bmatrix} \mathbf{Z}^{\mathrm{B}^k}_{y\mathrm{R}^k}(\overline{z}^k_1) \\ \vdots \\ \mathbf{Z}^{\mathrm{B}^k}_{y\mathrm{R}^k}(\overline{z}^k_{\mathrm{N}^k_z}) \end{bmatrix} (-\widehat{\mathbf{H}}^k_{\mathrm{B}}) = \widehat{\mathbf{Z}}^{\mathrm{B}^k}_{y\mathrm{R}}(-\widehat{\mathbf{H}}^k_{\mathrm{B}})
\end{aligned} \tag{8.91}$$

$z^k_{\mathrm{N}^k_z}$ is the last column at subport C^k or D^k in z_1-direction. The field vectors $\widehat{\mathbf{H}}_{\mathrm{A}^k}$ and $\widehat{\mathbf{H}}_{\mathrm{B}^k}$ are both determined from $\widehat{\mathbf{H}}_{\mathrm{A}}$ by the procedure described. These relations may be expressed by above:

$$\widehat{\mathbf{H}}^k_{\mathrm{A}} = \widehat{\mathbf{V}}^k_{\mathrm{AA}}\widehat{\mathbf{H}}_{\mathrm{A}} \qquad -\widehat{\mathbf{H}}^k_{\mathrm{B}} = \widehat{\mathbf{V}}^k_{\mathrm{BA}}\widehat{\mathbf{H}}_{\mathrm{A}} \tag{8.92}$$

The fields at all the subports C^k and D^k (subports k of ports C and D) are now collected in the common column vectors to obtain the total column vectors for $\widehat{\mathbf{E}}^{\mathrm{R}}_y$ at ports C and D, respectively.

$$\widehat{\mathbf{E}}^{\mathrm{R}}_y = \begin{bmatrix} \widehat{\mathbf{E}}^{\mathrm{A}^1}_{y\mathrm{R}^1} + \widehat{\mathbf{E}}^{\mathrm{B}^1}_{y\mathrm{R}^1} \\ \vdots \\ \widehat{\mathbf{E}}^{\mathrm{A}^k}_{y\mathrm{R}^k} + \widehat{\mathbf{E}}^{\mathrm{B}^k}_{y\mathrm{R}^k} \\ \vdots \end{bmatrix} = \begin{bmatrix} \widehat{\mathbf{Z}}^{\mathrm{A}^1}_{y\mathrm{R}^1}\widehat{\mathbf{V}}^1_{\mathrm{AA}} + \widehat{\mathbf{Z}}^{\mathrm{B}^1}_{y\mathrm{R}^1}\widehat{\mathbf{V}}^1_{\mathrm{BA}} \\ \vdots \\ \widehat{\mathbf{Z}}^{\mathrm{A}^k}_{y\mathrm{R}^k}\widehat{\mathbf{V}}^k_{\mathrm{AA}} + \widehat{\mathbf{Z}}^{\mathrm{B}^k}_{y\mathrm{R}^k}\widehat{\mathbf{V}}^k_{\mathrm{BA}} \\ \vdots \end{bmatrix} \widehat{\mathbf{H}}_{\mathrm{A}} = \widehat{\mathbf{Z}}^{\mathrm{A}}_{y\mathrm{R}}\widehat{\mathbf{H}}_{\mathrm{A}} \tag{8.93}$$

This is the upper part of the vector in eq. (8.83).

We obtain the tangential electric field component $\mathbf{E}_z(\overline{z}_k)$ from the electric and magnetic fields $\widehat{\mathbf{E}}(z_k)$, $\widehat{\mathbf{H}}(z_k)$ according to eq. (8.16), which can be written in discretised form as:

$$\mathrm{j}\widehat{\boldsymbol{\epsilon}}^k_{zz}\mathbf{E}^k_z(z_k) = [\widehat{\overline{\boldsymbol{D}}}{}^{\bullet}_{\overline{y}} \quad \widehat{\overline{\boldsymbol{D}}}{}^{\circ}_{\overline{x}}]\widehat{\mathbf{H}}(\overline{z}_k) - \mathrm{j}[\widehat{\boldsymbol{M}}{}^{\bullet}_y\widehat{\boldsymbol{\epsilon}}_{zy} \quad \widehat{\boldsymbol{M}}{}^{\circ}_x\widehat{\boldsymbol{\epsilon}}_{zx}]\widehat{\mathbf{E}}(\overline{z}_k) \tag{8.94}$$

For $\widehat{\mathbf{E}}(\overline{z}_k)$ and $\widehat{\mathbf{H}}(\overline{z}_k)$ we introduce eqs. (8.84) and (8.85), with the components in column n according to eq. (8.86). To calculate $\mathbf{E}_z$ at ports C^k and D^k, again by an extrapolation process, we need the values of $\mathbf{E}_z$ that belong to the last two (three) and the first two (three) columns of $\widehat{\mathbf{E}}_z$ in the cross-section at z^k, respectively. We obtain at the boundaries R^k:

$$\mathbf{E}^k_{z\mathrm{R}}(z_k) = -\mathrm{j}(\boldsymbol{\epsilon}^{\mathrm{R}^k}_{zz})^{-1}[\boldsymbol{U}_1 e^{-\boldsymbol{\Gamma}^k_1 \overline{z}_k} \quad \boldsymbol{U}_2 e^{-\boldsymbol{\Gamma}^k_2(\overline{d}_k - \overline{z}_k)}](\widehat{\boldsymbol{M}}^k_{\mathrm{H}})^{-1}\widehat{\mathbf{H}}^k_{\mathrm{AB}} \qquad (8.95)$$

where:

$$\begin{aligned} \boldsymbol{U}_1 &= \overline{\boldsymbol{D}}^{\bullet}_{\overline{y}} \boldsymbol{T}^{\mathrm{R^k}}_{21_x} + \overline{h}_x^{-1} \boldsymbol{T}^{\mathrm{R^k}}_{21_y} - \mathrm{j}\boldsymbol{M}^{\bullet}_y \boldsymbol{\epsilon}^{\mathrm{R^k}}_{zy} \boldsymbol{T}^{\mathrm{Rk}}_{11_y} - \mathrm{j}\boldsymbol{\epsilon}^{\mathrm{R^k}}_{zx} \boldsymbol{T}^{\mathrm{R^k}}_{11_x} \\ \boldsymbol{U}_2 &= \overline{\boldsymbol{D}}^{\bullet}_{\overline{y}} \boldsymbol{T}^{\mathrm{R^k}}_{22_x} + \overline{h}_x^{-1} \boldsymbol{T}^{\mathrm{R^k}}_{22_y} - \mathrm{j}\boldsymbol{M}^{\bullet}_y \boldsymbol{\epsilon}^{\mathrm{R^k}}_{zy} \boldsymbol{T}^{\mathrm{R^k}}_{12_y} - \mathrm{j}\boldsymbol{\epsilon}^{\mathrm{R^k}}_{zx} \boldsymbol{T}^{\mathrm{R^k}}_{12_x} \end{aligned}$$

The matrix in the brackets has the same form as that in eq. (8.94). The matrices $\boldsymbol{\epsilon}^{\mathrm{R^k}}_{zu}, u = x, y, z$ are diagonal matrices with the permittivities ε_{zu} for the column at the ports. In the case of $\boldsymbol{\epsilon}^{\mathrm{R^k}}_{zx}$, we assume that there is no permittivity step in the vicinity of the ports. Otherwise the permittivity values have to be taken into account in the corresponding eq. (8.88). The difference matrix $\overline{\boldsymbol{D}}^{\bullet}_{\overline{y}}$ and the interpolation matrix $\boldsymbol{M}^{\bullet}_y$ are used as usual.

The interpolation matrix $\boldsymbol{M}^{\circ}_x$ is replaced by the extrapolation matrices $\boldsymbol{T}^{\mathrm{R^k}}_{11,12_x}$ in the direction of the ports. Because the positions of the discretisation points differ from those in eq. (8.88), here the matrices are given by:

$$\begin{aligned} \boldsymbol{T}^{\mathrm{C}k}_{11,12_x} &= (3\boldsymbol{T}^k_{11,12_{x\mathrm{N}}} - 3\boldsymbol{T}^k_{11,12_{x\mathrm{N}-1}} + \boldsymbol{T}^k_{11,12_{x\mathrm{N}-2}}) \\ \boldsymbol{T}^{\mathrm{D}k}_{11,12_x} &= (3\boldsymbol{S}^k_{11,12_{x1}} - 3\boldsymbol{T}^k_{11,12_{x2}} + \boldsymbol{T}^k_{11,12_{x3}}) \end{aligned} \qquad (8.96)$$

The extrapolation matrices $\boldsymbol{T}^{\mathrm{R^k}}_{11,12_y}$ and $\boldsymbol{T}^{\mathrm{R^k}}_{21,22_x}$ are constructed analogously to eq. (8.88). The subscripts x and y correspond to the components. The extrapolation matrices $\boldsymbol{T}^{\mathrm{R^k}}_{21,22_y}$ are different because the difference operator $\overline{\boldsymbol{D}}^{\circ}_{\overline{x}}$ must be applied to the columns. After this application, the columns of the derivatives can be extrapolated with eq. (8.88), because they have to fulfil Neumann boundary conditions. The columns of the derivatives can be expressed with the central differences of the neighbouring columns. The field component $\mathbf{H}^k_{yn}$ itself has to fulfil Dirichlet conditions. By using central differences for this case, we obtain:

$$\begin{aligned} \boldsymbol{T}^{\mathrm{C}^k}_{21,22_y} &= \tfrac{1}{8}(10\boldsymbol{T}^k_{11,12_{y\mathrm{N}}} - \boldsymbol{T}^k_{11,12_{y\mathrm{N}\text{-}}}) \\ \boldsymbol{T}^{\mathrm{D}^k}_{11,12_y} &= \tfrac{1}{8}(10\boldsymbol{T}^k_{11,12_{y1}} - \boldsymbol{T}^k_{11,12_{y2}}) \end{aligned} \qquad (8.97)$$

$\overline{h}_x$ in eq. (8.95) is the normalised discretisation distance in x-direction. Eq. (8.97) can easily be checked. If $\mathbf{H}^k_y(x)$ has a linear behaviour, then we have e.g. $\boldsymbol{T}^k_{11,12_{y2}} = 2\boldsymbol{T}^k_{11,12_{y1}}$ and therefore $\boldsymbol{T}^{\mathrm{D}^k}_{11,12_y} = \boldsymbol{T}^k_{11,12_{y1}}$. The field component $\mathbf{E}_{z\mathrm{R}}^{k}$ in eq. (8.95) at ports $\mathrm{R}^k \equiv \mathrm{C}^k$ or $\mathrm{R}^k \equiv \mathrm{D}^k$ is now expressed analogously to eq. (8.89) by:

$$\mathbf{E}_{z\mathrm{R}}^{k}(z_k) = \widehat{\mathbf{Z}}_{z\mathrm{R}^k}^{\mathrm{AB}^k}(\overline{z}_k)\widehat{\mathbf{H}}_{\mathrm{AB}}^{k} \tag{8.98}$$

By dividing the matrix $\widehat{\mathbf{Z}}_{z\mathrm{R}^k}^{\mathrm{AB}k}(\overline{z}_k)$ into a left and right part $\mathbf{Z}_{z\mathrm{R}^k}^{\mathrm{A}k}(\overline{z}_k)$ and $\mathbf{Z}_{z\mathrm{R}^k}^{\mathrm{B}k}(\overline{z}_k)$ (according to the subvectors $\widehat{\mathbf{H}}_{\mathrm{A}}^{k}$ and $-\widehat{\mathbf{H}}_{\mathrm{B}}^{k}$ of $\widehat{\mathbf{H}}_{\mathrm{AB}}^{k}$), we can write instead of eq. (8.98):

$$\mathbf{E}_{z\mathrm{R}}^{k}(z_k) = \mathbf{Z}_{z\mathrm{R}^k}^{\mathrm{A}^k}(\overline{z}_k)\widehat{\mathbf{H}}_{\mathrm{A}}^{k} + \mathbf{Z}_{z\mathrm{R}^k}^{\mathrm{B}^k}(\overline{z}_k)(-\widehat{\mathbf{H}}_{\mathrm{B}}^{k}) \tag{8.99}$$

Eq. (8.99) must now be discretised at points $\overline{z}_i^k = (\overline{z}_k)_i$ of ports C^k and D^k. This is analogous to the procedure for eq. (8.90). The field parts $\mathbf{E}_{z\mathrm{R}^k}^{\mathrm{A}^k}$ and $\mathbf{E}_{z\mathrm{R}^k}^{\mathrm{B}^k}$ corresponding to $\widehat{\mathbf{H}}_{\mathrm{A}}^{k}$ and $-\widehat{\mathbf{H}}_{\mathrm{B}}^{k}$, respectively, can be expressed as:

$$\begin{aligned}
\mathbf{E}_{z\mathrm{R}^k}^{\mathrm{A}^k} &= \begin{bmatrix} \mathbf{Z}_{z\mathrm{R}^k}^{\mathrm{A}k}(\overline{z}_1^k) \\ \vdots \\ \mathbf{Z}_{z\mathrm{R}^k}^{\mathrm{A}^k}(\overline{z}_{\mathrm{N}_z^k}^k) \end{bmatrix} \widehat{\mathbf{H}}_{\mathrm{A}}^{k} = \widehat{\mathbf{Z}}_{z\mathrm{R}}^{\mathrm{A}^k}\widehat{\mathbf{H}}_{\mathrm{A}}^{k} \\
\mathbf{E}_{z\mathrm{R}^k}^{\mathrm{B}^k} &= \begin{bmatrix} \mathbf{Z}_{z\mathrm{R}^k}^{\mathrm{B}^k}(\overline{z}_1^k) \\ \vdots \\ \mathbf{Z}_{z\mathrm{R}^k}^{\mathrm{B}^k}(\overline{z}_{\mathrm{N}_z^k}^k) \end{bmatrix} (-\widehat{\mathbf{H}}_{\mathrm{B}}^{k}) = \widehat{\mathbf{Z}}_{z\mathrm{R}}^{\mathrm{B}^k}(-\widehat{\mathbf{H}}_{\mathrm{B}}^{k})
\end{aligned} \tag{8.100}$$

$z_{\mathrm{N}_z^k}^k$ is the last column at subport C^k or D^k in z_1-direction. The field vectors $\widehat{\mathbf{H}}_{\mathrm{A}^k}$ and $\widehat{\mathbf{H}}_{\mathrm{B}^k}$ are given in eq. (8.90). The fields at all the subports C^k and D^k (subports k of ports C and D) should now be collected in the common column vectors for $\widehat{\mathbf{E}}_z^{\mathrm{R}}$ at ports C and D, respectively.

$$\widehat{\mathbf{E}}_z^{\mathrm{R}} = \begin{bmatrix} \widehat{\mathbf{E}}_{z\mathrm{R}^1}^{\mathrm{A}^1} + \widehat{\mathbf{E}}_{z\mathrm{R}^1}^{\mathrm{B}^1} \\ \vdots \\ \widehat{\mathbf{E}}_{z\mathrm{R}^k}^{\mathrm{A}^k} + \widehat{\mathbf{E}}_{z\mathrm{R}^k}^{\mathrm{B}^k} \\ \vdots \end{bmatrix} = \begin{bmatrix} \widehat{\mathbf{Z}}_{z\mathrm{R}^1}^{\mathrm{A}^1}\widehat{\mathbf{V}}_{\mathrm{AA}}^{1} + \widehat{\mathbf{Z}}_{z\mathrm{R}^1}^{\mathrm{B}^1}\widehat{\mathbf{V}}_{\mathrm{BA}}^{1} \\ \vdots \\ \widehat{\mathbf{Z}}_{z\mathrm{R}^k}^{\mathrm{A}^k}\widehat{\mathbf{V}}_{\mathrm{AA}}^{k} + \widehat{\mathbf{Z}}_{z\mathrm{R}^k}^{\mathrm{B}^k}\widehat{\mathbf{V}}_{\mathrm{BA}}^{k} \\ \vdots \end{bmatrix} \widehat{\mathbf{H}}_{\mathrm{A}} = \widehat{\mathbf{Z}}_{z\mathrm{R}}^{\mathrm{A}}\widehat{\mathbf{H}}_{\mathrm{A}} \tag{8.101}$$

This is the lower part of the vector in eq. (8.83).

In an analogous manner, we can calculate $\widehat{\mathbf{Z}}_{y\mathrm{R}}^{\mathrm{B}}$ and $\widehat{\mathbf{Z}}_{z\mathrm{R}}^{\mathrm{B}}$ under the condition $\widehat{\mathbf{H}}_{\mathrm{A}} = \mathbf{0}$. In summary, the results may be written as follows:

$$\widehat{\mathbf{E}}_{\mathrm{CD}} = \begin{bmatrix} \widehat{\mathbf{E}}_{y_2}^{\mathrm{C}} \\ \widehat{\mathbf{E}}_{x_2}^{\mathrm{C}} \\ \widehat{\mathbf{E}}_{y_2}^{\mathrm{D}} \\ \widehat{\mathbf{E}}_{x_2}^{\mathrm{D}} \end{bmatrix} = \begin{bmatrix} \widehat{\mathbf{Z}}_{y\mathrm{C}}^{\mathrm{A}} & \widehat{\mathbf{Z}}_{y\mathrm{C}}^{\mathrm{B}} \\ \widehat{\mathbf{Z}}_{z\mathrm{C}}^{\mathrm{A}} & \widehat{\mathbf{Z}}_{z\mathrm{C}}^{\mathrm{B}} \\ \widehat{\mathbf{Z}}_{y\mathrm{D}}^{\mathrm{A}} & \widehat{\mathbf{Z}}_{y\mathrm{D}}^{\mathrm{B}} \\ \widehat{\mathbf{Z}}_{z\mathrm{D}}^{\mathrm{A}} & \widehat{\mathbf{Z}}_{z\mathrm{D}}^{\mathrm{B}} \end{bmatrix} \begin{bmatrix} -\widehat{\mathbf{H}}_x^{\mathrm{A}} \\ \widehat{\mathbf{H}}_y^{\mathrm{A}} \\ \widehat{\mathbf{H}}_x^{\mathrm{B}} \\ -\widehat{\mathbf{H}}_y^{\mathrm{B}} \end{bmatrix} = \widehat{\mathbf{z}}_{\mathrm{CD}}^{\mathrm{AB}}\widehat{\mathbf{H}}_{\mathrm{AB}} \tag{8.102}$$

The other matrix parameters in eq. (8.75) are obtained in an analogous way.

Now, to analyse a special device, the load impedances at the ports must be calculated. This can be done in the usual way. In the case of photonic crystals, the connecting waveguides are periodic structures. The procedure for this case is described in Chapter 5.

8.4.5 Steps of the analysis procedure

The analysis of devices is now carried out as follows: we start at the output of the device and transform the load impedance/admittance through the sections and interfaces to the input. At the input we obtain the fields from the input impedance and the input wave from the source. These fields are then transformed in the opposite direction through the section toward the output of the device.

For eigenmode calculation, we transform the impedances/admittances from the upper and lower sides of the cross-section towards a matching interface. The field matching there results in an indirect eigenvalue problem, e.g. for the effective permittivity ε_{re}. The fields in the whole device are then obtained by transforming the field from the matching interface through the layers towards the upper and lower boundaries (see Section 5.5.4).

8.5 NUMERICAL RESULTS

The algorithm can be used for microwave and millimetre-wave structures. With an analogous procedure, conformal antennas were analysed [9], [10]. Here we show further numerical examples, dealing in particular with anisotropic materials.

To validate the proposed algorithm, a microstrip junction on an anisotropic substrate has been analysed [11]. The substrate has a thickness $h = 0.635$ mm and the width of the microstrip is $w = 0.6$ mm. Fig. 8.7 shows scattering parameters for this structure with anisotropic and isotropic substrates. In the example of Fig. 8.8, the effective permittivity ε_{re} – whose square root is equal to the propagation constant normalised with the free space wave number k_0 – for a channel guide is drawn as a function of the normalised free space wave number k_0. The channel material is anisotropic with the permittivity-tensor given in the figure. In [12] the structure was completely surrounded by metallic walls. In our case we have an electric wall on one side and a magnetic wall on the other. In [12] the waveguide behaves like a hollow guide at lower frequencies. This is different from our device. Nevertheless, there is a good agreement between the results. The algorithm has also been used to analyse waveguides in other than Cartesian coordinates.

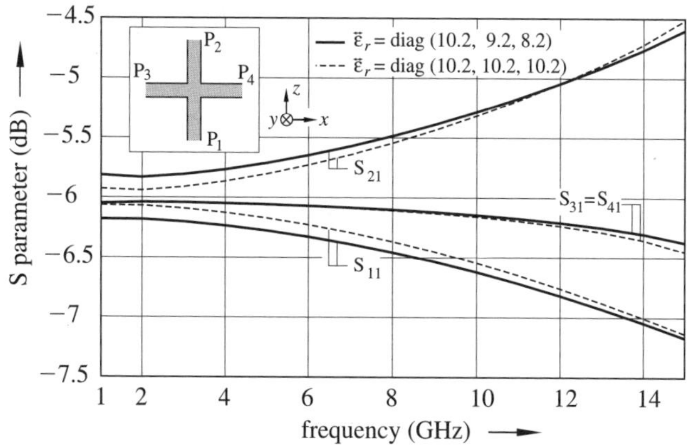

Fig. 8.7 Scattering parameters for a microstrip junction on an anisotropic structure (Reproduced by permission of Union Radio-Scientifique Internationale–International Union of Radio Science (URSI))

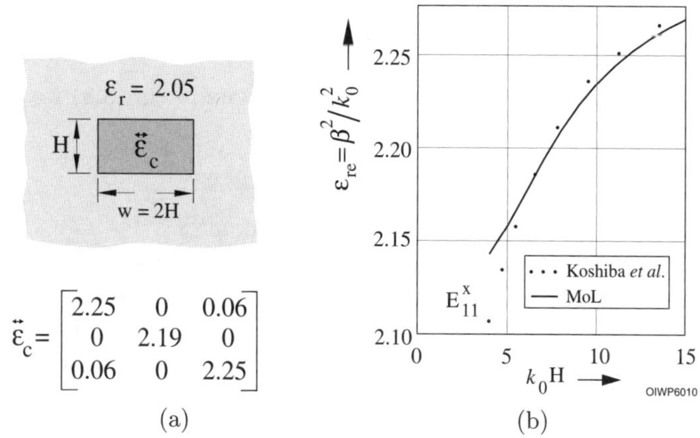

Fig. 8.8 Dispersion diagram (b) for a channel waveguide (a) • results from [12] (R. Pregla, 'Modeling of Optical Waveguide Structures with General Anisotropy in Arbitrary Orthogonal Coordinate Systems', *IEEE J. of Sel. Topics in Quantum Electronics*, vol. 8, pp. 1217–1224. © 2002 Institute of Electrical and Electronics Engineers (IEEE))

8.6 ANALYSIS OF WAVEGUIDE STRUCTURES IN SPHERICAL COORDINATES

8.6.1 Introduction

After dealing with arbitrary coordinates in the last sections, we will now show the analysis with spherical coordinates. This may be seen as a special case of the general one we examined before. Various structures, like conformal antennas on spherical surfaces or circular horns as primary feeds in e.g. Cassegrainian antenna systems, can be examined with such spherical coordinates. Therefore, they deserve a special treatment.

8.6.2 Generalised transmission line equations in spherical coordinates

The coordinate system is drawn in Fig. 8.9. We can give the following relation to Cartesian coordinates:

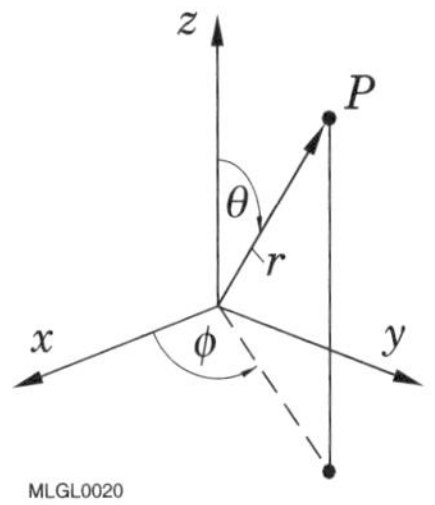

$$z = r\cos\theta$$
$$x = r\sin\theta\cos\phi$$
$$y = r\sin\theta\sin\phi$$

Fig. 8.9 Spherical coordinate system

From the general formula for the metric factors in eq. (8.3) we obtain:

$$g_{\mathrm{r}} = 1 \qquad g_\theta = r \qquad g_\phi = r\sin\theta \tag{8.103}$$

8.6.2.1 Material parameters in spherical coordinates

In this subsection we would like to represent the material parameters for anisotropic materials in spherical coordinates We assume that one of the crystal axes coincides with the z coordinate. The relation between the field components F_u in Cartesian ($u = x, y, z$) and F_v in spherical ($v = r, \theta, \phi$) coordinates is given by: (See Fig. 8.9.)

$$\begin{bmatrix} F_x \\ F_y \\ F_z \end{bmatrix} = \boldsymbol{G} \begin{bmatrix} F_r \\ F_\theta \\ F_\phi \end{bmatrix} \qquad \boldsymbol{G} = \begin{bmatrix} \sin\theta\cos\phi & \cos\theta\cos\phi & -\sin\phi \\ \sin\theta\sin\phi & \cos\theta\sin\phi & \cos\phi \\ \cos\theta & -\sin\theta & 0 \end{bmatrix} \tag{8.104}$$

The inverse matrix of $\boldsymbol{G}$ (describing the determination of the spherical components from the Cartesian ones) is just its transposition, i.e. $\boldsymbol{G}^{\mathrm{t}}$. Therefore, if a material has anisotropic behaviour that is described in Cartesian coordinates with the tensor $\overleftrightarrow{\boldsymbol{\nu}}_r^{\mathrm{C}}$ ($\boldsymbol{\nu} = \boldsymbol{\varepsilon}$ or $\boldsymbol{\mu}$), we obtain in spherical coordinates $\overleftrightarrow{\boldsymbol{\nu}}_r^{\mathrm{S}}$ by the relation:

$$\overleftrightarrow{\boldsymbol{\nu}}_r^{\mathrm{S}} = \boldsymbol{G}^{\mathrm{t}} \overleftrightarrow{\boldsymbol{\nu}}_r^{\mathrm{C}} \boldsymbol{G} \tag{8.105}$$

Even for uniaxial anisotropy, the tensor in spherical coordinates is not diagonal. Furthermore, all the elements in the tensor are functions of θ and ϕ.

We will demonstrate here only the transformation of the following tensor (in Cartesian coordinates):

$$\overleftrightarrow{\nu}_r^{\mathrm{C}} = \begin{bmatrix} \nu_{xx} & \mathrm{j}\kappa & 0 \\ -\mathrm{j}\kappa & \nu_{xx} & 0 \\ 0 & 0 & \nu_{zz} \end{bmatrix} \tag{8.106}$$

In spherical coordinates we obtain the following tensor:

$$\overleftrightarrow{\nu}_r^{\mathrm{S}} = \begin{bmatrix} \nu_{xx}\sin^2\theta + \nu_{zz}\cos^2\theta & (\nu_{xx}-\nu_{zz})\sin\theta\cos\theta & \mathrm{j}\kappa\sin\theta \\ (\nu_{xx}-\nu_{zz})\sin\theta\cos\theta & \nu_{xx}\cos^2\theta + \nu_{zz}\sin^2\theta & \mathrm{j}\kappa\cos\theta \\ -\mathrm{j}\kappa\sin\theta & -\mathrm{j}\kappa\cos\theta & \nu_{xx} \end{bmatrix} \tag{8.107}$$

Only, in this case the tensor elements do not depend on ϕ. For $\kappa = 0$ the two off-diagonal elements in the last column and row vanish. Even for diagonal matrices $\overleftrightarrow{\boldsymbol{\nu}}_r^{\mathrm{C}}$ with $\nu_{zz} \neq \nu_{xx}$, we obtain full matrices $\overleftrightarrow{\boldsymbol{\nu}}_r^{\mathrm{S}}$.

8.6.2.2 Curl operator for spherical coordinates in matrix notation

To obtain the required equations in spherical coordinates, it is possible to start from the general equations derived earlier in this chapter. However, sometimes it is easier to obtain optimal formulas by starting from basic equations. The curl operator in a spherical coordinate system may be written in matrix as:

$$\overline{\nabla} \times \vec{A} = \begin{bmatrix} 0 & -(\overline{r}s_\theta)^{-1}D_\phi & (\overline{r}s_\theta)^{-1}D_\theta s_\theta \\ (\overline{r}s_\theta)^{-1}D_\phi & 0 & -\overline{r}^{-1}D_{\overline{r}}\overline{r} \\ -\overline{r}^{-1}D_\theta & \overline{r}^{-1}D_{\overline{r}}\overline{r} & 0 \end{bmatrix} \begin{bmatrix} A_r \\ A_\theta \\ A_\phi \end{bmatrix} \tag{8.108}$$

where we introduce the following abbreviations and normalisations:

$$\overline{\nabla} = k_{\mathrm{o}}^{-1}\nabla \qquad \overline{r} = k_{\mathrm{o}} r \qquad s_\theta = \sin\theta \tag{8.109}$$

8.6.2.3 Propagation in r-direction

From the second and third equations of Ampère's law and from the third and second equations of the law of induction, we obtain:

$$-\mathrm{j}\begin{bmatrix} \epsilon_{r\theta} & \\ & \epsilon_{r\phi} \end{bmatrix} \begin{bmatrix} E_\theta \\ E_\phi \end{bmatrix} = \frac{1}{\overline{r}}\frac{\partial}{\partial\overline{r}}\overline{r}\begin{bmatrix} \widetilde{H}_\phi \\ -\widetilde{H}_\theta \end{bmatrix} - \frac{1}{\overline{r}}\begin{bmatrix} s_\theta^{-1}D_\phi \\ -D_\theta \end{bmatrix}\widetilde{H}_r \tag{8.110}$$

$$-\mathrm{j}\begin{bmatrix} \mu_{r\phi} & \\ & \mu_{r\theta} \end{bmatrix} \begin{bmatrix} \widetilde{H}_\phi \\ -\widetilde{H}_\theta \end{bmatrix} = \frac{1}{\overline{r}}\frac{\partial}{\partial\overline{r}}\overline{r}\begin{bmatrix} \widetilde{E}_\theta \\ \widetilde{E}_\phi \end{bmatrix} - \frac{1}{\overline{r}}\begin{bmatrix} D_\theta \\ s_\theta^{-1}D_\phi \end{bmatrix}E_r \tag{8.111}$$

Due to discretisation we will use later, we distinguish the material parameters in the three directions r, θ and ϕ. From the first equations we obtain:

$$\widetilde{H}_r = \mathrm{j}\mu_{rr}^{-1}(\overline{r}s_\theta)^{-1}[-D_\phi \quad D_\theta][\widehat{E}^r] \tag{8.112}$$

$$E_r = -\mathrm{j}\varepsilon_{rr}^{-1}(\overline{r}s_\theta)^{-1}[D_\theta \quad D_\phi][H^r] \tag{8.113}$$

where we define the following field vectors transverse to the r-direction:

$$[E^r] = \left[E_\theta, s_\theta E_\phi\right]^{\mathrm{t}} \qquad [H^r] = \left[s_\theta \widetilde{H}_\phi, -\widetilde{H}_\theta\right]^{\mathrm{t}} \tag{8.114}$$

By introducing the last two equations into the previous ones, we obtain:

$$\bar{r}\frac{\mathrm{d}}{\mathrm{d}\bar{r}}\bar{r}[E^r] = -\mathrm{j}[R_{\mathrm{H}}^{\mathrm{rS}}][H^r] - \mathrm{j}\bar{r}^2[\mu_{rt}][H^r] \tag{8.115}$$

$$\bar{r}\frac{\mathrm{d}}{\mathrm{d}\bar{r}}\bar{r}[H^r] = -\mathrm{j}[R_{\mathrm{E}}^{\mathrm{rS}}][E^r] - \mathrm{j}\bar{r}^2[\varepsilon_{rt}][E^r] \tag{8.116}$$

The matrices $[R_{\mathrm{E,H}}^{\mathrm{rS}}]$ and $[\mu_{rt}]$, $[\epsilon_{rt}]$ are defined by:

$$[R_{\mathrm{H}}^{\mathrm{rS}}] = \begin{bmatrix} D_\theta(\varepsilon_{rr}s_\theta)^{-1}D_\theta & D_\theta(\varepsilon_{rr}s_\theta)^{-1}D_\phi \\ D_\phi(\varepsilon_{rr}s_\theta)^{-1}D_\theta & D_\phi(\varepsilon_{rr}s_\theta)^{-1}D_\phi \end{bmatrix}$$

$$[R_{\mathrm{E}}^{\mathrm{rS}}] = \begin{bmatrix} D_\phi(\mu_{rr}s_\theta)^{-1}D_\phi & -D_\phi(\mu_{rr}s_\theta)^{-1}D_\theta \\ -D_\theta(\mu_{rr}s_\theta)^{-1}D_\phi & D_\theta(\mu_{rr}s_\theta)^{-1}D_\theta \end{bmatrix} \tag{8.117}$$

$$[\mu_{rt}] = \begin{bmatrix} s_\theta^{-1}\mu_{r\phi} & 0 \\ 0 & s_\theta\mu_{r\theta} \end{bmatrix} \qquad [\varepsilon_{rt}] = \begin{bmatrix} s_\theta\varepsilon_{r\theta} & 0 \\ 0 & s_\theta^{-1}\varepsilon_{r\phi} \end{bmatrix}$$

We combine the two first-order differential equation systems (8.115) and (8.116). By taking into account the fact that the products of the matrices $R_{\mathrm{E}}^{\mathrm{rS}}$ and $R_{\mathrm{H}}^{\mathrm{rS}}$ are zero, we obtain:

$$\bar{r}\frac{\partial}{\partial\bar{r}}\left(\bar{r}^2\frac{\partial}{\partial\bar{r}}\bar{r}[E^r]\right) + \bar{r}^2[S_{\mathrm{E}}^{\mathrm{rS}}][E^r] + \mathrm{j}2\bar{r}^3[\mu_{rt}][H^r] = [0\ 0]^{\mathrm{t}} \tag{8.118}$$

$$\bar{r}\frac{\partial}{\partial\bar{r}}\left(\bar{r}^2\frac{\partial}{\partial\bar{r}}\bar{r}[H^r]\right) + \bar{r}^2[S_{\mathrm{H}}^{\mathrm{rS}}][H^r] + \mathrm{j}2\bar{r}^3[\varepsilon_{rt}][E^r] = [0\ 0]^{\mathrm{t}} \tag{8.119}$$

The matrices $[S_{\mathrm{E,H}}^{\mathrm{rS}}]$ are given by:

$$[S_{\mathrm{E}}^{\mathrm{rS}}] = [R_{\mathrm{H}}^{\mathrm{rS}}][\varepsilon_{rt}] + [\mu_{rt}][R_{\mathrm{E}}^{\mathrm{rS}}] + \bar{r}^2[\varepsilon_{rt}][\mu_{rt}] \tag{8.120}$$

$$[S_{\mathrm{H}}^{\mathrm{rS}}] = [\varepsilon_{rt}][R_{\mathrm{H}}^{\mathrm{rS}}] + [R_{\mathrm{E}}^{\mathrm{rS}}][\mu_{rt}] + \bar{r}^2[\varepsilon_{rt}][\mu_{rt}] \tag{8.121}$$

8.6.2.4 TM$_r$ and TE$_r$ modes

An arbitrary electromagnetic field in a homogeneous region can be constructed as a superposition of its TM and TE parts [13]. We obtain these field parts in the following way:

TM$_r$ mode: $\widetilde{H}_r = 0$	TE$_r$ mode: $E_r = 0$
$\widetilde{H}_r = 0$ results in (see eq. (8.112)):	$E_r = 0$ results in (see eq. (8.113)):
$D_\phi E_\theta = D_\theta(s_\theta E_\phi) \longrightarrow [R_{\mathrm{E}}^{\mathrm{rS}}] = 0$	$D_\phi H_\theta = D_\theta(s_\theta H_\phi) \longrightarrow [R_{\mathrm{H}}^{\mathrm{rS}}] = 0$
As a further condition we obtain:	As a further condition we obtain:
$$\varepsilon_{r\phi} D_\phi \widetilde{H}_\phi = -\varepsilon_{r\theta} D_\theta(s_\theta \widetilde{H}_\theta) \quad (8.122)$$	$$\mu_{r\phi} D_\phi E_\phi = -\mu_{r\theta} D_\theta(s_\theta E_\theta) \quad (8.123)$$
Now eq. (8.116) reads as:	Now eq. (8.115) reads as:
$$\frac{\mathrm{d}}{\mathrm{d}\overline{r}} \overline{r}[H^r] = -\mathrm{j}\overline{r}[\varepsilon_{rt}][E^r] \quad (8.124)$$	$$\frac{\mathrm{d}}{\mathrm{d}\overline{r}} \overline{r}[E^r] = -\mathrm{j}\overline{r}[\mu_{rt}][H^r] \quad (8.125)$$
Introducing eq. (8.124) into eq. (8.115) results in:	Introducing eq. (8.125) into eq. (8.116) results in:
$$\overline{r}\frac{\mathrm{d}^2}{\mathrm{d}\overline{r}^2}\overline{r}[H^r] + [\varepsilon_{rt}][R_{\mathrm{H}}^{\mathrm{rS}}][H^r] + \overline{r}^2[p_{\varepsilon\mu}][H^r] = 0 \quad (8.126)$$	$$\overline{r}\frac{\mathrm{d}^2}{\mathrm{d}\overline{r}^2}\overline{r}[E^r] + [\mu_{rt}][R_{\mathrm{E}}^{\mathrm{rS}}][E^r] + \overline{r}^2[p_{\varepsilon\mu}][E^r] = 0 \quad (8.127)$$
with:	with:
$[p_{\varepsilon\mu}] = \mathrm{diag}(\varepsilon_{r\theta}\mu_{r\phi}, \varepsilon_{r\phi}\mu_{r\theta})$	$[p_{\varepsilon\mu}] = \mathrm{diag}(\varepsilon_{r\theta}\mu_{r\phi}, \varepsilon_{r\phi}\mu_{r\theta})$

Now we introduce the conditions (8.122) and (8.123) into eqs. (8.126) and (8.127) and obtain:

$$\overline{r}\frac{\mathrm{d}^2}{\mathrm{d}\overline{r}^2}\overline{r}[F^r] + [Q_{\mathrm{F}}^{\mathrm{rS}}][F^r] + \overline{r}^2[p_{\varepsilon\mu}][F^r] = 0 \tag{8.128}$$

with $F = H$ for TM$_r$ and $F = E$ for TE$_r$ modes. We obtain for the matrices $[Q_{\mathrm{E,H}}^{\mathrm{rS}}]$ (the material parameters are independent of ϕ and $D_\theta D_\phi = D_\phi D_\theta$):

$$[Q_{\mathrm{H}}^{\mathrm{rS}}] = \begin{bmatrix} \varepsilon_{r\theta} s_\theta D_\theta (\varepsilon_{rr} s_\theta)^{-1} D_\theta & \varepsilon_{r\theta} s_\theta D_\theta (\varepsilon_{rr} s_\theta)^{-1} D_\phi \\ 0 & \varepsilon_{r\phi}/(\varepsilon_{rr} s_\theta^2)(D_\theta(\varepsilon_{r\theta}/\varepsilon_{r\phi}) s_\theta D_\theta s_\theta + D_\phi^2) \end{bmatrix}$$

$$[Q_{\mathrm{E}}^{\mathrm{rS}}] = \begin{bmatrix} \mu_{r\phi}/(\mu_{rr} s_\theta^2)(D_\theta(\mu_{r\theta}/\mu_{r\phi}) s_\theta D_\theta s_\theta + D_\phi^2) & 0 \\ -\mu_{r\theta} s_\theta D_\theta (\mu_{rr} s_\theta)^{-1} D_\phi & \mu_{r\theta} s_\theta D_\theta (\mu_{rr} s_\theta)^{-1} D_\theta \end{bmatrix} \tag{8.129}$$

To solve eq. (8.128), we transform it into Bessel's equation. We should keep in mind the following relation:

$$\overline{r}\frac{\mathrm{d}^2}{\mathrm{d}\overline{r}^2}\overline{r}[F^r] = \frac{\mathrm{d}}{\mathrm{d}\overline{r}}\overline{r}^2\frac{\mathrm{d}}{\mathrm{d}\overline{r}}[F^r] \qquad [F^r] = [H^r] \text{ or } [E^r] \tag{8.130}$$

We assume $\varepsilon_{r\theta}\mu_{r\phi} = \varepsilon_{r\phi}\mu_{r\theta} = \varepsilon_r\mu_r$, replace $\sqrt{\mu_r\varepsilon_r}\ \overline{r} = t$ and transform the field components according to:

$$[F^r] = \sqrt{t}^{-1}[G_{\mathrm{F}}^r](t) \tag{8.131}$$

Now, because of the equality:

$$\sqrt{t}\frac{\mathrm{d}}{\mathrm{d}t}\left(t^2\frac{\mathrm{d}}{\mathrm{d}t}(\sqrt{t}^{-1}[G^r](t))\right) = t\frac{\mathrm{d}}{\mathrm{d}t}\left(t\frac{\mathrm{d}}{\mathrm{d}t}[G^r](t)\right) - \frac{1}{4}[G^r](t) \tag{8.132}$$

we obtain the following wave equations:

$$t\frac{\mathrm{d}}{\mathrm{d}t}\left(t\frac{\mathrm{d}}{\mathrm{d}t}[G^r_{\mathrm{E}}]\right) + \left([Q^{\mathrm{rS}}_{\mathrm{E}}] - \frac{1}{4}[I]\right)[G^r_{\mathrm{E}}] + t^2[G^r_{\mathrm{E}}] = 0 \tag{8.133}$$

$$t\frac{\mathrm{d}}{\mathrm{d}t}\left(t\frac{\mathrm{d}}{\mathrm{d}t}[G^r_{\mathrm{H}}]\right) + \left([Q^{\mathrm{rS}}_{\mathrm{H}}] - \frac{1}{4}[I]\right)[G^r_{\mathrm{H}}] + t^2[G^r_{\mathrm{H}}] = 0 \tag{8.134}$$

8.6.2.5 Wave equations for ϕ-independent fields

In the case of ϕ-independent fields we obtain uncoupled equations for the components E_ϕ and $\widetilde{H}_\phi$ (introduce $D_\phi = 0$ into eq. (8.128)):

$$\overline{r}\frac{\mathrm{d}^2}{\mathrm{d}\overline{r}^2}\overline{r}E_\phi + P^\mu_\theta E_\phi + \overline{r}^2\mu_{r\theta}\varepsilon_{r\phi}E_\phi = 0 \quad P^\mu_\theta = \mu_{r\theta}D_\theta(\mu_{rr}s_\theta)^{-1}D_\theta s_\theta \tag{8.135}$$

$$\overline{r}\frac{\mathrm{d}^2}{\mathrm{d}\overline{r}^2}\overline{r}\widetilde{H}_\phi + P^\varepsilon_\theta \widetilde{H}_\phi + \overline{r}^2\varepsilon_{r\theta}\mu_{r\phi}\widetilde{H}_\phi = 0 \quad P^\varepsilon_\theta = \varepsilon_{r\theta}D_\theta(\varepsilon_{rr}s_\theta)^{-1}D_\theta s_\theta \tag{8.136}$$

To solve the last two equations, we again transform them into Bessel's equation. We replace $\sqrt{\mu_{r\theta}\varepsilon_{r\phi}}\ \overline{r} = t$ or $\sqrt{\varepsilon_{r\theta}\mu_{r\phi}}\ \overline{r} = t$, respectively, and transform the field components according to:

$$E_\phi = \sqrt{t}^{-1}G^r_{\mathrm{E}}(t) \qquad H_\phi = \sqrt{t}^{-1}G^r_{\mathrm{H}}(t) \tag{8.137}$$

By using an analogous equality to that in eq. (8.132) we obtain the following wave equations:

$$t\frac{\mathrm{d}}{\mathrm{d}t}\left(t\frac{\mathrm{d}}{\mathrm{d}t}G^r_{\mathrm{E}}\right) + \left(P^\mu_\theta - \frac{1}{4}\right)G^r_{\mathrm{E}} + t^2 G^r_{\mathrm{E}} = 0 \tag{8.138}$$

$$t\frac{\mathrm{d}}{\mathrm{d}t}\left(t\frac{\mathrm{d}}{\mathrm{d}t}G^r_{\mathrm{H}}\right) + \left(P^\varepsilon_\theta - \frac{1}{4}\right)G^r_{\mathrm{H}} + t^2 G^r_{\mathrm{H}} = 0 \tag{8.139}$$

We can determine the other field components by:

TM$_r$ mode: $\widetilde{H}_\phi$	TE$_r$ mode: E_ϕ	
$\varepsilon_{r\theta}E_\theta = \dfrac{\mathrm{j}}{\overline{r}}\dfrac{\partial}{\partial\overline{r}}(\overline{r}\widetilde{H}_\phi)$	$\mu_{r\theta}\widetilde{H}_\theta = \dfrac{-\mathrm{j}}{\overline{r}}\dfrac{\partial}{\partial\overline{r}}(\overline{r}E_\phi)$	(8.140)
$\varepsilon_{rr}E_r = \dfrac{-\mathrm{j}}{\overline{r}s_\theta}\dfrac{\partial}{\partial\theta}(s_\theta\widetilde{H}_\phi)$	$\mu_{rr}\widetilde{H}_r = \dfrac{\mathrm{j}}{\overline{r}s_\theta}\dfrac{\partial}{\partial\theta}(s_\theta E_\phi)$	

8.6.2.6 Propagation in θ-direction

By using the abbreviations:

$$\overline{s}_\theta = \overline{r} s_\theta \qquad p_\varepsilon = (s_\theta \varepsilon_{r\theta})^{-1} \qquad p_\mu = (s_\theta \mu_{r\theta})^{-1} \tag{8.141}$$

we obtain from the first and third equations of Faraday's law and the law of induction:

$$-\mathrm{j}\overline{r}\begin{bmatrix}\overline{s}_\theta \varepsilon_{rr} & \\ & \overline{s}_\theta^{-1}\varepsilon_{r\phi}\end{bmatrix}\begin{bmatrix}E_r \\ \overline{s}_\theta E_\phi\end{bmatrix} = \frac{\partial}{\partial\theta}\begin{bmatrix}-\overline{s}_\theta \widetilde{H}_\phi \\ \widetilde{H}_r\end{bmatrix} - \begin{bmatrix}-D_\phi \\ D_{\overline{r}}\end{bmatrix}\overline{r}\widetilde{H}_\theta \tag{8.142}$$

$$-\mathrm{j}\overline{r}\begin{bmatrix}\overline{s}_\theta^{-1}\mu_{r\phi} & \\ & \overline{s}_\theta \mu_{rr}\end{bmatrix}\begin{bmatrix}-\overline{s}_\theta \widetilde{H}_\phi \\ \widetilde{H}_r\end{bmatrix} = \frac{\partial}{\partial\theta}\begin{bmatrix}E_r \\ \overline{s}_\theta E_\phi\end{bmatrix} - \begin{bmatrix}D_{\overline{r}} \\ D_\phi\end{bmatrix}\overline{r}E_\theta \tag{8.143}$$

We define the following field vectors:

$$[E^\theta] = [E_r \quad \overline{s}_\theta E_\phi]^{\mathrm{t}} \qquad [H^\theta] = [-\overline{s}_\theta \widetilde{H}_\phi \quad \widetilde{H}_r]^{\mathrm{t}} \tag{8.144}$$

Now the second equations of Faraday's law and the law of induction (which we had not used so far) give:

$$\mathrm{j}\mu_{r\theta} s_\theta \overline{r}\widetilde{H}_\theta = [-D_\phi \quad D_{\overline{r}}][E^\theta] \tag{8.145}$$

$$\mathrm{j}\varepsilon_{r\theta} s_\theta \overline{r} E_\theta = [D_{\overline{r}} \quad D_\phi][H^\theta] \tag{8.146}$$

We distinguish between ε_{rr}, $\varepsilon_{r\theta}$ and $\varepsilon_{r\phi}$, and between μ_{rr}, $\mu_{r\theta}$ and $\mu_{r\phi}$. This is because these quantities are discretised on different places in the case of inhomogeneous layers. Actually, we cannot introduce anisotropic material in this simple way. By using the transformation described in Section 8.6.2.1 we obtain material parameters that are functions of θ and ϕ. Therefore, an analytical solution in θ-direction is not possible. By introducing $\mathrm{j}\widetilde{H}_\theta$ and $\mathrm{j}E_\theta$ from eqs. (8.145) and (8.146), we obtain:

$$\frac{\partial}{\partial\theta}[H^\theta] = -\mathrm{j}[R_{\mathrm{E}}^\theta][E^\theta] = -\mathrm{j}\big([R_{\mathrm{E}}^{\theta\mathrm{S}}] + \overline{r}[\varepsilon_{rt}^{\theta\mathrm{S}}]\big)[E^\theta] \tag{8.147}$$

$$\frac{\partial}{\partial\theta}[E^\theta] = -\mathrm{j}[R_{\mathrm{H}}^\theta][H^\theta] = -\mathrm{j}\big([R_{\mathrm{H}}^{\theta\mathrm{S}}] + \overline{r}[\mu_{rt}^{\theta\mathrm{S}}]\big)[H^\theta] \tag{8.148}$$

We divide the matrices $[R_{\mathrm{E,H}}^\theta]$ into two parts. These matrices are:

$$[R_{\mathrm{E}}^{\theta\mathrm{S}}] = \begin{bmatrix}D_\phi p_\mu D_\phi & -D_\phi p_\mu D_{\overline{r}} \\ -D_{\overline{r}} p_\mu D_\phi & D_{\overline{r}} p_\mu D_{\overline{r}}\end{bmatrix} \qquad [\varepsilon_{rt}^{\theta\mathrm{S}}] = \begin{bmatrix}\overline{s}_\theta \varepsilon_{rr} & 0 \\ 0 & \overline{s}_\theta^{-1}\varepsilon_{r\phi}\end{bmatrix} \tag{8.149}$$

$$[R_{\mathrm{H}}^{\theta\mathrm{S}}] = \begin{bmatrix}D_{\overline{r}} p_\varepsilon D_{\overline{r}} & D_{\overline{r}} p_\varepsilon D_\phi \\ D_\phi p_\varepsilon D_{\overline{r}} & D_\phi p_\varepsilon D_\phi\end{bmatrix} \qquad [\mu_{rt}^{\theta\mathrm{S}}] = \begin{bmatrix}\overline{s}_\theta^{-1}\mu_{r\phi} & 0 \\ 0 & \overline{s}_\theta \mu_{rr}\end{bmatrix} \tag{8.150}$$

The products of the matrices $[R_{\mathrm{E}}^{\theta\mathrm{S}}]$ and $[R_{\mathrm{H}}^{\theta\mathrm{S}}]$ are always equal to zero.

8.6.2.7 Propagation in ϕ-direction

From the two first equations of Faraday's law and of the law of induction we obtain:

$$-\mathrm{j}\begin{bmatrix}\varepsilon_{rr} & \\ & \varepsilon_{r\theta}\end{bmatrix}\begin{bmatrix}E_r\\ E_\theta\end{bmatrix}=\frac{1}{\overline{r}s_\theta}\frac{\partial}{\partial\phi}\begin{bmatrix}\widetilde{H}_\theta\\ -\widetilde{H}_r\end{bmatrix}-\begin{bmatrix}(\overline{r}s_\theta)^{-1}D_\theta s_\theta\\ -\overline{r}^{-1}D_{\overline{r}}\overline{r}\end{bmatrix}\widetilde{H}_\phi \quad (8.151)$$

$$-\mathrm{j}\begin{bmatrix}\mu_{r\theta} & \\ & \mu_{rr}\end{bmatrix}\begin{bmatrix}\widetilde{H}_\theta\\ -\widetilde{H}_r\end{bmatrix}=\frac{1}{\overline{r}s_\theta}\frac{\partial}{\partial\phi}\begin{bmatrix}E_r\\ E_\theta\end{bmatrix}-\begin{bmatrix}\overline{r}^{-1}D_{\overline{r}}\overline{r}\\ (\overline{r}s_\theta)^{-1}D_\theta s_\theta\end{bmatrix}E_\phi \quad (8.152)$$

Again, we distinguish between ε_{rr}, $\varepsilon_{r\theta}$ and $\varepsilon_{r\phi}$, and between μ_{rr}, $\mu_{r\theta}$ and $\mu_{r\phi}$, because these quantities are discretised on different places in the case of inhomogeneous layers. As before, we cannot introduce anisotropic material in this simple way. By using the transformation described in Section 8.6.2.1 we obtain material parameters that are functions of θ and ϕ, so that an analytical solution in ϕ-direction is not possible. From the third equations of Faraday's law and the law of induction we obtain:

$$\mathrm{j}\mu_{r\phi}\overline{r}\widetilde{H}_\phi=\begin{bmatrix}D_\theta & -D_{\overline{r}}\end{bmatrix}\begin{bmatrix}E_r\\ \overline{r}E_\theta\end{bmatrix} \quad (8.153)$$

$$\mathrm{j}\varepsilon_{r\phi}\overline{r}E_\phi=\begin{bmatrix}D_{\overline{r}} & D_\theta\end{bmatrix}\begin{bmatrix}\overline{r}\widetilde{H}_\theta\\ -\widetilde{H}_r\end{bmatrix} \quad (8.154)$$

The equations above may be rewritten as:

$$-\mathrm{j}\begin{bmatrix}\overline{r}^2 s_\theta\varepsilon_{rr} & \\ & s_\theta\varepsilon_{r\theta}\end{bmatrix}\begin{bmatrix}E_r\\ \overline{r}E_\theta\end{bmatrix}=\frac{\partial}{\partial\phi}\begin{bmatrix}\overline{r}\widetilde{H}_\theta\\ -\widetilde{H}_r\end{bmatrix}+\mathrm{j}\begin{bmatrix}D_\theta s_\theta\\ -s_\theta D_{\overline{r}}\end{bmatrix}j\overline{r}\widetilde{H}_\phi \quad (8.155)$$

$$-\mathrm{j}\begin{bmatrix}s_\theta\mu_{r\theta} & \\ & \overline{r}^2 s_\theta\mu_{rr}\end{bmatrix}\begin{bmatrix}\overline{r}\widetilde{H}_\theta\\ -\widetilde{H}_r\end{bmatrix}=\frac{\partial}{\partial\phi}\begin{bmatrix}E_r\\ \overline{r}E_\theta\end{bmatrix}+\mathrm{j}\begin{bmatrix}s_\theta D_{\overline{r}}\\ D_\theta s_\theta\end{bmatrix}j\overline{r}E_\phi \quad (8.156)$$

We use the following normalisations:

$$\overline{r}_0=r_0k_0 \qquad D_{\overline{\theta}}=\partial/\partial(\overline{r}_0\theta) \qquad \partial\overline{\phi}=\partial(\overline{r}_0\phi) \qquad \overline{r}_n=\overline{r}/\overline{r}_0 \quad (8.157)$$

where r_0 is a suitable constant radius. For $r_0 \longrightarrow \infty$ we obtain the following limits (r has the same size in the associated structure):

$$\overline{r}_n=1 \qquad s_\theta=1 \qquad \overline{r}_0\partial\phi\rightarrow\partial\overline{z} \qquad \partial\overline{r}\rightarrow\partial\overline{t} \qquad \overline{r}_0\partial\theta\rightarrow\partial\overline{y} \quad (8.158)$$

As we can see, these formulas transform to Cartesian coordinates, and we could also e.g. use the developed programs for analysing 'Cartesian structures'. By defining:

$$[\widehat{E}^\phi]=\begin{bmatrix}E_r\\ \overline{r}_nE_\theta\end{bmatrix} \qquad [\widehat{H}^\phi]=\begin{bmatrix}\overline{r}_n\widetilde{H}_\theta\\ -\widetilde{H}_r\end{bmatrix} \quad (8.159)$$

and by the introduction of $\mathrm{j}\widetilde{H}_\phi$ and $\mathrm{j}E_\phi$ from eqs. (8.153) and (8.154), we obtain:

$$\frac{\partial}{\overline{r}_0\partial\phi}[H^\phi] = -\mathrm{j}[R_\mathrm{E}^\phi][E^\phi] = -\mathrm{j}\big([R_\mathrm{E}^{\phi\mathrm{S}}] + \widehat{s}_\theta[\varepsilon_{rt}^{\phi\mathrm{S}}]\big)[E^\phi] \tag{8.160}$$

$$\frac{\partial}{\overline{r}_0\partial\phi}[E^\phi] = -\mathrm{j}[R_\mathrm{H}^\phi][H^\phi] = -\mathrm{j}\big([R_\mathrm{H}^{\phi\mathrm{S}}] + \widehat{s}_\theta[\mu_{rt}^{\phi\mathrm{S}}]\big)[H^\phi] \tag{8.161}$$

with $\widehat{s}_\theta = \mathrm{diag}(s_\theta, s_\theta)$. We divide the matrices $[R_{\mathrm{E,H}}^\phi]$ into two parts. With the abbreviations $s_\theta^\varepsilon = s_\theta\varepsilon_{r\phi}^{-1}$ and $s_\theta^\mu = s_\theta\mu_{r\phi}^{-1}$ we obtain for these matrices:

$$[R_\mathrm{E}^{\phi\mathrm{S}}] = \begin{bmatrix} D_{\overline{\theta}}s_\theta^\mu D_{\overline{\theta}} & -D_{\overline{\theta}}s_\theta^\mu D_{\overline{r}} \\ -D_{\overline{r}}s_\theta^\mu D_{\overline{\theta}} & D_{\overline{r}}s_\theta^\mu D_{\overline{r}} \end{bmatrix} \qquad [\widehat{\varepsilon}_{rt}^{\phi\mathrm{S}}] = \begin{bmatrix} \overline{r}_n^2\varepsilon_{rr} & 0 \\ 0 & \varepsilon_{r\theta} \end{bmatrix} \tag{8.162}$$

$$[R_\mathrm{H}^{\phi\mathrm{S}}] = \begin{bmatrix} D_{\overline{r}}s_\theta^\varepsilon D_{\overline{r}} & D_{\overline{r}}s_\theta^\varepsilon D_{\overline{\theta}} \\ D_{\overline{\theta}}s_\theta^\varepsilon D_{\overline{r}} & D_{\overline{\theta}}s_\theta^\varepsilon D_{\overline{\theta}} \end{bmatrix} \qquad [\widehat{\mu}_{rt}^{\phi\mathrm{S}}] = \begin{bmatrix} \mu_{r\theta} & 0 \\ 0 & \overline{r}_n^2\mu_{rr} \end{bmatrix} \tag{8.163}$$

(Please note that $D_{\overline{r}}$ is normalised with k_0 but not with $\overline{r}_0$.) The products of the matrices $[R_\mathrm{E}^{\phi\mathrm{S}}]$ and $[R_\mathrm{H}^{\phi\mathrm{S}}]$ are always equal to zero.

By combining the two first-order equations we obtain the wave equations:

$$\frac{\mathrm{d}^2}{\overline{r}_0^2\mathrm{d}\phi^2}[E^\phi] - [Q_\mathrm{E}^\phi][E^\phi] = 0 \tag{8.164}$$

$$\frac{\mathrm{d}^2}{\overline{r}_0^2\mathrm{d}\phi^2}[H^\phi] - [Q_\mathrm{H}^\phi][H^\phi] = 0 \tag{8.165}$$

With the abbreviations $\widehat{p}_{\varepsilon\mu} = \mathrm{diag}(\varepsilon_{rr}\mu_{r\theta}, \varepsilon_{r\theta}\mu_{rr})$ and $\widehat{p}_{rs} = \mathrm{diag}(\overline{r}_n s_\theta, \overline{r}_n s_\theta)$, the matrices $[Q_{\mathrm{E,H}}^\phi]$ are given by:

$$-[Q_\mathrm{E}^\phi] = [R_\mathrm{H}^{\phi\mathrm{S}}]\widehat{s}_\theta[\varepsilon_{rt}^{\phi\mathrm{S}}] + \widehat{s}_\theta[\mu_{rt}^{\phi\mathrm{S}}][R_\mathrm{E}^{\phi\mathrm{S}}] + \widehat{p}_{rs}^2\widehat{p}_{\varepsilon\mu} \tag{8.166}$$

$$-[Q_\mathrm{H}^\phi] = [R_\mathrm{E}^{\phi\mathrm{S}}]\widehat{s}_\theta[\mu_{rt}^{\phi\mathrm{S}}] + \widehat{s}_\theta[\varepsilon_{rt}^{\phi\mathrm{S}}][R_\mathrm{H}^{\phi\mathrm{S}}] + \widehat{p}_{rs}^2\widehat{p}_{\varepsilon\mu} \tag{8.167}$$

The submatrices of $[Q_\mathrm{E}^\phi]$ and $[Q_\mathrm{H}^\phi]$ can now easily be calculated. We will use the results of this section in the analysis of conformal antennas on a sphere.

8.6.2.8 Differential equations for E_r and $\widetilde{H}_r$

In layers (or sections) with homogeneous and isotropic material, we may write $\mu_{r\theta} = \mu_{rr} = \mu_{r\phi} = \mu_r$ and $\varepsilon_{rr} = \varepsilon_{r\theta} = \varepsilon_{r\phi} = \varepsilon_r$. In that case, the submatrices $Q_{\mathrm{E}12}^\phi$ and $Q_{\mathrm{H}21}^\phi$ are zero and the equations for E_r and $\widetilde{H}_r$ decouple:

$$\frac{\mathrm{d}^2}{\overline{r}_0^2\mathrm{d}\phi^2}E_r + s_\theta D_{\overline{\theta}}s_\theta D_{\overline{\theta}}E_r + D_{\overline{r}}s_\theta D_{\overline{r}}\overline{r}_n^2 s_\theta E_r + (\overline{r}_n s_\theta)^2\varepsilon_r\mu_r E_r = 0 \tag{8.168}$$

$$\frac{\mathrm{d}^2}{\overline{r}_0^2\mathrm{d}\phi^2}\widetilde{H}_r + s_\theta D_{\overline{\theta}}s_\theta D_{\overline{\theta}}\widetilde{H}_r + D_{\overline{r}}s_\theta D_{\overline{r}}\overline{r}_n^2 s_\theta \widetilde{H}_r + (\overline{r}_n s_\theta)^2\varepsilon_r\mu_r \widetilde{H}_r = 0 \tag{8.169}$$

We denormalise these equations by multiplication with $\overline{r}_0^2$ and write them in the following form:

$$\frac{\mathrm{d}^2}{\mathrm{d}\overline{r}^2}(\overline{r}^2 E_r) + s_\theta^{-1} D_\theta s_\theta D_\theta E_r + s_\theta^{-2} D_\phi D_\phi E_r + \overline{r}^2 \varepsilon_r \mu_r E_r = 0 \tag{8.170}$$

$$\frac{\mathrm{d}^2}{\mathrm{d}\overline{r}^2}(\overline{r}^2 \widetilde{H}_r) + s_\theta^{-1} D_\theta s_\theta D_\theta \widetilde{H}_r + s_\theta^{-2} D_\phi D_\phi \widetilde{H}_r + \overline{r}^2 \varepsilon_r \mu_r \widetilde{H}_r = 0 \tag{8.171}$$

It should be mentioned that the operators (especially the differential ones) have to be applied to all terms of this operator. Therefore the expression $D_\theta s_\theta D_\theta (tE_r)$ is the same as $D_\theta(s_\theta D_\theta(tE_r))$. In the next step we replace $\sqrt{\varepsilon_r \mu_r}\overline{r} = t$ and transform, analogously to eq. (8.137):

$$t\frac{\mathrm{d}^2}{\mathrm{d}t^2}(t(tE_r)) + s_\theta^{-1} D_\theta s_\theta D_\theta (tE_r) + s_\theta^{-2} D_\phi D_\phi (tE_r) + t^2(tE_r) = 0 \tag{8.172}$$

$$t\frac{\mathrm{d}^2}{\mathrm{d}t^2}(t(t\widetilde{H}_r)) + s_\theta^{-1} D_\theta s_\theta D_\theta (t\widetilde{H}_r) + s_\theta^{-2} D_\phi D_\phi (t\widetilde{H}_r) + t^2(t\widetilde{H}_r) = 0 \tag{8.173}$$

We replace $t\frac{\mathrm{d}^2}{\mathrm{d}t^2}(tF)$ with $\frac{\mathrm{d}}{\mathrm{d}t}t^2\frac{\mathrm{d}}{\mathrm{d}t}F$ ($F = tE_r$ or $F = t\widetilde{H}_r$) and transform the field components according to:

$$tE_r = \sqrt{t}^{-1} G_\mathrm{E}(t) \qquad tH_r = \sqrt{t}^{-1} G_\mathrm{H}(t) \tag{8.174}$$

Due to the equality:

$$\sqrt{t}\frac{\mathrm{d}}{\mathrm{d}t}t^2\frac{\mathrm{d}}{\mathrm{d}t}(\sqrt{t}^{-1}G(t)) = t\frac{\mathrm{d}}{\mathrm{d}t}t\frac{\mathrm{d}}{\mathrm{d}t}G(t) - \frac{1}{4}G(t) \tag{8.175}$$

we obtain the following wave equations of Bessel type:

$$t\frac{\mathrm{d}}{\mathrm{d}t}t\frac{\mathrm{d}}{\mathrm{d}t}G_\mathrm{E} + \left(s_\theta^{-2} D_\phi D_\phi + s_\theta^{-1} D_\theta s_\theta D_\theta - \frac{1}{4}\right) G_\mathrm{E} + t^2 G_\mathrm{E} = 0 \tag{8.176}$$

$$t\frac{\mathrm{d}}{\mathrm{d}t}t\frac{\mathrm{d}}{\mathrm{d}t}G_\mathrm{H} + \left(s_\theta^{-2} D_\phi D_\phi + s_\theta^{-1} D_\theta s_\theta D_\theta - \frac{1}{4}\right) G_\mathrm{H} + t^2 G_\mathrm{H} = 0 \tag{8.177}$$

In Section 8.6.4 we will show how these differential equations can be solved for the r-direction. In the analysis process we must also calculate the other field components. This calculation will be shown as well.

8.6.3 Analysis of special devices – conformal antennas

8.6.3.1 Introduction

In this subsection we will show the analysis of conformal antennas on spheres. We have already dealt with such antennas on a cylindrical body in Section 4.7.1. Fig. 8.10 shows a circular microstrip resonator on a sphere which can act as a conformal antenna. The discretisation is performed in the θ-direction. Another example is drawn in Fig. 8.11. We assume propagation in

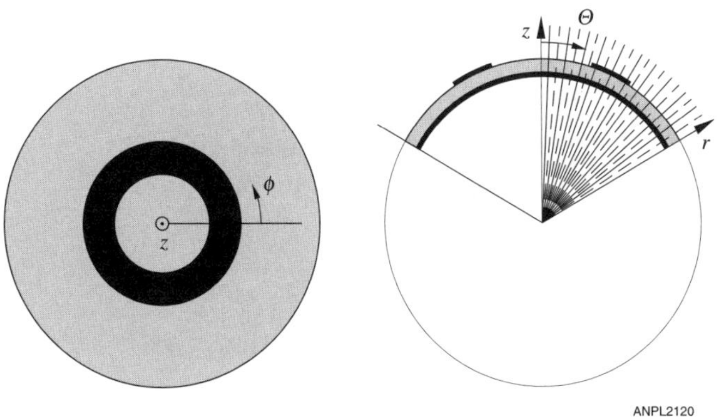

———	e	$E_r\ E_\phi\ H_\theta\ S_\phi$	$\varepsilon_{rr}\ \varepsilon_{\phi\phi}\ \mu_{\theta\theta}$
– – –	h	$H_r\ H_\phi\ E_\theta\ S_\theta$	$\varepsilon_{\theta\theta}\ \mu_{rr}\ \mu_{\phi\phi}$

Fig. 8.10 Circular microstrip resonator on a sphere

ϕ-direction. Therefore, we discretise the fields in r- and θ-directions. A more general cross-section is presented in Fig. 8.12, in which the discretisation lines are shown as points. Generally, the examined cross-sections can consist of various different layers.

Fig. 8.11 Conformal antenna on a sphere (R. Pregla, 'Efficient Analysis of Conformal Antennas with Anisotropic Material (0682)', in *AP 2000 Millennium Conference on Antennas and Propagation.* © 2000 European Space Agency (ESA))

8.6.3.2 Discretisation

The corresponding differential equations for this case (see Section 8.6.2.7) are now discretised. Fig. 8.12 shows the cross-section of a conformal antenna with the adequate discretisation points. As usual, the field components are

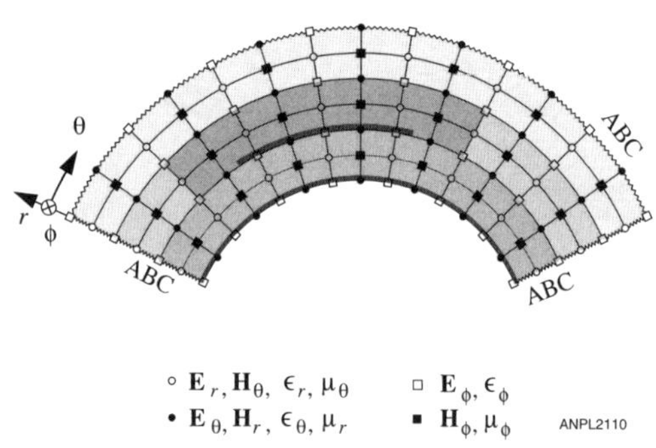

Fig. 8.12 Cross-section of a conformal antenna on a sphere

discretised on different positions. However, the positions of E_r and H_θ, as well as of E_θ and H_r, coincide. These are the field components with which the two parts of the Poynting vector in ϕ-direction – the assumed direction of wave propagation – are calculated. We collect the discretised field components in column vectors.

To order these components, we start with the inner point on the left side and go towards the outside of the first column (see Fig. 8.12). Then we continue with the second column and put these values below those of the first column etc. The column vectors are indicated by boldface; the subscript i represents the ith column. The collection of all these vectors in the total column vector is marked with a hat ($\widehat{\ }$). If we have N_θ° columns of $\circ$-discretisation points and N_r° points in each column, we obtain as the total number of discretisation points $\mathrm{N}_r^\circ \mathrm{N}_\theta^\circ$. If we have ideal metal in a cross-section we reduce the total number of discretisation points by the number of discretisation points in the metal area. In the case of non-ideal metal, the metal is modelled as a dielectric with complex permittivity. The numbers of $\bullet$ columns and rows are $\mathrm{N}_\theta^\bullet$ and $\mathrm{N}_r^\bullet$, respectively. Now, due to the discretisation, we transform the continuous values on the left into discrete ones on the right:

$$
\begin{aligned}
E_r, \widetilde{H}_r &\longrightarrow \widehat{\mathbf{E}}_r, \widehat{\widetilde{\mathbf{H}}}_r \\
E_\theta, E_\phi &\longrightarrow \widehat{\mathbf{E}}_\theta, \widehat{\mathbf{E}}_\phi \\
\widetilde{H}_\theta, \widetilde{H}_\phi &\longrightarrow \widehat{\widetilde{\mathbf{H}}}_\theta, \widehat{\widetilde{\mathbf{H}}}_\phi \\
\varepsilon_{rr}, \varepsilon_{r\theta}, \varepsilon_{r\phi} &\longrightarrow \widehat{\boldsymbol{\epsilon}}_r^\circ, \widehat{\boldsymbol{\epsilon}}_\theta^\bullet, \widehat{\boldsymbol{\epsilon}}_\phi^\square \\
\mu_{rr}, \mu_{r\theta}, \mu_{r\phi} &\longrightarrow \widehat{\boldsymbol{\mu}}_r^\bullet, \widehat{\boldsymbol{\mu}}_\theta^\circ, \widehat{\boldsymbol{\mu}}_\phi^\blacksquare \\
\overline{r}_n = \frac{\overline{r}}{\overline{r}_0} &\longrightarrow \widehat{\overline{\mathbf{r}}}_n^\circ, \widehat{\overline{\mathbf{r}}}_n^\bullet, \widehat{\overline{\mathbf{r}}}_n^\square, \widehat{\overline{\mathbf{r}}}_n^\blacksquare \\
\sin\theta &\longrightarrow \widehat{\mathbf{s}}_\theta^\circ, \widehat{\mathbf{s}}_\theta^\bullet, \widehat{\mathbf{s}}_\theta^\square, \widehat{\mathbf{s}}_\theta^\blacksquare
\end{aligned}
$$

$$\frac{\partial}{\partial \overline{r}} \longrightarrow \overline{h}_r^{-1} \boldsymbol{D}_{\overline{r}}^{\circ,\bullet} = \overline{\boldsymbol{D}}_{\overline{r}}^{\circ,\bullet}$$
$$\overline{r}_0^{-1} \frac{\partial}{\partial \theta} = \frac{\partial}{\partial \theta_n} \longrightarrow (\overline{r}_0 h_\theta)^{-1} \boldsymbol{D}_\theta^{\circ,\bullet} = \overline{\boldsymbol{D}}_{\theta_n}^{\circ,\bullet} = \overline{\boldsymbol{D}}_{\overline{\theta}}^{\circ,\bullet}$$
$$\theta_n = \overline{r}_0 \theta = \overline{\theta} \tag{8.178}$$

We collect the permittivities and permeabilities in the same order as the field components. However, we put them on the main diagonal of a diagonal matrix (not in a vector). The permittivities and permeabilities correspond to the field components, which are discretised on different points (see Fig. 8.12). Therefore, we have three different permittivity and permeability matrices, respectively, in the case of inhomogeneous material distribution in the cross-section. This is the reason why we distinguished between the three values even in the isotropic case. The derivatives are replaced by central differences. All differences in one row (part of circular line) or one column are collected in difference operator matrices $\boldsymbol{D}_\theta^{\circ,\bullet}$ and $\boldsymbol{D}_{\overline{r}}^{\circ,\bullet}$, respectively. The matrices $\boldsymbol{D}_\theta^{\circ}$, $\boldsymbol{D}_{\overline{r}}^{\bullet}$ and $\boldsymbol{D}_{\overline{r}}^{\circ}$, $\boldsymbol{D}_\theta^{\bullet}$ have to fulfil Dirichlet and Neumann boundary conditions, respectively, when there are electric walls. When there are magnetic walls the conditions have to be exchanged. Electric and magnetic walls have to be positioned at different places (analogous to Fig. 2.27). If required, absorbing boundary conditions (ABC) have to be introduced instead of Dirichlet walls. When there are inhomogeneous cross-sections without metallic subsections inside, the difference operators can be described by Kronecker products between the difference operators for the rows ($\boldsymbol{D}_{\overline{r}}^{\circ,\bullet}$) or columns ($\boldsymbol{D}_\theta^{\circ,\bullet}$) and unit matrices that have the order of the number of rows or columns. So we obtain:

$$\widehat{\boldsymbol{D}}_{\overline{r}}^{\circ,\bullet} = \boldsymbol{I}_\theta^{\circ,\bullet} \otimes \boldsymbol{D}_{\overline{r}}^{\circ,\bullet} \quad \widehat{\boldsymbol{D}}_\theta^{\circ,\bullet} = \boldsymbol{D}_\theta^{\circ,\bullet} \otimes \boldsymbol{I}_r^{\circ,\bullet}$$

$\boldsymbol{I}_r^{\circ,\bullet}$ and $\boldsymbol{I}_\theta^{\circ,\bullet}$ are unit matrices of the order $\mathrm{N}_r^{\circ,\bullet}$ and $\mathrm{N}_\theta^{\circ,\bullet}$, respectively. With a metallic subsection (analogous to Fig. 2.27) the cross-sections are divided by lines in the radial and circular directions according to the metallisation boundaries. The further procedure is analogous to that in the Cartesian case. The solution is also obtained analogously, and is therefore not repeated here. In contrast to the Cartesian case, we must here discretise the radii and the function $\sin\theta = s_\theta$. In the discretised points of a circular row we have a constant radius. In a radial column we have a constant θ. Hence we obtain:

$$\widehat{\overline{\boldsymbol{r}}}_n^{\circ,\bullet} = \boldsymbol{I}_\theta^{\circ,\bullet} \otimes \overline{\boldsymbol{r}}_n^{\circ,\bullet} \qquad \widehat{\boldsymbol{s}}_\theta^{\circ,\bullet} = \boldsymbol{s}_\theta^{\circ,\bullet} \otimes \boldsymbol{I}_r^{\circ,\bullet}$$

Analogous equations are valid for the quantities labelled with □ and ▪.

8.6.3.3 GTL equations and solutions

With the supervectors of the discretised field components:

$$\widehat{\mathbf{E}}^\phi = \left[\widehat{\mathbf{E}}_r^{\mathrm{t}}, (\widehat{\overline{\boldsymbol{r}}}_n^{\bullet} \widehat{\mathbf{E}}_\theta)^{\mathrm{t}}\right]^{\mathrm{t}} \qquad \widehat{\mathbf{H}}^\phi = \left[(\widehat{\overline{\boldsymbol{r}}}_n^{\circ} \widehat{\mathbf{H}}_\theta)^{\mathrm{t}}, -\widehat{\mathbf{H}}_r^{\mathrm{t}}\right]^{\mathrm{t}} \tag{8.179}$$

we write for eqs. (8.160) and (8.161) and eqs. (8.162) and (8.163), respectively:

$$\frac{\partial}{\overline{r}_0\partial\phi}\widehat{\mathbf{H}}^{\phi} = -\mathrm{j}\boldsymbol{R}_{\mathrm{E}}^{\phi}\widehat{\mathbf{E}}^{\phi} = -\mathrm{j}\big(\boldsymbol{R}_{\mathrm{E}}^{\phi\mathrm{S}} + \widehat{\mathbf{s}}_{\theta}\widehat{\boldsymbol{\epsilon}}_{rt}^{\phi\mathrm{S}}\big)\widehat{\mathbf{E}}^{\phi} \tag{8.180}$$

$$\frac{\partial}{\overline{r}_0\partial\phi}\widehat{\mathbf{E}}^{\phi} = -\mathrm{j}\boldsymbol{R}_{\mathrm{H}}^{\phi}\widehat{\mathbf{H}}^{\phi} = -\mathrm{j}\big(\boldsymbol{R}_{\mathrm{H}}^{\phi\mathrm{S}} + \widehat{\mathbf{s}}_{\theta}\widehat{\boldsymbol{\mu}}_{rt}^{\phi\mathrm{S}}\big)\widehat{\mathbf{H}}^{\phi} \tag{8.181}$$

with $\widehat{\mathbf{s}}_{\theta} = \mathrm{diag}(\widehat{\mathbf{s}}_{\theta}^{\circ}, \widehat{\mathbf{s}}_{\theta}^{\bullet})$. By using the abbreviations $\widehat{\mathbf{s}}_{\theta}^{\varepsilon} = \widehat{\mathbf{s}}_{\theta}^{\square}\widehat{\boldsymbol{\epsilon}}_{\phi}^{\square\,-1}$ and $\widehat{\mathbf{s}}_{\theta}^{\mu} = \widehat{\mathbf{s}}_{\theta}^{\blacksquare}\widehat{\boldsymbol{\mu}}_{\phi}^{\blacksquare\,-1}$, we obtain for these matrices:

$$\boldsymbol{R}_{\mathrm{E}}^{\phi\mathrm{S}} = \begin{bmatrix} -\widehat{\overline{\boldsymbol{D}}}_{\overline{\theta}}^{\circ\mathrm{t}}\,\mathbf{s}_{\theta}^{\mu}\widehat{\overline{\boldsymbol{D}}}_{\overline{\theta}}^{\circ} & \widehat{\overline{\boldsymbol{D}}}_{\overline{\theta}}^{\circ\mathrm{t}}\,\mathbf{s}_{\theta}^{\mu}\widehat{\overline{\boldsymbol{D}}}_{\overline{r}}^{\bullet} \\ \widehat{\overline{\boldsymbol{D}}}_{\overline{r}}^{\bullet\mathrm{t}}\,\mathbf{s}_{\theta}^{\mu}\widehat{\overline{\boldsymbol{D}}}_{\overline{\theta}}^{\circ} & -\widehat{\overline{\boldsymbol{D}}}_{\overline{r}}^{\bullet\mathrm{t}}\,\mathbf{s}_{\theta}^{\mu}\widehat{\overline{\boldsymbol{D}}}_{\overline{r}}^{\bullet} \end{bmatrix} \qquad \widehat{\boldsymbol{\epsilon}}_{rt}^{\phi\mathrm{S}} = \begin{bmatrix} \widehat{\overline{\mathbf{r}}}_{n}^{\circ 2}\widehat{\boldsymbol{\epsilon}}_{r}^{\circ} & \boldsymbol{0} \\ \boldsymbol{0} & \widehat{\boldsymbol{\epsilon}}_{\theta}^{\bullet} \end{bmatrix} \tag{8.182}$$

$$\boldsymbol{R}_{\mathrm{H}}^{\phi\mathrm{S}} = \begin{bmatrix} -\widehat{\overline{\boldsymbol{D}}}_{\overline{r}}^{\circ\mathrm{t}}\,\widehat{\mathbf{s}}_{\theta}^{\varepsilon}\widehat{\overline{\boldsymbol{D}}}_{\overline{r}}^{\circ} & -\widehat{\overline{\boldsymbol{D}}}_{\overline{r}}^{\circ\mathrm{t}}\,\widehat{\mathbf{s}}_{\theta}^{\varepsilon}\widehat{\overline{\boldsymbol{D}}}_{\overline{\theta}}^{\bullet} \\ -\widehat{\overline{\boldsymbol{D}}}_{\overline{\theta}}^{\bullet\mathrm{t}}\,\widehat{\mathbf{s}}_{\theta}^{\varepsilon}\widehat{\overline{\boldsymbol{D}}}_{\overline{r}}^{\circ} & -\widehat{\overline{\boldsymbol{D}}}_{\overline{\theta}}^{\bullet\mathrm{t}}\,\widehat{\mathbf{s}}_{\theta}^{\varepsilon}\widehat{\overline{\boldsymbol{D}}}_{\overline{\theta}}^{\bullet} \end{bmatrix} \qquad \widehat{\boldsymbol{\mu}}_{rt}^{\phi\mathrm{S}} = \begin{bmatrix} \widehat{\boldsymbol{\mu}}_{\theta}^{\circ} & \boldsymbol{0} \\ \boldsymbol{0} & \widehat{\overline{\mathbf{r}}}_{n}^{\bullet 2}\widehat{\boldsymbol{\mu}}_{r}^{\bullet} \end{bmatrix} \tag{8.183}$$

Like the operators in the continuous case, the products of the matrices $\boldsymbol{R}_{\mathrm{E}}^{\phi\mathrm{S}}$ and $\boldsymbol{R}_{\mathrm{H}}^{\phi\mathrm{S}}$ are always equal to zero.

By combining the two first-order equations we obtain the wave equations:

$$\frac{\mathrm{d}^2}{\overline{r}_0^2\mathrm{d}\phi^2}\widehat{\mathbf{H}}^{\phi} - \widehat{\mathbf{Q}}_{\mathrm{H}}^{\phi}\widehat{\mathbf{H}} = \mathbf{0} \qquad \frac{\mathrm{d}^2}{\overline{r}_0^2\mathrm{d}\phi^2}\widehat{\mathbf{E}}^{\phi} - \widehat{\mathbf{Q}}_{\mathrm{E}}^{\phi}\widehat{\mathbf{E}} = \mathbf{0} \tag{8.184}$$

Now we use the abbreviations $\widehat{\mathbf{p}}_{\varepsilon\mu} = \mathrm{diag}(\widehat{\boldsymbol{\epsilon}}_{r}^{\circ}\widehat{\boldsymbol{\mu}}_{\theta}^{\circ}, \widehat{\boldsymbol{\epsilon}}_{\theta}^{\bullet}\widehat{\boldsymbol{\mu}}_{r}^{\bullet})$ and $\widehat{\mathbf{p}}_{rs} = \mathrm{diag}(\widehat{\overline{\mathbf{r}}}_{n}^{\circ}\widehat{\mathbf{s}}_{\theta}^{\circ}, \widehat{\overline{\mathbf{r}}}_{n}^{\bullet}\widehat{\mathbf{s}}_{\theta}^{\bullet})$. The matrices $\boldsymbol{Q}_{\mathrm{E,H}}^{\phi}$ are given by:

$$-\boldsymbol{Q}_{\mathrm{E}}^{\phi} = \boldsymbol{R}_{\mathrm{H}}^{\phi\mathrm{S}}\widehat{\mathbf{s}}_{\theta}\widehat{\boldsymbol{\epsilon}}_{rt}^{\phi\mathrm{S}} + \widehat{\mathbf{s}}_{\theta}\widehat{\boldsymbol{\mu}}_{rt}^{\phi\mathrm{S}}\boldsymbol{R}_{\mathrm{E}}^{\phi\mathrm{S}} + \widehat{\mathbf{p}}_{rs}^{2}\widehat{\mathbf{p}}_{\varepsilon\mu} \tag{8.185}$$

$$-\boldsymbol{Q}_{\mathrm{H}}^{\phi} = \boldsymbol{R}_{\mathrm{E}}^{\phi\mathrm{S}}\widehat{\mathbf{s}}_{\theta}\widehat{\boldsymbol{\mu}}_{rt}^{\phi\mathrm{S}} + \widehat{\mathbf{s}}_{\theta}\widehat{\boldsymbol{\epsilon}}_{rt}^{\phi\mathrm{S}}\boldsymbol{R}_{\mathrm{H}}^{\phi\mathrm{S}} + \widehat{\mathbf{p}}_{rs}^{2}\widehat{\mathbf{p}}_{\varepsilon\mu} \tag{8.186}$$

The submatrices of $\boldsymbol{Q}_{\mathrm{E}}^{\phi}$ and $\boldsymbol{Q}_{\mathrm{H}}^{\phi}$ can now easily be determined.

This solution is analogous to the procedure for planar structures in the Cartesian case in Chapter 2 and will not be repeated here. Finally, we obtain the **y**- (or short-circuit admittance) matrix formulation for the fields between ports A and B:

$$\begin{bmatrix} \widehat{\widehat{\mathbf{H}}}_{\mathrm{A}} \\ -\widehat{\widehat{\mathbf{H}}}_{\mathrm{B}} \end{bmatrix} = \begin{bmatrix} \overline{\mathbf{y}}_1 & \overline{\mathbf{y}}_2 \\ \overline{\mathbf{y}}_2 & \overline{\mathbf{y}}_1 \end{bmatrix} \begin{bmatrix} \widehat{\widehat{\mathbf{E}}}_{\mathrm{A}} \\ \widehat{\widehat{\mathbf{E}}}_{\mathrm{B}} \end{bmatrix} \qquad \begin{aligned} \overline{\mathbf{y}}_1 &= \widehat{\widehat{\mathbf{Y}}}_0/\tanh(\widehat{\boldsymbol{\Gamma}}\overline{r}_0\Phi_{\mathrm{AB}}) \\ \overline{\mathbf{y}}_2 &= -\widehat{\widehat{\mathbf{Y}}}_0/\sinh(\widehat{\boldsymbol{\Gamma}}\overline{r}_0\Phi_{\mathrm{AB}}) \end{aligned} \tag{8.187}$$

By inverting this expression, we obtain the **z**- (or open-circuit impedance) matrix formulation:

$$\begin{bmatrix} \widehat{\widehat{\mathbf{E}}}_{\mathrm{A}} \\ \widehat{\widehat{\mathbf{E}}}_{\mathrm{B}} \end{bmatrix} = \begin{bmatrix} \widehat{\overline{\mathbf{z}}}_1 & \widehat{\overline{\mathbf{z}}}_2 \\ \widehat{\overline{\mathbf{z}}}_2 & \widehat{\overline{\mathbf{z}}}_1 \end{bmatrix} \begin{bmatrix} \widehat{\widehat{\mathbf{H}}}_{\mathrm{A}} \\ -\widehat{\widehat{\mathbf{H}}}_{\mathrm{B}} \end{bmatrix} \qquad \begin{aligned} \overline{\mathbf{z}}_1 &= \widehat{\widehat{\mathbf{Z}}}_0/\tanh(\widehat{\boldsymbol{\Gamma}}\overline{r}_0\Phi_{\mathrm{AB}}) \\ \overline{\mathbf{z}}_2 &= \widehat{\widehat{\mathbf{Z}}}_0/\sinh(\widehat{\boldsymbol{\Gamma}}\overline{r}_0\Phi_{\mathrm{AB}}) \end{aligned} \tag{8.188}$$

We introduce the abbreviation $\Phi_{\mathrm{AB}} = \Phi_{\mathrm{B}} - \Phi_{\mathrm{A}}$. The division, indicated by '/' stands for the multiplication with the inverse matrix. After a suitable normalisation, we have $\widehat{\widetilde{\mathbf{Z}}}_0 = \widehat{\widetilde{\mathbf{Y}}}_0 = \widehat{\mathbf{I}}$. The admittance/impedance transformation through homogeneous waveguide sections and through concatenations of waveguide sections can be performed as described in Section 2.5 and we do not go into details here.

8.6.4 Analysis of special devices – conical horn antennas

8.6.4.1 Introduction

In this section we use the results of Section 8.6.2.3 for the analysis of radiating problems that can be described in the spherical coordinate system. To predict the performance of a Cassegrainian antenna (see Fig. 8.13), we need a detailed knowledge of the characteristics of the electromagnetic field in the near-field region of the primary feed shown in Fig. 8.14. As the first step of the analysis, we consider the primary feed of the antenna system.

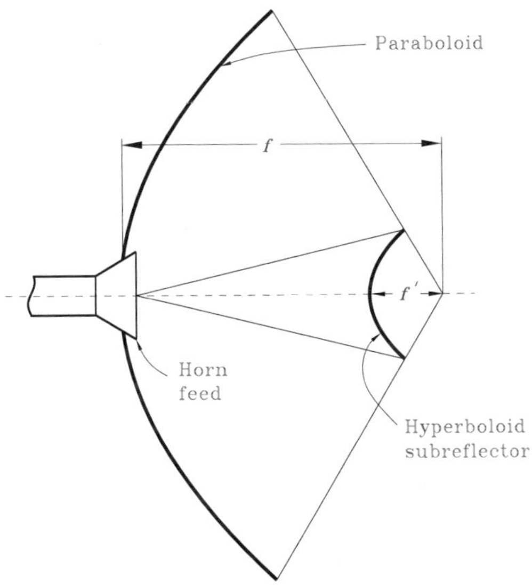

Fig. 8.13 Cassegrainian antenna (Reproduced by permission of Copernicus Gesellschaft mbH)

8.6.4.2 Discretisation

An arbitrary electromagnetic field in a homogeneous region can be constructed as a superposition of its TM and TE parts (see [13] and Section 8.6.2.3). For the special case $D_\phi = \partial/\partial\phi = 0$ (azimuthally independent fields), the TM problem can be described with the azimuthal magnetic field component $\widetilde{H}_\phi$

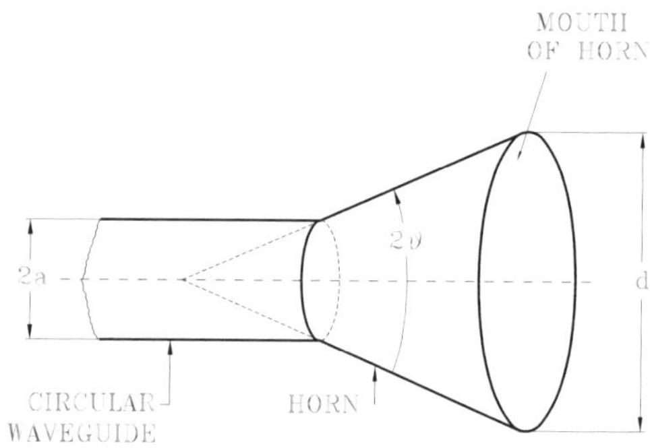

Fig. 8.14 Circular horn as primary feed element (Reproduced by permission of Copernicus Gesellschaft mbH)

and the TE problem can be described with the electric field component $\widetilde{E}_\phi$. In general, the ϕ dependence is described by:

$$E_\theta, \widetilde{H}_\phi \sim \sin(m\phi) \qquad E_\phi, \widetilde{H}_\theta \sim \cos(m\phi) \tag{8.189}$$

Therefore, we may write $D_\phi^2 = \partial^2/\partial\phi^2 = -m^2$.

Fig. 8.15 shows the horizontal section of the discretised computational window. Since we discretise only in θ-direction, the fields are determined analytically on lines in the radial direction. We use different line systems, which are shifted towards each other to satisfy the interface conditions in the direction of discretisation. In particular, for $\theta = 0$ the discretisation depends on the mode to be calculated. In the further analysis we will describe only the case shown in Fig. 8.15a. We obtain:

$$\begin{array}{rcll}
E_r, \widetilde{H}_r & \longrightarrow & \mathbf{E}_r, \widetilde{\mathbf{H}}_r & \text{column vectors} \\
E_\theta, E_\phi & \longrightarrow & \mathbf{E}_\theta, \mathbf{E}_\phi & \text{column vectors} \\
\widetilde{H}_\theta, \widetilde{H}_\phi & \longrightarrow & \widetilde{\mathbf{H}}_\theta, \widetilde{\mathbf{H}}_\phi & \text{column vectors} \\
\sin\theta & \longrightarrow & \mathbf{s}_\theta^\circ, \mathbf{s}_\theta^\bullet & \text{diagonal matrices} \\
\sqrt{\sin\theta} & \longrightarrow & \mathbf{w}^{\circ,\bullet} & \text{diagonal matrices} \\
\dfrac{\partial}{\partial\theta} & \longrightarrow & h_\theta^{-1}\mathbf{D}_\theta^{\circ,\bullet} = \overline{\mathbf{D}}_\theta^{\circ,\bullet} &
\end{array} \tag{8.190}$$

h_θ is the angular discretisation distance in θ-direction.

8.6.4.3 Solution of the wave equation

Now we combine the discretised fields into supervectors (indicated by a hat $(\widehat{\ })$) and normalise according to eq. (8.131) in discretised form. Then by discretising the derivatives with respect to θ with finite differences, the wave equations (8.133) and (8.134) are transformed to:

$$t\frac{\mathrm{d}}{\mathrm{d}t}\left(t\frac{\mathrm{d}}{\mathrm{d}t}\widehat{\mathbf{G}}_{\mathrm{E,H}}^r\right) + \left(\widehat{\boldsymbol{Q}}_{\mathrm{E,H}}^{\mathrm{rS}} - \frac{1}{4}\boldsymbol{I}\right)\widehat{\mathbf{G}}_{\mathrm{E,H}}^r + t^2\widehat{\mathbf{G}}_{\mathrm{E,H}}^r = 0 \tag{8.191}$$

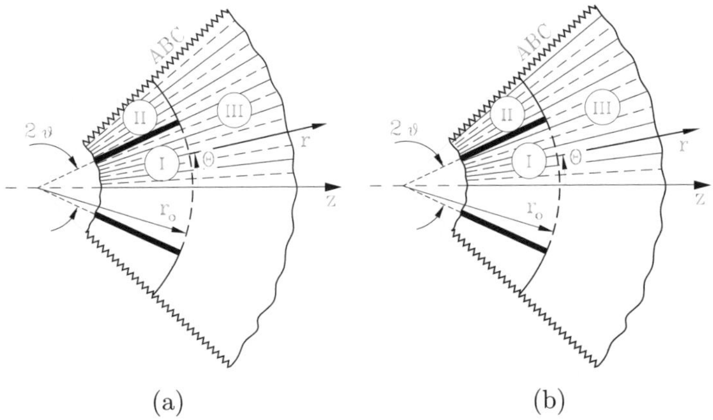

Fig. 8.15 Circular horn with discretisation lines (a) • —— E_r, E_ϕ, H_θ, s_θ ∘ - - - - H_r, H_ϕ, E_θ, s_θ: H_{0n}, H_{11}, E_{11} (b) —— H_r, H_ϕ, E_θ - - - - E_r, E_ϕ, H_θ: E_{0n} (Reproduced by permission of Copernicus Gesellschaft mbH)

The material is homogeneous and isotropic. Therefore, the matrices $\widehat{\boldsymbol{Q}}^{\mathrm{rS}}_{\mathrm{E,H}}$ read:

$$\widehat{\boldsymbol{Q}}^{\mathrm{rS}}_{\mathrm{H}} = \begin{bmatrix} \boldsymbol{Q}^{\mathrm{rS}}_{\mathrm{H11}} & \boldsymbol{Q}^{\mathrm{rS}}_{\mathrm{H12}} \\ \boldsymbol{0} & \boldsymbol{Q}^{\mathrm{rS}}_{\mathrm{H22}} \end{bmatrix} = \begin{bmatrix} \mathbf{s}^{\circ}_{\theta}\boldsymbol{D}^{\bullet}_{\theta}\mathbf{s}^{\bullet\,-1}_{\theta}\boldsymbol{D}^{\circ}_{\theta} & -m\mathbf{s}^{\circ}_{\theta}\boldsymbol{D}^{\bullet}_{\theta}\mathbf{s}^{\bullet\,-1}_{\theta} \\ \boldsymbol{0} & \mathbf{s}^{\bullet\,-2}_{\theta}(\boldsymbol{D}^{\circ}_{\theta}\mathbf{s}^{\circ}_{\theta}\boldsymbol{D}^{\bullet}_{\theta}\mathbf{s}^{\bullet}_{\theta} - m^2\boldsymbol{I}^{\bullet}) \end{bmatrix}$$

$$\widehat{\boldsymbol{Q}}^{\mathrm{rS}}_{\mathrm{E}} = \begin{bmatrix} \boldsymbol{Q}^{\mathrm{rS}}_{\mathrm{E11}} & \boldsymbol{0} \\ \boldsymbol{Q}^{\mathrm{rS}}_{\mathrm{E21}} & \boldsymbol{Q}^{\mathrm{rS}}_{\mathrm{E22}} \end{bmatrix} = \begin{bmatrix} \mathbf{s}^{\circ\,-2}_{\theta}(\boldsymbol{D}^{\bullet}_{\theta}\mathbf{s}^{\bullet}_{\theta}\boldsymbol{D}^{\circ}_{\theta}\mathbf{s}^{\circ}_{\theta} - m^2\boldsymbol{I}^{\circ}) & \boldsymbol{0} \\ -m\mathbf{s}^{\bullet}_{\theta}\boldsymbol{D}^{\circ}_{\theta}\mathbf{s}^{\circ\,-1}_{\theta} & \mathbf{s}^{\bullet}_{\theta}\boldsymbol{D}^{\circ}_{\theta}\mathbf{s}^{\circ\,-1}_{\theta}\boldsymbol{D}^{\bullet}_{\theta} \end{bmatrix} \tag{8.192}$$

The matrices $\boldsymbol{P}^{\theta}_{\mu,\varepsilon}$ in eqs. (8.135) and (8.136) read in discretised form:

$$\boldsymbol{P}^{\theta}_{\mu,\varepsilon} = \boldsymbol{D}^{\bullet}_{\theta}\mathbf{s}^{\bullet\,-1}_{\theta}\boldsymbol{D}^{\circ}_{\theta}\mathbf{s}^{\circ}_{\theta} \tag{8.193}$$

To obtain partly symmetric submatrices, a normalisation according to e.g.:

$$\begin{aligned} \mathbf{w}^{\circ,\bullet}\widetilde{\mathbf{H}}_{\phi} &= \widetilde{\mathbf{H}}_{n} \\ \mathbf{w}^{\circ,\bullet\,-1}\boldsymbol{D}^{\circ,\bullet}\mathbf{w}^{\circ,\bullet} &= \boldsymbol{D}^{\circ,\bullet}_{n} \\ \mathbf{w}^{\circ,\bullet}\boldsymbol{D}^{\circ,\bullet}\mathbf{w}^{\circ,\bullet\,-1} &= \boldsymbol{D}^{\circ,\bullet}_{n} \end{aligned}$$

is possible. However, we will not use this expression in order to avoid a further complication of the equations. Now we transform to the principle axes (diagonalisation):

$$\widehat{\mathbf{G}}^{r}_{\mathrm{H}} = \widehat{\boldsymbol{T}}_{\mathrm{H}}\widehat{\overline{\mathbf{G}}}^{r}_{\mathrm{H}} \qquad \widehat{\boldsymbol{T}}^{-1}_{\mathrm{H}}\widehat{\boldsymbol{Q}}^{\mathrm{rS}}_{\mathrm{H}}\widehat{\boldsymbol{T}}_{\mathrm{H}} = -\widehat{\boldsymbol{\lambda}}^{2}_{\mathrm{H}} \tag{8.194}$$

$$\widehat{\mathbf{G}}^{r}_{\mathrm{E}} = \widehat{\boldsymbol{T}}_{\mathrm{E}}\widehat{\overline{\mathbf{G}}}^{r}_{\mathrm{E}} \qquad \widehat{\boldsymbol{T}}^{-1}_{\mathrm{E}}\widehat{\boldsymbol{Q}}^{\mathrm{rS}}_{\mathrm{E}}\widehat{\boldsymbol{T}}_{\mathrm{E}} = -\widehat{\boldsymbol{\lambda}}^{2}_{\mathrm{E}} \tag{8.195}$$

If the matrices $\widehat{\boldsymbol{Q}}^{\mathrm{rS}}_{\mathrm{E,H}}$ have the form according to eqs. (8.192), we obtain:

$$\widehat{\boldsymbol{\lambda}}^2_{\mathrm{E,H}} = \begin{bmatrix} \boldsymbol{\lambda}^2_{\circ\mathrm{E,H}} & \boldsymbol{0} \\ \boldsymbol{0} & \boldsymbol{\lambda}^2_{\bullet\mathrm{E,H}} \end{bmatrix} \quad \widehat{\boldsymbol{T}}_{\mathrm{H}} = \begin{bmatrix} \boldsymbol{T}^{\circ}_{\mathrm{H}} & \boldsymbol{T}^{\circ\bullet}_{\mathrm{H}} \\ \boldsymbol{0} & \boldsymbol{T}^{\bullet}_{\mathrm{H}} \end{bmatrix} \quad \widehat{\boldsymbol{T}}_{\mathrm{E}} = \begin{bmatrix} \boldsymbol{T}^{\circ}_{\mathrm{E}} & \boldsymbol{0} \\ \boldsymbol{T}^{\bullet\circ}_{\mathrm{E}} & \boldsymbol{T}^{\bullet}_{\mathrm{E}} \end{bmatrix} \tag{8.196}$$

Then we obtain uncoupled equations for Bessel functions of order ν:

$$t\frac{\mathrm{d}}{\mathrm{d}t}\left(t\frac{\mathrm{d}}{\mathrm{d}t}\widehat{\overline{\mathbf{G}}}^{r}_{\mathrm{E,H}}\right) + (t^2\widehat{\boldsymbol{I}} - \widehat{\boldsymbol{\nu}}^2)\widehat{\overline{\mathbf{G}}}^{r}_{\mathrm{E,H}} = \mathbf{0} \quad \text{with} \quad \widehat{\boldsymbol{\nu}} = \left(\widehat{\boldsymbol{\lambda}}^2 + \frac{1}{4}\widehat{\boldsymbol{I}}\right)^{\frac{1}{2}} \tag{8.197}$$

$\boldsymbol{\lambda}^2_{\circ\mathrm{E,H}}$, $\boldsymbol{T}^{\circ}_{\mathrm{E,H}}$ and $\boldsymbol{\lambda}^2_{\bullet\mathrm{E,H}}$, $\boldsymbol{T}^{\bullet}_{\mathrm{E,H}}$ are eigensolutions of $\boldsymbol{Q}^{\mathrm{rS}}_{\mathrm{E,H11}}$ and $\boldsymbol{Q}^{\mathrm{rS}}_{\mathrm{E,H22}}$, respectively. The eigenvector submatrices $\boldsymbol{T}^{\bullet\circ}_{\mathrm{E}}$ and $\boldsymbol{T}^{\circ\bullet}_{\mathrm{H}}$ are obtained from these matrices as a solution of the system of equations:

$$\boldsymbol{Q}^{r\mathrm{S}}_{\mathrm{E22}}\boldsymbol{T}^{\bullet\circ}_{\mathrm{E}} - \boldsymbol{T}^{\bullet\circ}_{\mathrm{E}}\boldsymbol{\lambda}^2_{\circ} = -\boldsymbol{Q}^{r\mathrm{S}}_{\mathrm{E21}}\boldsymbol{T}^{\circ}_{\mathrm{E}} \qquad \boldsymbol{Q}^{r\mathrm{S}}_{\mathrm{H11}}\boldsymbol{T}^{\circ\bullet}_{\mathrm{H}} - \boldsymbol{T}^{\circ\bullet}_{\mathrm{H}}\boldsymbol{\lambda}^2_{\bullet} = -\boldsymbol{Q}^{r\mathrm{S}}_{\mathrm{H12}}\boldsymbol{T}^{\bullet}_{\mathrm{H}} \tag{8.198}$$

The kth column vectors in $\boldsymbol{T}^{\bullet\circ}_{\mathrm{E}}$ and $\boldsymbol{T}^{\circ\bullet}_{\mathrm{H}}$, respectively, are obtained from:

$$(\boldsymbol{Q}^{r\mathrm{S}}_{\mathrm{E22}} - \lambda^2_{\circ k}\boldsymbol{I})\boldsymbol{T}^{\bullet\circ(k)}_{E} = -\boldsymbol{Q}^{r\mathrm{S}}_{\mathrm{E21}}\boldsymbol{T}^{\circ(k)}_{\mathrm{E}} \qquad (\boldsymbol{Q}^{r\mathrm{S}}_{\mathrm{H11}} - \lambda^2_{\bullet k}\boldsymbol{I})\boldsymbol{T}^{\circ\bullet(k)}_{\mathrm{H}} = -\boldsymbol{Q}^{r\mathrm{S}}_{\mathrm{H12}}\boldsymbol{T}^{\bullet(k)}_{\mathrm{H}} \tag{8.199}$$

Inverting $\widehat{\boldsymbol{T}}_{\mathrm{E,H}}$ results in:

$$\widehat{\boldsymbol{T}}^{-1}_{\mathrm{E}} = \begin{bmatrix} \boldsymbol{T}^{\circ-1}_{\mathrm{E}} & \boldsymbol{0} \\ -\boldsymbol{T}^{\bullet-1}_{\mathrm{E}}\boldsymbol{T}^{\bullet\circ}_{\mathrm{E}}\boldsymbol{T}^{\circ-1}_{\mathrm{E}} & \boldsymbol{T}^{\bullet-1}_{\mathrm{E}} \end{bmatrix} \qquad \widehat{\boldsymbol{T}}^{-1}_{\mathrm{H}} = \begin{bmatrix} \boldsymbol{T}^{\circ-1}_{\mathrm{H}} & -\boldsymbol{T}^{\circ-1}_{\mathrm{H}}\boldsymbol{T}^{\circ\bullet}_{\mathrm{H}}\boldsymbol{T}^{\bullet-1}_{\mathrm{H}} \\ \boldsymbol{0} & \boldsymbol{T}^{\bullet-1}_{\mathrm{H}} \end{bmatrix} \tag{8.200}$$

The general solution of Bessel's equation is a linear combination of cylindrical functions:

$$\widehat{\overline{\mathbf{G}}} = \mathcal{C}^{(2)}_{\widehat{\boldsymbol{\nu}}}(t)\widehat{\mathbf{A}} + \mathcal{C}^{(1)}_{\widehat{\boldsymbol{\nu}}}(t)\widehat{\mathbf{B}} \tag{8.201}$$

$\mathcal{C}^{(1)}_{\widehat{\boldsymbol{\nu}}}$ and $\mathcal{C}^{(2)}_{\widehat{\boldsymbol{\nu}}}$ are diagonal matrices of the cylindrical functions. $\widehat{\boldsymbol{\nu}}$ is a diagonal matrix and gives the orders of the cylindrical functions. $\widehat{\boldsymbol{\nu}}$ is a function of the eigenvalues $\widehat{\boldsymbol{\lambda}}$. The eigenvalues are complex as a consequence of the absorbing boundary conditions. Therefore, $\widehat{\boldsymbol{\nu}}$ are also complex numbers. We proceed as described for the r-direction in the cylindrical coordinate system. Therefore, we now write a relation between the fields at ports A and B (at $r = r_{\mathrm{A}}$ and $r = r_{\mathrm{B}}$ or $t = t_{\mathrm{A}}$ and $t = t_{\mathrm{B}}$). From the general solution we obtain for the normalised radii $t = t_{\mathrm{A}}$ and $t = t_{\mathrm{B}}$ (see Fig. 4.18):

$$\begin{bmatrix} \widehat{\overline{\mathbf{G}}}_{\mathrm{FA}} \\ \widehat{\overline{\mathbf{G}}}_{\mathrm{FB}} \end{bmatrix} = \begin{bmatrix} J_{\widehat{\boldsymbol{\nu}}}(t_{\mathrm{A}}) & Y_{\widehat{\boldsymbol{\nu}}}(t_{\mathrm{A}}) \\ J_{\widehat{\boldsymbol{\nu}}}(t_{\mathrm{B}}) & Y_{\widehat{\boldsymbol{\nu}}}(t_{\mathrm{B}}) \end{bmatrix} \begin{bmatrix} \widehat{\mathbf{A}} \\ \widehat{\mathbf{B}} \end{bmatrix} \tag{8.202}$$

The derivatives at these places are determined as:

$$t_{\mathrm{A,B}}\frac{\mathrm{d}}{\mathrm{d}t}\begin{bmatrix} \widehat{\overline{\mathbf{G}}}_{\mathrm{FA}} \\ \widehat{\overline{\mathbf{G}}}_{\mathrm{FB}} \end{bmatrix} = \widehat{\boldsymbol{p}}^{-1}_{\boldsymbol{\nu}} \begin{bmatrix} \overline{\boldsymbol{r}}_{\widehat{\boldsymbol{\nu}}} & \frac{2}{\pi}\boldsymbol{I} \\ -\frac{2}{\pi}\boldsymbol{I} & \overline{\boldsymbol{q}}_{\widehat{\boldsymbol{\nu}}} \end{bmatrix} \begin{bmatrix} \widehat{\overline{\mathbf{G}}}_{\mathrm{FA}} \\ \widehat{\overline{\mathbf{G}}}_{\mathrm{FB}} \end{bmatrix} = \widehat{\overline{\boldsymbol{\Lambda}}}_{\mathrm{F}} \begin{bmatrix} \widehat{\overline{\mathbf{G}}}_{\mathrm{FA}} \\ \widehat{\overline{\mathbf{G}}}_{\mathrm{FB}} \end{bmatrix} \tag{8.203}$$

where t_A is valid for the upper line (matching surface A) and t_B for the lower line (matching surface B). $\boldsymbol{p}_\nu$, $\overline{\boldsymbol{r}}_\nu$ and $\overline{\boldsymbol{q}}_\nu$ are normalised cross-products (see formula 9.1.32 in [14]):

$$\begin{array}{llll}
\boldsymbol{p}_{\widehat{\boldsymbol{\nu}}} & = J_{\widehat{\boldsymbol{\nu}}}(t_\mathrm{A})Y_{\widehat{\boldsymbol{\nu}}}(t_\mathrm{B}) - J_{\widehat{\boldsymbol{\nu}}}(t_\mathrm{B})Y_{\widehat{\boldsymbol{\nu}}}(t_\mathrm{A}) & \widehat{\boldsymbol{p}}_{\widehat{\boldsymbol{\nu}}} = \mathrm{Diag}(\boldsymbol{p}_{\widehat{\boldsymbol{\nu}}}, \boldsymbol{p}_{\widehat{\boldsymbol{\nu}}}) & \\
\boldsymbol{r}_{\widehat{\boldsymbol{\nu}}} & = J'_{\widehat{\boldsymbol{\nu}}}(t_\mathrm{A})Y_{\widehat{\boldsymbol{\nu}}}(t_\mathrm{B}) - Y'_{\widehat{\boldsymbol{\nu}}}(t_\mathrm{A})J_{\widehat{\boldsymbol{\nu}}}(t_\mathrm{B}) & \overline{\boldsymbol{r}}_{\widehat{\boldsymbol{\nu}}} = t_\mathrm{A}\boldsymbol{r}_{\widehat{\boldsymbol{\nu}}} & \\
\boldsymbol{q}_{\widehat{\boldsymbol{\nu}}} & = J_{\widehat{\boldsymbol{\nu}}}(t_\mathrm{A})Y'_{\widehat{\boldsymbol{\nu}}}(t_\mathrm{B}) - J'_{\widehat{\boldsymbol{\nu}}}(t_\mathrm{B})Y_{\widehat{\boldsymbol{\nu}}}(t_\mathrm{A}) & \overline{\boldsymbol{q}}_{\widehat{\boldsymbol{\nu}}} = t_\mathrm{B}\boldsymbol{q}_{\widehat{\boldsymbol{\nu}}} & (8.204) \\
\boldsymbol{s}_{\widehat{\boldsymbol{\nu}}} & = J'_{\widehat{\boldsymbol{\nu}}}(t_\mathrm{A})Y'_{\widehat{\boldsymbol{\nu}}}(t_\mathrm{B}) - J'_{\widehat{\boldsymbol{\nu}}}(t_\mathrm{B})Y'_{\widehat{\boldsymbol{\nu}}}(t_\mathrm{A}) & \overline{\boldsymbol{s}}_{\widehat{\boldsymbol{\nu}}} = t_\mathrm{A}t_\mathrm{B}\boldsymbol{s}_{\widehat{\boldsymbol{\nu}}} & \\
\boldsymbol{W}_\mathrm{P} & = J_{\widehat{\boldsymbol{\nu}}}(t_\mathrm{P})Y'_{\widehat{\boldsymbol{\nu}}}(t_\mathrm{P}) - J'_{\widehat{\boldsymbol{\nu}}}(t_\mathrm{P})Y_{\widehat{\boldsymbol{\nu}}}(t_\mathrm{P}) & \overline{\boldsymbol{W}} = t_\mathrm{P}\boldsymbol{W}_\mathrm{P} = \dfrac{2}{\pi}\boldsymbol{I} &
\end{array}$$

$\mathcal{C}'_{\widehat{\boldsymbol{\nu}}} = \mathrm{d}\mathcal{C}_{\widehat{\boldsymbol{\nu}}}(t)/\mathrm{d}t$ is the derivative of the Bessel functions $\mathcal{C}_{\widehat{\boldsymbol{\nu}}}$ with respect to the argument t. In our case, the arguments $t_\mathrm{A,B}$ are scalar values according to:[2]

$$t_\mathrm{A} = \sqrt{\varepsilon_r\mu_r}\overline{r}_a = \sqrt{\varepsilon_r\mu_r}k_0 r_a \qquad t_\mathrm{B} = \sqrt{\varepsilon_r\mu_r}\overline{r}_b = \sqrt{\varepsilon_r\mu_r}k_0 r_a \qquad (8.205)$$

Now we introduce the GTL equation (8.125) on the left side of eq. (4.189):

TM_r mode: $\widetilde{H}_r = 0$	TE_r mode: $E_r = 0$
We redefine:	We redefine:
$[\widehat{H}^r_\mathrm{Hn}] = \sqrt{\frac{\mu_r}{\varepsilon_r}}[\widehat{H}^r]$ (8.206)	$[\widehat{E}^r_\mathrm{En}] = [\widehat{E}^r_\mathrm{E}]$ (8.207)
$[\widehat{E}^r_\mathrm{Hn}] = [s_\theta E_\theta, E_\phi]^\mathrm{t}$ (8.208)	$[\widehat{H}^r_\mathrm{En}] = \sqrt{\dfrac{\mu_r}{\varepsilon_r}}[H_\phi, -s_\theta H_\theta]^\mathrm{t}$ (8.209)
Now we discretise and transform:	Now we discretise and transform:
$\widehat{\mathbf{E}}^r_{\mathrm{H}n} = \boldsymbol{T}_\mathrm{H}\widehat{\overline{\mathbf{E}}}^r_{\mathrm{H}n}$ (8.210)	$\widehat{\mathbf{H}}^r_{\mathrm{E}n} = \boldsymbol{T}_\mathrm{E}\widehat{\overline{\mathbf{H}}}^r_{\mathrm{E}n}$ (8.211)
With $[\widehat{G}^r_\mathrm{Hn}] = \sqrt{t}[\widehat{H}^r_\mathrm{Hn}]$, eq. (8.124) reads:	With $[\widehat{G}^r_\mathrm{En}] = \sqrt{t}[\widehat{E}^r_\mathrm{En}]$, eq. (8.125) reads:
$\dfrac{\mathrm{d}}{\mathrm{d}t}(\sqrt{t}\,\widehat{\overline{\mathbf{G}}}^r_{\mathrm{H}n}) = -\mathrm{j}t\,\widehat{\overline{\mathbf{E}}}^r_{\mathrm{H}n}$ (8.212)	$\dfrac{\mathrm{d}}{\mathrm{d}t}(\sqrt{t}\,\widehat{\overline{\mathbf{G}}}^r_{\mathrm{E}n}) = -\mathrm{j}t\,\widehat{\overline{\mathbf{H}}}^r_{\mathrm{E}n}$ (8.213)

To obtain the left side of eq. (8.203) we construct the derivative:

$$\frac{\mathrm{d}}{\mathrm{d}t}(\sqrt{t}\,\widehat{\overline{\mathbf{G}}}) = \frac{1}{2\sqrt{t}}\widehat{\overline{\mathbf{G}}} + \sqrt{t}\frac{\mathrm{d}}{\mathrm{d}t}\,\widehat{\overline{\mathbf{G}}} \qquad (8.214)$$

[2]Please note the difference between $\boldsymbol{r}_{\widehat{\boldsymbol{\nu}}}$ (bold) and the coordinate r (not bold).

Now, using eqs. (8.212) and (8.213), respectively, we can write:

$$t_{\mathrm{A,B}}\frac{\mathrm{d}}{\mathrm{d}t}\begin{bmatrix}\widehat{\overline{\mathbf{G}}}^r_{\mathrm{FnA}}\\ \widehat{\overline{\mathbf{G}}}^r_{\mathrm{FnB}}\end{bmatrix} = -\mathrm{j}\sqrt{t_{\mathrm{A,B}}}t_{\mathrm{A,B}}\begin{bmatrix}\widehat{\overline{\mathbf{F}}}^{rd}_{\mathrm{FnA}}\\ \widehat{\overline{\mathbf{F}}}^{rd}_{\mathrm{FnB}}\end{bmatrix} - \frac{1}{2}\begin{bmatrix}\widehat{\overline{\mathbf{G}}}^r_{\mathrm{FnA}}\\ \widehat{\overline{\mathbf{G}}}^r_{\mathrm{FnB}}\end{bmatrix} \tag{8.215}$$

where F has to be replaced by E or H and F^d by H or E, respectively. Finally we now can write:

$$\sqrt{t_{\mathrm{A,B}}}t_{\mathrm{A,B}}\begin{bmatrix}\widehat{\overline{\mathbf{E}}}^r_{\mathrm{HnA}}\\ \widehat{\overline{\mathbf{E}}}^r_{\mathrm{HnB}}\end{bmatrix} = \mathrm{j}\left(\widehat{\overline{\boldsymbol{\Lambda}}}_{\mathrm{H}} + \frac{1}{2}\widehat{\boldsymbol{I}}\right)\begin{bmatrix}\widehat{\overline{\mathbf{H}}}^r_{\mathrm{HnA}}\\ \widehat{\overline{\mathbf{H}}}^r_{\mathrm{HnB}}\end{bmatrix} \tag{8.216}$$

$$\sqrt{t_{\mathrm{A,B}}}t_{\mathrm{A,B}}\begin{bmatrix}\widehat{\overline{\mathbf{H}}}^r_{\mathrm{EnA}}\\ \widehat{\overline{\mathbf{H}}}^r_{\mathrm{EnB}}\end{bmatrix} = \mathrm{j}\left(\widehat{\overline{\boldsymbol{\Lambda}}}_{\mathrm{E}} + \frac{1}{2}\widehat{\boldsymbol{I}}\right)\begin{bmatrix}\widehat{\overline{\mathbf{E}}}^r_{\mathrm{EnA}}\\ \widehat{\overline{\mathbf{E}}}^r_{\mathrm{EnB}}\end{bmatrix} \tag{8.217}$$

These formulas are analogous to the $\boldsymbol{z}$-matrix or $\boldsymbol{y}$-matrix formulas in other sections and we can treat them in an analogous manner.

8.6.4.4 Special solutions

For infinite outer regions, where the field is radiating, we should derive separate equations. This is also required e.g. in region II of Fig. 8.15. There, the field propagates towards the centre and is absorbed by the ABCs. In both cases we use Hankel functions as a solution. For the outward propagating fields the Hankel functions of the second kind $H^{(2)}_{\widehat{\nu}}$ are applied. The fields propagating towards the centre are described by Hankel functions of the first kind $H^{(1)}_{\widehat{\nu}}$. Instead of eq. (8.203) we obtain:

$$t\frac{\mathrm{d}}{\mathrm{d}t}\widehat{\overline{\mathbf{G}}}_{\mathrm{F}n} = tH'_{\widehat{\boldsymbol{\nu}}}(t)H^{-1}_{\widehat{\boldsymbol{\nu}}}(t)\widehat{\overline{\mathbf{G}}}_{\mathrm{F}n} \tag{8.218}$$

$H_{\widehat{\boldsymbol{\nu}}}$ is equal to $H^{(2)}_{\widehat{\boldsymbol{\nu}}}$ in the (infinite outer) region and to $H^{(1)}_{\widehat{\boldsymbol{\nu}}}$ for regions where we have only fields propagating to the centre. $H'^{(1,2)}_{\widehat{\boldsymbol{\nu}}}$ is again the derivative with respect to the argument t. Now we use the abbreviation:

$$\overline{\boldsymbol{\Lambda}}^{\mathrm{C}}_{\mathrm{E,H}} = t_{\mathrm{C}}H'_{\widehat{\boldsymbol{\nu}}}(t_{\mathrm{C}})H^{-1}_{\widehat{\boldsymbol{\nu}}}(t_{\mathrm{C}}) \tag{8.219}$$

We obtain instead of eqs. (8.216) and (8.217) for the radius $r = r_{\mathrm{C}}$ (or $t = t_{\mathrm{C}}$):

$$\sqrt{t_{\mathrm{C}}}\,t_{\mathrm{C}}\widehat{\overline{\mathbf{E}}}^r_{\mathrm{HnC}} = \mathrm{j}(\widehat{\overline{\boldsymbol{\Lambda}}}^{\mathrm{C}}_{\mathrm{H}} + \tfrac{1}{2}\widehat{\boldsymbol{I}})\widehat{\overline{\mathbf{H}}}^r_{\mathrm{HnC}} \tag{8.220}$$

$$\sqrt{t_{\mathrm{C}}}\,t_{\mathrm{C}}\widehat{\overline{\mathbf{H}}}^r_{\mathrm{EnC}} = \mathrm{j}(\widehat{\overline{\boldsymbol{\Lambda}}}^{\mathrm{C}}_{\mathrm{E}} + \tfrac{1}{2}\widehat{\boldsymbol{I}})\widehat{\overline{\mathbf{E}}}^r_{\mathrm{EnC}} \tag{8.221}$$

r_{C} could e.g. be identical to r_0 in Fig. 8.15.

8.6.4.5 Radiation from open horns

With the developed formulas we could e.g. determine the radiation from an open horn as shown in Fig. 8.15. By using impedances/admittances (which have to be defined in a suitable way), the procedure is analogous to that described for open-ending circular waveguides in Section 8.5.5. Therefore, we will not repeat the procedure here. The difference is that we only need to use the transformation matrix $\widehat{\boldsymbol{T}}_{\mathrm{H}}$ ($\widehat{\boldsymbol{T}}_{\mathrm{E}}$) for TM_r modes (TE_r modes).

8.6.5 Numerical results

To verify the derived algorithm, the mode E_{01} of the horn antenna was determined [15,16]. In Fig. 8.16 the absolute value of the azimuthal component H_φ is plotted versus the discretisation angle θ on the matching surface.

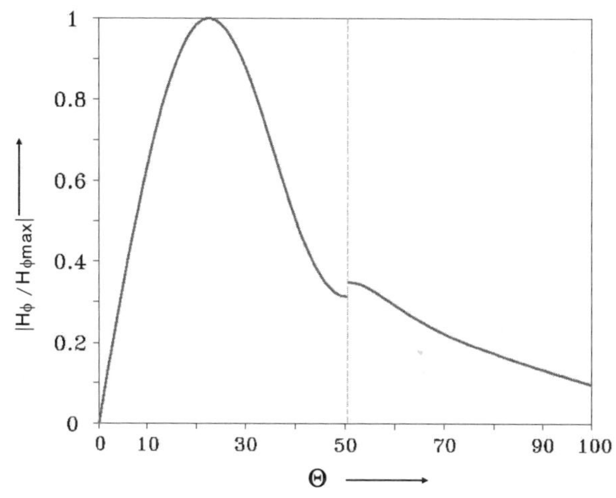

Fig. 8.16 H_φ component on the matching surface (Reproduced by permission of Copernicus Gesellschaft mbH)

It can be clearly seen that the azimuthal component of the magnetic field is not continuous at the position of the metal wall of the horn antenna, as expected. The absolute value of the E_θ component is shown versus the discretisation angle θ for the fundamental mode in Fig. 8.17. As expected, the electric field component possesses a pole at the metal wall of the horn. We found that the calculation of the Bessel functions must be carried out very carefully. Due to the relation between the complex eigenvalues and the complex orders of the Bessel functions in eq. (8.197), the order increases with the number of lines. So we had to utilise various different numerical algorithms [14] (like expansion in Airy functions (uniform expansion) and asymptotic expansions for large orders) to compute the cylinder functions.

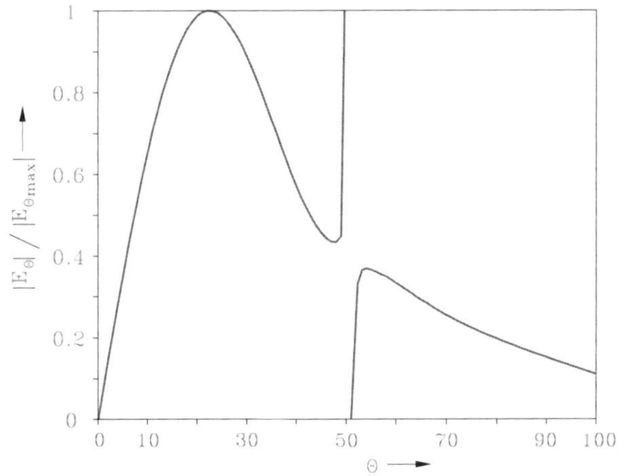

Fig. 8.17 E_θ component on the matching surface (Reproduced by permission of Copernicus Gesellschaft mbH)

8.7 ELLIPTICAL COORDINATES

In this section we will deal with another coordinate system for which we can give analytical expressions, at least in some special cases. The elliptical coordinates are ξ, η, z, ξ and η are given by:

$$\xi = \frac{r_1 + r_2}{2f} \qquad \eta = \frac{r_1 - r_2}{2f} \tag{8.222}$$

r_1 and r_2 are the distances to the foci of a family of conical ellipses and hyperbolas; $2f$ is the distance between foci (see Fig. 8.18).

$$a = f\xi \qquad b = f\sqrt{\xi^2 - 1} \qquad e = \frac{f}{a} \tag{8.223}$$

where a = semi-major axis, b = semi-minor axis, e = eccentricity.

The equations of families of confocal ellipses and hyperbolas are given by:

$$\frac{x^2}{\xi^2} + \frac{y^2}{\xi^2 - 1} = f^2 \qquad (1 < \xi < \infty) \tag{8.224}$$

$$\frac{x^2}{\eta^2} - \frac{y^2}{1 - \eta^2} = f^2 \qquad (-1 < \eta < 1) \tag{8.225}$$

For the relations between Cartesian and elliptical coordinates we have:

$$x = f\xi\eta \qquad y = f\sqrt{(\xi^2 - 1)(1 - \eta^2)} \tag{8.226}$$

The metric coefficients are given by:

$$g_\xi = f\sqrt{\frac{\xi^2 - \eta^2}{\xi^2 - 1}} \qquad g_\eta = f\sqrt{\frac{\xi^2 - \eta^2}{1 - \eta^2}} \qquad g_z = 1 \tag{8.227}$$

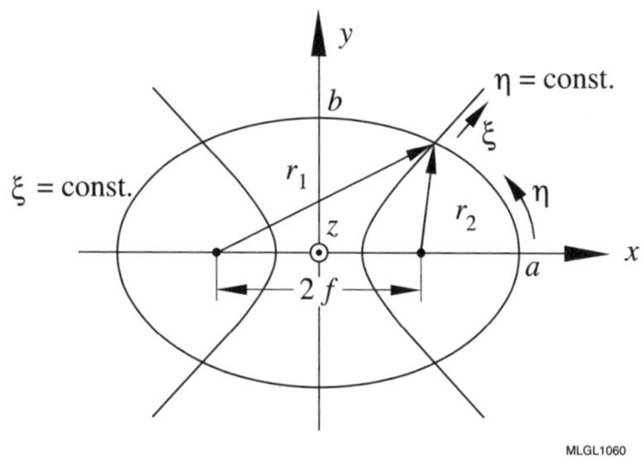

Fig. 8.18 Elliptical coordinate system: ξ and η may be also replaced by ξ° and η°

By using ξ° and η° instead of ξ and η according to:

$$\eta = \cos\eta^\circ \qquad \xi = \cosh\xi^\circ \tag{8.228}$$

(ξ and η in Fig. 8.18 should be replaced with ξ° and η°) we obtain:

$$\begin{array}{lll} x = f\cos\eta^\circ\cosh\xi^\circ & (0 < \xi^\circ < \infty) & a = f\cosh\xi^\circ_o \\ y = f\sin\eta^\circ\sinh\xi^\circ & (0 < \eta^\circ < 2\pi) & b = f\sinh\xi^\circ_o \end{array} \tag{8.229}$$

The metric coefficients are now given as:

$$g_t = g_{\xi^\circ} = g_{\eta^\circ} = f\sqrt{\sin^2\eta^\circ + \sinh^2\xi^\circ} = f\sqrt{\cosh^2\xi^\circ - \cos^2\eta^\circ} \qquad g_z = 1 \tag{8.230}$$

8.7.1 GTL equations for z-direction

Now we will give the equation for the z-direction. The coordinates should be assigned according to: $x_1 = \xi^\circ$, $x_2 = \eta^\circ$ and $x_3 = z$. We are only interested here in the uniaxial anisotropic case. From eq. (8.9) we obtain with $\nu_{11} = \nu_{22} = \nu_{rt}$, $\nu_{33} = \nu_{zz}$ and $\nu_{ik|i\neq k} = 0$:

$$[\overline{\varepsilon}_r] = \mathrm{diag}(\varepsilon_{rt}, \varepsilon_{rt}, g_t^2\varepsilon_{zz}) \qquad [\overline{\mu}_r] = \mathrm{diag}(\mu_{rt}, \mu_{rt}, g_t^2\mu_{zz}) \tag{8.231}$$

Here we define the supervectors according to:

$$[E^z] = \overline{g}_t[E_{\xi^\circ}, E_{\eta^\circ}]^{\mathrm{t}} \qquad [H^z] = \overline{g}_t[\widetilde{H}_{\eta^\circ}, -\widetilde{H}_{\xi^\circ}]^{\mathrm{t}} \tag{8.232}$$

When we use normalised coordinates, we must also normalise f, g_t, a and b with k_0, e.g. $\overline{g}_t = k_0 g_t$. The GTL equations are given by:

$$\frac{\partial}{\partial\overline{z}}[H^z] = -\mathrm{j}[R_{\mathrm{E}}^z][E^z] \qquad \frac{\partial}{\partial\overline{z}}[E^z] = -\mathrm{j}[R_{\mathrm{H}}^z][H^z] \tag{8.233}$$

and the operators $[R_{\mathrm{E}}^{z}]$ and $[R_{\mathrm{E}}^{z}]$ are determined as:

$$[R_{\mathrm{H}}^{x_3}] = \begin{bmatrix} \mu_{rt} + D_{\xi^\circ}\overline{\varepsilon}_{zz}^{-1}D_{\xi^\circ} & D_{\xi^\circ}\overline{\varepsilon}_{zz}^{-1}D_{\eta^\circ} \\ D_{\eta^\circ}\overline{\varepsilon}_{zz}^{-1}D_{\xi^\circ} & \mu_{rt} + D_{\eta^\circ}\overline{\varepsilon}_{zz}^{-1}D_{\eta^\circ} \end{bmatrix} \tag{8.234}$$

$$[R_{\mathrm{E}}^{x_3}] = \begin{bmatrix} \varepsilon_{rt} + D_{\eta^\circ}\overline{\mu}_{zz}^{-1}D_{\eta^\circ} & -D_{\eta^\circ}\overline{\mu}_{zz}^{-1}D_{\xi^\circ} \\ -D_{\xi^\circ}\overline{\mu}_{zz}^{-1}D_{\eta^\circ} & \varepsilon_{rt} + D_{\xi^\circ}\overline{\mu}_{zz}^{-1}D_{\xi^\circ} \end{bmatrix} \tag{8.235}$$

We use the abbreviations $\overline{\varepsilon}_{zz} = \overline{g}_t^2\varepsilon_{zz}$ and $\overline{\mu}_{zz} = \overline{g}_t^2\mu_{zz}$ here.
The longitudinal components are computed as:

$$E_z = -\mathrm{j}\overline{\varepsilon}_{zz}^{-1}[D_{\xi^\circ} \quad D_{\eta^\circ}][H^z] \qquad H_z = \mathrm{j}\overline{\mu}_{zz}^{-1}[-D_{\eta^\circ} \quad D_{\xi^\circ}][E^z] \tag{8.236}$$

8.7.2 GTL equations for ξ-direction

Now we examine the ξ-direction. The coordinates should be assigned according to: $x_1 = \eta^\circ$, $x_2 = z$ and $x_3 = \xi^\circ$. Here we define the transverse vectors according to:

$$[E^\xi] = [\overline{g}_t E_{\eta^\circ}, E_z]^{\mathrm{t}} \qquad [H^\xi] = [\widetilde{H}_z, -\overline{g}_t\widetilde{H}_{\eta^\circ}]^{\mathrm{t}} \tag{8.237}$$

As before, we deal only with the uniaxial anisotropic case. From eq. (8.9) we obtain with $\nu_{11} = \nu_{33} = \nu_{rt}$, $\nu_{22} = \nu_{zz}$ and $\nu_{ik|i\neq k} = 0$:

$$[\overline{\varepsilon}_r] = \mathrm{diag}(\varepsilon_{rt}, \overline{g}_t^2\varepsilon_{zz}, \varepsilon_{rt}) \qquad [\overline{\mu}_r] = \mathrm{diag}(\mu_{rt}, \overline{g}_t^2\mu_{zz}, \mu_{rt}) \tag{8.238}$$

The GTL equations are given by:

$$\frac{\partial}{\partial\xi^\circ}[H^\xi] = -\mathrm{j}[R_{\mathrm{E}}^{\xi^\circ}][E^\xi] \qquad \frac{\partial}{\partial\xi^\circ}[E^\xi] = -\mathrm{j}[R_{\mathrm{H}}^{\xi^\circ}][H^\xi] \tag{8.239}$$

and we have for the operators $[R_{\mathrm{E}}^{\xi^\circ}]$ and $[R_{\mathrm{H}}^{\xi^\circ}]$:

$$[R_{\mathrm{H}}^{\xi^\circ}] = [R_{\mathrm{H}}^{\mathrm{e}\xi^\circ}] + [\mu_{\mathrm{M}}^{\xi^\circ}] \qquad [R_{\mathrm{E}}^{\xi^\circ}] = [R_{\mathrm{E}}^{\mathrm{e}\xi^\circ}] + [\varepsilon_{\mathrm{M}}^{\xi^\circ}] \tag{8.240}$$

$$[R_{\mathrm{H}}^{\mathrm{e}\xi^\circ}] = \begin{bmatrix} D_{\eta^\circ}\varepsilon_{rt}^{-1}D_{\eta^\circ} & D_{\eta^\circ}\varepsilon_{rt}^{-1}D_{\overline{z}} \\ D_{\overline{z}}\varepsilon_{rt}^{-1}D_{\eta^\circ} & D_{\overline{z}}\varepsilon_{rt}^{-1}D_{\overline{z}} \end{bmatrix} \qquad [\mu_{\mathrm{M}}^{\xi^\circ}] = \begin{bmatrix} \overline{g}_t^2\mu_{zz} & \\ & \mu_{rt} \end{bmatrix} \tag{8.241}$$

$$[R_{\mathrm{E}}^{\mathrm{e}\xi^\circ}] = \begin{bmatrix} D_{\overline{z}}\mu_{rt}^{-1}D_{\overline{z}} & -D_{\overline{z}}\mu_{rt}^{-1}D_{\eta^\circ} \\ -D_{\eta^\circ}\mu_{rt}^{-1}D_{\overline{z}} & D_{\eta^\circ}\mu_{rt}^{-1}D_{\eta^\circ} \end{bmatrix} \qquad [\varepsilon_{\mathrm{M}}^{\xi^\circ}] = \begin{bmatrix} \varepsilon_{rt} & \\ & \overline{g}_t^2\varepsilon_{zz} \end{bmatrix} \tag{8.242}$$

The ξ-components are determined as:

$$\overline{g}E_{\xi^\circ} = -\mathrm{j}\varepsilon_{rt}^{-1}[D_{\eta^\circ} \quad D_{\overline{z}}][H^\xi] \qquad \overline{g}H_{\xi^\circ} = \mathrm{j}\mu_{rt}^{-1}[-D_{\overline{z}} \quad D_{\eta^\circ}][E^\xi] \tag{8.243}$$

The products of the matrices $[R_{\mathrm{E}}^{\xi^\circ}]$ and $[R_{\mathrm{H}}^{\xi^\circ}]$ result in:

$$[R_{\mathrm{E}}^{\xi^\circ}][R_{\mathrm{H}}^{\xi^\circ}] = [R_{\mathrm{E}}^{\mathrm{e}\xi^\circ}][\mu_{\mathrm{M}}^{\xi^\circ}] + [\varepsilon_{\mathrm{M}}^{\xi^\circ}][R_{\mathrm{H}}^{\mathrm{e}\xi^\circ}] + [\varepsilon_{\mathrm{M}}^{\xi^\circ}][\mu_{\mathrm{M}}^{\xi^\circ}] \tag{8.244}$$

$$[R_{\mathrm{H}}^{\xi^\circ}][R_{\mathrm{E}}^{\xi^\circ}] = [R_{\mathrm{H}}^{\mathrm{e}\xi^\circ}][\varepsilon_{\mathrm{M}}^{\xi^\circ}] + [\mu_{\mathrm{M}}^{\xi^\circ}][R_{\mathrm{E}}^{\mathrm{e}\xi^\circ}] + [\mu_{\mathrm{M}}^{\xi^\circ}][\varepsilon_{\mathrm{M}}^{\xi^\circ}] \tag{8.245}$$

Therefore, we obtain the following differential equations for E_z and $\widetilde{H}_z$ in homogeneous layers in the η°-direction:

$$\begin{aligned} D_{\xi^\circ} D_{\xi^\circ} \widetilde{H}_z + D_{\overline{z}} \mu_{rt}^{-1} D_{\overline{z}} \overline{g}_t^2 \mu_{zz} \widetilde{H}_z + D_{\eta^\circ} D_{\eta^\circ} \widetilde{H}_z + \varepsilon_{rt} \overline{g}_t^2 \mu_{zz} \widetilde{H}_z = 0 \\ D_{\xi^\circ} D_{\xi^\circ} E_z + D_{\overline{z}} \varepsilon_{rt}^{-1} D_{\overline{z}} \overline{g}_t^2 \varepsilon_{zz} E_z + D_{\eta^\circ} D_{\eta^\circ} E_z + \mu_{rt} \overline{g}_t^2 \varepsilon_{zz} E_z = 0 \end{aligned} \tag{8.246}$$

Expressions for the η components are obtained from eq. (8.239):

$$\mathrm{j}\overline{g}_t(\varepsilon_{rt}\mu_{rt} + D_{\overline{z}} D_{\overline{z}}) \begin{bmatrix} E_{\eta^\circ} \\ \widetilde{H}_{\eta^\circ} \end{bmatrix} = \begin{bmatrix} \mathrm{j} D_{\overline{z}} D_{\eta^\circ} & -\mu_{rt} D_{\xi^\circ} \\ \varepsilon_{rt} D_{\xi^\circ} & \mathrm{j} D_{\overline{z}} D_{\eta^\circ} \end{bmatrix} \begin{bmatrix} E_z \\ \widetilde{H}_z \end{bmatrix} \tag{8.247}$$

We assume now that the propagation in z-direction occurs according to $e^{-\mathrm{j}\sqrt{\varepsilon_{re}}\overline{z}}$. We replace $D_{\overline{z}}$ by $-\mathrm{j}\sqrt{\varepsilon_{re}}$ and obtain:

$$\begin{aligned} D_{\xi^\circ} D_{\xi^\circ} \widetilde{H}_z + D_{\eta^\circ} D_{\eta^\circ} \widetilde{H}_z + \overline{g}_t^2 \varepsilon_{dh} \widetilde{H}_z = 0 \quad & \varepsilon_{dh} = \varepsilon_{rt}\mu_{zz} - \varepsilon_{re}\mu_{zz}/\mu_{rt} \\ D_{\xi^\circ} D_{\xi^\circ} E_z + D_{\eta^\circ} D_{\eta^\circ} E_z + \overline{g}_t^2 \varepsilon_{de} E_z = 0 \quad & \varepsilon_{de} = \mu_{rt}\varepsilon_{zz} - \varepsilon_{re}\varepsilon_{zz}/\varepsilon_{rt} \end{aligned} \tag{8.248}$$

In this case we obtain for the η components:

$$\mathrm{j}\overline{g}_t \varepsilon_{re} \begin{bmatrix} E_{\eta^\circ} \\ \widetilde{H}_{\eta^\circ} \end{bmatrix} = - \begin{bmatrix} \sqrt{\varepsilon_{re}} D_{\eta^\circ} & -\mu_{rt} D_{\xi^\circ} \\ \varepsilon_{rt} D_{\xi^\circ} & \sqrt{\varepsilon_{re}} D_{\eta^\circ} \end{bmatrix} \begin{bmatrix} E_z \\ \widetilde{H}_z \end{bmatrix} \tag{8.249}$$

Eqs. (8.246) and (8.247) can be further simplified by separating the variables. With the ansatz $F_z = S(\eta^\circ)R(\xi^\circ)$ (where $F_z = E_z$ or $\widetilde{H}_z$), we obtain by using the method of separation of the variables:

$$D_{\eta^\circ} D_{\eta^\circ} S(\eta^\circ) \frac{1}{S} - \frac{\varepsilon_d \overline{f}^2}{2} \cos(2\eta^\circ) = -a^\circ \tag{8.250}$$

$$D_{\xi^\circ} D_{\xi^\circ} R(\xi^\circ) \frac{1}{R} + \frac{\varepsilon_d \overline{f}^2}{2} \cosh(2\xi^\circ) = a^\circ \tag{8.251}$$

or in the canonical form for Mathieu's differential equations:

$$\frac{\partial^2}{\partial \eta^{\circ 2}} S + (a^\circ - 2q\cos 2\eta^\circ) S = 0 \tag{8.252}$$

$$\frac{\partial^2}{\partial \xi^{\circ 2}} R - (a^\circ - 2q\cosh 2\xi^\circ) R = 0 \tag{8.253}$$

with $q = \varepsilon_{de}\overline{f}^2/4$ or $q = \varepsilon_{dh}\overline{f}^2/4$, respectively.

Besides, we see from eq. (8.248) that the propagation constant for hollow waveguides is given by $k_z = \sqrt{k_0^2 - k_c^2}$, where k_c is the cut-off wave number.

8.7.3 GTL equations for η-direction

Finally, we derive the GTL equations for the η-direction. Here, the coordinates are assigned according to: $x_1 = z$, $x_2 = \xi^\circ$, and $x_3 = \eta^\circ$. The transverse field vectors are defined as:

$$[E^\eta] = [E_z, \overline{g}_t E_{\xi^\circ},]^{\mathrm{t}} \qquad [H^\eta] = [\overline{g}_t \widetilde{H}_{\xi^\circ}, -\widetilde{H}_z]^{\mathrm{t}} \tag{8.254}$$

As before, we consider only the uniaxial anisotropic case. We obtain from eq. (8.9) with $\nu_{11} = \nu_{zz}$, $\nu_{22} = \nu_{33} = \nu_{rt}$ and $\nu_{ik|i\neq k} = 0$:

$$[\overline{\varepsilon}_r] = \mathrm{diag}(\overline{g}_t^2 \varepsilon_{zz}, \varepsilon_{rt}, \varepsilon_{rt}) \qquad [\overline{\mu}_r] = \mathrm{diag}(\overline{g}_t^2 \mu_{zz}, \mu_{rt}, \mu_{rt}) \tag{8.255}$$

The GTL equations read:

$$\frac{\partial}{\partial \eta^\circ}[H^\eta] = -\mathrm{j}[R_\mathrm{E}^{\eta^\circ}][E^\eta] \qquad \frac{\partial}{\partial \eta^\circ}[E^\eta] = -\mathrm{j}[R_\mathrm{H}^{\eta^\circ}][H^\eta] \tag{8.256}$$

and the matrices $[R_\mathrm{E}^{\eta^\circ}]$ and $[R_\mathrm{H}^{\eta^\circ}]$ are:

$$[R_\mathrm{H}^{\eta^\circ}] = [R_\mathrm{H}^{\mathrm{e}\eta^\circ}] + [\mu_\mathrm{M}^{\eta^\circ}] \qquad [R_\mathrm{E}^{\eta^\circ}] = [R_\mathrm{E}^{\mathrm{e}\eta^\circ}] + [\varepsilon_\mathrm{M}^{\eta^\circ}] \tag{8.257}$$

$$[R_\mathrm{H}^{\mathrm{e}\eta^\circ}] = \begin{bmatrix} D_{\overline{z}}\varepsilon_{rt}^{-1}D_{\overline{z}} & D_{\overline{z}}\varepsilon_{rt}^{-1}D_{\xi^\circ} \\ D_{\xi^\circ}\varepsilon_{rt}^{-1}D_{\overline{z}} & D_{\xi^\circ}\varepsilon_{rt}^{-1}D_{\xi^\circ} \end{bmatrix} \qquad [\mu_\mathrm{M}^{\eta^\circ}] = \begin{bmatrix} \mu_{rt} & \\ & \overline{g}_t^2\mu_{zz} \end{bmatrix} \tag{8.258}$$

$$[R_\mathrm{E}^{\mathrm{e}\eta^\circ}] = \begin{bmatrix} D_{\xi^\circ}\mu_{rt}^{-1}D_{\xi^\circ} & -D_{\xi^\circ}\mu_{rt}^{-1}D_{\overline{z}} \\ -D_{\overline{z}}\mu_{rt}^{-1}D_{\xi^\circ} & D_{\overline{z}}\mu_{rt}^{-1}D_{\overline{z}} \end{bmatrix} \qquad [\varepsilon_\mathrm{M}^{\eta^\circ}] = \begin{bmatrix} \overline{g}_t^2\varepsilon_{zz} & \\ & \varepsilon_{rt} \end{bmatrix} \tag{8.259}$$

We have for the η components:

$$\overline{g}E_{\eta^\circ} = -\mathrm{j}\varepsilon_{rt}^{-1}[D_{\overline{z}} \quad D_{\xi^\circ}][H^\eta] \qquad \overline{g}H_{\eta^\circ} = \mathrm{j}\mu_{rt}^{-1}[-D_{\xi^\circ} \quad D_{\overline{z}}][E^\eta] \tag{8.260}$$

We obtain for the products of the matrices $[R_\mathrm{E}^{\eta^\circ}]$ and $[R_\mathrm{H}^{\eta^\circ}]$:

$$[R_\mathrm{E}^{\eta^\circ}][R_\mathrm{H}^{\eta^\circ}] = [R_\mathrm{E}^{\mathrm{e}\eta^\circ}][\mu_\mathrm{M}^{\eta^\circ}] + [\varepsilon_\mathrm{M}^{\eta^\circ}][R_\mathrm{H}^{\mathrm{e}\eta^\circ}] + [\varepsilon_\mathrm{M}^{\eta^\circ}][\mu_\mathrm{M}^{\eta^\circ}] \tag{8.261}$$

$$[R_\mathrm{H}^{\eta^\circ}][R_\mathrm{E}^{\eta^\circ}] = [R_\mathrm{H}^{\mathrm{e}\eta^\circ}][\varepsilon_\mathrm{M}^{\eta^\circ}] + [\mu_\mathrm{M}^{\eta^\circ}][R_\mathrm{E}^{\mathrm{e}\eta^\circ}] + [\mu_\mathrm{M}^{\eta^\circ}][\varepsilon_\mathrm{M}^{\eta^\circ}] \tag{8.262}$$

Therefore, we obtain the following differential equations for E_z and $\widetilde{H}_z$ for homogeneous layers in ξ°-direction:

$$\begin{aligned} D_{\xi^\circ}D_{\xi^\circ}\widetilde{H}_z + D_{\overline{z}}\mu_{rt}^{-1}D_{\overline{z}}\overline{g}_t^2\mu_{zz}\widetilde{H}_z + D_{\eta^\circ}D_{\eta^\circ}\widetilde{H}_z + \varepsilon_{rt}\overline{g}_t^2\mu_{zz}\widetilde{H}_z &= 0 \\ D_{\xi^\circ}D_{\xi^\circ}E_z + D_{\overline{z}}\varepsilon_{rt}^{-1}D_{\overline{z}}\overline{g}_t^2\varepsilon_{zz}E_z + D_{\eta^\circ}D_{\eta^\circ}E_z + \mu_{rt}\overline{g}_t^2\varepsilon_{zz}E_z &= 0 \end{aligned} \tag{8.263}$$

These equations are identical to the expressions in eq. (8.246). Equations for the ξ components are obtained from eq. (8.256):

$$\mathrm{j}\overline{g}_t(\varepsilon_{rt}\mu_{rt} + D_{\overline{z}}D_{\overline{z}}) \begin{bmatrix} E_{\xi^\circ} \\ \widetilde{H}_{\xi^\circ} \end{bmatrix} = \begin{bmatrix} \mathrm{j}D_{\overline{z}}D_{\xi^\circ} & \mu_{rt}D_{\eta^\circ} \\ -\varepsilon_{rt}D_{\eta^\circ} & \mathrm{j}D_{\overline{z}}D_{\xi^\circ} \end{bmatrix} \begin{bmatrix} E_z \\ \widetilde{H}_z \end{bmatrix} \tag{8.264}$$

Eqs. (8.248) and (8.250)–(8.253) are also valid here.

8.7.4 Hollow waveguides with elliptic cross-section

The propagation of the modes in hollow waveguides can be described by knowing the cut-off wave number or cut-off wavelength. Therefore, we calculate the cut-off wave number normalised by the circumference of the

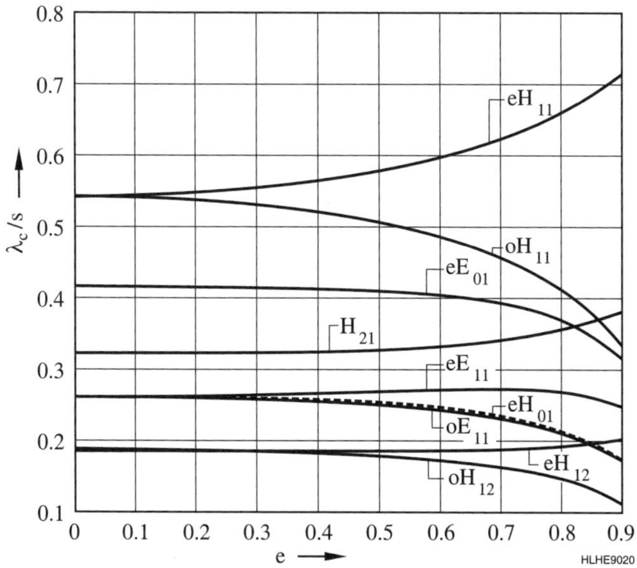

Fig. 8.19 Normalised cut-off wavelength λ_c in hollow waveguides with elliptic cross-section as a function of the eccentricity e

boundary ellipse $s = 4aE(e)$. E is the elliptic integral and e the eccentricity. For hollow waveguides and at cut-off we have $\varepsilon_{de,h} = 1$. Now we discretise eq. (8.248) for the coordinates ξ° and η°. We use N_ξ° and N_η° discretisation points in ξ°- and η°-direction, respectively. The equidistant discretised values of ξ° and η° in one row or column are collected in diagonal matrices $\boldsymbol{\xi}^\circ$ and $\boldsymbol{\eta}^\circ$. The discretised values of F_z are collected in a vector. We construct subvectors for constant η° values and start at the first discretised η° value. The discretised wave equation is given by:

$$\widehat{\boldsymbol{Q}}\widehat{\mathbf{F}}_z = \widehat{\mathbf{0}} \qquad \widehat{\boldsymbol{Q}} = \widehat{\overline{\boldsymbol{P}}}_{\eta^\circ} + \widehat{\overline{\boldsymbol{P}}}_{\xi^\circ} + \widehat{\overline{\mathbf{g}}}_t^2 \qquad \widehat{\mathbf{F}}_z = \widehat{\mathbf{E}}_z \quad \text{or} \quad \widehat{\mathbf{H}}_z \tag{8.265}$$

where:

$$\begin{aligned}
\widehat{\mathbf{F}}_z &= [\mathbf{F}_{z1}^{\mathrm{t}}, \mathbf{F}_{z1}^{\mathrm{t}}, \ldots, \mathbf{F}_{zN_{\eta^\circ}}^{\mathrm{t}}]^{\mathrm{t}} \\
\widehat{\overline{\boldsymbol{P}}}_{\xi^\circ} &= \boldsymbol{I}_{\eta^\circ} \otimes \overline{\boldsymbol{P}}_{\xi^\circ} \qquad \overline{\boldsymbol{P}}_{\xi^\circ} = -\overline{\boldsymbol{D}}_{\xi^\circ}^{\mathrm{t}} \overline{\boldsymbol{D}}_{\xi^\circ} \\
\widehat{\overline{\boldsymbol{P}}}_{\eta^\circ} &= \overline{\boldsymbol{P}}_{\eta^\circ} \otimes \boldsymbol{I}_{\xi^\circ} \qquad \overline{\boldsymbol{P}}_{\eta^\circ} = -\overline{\boldsymbol{D}}_{\eta^\circ}^{\mathrm{t}} \overline{\boldsymbol{D}}_{\eta^\circ} \\
\widehat{\overline{\mathbf{g}}}_t^2 &= \overline{f}^2(\boldsymbol{I}_{\eta^\circ} \otimes \cosh^2 \boldsymbol{\xi}^\circ - \cos^2 \boldsymbol{\eta}^\circ \otimes \boldsymbol{I}_{\xi^\circ})
\end{aligned} \tag{8.266}$$

The matrices $\boldsymbol{I}_{\xi^\circ}$ and $\boldsymbol{I}_{\eta^\circ}$ are unit matrices of the order N_ξ° and N_η°, respectively. For $\overline{f}$ we may write:

$$\overline{f} = \frac{\pi}{2} \frac{s}{\lambda} \frac{e}{E}$$

Due to the symmetry, only a quarter of the elliptic cross-section will be used for the analysis, where thc coordinates vary between $(0 < \eta^\circ < \pi/2)$ and $(0 < \xi^\circ < \xi_0^\circ)$ and where ξ_0° is given by $\cosh \xi_0^\circ = 1/e$. The following table summarises the boundary conditions (BC) that we have to introduce into the matrices $\boldsymbol{D}$ for some of the modes. The first condition is for $\xi^\circ = 0$ and $\eta^\circ = 0$ and the second one for $\xi^\circ = \xi_0^\circ$ and $\eta^\circ = \pi/2$, respectively. We use D for Dirichlet BC and N for Neumann BC.

mode	BC in $\boldsymbol{D}_{\eta^\circ}$	BC in $\boldsymbol{D}_{\xi^\circ}$
evenH_{11}	ND	NN
oddH_{11}	DN	DN
evenE_{11}	ND	ND
oddE_{11}	DN	DD
evenH_{01}	NN	NN
evenE_{01}	NN	ND

For a given value of e we vary λ/s until the adequate eigenvalue of the matrix $\widehat{\boldsymbol{Q}}$ is equal to zero. Fig. 8.19 shows the computed results for the cut-off wavelength in elliptical hollow waveguides for some different modes. For comparison of the results see [17]. The values for λ/s at $e = 0$ are obtained from the cut-off wave numbers of the circular hollow waveguides, which are given as: H_{11}: 0.54313, E_{01}: 0.41583, H_{21}: 0.32741, H_{01} and E_{11}: 0.26098, H_{12}: 0.187566.

References

[1] V. K. Tripathi, 'On the Analysis of Symmetrical Three-Line Microstrip Circuits', *IEEE Trans. Microwave Theory Tech.*, vol. MTT–25, no. 9, pp. 726–729, Sep. 1977.

[2] R. Pregla, 'Analysis of Planar Waveguides with Arbitrary Anisotropic Material', in *U.R.S.I Intern. Symp. Electromagn. Theo.*, Victoria, Canada, May 2001, pp. 425–427.

[3] L. B. Felsen and N. Marcuvitz (Eds.), *Radiation and Scattering of Waves*, IEEE Press, New York, USA, 1996.

[4] R. Pregla, 'Method of Lines for the Analysis of Multilayered Gyrotropic Waveguide Structures', *IEE Proc. -H, Microwave Antennas Propagation*, (Special Issue On Gyroelectric Waveguides And Their Circuit Application), vol. 140, no. 3, pp. 183–192, June 1993.

[5] R. Pregla, 'Efficient and Accurate Modeling of Planar Anisotropic Microwave Structures by the Method of Lines', *IEEE Trans. Microwave Theory Tech.*, vol. 50, pp. 1469–1479, June 2002.

[6] R. Pregla, 'Modeling of Optical Waveguide Structures with General Anisotropy in Arbitrary Orthogonal Coordinate Systems', *IEEE J. of Sel. Topics in Quantum Electronics*, vol. 8, pp. 1217–1224, Dec. 2002.

[7] R. Pregla, 'The Impedance/Admittance Transformation – An Efficient Concept for the Analysis of Optical Waveguide Structures', in *OSA Integr. Photo. Resear. Tech. Dig.*, Santa Barbara, USA, July 1999, pp. 40–42.

[8] R. Pregla, 'Analysis of Waveguide Junctions and Sharp Bends with General Anisotropic Material by Using Orthogonal Propagating Waves', in *ICTON Conf.*, Warsaw, Poland, 2003, vol. 5, pp. 116–121.

[9] R. Pregla, 'Efficient Modelling of Conformal Antennas', in *Proc. of the 1st European Workshop on Conformal Antennas*, Karlsruhe, Germany, 29. Oktober 1999, pp. 36–39.

[10] R. Pregla, 'Modeling of Waveguide Structures with General Anisotropy in Arbitrary Orthogonal Coordinate Systems', in *Proceedings of IGTE Symp.*, Graz, Austria, Sep. 2002.

[11] R. Pregla and L. Greda, 'Modeling of Waveguide Junctions with General Anisotropic Material by Using Orthogonal Propagating Waves', in *URSI-B Symp.*, Pisa, Italy, May 2004.

[12] M. Koshiba, K. Hayata and M. Suzuki, 'Approximate Scalar Finite-Element Analysis of Anisotropic Optical Waveguides With Off-Diagonal

Elements in a Permittivity Tensor', *IEEE Trans. Microwave Theory Tech.*, vol. MTT-32, pp. 587–593, 1984.

[13] R. F. Harrington, *Field Computing by Moment Methods*, chapter 3, Macmillan, New York, 1968.

[14] M. Abramowitz and I. A. Stegun, *Handbook of Mathematical Functions*, chapter 9, Dover Publ., New York, 1965.

[15] R. Pregla and D. Kremer, 'The Method of Lines for the Analysis of Dipoles and Horn Antennas', in *Progress In Electromagnetic Research Symp. (PIERS)*, Seattle, WA, USA, July 1995, p. 882.

[16] D. Kremer and R. Pregla, 'Hybrid Analysis of Conical Horn Antennas by the Method of Lines', in *Kleinheubacher Berichte (German URSI Symposium)*, vol. 39, 1995, pp. 411–417.

[17] N. Marcuvitz, *Waveguide Handbook*, P. Peregrinus Ltd., London, Great Britain, 1986.

Further Reading

[18] R. Pregla, 'A Generalized Algorithm for Analysis of Planar Multilayered Anisotropic Waveguide Structures by the Method of Lines', *Int. J. Electronics and Comm. (AEÜ)*, vol. 52, no. 2 , pp. 94-98, Mar. 1998, Corr.: vol. 53, no. 1 , Jan. 1999.

[19] R. Pregla, 'The Method of Lines as Generalized Transmission Line Technique for the Analysis of Multilayered Structures', *Int. J. Electronics and Comm. (AEÜ)*, vol. 50, no. 5, Sept. 1996, pp. 293–300.

[20] R. Pregla, 'Novel FD-BPM for Optical Waveguide Structures with Isotropic or Anisotropic Material', *Eur. Conf. on Integrated Optics 1999*, ECIO'99, Apr. 14-16 1999, Torino, Italy, pp. 55–58.

[21] R. Pregla, 'Novel Algorithms for the Analysis of Optical Fiber Structures with Anisotropic Materials', *Int. Conf. on Transparent Optical Networks*, June 1999, Kielce, Poland, pp. 49–52.

[22] R. Pregla, 'Efficient Analysis of Conformal Antennas with Anisotropic Material', *AP 2000 Millennium Conf. on Antennas and Propagation (paper 0682)*, Davos, Switzerland, 9–14 April 2000. ESA Publ. Division, ESTEC, NL-2200 AG Noordwijk, CD: ISBN 92-9092-776-3.

[23] A. Barcz, 'Analysis of the Wave Propagation in Planar Structures with Substrates of Magnetized Ferrites', PhD thesis, FernUniversität in Hagen, 2006.

CHAPTER 9

SUMMARY AND PROSPECT FOR THE FUTURE

In this book we have explained the basic principles of the Method of Lines, and we hope that readers will come to the conclusion that it is a powerful tool for solving electromagnetic field problems. To summarise, the following characteristics should be mentioned:

- Different problems are described in a unique way by the Generalised Transmission Line Equations, which can be formulated in all orthogonal coordinate systems and for arbitrary complex material tensors.
- The electromagnetic waves are described in a physically adequate way. This is done by transforming the impedances from the far end of the sources towards the source, then computing the fields in the opposite direction, i.e. outwards.
- The results have a high accuracy. The eigensolutions have monotonic convergence behaviour and the fields show a smooth distribution.
- Extensions of the algorithm have been developed for problems which cannot be solved in an ideal way, e.g. algorithms with discretisation lines of different lengths.

Even though many problems have been dealt with in this book, there are other ones which have not yet been solved. Some of these problems are listed below. This list cannot of course be complete.

- If waveguides of different forms are connected to each other, it might be necessary to use different coordinate systems to describe the wave propagation. A classical problem is the concatenation of the rectangular waveguide to a horn radiator. When using Cartesian and spherical coordinates there is a region that does not belong to any of these systems. In the Mode Matching Technique this problem is 'solved' by the so called 'Zwischenmedium' (intermediate medium) [1]. Alternative, more accurate solutions must be developed.
- Each numerical method has its own characteristics and is suited for specific problems. Complex structures consist of different regions. Therefore, for a region A the MoL might be best suited, whereas

region B could be better treated with a different method. Hence, a combination of different methods appears appropriate. Examples are the combination of the MoL with the Multiple Multipole Method [2], [3] and the combination of the MoL with a special FDFD described in [4] and in this book. There are many other problems where such a combination might be very helpful. Therefore, in future work dealing with structures of increasing complexity, such combinations of methods should be further developed.

- Importantly, for some coordinate systems adequate analytical or numerical solutions for the GTL equations have not yet been found.
- Some problems (like pulse propagation) are best solved in the time domain. A few papers dealing with the MoL in time domain are mentioned in the introduction [5]– [9]. A new approach was presented at the ICTON 07 in Rome [10]. However, further work is required to treat time domain problems in the MoL in a more efficient way.
- All problems described in this book are linear. However, nonlinear problems are also important. To the best knowledge of the author, only a few MoL papers so far, e.g. [11], have dealt with nonlinear problems. So more work in this area appears useful.
- The algorithms that were developed for solving Maxwell's equations could also be transformed into other fields in physics to solve adequate partial equations there. An example is the heat equation that has to be dealt with when lasers are examined. The heat equation for VCSELs was solved in [12]– [14].

As can be seen, there is still a lot more work to do in extending the MoL algorithms. The authors invite readers to participate in such work and wishes them best of luck in all their efforts.

References

[1] G. Piefke, *Feldtheorie III*, vol. 782 of *Hochschultaschenbücher*, B.I. Wissenschaftsverlag, Mannheim/Wien/Zürich, 1977, chap. 13.4.

[2] W. Pascher, 'Combination of the Method of Lines and the Generalized Multipole Technique', in 26th *European Microwave Conference*, Prague, 1996, pp. 452–456.

[3] W. Pascher, 'Interface Considerations in the Combination of the Method of Lines and the Generalized Multipole Technique', in *U.R.S.I Intern. Symp. Electromagn. Theor.*, Thessaloniki, Greece, May 1998, pp. 372–374.

[4] R. Pregla, 'Modeling of Optical Waveguides and Devices by Combination of the Method of Lines and Finite Differences of Second Order Accuracy', *Opt. Quantum Electron.*, vol. 38, no. 1–3, pp. 3–17, 2006, Special Issue on Optical Waveguide Theory and Numerical Modelling.

[5] S. Nam, S. El–Ghazaly, H. Ling and T. Itoh, 'Time-Domain Method of Lines', *Electron. Lett.*, vol. 24, no. 2, pp. 128–129, 1988.

[6] S. Nam, S. El–Ghazaly, H. Ling and T. Itoh, 'Time-Domain Method of Lines Applied to a Partially Filled Waveguide', in *IEEE MTT-S Int. Symp. Dig.*, 1988, vol. 2, pp. 627–630.

[7] S. Nam, H. Ling and T. Itoh, 'Time-Domain Method of Lines Applied to the Uniform Microstrip Line and its Step Discontinuity', in *IEEE MTT-S Int. Symp. Dig.*, 1989, vol. 3, pp. 997–1000.

[8] S. Nam, H. Ling and T. Itoh, 'Time-Domain Method of Lines Applied to Planar Guided Wave Structures', *IEEE Trans. Microwave Theory Tech.*, vol. 37, no. 5, pp. 897–901, 1989.

[9] S. Nam, H. Ling and T. Itoh, 'Characterization of Uniform Microstrip Line and its Discontinuities Using the Time-Domain Method of Lines', *IEEE Trans. Microwave Theory Tech.*, vol. 37, no. 12, pp. 2051–2057, 1989.

[10] J. Gerdes, 'Hybrid Time-Frequency Domain Eigenmode Propagation Analysis of Optical Waveguides Based on the Method of Lines', in *ICTON Conf.*, Rome, Italy, 2007, vol. 4, pp. 324–327.

[11] M. Bertolotti, M. Masciulli and C. Sibilia, 'MoL Numerical Analysis of Nonlinear Planar Waveguide', *J. Lightwave Technol.*, vol. 12, pp. 784–789, 1994.

[12] E. Ahlers, S. Helfert and R. Pregla, 'Accurate Analysis of Vertical Cavity Surface Emitting Laser Diodes (in German)', in *Deutsche Nationale U.R.S.I Konf.*, Kleinheubach, Oct. 1995, vol. 39, pp. 135–144.

[13] E. Ahlers, S. F. Helfert and R. Pregla, 'Modelling of VCSELs by the Method of Lines', in *OSA Integr. Photo. Resear. Tech. Dig.*, Boston, USA, 1996, vol. 6, pp. 340–343.

[14] O. Conradi, S. Helfert and R. Pregla, 'Comprehensive Modeling of Vertical-Cavity Laser-Diodes by the Method of Lines', *IEEE J. Quantum Electron.*, vol. 37, pp. 928–935, 2001.

Appendix A
DISCRETISATION SCHEMES AND DIFFERENCE OPERATORS

A.1 DETERMINATION OF THE EIGENVALUES AND EIGENVECTORS OF $\boldsymbol{P}$

The elements $\boldsymbol{\lambda}_k^2$ of the matrix $\boldsymbol{\lambda}^2$ are the eigenvalues, and the column vectors t_k of the transformation matrix $\boldsymbol{T}$ are the eigenvectors belonging to the matrix $\boldsymbol{P} = \boldsymbol{D}^{\mathrm{t}}\boldsymbol{D}$:

$$(\boldsymbol{P} - \lambda_k^2 \boldsymbol{I})t_k = \mathbf{0} \tag{A.1}$$

As $\boldsymbol{P}$ is a tridiagonal matrix, we obtain a second-order difference equation:

$$-t_{i-1}^{(k)} + (2 - \lambda_k^2)t_i^{(k)} - t_{i+1}^{(k)} = 0 \tag{A.2}$$

Only the first and the last equation in (A.1) are different from this form. With the substitution:

$$t_i^{(k)} = A_k e^{ji\varphi_k} + B_k e^{-ji\varphi_k} \tag{A.3}$$

we obtain the characteristic equation from (A.2):

$$\lambda_k^2 = 2(1 - \cos\varphi_k)$$

or:

$$\boxed{\lambda_k^2 = 4\sin^2(\varphi_k/2)} \tag{A.4}$$

This equation is valid for all boundary combinations. Only the φ_k depend on the boundary combinations.

To determine the φ_k and the A_k and B_k in (A.3), the first and the last equations are used. From the explanations in Section 2.2 it is known that the boundary conditions for two components e.g. E_z and H_z are dual to each other. Thus it is convenient to look at the solutions for two components (here generally named ϕ_1 and ϕ_2) together. For a given problem, the pair of boundary conditions is either:

I. ϕ_1: Dirichlet–Neumann (**DN**) ϕ_2: Neumann–Dirichlet (**ND**)

or:

II. ϕ_1: Dirichlet–Dirichlet (**DD**) ϕ_2: Neumann–Neumann (**NN**)

For an easy solution to the first and last equations (A.1), these equations are extended by fictitious quantities t_0, t_{N+1} or even t_{N+2}, which shall also obey the law (A.3). Thus eq. (A.2) is fulfilled as well.

I. Dirichlet–Neumann (DN) condition and Neumann–Dirichlet (ND) condition:
The number of lines is N for both components. The first and the last equations run as follows:

DN	**ND**
$t_0^{(k)} = 0$	$t_0^{(k)} - t_1^{(k)} = 0$
$-t_N^{(k)} + t_{N+1}^{(k)} = 0$	$t_{N+1}^{(k)} = 0$

or with (A.3):

$$\begin{bmatrix} 1 & 1 \\ e^{jN\varphi_k}(e^{j\varphi_k} - 1) & e^{-jN\varphi_k}(e^{-j\varphi_k} - 1) \end{bmatrix} \cdot \begin{bmatrix} A_k \\ B_k \end{bmatrix} = \mathbf{0} \qquad \Bigg| \qquad \begin{bmatrix} 1 - e^{j\varphi_k} & 1 - e^{-j\varphi_k} \\ e^{j(N+1)\varphi_k} & e^{-j(N+1)\varphi_k} \end{bmatrix} \cdot \begin{bmatrix} A_k \\ B_k \end{bmatrix} = \mathbf{0}$$

From the conditions for nontrivial solutions follows:

$$\boxed{\varphi_k = \frac{k - 1/2}{N + 1/2}\pi, k = 1, 2, \ldots, N} \tag{A.5}$$

and from eq. (A.3) with:

DN	**ND**
$A_k = -B_k$	$(1 - e^{j\varphi_k})A_k = -(1 - e^{-j\varphi_k})B_k$
	$e^{j\varphi_k/2}A_k = e^{-j\varphi_k/2}B_k$
$t_i^{(k)} = A_k \sin i\varphi_k$	$t_i^{(k)} = A_k \cos(i - 1/2)\varphi_k$

Thus the following transformation matrices result for $i, k = 1, 2, \ldots, N$:

$$\boxed{\begin{aligned} &\text{For } \mathbf{DN} \quad T_{\mathrm{DN}ik} = \sqrt{\frac{2}{N + 1/2}} \sin \frac{i(k - 1/2)\pi}{N + 1/2} \\ &\text{For } \mathbf{ND} \quad T_{\mathrm{ND}ik} = \sqrt{\frac{2}{N + 1/2}} \cos \frac{(i - 1/2)(k - 1/2)\pi}{N + 1/2} \end{aligned}} \tag{A.6}$$

The vectors are orthonormal, which means that:

$$\boldsymbol{T}^{\mathrm{t}}\boldsymbol{T} = \boldsymbol{I}$$

II. Dirichlet–Dirichlet (DD) condition and Neumann–Neumann (NN) condition:
The number of lines for the boundary condition DD is N, the number of lines for NN is $N+1$. The first and the last equations respectively run as follows:

DD	**NN**
$t_0^{(k)} = 0$	$t_0^{(k)} - t_1^{(k)} = 0$
$t_{N+1}^{(k)} = 0$	$t_{N+2}^{(k)} - t_{N+1} = 0$

or with (A.3):

$$\begin{bmatrix} 1 & 1 \\ e^{j(N+1)\varphi_k} & e^{-j(N+1)\varphi_k} \end{bmatrix} \cdot \begin{bmatrix} A_k \\ B_k \end{bmatrix} = \mathbf{0} \qquad \Bigg| \qquad \begin{bmatrix} 1 & 1 \\ e^{j(N+1)\varphi_k} & e^{-j(N+1)\varphi_k} \end{bmatrix} \cdot \begin{bmatrix} (1-e^{j\varphi_k})A_k \\ (1-e^{-j\varphi_k})B_k \end{bmatrix} = \mathbf{0}$$

From the condition for nontrivial solutions follows:

$$\varphi_k = \frac{k\pi}{N+1}$$
$$\text{For } \mathbf{DD} \quad k = 1, 2, \ldots, N$$
$$\text{For } \mathbf{NN} \quad k = 0, 1, 2, \ldots, N \tag{A.7}$$

and from eq. (A.3) with:

DD	**NN**
$A_k = -B_k$	$(1 - e^{j\varphi_k})A_k = -(1 - e^{-j\varphi_k})B_k$
	$e^{j\varphi_k/2}A_k = +e^{-j\varphi_k/2}B_k$
$t_i^{(k)} = A_k \sin i\varphi_k$	$t_i^{(k)} = A_k \cos(i - 1/2)\varphi_k$

we finally obtain the transformation matrices:

$$\text{For } \mathbf{DD} \quad T_{\mathrm{DD}i,k} = \sqrt{\frac{2}{N+1}} \sin \frac{ik\pi}{N+1} \qquad i, k = 1, 2, \ldots, N$$
$$\text{For } \mathbf{NN} \quad T_{\mathrm{NN}i,0} = \sqrt{\frac{1}{N+1}}$$
$$T_{\mathrm{NN}i,k} = \sqrt{\frac{2}{N+1}} \cos \frac{(i+1/2)k\pi}{N+1} \qquad i = 0, 1, \ldots, N \quad k = 1, 2, \ldots, N \tag{A.8}$$

In this case the eigenvectors are also orthonormal, which means again:

$$\mathbf{T}^{\mathrm{t}}\mathbf{T} = \mathbf{I}$$

The eigenvalues are equal in case I. In case II the first eigenvalue for the boundary conditions NN is equal to zero, whereas the other eigenvalues are equal to those for the boundary conditions DD. We can give the following relations:

$$\boldsymbol{\lambda}^2_{\mathrm{DN}} = \boldsymbol{\lambda}^2_{\mathrm{ND}} \tag{A.9}$$

$$\boldsymbol{\lambda}^2_{\mathrm{NN}} = \mathrm{blockdiag}(\boldsymbol{0}, \boldsymbol{\lambda}^2_{\mathrm{DD}}) \tag{A.10}$$

where we have generally set:

$$\boldsymbol{\lambda}^2 = \mathrm{diag}(\lambda_k^2) \tag{A.11}$$

A.1.1 Calculation of the matrices $\boldsymbol{\delta}$

The matrix $\boldsymbol{\delta}$ is defined as a product of the difference matrix $\boldsymbol{D}$ with the corresponding transformation matrix $\boldsymbol{T}$ from the right and with the transposed matrix $\boldsymbol{T}_{\mathrm{d}}^{\mathrm{t}}$ of the dual boundary value problem from the left:

$$\boldsymbol{\delta} = \boldsymbol{T}_{\mathrm{d}}^{\mathrm{t}} \boldsymbol{D} \boldsymbol{T} \tag{A.12}$$

Here the subscript d stands for 'dual'. Before $\boldsymbol{\delta}$ is determined, one result can be anticipated. The eigenvalue matrix $\boldsymbol{\lambda}^2$ results from the product:

$$\boldsymbol{\delta}^{\mathrm{t}} \boldsymbol{\delta} = \boldsymbol{T}^{\mathrm{t}} \boldsymbol{D}^{\mathrm{t}} \boldsymbol{T}_{\mathrm{d}} \boldsymbol{T}_{\mathrm{d}}^{\mathrm{t}} \boldsymbol{D} \boldsymbol{T} = \boldsymbol{T}^{\mathrm{t}} \boldsymbol{D}^{\mathrm{t}} \boldsymbol{D} \boldsymbol{T} = \boldsymbol{\lambda}^2 \tag{A.13}$$

Thus a relation between $\boldsymbol{\delta}$ and $\boldsymbol{\lambda}^2$ is presumed.

I. Boundary conditions **DD** *and* **NN**:

For the boundary conditions **DD** the difference matrix $\boldsymbol{D}_{\mathrm{DD}}$ is given by:

$$\boldsymbol{D}_{\mathrm{DD}} = \begin{bmatrix} 1 & & \\ -1 & \ddots & \\ & \ddots & 1 \\ & & -1 \end{bmatrix} \tag{A.14}$$

The corresponding $\boldsymbol{\delta}$ is:

$$\boldsymbol{\delta} = \boldsymbol{T}_{\mathrm{NN}}^{\mathrm{t}} \boldsymbol{D}_{\mathrm{DD}} \boldsymbol{T}_{\mathrm{DD}} \tag{A.15}$$

If we first multiply $\boldsymbol{T}_{\mathrm{NN}}^{\mathrm{t}}$ by $\boldsymbol{D}_{\mathrm{DD}}$, we find that the first row of the result vanishes as all elements of the first column vector in $\boldsymbol{T}_{\mathrm{NN}}$ are identical. The remaining part of the resulting matrix can be written as a product of $\boldsymbol{\lambda}_{\mathrm{DD}}$ and $\boldsymbol{T}_{\mathrm{DD}}$, where $\boldsymbol{\lambda}_{\mathrm{DD}}$ is the diagonal matrix formed by the positive square roots of $\boldsymbol{\lambda}^2_{\mathrm{DD}}$. Hence there is:

$$\boldsymbol{T}_{\mathrm{NN}}^{\mathrm{t}} \boldsymbol{D}_{\mathrm{DD}} = \begin{bmatrix} -\boldsymbol{0}- \\ \boldsymbol{\lambda}_{\mathrm{DD}} \boldsymbol{T}_{\mathrm{DD}} \end{bmatrix} = \begin{bmatrix} -\boldsymbol{0}- \\ \boldsymbol{\lambda}_{\mathrm{DD}} \end{bmatrix} \boldsymbol{T}_{\mathrm{DD}} \tag{A.16}$$

As $\boldsymbol{T}_{\rm DD}$ is a symmetric matrix, hence $\boldsymbol{T}_{\rm DD} = \boldsymbol{T}^{\rm t}_{\rm DD}$. Consequently $\boldsymbol{\delta}$ is given by:

$$\boldsymbol{\delta} = \begin{bmatrix} -\boldsymbol{0}- \\ \boldsymbol{\lambda}_{\rm DD} \end{bmatrix} \tag{A.17}$$

Thus $\boldsymbol{\delta}$ is a quasi-diagonal matrix.

For the boundary conditions NN the result is $-\boldsymbol{\delta}^{\rm t}$, because there is:

$$-\boldsymbol{\delta}^{\rm t} = \boldsymbol{T}^{\rm t}_{\rm DD}(-\boldsymbol{D}^{\rm t}_{\rm DD})\boldsymbol{T}_{\rm NN} \tag{A.18}$$

and $-\boldsymbol{D}^{\rm t}_{\rm DD}$ is the difference operator for the boundary conditions NN. With $\boldsymbol{\delta}$ according to eq. (A.17) there is:

$$\boldsymbol{\delta}^{\rm t}\boldsymbol{\delta} = \boldsymbol{\lambda}^2_{\rm DD} \quad \text{and} \quad \boldsymbol{\delta}\boldsymbol{\delta}^{\rm t} = \boldsymbol{\lambda}^2_{\rm NN} \tag{A.19}$$

For the conversion of equations containing $\boldsymbol{\delta}$ we hint at a useful relation. Because of the special structure of $\boldsymbol{\delta}$ according to eq. (A.17) not only:

$$\boldsymbol{\delta}\boldsymbol{\lambda}^2_{\rm DD} = \boldsymbol{\lambda}^2_{\rm NN}\boldsymbol{\delta} \quad \text{and} \quad \boldsymbol{\delta}^{\rm t}\boldsymbol{\lambda}^2_{\rm NN} = \boldsymbol{\lambda}^2_{\rm DD}\boldsymbol{\delta}^{\rm t} \tag{A.20}$$

is valid, which follows from eq. (A.19), but e.g. also:

$$\boldsymbol{\delta}\boldsymbol{\Gamma}^2_{y\rm DD} = \boldsymbol{\Gamma}^2_{y\rm NN}\boldsymbol{\delta} \quad \text{and} \quad \boldsymbol{\delta}^{\rm t}\boldsymbol{\Gamma}^2_{y\rm NN} = \boldsymbol{\Gamma}^2_{y\rm DD}\boldsymbol{\delta}^{\rm t} \tag{A.21}$$

Instead of $\boldsymbol{\Gamma}^2_y$ other diagonal matrices determined from $\boldsymbol{\lambda}^2$ and $\boldsymbol{\Gamma}_y$ can be used.

II. Boundary conditions DN and ND:

For the boundary conditions **DN** the difference matrix $\boldsymbol{D}$ is given by:

$$\boldsymbol{D}_{\rm DN} = \begin{bmatrix} 1 & & & \\ -1 & \ddots & & \\ & \ddots & \ddots & \\ & & -1 & 1 \end{bmatrix} \tag{A.22}$$

The corresponding $\boldsymbol{\delta}$ is:

$$\boldsymbol{\delta} = \boldsymbol{T}^{\rm t}_{\rm ND}\boldsymbol{D}_{\rm DN}\boldsymbol{T}_{\rm DN} \tag{A.23}$$

We find that:

$$\boldsymbol{\delta} = \text{diag}(\lambda_k) = \boldsymbol{\lambda}_{\rm DN} = \boldsymbol{\lambda}_{\rm ND} \tag{A.24}$$

λ_k are the positive square roots of λ^2_k. For the boundary conditions **ND** we obtain analogously to the above, $-\boldsymbol{\delta}^{\rm t}$.

A.1.2 Derivation of the eigenvalues of the Neumann problem from those of the Dirichlet problem

We assume that the solution of the problem:

$$\overline{\boldsymbol{D}}^{\mathrm{t}}\overline{\boldsymbol{D}}\boldsymbol{T}_{\mathrm{D}} = \boldsymbol{T}_{\mathrm{D}}\overline{\boldsymbol{\lambda}}_{\mathrm{D}}^{2}$$

is known. $\boldsymbol{D}$ is the difference operator of the Dirichlet problem of the form:

$$\boldsymbol{D} = \begin{bmatrix} 1 & & \\ -1 & \ddots & \\ & \ddots & 1 \\ & & -1 \end{bmatrix}$$

We have for the Neumann problem:

$$\overline{\boldsymbol{D}}\overline{\boldsymbol{D}}^{\mathrm{t}}\boldsymbol{T}_{\mathrm{N}} = \boldsymbol{T}_{\mathrm{N}}\overline{\boldsymbol{\lambda}}_{\mathrm{N}}^{2}$$

Because we may write the first equation above in the form:

$$\overline{\boldsymbol{D}}\overline{\boldsymbol{D}}^{\mathrm{t}}(\overline{\boldsymbol{D}}\boldsymbol{T}_{\mathrm{D}}) = (\overline{\boldsymbol{D}}\boldsymbol{T}_{\mathrm{D}})\overline{\boldsymbol{\lambda}}_{\mathrm{D}}^{2}$$

we obtain for the Neumann boundary conditions:

$$\overline{\boldsymbol{\lambda}}_{\mathrm{N}}^{2} = \begin{bmatrix} 0 & \\ & \overline{\boldsymbol{\lambda}}_{\mathrm{D}}^{2} \end{bmatrix} \quad \text{and} \quad \boldsymbol{T}_{\mathrm{N}} = [\vec{\mathbf{T}}_{o} \vdots \overline{\boldsymbol{D}}\boldsymbol{T}_{\mathrm{D}}\overline{\boldsymbol{\lambda}}_{\mathrm{D}}^{-1}]$$

where $\vec{\mathbf{T}}_o$ is the eigenvector for the eigenvalue 0. Further, it is valid:

$$\boldsymbol{\delta}_{\mathrm{D}} = \boldsymbol{T}_{\mathrm{N}}^{\mathrm{t}}\overline{\boldsymbol{D}}\boldsymbol{T}_{\mathrm{D}} \qquad \boldsymbol{\delta}_{\mathrm{N}}' = \boldsymbol{T}_{\mathrm{D}}^{\mathrm{t}}\overline{\boldsymbol{D}}^{\mathrm{t}}\boldsymbol{T}_{\mathrm{N}}$$

$$\boldsymbol{\delta}_{\mathrm{D}} = \begin{bmatrix} \vec{\mathbf{T}}_{o}^{\mathrm{t}} \\ \overline{\boldsymbol{\lambda}}_{\mathrm{D}}^{-1}\boldsymbol{T}_{\mathrm{D}}^{\mathrm{t}}\overline{\boldsymbol{D}}^{\mathrm{t}} \end{bmatrix} \quad \overline{\boldsymbol{D}}\boldsymbol{T}_{\mathrm{D}} = \begin{bmatrix} \vec{\mathbf{T}}_{o}^{\mathrm{t}}\overline{\boldsymbol{D}}\boldsymbol{T}_{\mathrm{D}} \\ \overline{\boldsymbol{\lambda}}_{\mathrm{D}}^{-1}\boldsymbol{T}_{\mathrm{D}}^{\mathrm{t}}\overline{\boldsymbol{D}}^{\mathrm{t}}\overline{\boldsymbol{D}}\boldsymbol{T}_{\mathrm{D}} \end{bmatrix} = \begin{bmatrix} \cdots & 0 & \cdots \\ \ddots & \overline{\boldsymbol{\lambda}}_{\mathrm{D}} & \\ & & \ddots \end{bmatrix}$$

The eigenvector $\vec{\mathbf{T}}_o$ will be calculated from the equation $\overline{\boldsymbol{D}}^{\mathrm{t}}\vec{\mathbf{T}}_o = \vec{\mathbf{0}}$. It is constant in case of equidistant discretisation, but has to be orthonormalised. Furthermore, we obtain:

$$\boldsymbol{\delta}_{\mathrm{N}}' = [\boldsymbol{T}_{\mathrm{D}}^{\mathrm{t}}\overline{\boldsymbol{D}}^{\mathrm{t}}\vec{\mathbf{T}}_{o} \vdots \boldsymbol{T}_{\mathrm{D}}^{\mathrm{t}}\overline{\boldsymbol{D}}^{\mathrm{t}}\overline{\boldsymbol{D}}\boldsymbol{T}_{\mathrm{D}}\overline{\boldsymbol{\lambda}}^{-1}] = \begin{bmatrix} \vdots & \ddots & \\ 0 & & \overline{\boldsymbol{\lambda}}_{D} \\ \vdots & & \ddots \end{bmatrix}$$

It should be noticed that $\boldsymbol{\delta}_{\mathrm{N}} = \boldsymbol{T}_{\mathrm{D}}^{\mathrm{t}}(-\overline{\boldsymbol{D}}^{\mathrm{t}})\boldsymbol{T}_{\mathrm{N}} = -\boldsymbol{\delta}_{\mathrm{N}}'$.

A.1.3 The component of ε_r at an abrupt transition

Here we would like to examine which value has to be chosen for the material parameters at an abrupt interface. We will examine the two-dimensional TM case corresponding to Section 2.2.4, particularly the discretisation of eqs. (2.28)–(2.30). As numerical investigations have shown, it is best to centre the interface between the points where the fields E_x and H_y are discretised. Looking at eqs. (2.28)–(2.30), we see that in this case ε_{zz} is positioned on the interface and we must find the best value at this place.

The principle behaviour of ψ_e at an interface is shown in Fig. A.1 [1]. ψ_e could be the magnetic field component H_y or a component of a vector potential. As we can also see, the boundary could be shifted with respect to the discretisation lines. This is indicated by the parameter p_μ [4,6]. We define the distance from the dual discretisation line system, labelled h here. First we assume that this value is zero ($p_\mu = 0$).

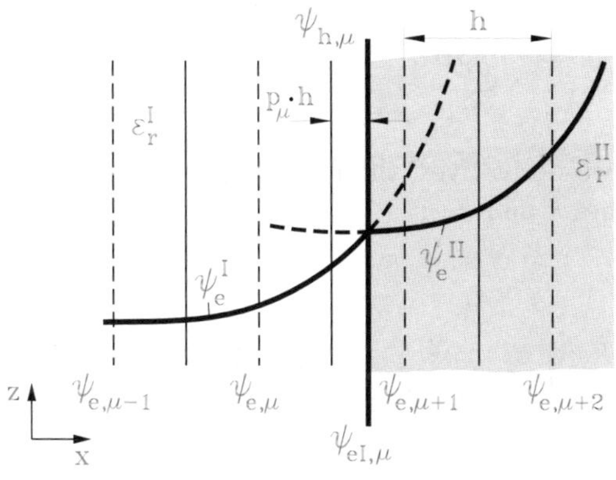

Fig. A.1 Behaviour of ψ_e at an abrupt transition from ε_{r1} to ε_{r2} (S. F. Helfert and R. Pregla, 'Finite Difference Expressions for Arbitrarily Positioned Dielectric Steps in Waveguide Structures', in *J. Lightwave Technol.* vol. 14, no. 10, pp. 2414–2421. © 1996 Institute of Electrical and Electronics Engineers (IEEE))

In what follows, we assume isotropic materials. The subscript zz indicates that the permittivity is determined at a different position than ε_{xx}. However, the results that we find can also be applied to the general case.

At the interfaces between two layers, H_y and E_z have to be continuous. The electric field component E_z can be determined from H_y as a derivative with respect to x multiplied with the inverse of ε_r(see eq. (2.12)). Therefore, the following conditions have to be fulfilled:

$$\psi_I^\mathrm{I} = \psi_\mathrm{I}^\mathrm{II} \tag{A.25}$$

$$\frac{1}{\varepsilon_{r1}} \frac{\partial \psi_\mathrm{I}^\mathrm{I}}{\partial x} = \frac{1}{\varepsilon_{r2}} \frac{\partial \psi_\mathrm{I}^\mathrm{II}}{\partial x} \tag{A.26}$$

Consequently, in Fig A.1 the ψ-curve is represented with a break on the interface, and the dashed curves are drawn without any break for the definition of the auxiliary quantities ψ_{k+1}^{I} and ψ_k^{II}. These auxiliary quantities are necessary to compute the second derivative on the lines k and k+1. From the incremental formula (Taylor series up to the linear term, applied on the interface line I), it is found that:

$$\begin{aligned} \psi_{k+1}^{\mathrm{I}} &\approx \psi_{\mathrm{I}} + \frac{\partial\psi^{\mathrm{I}}}{\partial x}\frac{h}{2} & \psi_k &\approx \psi_{\mathrm{I}} - \frac{\partial\psi^{\mathrm{I}}}{\partial x}\frac{h}{2} \\ \psi_{k+1} &\approx \psi_{\mathrm{I}} + \frac{\partial\psi^{\mathrm{II}}}{\partial x}\frac{h}{2} & \psi_k^2 &\approx \psi_{\mathrm{I}} - \frac{\partial\psi^{\mathrm{II}}}{\partial x}\frac{h}{2} \end{aligned} \tag{A.27}$$

Condition (A.25) was immediately taken into account in these equations. With (A.26) we find for the two derivatives on the interface line:

$$\begin{aligned} h\frac{\partial\psi^{\mathrm{I}}}{\partial x} &\approx \frac{2\varepsilon_{r1}}{\varepsilon_{r1}+\varepsilon_{r2}}(\psi_{k+1}-\psi_k) \\ h\,\frac{\partial\psi^{\mathrm{II}}}{\partial x} &\approx \frac{2\varepsilon_{r2}}{\varepsilon_{r1}+\varepsilon_{r2}}(\psi_{k+1}-\psi_k) \end{aligned} \tag{A.28}$$

With these two derivatives, the following are given by the special second derivatives on lines k and k+1:

Line k:

$$h^2\frac{\partial}{\partial x}\left(\frac{1}{\varepsilon_r}\frac{\partial\psi}{\partial x}\right) \approx \frac{2}{\varepsilon_{r1}+\varepsilon_{r2}}(\psi_{k+1}-\psi_k) - \frac{1}{\varepsilon_{r1}}(\psi_k-\psi_{k-1}) \tag{A.29}$$

Line k+1:

$$h^2\frac{\partial}{\partial x}\left(\frac{1}{\varepsilon_r}\frac{\partial\psi}{\partial x}\right) \approx \frac{1}{\varepsilon_{r2}}(\psi_{k+2}-\psi_{k+1}) - \frac{2}{\varepsilon_{r1}+\varepsilon_{r2}}(\psi_{k+1}-\psi_k) \tag{A.30}$$

We get the same results from eq. (A.29) and (A.30), if we take for ε_{zz} the value $\varepsilon_{zz} = (\varepsilon_{r1}+\varepsilon_{r2})/2$, that is the arithmetic mean of the two ε_r values. A closer inspection shows that we get the same result if we assume a linear change of ε_r at boundaries. As we know, for technological reasons, we never have perfect interfaces, but rather a smooth transition between two regions. Therefore, such an approximation is justified. If the boundary is not centered between two discretisation lines but is at a certain distance that can be described with the parameter p_μ (see Fig. A.1), we take the following value:

$$\varepsilon_r = \frac{\varepsilon_{r1}+\varepsilon_{r2}}{2} + p_\mu(\varepsilon_{r1}-\varepsilon_{r2}) \qquad -0.5 \le p_\mu \le 0.5$$

This rule is also valid for the perpendicular polarisation, and in the case of anisotropic materials. If the permittivity (or the permeability) is discretised on an interface between two layers, we assume a linear function. Then we take its average value (or the value corresponding to this linear behaviour). An improvement of these expressions can be found in [6] where second order expressions were derived.

A.1.4 Eigenvalues and eigenvectors for periodic boundary conditions

The eigenvalues and eigenvectors of the matrix $\boldsymbol{P}_n$ for periodic structure s are determined from the eigenvalues and eigenvectors of the difference matrix $\boldsymbol{D}_n$, which is given by eq. (A.88) for phase-normalised field quantities or potentials. $\boldsymbol{D}_n$ is a circulant matrix. Analogous to Section A.1, we find for its eigenvalues:

$$\boxed{\delta_{nk} = 2je^{j\varphi_k/2}\sin\frac{\varphi_k - \beta_x h}{2}} \tag{A.31}$$

$$\boxed{\varphi_k = \frac{2\pi}{N} \quad k; k = 1, 2 \ldots N} \tag{A.32}$$

As the eigenvector matrix, either $\boldsymbol{T}_e$ or $\boldsymbol{T}_h$ is chosen, which are Fourier matrices:

$$\boxed{T_{e\,ik} = \frac{1}{\sqrt{N}}e^{ji\varphi_k}; \quad T_{h\,ik} = \frac{1}{\sqrt{N}}e^{j(i+1/2)\varphi_k}} \tag{A.33}$$

The general phase in the eigenvector matrix is arbitrary. Because of the shifting of the discretisation lines by $h/2$, a phase shifting by $\varphi_k/2$ with respect to $\boldsymbol{T}_e$ is introduced in $\boldsymbol{T}_h$:

$$\boldsymbol{T}_h = \boldsymbol{T}_e\boldsymbol{S}_\varphi \tag{A.34}$$

with:

$$\boldsymbol{S}_\varphi = \mathrm{diag}(e^{j\varphi_k/2}) \tag{A.35}$$

As the matrix $\boldsymbol{P}_n$ is given as a product in eq. (A.89), its eigenvalues $\boldsymbol{\lambda}^2$ are products as well:

$$\boldsymbol{\lambda}^2 = \boldsymbol{\delta}_n\boldsymbol{\delta}_n^* \tag{A.36}$$

where:

$$\boldsymbol{\lambda} = \mathrm{diag}\left(2\sin\frac{\varphi_k - \beta_x h}{2}\right) \tag{A.37}$$

$$\boldsymbol{\delta}_n = j\boldsymbol{\lambda}\boldsymbol{S}_\varphi = j\boldsymbol{S}_\varphi\boldsymbol{\lambda} \tag{A.38}$$

In addition there is:

$$\begin{aligned} \boldsymbol{T}_{e,h}^{*t}\boldsymbol{D}_n\boldsymbol{T}_{e,h} &= \boldsymbol{\delta}_n \\ \boldsymbol{T}_h^{*t}\boldsymbol{D}_n\boldsymbol{T}_e &= j\boldsymbol{\lambda} \\ \boldsymbol{T}_e^{*t}\boldsymbol{D}_n^{*t}\boldsymbol{T}_h &= -j\boldsymbol{\lambda} \end{aligned} \tag{A.39}$$

A.1.5 Discretisation for non-ideal places of the boundaries

The equidistant discretisation has some advantages over the non-equidistant one. In particular, the convergence behaviour is monotonic. Therefore, the equidistant discretisation allows polynomial extrapolation for obtaining 'exact' values. If the discretisation scheme is predominated by one of the dimensions, not all of the other dimensions coincide with this scheme. For example, in the filter structure of Fig. 3.21 the choice of discretisation scheme is determined by the broadside of the waveguide. The width of the coupling slots does not coincide with this scheme in general. Nevertheless, it is possible to fulfil the boundary conditions. The method by which this can be done is shown in Fig. A.2. We assume that a field component F has to be discretised on the full lines and F has to fulfil the Dirichlet boundary condition on the metallic walls, which in the ideal case lie on lines 0 and $N+1$. The real positions may be described by the distance $p_l h$ from line 1 and $p_r h$ from line N, respectively. $p_{l,r}$ therefore may take the values $0 < p_{l,r} \leq 1$ (for e.g. $p_l = 0$ or $p_r = 0$, the number N should be reduced by 1, or in other words the first or last discretisation line for F in that region should be line 2 or $N-1$, respectively).

For $0.5 \leq p_{l,r} \leq 1$ and Dirichlet boundary conditions, the difference operator may be written in the form:

$$\boldsymbol{D}_1 = \begin{bmatrix} p_l^{-1} & & & \\ -1 & 1 & & \\ & \ddots & \ddots & \\ & & -1 & 1 \\ & & & -p_r^{-1} \end{bmatrix} \qquad \boldsymbol{D}_2 = \begin{bmatrix} -1 & 1 & & & \\ & \ddots & \ddots & & \\ & & \ddots & \ddots & \\ & & & -1 & 1 \end{bmatrix} \tag{A.40}$$

The second-order difference operator $\boldsymbol{P}_{\mathrm{D}}$ for the component F with Dirichlet boundary conditions is:

$$\boldsymbol{P}_{\mathrm{D}} = \boldsymbol{D}_2 \boldsymbol{D}_1 \tag{A.41}$$

where the operator $\boldsymbol{D}_2$ is obtained from $\boldsymbol{D}_1$ by $\boldsymbol{D}_2 = -\boldsymbol{D}_1^t$ with $p_{l,r} = 1$. The number M is given by $M = N + 1$.

We assume that a field component G has to fulfil the Neumann boundary conditions and is discretised on the dashed lines. The second-order difference operator $\boldsymbol{P}_{\mathrm{N}}$ for the component G with Neumann boundary conditions can be written in the form:

$$\boldsymbol{P}_{\mathrm{N}} = \boldsymbol{D}_1 \boldsymbol{D}_2 \tag{A.42}$$

Now we assume that the parameters $p_{l,r}$ are less than 0.5. (If we have a non-ideal position of the boundary on one side only, we need only construct the difference operator for this case in the way described here.)

The situation is sketched in Fig. A.3. The full lines for the component F with Dirichlet boundary conditions are now – very unusually – closer to the metallic boundaries than the dashed lines, and we have $M = N - 1$.

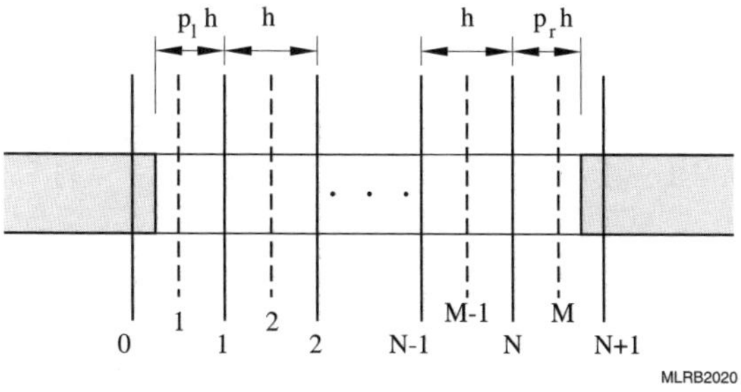

Fig. A.2 Discretisation for non-ideal locations of the boundaries ___ F component with Dirichlet boundary conditions - - - H component with Neumann boundary conditions

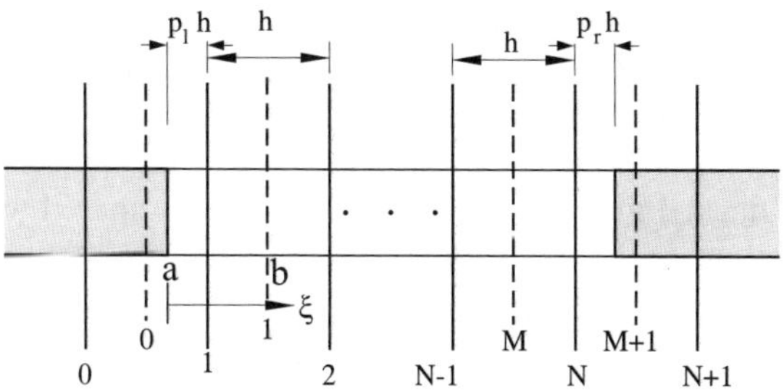

Fig. A.3 Discretisation for non-ideal locations of the boundaries ___ F component with Dirichlet boundary conditions - - - H component with Neumann boundary conditions

The difference operators $\boldsymbol{D}$ must be constructed separately for the components F and G. The difference operator $\boldsymbol{D}_1$ for the first derivative of F is the same as in eq. (A.40). We assume that the (first and last) values F_0' and F_{M+1}' are on the boundaries on the left and right sides, respectively. For the behaviour of $F'(\xi/h)$ on the left side we may write:

$$F'(\xi/h) = F_a' + c_1(\xi/h)^2$$

It follows that:

$$F''(\xi/h) = 2c_1(\xi/h)$$

From these two equations we obtain at $\xi/h = p_l$ (or the first full line 1):

$$F_1'' = A_l(F_b' - F_a')$$

with $F_b' = F'(p_l + 0.5)$ and $A_l = 2p_l/(p_l + 0.5)^2$. Therefore, we write for the difference operator $\boldsymbol{D}_2^{\mathrm{D}}$:

$$\boldsymbol{D}_2^{\mathrm{D}} = \begin{bmatrix} -A_l & A_l & & & \\ & -1 & 1 & & \\ & & \ddots & \ddots & \\ & & -1 & 1 & \\ & & & -A_r & A_r \end{bmatrix} \qquad \boldsymbol{D}_1^{\mathrm{N}} = \begin{bmatrix} -B_l & B_l & & & \\ -1 & 1 & & & \\ & & \ddots & \ddots & \\ & & & -1 & 1 \\ & & & -B_r & B_r \end{bmatrix} \tag{A.43}$$

In the case of component G, we have to fulfil Neumann boundary conditions. By an analogous procedure to that used before, we determine the first derivative of G at the first full line (or $\xi/h = p_l$) and obtain:

$$G_1' = G'(p_l) = B_l(G_2 - G_1)$$

with $B_l = p_l/(p_l + 1)$. The derivation for the right side is done analogously. The difference operator $\boldsymbol{D}_1^{\mathrm{N}}$ is given in eq. (A.43). The difference operator for obtaining the second derivative of G must have the same form as Neumann boundary conditions or $\boldsymbol{D}_2$ in eq. (A.40). This can easily be seen from Fig. A.3. The number of dashed lines is now reduced by one compared with the former case. The second-order difference operators are:

$$\boldsymbol{P}_{\mathrm{D}} = \boldsymbol{D}_2^{\mathrm{D}}\boldsymbol{D}_1 \qquad \boldsymbol{P}_{\mathrm{N}} = \boldsymbol{D}_2\boldsymbol{D}_1^{\mathrm{N}} \tag{A.44}$$

for Dirichlet and Neumann boundary conditions, respectively. It is important to point out that the eigenvectors of $\boldsymbol{P}_{\mathrm{D}}$ and $\boldsymbol{P}_{\mathrm{N}}$ in eq. (A.44) are not orthogonal! Therefore, we cannot simply determine their inverse by transposing. This is in contrast to the operators in eqs. (A.41)–(A.41), where we could write (after suitable normalisation) $\boldsymbol{T}^{-1} = \boldsymbol{T}^{\mathrm{t}}$.

It should be mentioned that for inhomogeneous layers the matrices of the material parameters must be introduced.

Note: if only first-order difference operators are required for the matching process, they must be different from those above in some cases. For $0.5 \leq p_{l,r} \leq 1$ we can use the operators in eq. (A.40). If we want to calculate the first derivative of F for $0.0 \leq p_{l,r} \leq 0.5$ we must use $\boldsymbol{D}_2$ in eq. (A.40). For the first derivative of G we must use $\boldsymbol{D}_1^{\mathrm{N}}$ in eq. (A.43). ($0.0 \leq p_{l,r} \leq 0.5$.)

A.2 ABSORBING BOUNDARY CONDITIONS (ABCs)

A.2.1 Introduction[1]

In inhomogeneous waveguide structures, part of the field of a propagating wave experiences scattering. This scattered field propagates to the walls and must be absorbed there to avoid reflections.

[1]Sections A.2.1–A.2.5 reproduced by permission of EMW Publishing (R. Pregla 'MoL-BPM Method of Lines Based Beam Propagation Method', in *Methods for Modeling and Simulation of Guided-Wave Optoelectronic Devices (PIER 11)*, W. P. Huang (Ed.), Progress in Electromagnetic Research, pp. 51–102, 1995).

For this task, absorbing boundary conditions (ABCs) as described in [2] are introduced in the MoL [3]. The ABCs are obtained by factorisation of the wave equation.

A.2.2 Factorisation of the Helmholtz equation

We will demonstrate the procedure of factorisation of the Helmholtz equation for the x-direction (see Fig. A.4). In the regions at the left and right sides the local wave equations are factorised [2], making use of the fact that the waves are supposed to obey Sommerfeld's radiation condition at the boundary.

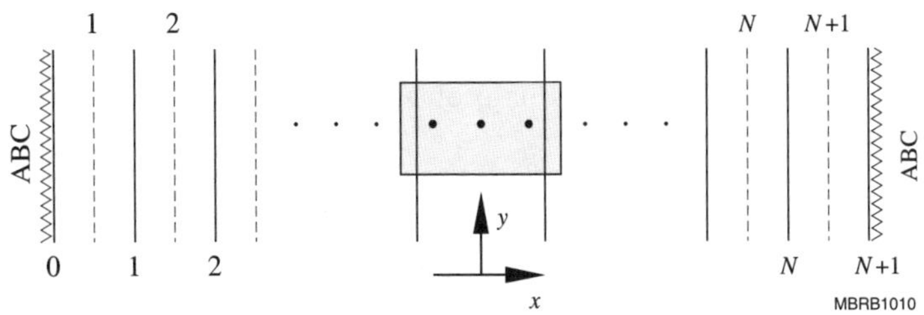

Fig. A.4 Absorbing boundary conditions (ABC) on the left and right sides of a waveguide (R. Pregla, 'MoL-BPM Method of Lines Based Beam Propagation Method', in *Methods for Modeling and Simulation of Guided-Wave Optoelectronic Devices (PIER 11)*, W. P. Huang, (Ed.), Progress in Electromagnetic Research, pp. 51–102. © 1995 EMW Publishing)

Using the operator L where:

$$L = \frac{\partial^2}{\partial \overline{x}^2} + \frac{\partial^2}{\partial \overline{y}^2} + \frac{\partial^2}{\partial \overline{z}^2} + \epsilon_r = D_{\overline{x}}^2 + D_{\overline{y}}^2 + D_{\overline{z}}^2 + \epsilon_r \tag{A.45}$$

$\overline{x}$, $\overline{y}$, $\overline{z}$ being Cartesian coordinates normalised with respect to the free space wave number, we write instead of the wave equation:

$$L\phi = L^+ L^- \phi = 0 \tag{A.46}$$

ϕ stands for any of the electric or magnetic field components. We obtain two separate equations as solutions for propagating fields in $+x$- and $-x$-directions:

$$L^+\phi = 0 \text{ and } L^-\phi = 0 \tag{A.47}$$

where:

$$L^+\phi = (D_{\overline{x}} + j\sqrt{\epsilon_r}\sqrt{1+S^2})\phi = 0 \tag{A.48}$$

$$L^-\phi = (D_{\overline{x}} - j\sqrt{\epsilon_r}\sqrt{1+S^2})\phi = 0 \tag{A.49}$$

$$S^2 = \epsilon_r^{-1}(D_{\overline{y}}^2 + D_{\overline{z}}^2) \tag{A.50}$$

The first equation describes wave propagation in $+x$-direction only and the second one in $-x$-direction only. The scattered field from the guide in Fig. A.4 propagates on the right side in $+x$-direction and on the left side in $-x$-direction. On both sides, only outgoing waves are allowed. Therefore, on the right side the field will be described by eq. (A.48) only and on the left side by eq. (A.49) only. Now the ABCs will be obtained from these two equations for the right and left boundary, respectively. The main problem in eqs. (A.48) and (A.49) is the appearance of the differential operators under the radical sign. Therefore, L^+ and L^- are nonlocal pseudodifferential operators and a direct numerical implementation is not possible.

A.2.3 Padé approximation

The usual way [2] to overcome this problem is to apply Padé approximation by the rational function:

$$\sqrt{1+S^2} \approx \frac{p_0 + p_2 S^2}{q_0 + q_2 S^2} \qquad |S^2| < 1 \tag{A.51}$$

With this approximation we can write eqs. (A.48) and (A.49) in the form:

$$\{\mp j[q_0 + q_2\epsilon_r^{-1}(D_{\bar{y}}^2 + D_{\bar{z}}^2)]D_{\bar{x}} + \sqrt{\epsilon_r}[p_0 + p_2\epsilon_r^{-1}(D_{\bar{y}}^2 + D_{\bar{z}}^2)]\}\phi = 0 \tag{A.52}$$

Introducing:

$$S^2 = \epsilon_r^{-1}(D_{\bar{y}}^2 + D_{\bar{z}}^2) = -1 - \epsilon_r^{-1}D_{\bar{x}}^2 \tag{A.53}$$

from the wave equation, we obtain:

$$\left\{\epsilon_r^{-1}D_{\bar{x}}^2 \pm j\left[\frac{q_0}{p_2} - \frac{q_2}{p_2}(1+\epsilon_r^{-1}D_{\bar{x}}^2)\right]\epsilon_r^{-\frac{1}{2}}D_{\bar{x}} + \left(1 - \frac{p_0}{p_2}\right)\right\}\phi = 0 \tag{A.54}$$

Discretising this equation on line 1 with sign minus and on line N with sign plus by using the Taylor series to obtain the potentials on the neighbouring lines gives:

$$\begin{aligned}\begin{Bmatrix}\phi_o \\ \phi_{N+1}\end{Bmatrix} &= \frac{2p_2 - \frac{1}{2}\bar{n}_a(1-q_2) + 3q_2\bar{n}_a^{-1} + (p_2 - p_0)\bar{n}_a^2}{p_2 + \frac{1}{3}\bar{n}_a(1-q_2) + q_2\bar{n}_a^{-1}}\begin{Bmatrix}\phi_1 \\ \phi_N\end{Bmatrix} \\ &\quad - \frac{p_2 - \bar{n}_a(1-q_2) + 3q_2\bar{n}_a^{-1}}{p_2 + \frac{1}{3}\bar{n}_a(1-q_2) + q_2\bar{n}_a^{-1}}\begin{Bmatrix}\phi_2 \\ \phi_{N-1}\end{Bmatrix} \\ &\quad - \frac{\frac{1}{6}\bar{n}_a(1-q_2) - q_2\bar{n}_a^{-1}}{p_2 + \frac{1}{3}\bar{n}_a(1-q_2) + q_2\bar{n}_a^{-1}}\begin{Bmatrix}\phi_3 \\ \phi_{N-2}\end{Bmatrix}\end{aligned} \tag{A.55}$$

with the abbreviation $\bar{n}_a = j\bar{h}\sqrt{\epsilon_r}$, $\bar{h} = k_0 h$ and where, without loss of generality, $q_0 = 1$ was introduced. We write these equations in the form:

$$\phi_0 = (1 - a_1)\phi_1 + b_1\phi_2 + c_1\phi_3 \tag{A.56}$$

$$\phi_{N+1} = (1 - a_N)\phi_N + b_N\phi_{N-1} + c_N\phi_{N-2} \tag{A.57}$$

where the constants $a_1, a_N, b_1, b_N, c_1, c_N$ are given by:

$$a_{1(N)} = 1 - \frac{2p_2 - \frac{1}{2}\bar{n}_a(1-q_2) + 3q_2\bar{n}_a^{-1} + (p_2 - p_0)\bar{n}_a^2}{p_2 + \frac{1}{3}\bar{n}_a(1-q_2) + q_2\bar{n}_a^{-1}} \tag{A.58}$$

$$b_{1(N)} = -\frac{p_2 - \bar{n}_a(1-q_2) + 3q_2\bar{n}_a^{-1}}{p_2 + \frac{1}{3}\bar{n}_a(1-q_2) + q_2\bar{n}_a^{-1}} \tag{A.59}$$

$$c_{1(N)} = -\frac{\frac{1}{6}\bar{n}_a(1-q_2) - q_2\bar{n}_a^{-1}}{p_2 + \frac{1}{3}\bar{n}_a(1-q_2) + q_2\bar{n}_a^{-1}} \tag{A.60}$$

The interpretation of eqs. (A.56) and (A.57) is as follows: the field outside the calculation window (linearly) depends only on the field inside the window, because we have sources only inside the window. It should be mentioned that the parameters ϵ_r, p_0, p_2 and q_2 can be different on both sides of the waveguide. Therefore, the absorbing boundary conditions on both sides of the waveguide are different in general.

The expansion coefficients are chosen from the three angles Θ_1, Θ_2 and Θ_3 with total absorption. For other angles, the numerical implementation is not exact. The reflection coefficient at the boundary is given by [2]:

$$r = \frac{\cos\Theta - q_2\cos\Theta\sin^2\Theta - p_0 + p_2\sin^2\Theta}{\cos\Theta - q_2\cos\Theta\sin^2\Theta + p_0 - p_2\sin^2\Theta} \tag{A.61}$$

Figure A.5 is a plot of the reflection coefficient r vs. angle of incidence Θ for different approximations and various choices of absorbing angles Θ_i. Therefore, with $r = 0$ for three chosen angles Θ_1, Θ_2 and Θ_3, we obtain the coefficients from:

$$\begin{bmatrix} 1 & -\sin^2\Theta_1 & \cos\Theta_1\sin^2\Theta_1 \\ 1 & -\sin^2\Theta_2 & \cos\Theta_2\sin^2\Theta_2 \\ 1 & -\sin^2\Theta_3 & \cos\Theta_3\sin^2\Theta_3 \end{bmatrix} \begin{bmatrix} p_0 \\ p_2 \\ q_2 \end{bmatrix} = \begin{bmatrix} \cos\Theta_1 \\ \cos\Theta_2 \\ \cos\Theta_3 \end{bmatrix} \tag{A.62}$$

with the solution:

$$\begin{aligned} q_2 &= (1 + \cos\Theta_1\cos\Theta_2 + \cos\Theta_1\cos\Theta_3 + \cos\Theta_2\cos\Theta_3)^{-1} \\ p_0 &= (\cos\Theta_1 + \cos\Theta_2 + \cos\Theta_3 + \cos\Theta_1\cos\Theta_2\cos\Theta_3)q_2 \\ p_2 &= (\cos\Theta_1 + \cos\Theta_2 + \cos\Theta_3)q_2 \end{aligned} \tag{A.63}$$

The solutions in eq. (A.63) were calculated by using MAPLE. The angles of total absorption can be chosen in such a way as to achieve an equiripple reflection behaviour in a specified region.

A.2.4 Polynomial approximations

In the case where $q_2 = 0$ (polynomial approximation) and there are two angles Θ_1 and Θ_2 of total absorption, we obtain the solution:

$$p_0 = \frac{1 + \cos\Theta_1\cos\Theta_2}{\cos\Theta_1 + \cos\Theta_2} \qquad p_2 = \frac{1}{\cos\Theta_1 + \cos\Theta_2} \tag{A.64}$$

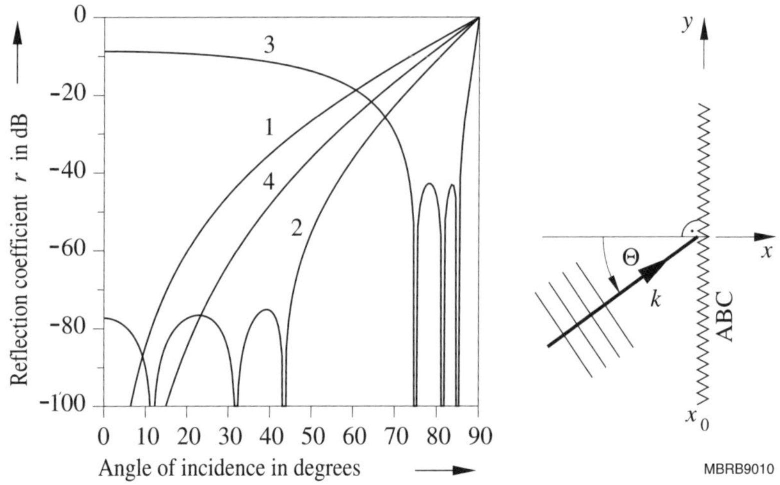

Fig. A.5 Reflection coefficient r at the boundary as a function of the angle of incidence Θ
(1) : (2.0) p_0=1.0 p_2=0.5 q_2=0.0 Θ_1=0° Θ_2=0°
(2) : (2.2) p_0=0.999730 p_2=0.808640 q_2=0.316570 Θ_1=11.7° Θ_2=31.9° Θ_3=43.5°
(3) : (2.2) p_0=0.464383 p_2=0.461243 q_2=0.930845 Θ_1=75.0° Θ_2=81.4° Θ_3=85.0°
(4) : (4.0) p_0=1.0 p_2=0.5 p_4= − 0.125 Θ_1=0° Θ_2=0° Θ_3=0°
(R. Pregla, 'MoL-BPM Method of Lines Based Beam Propagation Method', in *Methods for Modeling and Simulation of Guided-Wave Optoelectronic Devices (PIER 11)*, W. P. Huang (Ed.), Progress in Electromagnetic Research, pp. 51–102. © 1995 EMW Publishing)

By an alternative discretisation [8] the following coefficients are obtained:

$$a_{1(\mathsf{N})} = 1 - 2\frac{2p_2 + (p_2 - p_0)\overline{n}_a^2}{2p_2 + \overline{n}_a} \qquad b_{1(\mathsf{N})} = -\frac{2p_2 - \overline{n}_a}{2p_2 + \overline{n}_a} \qquad c_{1(\mathsf{N})} = 0 \tag{A.65}$$

To obtain optimal absorption in the vicinity of $\Theta = 0$ we can also use a Taylor expansion for the square root of fourth order:

$$\sqrt{1 + S^2} = 1 + \tfrac{1}{2}S^2 - \tfrac{1}{8}S^4 \tag{A.66}$$

The corresponding operator equation, analogous to eq. (A.54), can be factorised in the following form:

$$\{(D_{\overline{x}} \pm j\sqrt{\epsilon_r})^3(D_{\overline{x}} \pm j3\sqrt{\epsilon_r})\}\phi = 0 \tag{A.67}$$

The first factor produces a third-order reflection zero at $\Theta = 0$. Therefore, an improved behaviour of r is obtained (see curve 4 in Fig. A.5). For the second factor, a wave with a real angle of incidence does not exist.

Some care is necessary in using this approximation. For discretisation of this equation, a potential at one more discretisation point, as for eq. (A.54), is necessary.

A.2.5 Construction of the difference operator for ABCs

The first-order partial differential quotients are approximated by the potential differences of neighboured lines and are valid in the middles of the lines. In Fig. A.4 we have placed the ABCs on the full lines 0 and $N+1$ or at places where we otherwise introduce the Dirichlet boundary conditions. Therefore, in an analogous way to Section 1.2.2, using eqs. (A.56) and (A.57), we construct the difference operator in following way:

$$\bar{h}\frac{\partial\phi}{\partial\bar{x}} \longrightarrow \underbrace{\begin{array}{c|cccc|c} \phi_0 & \phi_1 & \cdots & \cdots & \phi_{\mathrm{N}} & \phi_{\mathrm{N}+1} \\ -1 & 1 & & & & \\ & -1 & \ddots & & & \\ & & \ddots & \ddots & & \\ & & & -1 & 1 & \\ & & & & -1 & 1 \end{array}}_{\boldsymbol{D}_{\bar{x}}} \longrightarrow \underbrace{\begin{bmatrix} a_1 & -b_1 & -c_1 & & \\ -1 & 1 & & & \\ & \ddots & \ddots & & \\ & & & -1 & 1 \\ & & c_{\mathrm{N}} & b_{\mathrm{N}} & -a_{\mathrm{N}} \end{bmatrix}}_{\boldsymbol{D}^a_{\bar{x}}} \begin{bmatrix} \phi_1 \\ \vdots \\ \phi_{\mathrm{N}} \end{bmatrix} \tag{A.68}$$

or:

$$\bar{h}\frac{\partial\phi}{\partial\bar{x}} \rightarrow \boldsymbol{D}^a_{\bar{x}}\boldsymbol{\phi} \tag{A.69}$$

The difference operator $\boldsymbol{D}^a_{\bar{x}}$ is obtained as a modified difference operator $\boldsymbol{D}_{\bar{x}}$. The results are valid on the dashed lines in Fig. A.4. For the derivative of these values, we only have to multiply the result in eq. (A.69) by the difference operator for Neumann boundary conditions, or $-\boldsymbol{D}^{\mathrm{t}}_{\bar{x}}$. The second-order partial differential quotient can now be approximated by:

$$\bar{h}^2\frac{\partial^2\phi}{\partial\bar{x}^2} \rightarrow -\boldsymbol{D}^{\mathrm{t}}_{\bar{x}}\boldsymbol{D}^a_{\bar{x}}\boldsymbol{\phi} \tag{A.70}$$

If we have another field component ψ which is discretised on the dashed lines, then in an analogous way we can construct the approximation for the first derivative using the difference operator $-\boldsymbol{D}^{\mathrm{t}}_{\bar{x}}$ and for the second derivative using $\boldsymbol{D}^{\mathrm{a}}_{\bar{x}}$. So we obtain:

$$\bar{h}\frac{\partial\psi}{\partial\bar{x}} \rightarrow -\boldsymbol{D}^{\mathrm{t}}_{\bar{x}}\boldsymbol{\psi} \qquad \bar{h}^2\frac{\partial^2\psi}{\partial\bar{x}^2} \rightarrow -\boldsymbol{D}^a_{\bar{x}}\boldsymbol{D}^{\mathrm{t}}_{\bar{x}}\boldsymbol{\psi} \tag{A.71}$$

In the case of x-dependant material parameters the matrix of the discretised parameters may have to be introduced between the difference operators.

A.2.6 Special boundary conditions (SBCs)

In the integrated microwave and optical circuits, electromagnetic eigenmode problems have to be solved by numerical methods, where the structure normally has to be surrounded by a box of metallic or magnetic walls, whereby box modes can occur and complicate the numerical evaluation. Since these boundaries must not have any influence on the field distribution, the distance chosen between the walls must be very large, particularly for a structure with weak confinement. For such a structure, the numerical effort is high.

It could be demonstrated that these disadvantages can be overcome by the introduction of special boundary conditions (SBCs) into the difference operators that will be obtained from the absorbing boundary conditions. The introduction of the special boundary conditions in the MoL is very advantageous because the SBCs can be placed closer to the guiding structure, whereby the discretised area and the corresponding matrices become smaller. With the SBCs we simulate an infinite computation window at the place where they are introduced, and box modes, which complicate the numerical evaluation, do not occur. The procedure to obtain the SBCs is completely analoguous to the case of ABCs. We assume propagation in z-direction according to $\exp -\mathrm{j}\sqrt{\varepsilon_{re}}\overline{z}$. Therefore, for $D_{\overline{z}}^2$ in eq. (A.45) we have to introduce $-\varepsilon_{re}$. The equations in the case of SBCs are formally identical to eqs. (A.48) and (A.49). Instead of $\overline{n}_a$ we have to introduce $\overline{n}_d$ into eqs. (A.58)–(A.60), where $\overline{n}_d$ is given by:

$$\overline{n}_d = -\overline{h}\sqrt{\varepsilon_{re} - \varepsilon_{\mathrm{R}}} \tag{A.72}$$

ε_{R} being the permittivity of the border where the SBCs are introduced. In the case of Fig. A.7b, we have on the upper side (lower side) $\varepsilon_{\mathrm{R}} = n_c^2$ ($\varepsilon_{\mathrm{R}} = n_s^2$). For the parameter combination $p_0 = 1$, $p_2 = 0.75$ and $q_2 = 0.25$ the reflection coefficient at the boundary is zero for perpendicular incidence.

The regions in which the SBCs are positioned have a lower refractive index than the effective index. For that very reason the coefficients a, b and c are real. That means that the SBCs describe the decaying character of the field in an infinite half-space beyond the boundary.

A.2.7 Numerical results

Let us demonstrate how well the absorbing boundary conditions work. In Fig. A.6 the propagation of a Gaussian beam in a homogeneous medium towards the boundary under an angle of 45 degrees with implemented ABCs is demonstrated. The figure shows the beam in different positions, beginning with the starting position. No reflections can be seen if the beam propagates through the boundary.

An open microstrip resonator, which consists of a rectangular microstrip patch on dielectric substrate, was examined in [8]. The results for the complex frequency were in good agreement with [7].

For the example of the weakly-guiding rib waveguide, shown in Fig. A.7a, we demonstrate the advantages of the introduction of SBCs opposed to

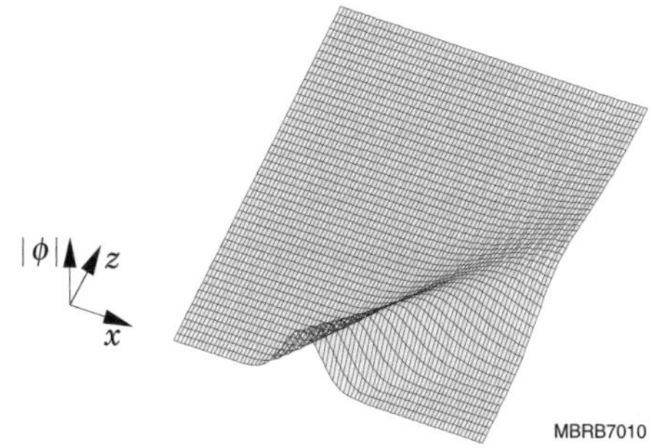

Fig. A.6 Propagation of a Gaussian beam towards the absorbing boundary on the right side under an angle of 45 degrees (R. Pregla, 'MoL-BPM Method of Lines Based Beam Propagation Method', in *Methods for Modeling and Simulation of Guided-Wave Optoelectronic Devices (PIER 11)*, W. P. Huang (Ed.), Progress in Electromagnetic Research, pp. 51–102. © 1995 EMW Publishing)

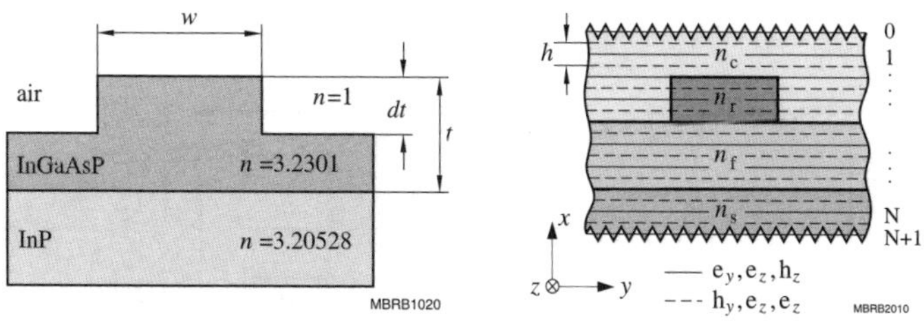

(a) (b)

Fig. A.7 Rib waveguide: cross-section (a) and discretisation lines (b) $d_t = 0.3\ \mu$m, $t = 1.3\ \mu$m, $w = 2.0\ \mu$m, $\lambda = 1.286\ \mu$m

the hitherto-used metallic or magnetic walls [9], [10]. The principle of discretisation is shown in Fig. A.7b. The fact that the SBCs can be positioned very near to the film is illustrated in Fig. A.8, where the normalised phase parameter $B = (n_{\text{eff}}^2 - n_s^2)/(n_f^2 - n_s^2)$ is plotted versus the distance d between the film and the wall in the substrate for the HE_{00} and EH_{00} mode. For analysis of the field, we consider a ridge waveguide in Fig. A.8. The field distributions in the symmetry plane of the rib waveguide for metallic walls at positions 1 and 2 are illustrated in Fig. A.9 for the EH_{00} mode. Here we can see that for an SBC already at position 2, where the SBC is very close to the film, the field distribution agrees very well with the reference distribution, whereas for the metallic walls the disturbance is substantial at position 2. The deviation from the reference distribution will be smaller only for very large distances between the metallic wall and the film. Adequate strong disturbances of the field distributions will be obtained by using magnetic walls. The reference distribution was obtained by using a metal wall at a distance of $d = 6\ \mu$m from the film.

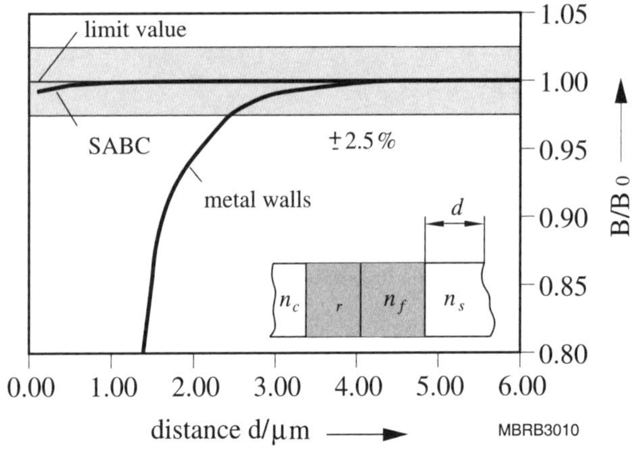

(a)

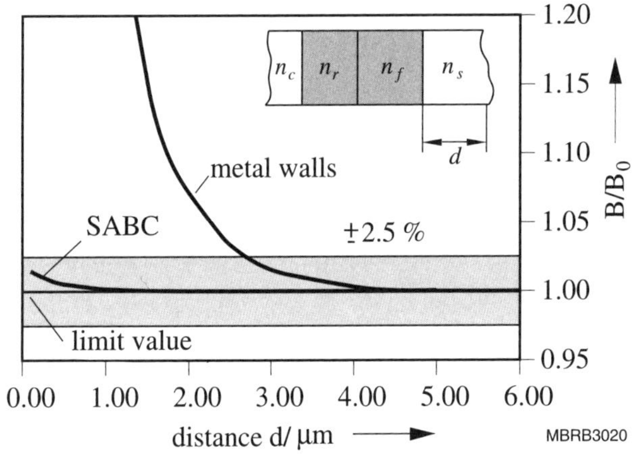

(b)

Fig. A.8 Convergence behaviour of the phase parameter B (a) HE_{00} mode with $B_0 = 0.18006416$ (b) EH_{00} mode with $B_0 = 0.14628822$ (Reproduced by permission of Copernicus Gesellschaft mbH)

Fig. A.10a shows the contour plot of the field distribution for the reference case. The diagrams in Fig. A.10b and c illustrate the contour plots for wall position 1 for metallic walls and for SBCs for the distance $d = 1\ \mu$m. Fig. A.10b clearly shows that the field distribution is strongly disturbed by the metallic wall, whereas with the SBC in Fig. A.10c the field distribution is identical to the reference distribution in Fig. A.10a.

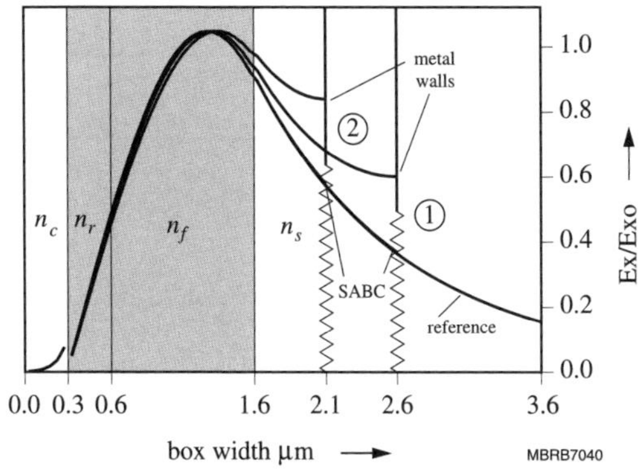

Fig. A.9 Normalised x component of the electric field for SBCs and metallic walls at positions 1 and 2 (Reproduced by permission of Copernicus Gesellschaft mbH)

The introduction of special absorbing boundary conditions to the Method of Lines is very advantageous because the SBCs can be placed closer to the guiding structure, whereby the discretised area and the corresponding matrices become smaller. With the SBCs we simulate an infinite computation window at the place where they are introduced, and box modes, which complicate the numerical evaluation, do not occur.

A.2.8 ABCs for cylindrical coordinates

Normalising the potential according to $\psi_N = \sqrt{\bar{r}}\psi$, the wave equation for ${\psi_N}^2$:

$$\left(\sqrt{\bar{r}}D_{\bar{r}}\frac{1}{\sqrt{\bar{r}}}\right)\left(\frac{1}{\sqrt{\bar{r}}}D_{\bar{r}}\sqrt{\bar{r}}\right)\psi_N + D_{\bar{z}}^2\psi_N + \bar{r}^{-2}D_\varphi^2\psi_N + \varepsilon_r\psi_N = 0 \quad \text{(A.73)}$$

where $D_{\bar{r}}$, D_φ and $D_{\bar{z}}$ are the differential operators with respect to the cylindrical coordinates, r, φ and z may be factorised in the following way:

$$\left(\sqrt{\bar{r}}D_{\bar{r}}\frac{1}{\sqrt{\bar{r}}} \pm \mathrm{j}\sqrt{\varepsilon_r}\sqrt{1+S^2}\right)\psi_N = 0 \quad \text{(A.74)}$$

$$S^2 = \frac{1}{\varepsilon_r}\left(\frac{1}{\bar{r}^2}D_\varphi^2 + D_{\bar{z}}^2\right) \quad \text{(A.75)}$$

which may be approximated by e.g.:

$$\left(\sqrt{\bar{r}}D_{\bar{r}}\frac{1}{\sqrt{\bar{r}}} + \mathrm{j}\sqrt{\varepsilon_r}\left(1+\frac{1}{2}S^2\right)\right)\psi_N = 0 \quad \text{(A.76)}$$

$$\left(\mathrm{j}2\sqrt{\varepsilon_r}\sqrt{\bar{r}}D_{\bar{r}}\frac{1}{\sqrt{\bar{r}}} - 2\varepsilon_r - \frac{1}{\bar{r}^2}D_\varphi^2 - D_{\bar{z}}^2\right)\psi_N = 0 \quad \text{(A.77)}$$

[2]Other forms of wave equations should be factorised in an analogous way.

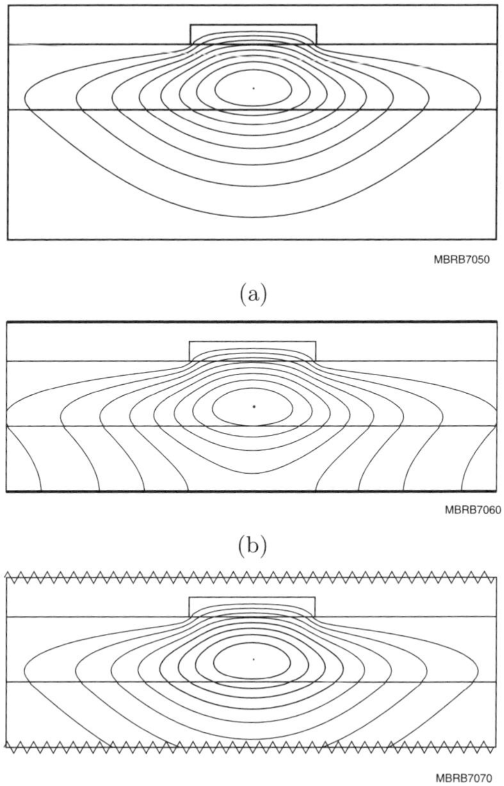

Fig. A.10 Contour plots of x component of the electric field (position 1: $d = 1$ μm) (a) metallic walls at $d = 6$ μm (reference contour plot) (b) metallic walls at position 1 (c) SBCs at position 1 (Reproduced by permission of Copernicus Gesellschaft mbH)

Summation of eqs. (A.73) and (A.77) yields:

$$\left[\left(\sqrt{\bar{r}}D_{\bar{r}}\frac{1}{\sqrt{\bar{r}}}\right)\left(\frac{1}{\sqrt{\bar{r}}}D_{\bar{r}}\sqrt{\bar{r}}\right) + \mathrm{j}2\sqrt{\varepsilon_r}\sqrt{\bar{r}}D_{\bar{r}}\frac{1}{\sqrt{\bar{r}}} - \varepsilon_r\right]\psi_N = 0 \tag{A.78}$$

or after factorising:

$$[(\sqrt{\bar{r}}D_{\bar{r}}\sqrt{\bar{r}}^{-1}) + \mathrm{j}\sqrt{\varepsilon_r}]\psi_N = 0 \tag{A.79}$$

The solution of this equation is:

$$\psi_N = e^{-\mathrm{j}\sqrt{\varepsilon_r}\bar{r}}A; \quad \psi = \frac{1}{\sqrt{\bar{r}}}e^{-\mathrm{j}\sqrt{\varepsilon_r}\bar{r}}A \tag{A.80}$$

Therefore, ψ has the form known from asymptotic expansion of a Hankel function of the second kind for large arguments. The operator D_r has to be discretised in the same way as in Cartesian coordinates. Even a Padé approximation may be used for the square root above.

A.2.9 Periodic boundary conditions

Besides Dirichlet, Neumann or absorbing boundaries, we sometimes need periodic boundary conditions. These are of particular importance when calculating band structures in photonic crystals. Let us take a look at Fig. A.11 for this purpose. We have a period length P, and N is the number of the lines within this period. We can write according to Floquet's theorem:

$$E_{yi+N} = s^{-2N} E_{yi}, H_{yi+N} = s^{-2N} H_{yi} \tag{A.81}$$

with:

$$s = e^{jk_x h/2}; s^{-2N} = e^{-jk_x Nh}$$

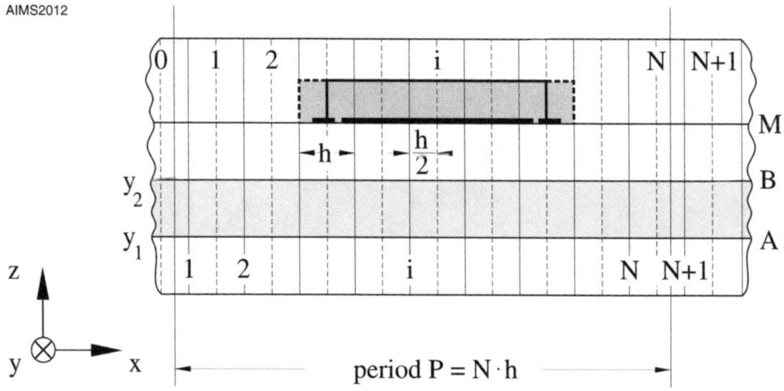

Fig. A.11 To the discretisation with periodic boundary conditions

If no propagation in x-direction is to take place, the value of k_x is made equal to zero. The difference operator D now acquires the form:

$$D = \begin{bmatrix} -1 & 1 & & \\ & \ddots & \ddots & \\ & & \ddots & 1 \\ s^{-2N} & & & -1 \end{bmatrix} \tag{A.82}$$

The modification of the operators presented in eq. (1.8) is very small and causes no problems.

In the analysis of periodic structures it is convenient to normalise the field with respect to the phase in the following way [5]:

$$e^{jk_x x} E_y; e^{jk_x x} H_y \tag{A.83}$$

These products are periodic functions in x. In discretised form we write:

$$\begin{aligned} E_y e^{jk_x x} &\longrightarrow \mathbf{S}_e \mathbf{E}_y \\ H_y e^{jk_x x} &\longrightarrow \mathbf{S}_h \mathbf{H}_y \end{aligned} \tag{A.84}$$

with:

$$\boldsymbol{S}_e = \mathrm{diag}(e^{jk_x ih}) \tag{A.85}$$

$$\boldsymbol{S}_h = \mathrm{diag}(e^{jk_x(i+\frac{1}{2})h}) \tag{A.86}$$

For the phase-normalised derivatives we obtain:

$$\begin{aligned} he^{jk_x x}\frac{\partial E_y}{\partial x} &\longrightarrow \boldsymbol{S}_h \boldsymbol{D} \boldsymbol{S}_e^* \boldsymbol{S}_e \mathbf{E}_y = \boldsymbol{D}_n \boldsymbol{S}_e \mathbf{E}_y \\ he^{jk_x x}\frac{\partial H_y}{\partial x} &\longrightarrow -\boldsymbol{S}_e \boldsymbol{D}^t \boldsymbol{S}_h^* \boldsymbol{S}_h \mathbf{H}_y = -\boldsymbol{D}_n^{*t} \boldsymbol{S}_h \mathbf{H}_y \end{aligned} \tag{A.87}$$

The phase-normalised difference operators D_n and P_n are given by:

$$\boldsymbol{D}_n = \boldsymbol{S}_h \boldsymbol{D} \boldsymbol{S}_e^* = \begin{bmatrix} -s & s^* & & \\ & \ddots & \ddots & \\ & & \ddots & s^* \\ s^* & & & -s \end{bmatrix} \tag{A.88}$$

and:

$$\boldsymbol{P}_n = \boldsymbol{D}_n \boldsymbol{D}_n^{*t} = \boldsymbol{D}_n^{*t} \boldsymbol{D}_n \tag{A.89}$$

Hence $\boldsymbol{P}_n$ is the same for both normalised field quantities.

A.3 HIGHER-ORDER DIFFERENCE OPERATORS [11]

A.3.1 Introduction[3]

Customarily the difference operators are derived from Taylor expansions that are truncated after the quadratic term [12, 13]. A fourth-order approximation for planar microwave structures is first investigated in [14]. The authors of [15] and [16] explicitly give higher-order approximations for the necessary first derivative. These authors claim that different transformation matrices are necessary for the second- and fourth-order approximations. It will be shown in this section that this is not the case. In fact, in [15] and [16] the same matrices are used in both cases. Only the amplitudes of the eigenvectors are changed. In this way the important orthogonality is lost unnecessarily. [17] shows how higher-order approximations can be combined with interface conditions at dielectric steps.

[3]Sections A3.1–A3.3 reproduced from R. Pregla, 'Higher Order Approximation for the Difference Operators in the Method of Lines', in *IEEE Microwave Guided Wave Lett.* vol. 5, no. 2, pp. 53–55. © 1995 Institute of Electrical and Electronics Engineers (IEEE).

This section presents a possibility for obtaining higher-order approximations by a recurrence relation. The order of this approximation is higher than the fourth-order approximation. Due to the special form, an exact degree cannot be given. The obtained results are valid not only in Cartesian coordinates but also in cylindrical ones [18–20]. By using higher-order approximations, it is possible either to minimise the computational effort or to get smaller calculation errors with the same computational effort. The author has obtained the idea for this development from the work of G. R. Hadley [21], who has derived wide-angle BPM approximations by recurrence relations.

A.3.2 Theory

To solve the wave equation in the MoL, one of the second-order differential operators $\partial^2\psi/\partial u^2$, where u stands for x, y, z, φ (or other suitable coordinates), has to be replaced by a suitable difference operator. From the Taylor series expansion for ψ on line i (ψ is the scalar component of the vector potential or one of the field components from which the total field is obtained) one obtains:

$$\left(2\frac{h^2}{2!}\frac{\partial^2\psi}{\partial u^2}+2\frac{h^4}{4!}\frac{\partial^4\psi}{\partial u^4}+2\frac{h^6}{6!}\frac{\partial^6\psi}{\partial u^6}+2\frac{h^8}{8!}\frac{\partial^8\psi}{\partial u^8}+\ldots\right)_i = \psi_{i-1}-2\psi_i+\psi_{i+1} \quad \text{(A.90)}$$

or, in another form:

$$h^2\frac{\partial^2}{\partial u^2}\left\{1+2\frac{h^2}{4!}\frac{\partial^2}{\partial u^2}\left[1+\frac{4!}{6!}h^2\frac{\partial^2}{\partial u^2}\left(1+\frac{6!}{8!}h^2\frac{\partial^2}{\partial u^2}(1\ldots)\right)\right]\right\}\psi_i = -P\cdot\psi_i \quad \text{(A.91)}$$

$-P\cdot\psi_i$ is an abbreviation for the right side of eq. (A.90). P describes the scheme of numbers $\{-1, 2, -1\}$.

Concentrating only on the operators, we obtain:

$$h^2\frac{\partial^2}{\partial u^2} = \frac{-P}{\left\{1+2\frac{h^2}{4!}\frac{\partial^2}{\partial u^2}\left[1+\frac{4!}{6!}h^2\frac{\partial^2}{\partial u^2}\left(1+\frac{6!}{8!}h^2\frac{\partial^2}{\partial u^2}(1\ldots)\right)\right]\right\}} \quad \text{(A.92)}$$

We can interpret eq. (A.92) as a recurrence relation that improves the approximation for the required difference operator.

Starting with the normally-used second-order approximation:

$$h^2\frac{\partial^2}{\partial u^2} \longrightarrow -\mathbf{P} \quad \text{(A.93)}$$

we obtain after the first recursion:

$$h^2\frac{\partial^2}{\partial u^2} \rightarrow -\mathbf{P}\left\{\mathbf{I}-\frac{2!}{4!}\mathbf{P}\left[\mathbf{I}-\frac{4!}{6!}\mathbf{P}\left(\mathbf{I}-\frac{6!}{8!}\mathbf{P}(\mathbf{I}-\ldots)\right)\right]\right\}^{-1} \quad \text{(A.94)}$$

If $\boldsymbol{P}$ has the eigenvalue matrix $\boldsymbol{\lambda}^2$, we obtain the new eigenvalue matrix $\boldsymbol{\lambda}_1^2$ by:

$$\boldsymbol{\lambda}_1^2 = \boldsymbol{\lambda}^2 \left\{ \boldsymbol{I} - \frac{1}{12}\boldsymbol{\lambda}^2 \left[\boldsymbol{I} - \frac{1}{30}\boldsymbol{\lambda}^2 \left(\boldsymbol{I} - \frac{1}{56}\boldsymbol{\lambda}^2 (\boldsymbol{I} - \ldots) \right) \right] \right\}^{-1} \tag{A.95}$$

To obtain all field components, we need the difference operator for the differential quotient of the first order as well. By a similar procedure, starting from the line between lines i and $i+1$ (that is, from the line system of the other component necessary for the expansion of the field or the dual boundary problem) we obtain, analogous to eq. (A.90):

$$h\frac{\partial \psi}{\partial u} + 2\frac{(h/2)^3}{3!}\frac{\partial^3 \psi}{\partial u^3} + 2\frac{(h/2)^5}{5!}\frac{\partial^5 \psi}{\partial u^5} + 2\frac{(h/2)^7}{7!}\frac{\partial^7 \psi}{\partial u^7} + \ldots = \psi_{i+1} - \psi_i \tag{A.96}$$

or, for the operator:

$$h\frac{\partial}{\partial u} = D\left\{1 + \frac{h^2}{2^2 3!}\frac{\partial^2}{\partial u^2}\left[1 + \frac{3!}{2^2 5!}h^2\frac{\partial^2}{\partial u^2}\left(1 + \frac{5! h^2}{2^2 7!}\frac{\partial^2}{\partial u^2}(1 + \ldots)\right)\right]\right\}^{-1} \tag{A.97}$$

D stands for the number scheme $\{-1,\ 1\}$. The dual problem must be introduced for the approximations of the second-order differential operators on the right-hand side. The higher-order approximations obtained in this way are appropriate to the higher-order approximations of the second-order differential quotient. For the δ-matrices (see Chapter 1 and appendices) we obtain:

$$\boldsymbol{\delta}_1 = \boldsymbol{\delta}\left\{\boldsymbol{I} - \frac{\boldsymbol{\lambda}^2}{4 \cdot 3!}\left[\boldsymbol{I} - \frac{\boldsymbol{\lambda}^2 3!}{4 \cdot 5!}\left(\boldsymbol{I} - \frac{\boldsymbol{\lambda}^2 5!}{4 \cdot 7!}\left(\boldsymbol{I} - \frac{\boldsymbol{\lambda}^2 7!}{4 \cdot 9!}(\boldsymbol{I} - \ldots)\right)\right)\right]\right\}^{-1} \tag{A.98}$$

If δ has the subscript e, λ^2 must also have the subscript e. If the order in the product of the braces and δ is changed, then λ^2 must obtain the subscript of the dual problem.

The two equations (A.95) and (A.98) could be rewritten in the following form:

$$\begin{aligned}\boldsymbol{\lambda}_1^2 &= \frac{\boldsymbol{\lambda}^2}{\boldsymbol{I} - \frac{2}{4!}\boldsymbol{\lambda}^2 + \frac{2}{6!}\boldsymbol{\lambda}^4 - \frac{2}{8!}\boldsymbol{\lambda}^6 + \frac{2}{10!}\boldsymbol{\lambda}^{10} - + \ldots} \\ &= \boldsymbol{\lambda}^2\boldsymbol{\lambda}^2\left\{2\boldsymbol{I} - 2\left[\boldsymbol{I} - \frac{1}{2!}\boldsymbol{\lambda}^2 + \frac{1}{4!}\boldsymbol{\lambda}^4 - + \ldots\right]\right\}^{-1} = \frac{\boldsymbol{\lambda}^2\boldsymbol{\lambda}^2}{2(\boldsymbol{I} - \cos\boldsymbol{\lambda})}\end{aligned} \tag{A.99}$$

$$\boldsymbol{\delta}_1 = \boldsymbol{\delta}\boldsymbol{\lambda}\left\{2\left[\frac{1}{1!}\left(\frac{\boldsymbol{\lambda}}{2}\right) - \frac{1}{3!}\left(\frac{\boldsymbol{\lambda}}{2}\right)^3 + \frac{1}{5!}\left(\frac{\boldsymbol{\lambda}}{2}\right)^5 - + \ldots\right]\right\}^{-1} = \frac{\boldsymbol{\delta}\boldsymbol{\lambda}}{2\sin\frac{\boldsymbol{\lambda}}{2}} \tag{A.100}$$

For the product of the δ matrices, we obtain:

$$\boldsymbol{\delta}_1^t \boldsymbol{\delta}_1 = \frac{\boldsymbol{\lambda}^2 \boldsymbol{\lambda}^2}{4 \sin^2 \dfrac{\boldsymbol{\lambda}}{2}} = \frac{\boldsymbol{\lambda}^2 \boldsymbol{\lambda}^2}{2(\boldsymbol{I} - \cos \boldsymbol{\lambda})} = \boldsymbol{\lambda}_1^2 \tag{A.101}$$

$\boldsymbol{\lambda}^2$ must have the same subscript as $\boldsymbol{\delta}$. If the order of the product on the left-hand side is changed, the subscript of $\boldsymbol{\lambda}^2$ must be that of the dual problem. Because of the identity in eq. (A.101), the new results fit in the scheme known for $\boldsymbol{\delta}$ and $\boldsymbol{\lambda}$ (see appendices). It should be mentioned that by setting the bracket in eq. (A.92) equal to 1, the result in [12] is obtained. The same result can be obtained by expanding the trigonometric functions in eqs. (A.96) and (A.98). From the results obtained for the first step, recurrence relations can be written in the following form:

$$\boldsymbol{\lambda}_{k+1}^2 = \boldsymbol{\lambda}^2 \boldsymbol{\lambda}_k^2 \{2(\boldsymbol{I} - \cos \boldsymbol{\lambda}_k)\}^{-1} \tag{A.102}$$

$$\boldsymbol{\delta}_{k+1} = \boldsymbol{\delta} \boldsymbol{\lambda}_k \left\{ 2 \sin \frac{\boldsymbol{\lambda}_k}{2} \right\}^{-1} \tag{A.103}$$

These recurrence relations are given for completeness. Numerically, the first step in eqs. (A.99) and (A.100) is sufficient. The new approximation for the eigenvalues does not change the eigenvectors, as can be seen from eq. (A.94). This is true for the fourth-order approximation, too.

A.3.3 Numerical results

To verify the equations, the convergence behaviour of the propagation constant of the quasi-TE mode for a rib waveguide is analysed. Fig. A.12 shows the propagation constant normalised with the extrapolated value against the discretisation distance. The comparison of the curves demonstrates the increase of exactness. Especially noteworthy is the improvement in comparison to the second-order approximation. Compared to the fourth-order approximation, the further improvement is small in this example. The main importance of the new approximation lies in the relations in eqs. (A.99)–(A.101). The product relation in eq. (A.101) is especially important. Using these relations for the approximation, the whole apparatus developed for the second-order approximation in the MoL can be used. This is not the case for the traditional fourth-order approximation. However, the result in eq. (A.101) gives us an idea for constructing new self-consistent approximations for the fourth-order, too. This can be done by using the square root of $\boldsymbol{\lambda}^2$ for $\boldsymbol{\delta}$ (and choosing the suitable sign: see appendices) or using the higher-order $\boldsymbol{\delta}$ and constructing the $\boldsymbol{\lambda}^2$ by a suitable product. For the case of a microstrip cross-section, the edge parameter value p = 0.2735 was found.

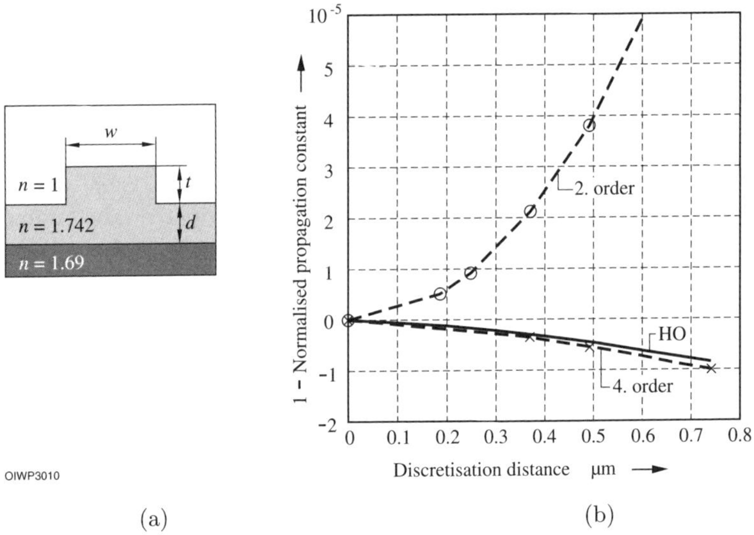

Fig. A.12 Convergence curves of the normalised propagation constant (b) for a rib waveguide of cross-section (a). Dimensions: $\lambda_0 = 1.55\mu\text{m}$, $k_0 d = 5, t/d = 1, w/d = 6$. Curve HO obtained by W.-D. Yang with eqs. (A.99) and (A.100)

A.4 NON-EQUIDISTANT DISCRETISATION

A.4.1 Introduction

In the non-equidistant discretisation scheme used previously with the MoL, the difference operators for the first and second derivatives were calculated from two and three consecutive lines, respectively [22]. However, as shown in [23], the accuracy of such formulation is low in the case of non-equidistant discretisation. For the equidistant discretisation, the difference operators for the first and second derivative have second-order accuracy. On the other hand, in the non-equidistant case both difference operators have only first- or zeroth-order accuracy. Thus, for the non-equidistant discretisation, the second derivative has an error, which cannot be reduced by reducing the distance between the lines. In this section we would like to describe optimal difference operators for the non-equidistant discretisation. These formulations are based on the ideas presented in [23]. We apply the ideas and show how they can also be used in the case of inhomogeneous material.

A.4.2 Theory

It can easily be shown that second-order accuracy is achieved if we approximate the first derivative in the middle between two discretisation lines. We assume a parabolic change of the field component f in the vicinity of the ith line at x_i (Fig. A.13) according to:

$$f = f_i + a_1(x - x_i) + a_2(x - x_i)^2 \tag{A.104}$$

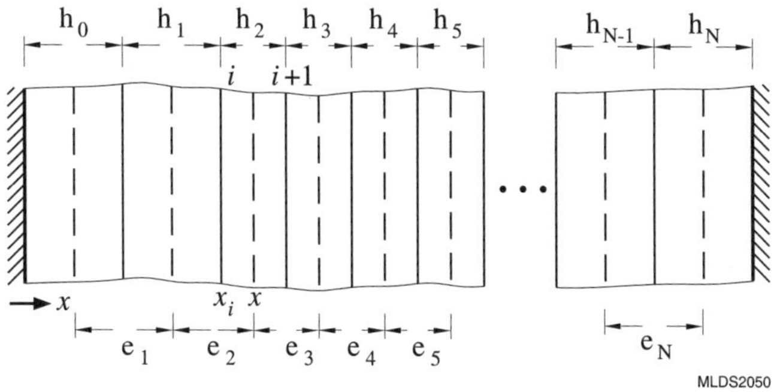

Fig. A.13 Discretisation scheme with non-equidistant lines. $e_i = \frac{1}{2}(h_{i-1} + h_i)$ (L. Greda and R. Pregla, 'Hybrid Analysis of Three-Dimensional Structures by the Method of Lines Using Novel Nonequidistant Discretization', in *IEEE MTT-S Int. Symp. Dig.*, pp. 1877–1880. © 2002 Institute of Electrical and Electronics Engineers (IEEE))

The field component $f = f_{i+1}$ on the line numbered $i + 1$ (i.e. at $x_{i+1} = x_i + h_i$) may be expressed as:

$$f_{i+1} = f_i + a_1 h_i + a_2 h_i^2 \tag{A.105}$$

From eq. (A.104) we obtain for the first derivative of f (indicated as f') at the position $x = x_i + h_i/2$ (i.e. in the middle of x_i and x_{i+1}):

$$f'(x_i + h_{i/2}) = a_1 + a_2 h_i \tag{A.106}$$

If we use eq. (A.105) and apply central difference we can approximate the derivative as:

$$\left.\frac{\partial f}{\partial x}\right|_{i+1/2} = \frac{f_{i+1} - f_i}{h_i} = a_1 + a_2 h_i \tag{A.107}$$

As we can see, the numerical approximation of the first derivative (A.107) is equal to the analytical value of a parabolic function (A.106). This proves the second-order accuracy of the approximation of the first derivative calculated in the middle of two lines. Therefore, in the Method of Lines (the proposed formulas can be used with other FD methods as well), the non-equidistant discretisation should be done as shown in Fig. A.13. It should be stressed that the lines for one of the field components (e.g. discretised on dashed lines) should be exactly in the middle of the other type of lines (solid). Thus we can write for the first derivative of the field component discretised on solid lines:

$$\frac{\mathrm{d}\mathbf{F}_{\mathrm{DD}}}{\mathrm{d}x} = \boldsymbol{h}^{-1}\boldsymbol{D}_{\mathrm{DD}}\mathbf{F} = \overline{\boldsymbol{D}}_{\mathrm{DD}}\mathbf{F} \tag{A.108}$$

where:

$$\boldsymbol{h} = \mathrm{diag}(h_0, h_1, \ldots, h_{N-1}, h_N)$$

$\boldsymbol{D}_{\mathrm{DD}}$ denotes the difference operator for central differences and Dirichlet boundary conditions. However, the other field component (discretised on solid lines) is not in the middle of dashed lines in the case of non-equidistant discretisation. For that reason, eq. (A.104) has only first-order accuracy in this case.

We want to approximate a continuous function $f(x)$ (as in Fig. A.14) with a quadratic polynomial f_{a}. For this purpose we take three consecutive lines x_{i-1}, x_i, x_{i+1} and the values of the function at these points $f(x_{i-1}) = f_{i-1}$, $f(x_i) = f_i$ and $f(x_{i+1}) = f_{i+1}$. Then we obtain according to [23]:

$$f_{\mathrm{a}}(x) = f_{i-1}\frac{(x-x_i)(x-x_{i+1})}{(x_{i-1}-x_i)(x_{i-1}-x_{i+1})} + f_i\frac{(x-x_{i-1})(x-x_{i+1})}{(x_i-x_{i-1})(x_i-x_{i+1})} + f_{i+1}\frac{(x-x_i)(x-x_{i-1})}{(x_{i+1}-x_i)(x_{i+1}-x_{i-1})} \tag{A.109}$$

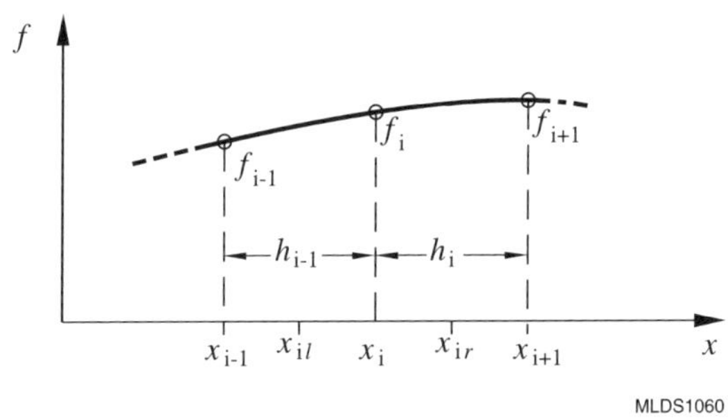

Fig. A.14 Quadratic approximation for interpolation (Reproduced by permission of Springer Netherlands)

The first derivative of $f(x)$ close to these three points, and especially for x between x_{i-1} and x_{i+1}, is approximately given by:

$$\frac{\mathrm{d}f(x)}{\mathrm{d}x} \approx f_{i-1}\frac{(x-x_i)+(x-x_{i+1})}{(x_{i-1}-x_i)(x_{i-1}-x_{i+1})} + f_i\frac{(x-x_{i-1})+(x-x_{i+1})}{(x_i-x_{i-1})(x_i-x_{i+1})} + f_{i+1}\frac{(x-x_i)+(x-x_{i-1})}{(x_{i+1}-x_i)(x_{i+1}-x_{i-1})} = \frac{\mathrm{d}f_{\mathrm{a}}(x)}{\mathrm{d}x} \tag{A.110}$$

This formula has second-order accuracy, or in other words: if the function f is of second-order, eq. (A.110) gives exact values. Because the first derivative is calculated from three consecutive lines, there are two possibilities for choosing these lines. To obtain the best accuracy, one should choose the two

points x_{i-1}, x_i and x_{i+1} we can write for the quadratic approximation at x:

$$\begin{aligned} f(x) = f(x_{\mathrm{i}-1})\frac{(x - x_{\mathrm{i}})(x - x_{\mathrm{i}+1})}{h_{i-1}(h_{i-1} + h_i)} + f(x_{\mathrm{i}})\frac{(x - x_{\mathrm{i}-1})(x - x_{\mathrm{i}+1})}{-h_{i-1}h_i} \\ f(x_{\mathrm{i}-1})\frac{(x - x_{\mathrm{i}})(x - x_{\mathrm{i}-1})}{h_i(h_{i-1} + h_i)} \end{aligned} \quad (A.114)$$

This approximation is especially useful for points that are positioned between x_{i-1}, x_i and x_{i+1}. Let us now determine the approximate values in the middle of x_{i-1} and x_i and of x_i and x_{i+1}, respectively. We obtain for a point left of x_i (at $x = x_{i\mathrm{l}} = x_{i-1} + h_{i-1}/2$; $p = h_i/h_{i-1}$):

$$f(x_{i\mathrm{l}}) = f(x_{i-1})\frac{1 + 2p}{4(1 + p)} + f(x_i)\frac{1 + 2p}{4p} - f(x_{i+1})\frac{1}{4p(1 + p)} \quad (A.115)$$

For the special case $h_{i-1} = h_i$ or $p = 1$ we obtain:

$$\begin{aligned} f(x_{i\mathrm{l}}) &= \tfrac{3}{8}f(x_{\mathrm{i}-1}) + \tfrac{3}{4}f(x_{\mathrm{i}}) - \tfrac{1}{8}f(x_{\mathrm{i}+1}) \\ &= \tfrac{1}{2}[f(x_{\mathrm{i}-1}) + f(x_{\mathrm{i}})] - \tfrac{1}{8}[f(x_{\mathrm{i}-1}) - 2f(x_{\mathrm{i}}) + f(x_{\mathrm{i}+1})] \end{aligned} \quad (A.116)$$

For a point to the right of x_i ($x = x_{i\mathrm{r}} = x_i + h_i/2$) we have:

$$f(x_{i\mathrm{r}}) = -f(x_{i-1})\frac{p^2}{4(1 + p)} + f(x_i)\frac{2 + p}{4} + f(x_{i+1})\frac{2 + p}{4(1 + p)} \quad (A.117)$$

For the special case $h_{i-1} = h_i$ or $p = 1$ we obtain:

$$\begin{aligned} f(x_{i\mathrm{r}}) &= -\tfrac{1}{8}f(x_{i-1}) + \tfrac{3}{4}f(x_{\mathrm{i}}) + \tfrac{3}{8}f(x_{\mathrm{i}+1}) \\ &= \tfrac{1}{2}[f(x_i) + f(x_{i+1})] - \tfrac{1}{8}[f(x_{\mathrm{i}-1}) - 2f(x_{\mathrm{i}}) + f(x_{\mathrm{i}+1})] \end{aligned} \quad (A.118)$$

In what follows, we give some examples of interpolation matrices. The values on a Dirichlet (D) boundary are equal to zero and therefore omitted. For a Neumann boundary the value of the first derivative is zero. Therefore, we assume identical values of the fields on the first (last) position inside the computational window and on the last (first) one outside this window. (We considered the boundary on the left (right) side.) The coefficient for the field outside is then added to the corresponding point inside the computational window.

With the assumptions above we can construct the extrapolation matrices. First we consider the case in which the third point for the derivative is on the left. In the matrix (A.119) we have Neumann conditions on the left and Dirichlet boundaries on the right. Therefore, '−1' is added to '6', resulting in '5' (left side), whereas the '3' on the right is omitted. For 'DN' eq. (A.119), we omit the '6' on the left due to the Dirichlet condition. Now we assume a linear behaviour of the field in the region of the boundary. The field on the position next to the last one outside the computational window is identical to the first inside this window, but with the opposite sign. Hence, we must multiply the coefficient by '-1' and then add this value to that of the first position inside the computational window. As we can see, we have to add '1' to '3' and obtain '4'. (The '-1' in the second row is simply omitted due to the Dirichlet condition.) At the right (Neumann) boundary, we do not need an additional interpolation, because the last row already gives the derivative on the last position in the computational window.

$$\boldsymbol{M}_{\mathrm{ND}} = \frac{1}{8}\begin{array}{c}-1\\ \\ \\ \\ \end{array}\begin{bmatrix} 6 & 3 & & & \\ -1 & 6 & 3 & & \\ & \ddots & \ddots & \ddots & \\ & & & -1 & 6 \end{bmatrix}\begin{array}{c} \\ \\ \\ 3\end{array} = \frac{1}{8}\begin{bmatrix} 5 & 3 & & \\ -1 & 6 & 3 & \\ & \ddots & \ddots & \ddots \\ & & -1 & 6 \end{bmatrix} \tag{A.119}$$

$$\boldsymbol{M}_{\mathrm{DN}} = \frac{1}{8}\begin{array}{cc}-1 & 6\\ & -1\\ & \\ & \end{array}\begin{bmatrix} 3 & & & \\ 6 & 3 & & \\ \ddots & \ddots & \ddots & \\ & -1 & 6 & 3 \end{bmatrix} = \frac{1}{8}\begin{bmatrix} 4 & & & \\ 6 & 3 & & \\ \ddots & \ddots & \ddots & \\ & -1 & 6 & 3 \end{bmatrix} \tag{A.120}$$

Analogously, we can construct the interpolation matrices, if the third point from the right side is used to approximate the first derivative. We obtain:

$$\boldsymbol{M}_{\mathrm{ND}} = \frac{1}{8}\begin{bmatrix} 3 & 6 & -1 & \\ & 3 & 6 & -1 \\ & & \ddots & \\ & & & 3 \end{bmatrix}\begin{array}{cc} & \\ & \\ -1 & \\ 6 & -1 \end{array} = \frac{1}{8}\begin{bmatrix} 3 & 6 & -1 & \\ & 3 & 6 & -1 \\ & & 3 & 6 \\ & & & 4 \end{bmatrix} \tag{A.121}$$

$$\boldsymbol{M}_{\mathrm{DN}} = \frac{1}{8}\begin{array}{c}3\\ \\ \\ \\ \end{array}\begin{bmatrix} 6 & -1 & & \\ 3 & 6 & -1 & \\ & \ddots & \ddots & \ddots \\ & & 3 & 6 \end{bmatrix}\begin{array}{c} \\ \\ \\ -1\end{array} = \frac{1}{8}\begin{bmatrix} 6 & -1 & & \\ 3 & 6 & -1 & \\ & \ddots & \ddots & \ddots \\ & & 3 & 5 \end{bmatrix} \tag{A.122}$$

The two possibilities can also be mixed. To obtain the values on the Neumann boundary, a row $[9\,{-1}\ldots]$ at the upper side or a row $[\ldots{-1}\ 9]$ at the lower side of the matrix has to be added. Such an interpolation procedure can also be used to extend [37].

A.4.4 Numerical results

To compare the accuracy of the old (see e.g. [22], [24]) and the new non-equidistant difference operators, the first and second derivatives of $\sin(x), x \in [0; \pi/2]$ and $e^{-x}, x \in [0; 1.5]$ functions were calculated. The $\sin(x)$ function was discretised with sinusoidal decrease of discretisation distance. For the e^{-x} function we used a geometrical increase of the discretisation distance. Additionally, an abrupt change of the discretisation distance between two lines was introduced. The results presented in Fig. A.16 show the deviation of the approximated derivatives and the analytical values. As we can see, the numerical error of the new proposed non-equidistant difference operators is much lower than that of the previously used scheme. In particular, if there are abrupt changes in the discretisation distance, which are frequently unavoidable, the new difference operators give results with considerably higher accuracy.

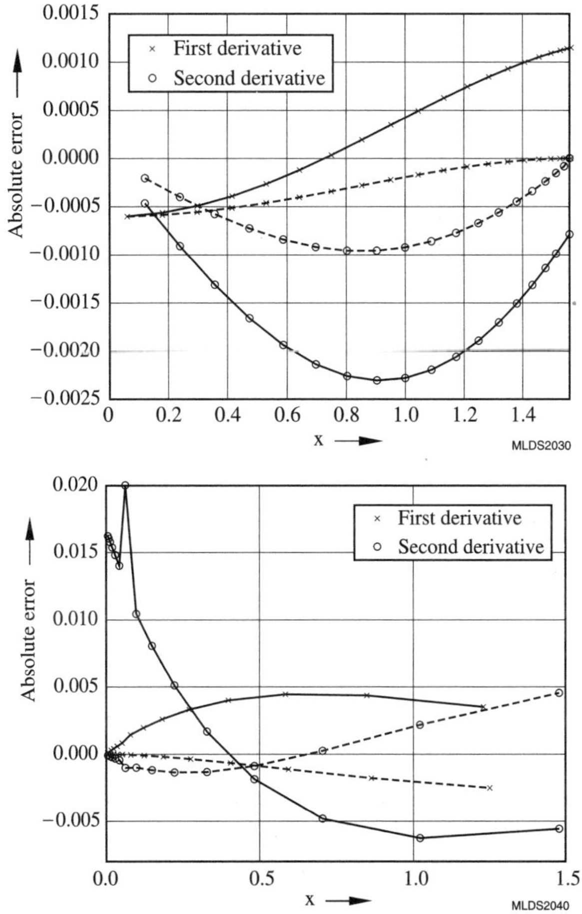

Fig. A.16 Absolute numerical error of the new and the old non-equidistant difference operators. Discretised functions: $\sin(x)$ (upper) and e^{-x} (lower). — old difference operators - - - - new difference operators (L. Greda and R. Pregla, 'Hybrid Analysis of Three-Dimensional Structures by the Method of Lines Using Novel Nonequidistant Discretization', in *IEEE MTT-S Int. Symp. Dig.*, pp. 1877–1880. © 2002 Institute of Electrical and Electronics Engineers (IEEE))

A.5 REFLECTIONS IN DISCRETISATION GRIDS

A.5.1 Introduction

It is well-known from beam propagation methods that numerical reflections occur in non-equidistant discretisation schemes. Therefore, special care is required when constructing the difference operators. The phenomenon can be studied by a simple 1D model, as shown in Fig. A.17. We will show that the reflections are much smaller by introducing difference operators that have second-order accuracy for the approximated function and its first derivative.

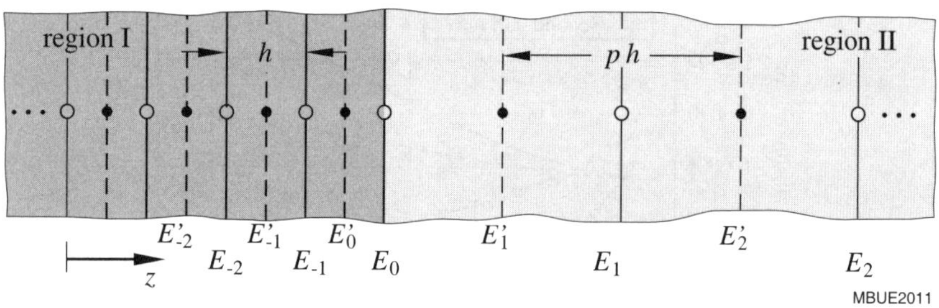

Fig. A.17 Concatenations of two discretisation schemes of different discretisation distances (Reproduced by permission of Springer Netherlands)

A.5.2 Dispersion relations

We study the internal reflections in discretisation schemes using a plane wave propagating in z-direction. This can be done with a 1D model, as shown in Fig. A.17.

The plane wave is described by the following transmission line equations:

$$\frac{\mathrm{d}}{\mathrm{d}z}E_{\mathrm{x}} = -\mathrm{j}\mu_{\mathrm{r}}k_0\widetilde{H}_{\mathrm{y}} \qquad \frac{\mathrm{d}}{\mathrm{d}z}\widetilde{H}_{\mathrm{y}} = -\mathrm{j}\varepsilon_{\mathrm{r}}k_0E_{\mathrm{x}} \tag{A.123}$$

The wave equation is obtained by combining these expressions. We obtain:

$$\frac{\mathrm{d}^2}{\mathrm{d}\overline{z}^2}E_{\mathrm{x}} + E_{\mathrm{x}} = 0 \qquad \overline{z} = \sqrt{\varepsilon_{\mathrm{r}}\mu_{\mathrm{r}}}k_0z \tag{A.124}$$

In what follows, we model the propagation in z-direction with finite differences. The discretisation points are shown in Fig. A.17, where we can see two regions in which we have different discretisation distances. All other parameters are identical. The wave propagates from the finer to the coarser mesh. The propagation inside regions I and II (for region I, introduce $p = 1$) can be described by the difference equation:

$$E_{n+1} - 2E_n + E_{n-1} + (p\overline{h})^2E_n = 0 \tag{A.125}$$

$\overline{h} = \sqrt{\varepsilon_r \mu_r} k_0 h$. Central differences were introduced into the above equations. Eq. (A.125) is valid in the homogeneous discretisation regions. For point 0 (see Fig. A.17) a separate equation must be written. With the ansatz $E_n = e^{-j\beta_{I,II} n} A$ we obtain the following dispersion relation:

$$\cos \beta_I = 1 - \frac{1}{2}\overline{h}^2 \qquad \cos \beta_{II} = 1 - \frac{1}{2}(p\overline{h})^2 \tag{A.126}$$

or with the relation $\cos(2\alpha) = 1 - 2\sin^2 \alpha$:

$$\sin \frac{\beta_I}{2} = \pm \frac{1}{2}\overline{h} \qquad \sin \frac{\beta_{II}}{2} = \pm \frac{1}{2}p\overline{h} \tag{A.127}$$

From these equations we can see that the description of the propagation with finite differences in the direction of propagation leads to a phase error. This error can be studied by a comparison with the exact propagation constant. The normalised exact propagation constant according to eq. (A.124) is equal to $\overline{\beta}_e = 1$. The approximate normalised propagation constant according to the finite differences is given by $\overline{\beta}_a = \beta_I/\overline{h}$, where β_I is obtained from eq. (A.127a). The relative error is defined as $(\overline{\beta}_e - \overline{\beta}_a)/\overline{\beta}_e$. This error is plotted as a function of the normalised discretisation distance in Fig. A.18.

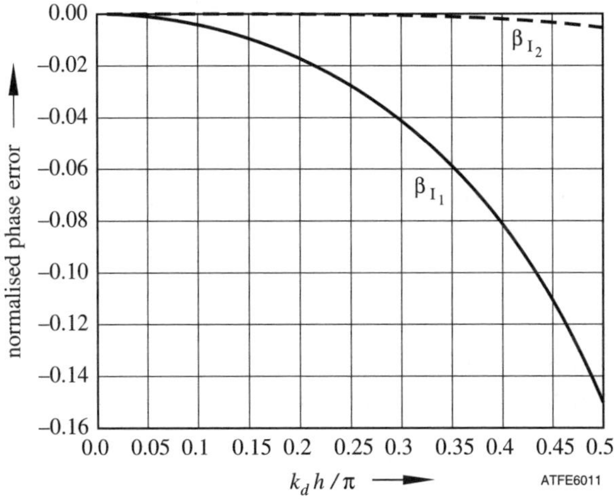

Fig. A.18 The error in the propagation constant as a function of the normalised discretisation distance $k_d h = \sqrt{\varepsilon_r \mu_r} k_0 h$ — results according to eq. (A.127) - - - results according to eq. (A.131)

Furthermore, eqs. (A.126a) and (A.127a) show that there is a limit for the maximum discretisation distance. If $\overline{h}$ exceeds the value 2, eqs. (A.126a) and (A.127a) do not have real solutions for β. (This was also shown in [25].) Propagation is not possible any longer. A more detailed solution is drawn in Fig. A.19. For values of $k_d\overline{h}$ greater than 2 the phase constant β_I is equal to π and the damping factor α starts to grow with increasing $k_d h$. This result shows that we must discretise very carefully. In some papers (see e.g. [26] and the comment in [27]) the infinite space is transformed into a finite region. If we discretise, the discretisation distance close to the border will automatically become greater than 2 after such a transformation. Furthermore, these investigations show that it is generally not possible to model radiation by such a coordinate transformation [34].

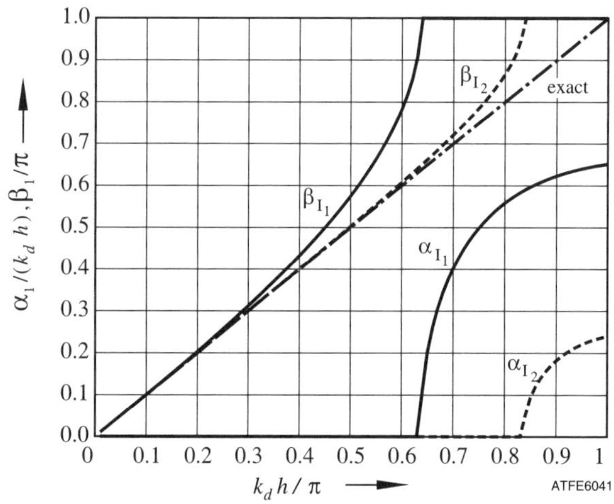

Fig. A.19 Real and imaginary parts of the propagation constant as a function of the normalised discretisation distance $k_d h = \sqrt{\varepsilon_r \mu_r} k_0 h$. — results according to eq. (A.127) - - - results according to eq. (A.131)

We can reduce the phase error and move the band edge to a higher frequency by introducing higher-order difference operators according to Section A.3. Using again the abbreviation P for the scheme [1 −2 1] and λ^2, associated with this scheme, we obtain $\lambda^2 = 2(\cos\beta - 1)$ or:

$$\cos\beta_{\rm I} = 1 + \tfrac{1}{2}\lambda^2 \qquad \text{(A.128)}$$

From eq. (A.125) follows (we assume $p = 1$) for the first value of λ^2: $\lambda^2 = -\overline{h}^2$. We obtain a more accurate value λ_1^2 for λ^2 by using eq. (A.99) in Section A.3:

$$\lambda_1^2 = \frac{\lambda^2 \lambda^2}{2(1 - \cos\lambda)} \qquad \text{(A.129)}$$

Now a more correct value for β_I is obtained from:

$$\cos\beta_{\rm I} = 1 + \frac{1}{2}\lambda_1^2 = 1 + \frac{\overline{h}^4}{4(1 - \cosh\overline{h})} \qquad \text{(A.130)}$$

By approximating the function $\cosh \overline{h}$ with a quadratic polynom, we obtain again the result given in eq. (A.126a). The expression analogous to eq. (A.127a) has the form:

$$\sin \frac{\beta_{\mathrm{I}}}{2} = \frac{\overline{h}^2}{2\sqrt{2(\cosh \overline{h} - 1)}} \tag{A.131}$$

The numerical results from these formulas are also included in Figs. A.18 and A.19. From Fig. A.18 it can be seen that the phase error is now much smaller. Fig. A.19 shows that the band edge has been shifted towards a higher frequency.

A.5.3 Reflections at discretisation transitions

In our further investigation we assume that $k_{\mathrm{d}}\overline{h}$ and $pk_{\mathrm{d}}\overline{h}$ are below the values for the bend edge (i.e. propagation is possible). The general solution of eq. (A.125) for the two regions I and II, respectively, is now given as:

$$E_n = e^{-\mathrm{j}\beta_{\mathrm{I}}n} E_{\mathrm{fI}} + e^{\mathrm{j}\beta_{\mathrm{I}}n} E_{\mathrm{bI}} \qquad E_n = e^{-\mathrm{j}\beta_{\mathrm{II}}n} E_{\mathrm{fII}} + e^{\mathrm{j}\beta_{\mathrm{II}}n} E_{\mathrm{bII}} \tag{A.132}$$

We assume that region II is infinite. The solution in this part is given by:

$$E_n = e^{-\mathrm{j}\beta_{\mathrm{II}}n} E_0 \qquad n \geq 0 \tag{A.133}$$

The derivative $\mathrm{d}E/\mathrm{d}\overline{z}$ in the first • point (labelled '1' in Fig. A.17–just after the change of discretisation) can now be given in the following forms, using the dispersion relation of eq. (A.127b):

$$\begin{aligned} E_1' &= (p\overline{h})^{-1}(E_1 - E_0) = (p\overline{h})^{-1}(e^{-\mathrm{j}\beta_{\mathrm{II}}} - 1)E_0 \\ &= -\mathrm{j}2(p\overline{h})^{-1} \sin \frac{\beta_{\mathrm{II}}}{2} e^{-\mathrm{j}\beta_{\mathrm{II}}/2} E_0 = -\mathrm{j}e^{-\mathrm{j}\beta_{\mathrm{II}}/2} E_0 \end{aligned} \tag{A.134}$$

Likewise, we obtain at the • point, labelled '2':

$$\begin{aligned} E_2' &= (p\overline{h})^{-1}(E_2 - E_1) = (p\overline{h})^{-1}(e^{-\mathrm{j}2\beta_{\mathrm{II}}} - e^{-\mathrm{j}\beta_{\mathrm{II}}})E_0 \\ &= e^{-\mathrm{j}\beta_{\mathrm{II}}} E_1' = -\mathrm{j}e^{-\mathrm{j}3\beta_{\mathrm{II}}/2} E_0 \end{aligned} \tag{A.135}$$

In region I we write the general solution in the form:

$$E_n = e^{-\mathrm{j}\beta_{\mathrm{I}}n} E_{\mathrm{f0}} + e^{\mathrm{j}\beta_{\mathrm{I}}n} E_{\mathrm{b0}} \qquad n \leq 0 \tag{A.136}$$

E_{f0} and E_{b0} are the phasors of the forward and backward propagating waves at $n = 0$ (i.e. at the interface), respectively. By introducing the reflection factor r at the interface (i.e. $n = 0$) we can write $E_{\mathrm{b0}} = rE_{\mathrm{f0}}$ and therefore:

$$E_0 = (1 + r)E_{\mathrm{f0}} \tag{A.137}$$

For the first derivatives (E_0', E_{-1}') at the • points 0 and -1, which we need in our calculations later on, we obtain with the help of eq. (A.136) and by using the dispersion relation in eq. (A.127a):

$$\begin{aligned} E_0' &= \overline{h}^{-1}(E_0 - E_{-1}) = \overline{h}^{-1}(1 + r - (e^{\mathrm{j}\beta_\mathrm{I}} + re^{-\mathrm{j}\beta_\mathrm{I}}))E_{\mathrm{f}0} \\ &= -\mathrm{j}2\overline{h}^{-1}\sin\frac{\beta_\mathrm{I}}{2}(e^{\mathrm{j}\beta_\mathrm{I}/2} - re^{-\mathrm{j}\beta_\mathrm{I}/2})E_{\mathrm{f}0} = -\mathrm{j}(e^{\mathrm{j}\beta_\mathrm{I}/2} - re^{-\mathrm{j}\beta_\mathrm{I}/2})E_{\mathrm{f}0} \end{aligned} \tag{A.138}$$

$$\begin{aligned} E_{-1}' &= \overline{h}^{-1}(E_{-1} - E_{-2}) = \overline{h}^{-1}((e^{\mathrm{j}\beta_\mathrm{I}} + re^{-\mathrm{j}\beta_\mathrm{I}}) - (e^{2\mathrm{j}\beta_\mathrm{I}} + re^{-2\mathrm{j}\beta_\mathrm{I}}))E_{\mathrm{f}0} \\ &= -\mathrm{j}2\overline{h}^{-1}\sin\frac{\beta_\mathrm{I}}{2}(e^{\mathrm{j}3\beta_\mathrm{I}/2} - re^{-\mathrm{j}3\beta_\mathrm{I}/2})E_{\mathrm{f}0} = -\mathrm{j}(e^{\mathrm{j}3\beta_\mathrm{I}/2} - re^{-\mathrm{j}3\beta_\mathrm{I}/2})E_{\mathrm{f}0} \end{aligned} \tag{A.139}$$

So far we have not fulfilled the wave equation in the ∘ point labelled '0'. Instead of eq. (A.125) we write more generally:

$$\overline{h}E_0'' + \overline{h}E_0 = 0 \qquad p\overline{h}E_0'' + p\overline{h}E_0 = 0 \tag{A.140}$$

By using the two-point formula in the case of non-equidistant discretisation (A.107) we write the second derivative as the difference of the first derivatives that we calculated before:

$$\overline{h}E_0'' = \overline{h}\frac{E_1' - E_0'}{\overline{h}/2 + p\overline{h}/2} = \frac{2}{p+1}(E_1' - E_0') \tag{A.141}$$

Better results (with much smaller reflections) can be obtained by the three-point formula (or quadratic approximation) given in [23] for the first derivative. As shown in Section A.4.2, this principle was extended for the first and second derivative in [28] by using the chain rule, and was verified in Section A.4.4 (see also [29]). With this principle, the second derivative as a function of the first ones in the ∘ point labelled '0' is given by:

$$\overline{h}E_0'' = \frac{1-p}{3+p}E_{-1}' + \frac{p-3}{p+1}E_0' + \frac{8}{(p+1)(p+3)}E_1' \tag{A.142}$$

Using E_2' instead of E_{-1}', we likewise obtain:

$$p\overline{h}E_0'' = \frac{-8p^2}{(p+1)(1+3p)}E_0' + \frac{3p-1}{p+1}E_1' + \frac{1-p}{1+3p}E_2' \tag{A.143}$$

For $p = 1$ (or equidistant discretisation) the formulas in eqs. (A.141)–(A.143) reduce to the two-point formula that was used previously for the homogeneous parts: $\overline{h}E_0'' = E_1' - E_0'$. This shows that the central differences give exact results in the middle of the discretisation points if the function has a quadratic behaviour. By substituting eqs. (A.141) and (A.142) into eq. (A.140) and using eqs. (A.134) and (A.137)–(A.139) we can determine the reflection coefficient. For a discretisation with the two-point formula at the transition we obtain:

$$\frac{2}{p+1}((1+r)e^{-j\beta_{\mathrm{II}}/2} - (e^{j\,\beta_{\mathrm{I}}/2} - re^{-j\beta_{\mathrm{I}}/2})) + j(1+r)\overline{h} = 0 \tag{A.144}$$

The quantity E'_{f0} has been cancelled out in all terms. Finally, we introduce the dispersion relation in eq. (A.127) and obtain for the reflection coefficient r_2 (subscript 2 for the two-point formula at the transition):

$$r_2 = \frac{e^{j\beta_{\mathrm{I}}/2} - e^{-j\beta_{\mathrm{II}}/2} - j\frac{1}{2}(p+1)\overline{h}}{e^{-j\beta_{\mathrm{I}}/2} + e^{-j\beta_{\mathrm{II}}/2} + j\frac{1}{2}(p+1)\overline{h}} = \frac{\cos\frac{\beta_{\mathrm{I}}}{2} - \cos\frac{\beta_{\mathrm{II}}}{2}}{\cos\frac{\beta_{\mathrm{I}}}{2} + \cos\frac{\beta_{\mathrm{II}}}{2}} = \frac{\left(\frac{\overline{h}}{2}\right)^2(p^2-1)}{\left(\cos\frac{\beta_{\mathrm{I}}}{2} + \cos\frac{\beta_{\mathrm{II}}}{2}\right)^2} \tag{A.145}$$

For the three-point formula we obtain in an analogous way:

$$\frac{1-p}{3+p}(e^{j3\beta_{\mathrm{I}}/2} - re^{-j3\beta_{\mathrm{I}}/2}) + \frac{p-3}{p+1}(e^{j\beta_{\mathrm{I}}/2} - re^{-j\beta_{\mathrm{I}}/2}) + \frac{8(1+r)}{(p+1)(p+3)}e^{-j\beta_{\mathrm{II}}/2} + j(1+r)\overline{h} = 0 \tag{A.146}$$

This expression can be simplified with the help of the dispersion relation in eq. (A.127). The reflection coefficient r_3 (subscript 3 for the three-point formula at the transition) is then given as:

$$\begin{aligned} r_3 &= \frac{(1-p^2)e^{j3\beta_{\mathrm{I}}/2} + (p^2-9)e^{j\beta_{\mathrm{I}}/2} + 8e^{-j\beta_{\mathrm{II}}/2} + j(p+1)(p+3)\overline{h}}{(1-p^2)e^{-j3\beta_{\mathrm{I}}/2} + (p^2-9)e^{-j\beta_{\mathrm{I}}/2} - 8e^{-j\beta_{\mathrm{II}}/2} - j(p+1)(p+3)\overline{h}} \\ &= \frac{\cos\frac{\beta_{\mathrm{I}}}{2}(p^2-1)\overline{h}^2 + 8\left(\cos\frac{\beta_{\mathrm{II}}}{2} - \cos\frac{\beta_{\mathrm{I}}}{2}\right) + j\frac{1}{2}(p^2-1)\overline{h}^3}{\cos\frac{\beta_{\mathrm{I}}}{2}(p^2-1)\overline{h}^2 - 8\left(\cos\frac{\beta_{\mathrm{II}}}{2} + \cos\frac{\beta_{\mathrm{I}}}{2}\right) - j\frac{1}{2}(p^2-1)\overline{h}^3} \end{aligned} \tag{A.147}$$

with $\cos\frac{\beta_{\mathrm{I}}}{2} = \sqrt{1-(\frac{\overline{h}}{2})^2}$ and $\cos\frac{\beta_{\mathrm{II}}}{2} = \sqrt{1-(\frac{p\overline{h}}{2})^2}$. The square roots can be approximated as $\sqrt{1+x} = 1+0.5x$. For $p=1$ the reflection coefficient is zero in both cases, as expected.

All quantities in eqs. (A.145) and (A.147) depend on the normalised discretisation distance $\overline{h}$ and the parameter p. Fig. A.20 shows the reflection coefficients as functions of the normalised discretisation distance $\overline{h}$ parameterised with p. We used for p the following values: 1.2, 2, 3, 5 and 10. In all cases the reflections are lower with the quadratic approximation in the three-point formula. The logarithmic diagram in Fig. A.21 shows that the transitions described by the three-point formula are always more then 20 dB better than those obtained with the two-point formula.

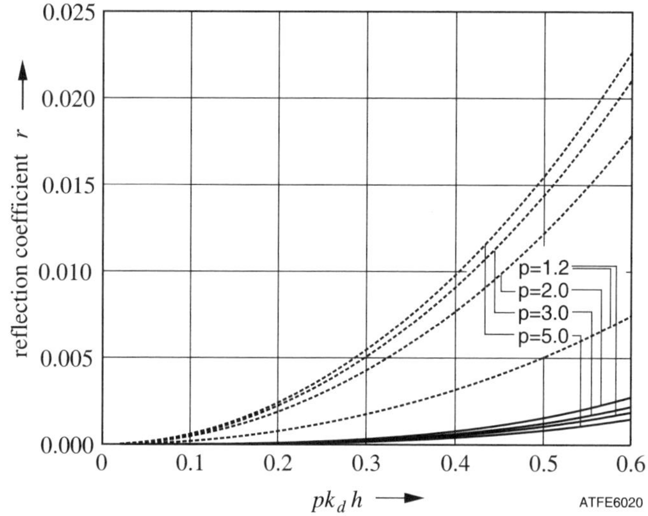

Fig. A.20 Reflection coefficients vs. normalised discretisation distance $\bar{h}$ with p as a parameter: $k_d h = \sqrt{\varepsilon_r \mu_r} k_0 h$. - - - two-point formula — three-point formula

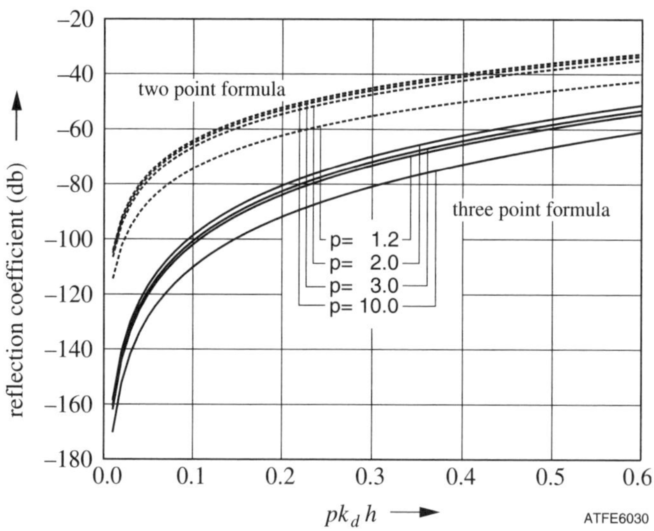

Fig. A.21 Reflection coefficients in dB as functions of the normalised discretisation distance $\bar{h}$ with p as parameter: $k_d h = \sqrt{\varepsilon_r \mu_r} k_0 h$ (Reproduced by permission of Springer Netherlands)

A.6 FIELD EXTRAPOLATION FOR NEUMANN BOUNDARY CONDITIONS

Sometimes we need the fields at the boundaries when we have Neumann boundary conditions. For this purpose, we consider an arbitrary function $\psi(\xi)$. Let ξ be a coordinate perpendicular to the boundary.

The first discretisation point in the vicinity of the boundary is positioned at distance ph from this boundary, and h is the discretisation distance for an equidistant discretisation with $0 < p < 1$ (see Fig. A.22).

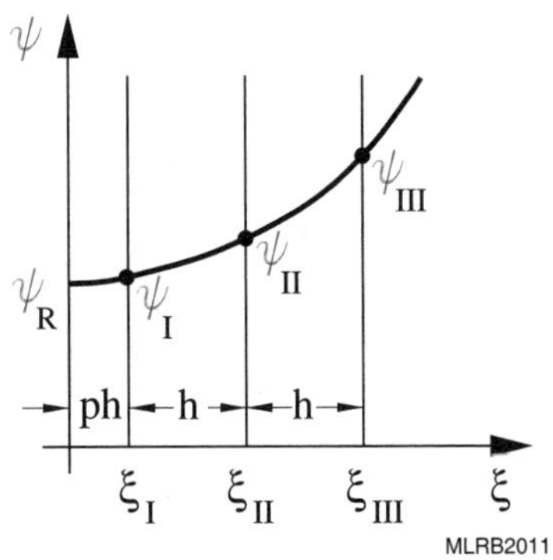

Fig. A.22 Field behaviour with Neumann boundary conditions

To obtain the field at this boundary, we must extrapolate from the neighbouring discretisation points. For the field we assume a quadratic function. By introducing:

$$\psi = \psi_{\mathrm{R}} + a\left(\frac{\xi}{h}\right)^2 \tag{A.148}$$

we obtain at the positions $\xi_{\mathrm{I}} = ph$ and $\xi_{\mathrm{II}} = (1+p)h$:

$$\psi_{\mathrm{I}} = \psi_{\mathrm{R}} + ap^2 \qquad \psi_{\mathrm{II}} = \psi_{\mathrm{R}} + a(1+p)^2 \tag{A.149}$$

Therefore, we can compute the field on the boundary as:

$$\begin{aligned} \psi_{\mathrm{R}} &= \frac{(1+p)^2}{1+2p}\psi_{\mathrm{I}} - \frac{p^2}{1+2p}\psi_{\mathrm{II}} = \alpha\psi_{\mathrm{I}} + \beta\psi_{\mathrm{II}} \\ p = \tfrac{1}{2} &\longrightarrow \psi_{\mathrm{R}} = \tfrac{9}{8}\psi_{\mathrm{I}} - \tfrac{1}{8}\psi_{\mathrm{II}} \\ p = 1 &\longrightarrow \psi_{\mathrm{R}} = \tfrac{4}{3}\psi_{\mathrm{I}} - \tfrac{1}{3}\psi_{\mathrm{II}} \end{aligned} \tag{A.150}$$

For the discretised field we have:

$$\boldsymbol{\psi} = \boldsymbol{T}\overline{\boldsymbol{\psi}} \tag{A.151}$$

Therefore, we obtain instead of eq. (A.150):

$$\psi_{\mathrm{R}}(y) = \mathbf{T}_{\Delta\mathrm{R}}\overline{\boldsymbol{\psi}}(y) \tag{A.152}$$

where $\mathbf{T}_{\Delta\mathrm{R}}$ depends on the position of the boundary (R $\equiv$ A: side where discretisation starts; R $\equiv$ B: side where discretisation ends). Explicitly, we have:

$$\mathbf{T}_{\Delta\mathrm{A}} = \alpha\mathbf{T}_1 - \beta\mathbf{T}_2 \quad \mathbf{T}_{\Delta\mathrm{B}} = \alpha\mathbf{T}_{\mathrm{N}} - \beta\mathbf{T}_{\mathrm{N}-1} \tag{A.153}$$

for the field on the start and end side of the discretisation, respectively. $\mathbf{T}_1$ and $\boldsymbol{T}_2(\mathbf{T}_{\mathrm{N}}$ and $\mathbf{T}_{\mathrm{N}-1})$ are the first (last) and the second (next to last) row vectors of $\boldsymbol{T}$.

We generalise this procedure and assume now that the function also contains a linear term. We need a further discretisation point. With:

$$\psi = \psi_{\mathrm{R}} + a\left(\frac{\xi}{h}\right) + b\left(\frac{\xi}{h}\right)^2 \tag{A.154}$$

we obtain on points $\xi_{\mathrm{I}} = ph$, $\xi_{\mathrm{II}} = (1+p)h$ and $\xi_{\mathrm{III}} = (2+p)h$:

$$\begin{aligned}
\psi_{\mathrm{I}} &= \psi_{\mathrm{R}} + pa + p^2 b \\
\psi_{\mathrm{II}} &= \psi_{\mathrm{R}} + (1+p)a + (1+p)^2 b \\
\psi_{\mathrm{III}} &= \psi_{\mathrm{R}} + (2+p)a + (2+p)^2 b
\end{aligned} \tag{A.155}$$

Therefore, ψ_{R} on the boundary is determined as:

$$\begin{aligned}
\psi_{\mathrm{R}} &= \alpha\psi_{\mathrm{I}} - \beta\psi_{\mathrm{II}} + \gamma\psi_{\mathrm{III}} \\
&= \tfrac{1}{2}(1+p)(2+p)\psi_{\mathrm{I}} - p(2+p)\psi_{\mathrm{II}} + \tfrac{1}{2}p(1+p)\psi_{\mathrm{III}} \\
p = \tfrac{1}{2} &\longrightarrow \psi_{\mathrm{R}} = \tfrac{1}{8}(15\psi_{\mathrm{I}} - 10\psi_{\mathrm{II}} + 3\psi_{\mathrm{III}}) \\
p = 1 &\longrightarrow \psi_{\mathrm{R}} = 3\psi_{\mathrm{I}} - 3\psi_{\mathrm{II}} + \psi_{\mathrm{III}}
\end{aligned} \tag{A.156}$$

The new row vectors are given by:

$$\mathbf{T}_{\Delta\mathrm{A}} = \alpha\mathbf{T}_1 - \beta\mathbf{T}_2 + \gamma\mathbf{T}_3 \qquad \mathbf{T}_{\Delta\mathrm{B}} = \alpha\mathbf{T}_{\mathrm{N}} - \beta\mathbf{T}_{\mathrm{N}-1} + \gamma\mathbf{T}_{\mathrm{N}-2} \tag{A.157}$$

A.7 ABOUT THE NATURE OF THE METHOD OF LINES

A.7.1 Introduction

In this section the nature of the Method of Lines is investigated [30]. The relation to the discrete Fourier transformation is pointed out. First, however, the connection between the shielded and the equivalent periodic structures must be established.

A.7.2 Relation between shielded structures and periodic ones

In analysing periodic structures and homogeneous dielectric layers with the Method of Lines it has been found that the matrices representing the eigenvalue problem of e.g. the propagation constant and current distribution have the Toeplitz structure. Toeplitz matrices are determined by their first row and column only and are therefore inverted quickly and easily [33]. Thus it may be convenient to analyse a shielded planar waveguide by converting it into an equivalent periodic structure in order to take advantage of this property. The foundation of the conversion is the image theory. Instead of the laterally shielded microstrip line in Fig. A.23a, the periodic structure of Fig. A.23b, for example, can be analysed. Naturally, the propagation constant in the x-direction must be set to zero. Additional attention must be paid to determine the odd mode of the periodic structure (the even mode corresponds to the mode in Fig. A.23a with magnetic side walls). The advantage of Toeplitz matrices in Fig. A.23b for homogeneous substrates is offset by the disadvantage of having a matrix twice as large. For large matrices, however, the advantage dominates over the disadvantage. The geometric conversion corresponds mathematically to the fact that matrices of the structure in Fig. A.23a are sums of Toeplitz and Hankel matrices. This matrix sum may be converted in a block Toeplitz matrix of twice the original size.

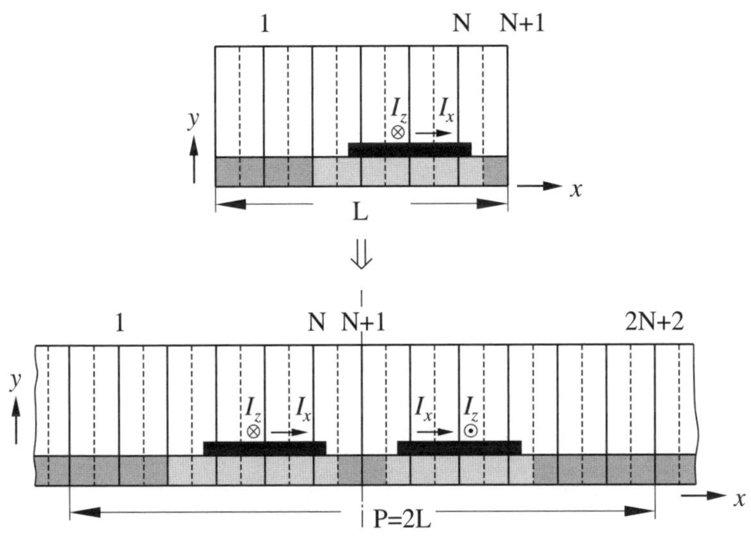

Fig. A.23 Relations between a shielded microstrip line and a periodic one

A.7.3 Method of Lines and discrete Fourier transformation

Now we are able to say something about the nature of the Method of Lines. In the field theory of planar structures, partial differential equations of the Sturm–Liouville type:

$$\varepsilon_r(x)\frac{\partial}{\partial \overline{x}}\left(\varepsilon_r^{-1}(x)\frac{\partial F}{\partial \overline{x}}\right) + \frac{\partial^2 F}{\partial \overline{y}^2} + (\varepsilon_r(x) - \varepsilon_{re})F = 0 \qquad \varepsilon_{re} = k_z^2/k_0^2 \quad \text{(A.158)}$$

must be solved. Metallic side walls must be taken into account by the corresponding boundary conditions. Because F, ε_r, p_ε and G with:

$$G = \varepsilon_r\frac{\partial}{\partial \overline{x}}\left(\varepsilon_r^{-1}\frac{\partial F}{\partial \overline{x}}\right) + \varepsilon_r F = \frac{\partial^2 F}{\partial \overline{x}^2} - p_\varepsilon\frac{\partial F}{\partial \overline{x}} + \varepsilon_r F \qquad p_\varepsilon = \frac{1}{\varepsilon_r}\frac{\partial \varepsilon_r}{\partial \overline{x}} \quad \text{(A.159)}$$

are periodic functions of period P, we use the following Fourier expansions:

$$F(x,y) = \sum_{n=-\infty}^{\infty} \widetilde{F}_n(y)e^{k_n x} \qquad k_n = \mathrm{j}2\pi n/P \quad \text{(A.160)}$$

$$G(x,y) = \sum_{n=-\infty}^{\infty} g_n(x)\widetilde{F}_n(y)e^{k_n x} \qquad g_n = k_n^2 - p_\varepsilon(x)k_n + \varepsilon_r(x) \quad \text{(A.161)}$$

The second expansion was written with the help of the first one. In the Mode Matching Technique these series are terminated at sufficiently high $n = N'$. In the Method of Lines we discretise, i.e. we take F only for N points $x = mP/N = x_m$ $(m = 0, 1, \ldots, N-1)$ altogether. With $F(x_m) = F_m$, $p_\varepsilon(x_m) = p_{\varepsilon m}$, $\varepsilon_r(x_m) = \varepsilon_{rm}$ and $G(x_m) = G_m$, we have:

$$F_m = \sum_{n=-\infty}^{\infty} \widetilde{F}_n e^{\widetilde{k}_n m} \qquad \widetilde{k}_n = \mathrm{j}2\pi n/N \quad \text{(A.162)}$$

$$G_m = \sum_{n=-\infty}^{\infty} \widetilde{g}_n\widetilde{F}_n e^{\widetilde{k}_n m} \qquad \widetilde{g}_n = k_n^2 - p_{\varepsilon m}k_n + \varepsilon_{rm} \quad \text{(A.163)}$$

With:

$$\overline{F}_i = \sum_{r=-\infty}^{\infty} \widetilde{F}_{i+rN} \qquad \overline{F}_i = \overline{F}_{i+N} \qquad \overline{G}_i = \overline{G}_{i+N} \quad \text{(A.164)}$$

$$\overline{G}_i = \sum_{r=-\infty}^{\infty} \widetilde{g}_{i+rN}\widetilde{F}_{i+rN} \qquad \widetilde{g}_{i+rN} = k_{i+rN}^2 - p_{\varepsilon m}k_{i+rN} + \varepsilon_{rm} \quad \text{(A.165)}$$

we get:

$$F_m = \sum_{i=0}^{N-1} \overline{F}_i e^{j2\pi m\frac{i}{N}} \qquad G_m = \sum_{i=0}^{N-1} \overline{G}_i e^{j2\pi m\frac{i}{N}} \quad \text{(A.166)}$$

The expressions in eq. (A.166) represent discrete Fourier transformations (DFTs). The inverse transformation of the left side equation is as follows (as we can immediately see from the orthogonal properties of the exponential function):

$$\overline{F}_i = \frac{1}{N} \sum_{m=0}^{N-1} F_m \; e^{-j2\pi i \frac{m}{N}} \tag{A.167}$$

So far no approximations have been made. Thus, if the $\overline{F}_i$ are known exactly, the values of the function F at the points x_m are also known exactly. The $\overline{F}_i$ must be determined from the differential equation according to eq. (A.158). Therefore, we must now look at G_m. Approximately, we introduce an effective $\overline{\lambda}_i^2$, which may be seen as eigenvalue:

$$\overline{G}_i = \overline{F}_i \overline{\lambda}_i^2 \tag{A.168}$$

Formally, we may write:

$$\overline{\lambda}_i^2 = \frac{\sum\limits_{r=-\infty}^{\infty} \widetilde{g}_{i+rN} \widetilde{F}_{i+rN}}{\sum\limits_{r=-\infty}^{\infty} \widetilde{F}_{i+rN}} \tag{A.169}$$

However, this expression cannot be used for its determination. By using eq. (A.168) we obtain the following DFT pair:

$$G_m = \sum_{i=0}^{N-1} \overline{\lambda}_i^2 \overline{F}_i \; e^{j2\pi m \frac{i}{N}} \qquad \overline{\lambda}_i^2 \, \overline{F}_i = \frac{1}{N} \sum_{m=0}^{N-1} G_m \; e^{-j2\pi i \frac{m}{N}} \tag{A.170}$$

By substituting eqs. (A.166a) and (A.170a) into eq. (A.158), we obtain the ordinary differential equations:

$$\frac{d^2 \overline{F}_i}{d\overline{y}^2} - (\varepsilon_{re} - \overline{\lambda}_i^2) \overline{F}_i = 0 \tag{A.171}$$

for $\overline{F}_i$. The $\overline{F}_i$, which are functions of $\overline{y}$, of course, can only be determined exactly if the values $\overline{\lambda}_i^2$ are known exactly. This is the only point where an approximation has to be used in the discretised representation.

A.7.4 Discussion

In order to get accurate results, $\overline{\lambda}_i^2$ must be determined accurately. In the Method of Lines the differential quotient, and so $\overline{\lambda}_i^2$, are determined from the corresponding difference quotient. According to the mean value theorem of WEIERSTRASS, there is a place in every discretisation interval where the difference quotient is exactly equal to the differential quotient. Therefore it is to be expected that these values are very accurate.

Thus monotonous curves, i.e. nonoscillating ones, are to be expected for the field quantities, such as the current distribution [30], which is presented for a microstrip line in Fig. A.24. This curve is drawn using the results for eleven lines for J_z on the strip. When the current distribution for nine, seven, and five lines is calculated, we notice that the corresponding points lie very close to the full line. For three lines only do the two outer points show a certain deviation. The current J_x is phase-shifted by $\pi/2$ with respect to J_z. Thus jJ_x/J_0 is a real quantity. This diagram and the whole section are a proof that in the Method of Lines the fields on the lines are determined very accurately and hence λ_i^2 are very well approximated. The normal component of the electric field at an interface is sketched in Fig. A.25 [38]. We can clearly recognise the step as required. Even if we discretise with very few lines, we find that the determined values of the field are on this curve. This shows the high accuracy with which the fields are modelled.

A.8 RELATION BETWEEN THE MODE MATCHING METHOD (MMM) AND THE METHOD OF LINES (MoL) FOR INHOMOGENEOUS MEDIA

The equivalence of the MoL and the variational method was shown in [31]. The Rayleigh–Ritz Method (RRM) is one of the algorithms used to solve such variational problems. In this section we show the particular relation between the MoL and RRM.

The analysis of waveguides with inhomogeneous media by the MMM in the special form of the Rayleigh–Ritz Method is described by Collin in [32]. For LSM modes, propagating in z-direction according to $\exp(-\gamma z)$, the following wave equation is valid (eq. (22) in Collin [32], p. 233):

$$\frac{\mathrm{d}}{\mathrm{d}\overline{x}}\frac{1}{\varepsilon_r}\frac{\mathrm{d}\phi_e}{\mathrm{d}\overline{x}} + \frac{1}{\varepsilon_r}(\varepsilon_r + \gamma^2 - q^2)\phi_e = 0 \qquad \text{(A.172)}$$

In general we must write $k_y^2 = -\partial^2/\partial y^2$ instead of q^2 (in [32], h^2). According to [32] we introduce as ansatz for the eigensolutions:

$$\phi_{ek} = \sum_{n=0}^{N} a_{nk} f_n(x) \qquad \text{(A.173)}$$

where $f_n(x)$ are the eigensolutions of the associated homogeneous region. The following general eigenvalue problem must be solved:

$$\sum_{n=0}^{N} a_{nk}(T_{sn} - \gamma_k^2 P_{sn}) = 0 \qquad s = 0, 1, 2, \ldots, N \qquad \text{(A.174)}$$

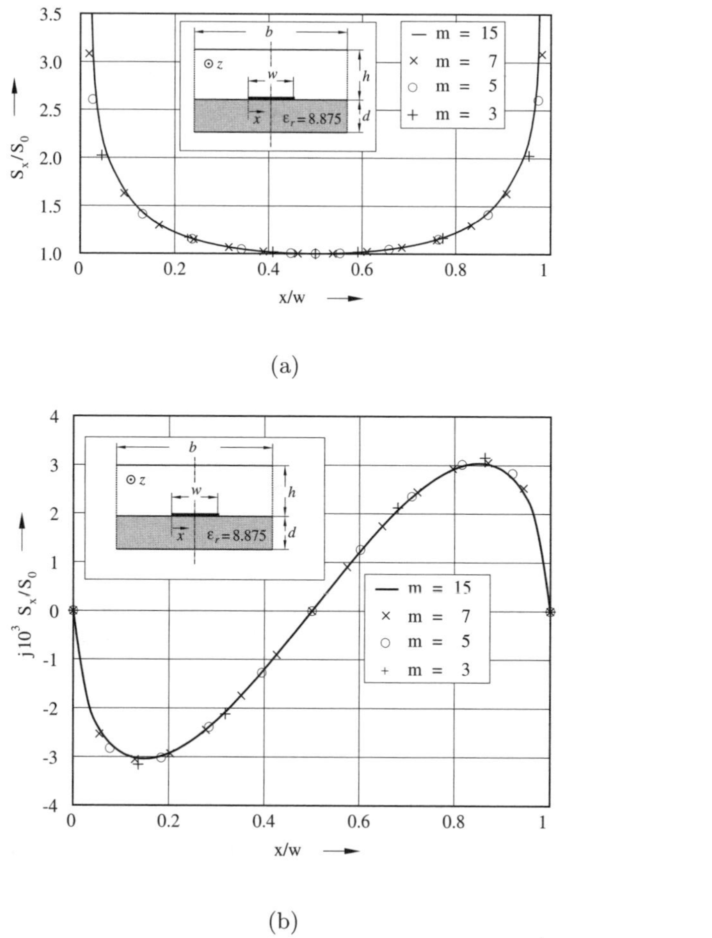

Fig. A.24 Current density distribution of a shielded microstrip line; (a) S_z (b) S_x. Dimensions: $w = 1$ mm, $d = 1$ mm, $b = 12$ mm, $h = 9$ mm; m is the number of discretisation points in half of the microstrip (Reproduced by permission of Elsevier)

with:

$$P_{sn} = \int_0^a \frac{1}{\varepsilon_r} f_n f_s \mathrm{d}x = P_{ns} \tag{A.175}$$

$$T_{sn} = \int_0^a \frac{1}{\varepsilon_r} \left[\frac{\mathrm{d}f_n}{\mathrm{d}\overline{x}} \frac{\mathrm{d}f_s}{\mathrm{d}\overline{x}} - (\varepsilon_r - h^2) f_n f_s \right] \mathrm{d}x = T_{ns} \tag{A.176}$$

If we replace ϕ_e with $\epsilon_{xx} E_x$ or H_y, eq. (A.172) is identical (analogous) to eqs. (2.29) and (2.30) in Section 2.2.4 before the discretisation. We would like to show the analogy of the above algorithm with the MoL. We discretise the

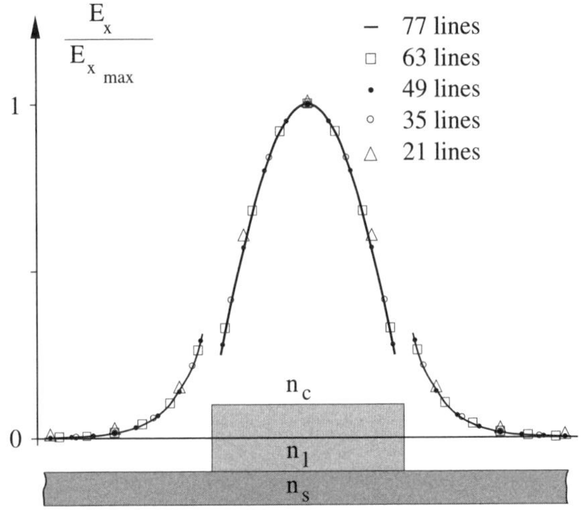

Fig. A.25 Normal component of the electric field at an interface (U. Rogge and R.Pregla, 'Method of Lines for the Analysis of Strip-Loaded Optical Waveguides', *J. Opt. Soc. Am. B*, vol. 8, no. 2, pp. 459–463. © 1991 Optical society of America (OSA))

above equations and write them according to the MoL. We obtain:

$$f_n \to \boldsymbol{T}_0 : \quad \text{eigenvectormatrix for the homogeneous region}$$

$$P_{sn} \to \boldsymbol{\Pi} : \quad \boldsymbol{\Pi} = \boldsymbol{T}_0^{\mathrm{t}} \boldsymbol{\epsilon}_{xx}^{-1} \boldsymbol{T}_0 \Longrightarrow \boldsymbol{\Pi}^{-1} = \boldsymbol{T}_0^{\mathrm{t}} \boldsymbol{\epsilon}_{xx} \boldsymbol{T}_0 = \overline{\boldsymbol{\epsilon}}_{xx} \tag{A.177}$$

$$T_{sn} \to \boldsymbol{\Theta} : \quad \boldsymbol{\Theta} = (\overline{\boldsymbol{D}} \boldsymbol{T}_0)^{\mathrm{t}} \boldsymbol{\epsilon}_{zz}^{-1} (\overline{\boldsymbol{D}} \boldsymbol{T}_0) - \boldsymbol{T}_0^{\mathrm{t}} (\boldsymbol{I} - q^2 \boldsymbol{\epsilon}_{xx}^{-1}) \boldsymbol{T}_0 \tag{A.178}$$

$$a_{nk} \to \boldsymbol{S} : \quad \text{eigenvectormatrix according eq. (A.174)}$$

$$\phi_{ek} \to \boldsymbol{T}^{\varepsilon} : \quad \boldsymbol{\phi}_e = \boldsymbol{T}^{\varepsilon} \overline{\boldsymbol{\phi}}_e \quad \boldsymbol{T}^{\varepsilon} = \boldsymbol{T}_0 \boldsymbol{S} \quad \text{(see eq. (A.173))} \tag{A.179}$$

$$\gamma_k^2 \to \boldsymbol{\Gamma}^2 : \quad \boldsymbol{\Gamma}^2 = \overline{\boldsymbol{\lambda}}_e^2 + q^2 \boldsymbol{I} \tag{A.180}$$

Now we obtain instead of the eigenvalue problem in eq. (A.174):

$$\boldsymbol{\Theta S} = \boldsymbol{\Pi S \Gamma}^2 \qquad \Rightarrow \qquad \boldsymbol{\Pi}^{-1} \boldsymbol{\Theta S} = \boldsymbol{S \Gamma}^2 \tag{A.181}$$

or:

$$\boldsymbol{\Pi}^{-1} \boldsymbol{\Theta} = \boldsymbol{T}_0^{\mathrm{t}} \boldsymbol{\epsilon}_{xx} \overline{\boldsymbol{D}}^{\mathrm{t}} \boldsymbol{\epsilon}_{zz}^{-1} \overline{\boldsymbol{D}} \boldsymbol{T}_0 - \boldsymbol{T}_0^{\mathrm{t}} \boldsymbol{\epsilon}_{xx} \boldsymbol{T}_0 + q^2 \boldsymbol{I} \tag{A.182}$$

$$\boldsymbol{T}_0^{\mathrm{t}} (\boldsymbol{\epsilon}_{xx} \overline{\boldsymbol{D}}^{\mathrm{t}} \boldsymbol{\epsilon}_{zz}^{-1} \overline{\boldsymbol{D}} \boldsymbol{T}_0 - \boldsymbol{\epsilon}_{xx} + q^2 \boldsymbol{I}) \boldsymbol{T}_0 \boldsymbol{S} = \boldsymbol{S \Gamma}^2 \tag{A.183}$$

With eqs. (A.179) and (A.180) we finally obtain:

$$(\boldsymbol{\epsilon}_{xx} \overline{\boldsymbol{D}}^{\mathrm{t}} \boldsymbol{\epsilon}_{zz}^{-1} \overline{\boldsymbol{D}} \boldsymbol{T}_0 - \boldsymbol{\epsilon}_{xx}) \boldsymbol{T}^{\varepsilon} = \boldsymbol{T}^{\varepsilon} \overline{\boldsymbol{\lambda}}_e^2 \tag{A.184}$$

So the equivalence is shown. Now, because the Rayleigh–Ritz Method has stationary behaviour – as shown in [32] – the MoL must also be stationary. Analogous behaviour can also be shown for the LSE (or TE) modes, with a simpler derivation.

A.9 RECIPROCITY AND ITS CONSEQUENCES

By definition, the reaction R of the field a with the sources b of a field b is given by:

$$R = \langle a, b \rangle = \iiint (\vec{E}_a \vec{J}_b - \vec{H}_a \vec{M}_b) \mathrm{d}V \tag{A.185}$$

With this definition, the reciprocity theorem can be written as:

$$\langle a, b \rangle = \langle b, a \rangle \tag{A.186}$$

In words:

> The reaction of the field a with the sources b of the field b is equal to the reaction of the field b with the sources a of the field a.

Let us now examine the situation at the general waveguide port of a circuit, which has a general source (see Fig. A.26). Usually it is impossible to work with the real sources. Therefore, we first determine the equivalent Huygens sources on the outer side of the dashed area. We obtain finite values for these sources only at the part of the dashed area that is equivalent to the waveguide port, because at the other parts the dashed area is on a magnetic or an electric wall. With $\vec{n} = \vec{e}_z$, the Huygens sources at the port are given by:

$$\vec{J}_\mathrm{H} = \vec{e}_z \times \vec{H} \qquad \vec{M}_\mathrm{H} = \vec{E} \times \vec{e}_z \tag{A.187}$$

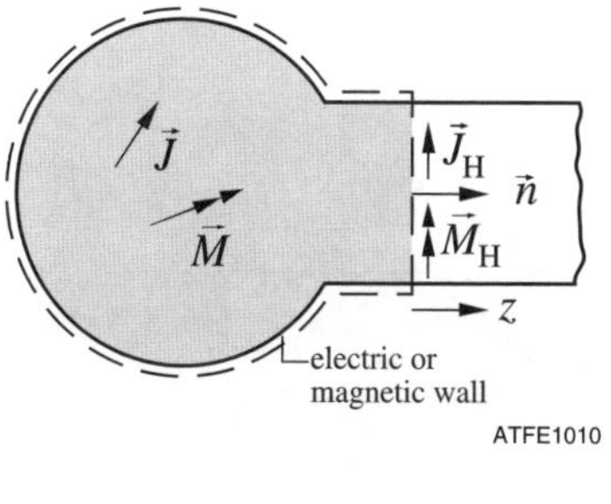

Fig. A.26 Definition of Huygens-equivalent sources at a circuit port

Now we can replace the material inside the volume of the real sources. By using material of ideal magnetic conductivity, the Huygens magnetic current source $\vec{M}_\mathrm{H}$ will be short-circuited and without effect. For $\vec{J}_\mathrm{H}$ we obtain:

$$\vec{J}_\mathrm{H} = \vec{e}_z \times (H_x \vec{e}_x + H_y \vec{e}_y) = -H_y \vec{e}_x + H_x \vec{e}_y = \vec{J}_b \tag{A.188}$$

We can determine the reaction of the field a given by:

$$\vec{E}_a = E_x^a \vec{e}_x + E_y^a \vec{e}_y \tag{A.189}$$

with the source b and obtain:

$$\langle a, b \rangle = \iint \vec{E}_a \vec{J}_b \mathrm{d}A = \iint (E_y^a H_x^b - E_x^a H_y^b) \mathrm{d}A \tag{A.190}$$

In discretised form we can write for the integral on the right-hand side:

$$\langle a, b \rangle = -\widehat{\mathbf{E}}^{at} \widehat{\boldsymbol{f}} \widehat{\mathbf{H}}^b = -\widehat{\mathbf{H}}^{bt} \widehat{\boldsymbol{f}} \widehat{\mathbf{E}}^a \tag{A.191}$$

$\widehat{\mathbf{E}}$ and $\widehat{\mathbf{H}}$ are the vectors of the discretised fields as defined throughout this book. $\widehat{\boldsymbol{f}}$ is a diagonal matrix with area elements, i.e. $\boldsymbol{f}_i = h_{xi} h_{yi}$. Let us now use the results in eq. (A.191) for a general two-port (see Fig. A.27), which could also be a period of a periodic structure. The fields at ports A and B are related by the open-circuit impedance matrix parameters. The magnetic fields $\widehat{\mathbf{H}}_\mathrm{A}$ and $\widehat{\mathbf{H}}_\mathrm{B}$ are now used as sources. $\widehat{\mathbf{E}}_\mathrm{B}$ and $\widehat{\mathbf{E}}_\mathrm{A}$ are the associated fields at the magnetic walls at the ports. We have at:

$$\text{port B:} \quad \widehat{\mathbf{E}}^a = \widehat{\mathbf{E}}_\mathrm{B} = \mathbf{z}_{21} \widehat{\mathbf{H}}_\mathrm{A} \quad \text{and} \quad \widehat{\mathbf{H}}^b = -\widehat{\mathbf{H}}_\mathrm{B}$$

$$\begin{aligned} \langle a, b \rangle &= -(-\widehat{\mathbf{H}}_\mathrm{B})^\mathrm{t} \widehat{\boldsymbol{f}}_\mathrm{B} \widehat{\mathbf{E}}_\mathrm{B} = \widehat{\mathbf{H}}_\mathrm{B}^\mathrm{t} \widehat{\boldsymbol{f}}_\mathrm{B} \mathbf{z}_{21} \widehat{\mathbf{H}}_\mathrm{A} \\ &= -\widehat{\mathbf{E}}_\mathrm{B}^\mathrm{t} \widehat{\boldsymbol{f}}_\mathrm{B} (-\widehat{\mathbf{H}}_\mathrm{B}) = \widehat{\mathbf{H}}_\mathrm{A}^\mathrm{t} \mathbf{z}_{21}^\mathrm{t} \widehat{\boldsymbol{f}}_\mathrm{B} \widehat{\mathbf{H}}_\mathrm{B} \end{aligned} \tag{A.192}$$

$$\text{port A:} \quad \widehat{\mathbf{E}}^b = \widehat{\mathbf{E}}_\mathrm{A} = -\mathbf{z}_{12} \widehat{\mathbf{H}}_\mathrm{B} \quad \text{and} \quad \widehat{\mathbf{H}}^a = \widehat{\mathbf{H}}_\mathrm{A}$$

$$\begin{aligned} \langle b, a \rangle &= -\widehat{\mathbf{E}}_\mathrm{A}^\mathrm{t} \widehat{\boldsymbol{f}}_\mathrm{A} \widehat{\mathbf{H}}_\mathrm{A} = \widehat{\mathbf{H}}_\mathrm{B}^\mathrm{t} \mathbf{z}_{12}^\mathrm{t} \widehat{\boldsymbol{f}}_\mathrm{A} \widehat{\mathbf{H}}_\mathrm{A} \\ &= -\widehat{\mathbf{H}}_\mathrm{A}^\mathrm{t} \widehat{\boldsymbol{f}}_\mathrm{A} \widehat{\mathbf{E}}_\mathrm{A} = \widehat{\mathbf{H}}_\mathrm{A}^\mathrm{t} \widehat{\boldsymbol{f}}_\mathrm{A} \mathbf{z}_{12} \widehat{\mathbf{H}}_\mathrm{B} \end{aligned} \tag{A.193}$$

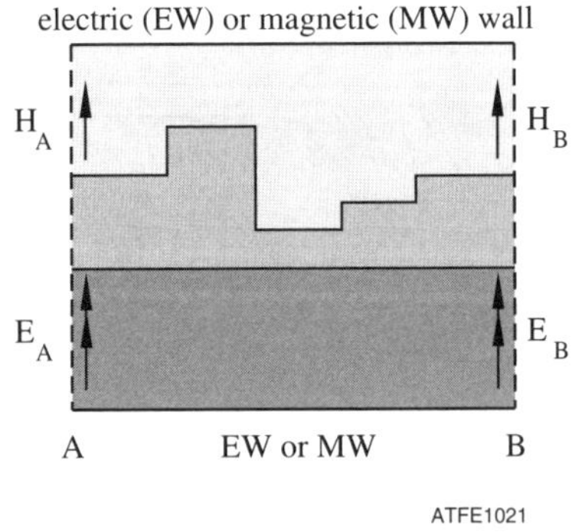

Fig. A.27 General two-port with sources at the ports

For reciprocity we must have:

$$\widehat{\boldsymbol{f}}_\mathrm{B} \mathbf{z}_{21} = \mathbf{z}_{12}^\mathrm{t} \widehat{\boldsymbol{f}}_\mathrm{A} \tag{A.194}$$

Depending on the discretisation scheme, we might be able to simplify eq. (A.194) further. In the case of equidistant discretisation, $\widehat{\boldsymbol{f}} = h_x h_y \boldsymbol{I}$ is a diagonal matrix with constant elements and can be omitted for $(h_x h_y)_{\rm A} = (h_x h_y)_{\rm B}$.

Analogously, we replace the material inside the volume containing the real sources (see Fig. A.26) with a material of ideal electric conductivity. The Huygens electric current source $\vec{J}_{\rm H}$ is short-circuited and therefore without effect. For $\vec{M}_{\rm H}$ we obtain:

$$\vec{M}_{\rm H} = (E_x \vec{e}_x + E_y \vec{e}_y) \times \vec{e}_z = E_y \vec{e}_x - E_x \vec{e}_y = \vec{M}_b \tag{A.195}$$

For the reaction of the field a given by:

$$\vec{H}_a = H_x^a \vec{e}_x + H_y^a \vec{e}_y \tag{A.196}$$

with the source b we obtain:

$$\langle a, b\rangle = -\iint \vec{H}_a \vec{M}_b {\rm d}A = -\iint (H_x^a E_y^b - H_y^a E_x^b) {\rm d}A \tag{A.197}$$

In discretised form we obtain for the integrals:

$$\langle a, b\rangle = \widehat{\mathbf{H}}^{at} \widehat{\boldsymbol{f}} \widehat{\mathbf{E}}^{b} = \widehat{\mathbf{E}}^{bt} \widehat{\boldsymbol{f}} \widehat{\mathbf{H}}^{a} \tag{A.198}$$

Let us introduce the equations for a two-port according to eq. (A.185b) into eq. (A.198). The electric fields $\widehat{\mathbf{E}}_{\rm A}$ and $\widehat{\mathbf{E}}_{\rm B}$ are used as sources. $\widehat{\mathbf{H}}_{\rm B}$ and $\widehat{\mathbf{H}}_{\rm A}$ are the associated fields at the electric walls at the ports. We have at:

$$\begin{aligned} \text{port A:}\quad \widehat{\mathbf{H}}^b = \widehat{\mathbf{H}}_{\rm A} &= \boldsymbol{y}_{12} \widehat{\mathbf{E}}_{\rm B} \quad \text{and} \quad \widehat{\mathbf{E}}^a = \widehat{\mathbf{E}}_{\rm A} \\ \langle b, a\rangle &= \widehat{\mathbf{H}}_{\rm A}^{\rm t} \widehat{\boldsymbol{f}}_{\rm A} \widehat{\mathbf{E}}_{\rm A} = \widehat{\mathbf{E}}_{\rm B}^{\rm t} \boldsymbol{y}_{12}^{\rm t} \widehat{\boldsymbol{f}}_{\rm A} \widehat{\mathbf{E}}_{\rm A} \\ &= \widehat{\mathbf{E}}_{\rm A}^{\rm t} \widehat{\boldsymbol{f}}_{\rm A} \widehat{\mathbf{H}}_{\rm A} = \widehat{\mathbf{E}}_{\rm A}^{\rm t} \widehat{\boldsymbol{f}}_{\rm A} \boldsymbol{y}_{12} \widehat{\mathbf{E}}_{\rm B} \end{aligned} \tag{A.199}$$

$$\begin{aligned} \text{port B:}\quad \widehat{\mathbf{H}}^a = -\widehat{\mathbf{H}}_{\rm B} &= \boldsymbol{y}_{21} \widehat{\mathbf{E}}_{\rm A} \quad \text{and} \quad \widehat{\mathbf{E}}^b = \widehat{\mathbf{E}}_{\rm B} \\ \langle a, b\rangle &= \widehat{\mathbf{H}}_{\rm B}^{\rm t} \widehat{\boldsymbol{f}}_{\rm B} \widehat{\mathbf{E}}_{\rm B} = \widehat{\mathbf{E}}_{\rm A}^{\rm t} \boldsymbol{y}_{21}^{\rm t} \widehat{\boldsymbol{f}}_{\rm B} \widehat{\mathbf{E}}_{\rm B} \\ &= \widehat{\mathbf{E}}_{\rm B}^{\rm t} \widehat{\boldsymbol{f}}_{\rm B} \widehat{\mathbf{H}}_{\rm B} = \widehat{\mathbf{E}}_{\rm B}^{\rm t} \widehat{\boldsymbol{f}}_{\rm B} \widehat{\boldsymbol{y}}_{21} \widehat{\mathbf{E}}_{\rm A} \end{aligned} \tag{A.200}$$

Reciprocity requires:

$$\widehat{\boldsymbol{f}}_{\rm B} \boldsymbol{y}_{21} = \boldsymbol{y}_{12}^{\rm t} \widehat{\boldsymbol{f}}_{\rm A} \tag{A.201}$$

References

[1] R. Pregla, M. Koch and W. Pascher 'Analysis of Hybrid Waveguide Structures Consisting of Microstrips and Dielectric Waveguides', *Proceedings of the 17th Europ. Microwave Conf.*, Rom 1987, pp. 927–932.

[2] T. G. Moore, J. G. Blaschak, A. Taflove and G. A. Kriegsmann, 'Theory and Application of Radiation Boundary Operators', *IEEE Trans. Antennas. Propagation*, vol. AP-36, pp. 1797–1812, Dec. 1988.

[3] R. Pregla, 'MoL-BPM Method of Lines Based Beam Propagation Method', in *Methods for Modeling and Simulation of Guided-Wave Optoelectronic Devices (PIER 11)*, W. P. Huang (Ed.), Progress in Electromagnetic Research, pp. 51–102. EMW Publishing, Cambridge, Massachusetts, USA, 1995.

[4] R. Pregla, 'The Method of Lines for the Unified Analysis of Microstrip and Dielectric Waveguides', *Electromagnetics*, vol. 15, no. 5, Sept–Oct. 1995, pp. 441–456.

[5] S. B. Worm, *Analysis of Planar Microwave Structures with Arbitrary Contour* (in German), PhD thesis, FernUniversität – Hagen, 1983.

[6] S. F. Helfert and R. Pregla, 'Finite Difference Expressions for Arbitrarily Positioned Dielectric Steps in Waveguide Structures', *J. Lightwave Technol.*, vol. 14, no. 10, pp. 2414–2421, Oct. 1996.

[7] S. Nam and T. Itoh, 'Calculation of Accurate Complex Resonant Frequency of an Open Microstrip Resonator Using the Spectral Domain Method', *J. Electrom. Waves and Appl.*, vol. 2, pp. 635–651, 1988.

[8] A. Dreher and R. Pregla, 'Full-Wave Analysis of Radiating Planar Resonators with the Method of Lines', *IEEE Trans. Microwave Theory Tech.*, vol. MTT-41, no. 8, pp. 1363–1368, Aug. 1993.

[9] D. Kremer and R. Pregla, 'The Method of Lines with Special Absorbing Boundary Conditions – Modelling of Active and Lossy Waveguides for the Integrated Optics', in *Deutsche Nationale U.R.S.I Konf.*, Kleinheubach, Germany, 1992, vol. 36, pp. 81–86.

[10] R. Pregla and D. Kremer, 'Method of Lines with Special Absorbing Boundary Conditions – Analysis of Weakly Guiding Optical Structures', *IEEE Microwave Guided Wave Lett.*, vol. 2, no. 6, pp. 239–241, 1992.

[11] R. Pregla, 'Higher Order Approximation for the Difference Operators in the Method of Lines', *IEEE Microwave Guided Wave Lett.*, vol. 5, no. 2, pp. 53–55, 1995.

[12] S. G. Michlin and Ch. L. Smolizki, *Näherungsmethoden zur Lösung von Differential- und Integralgleichungen*, Teubner, Leipzig, 1969.

[13] D. J. Jones, J. C. South and E. B. Klunker, 'On the Numerical Solution of Elliptic Partial Differential Equations by the Method of Lines', *J. Computat. Phys.*, vol. 9, pp. 496–527, 1972.

[14] U. Schulz, *The Method of Lines – A New Technique for the Analysis of Planar Microwave Structures* (in German), PhD thesis, FernUniversität – Hagen, Hagen, 1980.

[15] S. Xiao, R. Vahldieck, H. Jin and Z. Cai, 'A Modified MoL Algorithm with Faster Convergence and Improved Computational Efficiency', in *IEEE MTT-S Int. Symp. Dig.*, 1991, vol. 1, pp. 357–360.

[16] S. Xiao and R. Vahldieck, 'Full-Wave Characterization of Cylindrical Layered Multiconductor Transmission Lines Using the MoL', in *IEEE MTT-S Int. Symp. Dig.*, 1994, vol. 1, pp. 349–352.

[17] A. Kornatz and R. Pregla, 'Increase of the Order of Approximation and Improvement of the Interface Conditions for the Method of Lines', *J. Lightwave Technol.*, vol. 11, no. 2, pp. 249–251, Feb. 1993.

[18] Xu Yansheng, 'Application of Method of Lines to Solve Problems in the Cylindrical Coordinates', *Microw. Opt. Techn. Lett.*, vol. 1, pp. 173–175, 1988.

[19] Y. J. He and S. F. Li, 'Analysis of Arbitrary Cross-Sections Using the Method of Lines', *IEEE Trans. Microwave Theory Tech.*, vol. MTT–42, no. 1, pp. 162–164, Jan. 1994.

[20] R. Pregla and L. Vietzorreck, 'Calculation of Input Impedances of Planar Antennas with the Method of Lines', in *Progress in Electromagnetics Research Symp. (PIERS)*, Noordwijk, The Netherlands, 1994, pp. CD-ROM.

[21] G. R. Hadley, 'Multistep Method for Wide-Angle Beam Propagation', *Opt. Lett.*, vol. 17, no. 24, pp. 1743–1745, 1992.

[22] H. Diestel and S. B. Worm, 'Analysis of Hybrid Field Problems by the Method of Lines with Nonequidistant Discretization', *IEEE Trans. Microwave Theory Tech.*, vol. MTT-32, pp. 633–638, 1984.

[23] R. Gordon, J.-F. Lee and R. Mittra, 'A Technique for Using the Finite Difference Frequency Domain Method with a Non-Uniform Mesh', *AEÜ*, vol. 47, no. 3, pp. 143–148, 1993.

[24] R. Pregla and W. Pascher, 'The Method of Lines', in *Numerical Techniques for Microwave and Millimeter Wave Passive Structures*, T. Itoh (Ed.), pp. 381–446. J. Wiley Publ., New York, USA, 1989.

[25] S. Helfert, *Algorithmen zur Analyse periodischer Strukturen und zur Modellierung dünner Schichten*, PhD thesis, FernUniversität – Hagen, 1999.

[26] Xian Hua Yang and Lotfollah Shafai, 'Full Wave Approach for the Analysis of Open Planar Waveguides with Finite Width Dielectric Layers and Ground Planes', *IEEE Trans. Microwave Theory Tech.*, vol. MTT-42, no. 1, pp. 142–149, 1994.

[27] R. Pregla, 'Comments on "Full Wave Approach for the Analysis of Open Planar Waveguides with Finite Width Dielectric Layers and Ground Planes"', *IEEE Trans. Microwave Theory Tech.*, vol. 42, p. 1717, 1994.

[28] Ł. Gręda and R. Pregla, 'Hybrid Analysis of Three-Dimensional Structures by the Method of Lines Using Novel Nonequidistant Discretization', in *IEEE MTT-S Int. Symp. Dig.*, Seattle, USA, June 2002, pp. 1877–1880.

[29] L. Greda and R. Pregla, 'Modeling of Planar Microwave Filters', in *European Microwave Week*, Munich, Germany, 2003.

[30] R. Pregla, 'About the Nature of the Method of Lines', *AEÜ*, vol. 41, pp. 368–370, 1987.

[31] W. Hong and W.-X. Zhang, 'On the Equivalence Between the Method of Lines and the Variational Method', *AEÜ*, vol. 45, pp. 198–201, 1991.

[32] R. E. Collin, *Field Theory of Guided Waves*, McGraw-Hill, New York, USA, 1960.

[33] W. Pascher and R. Pregla, 'Full Wave Analysis of Complex Planar Microwave Structures', *Radio Sci.*, vol. 22, pp. 999–1002, 1987.

[34] F. Rellich, 'Über das Asymptotische Verhalten der Lösungen von $\Delta u + \lambda u = 0$ in Unendlichen Gebieten', *Jahresbericht der Deutschen Mathematischen Vereinigung*, vol. 53, no. 1, pp. 57–65, 1943.

[35] L. A. Greda and R. Pregla, 'New High-Accuracy Subgridding Technique for the Method of Lines', in *IEEE MTT-S Int. Symp. Dig.*, Fort Worth, USA, June 2004, pp. 1839–1842.

[36] L. A. Greda and R. Pregla, 'Novel Subgridding Technique for the Analysis of Optical Devices', *Opt. Quantum Electron.*, vol. 37, pp. 265–276, 2005, Special Issue on Optical Waveguide Theory and Numerical Modelling.

[37] R. Pregla, 'Modeling of Optical Waveguides and Devices by Combination of the Method of Lines and Finite Differences of Second Order Accuracy', *Opt. Quantum Electron.*, vol. 38, no. 1–3, pp. 3–17, 2006, Special Issue on Optical Waveguide Theory and Numerical Modelling.

[38] U. Rogge and R. Pregla, 'Method of Lines for the Analysis of Strip-Loaded Optical Waveguides', *J. Opt. Soc. Am. B*, vol. 8, no. 2, pp. 459–463, Feb. 1991.

Further Reading

[39] S. F. Helfert and P. Pregla, 'Modeling of Thin Dielectric Layers in Finite Difference Schemes', in *OSA Integr. Photo. Resear. Tech. Dig.*, Santa Barbara, USA, July 1999, pp. 253–255.

[40] S. F. Helfert and R. Pregla, 'Analysis of Thin Layers and Discontinuities', *Opt. Quantum Electron.*, vol. 31, pp. 721–732, 1999, Special Issue on Optical Waveguide Theory and Numerical Modelling.

[41] O. Conradi, S. Helfert and R. Pregla, 'Modification of the Finite Difference Scheme for Efficient Analysis of Thin Lossy Metal Layers in Optical Devices', *Opt. Quantum Electron.*, vol. 30, no. 14, pp. 369–373, 1998.

[42] D. Kremer and R. Pregla, 'The Method of Lines with Improved Special Absorbing Boundary Conditions', in *OSA Integr. Photo. Resear. Tech. Dig.*, Palm Springs, USA, 1993, pp. 41–44.

Appendix B
TRANSMISSION LINE EQUATIONS

B.1 TRANSMISSION LINE EQUATIONS IN FIELD VECTOR NOTATION

Here we would like to derive equations for describing field propagation in waveguides that have an analogous form to the well-known transmission line equations. We will assume waves propagating in z-direction. The material parameters have a special anisotropy as shown below. Let us write $\overleftrightarrow{\varepsilon}_r$ and $\overleftrightarrow{\mu}_r$ as:

$$\overleftrightarrow{\varepsilon}_r = \begin{bmatrix} \overleftrightarrow{\varepsilon}_{rt} & \\ & \varepsilon_{rz} \end{bmatrix} \qquad \overleftrightarrow{\varepsilon}_{rt} = \begin{bmatrix} \varepsilon_{r11} & \varepsilon_{r12} \\ \varepsilon_{r21} & \varepsilon_{r22} \end{bmatrix}$$

$$\overleftrightarrow{\mu}_r = \begin{bmatrix} \overleftrightarrow{\mu}_{rt} & \\ & \mu_{rz} \end{bmatrix} \qquad \overleftrightarrow{\mu}_{rt} = \begin{bmatrix} \mu_{r11} & \mu_{r12} \\ \mu_{r21} & \mu_{r22} \end{bmatrix} \tag{B.1}$$

This form includes materials such as magnetised gyromagnetic and gyroelectric media with respect to the z-axes. In this case the scalar permittivity and the scalar susceptibility have to be replaced by complex permittivity and complex susceptibility tensors, respectively.

Maxwell's equations read as follows:

$$\begin{aligned} \nabla \times \vec{\tilde{H}} &= \mathrm{j}k_0 \overleftrightarrow{\varepsilon}_r \vec{E} \qquad & \nabla \times \vec{E} &= -\mathrm{j}k_0 \overleftrightarrow{\mu}_r \vec{\tilde{H}} \\ \nabla(\overleftrightarrow{\varepsilon}_r \vec{E}) &= 0 & \nabla(\overleftrightarrow{\mu}_r \vec{\tilde{H}}) &= 0 \end{aligned} \tag{B.2}$$

We use $k_o = \omega\sqrt{\mu_o \varepsilon_o}$ for the wave number. The magnetic field is normalised with the wave impedance of free space $\eta_o = \sqrt{\mu_o/\varepsilon_o}$ according to $\vec{\tilde{H}} = \eta_o \vec{H}$. The fields may be decomposed in the part parallel to the z-direction and the transversal part (subscript t). The wave equations will be developed for the latter components. The same decomposition is done with the Nabla operator ∇. The derivation was demonstrated by e.g. D. M. Bolle [1]:

$$\vec{\tilde{H}} = \vec{\tilde{H}}_t + \tilde{H}_z \vec{e}_z \qquad \vec{E} = \vec{E}_t + E_z \vec{e}_z \tag{B.3}$$

$$\overline{\nabla} = \overline{\nabla}_t + \vec{e}_z \frac{\partial}{\partial \overline{z}} \qquad \overline{\nabla}_t = \vec{e}_u \frac{\partial}{\partial \overline{u}} + \vec{e}_v \frac{\partial}{\partial \overline{v}} \tag{B.4}$$

$\vec{e}_z$, $\vec{e}_u$ and $\vec{e}_v$ are the unit vectors in z-u- and v-directions, respectively. The coordinates are normalised according to $\overline{u} = k_0 u$, $\overline{v} = k_0 v$, $\overline{z} = k_0 z$,

$\overline{\nabla} = k_0^{-1}\nabla$:

$$\begin{aligned}&\left(\overline{\nabla}_t E_z - \frac{\partial}{\partial \overline{z}}\vec{E}_t\right)\times\vec{e}_z = -\mathrm{j}\overleftrightarrow{\mu}_{rt}\vec{\tilde{H}}_t \qquad && \mathrm{j}\varepsilon_{rz}E_z\vec{e}_z = \overline{\nabla}_t\times\vec{\tilde{H}}_t\\ &\left(\overline{\nabla}_t \tilde{H}_z - \frac{\partial}{\partial \overline{z}}\vec{\tilde{H}}_t\right)\times\vec{e}_z = \mathrm{j}\overleftrightarrow{\varepsilon}_{rt}\vec{E}_t \qquad && -\mathrm{j}\mu_{rz}\tilde{H}_z\vec{e}_z = \overline{\nabla}_t\times\vec{E}_t\end{aligned} \tag{B.5}$$

The two equations are multiplied by $\vec{e}_z\times$ from the left. We obtain:

$$\vec{e}_z\times\left(\left(\overline{\nabla}_t E_z - \frac{\partial}{\partial \overline{z}}\vec{E}_t\right)\times\vec{e}_z\right) = -\mathrm{j}\vec{e}_z\times\overleftrightarrow{\mu}_{rt}\vec{\tilde{H}}_t \tag{B.6}$$

$$\vec{e}_z\times\left(\left(\overline{\nabla}_t \tilde{H}_z - \frac{\partial}{\partial \overline{z}}\vec{\tilde{H}}_t\right)\times\vec{e}_z\right) = \mathrm{j}\vec{e}_z\times\overleftrightarrow{\varepsilon}_{rt}\vec{E}_t \tag{B.7}$$

By using the identity $\vec{a}\times(\vec{b}\times\vec{c}) = (\vec{a}\vec{c})\vec{b} - (\vec{a}\vec{b})\vec{c}$ we obtain for the left side of the first equation:

$$\begin{aligned}\vec{e}_z\times\left(\left(\overline{\nabla}_t E_z - \frac{\partial}{\partial \overline{z}}\vec{E}_t\right)\times\vec{e}_z\right) &= \overline{\nabla}_t E_z - \frac{\partial}{\partial \overline{z}}\vec{E}_t - \left(\vec{e}_z\left(\overline{\nabla}_t E_z - \frac{\partial}{\partial \overline{z}}\vec{E}_t\right)\right)\vec{e}_z\\ &= \overline{\nabla}_t E_z - \frac{\partial}{\partial \overline{z}}\vec{E}_t\end{aligned} \tag{B.8}$$

In an analogous way the second equation can be shortened. Therefore, we obtain:

$$\begin{aligned}\frac{\partial}{\partial \overline{z}}\vec{E}_t &= \mathrm{j}\overleftrightarrow{\mu}_{rt}(\vec{e}_z\times\vec{\tilde{H}}_t) + \overline{\nabla}_t E_z\\ \frac{\partial}{\partial \overline{z}}\vec{\tilde{H}}_t &= -\mathrm{j}\overleftrightarrow{\varepsilon}_{rt}(\vec{e}_z\times\vec{E}_t) + \overline{\nabla}_t\tilde{H}_z\end{aligned} \tag{B.9}$$

We introduce the right expressions in eq. (B.5) into these equations, which results in:

$$\frac{\partial}{\partial \overline{z}}\vec{E}_t = -\mathrm{j}[-\overleftrightarrow{\mu}_{rt}(\vec{e}_z\times\vec{\tilde{H}}_t) + \overline{\nabla}_t(\varepsilon_{rz}^{-1}\vec{e}_z(\overline{\nabla}_t\times\vec{\tilde{H}}_t))] \tag{B.10}$$

$$\frac{\partial}{\partial \overline{z}}\vec{\tilde{H}}_t = -\mathrm{j}[\overleftrightarrow{\varepsilon}_{rt}(\vec{e}_z\times\vec{E}_t) - \overline{\nabla}_t(\mu_{rz}^{-1}\vec{e}_z(\overline{\nabla}_t\times\vec{E}_t))] \tag{B.11}$$

From these equations we can obtain the generalised transmission line (GTL) equations in matrix notations. They are derived in Section 8.1. This form is much more suitable for our applications.

B.2 DERIVATION OF THE MULTICONDUCTOR TRANSMISSION LINE EQUATIONS

In this section we would like to show how the multiconductor transmission line equations can be derived. From eq. (B.5) we obtain for $E_z = 0$ and $H_z = 0$

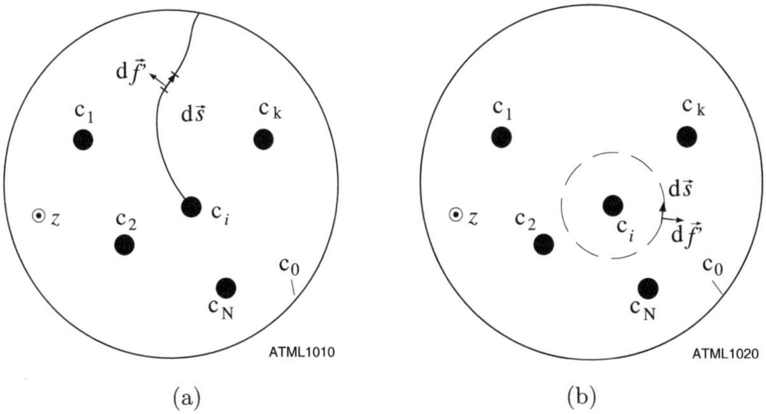

Fig. B.1 Cross-section of a multiconductor transmission line (a) with integration path of the electrical field (b) with closing area for integration according to eq. (B.20)

(TEM waves):

$$\overline{\nabla}_t \times \vec{E}_t = 0 \longrightarrow \vec{E}_t = -\text{grad}\ \phi \tag{B.12}$$

$$\overline{\nabla}_t \times \vec{\tilde{H}}_t = 0 \longrightarrow \vec{H}_t = -\text{grad}\ \psi \tag{B.13}$$

This can easily be seen from:

$$\left(\frac{\partial}{\partial x}\vec{e}_x + \frac{\partial}{\partial y}\vec{e}_y\right) \times \left(\frac{\partial \phi, \psi}{\partial x}\vec{e}_x + \frac{\partial \phi, \psi}{\partial y}\vec{e}_y\right) \equiv 0 \tag{B.14}$$

Therefore we obtain in this case, from eq. (B.9):

$$\frac{\mathrm{d}}{\mathrm{d}\overline{z}}\vec{E}_t = -\mathrm{j}\overleftrightarrow{\mu}_{rt}(-\vec{e}_z \times \vec{\tilde{H}}_t) \tag{B.15}$$

$$\frac{\mathrm{d}}{\mathrm{d}\overline{z}}\vec{\tilde{H}}_t = -\mathrm{j}\overleftrightarrow{\varepsilon}_{rt}(\vec{e}_z \times \vec{E}_t) \tag{B.16}$$

The combination results in (no variation of $\overleftrightarrow{\varepsilon}_{rt}$ and $\overleftrightarrow{\mu}_{rt}$ in z-direction):

$$\frac{\mathrm{d}^2}{\mathrm{d}\overline{z}^2}\vec{E}_t = \overleftrightarrow{\mu}_{rt}\overleftrightarrow{\varepsilon}_{rt}(\vec{e}_z \times (\vec{e}_z \times \vec{E}_t)) \rightarrow \left(\frac{\mathrm{d}}{\mathrm{d}\overline{z}^2} + \overleftrightarrow{\varepsilon}_{rt}\overleftrightarrow{\mu}_{rt}\right)\vec{E}_t = 0 \tag{B.17}$$

$$\frac{\mathrm{d}^2}{\mathrm{d}\overline{z}^2}\vec{H}_t = \overleftrightarrow{\varepsilon}_{rt}\overleftrightarrow{\mu}_{rt}(\vec{e}_z \times (\vec{e}_z \times \vec{H}_t)) \rightarrow \left(\frac{\mathrm{d}}{\mathrm{d}\overline{z}^2} + \overleftrightarrow{\varepsilon}_{rt}\overleftrightarrow{\mu}_{rt}\right)\vec{H}_t = 0 \tag{B.18}$$

From these equations it can easily be seen that the TEM field is propagating in z-direction with the propagation constant $\beta = k_0\sqrt{\varepsilon_r \mu_r}$ in the homogeneous and isotropic cases.

Let us now assume a multiconductor transmission line system in homogeneous isotropic medium (see Fig. B.1a). Integrating eq. (B.15) in the path from conductor c_i to conductor c_0 as shown in Fig. B.1a using:

$$(\vec{e}_z \times \vec{H}_t)\mathrm{d}\vec{s} = -(\vec{e}_z \times \mathrm{d}\vec{s})\vec{H}_t \quad \text{and} \quad \vec{e}_z \times \mathrm{d}\vec{s} = \mathrm{d}\vec{f}'$$

results in:

$$\frac{\mathrm{d}}{\mathrm{d}\bar{z}} \int_{c_i}^{c_0} \vec{E}_t \mathrm{d}\vec{s} = -\mathrm{j}\mu_r \int_{c_i}^{c_0} (\vec{e}_z \times \mathrm{d}\vec{s})\vec{\bar{H}}_t = -\mathrm{j}\mu_r \int_{c_i}^{c_0} \vec{\bar{H}}_t \mathrm{d}\vec{f}' \tag{B.19}$$

The integral on the left side is identical with the voltage U_i between the conductor c_i and the border. The integral on the right side results in the magnetic flux ϕ_i, which can be described by the currents and the inductance coefficients. After denormalisation we obtain:

$$\frac{\mathrm{d}}{\mathrm{d}z} U_i = -\mathrm{j}\mu_r\mu_0\omega \int_{c_i}^{c_0} \vec{H}_t \mathrm{d}\vec{f}' = -\mathrm{j}\omega\phi_i = -\mathrm{j}\omega \sum_{k=1}^{N} L'_{ik} I_k \tag{B.20}$$

The currents are assumed positive in z-direction. From eq. (B.20) we obtain in general:

$$\frac{\mathrm{d}}{\mathrm{d}z}[U] = -\mathrm{j}\omega[L][I] \tag{B.21}$$

Integrating eq. (B.16) on a closed path around conductor c_i (see Fig. B.1b) by using $(\vec{e}_z \times \vec{E}_t)\mathrm{d}\vec{s} = -(\vec{e}_z \times \mathrm{d}\vec{s})\vec{E}_t$ and $-\vec{e}_z \times \mathrm{d}\vec{s} = \mathrm{d}\vec{f}'$ Fig. B.1b), we obtain:

$$\frac{\mathrm{d}}{\mathrm{d}z} \oint \vec{H}_t \mathrm{d}\vec{s} = -\mathrm{j}\varepsilon\omega \oint (-\vec{e}_z \times \mathrm{d}\vec{s})\vec{E}_t = -\mathrm{j}\varepsilon\omega \oint \vec{E}_t \mathrm{d}\vec{f}' \tag{B.22}$$

According to Ampère's law, the result of the integral on the left side is equal to the current in conductor c_i. The integral on the right side results in the charge per unit length. So we have:

$$\frac{\mathrm{d}}{\mathrm{d}z} I_i = -\mathrm{j}\omega Q'_i$$

with:

$$Q'_i = C'_{ii} U_i + \sum_{k=1,\neq i}^{N} C_{ik}(U_i - U_k) \tag{B.23}$$

or in general:

$$\frac{\mathrm{d}}{\mathrm{d}z}[I] = -\mathrm{j}\omega[C'][U] \tag{B.24}$$

The combination of eqs. (B.21) and (B.24) results in:

$$\frac{\mathrm{d}^2}{\mathrm{d}z^2}[I] + \omega^2[C'][L'][I] = 0 \qquad \frac{\mathrm{d}^2}{\mathrm{d}z^2}[U] + \omega^2[L'][C'][U] = 0 \tag{B.25}$$

Since the field is propagating with the propagation constant β (see above), we obtain for the product of the matrices $[C']$ and $[L']$ the following expression:

$$[C'][L'] = [L'][C'] = \frac{\beta^2}{\omega^2} \tag{B.26}$$

References

[1] D.M. Bolle, Lectures on 'Microwave in Ferrites, Plasmas and Semiconductors', Technical University, Braunschweig, 1967.

Appendix C
SCATTERING PARAMETERS

By using the impedance/admittance transformation formulas we can calculate the input impedance of a concatenation structure starting at the end. In most cases, the load impedance matrix is the characteristic impedance. At the input of the structure we have a feeding waveguide. By suitable normalisation of the transformation matrices, we have the unit matrix as characteristic impedance $\overline{\mathbf{Z}}_0 = \boldsymbol{I}$. The load impedance of the input waveguide (the input impedance matrix of the structure) is given as $\overline{\mathbf{Z}}_A$. We then may write for the magnetic field at the input:

$$\overline{\mathbf{H}}_\mathrm{A} = 2(\overline{\mathbf{Z}}_\mathrm{A} + \boldsymbol{I})^{-1}\overline{\mathbf{E}}_\mathrm{Af} \tag{C.1}$$

$\overline{\mathbf{E}}_\mathrm{Af}$ is the vector of forward propagating modes. If we assume that the fundamental mode with unit amplitude is injected, then $\overline{\mathbf{E}}_\mathrm{Af}$ is given by:

$$\overline{\mathbf{E}}_\mathrm{Af} = [1, 0, \ldots, 0]^t \tag{C.2}$$

(The eigenmodes are ordered in such a way that the fundamental mode is located in the first column of $\overline{\boldsymbol{T}}$.) With eq. (C.1) and $\overline{\mathbf{E}}_\mathrm{A} = \overline{\mathbf{Z}}_\mathrm{A}\overline{\mathbf{H}}_\mathrm{A}$ we obtain:

$$\overline{\mathbf{E}}_\mathrm{A} = 2\overline{\mathbf{Z}}_\mathrm{A}(\overline{\mathbf{Z}}_\mathrm{A} + \boldsymbol{I})^{-1}\overline{\mathbf{E}}_\mathrm{Af} \tag{C.3}$$

The vector $\overline{\mathbf{E}}_\mathrm{A}$ contains the complex amplitude of all modes. The scattering coefficient $\mathbf{S}_{11}$ is now given by:

$$\mathbf{S}_{11} = (\boldsymbol{I} - \overline{\mathbf{Z}}_\mathrm{A})(\boldsymbol{I} + \overline{\mathbf{Z}}_\mathrm{A})^{-1} \tag{C.4}$$

The reflection coefficient of the fundamental mode is given in the above vector $\mathbf{S}_{11}$ as its first component. The absolute value of the transmittance S_{21} for a mode in the lossless case may be calculated from:

$$|S_{11}|^2 + |S_{21}|^2 = 1 \tag{C.5}$$

This formula does not give us the phase of the parameter S_{21}. From the field at the input of the structure we can calculate the field in the whole structure by using the derived field transfer equations. The stable and high accurate algorithm is described in Section 2.5.4. From the fields at the end, all the other scattering parameters can be obtained.

Appendix D
EQUIVALENT CIRCUITS FOR DISCONTINUITIES

The discontinuity under consideration is symmetric and is physically determined by the bend angle α. The electrical behaviour can be described by an equivalent circuit, i.e. a symmetric T or Π network as shown in Fig. D.1.

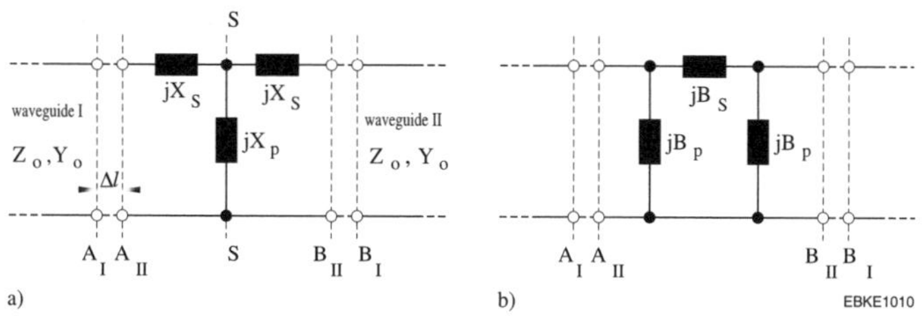

Fig. D.1 Symmetric T network (a) or symmetric Π network (b) as equivalent circuit for the bend discontinuity

To determine the circuit elements, two independent cases must be examined. We may use two waves for excitation, one from each side of the structure. If these two waves are in phase (even-mode excitation) the situation is the same as if a magnetic wall was introduced in the plane of symmetry S. The equivalent network must be open-circuited in plane S.

In the second case, the two waves have opposite phase (odd-mode excitation). An electric wall could be introduced in the plane of symmetry to obtain this behaviour. The equivalent circuit must be short-circuited in plane S. In the first case we obtain an input reflection coefficient $S_{11\text{o}}$ and input impedance $\mathrm{j}X_o = j(X_S + 2X_p)$ or input admittance $Y_\text{o} = jB_p$. In the second case the input reflection coefficient is $S_{11\text{s}}$, and the input impedance $jX_s = jX_S$ or the input admittance $Y_\text{s} = j(B_p + 2B_\text{S})$. These impedances (admittances) are connected with the reflection coefficients in the following way:

$$Y_\text{o} Z_0 = \frac{1 - S_{11\text{o}}}{1 + S_{11\text{o}}} = \frac{Z_0}{\mathrm{j}X_\text{o}} \qquad Y_\text{s} Z_0 = \frac{1 - S_{11\text{s}}}{1 + S_{11\text{s}}} = \frac{Z_0}{\mathrm{j}X_\text{s}} \tag{D.1}$$

The relation between $S_{11\text{o}}, S_{11\text{s}}$ and the scattering parameters is given by:

$$S_{11\text{o}} = S_{11} + S_{21} \qquad S_{11\text{s}} = S_{11} - S_{21} \tag{D.2}$$

We have to add to the reflection caused by the excitation at port I the transmission (with correct sign) added to the scattered portion S_{11} from the excitation of waveguide II.

We can also determine the equivalent elements from the $\mathbf{z}$- and $\boldsymbol{y}$-matrices with excitation as described above. Only the first components (representing the fundamental mode) of the vectors in the following table are different from zero. Therefore, only the first components of the matrices $\mathbf{z}$ and $\boldsymbol{y}$ are relevant.

Even mode excitation:	$\overline{\mathbf{E}}_{\mathrm{A}} = \overline{\mathbf{E}}_{\mathrm{B}} \qquad \overline{\mathbf{H}}_{\mathrm{A}} = -\overline{\mathbf{H}}_{\mathrm{B}}$
$\longrightarrow$	$\overline{Z}_{\mathrm{Aeven}} = \overline{\mathbf{z}}_{1_{11}} + \overline{\mathbf{z}}_{2_{11}} = \mathrm{j}X_s + \mathrm{j}2X_p$
$\longrightarrow$	$\overline{Y}_{\mathrm{Aeven}} = \overline{\boldsymbol{y}}_{1_{11}} + \overline{\boldsymbol{y}}_{2_{11}} = \mathrm{j}B_p$
Odd mode excitation:	$\overline{\mathbf{E}}_{\mathrm{A}} = -\overline{\mathbf{E}}_{\mathrm{B}} \qquad \overline{\mathbf{H}}_{\mathrm{A}} = \overline{\mathbf{H}}_{\mathrm{B}}$
$\longrightarrow$	$\overline{Z}_{\mathrm{Aodd}} = \overline{\mathbf{z}}_{1_{11}} - \overline{\mathbf{z}}_{2_{11}} = \mathrm{j}X_s$
$\longrightarrow$	$\overline{Y}_{\mathrm{Aodd}} = \overline{\boldsymbol{y}}_{1_{11}} - \overline{\boldsymbol{y}}_{2_{11}} = \mathrm{j}B_p + \mathrm{j}2B_s$
Equivalent elements:	$\mathrm{j}X_p = \overline{\mathbf{z}}_{2_{11}} \qquad \mathrm{j}X_s = \overline{\mathbf{z}}_{1_{11}} - \overline{\mathbf{z}}_{2_{11}}$
	$Y_s = -\overline{\boldsymbol{y}}_{2_{11}} \qquad Y_p = \overline{\boldsymbol{y}}_{1_{11}} + \overline{\boldsymbol{y}}_{2_{11}}$

Appendix E

APPROXIMATE METALLIC LOSS CALCULATION IN CONFORMAL STRUCTURES

The solution of the equation:

$$\frac{1}{\overline{r}}\frac{\partial}{\partial \overline{r}}\left(\overline{r}\frac{\partial \psi}{\partial \overline{r}}\right) + \varepsilon_r \psi = 0 \qquad \text{or} \qquad \overline{r}\frac{\partial}{\partial \overline{r}}\left(\overline{r}\frac{\partial \psi}{\partial \overline{r}}\right) + \varepsilon_r \overline{r}^2 \psi = 0 \tag{E.1}$$

is given with R_0 as one of the cylindrical functions of zeroth order by:

$$\psi = R_0(\sqrt{\varepsilon_r}\overline{r}) \tag{E.2}$$

There are two cases:

$$\begin{aligned} &\text{I.} \quad \psi = \widetilde{H}_z \to E_\phi = \mathrm{j}\varepsilon_\mathrm{d}^{-1}\frac{\partial}{\partial \overline{r}}\widetilde{H}_z = \mathrm{j}\varepsilon_r^{-1}\frac{\partial}{\partial \overline{r}}\widetilde{H}_z \\ &\text{II.} \quad \psi = E_z \to \widetilde{H}_\phi = -\mathrm{j}\varepsilon_r\varepsilon_\mathrm{d}^{-1}\frac{\partial}{\partial \overline{r}}E_z = -\mathrm{j}\frac{\partial}{\partial \overline{r}}E_z \end{aligned} \tag{E.3}$$

The first derivative of $R_0(x)$ with respect to x is given by $R_0'(x) = -R_1(x)$. Since ε_r has very large values in metal, we may approximate:

$$H_0^{(2)}(x) = \sqrt{\frac{2\mathrm{j}}{\pi x}}e^{-\mathrm{j}x} \qquad H_1^{(2)}(x) = \sqrt{\frac{2\mathrm{j}}{\pi x}}\mathrm{j}e^{-\mathrm{j}x} \tag{E.4}$$

Therefore, we have $H_1 = \mathrm{j}H_0$ and:

$$E_\phi = \sqrt{\varepsilon_r}^{-1}\widetilde{H}_z \qquad \text{with:} \qquad \varepsilon_r = -\mathrm{j}\frac{\kappa\eta_0}{k_0} \tag{E.5}$$

With $\eta_m = \dfrac{1+\mathrm{j}}{\sqrt{2}}\sqrt{\dfrac{k_0}{\kappa\eta_0}}$ we obtain:

$$E_\phi = \eta_m \widetilde{H}_z \qquad \widetilde{H}_\phi = -\sqrt{\varepsilon_r}E_z \to -E_z = \eta_m \widetilde{H}_\phi \tag{E.6}$$

Furthermore, we have:

$$\eta_m(\widetilde{H}_\phi \vec{e}_\phi + \widetilde{H}_z \vec{e}_z) = -\vec{e}_r \times (E_\phi \vec{e}_\phi + E_z \vec{e}_z) = -E_\phi \vec{e}_z + E_z \vec{e}_\phi$$

which means:

$$\eta_m \widetilde{H}_z = -E_\phi \qquad \eta_m \widetilde{H}_\phi = E_z \tag{E.7}$$

Analysis of Electromagnetic Fields and Waves R. Pregla

Using the following notations:

$$\vec{\widetilde{H}} = [\mathrm{j}\widetilde{H}_z \quad r\widetilde{H}_\phi]^{\mathrm{t}} \qquad \vec{E} = [rE_\phi \quad \mathrm{j}E_z]^{\mathrm{t}} \tag{E.8}$$

we may write with the help of eq. (E.7):

$$\begin{bmatrix} \mathrm{j}\widetilde{H}_z \\ \widetilde{H}_\phi \end{bmatrix} = \begin{bmatrix} I & \\ & r^{-1}I \end{bmatrix} \vec{\widetilde{H}} = -\mathrm{j}\eta_m^{-1} \begin{bmatrix} E_\phi \\ \mathrm{j}E_z \end{bmatrix} = Y_\mathrm{t} \begin{bmatrix} E_\phi \\ \mathrm{j}E_z \end{bmatrix} = Y_\mathrm{t} \begin{bmatrix} r^{-1}I & \\ & I \end{bmatrix} \vec{E} \tag{E.9}$$

Index

ABC, *see* boundary conditions, absorbing
abrupt transition, 3, 443
absorbing boundary conditions, 9, 153, 175, 289, 386, 449
admittance transformation, 3, 25, 76, 163, 280
admittance/impedance transformation, 3, 174, 222
  by special FDFD, 434
  concept, 110
  through circular layers/sections, 161, 168, 177
  through general layers/sections, 35
  through interfaces of layers/sections, 53
  through planar layers/sections, 22, 53
Ampère's law, *see* law, Ampère's
anisotropic, 3, 134, 142, 290, 313, 330, 381, 444
anisotropic material parameters, 149, 340
arbitrary coordinate, 401
azimuthal, 203

backward wave, 199
band diagram, 270, 294
bandgap structures, 270
Bessel functions, 75, 231, 369, 418
  modified, 161, 419
boundary conditions
  absorbing, 154, 175, 205, 230, 392, 455, 456
  Dirichlet, 6, 122, 261, 328, 358, 386, 387, 437
  Neumann, 6, 122, 261, 328, 358, 413, 437
  periodic, 9, 153, 175, 295, 331, 369, 445
  special, 454
Bragg-fibre, 173
Bragg-grating, 269, 271
branch waveguide coupler, 347
Brillouin zone, 294

Cartesian coordinate system, 82, 134, 381
central differences, 152, 261, 328, 399, 465
characteristic admittance, 96
characteristic impedance, 3, 82, 226, 277, 319, 370, 501
circular bend, 116, 164
circular waveguide, 77, 133
coaxial waveguide, 142, 199
concatenation, 3, 75, 222, 300, 320, 391, 501
conformal antenna, 1, 255, 401
convergence, 1, 43, 86, 210, 433, 446
coupled resonator filter, 89
cross-products, 161, 369, 419
crossed discretisation lines, 4, 77, 234, 300, 313, 347, 392

current density, 42, 224, 321
current distribution, 44, 250, 480
curved waveguides, 75
cut-off wavelength, 11, 75, 216, 375, 427
cylindrical coordinate system, *see* cylindrical coordinates
cylindrical coordinates, 4, 133, 348, 457

defect waveguide, 270
diaphragm, 86
dielectric waveguides, 375
  film waveguides, 22, 25
  inset dielectric guide, 41
  rib waveguide, 45, 164
difference matrix, 9, 398, 440
difference operator, 8, 75, 152, 328, 441
  higher order, 460
dipole antenna, 250
Dirichlet, 301, 387, 413, 442
Dirichlet condition, 77, 153, 154, 214, 387, 469
discontinuities, 85, 86, 105, 123, 133
discrete Fourier transformation, 45, 480, 481
discretisation, 4, 16, 80, 152, 160
  ϕ-direction, 78, 173
  r-direction, 80, 155
  line, 297
  one-dimensional, 26, 155, 289
  qualitative description, 5
  quantitative description, 7
  two-dimensional, 22, 152
discretisation line, 2
dispersion curve, 184, 330
  rib waveguide, 67

E-bend, 80
E-plane bend, *see* E-bend
E-plane corner, 102
E-plane junction, 75, 95
effective index, 25, 165, 309, 454
effective permittivity, 9, 184, 400
eigenmode, 4, 25, 77, 133, 347, 368, 389, 453
eigenvalue/eigenvector problem, 24, 96
electric wall, 8, 50, 77, 153, 340, 365, 386, 487, 503
elliptical coordinates, 218, 423
elliptical waveguide, 212

Faraday's law, *see* law, Faraday
fibre Bragg grating, 269
fibres, 133
  circular cross section, 211
  elliptical cross section, 217
  star cross section, 218
field extrapolation, 246
filter, 4, 15, 75, 89, 157, 269, 333, 446
finite difference method, 386
finite differences, 7, 22, 417, 472
  impedances, 148
  impedances/admittance transformation, 4, 66, 218
finite metallisation thickness, 3, 28
finite width, 184
Floquet impedance, 282
Floquet modal matrix, 276
Floquet mode, 228, 276, 277, 279, 280, 283
Floquet's theorem, 4, 274, 277, 331, 375, 459
forward propagation, 24, 225, 501

generalised port, 93, 241, 323, 393
generalised transmission line equations, *see* GTL equations
groove guide, 28, 42, 363
GTL equations, 4, 22, 23, 135, 139, 146, 257, 353, 434
  Cartesian coordinates, 381, 386, 390
  cylindrical coordinates, 134

general orthogonal coordinates, 4, 383, 386
gyrotropic material, 138, 384

H-bend, 79
H-plane bend, *see* H-bend
H-plane junctions, 98
Hankel functions, 204, 421
heat equation, 2, 434
Helmholtz equation, 6, 449
hexagonal lattice, 297
Huygen's source, 348

impedance transformation, 3, 55, 76, 110, 209, 285, 316, 320
interface, 4, 78, 84
impedance/admittance transformation with finite differences, *see* finite differences, impedance/admittance transformation
indirect eigenvalue problem, 28, 209, 333, 400
inset dielectric guide, 28, 40
interface condition, 38, 210, 315, 416, 460
interface with metallisation, 36, 38
interfaces, 56
interfaces between layers, 57
interfaces between sections, 171
interpolation, 63, 191, 468
discretisation schemes, 154
interpolation matrix, 49, 109, 214, 315, 383, 387, 469
isotropic materials, 16, 138, 443

junctions, 75, 112, 320, 391
E-plane rectangular, 76
general 3D, 95, 117
microstrip with rectangular waveguides, 340

Kronecker product, 50, 122, 153, 413

law
Ampère's, 19, 119, 135, 381, 498
Faraday's, 20, 120, 407
law of induction, 19, 120, 135, 404
layer, 3, 6, 16, 35, 133, 313, 347, 410, 443
homogeneous, 6, 28, 149, 179, 425
inhomogeneous, 3, 28, 151, 181, 407, 448
interface, 6, 57, 189, 209
linear field interpolation, 62
lines of varying length, 105, 211, 297, 299, 313
LSE modes, 78, 82
LSM modes, 78, 82, 484

magnetic boundary, 322
magnetic field, 199
magnetic wall, 2, 26, 77, 153, 289, 325, 351, 385, 453, 503
magnetised ferrites, 142, 143
magnetised plasma, 145, 199
magnetostatic field, 199
matching, 36, 59, 84, 99, 118, 161, 163, 171, 314, 400, 448
field matching, 56, 366, 400
impedance matching, 257
mode matching, 483
material tensor, 4, 30, 139, 386, 433
matrix partition, 40, 85, 224
Maxwell's equations, 4, 349, 434, 495
Maxwell's equations in matrix notation, 379
meander line, 15, 269, 331
metallic loss, 187, 505
metallisation, 2, 27, 56, 60, 84, 187, 313, 347
mode matching technique (MMT), 1, 45, 113, 347, 433, 481

monopole, 231, 233, 256
multiplication theorem, 172

Neumann condition, 96, 195, 356, 469
Neumann functions, 161
non-equidistant discretisation, 44, 210, 340, 463
non-linear material, 2

operator matrix, 24, 81, 236

Padé approximation, 450
patch antenna, 16, 231
period, 228, 270, 271, 280, 335, 459
  symmetric, 269, 271, 276
  unsymmetric, 270, 271, 283
periodic structures, 4, 269, 271, 276, 285, 288, 375, 400, 480
permeability, 19, 144, 379, 444
permittivity, 6, 19, 77, 78, 133, 298, 343, 379, 443, 495
photonic bandgap structures, *see* bandgap structures
photonic crystals, 9, 270, 293, 299, 301, 391, 458
planar circuit, 3, 45
planar structures, 4, 30, 189, 300, 414, 481
planar waveguide, 26, 173, 313, 322, 480
propagation constant, 2, 24, 96, 144, 272, 376, 401, 463, 497
properties, 1, 313, 379, 482

quadratic field interpolation, 63

radiation, 9, 26, 165, 421, 449
radiation loss, 172
reciprocity, 486
rectangular waveguide, 1, 4, 75, 112, 141, 330, 340, 363, 433
reflections in discretisation grids, 470
resonators, 133, 157, 333

scattering parameters, 4, 86, 100, 112, 255, 292, 330, 401, 501, 503
sharp bends, 105, 299, 313, 392
  microstrip, 330
  photonic crystal, 299, 306, 391
single carrier semiconductor, 199
slab waveguide, 25, 305
Smith chart, 4, 251, 253
special boundary conditions, *see* boundary conditions, special
spherical coordinate system, 402, 403
spherical coordinates, 5, 401, 403, 433
square lattice, 294, 296, 297
stable field transformation, 67
Sturm–Liouville, 21, 481
substrate, 2, 43, 184, 231, 297, 313, 333, 343, 401, 455
supermatrix, 96, 100
superstrate, 184
supervector, 21, 93, 153, 157, 323, 349, 381, 417
system equation, 33, 36, 38, 104

T-junction, 102, 113, 337
TE_z modes, 22, 78, 95, 138, 368
TE_{10} mode, 11, 78
TM_z modes, 78, 138, 368, 370, 405
transfer matrix, 25, 106, 180, 276, 316
transformation matrix, 42, 43, 84, 98, 244, 354, 391, 437
transformation matrix partition, 28, 76
transversal field vectors, 19
two-dimensional (2D), 22

variable thickness, 313
VCSEL, 2, 434
vector potentials, 21

wave admittance, 24, 25, 107, 365
wave impedance, 6, 24, 25, 107, 237, 281, 380, 495
waveguide, 2, 5, 16, 26, 75, 77, 88, 133, 271, 293, 313, 320, 347, 365, 379, 433, 446, 448, 495, 501, 504
  circular, 77, 133, 143, 192, 219, 222, 350, 421
  cross, 340, 348
  dielectric, 216, 375
  discontinuities, 3, 105, 123, 133, 219
  elliptic, 217
  planar, 2, 26, 173, 313, 480
  rectangular, 1, 75, 120, 141, 330, 340, 350
  ridge, 348, 368, 455
waveguide bends
  circular, 78, 141
  rectangular, 106, 112, 141
waveguide junctions, 93, 117, 322, 337, 391